Springer Series in Wood Science

Editor: T.E. Timell

M.H. Zimmermann
Xylem Structure and the Ascent of Sap (1983)

J.F. Siau
Transport Processes in Wood (1984)

R.R. Archer
Growth Stresses and Strains in Trees (1986)

W.E. Hillis
Heartwood and Tree Exudates (1987)

S. Carlquist
Comparative Wood Anatomy (1988)

L.W. Roberts/P.B. Gahan/R. Aloni
Vascular Differentiation and Plant Growth Regulators (1988)

C. Skaar
Wood-Water Relations (1988)

J.M. Harris
Spiral Grain and Wave Phenomena in Wood Formation (1989)

B.J. Zobel/J.P. van Buijtenen
Wood Variation (1989)

P. Hakkila
Utilization of Residual Forest Biomass (1989)

J.W. Rowe
Natural Products of Woody Plants (1989)

John W. Rowe (Ed.)

Natural Products of Woody Plants I

Chemicals Extraneous to the Lignocellulosic Cell Wall

With 244 Figures (both volumes)

Springer-Verlag Berlin Heidelberg GmbH

JOHN W. ROWE
USDA, Forest Service
Forest Products Laboratory
One Gifford Pinchot Drive
Madison, WI 53705-2398, USA

Series Editor:

T. E. TIMELL
State University of New York
College of Environmental, Science and Forestry,
Syracuse, NY 13210, USA

Cover: Transverse section of *Pinus lambertiana* wood. Courtesy of Dr. Carl de Zeeuw,
SUNY College of Environmental Science and Forestry, Syracuse, New York

ISBN 978-3-642-74077-0 ISBN 978-3-642-74075-6 (eBook)
DOI 10.1007/978-3-642-74075-6

Library of Congress Cataloging-in-Publication Data
Natural products of woody plants: chemicals extraneous to the lignocellulosic cell wall / John W. Rowe, ed. p. cm. – (Springer series in wood science)
ISBN 978-3-642-74077-0
1. Wood products. 2. Plant products. 3. Wood – Chemistry. I. Rowe, John W. II. Series. TS930.N38 1989

Originally published by Springer-Verlag Berlin Heidelberg New York in 1989
Softcover reprint of the hardcover 1st edition 1989

Typesetting: K+V Fotosatz GmbH, Beerfelden;

2131/3020-543210 – Printed on acid-free paper

With great pleasure and in grateful appreciation, this book is dedicated to

Oskar Jeger

Professor Emeritus, Swiss Federal Institute of Technology (ETH)

This outstanding organic chemist played a pivotal role in my professional development and instilled in me a sense of awe at the beauty and elegance of the natural organic chemicals that surround us all

JOHN W. ROWE

The Editor

Project Leader for Wood Extractives Research (retired), Forest Service, U.S. Department of Agriculture, Forest Products Laboratory, Madison, Wisconsin, USA

Educated at the Swiss Federal Institute of Technology (ETH), M.I.T., and the University of Colorado..

Former positions include: Research Chemist, Forest Products Laboratory (1957–1966). Author of numerous publications.

"Wood Salutes" Award (1975). Fellow, International Academy of Wood Sciences, American Institute of Chemists, American Association for the Advancement of Science. Chair Wisconsin Section (1968–1969), Alternate Counselor (1976–1978), American Chemical Society.

Research interests include terpenoid and steroid chemistry; tree extractives; wood chemistry; the potential of wood as a source of chemical intermediates and energy.

Preface

Wood as found in trees and bushes was of primary importance to ancient humans in their struggle to control their environment. Subsequent evolution through the Bronze and Iron Ages up to our present technologically advanced society has hardly diminished the importance of wood. Today, its role as a source of paper products, furniture, building materials, and fuel is still of major significance.

Wood consists of a mixture of polymers, often referred to as lignocellulose. The cellulose microfibrils consist of an immensely strong, linear polymer of glucose. They are associated with smaller, more complex polymers composed of various sugars called hemicelluloses. These polysaccharides are embedded in an amorphous phenylpropane polymer, lignin, creating a remarkably strong composite structure, the lignocellulosic cell wall.

Wood also contains materials that are largely extraneous to this lignocellulosic cell wall. These extracellular substances can range from less than 1% to about 35% of the dry weight of the wood, but the usual range is 2% – 10%. Among these components are the mineral constituents, salts of calcium, potassium, sodium, and other metals, particularly those present in the soil where the tree is growing. Some of the extraneous components of wood are too insoluble to be extracted by inert solvents and remain to give extractive-free wood its color; very often these are high-molecular-weight polyphenolics. Most of the extraneous components, however, can be removed with neutral, inert organic solvents or water to yield the *extractives*, a mixture of naturally occurring organic compounds. This large and diverse group is the subject of this book.

Woody tissue is capable of synthesizing a wide range of natural products, some of great complexity, sometimes in considerable quantity, and at times, in relatively pure form. Some of them, such as simple carbohydrates, phytosterols, non-alkaloidal nitrogenous compounds, and simple aliphatic, alicyclic, and aromatic compounds, are lignocellulosic precursors or are otherwise involved in the intermediary metabolism of the living cell. Among the more interesting materials, however, are the wide array of compounds that woody tissue has developed to protect itself from higher animals, insects, fungi, bacteria, and other decay-causing microorganisms. These protective compounds are found in especially high concentration in heartwood and bark and are often produced as exudates in response to wounds. Exudates are a widely diverse group, comprising oleoresins, kinos, balsams, latexes, gums, manna, and maple sugar.

Particularly significant to organic chemists in describing these compounds was W. E. Hillis's book "Wood Extractives and Their Significance to the Pulp and Paper Industry" (Academic Press, New York and London 1962, 513 pp.). His more recent, excellent book "Heartwood and Tree Exudates" (1987, 268 pp.) is part of the Springer Series in Wood Science, to which this book belongs.

Tree exudates were among the earliest items of trade in prehistoric times, with a wide varicty of uses. Amber was one of the prized items of trade in ancient times. Since then, extractives and exudates have found a wide array of uses, ranging from a variety of resins to dyes and perfume intermediates, that are discussed in detail in Chapter 10.

It is worth noting that products and chemical intermediates derived from these natural products are from a renewable resource with consequent advantages over competitive materials derived from petrochemicals. The significance of these compounds, however, does not stop here, for they also have a considerable effect on the utilization of wood. The odor, color, and decay resistance of wood are a function of the extractives. Ancient wooden temples, such as exist in the Far East, owe their durability to their preservative extractives. Wood utilization is adversely affected by some extractives. Understanding their chemistry has in the past led to control of these difficulties. Chapter 9 discusses the influence of extractives on wood properties and utilization.

Naturally-occurring compounds of wood have been a gold mine of considerable significance to the organic chemist. Challenges to the organic chemist started with how to isolate and purify these compounds, often a difficult task when a series of closely related compounds was present. The subsequent challenge of the proof of structure of these compounds became the life work of many of this century's most eminent chemists. Chapter 2 discusses this elegant work in detail. Determination of the biogenetic pathways led to a quantum leap in our understanding of the biochemical processes in plants. Studies of the properties of this wide array of natural products contributed greatly to our basic understanding of organic chemistry. The significant development of conformational analysis came mostly from a study of steroids and other compounds with a relatively rigid skeleton. Understanding of Wagner-Meerwein and other carbonium ion-like rearrangements resulted from studies of the rearrangements repeatedly encountered among the terpenoids.

Many of these natural products have been exciting targets for the synthetic organic chemist, including the interesting work of George Büchi, who synthesized patchouly alcohol before the correct structure was known. More recently, the emphasis has been on the preparation of synthetically-modified derivatives of the natural products. This has been particularly significant in the area of pharmaceuticals. An example of this approach was reported in the March 1, 1988 issue of the New York Times, when E. J. Corey and his co-workers reported the complete synthesis of the extremely complex diterpene ginkgolide B from the ancient ginkgo tree (J. Am. Chem. Soc. 110:649–651). Ginkgolide B has potential in treating asthma, toxic shock, graft rejection, Alzheimer's disease, and various circulatory disorders. Investigations on synthetic analogs are now under way. It is worth noting that almost half of all prescription drugs dispensed in the U.S. contain substances of natural origin. Even aspirin was first derived in part from willow bark. Current ethnobotanical studies, especially in the tropics, are of considerable importance for the identification of pharmaceutically effective, native remedies before the tropical forests are decimated, and native herbal practices are replaced by "modern" medical practices. In the future we may see increased emphasis on the isolation of the enzymes involved in the synthesis of natural products for the production of valuable extractives in vitro.

The chemistry of the many different types of these natural products is discussed in Chapters 4–8. Most of the forest products laboratories around the world have contributed significantly to this knowledge. As an increasing number of species has been studied, and as the list of known natural products has grown, chemotaxonomy (chemical systematics) has become of much interest under the tutelage of some of the contributors to this book, as well as scientists such as Erdtman, Swain, Hegenauer, Bate-Smith, and many others. This has led to a better understanding of tree genetics and to an improved classification of genera and species. The evolution of tree species can really be considered in part to be the continual acquisition of new enzymatic pathways. Chapter 3 discusses the evolution of natural products from a most interesting perspective. Future efforts in this direction will need to focus on the function of the enzymes controlling the level and composition of these natural products.

The eminent scientists who have contributed to this book document how, in the course of the last 50 years, we have seen an explosion of knowledge about the chemistry and biochemistry of these novel and important natural products derived from wood. We have, nevertheless, only scratched the surface. It is exciting to think of the knowledge that will be acquired during the next 50 years. This book will, I hope, prove to be of value to scientists seeking to expand their understanding of the array of fascinating natural products found in wood, and to stimulate their future research.

Madison, Wisconsin
October 1989

J. W. Rowe

Acknowledgments

This book would not have been possible without the dedicated, conscientious effort by the many outstanding scientists who have contributed to it. The U.S. Forest Products Laboratory contributed invaluable support services. However, the editor would not have been able to pull it all together without the indispensable, extensive services of Celeste Kirk, who, as redactor, spent untold hours helping the contributors to polish their manuscripts and put them into proper format for Springer-Verlag. Many of the contributors have expressed their gratitude for her efforts, and to these the editor adds his heartfelt thanks.

CELESTE HANSON KIRK

Ms. Kirk is a highly experienced editor. Prior to founding her own editing service, she was a technical editor at the U.S. Forest Products Laboratory, where she edited scientific papers for submission to over 60 different journals. She is a member of the Council of Biology Editors and Women in Communications.

Contents Volume I

Contents Volume II XXV

1 Introduction and Historical Background 1

1.1 Historical Uses of Extractives and Exudates 1
W.E. Hillis

1.1.1 Introduction 1
1.1.2 Major Uses of Extractives and Exudates 1
1.1.2.1 The Use of Durable Woods 2
1.1.2.2 Exudates 3
1.1.2.2.1 Varnishes 3
1.1.2.2.2 Lacquers 5
1.1.2.2.3 Gums 5
1.1.2.3 Tannins 5
1.1.2.4 Dyes 6
1.1.2.5 Perfumes 7
1.1.2.6 Rubber 8
1.1.2.7 Medicines 9
1.1.3 Lessons from History 10
References 12

1.2 Natural Products Chemistry – Past and Future 13
K. Nakanishi

1.2.1 Introduction 13
1.2.2 Isolation and Purification 16
1.2.3 Structure Determination 18
1.2.4 The Future of Natural Products Science 22
References 25

2 Fractionation and Proof of Structure of Natural Products 27
J. Snyder, R. Breuning, F. Derguini, and K. Nakanishi

2.1 Introduction 27
2.2 Novel Techniques and Recent Developments in Fractionation and Isolation 28
2.2.1 Countercurrent Chromatography 28
2.2.1.1 Coil Countercurrent Chromatography 29
2.2.1.2 Droplet Countercurrent Chromatography 36
2.2.1.3 Rotation Locular Countercurrent Chromatography 39
2.2.1.4 Centrifugal Partition Chromatography 44
2.2.1.5 Comparison of Partition Chromatographic Methods 46
2.2.2 Adsorption Chromatography 49

2.2.2.1 Ion-Pair Chromatography 50
2.2.2.2 Other New Methods of Column Chromatography 53
2.2.2.3 Supercritical Fluid Chromatography 55

2.3 Nuclear Magnetic Resonance Spectroscopy 60
2.3.1 Proton Nuclear Magnetic Resonance 60
2.3.1.1 Difference Decoupling 60
2.3.1.2 Difference NOE 61
2.3.1.3 Contact Shifts 64
2.3.1.4 Partial Relaxation 65
2.3.2 Carbon Nuclear Magnetic Resonance 66
2.3.2.1 J-Modulated Spin Echo 67
2.3.2.2 Insensitive Nuclei Enhanced by Polarization Transfer 68
2.3.2.3 Distortionless Enhancement by Polarization Transfer 69
2.3.2.4 Carbon-Proton Heteronuclear Coupling 70
2.3.2.5 Carbon-Proton Heteronuclear NOE 72
2.3.2.6 Deuterium Isotopic Shifts 73
2.3.3 Two-Dimensional NMR Spectroscopy 75
2.3.3.1 Two Dimensional J-Resolved Proton NMR Spectroscopy 75
2.3.3.2 Two Dimensional Correlation Spectroscopy 76
2.3.3.3 Two Dimensional-INADEQUATE (Incredible Natural Abundance Double Quantum Transfer Experiment) 83

2.4 Other Spectroscopic Techniques 85
2.4.1 Mass Spectrometry 85
2.4.1.1 Techniques That Enhance Sample Volatilization 86
2.4.1.2 Modern Techniques of Ionization/Desorption 88
2.4.1.3 Tandem Mass Spectrometry 91
2.4.2 Ultraviolet-Visible Spectroscopy 93
2.4.3 Infra-Red Spectroscopy 97
2.4.4 Circular Dichroism 101
2.4.4.1 The Nature of Circular Dichroism 103
2.4.4.2 The Additivity Relation in A Values 106

2.5 General Conclusions 109

References 109

3 Evolution of Natural Products 125
O. R. Gottlieb

3.1 Convergent Synthesis and the Origin of RNA-Based Life 125
3.2 Expansion of the Acetate, Mevalonate, and δ-Aminolevulinate Pathways in Bacteria and Algae 127
3.3 Expansion of the Shikimate Pathway in Terrestrial Plants 128
3.4 Phytochemistry and Plant Defense 133
3.5 Oxidation Levels of Angiospermous Micromolecules 137
3.6 Skeletal Specialization of Angiospermous Micromolecules 140
3.7 Quantification of Micromolecular Parameters 145
3.8 Phytochemical Gradients in Angiosperms 146
3.9 Future Perspectives 148
References 150

4 **Carbohydrates** 155
J. N. BeMiller

4.1 Introduction 155
4.2 Sucrose 156
4.3 Higher Oligosaccharides Related to Sucrose 157
4.4 Other Oligosaccharides 158
4.5 Monosaccharides 159
4.6 Alditols 159
4.7 Cyclitols 160
4.7.1 *myo*-Inositol 160
4.7.2 D-*chiro*-Inositol 161
4.7.3 Quebrachitol 161
4.7.4 D-Quercitol 162
4.7.5 Conduritol 162
4.7.6 Quinic Acid 162
4.8 Plant Glycosides 162
4.9 Starch 162
4.10 Extractable Polysaccharides 163
4.10.1 Arabinogalactans 164
4.10.1.1 Larch Arabinogalactans 165
4.10.1.2 Other Extractable, Nonexudate Arabinogalactans 165
4.10.2 Other Extractable Polysaccharides; The Pectic Polysaccharides ... 166
4.10.3 Exudate Gums 167
4.10.3.1 *Acacia* Gums 168
4.10.3.2 Exudate Gums of Other Rosales Genera 168
4.10.3.3 Gums of Combretaceae (Myrtiflorae) Genera 169
4.10.3.4 Exudate Gums of Anacardiaceae (Sapindales) 170
4.10.3.5 Exudate Gums of Families in the Orders Rutales, Parietales, and Malvales 170
4.10.3.6 Exudate Gums from Other Orders 171
4.10.3.7 Exudate Gums with Xylan Cores 172
References 172

5 **Nitrogenous Extractives** 179

5.1 Amino Acids, Proteins, Enzymes, and Nuccleic Acids 179
D. J. Durzan

5.1.1 Introduction 179
5.1.2 Composition 182
5.1.2.1 Free and Bound Amino Acids 183
5.1.2.2 Proteins and Enzymes 186
5.1.2.3 Nucleic Acids and Related Products 189
5.1.3 Factors Determining Composition 189
5.1.3.1 Genetics 189
5.1.3.2 Genetics × Environment 190
5.1.3.3 Growth and Development 190

5.1.3.4 Pathology 191
5.1.3.5 Impact of Humans 192
5.1.4 Conclusion 195
References 195

5.2 The Alkaloids 200
S.-I. SAKAI, N. AIMI, E. YAMANAKA, and K. YAMAGUCHI

5.2.1 Introduction 200
5.2.2 True Alkaloids 202
5.2.2.1 Alkaloids from Ornithine 202
5.2.2.1.1 Coca Alkaloids 202
5.2.2.1.2 Elaeocarpus Alkaloids 204
5.2.2.2 Alkaloids from Lysine 205
5.2.2.2.1 Punica Alkaloids 205
5.2.2.2.2 Lythraceae Alkaloids 206
5.2.2.2.3 Securinega Alkaloids 208
5.2.2.2.4 Cytisus Alkaloids 209
5.2.2.3 Alkaloids from Anthranilic Acid 210
5.2.2.3.1 Quinoline and Furoquinoline Alkaloids 210
5.2.2.3.2 Acridone Alkaloids 211
5.2.2.3.3 Evodia Alkaloids 212
5.2.2.3.4 Carbazole Alkaloids 213
5.2.2.4 Alkaloids from Nicotinic Acid (Celastraceae Alkaloids) 213
5.2.2.5 Alkaloids from Phenylalanine and Tyrosine 214
5.2.2.5.1 Benzylisoquinoline Alkaloids 214
5.2.2.5.2 Curare Alkaloids 217
5.2.2.5.3 Sinomenine 217
5.2.2.5.4 Aporphine-type Alkaloids 218
5.2.2.5.5 Berberine 218
5.2.2.5.6 Nitidine 219
5.2.2.5.7 Erythrina Alkaloids 221
5.2.2.5.8 Cephalotaxus Alkaloids 222
5.2.2.5.9 Ipecacuanha and Alangium Alkaloids 223
5.2.2.6 Alkaloids from Tryptophan 225
5.2.2.6.1 Calycanthus Alkaloids 225
5.2.2.6.2 Picrasma (Pentaceras) and Carboline Alkaloids 226
5.2.2.6.3 Rauwolfia Alkaloids 227
5.2.2.6.4 Tabernanthe Alkaloids 229
5.2.2.6.5 Ochrosia Alkaloids 231
5.2.2.6.6 Ervatamia Alkaloids 231
5.2.2.6.7 Uncaria-Mitragyna Alkaloids 232
5.2.2.6.8 Yohimbe Alkaloids 236
5.2.2.6.9 Cinchona Alkaloids 237
5.2.2.6.10 Guettarda Alkaloids 238
5.2.2.6.11 Strychnos Alkaloids 239
5.2.2.6.12 Gelsemium Alkaloids 241
5.2.2.6.13 Gardneria Alkaloids 243
5.2.2.6.14 Camptothecins 244
5.2.3 Pseudoalkaloids 246
5.2.3.1 Alkaloids from Polyketides 246
5.2.3.1.1 Pinidine 246

5.2.3.1.2 Galbulimima Alkaloids 246
5.2.3.2 Alkaloids from Mevalonate 248
5.2.3.2.1 Spiraea Alkaloids 248
5.2.3.2.2 Erythrophleum Alkaloids 248
5.2.3.2.3 Daphniphyllum Alkaloids 249
5.2.3.2.4 Apocynaceae Steroidal Alkaloids 250
5.2.3.2.5 Buxaceae Steroidal Alkaloids 250
References 251

6 Aliphatic and Alicyclic Extractives 259

6.1 Simple Organic Acids 259
Y. OHTA

6.1.1 Introduction 259
6.1.2 Organic Acids in the TCA and Glyoxylate Cycles 259
6.1.2.1 Citric Acid 260
6.1.2.2 Aconitic Acid 261
6.1.2.3 Isocitric Acid 261
6.1.2.4 α-Ketoglutaric Acid 261
6.1.2.5 Succinic Acid 262
6.1.2.6 Fumaric Acid 262
6.1.2.7 Malic Acid 262
6.1.2.8 Oxaloacetic Acid 263
6.1.2.9 Glyoxylic Acid 263
6.1.3 Other Metabolically Important Organic Acids 263
6.1.3.1 Glycolic Acid 263
6.1.3.2 Glyceric Acid 265
6.1.3.3 Pyruvic Acid 265
6.1.3.4 Malonic Acid 265
6.1.3.5 Shikimic Acid and Quinic Acid 266
6.1.4 Organic Acids of an End-Product Nature 266
6.1.4.1 Lactic Acid 266
6.1.4.2 Oxalic Acid 268
6.1.4.3 Tartaric Acid 268
6.1.4.4 Chelidonic Acid 269
6.1.4.5 Fluoroacetic Acid 269
References 270

6.2 Complex Aliphatic and Alicyclic Extractives 274
Y. OHTA

6.2.1 Introduction 274
6.2.2 γ-Lactones 275
6.2.3 δ-Lactones (2-Pyrones) 280
6.2.4 Cyanogenic Glycosides and Related Compounds 281
6.2.5 Highly Oxygenated Cyclohexanes 287
6.2.6 Cyclohexane Diols 289
6.2.7 Polycyclic Compounds 291
6.2.8 Miscellaneous 292
References 294

6.3 Fats and Fatty Acids 299
D. F. ZINKEL

6.3.1 Introduction 299
6.3.2 Fats as Food Reserves 299
6.3.3 Aliphatic Monocarboxylic Acids 300
6.3.3.1 Volatile Fatty Acids 301
6.3.3.2 Constituent Fatty Acids of Fats 301
References 302

6.4 Chemistry, Biochemistry, and Function of Suberin and Associated Waxes 304
P. E. KOLATTUKUDY and K. E. ESPELIE

6.4.1 Introduction 304
6.4.2 Waxes 304
6.4.2.1 Analysis of Plant Waxes 304
6.4.2.2 Composition of Suberin-Associated Waxes 306
6.4.2.2.1 Hydrocarbons in Suberin-Associated Waxes 307
6.4.2.2.2 Wax Esters 308
6.4.2.2.3 Free Fatty Alcohols 308
6.4.2.2.4 Free Fatty Acids 308
6.4.2.2.5 Polar Wax Components 309
6.4.2.2.6 Ferulic Acid Esters 309
6.4.2.2.7 Tabular Survey of Bark Wax Components 310
6.4.2.3 Biosynthesis of Wax Components 312
6.4.2.3.1 Very Long Fatty Acids 312
6.4.2.3.2 Fatty Alcohols 313
6.4.2.3.3 Wax Esters 313
6.4.2.3.4 Hydrocarbons and Derivatives 314
6.4.3 Suberin 316
6.4.3.1 Ultrastructure 316
6.4.3.1.1 Ultrastructural Characterization 316
6.4.3.1.2 Ultrastructural Identification of Suberin in Bark 317
6.4.3.2 Chemical Composition and Structure of the Polymer 323
6.4.3.2.1 Composition of the Aliphatic Portion 323
6.4.3.2.2 Phenolic Composition 326
6.4.3.2.3 Structure 333
6.4.3.3 Suberin Biosynthesis 337
6.4.3.3.1 Biosynthesis of the Aliphatic Monomers 337
6.4.3.3.1.1 ω-Hydroxylation of Fatty Acids 337
6.4.3.3.1.2 Oxidation of ω-Hydroxy Acids 337
6.4.3.3.1.3 Biosynthesis of Mid-Chain Oxygenated Suberin Monomers 339
6.4.3.3.2 Biosynthesis of the Aromatic Components of Suberin 341
6.4.3.3.3 Biosynthesis of Suberin from Aliphatic and Aromatic Monomers 342
6.4.3.4 Function of Suberin and Associated Waxes 343
6.4.3.4.1 Prevention of Water Loss 343
6.4.3.4.2 Suberization in Wound Healing 344
6.4.3.4.3 Suberization in Response to Stress 344
6.4.3.4.4 Suberization as a Means of Compartmentalization 345
6.4.3.5 Regulation of Suberization 346
6.4.3.6 Enzymatic Degradation of Suberin 347
References 349

7 Benzenoid Extractives ... 369

7.1 Monoaryl Natural Products ... 369
O. THEANDER and L. N. LUNDGREN

7.1.1 Introduction ... 369
7.1.2 Simple Phenols (C_6) ... 370
7.1.3 Phenolic Acids, Salicins and Other C_6-C_1 Compounds ... 371
7.1.3.1 Benzoic Acids and Related Compounds ... 371
7.1.3.2 Salicins and Related Compounds ... 371
7.1.4 Acetophenones and Other C_6-C_2 Compounds ... 373
7.1.5 Cinnamic Acids, Coumarins and Other Phenylpropanoids (C_6-C_3) 374
7.1.5.1 Cinnamic Acids ... 374
7.1.5.2 Coumarins ... 374
7.1.5.3 Other Phenylpropanoids ... 383
7.1.6 Miscellaneous Monoaryl Compounds ... 392
References ... 393

7.2 Gallic Acid Derivatives and Hydrolyzable Tannins ... 399
E. HASLAM

7.2.1 Introduction ... 399
7.2.2 Metabolism of Gallic Acid – General Observations ... 400
7.2.3 Biosynthesis of Gallic Acid ... 404
7.2.4 Metabolites of Gallic Acid ... 406
7.2.4.1 Simple Esters Occurrence and Detection ... 407
7.2.4.2 Depside Metabolites Group 2A ... 412
7.2.4.3 Metabolites Formed by Oxidative Coupling of Galloyl Esters Groups 2B and 2C, Ellagitannins ... 415
7.2.4.3.1 Hexahydroxydiphenoyl Esters ... 416
7.2.4.3.2 Dehydrohexahydroxydiphenoyl Esters ... 419
7.2.4.3.3 Group 2B Metabolites ... 419
7.2.4.3.4 Group 2C Metabolites ... 426
7.2.4.3.5 Postscript ... 429
7.2.5 The Interaction of Proteins with Metabolites of Gallic Acid ... 430
References ... 433

7.3 Lignans ... 439
O. R. GOTTLIEB and M. YOSHIDA

7.3.1 Introduction ... 439
7.3.2 Nomenclature and Numbering ... 441
7.3.3 Chemistry ... 501
7.3.4 Oligomeric Lignoids ... 505
References ... 505

7.4 Stilbenes, Conioids, and Other Polyaryl Natural Products ... 512
T. NORIN

7.4.1 Introduction ... 512
7.4.2 Stilbenes and Structurally Related Compounds ... 512

7.4.3 Conioids (Norlignans), Including Condensed and Structurally Related Compounds ... 517
7.4.4 Aucuparins and Structurally Related Biphenyls ... 520
7.4.5 Diarylheptanoids, Structurally Related Diarylheptanoids, and Bridged Biphenyls (Cyclophanes) ... 521
7.4.6 Miscellaneous Diaryl and Polyaromatic Compounds ... 525
7.4.7 Concluding Remarks ... 528
References ... 528

7.5 Flavonoids ... 533
J.B. Harborne

7.5.1 Introduction ... 533
7.5.2 Structural Types ... 536
7.5.2.1 Flavones and Flavonols ... 536
7.5.2.2 Flavonones and Flavononols ... 543
7.5.2.3 Chalcones and Aurones ... 546
7.5.2.4 Isoflavonoids and Neoflavonoids ... 548
7.5.3 Distribution ... 554
7.5.3.1 Distribution within the Plant ... 554
7.5.3.2 Patterns in Gymnosperm Woods ... 556
7.5.3.3 Patterns in Angiosperm Woods ... 558
7.5.3.3.1 Heartwood Flavonoids of the Anacardiaceae ... 559
7.5.3.3.2 Heartwood Flavonoids of the Leguminosae ... 562
7.5.4 Properties and Function ... 567
References ... 569

7.6 Biflavonoids and Proanthocyanidins ... 571
R.W. Hemingway

7.6.1 Introduction ... 571
7.6.2 Biflavonoids ... 571
7.6.2.1 Structural Variations of Biflavonoids ... 572
7.6.2.2 Distribution of Biflavonoids ... 577
7.6.2.3 Significant Properties of Biflavonoids ... 583
7.6.3 Proanthocyanidins ... 584
7.6.3.1 Flavan-3-ols ... 586
7.6.3.1.1 Structure of Flavan-3-ols ... 587
7.6.3.1.2 Distribution of Flavan-3-ols ... 589
7.6.3.1.3 Reactions of Flavan-3-ols ... 594
7.6.3.2 Flavan-3,4-diols ... 602
7.6.3.2.1 Structure of Flavan-3,4-diols ... 604
7.6.3.2.2 Distribution of Flavan-3,4-diols ... 608
7.6.3.2.3 Reactions of Flavan-3,4-diols ... 608
7.6.3.3 Oligomeric Proanthocyanidins ... 611
7.6.3.3.1 Structure and Distribution of Oligomeric Proanthocyanidins ... 612
7.6.3.3.1.1 Proquibourtinidins ... 613
7.6.3.3.1.2 Profisetinidins ... 613
7.6.3.3.1.3 Prorobinetinidins ... 619
7.6.3.3.1.4 Proteracacidins and Promelacacidins ... 619
7.6.3.3.1.5 Propelargonidins ... 619
7.6.3.3.1.6 Procyanidins ... 621

7.6.3.3.1.7 Prodelphinidins 629
7.6.3.3.2 Reactions of Oligomeric Proanthocyanidins 631
References 636

7.7 Condensed Tannins 651
L. J. PORTER

7.7.1 Introduction 651
7.7.2 Structure and Properties 652
7.7.2.1 Isolation and Purification 652
7.7.2.2 Elucidation of the Structure of Type 1 Proanthocyanidin Polymers 653
7.7.2.3 Structure of Type 2 Proanthocyanidin Polymers 660
7.7.2.4 Molecular Weight Distribution 661
7.7.2.5 Conformation and Solution Properties 664
7.7.2.6 Complexation 667
7.7.3 Distribution in Plants 669
7.7.3.1 Chemotaxonomic and Phylogenetic Significance 669
7.7.3.2 Distribution and Structural Variations within Plants 675
7.7.4 Metabolism 676
7.7.4.1 Biosynthesis 676
7.7.4.2 Seasonal Variation and Fate in Senescent Tissues 681
7.7.5 Role in Plants 682
7.7.5.1 Resistance to Insects 682
7.7.5.2 Resistance to Decay Fungi 683
7.7.5.3 Allelopathic Relationships 684
References 685

Contents Volume II

Contents Volume I ... XV

8 Isoprenoids ... 691

8.1 Terpenoids ... 691
SUKH DEV

8.1.1 Introduction ... 691
8.1.1.1 Nomenclature ... 692
8.1.1.2 Biosynthesis ... 692
8.1.2 Occurrence in Woody Plants ... 695
8.1.3 Classes: Distribution and Structural Types ... 696
8.1.3.1 Hemiterpenoids ... 696
8.1.3.2 Monoterpenoids ... 697
8.1.3.2.1 Distribution ... 697
8.1.3.2.2 Structural Types ... 698
8.1.3.2.3 Tropolones ... 711
8.1.3.3 Sesquiterpenoids ... 711
8.1.3.3.1 Distribution ... 711
8.1.3.3.2 Structural Types ... 712
8.1.3.4 Diterpenoids ... 746
8.1.3.4.1 Distribution ... 746
8.1.3.4.2 Structural Types ... 748
8.1.3.5 Sesterterpenoids ... 766
8.1.3.6 Non-Steroidal Triterpenoids ... 767
8.1.3.6.1 Distribution ... 767
8.1.3.6.2 Structural Types ... 768
8.1.3.7 Tetraterpenoids: Carotenoids ... 785
8.1.3.8 Polyterpenoids: Polyprenols ... 787
8.1.3.9 Meroterpenoids ... 788
8.1.4 Biological Role ... 789
References ... 791

8.2 Steroids ... 808
W. R. NES

8.2.1 Introduction ... 808
8.2.1.1 Meaning of the Terms "Steroid" and "Sterol" ... 808
8.2.1.2 Nomenclature ... 810
8.2.1.3 Biosynthesis of Basic Steroidal Structure ... 812
8.2.2 Sterols ... 815
8.2.2.1 Names, Structures, and Organismic Relationships ... 815
8.2.2.2 Identification of Sterols ... 818
8.2.2.3 Occurrence in Wood and Bark ... 819
8.2.2.4 Biosynthetic Origin of Individual Sterols ... 826

8.2.2.5 Function of Sterols ... 830
8.2.2.6 Phyletic and Phylogenetic Relationships ... 832
8.2.2.7 Industrial Utilization of Tree Sterols ... 833
8.2.3 Esters ... 834
8.2.4 Glycosides ... 835
8.2.5 Spiroketals (Saponins) ... 835
8.2.6 Ecdysteroids ... 836
8.2.6.1 Names, Structures, and Occurrence ... 836
8.2.6.2 Function of Plant Ecdysteroids ... 837
8.2.7 Cardiac Glycosides ... 837
References ... 839

9 The Influence of Extractives on Wood Properties and Utilization ... 843

9.1 Contribution of Extractives to Wood Characteristics ... 843
H. IMAMURA

9.1.1 Introduction ... 843
9.1.2 Color in Wood ... 844
9.1.2.1 Chemical Structure and Color ... 844
9.1.2.2 Color of Wood ... 844
9.1.2.3 Pigments Occurring in Wood ... 851
9.1.3 Odor in Wood ... 851
9.1.3.1 Volatile Components ... 851
9.1.3.2 Fragrant Components ... 852
9.1.3.3 Foul-Smelling Components ... 853
9.1.3.4 Removal of Foul Odors ... 855
9.1.3.5 Insect Attractants ... 855
9.1.4 Physical Properties ... 856
9.1.4.1 Wood Density and Strength ... 856
9.1.4.2 Other Physical Properties ... 859
References ... 859

9.2 Role of Wood Exudates and Extractives in Protecting Wood from Decay ... 861
J.H. HART

9.2.1 Introduction ... 861
9.2.1.1 Decay ... 861
9.2.1.2 Decay-Causing Organisms and Their Effect on Wood Structure ... 861
9.2.2 How Trees Defend Themselves Against Decay ... 863
9.2.2.1 Role of Wounds ... 863
9.2.2.2 Toxic Heartwood Components ... 867
9.2.2.2.1 Formation of Antimicrobial Compounds ... 867
9.2.2.2.2 Chemical Nature of Antimicrobial Compounds ... 868
9.2.3 Evaluation of Decay Resistance ... 870
9.2.3.1 Isolation and Evaluation of Compounds ... 870
9.2.3.2 Physiology of Decay Inhibition ... 872
9.2.3.3 Variation in Decay Resistance ... 874
9.2.3.3.1 Variation Between Tree Species ... 874
9.2.3.3.2 Variation Between Individuals of the Same Species ... 875
9.2.3.3.3 Variation Within an Individual Tree ... 876
References ... 878

9.3 Effect of Extractives on Pulping 880
W.E. HILLIS and M. SUMIMOTO

9.3.1 Introduction 880
9.3.2 Pulping Processes 881
9.3.2.1 Mechanical 882
9.3.2.2 Semichemical 883
9.3.2.3 Chemical 884
9.3.3 Pulpwood Quality 885
9.3.3.1 Effect of Extractives on Pulp Yield 885
9.3.3.2 Effect of Storage 885
9.3.4 Increased Consumption of Pulping Liquors 886
9.3.5 Effect on Pulping Processes 887
9.3.5.1 Reduced Penetrability of Liquors 887
9.3.5.2 Reduced Lignin Solubility 888
9.3.6 Effect on Equipment during Pulping 888
9.3.6.1 Wear and Corrosion 888
9.3.6.2 Blockage and Deposits 889
9.3.7 Pulp Properties 895
9.3.7.1 Color Changes Arising during Pulping and Bleaching 895
9.3.7.1.1 Mechanical Pulps 895
9.3.7.1.2 Chemical Pulps 897
9.3.7.2 Speck Formation and Pitch Problems during Pulping and Bleaching 901
9.3.7.3 Wettability 910
9.3.7.4 Sticking to Press Rolls 910
9.3.8 Spent Liquor Recovery 910
9.3.8.1 Concentration and Burning Difficulties 910
9.3.8.2 Foaming during Concentration and Oxidation 912
9.3.8.3 By-Product Recovery 912
9.3.9 Observations 912
References 914

9.4 Effect of Extractives on the Utilization of Wood 920
T. YOSHIMOTO

9.4.1 Introduction 920
9.4.2 Inhibition of Resin and Glue Curing by Extractives 920
9.4.2.1 Phenolic Resin Adhesives 922
9.4.2.2 Amino Resin Adhesives 923
9.4.3 Inhibition of Cement Hardening by Extractives 923
9.4.3.1 Effect of Extractives on Cement Hardening 923
9.4.3.2 Effect of Light Exposure on Wood Panels Used with Cement 924
9.4.4 Color Change by Light 925
9.4.4.1 Color Changes Upon Exposure to Light 925
9.4.4.2 Color Changes by Other Agents 927
References 929

9.5 Health Hazards Associated with Extractives 931
E.P. SWAN

9.5.1 Introduction 931
9.5.2 Toxic Extractives 932

9.5.2.1 Alkaloids and Amino Acids ... 932
9.5.2.2 Saponins and Glycosides ... 933
9.5.2.3 Quinones ... 933
9.5.2.4 Phenolics ... 933
9.5.2.5 Terpenes ... 933
9.5.3 Allergenic Extractives ... 934
9.5.3.1 Quinones ... 935
9.5.3.2 Alkyl Phenols ... 935
9.5.3.3 Terpenes ... 936
9.5.3.4 Phenolics ... 937
9.5.3.5 Other Allergenic Extractives ... 938
9.5.4 Carcinogenic Extractives ... 938
9.5.4.1 Early Work ... 938
9.5.4.2 Hausen's Contributions ... 938
9.5.4.3 Tannin ... 939
9.5.4.4 Other Carcinogenic Extractives ... 941
9.5.5 Hygiene and Safety ... 941
9.5.6 Discussion and Further Research ... 943
9.5.6.1 Discussion ... 943
9.5.6.2 Future Research ... 944
References ... 946

10 The Utilization of Wood Extractives ... 953

10.1 Naval Stores ... 953
D.F. Zinkel

10.1.1 Introduction ... 953
10.1.2 Naval Stores Sources ... 954
10.1.2.1 Gum Naval Stores ... 955
10.1.2.2 Wood Naval Stores ... 956
10.1.2.3 Sulfate (Kraft) Naval Stores ... 957
10.1.2.4 Potential New Sources ... 958
10.1.3 Turpentine ... 959
10.1.3.1 Pine Oil ... 961
10.1.3.2 Polyterpene Resins ... 961
10.1.3.3 Flavors and Fragrances ... 962
10.1.3.4 Insecticides ... 965
10.1.3.5 Miscellaneous Uses ... 966
10.1.4 Rosin ... 968
10.1.4.1 Paper Size ... 969
10.1.4.2 Polymerization Emulsifiers ... 970
10.1.4.3 Adhesives ... 970
10.1.4.4 Inks ... 971
10.1.4.5 Other Market Areas ... 972
10.1.5 Fatty Acids ... 973
10.1.5.1 Intermediate Chemicals ... 974
10.1.5.2 Protective Coatings ... 974
10.1.5.3 Other Uses ... 975
10.1.6 Miscellaneous Products ... 975
10.1.7 The Future for Naval Stores ... 977
References ... 977

10.2 Gums 978
J. N. BEMILLER

10.2.1 Introduction 978
10.2.2 Larch Arabinogalactan 979
10.2.2.1 Source 979
10.2.2.2 Production 980
10.2.2.3 Purification 980
10.2.2.4 Properties 981
10.2.2.5 Uses 982
10.2.3 Gum Arabic (Gum Acacia, Acacia Gum) 983
10.2.3.1 Source, Production, and Purification 983
10.2.3.2 Properties 983
10.2.3.3 Uses 984
10.2.4 Gum Karaya 984
10.2.5 Gum Tragacanth 985
10.2.6 Gum Ghatti 986
10.2.7 Predictions 986
References 987

10.3 Significance of the Condensed Tannins 988
L. J. PORTER and R. W. HEMINGWAY

10.3.1 Introduction 988
10.3.2 Tannins as Human and Animal Nutrition Factors 988
10.3.3 Pharmacological and Physiological Properties 991
10.3.4 Tannins as Insect, Mollusc, Bacterial, and Fungal Control Factors 992
10.3.5 Leather Tannage 993
10.3.5.1 Mechanisms of Vegetable Tanning 994
10.3.5.2 Vegetable Tanning Processes 997
10.3.5.3 Treatment of Vegetable Tanning Spent Liquors 1000
10.3.6 Condensed Tannins in Wood Adhesives 1002
10.3.6.1 Wattle Tannin-Based Particleboard Adhesives 1003
10.3.6.2 Wattle Tannin-Based Plywood Adhesives 1005
10.3.6.3 Wattle Tannin-Based Laminating Adhesives 1006
10.3.6.4 Other Wattle Tannin-Based Adhesives 1007
10.3.6.5 Conifer Bark and Related Tannins as Particleboard Adhesives ... 1008
10.3.6.6 Conifer Bark and Related Tannins as Plywood Adhesives 1011
10.3.6.7 Conifer Bark and Related Tannins in Cold-Setting Phenolic Resins 1014
10.3.7 Specialty Polymer Applications 1016
References 1018

10.4 Rubber, Gutta, and Chicle 1028
F. W. BARLOW

10.4.1 Introduction 1028
10.4.2 Historical Development of Natural Rubber Production 1029
10.4.2.1 *Hevea brasiliensis* 1029
10.4.2.2 Guayule 1031
10.4.3 Natural Rubber Production Processes 1032

10.4.3.1 Rubber Tree Growing 1032
10.4.3.2 Latex Collection 1034
10.4.3.3 Production of Latex Concentrate 1034
10.4.3.4 Production of Dry Rubber 1035
10.4.4 Packing and Market Grades 1037
10.4.4.1 Packing 1037
10.4.4.2 Grading 1037
10.4.5 Properties 1038
10.4.5.1 Natural Rubber 1038
10.4.5.2 Modified Natural Rubber 1040
10.4.5.2.1 Deproteinized Rubber 1040
10.4.5.2.2 Depolymerized Rubber 1040
10.4.5.2.3 Peptized Rubber 1040
10.4.5.2.4 Oil Extended Natural Rubber (OENR) 1041
10.4.5.2.5 MG Rubbers 1041
10.4.5.2.6 SP Rubbers 1041
10.4.6 Natural Rubber Utilization 1041
10.4.6.1 Vulcanization 1042
10.4.6.2 Transportation Items 1043
10.4.6.3 Mechanical Rubber Goods 1044
10.4.6.4 Footwear 1044
10.4.6.5 Miscellaneous Uses 1045
10.4.6.6 Health and Safety Factors 1045
10.4.7 Natural Rubber Economy 1046
10.4.8 Gutta Percha and Balata 1048
10.4.8.1 Production 1048
10.4.8.2 Properties and Uses 1049
10.4.9 Chicle 1049
References 1050

10.5 Other Extractives and Chemical Intermediates 1051
E. P. Swan

10.5.1 Introduction 1051
10.5.2 Conifer Extractives Utilization 1051
10.5.2.1 Balsams, Copals, Amber, and Other Products 1051
10.5.2.2 Cedar Wood Oils 1052
10.5.2.3 Non-Commercial Extractives 1053
10.5.2.3.1 Conidendrin 1053
10.5.2.3.2 Dihydroquercetin 1054
10.5.2.3.3 Juvabione and Related Insect Hormones 1054
10.5.2.3.4 Occidentalol 1054
10.5.2.3.5 Plicatic Acid 1054
10.5.2.3.6 Thujaplicins, Thujic Acid, and Methyl Esters 1055
10.5.2.3.7 Waxes 1055
10.5.3 Hardwood Extractives Utilization 1056
10.5.4 Prospects 1056
References 1058

10.6 Pharmacologically Active Metabolites 1059
C.W.W. BEECHER, N.R. FARNSWORTH, and C. GYLLENHAAL

10.6.1 Introduction 1059
10.6.2 Sources of Information 1059
10.6.3 Currently Used Drugs Produced in Wood 1061
10.6.3.1 Alkaloids 1061
10.6.3.1.1 Tropane 1061
10.6.3.1.2 Quinolizidine 1061
10.6.3.1.3 Phenylalanine Derivatives 1062
10.6.3.1.3.1 Simple Tyramine Derivatives 1062
10.6.3.1.3.2 Protoberberine 1062
10.6.3.1.3.3 Phthalideisoquinoline 1062
10.6.3.1.3.4 Benzo(c)phenanthridine 1062
10.6.3.1.3.5 Ipecac 1062
10.6.3.1.4 Tryptophane Derivatives 1062
10.6.3.1.4.1 Indole 1062
10.6.3.1.4.2 Quinoline 1063
10.6.3.2 Quinoids 1063
10.6.3.3 Lignans 1063
10.6.3.4 Triterpenes 1063
10.6.3.5 Pyrones 1063
10.6.3.6 Coumarins 1063
10.6.4 Potential Drugs Derived from Secondary Metabolites of Wood . 1063
10.6.4.1 Alkaloids 1064
10.6.4.1.1 Tropane 1064
10.6.4.1.2 Isoquinoline 1064
10.6.4.1.3 Indole Alkaloids 1065
10.6.4.1.4 Ansamacrolides 1065
10.6.4.1.5 Quinazoline 1065
10.6.4.1.6 Pyrazine Derivatives 1066
10.6.4.2 Quinoids 1066
10.6.4.3 Lignans 1066
10.6.4.4 Diterpenes 1068
10.6.4.5 Triterpenes 1069
10.6.4.6 Flavonoids 1069
10.6.5 Summary and Conclusions 1069
References 1120

11 The Future of Wood Extractives 1165
H.L. HERGERT

11.1 Introduction 1165

11.2 Requirements for Future Wood Extractives Ventures 1166
11.2.1 Low Investment Risk 1166
11.2.2 Good Sales Potential 1167
11.2.3 Inexpensive Raw Material 1168
11.2.4 Shared Capital Expense 1169
11.2.5 National Priority 1169
11.2.6 Realistic Research, Development, and Engineering 1169

11.3 Prospects for Existing Extractives-Based Industries 1170
- 11.3.1 Natural Rubber 1170
- 11.3.2 Rosin and Terpenes from Pine 1172
- 11.3.3 Carbohydrate Gums 1174
- 11.3.4 Tannins 1175

11.4 Failed Wood Extractives Ventures 1176

11.5 Future Directions for Industrially Oriented Extractives Research 1178
- 11.5.1 Control of Extractives Deposition 1179
- 11.5.2 Manipulation of Wood Growth by Chemicals 1180
- 11.5.3 New Techniques for Extractives Isolation 1182

11.6 Areas of Needed Basic Research 1184
- 11.6.1 Cambial Constituents: Growth Regulators 1185
- 11.6.2 Root Constituents: Role of Mycorrhizae 1186
- 11.6.3 Environmental Relationships 1186
- 11.6.4 Pharmacologically Active Compounds 1188
- 11.6.5 Phenolic Polymers 1189
- 11.6.6 Sites and Control Mechanisms of Biosynthesis 1190

11.7 Conclusions 1192

References 1193

Index of Plant Genera and Species 1197

Organic Compounds Index 1220

List of Contributors: Biographical Sketches

NORIO AIMI

Associate Professor, Faculty of Pharmaceutical Sciences, Chiba University, Chiba, Japan.
Born, 1938. Educated at Tokyo University.
Former positions include: Visiting Fellow, National Institutes of Health (1968–1969). Lecturer, Faculty of Pharmaceutical Sciences, Chiba University (1967–1969). Research Associate, Faculty of Pharmaceutical Sciences, Tokyo University (1965–1967).

FRED BARLOW

Consultant/Technical Writer, Stow, Ohio, U.S.A.
Educated at the University of Toronto.
Former positions include: Manager of Technical Service Laboratory, Ashland Chemical Company. Technical Service Specialist, Malaysian Rubber Bureau, U.S.A. Initiator of testing program for natural-rubber-treaded tires for Canadian Safety Authority and Malaysian Rubber Producers Research Association Laboratories. Author of numerous papers and a book.
Member, American Chemical Society Rubber Division; Northeast Ohio Rubber Group; American Society of Testing and Materials; and American Society for Quality Control.
Research interests include all aspects of rubber technology; statistical quality control; rubber physics; training.

CHRISTOPHER WILLIAM WARD BEECHER

Assistant Professor of Medicinal Chemistry and Pharmacognosy, College of Pharmacy, University of Illinois at Chicago, Chicago, Illinois, U.S.A.
Born, 1948. Educated at the University of Connecticut.
Member, Scientific Advisory Board, NAPRALERT; MEDFLOR; METMAP.
Research interests include genetic expression of secondary metabolism in plant tissue cultures; NAPRALERT database; plant taxonomy and morphogenesis; ethnobotany.

JAMES N. BEMILLER

Director, Whistler Center for Carbohydrate Research, and Professor, Department of Food Science, Purdue University, Lafayette, Indiana, U.S.A.
Educated at Purdue University.
Former positions include: Chairman, Department of Medical Biochemistry, Southern Illinois University, Carbondale (1979–1986); Professor, School of Medicine (1971–1976); Dean, College of Science (1976–1979); Chairman, Department of Chemistry and Biochemistry (1966–1971). Author of 17 books and 120 papers.
President, American Institute of Chemists. Past President, International Carbohydrate Organization. President, U.S. Advisory Committee for International Carbohydrate Symposia, Inc. General Chaiman, XIII International Carbohydrate Symposium.
Research interests include carbohydrates including gums and starches; methods in carbohydrate chemistry; food chemistry.

Reimar C. Breuning

Assistant Professor of Chemistry, Department of Chemistry, College of Chemistry, University of Hawaii, Honolulu, Hawaii, U.S.A.

Born 1948. Educated at the Ludwigs-Maximilian University, Munich.

Former positions include: Research Associate, Columbia University (1981–1985). Postdoctoral Fellow, Nagoya University (1979–1981).

Member, Gesellschaft Deutscher Chemiker; Phytochemical Society of Europe; American Chemical Society; American Association for the Advancement of Science.

Fadila Derguini

Associate Research Scientist, Department of Chemistry, Columbia University, New York, U.S.A.

Educated at the Université Pierre et Marie Curie, Paris.

Former positions include Chargée de Recherche, Centre National Recherche Scientifique, Ecole Normale Superieure, Paris (1982–1984) (on leave).

Research interests include organic synthesis and methodology; isolation and structural and bioorganic studies of retinal proteins.

Don J. Durzan

Professor, Department of Environmental Horticulture, University of California, Davis, California, U.S.A.

Educated at Cornell University.

Former positions include: Chair, Department of Pomology, University of California, Davis (1981–1987). Chair, Department of Special Studies/Senior Research Associate, Institute of Paper Chemistry, Appleton, Wisconsin (1977–1981). Member, Executive Committee, Biological Council of Canada (1975–1977). Senior Advisor to the Assistant Deputy Minister, Environmental Management Service, Ottawa (1975–1977). Head, Biochemistry Unit, Forest Ecology Institute, Environment Canada (1965–1975).

Research interests include biotechnologies based on growth and development of woody perennials; nitrogen metabolism and secondary products; expression of elite genetic traits in clonally propagated trees; mass propagation of conifer hybrids using cell and tissue culture methods.

Karl E. Espeli

Associate Entomologist, Department of Entomology, University of Georgia, Athens, Georgia, U.S.A.

Born, 1946. Educated at the University of Wisconsin, Madison.

Former positions include: Adjunct Associate Professor, Department of Entomology, Washington State University (1985–1986). Researcher, Institute of Biological Chemistry, Washington State University (1976–1985). Visiting Scientist, Protein Chemistry Department, Imperial Cancer Research Fund, London (1974–1976). Postdoctoral Fellow, Division of Biology, California Institute of Technology (1972–1974).

Damon Runyon Memorial Fellowship for Cancer Research (1972).

Research interests include natural products chemistry of insect/plant interactions.

Norman R. Farnsworth

Research Professor of Pharmacognosy, Department of Pharmacognosy and Pharmacology, and Director, Program for Collaborative Research in the Pharmaceutical Sciences, University of Illinois at Chicago, Chicago, Illinois, U.S.A.

Born, 1930. Educated at the University of Pittsburgh.

Former positions include: Head, Department of Pharmacognosy and Pharmacology, University of Illinois (1970–1982). Professor of Pharmacognosy, University of Pittsburgh (to 1970). Initiator, NAPRALERT database on natural products (1975). Member, World Health Organization Expert Advisory Panel on Traditional Medicine (1980–). Member, WHO Special Programme of Research Development and Research Training in Human Reproduction Steering Committee of Task Force on Plants for Fertility Regulation (1976–1987). Author of 400 papers.

Research Achievement Award in Natural Products, Academy of Pharmaceutical Sciences. Honorary doctorates, University of Paris; Uppsala University. Honorary memberships, American Society of Pharmacognosy, French Pharmacognosy Society.

Research interests include folklore medicine; evaluation, isolation and structure elucidation of biologically active natural products; drug development; medicinal plants.

Otto Richard Gottlieb

Professor of Organic Chemistry, Instituto de Quimica, Universidade de São Paulo, São Paulo, Brazil.

Educated at the Universidade Federal de Rio de Janeiro.

Former positions include: Professor, Researcher, Belo Horizonte, Basilia, Manaus, Recife, and Rio de Janiero. Natural products researcher in industry and government. Author of 500 papers and three books.

Anisio Teixeira Prize (1986). Science for Amazonia Medal (1978). Fritz Priegl Prize (1977). Member, Academia de Ciencias da America Latina; Academia Brasileira de Ciencias; Academia de Ciencias do Estado de São Paulo. Honorary doctorates, Universidade Federal de Alagoas; Universidade Federal da Paraiba; Universität Hamburg.

Research interests include phytochemistry of Brazilian plants; evolution of secondary metabolites in plant; chemical variability under environmental pressure; systematic approach to medicinal compounds from plants; chemical nomenclature.

Charlotte Gyllenhaal

Research Associate, Program for Collaborative Research in the Pharmaceutical Sciences, University of Illinois at Chicago, Chicago, Illinois, U.S.A.

Born, 1948. Educated at the University of Alabama.

Member, Society for Economic Botany.

Research interests include maintaining NAPRALERT natural products database, abstracting ethnomedical data and researching taxonomic and nomenclatural status of organisms; editing International Traditional Medicine Newsletter for the World Health Organization; supervising computer operations for a plant collection project in Southeast Asia for the National Cancer Institute.

Jeffrey B. Harborne

Head, Department of Botany, University of Reading, Reading, UK.

Born, 1928. Educated at Wycliffe College, University of Bristol.

Former positions include: Personal Professor, University of Reading (1976–1986). Visiting Professor, University of California, Santa Barbara (1977). Visiting Professor, University of Texas at Austin (1976). Research Fellow, Reader, University of Reading (1968–1976). Research Fellow, University of Liverpool (1965–1968). Biochemist, John Innes Institute (1955–1965). Author of 230 papers and 26 books.

First Silver Medal, Phytochemical Society of Europe (1986). Gold Medal in Botany, Linnean Society of London (1985). Vice Chair (1974–1976), Honorary Chair (1972–1974), Secretary (1967–1972), Phytochemical Society of Europe.

Research interests include ecological biochemistry; phytochemical methodology; plant chemosystematics; flavonoids.

JOHN HENDERSON HART

Professor, Departments of Botany & Plant Pathology and Forestry, Michigan State University, East Lansing, Michigan, U.S.A.

Born, 1936. Educated at Iowa State University, Ames.

Former positions include: Assistant and Associate Professor, Michigan State University (1963–1975). Visiting Scientist, U.S.D.A., Forest Service, Fort Collins, Colorado (1985). Visiting Scientist, Pacific Forest Research Centre, Victoria, British Columbia (1977–1978). Visiting Research Fellow, CSIRO, Melbourne, Australia (1970–1971).

EDWIN HASLAM

Head, Department of Chemistry, University of Sheffield, Sheffield, UK.

Educated at the University of Sheffield and Canterbury University.

Former positions include: Personal Chair, University of Sheffield (1980–1985). Visiting Professor, University of the South Pacific, Fiji (1984). Lecturer and Reader, Department of Chemistry, University of Sheffield (1957–1980). Hugh Kelly Senior Research Fellow, Rhodes University, Grahamstown (1975). Ramsay Memorial Fellow, Emmanuel College, University of Cambridge (1955–1957). Author/editor of 100 papers and four books.

Tate and Lyle Award, Phytochemical Society of Europe (1977).

Research interests include chemistry of vegetable tannins; the shikimate pathway; metabolic chemistry.

RICHARD W. HEMINGWAY

Project Leader for Processing of Southern Woods, Forest Service, U.S. Department of Agriculture, Southern Forest Experiment Station, Pineville, Louisiana, U.S.A.

Educated at the University of Michigan, Ann Arbor

Former positions include: Principle Research Scientist, Commonwealth Scientific and Industrial Research Organization, Melbourne, Australia (1967–1972). Visiting Scientist, Division of Scientific and Industrial Research, Wellington, New Zealand (1980).

Research interests include chemistry and utilization of condensed tannins, primarily from North American tree barks and agricultural residues.

HERBERT L. HERGERT

Senior Scientist, Repap Technologies, Philadelphia, Pennsylvania, U.S.A.

Educated at Oregon State University, Corvallis.

Former positions include: Vice President for Technical Marketing, ITT Rayonier (1980–1987); Vice President of Research (1973–1980); Research Chemist (1954–1973).

Research interests include pulping by-products and papermaking.

W. E. HILLIS

Honorary Research Fellow (retired), Commonwealth Scientific and Industrial Research Organization, and Senior Associate, Forestry Section, Melbourne University, Clayton, Victoria, Australia.

Educated at Melbourne University.

Former positions include: Chief Research Scientist, Division of Chemical & Wood Technology, CSIRO. Cecil Green Visiting Distinguished Professor, University of British Columbia (1983). Visiting Fellow, Department of Forestry, The Australian National University. Author of 170 papers and five books.

Stanley A. Clarke Medal (1986). Honorary Member (1986), Executive Board Member (1977–1983), International Union of Forestry Research Organizations. President (1978–1982), Fellow (1970), International Academy of Wood Science. Honorary Advisor, Research Institute of Chemical Processing and Utilization of Forest Products, Chinese Academy of Forestry, Nanjing, P.R. China.

Research interests include wood extractives and exudates and their effects on wood utilization; eucalypts for wood production; heartwood growth and conversion phenomena.

HIROYUKI IMAMURA

Professor of Wood Chemistry, Department of Wood Products, School of Agriculture, Kyushu University, Fukuoka City, Japan.

Born, 1927. Educated at the Kyushu University.

Former positions include: Professor of Tree Biochemistry, Gifu University, Gifu (1968–1982). Research Officer, Government Forest Experiment Station, Tokyo (1950–1968).

P. E. KOLATTUKUDY

Director, Biotechnology Center; Assistant Director, Ohio Agricultural Research and Development Center; and Professor, Departments of Biochemistry, Botany, and Plant Pathology, Ohio State University, Columbus, Ohio, U.S.A.

Born, 1937. Educated at Oregon State University, Corvallis.

Former positions include: Fellow and Director, Institute of Biological Chemistry, Professor of Biochemistry, Washington State University (1980–1986); Agricultural Chemist and Professor of Biochemistry (1973–1980); Associate Agricultural Chemist and Associate Professor (1969–1973). Assistant Biochemist, Connecticut Agricultural Experiment Station (1964–1969). Author of 234 publications.

WSU First President's Faculty Excellence Award. WSU Distinguished Faculty Address. Golden Apple Award, State of Washington. Outstanding Paper Award, American Oil Chemist Society. Member, Physiological Chemistry Study Section, National Institutes of Health (1984–1988).

Research interests include biochemistry; biophysics; plant physiology; molecular plant-microbe interactions.

LENNART N. LUNDGREN

Post-doctoral researcher, Organic Chemistry Division, Department of Chemistry, Swedish University of Agricultural Sciences, Uppsala, Sweden.

Born, 1942. Educated at the Swedish University of Agricultural Sciences.

Author of 15 papers.

Research interests include analytical and structural studies of low molecular natural products – mainly phenolic compounds; analytical and synthetic studies on herbicides.

KOJI NAKANISHI

Director, Suntory Institute for Bioorganic Research, Osaka, Japan; and Centennial Professor, Department of Chemistry, Columbia University, New York, U.S.A.

Born, 1925. Educated at Nagoya University.

Former positions include: Professor, Department of Chemistry, Columbia University (1969–1980). Director of Research, International Centre for Insect Physiology and Ecology, Kenya (1969–1977). Professor, Departments of Chemistry, Tohoku University, Sendai (1963–1969); Tokyo Kyoiku University (1958–1963). Author/editor of four books.

Paul Karrer Gold Medal, University of Zurich (1986). Alcon Award in Ophthalmology (1986). Research Achievement, American Society of Pharmacognosy (1985). Remsen Award, American Chemical Society (1981). Centenary Medal, British Chemical Society (1979). Chemical Society of Japan Award (1979). Ernst Günther Award, American Chemical Society (1978).

Research interests include isolation and structural and bioorganic studies of bioactive compounds; retinal proteins; spectroscopy.

WILLIAM R. NES (deceased, 1989)

W. L. Obold Professor of Biological Sciences, and Director, Institute for Population Studies, Drexel University, Philadelphia, Pennsylvania, U.S.A.

Born, 1926. Educated at the University of Virginia, Charlottesville.

Former positions included: Professor of Chemistry and Pharmaceutical Chemistry, University of Mississippi (1964–1967). Director, Steroid Training Program, Worcester Foundation for Experimental Biology, and Associate Professor of Biochemistry, Clark University (1958–1964). Staff member, National Institute of Arthritis and Metabolic Diseases (1951–1958). Post-graduate Fellow, Princeton University; University of Heidelberg; University of Wales. Fellow, Mayo Foundation (1950–1951).

Research Achievement Award, Drexel University. Legion of Merit Chapel of the Four Chaplains. Anna Fuller Award. Ad hoc member, Endocrinology Study Section, NIH (1987). Member, Physiological Chemistry Study Section, National Institutes of Health (1979–1981 and 1984–1987); Steroid Drug Panel, U.S. Food and Drug Administration (1974); and Metabolic Biology Panel, National Science Foundation (1966–1969). Author of 125 books and articles.

Research interests included stereochemical principles involved in biochemical pathways, enzymological mechanisms and functions of steroids now known to operate throughout nature.

TORBJÖRN NORIN

Professor and Head, Department of Organic Chemistry, Royal Institute of Technology, Stockholm, Sweden.

Born, 1933. Educated at the Royal Institute of Technology.

Former positions include: Head, Wood Chemistry Laboratory, and Professor, Wood Chemistry, Royal Institute of Technology (1966–1969). Director of Research, and Head, Department of Chemistry, Swedish Forest Products Laboratory (1966–1972). Acting Professor, Organic Chemistry, University of Umeå (1965). Assistant Professor, Organic Chemistry (1964). Post-doctoral Fellow, Oxford University (1961–1962). Author of 150 papers and two books.

Academician, Royal Swedish Academy of Engineering Sciences (1987). University of Helsinki Medal (1980). Norblad-Ekstrands Medal, Swedish Chemical Society (1966).

Member, Advisory Board for Ecology Swedish Natural Science Research Council (1979–1983). Scientific Advisor, Court of Appeal (Svea Hovratt) (1981–). Member, Advisory Board of Chemistry, Swedish Natural Science Research Council (1977–1983). Member, Board of Directors, Nobel Chemicals AB (1984–).

Research interests include synthetic organic chemistry, especially asymmetric synthesis and the use of enzymes in organic synthesis, with special reference to biologically active compounds such as insect pheromones and related chemical signals; natural products chemistry; wood chemistry.

YOSHIMOTO OHTA

Senior Researcher, Suntory Institute for Bioorganic Research, Osaka, Japan.

Educated at Osaka University.

Former positions include: Post-doctoral Fellow, University of Washington, Pullman (1972–1974). Research Chemist, The Institute of Food Chemistry (now Suntory Institute for Bioorganic Research) (1962–1972).

LAWRENCE JAMES PORTER

Scientist, Chemistry Division, Department of Scientific and Industrial Research, Petone, New Zealand.

Educated at Victoria University, Wellington.

Former positions include: Research Fellow, University of Sheffield (1975–1976). Visiting Scientist, U.S.D.A. Western Regional Research Center, Berkeley, California.

Nuffield Commonwealth Fellowship (1975).

Research interests include plant phenolic compounds; pioneering work on flavonoid chemistry of primitive plants; chemistry of polymeric condensed tannins; biological properties of condensed tannins.

SHIN-ICHIRO SAKAI

Professor of Pharmaceutical Chemistry, Faculty of Pharmaceutical Sciences, Chiba University, Chiba, Japan.

Born, 1930. Educated at University of Tokyo.

Former positions include Assistant Professor, Department of Pharmaceutical Sciences, Chiba University (1960–1962).

JOHN SNYDER

Assistant Professor, Department of Chemistry, Boston University, Boston, Massachussetts, U.S.A.

Born, 1951. Educated at the University of Chicago.

Former positions include: Research Fellow, Community for Scholarly Communications, P.R. China, National Academy of Sciences (1983). Post-doctoral Fellow, Department of Chemistry, Columbia University (1979–1983).

Dreyfus Young Investigator Award, Henry and Camille Dreyfus Foundation (1983). Member, American Chemical Society; American Society of Pharmacognosy.

Research interests include natural products chemistry; isolation and structural determination; synthesis.

Sukh Dev

INSA Research Professor, India Institute of Technology, New Dehli, India.
Born, 1924. Educated at the Indian Institute of Science, Bangalore.
Former positions include: Research Director, Malti-Chem Research Centre, Nandesari, Vadodara (1974–1988). Lecturer, Indian Institute of Science (1953). Head, Division of Organic Chemistry (Natural Products), National Chemical Laboratory, Pune (1960). Visiting Professor, Stevens Institute of Technology (1968); University of Georgia (1969); University of Oklahoma (1970–1971). Author of 280 publications, several books, and 50 patents.
Ernst Günther Award, American Chemical Society (1980). President, Indian Chemical Society (1978–1979). President, Organic Chemistry Division Committee, IUPAC (1987–1989). Fellow, Indian National Science Academy.
Research interests include natural products chemistry; non-benzenoid aromatic systems; organic reagents and reactions; photochemistry; new technology and product development.

Masashi Sumimoto

Professor, Department of Forest Products, Faculty of Agriculture, Kyushu University, Fukuoka, Japan.
Born, 1929. Educated at Osaka University.
Former positions include: Associate Professor, Kyushu University (1962–1976). Visiting Scientist, Royal Institute of Technology, Stockholm (1960–1961). Assistant Professor, Kagoshima University (1959–1960).
Research interests include behavior of wood extractives and lignin in pulping and bleaching.

Eric P. Swan

Research Scientist, Forintek Canada Corporation, and Adjunct Professor, Faculty of Forestry, University of British Columbia, Vancouver, British Columbia.
Born, 1931. Educated at McGill University, Montreal.
Former positions include: Staff Member, Research Division, Rayonier Canada (B.C.) Ltd., Vancouver, B.C.; Ohio State University, Columbus. Author of 50 publications.
Member, Chemical Institute of Canada; Phytochemical Society of North America; and Society of Sigma Xi.
Research interests include wood chemistry, especially of western red cedar bark, leaves and wood; heartwood formation especially of true firs; bark adhesives; HPLC analyses; mill identification of lumber.

Olof Theander

Professor of Organic Chemistry, Swedish University of Agricultural Sciences, Uppsala, Sweden.
Born, 1925. Educated at the Royal Institute of Technology, Stockholm.
Former positions include: Assistant Professor in Wood Chemistry, Royal Institute of Technology (1969–1972). Head, Carbohydrate Division, Swedish Forest Research Laboratory, Stockholm (1956–1972). Author of 240 articles and reviews.
Swedish Arrhenius Award in Chemistry (1974).
Research interests include analytical, structural, utilization, and synthetic studies of carbohydrates; phenols and isoprenoids in forestry and agricultural plants; and food products and modifications in cellulose- and food-technical processes.

KEIICHI YAMAGUCHI

Chief Chemist, DIC-Hercules Chemicals, Inc. Laboratory, Ichihara, Japan.
Born, 1949. Educated at Tokyo University.
Former positions include: Research Associate, Faculty of Pharmaceutical Sciences, Chiba University (1971–1986). Visiting Fellow, University of Virginia (1985–1986).

ETSUJI YAMANAKA

Manager, Research Department, Kotei Kasei Co., Ltd., Saitama, Japan.
Born, 1946. Educated at Tokyo University.
Former positions include: Research Associate, Faculty of Pharmaceutical Sciences, Chiba University (1978–1986). Visiting Fellow, University of Wisconsin-Milwaukee (1978–1987). Research Associate, Faculty of Pharmaceutical Sciences, Chiba University (1973–1976).

MASSAYOSHI YOSHIDA

Associate Professor, Institute of Chemistry, University of São Paulo, São Paulo, Brazil.
Born, 1939. Educated at São Paulo University.
Former positions include: Post-doctoral Fellow, National Research Council of Canada, Ottawa; Northeastern University, Boston. Author of 200 communications at congresses and 50 papers.
Research interests include chemical composition of Brazilian Lauraceae and Myristaceae, with emphasis on lignoids; phytochemistry; radiation chemistry; spectrometric techniques, such as ^{13}C NMR, ORD, and CD, in organic analysis.

TOMOTAKA YOSHIMOTO

Professor, Department of Forest Products, Faculty of Agriculture, University of Tokyo, Tokyo, Japan.
Educated at the University of Tokyo.

DUANE F. ZINKEL

Research Chemist, Forest Service, U.S. Department of Agriculture, Forest Products Laboratory, Madison, Wisconsin, U.S.A.
Educated at the University of Wisconsin, Madison.
Research interests include wood extractives, specializing in naval stores; analytical methods and natural product isolation and structure elucidation.

Chapter 1

Introduction and Historical Background

1.1 Historical Uses of Extractives and Exudates

W.E. HILLIS

1.1.1 Introduction

When woody plants once covered most of the earth's surface, they provided not only food, warmth, and shelter, but many of the needs of defense, medicines, culture, and simple pleasures. Some needs could be readily satisfied by using "minor forest products" such as resins for caulking boats, for use in torches, or for providing rigidity in the attachment of spear, arrow, or axe-heads to the shaft before the joints were finally lashed with fibrous material (24). Such resinous products could be provided by a number of sources and a few trials would reveal their effectiveness. Other materials, such as kinos or tannin extracts, when used as astringents for the treatment of diarrhea, would require a more sophisticated knowledge.

Most extracts and exudates were probably used locally and, as they were biodegradable, little evidence remains of their use; recorded information is sparse, widespread, and difficult to collect. In addition, their complex nature would largely prevent identifying their origin. With the change in needs and increasing performance requirements, many uses of minor forest products have decreased considerably in recent years. A history of their use could never be complete. Nevertheless, indications can still be obtained of the ways earlier people adapted or used extractives or exudates for their needs. This section attempts to summarize that situation with reference to major uses of the past.

1.1.2 Major Uses of Extractives and Exudates

Extractives are considered as the nonstructural, secondary metabolites of wood cells that can usually be removed with neutral solvents. Exudates are extracellular secondary metabolites that are formed by trees growing under certain conditions or after injury by fire, insect, fungal, or mechanical damage (18).

The absence of technical facilities for extraction and concentration of the extracts until recent times restricted the use of the nonstructural components of wood mainly to the exudates of trees. The exudates originate from injured cambium, or incised phloem, or sapwood and can dry to gelatinous or brittle solids on the bark. In many cases the exudate and the wood and bark extractives of the same species can have significantly different compositions, but they frequently served the same purpose. Consequently, they will be considered together in the following discussion. Although extractives in a concentrated form have been available to very few people, a variety of extractives has nevertheless been used. This situation may be exemplified by the volatilization of the perfumes in sandalwood by burning for ritual purposes, by soaking hides in crushed tissues rich in tannins, or by using woods naturally containing toxic components as arrowheads or in spears (3).

Extractives and exudates are considered together according to their use in various categories. The following examples have been arranged in an order that generally required an increasing sophistication of knowledge of the active components by the user. Although convenient for the present requirement, such an arrangement cannot provide precisely defined categories.

1.1.2.1 The Use of Durable Woods

There is little doubt the properties resulting from the presence of extractives in wood have been recognized for a long time. Teak (*Tectona grandis*, Verbenaceae) has long been valued for its very high resistance to fungal decay and insect and borer attack, and for its dimensional stability after drying. Teak wood was reputedly exported from India to Babylon and Yemen around 4000 B.C. (15). It was widely used for house-building in Southeast Asia and is still used for boat building. Teak is today the highest-priced timber in international trade. The export (in the 10th century B.C.) of the cedar wood from Lebanon (*Cedrus libani*, Pinaceae) to Jerusalem for building the temple of Solomon (II Chronicles 2, 16) was doubtless due not only to its size and medium density, but also partly to its durability (7).

The Chinese have appreciated the durability of wood for many centuries. The Imperial Palaces of Beijing were built almost entirely of *Persea nan-mu* (Lauraceae). Nan-mu was also frequently used for the buildings attached to the Ming tombs erected in the early 15th century. To satisfy the high demand for this wood, authorities found it necessary to transport logs for long distances, and in the year 1558, for example, 15712 logs (27 m in length and of large diameter) were taken from the Yunnan province, for over 3500 km, over mountain ranges and then by river and canal transport to Beijing (36). The present unchanged condition of those timbers indicate that careful observations must have previously been made to lead to the selection of this wood in a distant location. Similar buildings exist in Japan.

Other durable timbers, notably those of the family Cupressaceae, have been used for buildings in many countries and for other purposes, such as boats and coffins, and have survived microbiological and insect attack for centuries (18). The stave churches remaining in good condition in Norway after more than 800 years were built from logs of resinous Scots pine (*Pinus sylvestris*, Pinaceae) (2).

Ebony (*Diospyros* spp., Ebenaceae) was probably the most highly prized wood of the past; Egyptian royalty used ivory, gold, and ebony from Ethiopia and southern India for their furniture. The tomb of King Tut-Ankh-Amon, who died in 1350 B.C., contained a cedar box covered with ebony veneer inlaid with ivory. Ebony was also used as a tribute, and in the year 260 B.C., Ptolemy staged a procession in Alexandria, Egypt, of 600 ivory tusks and 2000 ebony logs from Ethiopia (23).

1.1.2.2 Exudates

In the past, most exudates of trees would have resulted from natural causes, although incision of the bark or other injuries could have been applied, as is now done. The composition of tree exudates is frequently different from that of extracts of woody tissues.

The most notable exudate is amber, prized by man since its discovery in Neolithic times (7000 to 8000 B.C.). Its reputed mystical healing and protective powers, as well as its esthetically appealing appearance – such as its use in old and modern jewelry – and its development into varnishes has resulted in its continued use to the present day. This is partly due to the large deposits of amber in eastern Baltic locations, which probably arose from *Araucaria* spp. (Araucariaceae). The discussions by Pliny in his *Historia Naturalis* (A.D. 77) (22) show that the possible origins of amber had been discussed in the literature before that time. Pliny was the first to propose that amber was a plant exudate (rather than of mineral origin), although he erroneously considered pine to be its origin. The history and detailed studies of amber have recently been reported (22). Ambers are fossilized resins containing nonvolatile terpenoid materials that have been progressively oxidized and polymerized to a point at which they can withstand chemical and microbiological attack (22). They occur most commonly in the Cretaceous and Tertiary strata (70 million to 2 million years ago) and no doubt originate from the huge prehistoric coniferous forests that then existed.

Records of the earliest use of other exudates are limited because of the difficulties of dating the material. An examination of a sarcophagus of the 19th Egyptian dynasty (1400 to 1300 B.C.) showed that an oleoresin or balsam had been used as a coating. The preservation of human bodies as mummies around 1000 years B.C. had involved the use of storax (from *Liquidambar orientalis*, Altingiaceae), mastic, Aleppo pine resin, and probably sandarac and amber (20). The exudate benzoin from *Styrax* spp. (Styracaceae) of Southeast Asia has been found in a sarcophagus dating from the 2nd century A.D. (23). The South American Incas used balsams (possibly from *Myroxylon balsamum*, Leguminosae) for embalming (20).

1.1.2.2.1 Varnishes

The principles of European varnish-making were first recorded in the 11th century; they consisted of heating the resin, probably amber, in linseed oil. Rembrandt and Leonardo da Vinci used amber varnishes as vehicles for their paints (27). The discovery of dammar resin by Europeans in the 17th century in Malaya and neighboring countries and their ready availability resulted in an extension of their applications. Originally they were used in resin candles for illumination, in the caulking of boats, and for preparing batik; then they became important for the preparation of varnishes, enamels, polishes, and in printing. Dammars originate from several genera of the Dipterocarpaceae (particularly *Shorea, Hopea,* and *Vatica* spp.); large quantities have been removed or washed from the soils formerly, or currently, covered by these forests. Dammar gathered from the soil is harder than that collected from the lumps (sometimes as large as 20 kg) in forks or on branches of trees which, in turn, is harder than that collected by cutting cavities into the sapwood of the trees. Usually the harder resins provide the harder and more durable varnishes (9, 14, 27).

Copals are harder resins than the dammars and are obtained in similar ways but mainly from *Copaifera demeusi* (Leguminosae) now, or previously, growing in the basin of the Congo (Zaire) river. *C. copallifera* and *Daniellia* spp. (Leguminosae) also supply significant amounts. Copals are used for the same purposes as the dammars; the export of 18000 tons in 1935 from the Congo indicates their former importance (27).

Kauri "gum" or the resinous exudate from *Agathis australis* (Araucariaceae) was first recorded by Captain Cook in November 1769. Lumps of the resin, some of very large size (30×15×15 cm), have been dug from the soil in many parts of the North Island of New Zealand (34). The peak annual production was 11116 tons in 1899 and altogether over 442000 tons were exported between 1847 and 1939 (11). In the early parts of the 20th century there were repeated warnings that the reserves of kauri resin would soon be depleted. In 1924 surveys conducted by the New Zealand government indicated that there were sufficient resources to meet global demands for well over 100 years. However, its replacement by synthetic resins resulted in only 228 kg being exported in 1982. Kauri provided the basis for high-quality varnishes, and the lower-quality grades were used for the manufacture of linoleum together with pine rosin.

Manila resins [sometimes called copal (5)] have been obtained by tapping the bark of *Agathis alba* (Araucariaceae) growing in Indonesia, the Philippines, and New Guinea, and have been used in recent times.

Mastic and sandarac, which are soluble in alcohol, were formerly used as varnishes, particularly in the Middle Ages, for the protection of oil and watercolor paintings or for coating metals. Sandarac is the resinous exudate of *Tetraclinis articulata* (Cupressaceae) and mastic is obtained from *Pistacia lentiscus* var. *chia* (Pistaciaceae). Mastic is still collected on the island of Chios in the Aegean; it was once the monopoly of Greek emperors. Its use to sweeten the breath and for dental purposes was first recorded in 400 B.C. (19).

The term "naval stores" has been in use since the 17th century when sailing ships used large quantities of tar and pitch from the coniferous forests of northern Europe. In recent years these materials have been obtained from tree stumps, as byproducts of the kraft pulping industry, and by tapping living trees. The United States leads in rosin production from tall oil from pulping; China and Portugal lead in production via tapping; and the USSR leads in production from stump wood (see also Chap. 10.1). More than one-half of the global production of over 1 million metric tons of rosin is produced by tapping (28) and is known as gum rosin.

1.1.2.2.2 Lacquers

The sap that exudes from the incised bark of *Rhus vernicifera* (Anacardiaceae) in China and Japan and of *Melanorrhoea usitata* (Anacardiaceae) of Burma contains urushiol. This and other components oxidize during drying at 20° to 30°C and at a humidity of 75 to 85%, producing the basis of the black lacquerware of the Orient (J. Kumanotani 1986, personal communication). Lacquerware has been discovered with its original color in Chinese burial mounds from the late Zhou and Han Dynasties (≈ 770 B.C. to 220 A.D.). Lacquerware imported over 1200 years ago into Japan is preserved in the "Sho-So-In" in Nara, and in that region the craft was further developed. At the present time, there is nothing to replace the beauty of the black lacquer articles, particularly those produced by the craftsmen of Japan. Lower-quality lacquerware is commonly used for everyday domestic use. There are several types of lacquerware. One of these is "Shunkei Nuri" in which the wooden base is stained yellow and then covered with transparent coatings to enhance the grain of the timber. This type of lacquerware has been in use since the 14th century. The lacquer is also colored with a vermilion pigment in China, and after preparation in thick layers it is intricately carved for decorative articles (18).

1.1.2.2.3 Gums

Gums have satisfied a number of needs with their property either to dissolve in water to form viscid solutions or to absorb water and form jellies or gelatinous pastes. Gum arabic, probably from North African *Acacia senegal* (Leguminosae) and other species, was used by the ancient Egyptians in the preparation of their colors for painting (19). The trade in gums for this purpose continued through the Middle Ages, and they are still used in the preparation of water colors. Gums have been used as the basis for adhesives for centuries and are still used industrially for this purpose and for imparting luster to various surfaces, as well as in inks and other uses. The Australian aborigines and others used the gums of *Acacia* spp. (25, 26) as a source of food; they are still used in confectionery, lozenges, etc.

A wide variety of gums exists with different compositions. Frequently the best exudation of gums is from trees in poor vigor or unhealthy condition or from trees growing in adverse climatic conditions. A large number, but not all, of *Acacia* species exude gum and they, as well as the exudate of *Astragalus gummifer* (Leguminosae), have been the main commercial source (18).

1.1.2.3 Tannins

Leather production is one of the world's oldest crafts; tannin-bearing pods of *Acacia nilotica* (Leguminosae) with partly processed and finished leathers have been found on a 5000-year-old tan-yard site in Upper Egypt. By about 600 B.C., tannage with vegetable tissues was common throughout the Mediterranean area – if not elsewhere (21, 39). Tanning was carried out by placing alternate layers of hide and crushed vegetable tissues (generally of oak origin), adding water, and allowing the hides to remain until a satisfactory leather was formed. In Australia, before contact with Europeans, it was the custom of some aboriginal people to partly fill animal skins with water in which was contained pieces of eucalypt kino (largely polymerized prodelphinidins) and to carry the skin on the back during their nomadic existence, with the agitation facilitating tanning.

It was not recognized until 1796 that there was an active principle in oak bark that combined with protein in the hide to form leather (35), and the word tannin was introduced to define a material that produces leather from hide. Pliny (30) mentions the blue-black iron tannate reaction as a method for identifying iron and that this reaction was the basis of one of the hair dyes used by ancient Egyptians. The same reaction has been the major source of ink until recent years. The ability of tannins to combine with iron has resulted in the preservation of iron objects that were in association with leather in archeological sites (21).

Mangrove tannin extract or "cutch" obtained from the barks of *Rhizophora* and *Bruguiera* spp. (Rhizophoraceae) was used for centuries in tropical countries to preserve cotton or linen clothing, sails, and fishing nets. Cutch and other tannin extracts were widely used for the latter purpose until recently, when the older fabrics were replaced by nylon.

Tannin extracts from various sources are still very largely used for the conversion of hides and skins to leather (see also Chap. 10.3). In 1959, the major sources were the bark of *Acacia mearnsii* (Leguminosae), the heartwoods of *Schinopsis balansae* (Anacardiaceae), *Castanea sativa* (Fagaceae) and *Eucalyptus wandoo* (Myrtaceae) and the cups and unripe acorns of *Quercus aegilops* (valonea) (Fagaceae). In 1959, a total of 434500 metric tons of these extracts was said to have been used globally but in 1982 this had purportedly fallen to 261,400 metric tons (but see also Chap. 10.3).

1.1.2.4 Dyes

The employment of natural dyestuffs dates back to antiquity and many of these probably arose from wood or bark. Most gave a dull brown or a yellow color, but the discovery of mordanting in ancient Hindustan and in the Orient resulted in brighter, more permanent colors. The first mordants were aluminum sulfate and iron sulfate. Knowledge of these spread westward to Egypt, Greece, Rome, and Europe. With the subsequent importation of log-wood, brazil wood, and other materials from the New World, the natural dyes became increasingly important until they were replaced by synthetic dyes (29).

The most notable wood sources of dyes have been a group of bright red woods under the general name of brazil or brasil. In about 1290 Marco Polo mentioned the use of *Caesalpinia sappan* (Leguminosae) in the Orient for this purpose. The wood subsequently became an important article of commerce in Europe during the Middle Ages and eventually became known as sappan. Later, the discovery of a similar bright red wood in eastern South America in large quantities, as well as its use as a dye by the native people, led to the naming of the region "Brazil" after the type of wood that had the color of glowing embers ("brasa" in Portuguese). Confusion exists concerning the scientific names of the exported wood, but it appears to have been *Caesalpinia echinata* (Leguminosae) (32, 33).

The first exploitable wealth found in Brazil was indeed brazil wood. Originally large quantities were present, and the demand for it led to the first three European settlements at Recife and elsewhere in 1502 (4). Uncontrolled felling led to the rapid destruction of the coastal forests and, by the middle of the 16th century, exports of the wood had fallen considerably, although the trade continued for another three centuries. At times it was the prize of piracy, and its sale was controlled by royal monopoly. With various mordants, bright reds to violets were obtained but the fugitive character of the colors reduced their use.

Logwood (*Haematoxylon campechianum*, Leguminosae) of tropical America was exported from about the middle of the 16th century. Subsequently, the wood extracts were also exported. In the latter half of the 19th century, the annual exports of the wood exceeded 100,000 tons. The major component of the extractives, namely hematoxylin, when oxidized, forms the dye hematin, which produces blue to black colors depending on the mordant of iron or copper salts or potassium dichromate.

Old fustic from the heartwood of *Chlorophora tinctoria* (Moraceae) of central America and young fustic from *Rhus cotinus* (Anacardiaceae) were used in the middle of the 17th century to produce a range of light-stable olive-yellow to gold shades depending on the mordants. Similar colors were obtained with quercitron bark (*Quercus discolor*, Fagaceae) of the USA.

Other woods were used as sources of dyes, but they were much less important than those mentioned. These other heartwoods included jakwood (*Artocarpus* spp., Moraceae) of India, Burma, Sri Lanka (Ceylon), and Indonesia, which provided a dye that, in conjunction with alum, gave the saffron color for the robes of Buddhist priests. Sanderswood (*Pterocarpus santalinus*, Leguminosae) from tropical Asia, camwood or barwood (*Baphia nitida*, Leguminosae) of the west coast of Africa, and narrawood (*Pterocarpus* spp.) of the Philippines were formerly exported to Europe and used to impart red to violet colors with different mordants.

1.1.2.5 Perfumes

Just as at the present time, ancient people prized various types of perfumes. Historical records show that some of these were obtained from wood or from tree exudates. Frankincense (from *Boswellia* spp., Burseraceae) and myrrh (from *Commiphora* spp., Burseraceae) were obtained from incisions in the bark of trees growing in eastern Mediterranean countries. These resinous exudates were traded by the Phoenicians and others more than 2000 years ago to provide perfumes, incense, and unguents (19), and by other agents in the 17th century B.C. (20) and their use continues today [see also (1)].

It is recorded that cinnamon barks were prescribed by Moses for the ritual of the Hebrew Tabernacle, and at the time of the Hebrew exodus from Egypt the barks were apparently originating from India. In about 300 B.C. the true cinnamon bark of *Cinnamomum zeylanicum* (Lauraceae) from southern India and Sri

Lanka entered the Mediterranean trade. The Portuguese captured and suppressed Ceylon in 1500 in order to control the best source of the bark; they were, in turn, followed by the Dutch and the English (6). The bark is still used for flavoring foods but now less so for perfumery purposes.

The use of the fragrant wood and oil of the Indian sandalwood species *Santalum album* (Santalaceae) is one of great antiquity. This species was traded from India through the Red Sea to Mediterranean countries in the 6th century A.D., and at the same time was imported by China (6). In 1511 the Portuguese occupied Malacca to establish a trading base to monopolize the spice and sandalwood trade that was carried out between China and other countries. Early in the 19th century the British had established plantations in Mysore, India, and used Singapore as their chief market.

Another *Santalum* species was discovered in Hawaii about 1790 and its export to China reached a peak between 1810 and 1825. This trade brought great wealth to the primitive people of the islands (once known as the Sandalwood Islands) until the resource was almost completely destroyed by 1840 and with it the economy of the islands (32). In 1843 reports reached Perth, Western Australia, of the high prices being paid in Singapore for sandalwood. At that stage, the colony Western Australia had a very large imbalance of trade but in clearing land for agriculture was burning its sandalwood (*Santalum spicatum*, Santalaceae) or using it for buildings or posts. Export was begun in 1846, and 2 years later 1335 tons earned more than one-half of the export income of the colony. With the government imposition of an export tax to raise more income, export ceased until 1857. Sometime about 1870, restrictions were placed on the export of sandalwood from India owing to the failure of the Mysore plantations. From 1872 to 1882 the export of sandalwood from Western Australia rose from 3942 to 9605 tons. For various reasons the trade declined but revived later. It reached an all-time peak in 1920 with 13945 tons exported, earning $467,162; the present-day equivalent is considerably higher. At that time the sapwood was removed before export; it is recorded that 14255 tons of sandalwood were cut in 1920 and 10839 tons in 1921. The amount cut has declined considerably in recent years, and the remaining material is found only in remote desert regions (37). The wood is used for incense and joss sticks in Buddhist rituals.

1.1.2.6 Rubber

The natives of tropical America were preparing articles from the rubbery exudates of *Hevea* spp. (Euphorbiaceae) and *Castilla* spp. (Moraceae) trees well before discovery by Europeans in the early 17th century. Following the developments of vulcanization and other treatments, an enormous exploitation of rubber trees, particularly of *Hevea brasiliensis*, growing in the delta of the Amazon River and in its upper reaches west of Manaus, took place from 1827 onward. Large fortunes were made by a few European entrepreneurs at the expense of the lives of many local and imported laborers. After some difficulties, rubber trees were cultivated in Malaysia in 1880, and subsequently in neighboring countries, from which most of the natural rubber of today is derived. Following severe competition from synthetic rubber, improved plantation techniques have ensured their continuation as a source of rubber (see also Chaps. 10.4 and 11.3.1). In 1980, over 3.8 billion metric tons of natural rubber were exported. Of this, 67% originated in Malaysia and Indonesia, and 7% in Brazil. Almost 6.5 billion tons of synthetic rubber were produced at the same time (18).

1.1.2.7 Medicines

Many plants have been used in the past for medicinal purposes and the search continues today for superior sources. Various extracts of trees and of exudates have been used for different complaints.

The use of the deep red or ruby red exudate of "Dragon's blood" from *Dracaena cinnabri* (Agavaceae) (or more recently *Daemonorops* spp., Palmae) as an astringent for the treatment of dysentery and diarrhea was recorded by Pliny (30) and early Greek writers. It has been found in tombs of that period (19). Its incorporation in violin varnishes as a dye was probably a more effective use.

"Cutch," the aqueous polyphenolic extract of the heartwood of *Acacia catechu* (Leguminosae) grown in India, was imported by the Chinese in the late 16th century or earlier; it was used as a medicine (and as a dye) in India early in that century. Other polyphenols and tannins have been used as astringents for the control of diarrhea. For example, malabar kino (an exudate from *Pterocarpus marsupium*, Leguminosae) and Bengal kino (from *Butea frondosa*, Leguminosae) were exported to Europe in the late 18th century. Their use and that of kinos from some eucalypt species (e.g., *Eucalyptus camaldulensis*, Myrtaceae) have been prescribed by some pharmacopoeias for this purpose until recently.

Mannitol is a major component of the exudates ("manna") of a range of trees including *Fraxinus ornus* (Oleaceae), *Olea europaea* (Oleaceae), *Alhagi camelorum* (Leguminosae), *Myoporum platycarpum* (Myoporaceae) (8, 12, 38). Mannitol was formerly used as a mild laxative. The manna that sustained the migration of the Israelites to Palestine in biblical times is reputed to have been composed of raffinose.

There are a few cases of effective medicines obtained from wood or bark for the cure of specific illnesses using western style medical treatment. Until recent times the bark of *Cinchona* spp. (Rubiaceae) provided quinine and hence control or cure of malaria. The first use of the bark by Europeans was recorded in 1636 in Peru with bark collected in the Andes. The subsequent exploitation of the trees almost led to their extinction until plantations in other tropical countries were established. *Cinchona calisaya* var. *ledgeriana* and *C. officinalis* provided the richest sources of quinine, and high-yielding hybrids were developed. This situation was not achieved without difficulty in smuggling the seeds during 1854 to 1859 from South America to the Orient, and together with the lack of germination of some seeds, this led to boom periods with inflated prices for the bark (6). Cinchona has lost much of its importance because of the synthesis of anti-malarials with greater effectiveness and fewer side effects than quinine.

Cavities in the trunks of *Dryobalanops aromatica* (Dipterocarpaceae) of Malaysia and Indonesia can contain crystals of camphor used for medicinal purposes. It was sold in the Mediterranean region 2000 years ago and was the first type of camphor to reach Europe. The Chinese were importing it from Malaysia before the 8th century, but subsequently replaced it with camphor from *Cinnamomum camphora* (Lauraceae), which was manufactured from old trees by the Chinese and Japanese before the 9th century (6). Natural camphor and camphor wood are still used, but camphor is now synthesized in increasing amounts. Also the wood of *Cedrus libani* (Pinaceae) has been used for medicinal purposes since biblical time (7, 10).

From the converse aspect, a number of woods can cause complaints and severe illness during processing, and the occupational risks from processing were defined in 1700 (3, 31). A number of woods cause dermatitis and general irritation for handlers (13), and some users are particularly sensitive to them. The irritant effects of different woods have been known in the wood-working trade for many years, but have been poorly recorded. The increased incidence of complaints following the introduction of new species of tropical timbers in recent years has been associated with the determination of the responsible active components. Constituents responsible for most of the observed cases of allergic contact dermatitis belong to benzo- and naphtho-quinones (13). These sensitizing quinones have been found in particular in *Dalbergia* and *Manchaerium* spp. (Leguminosae), *Tectona grandis* L. (Verbenaceae), *Cordia* spp. (Boraginaceae), *Paratecoma peroba* (Bignoniaceae), *Mansonia altissima* (Sterculiaceae), and *Diospyros celebica* (Ebenaceae). Some woods cause severe stomach complaints (*Erythrophleum letestui*, Leguminosae), others congestion of the lungs and bleeding nose (*Dysoxylum muelleri*, Meliaceae), asthma (*Thuja plicata*, Cupressaceae), nausea, cramps, irritation of mucous membranes (*Aspidosperma peroba*, Apocynaceae). *Cryptocarya pleurosperma* (Lauraceae) contains an alkaloid that causes several severe complaints, and other woods cause different problems (see also Chap. 9.5).

1.1.3 Lessons from History

From prehistoric times to the present day, extractives and exudates have provided a number of mankind's needs. Originally they were provided by local sources, and a large number of them were used for relatively simple purposes. Eventually, as international trade increased, and the sophistication of requirements and specifications rose, some materials were supplied in enlarged quantities when their superior properties became known. They became the objects of royal monopolies, trading nations were established by their exploitation, wars were fought over their supply, they enabled some countries or regions to become temporarily wealthy or to avoid bankruptcy, and they gave their names to countries. Eventually many exudates and extractives were replaced by other resources or by synthetic or alternative substances with superior properties that could be reliably supplied in uniform high qualities and at acceptable costs. In some cases, their ready availability, high qualities, and easy profit led to the destruction of the resources. Very few of the materials used 50 years ago are now used commercially. Previously the user sought the best natural product available and adapted his requirements to them. Today his precise needs can often be supplied from other sources.

In a world of diminishing resources per capita, there is a growing need for the development of renewable resources, such as extractives, exudates, and forest products. What conclusions regarding their use can be drawn from history? Unless the product can be supplied in the highest appropriate quality, with assurances of the maintenance of that quality and continued availability at acceptable costs, the product will be supplanted. Many of the woody plants producing these materials are slow-growing and hence costly to maintain. Furthermore, the collection and preparation of extractives and exudates are labor-intensive. Such activities, even in regions with increasing unemployment and low wages, will be fruitless unless the consumers' requirements are fulfilled.

The provision and use of these renewable resources can be increased by research and understanding. Consider rubber, for example. Vulcanization and other developments of rubber greatly increased its value to modern man. Biological research and strict merchandizing control have increased the supplies and quality of rubber from tropical rubber plantations so that it can and does compete with synthetics from nonrenewable resources. In another example, improved tapping methods and strict quality control have enabled China with its labor resources to increase its exports of resin and other naval stores products. The use of stimulants for resin production in the wood of certain pines will assist the maintenance of production of naval stores as a byproduct of chemical pulping. Plantations of selected high-yielding variants of tannin-bearing trees, improvements in the production and purification of the extracts, and development of uses such as adhesives will ensure the continued use of these extracts. With the decreasing availability of low-cost petrochemical intermediates and the increasing costs to control hazardous effluents of the industry, the production of minor forest products in developing nations may become viable in the future. However, unless adequate quantities of a material, such as a dammar, can be provided at low costs, in uniform quality, and with the supporting technical expertise for a particular need, developments are unlikely.

The history of the vicissitudes of the commercial use of extractives and exudates point to both short- and long-term needs for their increased understanding. The greater use of unfamiliar wood species to provide mankind's increasing needs of building materials, pulp and paper, feature timbers, and other products will encounter difficulties in the way of efficient utilization. Understanding the extractives responsible for certain adverse properties has in the past led to the control of resultant problems. On the other hand, examination of favorable properties of wood caused by extractives could provide understanding of the mechanism of that property and result in its enhancement or in development of alternatives of higher quality (16).

An exciting aspect of the study of extractives and exudates is the recent development in enzymology. These will help in the provision of answers to the questions frequently asked in the past of why and how the products extraneous to the lignocellulosic cell wall are formed. The different chemistry of the extractives and exudates from the same tree point to specific biological mechanisms. The different composition of the groups of compounds found in the oleoresin in the resin canals and in the adjacent parenchyma of conifers exemplifies the selectivity of these mechanisms. Even more dramatic examples of the detailed specificity of formation of individual compounds are shown in *Intsia bijuga* (Leguminosae) (and other species), wherein the vessels may be filled with a pure crystalline compound, but the adjacent parenchyma will contain a mixture of several compounds from different classes (17). When these mechanisms, which the tree readily controls, are understood, we will be closer to the formation by biological methods of the complex chemicals required by the consumer in a pure form without byproducts. This will require knowledge of the complex array of cell reactions in woody plants and the products they produce (18).

References

1 Abercrombie T J 1985 Arabia's frankincense trial. Nat Geogr 168:475–513

2 Aune P, Sack R L, Selberg A 1983 The stave churches of Norway. Sci Am 249(2):96–105

3 Bolza E 1980 Some health hazards in the forest products industries. Control 6(1):7–16

4 Brazilian Embassy 1976 Brazil – a Geography. Brazilian Embassy, London, 59 pp

5 Brown W H 1921 Minor products of Philippines forests. Bureau For Bull No 22, Manila Vol 2, 410 pp

6 Burkill I H 1935 A dictionary of the economic products of the Malay Peninsula. Vol 1–2. Crown Agents for the Colonies. London, 2402 pp

7 Edlin H L 1963 A modern sylva or a discourse of forest trees. 7. Cedars – Cedrus species. Quart J For 57:302–310

8 Flückiger F A 1894 Australische manna. Arch Pharm 232:311–314

9 Foxworthy F W 1922 Minor products of the Malay Peninsula. For Record (2):151–217

10 Hafizoglu H 1987 Studies on the chemistry of Cedrus libani. I. Wood extractives. Holzforschung 41:27–38

11 Harrison-Smith J L 1940–41 Kauri gum. NZ J For 4:284–292

12 Hatt H H, Hillis W E 1947 The manna of Myoporum platycarpum R.Br as a possible commercial source of mannitol. J Council Sci Ind Res (Aust) 20:207–224

13 Hausen B M 1981 Woods injurious to human health. Walter de Gruyter Berlin, 190 pp

14 Hedley Barry T 1936 The future of natural resins. J Oil Colour Chem Assoc 19(189):75–95

15 Hermann P 1952 Quoted by Sandermann W, Simatupang M H 1966 Zur Chemie und Biochemie des Teakholzes (Tectona grandis). Holz Roh-Werkst 24:190–204

16 Higuchi T (ed) 1985 Biosynthesis and biodegradation of wood components. Academic Press Orlando, 679 pp

17 Hillis W E 1977 Secondary changes in wood. In: Loewus F A, Runeckles V C (eds) Recent advances in phytochemistry 11:247–309

18 Hillis W E 1987 Heartwood and tree exudates. Springer Berlin, Heidelberg, New York, Tokyo, 268 pp

19 Howes F N 1949 Vegetable gums and resins. Chronica Botanica Co Waltham MA, 188 pp

20 Howes F N 1950 Age-old resins of the Mediterranean region and their uses. Econ Bot 4:1307–1316

21 Knowles E, White T 1958 The protection of metals with tannins. J Oil Colour Chem Assoc 41(1):10–23

22 Lagenheim J H 1969 Amber: A botanical inquiry. Science 163:1157–1169

23 Lucas A 1962 Ancient Egyptian materials and industries. 4th Edn. Rev by Harris J R. Arnold London, 323–324

24 Maiden J H 1894 Australian sandarac. Agric Gaz N S Wales 5:301–305

25 Maiden J H 1901 The gums, resins and vegetable exudations of Australia. J & Proc R Soc N S Wales 35:161–212

26 Maiden J H, Smith H G 1895 Contributions to a knowledge of Australian vegetable exudations. J & Proc R Soc N S Wales 29:393–404

27 Mantell C L, Kopf C W, Curtis J L, Rogers E M 1942 The technology of natural resins. John Wiley New York, 506 pp

28 Naval Stores Review 1986 Global rosin production lower in 1985. Naval Stores Review 96(2):13

29 Perkin A G, Everest A E 1918 The natural organic colouring matters. Longmans Green London, 655 pp
30 Pliny G S 1559 Historia Naturalis XXXIV, II, Col. 880 ed. Gelanii Venetiis. Quoted by Knowles and White 1958
31 Ramazzini B 1700 De morbis artificum diatriba. A Pazzini (ed) 1953 Tome Columbus, Chap 6 in Suppl. Quoted by E Bolza 1980.
32 Record S J, Hess R W 1943 Timbers of the New World. Yale Univ Press New Haven, 640 pp
33 Record S J, Mell C D 1924 Timbers of tropical America. Yale Univ Press New Haven, 610 pp
34 Reed A H 1954 The story of the kauri. A H and H W Reed Wellington NZ, 439 pp
35 Seguin A 1796 Rapport au comité de salut public sur les nouveaux moyens de tanner les cuirs proposés par le cit. Ann Chim Paris 20(15). Quoted by White 1957
36 Shaw N 1914 Chinese forest trees and timber supply. T Fisher Unwin London, 351 pp
37 Talbot L 1983 Sandalwood. For Focus W Aust 30 (Sept):21–31
38 Trabut M 1901 Sur la manne de l'olivier CR Acad Sci 132:225–226
39 White T 1957 Tannins – their occurrence and significance. J Sci Food Agric 8:377–385

1.2 Natural Products Chemistry – Past and Future

K. Nakanishi

1.2.1 Introduction

As is true of other fields of science, natural products chemistry is in a state of transition and is undergoing a profound change in character. Being intimately related to nature and natural phenomena, plants, animals, food, poisons, diseases, etc., the "chemical" studies in the natural products area can be traced back thousands of years. Traditional natural products chemistry necessarily started with the isolation and purification of a compound, followed by structure determination, and culminated in partial or total synthesis.

Professor Roger Adams once told me that when studying in Germany around 1915 he could read in one night the papers published in one year of the Journal of the American Chemical Society. It is hard to believe that such a fundamental concept as the tetravalency of carbon was forwarded only in 1857 (Kekulé), the benzene structure in 1865 (Kekulé), the tetrahedral structure of the saturated carbon atom in 1874 (van't Hoff, Le Bel), and the correct skeletal structure of steroids in 1932 (including X-ray) (Rosenheim/King; Wieland/Dane). On the other hand, basic investigations into optical activity had been completed by the mid-1850s (Pasteur), configurational correlation of glucose and some sugars were known by 1891 (Fischer), and structural and synthetic studies on purines, pyrimidines, and peptides had been conducted from 1880 through the mid-1910s (Fischer).

These and other developments are listed chronologically in Table 1.2.1, abstracted from Paul Karrer's *Organische Chemie* (7), the standard textbook in the pre-war era; it presents the important achievements in natural products chemistry up to 1938, when the field was dominated by studies carried out in the United Kingdom, Switzerland, France, and especially Germany.

Table 1.2.1. Important achievements in natural products chemistry up to 1938

Year	Discovery	Scientist (Nobel prize yr.)
1769	Crystalline tartaric acid from argol	Scheele
1772	Methane formation in marshes	Priestley
1773	Urea discovery	Rouelle
1775	Benzoic acid from gum benzoin	Scheele
1776	Uric acid from urine and urinary calculi	Scheele; Bergmann
1779	Glycerol from olive oil	Scheele
1780	Lactic acid from sour milk	Scheele
1785	Malic acid from apples	Scheele
1786	Oxidation of alcohol to acetic acid	Lavoisier
1789	General course of alcohol fermentation	Lavoisier
1797	Ether from alcohol by removal of water	Fourcroy
1805	Morphine from opium	Sertürner
1811	Glucose from starch and sulfuric acid	Kirchhoff
1815	Detection of optical activity	Biot
1818	Chlorophyll from leaves	Pelletier and Caventou
1819	Naphthalene discovery	Garden and Kidd
1820	Caffeine from coffee	Runge
1824	Synthesis of oxalic acid from cyanogen	Wöhler
1825	Benzene discovery in compressed oil gas	Faraday
1826	Hematin from blood	Tiedemann and Gmelin
1828	Synthesis of urea from ammonium cyanate	Wöhler
1831	Carotene from carrots	Wackenroder
1832	Anthracene from tar	Dumaas and Laurent
1834	Aniline, quinoline, pyrrole from tar	Runge
1838	*p*-Quinone from quinic acid	Wosskressensky
1841	Pure phenol from tar	Laurent
1846	Ether structure	Laurent
1847	Maltose from starch by diastase	Dubrunfaut
1848	Racemic acid resolution to tartaric acid	Pasteur
1852	Resolution of racemates by bases	Pasteur
1853	Synthesis of glycerides and fats	Berthelot
1857	Carbon tetravalency postulation	Kekulé
1858	Concept of modern structural formula	Kekulé
	Biochemical resolution of racemates	Pasteur
1859	Asymmetry concept and optical activity	Pasteur
1862	Crystalline hemoglobin	Hoppe-Seyler
1865	Benzene formula	Kekulé
1868	Lecithin from brain	Strecker
1870	Indigo synthesis from isatin	Baeyer (1905)
	Indigo synthesis from nitroacetophenone	Engler and Emmerling
1874	Tetrahedral arrangement around carbon	van't Hoff (1901); Le Bel
1876	Preparation of coumarin	W.H. Perkin, Sr.
1882	Synthetic polypeptides	Curtius
1883	Deduction of indigo formula	Baeyer
1891	Glucose configuration and synthesis	Fischer (1902)
1893	Camphor structure	Bredt
1899	Walden inversion	Walden
1904	Nicotine synthesis	Pictet
1906	Elution chromatography	Tswett
1907	Synthesis of octadecapeptide	Fischer
1909	Camphor total synthesis	Komppa
1912	Tannin structure	Fischer

Table 1.2.1 (continued)

Year	Discovery	Scientist (Nobel prize yr.)
1913	Preparation of pure anthocyanin	Willstätter (1915)
1923	Microanaylsis completed	Pregl (1923)
1925	Catechin structure, epicatechin synthesis	Freudenberg
1926	Medium-ring compound synthesis	Ruzicka (1939)
	Thyroxin structure and synthesis	Harrington
	Vitamin B_1 isolation	Jansen and Donath
	Irradiation of ergosterol to vitamin D	Windaus (1939); Rosenheim
1928	Vitamin C isolation	Szent-Györgyi (1937)
	Follicular hormone isolation	Butenandt; Doisy; Girard; Marriau
1930	Carotene and lycopene structure	Karrer (1937)
1931	Anthocyanin synthesis	Robinson (1947)
	Vitamin A structure	Karrer
1932	Testicular hormone isolation	Butenandt (1939)
	Cholesterol structure	Wieland (1925) and Rosenheim
1933	Vitamin B_2 isolation	Kuhn (1938)
	Vitamin C synthesis	Reichstein (1950); Haworth (1937); Hirst
1934	Auxin isolation	Kögl
	Corpus luteum hormone isolation, synthesis	Slotta; Hartmann and Wettstein; Butenandt; Fernholz
	Androsterone synthesis	Ruzicka (1939)
	Phthalocyanin discovery	Linstead
1935	Testosterone isolation	Laqueur
	Testosterone synthesis	Butenandt; Ruzicka; Wettstein
1936	Vitamin B_1 structure and synthesis	R.R. Williams; Windaus; Grewe
1937	Corticosterone isolation and structure	Reichstein; Kendall (1950)
1938	Vitamin E isolation, structure, synthesis	H.M. Evans; Fernholz; Karrer

During the 1940's, the center of organic chemistry shifted to the United States, where the emphasis has been on synthesis and physical organic chemistry. Traditional natural products chemistry and biosynthesis remained centered in Europe, and during the mid-1960's branched into Japan, which became another center. A possible reason that natural products chemistry, excepting synthesis, never entered the mainstream in American academic research is that research in this field requires a team effort, including collection of material, assay, and a variety of spectroscopic measurements, a requirement that is often extremely difficult to achieve for starting faculty members in a nonhierarchical system.

The entire field of natural products chemistry has been undergoing a dramatic metamorphosis in recent years, to such an extent that to many modern chemists the term itself carries the implication of being outdated. Instead, "bio-organic chemistry" has been in vogue since the 1980's, and it does cover a larger territory including biology and biochemistry. However, whether we call ourselves natural products chemists or bio-organic chemists, there is no doubt that the future in this area is becoming increasingly interdisciplinary and important. In the following, we discuss the current status of isolation and structure determination and the future of natural products chemistry; synthesis is not included.

1.2.2 Isolation and Purification

In any study in the area of natural products chemistry, isolation and purification are the mandatory first steps that one encounters and must accomplish. Unfortunately, however, this phase is often handled too casually and without the realization that success or failure of a project is often solely determined by this step. Perhaps this attitude reflects the days when we were dealing with much larger quantities of "static" compounds that could be isolated readily by extraction with organic solvents and recrystallization. Or perhaps it is because most chemists, excepting professional isolation chemists or chromatographers, regard purification as a trial and error art that lacks logic or innovation.

Even when an air-sensitive compound is isolated by ingenious manipulations, or a hormone is isolated in minuscule amounts from tons of starting material and after years of frustrating bioassays, the purification protocol usually is only applicable to that particular case and lacks generality. Unlike sequences or individual steps in synthesis or structure determination, which others can follow or appreciate, it is not easy to realize the significance of a specific isolation step, and it therefore does not register in one's mind. We can outline synthetic or structure determination sequences, but seldom do we recall an isolation procedure. A difficult isolation is usually presented at symposia in one or two slides and is seldom followed by discussions.

Structure determination has nowadays become quite routine, but isolation and purification steps will never be routine. This is especially true in dealing with "dynamic" natural products that exist in only minute quantities or may only have a fleeting existence; these have to be isolated after some derivatization and yet must be monitored by a bioassay.

Undoubtedly, exciting topics in the future will require expert and imaginative handling in the first crucial isolation step. The spectacular rediscovery of column chromatography by Kuhn during studies on carotenes (1931) led to the era of alumina chromatography. Elution chromatography was discovered in 1906 by the botanist Tswett, who showed that the petroleum ether extract of green leaves could be separated into bands of chlorophyll and yellow pigments when passed through adsorbent chalk; however, it attracted little attention until Kuhn's reintroduction. This was the prevalent technique for purification, apart from recrystallization, until discovery of paper chromatography in 1944 (Martin and Synge; Nobel prize 1952), followed by thin-layer chromatography and counter-current chromatography (Craig). This in turn was followed by introductions of high pressure liquid chromatography or HPLC (now called high-performance liquid chromatography), numerous variations in chromatographic techniques and packing material, including those used primarily in biochemistry – such as electrophoresis. The infinite variations in methods, equipment, solvent systems, etc., make it impossible to keep up with the progress and this presumably is another factor contributing to the prevailing lack of awareness.

It may even be stated that future exciting problems in characterization of bioactive factors rest only with those that present a challenging problem in isolation, whether due to the micro-quantity, sensitivity to air, light, or moisture, etc., the difficulty in assay, transient existence, or inherent difficulty in isolation as exemplified by sticky detergent-like compounds. The crucial nature of the isolation process is well-represented by enormous efforts directed toward isolations of neuropeptides and neurotransmitters, inherent elicitors of phytoalexins, inducers of plant and animal cell differentiation, and numerous hormones. Isolations are monitored by an assay, usually biological, and here again novel concepts for assays are an important factor that determines the value of a project. In a different context, successful separations of the cleavage products of proteins and nucleic acids were major factors responsible for the launching of genetic engineering. It is to be noted that many of these factors are hydrophilic and insoluble or only slightly soluble in organic solvents.

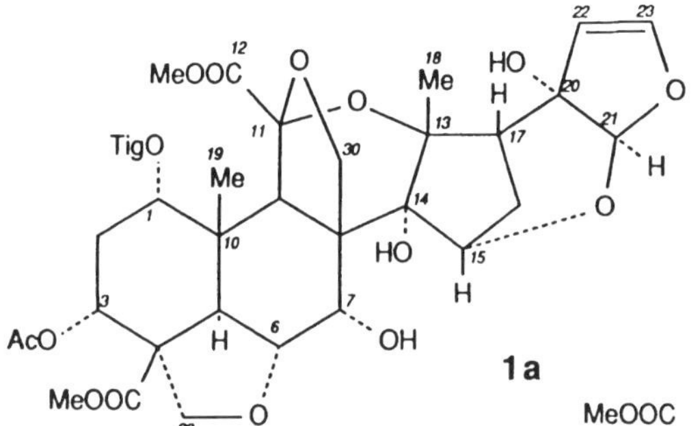

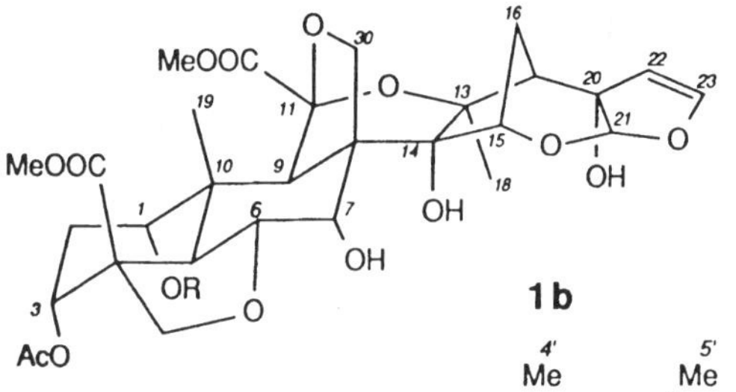

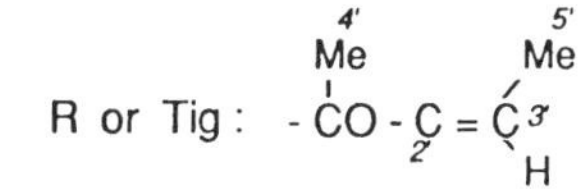

Except in industry, where research is often divided into professional teams, it is crucial for all natural products chemists to strive to be aware of the complicated art of isolation/purification. The following incidence does not belong to the category of highly active natural products occurring in minute quantities. Rather, it concerns a classical plant product, azadirachtin, for which partial structures were published by Morgan and coworkers in 1972 (3). It occurs in the neem tree *Azadirachtas indica*, and is the most potent insect antifeedant known. The finding that the action of azadirachtin and many other less potent congeners (18) is systemic, that the compounds are nontoxic, nonmutagenic, and biodegradable, and that they affect many insect species, has made the tree a focus of worldwide attention. Incorporating the results of earlier studies (3) and applying a new ^{13}C NMR technique, we proposed structure **1** in 1975 (25). However, since the structural evidence was not convincing, we used this molecule over the years as a sample for applying new NMR pulse sequences, but results were ambiguous. Members of our group who worked on azadirachtin were competent scientists who were basically spectroscopists rather than natural products chemists; consequently, despite the availability of sufficient plant extract, there was reluctance in performing extraction for more material and chemical modifications to prepare samples suited for NMR. This peculiar situation continued until Dan Schroeder, a post-doctoral fellow with natural products background, volunteered to do the isolation; he soon devised an improved isolation scheme and produced 10 g of microcrystalline azadirachtin (17). By the time we had arrived at a revised structure, Kraus informed us of their structure **2** (8), which was confirmed by Ley's group by X-ray (2). The full papers from the three groups have now been published back-to-back with a preceding summary article by Taylor (19). The azadirachtin problem certainly taught us a lesson in isolation.

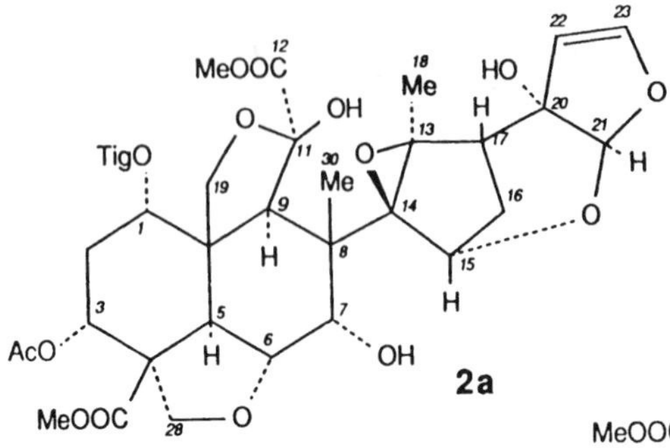

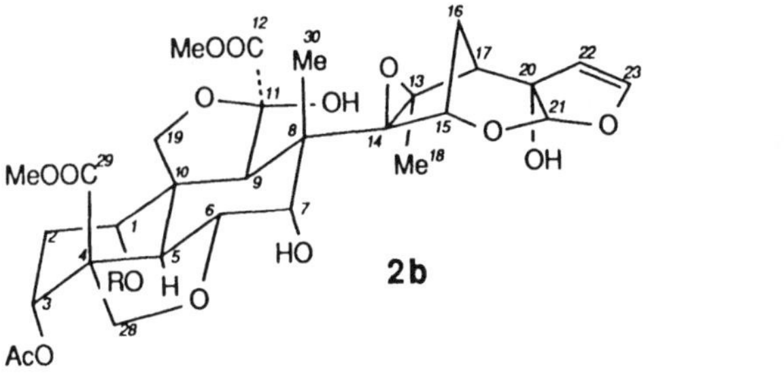

1.2.3 Structure Determination

Elucidation of complex terpenoid, alkaloid, and miscellaneous structures by chemical reactions, coupled with rationalizations of reactions, skeletal rearrangements, and biogenetic concepts (16) occupied the best brains and hands of natural products chemists up to the early 1940's. Woodward (Nobel prize 1965) published his short papers on the empirical additivity relations in the absorption maxima of polyenes and enones (22, 23, 24), [expanded by Fieser and Fieser (6)], the revolutionary Beckman DU spectrophotometer was introduced around this time, and commercial self-recording infrared spectrometers became available in 1949. These were major factors that led organic chemists to realize that spectroscopy provided data indispensable for routine research, and that it was a subject far more approachable than generally regarded. Nuclear magnetic resonance spectroscopy (NMR) and mass spectrometry (MS) were both incorporated into organic chemistry in the late 1950's. Although optical rotatory dispersion (ORD) and circular dichroism (CD) were used for chiroptical measurements in the 19th century, introduction of the Bunsen burner made available the 589 nm sodium D-line and led to the use of $[\alpha]_D$ values, which have no merit other than being physical constants expressing rotation signs. ORD and CD together with self-recording instruments were reintroduced into the organic community by C. Djerassi and by L. Velluz and M. Legrand, respectively, in the mid-1950's. The concept of conformational analysis was established around this time and made significant contributions to structural chemistry and the understanding of reactions (Barton and Hassel; Nobel prize 1969).

NMR, already the major physical tool by the mid-1970's, will continue its dominance in structural chemistry for some time to come, due to continued dramatic developments exemplified by introduction of the Fourier transform (FT) technique, new pulse sequences, solid state NMR, single-cell measurements, and NMR imaging in medicine. The popularity of IR peaked in the 1950's when most of the characteristic frequencies known today were discovered, but NMR replaced IR to the extent that in many cases chemists only used IR for detection of carbonyl functions. However, with the general availability of FTIR since the mid-1970's, IR is experiencing a renaissance; another technique in vibrational spectroscopy, particularly resonance Raman spectroscopy (RR), has become popular, and the field of vibrational spectroscopy has embarked on an extremely promising future. IR, RR, and a variety of other spectroscopic methods based on lasers are now available to study fast reactions exemplified by light energy conversions on a time scale never before attainable (lower picosecond range).

Because of its continued dramatic improvements, MS has become so professionalized that, except for mass spectroscopists, it is impossible to keep up with new developments. Rapid developments are seen in methods of sample volatilization, range of measurable molecular ion peaks (currently up to around 20,000), sensitivity (picogram level or less), tandem MS/MS, FTMS, etc. MS is the most expensive of the common physical tools used in natural products chemistry, is the most difficult to maintain properly, and is frequently nonroutine with regard to sample measurement; it is also the only destructive physical method, which prohibits remeasurement. Nevertheless, MS is indispensable.

With respect to X-ray crystallography, improvements in instrumentation are now enabling organic chemists to carry out their own measurements to ascertain structures, for example, of synthetic intermediates. The most exciting advance, however, is seen in the field of biopolymers, where entire structures of enzyme-substrate complexes, clusters of proteins and reactive centers, etc., have been elucidated. In most cases, such achievements lead to quantum jumps in our understanding of intricate biochemical processes by defining the three-dimensional structures of all molecules involved.

The physical tool that has tremendous potentialities but which is totally under-utilized is CD. ORD is used only for specific purposes these days because ORD curves, which basically give information identical to CD, are more complex. A barrier, largely psychological, exists in the daily use or investigations of chiroptical properties, and it is not uncommon for most organic chemists to adopt a resigned attitude toward analyzing CD data. This unfortunate situation is probably due to the empiricism seen in the handling of many CD curves and to the gap between data and theory – the coupled oscillator theory being an exception. A characteristic feature of chiroptical data is that, most commonly, they are used for determining the absolute configuration and, this being the case, there is only one out of two possibilities. If the underlying principles leading to CD Cotton effects are not well-defined, as is often the case, the solution to the problem is similar to coin flipping. Also, two wrong deductions will lead to the correct solution, while three wrong deductions will give an erroneous result; such examples can be seen in the literature. The art of CD data analysis is clearly different from that of NMR or MS where one can examine a chart for days, and slowly but logically elucidate structural moieties. CD thus offers an intriguing and worthwhile target for serious fundamental research in organic chemistry, physical chemistry, and biophysics, particularly in the direction of filling the gap between data and theory.

In the pre-spectroscopic period of structure determination, there were elements of drama and glamour, and a keen intuitive mind was necessary to elucidate the reactions correctly. Presence of a hydroxyl group had to be deduced from microanalysis (Pregl; Nobel prize 1923), followed by reactions such as dehydration, oxidation, or acetylation, and another elemental analysis. It should be remembered that each microanalysis is preceded by recrystallizations to obtain analytically pure samples. Extensive chemical reactions were necessary if the group was sterically hindered and unreactive – a typical case being the 11-keto function in cortisone [Kendall, 1936–37 (6 p 606)]. If an oxygen function is extremely sluggish to react, it would customarily be assigned to an ether group. The number of methyl groups were derived from Kuhn-Roth oxidations, and gross skeletal structures were deduced from zinc dust distillation or selenium dehydrogenation. It was not unusual to see an entire paper dedicated to the clarification of any such single aspect.

Introduction of IR dramatically changed the entire picture – e.g., presence of a saturated keto function in cortisone became immediately obvious from a single IR spectrum taken in 30 minutes. The first extensive diagnostic studies using IR were performed in connection with the structure of penicillin, a compound considered to be of extreme importance to the military during World War II. Massive collaborative investigations were carried out by American and British chemists between 1943 and 1946, involving 39 institutions (5), including Cambridge, Manchester, Oxford, Cornell, Harvard, Michigan, Rockefeller, and Stanford Universities. Exhaustive IR studies on models led to the conclusion that the 1690 cm^{-1} could best be reconciled with the unprecedented β-lactam-amide formula (5 p 382–414), a result confirmed by X-ray in 1946 [Crowfoot (5 p 310–366); Nobel prize 1964]. Structure determination was affected even more dramatically with the arrival of NMR. Numbers of methyl groups, acetyl groups, methoxyl groups, etc., are counted directly from the spectrum, and it becomes unnecessary to look for olefinic out-of-plane bending bands in the IR.

Structure determinations of most molecules have become routine and the element of excitement it had two decades ago may be largely gone. However, handling of spectroscopic data still requires careful and detailed attention. Results of total syntheses of natural products are absolute because they can be checked by comparison; structures determined by X-ray are likewise usually final. In contrast, structures derived from spectroscopy represent the most plausible, but one is never absolutely certain about their correctness. *Spectroscopic data can be consistent with a structure but they can never prove it.*

A single misinterpretation or oversight of an NMR signal can lead to a grossly incorrect structure. The following is an embarrassing example concerning a simple structure. Specionin, isolated in trace quantities from the ethanol extract of the leaves of the *Catalpa speciosa* tree, was assigned structure **3** on the basis primarily of NMR studies (4). It is a potent antifeedant against the Eastern spruce budworm (a major forest pest in northeast America) and has attracted the interest of several synthetic chemistry groups, who have shown (20) the structure to be **4** rather than **3**. In the structural studies, we had neglected to determine the class of hydroxyl functions, a routine first step that should have been done by ^{1}H-NMR in a solvating solvent (not the fault of the graduate student performing his first NMR measurements and structure determination).

Ginkgolides are bitter substances contained in leaves and root barks of the ginkgo tree *Ginkgo biloba* ("fossil tree"), which has remained unchanged for several millions of years; these trees were preserved in Chinese temples before becoming widespread throughout the world in the 20th century. Structural studies, including discovery of the now familiar nuclear Overhauser effect (NOE), led successfully to the esthetically pleasing cage structures **5–8** for the three ginkgolides

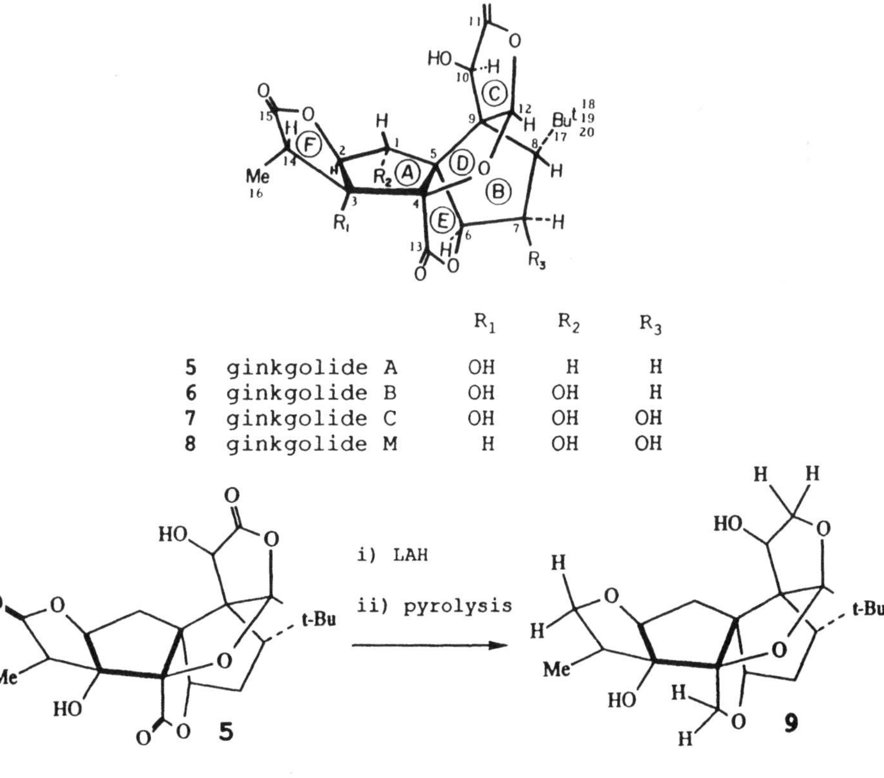

		R_1	R_2	R_3
5	ginkgolide A	OH	H	H
6	ginkgolide B	OH	OH	H
7	ginkgolide C	OH	OH	OH
8	ginkgolide M	H	OH	OH

GA, GB and GC (9–14,21). Although ginkgolides are currently attracting wide attention as platelet aggregating factor antagonists (15), they were devoid of any known activity when our studies were carried out in 1964 through 1966. Nevertheless, extensive investigations, involving the preparation of over 50 derivatives, were conducted by a team headed by Maruyama, because of the unique structures; they are incredibly stable, and crystals can be recovered after boiling off a nitric acid solution. The t-Bu group (probably still the only known case in nature) appearing as a 9H NMR singlet was verified by production of pivalic acid upon Kuhn-Roth oxidation. The nature of all hydroxyl functions were also clarified by NMR taken in DMSO-d_6 before and after addition of D_2O.

Out of desperation to secure any structural information, GA **5** was reduced with lithium aluminum hydride (LAH), but, as expected this gave only a thick oil, the product being an octaol. An undergraduate left this in a drying oven that unknown to him had a broken thermostat, and so was horrified to return from a group baseball game to find that the syrup had become scorched tar. However, chromatography of the tar yielded crystals of GA triether **9**, the most important of all derivatives ! Amazingly, the octaol had recycled back to give the original cage structure, but now the three lactones had been replaced by three tetrahydrofurans. The NMR of ginkgolides was uninformative since the proton systems are isolated by quaternary carbons, oxygens and lactones; the NMR of the triether was far more complex because of the six extra protons but contained more information. It was during the scrutinizing NMR measurements carried out by Woods and Miura that they found irradiation of the t-Bu peak led to 10% to 33% enhancements in the area of certain proton signals. Instead of dismissing this as an integrator error, they repeated the measurements numerous times until they were convinced of the increments. Woods was an uncompromising person and this attitude had led to the discovery of NOE (21) which was unknown at that time; soon after we had interpreted the results correctly, the first NOE paper dealing with β,β-dimethylacrylic acid appeared in the literature (1).

The crucial finding of NOE allowed us to put together all available structural information into a logical structure within a short period. During these chemical and spectroscopic studies we were constantly thinking about what the ultimate structures of ginkgolides might be. We derived great enjoyment from these studies but they were the last of the "romantic" and classical structure determinations we carried out.

1.2.4 The Future of Natural Products Science

Until recently it was customary for the primary targets in natural products chemistry to be isolation and structure determination, which in many cases was followed by a total synthesis. Without doubt it is more economical in terms of both time and input to determine structures by X-ray, particularly if they represent a new skeleton with multiple chiral centers. However, many are not crystallizable due to their minute quantity or inherent noncrystallinity, and it is cases like these that force one to develop new applications or twists in spectroscopy that may lead to a breakthrough methodology. This trend for developing new methods in structural studies is similar to but unfortunately not as common as that seen in synthesis, where the major effort is to devise new and general reagents or reactions as well as to complete the synthesis. A conceptually innovative approach is always a prime target in a synthesis; the chemical community is not interested in a synthesis paper that describes a synthesis by routine combinations of known reactions and concepts. In contrast, in structural studies, the focus of interest frequently rests only in the final structure, the process being neglected. This prevailing difference in attitude between synthetic and structural natural product chemists can be attributed to differences in the two fields. Namely, although synthesis has developed to such an extent that most compounds can be made by nonsynthetic chemists, possibilities for further development are still boundless. In structure determination we have been overly distracted by the spectacular development in instrumentations and the approach has become too mechanical or technical, often with little space left for intellect. This is what is meant by the comment "structure determination has become routine."

It should be stressed that in tackling future topics, a constant effort to develop imaginative approaches in isolation and structure determination is indispensable and, in collaboration with scientists of other disciplines, it is a role that the natural products chemists have to fulfill. Analysis of a complex structure by 500 MHz NMR employing a new two-dimensional pulse sequence cannot be regarded as being imaginative if it is a straight application. Topics are challenging because they resist solution by routine methods. The era in which natural products chemists were engaged in the isolation and structural investigation of a static compound from a natural source has mostly come to an end.

We should look at the field from a new angle and attempt to define its future role. The traditional classification of various disciplines of science – biology, chemistry, biochemistry, physics, physiology, medicine, etc. – is becoming less and less distinct with the advancement in scientific concepts, experimental methods, instrumentation, and technology. Scientific research is thus facing a period in which it is not only multidisciplinary but is almost becoming *nondisciplinary.* Collaborative efforts among scientists with broad and overlapping interests are essential as research moves into intricate topics dealing with the nature of life. Nature has been divided into many branches, and over the years each branch has flourished and become deeply specified, a tendency which continues to intensify. However, no such division exists in nature itself, and from the viewpoint of natural products chemistry, the finely-tuned division has left wide gaps between disciplines; these gaps abound in important problems waiting to be recognized and solved. Big gaps exist between chemistry and the fields of biology, biochemistry, pharmacology, physics, etc.; the same holds among other disciplines.

Because of their training, organic chemists have the clearest concepts regarding molecular structures, configuration, conformation, reactivities, and methods of synthesis. However, it is almost impossible for the chemists to realize or detect where exciting problems relating to life science exist. A biologist with a keen observational mind may take a natural phenomenon for granted, not knowing what modern natural products chemistry is capable of achieving or not realizing that a behavior may be induced by a compound. Similarly, biochemists may produce and analyze complex data pertaining to some toxicity but without placing the discussions on a structural basis. It is precisely in such areas that interdisciplinary collaboration, which starts from interdisciplinary conversation, is needed. A decade ago, it was hardly possible for the natural products chemists (or bio-organic chemists or natural products scientists) to contribute to the solution of such problems. Advancements in the field have made it possible for natural products chemists to participate in such topics and to put them on a clearer molecular structural basis.

A broad general area that could be considered a prime target for future research is clarification of the mode of action of physiologically or biologically active compounds ("factors") on a structural basis. And probably almost invariably, this will involve clarification of the interaction between the factor and its biopolymeric receptor. The plant hormones auxins, gibberellins, cytokinins, abscisic acid, ethylene, and brassinolides have been known for some time, but despite their importance as plant growth regulators and in biotechnology, our knowledge regarding their mode of action, receptors, etc., is extremely limited. To understand these will be very challenging; the problems are not unrelated to attempting to understand a small portion about life from a molecular structural viewpoint. The same applies to any bioactive natural products. Establishment of a structure is merely a starting point for more important questions to be answered. What factors are involved in inducing biosynthesis and how do they do it? How does a bioactive substance exert its activity? Understanding these on concrete structural terms constitutes immense problems for which we have as yet no answer for any compound. There are also such problems as elucidation of mechanisms of nitrogen fixation, photosynthesis, and cell differentiation. Many aspects of such extraordinarily complex phenomena have been clarified, but they remain as mysterious as ever.

In many cases, the more we know through studies performed by specific groups in various fields, the more enigmatic the sequence of events becomes. Science is still far from reproducing any natural life-related phenomena. Many of these interdisciplinary-type studies are being performed by biophysicists, biochemists, and molecular biologists. It has been difficult for natural products chemists to enter the field because their science has been mostly centered around exact structures of small molecules; however, the time has come in which they can discuss the interaction of small bioactive molecules with folded proteins or nucleic acids and their conformational changes, and to follow the course of these interactions in a time-resolved manner. Natural products chemists are also capable of synthesizing tailored molecules that will clarify specific aspects of a function. They are also the scientists who have the technique to properly isolate and determine the structures of factors that are extremely unstable, that only have a transient existence, or that exist in minuscule amounts, most of which would have been impossible to handle a decade ago. Natural products chemists are capable of clarifying detailed structural and conformational aspects of a life science-related phenomenon. This will enable scientists in other disciplines to have a clearer image, which in turn will advance the field further. The multidisciplinary approach and development of natural products science, and the role the bio-organic chemists is expected to perform have never been more unique, challenging, and exciting.

References

1 Anet F A L, Bourn A J R 1965 Effect of steric compression on coupling constants. J Am Chem Soc 87:5249–5250
2 Broughton H B, Ley S V, Slawin A M Z, Williams D J, Morgan E D 1986 X-Ray crystallographic structure determination of detigloyldihydroazadirachtin and reassignment of the structure of the limonoid insect antifeedant azadirachtin. J Chem Soc Chem Commun 46–47
3 Butterworth J H, Morgan E D, Percy G R 1972 The structure of azadirachtin; the functional groups. J Chem Soc Perkin Trans I 2445–2450
4 Chang C C, Nakanishi K 1983 Specionin, an iridoid insect antifeedant from Catalpa speciosa. J Chem Soc Chem Commun 605–606
5 Clarke H T, Johnson J R, Robinson R (eds) 1949 The chemistry of penicillin. Princeton Univ Press Princeton, 1094 pp
6 Fieser L F, Fieser M 1959 Steroids. Reinhold New York, 945 pp
7 Karrer P 1950 Organische Chemie. 4th English ed. Elsevier Amsterdam
8 Kraus W, Bokel M, Klenk A, Pohnl H 1985 The structure of azadirachtin and 22,23-dihydro-23β-methoxyazadirachtin. Tetrahedron Lett 26:6435–6438
9 Maruyama M, Terahara A, Itagaki Y, Nakanishi K 1967 The ginkgolides. I. Isolation and characterization of the various groups. Tetrahedron Lett 299–302
10 Maruyama M, Terahara A, Itagaki Y, Nakanishi K 1967 The ginkgolides. II. Derivation of partial structures. Tetrahedron Lett 303–308
11 Maruyama M, Terahara A, Nakadaira Y, Woods M C, Nakanishi K 1967 The ginkgolides. III. The structure of the ginkgolides. Tetrahedron Lett 309–313
12 Maruyama M, Terahara A, Nakadaira Y, Woods M C, Takagi Y, Nakanishi K (1967) The ginkgolides. IV. Stereochemistry of the ginkgolides. Tetrahedron Lett 315–319
13 Nakanishi K 1967 The ginkgolides. Pure and Appl Chem 14:89–113
14 Nakanishi K, Goto T, Ito S, Naatori S, Nozoe S (eds) 1974 Natural products chemistry. Vol I. Kodansha Tokyo/Academic Press New York, 295–300
15 Nunez D, chignard M, Korth R, Le Couedich J P, Norel X, Spinnewyn B, Braquet P, Benveniste J 1986 Specific inhibition of PAF-acether-induced platelet activation by BN 52021 and comparison with the PAF-acether inhibitors kadsurenone and CV 3988. Eur J Pharmacol 123:197–205
16 Robinson R 1955 The structural relations of natural products. Clarenden Press Oxford, 144 pp
17 Schroeder D, Nakanishi K 1987 A simplified isolation procedure for azadirachtin. J Nat Prod (1987) 50:241–244
18 Taylor D A H 1984 The chemistry of the limonoids from Meliaceae. In: Herz W, Grisebach H, Kirby G W (eds) Progress in the chemistry of natural products. Vol 45. Springer, Berlin Heidelberg New York Tokyo, 1–102
19 Taylor D A H 1987 Azadirachtin. Tetrahedron 43:2779–2787
20 Van der Eycken E, De Bruyn A, Van der Eycken J, Callant P, Vandewalle M 1986 Iridoids: The structure elucidation of specionin based on chemical evidence and ^{1}H NMR analysis. Tetrahedron 42:5385–5396
21 Woods M C, Miura I, Nakadaira Y, Terahara A, Maruyama M, Nakanishi K 1967 The ginkgolides. V. Some aspects of their NMR spectra. Tetrahedron Lett 321–326
22 Woodward R B 1941 Structure and absorption spectra of α,β-unsaturated ketones. J Am Chem Soc 63:1123–1126
23 Woodward R B 1942 Structure and absorption spectra. III. Normal conjugated dienes. J Am Chem Soc 64:72–75
24 Woodward R B 1942 Structure and absorption spectra. IV. Further observations on α,β-unsaturated ketones. J Am Chem Soc 64:76–77
25 Zanno P, Miura I, Nakanishi K, Elder D L 1975 Structure of the insect phagorepellent azadirachtin. Application of PRFT/CWD carbon-13 nuclear magnetic resonance. J Am Chem Soc 97:1975–1977

Chapter 2

Fractionation and Proof of Structure of Natural Products

J. SNYDER, R. BREUNING, F. DERGUINI, and K. NAKANISHI

2.1 Introduction

The past few years have witnessed a veritable explosion in the development of new methods for the isolation and structure determination of natural products. This is true in particular for the applications of novel separation technologies and modern spectroscopic techniques for structural analysis. In many instances, the discovery of these new methods was spawned by persistent, formidable obstacles in the preparation of pure substances from complex mixtures or in the microscale determination of complex structures. This is exemplified by the application of the benzoate chirality method for determining linkage sites of sugars in branched glycosides (236, 237) or the development of new countercurrent chromatography techniques for the separation of labile compounds like the tunichromes (53, 54). In other situations, an existing technique of theoretical interest was tailored to solve structural problems, as was the case with the application of two dimensional nuclear magnetic resonance in determining chemical structure.

In the realm of natural products isolation, with its "classical" extraction schemes and chromatographic techniques – gas chromatography, liquid chromatography, high performance LC, and thin layer chromatography, to name a few – new methodology was developed as research began to focus on molecules that had eluded previous investigators. These new methods can accommodate large molecular size, lack of mobility in commonly used adsorbents, sensitivity to decomposition, or lack of sufficient quantities for detection and analysis. Thus, for example, the revitalization of liquid-liquid chromatography in the form of coil countercurrent chromatography, droplet countercurrent chromatography, rotation locular countercurrent chromatography, and centrifugal partition chromatography has been successful in the isolation of compounds whose routine isolation by conventional column chromatographic techniques proved difficult and time-consuming.

It has always been said in physical organic chemistry that a reaction mechanism is never proven, only conceivable alternatives are eliminated. It can also be said that spectroscopic structure determination doesn't prove a new structure, but only eliminates other conceivable alternatives. Only X-ray crystallography can prove a structure. Spectroscopy provides a means of examining the physical properties of a compound that are affected by its chemical bonding characteristics and relative spatial orientations of the constituent atomic units. Advances in spectroscopic structure determination reflect advances in instrumentation that enhance sensitivity and resolution, and the development of new techniques for extracting maximum spectral information. The combined features of enhanced sensitivity and enhanced resolution enable weak interatomic interactions to be sampled and cleanly observed.

It is not the purpose of this section to give an account of all available isolation and structure determination techniques, but rather to review those developed in the past few years that have made important contributions to methodology in natural products chemistry. Emphasis will be on recent applications, with references given for further considerations.

2.2 Novel Techniques and Recent Developments in Fractionation and Isolation

2.2.1 Countercurrent Chromatography

Countercurrent chromatography (CCC) (40, 160, 175, 180, 255) is a form of liquid-liquid partition chromatography that does not utilize a solid support. Most modern countercurrent chromatographic techniques are not truly "countercurrent," for usually only one of the two phases is mobile.[1] However, the new countercurrent chromatographic techniques have advantages over the older countercurrent distribution or Craig-tube method (69) from which they have evolved. Modern countercurrent chromatography consumes considerably less solvent, time, and effort than the older technique, and avoids the problem of emulsion formation. Moreover, these new instruments come in handy sizes and offer several other advantages that make them attractive alternatives to column chromatography.

Basically, there are four new methods of commercially available countercurrent chromatography: droplet countercurrent chromatography (159, 162, 376), rotation locular countercurrent chromatography (359), several types of coil countercurrent chromatography (170, 176, 177, 183, 185) and centrifugal partition chromatography (51, 52, 172, 174, 178, 276). All are forms of liquid-liquid partition chromatography, as solutes are transported via partitioning between a liquid mobile phase and a liquid stationary phase. Consequently, the solute that partitions to a greater extent into the mobile phase will be eluted faster from the instruments.

Interest in countercurrent chromatography was revived in an effort to overcome problems frequently encountered in adsorption or exclusion chromatography, which uses a solid support. Most prominent among these problems was the difficulty in resolving mixtures of sensitive or highly polar compounds. Irreversible adsorption onto the solid phase, solid phase catalyzed decomposition, especially with silica gel, and severe tailing are obstacles to the purification of such molecules.

Liquid-liquid partition chromatography in the absence of a solid phase avoids these problems.[2] Furthermore, as the mechanism of solute transport is partition-

[1] A true *countercurrent* system (called "dual countercurrent system") has been introduced only very recently. In this system, one of the two mobile phases is a foam (175).

[2] Partitioning also plays a major role in chromatographic separations with a solid support, such as thin layer or column chromatography, when a liquid phase is bound to the solid adsorbent, forming

ing into the mobile phase, relative partitioning behavior controls the separation of different solute molecules. Since relative partitioning behavior can be significantly different than relative adsorption behavior, compounds that do not separate well with adsorption chromatography may pose no difficulty for liquid-liquid partition chromatography. Solute molecules that do not elute from the instrument but remain dissolved in the stationary liquid phase are routinely recovered by displacing the stationary phase from the columns. Such recovery of immobile solute is a major advantage of countercurrent chromatography. The number of theoretical plates available with modern instruments is often sufficient to enable final purification to be achieved without resorting to a subsequent procedure. Consequently, modern countercurrent chromatography nicely complements adsorption and exclusion chromatography, particularly reversed phase HPLC, in the isolation of polar compounds.

2.2.1.1 Coil Countercurrent Chromatography

Coil countercurrent chromatography (coil-CCC) is based on two systems, both developed by Ito (170, 176, 177, 183, 185). The simpler of the two is the hydrostatic equilibrium system, HSES (Fig. 2.1a), which employs a stationary coil. While

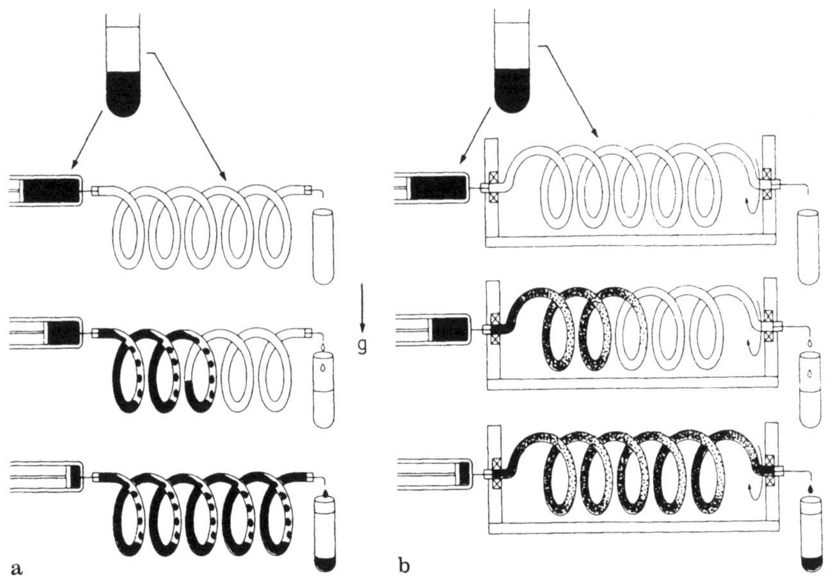

Fig. 2.1. Principle of the two basic coil-CCC systems. **a** Hydrostatic equilibrium system, HSES; **b** Hydrodynamic equilibrium system, HDES (170)

[2] (continued) a liquid stationary phase. An example would be the use of an organic eluent saturated with water with silica gel as the adsorbent. Here the water is bound to the silica gel acting as a stationary phase with partitioning of the solute between the mobile organic phase and the bound water strongly influencing the retention behavior. Nonetheless, an interaction between the silica gel and the solute still occurs and frequently leads to the above-mentioned difficulties.

of little practical utility, HSES illustrates the principle of coil-CCC. The coil is initially filled with the stationary phase. The second, and in this example, heavier phase is slowly applied by a pumping or syringe mechanism, passing through the stationary phase in the descending turn of each coil and displacing the stationary phase in each ascending turn, until the mobile phase emerges from the end of the coil. Should the lighter phase be selected as the mobile phase, it would displace the stationary phase from the descending turns and pass through the stationary phase in the ascending turns. Partitioning of the solute occurs at the interface of these two phases.

In the second form of coil-CCC, the hydrodynamic equilibrium system, HDES, a coil similar to that in HSES is rotated about its own axis, creating an "Archimedian screw" phenomenon (Fig. 2.1b). As a glass bead or air bubble in a rotating coil filled with water will traverse the length of the coil and emerge from one end, so will an equilibrated mobile phase, continuously applied by pumping, pass through a coil filled with an immiscible stationary phase.[3] Once the mobile phase has been applied and emerges from the coil terminus, hydrodynamic equilibrium has been attained and solutes may be introduced at the "head" of the coil.

The HDES is considerably more efficient than the HSES due to the greater interfacial area of the two phases and the greater degree of mixing; efficiencies of up to 1000 theoretical plates have been reported. Furthermore, HDES utilizes the entire coil for partitioning, whereas in the HSES about 50% of the coil is occupied by mobile phase only, leading to band broadening (Fig. 2.2). The HSES, however, invariably retains 50% of the stationary phase, whereas the HDES retention of stationary phase – particularly in centrifugal schemes described below – varies with the rotational speed and the solvents employed (169, 170, 171).

A number of centrifugal coil-CCC systems, both planetary and nonplanetary, have been devised employing HDES (181, 184, 376). All of these systems utilize a centrifugal force to accelerate the flow of the mobile phase through the coil. Continuous improvement in the design of the centrifugal systems has stabilized the retention of stationary phase once a critical rotation rate has been surpassed (179). Under such conditions, the heavier of the phases tends to the outside of each turn while the lighter of the phases tends to the inside. Fluctuations in the centrifugal force produce a thorough mixing at the interface of the two phases (170).

All of the centrifugal coil-CCC systems have been described in detail (170, 176, 177, 183, 185) (Fig. 2.3); the horizontal flow-through centrifuge is perhaps the most versatile (173). In this apparatus, two coils are mounted on opposite sides of the central axis, about which they maintain a synchronous planetary motion – that is, the coils revolve about their own axes at a rate equal to the rate of rotation about the central axis of the apparatus. The mode of synchronous planetary rotation, however, may be different for each coil. One coil may rotate on its axis counter to the direction of rotation of the apparatus (Fig. 2.3, Scheme

[3] As in all forms of countercurrent chromatography, the mobile and stationary phases should be equilibrated prior to use to prevent variation of the composition of two phases due to equilibration in the instrument. In some instances, however, this in situ equilibration can be utilized as a gradient-type technique to separate otherwise unresolvable compounds.

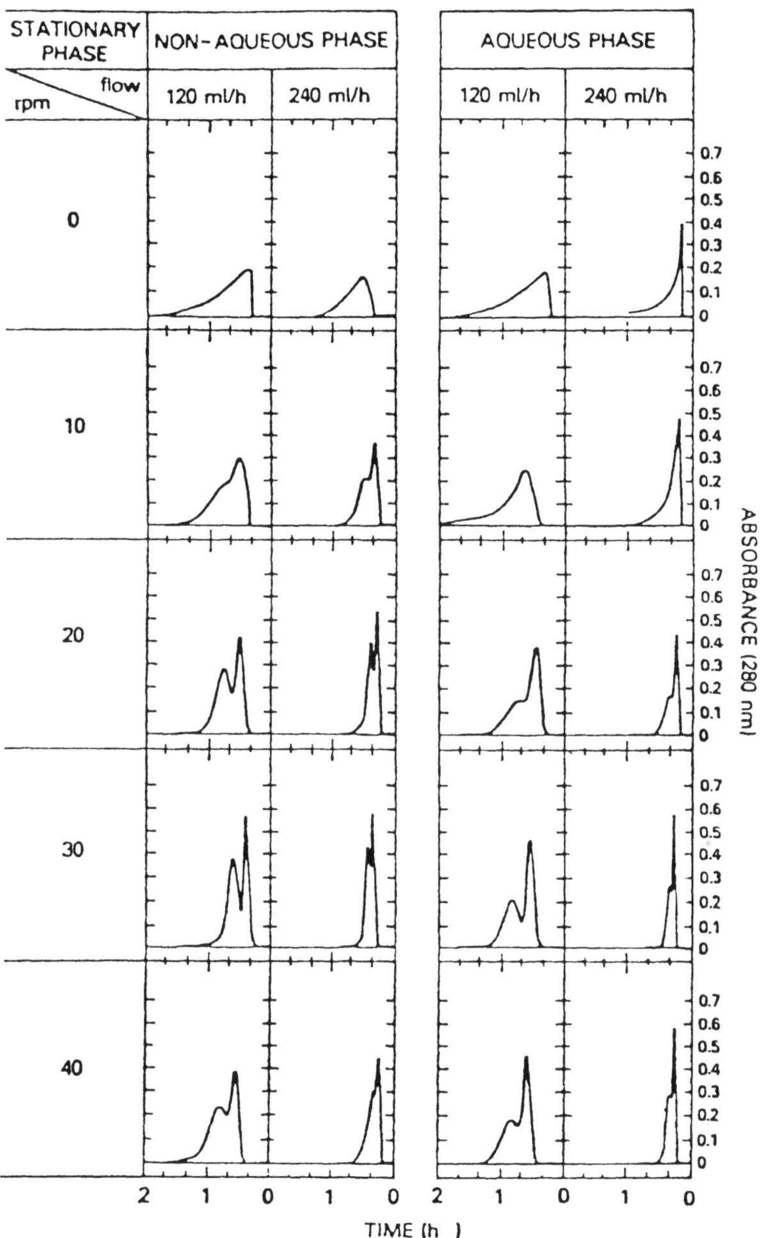

Fig. 2.2. Effects of coil revolution speed and flow rate upon separation efficiency of the dinitrophenyl amino acids, DNP-Glu and DNP-Ala. HSES mode is represented by rpm = 0; other separations represent HDES mode (169)

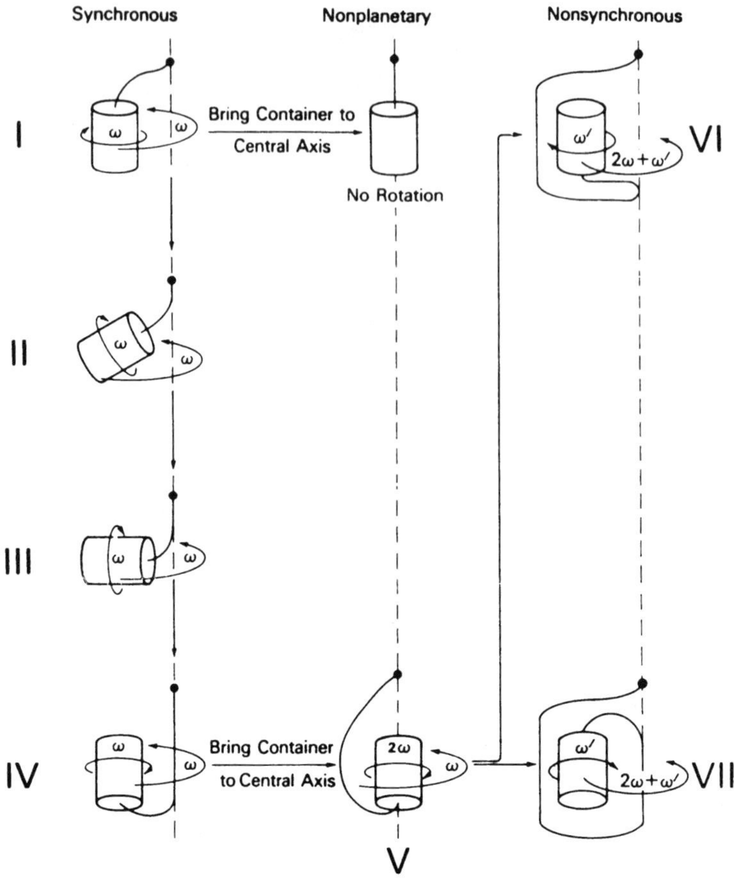

Fig. 2.3. Synchronous (I–IV), nonplanetary (V), and nonsynchronous modes of coil-CCC (170)

I), while the other coil rotates on its axis in the same direction as the rotation of the apparatus (Fig. 2.3, Scheme IV). These two different types of synchronous planetary motion generate a different distribution of centrifugal force vectors (Fig. 2.4), with a different distribution of the two phases in the coil. After exhaustive investigations into the optimization of resolution capabilities of all the centrifugal systems, Ito and coworkers found the synchronous planetary mode shown in Scheme I (Figs 2.3 and 2.4) as the best system for analytical work, whereas the

Fig. 2.4. Analysis of acceleration of synchronous planetary coil-CCC from Schemes I and IV, Fig. 2.3. **a** Diagram of synchronous planetary motion; **b** coordinate system for analysis; **c** orbits of arbitrary points; **d** distribution of centrifugal force vectors (170)

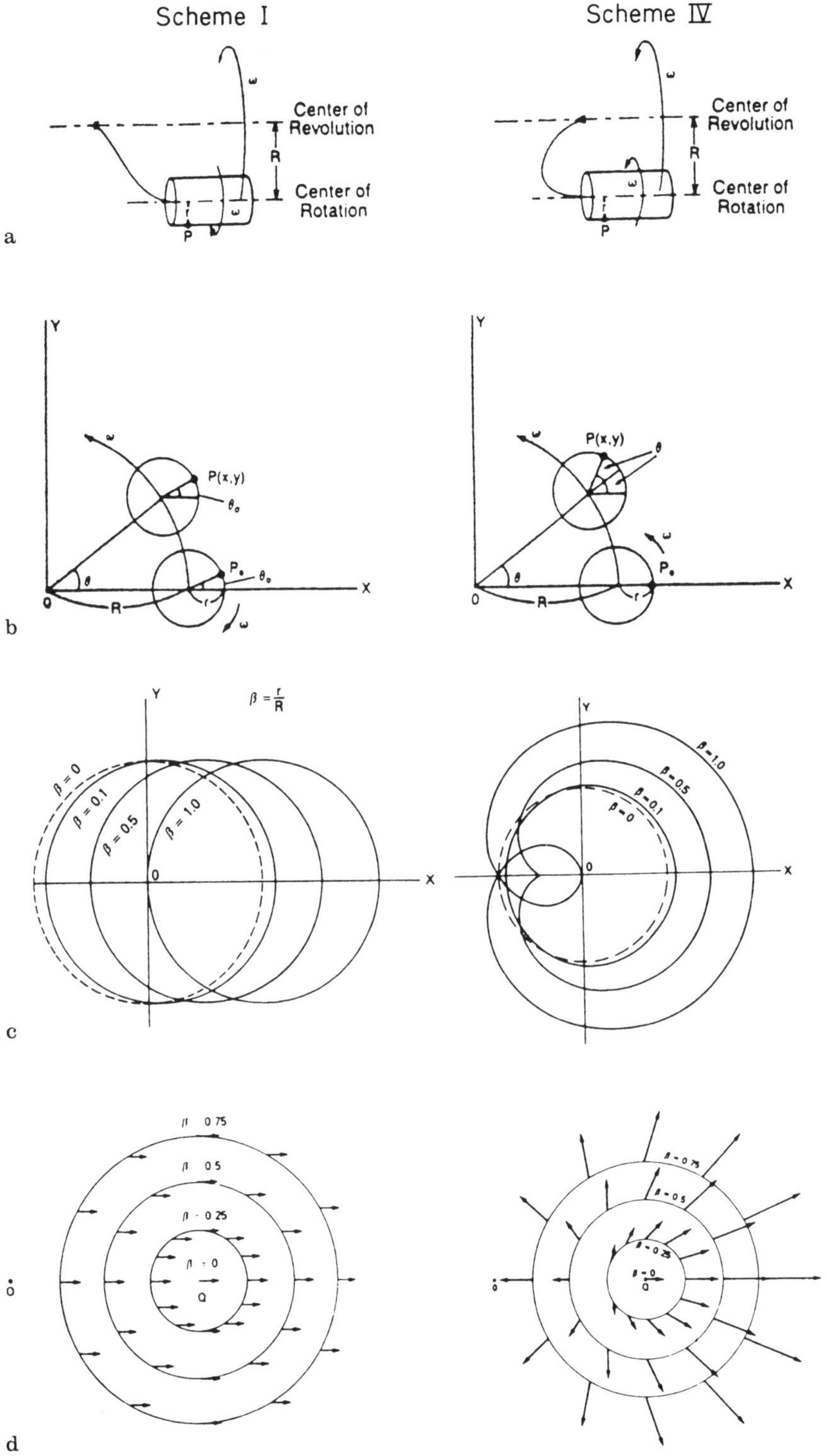
Scheme I
Scheme IV
Center of Revolution
Center of Rotation
R
ω
P
a
b
c
d
P(x,y)
θ_o
P_o
$\beta = \frac{r}{R}$
β = 0
β = 0.1
β = 0.5
β = 1.0
β 0.75
β 0.5
β 0.25
β = 0
O
X
Y

synchronous mode shown in Scheme IV (Figs 2.3 and 2.4), is best for preparative work.[4]

With a capability of 1000 theoretical plates as well as the flexibility in selecting any two-phase solvent systems, coil-CCC is well suited for the isolation of various types of natural products and other materials. Stationary phases containing a dissolved solute, such as an aqueous buffer or salt solution, are acceptable, as is the application of gradient-type mobile-phase systems, though such a gradient may lead to the slow elution of the stationary phase from the coil. The effects of parameters such as rotational speed and flow rate upon stationary phase retention and partition efficiency have been examined in detail. While the effects of such parameters undoubtedly vary with the biphasic solvent system and the nature of the solute to be purified, sufficient groundwork has been reported to enable extrapolation to the problem of interest in terms of selecting a solvent system, optimal rotation speed, and flow rate. A procedure for selecting a solvent system for coil-CCC has been reported (65) that is very similar to that described for RLCC and DCCC (see next two sections). These selection procedures are obviously interchangeable for the various countercurrent techniques.

To date, the number of reports on the application of coil-CCC to the isolation of natural products is still limited. The variety of structural types exemplified in these reports, however, indicates the versatility of the technique. Initial examples dealt mainly with DNP-amino acids (53, 54, 162, 171, 173, 178, 181, 184, 236, 237, 329). Recently, successful resolutions of many other compounds have been reported: Mandava and Ito demonstrated the analytical utility of coil-CCC in the separation of various plant hormones, including the indole auxins (Fig. 2.5), gibberellins, and cytokinins (248, 249, 250). Other results include the chromatography of dipeptides, gramicidins, purines, pyrimidines, and other molecules (66, 184, 208, 209, 239). All these separations employed the centrifugal planetary coil in the synchronous mode with sample loads of up to 1 g (176, 255).

A non-synchronous centrifugal planetary coil system has also been reported (182, 376). This countercurrent apparatus is similar to that used for the synchronous mode of coil-CCC, but has the added sophistication of an adjustable revolution rate of the coil; hence, the rate of revolution of the coil need not be synchronous to the rate of planetary rotation of the apparatus. (This method is occasionally referred to as planetary coil-CCC (PCCC).) Such flexibility is advantageous when the rate of coil revolution must be reduced to prevent emulsification of the two phases, which would result in transport of the stationary phase from the coil. This situation arises when there is low interfacial tension, as frequently occurs when separating macromolecules or partitioning intact cells with aqueous polymer phase systems. For instance, the nonsynchronous mode of centrifugal PCCC was successfully applied to the resolution of mammalian erythrocytes (183). This ability of aqueous solutions of two different polymers to form a two-phase liquid system allows for liquid-liquid partitioning between two aqueous phases (391).

---

[4] Instruments with a counterweight in place of one of the coils in the horizontal flow through coil-CCC serve as prototypes and are still the only commercially available coil-CCC instruments to date, though instruments with coils mounted both sides of the center axis will be available soon. See reference (181).

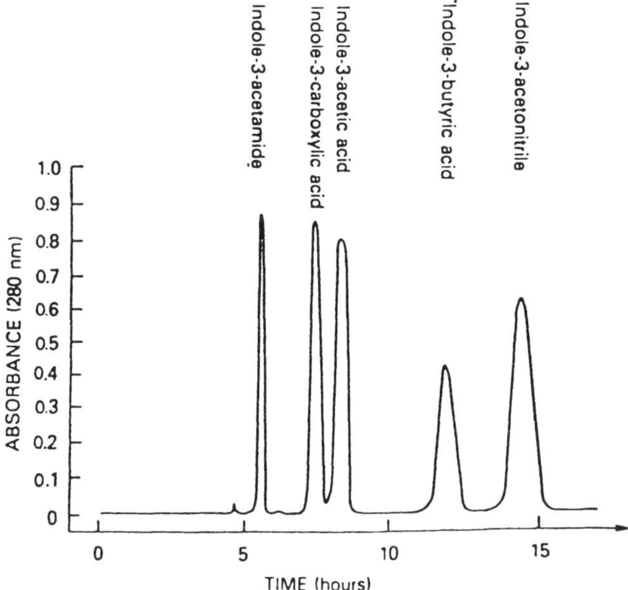

Fig. 2.5. Separation of indole compounds by coil-CCC in hexane-ethyl acetate-methanol-water (0.6 : 1.4 : 1 : 1), mobile phase : aqueous phase (248)

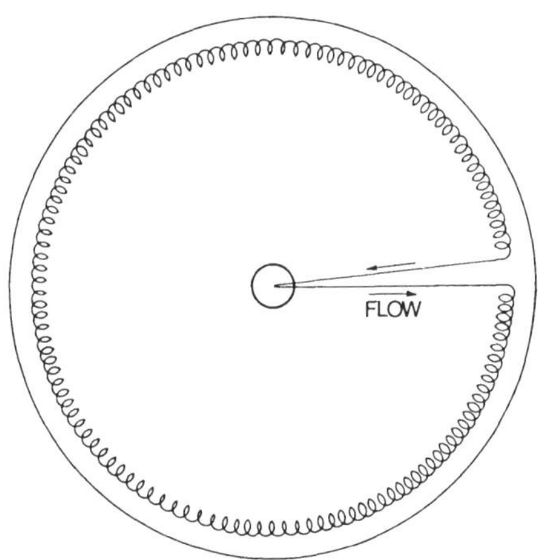

Fig. 2.6. Plan view of toroidal coil-CCC showing right-handed helix with clockwise flow and rotation (371)

The level of efficiency in these systems is comparatively less than that resulting from nonpolymer systems using an aqueous and an organic phase, due to the much slower rate of partitioning with a polymer. This is a consequence of the lack of significant interaction between the polymers of the two phases. Such systems have been utilized on the nonsynchronous coil-CCC (145, 233), in centrifugal partition chromatography (see below), and in a new related technique, toroidal CCC (101, 371) for the separation of cell populations as well as cell organelles and membranes. Toroidal CCC is a relatively simple CCCC technique consisting of a coil arranged circumferentially in a disc, the center of which marks the central axis of a centrifugal field (Fig. 2.6). The same principle of operation as described for the coil-CCC applies for the toroidal CCC.

2.2.1.2 Droplet Countercurrent Chromatography

Droplet countercurrent chromatography (DCCC) is currently the most widely used form of liquid-liquid chromatography without a solid support, because of its high efficiency and its excellent reproducibility in terms of instrument operation and separation results (139, 156, 157, 159, 162, 376). DCCC functions by pumping a mobile phase in the form of droplets through an immiscible stationary phase (Fig. 2.7). If the droplet phase is lighter than the stationary phase, we speak of "ascending mode," whereby the droplets rise through the stationary phase. The term "descending mode" is used if the mobile droplet phase is heavier. Partitioning of the solute occurs at the interface of the droplet surface, leading to separation of the solute components. The only restriction in the selection of solvent systems is that the mobile phase must pass through the stationary phase without displacement of the stationary phase. Usually this necessitates the formation of uniform droplets whose diameter is less than the inner diameter of the column, although wide-bore columns have helped to overcome this limitation. While the restriction of droplet formation limits the applicability of DCCC somewhat, recent results show that a sufficiently wide range of solvent system polarities are acceptable to enable separation of a great variety of natural products (139, 156, 157, 255, 296).

The first reported application of DCC utilized 300 tubes, 1.8 mm × 60 cm, to separate DNP-amino acids with a solvent system of chloroform: acetic

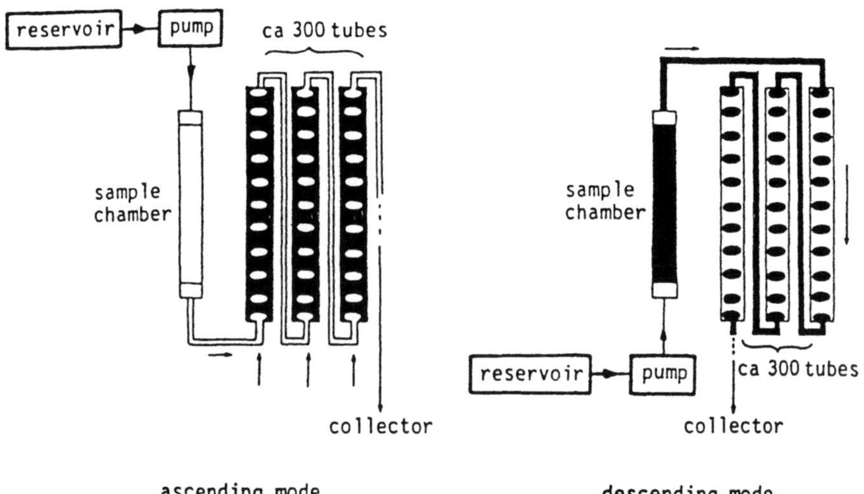

Fig. 2.7. Principle of operation of DCCC in two modes (156)

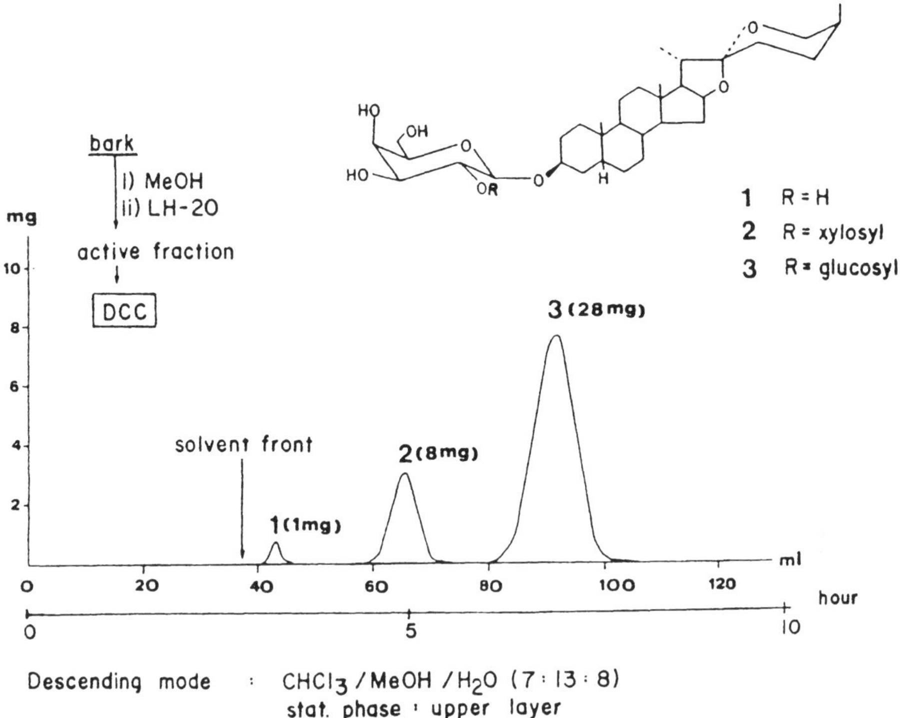

Fig. 2.8. DCCC separation of *Cornus florida* saponins (**1**, **2**, **3**), using $CHCl_3$–MeOH–H_2O (7:13:8, v/v), descending mode of operation (161)

acid:0.1N HCl (2:2:1), in which the lower layer was the mobile phase (296). Since then, numerous separations of natural products have been reported, particularly of saponins, tannins, and other polar molecules. Relatively nonpolar molecules have also been resolved applying nonaqueous, biphasic systems (31, 81, 158). Droplet countercurrent chromatography has proven to be the method of choice for the isolation of many saponins such as the molluscicidal principles of *Cornus florida* (161) (Fig. 2.8), as well as for the isolation of ecdysteroids from insects like *Bombyx mori* (221–223) (Fig. 2.9).

Several procedures may be used in selecting a solvent system for the DCCC. A method based upon thin-layer chromatography (TLC) on silica gel of the solutes has been described (66, 156). The TLC plate is developed with the equilibrated organic phase of an aqueous/organic biphasic system, whereby best results are obtained when the solutes show an R_f value between 0.4 and 0.7. The DCCC separation will then be run in "ascending mode" and mobile phase is chosen according to R_f values as shown in Fig 2.10. Since the organic phase is saturated with water, which is subsequently adsorbed onto the silica gel, the chromatographic behavior of the solutes is strongly influenced by partitioning between the organic eluent and the adsorbed water. Interaction of the solutes with the silica gel will still occur, however, so the order of elution from the DCCC will not necessarily reflect the order of decreasing R_f values. Similarly, the aqueous phase can

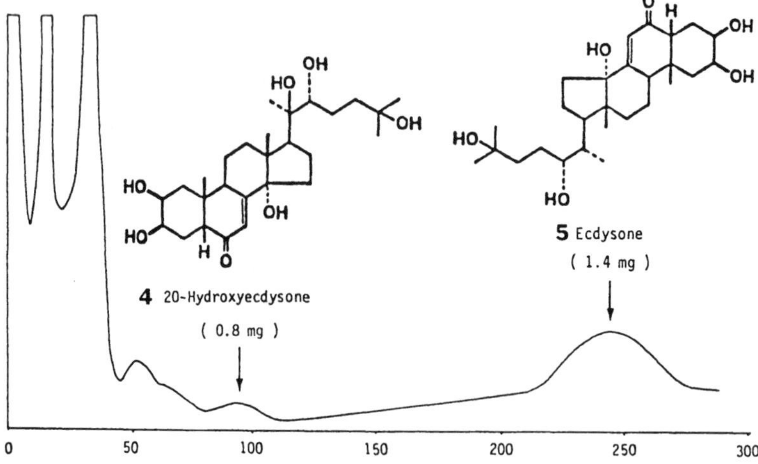

Fig. 2.9. DCCC separation of ecdysteroids from *Bombyx mori* (silkworm) ethyl acetate extract (0.65 g), using $CHCl_3-MeOH-H_2O$ (13:7:4, v/v), ascending mode, 1.0 ml per fraction, detection at 254 nm (221)

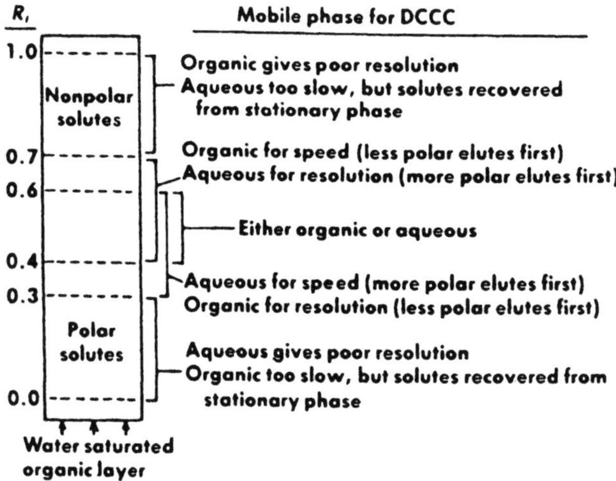

Fig. 2.10. TLC method for selecting a solvent system for DCCC that is also applicable to other CCC techniques (66)

be tested on C-8 reversed phase TLC plates, and a good separation obtained here would suggest a DCCC-run in "descending mode."

The number of biphasic systems applied successfully on DCCC has grown so extensively that the requirement of finding a solvent system suitable for droplet formation is no longer a limitation. Basically, all appropriate systems require two immiscible liquids together with the addition of at least one mutually miscible solvent, usually a water-soluble alcohol, in order to decrease the polarity difference between the two layers. The addition of this modifier allows for better partitioning of the solute molecules, which otherwise may be soluble in only one of the two phases. It also decreases the interfacial tension, thus allowing for suitable droplet formation. The advent of wide-bore glass columns – 2.7 and 3.4 mm i.d. – has broadened the applications of DCCC in several areas. First, the rate of elution can be significantly increased without displacing the stationary phase, thereby reducing the separation time. Second, greater quantities of solute may be charged onto the instrument without loss of resolution or an increase of analysis time. Most importantly, nonaqueous solvent systems are suitable for wider columns, which on the narrow-bore 2-mm columns would not form suitable droplets.

2.2.1.3 Rotation Locular Countercurrent Chromatography

Rotation locular countercurrent chromatography (RLCC) is perhaps the most obvious extension of classic countercurrent distribution. Originally proposed by Signer et al. (350), the apparatus employed for RLCC was later developed by several groups prior to the availability of commercial instruments (7, 40, 175, 180, 255, 351, 401). The prototypes of RLCC covered a range of efficiencies in terms of number of theoretical plates depending upon the capacities of the instrument. While less efficient than the planetary coil-CCC or the DCCC, these models demonstrated the utility of this new chromatographic technique.

In selecting the proper solvent system for a RLCC separation, another screening procedure of biphasic systems utilizing silica gel TLC can be employed (359); this method also works for DCCC. The TLCs of the sample are developed with each of the two phases; those systems giving R_f values greater than 0.8 in one of the phases and between 0.2 and 0.5 in the other phase are potentially applicable systems. Final selection of the solvent system is based on a test distribution of the sample mixture, 5 to 10 mg, between the two phases of the solvent system, 5 to 10 ml of each phase. A system in which 15% to 25% of the sample is distributed into one of the two phases is optimal with regard to resolution and length of time necessary to complete the chromatography. The phase containing the higher percentage of the sample mixture serves as the stationary phase. The amount of sample in each of the two phases after partitioning is usually determined by weighing the residues after evaporation when dealing with unknown compounds. In separation of known compounds with well-defined chromophores, absorbance readings may be taken of the two phases to determine the partition coefficients (65). Following the separation of the two phases, TLCs may be employed to examine the nature and selectivity of the partition.

The RLCC apparatus is constructed from 16 columns, each separated into 37 compartments or loculi by Teflon discs perforated in the center to allow the solvent to flow between the compartments. These columns are mounted cylindrically about a rotational axis and are connected in series with 1-mm Teflon tubing. The solvent serving as the stationary phase is loaded in the machine, expelling air, and the columns are inclined to an angle of 20° to 30°. In the ascending mode (Fig. 2.11), the lighter mobile phase is applied with a pump to the bottom of the first column and rises to the uppermost corner of the first loculus, displacing the stationary phase. As its volume increases, the mobile phase displaces the stationary phase, attains the level of the hole in the disc, and enters the second loculus, which is slightly elevated from the first loculus by the inclination of the columns. This

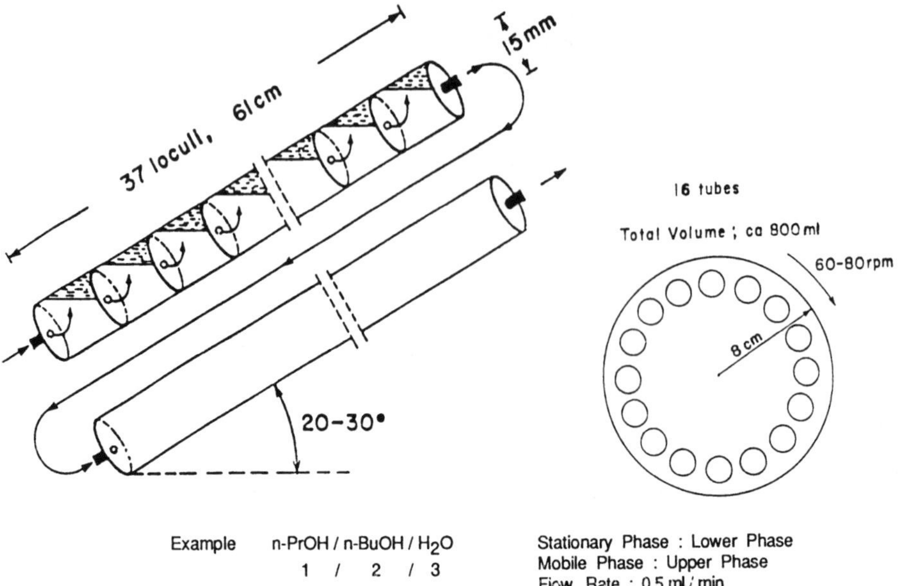

Fig. 2.11. Principle of operation of RLCC in ascending mode (359)

process of the mobile phase rising to the uppermost corner of each loculus, displacing the stationary phase, and passing on to the next loculus continues through all 16 columns, being directed to the bottom of successive columns by Teflon tubing.

When the apparatus is charged with the mobile phase, the sample is loaded into the injector loop and transferred to the first column.[5] Within each loculus, the sample is partitioned between the mobile and stationary phases whose interface is constantly renewed by the rotation of the apparatus at 60 to 80 rpm. This rotation speed does not allow excessive agitation to occur; this could result in the formation of emulsions, a problem commonly encountered in countercurrent distribution.

Should the lighter of the two phases dissolve the larger percentage of the sample, the descending mode of operation, with the heavier phase as the mobile phase, is used. The principle is the same as in the ascending mode, but the direction of application of the mobile phase is reversed. In the descending mode, the mobile phase is applied to the top of each column and drops to the lowermost corner of each loculus (Fig. 2.12).

Results on the early home-built models demonstrated the wide range of solvent systems applicable on RLCC. Using an aqueous methanol (5% water), petroleum ether system on an RLCC with 187 loculi, a mixture of chlorophylls (7) was resolved as well as a mixture of carotenoids (133) that had been extremely difficult to resolve by any other means – including countercurrent distribution, which had

[5] The sample is loaded into an injector loop as a solution in either the stationary or the mobile phase or in a biphasic mixture of both.

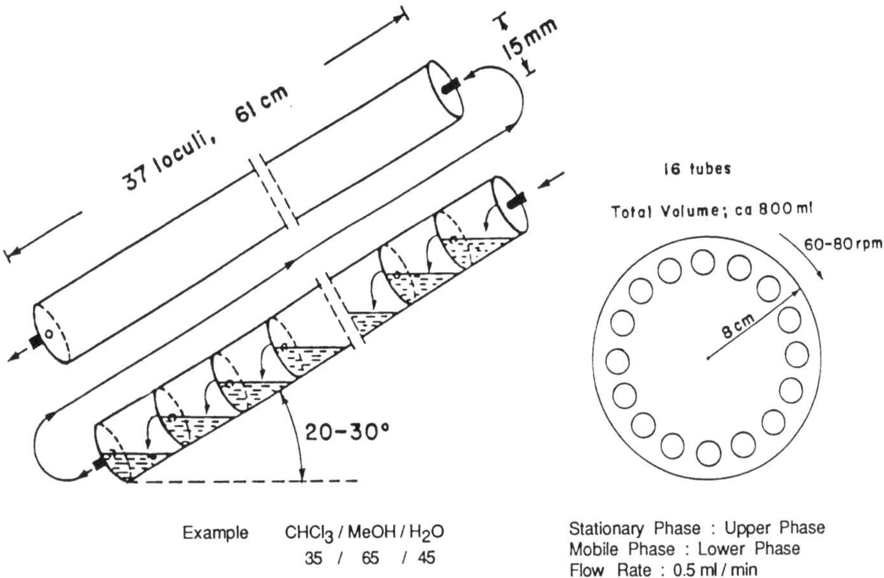

Fig. 2.12. Principle of operation of RLCC in descending mode (359)

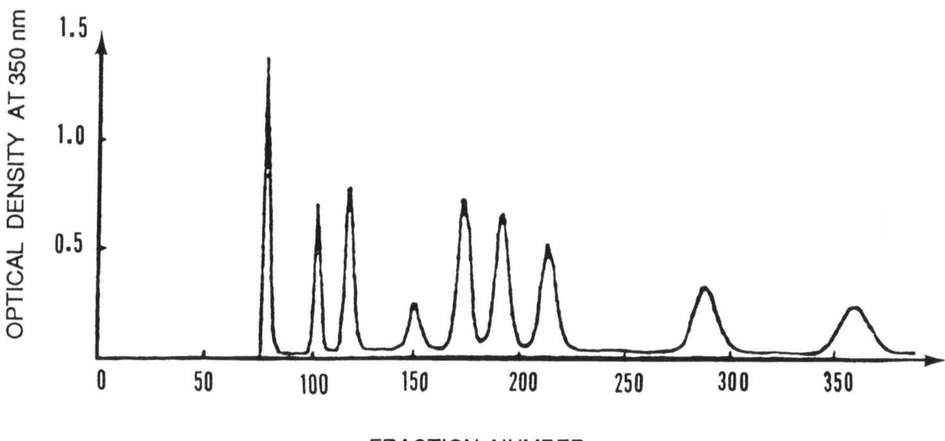

Fig. 2.13. Separation of DNP-amino acids on RLCC, using $CHCl_3$-glacial HOAc-0.1 N HCl (aq.) (2:2:1, v/v), ascending mode. Peaks from left to right: *N*-DNP-L-arginine, *N*-DNP-L-aspartic acid, *N*-DNP-D, L-glutamic acid, *N,N'*-diDNP-L-cystine, *N*-DNP-alanine, *N*-DNP-L-alanine, *N*-DNP-L-proline, *N*-DNP-L-valine, and *N*-DNP-L-leucine. Column efficiency is ≈3000 T.P.; total elution time is 70 h (180)

led to emulsion formation. Ito and Bowman reported the separation of DNP-amino acids on a machine with 5,000 loculi and an efficiency of about 3000 theoretical plates, or 60% partition efficiency per compartment (Fig. 2.13) (180).

The application of RLCC to the fractionation and isolation of natural products has also been described. Of particular interest was the isolation of chromo-

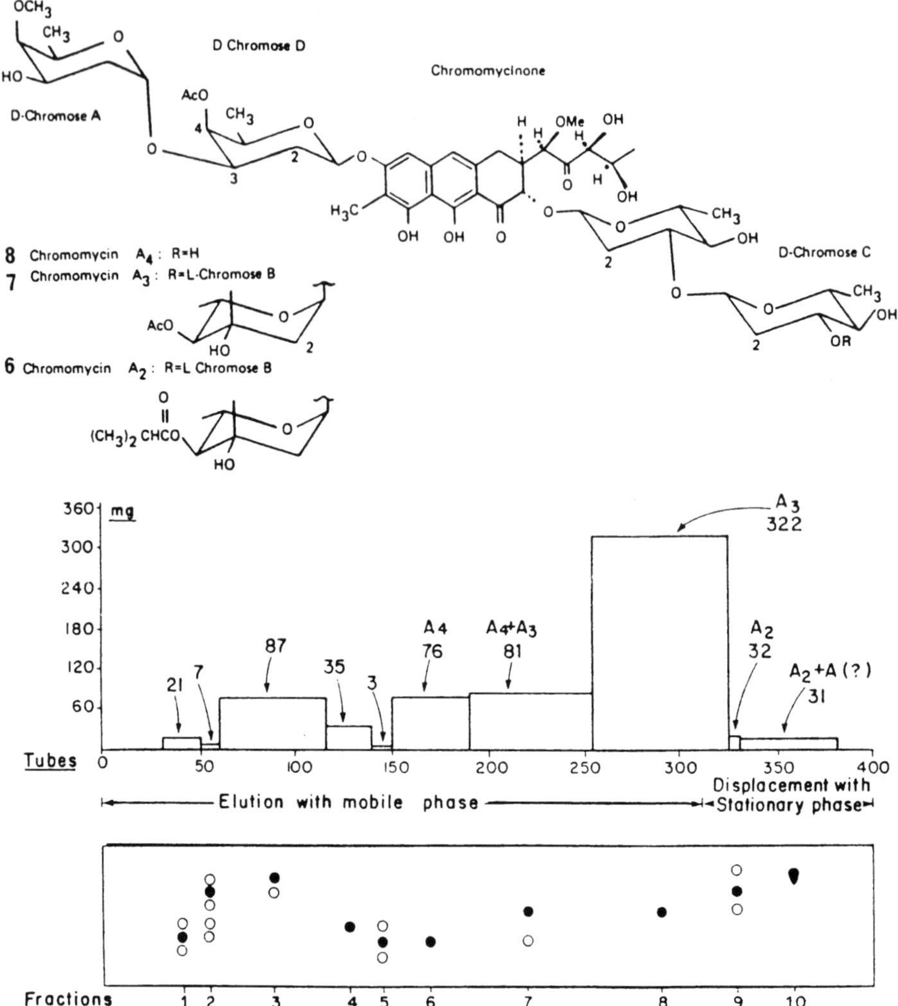

Fig. 2.14. (Upper) Structures of the antitumor antibiotics, chromomycins A2, A3, and A4 (**6, 7, 8**). (Lower) Separation of the chromomycins from crude extract (700 mg) on RLCC, $CHCl_3-MeOH-EtOAc-H_2O$ (2:2:4:1, v/v), ascending mode; flow rate, 0.5 ml/min, 6 ml per tube; total time, 2.5 days; material balance, 97%. TLCs of the fractions developed in 1% oxalic acid in EtOAc shown below each fraction (359)

mycins A_2, A_3, and A_4, anti-tumor antibiotics, from the crude culture extract of *Streptomyces griseus* No. 7 (359). RLCC proved to be a great improvement over the original isolation, which used chromatography on silica gel with 1% oxalic acid in ethyl acetate as the eluent (272, 273). The chromomycins were readily separated from early eluting impurities with similar R_f values, which had caused difficulties in the original isolation (Fig. 2.14).

Two interesting uses of a partition-chromatographic technique are illustrated by the application of RLCC to the resolution of the stereoisomers of the thioxan-

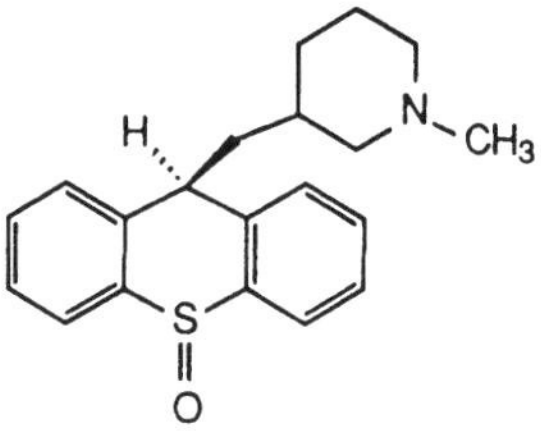

9, thioxanthene oxide

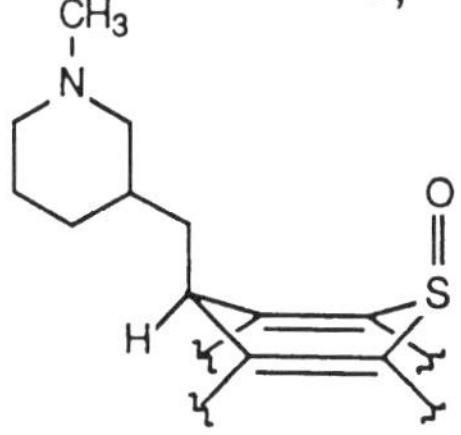

9a

$K_a = 1.04 \times 10^{-10}$

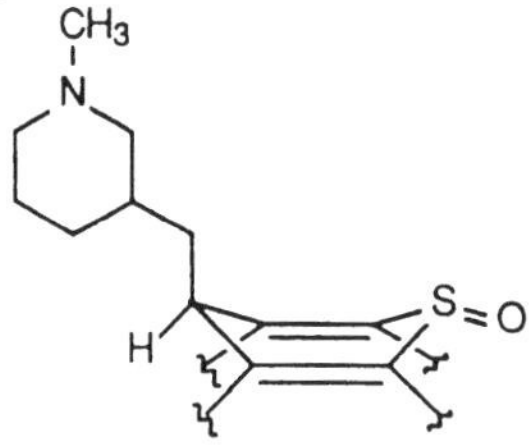

9b

$K_a = 1.73 \times 10^{-10}$

$$K_a = \frac{[B][H^+]}{[BH^+]}$$

Fig. 2.15. Structure of thioxanthene oxides (**9**, **9a**, **9b**) separated by RLCC (269) enantiomers with (*R*, *R*)-5-nonyltartrate

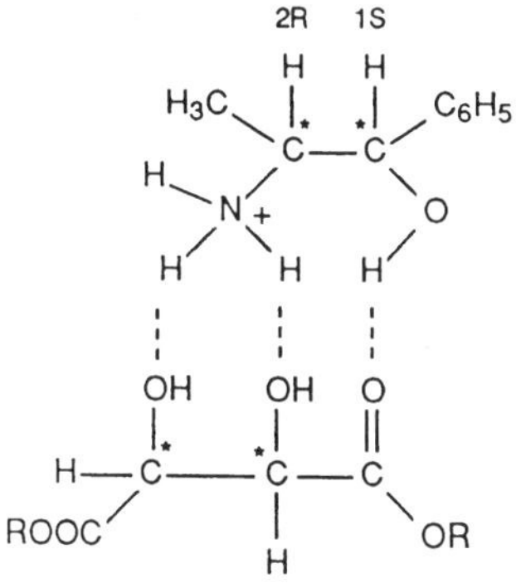

10a

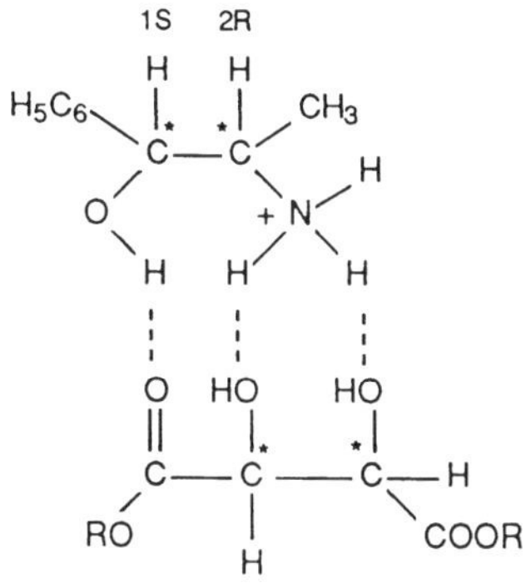

10b

R = 5-nanyl.

norephedrine = $H_5C_6-CH(OH)-CH(NH_2)-CH_3$

Fig. 2.16. Structure of diastereotopic complexes of norephedrine enantiomers (**10a**, **10b**) with (*R*, *R*)-5-nonyltartrate (82)

thene oxide (**9**) (Fig. 2.15) (269) and to the chiral resolution of norephedrine (Fig. 2.16) (82). The separation of **9a** and **9b** took advantage of the different pK_b's of these two stereoisomers, and employed an aqueous phase buffered to a pH at which one of the two stereoisomers was significantly more protonated than the other. The solvent systems suitable in this example were benzene-hexane (1:4) aqueous borate buffer (pH 8.73 or 9.43).

For the chiral resolution of racemic norephedrine, an aqueous phase containing 0.5M sodium hexafluorophosphate at pH 4 was utilized as the stationary phase with a mobile phase of 1,2-dichloroethane containing 0.3M (*R,R*)-5-nonyltartrate. The racemic norephedrine forms diastereotopic complexes **10a** and **10b**, which have different partitioning characteristics, whereby the nonyl-residues maintain the solubility in the organic phase. Such applications of a chiral template for the resolution of racemic mixtures via liquid-liquid chromatography should have a broad range of possibilities.

2.2.1.4 Centrifugal Partition Chromatography

A new, centrifugal partition-chromatographic (CPC) technique was recently introduced, operating upon similar basic principles as the DCCC described above (51, 52, 172, 174, 178, 276). In this instrument, however, the columns are literally channels (3 × 40 mm) drilled into a series of 12 polychlorotrifluoroethylene blocks or cartridges (54 columns per cartridge). The columns within each cartridge are connected by a fine channel that guides the mobile phase from the outlet of one column to the inlet of the next. The cartridges themselves are connected by Teflon tubing and are arranged around the rotor of a centrifuge with the longitudinal axis of the columns parallel to the centrifugal force vector (Fig. 2.17a, b) (332). A leak-free flow transition from the still to the moving parts of the CPC unit is provided through a rotary seal based upon the rotor-stator principle. It is

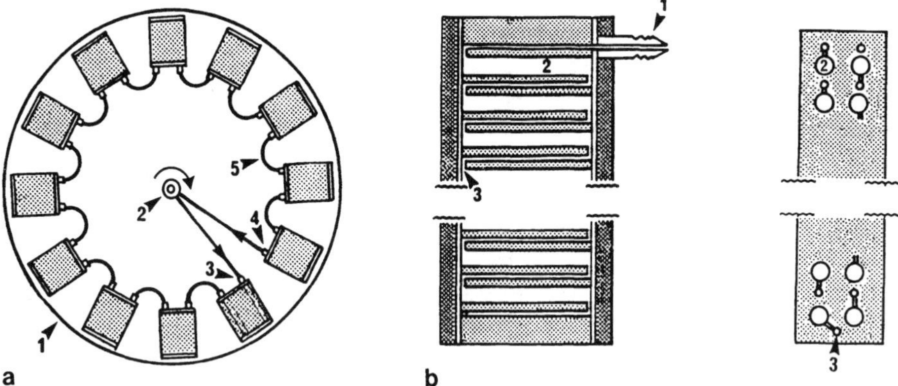

Fig. 2.17. a Arrangement of separation cartridges (*3*) in the rotor (*1*) of a CPC instrument; (*2*) rotary seal joint; (*4*) and (*5*) teflon tubing connections. **b** Side and frontal cross section of a separation cartridge in a CPC instrument: (*1*) connectors for solvent in- and outlet; (*2*) separation columns; (*3*) connecting channels (332)

made up of a static ceramic disk pressed against a moving graphite part and can withstand pressures of up to 65 atmospheres.

Actual droplet formation is not necessary as long as the mobile phase can pass through the stationary phase without displacing it. As a matter of fact, both aqueous and nonaqueous biphasic systems have been used (51, 332). Recent results reveal that a variety of natural products such as saponins, saturated fatty acids, and biomacromolecules – the latter on a biphasic aqueous polymer system – can be separated (332). Because of its comparatively rapid and non-destructive action, CPC has recently been used in the isolation of the tunichromes, the rather sensitive phenolic peptides from the vanadium-collecting sea squirt *Ascidia nigra* (53, 54) and of 11-*cis*- and *trans*-retinal from a complex photoisomerization mixture (51).

After filling the cartridges with stationary phase, the rotor is accelerated to 500 to 2000 rpm. If the stationary phase is the heavier one ("ascending mode"), the mobile phase is pumped "through" the stationary phase against the centrifugal force, which keeps the former in place and prevents it from being displaced. In "descending mode" the heavier mobile phase is pumped from the opposite side and accelerated by centrifugal action "through" the lighter stationary phase towards the outer diameter of the rotor (Fig. 2.18) (52). The rotational speed and thereby the centrifugal force can be varied to minimize the "bleeding" of the stationary phase by displacement. The flow rate of the mobile phase seems to have only minor influence on the resolution, which is rather dependent on the rotational speed and resulting back pressure (Figs. 2.19 and 2.20) (332).

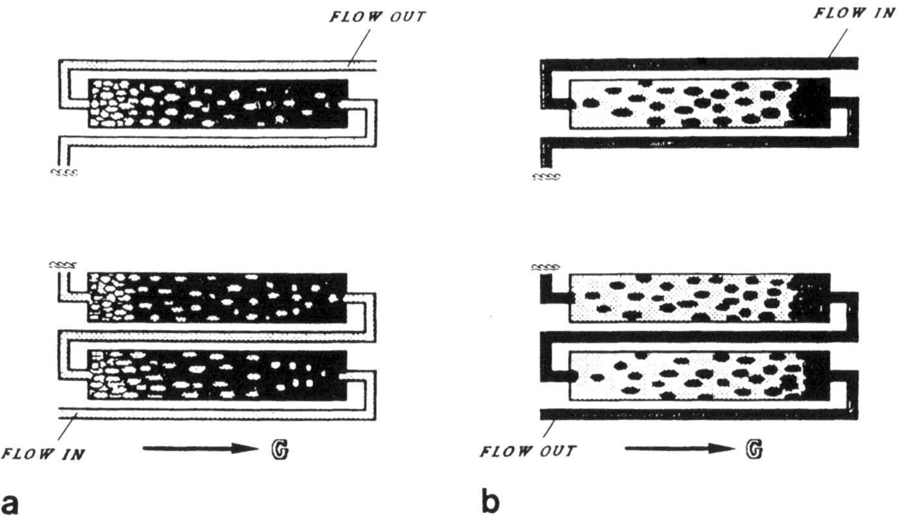

Fig. 2.18. Schematic flow pattern in the co-centrifugal columns of a CPC cartridge. **a** "Ascending" mode; **b** "descending" mode (52)

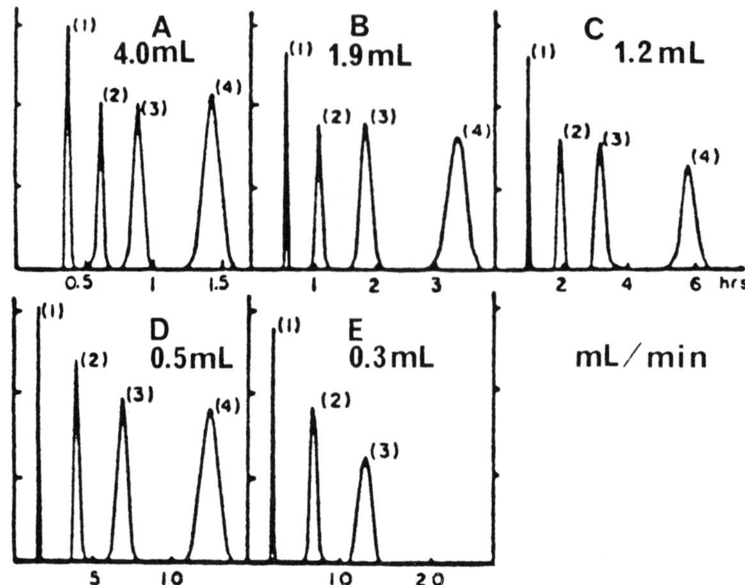

Fig. 2.19. CPC-separation of DNP-amino acids, using $CHCl_3$-0.1 M HCl–HOAc (2:1:2). Stationary phase: lower phase. Peaks (*1*) *N*-DNP-L-ornithine HCl, 4.0 mg, K = 50; (*2*) DNP-L-threonine, 4.0 mg, K = 2.0; (*3*) *N, N'*-di-DNP-L-cystine, 8.0 mg, K = 0.92; (*4*) DNP-L-proline, 12.0 mg, K = 0.54; rotational speed, 700 rpm. Note the time difference between these separations and the comparable RLCC chromatogram in Fig. 2.13 (332)

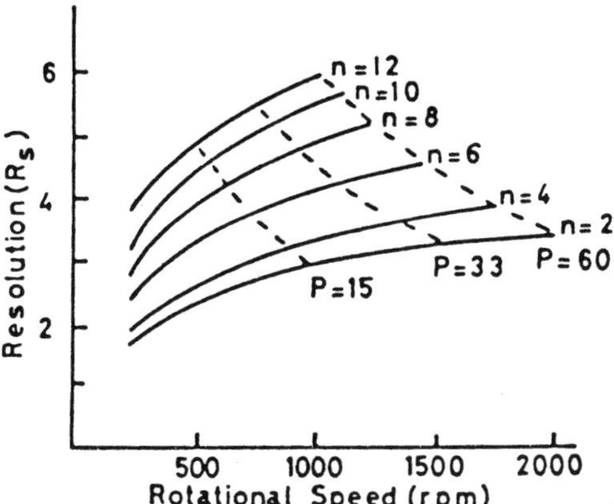

Fig. 2.20. Rotational speed vs resolution (Rs): solid lines correspond to number of cartridges used; dotted lines show the corresponding back pressure (332)

2.2.1.5 Comparison of Partition Chromatographic Methods

For the separation of solute molecules, all three methods of modern counter current chromatography have general advantages over adsorption and exclusion chromatography described in the introduction. There are relative advantages and disadvantages in the application of each of these forms of CCC so that no single method can optimally cover all the applications of the other three, although considerable overlap does exist. A particular variation of coil-CCC may have uniquely suitable applications that cannot be as adequately handled by another CCC technique. Given the same basic principle of separation, however, the results obtained on one instrument can usually be extrapolated to another.

The greatest number of theoretical plates is attained either with DCCC or the coil methods, both routinely providing ≥ 1000 theoretical plates. The commercial RLCC only provides about 300 theoretical plates, and data for CPC indicate numbers around 700. A new cartridge type, "W-250," was developed very recently with 200 columns per cartridge and a proposed theoretical plate number of 2000 for six cartridges, but no performance data are available yet. On the other hand, greater amounts can be routinely applied to the RLCC relative to the other instruments even with the advent of wider-bore tubing for the DCCC. A CPC pilot plant scale unit is available, however, at considerable cost. Moreover, there appear to be no limitations to the solvent systems that can be applied to RLCC and CPC, other than that they form two immiscible layers. This characteristic is not true of DCCC or some of the coil-CCC methods, which can be more restricted. Solvents should form droplets of some kind to be used on the DCCC, and solvents should form a stable stationary phase – that is, one that does not "bleed" from the coil under the centrifugal force – to be used on the coil-CCC. The relatively slow rate of rotation of the RLCC prevents emulsion formation, which can be a problem in coil-CCC, though the nonsynchronous coil-CCC overcomes this difficulty with an adjustable rotation rate. The RLCC, however, usually requires more solvent to complete the chromatography than does DCCC or CPC. For a CPC separation, on the other hand, droplet or emulsion formation in the two-phase solvent system does not matter at all, since continuous demixing takes place under the strong centrifugal force applied.

In a direct comparison of RLCC and DCCC (Fig. 2.21), two molluscicidal iridoid glycosides were isolated from *Olea europaea* (220). The RLCC isolation was considerably faster than the DCCC (3 days compared to 5) and used about the same quantity of eluent. On the other hand, the previously-mentioned separation of chromomycins took only 4.5 hours with CPC compared to 2.5 days with RLCC (52).

A survey of the literature readily leads to the conclusion that coil-CCC, RLCC, and CPC can accommodate a broader range of solute types than DCCC (330, 331). The former methods can be applied for the isolation of both natural products and biopolymers. There are no reports, however, about DCCC isolations of higher molecular-weight biopolymers.

While competing and somewhat conflicting claims are made for the coil-CCC methods and the DCCC, it seems safe to project that the range of solvents compatible with each instrument, as well as the time factor necessary to complete the chromatography will remain comparable. DCCC suffers somewhat from the limitation of solvent selectivity, since the mobile phase must form droplets or at least pass through the stationary phase in a droplet-like stream to avoid stationary phase displacement, and a limitation of solvent selectivity has been mentioned for the coil-CCC due to the tendency of the stationary phase to gradually be transported from the apparatus.

Countercurrent chromatography is playing an increasingly valuable role in modern isolation methods, complementing both the conventional adsorption

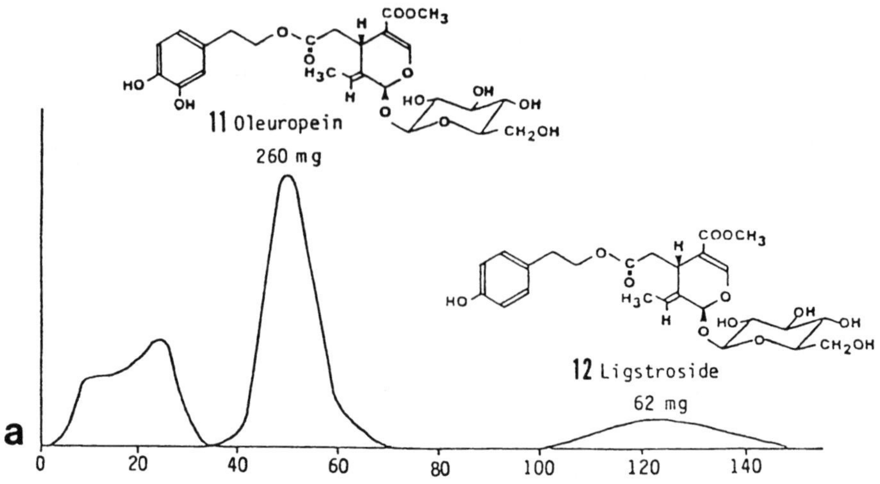

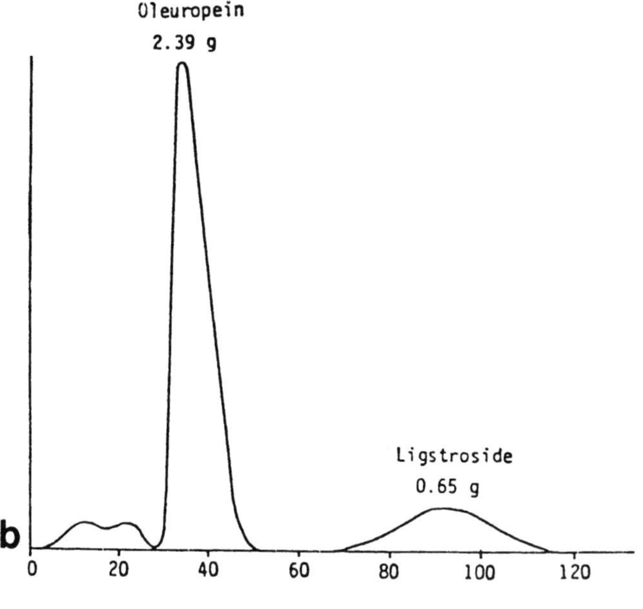

Fig. 2.21. Comparison of RLCC and DCCC for the isolation of molluscicides from *Olea europaea* methanol extract (220). Both methods employ $CHCl_3-MeOH-H_2O$ (13 : 7 : 4, v/v) solvent system in ascending mode. **a** RLCC separation, 1.4 ml per fraction, 3 days; **b** DCCC, 1.5 ml per fraction, 5 days

chromatography and molecular exclusion methods. Thus, applying both countercurrent chromatography and adsorption chromatography to an isolation scheme may greatly facilitate the otherwise difficult purification of many natural products and other compounds. The isolation of synthetic peptides illustrates this cooperative scheme using coil-CCC and HPLC (313). The greater solute capacity of RLCC makes it ideally suited for early fractionation steps in an isolation scheme, particularly for more polar fractions. The greater resolution capabilities of coil-CCC and/or DCCC are ideal for subsequent final purification. These latter two techniques effectively complement reversed phase HPLC for the isolation of many polar compounds.

Finally, the centrifugal partition chromatograph is an instrument with a tremendous amount of potential – and a relatively high price tag. The resolution of the chromomycins with the RLCC can be surpassed by CPC and in much less time (52). Any aqueous and non-aqueous biphasic solvent systems as well as aqueous polymer based biphasic systems can be employed while the sample capacity seems to equal that of coil-CCC and DCCC.

2.2.2 Adsorption Chromatography

Adsorption chromatography is unquestionably the major isolation technique applied in natural products chemistry and will probably remain so. The growth of normal and reversed phase high-performance (or formerly, "high-pressure") liquid chromatography (HPLC), the adaptation of bonded-phase packings to flash and medium-pressure liquid chromatography, as well as the increasing diversification of the types of chemically bonded phases, including chemically bonded chiral templates for chiral resolution (74, 297, 298, 310) have led to an arsenal of separation techniques capable of handling any isolation problem. Thorough treatises on all aspects of HPLC are available, addressing issues ranging from general overviews of applications and optimization of operating parameters to detailed examinations of specific stationary phases and separation problems associated with specific classes of solutes (138, 155, 218, 219, 326, 360). A review of even the most recent literature reporting new applications of adsorption chromatography and HPLC is beyond the scope of this chapter and the reader is referred to several excellent discussions as well as the expanding volume of journal literature specifically dedicated to this field.

Increasing sophistication in the practice of adsorption HPLC has created relatively new techniques, such as ion-pair chromatography, which incorporates partitioning phenomena as well as adsorption behavior in the separation mechanism. The influence of partitioning in the separation of natural products by molecular exclusion chromatography has also been reported (218, 270). Technology developed by HPLC application as well as by gas chromatography (GC) has been adapted to the development of supercritical fluid chromatography (SFC), making available a new approach to adsorption chromatography. Because these techniques are not too well documented, they will be treated in more detail in the following sections. New technology in manufacturing uniform adsorbent particles with controlled pore sizes up to 1000 Å allowed HPLC to enter biochemistry and compete successfully with established size-exclusion techniques for the separation of biological macromolecules in heretofore unknown short analysis times (63, 92).

Micronization in the field of column hardware and packing materials led to the development of microbore and short columns; the latter allow separation times sometimes to be shortened by a factor of 10 (83, 385). The advent of microbore columns created a whole new methodology in HPLC with detection limits lowered into the femto-mole range (224, 260, 291). Special micro-metering pumps give pulseless flow rates down to μl/min, and allow direct interfacing of the LC apparatus to a mass or infrared spectrometer. They also open the field of fused silica columns for HPLC use, and it might be not too optimistic to expect theoretical plate numbers of ≥500000 per LC column for the near future.

2.2.2.1 Ion-Pair Chromatography

As practiced today, ion-pair chromatography (147), which is also called "dynamic ion exchange chromatography" and "paired ion chromatography," can be viewed as a marriage of ion exchange chromatography and adsorption chromatography employing HPLC. It is an extremely versatile technique originally used to separate inorganic ions (50) and adapted for the resolution of charged as well as neutral organic molecules. While the majority of the reported applications fall into the area of analytical and bio-analytical chemistry (3, 238, 393), as more attention is focused on polar and charged, water-soluble natural products, the advantages of ion-pair chromatography for resolving such molecules should make this technique an indispensable tool in natural products chemistry as well.

A separation using ion-pair chromatography is based upon the resolution of solute molecules undergoing dynamic complexation with an added counter ion. Either a normal phase (90, 333) (silica gel, alumina, or cellulose, for example) or, more frequently, a reversed phase (41, 42, 89, 219) (e.g., C-18 or C-8) may be used. In general, the counter ion is initially adsorbed onto the stationary phase as well as being present in low concentration in the eluent. For reversed-phase ion-pair chromatography, the counter ion is usually relatively hydrophobic, whereas a normal-phase support such as silica gel requires adsorption of an aqueous solution of the counter ion, which gives rise to liquid-liquid partition characteristics for the separation.

While the precise nature of the mechanism of separation of compounds by ion-pair chromatography has been thoroughly examined and is still the subject of considerable debate (146, 362, 369), the basic features can be summarized by a series of equilibria representing the partitioning and/or adsorption of the ions and the complex between the mobile and stationary phases. The equilibria pertinent to reversed phases (Fig. 2.22) account for the possibility of ion-pair formation in either the mobile or stationary phases as well as the partitioning of the ions and the complex between the two phases. When normal phases are considered, equilibria accounting for appropriate interactions with the normal-phase solid support as well as with the stationary aqueous phase bound to the solid support must also be included. The relative influence of these dynamic processes depends upon the nature of the solute, counter ion, mobile phase, and stationary

S = solute I = complex ion

m = mobile phase s = stationary phase

$$S_{(m)} + I_{(m)} \rightleftharpoons SI_{(m)}$$

$$SI_{(m)} \rightleftharpoons SI_{(s)}$$

$$S_{(m)} \rightleftharpoons S_{(s)}$$

$$I_{(m)} \rightleftharpoons I_{(s)}$$

$$S_{(s)} + I_{(s)} \rightleftharpoons SI_{(s)}$$

$$S_{(m)} + I_{(s)} \rightleftharpoons SI_{(s)} \text{ or } SI_{(m)}$$

$$S_{(s)} + I_{(m)} \rightleftharpoons SI_{(s)} \text{ or } SI_{(s)}$$

Fig. 2.22. Equilibria that determine the retention and separation of a solute in ion-pair chromatography

phase. Other parameters that must be considered are the concentration of counter ion, flow rate, pH of the aqueous phase, ionic strength of the eluent and others, all of which have been the subject of considerable research (71, 77).

The fundamental characteristic of ion-pair chromatography is that the addition of the counter ion enhances the retention of the solute, without which the solute would either move with the solvent, in the case of reversed-phase support, or be completely retained, in the case of normal-phase support (or at least experience severe tailing and poor resolution). The enhanced retention is a consequence of the partitioning of the ion pair into the stationary phase subsequent to partitioning of the ions into the stationary phase. Thus, the equilibrium constants defined in Fig. 2.22, which are unique for each particular solute, counter ion, and stationary and mobile phase, are the factors that define the retention of a particular solute, the column efficiency, and hence the efficiency of the separation.

Parameters optimized for a specific separation problem include the pH of the aqueous phase, the addition of organic modifiers to the mobile solution, the concentration of the counter ion, and, to a lesser extent, the ionic strength of the aqueous phase. Resolution may be very sensitive to the pH of the aqueous phase, especially when solutes or counter ions with pK_a's between 2 and 10 are to be separated. Under such conditions, protonation-deprotonation equilibria are accessible within the pH limitations of most HPLC columns, including reversed phases. Such equilibria provide another solute-differentiation mechanism that can be exploited for enhancement of resolution. Under such conditions, the dynamic equilibria, which define the interactions of the solute with both the stationary phase and the counter ion, must include the behavior of both the protonated and unprotonated solute.

The capacity ratio of a solute is dependent upon the concentration of counter ion adsorbed on the stationary phase as well as on the concentration in the mobile phase when the stationary phase is saturated with adsorbed counter ion. As it is necessary to maintain a steady-state concentration of counter ion in the mobile phase, especially when the counter ion is UV-absorbing, the stationary phase is saturated with counter ion prior to application of the solute mixture. The concentration of counter ion that can be adsorbed onto the stationary phase in reversed-phase ion-pair chromatography is dependent upon the concentration of the bonded phase units – that is, upon the percentage of silylation of Si-OH groups (386) as well as other factors, such as the ionic strength and pH of the mobile phase and the lipophilicity of the counter ion.

The addition of organic modifiers can produce somewhat drastic effects upon the partitioning equilibria of both ions and ion pairs and hence upon the relative capacity ratio of the solute components. Variation of the concentration of 1-pentanol in chloroform as the mobile phase during the separation of alkylammonium ions with naphthalene 2-sulfonate as the counter ion proved to be a powerful tactic in enhancing resolution (70). Increasing the ionic strength too much leads to reduced retention of the solute as the added ions compete with the solute for pairing sites (210).

The variety of applications of ion-pair chromatography illustrates the versatility of this technique in separation science. Early applications utilized a normal-phase solid support and explored the resolution of amines and alkyl ammonium ions with an anionic counter ion such as picrate or naphthalene 2-sulfonate (41, 42, 89, 90, 219). These results prompted investigations into the suitability of ion-pair chromatography as a technique for use in pharmaceutical analysis such as the determination of ergot alkaloids (333).

With normal-phase supports, both the adsorbed water and the solid support can function as interactive stationary phases, which can hinder reproducibility and efficiency compared to reversed phases. Difficulties were encountered with silica gel as a solid support when the surface area of the silica gel was too high (70). Moreover, tailing and lack of resolution tended to hamper silica gel-based ion-pair separation when the solute mixture encompassed too great a variety of components. Column deterioration also restricted the aqueous phase flexibility with regard to pH and ion character. Using more inert supports such as ethanolized cellulose or diatomaceous earth also had drawbacks. Ethanolized cellulose is difficult to adapt to the high pressures of HPLC due to its compressibility, and diatomaceous earth gave poor peak symmetry. The more recent applications of ion-pair chromatography therefore successfully employ silica-based bonded-phase supports.

The types of molecules separated by reversed-phase ion-pair chromatography include compounds that were heretofore troublesome on HPLC, such as nucleotides (102, 211, 336), antibiotics in serum (302), inorganic ions (392), as well as neutral organic molecules with weak chromophores such as aliphatic alcohols and ketones (119), to name only a few. These latter compounds are thought to form associations with an added strongly absorbing counter ion – such as, methylene blue – or merely to non-associatively concentrate the dye in the same chromatographic zone. The dye allows thereby the detection of the non-chromophoric molecules. In the same manner, monosaccharides were separated with methylene blue as counter ion and borate buffer as mobile phase (120). In this case, the borate anion complexes with the sugars and this complex is thought to form subsequently an ion pair with the methylene blue. For the separation of zwitterionic molecules such as nucleotides, amino acids, or small peptides on reversed-phase columns, the addition of zwitterionic pairing reagents – e.g.,

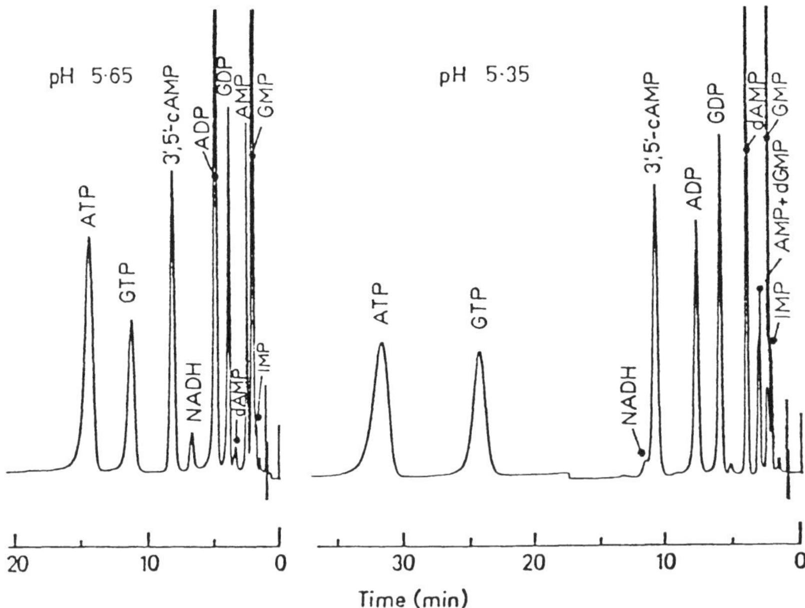

Fig. 2.23. Separation of nucleotides by zwitterion-pair chromatography, showing pH effect. Packing material, ODS-Hypersil; eluent, H_2O–MeOH (88:12, v/v), 75 mM in phosphate and 1.25 mM in 11-amino undecanoic acid (212)

11-aminoundecanoic acid – yielded excellent separations (212). As separation mechanism, a quadrupolar ion interaction or "zwitterion pairing" was suggested (210), which explains the strong sensitivity toward pH changes (Fig. 2.23) (212). The same quadrupolar interactive mechanism of zwitterion pairing has been subsequently exploited for the chiral resolution of tryptophan and glycyl phenylalanine enantiomers by using optically pure L-leucyl-L-leucyl-L-leucine as a zwitterionic counter ion (213).

As noted earlier, the addition of a counter ion with a strong chromophore allows for the extremely sensitive detection of non-absorbing solutes in either normal- or reversed-phase applications. The advent of electrochemical detectors, however, increases the list of counter ions appropriate for analytical detection considerably. Ion-pair chromatography therefore not only provides a new dimension for the use of HPLC in the isolation of natural products, it is also a new methodology for resolving and detecting trace amounts of compounds such as various metabolites in biological fluids that were previously elusive.

2.2.2.2 Other New Methods of Column Chromatography

A type of chromatography recently reported, termed micellar liquid chromatography or "soap chromatography," utilizes micelles as a component of the mobile phase. The partitioning of solute components into the micelles can be used as a selective transporting mechanism for the separation of neutral and ionic molecules with normal, reversed-phase, or even molecular exclusion gels as solid support.

The fundamentals of micellar chromatography are similar to those of ion-pair chromatography with the formation of a solute-micelle complex replacing that of an ion pair. The surfactant concentration is maintained above the critical micelle concentration (CMC) in the mobile phase so that micelles, either cationic as with cetyltrimethylammonium bromide, or anionic as with sodium dodecylsulfate, are present. The separation of solute components thus reflects a composite of their partitioning characteristics between the micelles and the mobile phase, between the micelles and the stationary phase, and between the stationary phase and the mobile phase.

The theoretical and parameter-optimization aspects have been examined and reveal that micellar chromatography can uniquely complement ion-pair chromatography (11). Thus, while ion-pair chromatography allows retention and ultimately separation of solute components – which would not otherwise be retained and, hence, not resolved – micellar chromatography reduces the retention of solute components that otherwise would not be transported via partitioning into the mobile micelles (10, 130, 402).

Two types of micellar chromatography are possible. For neutral solutes, a normal micelle with nonpolar interior and ionic surface, in an aqueous mobile phase against a reversed-phase solid support is used (10, 130, 402). The solute components, which may not be water soluble and would be adsorbed irreversibly on the reversed-phase support, are transported by partitioning into the nonpolar interior of the micelle. This type of micellar chromatography has been successful in separating pesticides, polynuclear aromatic compounds, and other model solutes (10, 130, 402).

For polar or charged solute components, reversed micelles, ionic interior with entrapped water molecules and a non-polar exterior, in an organic solvent mobile phase against a normal phase solid support is used. In this case, the solute molecules, such as nucleosides and amino acids, which normally would not be transported by an organic solvent, are mobilized by partitioning into the interior of the reversed micelle (12).

One major drawback to "soap chromatography" is the slow rate of mass transfer across the micellar interface. Poor "wetting" of the reversed-phase support by an aqueous mobile phase also slows mass transfer at the mobile phase/stationary phase interface and, hence, reduces column efficiency. It has been recently shown that adding an appropriate organic modifier – 1-propanol, for example – to the mobile aqueous phase in as low a concentration as 3% greatly enhances the mass transfer and increases column efficiency to a level comparable to that attained with adsorption chromatography (86).

Even gradients of micelle concentration can be utilized to facilitate difficult separations on reversed-phase supports (230). Once the CMC has been surpassed, further increasing the surfactant concentration increases the micelle concentration, but not the concentration of the free surfactant in the mobile phase. Therefore, the concentration of surfactant adsorbed into the stationary phase also remains constant; there is no need for a lengthy equilibration interval, which would make gradient chromatography unpredictable and irreproducible, as is unfortunately the case in ion-pair chromatography.

One particularly noteworthy application of micellar LC is the resolution of tRNAs using gel filtration (8). In this example, Sephadex G-100-120 gel, which retains tRNA and micelle-sized molecules inside the gel pores, was used. Upon incorporation of the tRNA inside the micelles, formed from cetyltrimethylammonium chloride, the size of the micelle inclusion complex prevented retention by the gel. Separation based mainly upon the relative partitioning of the tRNAs into the micelles was realized.

Micellar chromatography shows high selectivity in the resolution of model solutes, and also offers the possibility of room temperature phosphorescence detection (9, 397). This latter phenomenon is a consequence of micelle stabilization of the triplet state of photo-excited solute components, allowing phosphorescence as well as fluorescence to be observed. A unique application is the modification of a post-column reactor to create a micellar solution after non-micellar chromatography, by adding the necessary surfactant. Following the micelle formation, phosphorescence and fluorescence can be used for detection.

Using the same principle as micellar LC, a higher selectivity in solute transport has been achieved using aqueous cyclodextrin solutions as a mobile phase on polyamide TLC (149). In particular, the separation of *o*-, *m*-, and *p*-substituted phenols and benzoic acids proved to be trivial due to the different binding constants caused by the different substitutions. The greater the solute-cyclodextrin binding constant, the larger the Rf value. Presumably HPLC conditions will soon be tested for this highly selective separation procedure. Concomitantly, cyclodextrins were chemically bound to silica gel supports, thus giving a unique new group of stationary phases for HPLC. One of the latest applications was the separation of steroid epimers and isomers on a β-cyclodextrin column with water/acetonitrile mixtures as mobile phases (205).

2.2.2.3 Supercritical Fluid Chromatography

While chromatography employing either a gas or a liquid as the mobile phase is well known in the separation and isolation of natural products, chromatography employing a supercritical fluid (SFC) as the eluent is just being developed. A supercritical fluid is a gas maintained above its critical temperature under sufficiently high pressure so it exists in the fluid state. The extraordinary solubilizing ability of a supercritical fluid was noted more than one hundred years ago when Hannay and Hogarth observed the high solubility of potassium iodide in supercritical ethanol (140). Since then, numerous applications, particularly for industrial purposes, employing a supercritical fluid as an extraction medium have been reported. The extraction of caffeine from coffee by supercritical carbon dioxide (404, 405) is perhaps the best known, but by no means the only, example. Extraction with supercritical fluids is also widely employed in the deasphaltization of petroleum (268) and in the extraction of essential oils from spices (165).

The first application of a supercritical fluid as a chromatographic solvent was reported in the separation of thermally labile porphyrins using supercritical dichlorodifluoromethane as the eluent under adapted GC conditions, but at a much lower temperature, which avoided decomposition (194, 207). This and other (116, 190, 345, 346, 347, 348) early work on utilizing a supercritical fluid as an eluent in chromatography demonstrated the potential value of SFC in the isolation of compounds – especially thermally labile ones not suited for GC – and also pointed out some of the technical difficulties. For example, it was soon noted that many of the common stationary liquid phases used in gas chromatography could not be applied or had limited applications in SFC because they were solubilized by the supercritical fluid and transported off the column (263). Thus, Klesper noted that the column used to separate the porphyrins, 15% Kel-F on Chromosorb, could not be reused (194, 207).

While interest both in the physico-chemical aspects of SFC and in developing instrumentation for practical applications was generated by these early results (22, 206, 234), the rapid advent of high pressure liquid chromatography completely overshadowed progress in SFC, which remained more fundamental in nature. Relatively slow progress in SFC was also a consequence of problems inherent in this new chromatographic technique. The more obvious difficulties of operating the entire chromatographic procedure from injection to detection under high pressures, at times as high as 2000 atmospheres, were overcome by more or less adapting instrumentation from HPLC (117, 128, 191). The initial problem of solute precipitation upon emerging from the column, observed in the early adaptations of the GC flame ionization detector and caused by loss of the supercritical fluid state of the eluent, was overcome by advanced technology in the design of the nozzle at the column terminus (99) as well as by the advent of UV and fluorimetric detection prior to emergence of the solute from the column (100, 234, 303).

Due to its low critical temperature and relatively inert nature, carbon dioxide ($T_c = 31.04\,°C$) is the most studied eluent for SFC. The phase diagram of carbon dioxide (Fig. 2.24), with regions for different methods of separation denoted, gives a general overview of the relative conditions of temperature and pressure employed in various chromatographic techniques (LC, GC, and SFC). A recent study on the utility of supercritical xenon in SFC emphasizes its total spectral transparency from the vacuum UV to the NMR region. Together with its good solubilizing power, supercritical xenon would thus make an ideal, if costly, mobile phase especially if a Fourier-Transform infra-red spectrometer (Sect. 2.4.3) was used as detector (107).

Supercritical fluid chromatography can be considered as intermediate between GC and HPLC in terms of applications. Its efficiency with regard to height equivalent of a theoretical plate (HETP) and speed of analysis are comparable to GC due to the low viscosity and high diffusivity of the fluid state (292). However, given the relatively low critical temperatures of the eluents commonly employed in SFC, operating temperatures are substantially lower than those used for GC separation. Consequently, thermally labile compounds that decompose under GC conditions may be separated using SFC. More significant is the greater solubilizing power of supercritical fluid relative to a gas. Therefore, non-volatile compounds of higher molecular weight that cannot be mobilized on GC can be separated with SFC.

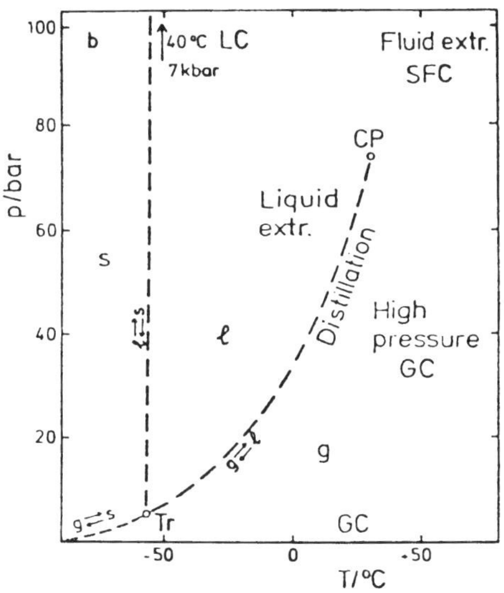

Fig. 2.24. Phase diagram of pure carbon dioxide (CO): p = pressure, T = temperature, CP = critical point, Tr = triple point, g = gas, l = liquid, s = solid. (U. van Wassen, I. Swaid, G. M. Schneider 1980 *Angew. Chem. Int. Ed. Eng. 19*, 575)

The main advantage of SFC over HPLC lies in the reduced analysis time. The phase diagrams of binary systems of a solute and an eluent such as carbon dioxide are dependent upon the nature of the solute and eluent and can still not be predicted (335). The solubility of a particular solute in a supercritical phase is extremely sensitive to the density of the fluid and therefore to the pressure of the system. Consequently, isothermal compression is the most important mechanism for bringing about mutual solubility. Furthermore, the density of the fluid required to achieve a single phase in a binary system is also sensitive to the nature, including molecular weight, of the solute. Thus, a solute has its own, frequently unique "threshold of partitioning," corresponding to the particular density of the fluid required to bring about complete dissolution in the mobile phase (345). Consequently, pressure programming is an integral part of SFC separations.

Because the density of the fluid is also dependent upon the temperature of the system, this parameter is significant as well. When operating above critical conditions, increasing the temperature initially reduces the density of the fluid and therefore the mobilities of the solutes; but further increasing the temperature may increase, decrease, or have no effect upon the partitioning of the solute into the fluid phase (129). Therefore, negative temperature progamming is sometimes employed along with pressure programming.

Temperature programming and particularly pressure programming provide an added dimension, along with the relative strengths of interactions of solute molecules with both the mobile fluid phase and the stationary phase, which contribute to the separation of compounds by SFC. Thus, SFC may prove to be more selective than HPLC in certain situations as the relative dominance of partition or adsorption interaction can be altered by pressure programming. Pressure is unquestionably the single most important parameter in SFC, and without pressure programming, some solutes would not be eluted.

Supercritical fluid chromatography employing carbon dioxide as an eluent is frequently operated in a temperature range up to 100 °C (110–115, 316). Pressures that conform to an eluent density of 200 to 500 times greater than that of the gas are usually employed. Other eluents frequently utilized besides carbon dioxide include ammonia ($T_c = 132.4\,°C$), nitrous oxide ($T_c = 36.5\,°C$), ethane ($T_c = 32.2\,°C$), ethylene ($T_c = 9.21\,°C$), propane ($T_c = 96.7\,°C$), pentane ($T_c = 196.5\,°C$), and chlorotrifluoromethane ($T_c = 28.9\,°C$) (129). Low concentrations of a moderator, usually a low molecular-weight alcohol such as methanol, ethanol, or isopropanol can be added to enhance resolution (314).

The increased solubilizing powers of a supercritical fluid cannot be entirely explained on the basis of an increased density of the fluid phase. Thus, supercritical carbon dioxide is a suitable solvent for SFC in terms of mobilizing a solute, but supercritical argon and nitrogen are not (116).

A "solvation" type interaction also contributes. Lewis acid-base interactions and co-complexation undoubtedly enhance the solvent qualities of carbon dioxide while hydrogen-bonding also contributes to those of ammonia (116).

The types of solute molecules that have been successfully separated by SFC as well as the number of eluents used are increasing. Since the initial separation of metalloporphyrins (194, 207), various types of solutes have been examined. The most common model solutes for investigating the physico-chemical and operational aspects of SFC were polycyclic aromatic compounds and polystyrene oligomers with carbon dioxide as the eluent. However, compounds as polar as nucleosides, amino acids, sugars, di- and tripeptides, and disaccharides have been resolved with SFC using supercritical ammonia. Many other solutes, such as polyethylene glycol oligomers (263), alkyl bromides (234), and natural products such as ubiquinones (110) have been separated with carbon dioxide. Figure 2.25 illustrates the separation of polycyclic aromatics and polystyrene oligomers (348). The separation of polystyrenes with molecular weights ≥ 2000 can be performed under a variety of conditions. These results represent a major improvement over gel permeation and other methods of analyzing polystyrenes. Supercritical ethylene has been used to separate the oleic acid glycerides (307).

The sensitivity of solute partitioning to the density of the fluid and ultimately to the operating pressure of the system was initially a source of difficulty in early work, as pressure gradients across the columns caused precipitation of the solute prior to emergence from the column. Recent results, however, indicate that both capillar columns and HPLC packed columns (4.6 mm i.d., packed with 3-, 5-, or 10-µm particles, including both normal and reversed phases) can be employed in SFC without a detrimental pressure drop (99, 100, 293, 294, 303, 355, 356, 357).

While research employing SFC is still in the physico-chemical and operational stages, the evolution of the technique is rapidly progressing (303–305). Commercial instruments are now available, but the application of SFC to research problems is still uncommon. One area that is particularly promising is the application of SFC for assaying compounds that are extracted industrially by supercritical fluids. For example, SFC can be used as an analytical tool to quantitate caffeine (114), the oligomer content of polymerizations (111), and fats and oils in foods or petrochemical products (316).

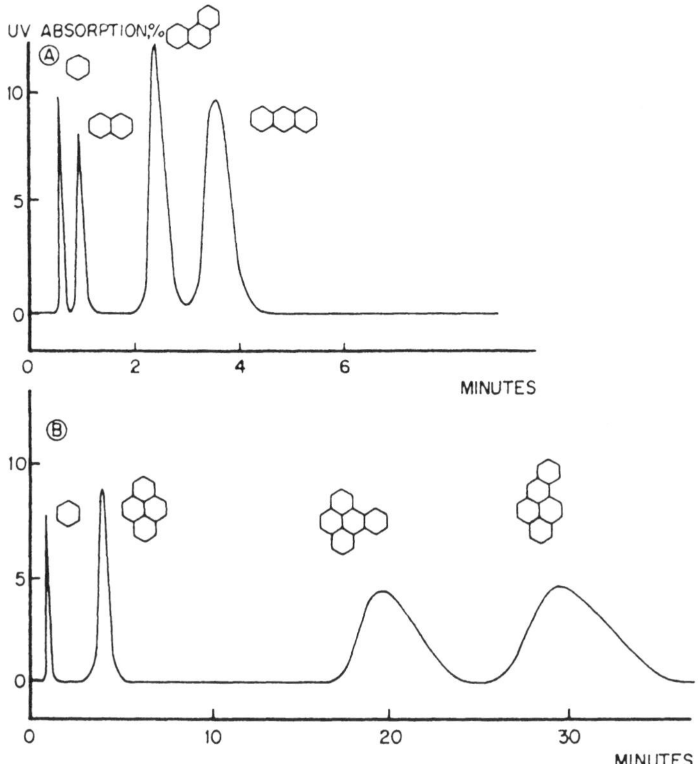

Fig. 2.25. Separation of polycyclic aromatics on alumina column with super-critical *n*-pentane as eluent, at 200 °C (348)

Another application that has tremendous potential is the interfacing of SFC with mass spectrometry (315, 355–357). The interfacing of gas chromatography with mass spectrometry is a well-known, widely used technique whose application is somewhat limited by the nature of the solute. Liquid chromatography interfaced with mass spectrometry still has some difficulties in solvent removal prior to ionization of the solute. Interfacing of SFC with MS does not suffer from these drawbacks and nicely complements GC-MS.

Undoubtedly, future research on SFC will expand its range of applications. While carbon dioxide remains the most widely used SFC eluent, investigations into the physico-chemical properties of other potential eluents should render more predictable separation results and increase the utility of SFC (e.g., 107). Applications to the isolation of natural products may prove uniquely suitable in difficult separations where pressure programming may complement relative adsorption to provide increased resolution. In this connection it is interesting to note a recent report on the use of liquid carbon dioxide as a solvent for TLC under sub-critical conditions for the separation of polycyclic aromatics (290). Resolution of the solute was good, but with a different order of elution compared to the use of hexane as mobile phase.

2.3 Nuclear Magnetic Resonance Spectroscopy

Advances in the major areas of spectroscopy have carried structural analysis to the point at which sub-milligram structure determinations have become routine. Progress in extracting the maximum spectral information has reached such sophistication that the necessity for chemical degradation becomes less and less frequent. Research problems previously surmountable only by tedious and prolonged methodology have now been rendered time-efficient by state-of-the-art instrumentation. While no single form of spectroscopy is currently capable of resolving all structural problems, nuclear magnetic resonance spectroscopy (NMR) is probably the technique of paramount importance for use in structural analysis of natural products.

Nuclear magnetic resonance spectroscopy (e.g., 6, 29, 30, 342) has progressed at a vigorous pace since the introduction of the first commercial spectrometers in the early 1960s, and closely followed the development of super-conducting magnets capable of maintaining stable magnetic fields of considerably higher field strength than that available from electromagnets. Furthermore, the increasing sophistication of microprocessors for signal acquisition and data manipulation, coupled with a deeper understanding of magnetic resonance phenomena by NMR investigators, have led to the discovery of invaluable methods, such as the development of two-dimensional NMR.

2.3.1 Proton Nuclear Magnetic Resonance

2.3.1.1 Difference Decoupling

One of the most useful proton nuclear magnetic resonance (^{1}H-NMR) one-dimensional (a single-frequency axis) techniques to be applied in structure determination is difference spectroscopy. This approach involves subtracting a spectrum obtained under the conditions of some perturbation from a normal, unperturbed spectrum. The resulting difference spectrum reveals the spectral changes caused by the perturbation, whereas those signals that remain unaffected are nulled. The perturbation can be a decoupling irradiation, a nuclear Overhauser enhancement, or even a population transfer experiment (135, 328).

Difference decoupling is a valuable technique for uncovering buried multiplets and assigning coupling partners (229). An example is shown in Fig. 2.26 (282). Cholic acid has three carbinol protons – 3-H, 7-H, and 12-H – two of which appear at 3.94 and 3.78 ppm. The assignments can be made in the following manner by revealing their adjacent protons, which are buried in the broad absorptions at 1.3 to 2.4 ppm (see expanded Spectrum B in Fig. 2.26). Spectrum C is obtained upon decoupling the 3.78-ppm peak and taking the difference; only the peaks that have undergone changes appear here. These peaks are a doublet of doublets (dd, J = 12, 5 Hz) at 1.94 ppm and a triplet (t, J = 12 Hz) at 1.53 ppm, both representing one proton each. It is only 12-H that has two adjacent protons; the other has four (3-H) and three (7-H) protons. Furthermore, the coupling pattern of the

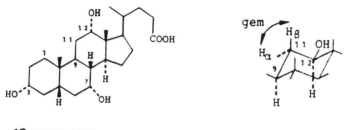

13 CHOLIC ACID

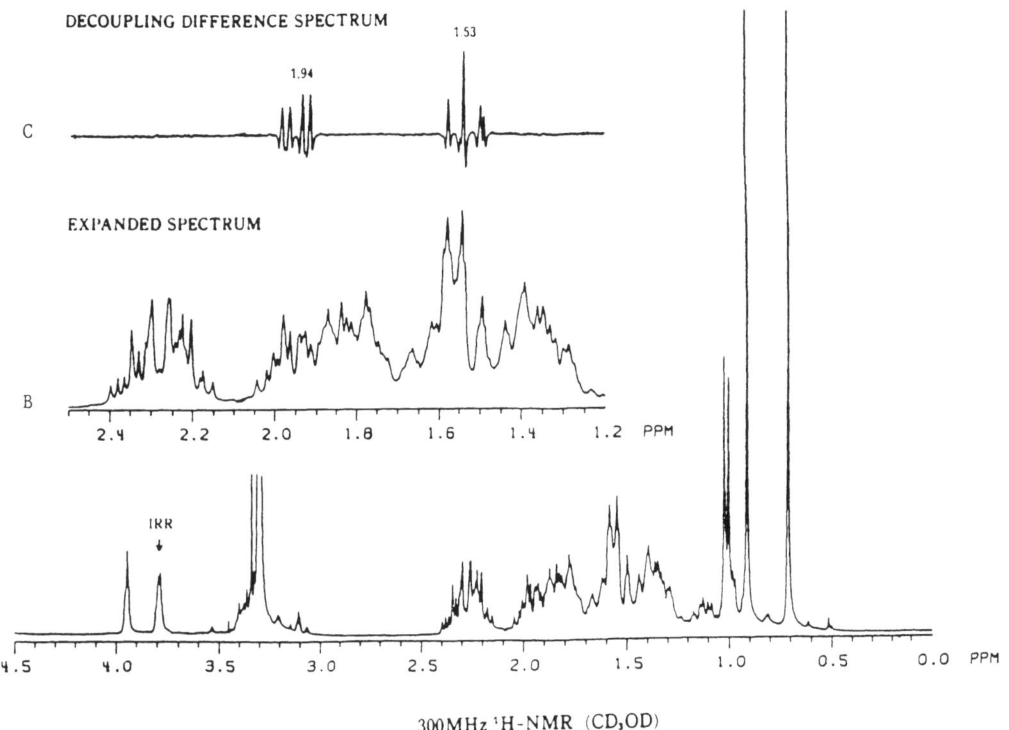

Fig. 2.26. Difference decoupling spectra of cholic acid (282)

12-H decoupled peaks in Fig. 2.26C shows that 1.53 t is due to axial 11β-H and the 1.94 dd is due to equatorial 11α-H. It is clear that such results could not have been obtained from straight decoupling experiments.

2.3.1.2 Difference NOE

Nuclear Overhauser enhancements (NOEs) are a consequence of the dipole-dipole relaxation mechanism that operates through space (as opposed to the

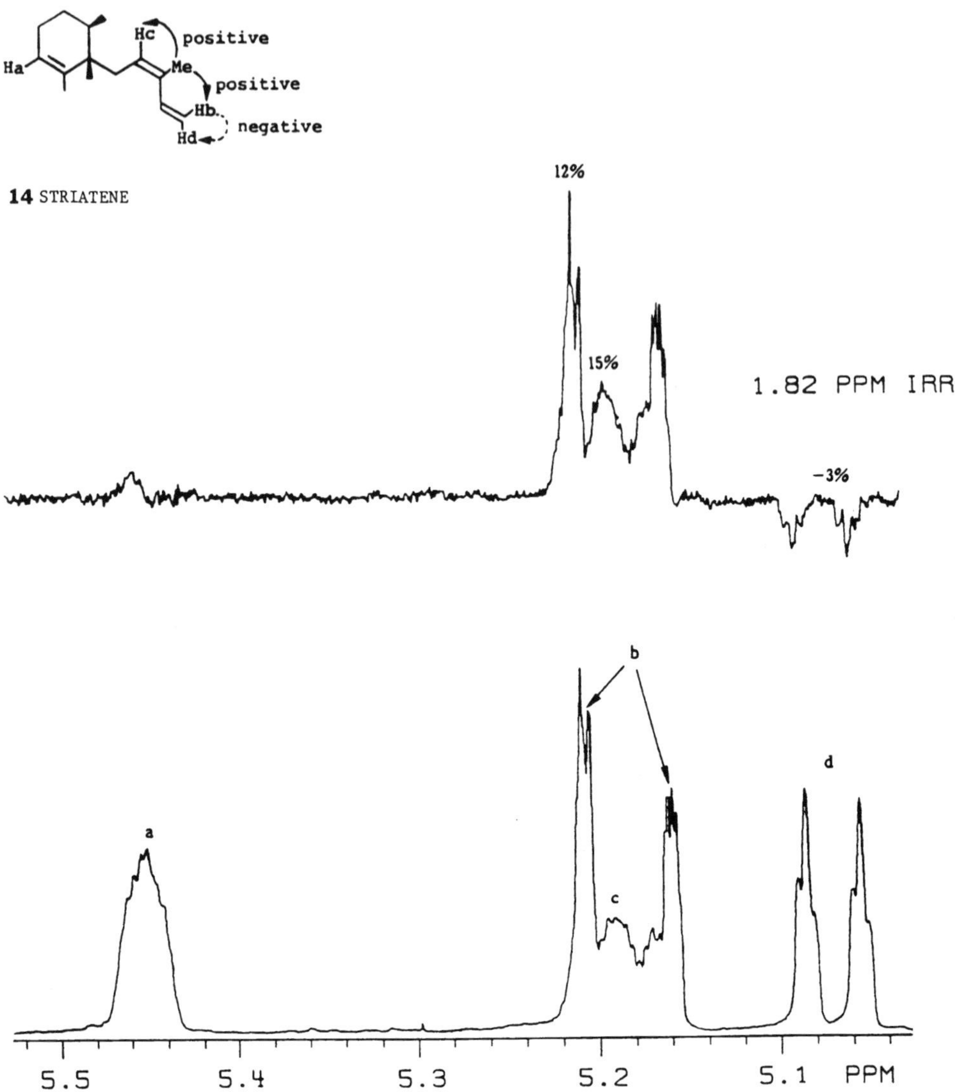

Fig. 2.27. NOE difference spectrum of striatene (282)

through-bond scalar coupling mechanism responsible for proton multiplicities). An excellent theoretical treatment is given by Noggle and Schirmer (287). The application of difference spectroscopy for the observation of NOEs is a powerful tool for identifying internuclear, through-space interactions and, hence, for defining relative stereochemistry (61, 64, 135, 202, 203, 214, 216, 217, 229, 328). The main advantage of the difference technique is that it enables small enhancements (NOEs $\leq 1\%$ have been reported) to be unmasked from the midst of signal congestion that would have been undetected without the subtraction of unaltered signals. Such a sensitivity of NOE detection is far beyond the capabilities of simple integration.

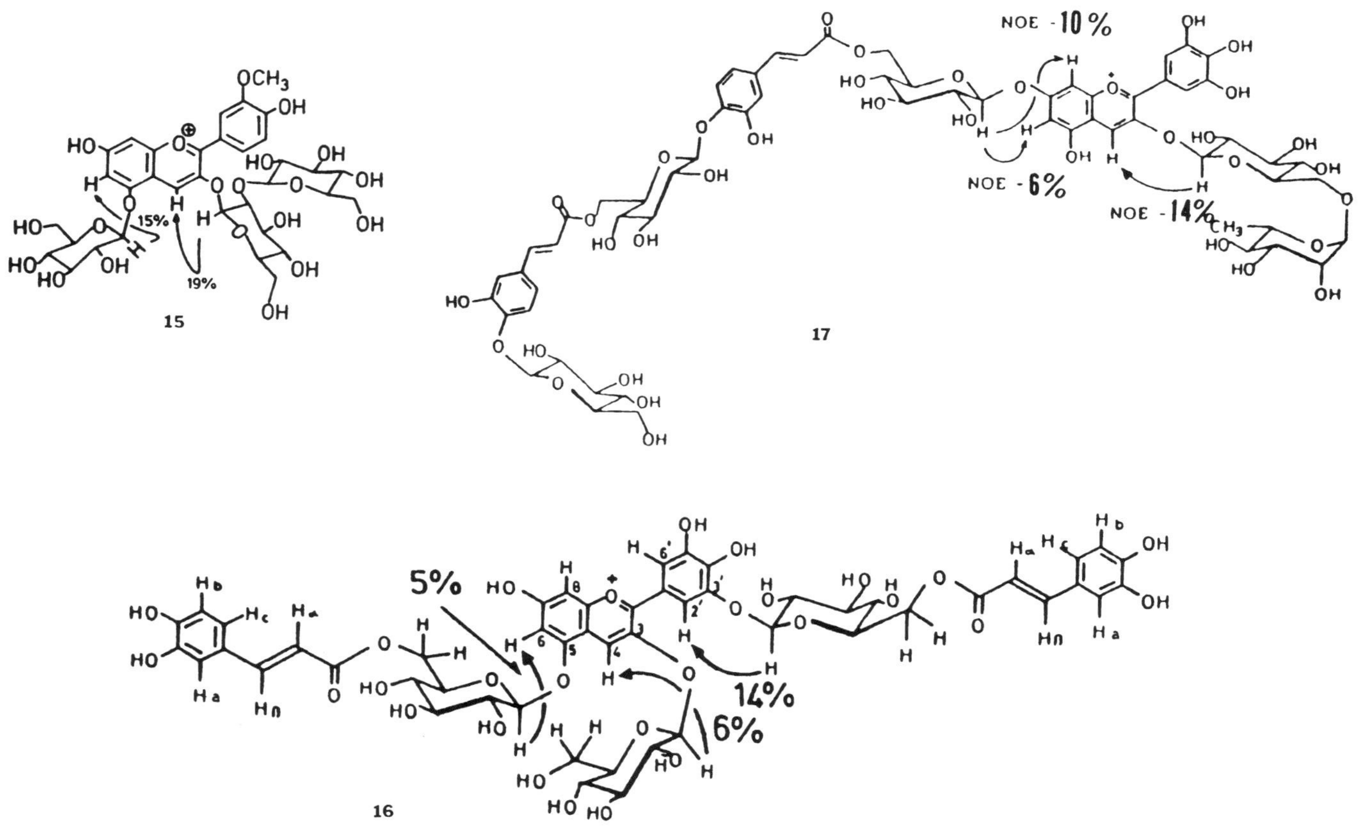

Fig. 2.28. NOEs observed on aglycone protons upon irradiation of sugar anomeric protons, which establish sugar linkage sites to aglycone: **15**, trisdeacyl heavenly blue anthocyanin, from blue petals of morning glory, *Ipomoea* spp (125); **16**, gentiodelphin, from the petals of *Gentiana makinoi* (126); **17**, platyconin A, from petals of *Platycoden grandiflorum* (127)

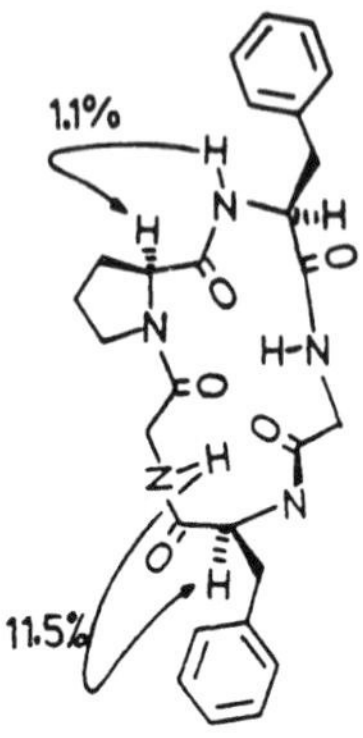

1.1%

11.5%

18

Fig. 2.29. NOEs that cross peptide bonds in cyclo (Pro-Phe-Gly-Phe-Gly) (229)

Figure 2.27 (282, 373) depicts the normal 360 MHz spectrum of striatene (lower trace) and the difference NOE resulting from irradiating the Me group at 1.82 ppm. The two adjacent olefinic protons, H_c (doublet of multiplets) and H_b (multiplet), appear as positive peaks because their intensities are enhanced upon decoupling of Me. In contrast, H_d appears as two negative multiplets because of reduction in intensity (−3%). This is due to the fact that, when three protons are aligned closely in space and in a sequential manner ($H_a - H_b - H_c$), irradiation of H_a will result in positive and negative NOEs on H_b and H_c, respectively. This is because the increased population of the excited spin state of H_a leads to an increased ground state population of H_b, which, in turn, induces the excited-state population of H_c to decrease (i.e., reduced peak intensity or negative NOE).

Recently there have been several reports on the determination of sugar linkage sites in saponins and other glycosides using NOEs from the anomeric proton to protons on the aglycone (125–127) (Fig. 2.28). In each example, an enhancement observed upon saturation of an anomeric proton located the sugar linkage site *ortho* to the enhanced aromatic proton. Other important applications of difference NOEs include conformation and sequencing studies of peptides by observing NOEs from the amidic proton of the peptide amide linkage to the α-proton of the subsequent amino acid (Fig. 2.29) (229, 323). Although the amidic proton of peptides is an exchangeable proton, its rate of exchange in peptides is usually quite slow, therefore caution must be exercised to avoid misinterpretation due to enhancement transfer.

2.3.1.3 Contact Shifts

With the advent of higher magnetic fields from superconducting magnets and advanced two-dimensional techniques to enhance resolution, older methods of signal resolution such as lanthanide-induced contact shifts (118, 201, 319, 349) and even solvent-induced shifts (5, 148, 186, 306) are less commonly employed. One application of contact-shift reagents (complexing metals that induce chemical shifts in nearby protons) that remains prominent is the use of chiral-shift reagents

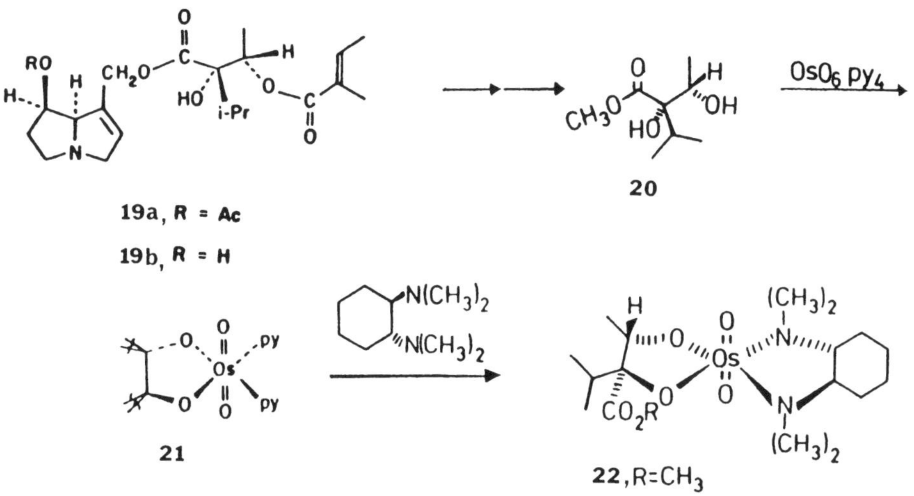

Fig. 2.30. Formation of chiral osmium complex (**22**) with chiral diol and chiral *trans-N, N, N′, N′*-tetramethyl-1,2-cyclohexanediamine (318)

for the determination of absolute stereochemistry (103, 121, 122, 193, 320, 398, 399).

The use of osmate (VI) complexes of chiral *trans-N,N,N′,N′*-tetramethyl-1,2-cyclohexanediamine for the determination of glycol absolute stereochemistry through formation of the osmate esters has been recently reported (318). Formation of the chiral complex [Fig. 2.30, (**22**)] allows stereochemistry of the original diol to be assigned by determination of the chemical shifts of the *N*-methyl groups. In this example, the chirality of the α-isopropyl-α,β-dihydroxybutyric acid moiety of the pyrrolizidine alkaloids (Fig. 2.30, **19a** and **19b**) isolated from *Myosotis scorpioides* was determined following alkaline hydrolysis and formation of its methyl ester. Comparison of the chemical shifts of the *N*-methyl groups with the known methyl ester of viridifloric acid (**20**) confirmed the chirality. The conversion of other natural products to diols, such as stereoselective glycol formation from olefins with osmium tetroxide, expands the applications of this technique.

2.3.1.4 Partial Relaxation

Spectral resolution enhancement can be achieved by partial relaxation, which capitalizes on the different relaxation times, T_1, of the protons in a molecule (324). The partial relaxation experiment usually employs the inversion recovery pulse sequence [$180°-\tau-90°-$acquisition] (390), though progressive saturation (106) can also be employed.

A series of partially relaxed spectra obtained by the inversion recovery pulse sequence with increasing delay time τ (in seconds) is shown as stack runs in Fig. 2.31 (282). At first, all peaks are negative; they then gradually become positive with τ, protons with shorter T_1 becoming positive earlier than those with longer

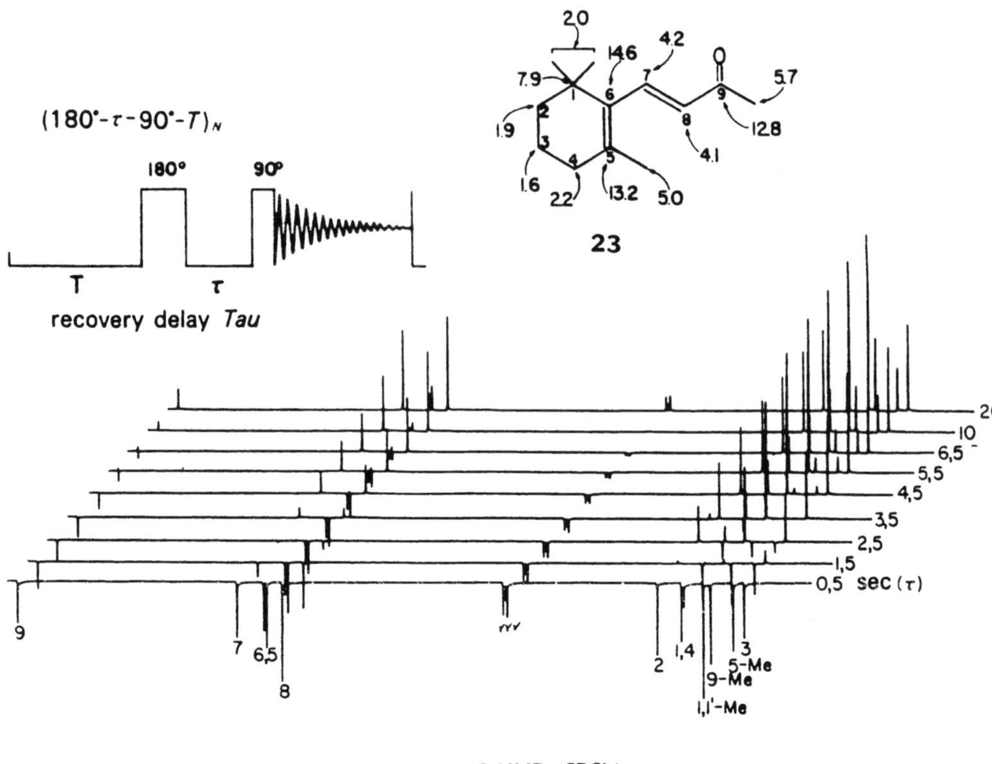

62.5MHz ^{13}C-NMR ($CDCl_3$)

Fig. 2.31. Inversion recovery spectra of β-ionone (**23**) (282)

T_1 values. From plots such as the one shown, the T_1 relaxation times of protons can be readily measured. The relaxation times of various protons in organic molecules usually increase in the order of methylene, methine, methyl, and quaternary. This is seen in the relaxation time of the protons in β-ionone.

By selecting the proper time interval, τ, resonances can be selectively nulled, thereby uncovering other overlapped resonances, which may appear as negative or positive signals depending upon their T_1's. Although the signal intensity may be reduced, the coupling constants can be measured. Furthermore, decoupling can also be employed during the partial relaxation pulse sequence to assign coupling partners (43).

2.3.2 Carbon Nuclear Magnetic Resonance

Inasmuch as establishing the carbon framework of a molecule overcomes a major obstacle to structure determination, carbon nuclear magnetic resonance (^{13}C-NMR) spectroscopy is potentially the most powerful spectroscopic implement for natural products characterization. The broad range of chemical shifts are valuable in determining the presence of different functional groups (49, 84, 85, 235, 321,

368, 394–396), while the dipolar and scalar coupling interactions provide the mechanism for establishing the carbon framework. The availability of high-field NMR instruments and improved sensitivity of the instrumentation in general, have enabled the extraction of previously inaccessible data on sub-milligram samples. Several reviews of new techniques in ^{13}C-NMR spectroscopy have appeared (132, 344, 381).

After attaining the chemical shift information of the carbon signals via broad-band proton decoupling, establishing the number of protons attached to each carbon is usually the next step. Formerly, single frequency off-resonance decoupling (SFORD) employing a proton decoupling irradiation frequency centered considerably upfield from any resonances, usually $\delta = -2$ to -3 ppm, with residual irradiation power only extending over the proton resonance frequency range was used (72, 131, 134, 225, 317). The signal broadening due to long-range couplings is eliminated and the size of multiplet splittings due to ^{1J}13-C, 1-H couplings is reduced. The number of protons attached to each carbon can then be determined from the resultant multiplicities of each signal – e.g., singlet, double, triplet, or quartet.

Difficulties are encountered in establishing carbon multiplicities with the SFORD technique when signal overlap becomes severe, or when only small quantitites of sample are available, thus dividing the signal intensity between the lines of the multiplet. New techniques that overcome the limitations of SFORD utilizing the spin echo are the J-modulated spin-echo (231) or the Attached Proton Test (312), the Insensitive Nuclei Enhanced by Polarization Transfer (56, 79, 104), and the Distortionless Enhancement by Polarization Transfer (80).

2.3.2.1 J-Modulated Spin Echo

Fourier Transform now produces a broad-band proton-decoupled spectrum with quaternary and methylene carbons appearing as positive singlets while methine and methyl carbons appear as negative singlets. The J-modulated spin echo ^{13}C-NMR of brevetoxin C, $C_{49}H_{69}O_{14}Cl$, a red-tide neurotoxin (123), is shown in Fig. 2.32. This compound has a total of 22 C-O bonds in the region of 55 to 90 ppm and it is difficult to differentiate the methine and quaternary carbons attached to oxygen in an off-resonance decoupled spectrum; however, in the spin echo spectrum, the two types of carbons can be readily differentiated and counted. The chemical shift differences between quaternary and methylene, and between methine and methyl carbons and data regarding the structure of the molecule are frequently sufficient to distinguish these two sets of signals with the same phase, and hence assign the multiplicities of all carbon signals. In some instances, particularly in the upfield aliphatic region, some ambiguity may exist in assigning a negative resonance as a doublet or quartet, or a positive resonance as a singlet or triplet. Relatively simple techniques described later, such as selective population transfer, can be used to clear up such uncertainties.

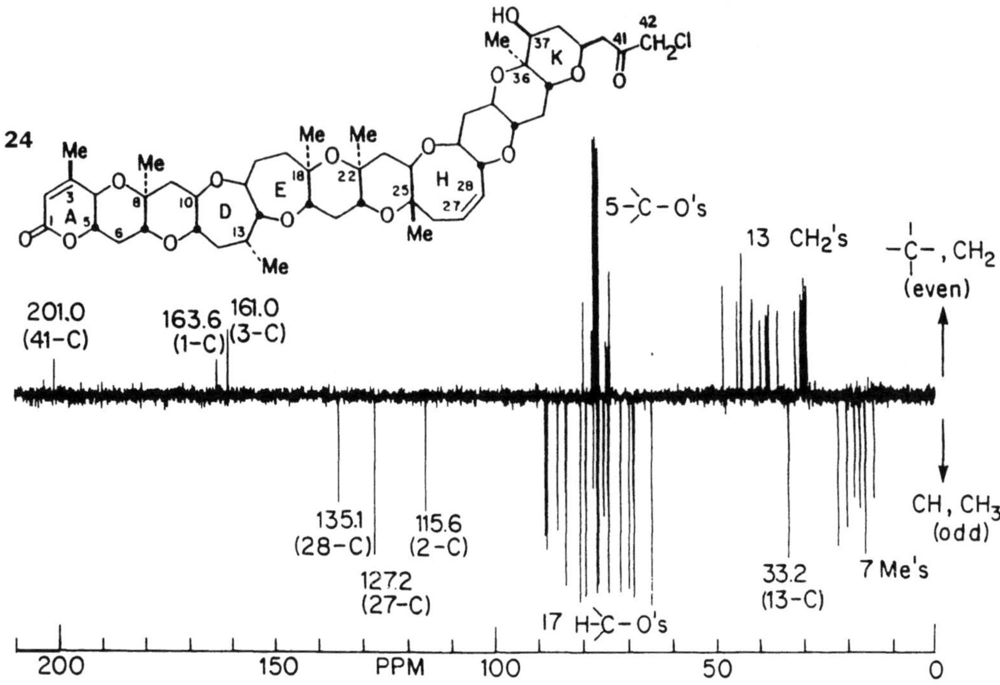

Fig. 2.32. J-modulated spin echo ^{13}C-NMR of 7 mg of brevetoxin C (**24**) ($C_{49}H_{69}O_{14}Cl$), 62.9 MHz, $CDCl_3$ (123)

2.3.2.2 Insensitive Nuclei Enhanced by Polarization Transfer

Another technique for determining carbon multiplicities similar to the J-modulated spin echo, but with added enhancement due to a polarization transfer from the protons to ^{13}C nuclei, is INEPT (Insensitive Nuclei Enhanced by Polarization Transfer) (56, 79, 104). The polarization transfer is capable of a roughly four-fold enhancement of the carbon signal intensity.

The INEPT pulse sequence represents a major advance in spectral editing for the determination of carbon multiplicity, but there are some minor drawbacks to the technique. The interval Δ in the INEPT pulse sequence is defined as a function of ^{1J}C-13, H-1, which is not invariant. Consequently, simultaneous, optimal refocusing of all signals is not possible, resulting in residual magnetization in the editing sequence. Signals are therefore not completely nulled in the resultant spectra. In addition, in the coupled INEPT spectra, phase distortion can be severe, and as indicated in the previous section, the relative intensities of the component transitions of triplets and quartets are greatly altered in comparison with their usual ratios.

2.3.2.3 Distortionless Enhancement by Polarization Transfer

A relatively new pulse sequence, called DEPT (Distortionless Enhancement by Polarization Transfer) (80), has been devised to overcome some difficulties of INEPT. In this pulse sequence, the pulse θ is varied to generate a series of subspectra, which are then combined in appropriate ratios to optimize spectral editing.

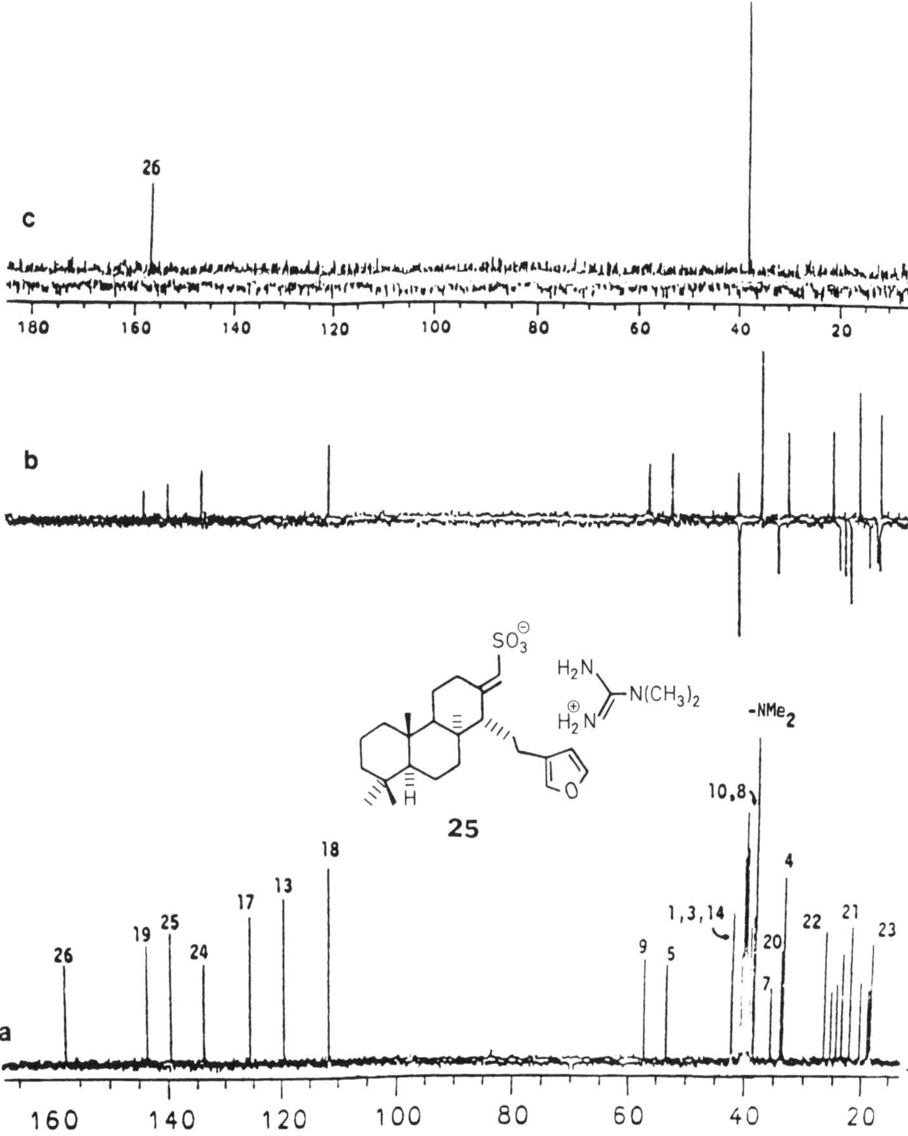

Fig. 2.33. Carbon-13 spectra of suvanine (**25**) in DMSO-d_6. **a** Broad-band proton decoupling; **b** DEPT spectra ($\theta = 135$) with broad-band proton decoupling; **c** selective INEPT sequence with single frequency irradiation of 6H singlet at $\delta = 2.94$ ($-NMe_2$) (251). (L. V. Manes, P. Crews, MR. Kernan, D. J. Faulkner, F. R. Fronczek, R. D. Gandour 1988 *J. Org. Chem. 53*, 570)

An elegant combination of the DEPT and long-range selective INEPT pulse sequences was recently applied to solve the structure of suvanine, a marine sesquiterpene (251). The multiplicities of the carbon signals were assigned with the DEPT sequence (Fig. 2.33 b). A selective INEPT sequence with irradiation of the guanidinium methyls only, $\delta = 2.94$ (6H,s), rather than broad-band proton saturation, was utilized. The 1/J value for determining the time interval, Δ, in the INEPT sequence was set to 0.167 s to detect modulation due to long-range coupling to the irradiated methyl protons. This 1/J value corresponds to long-range coupling of 6 Hz. In spectrum C (Fig. 2.33c), only the signals at $\delta = 156.7$ and $\delta = 37.7$, corresponding to the guanidinium quaternary carbon and the guanidinium methyl carbons, appear. This experiment supported the structure that suggests that these methyl protons would be long-range coupled to only the guanidinium carbon.

2.3.2.4 Carbon-Proton Heteronuclear Coupling

Considerable structural information can be obtained by examination of one-, two-, and three-bond ^{13}C-^{1}H coupling constants. To a close approximation, the coupling constants are determined by the contact interaction between nuclear and electronic spins (Fermi contact term) (23, 253, 254, 275). In general, as the s-character of the carbon hybridization increases, the contact term and, hence, the coupling constants increase. Electronegative substituents and angle strain, as well as unsaturation, increase the s-character of the carbon orbitals bonding to hydrogen, therefore increasing $^{1}J_{C\text{-}13,\,H\text{-}1}$ values.

Signal broadening due to long-range coupling and severe overlap in proton-undecoupled carbon spectra hinder determination of $^{1}J_{C\text{-}13,\,H\text{-}1}$ values. Furthermore, as these couplings are not manifestations of interactions that cross carbon-carbon bonds, they rarely are used to establish carbon skeletons directly. One-bond couplings can be invaluable, however, in suggesting the presence of certain structural units. Cyclopropyl units, for example, have upfield-shifted signals with

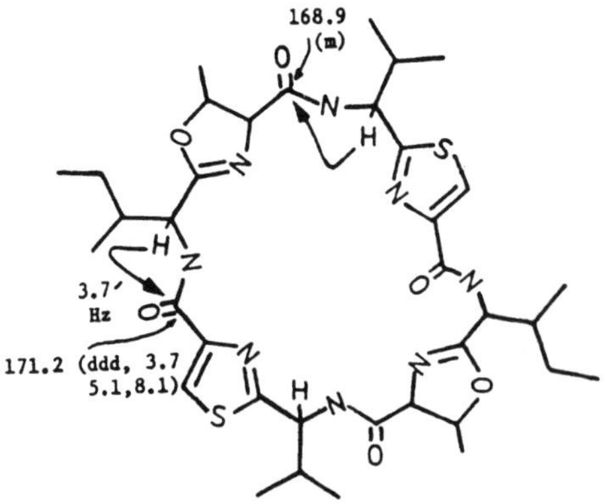

Fig. 2.34. Long-range couplings used to assign peptide sequence in ascidiacyclamide (**26**) (137)

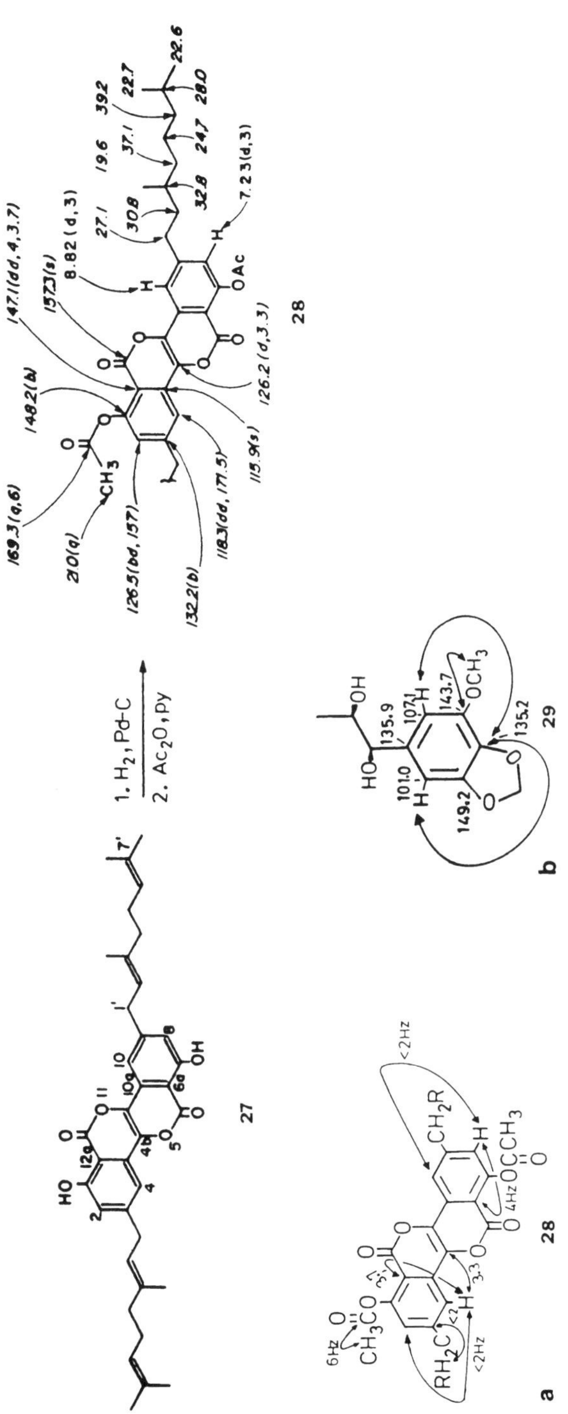

Fig. 2.35. Two- and three-bond long-range couplings observed in diacetyloctahydrocastanaguyone (**28**) (**a**) (358), and the phenyl propanoid (**29**) (**b**) isolated from *Daucus carota*

large ${}^{1}J_{C\text{-}13,\,H\text{-}1}$ values on the order of 160 Hz. The anomeric state of sugars can also be determined from the ${}^{1}J_{C\text{-}13,\,H\text{-}1}$ values (78, 91).

Long-range carbon-proton coupling constants, ${}^{2}J_{C\text{-}13,\,H\text{-}1}$ and ${}^{3}J_{C\text{-}13,\,H\text{-}1}$, are usually more valuable as a structural tool, for they represent interactions that cross carbon-carbon bonds (59, 94, 137, 141, 152, 271, 289). Long-range carbon-proton couplings across peptide bonds can aid in oligopeptide sequencing, as exemplified by the cyclic peptide ascidiacyclamide (Fig. 2.34) (137). Frequently, knowledge of long-range coupling interactions is irreplaceable in assigning quaternary carbon signals.

Long-range selective proton decoupling is especially valuable in assigning the carbon resonances of substituted aromatics (Fig. 2.35). The three-bond coupling observed between H-4 and 4b-C (${}^{3}J_{C\text{-}13,\,H\text{-}1} = 3.3$ Hz) was crucial in establishing the biisocoumarin skeleton of diacetyloctahydrocastanaguyone, (**28**) (358), a derivative of castanaguyone, (**27**), isolated from *Zanthoxylum fagara*. Other long-range couplings observed in **28** are also given. The long-range couplings of the phenyl propanoid, (**29**), isolated from *Daucus carota* flowers, enabled the assignment of the aromatic carbon resonances. Prior to the advent of long-range selective proton decoupling, such assignments were frequently and often erroneously based on calculations from substituent effects.

2.3.2.5 Carbon-Proton Heteronuclear NOE

The application of carbon-proton heteronuclear NOEs through gated decoupling of the protons is a well-known technique to enhance the signal-to-noise ratio in the ${}^{13}C$ spectrum. A maximum three-fold increase in signal intensity could be observed if the ${}^{13}C$ nuclei relaxed entirely through dipolar coupling with neighboring protons (287). The observation of selective dipole-dipole heteronuclear coupling, heteronuclear NOEs, can also be a valuable tool in structure determination and resonance assignment, particularly when the observed coupling occurs between a proton and carbon nucleus separated by two or three bonds. As with heteronuclear selective population transfer experiments, the adequacy of low levels of irradiation power for inducing NOEs in the carbon signals is a major advantage.

Examples of selective heteronuclear NOEs used in structure determination and signal assignment are shown in Fig. 2.36 (2, 339, 383). The correct assignments of the carbonyl carbon resonances of citraconic anhydride, (**30**), were made from the observed NOEs upon irradiation of the vinyl proton and the vinylic methyl group (383). In pentalenolactone G, (**31**), thought to be a biosynthetic precursor of pentalenolactone, the location of the gem-dimethyl at the C-2 and the methylene carbon at C-3, rather than the gem-dimethyl at C-3 and the methylene at C-2, was shown by enhancement of the ketonic carbonyl upon irradiation of both the methyl and methylene protons (339). A final example shows the use of heteronuclear NOEs to establish the structure of a quinone-diazoalkane adduct (**32**) (2). Enhancements of the carbon signals at $\delta = 90.0$, 90.5, and 195.0, upon irradiation of the one-proton singlet at 2.37, H-3a, supported the structure of the adduct as **32**.

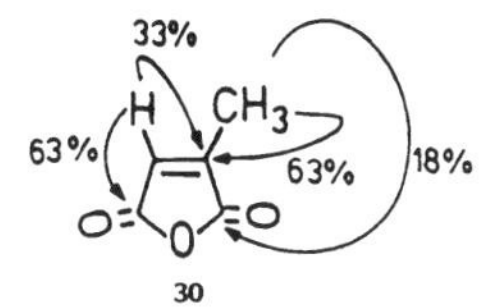

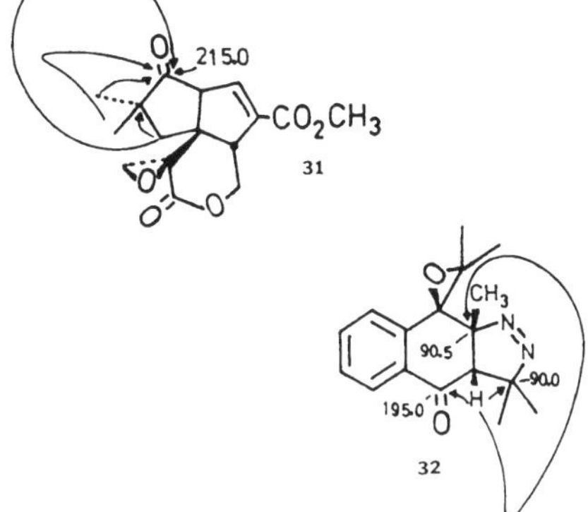

Fig. 2.36. ^{13}C-^{1}H Heteronuclear NOEs observed in citraconic anhydride (**30**) (383); pentalenolactone G (**31**) (339); and quinone diazoalkane adduct (**32**) (2)

2.3.2.6 Deuterium Isotopic Shifts

Substitution of deuterium for hydrogen induces a small upfield shift in the ^{13}C-NMR spectrum of directly-bonded carbons as well as increasingly smaller upfield shifts in the carbons two and three bonds away (142). Theoretical treatments of the isotope effect in NMR spectroscopy have appeared (55, 187, 188).

The best method for observing the deuterium isotopic shift in the carbon spectrum is the differential shift technique using coaxial NMR tubes (142). This technique utilizes an inner and outer tube of equal volume. One tube contains the sample dissolved in the deuterated solvent, such as D_2O or methanol-d_4, whereas the other contains the corresponding protio-solvent. Those carbons experiencing isotopic shifts appear as double resonances in the broad-band proton-decoupled spectrum, one signal originating from the protio-sample and the isotopically shifted resonance from the deuterium-exchanged sample.

Analysis of the deuterium isotopic shifts in the ^{13}C-NMR spectrum has been particularly useful for sugar and oligosaccharide structural studies and spectral assignments (18, 151). Application of the differential isotope shift technique was used to study the tautomers of psicose, (**33a**–**33d**), in solution (Fig. 2.37a) (384). Signal assignments were made by comparison of calculated and observed cumulative isotope shifts. This method distinguished the C-1 and C-6 resonances.

Deuterium isotopic shifts in the ^{13}C-NMR spectrum under partially exchanged conditions were used to assign the resonances of cellobiose (**34**) (Fig. 2.37b) (62). In this example, a 50% pre-exchanged sample of cellobiose was examined in DMSO-d_6, a solvent yielding slow-exchange conditions on the NMR time scale. Carbons experiencing only a single isotopic shift, visible upon exchange up to three bonds away (C-1, C-6, and C-6′) appeared as doublets, with upfield iso-

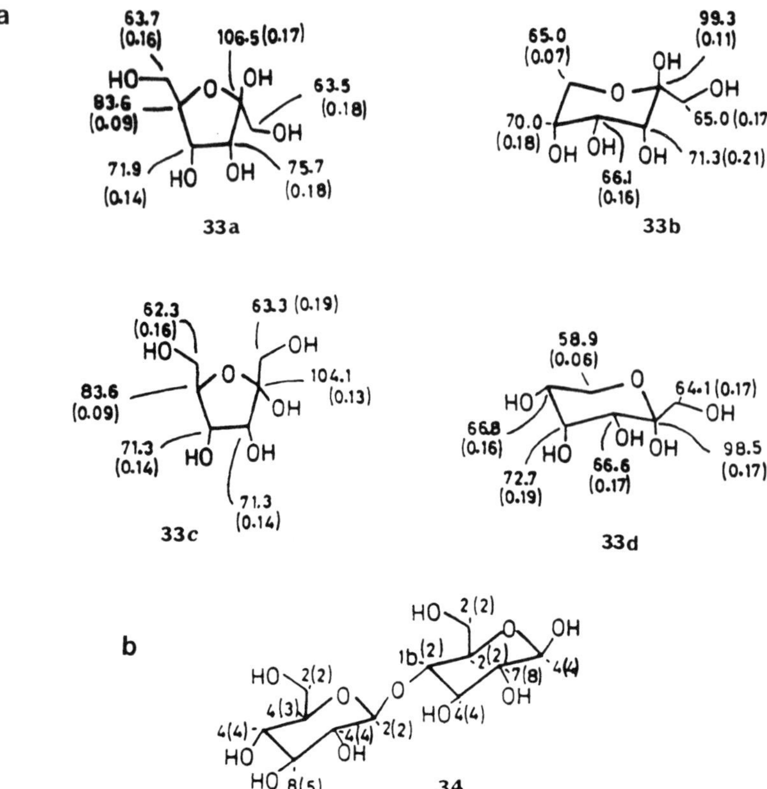

Fig. 2.37. a Differential deuterium isotope shifts for ^{13}C-NMR assignments for the tautomers of psicose (**33a**–**33d**) (384). Figure shows the carbon chemical shifts recorded in D_2O; numbers in parentheses represent the differential induced shift caused by deuterium exchange. **b** Number of carbon signals predicted for cellobiose (**34**) in DMSO-d_6 under partial exchange; numbers in parentheses represent number of observed signals (62)

Fig. 2.38. Structures **35** and **36** distinguished by loss of NOE to carbonyl carbon at $\delta = 188$ upon deuterium exchange (341)

topic shifts of −0.020, −0.110, and −0.113 ppm, respectively. Carbons with two possible isotopic shifts appear as four-line resonances (from the HH, HD, DH, and DD combinations) when all four isotope effects are of different magnitude (C-1, C-3, and C-2), or as three-line resonances if two isotope effects are of the same magnitude (C-5′). Three possible isotope effects produce a pattern of eight lines when the magnitudes are different.

In addition to isotope-induced shifts of NMR signals in the ^{13}C spectrum, reduction of signal intensity due to the loss of NOEs from exchangeable protons upon deuterium exchange can also be useful in structure assignment. The stereochemistry of the exocyclic double bond in **35** (Fig. 2.38) was resolved by determining the enhancement factor of the carbonyls with broad-band proton decoupling before and after deuterium exchange (341). A reduction in the intensity of the vinylogous amide carbonyl ($\delta = 188$) after deuterium exchange, but not in the amide carbonyl ($\delta = 166$) supported structure **35** rather than **36**.

2.3.3 Two-Dimensional NMR Spectroscopy

Two-dimensional NMR spectroscopy, first suggested by Jeener in 1971[6], encompasses several pulse sequences capable of selectively examining various coupling or exchange interactions by allowing the desired interaction to modulate time-dependently the resonance signals in a second time dimension (24). Basically, any one-dimensional multi-pulse sequence can be used to generate a two-dimensional spectrum by incrementing one of the time intervals prior to acquisition, thus creating a second time domain. The evolution of the magnetization vectors during this interval, which is controlled by some type of internuclear interaction, is time-dependent in the second dimension. The first dimension usually remains the normal FID time domain, which is Fourier Transformed to produce the chemical shift frequency domain.

By incrementing the interval of the second time dimension, a series of information-rich spectra is produced that contains the chemical shift frequency in one dimension and either the interaction frequency or interaction correlation in the second dimension. The projection of the frequency of these interactions, in the form of time-dependent intensity modulation or magnetization transfers along a second axis creates the two-dimensional spectrum. Fourier Transformation along this second dimension results in a tremendous resolution of spectral information. Several reviews of applications of two-dimensional NMR have appeared (37, 48, 105, 259, 277, 377). Only those experiments in routine use or of high potential value for natural products chemists are presented.

2.3.3.1 Two-Dimensional J-Resolved Proton NMR Spectroscopy

The two-dimensional J-resolved spectrum is extremely useful for resolving signals that are severely overlapped in a normal one-dimensional spectrum, allowing determination of both chemical shift frequencies and coupling constants that may otherwise be arduous to ascertain (15, 46, 136, 278, 280).

The two-dimensional spectrum may be plotted as a stacked plot, as shown for the insect antifeedant trichilin A, isolated from *Trichilia roka* (Fig. 2.39) (282, 284). Individual multiplicities were examined as cross-sections of the stacked plot

[6] Jeener, J. 1971. Presented at Ampere International Summer School. Basko Polje, Yugoslavia, unpublished.

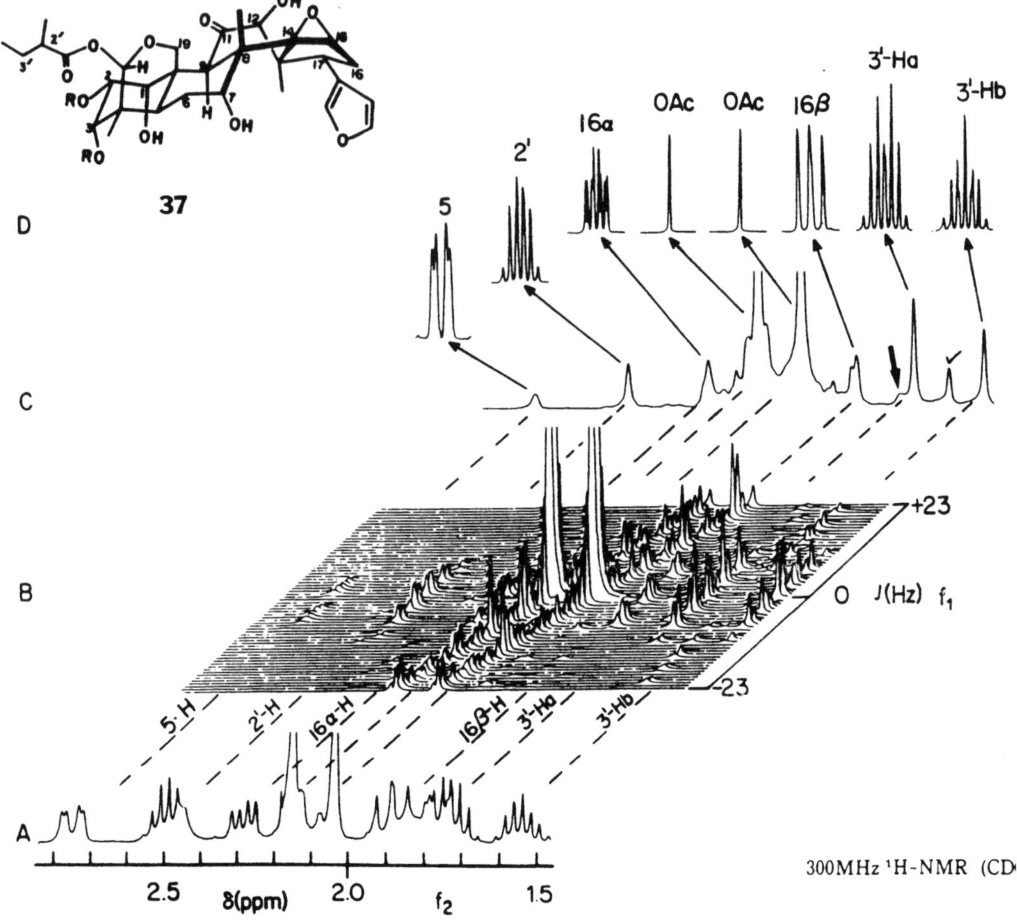

Fig. 2.39. J-resolved spectrum of trichilin A (**37**) (282)

to determine accurately coupling constants. These cross-sections are computer files of the Fourier Transform from the t_1 time domain. Other examples of the application of 2D-J spectroscopy have appeared in the literature (232).

2.3.3.2 Two-Dimensional Correlation Spectroscopy

Correlation spectroscopy is a unique method for mapping connectivity via spin coupling or exchange interactions (Jeener[6], 14). The spin-coupling interactions may be scalar or dipolar couplings. Signals in the 2D-spectrum indicating such interactions between nuclei A and B occur as cross-peaks at δ_A, δ_B. The pulse sequences employed in correlation spectroscopy feature a mixing pulse that transfers magnetization between the nuclei experiencing coupling. Spectral data from

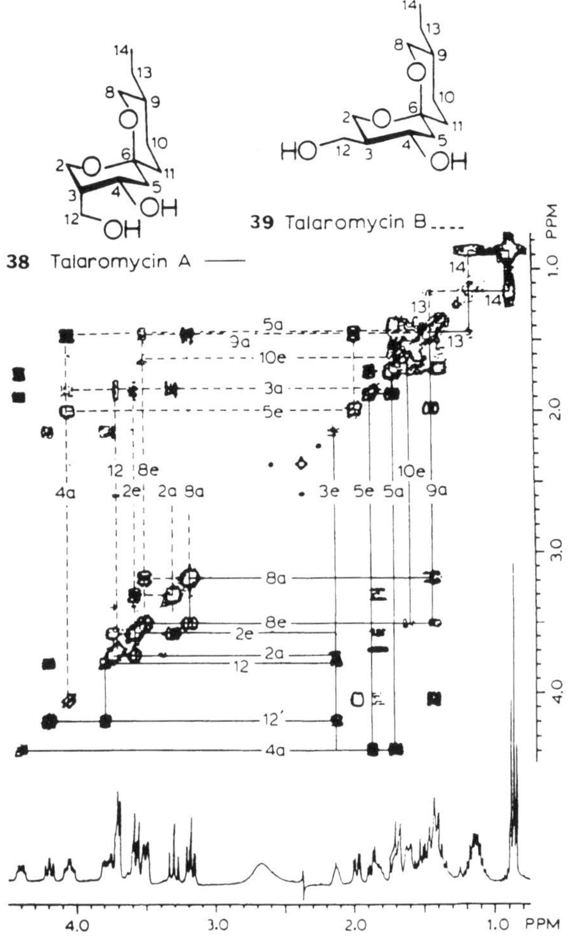

Fig. 2.40. Contour plot of the COSY spectrum of a mixture of the two diastereomeric talaromycins A and B (**38** and **39**). The lines on the spectrum connect each isomer individually (239)

2D-correlation experiments are almost invariably presented in a contour plot. Homonuclear correlation spectroscopy, in the form of COSY (proton correlation spectroscopy) (25, 28) or SECSY (2D-spin echo-correlated spectroscopy) (279) pulse sequences, is routinely employed in modern natural products structure determination (198, 328).

The recent example of talaromycins A and B (Fig. 2.40) produced by the toxicogenic fungus *Talaromyces stipitatus* illustrate the strategic utilization of sophisticated spectroscopic techniques in structure determination (239). These two toxins were isolated as a mixture and were extremely difficult to resolve. Due to the loss of water in the positive CI-MS, the molecular ions of these diastereomers were detected only by NCI-MS, with deuterium exchange NCI-MS confirming the presence of two exchangeable protons. The coupling connectivities of the individual diastereomers were mapped with a COSY spectrum (Fig. 2.40). The purified compounds were ultimately separated after formation of their phenyl boronic acid esters (Fig. 2.41).

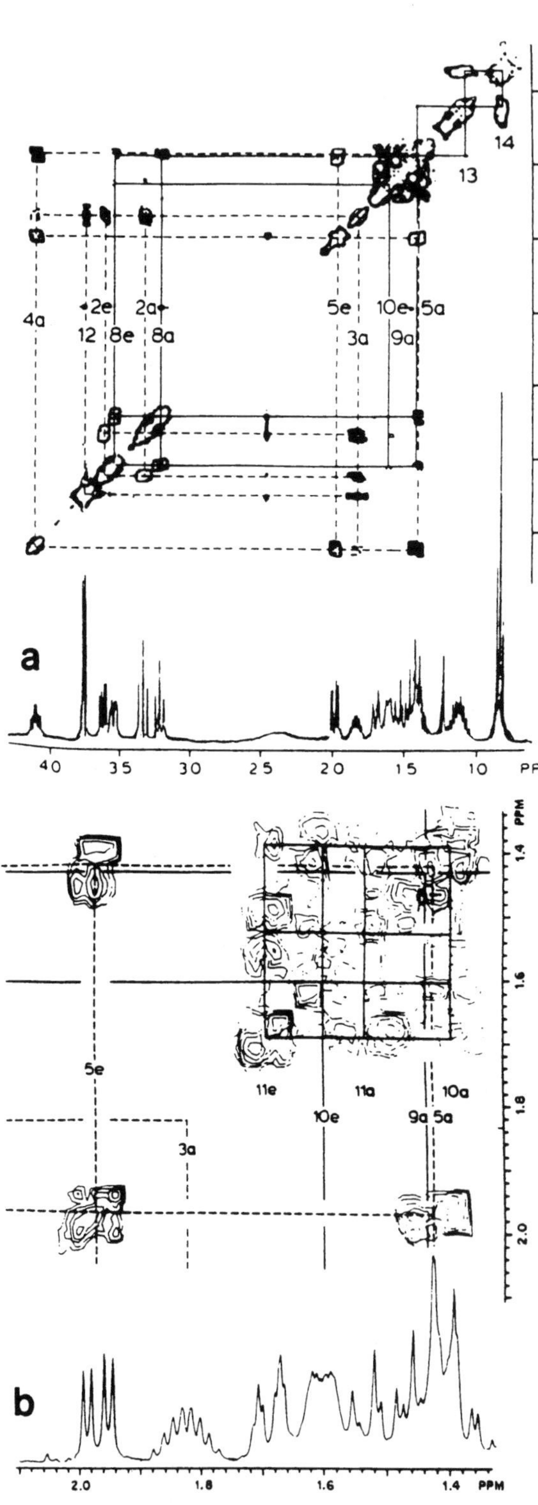

Fig. 2.41. a COSY spectrum of talaromycin B (**39**) after purification as phenylboronic acid ester; **b** expansion of the high-field region of the same spectrum. Solid and dashed lines distinguish the two isolated spin systems (239)

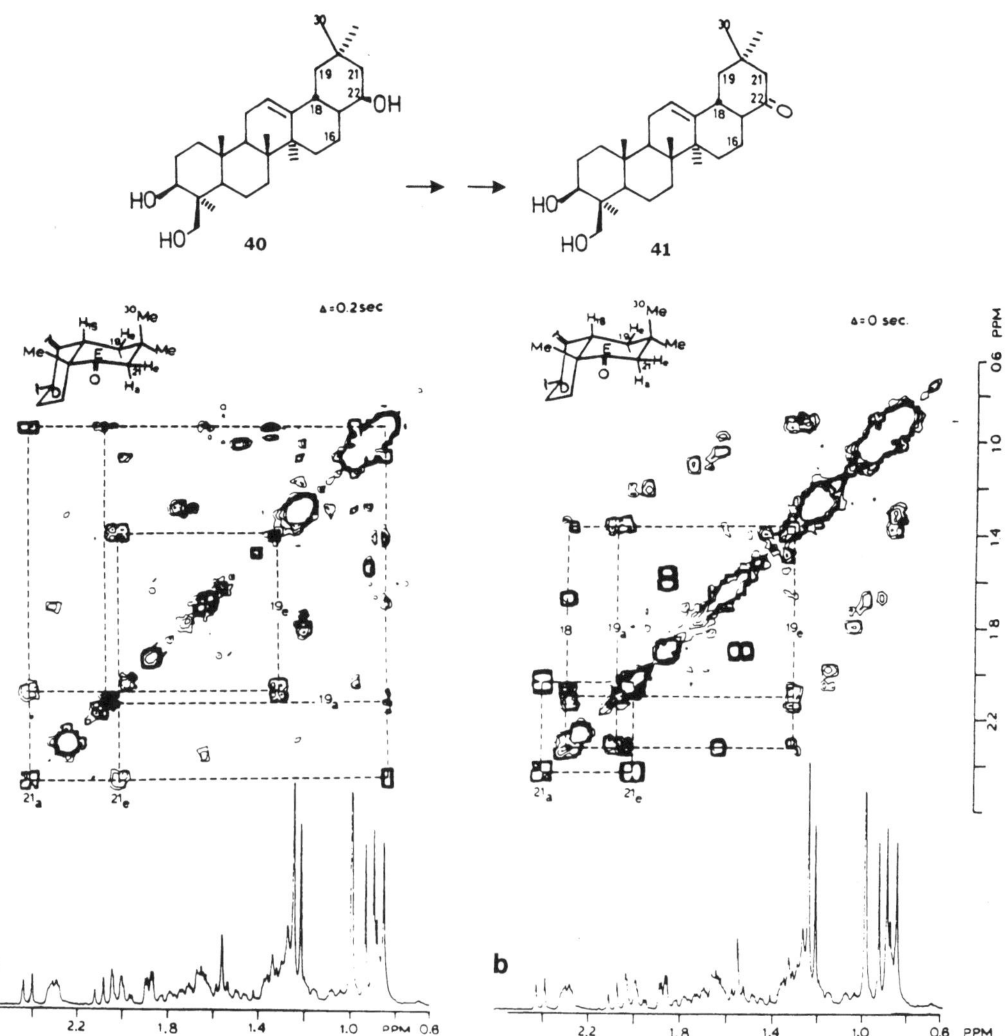

Fig. 2.42. a Long-range COSY spectrum of derivative **41** from hausterium-inducing component (**40**), isolated from *Lespedeza sericea*, showing four-bound coupling from H-21 (eq) to 30-CH_3 and H-19 (eq), utilizing a delay $\Delta = 0.02$ sec; **b** COSY spectrum without delay, $\Delta = 0.0$ (363)

The compound that induces hausterium growth in cultures of the plant parasite, *Agalinis purpurea*, isolated from the roots of *Lespedeza sericea*, was shown to have structure **41** (Fig. 2.42), based on long-range couplings observed in the ketoacetonide derivative (**40**) (363). Without resolving these long-range couplings, the location of the hydroxyl group on C-22 could not be distinguished from possible C-16 or C-21 hydroxyls. The COSY 2D-NMR spectrum (Fig. 2.42a) with a delay Δ of 0.2 s before and after the 90° mixing pulse of A showed four-bond coupling between the equatorial proton a to the carbonyl and both the axial 30-CH_3 and the 19-H equatorial proton, thus placing the original hydroxyl group at C-22.

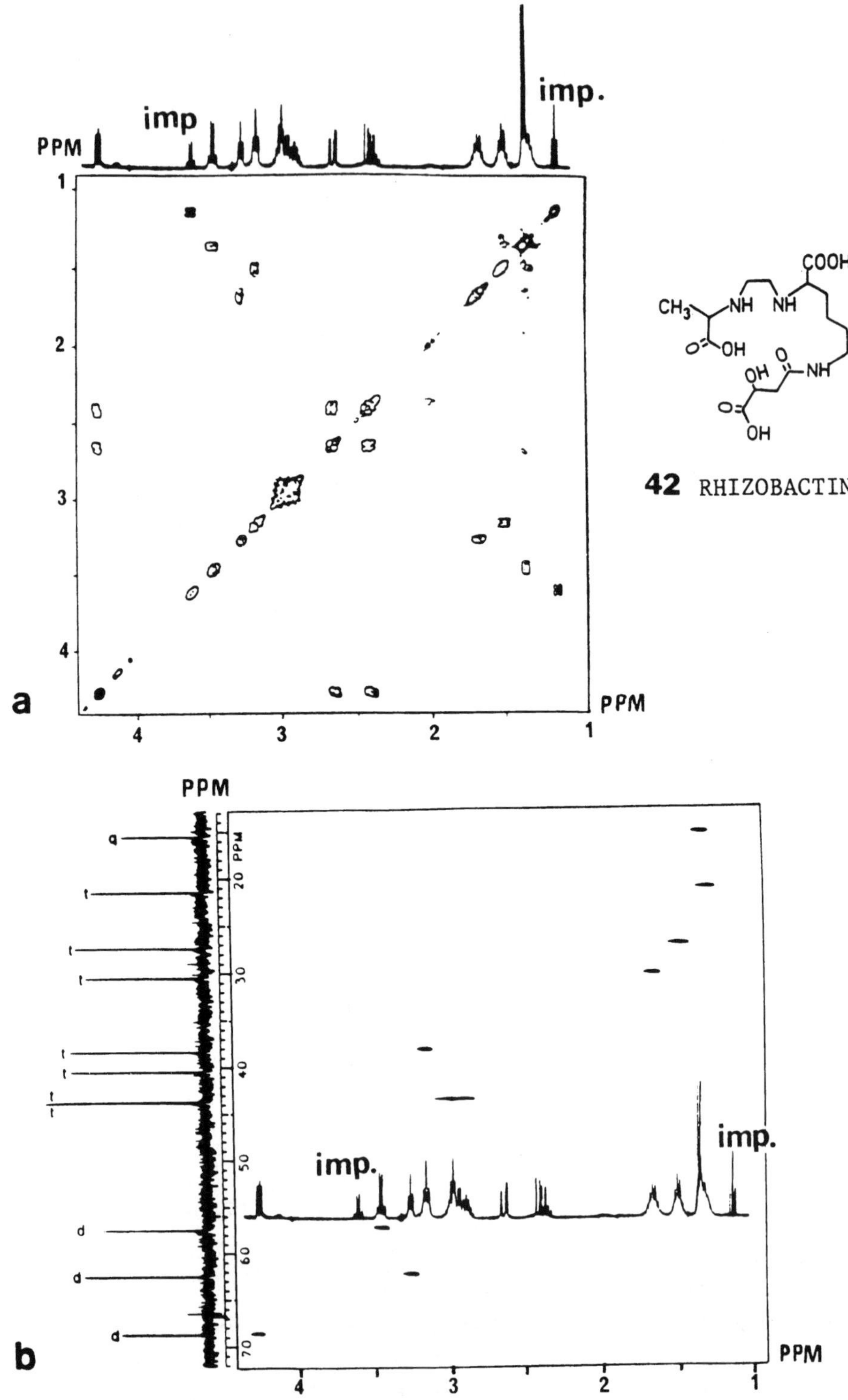
imp
imp.
PPM
a
PPM
COOH
CH3
NH
NH
OH
OH
NH
OH
42 RHIZOBACTIN
PPM
b
imp.
imp.
PPM

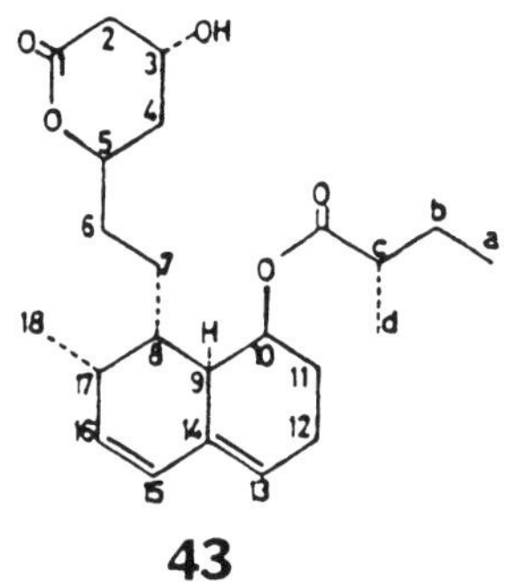

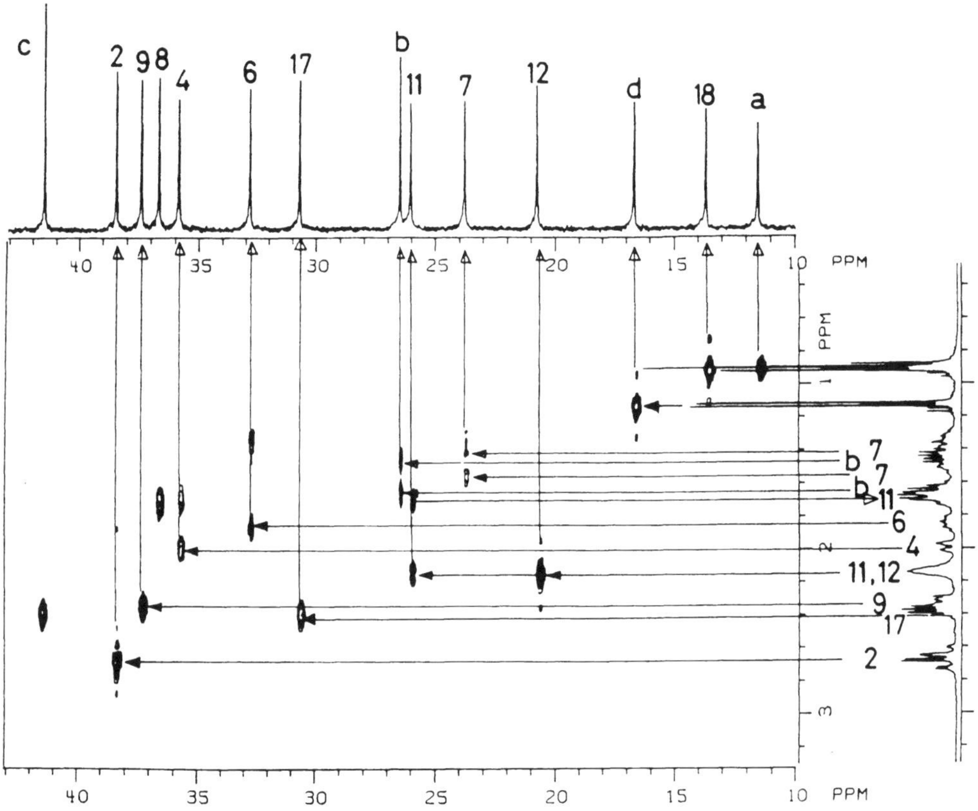

90MHz ^{13}C-NMR ($CDCl_3$)

Fig. 2.44. C-H COSY spectrum of compactin (**43**) (282)

Fig. 2.43a. Contour plot of the homonuclear COSY spectrum (HOMCOR) of rhizobactin (**42**) revealing the coupling connectivities of the four separate units that compose the molecule plus an impurity; **b** contour plot of the heteronuclear ^{13}C-^{1}H COSY (HECTOR) showing carbon and proton assignments. Both spectra were run in D_2O, pD = 8.4 (354)

These long-range couplings were not detected in the COSY spectrum without the delay Δ inserted in the pulse sequence (Fig. 2.42b).

Another excellent example of the application of NMR techniques in the structure determination of natural products is the structural resolution of rhizobactin, a siderophore isolated from *Rhizobium meliloti* (354). The proton homonuclear 2D-correlated spectrum revealed the four separate coupled units, whereas the ^{13}C-^{1}H heteronuclear 2D-correlated spectrum established the assignments (Fig. 2.43). Together, these spectra revealed that rhizobactin is composed of one unit each of ethylene diamine, alanine, lysine, and L-malic acid. The sequencing of

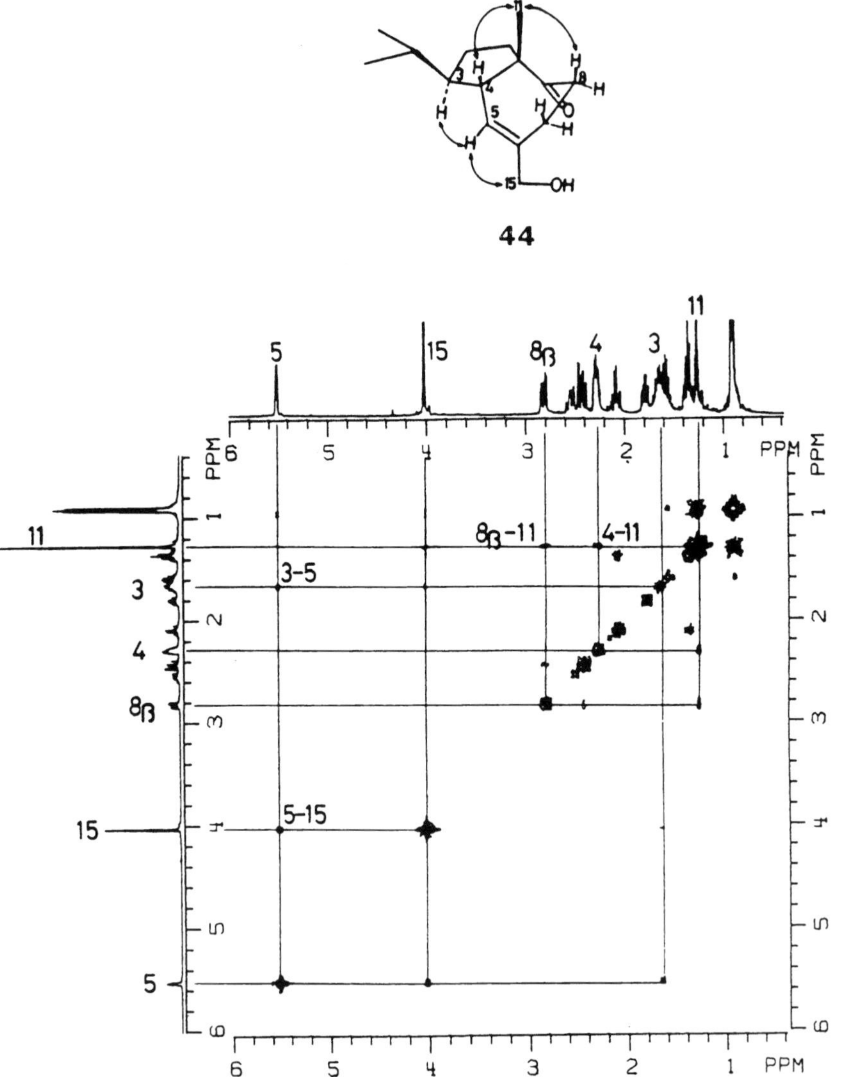

Fig. 2.45. NOESY spectrum of aphanamol (**44**) (282)

these moieties was established by the pD dependencies of the ^{13}C- and ^{1}H-chemical shifts.

Figure 2.44 (282) shows the enlarged high-field region of a C-H COSY spectrum of compactin. Clearly, the signals of C-7, 11, b (side-chain), 6, and 4 each have two cross-peaks; they are methylene carbons linked to the two protons correlated through the cross-peaks. The 20.8 ppm signal (C-12) is known to be due to CH_2 (from DEPT) but it has only one cross-peak; this is because the two 12-H's have the same chemical shifts (and are also overlapping with one of the 11-H's).

Cross-relaxation through dipolar coupling interactions (NOE) can be resolved and mapped in a fashion similar to scalar coupling interactions, thereby providing a means for establishing through-space connectivity (189, 228, 267). The use of the 2D-NOE experiment as a means for examining conformation structure of biomacromolecules is becoming routine. Both peptides (197, 199) and oligonucleotides (334) have been sequenced, and their conformations have been analyzed by 2D-NOE spectroscopy.

Figure 2.45 (282) shows the 2D-NOE spectrum of aphanamol-I. The proton assignments were accomplished by the COSY technique. The 2D-NOE method was then used to determine the conformation of the molecule. The cross-peak 4 – 11 indicates the 5- and 7-membered rings are *cis*-linked; cross-peak 3 – 5 shows the isopropyl group to have a β-configuration; and the $8\beta - 11$ peak suggests that the 7-membered ring adopts the conformation shown.

Since the 2D NOE detects the migration of spin density from one state to another, caution must be exercised in assigning off-diagonal signals that could arise from a spin-density transfer due to chemical exchange rather than dipolar coupling. The same pulse sequence is utilized to detect chemical exchange phenomena (44, 164, 246, 267). For example, the 2D-NOE spectrum of dimethylformamide shows an off-diagonal element connecting the two methyl groups. Such a spin-density transfer between the two methyl groups could be due to a dipolar relaxation mechanism or, more likely, chemical exchange via rotation about the carbon nitrogen bond. Studies of intramolecular chemical exchange utilizing 2D NMR can be very useful in the investigation of peptide conformations (200) and in the study of protonation sites (152). 2D NMR can also be used to study the kinetics of exchange processes (163).

2.3.3.3 Two-Dimensional INADEQUATE (Incredible Natural Abundance Double Quantum Transfer Experiment)

The 2D-INADEQUATE technique clarifies the connectivities of all carbon atoms mechanically and is therefore extremely powerful (26, 27, 47, 245, 286, 309, 322, 338). It requires close to 100 mg of sample, the pulse sequence is complex, and measurement time is long. These drawbacks, however, can be overcome by using ^{13}C-enriched samples. The technique is exemplified in Fig. 2.46 (282) with menthol. Thus, cross-peaks across the diagonal line interconnect linked carbons to give C-C connectivities: j is connected to i; i is connected to e; i is connected to d; and j is connected to h. This clarifies the moiety h-j-i-d (and e). Similarly, connectivities e-b, e-a, and d-g lead to moiety g-d-i-e-a (and b). Finally, additional connectivities h-f, f-c, and f-g lead to the entire menthol skeleton.

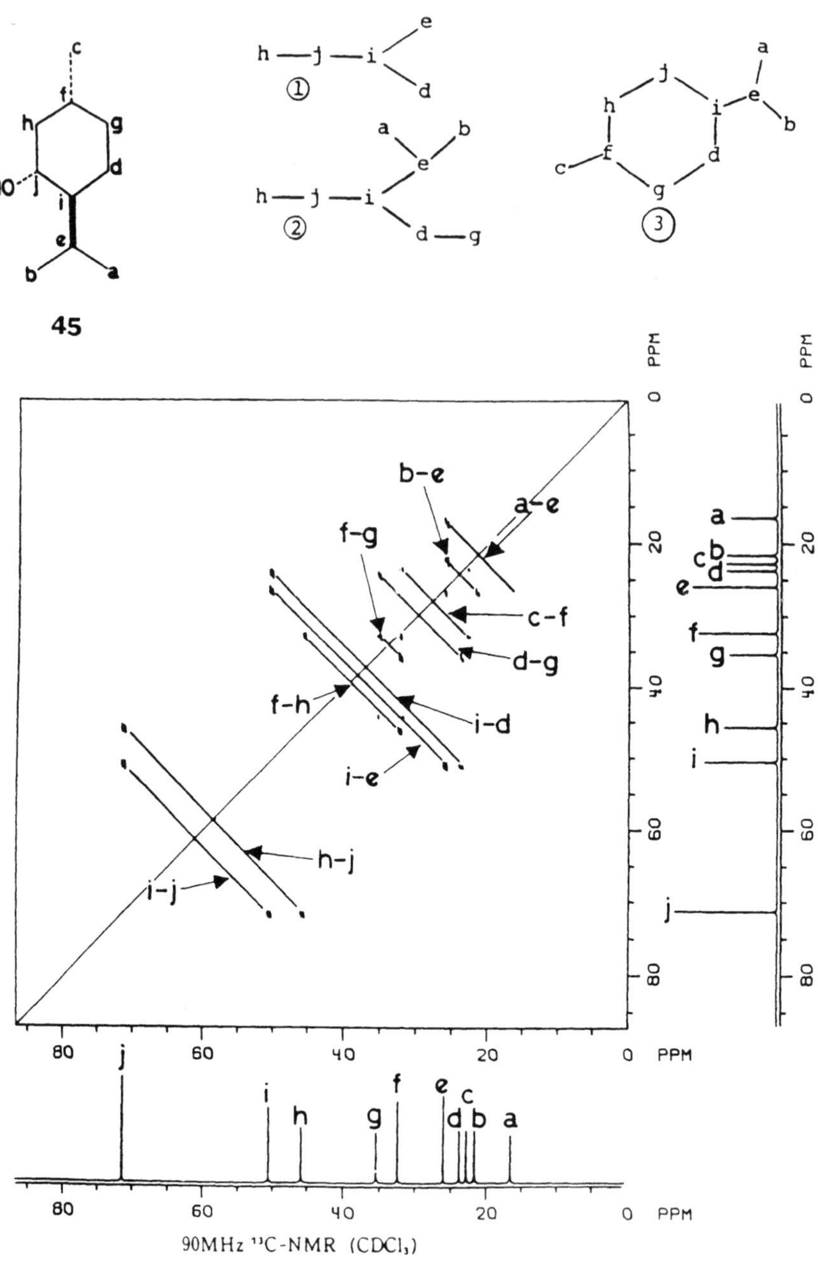

Fig. 2.46. 2D INADEQUATE spectrum of L-menthol (**45**) and (partial) structures showing C-C connectivities (282)

2.4 Other Spectroscopic Techniques

2.4.1 Mass Spectrometry

Mass spectrometry (261) is the second most important spectroscopic technique utilized in the determination of chemical structures. High resolution mass spectrometry is replacing combustion analysis as the method of choice for verifying the molecular formula of newly isolated compounds. Several advances such as field desorption and fast atom bombardment have enabled nonvolatile compounds of increasingly higher molecular weight to be analyzed by mass spectrometry without chemical derivatization. Equally important are the advances that interface a separation method with mass spectrometry, thereby allowing for rapid identification of known compounds by comparison with a library of known mass spectra. Such "hyphenated" techniques have advanced far beyond the original GC-MS to include LC-MS (13, 109, 258) and even MS-MS (262), which retain the extremely high sensitivity that is a major advantage of mass spectrometry. This section will discuss several recent developments in mass spectrometry that have or that will soon come into common usage in the determination of the structure of natural products.

Basically, mass spectrometry requires that the sample be ionized. The more familiar mass spectrometric techniques – electron impact and negative and positive chemical ionization – involve ionization of the sample in the gas state. Samples that cannot be volatilized, such as high-molecular-weight and ionic compounds, have to be derivatized to enhance their volatility. Such derivatization reactions usually produce molecules of even higher molecular weight, which can lead to reduced sensitivity in their detection or even go beyond the detection limits of the instrument. Furthermore, derivatization reactions of highly functionalized molecules can produce several products with varying degrees of derivatization due to incomplete reaction. Identification of molecular ions and fragmentation patterns can be quite confusing under such conditions.

The recent innovations in mass spectrometry involve methods for enhancing the vaporization, more accurately termed "desorption," of relatively nonvolatile samples without derivatization. Even more significant are those advances that now allow for direct ionization of the sample to occur on the probe with subsequent desorption of the ions. As the efficiency of these new techniques improves, arguments begin to arise as to whether ionization of the sample components actually occurs prior to or subsequent to desorption (34, 35, 153, 154). Such arguments, however, need not concern us here.

Molecular weights of >4000 for some biomacromolecules and of ≤ 10000 for polyethylenes can now be routinely measured, and detection limits are pushed forward on an almost weekly basis (96, 256). Within this range, it has been pointed out that the meaning of "molecular weight" itself will have to be reconsidered as the statistical significance of low natural abundance isotopes such as ^{2}H and ^{13}C can no longer be ignored (97). In effect, a statistical distribution of "molecular weights" will be found that accounts for the presence of these and other isotopes.

2.4.1.1 Techniques That Enhance Sample Volatilization

Electron impact mass spectrometry (EI) and chemical ionization mass spectrometry (CI) remain the most commonly employed mass spectrometric techniques due to their relatively low cost and simplicity. Negative chemical ionization mass spectrometry (NCI) is used less often, but remains useful.

In EI-MS, the sample, following vaporization, is exposed to a beam of energetic electrons in the order of 70 eV. The electron bombardment creates positive ions from the sample, which are subsequently mass analyzed. Spectra obtained from EI-MS are especially useful in providing information of fragmentation patterns, though molecular ions (M^+) can be recorded for more stable compounds. High molecular-weight and nonvolatile compounds, however, frequently show no molecular ion peaks, thus limiting the information obtainable from EI-MS (261).

In CI-MS, a reagent gas is bombarded with the electron beam under high pressure to produce reagent ions. Methane and isobutane are the most common reagent gases, though ammonia is also useful. Typical reagent ions produced using methane as the reagent gas are shown in Table 2.1. The reagent ions CH_5^+, $C_2H_5^+$, and $C_3H_5^+$ react with the sample through several possible mechanisms. Proton transfer, the most common mode of reaction, produces the $(M+1)^+$ ion used to determine the molecular weight. The reagent ions may also interact with the sample forming molecular association that produce $(M+17)^+$ and $(M+29)^+$ peaks. These peaks, when present, are very useful as markers for confirming the $(M+1)^+$ peak. Hydride transfer from the sample to the $C_2H_5^+$ reagent ion is also a possibility, particularly with hydrocarbons.

CI-MS is most widely used in confirming a molecular weight, for fragment ions are considerably less abundant than in EI-MS. The reagent gas can be changed to enhance the intensity of the $(M+1)^+$ peak and to reduce fragmentations. Isobutane, for example, is frequently utilized to observe the $(M+1)^+$ ion with little fragmentation. For more polar molecules, ammonia can be employed, usually producing an abundant $(M+18)^+$ peak in addition to the $(M+1)^+$. The molecular weight of an unknown natural product might be confirmed with CI-MS by comparing the spectrum obtained with methane or isobutane as reagent gas with the spectrum obtained with ammonia. The number of exchangeable protons in a sample is frequently determined by comparing the CI-MS spectrum using ammonia as the reagent gas with the spectrum obtained with ammonia-d_3 as the reagent gas (166). Under the latter conditions, exchangeable protons are replaced by deuterons with a resultant increase in molecular weight.

Table 2.1 Typical reagent ions produced using methane as the reagent gas in CI-MS (98)

Reactants		Products	Intensities
CH_4+e^-	→	$CH_4^{\bullet+}$, CH_3^+, $CH_2^{\bullet+}$, etc.	
$CH_4^{\bullet+}+CH_4$	→	$CH_5^+ + CH_3^\bullet$	47%
$CH_3^+ + CH_4$	→	$C_2H_5^+ + H_2$	41%
$CH_2^{\bullet+} + CH_4$	→	$C_2H_3^+ + H_2 + H^\bullet$	
$C_2H_3^+ + CH_4$	→	$C_3H_5^+ + H_2$	6%

With NCI-MS (73), negative ions are generated from the sample with various reagent gases, such as an isobutane-methylene chloride-oxygen mixture. Ionization can occur by several mechanisms, such as chloride ion addition. NCI is much more selective in producing ions from the sample, having a greater sensitivity for specific classes of compounds depending upon the exact nature of the reagent gas mixture. Furthermore, as the negatively ionized sample can readily regain stability by the simple loss of an electron, fragmentations are greatly reduced compared to positive CI. The high selectivity and reduced fragmentations make NCI useful in analyzing environmental samples for halogenated hydrocarbons. Natural products that fragment easily even with positive ion CI – such as hydroxylated compounds that readily lose water – may have their molecular ions successfully recorded with NCI (257).

These three techniques, EI-MS, CI-MS, and NCI-MS, along with field ionization (FI) (32), a technique little used in natural products chemistry today, all require ionization of the sample in the gas state. This limits the utility of such MS methods for analyzing the increasingly complex natural products of higher molecular weight being isolated today. Heating the probe to increase the amount of sample in the vapor state often leads to decomposition prior to vaporization because of thermal instability. The resulting spectrum contains information concerning the fragmentation only, with little or no molecular ion information.

It has been shown many times that the rate of heating is critical in mass spectral analysis (39). A rapid rate of heating can induce vaporization prior to decomposition even when the energy necessary for thermal decomposition is less than the heat of vaporization. This can only occur when sufficient thermal energy is supplied to overcome intermolecular interactions at a rate rapid enough so that the excited vibrational states are not attained by all sample molecules. In other words, one shoots past the decomposition temperature at a sufficiently rapid rate so that the intermolecular bonds that prevent vaporization (interactions on the external face of the molecule) are broken before the sample absorbs sufficient internal energy to completely decompose. The relative intensity of the molecular ion peak $(M+1)^+$ of thyrotropin-releasing hormone (= TRH) (m/e = 363) (Fig. 2.47) increases at a greater rate than the relative intensity of the fragment ion (m/e = 235) with increasing temperature (39). A non-interactive material, such as Teflon, on the sample holder or probe can also greatly enhance the amount of sample in the vapor state and improve the spectra (38). With an extended probe to place the sample as close as possible to the electron beam of the reagent gas (in CI-MS), it has been possible to observe the molecular and fragment ions of echinomycin (MW = 1101.27) in EI-MS, where normal EI and CI conditions had failed (75).

Many natural products, however, such as peptides – especially those containing arginine – oligosaccharides, nucleotides, and others with high molecular weights, cannot be successfully analyzed by these more common methods. More sophisticated techniques of ionization have therefore been developed, capable of analyzing compounds with molecular weights of 1000 to 5000. These techniques have made the question of whether ionization occurs prior to or subsequent to desorption somewhat moot.

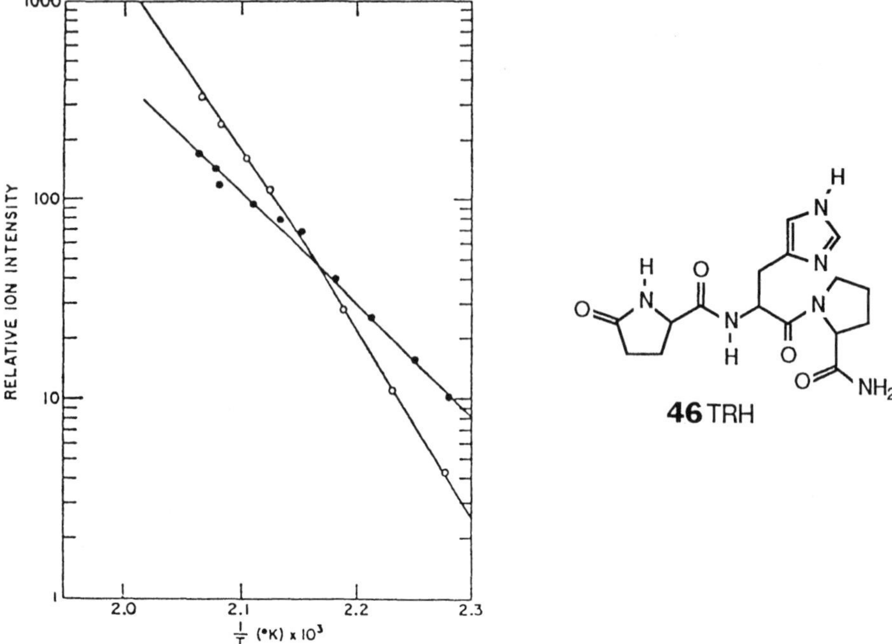

Fig. 2.47. Relative intensity of m/e 363 (○) $(M^+ + H)^+$ parent ion and m/e 235 (●) fragment ion in the mass spectrum of TRH (PCA–HIS–Pro–NH_2) (**46**) as a function of temperature (here as 1/T). Spectra were obtained by evaporating the solid sample from a copper probe in a collision chamber made from Teflon (39)

2.4.1.2 Modern Techniques of Ionization/Desorption

During the past decade, numerous techniques have been described for mass spectral analysis of nonvolatile and high-molecular-weight organic compounds. In field desorption mass spectrometry (FD-MS) (33, 36), the sample is deposited on an anode covered with carbonaceous microneedles, termed an "emitter" and placed in an electrostatic field. Ionization occurs on the surface and the ions are subsequently desorbed. The electrostatic field is thought to aid in the transfer of electrons from the sample to the emitter with the carbonaceous microneedles aiding desorption via a Coulombic repulsion. The emitter can also be heated to assist in the desorption. Because the field desorption energy necessary to remove the generated ions from the emitter is considerably less than the required sublimation or vaporization energy, spectra revealing molecular ions can be obtained of compounds that otherwise would undergo thermal decomposition prior to vaporization under EI-MS or CI-MS conditions (17, 75).

FD-MS has been shown to be an excellent method for recording the mass spectra of compounds showing both molecular and fragment-ion formation (1, 76, 88, 95, 108, 161, 372). Compounds that do not produce good mass spectra due to instability even under field desorption conditions – such as zwitterions – can be mixed with an acid such as *p*-toluenesulfonic acid prior to decomposition on the emitter to produce good results (196).

Recently it has been shown that the electrostatic field of field desorption is in many cases not necessary to induce ionization and can be replaced by the electron beam used in EI-MS or the reagent gas used in CI-MS. These new methods are termed desorption or direct electron impact (DEI) (361) and desorption chemical ionization (DCI) (167). The same emitter or activated probe used in FD-MS is employed, but placed close to the reagent gas or electron beam. In DCI-MS, which is more widely used than DEI-MS, the ionization mechanisms, deduced from the observed mass spectrum, are the same as observed for CI-MS. As the rate of desorption has also been shown to be independent of the reagent gas used, desorption of neutral molecules from the emitter followed by ionization is thought to be the dominant sequence of events leading to the mass spectrum (67). Nonvolatile or relatively unstable compounds that do not give good spectra with EI-, CI-, or FD-MS, such as guanosine and arginine, can be successfully analyzed by DCI-MS.

Recently, the application of a laser pulse to desorb polar, nonvolatile bioorganic molecules, such as oligosaccharides and glucosides, has been reported (311). Variation of the laser wavelength had no effect on the spectrum of sucrose, so it was concluded that desorption is a consequence of the near-spontaneous intense heating. Positive and negative ions are desorbed as are neutral molecules (68) and can be subsequently treated with a reagent gas to produce a CI-type spectrum, or bombarded with an electron beam for an EI-type spectrum. Using a pulsed CO_2-laser for desorption, cationized species such as $(M+Na)^+$ dominate with the electron beam off, but ionized species such as $(M+H)^+$ dominated with the electron beam on. Thus it was concluded that, without the electron beam, cationization prior to desorption dominates the spectrum in laser-desorption mode, but neutral desorption followed by ionization dominates with the electron beam on (366).

Several recently developed techniques to enhance sample desorption capitalize on the transfer of kinetic energy to the sample through molecular collisions to produce desorbed positive and negative ions as well as desorbed neutral molecules. The more useful of these "bombardment" techniques include secondary ion mass spectrometry (SIMS), fast atom bombardment mass spectrometry (FAB-MS), and californium-252 plasma desorption mass spectrometry (252 Cf-MS). The sample can be deposited on an inert surface and directly bombarded in the solid state, or more commonly it is dissolved (or suspended) in an inert and nonvolatile medium such as glycerol.

In SIMS, the sample is bombarded with a beam of ions generated from an inert gas, usually Ar^+ or Xe^+, with a kinetic energy of $\geqslant 3$ keV (21). Fast atom bombardment is an extension of SIMS, in which the beam of argon ions is accelerated to a kinetic energy of up to 10 keV and focused into a chamber of neutral argon (19, 20, 370). In this chamber, charge exchange occurs producing a beam of high-energy neutral argon atoms: Ar^+ (beam) + Ar^0 (chamber) $\rightarrow$ Ar^0 (beam) $+Ar^+$ (chamber). Ions are removed from the beam by electrostatic deflection. Both SIMS and FAB-MS are very good methods for recording mass spectra of high-molecular-weight and nonvolatile organic molecules. Numerous examples of both have appeared in the literature over the last few years (96, 97, 162, 164, 295, 300, 327, 403).

The utility of ^{252}Cf-MS (242, 243) is based on the fact that 3% of the radioactive decay of ^{252}Cf is spontaneous fission, the remaining 97% being α-decay. More than 40 ion pairs have been reported as the products of the spontaneous fission, such as $^{142}Ba^{+22}$ and $^{108}Te^{+22}$ with kinetic energies of 79 and 104 MeV, respectively. The average charge of a ^{252}Cf fragment is +20, with an average kinetic energy of about 10 MeV. These highly charged fragments bombard the solid sample deposited on nickel foil. As the production of a fragment pair from spontaneous fission generates particles traveling in opposite directions, time-of-flight mass spectrometers are used, with one fragment bombarding the sample, and its fission partner travelling in the opposite direction utilized as a precise time marker. Thus, it was possible to record the ions of palytoxin, the largest non-proteinaceous natural product (MW = 2681 daltons) (244), and of a protected nucleotide (as $(M+Na)^+$, m/e = 6301 ± 3) together with its dimer (with 12637 ± 10) (264). However, the translational energy spread occurring in ionization and ion dissociation considerably reduces the sensitivity and resolution of time-of-flight instruments at such high molecular weights (58) and directs the great potential of ^{252}Cf-plasma desorption rather towards tasks like the fragment-based sequence analysis of peptides (265). These energy-spread phenomena can be overcome with a relatively new instrument, the Fourier-Transform mass spectrometer FT-MS, where molecular fragmentations are kept "alive" for several magnitudes of time longer in circular motion within a cyclotron chamber. Since their cyclotron frequencies are directly proportional to their mass, measurement of this frequency produces values of unsurpassed resolution (usually > 1/100000 in accuracy). By the new ionization technique of pulsed Cs^+ desorption ionization, the ion of a cesium iodide cluster can be observed at exactly 16241 k daltons, marking the present detection limit (4). These "bombardment" techniques – SIMS, FAB-MS, ^{252}Cf plasma desorption Ms, and Cs^+ desorption ionization – all transfer mostly translation energy to the sample rather than vibrational energy. Fragmentation prior to desorption is therefore minimized and molecular ions over 3000 amu become readily observable. Another major advantage of these techniques, as well as of laser desorption mass spectrometry, is the desorption of both negative and positive ions as well as neutral molecules. Either of the ions can be selectively observed, or the desorbed neutrals can be treated with a reagent gas of electron beam to produce CI-MS- or EI-MS-type spectra.

An often observed phenomenon with bombardment and desorption techniques, including laser desorption, is the production of "cationized" species by complexation of the intact sample molecule with an ubiquitous alkali metal cation, usually Na^+ or K^+ (247, 285). These species give rise to $(M+Na)^+$ and $(M+K)^+$ peaks as well as some cationized fragment peaks observed in the mass spectrum. While some confusion can arise due to the appearance of these peaks, it has long been realized that they can significantly aid in the assignment of the molecular ion. Consequently, samples are frequently doped with an alkali metal, using LiCl, NaCl, or KCl, to enhance the intensity of the peaks due to cationized species (19, 311). Furthermore, cationization aids in the desorption of the sample similar to the effect noted previously by adding *p*-toluenesulfonic acid to a zwitterion sample for FD-MS (196).

2.4.1.3 Tandem Mass Spectrometry

Tandem mass spectrometry, the interfacing of two mass spectrometers, is a powerful method of obtaining structural information from a mixed sample on a selected compound with minimal interference due to unwanted chemical noise from other constituents (262). Any ionization technique, such as electron impact, positive or negative chemical ionization, field desorption, fast atom bombardment, and others, can be used to generate the initial array of molecular and fragment ions. Targeted primary ions are directed into the ionization chamber of the second mass spectrometer via a mass selection process that filters out other unwanted ions. If the selected ions have sufficient energy, they undergo spontaneous fragmentation to produce a daughter spectrum. Frequently, however, the selected ions must collide with a gas such as nitrogen to induce fragmentation in a process known as collisionally activated or collisionally induced decomposition, CAD or CID.

a MASS SPECTRUM

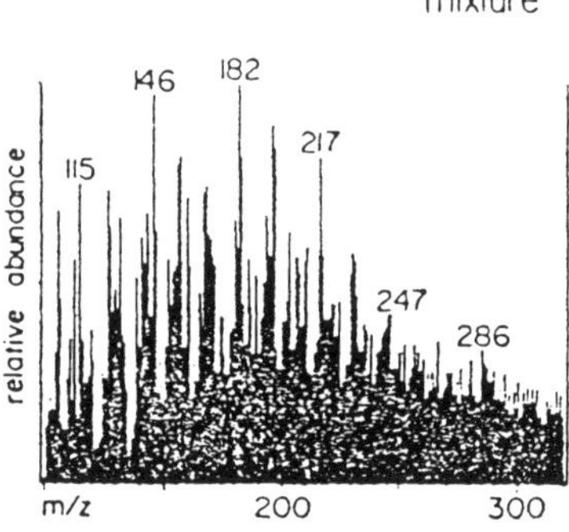

b DAUGHTER SPECTRUM OF 288⁺

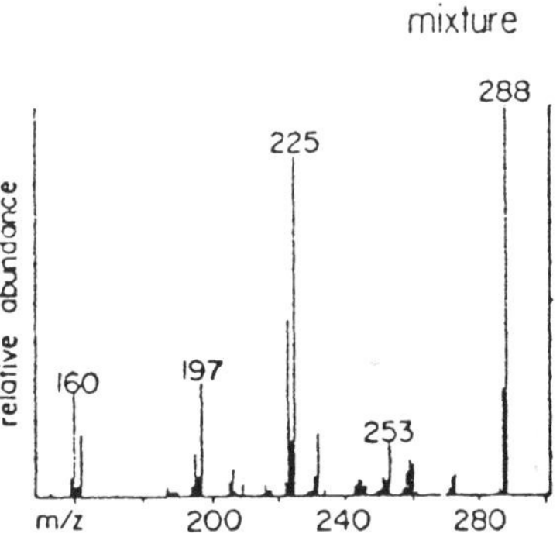

c DAUGHTER SPECTRUM OF 288⁺

1,2,4-trichlorodibenzo-p-dioxin

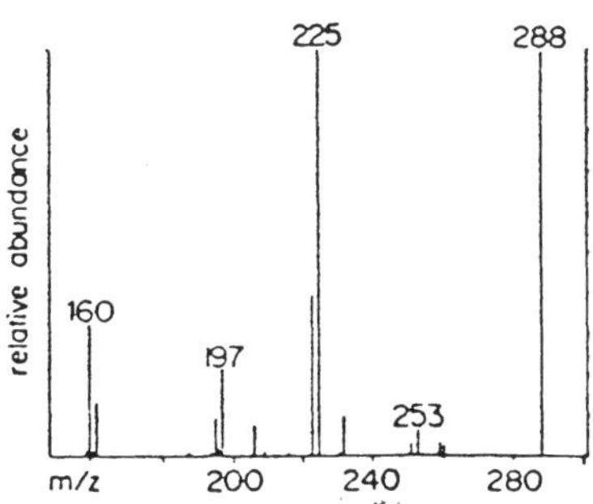

Fig. 2.48. a EI-MS of coal extract; **b** MS/MS daughter spectrum of same mixture with m/e 288^+ ion selected; **c** EI-MS of authentic 1,2,4-trichlorobenzo-*p*-dioxin (353)

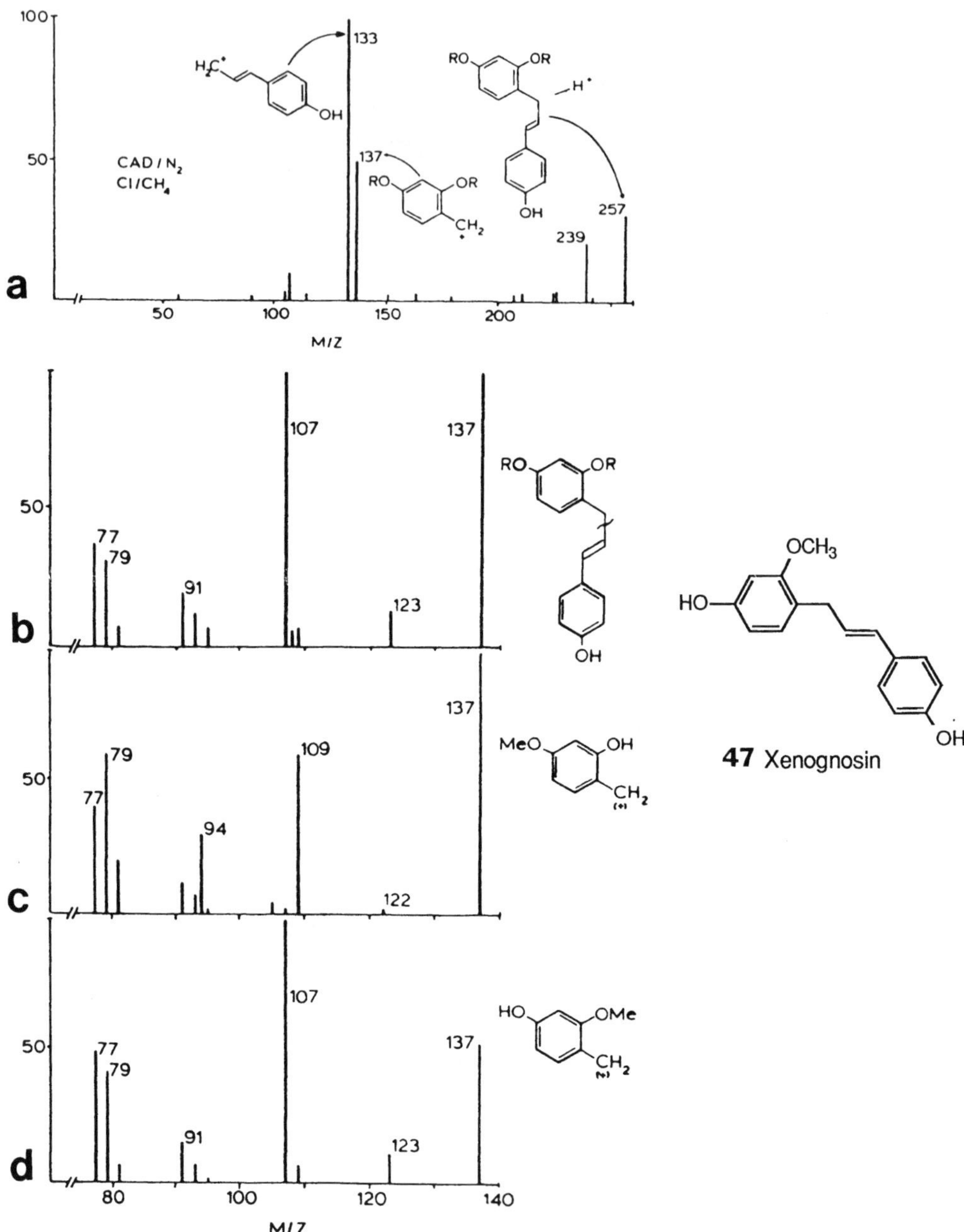

Fig. 2.49. MS/MS study of xenognosin (**47**). **a** CI-MS (CH_4 as reagent gas)/CAD (N_2) of (M+1)-ion; **b** CAD (N_2) daughter spectrum of selected m/c 137 ion from (**a**); **c** EI-MS of 2-hydroxy-4-methoxy benzylalcohol; **d** EI-MS of 4-hydroxy-2-methoxy benzylalcohol (240)

Tandem mass spectrometry can be operated in different scanning modes depending upon the information being sought (325). In the most frequently used scanning mode, the first mass spectrometer serves as a separator to select a particular ion, the fragmentation of which is scanned in the second mass spectrometer to produce the daughter spectrum. Tandem mass spectrometry operated in this scanning mode is a particularly elegant analytical technique for establishing the presence of targeted molecules without a lengthy, tedious purification procedure. Figure 2.48 shows the detection of 1,2,4-trichlorobenzo-*p*-dioxin in a sample of coal extract. The daughter spectrum produced from the targeted 288^+ parent ion matches the spectrum of the authentic dioxin (353).

The daughter spectrum produced from a selected parent ion may be crucial in distinguishing isomeric structures that would not be separated by the mass selection. The substitution pattern of the methoxyl-bearing aromatic part of the hausterial growth inducer xenognosin, isolated from gum tragacanth, was confirmed as 2-methoxy-4-hydroxyphenyl tandem mass spectrometry (240). The daughter spectrum produced the 137^+ parent ion of 2-methoxy-4-hydroxybenzyl alcohol and not 4-methoxy-2-hydroxybenzyl alcohol (Fig. 2.49). The small amount of isolated sample (400 μg) had precluded other methods of structure assignment.

Alternatively, the primary ions may be scanned with the second mass analyzer held constant to produce a parent spectrum. This type of tandem mass spectrometry reveals all the reactant ions giving rise to a particular fragment ion. Kondrat et al. (215) have demonstrated applications of this technique to the identification of biosynthetically related natural products that fragment to a common ion.

2.4.2 Ultraviolet-Visible Spectroscopy

Ultraviolet-visible (UV-Vis) spectroscopy (281, 337) is one of the oldest spectroscopic techniques and hence one of the first to be applied to structure determination. The substitution pattern of many aromatic natural products – flavonoids being an excellent example (241) – are still determined by UV-Vis spectroscopy including acid-based-, and aluminum trichloride-induced shifts of absorbance bands (227, 288, 343, 389). Adduct formation of the purine and pyrimidine bases are also investigated by UV-Vis spectroscopy (352), and solvent-induced shifts are often useful tools of structure determination for many other classes of compounds. There remains, however, a wealth of uncovered information in the UV-Vis spectra of natural products masked by broad, overlapping bands with little or no fine structure.

Application of derivative spectroscopy (375), performed by incrementing wavelength shifts to generate the derivative function d (Abs)/d(λ) with the aid of a microprocessor, provides a mechanism of abstracting a greater amount of information from the UV-Vis spectrum. Maxima are alternate positive and negative extrema in the even-order derivatives (zero-order, d^2, d^4,...), and go to zero in the odd-order derivatives (Fig. 2.50). Shoulders of inflection points become extrema in the odd-order derivatives. Terrace points produced by the overlap of two bands also appear as extrema in the odd-order derivatives. The overall effect on the de-

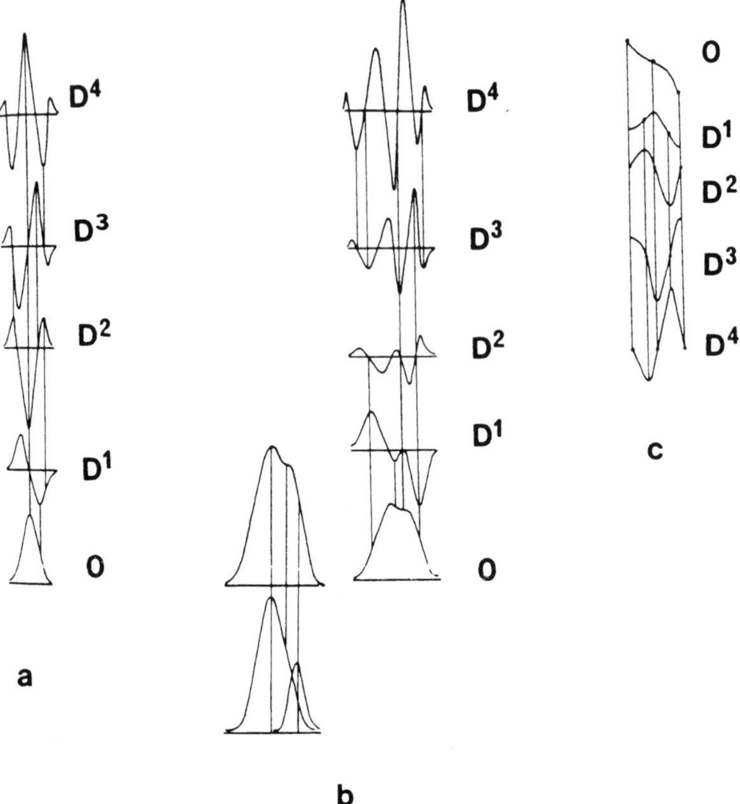

Fig. 2.50. UV-VIS derivative spectroscopy. **a** Zero to fourth derivatization of Gaussian-shaped absorption band; **b** derivative spectra of two superimposed Gaussian bands; **c** derivatives of a "shoulder" (375)

rivative spectrum is to increase resolution with the emergence of extrema in the higher order spectra corresponding to overlapped bands in the zero-order spectrum. Furthermore, derivative spectroscopy can be used as a base-line correction.

Some caution must be exercised in derivative spectroscopy, as artifacts arise due to the interaction of the side-wings (375). Furthermore, in the higher order spectra, the amplitude of a band does not reflect its amplitude in the original zero-order spectrum. Resolution of a particular band also depends upon the band shape, with Lorentzian curves resolving better than Gaussian curves.

The potential of UV-Vis derivative spectroscopy has not been fully exploited. Much of the spectroscopic analysis of aromatic compounds concerns the precise location of shoulders and maxima of hidden bands, a task ideally suited for derivative spectroscopy. The higher order derivative spectra can provide a fingerprint analysis that can distinguish compounds with the same chromophore, such as ketones (266). Isomeric compounds, such as *o*- and *p*-xylene, which cannot be differentiated by their zero-order spectra, can be quantitatively analyzed in higher order spectra (375). The technique of differential second-derivative UV (SEDUV)

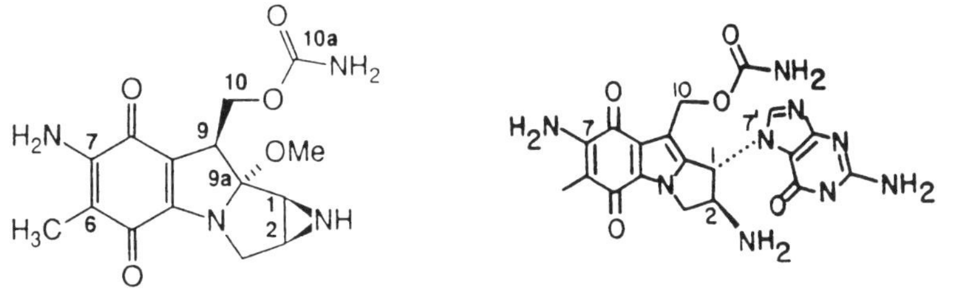

48 49

Fig. 2.51. Mitomycin C (**48**) and guanine adduct (**49**) obtained under acidic conditions (388)

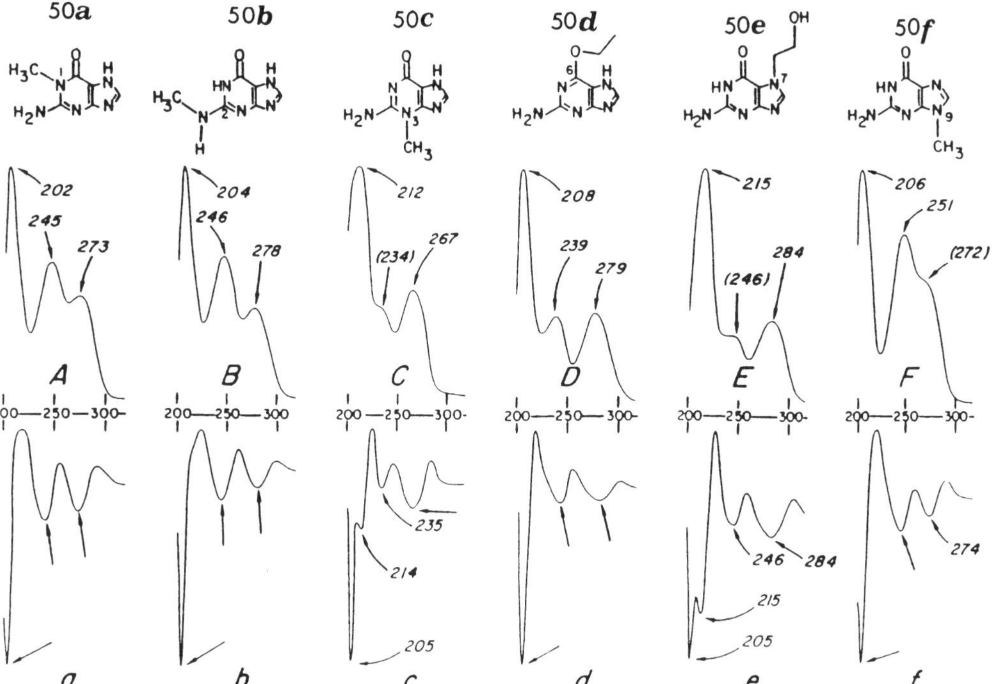

Fig. 2.52. Normal and second-derivative UV spectra of alkylguanines (**50a**–**50f**) (388)

has recently been used in the micro-scale structure determination of adducts obtained upon treatment of the clinically used anticancer agent mitomycin C (**48**) (Fig. 2.51) with the dinucleoside phosphate d(GpC) under acidic conditions (379, 388); the same extremely complex mixture of adducts is also obtained from the reaction of **48** with nucleic acids (378). The structure of one of the adducts is represented by **49** (Fig. 2.51).

^{1}H-NMR clarified the structure of the mitosene moiety but gave very limited information on the guanine portion due to the lack of usable proton signals other than 8′-H. The use of UV in determining the point of attachment of the mitosene to the guanine moiety was therefore explored. The UV spectra of alkylated

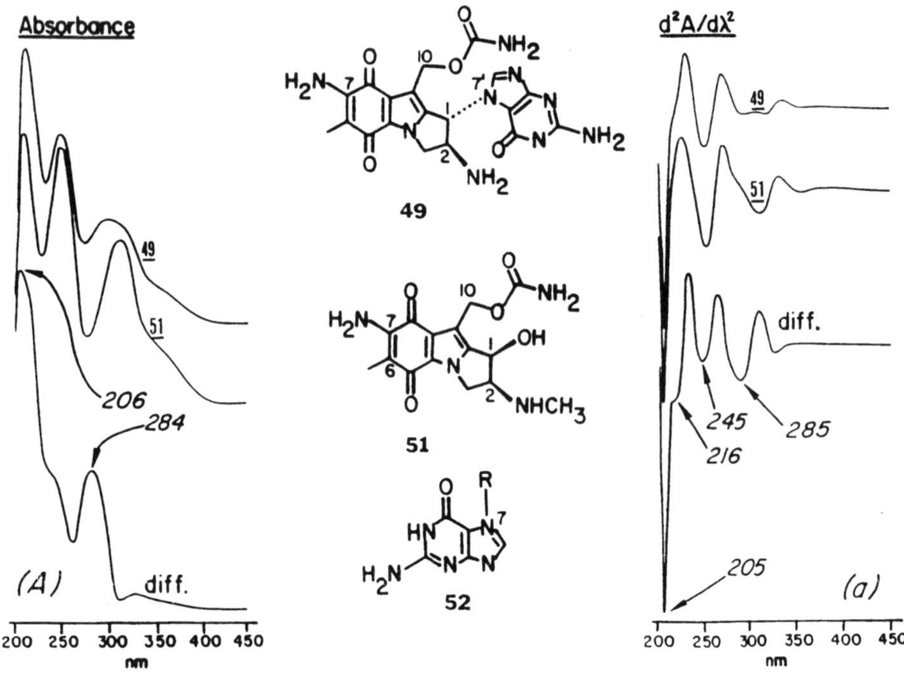

Fig. 2.53. Differential UV spectra, normal (**A**) and second derivative (**a**). UV: 0.1 M, K phosphate, pH 7.0; $d^2\lambda/d\lambda^2$:$\Delta\lambda = 6$ nm (388)

guanines (Fig. 2.52) do differ according to the substitution site but are not sufficiently characteristic to serve as fingerprints, especially when subtraction of strongly absorbing chromophores is involved. In contrast, the SEDUV spectra (Fig. 2.52a through f) show greatly enhanced resolution, as seen by clarification of shoulders (compare Fig. 2.52 C/c, E/e, F/f) and resolution of overlapping bands into two peaks (Fig. 2.52 C/c, E/e). Increase in derivative order leads to further enhanced resolution but also to reduce S/N (signal to noise ratio) and, hence, the second derivative is the best compromise in most cases.

The absorption spectrum of M-guanine A adduct (**49**) (Fig. 2.53) is a summation of mitosene bands at 245 (ε 14000), 350 (3200), and 530 nm (700) and substituted guanine bands at 204 to 215 (ε 15000), 234 to 246 (10000) and 267 to 284 nm (8000). Computer-assisted subtraction of the spectrum of model mitosene (**51**) from adduct **49** using the mitosene 350 nm absorbance as the normalizing wavelength leads to two (**51** subtracted from **49**) different spectra (Fig. 2.53A in the normal mode and Fig. 2.53a in the SEDUV mode). Comparison of the difference spectra with authentic sets (Fig. 2.52 A–F and a–f) shows that they match the spectra of *N7* (hydroxyethyl)guanine, (**52**) (Fig. 2.52 E/e). However, the agreement between the difference curve in Fig. 2.53A with maxima at 206/284 nm and the reference curve in Fig. 2.52E with maxima at 215/284 nm is not convincing. In contrast, agreement between the SEDUVset is excellent: 205/216/245/285 nm (Fig. 2.53a) and 205/215/246/284 nm (Fig. 2.52e).

Other promising applications of UV-Vis derivative spectroscopy appear for analytical and bio-organic chemistry through its ability to resolve hidden bands, in order to analyze complex mixtures. It has thus been possible to distinguish the phenylalanine, tyrosine, and, to a limited extent, tryptophan contributions to the aromatic absorbance region of peptides utilizing the fourth derivative (87, 301). The fourth derivative is especially suited for the examination of peptides as it reveals positive maxima with greatly enhanced resolution (greater resolution than observable in second- or third-order derivative spectra), while at the same time being more selective for narrow bands (57). The amplitude of a band observed in the fourth-derivative spectrum roughly parallels an inverse fourth-power dependence upon the band half-width. With fourth-derivative spectra, it has been possible to investigate the local environment of these aromatic amino acids in various peptides by precisely determining their absorbance maxima.

2.4.3 Infra-Red Spectroscopy

Infra-red spectroscopy, along with UV-Vis spectroscopy, represents another old spectroscopic technique once quite prominent in structure determination studies (281, 283). Since the advent of the more sophisticated NMR and MS techniques, however, IR has been used largely to locate carbonyl stretching frequencies and stretching frequencies of exchangeable hydrogens, and, to a lesser extent, to identify substitution patterns of aromatic systems. Infra-red spectra are routinely recorded for the sake of completeness, but rarely exploited for their full potential in structure determination. More sophisticated applications of IR spectroscopy in biochemical studies, such as the investigation of peptide conformation and binding studies, are nevertheless frequent (308, 364, 365).

The availability of infra-red spectrometers with Fourier-Transform capability has revitalized the field of infra-red spectroscopy and led to some exciting applications. With regard to natural products structure determination, Fourier Transform infra-red spectroscopy (FT-IR) has distinct advantages. The technique is extremely sensitive, and spectra can be recorded on submicrogram quantities, well below the sensitivity limits of NMR. Equally important is the ability to record high-resolution spectra of small amounts of samples in the solid state as well as in solution. Thus, the sample need not be soluble in a readily available deuterated solvent as is the case with NMR. The possibility of spectral storage enables the contributions from the solvent to be subtracted from the spectrum of highly dilute samples.

The characteristics of a high sensitivity and high resolution coupled with spectral storage capabilities make FT-IR an ideal analytical tool. The molecular uniqueness of an IR spectrum can be exploited for the rapid identification of known constitutents of an extract when interfaced with GC or LC (93, 168).

Difference FT-IR spectroscopy has given natural products chemists and bio-organic chemists another new tool in structure analysis that is extremely sensitive. Selective examination of specific bond vibrations through difference spectroscopy provides structural information that is not dependent upon an abundant nucleus possessing nuclear spin. Difference FT-IR spectroscopy offers the opportunity to

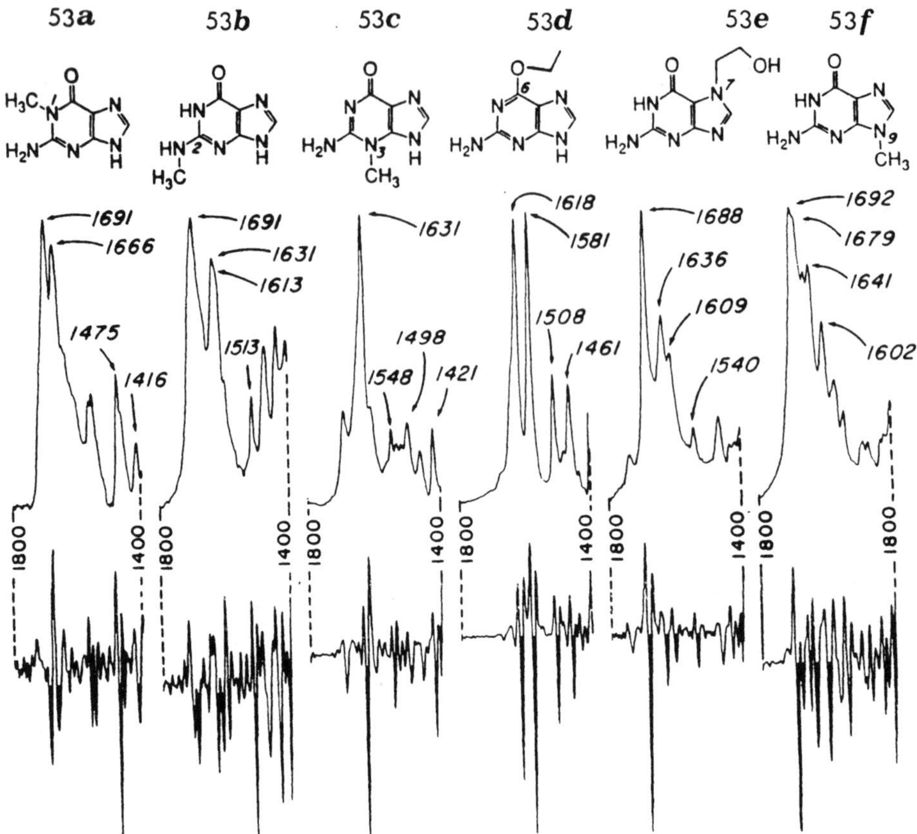

Fig. 2.54. Normal and second-derivative FT-IR spectra of alkylguanines (**53a–53f**) (378, 388)

select any desired spectrum as a control or baseline – such as the spectrum of the solvent. After recording the sample spectrum, subtraction yields the difference spectrum that reveals structural changes relative to the control.

As in the case of UV, the technique of differential second-derivative FT-IR (SEDIR) is exemplified by the determination of the site of addition of mitomycin C to the dinucleoside phosphate d(GpC) (378, 379, 388).

We previously showed that differential FT-IR using a lyophilized KBR technique offered a powerful method for determining substitution positions on purine bases (195). The resolution is greatly enhanced and shoulders became clear bands when solution spectra taken in an attenuated total reflectance CIRCLE microcell are recorded as second derivatives. Fig. 2.54 shows regular absorbance spectra of substituted guanines together with the corresponding second-derivative spectra (378, 388).

The 1800 to 1400 cm^{-1} region FT-IR of adduct **54** (Fig. 2.55A) (378) consists of bands due to the mitosene and guanine moieties. The mitosenes exemplified by **55** (Fig. 2.55A) exhibit a $\approx 1722\ cm^{-1}$ band due to the carbamate group, but this region is completely transparent in guanines. Weighted subtraction of the

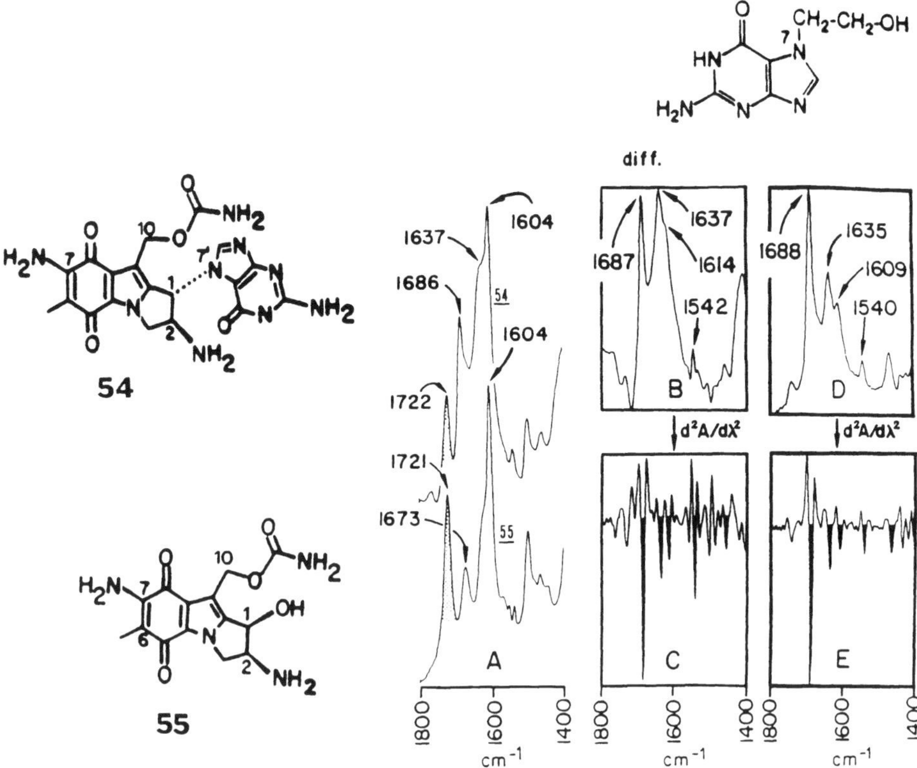

Fig. 2.55. Differential normal and second-derivative FT-IR spectra (378)

spectrum of **55** from that of **54** using the 1722 cm^{-1} band as subtraction marker thus produces the difference spectrum (Fig. 2.55C), which is composed of guanine-related bands. The peaks in curve B compare most closely with the FT-IR of *N*7-(hydroxy-ethyl)guanine (curve D) (see also Fig. 2.54), but the conclusion is not unambiguous. In contrast, a comparison of the differential SEDIR curve in Fig. 2.55C with curves in Fig. 2.54 shows clearly that the blacked-out peaks in Fig. 2.55C match only those of the *N*7-(hydroxy-ethyl)guanine (see also Fig. 2.55E). The method has been used for determining structures of other adducts as well (380).

The above-mentioned examples demonstrate that differential SEDUV and differential SEDIR offer extremely powerful micromethods for structural studies of molecules containing two (or more) nonconjugated moieties.

Subtraction spectra have been investigated as a method for studying the degree of branching and the branching types in dextrans (340). In this example, a linear dextran is used as the control or baseline spectrum, which is subtracted from the spectrum of the branched dextran. The difference spectrum contains a band at approximately 1090 cm^{-1}, whose intensity appears to be a function of the number of branching sites in the dextran. Other bands in the difference spectrum reflect the individual linkage sites. An FT-IR analysis for the determination of

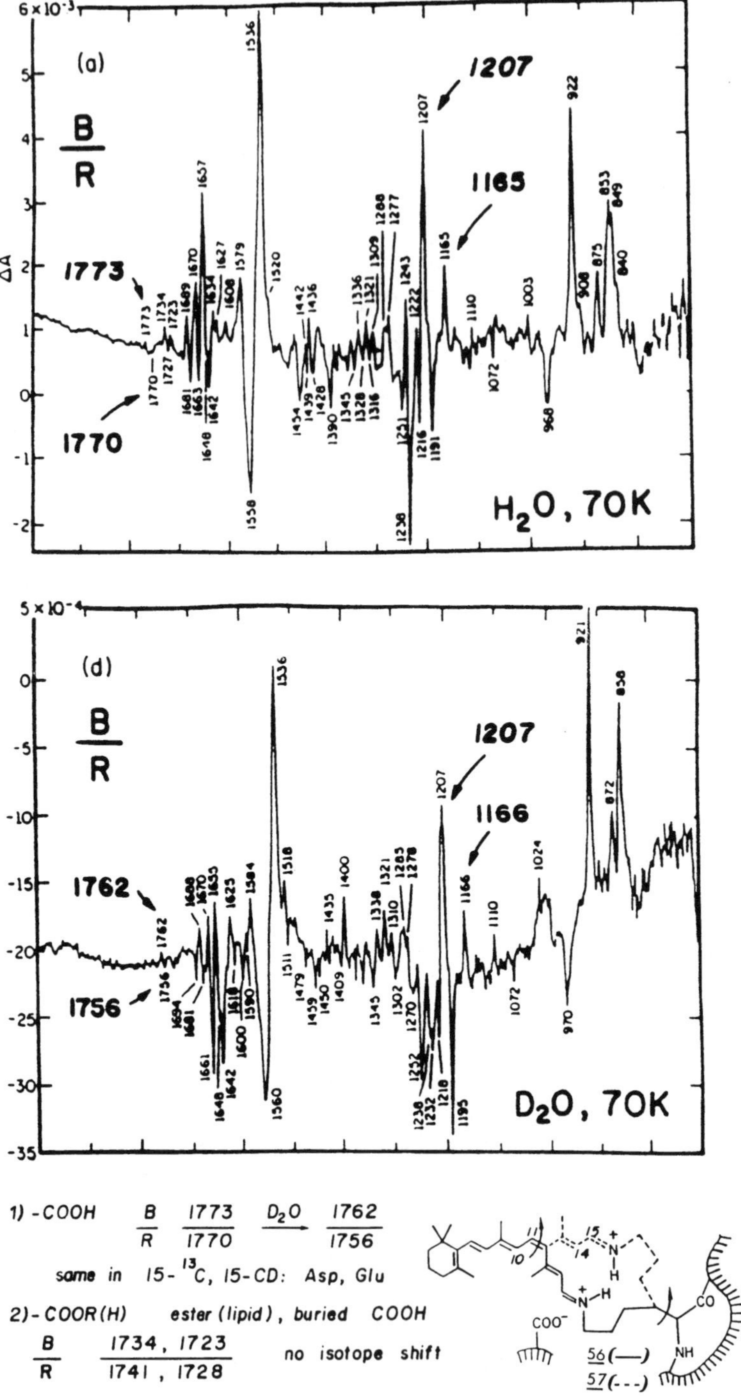

Fig. 2.56. Differential FT-IR of rhodopsin (**56**) and bathorhodopsin (**57**)

degree and type of branching has tremendous potential in solving this difficult problem.

Studies of protein binding sites and membrane behavior have been greatly aided by FT-IR spectroscopy in enabling structural investigation of aqueous systems. The baseline spectrum in this situation is the spectrum of water. The effect of temperature on membrane structure as followed by difference FT-IR in an aqueous system has been reported (252). The membrane spectrum is recorded at the control temperature. Spectra at increasingly higher temperatures are also recorded, and each difference spectrum reveals the temperature-induced changes undergone by the membrane.

A final example of FT-IR applications in bio-organic chemistry is the study of retinal binding in rhodopsin (**56**) (Fig. 2.56) (16).

Rhodopsin, the pigment responsible for vision, is made up of a glycoprotein (MW ≈ 40000) and the chromophore 11-*cis*-retinal. Upon absorption of light, the 11-*cis* chromophore photoisomerizes to all-*trans* (shown in Fig. 2.56 by solid and dotted lines in the retinal structure) to yield the primary photoproduct called bathorhodopsin; the retinal molecule is linked to the terminal amino group of a lysine residue of the protein. The difference spectra shown were measured by making a film of rhodopsin on an IR transparent window, cooling to 70 K, and measuring the FT-IR of rhodopsin (R). The sample was then irradiated to convert the pigment into bathorhodopsin (B), the FT-IR measured, R spectrum computer-subtracted from B spectrum; at 70 K temperature, the conversion is quenched at B intermediate and does not proceed further. The positive peaks are those that are stronger in B whereas negative peaks are those stronger in R. The difference spectra reflect differences arising from both the protein and the chromophore moieties. Although a detailed analysis is not appropriate here, two points can be readily recognized:

(1) The 1773 (B) and 1770 (R) bands shift to 1762 and 1756, respectively, in deuterated medium. When rhodopsin is reconstituted from retinal labeled at C-15 with ^{13}C or with ^{2}H, these bands do not shift positions. They are therefore due to the free carboxylic groups of either aspartic or glutamic acid residues rather than the chromophore. The difference spectra show that the environment of a carboxylic group changes slightly in going from R to B.

(2) B spectra show weak bands at 1734 and 1723 (not indicated in Fig. 2.56) whereas R spectra show weak bands at 1741 and 1728 cm^{-1}. Since these bands undergo no isotope shifts, they can be assigned to buried carboxylic groups or ester groups of the lipid.

2.4.4 Circular Dichroism

Of all the physical methods commonly in use in chemistry and biochemistry, circular dichroism (CD) is probably the most underused. The method has great potential but probably due to the still-empirical nature of many of the rules, there tends to exist a large psychological barrier to its daily use. These empirical rules, the most well-known of which is the octant rule (274), are valid for particular sets of compounds and yield invaluable information (192) if applied with care. In the

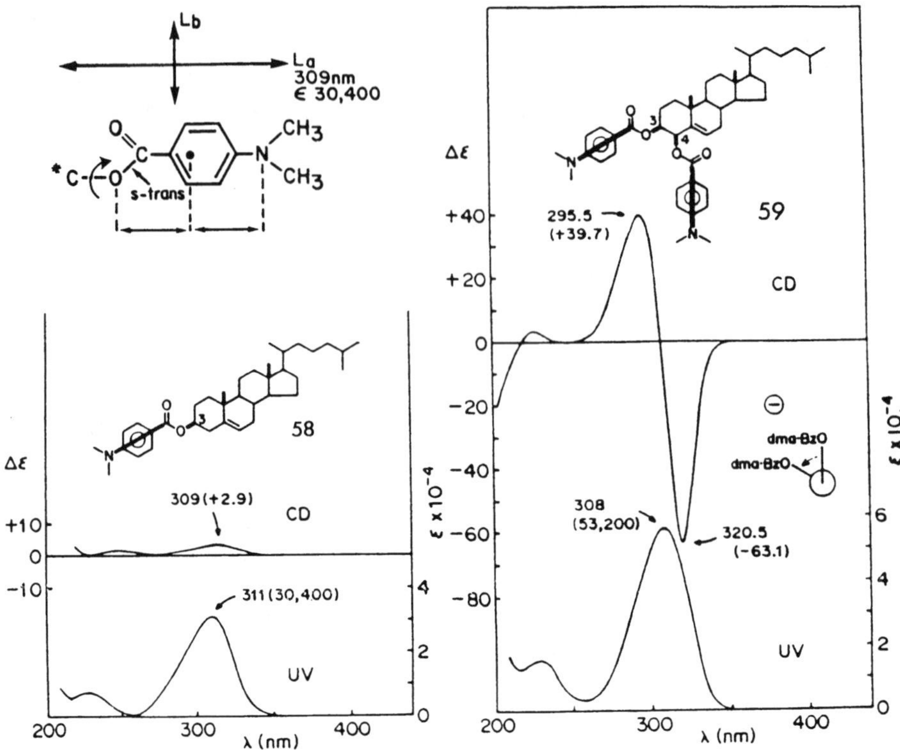

Fig. 2.57. Isolated and interacting chromophores, cholesterol 3-*p*-dimethylamino-benzoate **(58)** and steroidal 3,4-dibenzoate **(59)**

following, we will describe only the exciton chirality method (144), which is based on nonempirical theories of Kuhn (226) and Kirkwood (204); in many cases, it is simpler than the X-ray Bijvoet method because it is sensitive and does not require crystals.

The CD of cholesterol 3-*p*-dimethylaminobenzoate exhibits a Cotton effect (CE) of moderate intensity ($\Delta\varepsilon$+2.9 at 309 nm) – i.e., a wavelength close to its λ_{max} of 311 nm (ε 30400) (Fig. 2.57).

These maxima are due to the L_a transition that runs in the longitudinal direction. When two such chromophores are located nearby as in the steroidal 3,4-dibenzoate, the UV does not change except that its amplitude is doubled (it is the integrated area rather than the ε that is doubled); however, the CD changes drastically and gives rise to a bisignate or so-called "split CD" with intense extrema centered at its λ_{max} of 308 nm – i.e., at 320.5 nm ($\Delta\varepsilon$ −63.1) and 295 nm ($\Delta\varepsilon$ +39.7). We shall define the difference between the two extrema $\Delta\varepsilon$ values as "amplitude," or A, and assign to it a positive or negative sign depending on whether the longer wavelength split CD is positive or negative, respectively. In the case shown (Fig. 2.57), the A value is −102.8, not that the very large A value enables one to make measurements with only several micrograms of sample. The negative sign of A reflects the chirality between the two β C–O bonds at carbons

3 and 4. Strictly speaking it reflects the sense of screwness between the two L_a transition moments, but, because all ester bonds adopt an s-*trans* conformation (Fig. 2.57), the direction of the L_a transition roughly parallels that of the C–O bond. Thus, despite the flexible conformation around the C–O bond, the chirality between the L_a transitions approximate the chirality between the two C–O bonds, which, in the example shown, is counterclockwise or negative.

2.4.4.1 The Nature of Circular Dichroism

The split CD is caused by two interacting or coupled chromophores and is based on the non-empirical coupled oscillator and group polarizability theories of Kuhn (226) and Kirkwood (204), respectively. When two chromophores interact, the excited level splits into two (Fig. 2.58), one red-shifted and the other blue-shifted from the non-interacting level. The two split excited levels do not manifest themselves in the UV since the signs are always the same or postitive. However, in CD, the signs of split levels are opposite; when the transition moments of the two

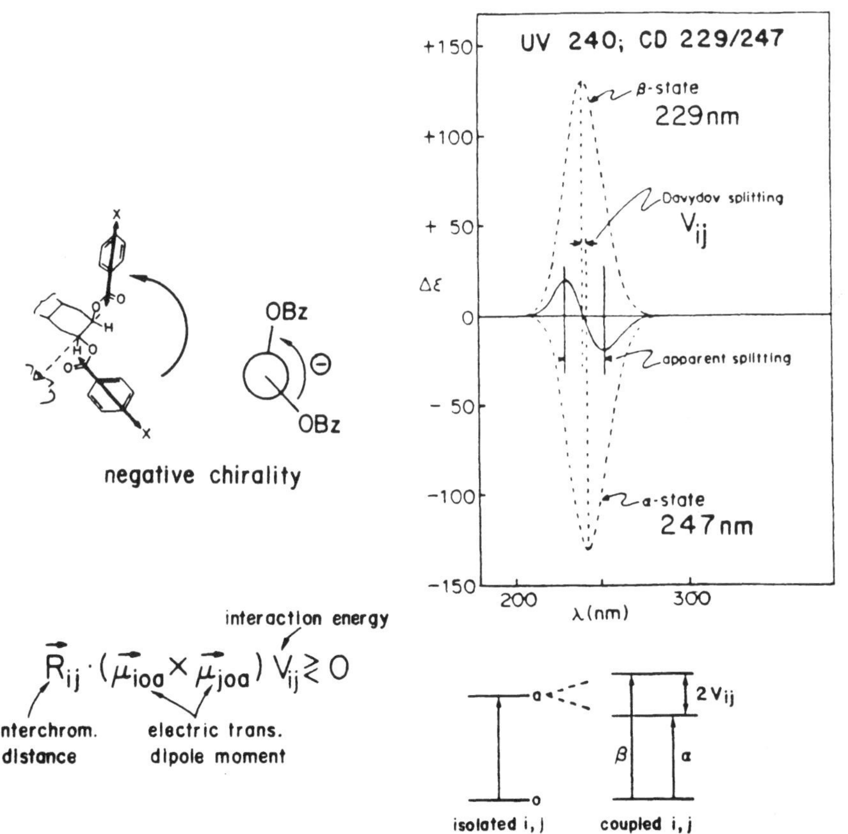

Fig. 2.58. Negatively coupled system

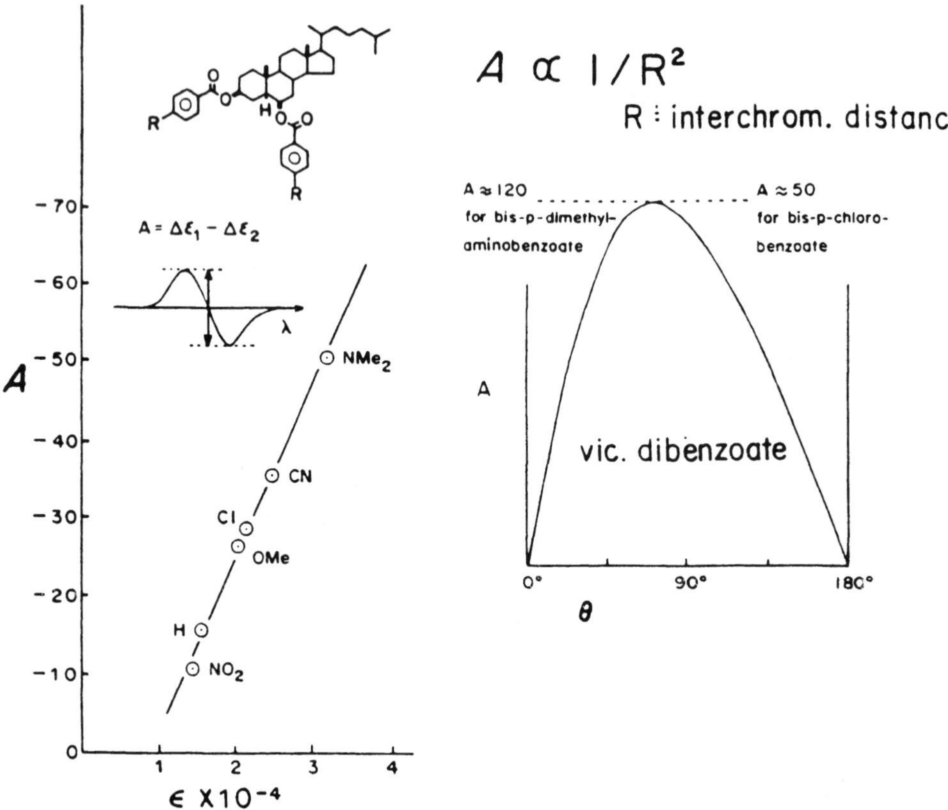

Fig. 2.59. Dependencies of split CD A values on ε, interchromophoric distance, and dihedral angle (143, 144)

chromophores constitute a negative chirality, the red-shifted transition α is negative and blue-shifted transition β is positive, thus giving rise to the split CD shown by the solid line. The situation is reversed in a system with positive chirality. The technique for determining absolute configurations, conformations, structures, etc., from the signs of split CD curves has been named the exciton chirality method (143, 274) and has been applied to numerous systems. Several practical aspects of this technique are summarized in the following (274):

a) Any chromophore with intense ε can be used, provided the direction of transition is known.
b) The split CD can be calculated from the interchromophoric distance and UV-Vis oscillator strength; the agreement between experiment and theory is excellent.
c) There is a linear relation between the UV ε and CD $\Delta\varepsilon$ values (Fig. 2.59); the stronger the ε is, the larger the absolute A value. In the example shown, the A is only -10 in *p*-nitrobenzoate but becomes -50 in *p*-dimethylamino-benzoate. Thus, chromophores with large ε values have greater sensitivity.
d) The A value is inversely proportional to the square of interchromophoric distance (Fig. 2.59).

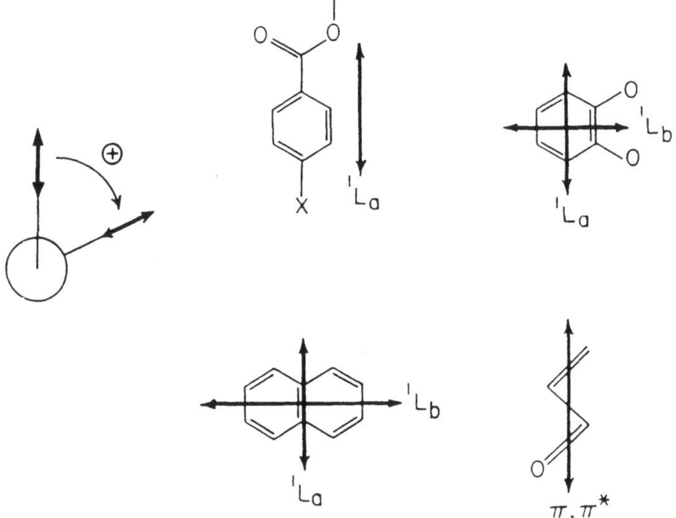

Fig. 2.60. Common chromophores used in the exciton chirality method; any chromophore with large ε and known direction of μ can be used

e) An A value of 26 is still observed between the 3β (eq) and 15β (ax) *p*-dimethyl-amino-benzoates in a saturated D-homo-steroid where the distance is ≤ 13 Å (60).
f) The A is maximal around dihedral angles of 70° and zero at 0° or 180° as calculated for *vic*-dibenzoates (Fig. 2.59).
g) The sign of the split CD is still retained when the separation in maxima of interacting chromophores is 80 nm. In a steroid, a 3β-benzoate with no *p*-substituent (λ_{max} 230 nm) is coupled sufficiently strongly to a 4β-*p*-dimethyl-amino-benzoate (λ_{max} 310 nm) to give a 312 nm CD with $\Delta\varepsilon$ −3.7, a sign in agreement with the screwness of the $3\beta/4\beta$ benzoates.
h) Any chromophore can be used, provided it absorbs strongly and the direction of the electric transition moment is well-defined (Fig. 2.60). Since the signs of split CD curves resulting from interacting chromophores represent the absolute sense of twist between coupled transition moments, their directions necessarily have to be well-defined in order to determine absolute configurations. The direction of transition moments of some representative chromophores are depicted in Fig. 2.60. The split CD resulting from spatial interaction of these chromophores has been used to determine the absolute configurations of numerous types of natural and unnatural products (144).
i) One can use interactions between existing chromophores in the substrate, introduce a chromophore that interacts with an existing chromophore, or introduce multiple chromophores that do not interact with the original chromophore. In the mitomycin derivative (Fig. 2.61), the mitosene chromophore absorbs at 245, 309, 350, and 530 nm. The β-naphthoate and cinnamate chromophores introduced at C-1 and C-2 still couple with the mitosene group; however, in the case of *p*-dimethylaminocinnamate (381 nm (ε 30400)), its long wave-

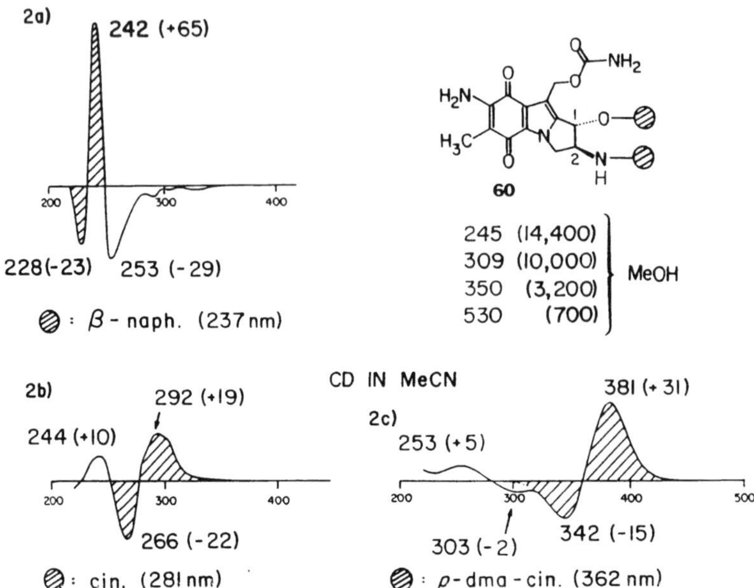

Fig. 2.61. A new chromophore, *p*-dimethylaminocinnamate and absolute configuration of mitomycin **(60)** (387). CD data are shown in nm ($\Delta\varepsilon$) and UV data in nm (ε)

length, λ_{max}, is separated from the first two intense maxima and gives rise to clearly split CD curves. The positive chirality is in accord with the recently reversed absolute configuration of mitomycin (150). This new chromophore should be promising because of its intense long wavelength maximum (387).

2.4.4.2 The Additivity Relation in A Values

When the *p*-bromobenzoates of over 40 hexopyranoses were prepared it was found that amplitudes of tri- and tetrabenzoates could be expressed by the sum of component three and six dibenzoate units, respectively (237). Thus the *A* values of various di-*p*-benzoate units (Fig. 2.62) can be regarded as constants, independent of other non-interacting groups or substituents such as CH_2, OH, OMe, OAc, etc. (depicted as "unsubstituted" carbon). In the case of D-galactose-2,3,4,6-tetrabenzoate, the observed *A* is +100, whereas the *A* value as estimated from the six dibenzoate units is +112 (Fig. 2.63). This additivity relation enables one to calculate the *A* value of interacting benzoates from standard dibenzoate values (Fig. 2.62) (124); it is to be noted that the values merely reflect the spatial disposition of coupled benzoate groups and are independent of other non-interacting groups. It is therefore applicable to unknown hexopyranose benzoates. Furthermore, the large *A* values make it possible to carry out measurements with μg quantities and also require no reference sample. The additivity relation is valid in congested caged systems, such as the tricothecene skeleton (299), and also between enones, benzoates, etc. (367).

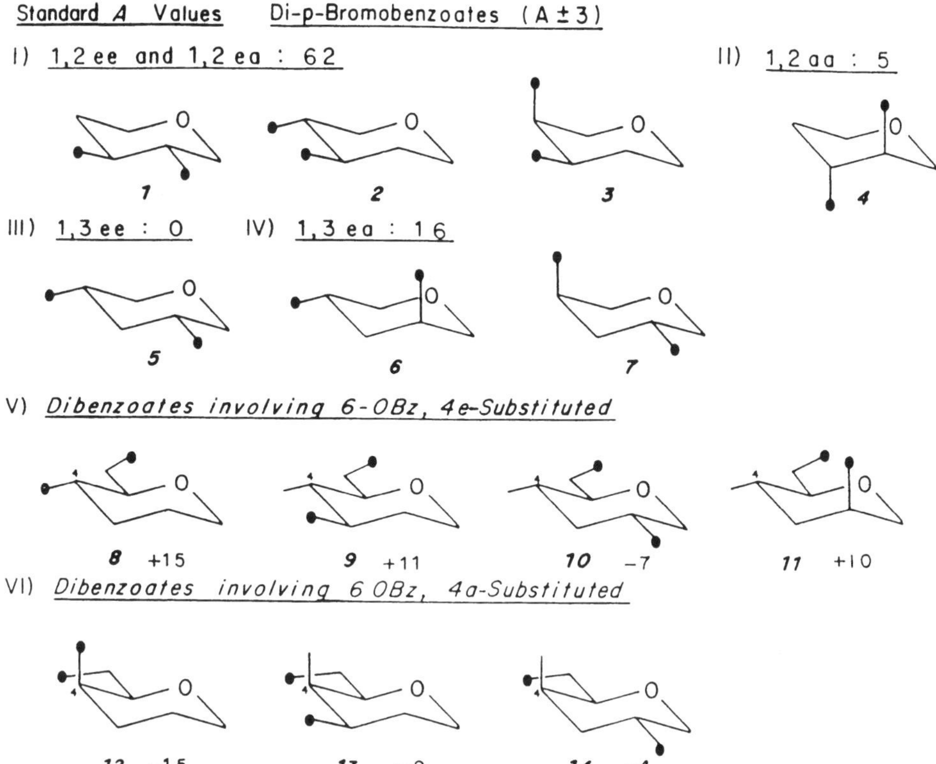

Fig. 2.62. *A* values of nexopyranose dibenzoates. The benzoates are denoted by black circles. The "unsubstituted" carbons can be any group that does not couple with the benzoate

Two important findings have been made recently. One is that the additivity relation has been found to be valid for the entire CD curve in multichromophore systems (400), as exemplified (Fig. 2.64) for α-methyl galactoside 2,3-di-*p*-bromobenzoate and 4,5-dimethoxycinnamate. The *p*-bromobenzoate (244.5 nm, ε 19500 in EtOH) and *p*-methoxycinnamate (311 nm, ε 24000 in MeCN) were selected as the interacting chromophores; it was sought to check whether sugars substituted with such groups would give rise to CD curves characteristic of the substitution pattern. If this were the case, the free hydroxyls in an oligosaccharide could be tagged by one chromophore, while those engaged in glycosidic bonds were tagged by the other chromophore; CD measurement of each sugar derivative would then disclose the substitution pattern as well as the sugar identity – i.e., the CD curves would be characteristic for each sugar derivative. Studies showed that indeed this is the case; 60 examples have shown that the additivity rule in CD curves holds without exception.

Thus for the case shown (Fig. 2.64), the calculated curve was made by adding the six experimental component CD curves (B and C represent, respectively, bromobenzoate and methoxycinnamate): 2B/3B, 4C/6C, 2B/4C, 3B/4C, and 3B/6C. Excellent agreement is seen throughout the entire CD range. This additiv-

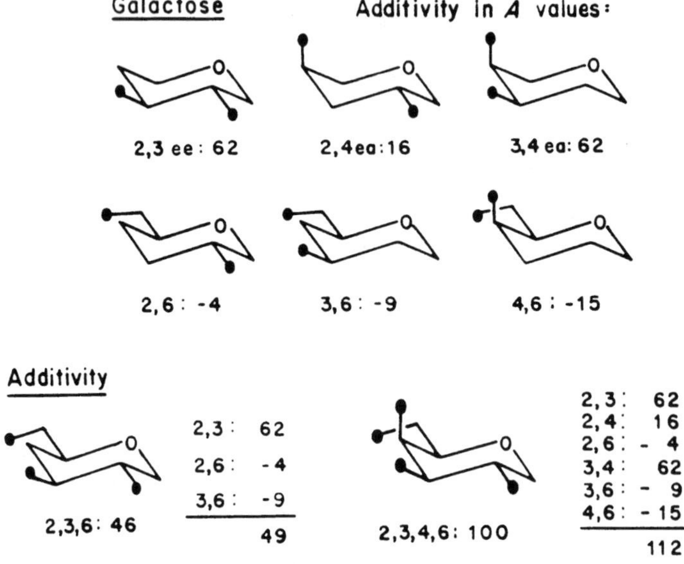

Fig. 2.63. A values of the six di-p-bromobenzoate units in D-galactose, and examples of observed and calculated A values (237)

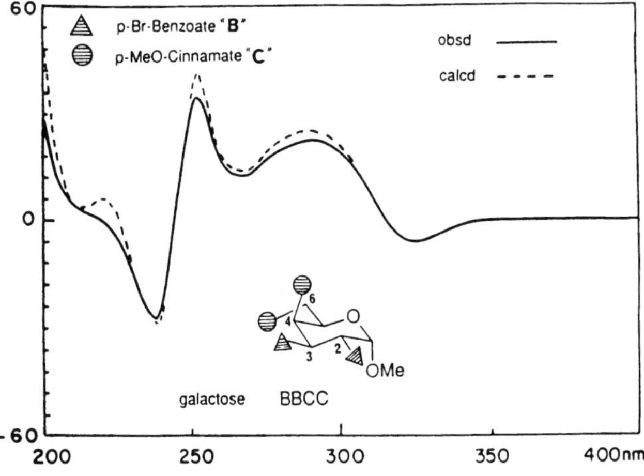

Fig. 2.64. Bichromophoric additivity in the CD curve of a hexopyranose tetraacylate

ity in CD curves is currently being exploited in a microscale structure determination method for oligosaccharides; besides, it should lead to important and practical consequences in general circular dichroism.

The other important finding was that the additivity in A values holds also for the benzylate group. This has been explored in detail with the p-phenylbenzyl group in the sugar benzylates of glucose, mannose, and galactose; again no exception has been found (374). This chromophore has the further advantage that it can be directly oxidized with ruthenium tetroxide to the p-phenylbenzoates, which give A values of five-fold intensity – i.e., the sensitivity is increased five-fold. The free hydroxyls in oligosaccharides can thus be benzylated and the perbenzylates submitted directly to methanolysis (unlike the benzoates, which undergo acyl migration under methanolysis conditions). The phenylbenzylate additivity and oxidation to phenylbenzoates, which also show additivity, form the basis of further methods currently under investigation for oligosaccharide structure determinations.

2.5 General Conclusions

We have attempted to cover the current status of isolation techniques and various spectroscopic methods used in modern natural products chemistry. Not surprisingly, the field is advancing very rapidly, and this article is bound to become rather obsolete in the not too distant future. However, it may serve as a convenient reference.

Quite a few of the techniques described are not routine, and even if the results are published in papers or depicted in figures, they are frequently the outcome of many frustrating attempts.

As techniques become more sophisticated, it becomes increasingly difficult to maintain each instrument in top condition. Despite this unavoidable difficulty, and despite the nonroutine nature of some measurements, it is important to know what can be achieved by present-day techniques. We hope that this article may serve to present an overview of modern techniques.

References

1 Ahmad V U, Najmus-Saquib Q, Usmaighani K, Fochs W, Voelter W 1980 New terpenoid from *Primula denticulata*. Z Naturforsch 35b:511 – 512

2 Aldersley M F, Dean F M, Mann B E 1983 The use of ^{13}C-{^{1}H selective} nuclear Overhauser enhancement experiments in the determination of the structure of a highly crowded quinone-diazoalkyl epoxide adduct. J Chem Soc Chem Commun 107 – 108

3 Al-Kaysi H N, Small L D, Murray W J, Haddadin M 1984 Quantitative analysis of quinidine analogs using ion-pairing HPLC. J Chromatogr 22:80 – 83

4 Amster I J, McLafferty F W, Castro M E, Russell D H, Cody R B Jr, Ghaderi S 1986 Detection of mass 16 241 ions by Fourier-transform mass spectrometry. Anal Chem 58:483 – 485

5 Ando I, Webb G A 1981 Some quantum chemical aspects of solvent effects on NMR parameters. Org Magn Res 15:111 – 130

6 Ando I, Webb G A 1983 Theory of NMR parameters. Academic Press New York, 217 pp

7 Arm H, Grob E C, Signer R 1966 Die Auftrennung der Chlorophylle durch Gegenstromextraktion. Helv Chim Acta 49:851 – 854

8 Armstrong D W, Fendler J H 1977 Differential partitioning of tRNAs between micellar and aqueous phases: a convenient gel filtration method for separation of tRNAs. Biochim Biophys Acta 478:75 – 80

9 Armstrong D W, Hinze W L, Bui K H, Singh H N 1981 Enhanced fluorescence and room temperature liquid phosphorescence detection in pseudophase liquid chromatography (PLC). Anal Lett 14:1659 – 1667

10 Armstrong D W, McNeely M 1979 Use of micelles in the TLC separation of polynuclear aromatic compounds and amino acids. Anal Lett 12:1285 – 1291

11 Armstrong D W, Nome F 1981 Partitioning behavior of solutes eluted with micellar mobile phases in liquid chromatography. Anal Chem 53:1662 – 1666

12 Armstrong D W, Terrill R Q 1979 Thin layer chromatographic separation of pesticides, decachlorobiphenyl, and nucleosides with micellar solutions. Anal Chem 51:2160–2163

13 Arpino P J, Guichon G 1979 LC/MS coupling. Anal Chem 51:682A–684A, 688A, 690A, 692A, 697A, 698A

14 Aue W P, Bartholdi E, Ernst R R 1976 Two-dimensional spectroscopy. Application to nuclear magnetic resonance. J Chem Phys 64:2229–2246

15 Aue W P, Karhan J, Ernst R R 1976 Homonuclear broad-band decoupling and two-dimensional J-resolved NMR spectroscopy. J Chem Phys 64:4226–4227

16 Bagley K A, Balogh-Nair V, Croteau A A, Dollinger G, Ebrey T G, Eisenstein L, Hong M K, Nakanishi K, Vittitow J 1985 Fourier-transform infrared difference spectroscopy of phodopsin and its photoproducts at low temperature. Biochem 24:6055–6071

17 Baldwin M A, McLafferty F W 1973 Direct chemical ionization of relatively involatile samples. Application to underivatized oligopeptides. Org Mass Spec 7:1353–1356

18 Balza F, Cyr N, Hamer G K, Perkin A S, Koch H J, Stuart R S 1977 Applications of catalytic, hydrogen-deuterium exchange in ^{13}C-NMR spectroscopy. Carbohyd Res 59:C7–C11

19 Barber M, Bordoli R S, Elliott G J, Sedgwick R D, Taylor A N 1982 Fast atom bombardment mass spectrometry. Anal Chem 54:645A–646A, 649A–650A, 653A, 655A, 657A

20 Barber M, Bordoli R S, Sedgwick R D, Tyler A N 1981 Fast atom bombardment of solids (FAB): A new ion source for mass spectrometry. J Chem Soc Chem Commun 325–327

21 Barber M, Vickerman J C, Wolstenholme J 1980 Secondary ion mass spectra of some simple organic molecules. J Chem Soc Faraday Trans I 76:549–559

22 Bartman D, Schneider G M 1973 Experimental results and physiochemical aspects of supercritical fluid chromatography with carbon dioxide as the mobile phase. J Chromatogr 83:135–145

23 Baum M W, Guenzi A, Johnson C A, Mislow K 1982 The dependence of ^{13}C-^{1}H coupling constants on C–C–C bond angles. Tetrahedron Lett 23:31–34

24 Bax A D 1982 Two-dimensional nuclear magnetic resonance in liquids. Delft University Press D Reidel Boston, 208 pp

25 Bax A, Freeman R 1981 Investigation of complex networks of spin-spin coupling by two-dimensional NMR. J Magn Res 44:542–561

26 Bax A, Freeman R, Frenkiel T A 1981 An NMR technique for tracing out the carbon skeleton of an organic molecule. J Am Chem Soc 103:2102–2104

27 Bax A, Freeman R, Kempsell S P 1980 Natural abundance ^{13}C-^{13}C coupling observed via double-quantum coherence. J Am Chem Soc 102:4849–4851

28 Bax A, Freeman R, Morris G 1981 Correlation of proton chemical shifts by two-dimensional Fourier transform NMR. J Magn Res 42:164–168

29 Becker E D 1980 High resolution NMR theory and chemical applications. 2nd edn. Academic Press New York, 354 pp

30 Becker E D, Ferretti J A, Gambhir P N 1979 Selection of optimum parameters for pulse Fourier Transform nuclear magnetic resonance. Anal Chem 51:1413–1420

31 Becker H, Reichling J, Hsieh W C, Wei C 1982 Water-free solvent system for droplet countercurrent chromatography and its suitability for the separation of nonpolar substances. J Chromatogr 237:307–310

32 Beckey H D 1971 Field ionization mass spectrometry. Pergamon Press New York, 359 pp

33 Beckey H D 1977 Principles of field ionization and field desorption mass spectrometry. Pergamon Press New York, 348 pp

34 Beckey H D 1979 Letters to the editor. Final note by Professor H D Beckey. Org Mass Spec 14:292

35 Beckey H D, Rollgen F W 1979 Field desorption 'without fields'? Org Mass Spec 14:188–190

36 Beckey H D, Schulten H R 1975 Field desorption mass spectrometry. Angew Chem Int Ed 14:403–415

37 Benn R, Gunther H 1983 Modern pulse methods in high-resolution NMR spectroscopy. Angew Chem Int Ed 22:350–380

38 Beuhler R J, Flanagan E, Greene L J, Friedman L 1972 Volatility enhancement of thyrotropin releasing hormone for mass spectrometric studies. Biochem Biophys Res Commun 46:1082–1088

39 Beuhler R J, Flanagan E, Greene L J, Friedman L 1974 Proton transfer mass spectrometry of peptides. A rapid heating technique for underivatized peptides containing arginine. J Am Chem Soc 96:3990–3999

40 Bhushan Mandava N (ed) 1984 Countercurrent Chromatography. J Liq Chromatogr Special Issue 7:227–431

41 Bidlingmeyer B A 1980 Separation of ionic compounds by reversed-phase liquid chromatography: an update of ion-pairing techniques. J Chromatogr 18:525–539
42 Bidlingmeyer B A 1983 Reversed-phase ion-pair liquid chromatography. LC Mag 1:344–349
43 Bock K, Meyer B, Thiem J 1978 Structural elucidation of diastereoisomeric 1,3,6-trioxacyclooctane systems by simultaneous relaxation and double resonance experiments. Angew Chem Int Ed 17:447–449
44 Bodenhausen G, Ernst R R 1981 The accordian experiment, a simple approach to three-dimensional NMR spectroscopy. J Magn Res 45:367–373
45 Bodenhausen G, Ernst R R 1982 Direct determination of rate constants of slow dynamic processes by two-dimensional 'accordian' spectroscopy in nuclear magnetic resonance. J Am Chem Soc 104:1304–1309
46 Bodenhausen G, Freeman R, Niedermeyer R, Turner D L 1977 Double Fourier transformation in high-resolution NMR. J Magn Res 26:133–164
47 Bolte P, Klessinger M, Wilhelm K 1984 ^{13}C-NMR INADEQUATE spectrum with broad band decoupling by supercycles. Angew Chem Int Ed 23:152–153
48 Bolton P H 1983 In: Laszlo P (ed) NMR of newly-accessible nuclei. Academic Press New York, 21–52
49 Breitmaier E, Voelter W 1978 ^{13}C-NMR Spectroscopy. Verlag Chemie Deerfield Beach FL, 401 pp
50 Brinkman U A Th, Devries G, Van Dalen E 1966 Chromatographic techniques using liquid anion exchangers. I. HCl systems. J Chromatogr 22:407–424
51 Bruening R C, Derguini F, Nakanishi K 1986 Preparative scale isolation of 11-cis-retinal from isomeric retinal mixture by centrifugal partition chromatography. J Chromatogr 357:340–343
52 Bruening R C, Derguini F, Nakanishi K Centrifugal partition chromatography in natural products chemistry. In: Mandava B, Ito Y (eds) Modern countercurrent chromatography. Marcel Dekker New York, in press
53 Bruening R C, Oltz E M, Furukawa J, Nakanishi K, Kustin K 1985 Isolation and structure of tunichrome B-1, a reducing blood pigment from the tunicate *Ascidia nigra L.* J Am Chem Soc 107:5298–5300
54 Bruening R C, Oltz E M, Furukawa J, Nakanishi K, Kustin K 1986 Isolation of tunichrome B-1, a reducing blood pigment of the sea squirt, *Ascidia nigra*. J Nat Prod 49:193–204
55 Buckingham A D, Urland W 1975 Isotope effects on molecular properties. Chem Rev 75:113–117
56 Burum D P, Ernst R R 1980 Net polarization transfer via a J-ordered state for signal enhancement of low-sensitivity nuclei. J Magn Res 39:163–168
57 Butler W 1979 Fourth derivative spectra. Methods Enzym 56:501–515
58 Chait B T, Field F H 1985 A study of the metastable fragmentation of ions produced by Californium-252 fission fragment bombardment of bovine insulin. Int J Mass Spectrom Ion Proc 65:169–180
59 Che C T, Cordell G A, Fung H H S 1983 Aristolindquinone – A new naphthoquinone from *Aristolochia indica* L. (Aristolochiaceae). Tetrahedron Lett 24:1333–1336
60 Chen S M L, Harada N, Nakanishi K 1974 Long range effect in the exciton chirality method. J Am Chem Soc 96:7352–7354
61 Cheung H T A, Watson T R 1980 Stereochemistry of the hexosulose in cardenolide glycosides of the Asclepiadaceae. J Chem Soc Perkin Trans I 2162–2168
62 Christofides J C, Davies D B 1983 A method to assign ^{13}C-NMR signals of compounds with exchangeable hydrogen atoms using two and three bond isotope effects: cellobiose. J Chem Soc Chem Commun 324–326
63 Cohen K A, Grillo S A, Dolan J W 1985 A comparison of protein retention and selectivity on large-pore reversed-phase HPLC columns. LC Mag 3:37–40
64 Conca E, De Bernardi M, Fronza G, Girometta M A, Mellerio G, Vidari G, Vita-Finzi P 1981 Fungal metabolites 10. New chromenes from *Lactarius fuliginosus* Fries and *Lactarius picinus* Fries. Tetrahedron Lett 43:4327–4330
65 Conway W D, Ito Y 1984 Solvent selection for countercurrent chromatography by rapid estimation of partition coefficients and application to polar conjugates of *p*-nitrophenol. J Liq Chromatogr 7:275–289
66 Conway W D, Ito Y 1984 Recent applications of countercurrent chromatography. LC Mag 2:368–376
67 Cotter R J 1980 Mass spectrometry of nonvolatile compounds: desorption from extended probes. Anal Chem 52:1589A, 1591A, 1594A, 1598A, 1600A, 1602A, 1604A, 1606A

68 Cotter R J 1980 Laser desorption chemical ionization mass spectrometry. Anal Chem 52:1767–1770
69 Craig L C 1944 Identification of small amounts of organic compounds by distribution studies. II. Separation by countercurrent distribution. J Biol Chem 155:519–534
70 Crommen J 1980 Ion-pair chromatography in the low concentration range by use of highly absorbing counter ions. III. High performance liquid chromatography of quaternary alkyammonium ions as ion pairs with naphthalene-2-sulfonate, using silica support of low surface area. J Chromatogr 193:225–234
71 Crommen J, Fransson B, Schill G 1977 Ion-pair chromatography in the low concentration range by use of highly absorbing counter ions. J Chromatogr 142:283–297
72 Dalling D K, Grant D M 1967 Carbon-13 magnetic resonance. IX. The methyl cyclohexanes. J Am Chem Soc 89:6612–6622
73 Daugherty R C 1981 Negative chemical ionization mass spectrometry. Anal Chem 53:625A, 629A, 630A, 632A, 634A, 636A
74 Davankov V A 1980 Resolution of racemates by ligand exchange chromatography. Adv Chromatogr 18:139–195
75 Dell A, Williams D H, Morris H R, Smith G A, Feeney J, Roberts G C K 1975 Structure revision of the antibiotic echinomycin. J Am Chem Soc 97:2497–2502
76 Demagos G P, Baltas W, Hofle G 1981 New anthraquinones and anthraquinone glycosides from *Morinda lucida*. Z Naturforsch 36B:1180–1184
77 Denkert M, Hackzell L, Schill G, Sjogren E 1981 Reversed-phase ion-pair chromatography with UV-absorbing ions in the mobile phase. J Chromatogr 218:31–43
78 Dill K, Hardy R E, Lacombe J M, Pavia A A 1982 ^{13}C-NMR spectral study of 3-0-α and β-D xylopyranosyl-L-serine and -threonine. Carbohyd Res 101:330–334
79 Doddrell D M, Pegg D T 1980 Assignment of proton-decoupled carbon-13 spectra of complex molecules by using polarization transfer spectroscopy. A superior method to off-resonance decoupling. J Am Chem Soc 102:6388–6390
80 Doddrell D M, Pegg D T, Bendall M R 1982 Distortionless enhancement of NMR signals by polarization transfer. J Magn Res 48:323–327
81 Domon B, Hostettmann M, Hostettmann K 1982 Droplet countercurrent chromatography with nonaqueous solvent systems. J Chromatogr 246:133–135
82 Domon B, Hostettmann K, Kovacevic K, Prelog V 1982 Separation of the enantiomers of (±) norephedrine by rotation locular counter-current chromatography. J Chromatogr 250:149–151
83 Dong M W, Gant J R 1984 Short three-micron columns. Applications in high-speed liquid chromatography. LC Mag 2:294–298, 300–303
84 Dorman D E, Angyal S J, Roberts J D 1970 Nuclear magnetic resonance spectroscopy. Carbon-13 spectra of some inositols and their 0-methylated derivatives. J Am Chem Soc 92:1351–1354
85 Dorman D E, Roberts J D 1970 Nuclear magnetic resonance spectroscopy. Carbon-13 spectra of some pentose and hexose aldopyranoses. J Am Chem Soc 92:1355–1361
86 Dorsey J G, De Echegaray M T, Landy J S 1983 Efficiency enhancement in micellar liquid chromatography. Anal Chem 55:924–928
87 Dunach M, Sabes M, Padros E 1983 Fourth-derivative spectrophotometry analysis of tryptophan environment in proteins: application to melittin, cytochrome C and bacteriorhodopsin. Eur J Biochem 134:123–128
88 Edgar J A, Frahn J L, Cockrum P A, Anderson N, Jago M V, Culvenor C C J, Jones J L, Murray K, Shaw K J 1982 Corynetoxins, causative agents of annual ryegrass toxicity: their identification as tunicamycin group antibiotics. J Chem Soc Chem Commun 222–224
89 Eksborg S, Lagerstrom P O, Modin R, Schill G 1973 Ion-pair chromatography of organic compounds. J Chromatogr 83:99–110
90 Eksborg S, Schill G 1973 Ion-pair partition chromatography of organic ammonium compounds. Anal Chem 45:2092–2100
91 Endo K, Takahashi K, Abe T, Hikino H 1981 Structure of forsythoside A, an antibacterial principle of *Forsythia suspensa* leaves. Heterocycles 16:1311–1314
92 Engelhardt H, Mueller H 1984 Optimal conditions for the reversed-phase chromatography of proteins. Chromatographia 19:77–84
93 Erickson M 1979 Gas chromatography/Fourier transform infrared spectroscopy applications. Appl Spectrosc Rev 15:261–325
94 Ewing D F 1975 Two-bond coupling between protons and carbon-13. Ann Rep NMR Spec 6A:389–437

95 Fales H M, Mine G W A, Winkler H U, Beckey H D, Damico J N, Borton R 1975 Comparison of mass spectra of some biologically important compounds as obtained by various ionization techniques. Anal Chem 47:207–219
96 Fenselau C 1982 Mass spectrometry of middle molecules. Anal Chem 54:105A, 196A, 108A, 110A, 112A, 114A
97 Fenselau C 1984 Fast atom bombardment and middle molecule mass spectrometry. J Nat Prod 47:215–225
98 Field F H, Munson M S B 1965 Reaction of gaseous ions. XIV. Mass spectrometric studies of methane at pressures to 2 torr. J Am Chem Soc 87:3289–3294
99 Fjeldsted J C, Kong R C, Lee M L 1983 Capillary supercritical-fluid chromatography with conventional flame detectors. J Chromatogr 279:449–455
100 Fjeldsted J C, Richter B E, Jackson W P, Lee M L 1983 Scanning fluorescence detection in capillary supercritical fluid chromatography. J Chromatogr 279:423–430
101 Flanagan S D, Johansson G, Yost B, Ito Y, Sutherland I A 1984 Toroidal coil countercurrent chromatography in the affinity partitioning of nicotinic cholinergic receptor enriched membranes. J Liq Chromatogr 7:385–402
102 Floyd T R, Crowther J B, Hartwick R A 1985 HPLC of nucleic acids: trends in the separation of synthetic single-stranded oligonucleotides and double-stranded DNA fragments. LC Mag 3:508–512, 516–520
103 Fraser R R, Pettit M A, Mislow M 1972 Separation of nuclear magnetic resonance signals of internally enantiotropic protons using a chiral shift reagent. The deuterium isotope effect on geminal proton-proton coupling constants. J Am Chem Soc 94:3253–3524
104 Freeman G A, Freeman R 1979 Enhancement of nuclear magnetic resonance signals by polarization transfer. J Am Chem Soc 101:760–762
105 Freeman R 1980 Nuclear magnetic resonance spectroscopy in two frequency dimensions. Proc R Soc London Ser A 373:149–178
106 Freeman R, Hill H D W 1971 Fourier transform study of NMR spin-lattice relaxation by progressive saturation. J Chem Phys 54:3367–3377
107 French S B, Novotny M 1986 Xenon, a uique mobile phase for supercritical fluid chromatography. Anal Chem 58:164–166
108 Fujimoto D, Hirama M, Iwahita T 1982 Histidinoanaline, a new crosslinking amino acid, in calcified tissue collagen. Biochem Biophys Res Commun 104:1102–1106
109 Games D E 1983 High-performance liquid chromatography/mass spectrometry (HPLC/MS). Adv Chromatogr 21:1–39
110 Gere D R 1983 Hewlett-Packard Appl Note 800-1. Hewlett-Packard Palo Alto CA
111 Gere D R 1983 Hewlett-Packard Appl Note 800-2. Hewlett-Packard Palo Alto CA
112 Gere D R 1983 Hewlett-Packard Appl Note 800-3. Hewlett-Packard Palo Alto CA
113 Gere D R 1983 Hewlett-Packard Appl Note 800-5. Hewlett-Packard Palo Alto CA
114 Gere D R 1983 Hewlett-Packard Appl Note 800-6. Hewlett-Packard Palo Alto CA
115 Gere D R, Stark T J, Tweeten T N 1983 Hewlett-Packard Appl Note 800-4. Hewlett-Packard, Palo Alto CA
116 Giddings J C, Myers M N, King J W 1969 Dense gas chromatography at pressures to 2000 atmospheres. J Chromatogr 7:276–283
117 Giddings J C, Myers M N, McLaren L, Keller R A 1968 High pressure gas chromatography of nonvolatile species. Science 162:67–73
118 Glockhart 1976 Lanthanide shift reagents in nuclear magnetic resonance spectroscopy. CRC Critical Reviews in Anal Chem 6:69–130
119 Gnanasambandan T, Freiser H 1982 Separation of aliphatic alcohols by paired ion liquid chromatography. Anal Chem 54:1282–1285
120 Gnanasambandan T, Freiser H 1982 Paired ion chromatography of certain monosaccharides. Anal Chem 54:2379–2380
121 Goering H L, Eikenberry J N, Koermer G S 1971 Tris[3-(trifluoromethylhydroxymethylene)-δ-camphorato]europium (III). A chiral shift reagent for direct determination of enantiomeric compositions. J Am Chem Soc 93:5913–5914
122 Goering H L, Eikenberry J N, Koermer G S, Lattimer C J 1974 Direct determination of enantiomeric compositions with optically active nuclear magnetic resonance lanthanide shift reagents. J Am Chem Soc 96:1493–1501

123 Golik J, James J C, Nakanishi K, Lin Y Y 1982 The structure of brevetoxin-C. Tetrahedron Lett 23:2535–2538
124 Golik J, Liu H W, Dinovi M, Furukawa J, Nakanishi K 1983 Characterization of methyl glycosides at the pico- to nano-gram level. Carbohyd Res 118:135–146
125 Goto T, Kondo T, Imagawa H, Takase S, Atobe M, Muira I 1981 Structure confirmation of tris-deacyl heavenly blue anthocyanin, an alkaline hydrolysis product of heavenly blue anthocyanin obtained from flowers of morning glory "heavenly blue." Chem Lett 883–886
126 Goto T, Kondo T, Tamura H, Imagawa H, Iino A, Takeda K 1982 Structure of gentiodelphin, an acylated anthocyanin isolated from *Gentiana makinoi*, that is stable in dilute aqueous solution. Tetrahedron Lett 23:3695–3698
127 Goto T, Kondo T, Tamura H, Kawahori K, Hattori H 1983 Structure of platyconin, a diacylated anthocyanin isolated from the chinese bell-flower *Platycondon grandiflorum*. Tetrahedron Lett 24:2181–2184
128 Gouw T H, Jentoft R E 1972 Supercritical fluid chromatography. J Chromatogr 68:203–223
129 Gouw T H, Jentoft R E 1975 Practical aspects in supercritical fluid chromatograph. Adv Chromatogr 13:1–40
130 Graham J A, Rogers L B 1980 Effects of surfactants on capacity factors of nonionic solutes in reversed-phase liquid chromatography. J Chromatogr 18:614–621
131 Grant D M, Paul E G 1964 Carbon-13 magnetic resonance. II. Chemical shift data for the alkanes. J Am Chem Soc 86:2984–2990
132 Gray G A 1984 New NMR experiments in liquids. In: Randall J C (ed) NMR and Macromolecules. ACS Symposium Series 247. Am Chem Soc Washington DC 97–118
133 Grob E C, Pfander H, Leuenberger U, Signer R 1971 Separation of carotinoid mixtures by countercurrent extraction. Chimia 25:332–333
134 Hagaman E W 1976 Pattern recognition of geminal proton nonequivalence and second-order coupling in ^{13}C single-frequency off-resonance decoupled spectra. Assignment criteria and structure elucidation. Org Magn Res 8:389–398
135 Hall L D, Sanders J K M 1980 Complete analysis of ^{1}H NMR spectra of complex natural products using a combination of one- and two-dimensional techniques. 1-Dehydrotestosterone. J Am Chem Soc 102:5703–5711
136 Hall L D, Sukumar S 1979 Applications of homonuclear two-dimensional J spectroscopy: an alternative to broad band heteronuclear and homonuclear decoupling. J Am Chem Soc 101:3120–3121
137 Hamamoto Y, Endo M, Nakagawa M, Nakanishi T Mizukawa K 1983 A new cyclic peptide, ascidiacylamide, isolated from Ascidian. J Chem Soc Chem Commun 323–324
138 Hancock W S, Spanow J T 1984 HPLC analysis of biological compounds, a laboratory guide. Marcel Dekker New York, 361 pp
139 Hanke F J, Kubo I 1985 Increasing the speed of droplet countercurrent chromatography separations. J Chromatogr 329:395–398
140 Hannay J B, Hogarth J 1879 On the solubility of solids in gases. Proc R Soc London 29:3244–3326
141 Hansen P E 1981 Carbon-hydrogen spin-spin coupling constants. Prog Nuc Magn Res Spec 14:175–296
142 Hansen P E 1983 Isotope effects on nuclear shielding. Ann Rep NMR Spec 15:105–234
143 Harada N, Nakanishi K 1972 The exciton chirality method and its application to configurational and conformational studies of natural products. Acc Chem Res 5:257–263
144 Harada N, Nakanishi K 1982 Circular dichroic spectroscopy – exciton coupling in organic stereochemistry. Tokyo Kagaku Dojin Tokyo, 460 pp
145 Harris J M, Case M G, Snyder R S, Chenault A A 1984 Cell separations on the countercurrent chromatograph. J Liq Chromatogr 7:419–431
146 Hearn M T W 1980 Ion-pair chromatography on normal- and reversed-phase systems. Adv Chromatogr 18:59–100
147 Hearn M T W (ed) 1985 Ion-pair chromatography: Theory and biological and pharmaceutical applications. Marcel Dekker New York, 294 pp
148 Herz W, Kumar N 1981 Heliangolides from *Helianthus maximiliani*. Phytochemistry 20:93–98
149 Hinze W L, Armstrong D W 1980 Thin layer chromatographic separation of *ortho-*, *meta-*, and *para*-substituted benzoic acids and phenols with aqueous solutions of α-cyclodextrin. Anal Lett 13:1093–1104

150 Hirayama N, Shirahata K 1983 Revised absolute configuration of mitomycin C. X-ray analysis of 1-N-(*p*-bromobenzoyl) mitomycin C. J Am Chem Soc 105:7199–7200
151 Ho S C, Koch H J, Stuart R S 1978 The effect of 0-deuteration upon the proton-decoupled, ^{13}C-NMR spectra of carbohydrates. Carbohyd Res 64:251–256
152 Hofle G 1977 ^{13}C-NMR-Spektroskopie chinoider Verbindungen-II. Substituierte 1,4-Naphthochinone und Anthrachinone. Tetrahedron 33:1963–1970
153 Holland J F 1979 Letters to the editor. Comments on the paper by H.D. Beckey and F.W. Röllgen ('Field desorption "without fields"?'). Org Mass Spec 14:188, 291
154 Holland J F, Soltmann B, Sweeley C C 1976 A model for ionization mechanisms in field desorption mass spectrometry. Biomed Mass Spec 3:340–345
155 Horvath C (ed) 1980, 1980, 1983 High-performance liquid chromatography: advances and perspectives. Vols 1 (330 pp), 2 (340 pp), 3 (230 pp). Academic Press New York
156 Hostettmann K 1980 Droplet countercurrent chromatography and its application to the preparative scale separation of natural products. Planta Med 39: 1–18
157 Hostettmann K 1983 Droplet countercurrent chromatography. Adv Chromatogr 21:165–186
158 Hostettmann K, Appolonia C, Domon B, Hostettmann M 1984 Droplet-countercurrent chromatography – new applications in natural products chemistry. J Liq Chromatogr 7:231–242
159 Hostettmann K, Hostettmann M, Marston A 1984 Isolation of natural products by droplet countercurrent chromatography and related methods. Nat Prod Rep 1:471–484
160 Hostettmann K, Hostettmann M, Marston A 1986 Preparative chromatography techniques – Applications in natural products isolation. Springer Berlin Heidelberg New York Tokyo, 139 pp
161 Hostettmann K, Hostettmann M, Nakanishi K 1978 Molluscicidal saponins from *Cornus florida* L. Helv Chim Acta 61:1990–1995
162 Hostettmann K, Hostettmann-Kaldas M, Nakanishi K 1979 Droplet countercurrent chromatography for the preparative isolation of various glycosides. J Chromatogr 170:355–361
163 Huang Y, Bodenhausen G, Ernst R R 1981 Use of spy nuclei for relaxation studies in nuclear magnetic resonance. J Am Chem Soc 103:6988–6989
164 Huang Y, Macura S, Ernst R R 1981 Carbon-13 exchange maps for the elucidation of chemical exchange networks. J Am Chem Soc 103:5327–5333
165 Hubert P, Vitzthum O G 1978 Fluid extraction of hops, spices, and tobacco with supercritical gases. Angew Chem Int Ed 17:710–715
166 Hunt D F, McEwen C N, Upham R A 1971 Chemical ionization mass spectrometry. II. Differentiation of primary, secondary, and tertiary amines. Tetrahedron Lett 4539–4542
167 Hunt D F, Shabanowitz J, Botz F W 1977 Chemical ionization mass spectrometry of salts and thermally labile organics with field desorption emitters as solid probes. Anal Chem 49:1160–1163
168 Idstein H, Herres W, Schreier P 1984 High-resolution gas chromatography-mass spectrometry and Fourier transform infrared analysis of cherimoya (*Annona cherimolia* Mill.) volatiles. J Agric Food Chem 32:383–389
169 Ito Y 1980 Preparative countercurrent chromatography with a slow rotating glass coil. J Chromatogr 196:295–301
170 Ito Y 1981 Countercurrent chromatography. J Biochem Biophys Methods 5:105–129
171 Ito Y 1981 New continuous extraction method with a coil planet centrifuge. J Chromatogr 207:161–169
172 Ito Y 1981 Efficient preparative countercurrent chromatography with a coil planet centrifuge. J Chromatogr 214:122–125
173 Ito Y 1982 Countercurrent chromatography. Trends Biochem Sci 7:47–50
174 Ito Y 1983 Countercurrent chromatography. Methods Enzym 91:335–351
175 Ito Y 1985 Foam countercurrent chromatography based on dual countercurrent system. J Liq Chromatogr 8:2131–2152
176 Ito Y 1986 High-speed countercurrent chromatography. CRC Critical Rev in Anal Chem 17:65–143
177 Ito Y, Aoki E, Kimura E, Nunogaki K, Nunogaki Y 1969 New micro liquid-liquid partition techniques with the coil planet centrifuge. Anal Chem 41:1579–1584
178 Ito Y, Bhatnagar R 1981 Preparative countercurrent chromatography with a rotating coil assembly. J Chromatogr 207:171–180
179 Ito Y, Bhatnagar R 1984 Improved scheme for preparative countercurrent chromatography with a rotating coil assembly. J Liq Chromatogr 7:257–273

180 Ito Y, Bowman R L 1970 Countercurrent chromatography: Liquid-liquid partition chromatography without solid support. J Chromatogr Sci 8:315–323
181 Ito Y, Bowman R L 1977 Horizontal flow-through coil planet centrifuge without rotating seals. Anal Biochem 82:63–68
182 Ito Y, Bramblett G T, Bhatnagar R, Huberman M, Leive L, Cullinane L M, Groves W 1983 Improved nonsynchronous flow-through coil planet centrifuge without rotating seals: principles and applications. Sep Sci Technol 18:33–48
183 Ito Y, Conway W D 1984 Development of countercurrent chromatography. Anal Chem 56:534A–552A
184 Ito Y, Sandlin J, Bowers W G 1982 High-speed preparative countercurrent chromatography with a coil planet centrifuge. J Chromatogr 244:247–258
185 Ito Y, Weinstein M A, Aoki I, Harada R, Kimura E, Nunogaki K 1966 The coil planet centrifuge. Nature 212:985–987
186 Jallali-Heravi M, Webb G A 1980 A theoretical study of solvent effects on the ^{13}C chemical shifts of some polar molecules. Org Magn Res 13:116–118
187 Jameson C J 1977 The isotope shift in NMR. J Chem Phys 66:4983–4988
188 Jameson C J 1981 Effects of intermolecular interactions and intramolecular dynamics on nuclear resonance. Bull Magn Res 3:3–28
189 Jeener J, Meier B H, Bachmann P, Ernst R R 1979 Investigation of exchange processes by two-dimensional NMR spectroscopy. J Chem Phys 71:4546–4553
190 Jentoft R E, Gouw T H 1970 Pressure-programmed supercritical fluid chromatography of wide molecular weight range mixtures. J Chromatogr 8:138–142
191 Jentoft R E, Gouw T H 1972 Apparatus for supercritical fluid chromatography with carbon dioxide as the mobile phase. Anal Chem 44:681–686
192 Kagan H B (ed) 1977 Stereochemistry fondamentale and methods. Vol 2: Determination of configurations by dipole moments, CD or ORD. Georg Thieme Stuttgart, 198 pp
193 Kainosho M, Ajisaka K, Pirkle W H, Beare S D 1972 The use of chiral solvents or lanthanide shift reagents to distinguish meso from d or l diastereomers. J Am Chem Soc 94:5924–5926
194 Karayannis N M, Corwin A H, Baker E W, Klesper E, Walter J A 1968 Apparatus and materials for hyperpressure gas chromatography of nonvolatile compounds. Anal Chem 40:1736–1739
195 Kasai H, Nakanishi K, Traiman S 1978 Two micromethods for determining the linkage of adducts formed between polyaromatic hydrocarbons and nucleic acid bases. J Chem Soc Chem Commun 798–800
196 Keough T, Stefano A J 1981 Acid enhanced field desorption mass spectrometry of zwitterions. Anal Chem 53:25–29
197 Kessler H 1982 Conformation and biological activity of cyclic peptides. Angew Chem Int Ed 21:512–523
198 Kessler H, Bermel M, Damm I 1982 Peptide conformations, NMR investigations of cyclic hexapeptides containing the active sequence of somatostatin. Tetrahedron Lett 23:4685–4688
199 Kessler H, Bermel M, Friedrich A, Frach G, Hall W E 1982 Peptide conformation. 17-*cyclo*-(L-Pro-L-Pro-D-Pro). Conformational analysis by 270- and 500-MHz one- and two-dimensional ^{1}H NMR spectroscopy. J Am Chem Soc 104:6297–6304
200 Kessler H, Schuck R, Siegmeier R 1982 Analysis of the conformational equilibrium of cyclo[Pro-NBGly$_2$] by means of two-dimensional NMR spectroscopy. J Am Chem Soc 104:4486–4487
201 Kime K A, Sievers R E 1977 A practical guide to uses of lanthanide NMR shift reagents. Aldrichimica Acta 10:54–62
202 Kinashi H, Someno K, Sakaguchi K, Higashijima T, Miyazawa T 1981 Alkaline degradation products of concanamycin A. Tetrahedron Lett 22:3857–3860
203 Kinashi H, Someno K, Sakaguchi K, Miyazawa T 1981 Structure of concanamycin A. Tetrahedron Lett 22:3861–3864
204 Kirkwood J G 1937 The theory of optical rotatory power. J Chem Phys 5:479–491
205 Kirschbaum J, Kerr L 1986 Separation of steroid epimers and isomers using cyclodextrin HPLC columns. LC Mag 4:30–32
206 Klesper E 1978 Chromatography with supercritical fluids. Angew Chem Int Ed 17:738–746
207 Klesper E, Corwin A H, Turner D A 1962 High pressure gas chromatography above critical temperatures. J Org Chem 27:700–701
208 Knight M, Ito Y, Peters P, diBello C 1985 Rapid purification of bombesin by countercurrent chromatography on the multi-layer coil planet centrifuge. J Liq Chromatogr 8:2281–2291

209 Knight M, Kask A M, Tamminga C A 1984 Purification of solid-phase synthesized peptides on the coil planet centrifuge. J Liq Chromatogr 7:351 – 362
210 Knox J H, Jurand J 1981 Zwitterion-pair chromatography of nucleotides and related species. J Chromatogr 203:85 – 92
211 Knox J H, Jurand J 1981 Mechanism of zwitterion-pair chromatography. I. Nucleotides. J Chromatogr 218:341 – 354
212 Knox J H, Jurand J 1981 Mechanism of zwitterion-pair chromatography. II. Ampicillin I. J Chromatogr 218:355 – 363
213 Knox J H, Jurand J 1982 Separation of optical isomers by zwitterion-pair chromatography. J Chromatogr 234:222 – 224
214 Kobayashi M, Yasuzawa T, Kobayashi Y, Kyogoka Y, Kitagawa I 1981 Alcyonolide, a novel diterpenoid from a soft coral. Tetrahedron Lett 22:4445 – 4448
215 Kondrat R W, McClusky G A, Cooks R G 1978 Multiple reaction monitoring in mass spectrometry/mass spectrometry for direct analysis of complex mixtures. Anal Chem 50:2017 – 2021
216 Kotovych G, Aarts G H M 1982 Application of the nuclear Overhauser effect difference experiment: assignment of the configuration of carboprostacyclin. Org Magn Res 18:77 – 81
217 Kraus W, Kypke K, Bokel M, Grimminger W, Sawitzki G, Schwinger G 1982 Surenolactone, a novel tetranortriterpenoid A/B dilactone from *Toona sureni* [Blume] Merrill (Meliaceae). Justus Liebigs Ann Chem 87 – 98
218 Kremmer T, Boross L 1979 Gel chromatography. John Wiley New York, 290 pp
219 Krstulovic A M, Brown P R 1982 Reversed phase high-performance liquid chromatography; theory, practice, and biomedical applications. John Wiley New York, 296 pp
220 Kubo I, Matsumoto A 1984 Molluscicides from olive Olea europaea and their efficient isolation by countercurrent chromatographies. J Agric Food Chem 32:687 – 688
221 Kubo I, Matsumoto A, Asano S 1985 Efficient isolation of ecdysteroids from the silkworm, *Bombyx mori* by droplet countercurrent chromatography. Insect Biochem 15:45 – 47
222 Kubo I, Matsumoto A, Ayafor J F 1984 Efficient isolation of a large amount of 20-hydroxyecdysone from *Vitex madiensis* (Verbenaceae) by droplet countercurrent chromatography. Agric Biol Chem 48:1683 – 1684
223 Kubo I, Matsumoto A, Hanke F J, Ayafor J F 1985 Analytical droplet countercurrent chromatography isolation of 20-hydroxyecdysone from *Vitex thyrsiflora* (Verbenacae). J Chromatogr 321:246 – 248
224 Kucera P (ed) 1984 Microcolumn high performance liquid chromatography. Elsevier Amsterdam, 302 pp
225 Kuhlman K F, Grant D M 1968 The nuclear Overhauser enhancement of the carbon-13 magnetic resonance spectrum of formic acid. J Am Chem Soc 90:7355 – 7357
226 Kuhn W 1930 The physical significance of optical rotatory power. Trans Faraday Soc 26:293 – 308
227 Kuiper J, Labadie R P 1983 Polyploid complexes within the genus *Galium*. 2. Galipuenylin, a new A-ring prenylated anthraquinone of *Galium album*. Planta Med 48:24 – 26
228 Kumar A, Ernst R R, Wurthrich K 1980 A two-dimensional nuclear Overhauser enhancement (2D NOE) experiment for the elucidation of complete proton-proton cross-relaxation networks in biological macromolecules. Biochem Biophys Res Commun 95:1 – 6
229 Kuo M C, Gibbons W A 1979 Total assignments, including four aromatic residues and sequence confirmation of the decapeptide tyrocidine A using different double resonance. J Biol Chem 254:6278 – 6287
230 Landy J S, Dorsey J G 1984 Rapid gradient capabilities of micellar liquid chromatography. J Chromatogr 22:68 – 70
231 Le Cocq C, Lallemand J Y 1981 Precise carbon-13 NMR multiplicity determination. J Chem Soc Chem Commun 150 – 152
232 Leibfritz D, Haupt E, Feigel M, Hull W E, Weber W D 1982 Structure elucidation of 9α-hydroxyestrone methyl ether by two-dimensional proton FT-NMR and ^{13}C NMR spectroscopy. Justus Liebigs Ann Chem 1971 – 1981
233 Leive L, Cullinane L M, Ito Y, Bramblett G 1984 Countercurrent chromatographic separation of bacteria with known differences in surface lipopolysaccharides. J Liq Chromatogr 7:403
234 Lentz H, Frank E U 1978 Phase equilibria and critical curves of binary ammonium-hydrocarbon mixtures. Angew Chem Int Ed 17:728 – 730
235 Levy G C, Nelson G L 1972 Carbon-13 nuclear magnetic resonance for organic chemists. John Wiley New York, 240 pp

236 Liu H W, Nakanishi K 1981 Additivity relation found in the amplitudes of exciton circular dichroism curves of pyranose benzoates. J Am Chem Soc 103:5591–5593
237 Liu H W, Nakanishi K 1982 Pyranose benzoates. An additivity relation in the amplitudes of exciton-split CD curves. J Am Chem Soc 104:1178–1185
238 Lunte C E, Kissinger P T 1983 Determination of pterins in biological samples by liquid chromatography/electrochemistry with a dual-electrode detector. Anal Chem 55:1458–1462
239 Lynn D G, Phillips N J, Hutton W C, Shabanowitz J, Fennel D I, Cole F J 1982 Talaromycins: application of homonuclear spin correlation maps to structure maps. J Am Chem Soc 104:7319–7322
240 Lynn D G, Steffens J C, Kamat V S, Graden D W, Shabanowitz J, Riopel J L 1981 Isolation and characterization of the first host recognition substance for parasitic angiosperms. J Am Chem Soc 103:1868–1870
241 Mabry T J, Markham K R, Thomas M B 1970 Systematic identification of flavonoids. Springer Berlin Heidelberg New York Tokyo, 354 pp
242 Macfarlane R D 1983 Californium-252 plasma desorption mass spectrometry, large molecules, software, and the essence of time. Anal Chem 55:1247A, 1248A, 1250A, 1255A, 1256A, 1258A, 1260A, 1262A, 1264A
243 Macfarlane R D, Torgerson D D 1976 Californium-252 plasma desorption mass spectrometry. Science 191:920–925
244 Macfarlane R D, Uemura D, Ueda K, Hirata Y 1980 ^{252}Cf plasma desorption mass spectrometry of palytoxin. J Am Chem Soc 102:875–876
245 MacKenzie N A, Baxter R L, Scott A I, Fagerness P E 1982 Uniformly ^{13}C-enriched substrates as NMR probes for metabolic events in vivo. Application of double quantum coherence to a biochemical problem. J Chem Soc Chem Commun 145–147
246 Macura S, Wurthrich K, Ernst R R 1982 Separation and suppression of coherent transfer effects in two-dimensional NOE and chemical exchange spectroscopy. J Magn Res 46:269–282
247 Mahato S B, Sahu N P, Ganguly A N, Miyahara K, Kawasaki T 1981 Steroidal glycosides of *Tribulus terrestris* Linn. J Chem Soc Perkin Trans I 2405–2410
248 Mandava N B, Ito Y 1982 Separation of plant hormones by countercurrent chromatography. J Chromatogr 247:315–325
249 Mandava N B, Ito Y 1984 Plant hormone analysis by countercurrent chromatography. J Liq Chromatogr 7:303–322
250 Mandava N B, Ito Y, Ruth J M 1985 Separation of S-triazine herbicides by countercurrent chromatography. J Liq Chromatogr 8:2221–2238
251 Manes L V, Naylor S, Crews P, Bakus G J 1985 Suvanine, a novel sesquiterpene from an *Ircinia* marine sponge. J Org Chem 50:284–286
252 Mantsch H H, Cameron D G, Umemura J, Casal H L 1980 Fourier transform infrared spectroscopy of aqueous systems: applications to the study of biological membranes. J Mol Struc 60:263–268
253 Marchand A P 1982 Stereochemical applications of NMR studies in rigid biocyclic systems. Verlag Chemie Deerfield Beach FL, 231 pp
254 Marshall J L 1983 Carbon-carbon and carbon-proton NMR couplings: Applications to organic stereochemistry and conformationals analysis. Verlag Chemie Deerfield Beach FL, 241 pp
255 Martin D G, Biles C, Peltonen R E 1986 Countercurrent chromatography in the fractionation of natural products. Am Lab 18:21–22, 24–26
256 Matsuda T, Matsuda H, Katakuse I 1979 Use of field desorption mass spectra of polystyrene and polypropylene glycol as mass references up to mass 10,000. Anal Chem 51:1329–1331
257 McCormick J D R, Morton G O 1982 Identity of cosynthetic factor 1 of *Streptomyces aureofaciens* and fragment FO from coenzyme F420 of *Methanobacterium* species. J Am Chem Soc 104:4014–4015
258 McFadden W H 1979 Interfacing chromatography and mass spectrometry. J Chromatogr 17:2–16
259 McFarlane W, Rycroft D S 1979 Magnetic multiple resonance. Ann Rep NMR Spec 9:319–414
260 McGuffin V L 1984 Novel applications of microcolumn liquid chromatography. LC Mag 2:282–288
261 McLafferty F W 1980 Interpretation of mass spectra. 3rd edn. Univ Science Books Mill Valley CA, 303 pp
262 McLafferty F W (ed) 1983 Tandem mass spectrometry. John Wiley New York, 506 pp

263 McLaren L, Myers M N, Giddings J C 1968 Dense-gas chromatography of nonvolatile substances of high molecular weight. Science 159:197–199

264 McNeal C J, Macfarlane R D 1981 Observation of a fully protected oligonucleotide dimer at m/z 12637 by ^{252}Cf-plasma desorption mass spectrometry. J Am Chem Soc 103:1609–1610

265 McNeal C J, Narang S A, Macfarlane R D, Hsiung H M, Broussean R 1980 Sequence determination of protected oligodeoxyribonucleotides containing phosphotriester linkages by californium-252 plasma desorption mass spectrometry. Proc Nat Acad Sci USA 77:735–739

266 Meal L 1983 Identification of ketones by second derivative ultraviolet spectrometry. Anal Chem 55:2448–2450

267 Meier B H, Ernst R R 1979 Elucidation of chemical exchange networks by two-dimensional NMR spectroscopy: The heptamethylbenzenonium ion. J Am Chem Soc 101:6441–6442

268 Messmore H E 1943 United States Pat 2420185

269 Michaelis W, Schindler O, Signer R 1966 Trennung und stereochemische Zuordnung von zwei substituierten isomeren Thioxanthenoxiden. Helv Chim Acta 49:42–53

270 Miller T W, Chaiet L, Cole D J, Cole L J, Flor J E, Goegelman R T, Gullo V P, Joshua H, Kempf A J, Krellwitz W R, Monaghan R L, Ormond R E, Wilson K E, Albers-Schonberg G, Putter T 1979 Avermectins, new family of potent anthelmintic agents: isolation and chromatographic properties. Antimicrobial Agts Chemother 15:368–371

271 Muira I, Hostettmann K, Nakanishi K 1978 ^{13}C-NMR of naturally occurring xanthone aglycones and glycosides. Nouv J Chim 2:653–657

272 Miyamoto M, Kawamatsu Y, Kawashima K, Sinohara M, Nakanishi K 1964 Chromomycinone, the aglycone of chromomycin A_3. Tetrahedron Lett 2355–2365

273 Miyamoto M, Morita K, Kawamatsu Y, Nakanishi K 1967 Chromomycin A_2, A_3, and A_4. Tetrahedron 23:421–437

274 Moffitt W, Woodward R B, Moscowitz A, Klyne W, Djerassi C 1961 Structure and the optical rotatory dispersion of saturated ketones. J Am Chem Soc 83:4013–4018

275 Mooney E F, Winson P H 1969 Carbon-13 nuclear magnetic resonance spectroscopy. Carbon-13 chemical shifts and coupling constants. Ann Rev NMR Spec 2:153–218

276 Murayama W, Kobayashi T, Kosage Y, Yano H, Nunogaki Y, Nunogaki K 1982 A new centrifugal countercurrent chromatograph and its application. J Chromatogr 239:643–649

277 Nagayama K 1980 Two-dimensional NMR spectroscopy: an application to the study of flexibility of protein molecules. Adv Biophys 14:139–204

278 Nagayama K, Bachmann P, Wurthrich K, Ernst R R 1978 The use of cross-sections and of projections in two-dimensional NMR spectroscopy. J Magn Res 31:133–148

279 Nagayama K, Kumar A, Wurthrich K, Ernst R R 1980 Experimental techniques of two-dimensional correlated spectroscopy. J Magn Res 40:321–334

280 Nagayama K, Wurthrich K, Bachman P, Ernst R R 1977 Two-dimensional J-resolved ^{1}H-NMR spectroscopy for studies of biological macromolecules. Biochem Biophys Res Commun 78: 99–105

281 Nakanishi K, Goto T, Ito S, Natori S, Nazoe S (eds) 1974 Natural products chemistry, vol 1. Academic Press New York, 562 pp

282 Nakanishi K, Kusumi T, Iwashita T, Naoki H 1986 FT-NMR with charts. Kodansha Tokyo, 195 pp

283 Nakanishi K, Solomon P H 1977 Infrared absorption spectroscopy, 2nd edn. Holden-Day San Francisco, 287 pp

284 Nakatani M, James J C, Nakanishi K 1981 Isolation and structures of trichilins, antifeedants against southern army worms. J Am Chem Soc 103:1228–1230

285 Nakayama H, Furihata K, Seto H, Otake N 1981 Structure of monazomycin, a new ionophorous antibiotic. Tetrahedron Lett 22:5217–5220

286 Neszmelyi A, Lukacs G 1981 Natural abundance ^{13}C-^{13}C coupling constants for carbon connectivity pattern determination and N.M.R. spectral analysis: patchoulol. J Chem Soc Chem Commun 999–1001

287 Noggle J H, Schirmer R E 1971 The nuclear Overhauser effect. Academic Press New York, 266 pp

288 Nomura T, Fukai T 1978 Kuwanon E, a new flavanone derivative from the root bark of the cultivated mulberry tree (*Morus alba* L.). Heterocycles 9:1295–1300

289 Nomura T, Fukai T, Hano Y, Uzawa J 1981 Structure of sanggenon C, a natural hypotensive Diels-Alder adduct from chinese crude drug 'sang-bái-pí' (morus root barks). Heterocycles 16:2141–2148

290 Nota G, Improta C, Cannata A 1984 Liquid carbon dioxide as a chromatographic eluent. Preliminary thin-layer chromatographic experiments. J Chromatogr 285:194–197

291 Novotny M 1985 Microcolumn liquid chromatography. LC Mag 3:876–886

292 Novotny M, Bertsch W, Zlatkis A 1971 Temperature and pressure effects in supercritical-fluid chromatography. J Chromatogr 61:17–28

293 Novotny M, Springston S R 1983 Fundamentals of column performance in supercritical fluid chromatography. J Chromatogr 279:417–422

294 Novotny M, Springston S R, Peadon P A, Fjeldsted C, Lee M L 1981 Capillary supercritical fluid chromatography. Anal Chem 53:407A, 408A, 410A, 412A, 414A

295 Nozoe S, Koike Y, Tsuji E, Kusano G, Seto H 1983 Isolation and structure of gymnoprenols, a novel type of polyisoprenepolyols from *Gymnopilus spectabilis*. Tetrahedron Lett 24:1731–1734

296 Ogihara Y, Inoue O, Otsuka H, Kawai K, Tanimura T, Shibata S 1976 Droplet countercurrent chromatography for the separation of plant products. J Chromatogr 128:218–223

297 Okamoto Y, Suzuki K, Ohta K, Harada K, Yuki H 1979 Optically active poly(triphenylmethyl methacrylate) with one-handed helical conformation. J Am Chem Soc 101:4763–4765

298 Okamoto Y, Yuki H 1981 Chromatographic resolution of enantiomers having aromatic group by optically active poly(triphenyl methyl methacrylate). Chem Lett 835–838

299 Oltz E M, Nakanishi K, Yagen B, Corley D G, Rottinghaus G E, Tempesta M S 1986 CD additivity in trichothecene benzoates: application as a microanalytical method for trichothecene characterization. Tetrahedron 42:2615–2624

300 Onda M, Konda Y, Hatano A, Hata T, Onura S 1983 Structure of carcinophilin. 3. Structure elucidation by nuclear magnetic resonance spectroscopy. J Am Chem Soc 105:6311–6312

301 Padros E, Morrows A, Manosa J, Dunach M 1982 The state of tyrosine and phenylalanine residues in proteins analyzed by fourth-derivative spectrophotometry: Histone H1 and Ribonuclease A. Eur J Biochem 127:117–122

302 Patel C P 1985 A comparison of Ames TDA, Syva Emid, and HPLC methods in the determination of serum gentamicin. LC Mag 3:148, 150, 152

303 Peaden P A, Fjeldsted J C, Lee M L, Springston S R, Novotny M 1982 Instrumental aspects of capillary supercritical fluid chromatography. Anal Chem 54:1090–1093

304 Peaden P A, Lee M L 1982 Supercritical fluid chromatography: methods and principles. J Liq Chromatogr 5 (Suppl 2):179–221

305 Peaden P A, Lee M L 1983 Theoretical treatment of resolving power in open tubular column supercritical fluid chromatography. J Chromatogr 259:1–16

306 Pelter A, Warren R, Usmani J N, Ilyas M, Rahman W 1969 The use of solvent induced methoxy shifts as a guide to hinokiflavone structure. Tetrahedron Lett 4259–4263

307 Peter S, Brunner G 1978 The separation of nonvolatile substances by means of compressed gases in countercurrent processes Angew Chem Int Ed 17:746–750

308 Peticolas W L 1979 Low frequency vibrations and the dynamics of proteins and polypeptides. Methods Enzym 61:425–458

309 Pinto A C, Goncalves M L A, Filho R B, Neszmelyi A, Lukacs G 1982 Natural abundance ^{13}C-^{13}C coupling constants observed via double quantum coherence: structure elucidation of velloziolide, a diterpene with a novel skeleton. J Chem Soc Chem Commun 293–295

310 Pirkle W H, Welch C J 1984 Chromatographic separation of the enantiomers of acylated amines on chiral stationary phases. J Org Chem 49:138–140

311 Posthumus M A, Kistemaker P G, Meuzelaar H L C 1978 Laser desorption-mass spectrometry of polar nonvolatile bio-organic molecules. Anal Chem 50:985–991

312 Pratt S T, Shoolery J N 1982 Attached proton test for carbon-13 NMR. J Magn Res 46:535–539

313 Putterman G J, Spear M B, Perini F 1984 Synergistic use of countercurrent chromatography and high performance liquid chromatography for the purification of synthetic peptides. J Liq Chromatogr 7:341–350

314 Randall L G 1983 Choosing a modifier in carbon dioxide-based chromatography. Results of a preliminary modifier selectivity survey. Hewlett-Packard Tech Paper No 102. Hewlett-Packard Palo Alto CA, 16 pp

315 Randall L G, Wahrhaftig A L 1978 Dense gas chromatograph/mass spectrometer interface. Anal Chem 50:1703–1705

316 Rawdon M G, Norris T A 1984 Supercritical fluid chromatography as a routine analytical technique. Am Lab 16:17–20, 22–23

317 Reich H J, Jautelat M, Messe M T, Weigert F J, Roberts J D 1969 Nuclear magnetic resonance spectroscopy. Carbon-13 spectra of steroids. J Am Chem Soc 91:7445–7454
318 Resch J F, Meinwald J 1981 Use of osmate (VI) ester *trans-N, N, N'*, N'-tetramethyl-1,2-cyclohexanediamine complexes for determination of glycol stereochemistry. Tetrahedron Lett 22:3159–3162
319 Reuben J 1973 Paramagnetic lanthanide shift reagents in NMR spectroscopy: principles, methodology and applications. Prog Nuc Magn Res Spec 9:1–70
320 Rinaldi P 1982 The determination of absolute configuration using nuclear magnetic resonance techniques. Prog Nuc Magn Res Spec 15:291–352
321 Roberts J D, Weigert F J, Kroschwitz J I, Reigh H J 1970 Nuclear magnetic resonance spectroscopy. Carbon-13 chemical shifts in acyclic and alicyclic alcohols. J Am Chem Soc 92:1338–1347
322 Robinson J A, Turner D L 1982 Total assignment of the carbon-13 NMR spectrum of monensin by two-dimensional correlation spectroscopy. J Chem Soc Chem Commun 148–151
323 Roques B P, Rao R, Marion D 1980 Use of nuclear Overhauser effect in the study of peptides and proteins. Biochimie 62:753–773
324 Roth K 1979 Selektive Unterdrückung von NMR Signalen mit Hilfe von T_1 Pulssequenzen. Org Magn Res 12:271–273
325 Roush R A, Cooks R G 1984 Characterization of alkaloids and other secondary metabolites by multiple stage mass spectrometry. J Nat Prod 47:197–214
326 Runser D J 1981 Maintaining and troubleshooting HPLC systems. John Wiley New York, 176 pp
327 Samain D, Cook J C, Rinehart K L 1982 Structure of scopafungin, a potent nonpolyene antifungal antibiotic. J Am Chem Soc 104:4129–4141
328 Sanders J K M, Mersh J D 1982 Nuclear magnetic double resonance: the use of different spectroscopy. Prog Nuc Magn Res Spec 15:353–400
329 Sandlin J L, Ito Y 1984 Gram-quantity separation of DNP(dinitrophenyl) amino acids with multi-layer coil countercurrent chromatography (CCC). J Liq Chromatogr 7:323–340
330 Sandlin J L, Ito Y 1985 Large-scale preparative countercurrent chromatography for separation of polar compounds. J Chromatogr 348:131–139
331 Sandlin J L, Ito Y 1985 Large-scale preparative countercurrent chromatography with a coil planet centrifuge. J Liq Chromatogr 8:2153–2171
332 Sanki Engineering 1983 Continuous flow extraction and fractionation system. Centrifugal countercurrent chromatograph. Technical Manual CPC-N90 CPC-A90. Sanki Engineering, 50 pp
333 Santi W, Huen J M, Frei R W 1975 High-speed ion-pair partition chromatography in pharmaceutical analysis. J Chromatogr 115:423–436
334 Scheek R M, Russo N, Boelens R, Kapstein R, Von Boom J H 1983 Sequential resonance assignments in DNA ^1H NMR spectra by two-dimensional NOE spectroscopy. J Am Chem Soc 105:2914–2916
335 Schneider G M 1978 Physiochemical principles of extraction with supercritical gases. Angew Chem Int Ed 17:716–727
336 Schwenn J D, Jender H G 1980 Reversed-phase high-performance liquid chromatography of adenine nucleotides: application to the kinetics of an adenosine 3'-phosphate 5'-sulfatophosphate sulfotransferase from plants. J Chromatogr 193:285–290
337 Scott A I 1962 Interpretation of the ultra violet spectra of natural products. Pergamon Press New York, 500 pp
338 Seo S, Uomori A, Yoshimura Y, Takeda K 1983 Stereospecificity in the biosynthesis of phytosterol side chains: ^{13}C NMR signal assignments of C-26 and C-27. J Am Chem Soc 105:6343–6344
339 Seto H, Sasaki T, Yonehara H, Uzawa J 1978 Studies on the biosynthesis of pentalenolactone. Part I. Application of long range selective proton decoupling (LSPD) and selective $^{13}C\text{-}\{^1H\}$ NOE in the structural elucidation of pentalenolactone G. Tetrahedron Lett 923–926
340 Seymour F R, Julian R L 1979 Fourier-transform, infrared difference-spectroscopy for structural analysis of dextrans. Carbohyd Res 74:63–75
341 Shapiro M J, Kolpak M X, Lemke T L 1984 Structure elucidation using signal intensity effects in carbon-13 nuclear magnetic resonance. J Org Chem 49:187–189
342 Shaw D 1984 Fourier transform NMR spectroscopy, 2nd edn. Elsevier Amsterdam, 304 pp
343 Shirataki Y, Komatsu M, Yokoe I, Manaka A 1981 Studies on the constituents of *Sophora* species. XVI. Constituents of the root of *Euchresta japonica* Hood. f. ex Regel (1). Chem Pharm Bull 29:3033–3036
344 Shoolery J N 1984 Recent developments in carbon-13 and proton-NMR. J Nat Prod 47:226–259

345 Sie S T, Rijinders G W A 1966 High-pressure gas chromatography and chromatography with supercritical fluids. I. The effect of pressure on partition coefficients in gas-liquid chromatography with carbon dioxide as a carrier gas. Sep Sci 1:459–490
346 Sie S T, Rijinders G W A 1967 High pressure gas chromatography and chromatography with supercritical fluids. II. Permeability and efficiency of packed columns with high-pressure gases as mobile fluids under conditions of incipient turbulence. Sep Sci 2:699–727
347 Sie S T, Rijinders G W A 1967 High-pressure gas chromatography and chromatography with supercritical fluids. III. Fluid-liquid chromatography. Sep Sci 2:729–753
348 Sie S T, Rijinders G W A 1967 High-pressure gas chromatography and chromatography with supercritical fluids. IV. Fluid-solid chromatography. Sep Sci 2:755–777
349 Sievers R E (ed) 1973 Nuclear magnetic shift reagents. Academic Press New York, 410 pp
350 Signer R, Allemann K, Koehli E, Lehmann W, Mayer H, Ritschard W 1956 A laboratory apparatus for multiple partition of substances between two immiscible fluids. Dechema Monograph 27:32–44
351 Signer R, Arm H 1967 Eine vielstufige Verteilungsapparatur. Helv Chim Acta 50:46–53
352 Singer B 1975 Chemical effects of nucleic acid alkylation and their relation to mutagenesis and carcinogenesis. Prog Nuc Acid Res Mol Biol 15:219–284
353 Singleton K E, Cooks R G, Wood K V 1983 Utilization of natural isotopic abundance ratios in tandem mass spectrometry. Anal Chem 55:762–764
354 Smith M J, Shoolery J N, Schwyn B, Holden I, Neilands J B 1985 Rhizobactin, a structurally novel siderophore from *Rhizobium meliloti*. J Am Chem Soc 107:1739–1743
355 Smith R D, Felix W D, Fjeldsted J C, Lee M L 1982 Capillary column supercritical fluid chromatography/mass spectrometry. Anal Chem 54:1883–1885
356 Smith R D, Felix W D, Fjeldsted J C, Lee M L 1984 Capillary supercritical fluid chromatography. Anal Chem 56:619A, 620A, 622A, 624A, 627A, 628A
357 Smith R D, Fjeldsted J C, Lee M L 1982 Direct fluid injection interface for capillary supercritical fluid chromatography/mass spectrometry. J Chromatogr 247:231–243
358 Snyder J, Nakanishi K, Chaverria G, Leal Y, Ochoa C C, Dominguez X A 1981 The structure of castanaguyone a biisocoumarin plant product. Tetrahedron Lett 22:5015–5018
359 Snyder J K, Nakanishi K, Hostettmann K, Hostettmann M 1984 Applications of rotation locular countercurrent chromatography in natural products isolation. J Liq Chromatogr 7:243–256
360 Snyder L R, Kirkland J J 1979 Introduction to modern liquid chromatography. John Wiley New York, 863 pp
361 Soltmann B, Sweeley C C, Holland J F 1977 Electron impact ionization mass spectrometry using field desorption activated emitters as solid sample probes. Anal Chem 49:1164–1166
362 Sorel R H A, Hulshoff A 1983 Dynamic anion-exchange chromatography. Adv Chromatogr 21:87–129
363 Steffens J C, Roark J L, Lynn D G, Riopel J L 1983 Host recognition in parasitic angiosperms: use of correlation spectroscopy to identify long-range coupling in an haustorial inducer. J Am Chem Soc 105:1669–1671
364 Stephens R M 1976 Structural investigations of peptides and proteins. Part III: Conformation and interaction of peptides and proteins in solution. 5. Infrared spectroscopy. Amino Acids Pept Protein 8:217–221
365 Stephens R M 1979 Structural investigations of peptides and proteins. Part III: Conformation and interaction of peptides and proteins in solution. 4. Infrared spectroscopy. Amino Acids Pept Protein 10:237–241
366 Stoll R, Rollgen W R 1979 Laser desorption mass spectrometry of thermally labile compounds using a continuous wave CO_2 laser. Org Mass Spec 14:642–645
367 Stonard R, Trainor D A, Nakatani M, Nakanishi K 1983 Additivity relation in the amplitudes of exciton-split circular dichroism curves arising from interactions between different chromophores and its application in structural studies. J Am Chem Soc 105:130–131
368 Stothers J B 1972 Carbon-13 NMR spectroscopy. Academic Press New York, 559 pp
369 Stranahan J J, Deming S N 1982 Mechanistic interpretations and simulations of induced peaks in liquid chromatography. Anal Chem 54:1540–1546
370 Surman D J, Vickerman J C 1981 Fast atom bombardment quadrupole mass spectrometry. J Chem Soc Chem Commun 324–325

371 Sutherland I A, Heywood-Waddington D, Peters T J 1984 Toroidal coil countercurrent chromatography: a fast simple alternative to countercurrent distribution using aqueous two phase partition. J Liq Chromatogr 7:363–384
372 Tachibana K, Scheuer P J, Tsukitani Y, Kikuchi H, Van Engen D, Clardy J, Gopichand Y, Schmitz F J 1981 Okadaic acid, a cytotoxic polyether from two marine sponges of the genus *Halichodria*. J Am Chem Soc 103:2469–2471
373 Takeda R, Mori R, Hisose Y 1982 Structural and absolute configurational studies of striatene, striatol and β-monocyclonerolidol, three sesquiterpenoids from the liverwort *Ptychanthus striatus* (Lehm. et Lindemb.) Nees. Chem Lett 1625–1628
374 Takeda R, Zask A, Nakanishi K, Park M H 1987 Additivity in split cotton effect amplitudes of p-phenylbenzyl ethers and p-phenylbenzoates. J Am Chem Soc 109:914–915
375 Talsky G, Maryring L, Kruezer H 1978 High-resolution, higher-order UV/Vis derivative spectrometry. Angew Chem Int Ed 17:785–799
376 Tanimura T, Pisano J J, Ito Y, Bowman R L 1970 Droplet countercurrent chromatography. Science 169:54–56
377 Terpstra D 1979 Two-dimensional Fourier transform carbon-13 NMR. In: Levy G C (ed) Topics in carbon-13 NMR spectroscopy, vol 3. John Wiley New York, 62–78
378 Tomasz M, Lipman R, Lee M S, Verdine G L, Nakanishi K 1987 Reaction of acid activated mitomycin C with calf thymus DNA and model guanines: Elucidation of the base-catalysed degradation of N7-alkylguanine nucleosides. Biochem 26:2010–2027
379 Tomasz M, Lipman R, Verdine G L, Nakanishi K 1985 Nature of the destruction of deoxyguanosine residues by mitomycin C activated by mild acid pH. J Am Chem Soc 107:6120–6121
380 Tomasz M, Lipman R, Verdine G L, Nakanishi K 1986 Reassignment of the guanine-binding mode of reduced mitomycin C. Biochem 25:4337–4344
381 Turner C J 1984 Multipulse NMR in liquids. Prog Nuc Magn Res Spec 16:311–370
382 Ulrich E L, Westler W M, Markley J L 1983 Reassignments in the ^{1}H NMR spectrum of flavin adenine dinucleotide by two-dimensional homonuclear chemical shift correlation. Tetrahedron Lett 24:473–476
383 Uzawa J, Takeuchi S 1978 Application of selective ^{13}C-{^{1}H} nuclear Overhauser effects with low-power ^{1}H irradiation in carbon-13 NMR spectroscopy. Org Magn Res 11:502–506
384 Valentine K M, Doner L W, Pfeffer P E 1981 ^{13}C-NMR resonance assignments of psicose by the differential isotope-shift technique. Carbohyd Res 96:293–298
385 Van Der Wal S J 1985 Practical aspects of high-speed liquid chromatography. LC Mag 3:488, 490, 492, 494, 496
386 Van Lancher M A, Nelis H J C F, Leenheer A P 1983 Reversed-phase ion-pair chromatography of anthracyclines. J Chromatogr 254:45–52
387 Verdine G L, Nakanishi K 1985 p-Dimethylaminocinnamate, a new red-shifted chromophore for use in the exciton chirality method. Its application to mitomycin C. J Chem Soc Chem Commun 1093–1095
388 Verdine G L, Nakanishi K 1985 Use of differential second-derivative UV and FTIR spectroscopy in structure studies of multichromophoric compounds. J Am Chem Soc 107:6118–6120
389 Vernes B, Chari V M, Wagner H 1981 Structure elucidation and synthesis of flavonol acylglycosides. III. The synthesis of Tiliroside. Helv Chim Acta 64:1964–1967
390 Vold R L, Waugh J S, Klein M P, Phelps D E 1968 Measurements of spin relaxation in complex systems. J Chem Phys 48:3831–3832
391 Walter H, Krob E J, Brooks D E 1976 Membrane surface properties other than charge involved in cell separation by partition in polymer aqueous two-phase systems. Biochemistry 15:2959–2964
392 Walters F H 1985 Recent advances in inorganic liquid chromatography. LC Mag 3:1056–1061
393 Warsh J J, Chiu A, Godse D D 1982 Simultaneous determination of norepinephrine, dopamine and seratonin in rat brain regions by ion-pair liquid chromatography on octy silane columns and amperometric detection. J Chromatogr 228:131–141
394 Wehrli F W, Nishida T 1979 The use of carbon-13 nuclear magnetic resonance spectroscopy in natural products chemistry. Fortschr Chem Org Naturst 36:1–229
395 Wehrli F W, Wirthlin T 1978 Interpretation of carbon-13 NMR spectra. Heyden Press Philadelphia, 310 pp

396 Weigert J F, Roberts J D 1970 Nuclear magnetic resonance spectroscopy. Carbon-13 chemical shifts in cycloalkanones. J Am Chem Soc 92:1347–1351
397 Weinberger R, Yarmchuk D, Cline Love L J 1982 Liquid chromatographic phosphorescence detection with micellar chromatography and postcolumn reaction modes. Anal Chem 54:1552–1558
398 Whitesides G M, Lewis D W 1970 Tris[3-(*tert*-butylhydroxymethylene)-δ-camphorato]europium (III). A reagent for determining enantiomeric purity. J Am Chem Soc 92:6979–6980
399 Whitesides G M, Lewis D W 1971 The determination of enantiomeric purity using chiral lanthanide shift reagents. J Am Chem Soc 93:5914–5916
400 Wiesler W T, Vázquez J T, Nakanishi K 1986 Additivity in complex CD curves of multichromophoric systems. J Am Chem Soc 108:6811–6813
401 Winistorfer P, Kovats E Sz 1967 Use of hydrotopic solutions for countercurrent separations. J Gas Chromatogr 5:362–372
402 Yarmchuk P, Weinberger R, Hirsch R F, Cline Love L J 1982 Selectivity in liquid chromatography with micellar mobile phases. Anal Chem 54:2233–2238
403 Yokota T, Arima M, Takahashi N 1982 Castasterone, a new phytosterol with plant-hormone potency, from chestnut insect gall. Tetrahedron Lett 23:1275–1278
404 Zosel K L 1970 DAS 2005293
405 Zosel K L 1978 Separation with supercritical gases: Practical applications. Angew Chem Int Ed 17:702–709

Added Note

In the time that has elapsed since this chapter was written, the field of NMR spectroscopy has witnessed a veritable explosion in the development of new software and hardware, and in the applications of new pulse sequences for structure determination. Indeed, a new dimension has literally been added with the successful use of 3D-NMR in structural problems (2, 3, 5, 6). The material on NMR spectroscopy and its applications to structure determination presented in this chapter is intended only to serve as an introduction for the uninitiated. If one wishes to further pursue the topic, the reader is referred to an excellent recent review of two dimensional NMR techniques written by Horst Kessler and coworkers (4). For a fundamental dissertation on the theory of modern NMR spectroscopy, the reader is referred to a truly remarkable book by Ernst, Bodenhausen and Wokaun (1).

1 Ernst R R, Bodenhausen G, Wokaun A 1987 Principles of Nuclear Magnetic Resonance in One and Two Dimensions. Clarendon Press, Oxford
2 Fesik S W, Gampe R T, Zuiderweg E R P 1989 J Am Chem Soc 111:770
3 Griesinger C, Sorensen O W, Ernst R R 1987 J Am Chem Soc 109:7227
4 Kessler H, Gehrke M, Griesinger C 1988 Angew Chem Int Ed Engl 27:490
5 Marion D, Kay L E, Sparks S W, Torchia D A, Bax A 1989 J Am Chem Soc 111:1515
6 Vuister G W, de Waard P, Boelens R, Vliegenhart J F G, Kaptein R 1989 J Am Chem Soc 111:772

Chapter 3

Evolution of Natural Products*

O. R. GOTTLIEB

3.1 Convergent Synthesis and the Origin of RNA-Based Life

The evolution of natural products on earth has fascinated scientists and laymen alike for many years. A vast literature has been accumulated concerning the hypothetical interface of the prebiotic (chemical) and the organismic (biochemical) phases (50, 60). Nevertheless the entire concept, including the validity of its experimental basis, has more recently been questioned. Indeed the concept introduces a problem of the hen-and-egg type. Which ones came first – the complex molecules with genetic and enzymatic potential or the molecules generated by such catalytic macromolecules? Cairns-Smith (10) gave the sole reasonable answer: genes and enzymes. According to this author the most central molecules of life are the same in all organisms on earth today. Hence all life has descended from a common ancestor in which the central biochemical system was already fixed. That it should have remained fixed for so long is surely because of the critical interdependence of all components of the central highly complex machinery. Hence the ancestor must be situated at a quite high position of the evolutionary tree, preceded by simpler forms in which chemical reactions were catalyzed initially by geochemical genetic material such as clay crystals and metal ions. In the resulting progressively more sophisticated system "genetic takeover" must have occurred, the inorganic material having been gradually replaced by an organic one, preferentially endowed with information-carrying capacity *and* catalytic activity. So far only one macromolecule is known to possess such a double capacity: RNA (12). However, direct synthesis of RNA is an improbable event. The enzyme catalyzing its in vitro formation is far too complex to have had a clay template analogue on primitive earth (10).

Cairns-Smith's enigma stimulates consideration of the intermediate forms of the genetic takeover in which inorganic and organic materials co-exist. Both types of catalysts in separation may be inefficient. Together each one may be responsible for the partial synthesis of a complex molecule. To join these parts into macromolecules such as RNA, it would conceivably suffice to rely on a possibly pre-existing general dehydration catalyst. McKey (58) also proposed a double pathway theory in order to rationalize the biosynthesis of indole-iridoid alkaloids in species of the Gentianiflorae [sensu Dahlgren (18)]. Thus the same concept may justify the synthesis of complex molecules at the two extremes of the organismic and molecular spectra: macromolecules (RNA, etc.) in the most primitive pro-biotic

---

* This chapter is dedicated to the memory of Prof. Dr. Rolf Dahlgren (1932–1987), Botanical Museum, University of Copenhagen, who courageously broke away from tradition, introducing chemical data not only for refinement but as decisive criteria in the construction of his system of angiosperm classification.

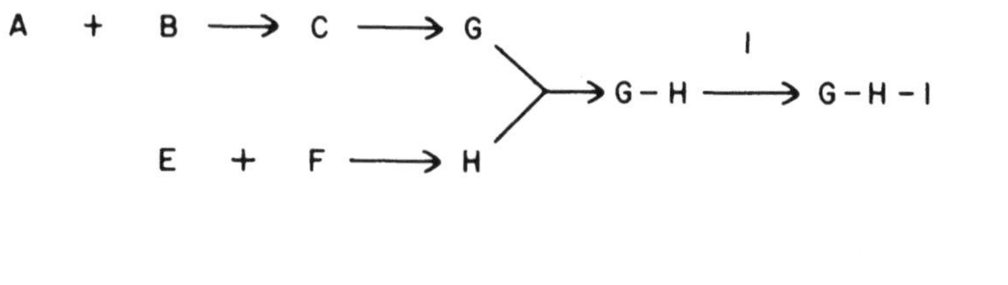

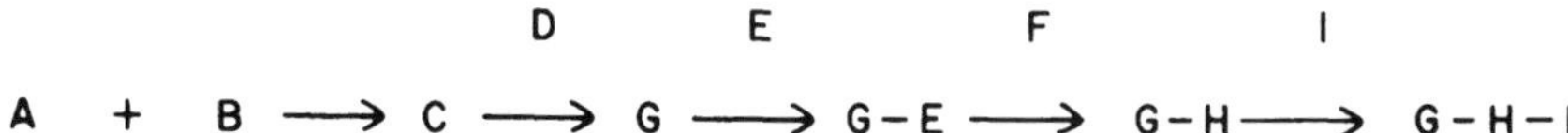

Fig. 3.1. The two strategies of chemical synthesis – convergent synthesis (**above**) and linear synthesis (**below**) – are exemplified. Molecules and molecular fragments are represented by capital letters (11)

unicorpuscular forms, and micromolecules (indole-iridoid alkaloids) in the highly developed pluricellular plants. Hence there is no reason to believe that it does not apply to intermediate organisms as well.

The organic chemist will find this hypothesis reasonable. Indeed he recognizes the potential advantages of a convergent synthesis (Fig. 3.1). "If, for example, a molecule consists of two major fragments, G and H, and a side chain, I, it is more efficient to synthesize G and H separately and then combine them, rather than make G first and build H upon it, step by step. The overall yield in a synthetic sequence is the product of the yields of the individual steps so total yield tends to decrease with an increasing number of steps in a sequence. A linear sequence maximizes the number of steps to which the original starting materials must be subjected. A convergent synthesis, in contrast, allows one to build up separate fragments and then combine them; the number of steps to which each set of starting materials is subjected is thus decreased" (11).

"Convergent synthesis" in the former paragraph is used to describe a certain strategy of synthetic organic chemistry. The term convergence, used by biologists, has a totally different meaning, being defined as the resemblance of structure due to adaptation to similar environmental conditions that have occurred in different evolutionary lines.

The formation of new, more effective molecular systems by the interaction of two (or more) precursors, each of them already endowed with its own evolutionary history, has an analogy on the level of organisms. According to the endosymbiotic theory defended by Margulis (53), several classes of cellular organelles – mitochondria, plastids, and even perhaps flagella – had once been free-living bacteria that were acquired symbiotically and in a certain sequence by host prokaryotes.

Having recognized the possibility of forming molecular and organismic systems by adding part-structures, it is most important to consider the peculiarity that the characteristics of the whole cannot be deduced from the knowledge of the components. The appearance of new and unexpected characteristics in wholes has been designated as emergence. Although yet a more philosophical than practical concept, emergence includes the most striking features of adaptive machinery (56).

This is then "the greatest show on earth": At the start of RNA-based life, the first and presumably simplest organism had stored in itself all the necessary information to construct you and me if – but what an if! – during close to 4 billion years the environmental conditions, including catastrophic events (20), had evolved precisely as they did.

3.2 Expansion of the Acetate, Mevalonate, and δ-Aminolevulinate Pathways in Bacteria and Algae

In order to formulate the sequences of organic reactions that occurred in organisms, and hence to contribute to the understanding of nature, only living forms can, of course, be analyzed. However, extrapolation of the data to past organisms is a valid procedure because of fossil links, fixedness of the central biochemical, uniformly RNA-based theme and the implied (10) use of CO_2 via photosynthesis as the carbon source even in the most primitive organisms.

The common ancestor to all extant organisms, were it alive today, would be placed in the genus *Clostridium* (13). It may be surprising, but it is nevertheless true: Clostridia contain saturated fatty acids, squalene, proteins including phenylalanine and tryptophane, tetrapyrrolic corrinoids and nucleosides composed of pyrimidines, purins, and sugars – all well known components of modern organisms. Evolutionary advance of the major primitive prokaryotic lineages of the kingdom Monera [sensu Whittaker (85)], was accompanied by the development of several fundamental pathways based on acetate (saturated and monoolefinic fatty acids – double-bond formation involving dehydration), mevalonate (squalene and hopanoids – the latter formed by an acid-catalyzed cyclization, not, as the more recent steroids, by an epoxide-mediated one – and carotenoids and arylcarotenoids), shikimate (menaquinones and ubiquinones), and δ-aminolevulinate (hemes). Mechanistically all these reactions can be explained by condensations and do not involve oxidations.

On the primitive earth most of the oxygen liberated by Cyanobacteria through the classical reaction (86)

$$nCO_2 + nH_2O \rightarrow (CH_2O)_n + nO_2$$

and other sources did not stay in the atmosphere but was used up by dissolved ferrous salts and sulfur-oxidizing bacteria. Only after most of the ferrous and sulfide ions had been deposited, under the form of ferric oxides and sulfates, respectively, did atmospheric oxygen rapidly reach present levels (25). For the organisms that had become adapted to the oxidizing environment by evolution of a protective system, the abundance of oxygen in the atmosphere even became an asset. Under anoxic conditions, energy had to be liberated from the photosynthesized organic matter by fermentation. Respiration, which becomes possible in the presence of oxygen, releases (through inversion of the reaction above) far more energy than fermentation (70) and counteracts the oxidative degradation of matter. The accumulation of substantial amounts of oxygen in the atmosphere from about 1.8 billion years ago onward is documented in the structure of the constituents of all subsequently evolved prokaryotic organisms. The synthesis of such structures by expanded acetate (polyunsaturated fatty acids), mevalonate (steroids, polyprenols), shikimate, and δ-aminolevulinate pathways, manifestedly requires not only condensation reactions but also oxidation reactions.

About 1.2 billion years ago, another interface appeared with the formation of cells containing double-membrane-bound organelles. The rationalization of the various events is controversial. The leading endosymbiotic theory was mentioned above. In other words, analogously to the primordial prokaryotic cell, a set of autoreplicable chemical systems selected and held together by emergence, the primitive eukaryotic cell is envisaged as a set of parts each formerly endowed with an independent chemistry. Here again the association of parts as different as possible may have been oriented by the formation of a system of highest emergence value in face of the existing ambient conditions.

Cyanobacteria, which dominated the earth from 3.9 to 0.6 billion years ago, contain, besides phycobilins, typical tetrapyrrolic pigments of bacteria, also chlorophyll a and β-carotene. In *Prochloron* (51), a sister group of Cyanobacteria, phycobilins are absent and the two other pigments are complemented by chlorophyll b, establishing itself for the first time the triad of pigments that occurs in all more modern green organisms of the kingdoms Protista and Plantae [sensu Whittaker (85)]. Indeed Prochloron is thought to represent the prokaryotic counterpart of the chloroplasts of green algae and plants (54). Whatever one's opinion on this matter, the fact remains that the chemical machinery of prokaryotes also operates in the eukaryotic algae and was there further expanded by the accretion of polyketides on the acetate pathway and of sesquiterpenes on the mevalonate pathway. Again, in spite of the profound disjunction that separates Monera and Protista, both kingdoms seem to be linked by a chemical continuity.

Still more recently, perhaps 0.6 billion years ago, one more interface appeared with the development of multicellular eukaryotes with differentiated cells. Their three kingdoms represent the three great ecological strategies for larger organisms: production (plants), absorption (fungi), and ingestion (animals). With respect to the consequences of production, one already suspects that chemical compounds should arise through the expansion of ancestral metabolic pathways (Fig. 3.2), in spite of the fact that morphogenesis, the process by which cells are organized into tissues, organs, and whole organisms, is a central unsolved problem in biology (14). We shall follow the evolution of chemical phenomena from this point to the flowering plants, in an attempt to understand the development of lignoid chemistry.

3.3 Expansion of the Shikimate Pathway in Terrestrial Plants

The ultraviolet part of solar radiation is capable of promoting not only the formation but also the cleavage of chemical bonds. Hence the organic compounds

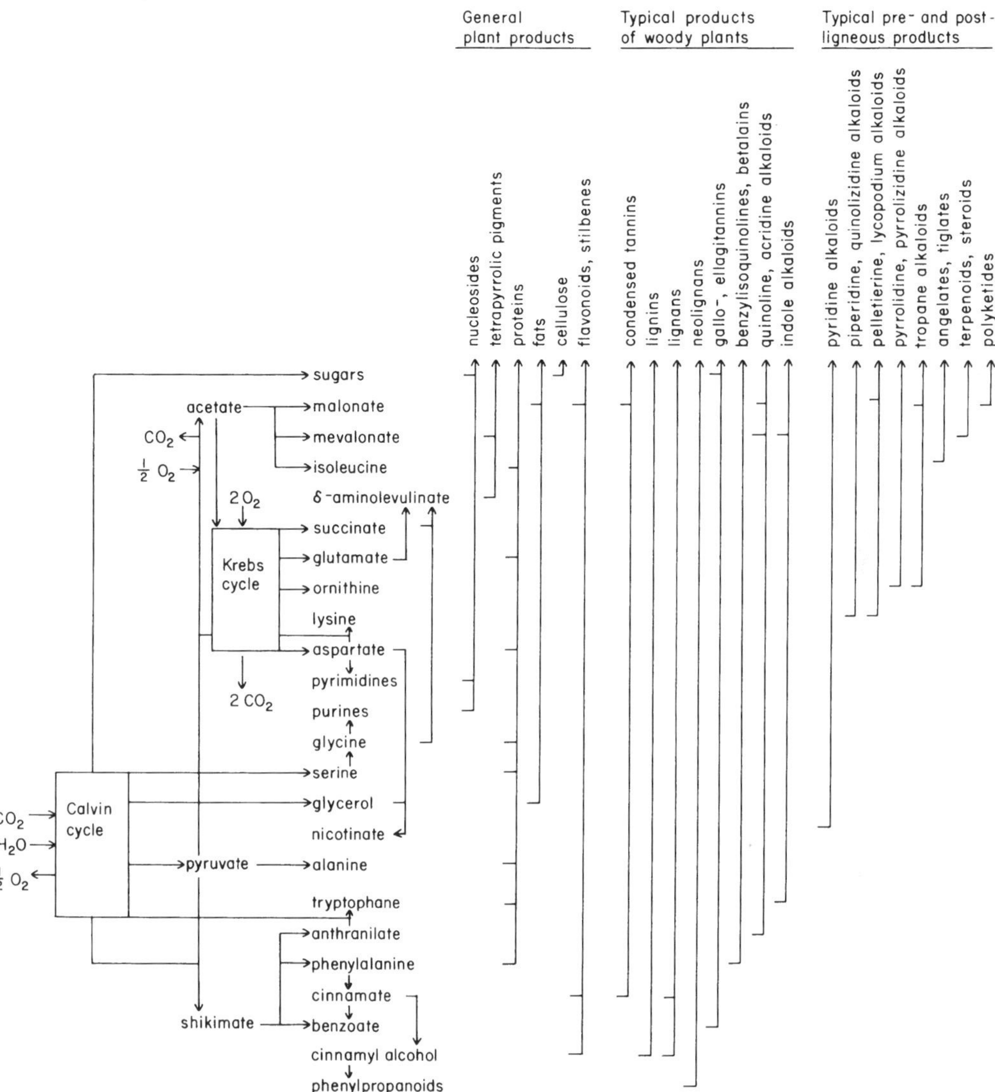

Fig. 3.2. Fundamental biosynthetic pathways leading through primary metabolites to macromolecular and micromolecular natural products

of the primordial forms and organisms were viable only in aquatic media, protected by the UV-absorbing oxygen of water, or in the internal cavities of hydrated soil, protected by the light-excluding walls of the solid material; the surface of the dry land remained sterile. Atmospheric oxygen led to the development of a stratospheric layer of ozone which, less than 1 billion years ago, had acquired sufficient thickness to shield the earth from the major portion of the short-wave (<300 nm) range of the solar spectrum. This was the time of separation into multicellular plants, fungi, and animals, each of these kingdoms having developed a different defensive system against excess ultraviolet radiation. The permanent regeneration of burnt cells of the animal skin, for instance, does not operate in plants.

Chlorophytes constitute a major group of algae within which evolutionary lineages have led from unicellular forms to multicellular organisms. One of these lineages encompasses the Charales, which are considered ancestral to land plants (59). However, the aqueous habitat confers protection against solar radiation to algae. Thus, about 0.5 billion years ago, when invading humid environments of the land surface, possibly through primitive members of the bryophytes (mosses), the plant phylum must have solved the problem of the UV screen. Chemically Chlorophyta and Bryophyta have much in common. The photosynthetic systems of both divisions contain chlorophylls a and b, as well as carotenoids such as β-carotene and xanthophylls. Both store starch and possess cellulosic cell walls, characteristics that continue to prevail in all more modern plant groups. Furthermore, algae and bryophytes lack true lignins. Nevertheless, the possibility of a direct evolutionary connection of green algae and modern bryophytes is not supported by steroid and triterpenoid data (77). Indeed the presence of highly oxygenated flavonoids (33, 55) as well as of isoflavones (1) shows that the modern bryophytes cannot represent the primitive (extinct) tracheophytes from which they must have descended in parallel to the more advanced cormophytes.

The most conspicuous chemical difference between aquatic protists and terrestrial plants lies in the exploitation of the shikimate pathway. In all forms of algae, this is limited to the production of phenylalanine and tyrosine, already incorporated into proteins since primitive bacteria, and of derivatives of a few other pretyrosine molecules (Figs. 3.3 and 3.4). Only with the primitive tracheophytes, as

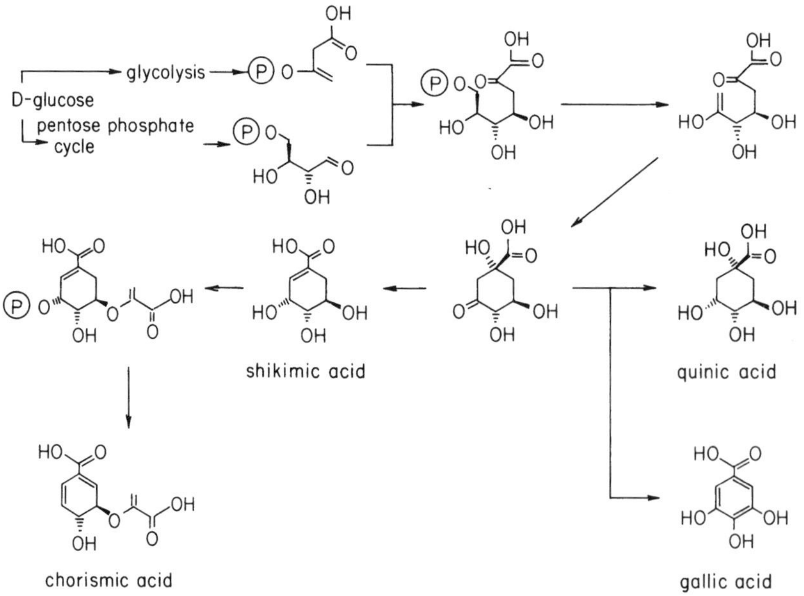

D-glucose
glycolysis
pentose phosphate cycle
shikimic acid
quinic acid
chorismic acid
gallic acid

Fig. 3.3 Initial steps of the shikimate pathway

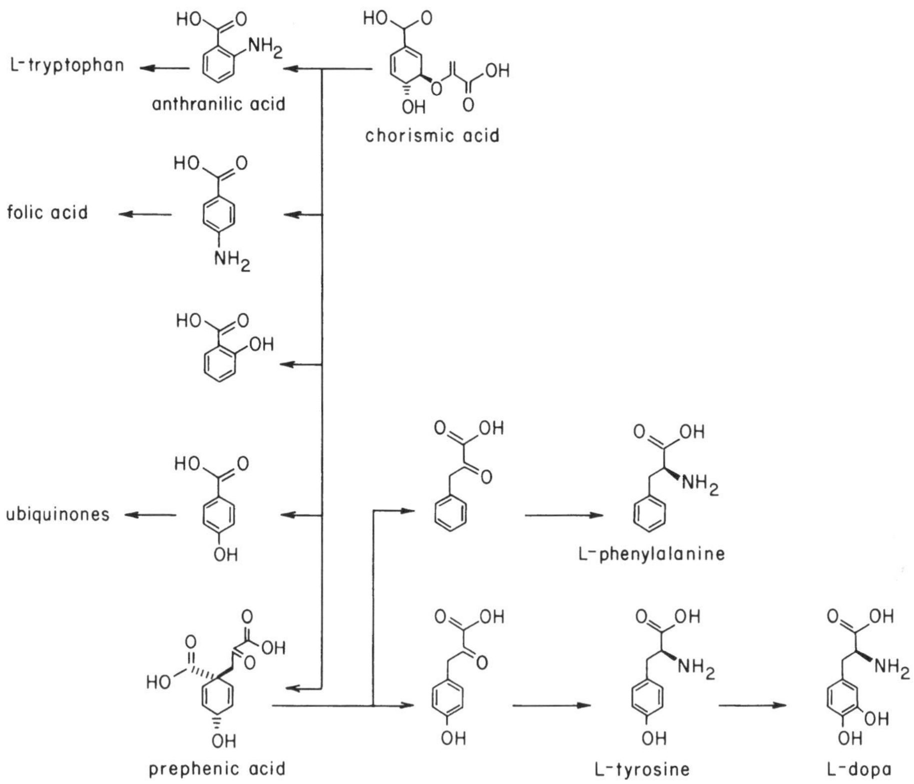

Fig. 3.4. Intermediate steps of the shikimate pathway

indicated by the analysis of bryophytes, is a post-tyrosine chemistry, based on cinnamic acid, initiated (Fig. 3.5). Deamination of aromatic amino acids to cinnamic acids may have occurred sporadically in other groups and certainly occurs in fungi. The special importance that cinnamoyl units acquired in terrestrial plants, however, is possibly due to their activation by coenzyme A. Polyketides, compounds formed by the condensation of acetyl-CoA as starter unit and malonyl-CoA for chain extension (Fig. 3.6, R = Me), dominate the allelochemistry of aerobic bacteria and of algae. The replacement in this process of the starter unit by cinnamoyl-CoA in bryophytes led to flavones and flavonols (Fig. 3.6, R = St). Flavonoids, such as chalcones, aurones, flavones, and flavonols, absorb UV light and should act as photoscreens (57) in all divisions of terrestrial plants. To the light-absorbing flavonoids also belong the anthocyanidins, which, present in the great majority of Plantae, supply an additional element of color to the landscape.

Apart from UV radiation, there exist other phenomena in an aerial environment that require adaptation of plant life: one concerns the relatively low humidity, and another the feeding pressure of herbivores. As in all chemical phenomena at the water/land interface, cinnamates are involved. Resistance to desiccation is provided by the water-impervious coating with "waxes" containing cutin in the aerial parts and suberin in the underground parts and at wound surfaces. Both

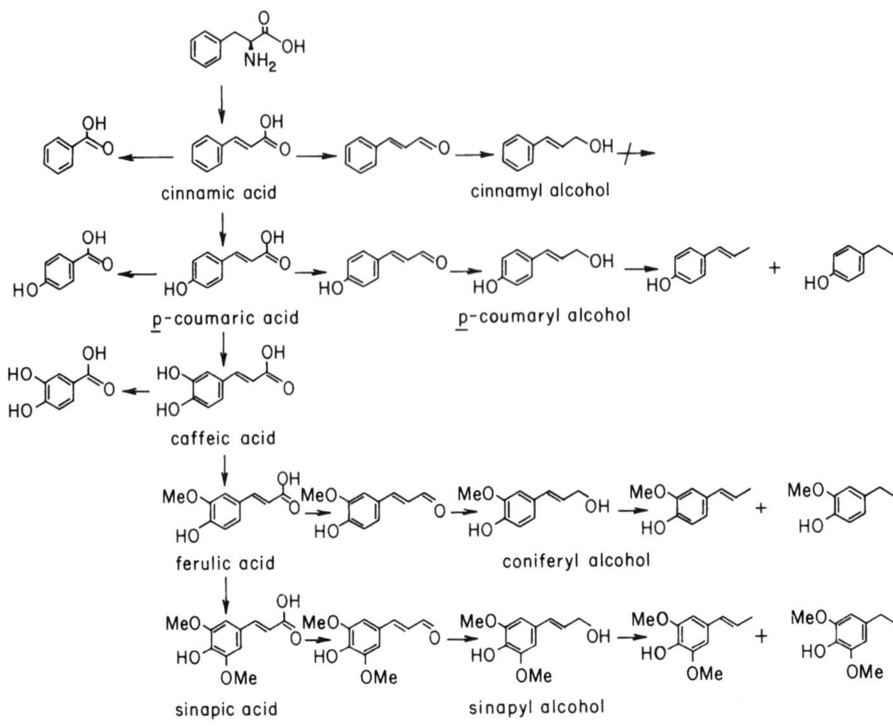

Fig. 3.5. Post-amino acid steps of the shikimate pathway

materials, cutin and suberin, are polymeric *p*-coumarates and ferulates of ω-hydroxy C_{16}- and C_{18}-fatty acids, which additionally constitute important barriers between the plants and their environments (47).

The aqueous habitat of algae not only confers protection against solar radiation, desiccation, and to some extent insect predation, but also favors buoyancy and contact of all organs with water and nutrients. This avoids the difficulties inherent in growth on dry land: the possibility of suffocation and shading by a horizontal mass of organisms, and the necessity of conducting water and soluble nutrients to organs. Resistance to gravity and opening of tracheids and vessels require the acquisition of rigidity. Algae are flexible as a consequence of their chemical composition. Sugars and amino acids can only form flexible macromolecules because the monomers are linked through heteroatoms (oxygen in cellulose and in pectins, major cell wall components of algae and of land plants, nitrogen in protein) with nonbonding orbitals. Sugars and amino acids cannot be polymerized by linkage of C-C bonds and cross links that would impart rigidity. New pathways leading to rigid polymers were required in land plants. It was first observed by Bate-Smith and Metcalfe (5) that lignins, the polymers that harden the cellulosic tissue of vascular plants, and condensed tannins, the polymers that function as feeding deterrents, frequently occur together. Indeed, the syntheses of both materials require reductive power (Figs. 3.7 and 3.8). Besides condensations (e.g., to flavonoids), reductions are also activated by the thiol ester function of

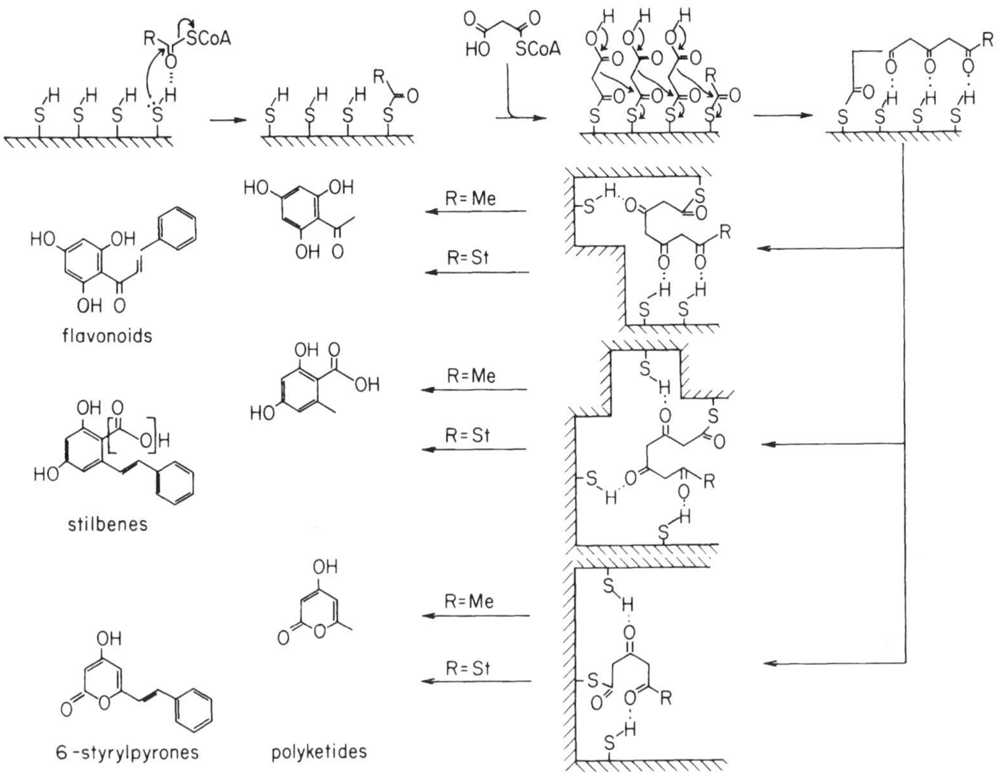

Fig. 3.6. Idealized biosynthetic scheme [adapted in part from ref. (52)] for some acetate-derived [chain initiation by acetate (R = methyl represented by Me), chain continuation by malonate] and for some cinnamate-acetate-derived [chain initiation by cinnamate (R = styryl represented by St), chain continuation by malonate] classes of secondary metabolites. The enzyme surface, through its thiol terminals, is postulated to bring the reactants into appropriate vicinity for chain formation. Specific forms of the enzyme surfaces are idealized to lead either to flavonoids, stilbenes, and 6-styrylpyrones or to three types of polyketides according to the nature of R

cinnamoyl-CoA. The resulting cinnamyl alcohols (Fig. 3.5) polymerize oxidatively into lignins (Figs. 3.8 and 3.9).

It is instructive to consider that the reductive sequences must be absent from mosses. Not only do these plants lack proanthocyanidins (79), but the structure of the lignin-like compound that encrusts the cell walls of *Sphagnum* is based on a cinnamic acid (82).

3.4 Phytochemistry and Plant Defense

Organisms are subject to attack by predators and may produce allelochemicals for protection. Algae, for example, contain defensive polyketides and sesquiterpenoids. The former are generated by the condensation of acetate units, a process

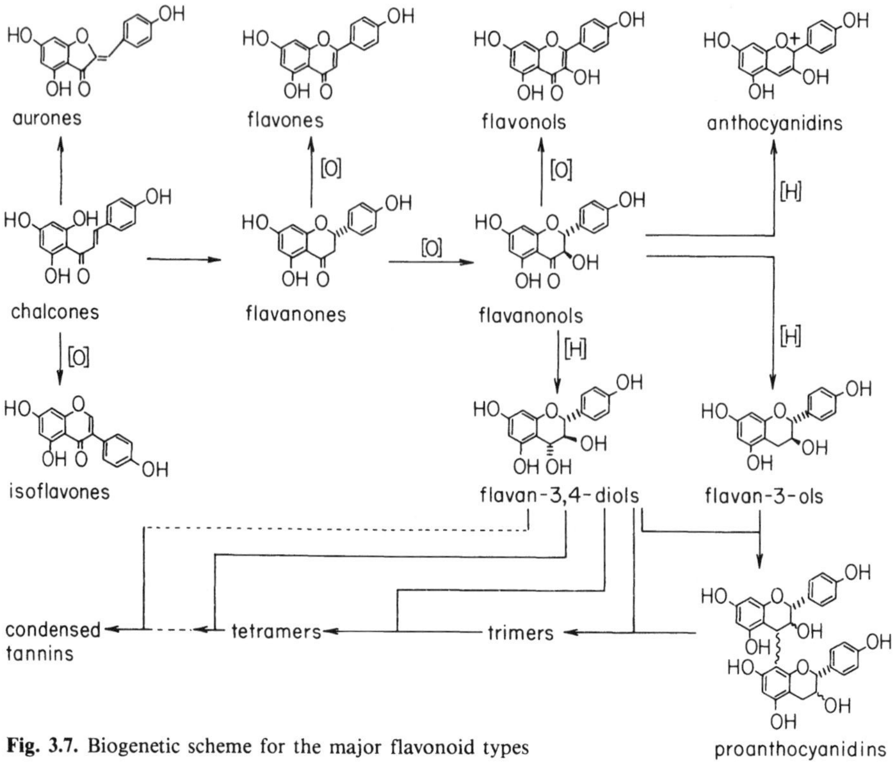

Fig. 3.7. Biogenetic scheme for the major flavonoid types

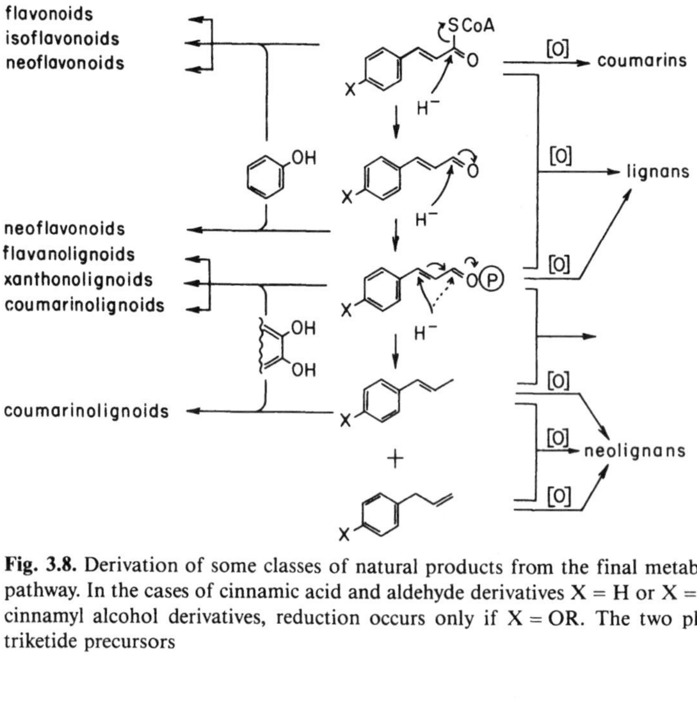

Fig. 3.8. Derivation of some classes of natural products from the final metabolites of the shikimate pathway. In the cases of cinnamic acid and aldehyde derivatives X = H or X = OR. In the case of the cinnamyl alcohol derivatives, reduction occurs only if X = OR. The two phenolic units represent triketide precursors

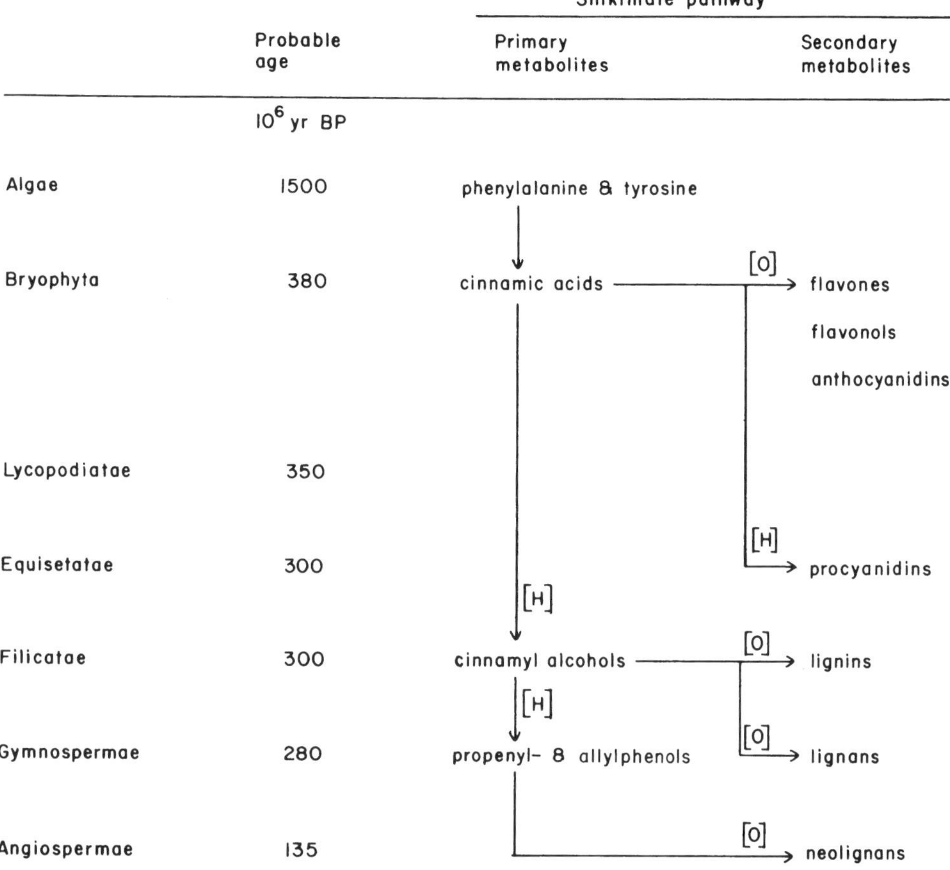

Fig. 3.9. Principal groups of living green organisms, listed from top to bottom according to their probable geological time of origin (78), and their major classes of shikimate-derived metabolites featured cumulatively from top to bottom – e.g., lignins occur in Filicatae, Gymnospermae, and Angiospermae. Note that some extinct representatives of Lycopodiatae and Equisetatae were strongly lignified

that leads primarily to fatty acids, and the latter are generated by the condensation of mevalonate units, a process that leads primarily to steroids. The products of secondary metabolism, structurally very variable micromolecules, supposedly with an ecological role, arise by extension of the primary routes to micro- and macromolecules of intrinsic value to their producer. Analogously, in ligneous plants, not only is the shikimate pathway used in the synthesis of cinnamyl alcohols, precursors of the lignins whose intrinsic value was examined above, but the evolutionary sequence continues with the transformation of post-tyrosine products into allelochemicals. This connection between intrinsic and ecological function (29) has clear evolutionary connotations. Thus in the most ancient *surviving* tracheophytes (Fig. 3.9), Lycopodiatae (lycopods) and Equisetatae (horsetails), in which lignification is absent or very slight, common secondary metabolites consist in the lysine- and acetate-derived *Lycopodium* alkoloids and in the mevalonate-derived steroid saponins, respectively, as well as in nicotin. The flavonoid theme in Lycopodiatae is expressed by flavones and bisflavones (as in Gymnospermae), whereas in Equisetatae mainly flavonols and only secondarily flavones abound. Here proanthocyanidins start to appear, whereas in Filicatae (ferns) over 92% of all examined genera contain them (79). Ferns also show considerable lignification. Their major and highly characteristic allelochemical groups nevertheless are acetate-derived methylenebisphloroglucinols and mevalonate-derived sesquiterpenoid indanones and triterpenoid ecdysones.

In conifers, the presence of condensed tannins, detected in 74% of the examined genera, remains important, and lignification of tissue attains its maximum. Mevalonate-derived products, such as mono- and sesquiterpenoids in essential oils, diterpenoids in balsams and resins, and triterpenoids such as ecdysones, play an important allelochemical role. In addition, flavones, including bisflavones and C-glycoflavones, a few flavonols (including myricetin), and stilbenes are widespread in conifers. Flavonoid formation preceded that of lignins evolutionarily, hence flavonoids have no direct connection with the woody habit, being simply characteristic of land plants. A direct link between lignification and secondary metabolism is established by the involvement of cinnamyl alcohols in lignin formation, but also by the synthesis of micromolecules, either through their oxidative transformation into lignans or through further reduction along the primary shikimate pathway to propenyl- and allylphenols (Fig. 3.8). Gymnosperms thus attained a climax in the expansion of the shikimate pathway and one is entitled to wonder why this broader synthetic capacity was so badly exploited for micromolecular variation. True, we have no chemical knowledge of now extinct but formerly such important gymnospermous groups as seed ferns, Nilssoniales, and Bennettitatae, and can only extrapolate data for surviving taxa. Considering their vast geographical distribution, their long period of domination of the land flora (63) or even only their considerable body size, the number of known biosynthetic classes of secondary metabolites from gymnosperms is, relative to angiosperms, exceedingly small. The probable reason for this chemical difference between gymnosperms and the equally highly lignified (with exception of the more advanced taxa) angiosperms lies in their defense mechanisms.

It is widely agreed that selection of secondary metabolites by, and coevolution with, phytophagous and pathogenic agents are the prime factors in shaping the diversity of natural products (21, 64). However, the major defense mechanism in pteridophytes and gymnosperms is still unspecialized, based on lignins and condensed tannins, which make plant tissue – carbohydrates and proteins – mechanically and enzymatically indigestible. In contrast, the defenses in angiosperms became more specialized. Small amounts of toxic micromolecules gradually replaced condensed tannins, detected only in 54% of the examined genera, and lignins. The unspecialized defense of pteridophytes and gymnosperms can only be breached with difficulty. In the case of the specialized defense in angiosperms, however, selection can lead to the introduction of detoxification mechanisms for action on particular micromolecules. Selection entails a shift first from one compound to another within the same biosynthetic group of secondary metabolites and, after exhaustion of the possibilities in this group, to another (15, 49). This replacement of general defense by specialized defense is exemplified by the positive relationship of diminishing lignosity and content of condensed tannins with benzylisoquinoline alkaloid diversity in the families belonging to the Magnoliiflorae-Ranunculiflorae group of superorders (33).

Thus we may arrive at the postulate that the enormous diversity of angiospermous micromolecules is due to interaction with (as opposed to simple repulsion of) pathogens and herbivores. The result of the frequently mutualistic association of flowering plants mostly, but not exclusively, with insects was the explosion of natural products chemistry that conditioned both biological systems to dominate the earth for the last 100 million years. Nevertheless, we have not yet discussed how this chemical diversification came to pass: was it haphazard or did it follow any recognizable trend? There is only one practical approach to an answer to this question: the inspection of the structures of the several tens of thousands of known natural compounds in relation to recognized morphologically homogeneous taxa such as Dahlgren's superorders (18). Analysis of the data reveals micromolecules of flowering plants to be endowed with two dynamic characteristics: a broad range of oxidation-reduction levels and the derivation from the initial (Fig. 3.3), intermediate (Fig. 3.4), and final (Figs. 3.5 and 3.8) primary metabolites of the shikimate pathway. Our theoretical working hypotheses on such matters (33) will now be examined in the light of comparative phytochemical evidence.

3.5 Oxidation Levels of Angiospermous Micromolecules

Secondary metabolites such as flavonoids, lignoids, polyketides, terpenoids, steroids, and alkaloids found in gymnosperms may also occur in angiosperms, where they are frequently accompanied by more highly oxidized, and occasionally by less highly oxidized, derivatives of the same biogenetic group. This observation was quantified for flavonoids. Averaging the data for the oxidation patterns of all known flavonoids, both the triketide and the cinnamate-derived moieties appear at conspicuously lower oxidation states for gymnospermous than for angiospermous compounds. Practically all known flavonoids of pteridophytes and gymnosperms have 5, 7-dioxygenated A-rings, suggesting the direct condensation of a cinnamate unit with the triketide precursor to have taken place. The presence of additional oxygen atoms, introduced by oxidation at carbons 6 or 8, or both, is common only in angiosperms (and in Bryophyta). The cinnamate unit of flavonoids in pteridophytes and gymnosperms is predominantly oxygenated at 4′ and to a progressively lesser extent at 3′, 4′ and at 3′, 4′, 5′. The analogous unit in angiosperms is commonly oxygenated at 3′, 4′ and to a progressively lesser extent at 4′ and 3′, 4′, 5′. The well known high relative contribution of coniferyl (4-hydroxy-3-methoxycinnamyl) alcohol versus sinapyl (4-hydroxy-3, 5-dimethoxycinnamyl) alcohol in the synthesis of fern and conifer versus flowering plant lignins (69) is viewed as further evidence for the higher oxidation level in angiosperms with respect to cinnamate-derived moieties. The Chlamydospermae are considered to occupy an intermediate phylogenetic position among the gymnosperms and the angiosperms. Indeed the lignin types of *Welwitschia* and the gymnosperms are closely related; the same is true for *Ephedra, Gnetum*, and the angiosperms. Moreover, lignans from gymnosperms stem commonly from hydroxymethoxycinnamyl alcohol, whereas in the angiosperms lignans and neolignans are frequently derived from hydroxydimethoxycinnamyl alcohol as well.

With regard to polyketides, accumulation of highly oxidized derivatives (e.g., polyacetylenes) is restricted to angiosperms. Terpenoids, as highly oxidized as the monoterpenoid iridoids, the sesquiterpenoid lactones, the diterpenoid tanshinones, royleanones and coleones, the triterpenoid cucurbitacins and limonoids, and the steroidal withaferins, are all known only from the angiosperms. The formation of indole (tryptophane-monoterpenoid) alkaloids as well as of benzylisoquinoline alkaloids involves oxidative steps and is also mostly restricted to the angiosperms. The transformation of the ubiquitous protein amino acids into cyanogenic glycosides also requires oxidative steps. Thus it is not surprising that the cyanogens of pteridophytes and gymnosperms should derive only from phenylalanine and tyrosine respectively (41), while the cyanogens of angiosperms derive additionally from several other precursors such as valine, isoleucine, and leucine.

The evidence seems to suggest that the origin of the angiosperms coincided with a substantial amplification of the oxidation-reduction potential of enzymatic

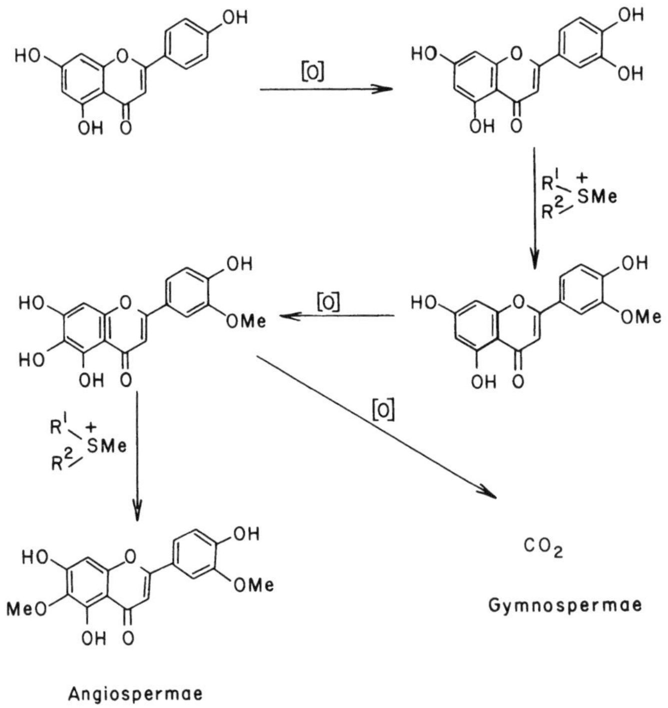

Fig. 3.10. Gymnosperms are endowed with a sole type of *O*-methyltransferase. This was shown to act exclusively on the 3′-hydroxyl (45). Any additionally hydroxylated flavonoid thus may possess a very short half-life. In angiosperms, other *O*-methyltransferases direct *O*-methylation to other hydroxyls, protecting the flavonoids against oxidative degradation

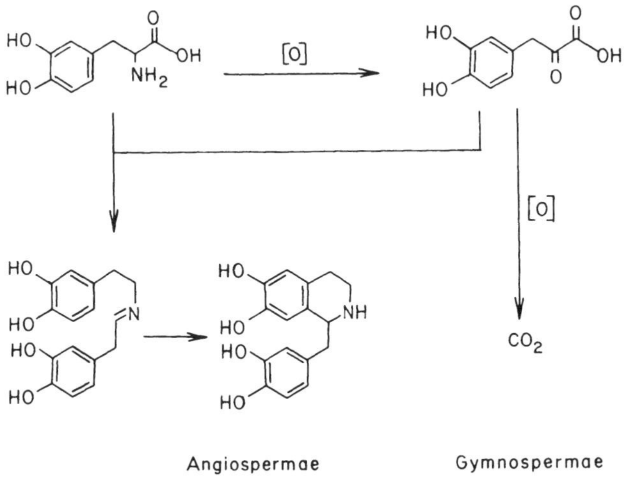

Angiospermae Gymnospermae

Fig. 3.11. In gymnosperms, arylpyruvates may suffer oxidative degradation. In angiosperms, condensation of arylpyruvates with amino acids leads to presumably more stable benzylisoquinoline alkaloids

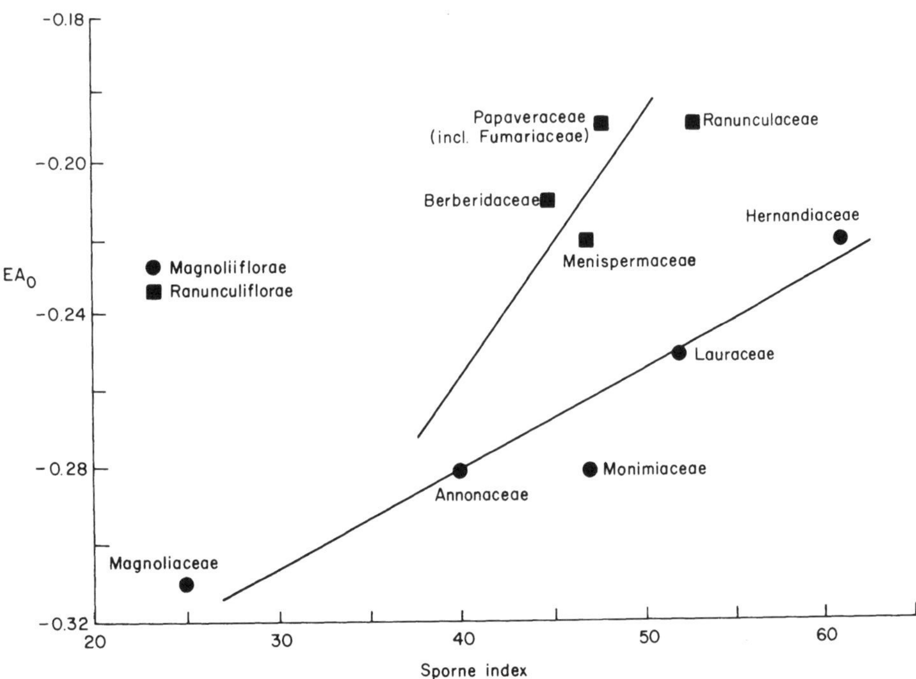

Fig. 3.12. Correlation of the evolutionary advancement parameters relative to the oxidation levels (EA_0) of benzylisoquinoline alkaloids (Barreiros, Kaplan, Gottlieb unpublished data) and Sporne's (76) evolutionary advancement indices of families belonging to the superorders Magnoliiflorae (circles) and Ranunculiflorae (squares) [sensu Dahlgren (18)]. The EA_0 values represent the weighted means of the oxidation levels of all benzylisoquinoline alkaloids known to occur in the family. The genus is taken as the fundamental taxonomic unit – i.e., the oxidation level of one compound present in *n* genera enters the list of oxidation levels *n* times in order to heighten the significance of the breadth of the compound's distribution

catalysts for the formation of micromolecules. Alternatively, however, gymnosperms may also produce highly oxidized molecular species. Since such compounds are rarely isolated, they must have a very short half-life. Indeed, with respect to flavonoids, gymnosperms possess only one protective device: An *O*-methyltransferase that acts specifically on the hydroxyl at position 3′ (45) (Fig. 3.10). Angiosperms possess this and other regiospecific *O*-methyl transferases. Similarly, amines and carbonyl compounds may be easily degraded in gymnosperms. Stabilization of these classes of metabolites via condensation to imines at the gymnosperm/angiosperm interface may explain why only the latter division of plants produces alkaloids in abundance (Fig. 3.11).

But what about the angiosperms themselves? Does the origin of new evolutionary lines within this plant division follow the same trend? Indeed, this seems to be the case for taxa of higher hierarchic levels – e.g., superorders, possibly also orders,and sometimes even families (e.g., Fig. 3.12). Just as for the origin of the angiosperms themselves, the radiation of each major plant group coincided with an abrupt amplification of the oxidation level of the compounds of one of its usually highly diversified biosynthetic groups of secondary metabolites. In contrast, correlation of morphological evidence for taxa of lower hierarchic levels, such as genera or species, with some micromolecular groups indicates the reversal of the trend – i.e., within each lineage of plants, evolutionary diminution of the oxidation level may occur. The line between "higher" and "lower" levels is of course not always easy to draw and will vary from taxon to taxon. These concepts were deduced upon consideration of flavonoids (9, 32, 35), benzylisoquinoline alkaloids (27, 30), indole alkaloids (8), quinolizidine alkaloids (65, 66, 67, 68), iridoids (46), sesquiterpene lactones (22, 23), polyacetylenes (26), gallic and ellagic acids (87), quassinoids and limonoids (72, 73), neolignans and stryrylpyrones (36, 37), flavonoids, quinolizidines, and non-protein amino acids (31), as well as the blue colored complex of flowers (38, 39). Nevertheless before the evidence can be summarized it is necessary to refer to another dynamic micromolecular character, namely skeletal specialization.

3.6 Skeletal Specialization of Angiospermous Micromolecules

Amplification of the oxidation-reduction potential is not the only cause for diversity of secondary metabolites in angiosperms. In order to understand the phenomenon more fully it is necessary to remember also the best known of the evolutionary trends: the gradual substitution of woody forms by herbaceous ones that operate in several lineages. Reduced utilization for the production of lignins and condensed tannins should cause initially a surplus of cinnamate, raw material for the biosynthesis of allelochemicals of primitive angiosperms. The continuing decrease in the importance of the shikimate pathway and the connection between primary and secondary metabolism requires that the transition from the woody to the herbaceous habit be accompanied in the long run by a curtailment of shikimate precursors for the biosynthesis of allelochemicals. Let us look now into the evidence for these postulates.

Table 3.1 registers the occurrence of biogenetic groups of compounds, considered to be chemosystematically relevant, in superorders of angiosperms. The first four superorders of dicotyledons, jointly with the monocotyledons, form the magnolialean block and the remaining dicotyledonous superorders form the rosiflorean block [sensu Kubitzki and Gottlieb (34, 48)]. The underlying basic concept that the deepest disjunction in the angiosperms does not lie between the monocots and the dicots was pioneered on chemical grounds (presence/absence of gallo- and ellagitannins) by Bate-Smith (2, 3) and forcefully driven home on morphological grounds by Huber (43, 44). In Table 3.1, the superorders are juxtaposed as far as possible in chemically similar groups. The position of a superorder within its group obeys, from top to bottom, the sequence of increasing values of mean morphological advancement indices for families. With the inherent limitations of a linear arrangement, the groups are placed in increasing order of the morphological advancement indices of their superorders.

For shikimate-derived biogenetic classes distinction is made in Table 3.1 between "inherited", "general", and "special" characters. The first one, represented by lignins, condensed tannins and lignans, occur in most angiosperms, just as in all possible ancestral tracheophytes. "General" characters, such as neolignans, tyrosine-derived alkaloids, and cyanogens, as well as pyrones, though accumulated chiefly in the magnolialean block, may also appear sporadically in the rosiflorean block. Inversely, the typical micromolecular classes of the rosiflorean block – including not only the "special" shikimate-derived classes, such as gallic acid derivatives, phenylalanine-derived cyanogens, anthranilate-derived alkaloids, indole alkaloids, but also tropane alkaloids, pyrrolizidine alkaloids, glucosinolates, iridoids, limonoids, cucurbitacins, etc. – occur not at all or only extremely rarely in the magnolialean block.

Further evidence revealing the advanced evolutionary status of the rosiflorean group concerns biosynthetic pathways. Simple pathways, involving the acetate, mevalonate, amino acid, or shikimate routes operate separately in the biosynthesis of individual micromolecules in all possible ancestors and in both angiospermous blocks. Apart from polyketides (condensation of acetates) and terpenoids (condensation of mevalonate-derived units), this primitive condition is exemplified by lignans (two cinnamyl alcohols), neolignans (two propenyl- or allylphenols), bisflavonoids (two flavones), methylenebisphloroglucinols (two methylated phloroglucinols), terphenyl derivatives, benzylisoquinoline alkaloids, betacyanins (two phenylalanines or tyrosines in all three), and *Calycanthus* alkaloids (two tryptophanes). Mixed pathways, involving combinations of these routes (for instance, the shikimate plus the mevalonate route or the mevalonate plus the amino acid route), which are relatively rare in the synthesis of individual compounds in pre-angiosperms and the magnolialean block, are common in the rosiflorean block. Betaxanthins (tyrosine plus ornithine), anthranilate-derived alkaloids (anthranilate plus, alternatively, polyketides, acetate and mevalonate, phenylalanines, or proline), esters of angelic, tiglic, or other branched acids with four to seven carbon atoms (amino acids plus alcohols based on terpenoids, pyrrolizidine alkaloids, etc.), prenylated phenolics (mevalonate-derived units plus coumarins, acetophenones, flavonoids, etc.), pyrrolizidine alkaloids (ornithine plus acetate), and indole alkaloids (tryptophane plus mevalonate) exemplify these advanced situations.

Table 3.1. Diversification[a] of secondary metabolites in superorders of angiosperms characterized by the mean Sporne indices (86) for their families

Compounds	Magnolialean Block					Rosiflorean Block																			
	Mag	Nym	Ran	Car	Monocots	Ros	Pod	Myr	Pro	Mal	The	Vio	Pri	Fab	Pol	Rut	San	Ara	Ast	Bal	Sol	Cor	Loa	Gen	Lam
Sporne index	43	57	46	54	–	42	48	51	57	41	42	47	53	48	52	52	56	57	61	67	60	50	64	65	72
Shikimate-derived compounds																									
Inherited																									
lignins	V	R	C	C	R	V	–	V	–	V	–	–	–	V	–	V	–	C	R	–	R	V	–	C	C
condensed tannins	V	C	C	R	C	V	–	C	–	V	–	–	–	V	–	V	–	–	–	–	–	C	–	–	–
lignans	C	C	C	–	–	R	–	R	–	C	–	R	–	–	–	V	R	C	C	–	–	C	–	V	V
General																									
neolignans	V	R	–	–	R	–	–	R	–	–	–	–	–	–	–	C	–	R	–	–	–	–	–	–	R
benzylisoquinoline alkaloids	V	R	V	–	R	R	–	–	–	C	–	–	–	C	–	C	R	R	R	–	R	R	–	R	–
other phenylalanine-derived alkaloids	–	–	R	C	V	–	–	–	–	–	–	–	–	–	–	–	–	–	–	–	–	R	–	–	–
betalains	–	–	–	C	–	–	–	–	–	–	–	–	–	–	–	–	–	–	–	–	–	–	–	–	–
tyrosine-derived cyanogens	C	C	C	C	C	R	–	–	–	R	–	–	–	R	–	R	R	–	–	–	R	R	–	R	–
Special																									
gallo- and ellagitannins	–	R	–	–	–	V	V	V	–	V	V	C	V	R	V	C	C	R	–	–	–	C	–	–	–
myricetin	R	R	R	–	–	V	–	V	V	V	V	C	V	C	V	C	C	–	–	–	–	C	–	–	–
isoflavonoids	R	–	–	R	R	R	–	–	–	R	–	–	–	V	–	–	–	–	R	–	–	–	–	–	–
phenylalanine-derived cyanogens	–	–	–	–	R	C	–	C	–	–	–	–	R	R	–	R	R	–	R	–	–	–	–	R	R
anthranilate-derived alkaloids	–	–	–	–	–	–	–	–	–	–	–	–	–	–	–	V	–	–	–	–	–	–	–	–	–
Acetate-derived compounds																									
esters of C_4-C_7 branched acids	R	–	–	–	–	–	–	–	–	R	–	–	R	–	–	C	R	C	V	–	C	C	–	R	R
prenylated compounds	–	–	C	–	C	R	–	–	V	C	V	R	–	V	–	V	–	V	V	–	C	–	–	C	C
iridoids	–	–	–	–	–	R	–	–	–	–	–	–	–	–	–	–	–	–	–	–	–	V	V	V	V
indolo-iridoid alkaloids	–	–	–	–	–	–	–	–	–	–	–	–	–	–	–	–	–	–	–	–	–	–	–	V	–
steroid saponins	–	–	–	–	V	–	–	–	–	–	–	–	–	R	–	C	–	–	R	–	C	R	–	–	C
sesquiterpene lactones	R	–	–	–	–	–	–	–	–	–	–	–	–	–	–	–	–	–	V	–	–	–	–	–	–
polyacetylenes	R	–	–	–	–	–	–	–	–	R	–	–	–	R	–	R	C	C	V	–	–	–	–	R	R

[a] V = very common; C = common; R = rare

Tentatively, given the paucity of chemical data for some of the superorders, three groups can be discerned in Table 3.1: The magnolialean group for which shikimate-derived compounds of the "inherited" and the "general" types predominate vastly over the shikimate-derived compounds of the "special" type, as well as over the acetate-derived compounds; one rosiflorean group in which shikimate-derived compounds of the "special" type (mostly gallo- and ellagi-tannins) seem to repress the formation not only of the shikimate compounds of "general" type, but also of the acetate-derived compounds; and another rosiflorean group in which acetate-derived compounds predominate vastly over shikimate-derived compounds of the "general" and the "special" (gallo- and ellagi-tannins absent!) types.

Within this perspective, the data of Table 3.1 are consistent with the theory above, according to which the shikimate pathway was envisaged to have reached its extreme primary metabolites (Figs. 3.5 and 3.8) prior to the origin of the flowering plants (Fig. 3.13). With the amplification of the oxidation potential of the plant cell oxidative dimers of these allyl- and propenyl-phenols, the neolignans (Fig. 3.8) were formed in Magnoliiflorae, generally considered to represent the

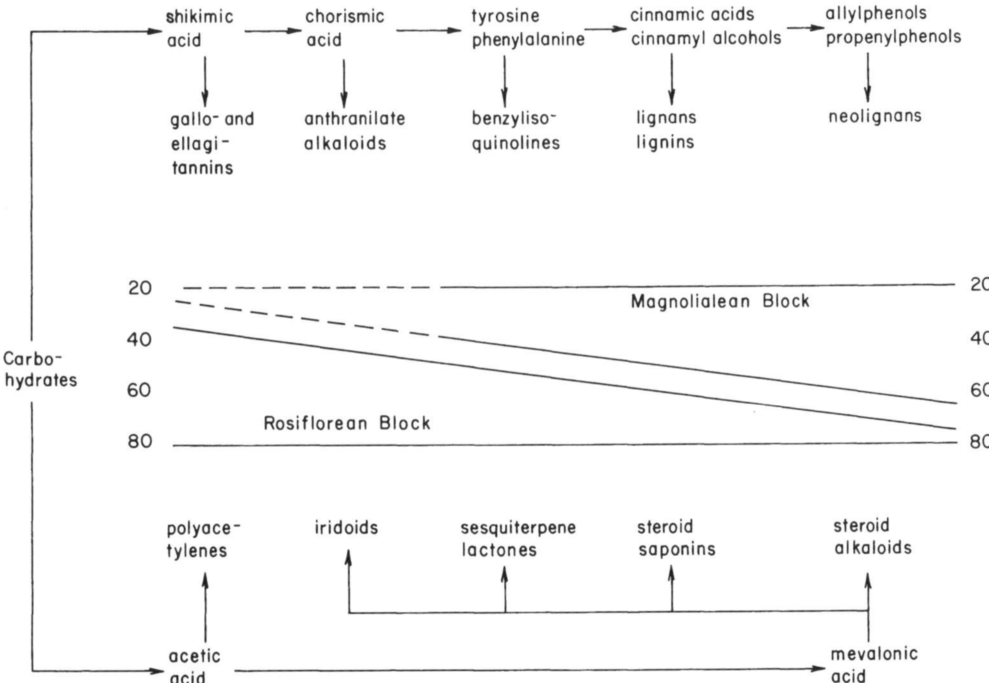

Fig. 3.13. Biosynthetic derivation and occurrence in the two blocks of angiosperms of systematically significant biogenetic groups of micromolecules – e.g., neolignans are common in the more primitive (as featured by the scale of Sporne indices) magnolialean plant groups; gallo- and ellagi-tannins are more common in the primitive rosiflorean groups; and steroid alkaloids are more common in the advanced rosiflorean groups

most primitive angiosperms. The abbreviation of the final steps of the shikimate pathway would result in the accumulation of phenylalanine and tyrosine (Figs. 3.4 and 3.5). The interaction of such metabolites, again through reactions involving oxidation, leads to benzylisoquinolines, presumably the first alkaloidal types of angiosperms to condition their co-evolution with herbivores and especially with mammals (21, 64, 78). The biosynthetic complexity of these alkaloids increases from the simplest aporphines of Magnoliiflorae in three directions to the berberines, morphines, and thebaines of Ranunculiflorae, to the phenylethylisoquinolines and the benzyl phenylethyl amines of Liliiflorae, and to the betalains of Caryophylliflorae (which also produce mescalines and β-carbolines). Further shortening of the shikimate pathway with respect to the synthesis of secondary metabolites leads to the accumulation of 5-dehydroshikimic acid (Figs. 3.4), the precursor of gallic and ellagic and acid-derived tannins, and of chorismic acid (Fig. 3.3 and 3.4), the precursor of anthranilate and of tryptophane-derived alkaloids.

With gallic acid, the possibilities of diversifying the production of allelochemicals through gradual curtailment of the shikimate pathway seem to be exhausted, and a switch-over to the exploration of the acetate pathway occurs in the most highly developed superorders of the rosiflorean block (Fig. 3.13). This statement refers to the metabolites of the shikimate pathway as precursors of micromolecules. A complete suppression of the shikimate pathway does not occur because the aromatic amino acids continue to be produced. Besides, cinnamate or cinnamyl alcohol surplus from reduced consumption for the production of lignins may account for a certain abundance of lignans (Table 3.1), coumarins, and even flavonoids, although the latter are subjected to selective forces different from those governing the presence of other secondary metabolites in herbaceous forms. One instructive case in point is provided by the Araliales, which contain the closely allied families Araliaceae and Apiaceae. While the latter are mainly herbaceous and make ample use of cinnamate in the synthesis of coumarins, no such substances have been detected so far in the predominantly woody counterpart, the Araliaceae. Nevertheless, in these and especially in even more highly advanced angiosperm groups, shikimate-derived secondary metabolites play a relatively minor role. In these lineages the full potential of mevalonate utilization has led to steroid alkaloids, iridoids, iridoid alkaloids, and sesquiterpene lactones, while the utilization of acetate has led to polyacetylenes.

Thus gradual curtailment of the shikimate pathway for the production of primary precursors to biosynthetic groups of secondary metabolites constitutes a general trend observable through the analysis of angiosperm taxa of high hierarchic rank. It is superimposed within each biogenetic set of compounds, and hence in the case of taxa of lower hierarchic rank, by a diversificatory trend compatible with the enhanced biosynthetic power emanating from the increase in oxidation-reduction potential.

3.7 Quantification of Micromolecular Parameters

The skeletal diversification of compounds can be quantified by a topological method. In its present form the method consists of counting for each compound the number of bonds that must be established and broken upon relating the compound to the general precursor of the compound's biosynthetic group. An example of the concommitant correlation of both statistical parameters (oxidation level and skeletal specialization) to a phylogenetic problem follows.

Oxidation levels and skeletal specializations of sesquiterpene lactones were calculated and used in the determination of evolutionary advancement parameters EA_0 (relative to oxidation of the compounds) and EA_s (relative to specialization of the compounds) for tribes of Asteraceae. The resulting plot (Fig. 3.14), though compatible with Wagenitz's clustering of these tribes on morphological grounds (83), originally lacked one piece of information: the evolutionary polarity. The problem was solved tentatively by calculation, for each of the tribes, of parameters revealing the relative importance of glycosylation versus methylation of their flavonoids. The rationale for the usefulness of these data hinges on con-

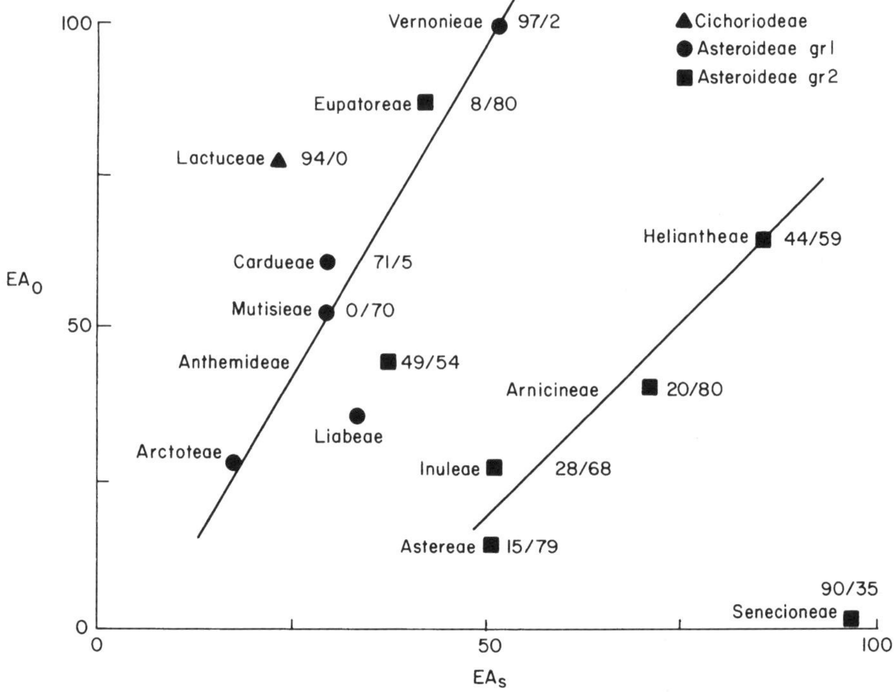

Fig. 3.14. Correlation of EA_0 and EA_s parameters (given in percentage of highest minus lowest observed values) for sesquiterpene lactones from tribes of Asteraceae. These tribes belong to the subfamilies Cichorioideae (triangle), Asteroideae group 1 (circles) and Asteroideae group 2 (squares) [sensu Wagenitz (83)]. Each tribe is characterized additionally by its EA_G/EA_M parameters expressing the percentage, with respect to the total number of reported flavonoids, of the number of *O*-glycosylated flavonoids versus *O*-methylated flavonoids

sideration of metabolic costs and effectiveness of deterrence against herbivores. Based on this concept the diversion of carbohydrates from their essential physiological role to the protection of the phenolic hydroxyl of secondary metabolites must indeed be a primitive feature.

Verification of the validity of this postulate is of great importance. Flavonoids are ubiquitous in plants and their structures and distribution have been studied extensively. They would indeed be excellent general markers of evolutionary polarity.

3.8 Phytochemical Gradients in Angiosperms

Natural products chemists have endeavored, during the past few decades, to contribute to the classification of plants. The potentialities of the approach were revealed by Erdtman (24) for gymnosperms and by Hegnauer (40) for angiosperms, to cite only two major exponents of the field. Indeed, chiefly on the level of the family and below, chemists have achieved some convincing successes (28) and morphologists have reluctantly (42) started to apply this old-new tool even to the family level and above. Three of the latest systems of angiosperm classification (16, 80, 81) pay some attention to chemical criteria for the placement of taxa, and the fourth one (18) relies more heavily upon them, especially for the realignment of many sympetalous families into "iridoid" and "sesquiterpene lactone/polyacetylene" blocks.

However, the presence or absence of compounds as classificatory criteria may be misleading. The complete range of precursors of natural products was present in the angiosperms from the start, and the facts – e.g., that benzylisoquinoline alkaloids and pyrones appear in Lauraceae, Rutaceae, and Asteraceae; that phenylethylisoquinoline alkaloids appear in Cephalotaxaceae, Liliaceae, and Aquifoliaceae, or that esters of angelic and tiglic acids appear in Schisandraceae, Euphorbiaceae, and Solanaceae (Table 3.1) – are not necessarily an indication of close relationship. Such direct application of micromolecular data is useful only on lower hierarchic level within taxa that are also morphologically homogeneous. Even in this favorable circumstance only versatility of structural variation confers on a biosynthetic group of micromolecules the status of a valid chemosystematic marker, lessening the ever-present danger of erroneous interpretations due to analogies produced by convergent evolution.

All this means, of course, that emphasis on the structure of accumulated metabolites must be shifted to something else. Birch (6) suggested considering alterations of biosynthetic pathways. However, it is usually impractical to inquire into the biochemical history of a constituent before daring to think about its significance, physiologically, systematically, ecologically or otherwise (Bate-Smith, 1983, personal communication). Even more seriously, no procedures exist to assess the meaning of such alterations with plant evolution (7). And finally we should not lose sight of phenetic definability (morphologic, micromolecular, etc.) as one of the goals in the organization of affinity groups into formal taxa (17). Phylogenetic classifications based on concealed evidence (amino acid sequence data for macromolecules, biosynthetic pathways for micromolecules) may deviate to a considerable extent from a phenetically useful classification (19).

Bate-Smith (3, 4) made the first step in the right direction by drawing attention to the dynamic distribution of gallo- and ellagi-tannins as well as of flavonols. The next step consists in the quantitative measurement of the evolutionary trends referred to above in order to demonstrate micromolecular continuities and disjunctions. In all our work up to 1982 (33) compounds were characterized by oxidation level (O) and by a parameter involving biogenetic pathways (S). While the O values were then and continue to be easily accessible, the S values (authentic "concealed evidence") were difficult to produce, more often than not requiring hypothetical decisions on biosynthetic sequence and tedious determinations of the frequency of the compound's occurrence. In consequence of this last requirement, the S values were dated and the frequency had to be re-calculated with each fresh report on the compound. In all our more recent papers, the determination of skeletal specialization (S) requires only the recognition of the compound's biogenetic affiliation, a simple calculation independent of prior or posterior registries of occurrence and the determination of the mean O and S values for the taxon. (Mean values of a numerical parameter are less prone to change with the incorporation of new data into the system, and thus produce more stable and meaningful results than extreme values.)

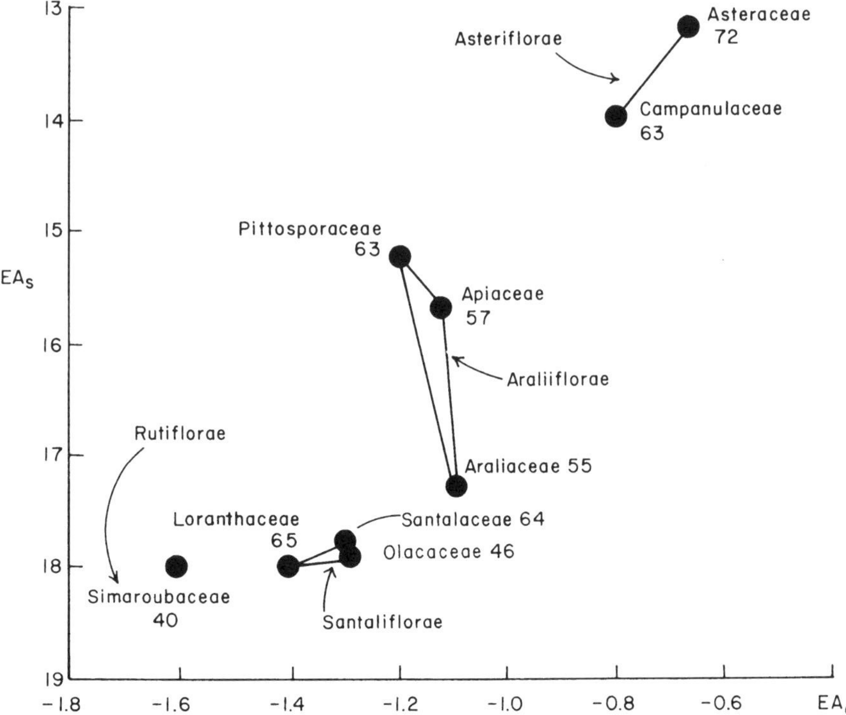

Fig. 3.15. Correlation of polyacetylene based EA_s and EA_0 values for families of dicotyledons characterized by their Sporne indices. The EA_s parameter for skeletal specialization is here expressed simply by the mean number of carbon atoms of the polyacetylene molecules

The existence of micromolecular gradients in Angiospermae with a life span of a mere 100 million years is no absurdity. Indeed the entire introduction of this chapter was dedicated to a demonstration of chemical continuity in organisms spanning several billions of years in permanently varying environments. Examples of such gradients based on micromolecular EA_0 and EA_s parameters have been reported for benzylisoquinoline alkaloids from Magnoliiflorae in several alternative directions to Ranunculiflorae, to Nymphaeiflorae, to Caryophilliflorae and to the monocotyledons (30); for quassinoids and limonoids from Simaroubaceae and Meliaceae to Rutaceae and Cneoraceae (72); for iridoids from Cornales (Corniflorae) to Gentianiflorae and from Ericales (also Corniflorae) to Lamiiflorae (46); for sesquiterpene lactones in angiosperms (22) and specially in Asteriflorae (23); and for polyacetylenes from Rutiflorae and Santaliflorae over Araliiflorae to Asteriflorae (26) (Fig. 3.15).

3.9 Future Perspectives

The search for chemical gradients must of course be continued in angiosperms and extended to other groups of organisms. The resulting chemical networks, superimposed on systems based on morphological features have been illustrated by a circular scheme (Fig. 3.16) of the type pioneered by Sporne for the representation of angiosperms (74). Such networks may become powerful tools in the selection of taxa with predictively useful novel chemicals or in ecological and ecogeographical studies. The latter subject merits an important comment. Small endemic plant populations often possess aberrant chemical composition. Thus, omitting the respective data from the analysis of the widespread representatives of the same taxon will lead to closer chemical/morphological correlations. This topic pertains to the study of chemical "replacement characters" in the same or closely related taxa, a fascinating though chemically little-known subject. Indeed there is no reason to believe that both the chemical and the morphological data sets always evolve in synchronous fashion. In one lineage of plant taxa – characterized by morphology or anatomy – trends in chemical compositions will be observed, and – vice versa – if chemical characters are used – only trends in morphological changes may be observed. At present our knowledge of botanical forms is overwhelmingly superior to our knowledge concerning botanical chemistry. Hence it is the former situation that prevails. However as early as 1966 Turner [cited in ref. (71)] was of the opinion that "there is certain to come a day when the present gap between organismally centered systematics and molecularly centered systematics becomes first a blur, then a conceptual framework from which will grow remarkable insights into the whole of biological evolution." Although certainly correct in outline, the verification of this assumption must wait for the development of appropriate methodology for the comparison of chemical and morphological data. Indeed part of the present chapter is dedicated to this important purpose.

Biologists usually claim that chemical data are precise and hence difficult to compare with morphological or, in general terms, biological data, assumed to be

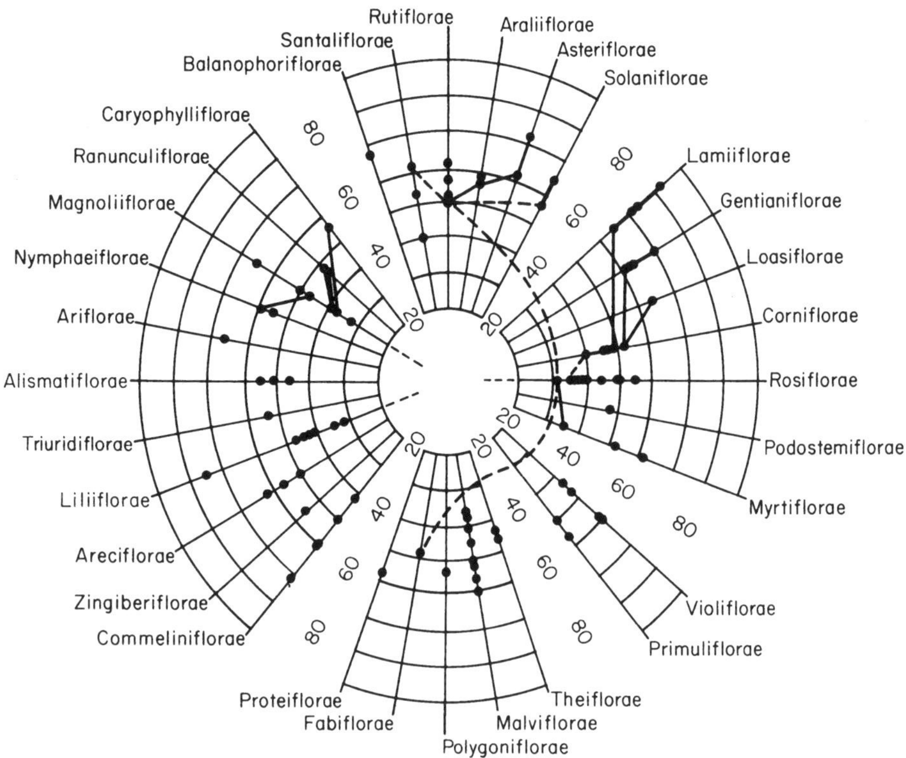

Fig. 3.16. Representation of the angiospermous magnolialean block (sector on the left) and rosiflorean block (four sectors on the right) with their superorders (rays) and orders (dots), the latter placed on a 0- to 100-point scale according to the mean Sporne indices of their dicotyledonous families (76) and the artificially modified (34) mean Sporne indices of their monocotyledonous families (74). The dots connected by solid lines represent orders for which the existence of chemical relationships, mostly in form of chemical gradients (increasing oxidation levels and skeletal specializations of the contained micromolecules), were recognized. Connections by broken lines are more hypothetical and refer in most cases to the frequent prenylation of phenolics. The broken lines in the center indicate possible accesses to the most primitive orders

much more variable and hence difficult to define (42). This point of view is debatable. For the great majority of cases, chemical data are based on the analysis, with varying proficiency, or one sample, while morphological data are based on the examination of many samples. It is predicted that when a comparable number of chemical analyses becomes available, chemistry of a plant group will appear to be as variable as morphology. We have already started to define taxa by the ranges of EA_0 and EA_s values rather than only by the averages or one of the extremes of these parameters.

Furthermore, as demonstrated by Sporne's work (74, 75, 76), the quantification of morphological data is possible at least for families of angiosperms. It is hoped that statistical approaches will become available for all hierarchic ranks and all plant groups. Indeed, according to Niklas (61), in order to simulate plant evolution, one must develop mathematical (i.e., statistical) techniques for quantifying the competitive advantages offered by various features. It is possible, for example, to quantify the factors involved in the trade-off between presenting large areas of photosynthetic tissue to the sun and bearing the resulting mechanical stresses.

In conclusion then, the path to lignification was superimposed in plants on the other metabolic routes about 400 million years before the present age (62). Having passed through a climactic stage about 150 to 100 million years ago, it shows unmistakable signs of an evolutionary diminution. How far this process would normally lead is irrelevant, since worldwide deforestation is being dramatically accelerated by man. Artificially planted forests may alleviate some of the consequent ecological and industrial problems but, because so very few species are involved, that will not slow down the loss of the vast majority of arboreal species. Now, as here envisaged, the processes of lignification and of production of natural products of woody plants are evolutionarily and biosynthetically connected, and hence we are witnessing at present the loss of invaluable chemical and biological models – models on which, lest we forget, scientific organic chemistry and hence the entire industrial age was and continues to be built.

Indeed, as this book demonstrates, the tropical rain forest, which harbors most woody plants, is a memory bank from which we have retrieved so far only a proportionally minute and highly fragmentary part of the contained information. In the words of Myers [cited in ref. (84)], if present patterns of converting the tropical rain forest persist, it may be the worst biological debacle since life's first emergence on the planet 3.6 billion years ago.

References

1 Anhut S, Zinsmeister H D, Mues R, Barz W, Mackenbrock K, Köster K, Markham K R 1984 The first identification of isoflavones from a bryophyte. Phytochemistry 23:1073–1075

2 Bate-Smith E C 1972 Chemistry and phylogeny of the angiosperms. Nature 269:353–354

3 Bate-Smith E C 1973 Systematic distribution of ellagitannins in relation to the phylogeny and classification of the angiosperms. In: Bendz G, Santesson J (eds) Chemistry in botanical classification. Academic Press London, 93–102

4 Bate-Smith E C 1984 Age and distribution of galloyl esters, iridoids and certain other repellents in plants. Phytochemistry 23:945–950

5 Bate-Smith E C, Metcalfe C R 1957 Leuco-anthocyanins. 3. The nature and systematic distribution of tannins in dicotyledonous plants. J Linn Soc Bot 55:669–705

6 Birch A J 1963 Biosynthetic pathways. In: Swain T (ed) Chemical plant taxonomy. Academic Press London, 141–166

7 Birch A J 1973 Biosynthetic pathways in chemical phylogeny. In: Bendz G, Santesson J (eds) Chemistry in botanical classification. Academic Press London, 261–270

8 Bolzani V da S, da Silva M F das G F, da Rocha A I, Gottlieb O R 1984 Indole alkaloids as systematic markers of the Apocynaceae. Biochem Syst Ecol 12:159–166

9 Cagnin M A H, Gottlieb O R 1978 Isoflavonoids as systematic markers. Biochem Syst Ecol 6:225–238

10 Cairns-Smith AG 1985 The first organisms. Sci Am 252(6):74–82

11 Carey F A, Sundberg R J 1984 Advanced organic chemistry. Part B: Reactions and synthesis. 2nd edn. Plenum Press New York, 570

12 Cech T R 1986 RNA as an enzyme. Sci Am 255(3):76–84

13 Chapman D J, Ragan M A 1980 Evolution of biochemical pathways: Evidence from comparative biochemistry. Ann Rev Plant Physiol 31:639–678
14 Clark R L, Stech T L 1979 Morphogenesis in Dictyostelium: an orbital hypothesis. Science 204:1163–1168
15 Cronquist A 1977 On the taxonomic significance of secondary metabolites in angiosperms. In: Kubitzki K (ed) Flowering plants. Evolution and classification of higher categories (Plant Syst Evol, Suppl 1). Springer Wien, 179–189
16 Cronquist A 1981 An integrated system of classification of flowering plants. Columbia University Press New York, 2162 pp
17 Cronquist A 1983 Some realignments in the dicotyledons. In: Ehrendorfer F, Dahlgren R (eds) New evidence of relationships and modern systems of classification of the angiosperms. Nord J Bot 3:75–83
18 Dahlgren R 1980 A revised system of classification of the angiosperms. J Linn Soc Bot 80:91–124
19 Dahlgren R 1983 The importance of modern serological research for angiosperm classification. In: Jensen U, Fairbrothers D E (eds) Proteins and nucleic acids in plant systematics. Springer Berlin Heidelberg New York Tokyo, 371–394
20 Eccles J C 1979 The human mystery. Springer, Berlin Heidelberg New York Tokyo, 57
21 Ehrlich P R, Raven P H 1965 Butterflies and plants: A study in co-evolution. Evolution 18:586–608
22 Emerenciano V de P, Kaplan M A C, Gottlieb O R 1985 Evolution of sesquiterpene lactones in angiosperms. Biochem Syst Ecol 13:145–166
23 Emerenciano V de P, Kaplan M A C, Gottlieb O R, Bonfanti M R de M, Ferreira Z S, Comegno L M A 1986 Evolution of sesquiterpene lactones in Asteraceae. Biochem Syst Ecol 14:585–589
24 Erdtman H 1973 Molecular taxonomy. In: Miller L P (ed) Phytochemistry. Vol III. Inorganic elements and special groups of chemicals. Van Nostrand Reinhold New York, 327–350
25 Fabian P 1984 Atmosphäre und Umwelt. Springer, Berlin Heidelberg New York Tokyo, 24
26 Ferreira Z S, Gottlieb O R 1982 Polyacetylenes as systematic markers in dicotyledons. Biochem Syst Ecol 10:155–160
27 Ferreira Z S, Gottlieb O R, Roque N F 1980 Chemosystematic implications of benzyltetrahydroisoquinolines in Aniba species. Biochem Syst Ecol 8:51–54
28 Gershenzon J, Mabry T J 1983 Secondary metabolites and the higher classification of angiosperms. In: Ehrendorfer F, Dahlgren R (eds) New evidence of relationships and modern systems of classification of the angiosperms. Nord J Bot 3:5–34
29 Gomes C M R, Gottlieb O R 1978 The evolution of structural biopolymers and secondary metabolites is connected ? Rev Bras Bot 1:41–45
30 Gomes C M R, Gottlieb O R 1980 Alkaloid evolution and angiosperm systematics. Biochem Syst Ecol 10:81–87
31 Gomes C M R, Gottlieb O R, Gottlieb R C, Salatino A 1981 Chemosystematics of the Papilionoideae. In: Polhill R M, Raven P H (eds) Advances in legume systematics, Part 2. Royal Botanic Gardens Kew, 465–488
32 Gomes C M R, Gottlieb O R, Marini-Bettòlo G B, Delle Monache F, Polhill R M 1981 Systematic significance of flavonoids in Derris and Lonchocarpus (Tephrosieae). Biochem Syst Ecol 9:129–147
33 Gottlieb O R 1982 Micromolecular evolution, systematics and ecology, an essay into a novel botanical discipline. Springer Berlin Heidelberg New York Tokyo, 170 pp
34 Gottlieb O R 1984 Phytochemistry and the evolution of angiosperms. Anais Acad Brasil Ciên 56:43–50
35 Gottlieb O R, Guajardo T E, Young M C M 1982 Evolution of flavonoids in Embryobionta. In: Fárkas L, Gábor M, Kállay F, Wagner H (eds) Flavonoids and bioflavonoids 1981. Elsevier Amsterdam, 227–244
36 Gottlieb O R, Kubitzki K 1981 Chemogeography of Aniba (Lauraceae). Plant Syst Evol 137:281–289
37 Gottlieb O R, Kubitzki K 1981 Chemosystematics of Aniba (Lauraceae). Biochem Syst Ecol 9:5–12
38 Gottsberger G, Gottlieb O R 1980 Blue flowers and phylogeny. Rev Bras Bot 3:79–83
39 Gottsberger G, Gottlieb O R 1981 Blue flower pigmentation and evolutionary advancement. Biochem Syst Ecol 9:13–18

40 Hegnauer R 1962–1973 Chemotaxonomie der Pflanzen. Vols 1–6. Birkhäuser Basel
41 Hegnauer R 1977 Cyanogenic compounds as systematic markers in Tracheophyta. In: Kubitzki K (ed) Flowering plants. Evolution and classification of higher categories (Plant Syst Evol, Suppl 1). Springer Wien, 191–209
42 Heywood V H 1973 The role of chemistry in plant systematics. Pure Appl Chem 34:355–375
43 Huber H 1977 The treatment of monocotyledons in an evolutionary system of classification. In: Kubitzki K (ed) Flowering plants. Evolution and classification of higher categories (Plant Syst Evol, Suppl 1). Springer Wien, 285–298
44 Huber H 1982 Die zweikeimblättrigen Gehölze im System der Angiospermen. Mitt Bot Staatssamml München 18:59–78
45 Ibrahim R K, De Luca V, Jay M, Voirin B 1982 Polymethylated flavonol synthesis is catalyzed by distinct *O*-methyltransferases. Naturwissenschaften 69:41–42
46 Kaplan M A C, Gottlieb O R 1982 Iridoids as systematic markers in dicotyledons. Biochem Syst Ecol 10:329–347
47 Kolattukudy P E 1980 Biopolyester membranes of plants. Science 208:990–1000
48 Kubitzki K, Gottlieb O R 1984 Micromolecular patterns and the evolution and major classification of angiosperms. Taxon 33:375–391
49 Kubitzki K, Gottlieb O R 1984 Phytochemical aspects of angiosperm origin and evolution. Acta Bot Neerl 33:457–468
50 Kuhn H, Waser J 1981 Molecular self-organization and the origin of life. Angew Chem Int Ed 20:500–520
51 Lewin R A 1981 *Prochloron* and the theory of symbiogenesis. In: Fredrick J F (ed) Origins and evolution of eukaryotic intracellular organelles. Ann New York Acad Sci 361:325–329
52 Manitto P 1981 Biosynthesis of natural products. Ellis Horwood Chichester, 176
53 Margulis L 1981 Symbiosis in cell evolution – Life and its environment on the Earth. Freeman, San Francisco, 419 pp
54 Margulis L, Schwartz K W 1982 Five kingdoms – An illustrated guide to the phyla of life on Earth. Freeman San Francisco, 338 pp
55 Markham K R, Porter L J 1978 Chemical constituents of the bryophytes. Prog Phytochem 5:181–272
56 Mayr R 1982 The growth of biological thought – diversity, evolution and inheritance. Harvard University Press Cambridge, 63–64
57 McClure J W 1975 Physiology and function of flavonoids. In: Harborne J B, Mabry T J, Mabry H (eds) The flavonoids. Academic Press London, 970–1055
58 McKey D 1980 Origins of novel alkaloid types: a mechanism for rapid phenotypic evolution of plant secondary compounds. Am Nat 115:754–759
59 Melkonian M 1982 Structural and evolutionary aspects of the flagellar apparatus in green algae and land plants. Taxon 31:255–265
60 Miller S L, Orgel L E 1974 The origins of life on the earth. Prentice-Hall Englewood Cliffs, 229 pp
61 Niklas K J 1986 Computer-simulated plant evolution. Sci Am 254(3):68–75
62 Niklas K J, Pratt L M 1980 Evidence for lignin-like constituents in early Silurian (Llandoverian) plant fossils. Science 209:396–397
63 Niklas K J, Tiffney B H, Knoll A H 1983 Patterns in vascular land plant diversification. Nature 303:614–616
64 Rhoades D F (1979) Evolution of plant chemical defense against herbivores. In: Rosenthal G A, Janzen D H (eds) Herbivores, their interaction with secondary metabolites. Academic Press New York, 3–54
65 Salatino A, Gottlieb O R 1980 Quinolizidine alkaloids as systematic markers of the Papilionoideae. Biochem Syst Ecol 8.133–147
66 Salatino A, Gottlieb O R 1981 A chemo-geographical perspective of the evolution of quinolizidine bearing Papilinoideae. Rev Bras Bot 4:83–88
67 Salatino A, Gottlieb O R 1981 Quinolizidine alkaloids as systematic markers of the Genisteae. Biochem Syst Ecol 9:267–273
68 Salatino A, Gottlieb O R 1983 Chemogeographical evolution of quinolizidines in Papilionoideae. Plant Syst Evol 143:167–174
69 Sarkanen K V, Ludwig C H 1971 Lignins. Occurrence, formation, structure, and reactions. John Wiley New York, 916 pp

70 Schopf J W 1978 The evolution of the earliest cells. Sci Am 239:110–138

71 Seaman F C 1982 Sesquiterpene lactones as taxonomic characters in the Asteraceae. Bot Rev 48:121–595

72 da Silva M F das G F, Gottlieb O R 1987 Evolution of quassionoids and limonoids in the Rutales. Biochem Syst Ecol 15:85–103

73 da Silva M F das G F, Gottlieb O R, Dreyer D L 1984 Evolution of limonoids in Meliaceae. Biochem Syst Ecol 12:299–310

74 Sporne K R 1974 The morphology of the angiosperms. The structure and evolution of flowering plants. Hutchinson University Library London, 207 pp

75 Sporne K R 1976 Character correlations among angiosperms and the importance of fossil evidence in assessing their significance. In: Beck C B (ed) Origin and early evolution of angiosperms. Columbia University Press New York, 312–329

76 Sporne K R 1980 A re-investigation of character correlations among dicotyledons. New Phytol 85:419–449

77 Suire C, Asakawa Y 1979 Chemotaxonomy of bryophytes: a survey. In: Clarke G C S, Duckett J G (eds) Bryophyte systematics. Academic Press London, 447–477

78 Swain T 1977 Secondary compounds as protective agents. Ann Rev Plant Physiol 28:479–501

79 Swain T 1979 Tannins and lignins. In: Rosenthal G A, Janzen D H (eds) Herbivores, their interaction with secondary metabolites. Academic Press New York, 657–682

80 Takhtajan A 1980 Outline of the classification of flowering plants. Bot Rev 46:226–359

81 Thorne R F 1981 Phytochemistry and angiosperm phylogeny, a summary statement. In: Young D A, Seigler D S (eds) Phytochemistry and angiosperm phylogeny. Praeger New York, 233–295

82 Tutchek R 1975 Isolation and characterization of the p-hydroxy-β-(carboxymethyl)-cinnamic acid (sphagnum acid) from the cell wall of Sphagnum magellanicum Brid. Z Pflanzenphysiol 76:353–365

83 Wagenitz G 1976 Systematics and phylogeny of the Compositae (Asteraceae). Plant Syst Evol 125:29–46

84 White P T, Blair J C 1983 Nature's dwindling treasures: Rain forests. Nat Geogr 163[1]:2–47

85 Whittaker R H 1969 New concepts of kingdoms of organisms. Science 163:150–160

86 Wolken J J 1975 Photoprocesses, photoreceptors and evolution. Academic Press New York, 317 pp

87 Wolter-Filho W, da Rocha A I, Yoshida M, Gottlieb O R 1985 Ellagic acid derivatives from Rhabdodendron macrophyllum. Phytochemistry 24:1991–1993

Chapter 4

Carbohydrates

J. N. BeMiller

4.1 Introduction

Carbohydrates are found in all living organisms. Indeed, they are the most abundant of the natural organic compounds. It is estimated that well over half the organic carbon on earth is in the form of carbohydrates, the great majority of it in plants. Almost three-fourths of the dry weight of plants is carbohydrate, most of which is in cell walls (structural components). In higher land plants, these carbohydrate components of the cell wall are cellulose, the hemicelluloses, and the pectic substances. The subject of this chapter is the carbohydrates other than those that are constituents of primary or secondary cell walls.

Carbohydrates (saccharides) occur in several forms. Few monosaccharides (sugars that cannot be broken down further by hydrolysis) are found free in nature, and these few are usually present in only small amounts. Most monosaccharides occur in glycosidic combinations, most often with either more of the same sugar or different sugars in the form of polymers (polysaccharides). Less frequently, except in the case of sucrose, they are joined together in short chains of 2–8 sugar units called oligosaccharides. Mono- and oligosaccharides may also be glycosidically linked to non-sugar organic residues. Derivatives of simple sugars are also present in plants. Each of these classes will be discussed.

In presenting the carbohydrates extraneous to the lignocellulosic cell walls of woody plants, emphasis will be on occurrence rather than amount because, not only is there insufficient space for a discussion of the latter topic but, in many cases, detailed information is not available. The carbohydrate composition of woody and other plants cannot be defined precisely for a given species or even for any given plant because chemical composition varies with plant part, type of wood, geographic location, season, climate, and soil conditions.

Because so little information on the non-cell wall carbohydrates of woody tissues is available, what is reported here can only be used as an indication of what might be found in a particular tissue and species. In this area, unreported often means not looked for, undetected, or not identified.

The occurrence, translocation, anabolism, and catabolism of carbohydrates related to the bioenergetics of plants have been reviewed (141). The energy sources stored for growth in living cells within woody tissue include mono- and oligosaccharides. The mono- and oligosaccharides are usually determined together in saps and in extracts of various tissues. They are most often extracted with hot 80% ethanol; 75% methanol, water, aqueous acetone, and other concentrations of ethanol have also been used.

Sucrose is the primary translocated photosynthate; other saccharides may also serve in the translocation of chemical energy.

Both sucrose and the higher oligosaccharides may function as soluble reserve substances. There is evidence of an active turnover of these carbohydrates that is related to biosynthetic demands (for example, requirement for monoterpene biosynthesis) (178) and/or fluctuations in composition of sieve-tube assimilates (see, for example, 166). As a result, there is an appreciable within-tree variation in the kind and amount of sap carbohydrates at any given time and within short periods of time (see, for example, 178). The content of any given soluble carbohydrate and the soluble carbohydrates as a whole are influenced by the season, rainfall, atmospheric temperature, sunlight, nutrition, and degree of foliation (see, for example, 50, 149, 166) and vary with plant part (see, for example, 79, 149).

Mono- and oligosaccharides (Sects. 4.2 through 4.5) and other carbohydrates (Sects. 4.8 through 4.10) have been extracted from bark, trunkwood, sapwood, root tissue, and saps of both angiosperms and gymnosperms. The alditols (Sect. 4.6) and cyclitols (Sect 4.7) are, or arise from, intermediates in metabolism; D-glucitol (Sect. 4.6) is involved in cryoprotection. The cyclitol methyl ethers (Sect. 4.7) are considered to be secondary plant substances. Much more is known about the occurrence and function of these carbohydrates in foliage, fruit, and seeds than in woody tissue, primarily because of the greater ease of working with non-woody tissues.

No claim is made for the inclusiveness of this chapter as far as specific species are concerned. Rather, what is included is meant to be indicative of the types of carbohydrates found extraneous to the primary and secondary cell walls, and the orders and families in which particular carbohydrates are found. Examples given here are from the more recent literature for the most part. Although the more recent investigations used better methods of analysis, it should be recognized that contemporary methods of carbohydrate analysis have yet to be applied to examination of the extractable carbohydrates of woody tissues. To confirm the presence of extractable carbohydrates in woody tissue, it is recommended that a cross search by compound (number and/or name) and genus be conducted and that the papers uncovered in this search be scrutinized for occurrence in a specific tissue.

4.2 Sucrose

By far the most abundant of the naturally occurring oligosaccharides is the disaccharide sucrose (*1*), which is the only economically important low-molecular-weight carbohydrate obtained from woody tissue. Sucrose [α-D-glucopyranosyl-(1→2)-β-D-fructofuranose] is not stored in wood, but it has been found in all saps that have been examined. It is the primary carbohydrate of most saps, being the only detectable sugar in some, although some contain glucose and fructose and are almost devoid of sucrose – e.g. white birch (*Betula alba*) (173).

It is well known that, in the spring, the sap of the sugar maple (*Acer saccharum*) contains 3% – 5% sucrose. Other *Acer* species generally contain less (1% – 3.7%). In all *Acer* species, approximately 95% of the dry matter and 99.9% of the total sap sugar is sucrose (121, 175). For a discussion of sap extraction from *Acer saccharum* and measurement of soluble sugars, see Gregory and Hawkins

(86). Other prolific sap producers are linden (*Tilia* sp.) and dogwood (*Cornus* sp.).

Sap availability in several European tree species has been investigated (77, 103). Sugar analysis of these saps revealed that sucrose was the only sugar present in sycamore maple (*Acer pseudoplatanus*) sap. Sucrose has also been extracted from slippery elm (*Ulmus fulva*) bark (51).

Seasonal variation in the sucrose content of willow (Salix) has been investigated (148). Maximum concentration occurs in late winter and is in the range of 3%–5%. Seasonal variation in the sugar composition of not only saps, but also wood cores, of both angiosperms and gymnosperms (see 93) has been determined. Perhaps the highest sucrose content yet reported in woody plants is that of the phloem sap of the Australian shrub *Grevillea leucopteris* which contains 14%–17% (92). The function and metabolism of sucrose in plants (29, 69, 78, 81, 104, 130, 138), the regulation of sucrose levels in plant cells (139), and the translocation of sucrose and related oligosaccharides (82) have been reviewed.

4.3 Higher Oligosaccharides Related to Sucrose

Much less effort has been directed toward investigations of the oligosaccharides of woody plants than toward the oligosaccharides of nonwoody plants. Issues that need to be addressed are taxonomic distribution, compartmentation within the plant and within cells, function(s), seasonal variation, anabolism, catabolism, and regulatory mechanisms. However, some information is available.

The trisaccharide raffinose (**2**) [*O*-α-D-galactopyranosyl-(1→6)-α-D-glucopyranosyl-(1→2)-β-D-fructofuranose)] is the second most abundant oligosaccharide in the plant kingdom and, like sucrose, may be ubiquitous – i.e. it may yet be found in plants not known to contain it when all tissues are carefully analyzed at all stages of development and in all seasons.

Raffinose was first crystallized from an extract of the tissues of *Eucalyptus manna* in 1843 (101). Recently, it has been reported to be present in the phloem and cambial saps of *Eucalyptus regnans* (166), *Pinus silvestris* (114), *Picea abies* (114), and *Abies alba* (114); the trunkwood of *Acer pseudoplatanus, Betula verrucosa, Tilia cordata*, and *Picea abies* (93), from which it almost vanished in the summer months; in the manna of *Fraxinus* spp. (ash) (168) along with D-mannitol (132), the major component; in the manna of *Eucalyptus* spp. (173) along with other sugars such as D-glucose, D-fructose, and L-arabinose; in the branchwood and branch sap of *Betula pendula* as a function of catkin growth and blossoming (149); and in the roots of Scots pine (*Pinus sylvestris*) (109). The effect of temperature on the accumulation of raffinose and related sugars (D-glucose, D-fructose, D-galactose, sucrose, stachyose, verbascose) in the bark of *Syringa vulgaris* and *Betula alba* has been examined (98).

The tetrasaccharide stachyose (**3**) [α-galactopyranosyl-(1→6)-α-D-galactopyranosyl-(1→6)-α-D-glucopyranosyl-(1→2)-β-D-fructofuranose] seems to be almost as widely distributed as raffinose (98), but is present in even lower concentrations. Although raffinose and stachyose occur in all parts of plants, they are segregated into storage tissues and leaves, for the most part. Low concentrations have been found in saps (see, for example, 166) and traces in wood, indicating that they may function both in the translocation of photosynthates and as soluble reserve compounds. The concentration of stachyose exceeded that of raffinose in the symplast of *Betula pendula* at 0 °C (149).

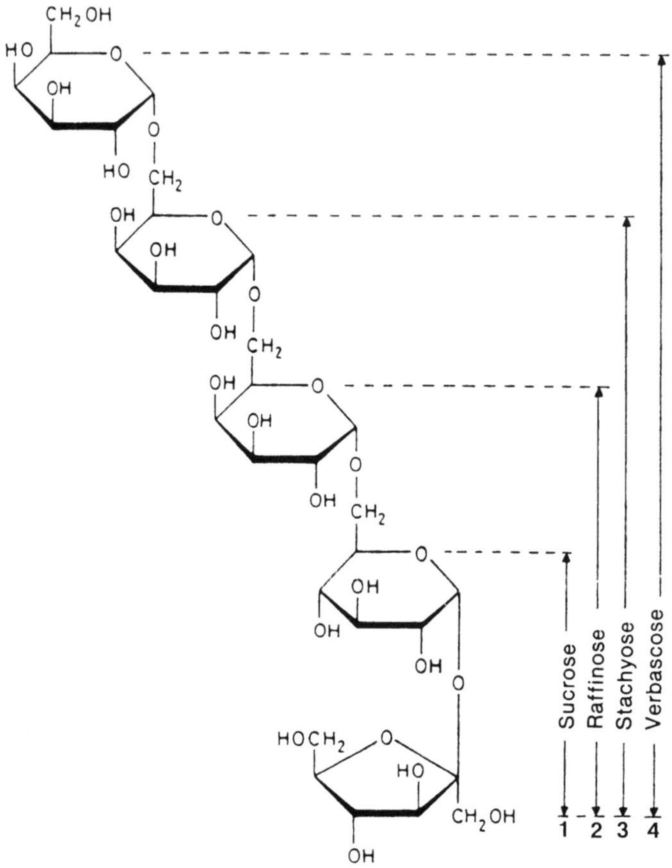

The next higher oligomer, verbascose (**4**), is generally found in trace amounts only in tissues in which raffinose accumulates to a high concentration.

4.4 Other Oligosaccharides

Maltose [α-D-glucopyranosyl-(1→4)-D-glucopyranose] has been found in the trunkwood of *Acer pseudoplatanus* and *Picea abies* (93) and in the symplast of *Betula pendula* at 0 °C (149).

The rare trisaccharide melezitose [*O*-α-D-glucopyranosyl-(1→3)-β-D-fructo-furanosyl-(1→2)-α-D-glucopyranose] occurs in the manna of Douglas-fir (*Pseudotsuga menziesii*) (95), Virginia pine (*Pinus virginiana*), and *Alhagi camelorum* as a result of insect attack.

4.5 Monosaccharides

D-Glucose and D-fructose are also found, along with sucrose, in the sapwood of both angiosperms and gymnosperms. Although sucrose is often the major component, the monosaccharides predominate in some cases. In an investigation of the spring sap of several European tree species (77), it was found that, while sucrose was essentially the only sugar present in saps of sycamore maple and other maples, D-glucose and D-fructose predominated in the saps of other trees. Their concentrations in the spring sap of the Finnish birches *Betula pendula, B. pubescens*, and *B. pendula* var. *carelica* were about equal and 4 to 7 times that of sucrose (103); the differences were even more pronounced in the North American gray birch (*Betula populifolia*) (50). These monosaccharides are also present in the phloem and cambial saps of *Eucalyptus regnans* (166), *Pinus sylvestris* (114), *Picea abies* (114), and *Abies alba* (114).

In the phloem sap of *Betula pendula* at 0 °C, sugars found (in order of concentration) were D-fructose > sucrose > stachyose > D-glucose > maltose > raffinose, while in the xylem sap the order was D-fructose > D-glucose > sucrose > > stachyose (149). The ubiquitous occurrence of D-glucose and D-fructose is exemplified by its presence in sap (149, 150), bark (179), roots (109), and trunkwood (93).

D-Galactose has been found in Siberian larch (*Larix sibirica*) wood (28), the spring sap of birches (*Betula* spp.) (103, 149), the xylem of Scots pine (*Pinus sylvestris*) (27), and red beech (*Fagus sylvatica*) sap (150). D-Mannose has been found in saps (149, 150). Both may be more widespread than indicated.

Of the pentoses, L-arabinose has been found in the heartwood and sapwood of many species of both angiosperms and gymnosperms (7, 27, 73, 160). The nutritive sap of pine is reported to contain a small amount of D-xylose (112). Traces of D-xylose were also found in the symplast of *Betula pendula* (149).

Traces of L-rhamnose and L-fucose were found in the xylem of *Pinus sylvestris* (27). These sugars, along with L-arabinose and D-galactose with which they were found, could have arisen from the breakdown of pectic substances (Sect. 4.10.2) or phenolic glycosides (Sect. 4.8). Traces of the D-ribose have been found in both the symplast and apoplast of *Betula pendula* (50).

4.6 Alditols

Alditols are acyclic, polyhydric alcohols. They are classified according to the number of carbon atoms they contain. Members of the tetritol (4 carbon atoms) to octitol (8 carbon atoms) families are known to occur in plants, with the hexitols being the most common. The distribution of alditols and anhydrohexitols and their use in taxonomy has been reviewed (137).

Several workers have reported correlations between tracheal sap concentrations of D-glucitol (sorbitol) and cold hardiness of apple (*Malus domestica*) trees (88, 96, 176). Throughout the dormant period, sap concentrations of sorbitol generally increase during periods of subfreezing temperatures and decrease during warmer periods. Early increases coincide with the initiation of leaf senescence. Sap sorbitol concentration reaches a maximum in late winter, then decreases in the spring. It is suggested that sorbitol in xylem sap during growth represents return transport from the roots and that the amount of sorbitol in the return transport reflects the ratio of leaf area to assimilate demand by the tree (88).

D-Mannitol, discovered in the manna of ash (*Fraxinus* sp.) in 1806, is the most widely distributed alditol in both higher and lower plants (137). Although it is particularly prevalent in lower plants, especially in combined forms, it has been found in more than 50 diverse families of higher plants. It has recently been reported to be present in the free form in *Pinus radiata* root extracts (68). It is the principal component of the manna of *Fraxinus ornus* (ash) (168).

Galactitol (dulcitol), first discovered in "manna of Madagascar" in 1850, has been extracted from various tissues of several woody plants (137).

4.7 Cyclitols

Cyclitols are polyhydroxycycloalkanes and -alkenes. The most abundant of these carbocyclic compounds are the hexahydroxycyclohexanes, commonly called inositols, and their methyl ethers. The nomenclature of cyclitols is problematic; several systems have been proposed and used. The rules used here are those of IUPAC and IUB (97).

Cyclitols are widely distributed in nature, albeit never in large quantities. The patterns of distribution and the biochemistry of cyclitols and their conjugates and derivatives are so interesting and important that their occurrence (23, 119, 140, 151), isolation (151), determination (23, 151), taxonomic significance (137), chemistry (23, 25, 151), biochemistry (24, 118–120, 151) and physiological importance (24, 151) have been frequently reviewed and discussed. Metabolic relationships between isomeric inositols and their methyl ethers, the amounts found in some plant tissue, and the fluctuations encountered suggest a storage function (see 119).

4.7.1 *myo*-Inositol

myo-Inositol, a hexahydroxycyclohexane, is so common in plants that it is generally regarded as ubiquitous; in fact it is likely that it is present in all living cells, where it is converted into D-glucuronic acid and subsequently into D-xylose, D-galacturonic acid, L-arabinose, and L-ascorbic acid. It accumulates in the cambial saps of *Pinus sylvestris, Picea abies*, and *Abies alba* (114); in the phloem and cambial saps of *Eucalyptus regnans* (166); in the spring sap of birches (*Betula* spp.) (103); and in the bark of *Picea abies, Pinus nigra*, and *Acer campestre* (71).

myo-Inositol is most often found as ester, ether, and/or glycoside derivatives. Phytic acid, the hexakis-*O*-phosphate monoester of *myo*-inositol occurs in most, if not all, higher plants, but it is yet to be demonstrated that it is a component of woody tissue. It is recovered from plant tissues as phytin, the mixed calcium-magnesium salt. Lower phosphate esters of *myo*-inositol arise from dephosphorylation of phytic acid. Although they too are common in plant tissues, neither have they been reported to occur in woody tissue. Likewise, phosphatidylinositol, a normal constituent of plant phospholipids, has not been shown to occur in woody tissue. Interest in all these compounds has increased dramatically in recent years because of their role in biochemical regulatory processes.

Methyl ethers of several cyclitols occur as secondary substances in many higher plants. Attempts to use their occurrence for chemical taxonomy have been made (137). Sequoyitol (5-*O*-methyl-*myo*-inositol) is found in all families of gymnosperms, and only in gymnosperms, although its distribution appears to be sporadic (137). It has been found in wood, bark, and cambial sap (114). D-Bornesitol (1-*O*-methyl-*myo*-inositol) is present in the latex of trees in the Apocynaceae family and in the wood of *Sarcocephalus diderrichii* (see 137). Dambonitol (1,3-di-*O*-methyl-*myo*-inositol) has been found in Gabon (*Castilloa elastica*) rubber latex (see 137).

Galactinol [1-α-galactinol, 1L-1-(*O*-α-D-galactopyranosyl)-*myo*-inositol] is the most common glycoside of *myo*-inositol. Galactinol is a galactosyl unit donor for the biosynthesis of oligosaccharides in the raffinose series and occurs in all plants that produce raffinose and stachyose. Its concentration in the phloem and cambial saps of *Eucalyptus regnans* fluctuates in a way consistent with its role as a galactosyl donor during periods of rapid growth (and of *myo*-inositol as a galactosyl acceptor) (166).

4.7.2 D-*chiro*-Inositol

D-*chiro*-Inositol (D-inositol) was isolated from the heartwood of *Pinus lambertiana* (47) and detected in the cambial sap of *Pinus sylvestris, Picea abies*, and *Abies alba* (114). It will undoubtedly be found to be present in other coniferous saps and woods, especially those containing D-pinitol, when modern analytical techniques are used.

D-Pinitol (1D-3-*O*-methyl-*chiro*-inositol) is the most widely distributed inositol ether. It has been found in six gymnosperm families and 13 angiosperm families (137). In gymnosperms, it is a component of the resin of *Pinus lambertiana* (see 137); the bark of *Picea abies, Pinus nigra* (71), *Pinus halepensis*, and *Schinus molle* (72) (where its content has been determined as a function of season); the cambial sap of *Pinus silvestris, Picea abies*, and *Abies alba* (114); and the wood of *Pinus* spp. (see 6, 137) and *Sequoia sempervirens* (153). It also occurs in various plant parts, including latex and wood (*Acacia mollissima*), of angiosperms (see 137).

4.7.3 Quebrachitol

Quebrachitol (1L-2-*O*-methyl-*chiro*-inositol) is present in several woods; the barks of the quebracho tree (*Aspidosperma quebracho*), *Haplophyton cimi-*

cidum, and *Acer campestre* (see 71, 137); and as a constituent of Borneo and *Hevea brasiliensis* latexes (see 137). Quebrachitol occurs in maple sap in a concentration greater than that of D-glucose plus D-fructose (167).

4.7.4 D-Quercitol

D-Quercitol, a pentahydroxycyclohexane, is found in the bark of *Quercus suber* and *Tiliacora acuminata* and roots of *Cissampelos pareira, Legnephora moorii* and *Cyclea burmanni* (see 137). It appears to be uniformly present in the bark of oak (*Quercus* spp.).

4.7.5 Conduritol

Conduritol, a tetrahydroxyclyclohexene, has been found only in the bark of *Gonolubus condurango* (see 39).

4.7.6 Quinic Acid

Quinic acid, a carboxylated tetrahydroxycyclohexane, is a secondary plant substance formed from dehydroquinic acid, an intermediate in the shikimic acid pathway, the metabolic pathway leading to aromatic compounds. It, therefore, is ubiquitous in living plant cells. It has been isolated from cinchona bark and detected in the cambial sap of conifers (114).

4.8 Plant Glycosides

The phenolic compounds of woody plant tissues are described and discussed in Chap. 7. They are mentioned here only to point out that they are present most often in living cells as glycosides. Glycosides of flavones, flavanols (3-hydroxyflavones), flavanones, flavanolols (2,3-dihydroflavanols), stilbenes and saponins, gallic acid derivatives, and condensed tannins are all common constituents of woody tissues (see Chap. 7). Anthocyanidin glycosides are less commonly found in wood. Both true glycosidies and *C*-glycosyl compounds ("*C*-glycosides") are known to occur in plants. Seasonal variations of phenolic glycosides in the cambial sap of woods have been determined (169).

4.9 Starch

Starch is the most common extraneous polysaccharide of wood. It is found in angiosperms in the living, ray and longitudinal, parenchymatous cells of sapwood. It almost never occurs in heartwood. There has been some uncertainty about the amounts present because the starch content undergoes wide seasonal variations, and methods for its determination have been inadequate. The contemporary method of starch analysis (170) has been applied to shoots of dormant jack pine (*Pinus banksiana*), white spruce (*Picea glauca*), and bigtooth aspen (*Populus grandidentata*) and roots of dormant red oak (*Quercus rubra*) (87). Starch contents ranged from ~15% in shoot to ~140% in root tissue.

The dependence of the amount of starch in woody tissues of various trees on season (see, for example, 72, 110), sunlight (see, for example, 122), foliation (see, for example, 178), nutrients (see, for example, 122), and the demand for monoterpene biosynthesis to repel insect attack (178) has been investigated; as expected, the starch content is directly proportional to the availability of translocatable carbohydrates from photosynthesis and inversely proportional to energy demands for anabolic processes.

The xylem ray cells of the sugar maple (*Acer saccharum*) contain ~10% starch (172). In birch (*Betula* spp.) stored starch is converted into sucrose when the ambient temperature drops to ~5 °C (50), perhaps to provide cryoprotection.

In general, the amount of starch, which may be relatively high in living parenchymatous cells, is very small compared to the total mass of the wood and is dependent on the number of living cells and their metabolic state. In trees that are affected by seasons, there are two maxima and two minima (131). Starch accumulates in the summer. The first maximum occurs in the fall before the deciduous trees shed their leaves. The first minimum occurs in late winter. Starch is absent in late winter in linden (*Tilia*), birch (*Betula*), and some conifers; in oak (*Quercus*), maple (*Acer*), elm (*Ulmus*), and ash (*Fraxinus*) the starch content of wood decreases and the content of bark drops to zero. The second maximum occurs in early spring prior to budding. The second minimum is correlated with growing shoots (late spring).

A notable exception as far as starch storage is concerned is the sago palm, the pith of which contains ~40% starch (on average) (66). To obtain the starch, 8-year-old or older trees are cut. A long strip of bark is removed from cut trunks (in native production) or the trunks are split (in commercial production) to expose the pith, which is scraped or rasped out. A single tree will yield 600 to 800 lb (270–360 kg) of pith. Sago starch [200 to 400 lb (90–180 kg)] is obtained by kneading the pith with water. Fourteen species of eight genera of sago palms, which grow in many tropical areas, are used by natives to obtain sago starch. *Metroxylon* spp. and *Arenga* spp. in the Eastern Hemisphere and *Mauritia* spp. in the Western Hemisphere are used for the commercial production of sago starch (66).

The other commercial importance of starch in woody tissue relates to the correlation between the presence of starch in felled timber and susceptibility to attack by insects and fungi (177).

4.10 Extractable Polysaccharides

The most abundant polysaccharide of woody tissue is cellulose. Next in overall amount is the class of structurally diverse substances called hemicelluloses, which are also involved in the construction of secondary cell walls. These polysaccharides are associated with lignin, both through chemical linkages and physical interactions, and are outside the coverage of this chapter.

Other polysaccharides are less associated and can be obtained by simple extractions of the tissue. The few known examples are given here. It must be recognized that only a few species have been investigated for extractable polysaccharides.

4.10.1 Arabinogalactans

The arabinogalactans are L-arabinosyl-substituted, branched galactans. Those extractable from woody tissue are type-II arabinogalactans (162). The occurrence, structures, and functions of this complex, interesting, and ubiquitous family of polysaccharides have been reviewed several times (see 65, 80, 162). They may be neutral or acidic polysaccharides. The ratio of D-galactopyranose to L-arabinofuranose, the major monosaccharides, varies between 10:90 and 85:15, with most containing more D-galactose than L-arabinose (80). Some L-arabinofuranosyl units, the most abundant form of L-arabinose, are attached singly to D-galactopyranosyl units; others may be present in short chains. L-Arabinosyl units at nonreducing termini may be in the pyranose ring form. Other neutral sugar units that may be present in arabinogalactans are L-rhamopyranosyl (up to 11%), D-mannopyranosyl (up to 16%), D-xylopyranosyl (up to 7%) and D-glucopyranosyl (up to 4%) units. Some arabinogalactans are acidic polysaccharides containing units of D-glucopyranosyluronic acid and/or 4-*O*-methyl-D-glucopyranosyluronic acid (up to 28%) or D-galactopyranosyluronic acid and/or 4-*O*-methyl-D-galactopyranosyluronic acid (up to 26%) units. Uronic acid units are found as nonreducing termini of the galactan backbone (80).

Arabinogalactans form a family of polysaccharides with a branched framework made up predominately of (1→3)-linked β-D-galactopyranosyl units with varying amounts of (1→6)-linked β-D-galactopyranosyl units. Their structures are often complex, with an apparent random distribution of side chains in at least some. They often occur as a mixture of components that differ widely in both molecular weight and chemical structure.

Most type-II arabinogalactans contain some protein. The amount is usually between 2% and 10%, but can be as high as 59% (65, 80). The polypeptide portions of some of the arabinogalactan-proteins are characterized by high hydroxyproline contents; they usually also have elevated amounts of proline and serine. Examples are the polypeptides of the polysaccharide-protein of the exudate gums from *Acacia* (Sect. 4.10.3.1), *Astragalus*, and *Prosopis* (Sect. 4.10.3.2) species (16–18). However, the polypeptide portions of other arabinogalactan-proteins – for example, those of the exudate gums from *Combretum*, *Terminalia*, and *Anogeissus* species (Sect. 4.10.3.3) – do not have especially high contents of hydroxyproline (18, 19).

All parts of higher plants (trunks, stems, roots, leaves, seeds, flowers) contain arabinogalactans and arabinogalactan-proteins, in some cases in considerable amounts. Their role in construction of the primary cell wall has been discussed (3, 4, 111).

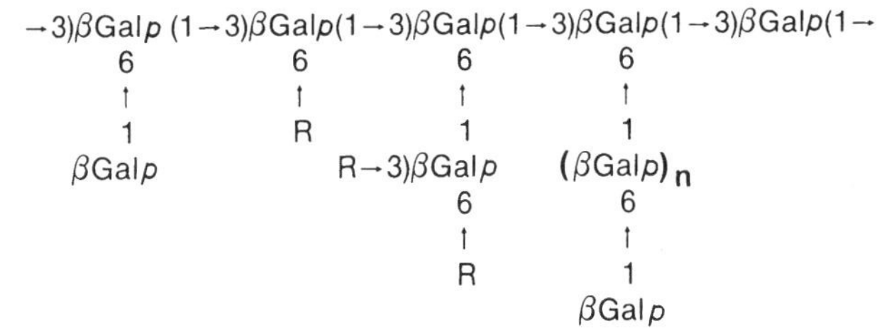

```
→3)βGalp (1→3)βGalp(1→3)βGalp(1→3)βGalp(1→3)βGalp(1→
   6          6          6          6
   ↑          ↑          ↑          ↑
   1          R          1          1
 βGalp              R→3)βGalp    (βGalp)n
                         6          6
                         ↑          ↑
                         R          1
                                  βGalp
```

R = one of βGalp(1→, LAraf(1→, βLArap(1→3)LAraf(1→ , βGlcpA(1→

Fig. 4.1. The main structural features of larchwood arabinogalactan

4.10.1.1 Larch Arabinogalactans

The water-extractable arabinogalactans are most abundant in the larches (*Larix* spp.). The most extensively studied has been that of the heartwood of the western larch (*Larix occidentalis*). The source, production, properties, and potential uses of this polysaccharide are reviewed in Chap. 10.2.2.

The structural information on larchwood arabinogalactan has been reviewed (162). Its main structural features are represented by the structure in Fig. 4.1.

Units such as these (MW ~ 2200 daltons) are joined by sugar units vulnerable to periodate oxidation to give polymers with a wide range of molecular weights of from 16–18000 to <100000 (49, 60).

It has been reported that the ratio of galatose to arabinose in the arabinogalactan of *Larix occidentalis* increases and that the number-average molecular weight of the polymer decreases with increasing age of the wood (67, 154).

The arabinogalactans of other larch species have been less extensively examined (see 162). Recent investigations of the arabinogalactan of Siberian larch (*Larix sibirica*) wood (28) have revealed a trimodal molecular weight distribution and chemical combination with protein and lignin (26). A positive relation between the percentage of heartwood and the content of arabinogalactan was found in *Larix gmelini* (108). All larch arabinogalactans occur in the lumen of tracheids and ray cells as an amorphous mass of extracellular material (67, 154).

4.10.1.2 Other Extractable, Nonexudate Arabinogalactans

Arabinogalactans similar to the larchwood arabinogalactan occur in smaller amounts in the heartwood of spruces (*Picea* spp.), pines (*Pinus* spp.), and firs (*Abies* spp.) (162), but their structures have not been subjected to the same rigorous examination given to larch arabinogalactans because of their lack of potential commercial importance. The concentration of arabinogalactan in the xylem of Scots pine (*Pinus sylvestris*) is at a maximum during the cell elongation phase (27).

Sap of the sugar maple (*Acer saccharum*) yields an arabinogalactan with a branched structure that includes an L-rhamnopyranosyl end group linked (1→6) to a D-galactopyranosyl main-chain unit (2). An arabinogalactan from the cambial zone of *Populus tremuloides* (aspen) contains D-galactosyl, L-arabinosyl, L-rhamnosyl, and D-glucosyluronic acid units. It has a main chain of (1→3)-linked β-D-galactopyranosyl units that is heavily branched at *O*-6. It occurs as the polysaccharide portion of an arabinogalactan-protein (154).

The latter two examples indicate that extractable arabinogalactans are not limited to conifers. However, the extractable arabinogalactans of angiosperms are more similar to the exudate gums (Sect. 4.10.3) than to the extractable arabinogalactans of the woody tissue of conifers.

4.10.2 Other Extractable Polysaccharides; The Pectic Polysaccharides

Larix spp. (108), other gymosperms, and angiosperms (see 99, 100, 155–158) contain polysaccharides that can be extracted without delignification, i.e. that can be extracted without prior conversion to holocellulose – but few have been investigated. Generally, the mixtures of polysaccharides that are extracted with hot water or, more commonly, with hot aqueous solutions of calcium and magnesium ion-chelating agents such as ethylenediamine tetraacetate, ammonium oxalate, and sodium hexametaphosphate are called pectic substances or pectic polysaccharides.

Extraction of maritime pine (*Pinus pinaster*) sawdust with boiling water yields a highly branched arabinan, a galactan, an arabinoglucuronoxylogalactan, a galactoglucomannan, and a polysaccharide containing D-galactose, D-xylose, L-rhamnose and a uronic acid (142–144).

The inner bark of slippery elm (*Ulmus fulva*) contains ~16% (dry wt basis) of a branched polysaccharide composed of D-galactosyl, 3-*O*-methyl-D-galactosyl, D-galactosyluronic acid, and L-rhamnosyl units (51, 83, 90, 174). The main chain of this polysaccharide is a rhamnogalacturonan composed of alternating *O*-4 substituted D-galactopyranosyluronic acid units and *O*-2 substituted L-rhamnopyranosyl units (51). Polysaccharide has also been isolated by water extraction of the bark of *Hydrangea particulata* (113).

Broadly defined, these polysaccharides, along with the arabinogalactans, comprise the pectic polysaccharides; strictly defined, pectic polysaccharides are only those glycans whose principal structural component is a D-galacturonan chain. (There is no generally agreed upon definition. Arabinans, galactans, and rhamnogalacturonans are often included as pectic polysaccharides because they are found in close physical association with the galacturonans and are extracted with them.)

Pectic polysaccharides comprise ~35% of the primary cell wall of dicots. The middle lamella, which lies between the primary walls of adjacent cells, is particularly rich in pectic polysaccharides. The amount of the pectic polysaccharides that can be extracted varies from 0.5%–1.5% from heavily lignified tissues to 15%–30% from parenchymatous and meristematic tissues; in either case, only 50%–90% of the total pectic polysaccharides can be removed by direct extraction (see 152).

The pectic polysaccharides of woody dicots have been divided into the following categories (70): rhamnogalacturonans I (the primary backbone of the pectic polymers), rhamnogalacturonans II (very complex structurally; contain a large number and variety of terminal glycosyl units), homogalacturonan, arabinans (highly branched; L-arabinose largely as L-arabinofuranosyl units; isolated from white willow (*Salix alba*) (105), *Rose glauca* bark (102), aspen (*Populus tremuloides*) bark (100), and *Pinus pinaster* sawdust (142–144)), galactans (primarily (1→4)-linked β-D-galactans; found in *Rosa glauca* bark (102), white willow (171), tension wood of beech (*Fagus sylvatica*) (128), and *Pinus pinaster* sawdust (142–144)), and arabinogalactans. Although these polysaccharides are widespread and one or more of this group are found in all higher plant tissues, few studies of the presence and structure of those that occur in woody tissue have been reported. What is known about them has been obtained largely from investigations of nonwoody plant tissues and, especially in recent years, analyses of cell-suspension cultures. Indeed, all classes of pectic polysaccharides listed above have been found or indicated to be present in cultures of sycamore maple cells (see 70), an indication of their presence in the primary cell wall of its tissues.

A water extract of the cambial tissue of *Populus tremuloides* (aspen) contained galacturonans (40%), arabinogalactan (12%), xyloglucan (6%), xylan (10%), glucomannan (1%), and 4-*O*-methylglucuronoxylan (155–158). The bark of the same tree contains arabinan and 4-*O*-methylglucurononxylan (99, 100). A water extract of the cambial tissue of *Tilia americana* (basswood) contained the same polysaccharides, with more xyloglucan and less xylan (155–158).

Water extracts of the wood of *Pinus taeda* (loblolly pine) and *Pinus radiata* contained a complex mixture of polysaccharides that could be divided into two fractions – one rich in L-arabinose and one rich in D-mannose (54).

In a study of *Picea abies, Pinus sylvestris, Fagus sylvatica*, and *Quercus robur*, it was found that the amount of soluble saccharides in the phloem was substantially higher than that in the xylem and that pectic substances predominated (73).

4.10.3 Exudate Gums

Of the plant gums that have been subject to structural characterization, the majority are branch-on-branch, substituted arabinogalactans (30, 55, 56, 159, 162, 174). Examples of this type of polysaccharide have been found in 14 of the 92 orders of angiosperms and three orders of gymnosperms (162). The exudate gums appear to be metabolic end products, i.e. they are not reserve food materials. It is generally assumed that the exudates, because they are produced in response to various injuries, seal off the injured tissue, prevent the loss of moisture, and protect against infection – i.e. production of exudates and exudate polysaccharides is a defense mechanism. Most exudate gums are acidic, substituted, branched galactans. The structural information for most is difficult to interpret because of the complexity of their structures and because most, if not all, preparations contain a spectrum of related molecules. Most, if not all, are polysaccharide proteins (18). A classification of exudate gums similar to that used by Stephen (162) will be followed here.

4.10.3.1 Acacia Gums

Exudate gums from more than 60 species of *Acacia* (order Rosales, Family Leguminosae, subfamily Mimosoidae) have been examined (159, 162, 174); but taxonomic classification based on structures has not been achieved because of the many species of *Acacia*, the uncertain assignment of a particular collection of tears to a particular species, the variations due to geographic location, season, and climate, and the finding that gum samples collected from different branches of a single tree (*Acacia pycantha*) vary, at least, in average molecular weights (1).

A discussion of the evidence for the structures of commercial *Acacia* gums, particularly *Acacia senegal* gum, commonly called gum arabic (Sect. 10.2.3), has appeared (85). All *Acacia* gums are polydisperse – i.e. they are mixtures of components with a distribution of molecular weights and chemical structures, as are most plant polysaccharides except some of the very few that are linear homopolymers and, hence, only polymolecular. Molecular weights of the components of *Acacia* gums from different sources vary from ~ 6000 to $>10^6$ daltons (162). As with the larch arabinogalactan (60), a single Smith degradation often reduces these polymolecular mixtures into single components, indicating that the native polysaccharide molecules are composed of various multiples of a basic, but somewhat variable, building unit. A single Smith degradation of gums from the following sources yields the products indicated: *Acacia cyclops*, MW 2000 (162); *A. implexa*, MW 8000 (162); *A. filicifolia*, MW 2100 (59); *A. podalyriaefolia*, MW 2100 (48); *A. mearnsii*, MW 6000 (63); *A. elata*, MW 2100 and 4400 (48). Other gums are more periodate resistant, e.g. those from *Acacia salinga* (58), *A. longifolia* (61), and *A. senegal* (162). The latter gives equal amounts of units of ~ 16000 and ~ 30000 daltons after a two-stage Smith degradation (162). It is suggested that, even with this gum, blocks of periodate oxidation-immune galactosyl residues no larger than 13 units in size may be the basic building blocks (162). All *Acacia* gums may be polysaccharide-proteins (see, for example, 20, 65). Some contain as much as 9% protein (65).

4.10.3.2 Exudate Gums of Other Rosales Genera

The most extensively studied exudate gum of the genus *Prosopis*, which belongs to the same subfamily (Mimosoideae) as *Acacia*, is mesquite (*Prosopis juliflora*) gum (30, 45, 46, 76, 174). Its structure is that of a highly branched, acidic arabinogalactan resembling *Acacia* gums, as are the structures of *Prosopis chilensis* and *P. glandulosa* gums (62, 162). The latter two gums are converted into a single subunit of ~ 6000 daltons upon Smith degradation (62) and thus resemble *Acacia mearnsii* gum. *Prosopis* gums also contain protein (16).

Some exudate gums of the genus *Albizia*, also in the Mimosoideae subfamily, contain the same monosaccharide units, viz., D-galactopyranosyl, L-arabinofuranosyl, D-glucopyranosyluronic acid, 4-*O*-methyl-D-glucopyranosyluronic acid, and L-rhamnopyranosyl units, as do *Acacia* and *Prosopis* gums. Others – e.g. those from *Albizia zygia* (Macbride) (*A. brownei*, West African walnut), *A. sericocephela*, and *A. glaberrima* – contain 6%–13% D-mannopyranosyl units

(10, 74, 162). Often, *Albizia* gums contain few, if any, L-rhamnopyranosyl units, but the reported amount of L-rhamnose in *A. glaberrima* gum is 28% (10, 162).

Gum tragacanth (Chap. 10.2.5) is obtained from various species of *Astragalus* (Leguminosae, subfamily Lotoidea), primarily *A. gummifer.* It is a mixture of polysaccharides; 60%–70% of the gum is an acidic, water-swellable polysaccharide called tragacanthic acid (formely called bassorin) that is composed of α-D-galactopyranosyluronic acid, β-D-xylopyranosyl, β-D-galactopyranosyl, and α-L-fucopyranosyl units. (It also contains a small amount of L-rhamnose.) The proposed structure is that of a branched galacturonan in which the main pectic acid chain is substituted at *O*-3 with single β-D-xylopyranosyl units or with β-D-xylopyranosyl units substituted at *O*-2 with β-D-galactopyranosyl or α-L-fucopyranosyl units (31). It has been reported that the acidic fraction of the exudate gum of *Astragalus microcephalus* can be fractionated (116) into four components, some of which are alkali-soluble and some of which are alkali-insoluble (115). The neutral fraction of the gum is a mixture of highly branched arabinogalactans that are unique in having a high arabinose: galactose ratio (as much as 4:1) and which contain small amounts of D-xylose (32, 115, 117); some of these polysaccharides may be slightly acidic (127). An arabinoxylan has been isolated from *A. microcephalus* gum (115). The polypeptide portion of *Astragalus* gums has been examined (17).

Exudate gums from *Prunus cerasus* (cherry), *P. virginiana* (cherry), *P. insitia* (damson plum), *P. domestica* (egg plum), *P. persica* (peach) (106, 107), *P. armeniaca* (apricot), *P. amygdalus* (almond), and *P. spinosa* (blackthorn) (Rosaceae) have been investigated (see 30, 159, 162, 174). They are composed of D-galactosyl, L-arabinosyl, D-xylosyl, D-glucuronosyl, and 4-*O*-methyl-D-glucuronosyl units, and smaller amounts of D-mannosyl and L-rhamnosyl units, the proportions varying even within a single species. The same general structure of a main chain of (1→6)-linked β-D-galactopyranosyl units substituted frequently at *O*-3 with short side chains is indicated (162). The exudate gums of *Virgilia oroboides* and *V. divaricata* are also arabinogalactans (see 159, 162). They contain D-glucopyranosyluronic acid, D-xylopyranosyl, and D-mannopyranosyl units.

4.10.3.3 Gums of Combretaceae (Myrtiflorae) Genera

The most extensively studied exudate gum from this plant family is that from *Anogeissus latifolia* which is known as gum ghatti or Indian gum (Chap. 10.2.6). This polysaccharide has a main chain of alternating β-D-glucopyranosyluronic acid units substituted at *O*-4 and D-mannopyranosyl units substituted at *O*-2 (30, 162). The D-mannopyranosyl units constitute double branch points. The side chains contain L-arabinopyranosyl, L-arabinofuranosyl, D-galactopyranosyl, and D-glucopyranosyluronic acid units.

The gum from *Anogeissus leiocarpus* has been fractionated into two components; the more abundant of these has been studied and shown to be similar in structure (but not identical) to the *A. latifolia* gum polysaccharide (30, 33, 34, 43, 162).

All exudate gums examined from *Combretum* species contain D-glucopyranosyluronic acid, 4-*O*-methyl-D-glucopyranosyluronic acid, D-galactopyranosyluronic acid, D-galactopyranosyl, L-arabinofuranosyl, L-arabinopyranosyl, and L-rhamnopyranosyl units; some also contain D-xylopyranosyl and/or D-mannopyranosyl units. However, there is wide variation in the proportions of these monosaccharide units. Investigations of the structures of the exudate gums of *Combretum leonense, C. hartmannianum, C. elliottii*, and *C. salicifolium* indicate structural similarity with the exudate gums of *Anogeissus* species (see 30, 162). Little is known about exudate gums from *Terminalia* species (see 162). The polypeptide portions of gums from all three genera have been investigated (19).

4.10.3.4 Exudate Gums of Anacardiaceae (Sapindales)

Exudate gums produced by several genera in the family Anacardiaceae have been examined (see 162). The most extensively investigated is that of *Anacardium occidentale* (the cashewnut of cashew tree) (30, 159, 174). This gum has a main chain of (1→3)-linked β-D-galactopyranosyl units interspersed with (1→6) linkages. Most main chain units carry a branched side chain at *O*-6. The side chains are composed of (1→3)- and (1→6)-linked D-galactopyranosyl units to which are attached L-arabinofuranosyl, L-arabinopyranosyl, L-rhamnopyranosyl, and D-glucopyranosyluronic acid units (8). The polypeptide portion of mu'gongo (mongo, mungango) gum from *Sclerocarya* spp. has been examined (9).

Gum jeol is obtained as an exudate of *Lannea coromandelica* (also known as *Odina wodier*). It is similar to the exudate gums of *Lannea humilis* and *L. schimperi* (12, 14, 15, 52, 53). These polysaccharides have structural features in common with the *Anacardium* gums. Relatively little structural information on the exudate gum of *Spondias cytheria* (West Indian gold apple) or on gums of other *Spondias* species exists (see 159, 162).

4.10.3.5 Exudate Gums of Families in the Orders Rutales, Parietales, and Malvales

Trees of the genus *Boswellia* (Burseraceae) exude gum frankincense, a mixture of a resin and an acidic polysaccharide. The acidic polysaccharides of the exudates of *Boswellia papyrifera* and *B. carteri* are indicated to be substituted, branched galactans (see 159, 162, 174). The polysaccharide obtained from gum myrrh, which exudes from the damaged bark of *Commiphora myrrha* (Burseraceae) (159, 174), is difficult to isolate because of its association with protein (162).

The most extensively studied citrus (Rutaceae) gum is that exuded from *Citrus limon* (lemon) (159, 162, 174). The indications are that it has a branch-on-branch galactan framework (see 162). *Citrus maxima* (grapefruit) and *C. sinensis* (orange) trees produce less exudate, and their exudate gums have been less extensively studied (159, 162, 174). The polypeptide portion of bhirra gum from *Chloroxylan* spp. (Rutaceae) has been examined (9).

The exudate gum of *Khaya senegalensis* (African mahogany) (Meliaceae) contains two polysaccharides. The major component is an acetylated branched polymer of complex structure that has some of the structural features of *Sterculia setigera* and *Cochlospermum gossypium* gum polysaccharides (40). The minor component is a highly branched, acidic arabinogalactan with some structural features similar to those of the exudate polysaccharides of *Acacia senegal* and *A. pycnantha* (41). The exudate polysaccharide of *Khaya grandifolia* has a structure similar to that of the major component of *K. senegalensis* gum (39, 159).

Ketha gum, which exudes from the bark of *Feronia elephantum* (Rutaceae, Rutales) is reported to have properties similar to those of the *Acacia* gums (159). The polypeptide portion of a gum from trees of *Azadirachta* spp. (Meliaceae) has been examined (9). Neem gum (*Azadirachta indica*) (159) contains 35% protein (13).

Exudates of *Sterculia urens* (Sterculiaceae, Malvales) are known as gum karaya (*Sterculia* gum, Indian tragacanth, Indian gum, Kadayo gum, Katilo gum, Kullo gum, Kuteera gum, Kutira gum, Mucara gum) (Sect. 10.2.4). Exudates from other *Sterculia* species, *Cochlospermum gossypium*, and *Cochlosperum kunth* (*Cochlospermacea, Parietales*) are also so designated (84). The *Sterculia urens* exudate polysaccharide is acetylated; it appears to be composed of two kinds of interconnected branched chains – one with a main chain structure of →4)αGalpA(1→2)LRhap(1→ and the other with a main chain structure of →4)αGalpA(1→4)αGalp(1→ (136). The polysaccharide of the *Sterculia urens* exudate appears to have several structural features that are similar to those of the exudate gum polysaccharides of *S. setigera* (37, 89, 94), *S. caudata* (≡*Brachychiton diversifolium*) (36, 91), and *Cochlospermum gossypium* (37, 38).

4.10.3.6 Exudate Gums from Other Orders

Other polysaccharides that are extraneous to the lignocellulosic cell wall and of the arabinogalactan type are the following: the exudate gum of *Cussonia spicata* (Araliaceae) (64); the exudate gum of *Crevillea robusta* (silk oak) (Proteaceae, Proteales) (11, 165); the exudate gum of *Brabeium stellatifolium* ("wild almond") (164); the exudate found only occasionally on the bark of *Hakea acicularis* (Proteaceae, Proteales) (159) trees introduced into South Africa from Australia (161); oleoresin gums from *Araucaria bidwillii* (21, 22, 35, 42, 44) and *A. klinkii* (Araucariaceae, Coniferae) (57); extracts of *Encephalartos longifolius* (a cycad) (163).

A group of acidic polysaccharides can be isolated from the stems of members of the Cactaceae family by cold water extraction. All are present in the parenchyma of these plants. The most studied have been those of *Opuntia* species. The (unfractionated) polysaccharide obtained from *Cereus peruvianus* contains 44% uronic acid and that from *Wigginsia erinacea* 51% (145); polysaccharides extracted or exuded from *Opuntia fulgida* (Cholla gum), *O. ficus-indica* (a plant with fleshy, modified stems), *O. aurantiaca, O. monacantha,* and *O. nopalea-coccinillifera* contain less (11%–25%) (5, 123–126, 129, 133–135, 145). Each of these extracts of *Opuntia* species contains a group of closely related, complex, highly branched polysaccharides with a backbone chain composed of *O*-4 substituted D-galactopyranosyluronic acid and *O*-2 substituted L-rhamnopyranosyl units. The main chain is substituted with galacto-oligosaccharides, primarily

(1→6)-linked, which in turn are substituted with oligosaccharides containing β-D-xylopyranosyl, α-L-arabinofuranosyl, and β-D-galactopyranosyl units (126).

The polypeptide portion of shahjan gum from *Moringa* spp. (Moringaceae, Rhoedales) has been examined (9).

4.10.3.7 Exudate Gums with Xylan Cores

Much less common than exudate gums with galactan cores are those with xylan cores. Those known to have xylan cores, to which are attached short branches, are sapote gum from *Sapota achras* (sapodilla) (Sapotaceae) (75, 159) and the exudate polysaccharide of *Rhizophora mangle* (mangrove) (Rhizophoraceae) (146, 147). Chagual gum exudes from *Puya* spp. following insect damage (159). An arabinoxylan has been isolated from *Astragalus microcephalus* gum (115).

References

1 Adam J W H, Churms S C, Stephen A M, Streefkerk D G, Williams E H 1977 Structural relationships of whole Acacia pycnantha gum and a component of low molecular weight. Carbohyd Res 54:304–307

2 Adams G A, Bishop C T 1960 Constitution of an arabinogalactan from maple sap. Can J Chem 38:2380–2386

3 Albersheim P 1974 The primary cell wall and control of elongation growth. Ann Proc Phytochem Soc 10 (Plant Carbohyd Biochem):145–164

4 Albersheim P 1975 The walls of growing plant cells. Sci Am 232(4):81–95

5 Amin E S, Awad O M, El-Sayed M M 1970 The mucilage of Opuntia ficus-indica Mill. Carbohyd Res 15:159–161

6 Anderson A B 1953 Pinitol from sugar pine stump wood. Ind Eng Chem 45:593–596

7 Anderson A B, Erdtman H 1949 L-Arabinose from heartwood of western red cedar (Thuja plicata). J Am Chem Soc 71:2927–2928

8 Anderson D M W, Bell P C 1975 Structural analysis of the gum polysaccharide from Anacardium occidentale. Anal Chim Acta 79:185–197

9 Anderson D M W, Bell P C, Gill M C L, McDougall F J, McNab C G A 1986 Studies of uronic acid materials. 83. The gum exudates from Chloroxylon swietenia, Sclerocarya caffra, Azadirachta indica and Moringa oleifera. Phytochemistry 25:247–249

10 Anderson D M W, Cree G M, Marshall J J, Rahman S 1966 Studies on uronic acid materials. XIII. The composition of gum exudates from Albizia sericocephala and Albizia glaberrima with an appendix on botanical nomenclature in the genus Albizia. Carbohyd Res 2:63–69

11 Anderson D M W, dePinto G L 1982 Studies of uronic acid materials. Gum exudates from the genus Grevillea (Protaceae). Carbohyd Polym 2:19–24

12 Anderson D M W, Hendrie A 1970 Uronic acid materials. 35. Analytical study of gum exudates from some species of the genus Lannea. Phytochemistry 9:1585–1588

13 Anderson D M W, Hendrie A 1971 The proteinaceous gum polysaccharide from Azadirachta indica A. Juss. Carbohyd Res 20:259–268

14 Anderson D M W, Hendrie A 1972 The structure of Lannea humilis gum. Carbohyd Res 22:265–279

15 Anderson D M W, Hendrie A 1973 The structure of Lannea coromandelica gum. Carbohyd Res 26:105–115

16 Anderson D M W, Howlett J F, McNab C G A 1985 Studies of uronic acid material. 79. The amino acid composition of gum exudates from Prosopis species. Phytochemistry 24:2718–2720

17 Anderson D M W, Howlett J F, McNab C G A 1985 Studies of uronic acid materials. 81. The amino acid composition of the proteinaceous component of gum tragacanth (Asiatic Astragalus spp.). Food Addit Contam 2:231–235

18 Anderson D M W, Howlett J F, McNab C G A 1986 Studies of uronic acid materials. 89. The hydroxyproline content of gum exudates from several plant genera. Phytochemistry 26:309–311
19 Anderson D M W, Howlett J F, McNab C G A 1987 Studies of uronic acid materials. 88. Amino acid composition of gum exudates from some African Combretum, Terminalia and Anogeissus species. Phytochemistry 26:837–839
20 Anderson D M W, McDougall F J 1987 The composition of the proteinaceous gums exuded by Acacia gerrardii and Acacia goetzii subsp. goetzii. Food Hydrocolloids 1:327–331
21 Anderson D M W, Munro A C 1969 An analytical study of gum exudates from the genus Araucaria Jussieu (Gymnospermae). Carbohyd Res 11:43–51
22 Anderson D M W, Munro A C 1969 Presence of 3-*O*-methylrhamnose in Araucaria resinous exudates. Phytochemistry 8:633–634
23 Anderson L 1972 The cyclitols. In: Pigman W, Horton D (eds) The carbohydrates, 3rd ed, Vol 1A. Academic Press New York, 519–579
24 Anderson L, Wolter K E 1966 Cyclitols in plants: Biochemistry and physiology. Ann Rev Plant Physiol 17:209–222
25 Angyal S J, Anderson L 1959 The cyclitols. Adv Carbohyd Chem 14:135–212
26 Antonova G F, Maierle S N 1974 Characteristics of the composition of arabinogalactan polysaccharide produced in Siberian larchwood. Probl Fiziol Biokhim Drev Rast 1:4–5
27 Antonova G F, Malyutina E S, Shebeko V V 1983 Study of water-soluble polysaccharides in developing xylem of Scotch pine. Fiziol Rast 30:151–157
28 Antonova G F, Usov A I 1984 Structure of an arabinogalactan from Siberian larch (*Larix sibirica* Ladeb.) wood. Bioorg Khim 10:1664–1669
29 ap Rees T 1974 Pathways of carbohydrate breakdown in higher plants. In: Northcote D H (ed) Biochemistry I, vol 11. Butterworth London, 89–127
30 Aspinall G O 1969 Gums and mucilages. Adv Carbohyd Chem Biochem 24:333–379
31 Aspinall G O, Baillie J 1963 Gum tragacanth. Part I. Fractionation of the gum and the structure of tragacanthic acid. J Chem Soc 1702–1714
32 Aspinall G O, Baillie J 1963 Gum tragacanth. Part II. The arabinogalactan. J Chem Soc 1714–1721
33 Aspinall G O, Carlyle J J 1969 Anogeissus leiocarpus gum. IV. Exterior chains of leiocarpan A. J Chem Soc (C) 851–856
34 Aspinall G O, Carlyle J J, McNab J M, Rudowski A 1969 Anogeissus leiocarpus gum. II. Fractionation of the gum and partial hydrolysis of leiocarpan A. J Chem Soc (C) 840–845
35 Aspinall G O, Fairweather R M 1965 Araucaria bidwillii gum. Carbohyd Res 1:83–92
36 Aspinall G O, Fraser R N 1965 Plant gums of the genus Sterculia. II. Sterculia caudata gum. J Chem Soc 4318–4325
37 Aspinall G O, Fraser R N, Sanderson G R 1965 Plant gums of the genus Sterculia. III. Sterculia setigera and Cochlospermum gossypium gums. J Chem Soc 4325–4329
38 Aspinall G O, Hirst E L, Johnston M J 1962 The acidic sugar components of Cochlospermum gossypium gum. J Chem Soc 2785–2789
39 Aspinall G O, Hirst E L, Matheson N K 1956 Plant gums of the genus Khaya. The structure of Khaya grandifolia gum. J Chem Soc 989–997
40 Aspinall G O, Johnston M J, Stephen A M 1960 Plant gums of the genus Khaya. II. The major component of Khaya senegalensis gum. J Chem Soc 4918–4927
41 Aspinall G O, Johnston M J, Young R 1965 Plant gums of the genus Khaya. III. The minor component of Khaya senegalensis gum. J Chem Soc 2701–2710
42 Aspinall G O, McKenna J P 1968 Araucaria bidwillii gum. II. Further studies on the polysaccharide components. Carbohyd Res 7:244–254
43 Aspinall G O, McNab J M 1969 Anogeissus leiocarpus gum. III. Interior chains of leiocarpan A. J Chem Soc (C) 845–851
44 Aspinall G O, Molloy J A, Whitehead C C 1970 Araucaria bidwillii gum. III. Partial acid hydrolysis of the gum. Carbohyd Res 12:143–146
45 Aspinall G O, Whitehead C C 1970 Mesquite gum. I. The 4-*O*-methylglucuronogalactan core. Can J Chem 48:3840–3849
46 Aspinall G O, Whitehead C C 1970 Mesquite gum. II. The arabinan peripheral chains. Can J Chem 48:3850–3855

47 Ballou C E, Anderson A B 1953 On the cyclitols present in sugar pine (Pinus lambertiana Dougl.). J Am Chem Soc 75:648–650

48 Bekker P I, Churms S C, Stephen A M, Woolard G R 1972 Structural features of the gums of Acacia podalyriaefolia and Acacia elata. Standardization of the Smith degradation procedure. J S Afr Chem Inst 25:115–130

49 Belue G P, McGinnis G D 1974 High-pressure liquid chromatography of carbohydrates J Chromatogr 97:25–31

50 Beveridge T, Bruce K, Kok R 1978 Carbohydrate and mineral composition of gray birch syrup. Can Inst Food Sci Technol J 11:28–30

51 Beveridge R J, Stoddart J F, Szarek W A, Jones J K N 1969 Some structural features of the mucilage from the bark of Ulmus fulva (slippery elm mucilage). Carbohyd Res 9:429–439

52 Bhattacharyya A K, Rao C V N 1964 Gum jeol: The structure of the degraded gum derived from it. Can J Chem 42:107–112

53 Bhattacharyya AK, Mukherjee AK 1964 Structural studies of the gum jeol polysaccharide. Bull Chem Soc Japan 37:1425–1429

54 Brasch D J, Jones J K N, Painter T J, Reid P E 1959 Structure of some water-soluble polysaccharides from wood. Proc 2nd Cellulose Conf 3–15

55 Caffini N O, Priolo de Lufrano N S 1974 Composition of gums and mucilages of higher plants. Phytochemical revision. Rev Farm (Bueños Aires) 116:95–105

56 Caffini N O, Priolo de Lufrano N S 1975 Composition of gums and mucilages from higher plants. Phytochemical review. Rev Farm (Bueños Aires) 117:38–45

57 Churms S C, Freeman B H, Stephen A M 1973 Composition of the polysaccharide fraction of Araucaria klinkii gum. J S Afr Chem Inst 26:111–117

58 Churms S C, Merrifield E H, Miller C L, Stephen A M 1979 Some new aspects of the molecular structure of the polysaccharide gum from Acacia salinge (syn. cyanophylla). S Afr J Chem 32:103–106

59 Churms S C, Merrifield E H, Stephen A M 1977 Structural features of the gum exudates from some Acacia species of the series Phyllodineae and Botryocephalae. Carbohyd Res 55:3–10

60 Churms S C, Merrifield E H, Stephen A M 1978 Regularity within the molecular structure of arabinogalactan from Western larch (Larix occidentalis). Carbohyd Res 64:C1–C2

61 Churms S C, Merrifield E H, Stephen A M 1981 Chemical and chemotaxonomic aspects of the polysaccharide gum from Acacia longifolia. S Afr J Chem 34:8–11

62 Churms S C, Merrifield E H, Stephen A M 1981 Smith degradation of gum exudates from some Prosopis species. Carbohyd Res 90:261–267

63 Churms S C, Merrifield E H, Stephen A M, Stephen E W 1978 Evidence for sub-unit structure in the polysaccharide gum from Acacia mearnsii. S Afr J Chem 31:115–116

64 Churms S C, Stephen A M 1971 Structural aspects of the gum of Cussonia Thunb. (Araliaceae). Carbohyd Res 19:211–221

65 Clarke A E, Anderson R L, Stone B A 1979 Form and function of arabinogalactans and arabinogalactan proteins. Phytochemistry 18:521–540

66 Corbishley D A, Miller W 1984 Tapioca, arrowroot, and sago starches: Production. In: Whistler R L, BeMiller J N, Paschall E F (eds) Starch: Chemistry and technology, 2nd ed. Academic Press Orlando, 469–478

67 Côté W A Jr, Simson B W, Timell T E 1967 Studies on larch arabinogalactan. II. Degradation within the living tree. Holzforschung 21:85–88

68 Cranswick A M, Zabkiewicz J A 1979 Quantitative analysis of monosaccharides, cyclitols, sucrose, quinic and shikimic acids in Pinus radiata extracts on a glass support-coated open tubular capillary column by automated gas chromatography. J Chromatogr 171:233–242

69 Davies D R 1974 Some aspects of sucrose metabolism. Ann Proc Phytochem Soc 10 (Plant Carbohyd Biochem): 145–164

70 Dey P M, Brinson K 1984 Plant cell-walls. Adv Chem Biochem 42:265–382

71 Diamantoglou S 1974 Variation of cyclitol content in vegetative parts of higher plants. Biochem Physiol Pflanzen 166:511–523

72 Diamantoglou S 1980 Carbohydrate content and osmotic conditions of leaves and bark of Pinus halepensis Mill. and Schinus molle L. throughout a year. Ber Deutsch Bot Ges 93:449–457

73 Dietrichs H H, Garves K, Behrensdorf D, Sinner M 1978 Studies on the carbohydrates of barks of native trees. Holzforschung 32:60–67

74 Drummond D W, Percival E 1961 Structural studies on the gum exudate of Albizzia zygia (Macbride). J Chem Soc 3908–3917
75 Dutton G G S, Kabir S 1973 Structural studies of sapote (Sapota achras) gum. Carbohyd Res 28:187–200
76 Dutton G G S, Unrau A M 1963 Periodate oxidation of mesquite gum. Can J Chem 41:1417–1423
77 Essiamah S K 1980 Spring sap of trees. Ber Deutsch Bot Ges 93:257–267
78 Feingold D S, Avigad G 1980 Sugar nucleotide transformations in plants. In: Preiss J (ed) The biochemistry of plants, vol 3. Academic Press New York, 102–107
79 Ferguson A R, Eiseman J A, Dale J R 1981 Xylem sap from Actinidia chinensis: Gradients in sap composition. Ann Bot (London) 48:75–80
80 Fincher G B, Stone B A, Clarke A E 1983 Arabinogalactan-proteins: Structure, biosynthesis, and function. Ann Rev Plant Physiol 34:47–70
81 Gander J E 1976 Mono- and oligosaccharides. In: Bonner J, Varner J E (eds) Plant biochemistry, 3rd ed. Academic Press New York, 337–339
82 Giaquinta R T 1980 Translocation of sucrose and oligosaccharides. In: Preiss J (ed) The biochemistry of plants, vol 3. Academic Press New York, 271–320
83 Gill R E, Hirst E L, Jones J K N 1946 Mucilage from the bark of Ulmus fulva. II. Sugars formed in the hydrolysis of the methylated mucilage. J Chem Soc 1025–1029
84 Glicksman M 1983 Gum karaya (Sterculia gum). In: Glicksman M (ed) Food hydrocolloids, vol II. CRC Press Boca Raton FL, 39–47
85 Glicksman M, Sand RE 1973 Gum arabic. In: Whistler R L, BeMiller J N (eds) Industrial gums. Academic Press New York, 197–263
86 Gregory R A, Hawley G J 1983 Sap extraction and measurement of soluble sap sugars in sugar maple. Can J For Res 13:400–404
87 Haissig B E, Dickson R E (1979) Starch measurement in plant tissue using enzymatic hydrolysis. Physiol Plant 47:151–157
88 Hansen P, Grausland J 1978 Levels of sorbitol in bleeding sap in relation to leaf mass and assimilate demand in apple trees. Physiol Plant 42:129–133
89 Hirst E L, Hough L, Jones J K N 1949 The structure of Sterculia setigera gum. I. An investigation by the method of paper partition chromatography of the products of hydrolysis of the gum. J Chem Soc 3145–3151
90 Hirst E L, Hough L, Jones J K N 1951 Constitution of the mucilage from the bark of Ulmus fulva (slippery elm mucilage). III. The isolation of 3-monomethyl D-galactose from the products of hydrolysis. J Chem Soc 323–325
91 Hirst E L, Percival E, Williams R S 1958 The structure of Brachychiton diversifolium gum (Sterculia caudata). J Chem Soc 1942–1950
92 Hocking P J 1983 The dynamics of growth and nutrient accumulation by fruits of Grevillea leucopteris Meissn., a proteaceous shrub, with special reference to the composition of xylem and phloem sap. New Phytol 93:511–529
93 Höll W 1981 A thin-layer chromatographic demonstration of the annual fluctuation of soluble sugars in the trunkwood of three angiosperms and one gymnosperm. Holzforschung 35:173–175
94 Hough L, Jones J K N 1950 The structure of Sterculia setigera gum. II. An investigation by the method of paper partition chromatography of the products of hydrolysis of the methylated gum. J Chem Soc 1199–1203
95 Hudson C S, Sherwood S F 1918 The occurrence of melezitose in a manna from the Douglas fir. J Am Chem Soc 40:1456–1460
96 Ichiki S, Yamaya H 1982 Sorbitol in tracheal sap of dormant apple (Malus domestica Borkh) shoots as related to cold hardiness. Plant Cold-Hardiness Freezing Stress: Mech Crop Implic (Proc Int Plant Cold-Hardiness Seminar) 2:181–187
97 IUPAC Commission on the Nomenclature of Organic Chemistry and the IUPAC-IUB Commission on Biochemical Nomenclature 1974 Nomenclature of cyclitols. Recommendations (1973) Pure Appl Chem 37:283–297. In: Fasman GD (ed) Handbook of biochemistry and molecular biology. Lipids, carbohydrates, steroids, 3rd ed. CRC Press Cleveland OH, 89–99
98 Jeremias K 1962 The influence of temperature on the accumulation of raffinose sugar. Ber Deutsch Bot Ges 75:313–322
99 Jiang K S, Timell T E 1972 Polysaccharides in the bark of aspen (Populus tremuloides). I. Isolation and constitution of a 4-*O*-methylglucuronoxylan. Cellul Chem Technol 6:493–498

100 Jiang K S, Timell T E 1972 Polysaccharides in the bark of aspen (Populus tremuloides). II. Isolation and structure of an arabinan. Cellul Chem Technol 6:499–502
101 Johnston J F W 1843 Phil Mag 23:14–18 reported by Dey P M 1980 Biochemistry of α-D-galactosidic linkages in the plant kingdom. Adv Carbohyd Chem Biochem 37:283–372
102 Joseleau J-P, Chambat G, Vignon M, Barnoud F 1977 Chemical and ^{13}C n.m.r. studies on two arabinans from the inner bark of young stems of Rosa glauca. Carbohyd Res 58:165–175
103 Kallio H, Ahtonen S, Raulo J, Linko R R 1985 Identification of the sugars and acids in birch sap. J Food Sci 50:266–267
104 Kandler O, Hopf H 1980 Metabolism and function of oligosaccharides. In: Preiss J (ed) The biochemistry of plants, vol 3. Academic Press New York, 221–270
105 Karácsonyi S, Toman R, Janeček F, Kubračkova M 1975 Polysaccharides from the bark of the white willow (Salix alba L.): Structure of an arabinan. Carbohyd Res 44:285–290
106 Kardošová A, Rosík J, Kubala J 1978 Neutral oligosaccharides from the enzyme hydrolysate of the polysaccharide of peach-tree gum (Prunus persica (L.) Batsch). Coll Czech Chem Commun 43:3428–3432
107 Kardošová A, Rosík J, Kubala J, Kovácik V 1979 Structure of neutral oligosaccharides from the enzyme hydrolysate of polysaccharide of peach gum (Prunus persica (L.) Batsch). Coll Czech Chem Commun 44:2250–2254
108 Kasapenko L F 1983 Variation in the polysaccharide content of Larix gmelini wood. Prodovol Kormovye Resur Lesor Sib, 114–121
109 Kaverzina L N, Prokushkin S G, Degermendzhi N N 1981 Composition and dynamics of root exudates in scotch pine. Lesovedenie 32–38
110 Kazaryan V V, Zakaryan S O, Oganesyan L N 1982 Seasonal dynamics of carbohydrates, nitrogen and phosphorus compounds in the roots of woody newly acclimatized plants. Biol Zh Avm 35:667–672
111 Keegstra K, Talmadge K W, Bauer W D, Albersheim P 1973 The structure of plant cell walls. III. A model of the walls of suspension-cultured sycamore cells based on the interconnections of the macromolecular components. Plant Physiol 51:188–196
112 Klason P 1929 Constitution of pine wood lignin. VIII. The nutritive sap of the pine. Ber 62B:635–639
113 Komatsu S, Ueda H 1925 The constitution of polysaccharides. III. On plant mucilage I. Mem Coll Sci Kyoto Imp Univ 8:51–57
114 Kretz A 1973 Sugars, cyclitols, and organic acids in the cambial sap of Pinus sylvestris L., Picea abies Karst., and Abies alba Mill. Planta (Berlin) 110:1–14
115 Kuliev V B, Kasumov K N 1982 Polysaccharide of the gums of Astragalus microcephalus Willd. from Azerbaijan. Rastit Resur 18:390–394
116 Lee S H 1961 The new polysaccharides of gum tragacanth. I. Isolation of the polysaccharide and monosaccharide components. Seoul Univ J Ser D 10:47–54
117 Lee S H 1962 New polysaccharide of gum tragacanth. II. The chemical structure of polysaccharide C. J Korean Agr Chem Soc 3:25
118 Loewus F, Chen M-S, Loewus M W 1980 The *myo*-inositol pathway to cell wall polysaccharides. In: Loewus F (ed) Biogenesis of cell wall polysaccharides. Academic Press New York, 1–27
119 Loewus F A, Loewus M W 1980 *myo*-Inositol: Biosynthesis and metabolism. In: Preiss J (ed) The biochemistry of plants, vol 3. Academic Press New York, 43–76
120 Loewus F A, Loewus M W 1983 *myo*-Inositol: Its biosynthesis and metabolism. Ann Rev Plant Physiol 34:137–161
121 Marvin J W 1957 Investigations of the sugar content and flow mechanism of maple sap. Tappi 40:209–216
122 McDonald A J S, Ericsson A, Lohammar T 1986 Dependence of starch storage on nutrient availability and photon flux density in small birch (Betula pendula Roth). Plant Cell Environ 9:433–438
123 McGarvie D, Parolis H 1979 The mucilage of Opuntia ficus-indica. Carbohyd Res 69:171–179
124 McGarvie D, Parolis H 1979 The mucilage of Opuntia ficus-indica. 2. The degraded polysaccharide. J Chem Soc Perkin Trans I 1464–1466
125 McGarvie D, Parolis H 1981 The acid-labile, peripheral chains of the mucilage of Opuntia ficus-indica. Carbohyd Res 94:57–65
126 McGarvie D, Parolis H 1981 The mucilage of Opuntia aurantiaca. Carbohyd Res 94:67–71

127 Meer G, Meer W A, Gerard T 1973 Gum tragacanth. In: Whistler R L, BeMiller J N (eds) Industrial gums. Academic Press New York, 289–299
128 Meier H 1962 Studies on a galactan from tension wood of beech (Fagus sylvatica L.). Acta Chem Scand 16:2275–2283
129 Moyna P, DiFabio J L 1978 Composition of Cactaceae mucilages. Planta Med 34:207–210
130 Nikaido H, Hassid W Z 1971 Biosynthesis of saccharides from glycopyranosyl esters of nucleoside pyrophosphates ("sugar nucleotides"). Adv Carbohyd Chem Biochem 26:351–483
131 Nikitin N I 1962 The chemistry of cellulose and wood. (Translation from Khim Drev Tsellyul by J. Schmorak, Israel Program Scientific Trans, Jerusalem, 1966)
132 Onuki M 1933 Chemical constitution of stachyose. J Agr Chem Soc Japan 9:90–111
133 Parikh V M, Jones J K N 1966 Cholla gum. I. Structure of the degraded cholla gum. Can J Chem 44:327–333
134 Parikh V M, Jones J K N 1966 Cholla gum. II. Structure of the undegraded cholla gum. Can J Chem 44:1531–1539
135 Paulsen B S, Lund P S 1979 Water-soluble polysaccharides of Opuntia ficus-indica cv "Burbank's Spineless". Phytochem 18:569–571
136 Phillips G O, Pass G, Jeffries M, Morley R G 1980 Use and technology of exudate gums from tropical sources. In: Neukom H, Pilnik W (ed) Gelling and thickening agents in foods. Forster Zürich, 135–161
137 Plouvier V 1963 Distribution of aliphatic polyols and cyclitols. In: Swain T (ed) Chemical plant taxonomy. Academic Press London, 313–336
138 Pontis H G 1977 Riddle of sucrose. Int Rev Biochem 13:79–117
139 Pontis H G, Salerno G L 1980 Regulation of sucrose levels in plant cells. In: Marshall J J (ed) Mechanisms of saccharide polymerization and depolymerization. Academic Press New York, 31–42
140 Posternak T 1966 The cyclitols. Holden-Day San Francisco
141 Preiss J (ed) 1980 The biochemistry of plants, vol 3. Carbohydrates: structure and function. Academic Press, New York
142 Roudier A J, Eberhard L 1963 Investigation of the hemicelluloses of maritime pine of swamps (Pinus maritima). III. Polyosides extracted from the wood with boiling water. Constitution of a glucuronoarabinoxylogalactan among them. Bull Soc Chim Fr 844–850
143 Roudier A J, Eberhard L 1965 Hemicelluloses of wood of the maritime pine of Landes. IV. Polysaccharides extracted by boiling water. Structure of an arabinan present in the extract. Bull Soc Chim Fr 460–464
144 Roudier A J, Eberhard L 1967 Hemicelluloses from maritime pine of swamps (Pinus maritima) VI. Polysaccharides extracted from this wood by hot water. Structure of a galactoglucomannan present among them. Bull Soc Chim Fr 1741–1745
145 Saag L M K, Sanderson G R, Moyna P, Ramos G 1975 Cactaceae mucilage composition. J Sci Fd Agr 26:993–1000
146 Sarkar M, Rao C V N 1974 Graded hydrolysis of the gum from Rhizophora mangle. Indian J Chem 11:1129–1133
147 Sarkar M, Rao C V N 1975 Structure of degraded mangle gum. Carbohyd Res 41:163–174
148 Sauter J J 1980 Seasonal variation of sucrose content in the xylem sap of Salix. Z Pflanzenphysiol 98:377–391
149 Sauter J J, Ambrosius T 1986 Changes in the partitioning of carbohydrates in the wood during bud break in *Betula pendula* Roth. J Plant Physiol 124:31–43
150 Schwalbe C G, Ender W 1931 Wood saps. I. Sap of the red beech (Fagus sylvatica). Cellulosechemie 12:316–318
151 Sebrell W H Jr, Harris R S (eds) 1971 Inositols. In: The vitamins, 2nd ed., Vol III. Academic Press New York, 340–415
152 Selvendran R R 1985 Developments in the chemistry and biochemistry of pectic and hemicellulosic polymers. J Cell Sci (Suppl) 2:51–88
153 Sherrard E C, Kurth E F 1928 Occurrence of pinite in redwood. Ind Eng Chem 20:722–723
154 Simson B W, Côté A W Jr, Timell T E 1968 Larch arabinogalactan. IV. Molecular properties. Svensk Papperstidn 71:699–710

155 Simson B W, Timell T E 1978 Polysaccharides in cambial tissues of Populus tremuloides and Tilia americana. I. Isolation, fractionation, and chemical composition of the cambial tissues. Cellul Chem Technol 12:39–50
156 Simson B W, Timell T E 1978 Polysaccharides in cambial tissues of Populus tremuloides and Tilia americana. II. Isolation and structure of a xyloglucan. Cellul Chem Technol 12:51–62
157 Simson B W, Timell T E 1978 Polysaccharides in cambial tissues of Populus tremuloides and Tilia americana. III. Isolation and constitution of an arabinogalactan. Cellul Chem Technol 12:63–77
158 Simson B W, Timell T E 1978 Polysaccharides in cambial tissues of Populus tremuloides and Tilia americana. IV. 4-*O*-Methylglucuronoxylan and pectin. Cellul Chem Technol 12:79–84
159 Smith F, Montgomery R 1959 The chemistry of plant gums and mucilages. Van Nostrand-Reinhold Princeton NJ
160 Smith L V, Zavarin E 1960 Free mono- and oligosaccharides of some California conifers. Tappi 43:218–221
161 Stephen A M 1956 The composition of Hakea acicularis gum. J Chem Soc 4487–4490
162 Stephen A M 1983 Other plant polysaccharides. In: Aspinall G O (ed) The polysaccharides, vol 2. Academic Press New York, 97–193
163 Stephen A M, de Bruyn D C 1967 The gum exudate of Encephalartos longifolius Lehm. (female) (family Cycadaceae). Carbohyd Res 5:256–265
164 Stephen A M, Van der Bijl P 1974 Structural studies of Brabeium stellatifolium gum by methylation analysis and Smith degradation. J S Afr Chem Inst 27:37–43
165 Stephen A M, van der Bijl P 1975 Molecular structure of Grevillea robusta gum. J S Afr Chem Inst 28:43–46
166 Stewart C M, Melvin J F, Ditchburne N, Tham S H, Zerdoner E 1973 The effect of season of growth on the chemical composition of cambial saps of Eucalyptus regnans trees. Oecologia 12:349–372
167 Stinson E E, Dooley C J, Purcell J M, Ard J S 1967 Quebrachitol – a new component of maple sap and sirup. J Agr Food Chem 15:394–397
168 Tanret C 1902 Bull Soc Chim (3) 27:947
169 Terazawa M, Miyake M 1984 Phenolic compounds in living tissue of woods. II. Seasonal variations of phenolic glycosides in the cambial sap of woods. Mokuzai Gakkaishi 30:329–334
170 Thivend P, Mercier C, Guilbot A 1972 Determination of starch with glucoamylase. Methods Carbohyd Chem 6:100–105
171 Toman R, Karácsonyi S, Kováčik V 1972 Polysaccharides from the bark of the white willow (Salix alba L.): Structure of a galactan. Carbohyd Res 25:371–378
172 Wallner W E, Gregory R A 1980 Relationship of sap sugar concentrations in sugar maple to ray tissue and parenchyma flecks caused by Phytobia setosa. Can J For Res 10:312–315
173 Wehmer C 1929 Die Pflanzenstoffe. 2nd ed., vol I. 1931 vol II. 1935 Supplement. G Fischer Jena
174 Whistler R L, Smart C L 1953 Polysaccharide chemistry. Academic Press New York, 331–332
175 Willets C O, Hills C H 1976 Maple sirup producers manual. US Dept Agr, Agr Handbook No. 134
176 Williams M W, Raese J T 1974 Sorbitol in tracheal sap of apple as related to temperature. Physiol Plant 30:49–52
177 Wise L E 1952 Miscellaneous extraneous components of wood. In: Wise L E, Jahn E C (eds) Wood chemistry, 2nd ed. Reinhold New York, 644 pp
178 Wright L C, Berryman A A, Gurusiddaiah S 1979 Host resistance to the fir engraver bettle, Scolytus ventralis (Coleoptera: Scolytidae). 4. Effect of defoliation on wound monoterpene and inner bark carbohydrate concentrations. Can Entomol 111:1255–1262
179 Yoshimoto T, Saburi Y, Samejima M, Taoka S, Ogawa T 1984 Fatty acids, terpenes, and sugars in extracts from coniferous barks. Mokuzai Gakkaishi 30:335–339

Chapter 5

Nitrogenous Extractives

The aim of Chapter 5 is to integrate in a general way the different trends in the literature and indicate where the reader can find specific details concerning the chemistry of nitrogenous extractives such as amino acids, proteins, nucleic acids, and alkaloids, and their products that occur in and beyond the lignocellulosic cell wall. Since most of the literature is covered mainly to 1983, a few key references have been added to bring selected topics up to date. The expectation is that with the further development of useful experimental systems we would be able to explain eventually why and where such compounds occur and how they can be used to our benefit.

The chemistry of cell walls usually reveals the presence of traces of nitrogenous extractives – e.g., 0.04% to 0.17% N for elemental compositions (56, 117, 121). Compounds that contain this nitrogen are now believed to have great significance in the responsiveness of woody plants to a range of biotic and abiotic stimuli. Because the development of cell walls involves proteins (structural, enzymatic, reserves) synthesized from simple nitrogen sources, and proteins that are degraded to the free amino acids, nitrogenous compounds are considered to occur in the extracellular environment.

Another major nitrogenous constituent in trees is chlorophyll, but this is essentially absent from woody tissues. The traces of nitrogen-containing compounds in woody tissues often represent intermediary metabolites and translocated forms (Sect. 5.1), and compounds more typical of end-products of metabolism such as the alkaloids (Sect. 5.2). Alkaloids originate from acetate-malonate pathway, the mevalonate pathway, and various amino acid pathways.

5.1 Amino Acids, Proteins, Enzymes, and Nucleic Acids

D. J. DURZAN

5.1.1 Introduction

The natural occurrence and the role of free and bound amino acids, proteins, nucleic acids, and related products (e.g., 8, 42, 116) are described in relation to factors determining the composition of the lignocellulosic cell wall. The biosynthesis of extraneous nitrogenous compounds involves a sequence of events that stem from molecular reactions inside cells, on the surfaces of cells, and in the soluble component of plant cell walls.

Usually growth and development are integrated by linkages of daughter cells composing tissues and organs that evolve during the life cycle of the woody plant.

Partial integration is achieved internally through plasmodesmata (cytoplasmic connections between cells). Hence, the external release of compounds during assimilation, as responses to stress, and for the development of lignocellulosic structures is of considerable interest. For the latter, the unusual release of nitrogenous compounds must eventually relate to the well-being of plants, the origin of secondary meristems, and the amplified production of wood, fiber, and extractives.

The release of soluble materials by cells to the external environment is usually mediated by the developing surface of the plasma membrane. Hence, the study of the allocation of metabolites requires a better understanding of the origin, irritability, and activities of live cells. Once compounds are released or discharged, further chemical modifications may occur spontaneously or enzymatically. Fortunately, this complex process can now be studied in cells with or without cell walls – i.e., as protoplasts or under aseptic conditions.

The unusual appearance of extraneous nitrogenous compounds along the various phases of the life cycle, particularly for perennial woody species, are elicited by numerous genetic and environmental factors. The main factors contributing to the release of these compounds, however, involve stress or injury from insects, disease, and the impact of human activities.

The occurrence of nitrogenous compounds, particularly in cells in mature parts of the tree, may represent physical, chemical, and biological changes over the life cycle. For dead woody cells that compose the vascular system, the nitrogenous extractives may reflect the recent past history of the cells and the enzymes that have survived cellular differentiation. Labile enzymes are present in the secondary walls of differentiated reaction wood. Peroxidase, which is a universally stable enzyme in cell walls, cannot be detected histochemically in walls of dead cells (120).

Historically, nitrogenous extractives are often described in terms of their natural occurrence, chemistry, and taxonomy. This was true at least until cells and tissues could be brought into the laboratory and controlled under more convenient cultural conditions. Much of the recent scattered literature therefore combines chemical analysis and cell and tissue culture methods with variables associated with experimental systems. For the woody perennials, greater emphasis is now placed on cell wall synthesis, problems of cellular differentiation, soluble compartments in the cell wall, and a wide range of specialized topics (Tables 5.1.1 – 5.1.3).

For recent methods dealing with the in vitro cultivation of cells of trees, reference may be made to the books by Bonga and Durzan (13, 14). The application of cell and tissue culture in pathology has been covered by Helgeson and Deverall (61) and in a series edited by Sharp et al. (122). The harvesting of profitable products from plant tissue cultures has been evaluated by Curtin (29). Metabolites liberated by organs – e.g., roots of white pine (*Pinus strobus*) – may be studied with carbon-14 using the culture methods of Slankis et al. (125).

For the study of the soluble compartments of cell walls, the recent methods described by Terry and Bonner (131) and by Cosgrove and Cleland (25, 26) are gaining considerable interest but these methods are not without their pitfalls.

The production of extracellular compounds through refinements in plant biotechnology is a major current trend (29, 127). The opportunity here is that cell

Table 5.1.1 Annotated chronological summary: studies on formation of materials containing amino acids by natural occurrence or in cell and tissue cultures

Amino acid, protein, etc.	Species and remarks	References
Free amino acids	Alcohol extracts of wood (several species)	52
21 amino acids in "skin substance"	In glycoprotein component cell wall fibers	92
Proline	Postulated as xylogenic factor during induction of wound vessel differentiation in *Coleus*	114
Secretion of amino acids	Transfer cells in sieve elements of legumes	109
Phosphatase	Role of polysaccharides, non-secretion, and origin of enzymes during penetration of lateral root meristems in *Pisum sativum* and *Zea mays*	5
Bound amino acids and reaction products	Amino acid composition of wood proteins of *Pinus* and *Eucalyptus* spp.	121
Cell wall proteins and extensin	Review of literature	78
Extracellular polysaccharide with 5% to 10% protein	From cell suspensions of bush bean (*Phaseolus vulgaris*)	83
Lectins	As tools for studying polysaccharides, glycoproteins, and cell surfaces	88
	Lectins in higher plants	87
	Specificity in *Rhizobium*-legume root nodule symbiosis	12
β-Lectins	Arabinogalactan-proteins precipitated by Yariv β-glycosyl artificial antigens	69
Proteins among polysaccharides and lignin-like substances	Deposition beneath cultures of shoot apical meristems	115
Cell surface proteins	Incompatibility in forest trees	59
	Incompatibility and pollen stigma interactions	63
β-Hydroxy-phenylalanine and β-hydroxytyrosine	Glycoprotein linkages in plant cells	86
Glycoprotein (chemotactin)	Suggests chemotaxis of *Rhizobium* sp. to root hairs under influence of Chemotacin (*Lotus corniculatus*)	28
Nucleotide diphosphate sugars	View that lignin-carbohydrate complex is formed by transglycosylation from nucleotide diphosphate sugars (UDGP) to primary OH group at the C_γ of the lignin side chain in cell wall	65
Nitrogenous compounds	Limited role in nonspecific defence in bark and wood during wounding, insect and pathogen attack	101
Lectin (chemotactin)	Lectin immobilizes *Rhizobium* sp.	33
	Review	6
Arabino-3,6-galactan proteins	In gum exudates and callus cells in culture review	19
Proteins and glycoproteins	Pollen-stigma interactions and surface components in *Gladiolus*	20
Protein-rich polysaccharide slime	secretion from ovary glands in *Aptenia cordifolia*	73
Secretory arabino-galactan-protein	In cell walls of *Phaseolus*	139
	Chemistry of recognition	91
Hydroxyproline-rich proteins	Levels in water-extracted cell walls of *Pisum sativum* is increased by ethylene	132

Table 5.1.1 (continued)

Amino acid, protein, etc.	Species and remarks	References
Lignin-protein complex	In cell walls of callus from *Pinus elliottii*	146, 140
Isodityrosine	New cross-linking amino acid from cell-wall glycoprotein of potato (*Solanum tuberosum*) calli	48
Lectin	Content in extracellular medium is significantly higher for cultures of the winged bean (*Psophocarpus tetragonolobus*) and for tissues grown in vitro in general	96
Protein glycosylation	Review for higher plants	32
Cell wall digesting enzymes	Induction of ethylene biosynthesis by enzymes in tobacco (*Nicotiana tabacum*) leaf discs	4
Salt-soluble protein with enzymatic activity	Release during ripening of normal and mutant tomato (*Lycopersicon esculentum*) cell walls	66
Glycoprotein cell fusion hormone	In filaments of red alga (*Griffithsia pacifica*)	143
Abiotic elicitor (ribonuclease)	To induce phytoalexin in french bean (*Phaseolus vulgaris*) and show intercellular transmission of elicitation response	37
Glycoproteins in somatic polyembryogenesis	Acetocarmine-reactive material in embryonal cells in mucilage around embryonal suspensor masses of loblolly pine (*Pinus taeda*)	58

walls or cells themselves could be used as reaction vessels or surfaces to produce specific, commercially important, extracellular products, which the chemist cannot match with current engineering technologies. The use of binding methods and carriers is aimed at achieving reactors of high activity, high conversion yields, and high fluxes, with long stability of the enzyme and matrix. This topic is quite broad. A few selected references to special topics such as immobilized enzymes and production of single cell proteins are given in Table 5.1.3.

5.1.2 Composition

The range of naturally occurring nitrogenous extractives in woody plants is reviewed by Gottlieb in Chap. 3 of this volume and is prerequisite to this chapter. Larsen (80) has discussed the physical and chemical properties of the amino acids. Durzan and Steward (42), Higgins and Payne (64), and Miége (99) have reviewed the nitrogen metabolism of the free amino acids, peptides, proteins and nucleic acids in higher plants.

The production of extracellular nitrogenous compounds normally proceeds through several levels – molecular, subcellular, and cellular – and extends into the physiological and biological performance. This process usually involves the amplified production of natural products such as the plant growth regulators (47), alkaloids (Sakai et al., Sect. 5.2), a variety of osmolytes (148), and proteinaceous substances such as the lectins (87, 88).

Lectins are sugar-binding proteins or glycoproteins of nonimmune origin. They agglutinate cells and precipitate glycoconjugates. Lectins may have two binding sites and show some toxic or hormone-like activities. They normally do not exhibit enzymatic activities toward carbohydrates to which they bind. Lectins also do not require free glycosidic hydroxyl groups on these sugars for their binding.

Our understanding of the range of these and other substances should be improved by developing new ways of recovering extracellular products. Using a novel method for removing low molecular weight components from the free space of the cell wall, Cosgrove and Cleland (25, 26) have estimated that in the free space of growing stem tissues, 25% of the solute is inorganic electrolytes. The other 75% constitutes organic nonelectrolytes with an average size similar to that of glucose.

Osmotic pressures in the wall space are believed to explain the observation that nontranspiring plants have negative water potentials. Cosgrove and Cleland conclude that the internal gradient in water potential from the xylem to the epidermis, which sustains cell enlargement, is small. Auxin does not alter the hydraulic conductance of stem tissue either at the cellular or whole tissue level. How extracellular nitrogenous compounds contribute to or modify this response remains unknown.

5.1.2.1 Free and Bound Amino Acids

Generally, amino acids, proteins, and their products are synthesized by enzymes according to current dogma (DNA makes mRNA makes protein). The occurrence of extracellular proteins involves the dynamics of cellulose and callose biosynthesis and cell wall development (9, 106). Against this background, the emergence of a pattern in the cell walls of woody plants may be studied (62). At the cellular levels, macromolecular biosynthesis is commonly believed to involve the dynamics of membrane transformation and flow (100, 76).

Components of membrane transformation systems include the nucleus with its DNA, nonhistone chromosomal proteins, and the derived informational molecules (e.g., mRNA) for protein synthesis. The mRNA moves into the cytoplasm through the transformation of membranes. For example, the rough endoplasmic reticulum contains ribosomal RNA with its templates for protein synthesis. The system evolves with cytoplasmic organelles and substrates to influence structure and function. During this process the vacuoles, microtubules, endoplasmic reticulum, subcellular organelles, and plasmalemma may release specific metabolic nitrogenous products (conjugates, enzymes) internally or externally. Some of these may be involves in the biosynthesis of cutin and suberin (76).

The processing of protein usually involved synthesis, segregation, transport, concentration, storage, and discharge at the membrane (108). Much of the current view on the control of this process implicates plant growth regulators such as the auxins and cytokinins, which control cell division and expansion during growth and development of the cell wall (77, 106). Amino acid conjugates of natural and synthetic growth regulators are now commonly found (42).

Free amino acids have been detected in alcohol extracts of wood of various species (52, 98). Bound amino acids in sapwood, heartwood, and reaction wood have been reported by Scurfield and Nicholls (121). The amounts of tyrosine in the wood of *Pinus radiata* seedlings are incompatible with the belief that tyrosine determines the intensity of staining of reaction wood cell walls with Millons reagent.

Tyrosine is the source of *p*-hydroxybenzaldehyde when milled wood lignin is oxidized with nitrobenzene. The loss of more than 20% of individual amino acids in the presence of milled wood lignin and a number of sugars and glucans was demonstrated. Numerous unidentified compounds were detected; some may arise as reaction products among amino acids and sugars (e.g., 57).

When woody plant cells are grown in culture and exploited for their products, the medium in which the cells grow is commonly "conditioned" by the excretion or release of a wide variety of compounds (Fig. 5.1.1). In jack pine (*Pinus banksiana*), these are free amino acids and amino sugars (glucosamine) (41). Other compounds may represent substrates such as *S*-adenosylmethionine for the methylation of hemicellulose (70), or methionine derivatives for the formation of ethylene, ammonia, HCN, and asparagine during aging and ripening processes (149).

Extracellular free amino acids are usually those commonly found in protein. Some non-protein amino acids, such as γ-aminobutyric acid and β-amino acids, also occur. Low levels of β-amino acids are usually products of nucleic acid degradation.

Cyanogenic glucosides and conjugates between amino acids, and phenols or between amino acids and growth regulators may be found at the low levels but the role of these compounds as indicators of the physiological state of cells remains to be worked out.

Recently, the physiological roles of extracellular components have been studied. Biologically active cell-wall fragments in sycamore maple (*Acer pseudoplatanus*) cells have been related to amino acid transport and protein synthesis (50, 147) and to the control of morphogenesis (134; Sect. 5.1.3.3).

Acid-released sycamore maple cell-wall fragments inhibit cellular growth. High fragment concentrations killed the cultured cells (50). At sublethal doses, wall fragments inhibited the transmembrane movement of amino acids within 1 minute after treatment. After 1 hour of treatment, amino acid transport was restored. Total protein synthesis was suppressed independently of the effect on amino acid transport. Results were related to the mechanism of killing plant cells by pectic enzymes and to the defense responses shown by plants to pathogens.

Free arabinose and galactose, which are often associated with hydroxyproline-rich proteins, are found in the free space of cell walls (132). Levels of proteins and free hydroxyproline are increased by ethylene treatment of pea (*Pisum sativa*) stem segments. However, the total amount of hydroxyproline in ethanol-insoluble polymers after extraction of the free space with water and Ca^{2+} was not influenced by ethylene. Terry et al. (132) propose that the response to ethylene, which is now known to be derived from methionine, can be divided into two components. One requires changes in cellulose microfibrils of the cell wall, which result in reorientation of the plane of cell expansion. The other involves a change in the hemicellulosic xyloglucan, which inhibits extension growth of these cells.

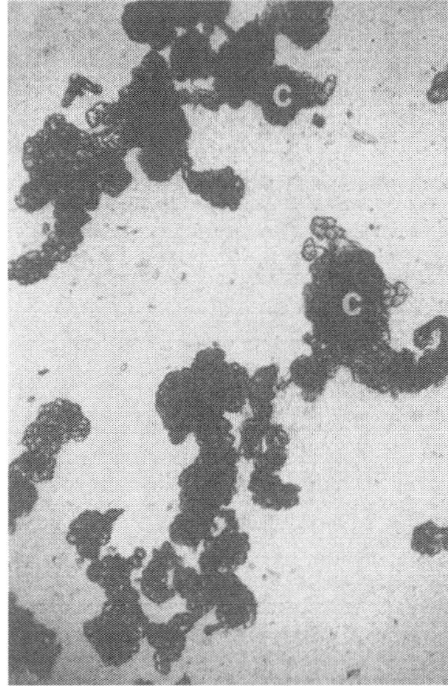

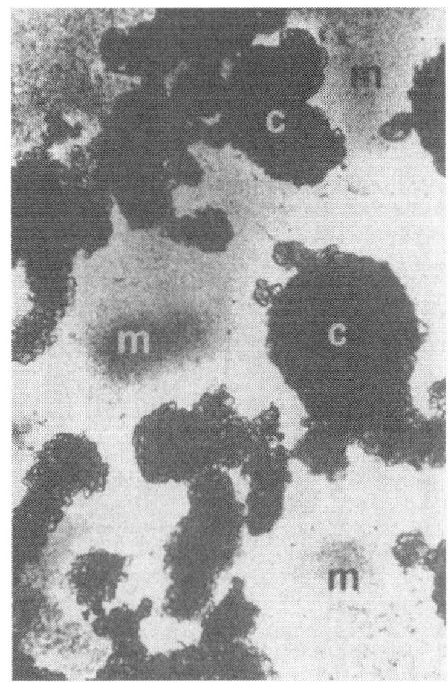

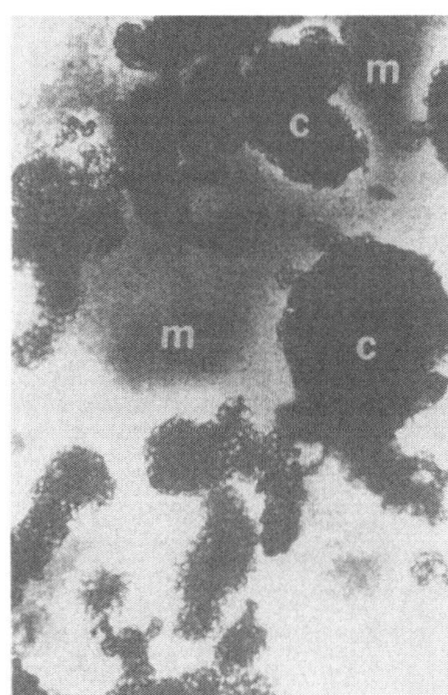

Fig. 5.1.1. Conditioning of the medium by growing clusters of cells (*c*) from aseptic cell suspension cultures of Douglas-fir (*Pseudotsuga menziesii*). During this process a variety of exogenous materials containing nitrogenous substances are released to condition the medium. Figures A through C show the growth of cell clusters and the release of exogenous water-insoluble materials (*m*) deposited on plastic surfaces. Eventually this material engulfs the growing tissues. Clumps (×31) after 4 (A), 11 (B), and 20 days (C). Similar substances from excised apical domes of conifer buds have been trapped on Millipore filters by Romberger and Tabor (115). Generally, exogenous factors are chemically complex (lectins, glycoproteins, etc.) and usually contain a limited range of bound protein amino acids. Analysis of depositions (*m*) shows them to contain bound amino acids, lignin, and polysaccharides. The ability of cells to produce (*m*) is a function of many stimuli. In most instances their production increases with sucrose levels in the medium. Production is believed to involve a balance of naturally occurring cytokinins and auxins or other growth regulators, either natural or synthetic

Nitrogenous compounds also occur as root exudates (125). These compounds provide nutrients for rhizosphere microbes and root symbionts. Nitrogenous compounds also play a role in drought tolerance, in salinity, in competition among higher plants, and in attack by root pathogens (1, 30, 126). The role of allelopathic compounds – i.e., those involved with competition between plants – has been reviewed by Putnam (111).

5.1.2.2 Proteins and Enzymes

Hydroxyproline-rich glycoproteins are among the most prominent proteins in cell walls. The hydroxylation of proline is one of over 100 post-translational modifications that can occur on amino acid residues in protein (137). As a result of protein turnover (31) and cell stress, sometimes these amino acids may be found in extracellular matrices.

Plants contain at least three classes of hydroxyproline-rich glycoproteins: 1) the cell wall proteins, 2) the arabinogalactan proteins, and 3) lectins. According to Cooper and Varner (23) the hydroxyproline-rich glycoprotein of the cell wall of carrots is secreted from the cytoplasm as a soluble monomer that slowly becomes insolubilized.

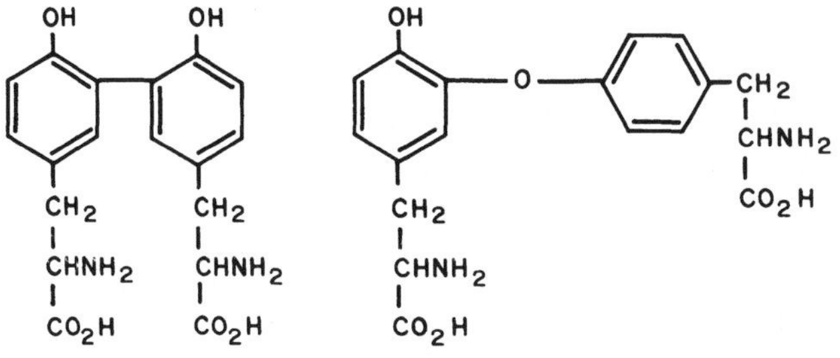

dityrosine　　　　**1** isodityrosine

Isoditrosine (**1**) is found in the wall during this insolubilization and could serve as a protein-protein cross link. The insolubilization of the glycoprotein is inhibited by peroxidase inhibitors and free radical scavengers, the most effective being L-ascorbate. Fry (49) believes that the cross-linked protein network is under the control of an extracellular peroxidase/ascorbic acid oxidase system. Isodityrosine, a cross-linked amino acid in cell-wall glycoprotein, is considered to be formed outside the cell wall by peroxidase (48). Cross-linking of matrix polymers in the growing cell walls of angiosperms has recently been summarized by Fry (49).

Arabinoglycoproteins occur in seeds in cell and tissue cultures (19), and in cambial tissue. An example of the extraneous materials containing bound amino acids and deposited on a plastic dish by Douglas-fir (*Pseudotsuga menziesii*) cells growing into sphaeroblasts is shown in Fig. 5.1.1 (cf., 13 p 36–71).

With cells and tissues in suspension culture, the matrix of deposited materials extracellular to the cell wall often stains for lignin and contains tannins and glycoproteins of yet undefined composition (82). This material binds with synthetic growth regulators introduced into the medium. The chemical nature of these extraneous materials is a strong function of the tissue or cell explant source and of the conditions of culture (40).

In this context it is interesting to note an unsuccessful attempt to apply a lactoperoxidase-iodide surface-labeling procedure to excised segments of *Cucurbita* (112). The aim was to study phytochrome and monitor the distribution of the plasma membrane during subsequent cell fractionation. The reason for this fail-

Table 5.1.2. Annotated chronological summary: Studies on production of materials containing amino acids related to disease and pathology (phytoalexins and elicitors)

Amino acid, protein, etc.	Species and remarks	References
[^{14}C]-Leucine	Effect of acid-released cell wall fragments on amino acid transport and protein synthesis in sycamore maple (*Acer pseudoplatanus*) cell suspensions	50
Enzyme production	Bacteria from water-stored Sitka spruce (*Picea sitchensis*)	46
Proteins	Review on biochemistry of pathogenesis in plants	2
	Review on biochemical mechanisms of disease resistance	7
Extracellular pectic enzymes	*Macrophonia phaseolina* (root-rot of *Sesamum indicum*)	38
Hydrolytic enzymes	Suction marks in nutritive cells of a gall on leaves of *Acer* caused by a stinging parasite	119
Enzymes in lignin degradation	Activity in *Phanerochaete chrysosporium*	72
	Hydrolytic and oxidative enzymes in *Sporotrichum*	43
Proteinase inhibitor (inducing activity)	In sycamore maple (*Acer pseudoplatanus*) and tomato (*Lycopersicon esculentum*) cell	118
	Activity resides in oligosaccharides enzymatically released from wall	10
Polygalacturonase	Elicitor from *Rhizopus* for casbene synthetase activity	81
Cell wall degrading enzymes	Isolates of *Ophiostoma ulmi*	129
Cell wall fragments	Inhibition of [^{14}C]-leucine incorporation into cell walls of sycamore maple (*Acer pseudoplatanus*)	147
Enzyme to depolymerize lignin	Isolated and purified from white-rot fungus *Phanerochaete chrysosporium*	133

ure was the presence of an extracellular component that was heavily labeled by the iodinating system and that becomes associated with cellular membranes upon homogenization. The extracellular constituent believed to be a phloem exudate protein was found to mimic the expected behavior of an iodinated plasma membrane.

Proteins and their derivatives, especially the lectins, extracellular to the cell wall, are now a common observation (Tables 5.1.1 and 5.1.2). Lectins from the bark of black locust (*Robinia pseudoacacia*) have been isolated by specific absorption on formaldehyde-fixed human erythrocytes and eluted with a borate solution (68). The lectin is homogeneous on disc electrophoresis and yields three bands on isoelectric focusing. It has a molecular weight of $\cong$110000 and consists of two subunits (MW 29000 and 31500). Its pI is 5.9 and it contains large amounts of aspartic acid, threonine, and serine, no cysteine, and very little methionine. Covalently bound neutral sugar constitutes 7.2% of the structure and glucosamine constitutes 0.47%.

Lectins seem to have several functions (45, 87). By contrast, other proteins, such as the reserve or storage proteins, are rarely found except under highly heterotrophic nutritional conditions. These occur in the ovule, where reserve proteins are being digested for the benefit of the developing embryo (Fig. 5.1.2).

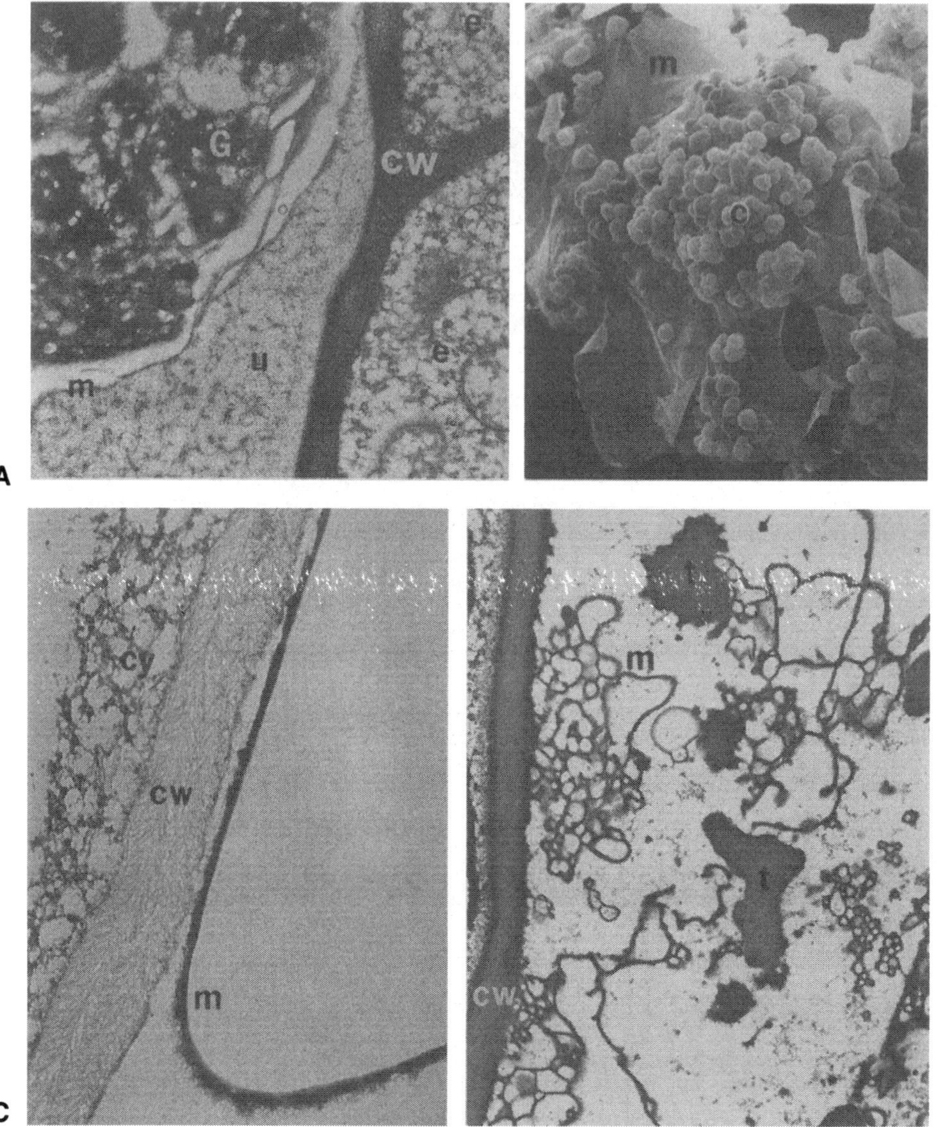

Fig. 5.1.2. Structural bases for the production of extracellular compounds. A) Electron micrograph of a thin section of cells at the surface of developing embryos (*e*) of Douglas-fir (*Pseudotsuga menziesii*). Cell walls (*CW*) are bound by a membrane (*m*). A fibrillar matrix of undefined materials (*u*) accumulates between the cell wall and membrane. The matrix is believed to include nourishment derived from digested cells (*G*) of the female gametophyte derived from the mother tree (×12000). A similar membrane is seen in cells grown in cell suspension cultures. B) Scanning electron micrograph of a cell cluster in vitro with an electron-dense membrane (*m*) over cells (*c*) (×2000). C) Transmission electron micrograph of a thin section a Douglas-fir cell (*cy* = cytoplasm; *cw* = cell wall) with a membrane (*m*) (×33000). D) Thin section (×10000) showing membrane development on the wall of a cell (*cw*) after 20 days in culture (*t* = tannin). All of these substances and structures can slough off into the culture medium or into intercellular spaces and growing clumps. This process contributes to a wide variety of substances now recognized as being produced extracellular to the lignocellulosic cell wall. The transformation and flow of membrane and subcellular organelle may provide the structural basis for the biosynthetic responsiveness of cells (76, 100) and to the unusual features of cells such as warts and membranes of yet undefined composition and structure (84)

The composition and description of cell wall proteins and lectins have been reported by Lis and Sharon (87, 88), Clarke et al. (19, 20), and Fincher et al. (45). A novel method to isolate proteins has been described by Lamport and Epstein (79). $CaCl_2$ was used to elute a column of packed tomato (*Lycopersicon esculentum*) cells to isolate the precursors of extensin.

The challenge remains that, for such molecular systems to become useful, considerably more work must be done before their value can be assessed fully. For proteins, this will include a better understanding of post-translational reactions. This will also depend on the recent progress related to the carbohydrate chemistry and models of the plant cell wall (e.g., 34, 39, 95).

Examples of enzymes reported as extracellular to the cell wall are listed in Tables 5.1.1 and 5.1.2. Most studies are based on the turnover of cell walls during seed germination and fruit ripening (77). In these situations, most is known about the cleavage of cell wall polysaccharides.

Information of enzymes can be misleading. Care is always needed to develop reliable compositional and structural descriptions of cell wall changes and measurements of enzyme activity using the relevant polymers, inhibitors, or substances involved in these reactions.

As for lignin biochemistry, a significant development has been the isolation and purification of an enzyme from a white rot fungus. This enzyme is able to depolymerize lignin (Table 5.1.2). The enzyme is a heme protein and its activity is inhibited by scavengers of active oxygen species.

5.1.2.3 Nucleic Acids and Related Products

Examples of nucleic acids usually represent pathological situations – e.g., virus suspected in blackline disease of walnuts (*Juglans nigra*). Roles of extracellular calcium, ATPases and nucleotides are uncertain. No good evidence exists that nucleic acids or nucleotides are found extracellular to cell walls except under stress or injury (Sect. 5.1.3.4).

Nevertheless, to the molecular geneticist, the field of lignin biodegradation is now open for study through the use of enzymes and nucleic acids. Gene sequences of cellobiohydrolase and the cloning of hydrolytic cellulases have been reported (Sect. 5.1.3.5).

5.1.3 Factors Determining Composition

5.1.3.1 Genetics

Genetic controls over extracellular nitrogenous compounds are seen in 1) cell-to-cell recognition in situ and in vitro (19, 20, 45), 2) specific and nonspecific responses to insects and disease (101), and 3) xylem differentiation. In the latter, genetics and metabolism interact to provide regulatory and feed-back systems and an environmental sensing apparatus (9).

More direct evidence for the relation of genetics and extracellular compounds is based on mutations in crop plants. Mutations that involve the nonripening of tomato-cell walls (*Lycopersicon esculentum*) have been used to study the release of a salt-soluble protein (66). During ripening of normal tomatoes, salt-soluble protein increases. Work with normal and mutant cells suggests that a metabolically active protein, removable by strong salt solutions, cellulase, or polygalacturonase (PG), remains attached to the cell walls of the fruit until late in ripening. The unusual amount of protein attached to cell walls of mutant fruit reflects the absence of some or all of the isoenzymes of PG that are associated with normal ripening.

Since 1983, encouraging progress has been witnessed in understanding genes for the extensins. Varner (140) has reviewed the architectural reasons for the abundance of extensins in dicot cell walls. The biochemistry and genetics of cellulose degradation is a subject of a recent symposium (150).

5.1.3.2 Genetics × Environment

Nitrogenous compounds in rain falling through a forest canopy are well-known and may be of abiotic origin (94, 136). With the increasing prevalence of acid rain or acid depositions, we may expect more interest in the release of nitrogenous compounds by trees prone to this environmental hazard. In many species, nitrogenous compounds of biotic origin are found in pollination droplets. Acid deposition, rain, fog, and changes in weather can affect tree survival, fertilization, and fruit and seed set (107, 141).

Competition among plants, as illustrated by the release of allelopathic nitrogenous compounds, is a function of the interaction of gene products released into the environment. This topic has been reviewed by Putnam (111).

Inside forest trees under water stress, the plugging of sieve-plate pores with *P*-protein is an almost instantaneous reaction to pressure release in active sieve tubes. *P*-protein plugs, or slime plugs, are now thought of as the first line of defense against the loss of assimilates. Wound callose further strengthens the cell's defenses but at variable rates (44; cf., Sect. 5.1.3.4).

The transport processes in wood have been reviewed by Siau (124). He has shown that permeability and steady-state moisture movement relate to the structure and chemical composition of wood.

5.1.3.3 Growth and Development

Cell wall fragments released by *endo*-3-1,4-polygalacturonase treatment of isolated walls or prepared by base-catalyzed solubilization and partial degradation can control morphogenesis in tobacco (*Nicotiana tabacum*) explants (134). These oligosaccharins (oligosaccharides with regulatory activity) at 10^{-8} and 10^{-9} M induce tobacco explants to form vegetative buds instead of flowers or callus, or flowers instead of vegetative buds. Other oligosaccharins induce roots instead of buds. This activity being a response to crude preparations may yet involve nitrogenous compounds (2).

Membranes of undefined composition are occasionally seen over extracellular exudates of cells in situ and in vitro. For example, in the developing immature embryo of Douglas-fir (*Pseudotsuga menziesii*), cells adjacent to the nutritive haploid female gametophyte, which is being digested, are covered with a membrane. Under the membrane, fibrils can be viewed by the electron microscope (Fig. 5.1.2A). The membrane appears to originate with the wall of the mother cell during cell division. Around the membrane, fibrils are contained in a mucilaginous fluid rich in free amino acids and, undoubtedly, other extracellular nitrogenous compounds (58).

Occasionally the membrane is seen attached to cells in suspension culture (Fig. 5.1.2B). This membrane is associated with a variety of unusual structures containing phenolic residues and tannin. The role of these membranes or "skins" in osmotic phenomena, filtration, gas exchanges, lectin and enzyme partitioning, and energy transduction at or on surfaces is unknown. Lectins are associated usually with juvenile tissues, such as embryos and seeds, rather than with mature tissues (19, 20). However, lectins that bind manganese and zinc have been isolated from the bark of *Robinia pseudoacacia* (68).

It is possible that a proteinaceous residue derived from the breakdown of cytoplasmic membranes contributes to the membrane that covers the lumen surface of xylem cells in softwoods and hardwoods (84). Deposited nitrogenous constituents may be eluted subsequently by the transpiration stream (27). In *Pistacia vera*, zinc deficiency contributes to the immobility of the nitrogenous storage compound arginine where it builds up inside leaves (Durzan 1985, unpublished). The redistribution of amino acids is mediated by lignified transfer cells (105, 109).

During the life cycle of trees, hormone action involving the release of hormones from conjugates may involve the soluble and insoluble compartments of the cell wall (21). How oligosaccharins derived from cell walls contribute and relate to the control of morphogenesis and extracellular nitrogenous compounds remains unknown (3).

5.1.3.4 Pathology

The role of extracellular nitrogenous compounds in pathological situations has been discussed by Agrios (1) and Smith (126). Recent trends emphasize that antimicrobial compounds of low molecular weight and containing nitrogen may be synthesized by and accumulate in or external to the lignocellulosic cell wall after exposure to microorganisms (Table 5.1.2). These compounds are called phytoalexins (110).

Phytoalexins also accumulate after exposure to other compounds called elicitors. Elicitors may originate as a product of the invading pathogen or from exposure to other plants or the activities of humans (Sect. 5.1.3.6). The latter are termed abiotic if they represent elicitors such as detergents, heavy metals, UV light, etc.

Van der Molen et al. (138) presented evidence that occlusion of the vascular system, which occurs during the wilting of plants, is a direct consequence of ethylene action. They propose that a common signal for the formation of vascular gels in pathogen and in abiotic stress situations is elevated by ethylene production. The effect of the pathogen-produced, wall-degrading enzymes on ethylene production is apparently exerted before the production of amino-cyclopropane carboxylic acid (ACC). The latter is a precursor for ethylene and is derived from methionine.

Extracellular alarmones produced by stressed bacteria may be deposited on cell walls. Alarmones are signal molecules (e.g., 5-amino-4-imidazolecarboxamide riboside 5′-triphosphate) coupled to a particular stress in the bacterium and serve to rebalance the metabolic economy of the cell (11, 128). By contrast, siderphores are iron-specific ligands produced by microorganisms as scavenging agents to combat low iron stress (104). These substances contain nitrogen and may be found extracellular to lignified root cells. The infection of plants is often characterized by a hypersensitive reaction characterized by the appearance of necrotic lesions with lignification, which limits the spread of the disease. The increase in *O*-methyltransferase activity during the hypersensitive response of leaves to tobacco mosaic virus infection, is attributed to stimulation of the rate of de novo synthesis of the enzymes themselves.

The study of the chemistry of lignin breakdown products has progressed through an understanding of the physiology of wood-rotting fungi. A white-rot fungus has been used to analyze enzyme systems in lignin degradation through molecular genetics and biotechnologies of fundamental and commercial interest. The lignolytic activity of *Phanerochaete chrysosporium* is not inducible and is evolved by the exhaustion of available carbon, nitrogen, or sulfur in the growth medium (cf., 72) (Table 5.1.2).

5.1.3.5 Impact of Humans

With concerns about energy and food shortages and waste products, the interest in extracellular enzymes and immobilized or encapsulated cells for industrial processes had increased significantly (Table 5.1.3) (e.g., 17, 35, 97). Furthermore, since wood is the world's most abundant readily available renewable organic material, the newer biotechnologies hold promise to expanding the utilization of wood. The coupling of enzyme technology with genetic engineering technology for lignin and lignocellulose conversions seems inevitable (71). If "ligninases" can be produced in large quantity, lignin research would move rapidly into the biotechnological age.

Indeed progress is encouraging with the cloning of *Clostridium* genes responsible for cellulose degradation (24) and the cloning and expression an endocellulase gene in *E. coli* (22). However, enzymes catalyze reactions under mild conditions. This attribute is not always ideal for industrial application. Enzyme instability and susceptibility to organic solvents are limitations that are being removed by immobilization (135). In some cases, product collection and harvesting may be improved by the encapsulation of cells. As for woody perennials (cf., Fig. 5.1.3), we may still have difficulty in finding cells with the necessary characteristics, reactions, and molecular specificity for industrial use. As for the basic chemical reaction of fibers, information on the grafting of reactions to lignocellulosic fibers is increasing (67).

Table 5.1.3. Annotated chronological summary: Studies on the production of materials containing amino acids and general examples of biotechnologies involving extracellular proteins and cells

Amino acid, protein, etc.	Species and remarks	References
Enzymes	Linked for immunosorbent assays (ELISA test)	103
	Application to plant cells	75
Antibodies	Column of antibodies for isolation of gibberellins	51
Enzyme technology	Relation to renewable resources	53
Enzymes, hormones	Microencapsulation in biodegradable semipermeable membranes	18
Single cell protein	Feed-grade protein from carbohydrate cabohydrate sources	145
Lectins from *Robinia* bark	Isolated on formaldehyde-fixed human erythrocytes	68
Microbial cells with high urease activity	Entrapped in cellulose triacetate fibres	54
Enzymes	Absorbed on immobilized tannin	142
Fungal cellulose	Production by *Trichoderma reesei* in fermentors	55
Living cells	Microencapsulation of animal cells	85
Microbial single cell protein production	Review of literature on photosynthetic and non-photosynthetic organisms	89, 90
Immobilization enzymes	β-galactosidase on epoxy-activated acrylic beads	60
Ionic control of immobilized acid phosphatase	Kinetics of enzyme bound to cell walls of sycamore maple (*Acer pseudoplatanus*) and Donnan effects	113
Artificial enzymes	Selective synthesis of amino acids	15
Immobilized enzymes and cells	Review of literature	74
Hydrolytic cellulases	Cloning from *Trichoderma reesei*	123, 130
N-containing phytohormones	Monoclonal antibodies and immunoassays for plant growth regulators	144

While the contribution of extracellular nitrogenous compounds is uncertain, preparations of poplar (*Populus*) sp. leaves have been used as a high-energy feed supplement for livestock (e.g., 36). Some preparations may contain toxic extracellular compounds. Lectins from *Robinia pseudoacacia* are toxic to mammalian cells, and have been explored as therapeutic agents in cancer patients (cf., 81, 88).

One example related to forest trees is the possible use of extracellular substances to establish a synergistic relation between free-living nitrogen-fixing bacteria and a nonleguminous plant. This technique was initially applied to corn (*Zea mays*) (16). In this approach, ammonia-excreting strains of *Azotobacter vinelandii* provide fixed nitrogen to plants, and the plants supply carbon to the bacteria. Some scientists have speculated that N-fixing bacteria might someday be attached in grooves of xerophytic needles of tall coniferous trees. The needles act as cloud collectors in the early morning hour. The moisture collected would provide a medium for soluble nitrogen released by bacteria to stimulate tree growth, especially in N-poor soils.

The addition of an arbitrary unit of nitrogen to the nitrogen cycle in nature is estimated to create an 800-fold flow in N cycling (93). It is no wonder that

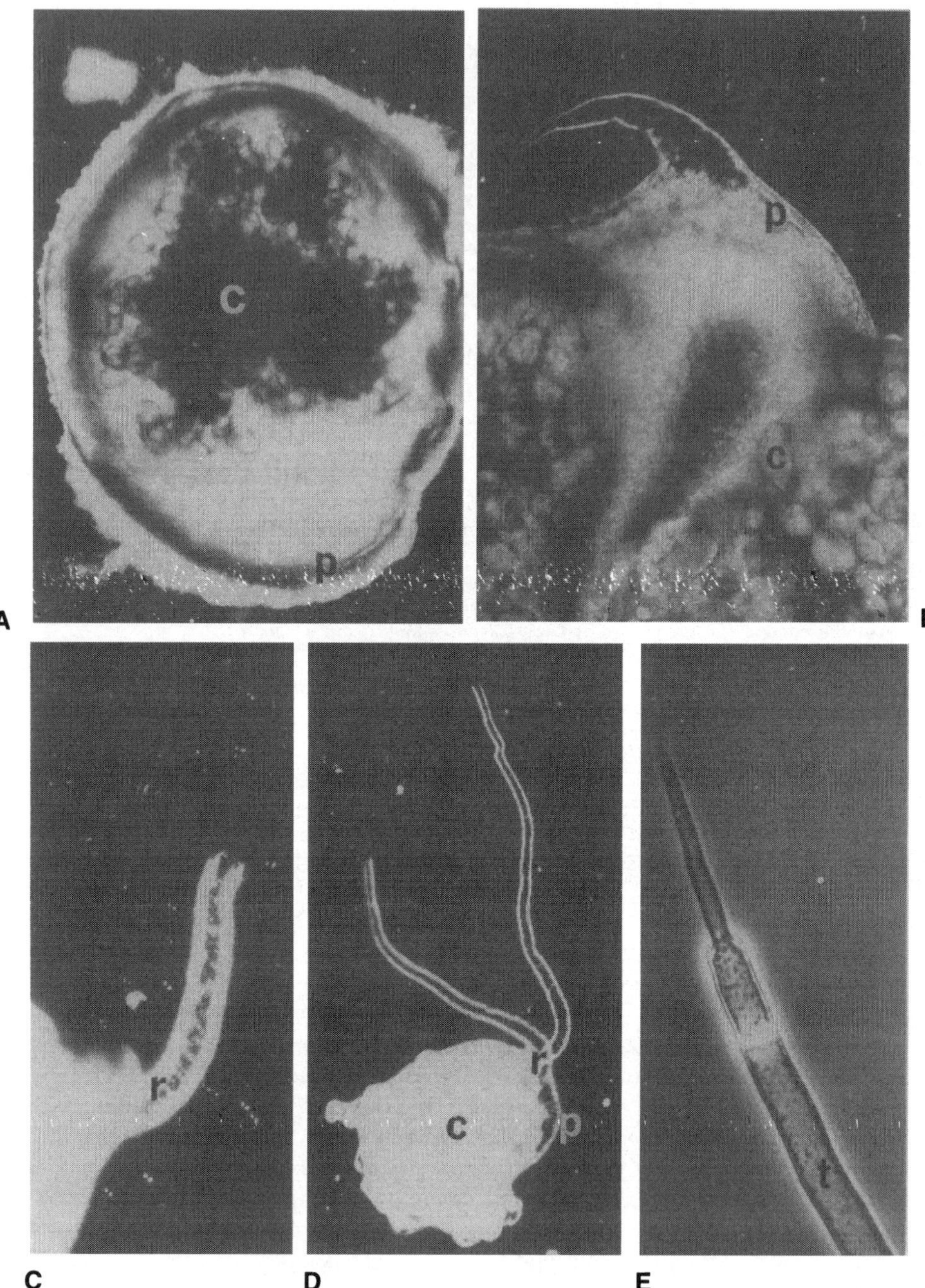

Fig. 5.1.3. Encapsulation of a cell clump of Douglas-fir (*Pseudotsuga menziesii*) in liquid medium (102) with soluble 0.1 mM polyvinylpyrrolidone. A) Encapsulated cell clump (*c*) in medium (*p*) (×60). B) When the encapsulated cells are exposed to pressure by tapping a glass cover slip, the encapsulated material will protrude and rapid growth will occur. C), D), and E) Even if the tube is ruptured (*r*) elongation will still occur. Tube growth after 1 (B, ×120), 3 (C, ×120), 18 (D, ×30), and 24 (E, ×120) minutes. The tubes (*t*) in C and E contain numerous extracellular particles

the nitrogen released and fixed by plants has been kept minimal and under control.

5.1.4 Conclusion

We are on the verge of an explosion in our knowledge of the role and use of extracellular nitrogenous compounds. The release from cells of these compounds is difficult to attribute solely to any single agent. Their occurrence results from interactions of numerous sources. Interactions must take into account inherent differences that arise from morphology at the source – the variety within the species, from its state of development, and from the environment during cultivation. Moreover, these factors vary greatly from one specific source to another.

References

1 Agrios G N 1969 Plant pathology. Academic Press New York, 629 pp
2 Albersheim P, Anderson-Procity A J 1975 Carbohydrates, proteins, cell surfaces, and biochemistry of pathogenesis. Ann Rev Plant Physiol 26:31–52
3 Albersheim P, Darvill AG 1985 Oligosaccharins. Amer Sci 253:58–64
4 Anderson J D et al 1982 Induction of ethylene biosynthesis in tobacco leaf-disks by cell-wall digesting enzymes. Biochem/Biophysiol Rev 107:588–596
5 Ashford A D, McCully M E 1970 Localization of naphthol as – B1 phosphatase – activity in lateral and main root meristems of pea and corn. Protoplasma 70:441
6 Bauer W D 1981 Infection of legumes by Rhizobia: Review. Ann Rev Plant Physiol 32:407–449
7 Bell A A 1981 Biochemical mechanisms of disease resistance. Ann Rev Plant Physiol 32:21–81
8 Bell E A, Charlwood B V 1980 Secondary plant products. Springer Heidelberg, 674 pp
9 Berlyn G P 1979 Physiological control of differentiation of xylem elements. Wood Fiver 11:109–126
10 Bishop P D, Makus D J, Pearce G, Ryan C A 1981 Proteinase inhibitor-inducing factor activity in tomato leaves resides in oligosaccharides enzymically released from cell walls. Proc Nat Acad Sci 78:3536–3540
11 Bochner B R, Ames B N 1982 ZTP (5-amino 4-imidazole carboxamide riboside 5′-triphosphate): A proposed alarmone for 10-formyltetrahydrofolate deficiency. Cell 29:929–937
12 Bohlool B B, Schmidt E L 1974 Lectins – possible basis for specificity in rhizobium-legume root nodule symbiosis. Science 185:269–271
13 Bonga J M, Durzan D J (eds) 1982 Tissue culture in forestry. Martinus Nijhoff The Hague, 420 pp
14 Bonga J M, Durzan D J (eds) 1987 Cell and tissue culture in forestry, vols 1–3. Martinus Nijhoff Dordrecht
15 Breslow R 1982 Artificial enzymes. Science 218:532–537
16 Brill W 1981 Agricultural microbiology. Sci Am 245:199–215
17 Bungay H R 1982 Overview of new biomass industries. In: Priorities in biotechnology research for international development. BOSTID, Nat Res Coun, National Academy Press Washington DC, 159–176
18 Chang T M S 1976 Biodegradable semipermeable microcapsules containing enzymes, hormones, vaccines, and other biologicals. J Bioeng 1:25–31
19 Clarke A E, Anderson R L, Stone B A 1979 Form and function of arabinogalactan-proteins – Review. Phytochemistry 18:521–540
20 Clarke A E, Clarke A, Gleeson P, Harrison S, Knox R B 1979 Secreted carbohydrates of the female reproductive tissues of gladiolus. (I). The stigma surface. Proc Nat Acad Sci 76:3358–3362

21 Cohen J D, Bandurski R S 1982 Chemistry and physiology of the bound auxins. Ann Rev Plant Physiol 33:403–430
22 Collmer A, Wilson D B 1983 Cloning and expression of a *Thermomonospore* YX endocellulase gene in *E. coli.* Bio/Technology 1:594–601
23 Cooper J B, Varner J E 1983 Insolubilization of hydroxyproline-rich cell wall glycoprotein in aerated carrot root slices. Biochem Biophys Res Comm 112:161–167
24 Cornet P, Millet J, Beǵuin P, Aubert J P 1983 Characterization of two cell (cellulase degradation) genes of Clostridium thermocellum coding for endoglucanases. Bio/Technology 1:589–594
25 Cosgrove D J, Cleland R E 1983a Solute in the free space of growing stem tissues. Plant Physiol 72:326–331
26 Cosgrove D J, Cleland R E 1983b Osmotic properties of pea internodes in relation to growth and auxin action. Plant Physiol 72:332–338
27 Cowling E B, Merrill W 1966 Nitrogen in wood and its role in wood deterioration. Can J Bot 44:1539–1554
28 Currier W W, Strobel G 1977 Chemotaxis of *Rhizobium* spp to a glycoprotein produced by a birdsfoot – trefoil roots. Science 196:434–436
29 Curtin M E 1983 Harvesting profitable products from plant tissue culture. Bio/Technology 1:649–657
30 Dandekar A M, LeRudlier D, Smith L T, Jakowec M W, Gong L S, Valentine R C 1983 *Rhizobium* as vectors for genetic engineering of salinity and drought tolerant legumes. In: Goldberg R B (ed) Plant molecular biology, vol 12. A R Liss Inc New York, 277–289
31 Davies D D 1982 Physiological aspects of protein turnover. In. Boulter D, Parthier B (eds) Nucleic acids and proteins in plants. I: Structure, biochemistry and physiology of proteins. Springer Heidelberg, 189–228
32 Delmer D P 1987 Cellulose biosynthesis. Ann Rev Plant Physiol 38:259–290
33 Dazzo F B 1978 Adsorption of microorganisms to roots and other plant surfaces. In: Britton G, Marshall K C (eds) Adsorption of microorganisms to surfaces. Wiley New York, 439 pp
34 Delmer D P 1982 Biosynthesis of cellulose. Carbohyd Chem 14:105–153
35 Demain A L 1981 Industrial microbiology. Science 214:987–995
36 Dickson R E, Larson R R 1977 Muka from *Populus* leaves: A high energy feed supplement for livestock. In: Proc TAPPI Conf For Biol Wood Chem. TAPPI Atlanta, 95–99
37 Dixon R A, Dey P M, Lawton M A, Lamb C W 1983 Phytoalexin induction in french bean. Intercellular transmission of elicitation in cell suspension cultures and hypocotyl sections of *Phaseolus vulgaris.* Plant Physiol 71:251–256
38 Dube H C, Gour H N 1976 Product of pectic enzymes by *Curvularia lunata* causing leaf-spot of cotton. Proc Ind Nat Sci Acad B 41:480–485
39 Dugger E M, Bartnicki-Garcia S (eds) 1984 Structure, function and biosynthesis of plant cell walls. Proc 7th Ann Symp Botany. Am Soc Plant Physiol Rockwille MD, 524 pp
40 Durzan D J (1984 Explant source: Juvenile vs. adult phase. In: Evans D A, Sharp W R, Ammirato P V, Yamada Y (eds) Handbook of plant cell culture. Plant tissue culture. MacMillan New York, 2:471–503
41 Durzan D J, Chalupa V 1976 Growth and metabolism of cells and tissue of jack pine (*Pinus banksiana*). 6. Free nirogenous compounds in cell suspension cultures of jack pine as affected by light and darkness. Can J Bot 54:496–506
42 Durzan D J, Steward F C 1984 Metabolism of organic nitrogenous compounds. In: Steward F C, Bidwell R G S (eds) Plant physiology, an advanced treatise, vol 8. Academic Press New York, 55–265
43 Eriksson K-E 1978 Enzyme mechanisms involved in cellulose hydrolysis by root fungus *Sporotrichum pulverulentum.* Biotechnol/Bioeng 29:317–332
44 Evert R F 1982 Sieve-tube structure in relation to function. BioScience 32:789–795
45 Fincher G B, Stoe B A, Clarke A E 1983 Arabinogalactan-proteins: Structure, biosynthesis, and function. Ann Rev Plant Physiol 34:47–70
46 Fogerty W M, Ward O P 1972 Enzyme production by bacteria isolated from water-stored sitka spruce (*Picea sitchensis*). J Appl Bacteriol 35:685–689
47 Fry S C 1979 Phenolic components of the primary cell wall and their possible role in the hormonal regulation of growth. Planta 157:111–123

48 Fry S C 1982 Isodityrosine, a new cross-linking amino acid from plant cell-wall glycoprotein. Biochem J 204:439–445

49 Fry S C 1986 Cross-linking of matrix polymers in the growing cell walls of angiosperms. Ann Rev Plant Physiol 37:165–186

50 Fry S C, Darvill A G, Albersheim P 1983 Amino acid transport and protein synthesis, possible primary targets of biologically active cell wall fragments. In: Jackson M B (ed) Interactions between nitrogen and growth regulators in the control of plant development. Monog. 9. British Plant Growth Regulator Group Wantage Oxfordshire, 33–45

51 Fuchs Y, Gertwan E 1974 Insoluble antibody column for isolation and quantitative-determination of gibberellins. Plant Cell Physiol 15:629–633

52 Fukuda T 1963 Studies on the chemical composition of woods. I. On the amino acids. J Japan Wood Res Soc 9:116–170

53 Gainer J L (ed) 1976 Enzyme technology of renewable resources. Proc NSF (RANN) Nat Sci Found Washington D C, 133 pp

54 Ghose T K, Kannan V 1979 Studies on fibre-entrapped whole microbial cells in urea hydrolysis. Enz Microb Technol 1:47–50

55 Ghose T K, Sahai V 1979 Production of cellulases by *Trichoderma reesei* QM-9414 in fed-batch and continuous flow-culture with cell recycle. Biotechnol/Bioeng 21:283–296

56 Greaves C, Schwartz H 1952 The chemical utilization of wood. Canada Dept Resour Dev Ottawa, 28 pp

57 Greenstein J P, Winitz M 1961 Chemistry of the amino acids, vol. 1. Wiley New York, 672 pp

58 Gupta P K, Durzan D J 1987 Biotechnology of somatic polyembryogenesis and plantlet regeneration in loblolly pine. Bio/Technology 5:147–151

59 Hagman M 1975 Incompatibility in forest trees. Proc R Soc London B 188:313–326

60 Hannibal-Friedrich O 1980 Immobilization of beta-galactosidase, albumin, and gamma-globulin on epoxy-activated acrylic bead. Biotechnol/Bioeng 22:157–175

61 Helgeson J P, Deverall B J 1983 Use of tissue culture and protoplasts in plant pathology. Academic Press New York, 194 pp

62 Heslop-Harrison J 1968 The emergence of pattern in the cell walls of higher plants. Dev Biol Suppl 2:118–150

63 Heslop-Harrison J 1975 Incompatibility and the pollen-stigma interaction. Ann Rev Plant Physiol 26:403–425

64 Higgins C V, Payne J W 1982 Plant peptides. In. Boulter D, Parthier B (eds) Nucleic acids and proteins in plants. I: Structure, biochemistry and physiology of proteins. Springer Berlin, 438–458

65 Higuchi T, Shimada M, Yamasaki T 1977 Biochemistry of lignin formation in wood. In: Proc TAPPI Conf For Biol Wood Chem. TAPPI Atlanta, 31–36

66 Hobson G E, Richardson C, Gillham D J 1983 Release of protein from normal and mutant tomato cell walls. Plant Physiol 71:635–638

67 Hon D N-S (ed) 1982 Graft copolymerization of lignocellulosic fibers. ACS Symp Ser 187. Am Chem Soc Washington DC, 381 pp

68 Hořejší V, Haškovec C, Kocourek J 1978 Studies on lectins. XXXVIII. Isolation and characterization of the lectin from black locust bark (*Robinia pseudoacacia* L). Biochim Biophys Acta 532:98–104

69 Jermyn M A, Yeow Y M 1975 Class of lectins present in the tissues of seed plants. Aust J Plant Physiol 2:501–531

70 Kauss H 1974 Biosynthesis of pectin and hemicelluloses. In: Pridham J B (ed) Plant carbohydrate biochemistry. Academic Press New York, 191–205

71 Kirk T K 1983 Biotechnology in utilization of wood. BioTechnology 1:666–668

72 Kirk T K, Schultz E, Connors W J, Lorenz L F, Zeikus J G 1978 Influence of culture parameters on lignin metabolism by *Phanerochaete chrysosporium*. Arch Microbiol 117:277–285

73 Kristen U 1980 Changes in exudate composition during the development of the ovary glands in *Aptenia cordifolia*. I. The quantitative saccharide-protein-relations and the protein components. Z Pflanzenphysiol 96:239–249

74 Klibanov A M 1983 Immobilized enzymes and cells as practical catalysts. Science 219:722–727

75 Knox R B, Clarke A E 1978 Localization of proteins and glycoproteins by binding to labelled antibodies and lectins. In: Hall J L (ed) Electron microscopy and cytochemistry of plant cells. Elsevier Amsterdam, 149–185

76 Kolattukudy P E 1981 Structure, biosynthesis, and biodegradation of cutin and suberin. Ann Rev Plant Physiol 32:539–567
77 Labavitch J M 1981 Cell wall turnover in plant development. Ann Rev Plant Physiol 32:385–406
78 Lamport D T A 1970 Cell wall metabolism. Ann Rev Plant Physiol 21:235–270
79 Lamport D T A, Epstein L 1983 A new model for the primary cell wall: A concatenated extensin-cellulose network. Preprint for Botany 221. Univ California Davis, 17 pp
80 Larsen P O 1980 Physical and chemical properties of amino acids. In: Stumpf P K, Conn E E (eds) Biochemistry of plants. Academic Press New York, 225–269
81 Lee S C, West C A 1981 Polygalacturonase from rhizopus-stuconifer, an elicitor of casbene synthetase-activity in caster bean (*Ricinus communis* L.) seedlings. Plant Physiol 67:633–639
82 Leppard G G, Colvin J R, Rose D, Martin S M 1971 Lignofibrils on the external cell wall surface of cultured plant cells. J Cell Biol 50:63–80
83 Liau D F, Boll W G 1972 Extracellular polysaccharide from cell suspension cultures of bush bean (*Phaseolus vulgaris* cv contender). Canadian J Bot 50:2031
84 Liese W 1965 The warty layer. In: Côté W A (ed) Cellular ultrastructure of woody plants. Syracuse University Press Syracuse, 251–269
85 Limm F, Sun A M 1980 Microencapsulated islets as bioartificial endocrine pancreas. Science 210:908–910
86 Lin T A, Kolattukudy P E 1976 Evidence for novel linkages in a glycoprotein involving beta-hydroxyphenylalanine, and beta hydroxytosine. Biochem Biophys Res Comm 72:243–250
87 Lis H, Sharon N 1981 Lectins in higher plants. In: Marcus A (ed) The biochemistry of plants. Proteins and nucleic acids. Academic Press New York, 6:371–447
88 Lis H, Sharon N 1986 Lectins as molecules and as tools. Ann Rev Biochem 55:55–67
89 Litchfield J H 1980 Microbial protein production. BioScience 30:387–396
90 Litchfield J H 1983 Single-cell proteins. Science 219:740–746
91 Loewus F A, Ryan C A (eds) 1981 The phytochemistry of cell recognition and cell surface interactions, vol 15. Recent advances in phytochemistry, Plenum Press New York, 277 pp
92 Lüdtke M, Lerch B 1968 Über die Aminosäuren der Hautsubstanz Pflanzlicher Faserzeller. Holzforschung 22:1–8
93 McCarl B, Raphael D, Stafford E 1975 The impact of man on the world nitrogen cycle. J Environ Managem 3:7–19
94 McKee H S 1962 Nitrogen metabolism in plants. Oxford University Press. Oxford, 728 pp
95 McNeil M, Darvill A G, Fry S C, Albersheim P 1984 Structure and function of the primary cell walls of plants. Ann Rev Biochem 53:625–663
96 Meimeth T, Tran Thanh Van K, Marcotte J-L, Harth Trinn T, Clarke A E 1982 Distribution of lectins in tissues derived from callus and roots of *Psophocarpus tetragonolobus* (winged bean). Plant Physiol 70:579–584
97 Menon M G K 1980 Production and availability of biomass. A state of the art report. Dep Sci Technol New Delhi, 27 pp
98 Merrill W, Cowling E B 1966 Role of nitrogen in wood deterioration: Amounts and distribution of nitrogen in tree stems. Can J Bot 44:1555–1580
99 Miège M N 1982 Protein types and distribution. In: Boulter D, Parthier B (eds) Nucleic acids and proteins in plants. I. Structure, biochemistry and physiology of proteins. Springer Heidelberg, 291–345
100 Mühlethaler K 1975 The ultrastructural of cells. In: Davies P (ed) Historical and current aspects of plant physiology: A symposium honoring F.C. Steward. NY State Coll Agr Life Sci Cornell Univ Ithaca, 226–242
101 Mullick D B 1977 The non-specific nature of defense in bark and wood during wounding, insect and pathogen attack. Rec Adv Phytochem 11:395–441
102 Murashige T, Skoog F 1962 A revised medium for rapid growth and bioassays with tobacco tissue cultures. Physiol Plant 15:473–479
103 Nakane P K, Pierce G B 1967 Enzyme-labeled antibodies – preparation and application for localization of antigens. J Histochem Cytochem 14:929
104 Neilands J B, Leong S A 1986 Siderophores in relation to plant growth and disease. Ann Rev Plant Physiol 37:187–208
105 Newcombe W, Peterson R L 1979 The occurrence and ontogeny of transfer cells associated with lateral roots and root nodules in Leguminosae. Can J Bot 57:2583–2602

106 Northcote D H 1982 Macromolecular aspects of cell wall differentiation. In: Boulter D, Parthier B (eds) Nucleic acids and proteins in plants. I. Structure, biochemistry and physiology of proteins. Springer Heidelberg, 637–655

107 O'Sullivan D A 1985 European concern about acid rain is growing. Chem Eng News 63(4):12–18

108 Palade G 1975 Intracellular aspects of the process of protein synthesis. Science 189:347–358

109 Pate J S 1983 Distribution of metabolites. In: Steward F C (ed) Plant physiology, a treatise. Academic Press New York, 335–401

110 Paxton J 1980 A new working definition of the term "phytoalexin." Plant Dis 64:734

111 Putnam A R 1983 Allelopathic chemicals. Nature's herbicides in action. Chem Eng News 61(14):34–45

112 Quail P H, Browning A H 1977 Failure of lactoperoxidase to iodinate specifically the plasma membrane of *Cucurbita* tissue segments. Plant Physiol 59:759–766

113 Ricard J, Noat G, Crasnier M, Job D 1981 Ionic control of immobilized enzymes: Kinetics of acid phosphatase bound to plant-cell walls. Biochemistry J 195:357–367

114 Roberts L W, Baba S 1968 Effects of proline on wound vessel member formation. Plant Cell Physiol 9:353

115 Romberger J A, Tabor C A 1975 The *Picea abies* shoot apical meristem in culture. II. Deposition of polysaccharides and lignin-like substances beneath cultures. J Bot 62:610–617

116 Rosenthal G A 1982 Plant nonprotein amino and imino acids. Academic Press New York, 273 pp

117 Rowe J W, Connor A H 1979 Extractives in eastern hardwoods – a review. USDA For Serv Gen Tech Rep FPL-18, 67 pp

118 Ryan C A, Bishop P, Pearce G, Darvill A G, McNeil M, Albersheim P 1981 A sycamore cell-wall polysaccharide and a chemically reacted tomato leaf polysaccharide possess similar proteinase inhibitor-inducing activities. Plant Physiol 68:616–618

119 Schmeits T G J, Sassen M M A 1978 Suction marks in nutrition cells of gall on leaves of Acer pseudoplatanus L. caused by *Eriophytes macrorrynchus-typicus* NAL. Acta Bot Neerl 27:27–33

120 Scurfield G 1967 The histochemistry of reaction wood differentiation in *Pinus radiata* D. Don. Aust J Bot 18:377–392

121 Scurfield G, Nicholls P W, 1970 Amino acid composition of wood proteins. J Expt Bot 21:857–868

122 Sharp W R, Evans D A, Ammirato P V, Yamada Y (eds) 1984 Handbook of plant cell culture, vol 2. Crop Species. MacMillan New York, 644 pp

123 Shoemaker S, Schweikart V, Ladner M, Gelfand D, Kwok S, Myambok K, Innis M 1983 Molecular cloning of exocello biohydrocase-1 derived from *Trichoderma reesei* strain-L27. BioTechnology 1:691–696

124 Siau J F 1983 Transport processes in wood. Springer Heidelberg, 245 pp

125 Slankis V, Runeckles V C, Krotkov G 1964 Metabolites liberated by roots of white pine (*Pinus strobus* L.) seedlings. Physiol Plant 17:301–313

126 Smith W H 1970 Tree pathology: A short introduction. Academic Press New York, 309 pp

127 Staba J E 1980 Plant tissue culture as a source of biochemicals. CRC Press Boca Raton, 304 pp

128 Stephens J C, Artz S W, Ames B N 1975 Guanosine 5′-diphosphate 3′-diphosphate (ppGpp): positive effector for histidine operon transcription and general signal for amino acid deficiency. Proc Nat Acad Sci USA 72:4389–4393

129 Svaldi R, Elgersma D M 1982 Further studies on the activity of cell wall degrading enzymes of aggressive and non-aggressive isolates of *Ophiostoma ulmi*. J Forest Path 12:29–36

130 Teeri T, Salovuori I, Knowles J 1983 The molecular cloning of the major cellulase gene from *Trichoderma reesei*. BioTechnology 1:696–699

131 Terry M E, Bonner B A 1980 An examination of centrifugation as a method of extracting an extracellular solution from peas, and its use for the study of IAA-induced growth. Plant Physiol 66:321–325

132 Terry M E, Rubenstein B, Jones R L 1981 Soluble cell wall polysaccharides released from pea stems by centrifugation. Plant Physiol 68:538–542

133 Tien M, Kirk TK 1983 Lignin-degrading enzyme from the hymenomycete *Phanerochaete chrysosporium* Burds. Science 221:661–663

134 Tran Thanh Van K, Toubart P, Cousson A, Darvill A G, Gollin D J, Chelf P, Albersheim P 1985 Oligosaccharins can control morphogenesis in tobacco explants. Nature 314:615–617

135 Trevan M D 1980 Immobilized enzymes. An introduction and application in biotechnology. Wiley London
136 Tukey H B 1970 The leaching of substances from plants. Ann Rev Plant Physiol 21:305–324
137 Uy R, Wold F 1977 Post-translational covalent modification of proteins. Science 198:890–896
138 Van der Molen G E, Labavitch J M, Strand L L, DeVay J E 1983 Pathogen-induced vascular gels: Ethylene as a host intermediate. Physiol Plant 59:573–580
139 Van Holst G J, Klis F, M, Bouman F Stegwee D 1980 Changing cell wall composition in hypocotyls of dark-grown bean seedlings. Planta 149:209–212
140 Varner J E 1987 Genes for the extensions, an abundant class of proteins. In: Breuning J, Harada J, Kosuge T, Hollaender A (eds) Tailoring genes for crop improvement. Plenum Press New York, 7–11
141 Waldman J M, Munger J W, Jacob D J, Flagan R C, Morgan J J, Hoffmann M R 1982 Chemical composition of acid fog. Science 218:677–679
142 Watanabe T, Fujimura M 1979 Characteristics and applications of tannin immobilized by covalent binding to aminohexyl cellulose. J Appl Biochem 1:28–36
143 Watson B A, Waaland S D 1983 Partial purification and characterization of a glycoprotein cell fusion hormone from *Griffithsia pacifica*, a red alga. Plant Physiol 71:327–332
144 Weiler E W 1984 Immunoassay of plant growth regulators. Ann Rev Plant Physiol 35:85–95
145 Wells J J 1977 Single cell protein and the protein economy. In: Hickson J L (ed) Sucrochemistry. ACS Symp Ser 41. Am Chem Soc Washington D C, 297–312
146 Whitmore F W 1982 Lignin-protein complex in cell walls of *Pinus elliottii*: Amino acid constituents. Phytochemistry 21:315–318
147 Yamazaki N, Fry S C, Darvill A G, Albersheim P 1983 Host-pathogen interactions. XXIV. Fragments isolated from suspension-cultured sycamore cell walls inhibit the ability of cells to incorporate [^{14}C]-leucine into proteins. Plant Physiol 72:864–869
148 Yancey P H, Clark M E, Hand S C, Bowlus R D, Somero G N 1982 Living with water stress: Evolution of osmolytic systems. Science 217:1214–1222
149 Yang S F, Hoffman N E 1984 Ethylene biosynthesis and its regulation in higher plants. Ann Rev Plant Physiol 35:155–189
150 Aubert, J-P, Begin P, Millet J 1988 Biochemistry and genetics of cellulose degradation. Academic Press London, 428 pp

5.2 The Alkaloids

S. I. SAKAI, N. AIMI, E. YAMANAKA, and K. YAMAGUCHI

5.2.1 Introduction

Alkaloids form the most important part of the large group of nitrogeneous secondary metabolites. The nitrogen comes either from the amino group of amino acids or from simple nitrogen-containing molecules such as ammonia, methyl- or ethylamine, and other analogous building blocks.

A definition that distinguishes alkaloids from other nitrogeneous secondary products is always accompanied by certain difficulties. As is well known, the original meaning of alkaloid is "alkali-like substance from nature." It is true that many of the alkaloids are basic, but there are also many alkaloids that are not basic. In these compounds, nitrogens may be in either an amide or an urethane moiety. Some molecules may have acidic functional groups, such as carboxyl or phenol groups, which cancel the nitrogen basicity and give a neutral or even an acidic character to the molecule. When nitrogen forms a pyrrole ring, this atom no longer shows basicity because of the aromaticity of the pyrrole.

Table 5.2.1. Representative families of alkaloid-containing woody plants

Gymnospermae	Angiospermae
Coniferales	Dicotyledoneae
Pinaceae	Archichlamydeae
Cephalotaxaceae	Magnoliales
Taxaceae	Magnoliaceae
Gnetales	Annonaceae
Ephedraceae	Calycanthaceae
	Lauraceae
	Hernandiaceae
	Ranunculales
	Ranunculaceae (herbaceous)
	Berberidaceae
	Menispermaceae
	Nympheraceae (herbaceous)
	Aristrochiales
	Aristrochiaceae
	Papaverales
	Papaveraceae (herbaceous)
	Rosales
	Leguminosae
	Rosaceae
	Geraniales
	Erythroxylaceae
	Euphorbiaceae (Daphniphyllaceae)
	Rutales
	Rutaceae
	Simaroubaceae
	Meliaceae
	Celastrales
	Celastraceae
	Buxaceae
	Icacinaceae
	Rhamnales
	Rhamnaceae
	Malvales
	Elaeocarpaceae
	Myrtiflorae
	Lythraceae
	Punicaceae
	Umbelliflorae
	Alangiaceae
	Nyssaceae
	Garryaceae
	Sympetalae
	Gentianales
	Loganiaceae
	Apocynaceae
	Rubiaceae
	Tubiflorae
	Convolvulaceae (herbaceous)
	Boraginaceae
	Solanaceae

Very often the definition of alkaloids is associated with their strong biological activities. Here again, however, we know a great many examples of alkaloids that do not show any recognizable physiological activities.

On the basis of the above considerations, we define the alkaloids here as a group of low-molecular weight nitrogeneous compounds that are apparently different from simple amines, amides, small peptides, and other nitrogeneous metabolites directly related to primary metabolism.

The plant parts in which the alkaloids are synthesized is an interesting subject. Very often alkaloids are abundant in the roots, stem bark, leaves, and seeds. Generally the alkaloid content is low in the woody parts of the plants. There is a large difference in alkaloid structures depending on the part of the plant where they are found. The important plant families that include alkaloid-producing woody plants are shown in Table 5.2.1 (5.2.2.6.7). The role of alkaloids in the living plants remains unclear, but this is not a problem specific to alkaloids. Alkaloids should probably be treated as members of a wide variety of secondary products such as phenolics, terpenoids, various kinds of glycosides, and others.

Alkaloids (26, 64, 127) are classified into two main groups based on their biogenetic origin: the true alkaloids and the pseudoalkaloids. The former are the alkaloids that are biosynthesized directly from amino acids. The latter have carbon skeletons derived from nonnitrogeneous biosynthetic precursors; they combine with a simple amine at some stage of the biosynthesis to form nitrogen-containing molecules. This distinction means that the nitrogen atoms of true alkaloids are directly derived from the amino group of an amino acid. Therefore the carbon skeleton surrounding the nitrogen atom has its origin in the parent amino acid, although many additional building blocks may join from the other biological sources. Typical of the pseudoalkaloids are steroidal alkaloids, in which ammonia or its biological equivalent is incorporated into the steroidal framework of mevalonate origin. Sometimes ammonia is replaced with simple alkyl derivatives, such as methylamine or ethylamine, to form pseudoalkaloids having *N*-methyl or *N*-ethyl groups. Most pseudoalkaloids in the higher plants are terpenoids. In very limited cases the fundamental carbon skeletons are formed through the acetate-malonate pathway or the shikimate pathway.

The above criteria will be adopted in this chapter to classify alkaloids. It is important to note that these classifications are concerned only with the main skeletal part or the part of interest in the molecules. Almost all alkaloids have some sort of additional building blocks, C-1 units of various origins being the smallest members.

5.2.2 True Alkaloids

5.2.2.1 Alkaloids from Ornithine

5.2.2.1.1 Coca Alkaloids

For centuries, inhabitants of Peru, Bolivia, and northern Argentina have chewed the leaves of the tree *Erythroxylum coca* to stimulate quick recovery from fatigue.

This habit is reported to go back to as early as the era of the Inca Empire. It was in the leaves of this tree of the Erythroxylaceae family that cocaine was found. It is one of the major coca alkaloids (24, 45, 163).

Erythroxylum coca is cultivated in a number of countries, including Peru, Bolivia, Colombia, Indonesia, and Sri Lanka. A second species of commercial importance is *Erythroxylum novogranatens.* Although the content of cocaine is highest in the leaves and the extraction is almost always done by using the leaves as the source material, cocaine and other coca alkaloids are also found in the bark. Representative coca alkaloids (**1**, **2**, **3**, **4**) are shown. All are derivatives of ecgonine (2-carboxytropanol).

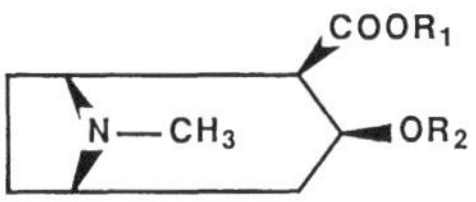

$COOR_1$, $N—CH_3$, OR_2

	R_1	R_2	
cocaine	CH_3	$—C(=O)—C_6H_5$	1
cinnamoyl-cocaine	CH_3	$—C(=O)—CH{=}CH\text{-}C_6H_5$	2
benzoylecgonine	H	$—C(=O)—C_6H_5$	3
methylecgonine	CH_3	H	4

Cocaine has lasting local anesthetic activity and also functions as a central nervous system stimulant. At the same time it causes narcosis, and the abuse of cocaine leads to addiction.

The bark of *Erythroxylum ellipticum*, a tall Australian tree that reaches a height of 35 feet, contains 0.32% crude alkaloids, from which tropine 3,4,5-trimethoxycinnamate and tropine benzoate have been isolated (83). These alkaloids are derivatives of tropine (3α-hydroxytropane), thus differing from cocaine-type alkaloids in that the latter are pseudotropine (3β-hydroxytropane) derivatives.

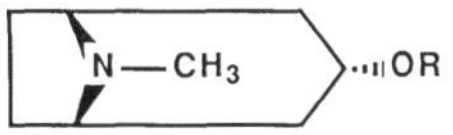

$N—CH_3$, OR

	R	
3,4,5-trimethoxy-cinnamoyltropine	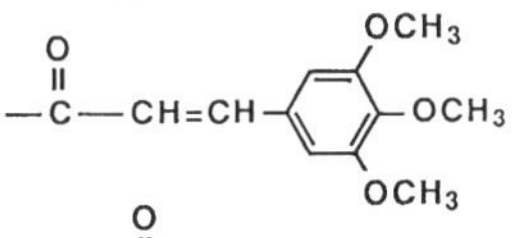 $—C(=O)—CH{=}CH—C_6H_2(OCH_3)_3$	5
benzoyltropine	$—C(=O)—C_6H_5$	6

Similar types of tropine-related alkaloids have been found in the various Solanaceous plants, most of them herbaceous. Well-known genera that contain these alkaloids are *Atropa, Hyoscyamus, Datura*, and *Scopolia.* An Australian plant, *Duboisia myopoloides*, is an example of a woody Solanaceous plant that contains tropane alkaloids. This plant contains scopolamine and other alkaloids, mainly in the leaves.

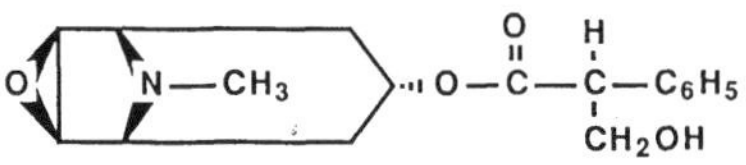

O, $N—CH_3$, $O—C(=O)—CH(CH_2OH)—C_6H_5$ scopolamine 7

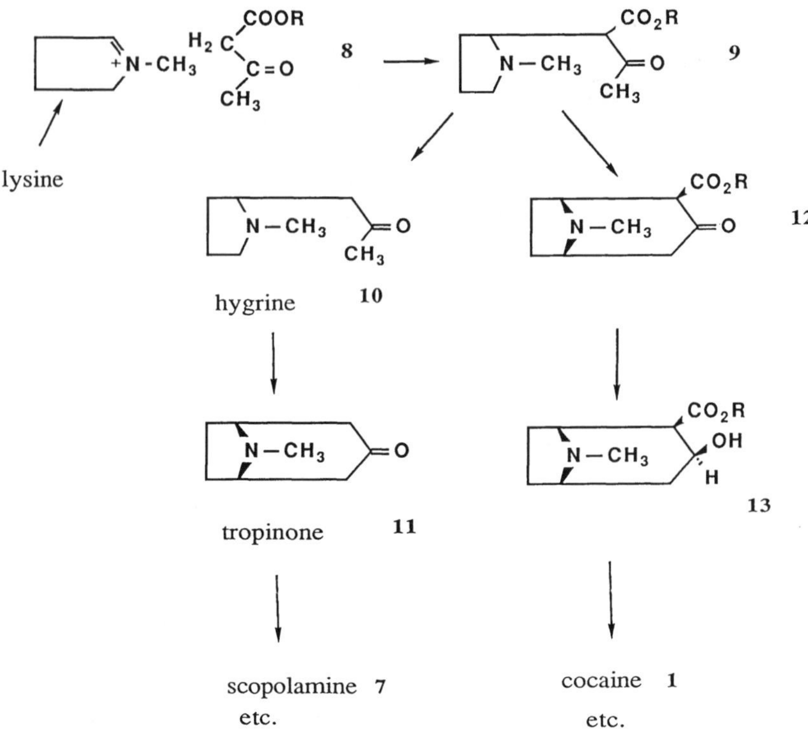

Fig. 5.2.1. The biogenetic route to cocaine and other related alkaloids

The general biogenetic route of cocaine and other related alkaloids is shown in Fig. 5.2.1. The fundamental skeleton is formed by the combination of lysine-ornithine derived pyrrolidine and the three- or four-carbon moiety of acetate-malonate origin. The acyl part of scopolamine has been demonstrated to be formed from phenylalanine through deamination and skeletal rearrangement (63).

5.2.2.1.2 Elaeocarpus Alkaloids

The family Elaeocarpaceae belongs to Malvales, and the plants of this family are all trees or bushes. The genus *Elaeocarpus* comprises about 250 species that occur mostly in tropical areas. Such species as *Elaeocarpus polydactylus, E. dolichostylis, E. sphaericus, E. densifolius*, and *E. altisectus* are all large rain-forest trees of New Guinea and contain a group of alkaloids (82).

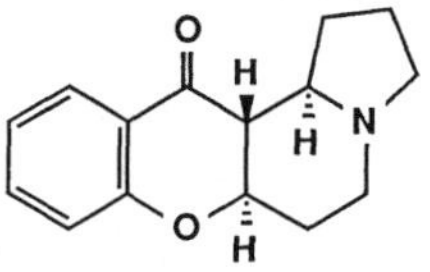

elaeocarpine **14** 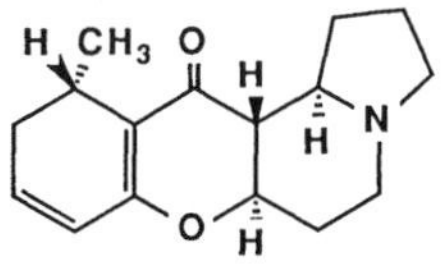 elaeocarpiline **15** 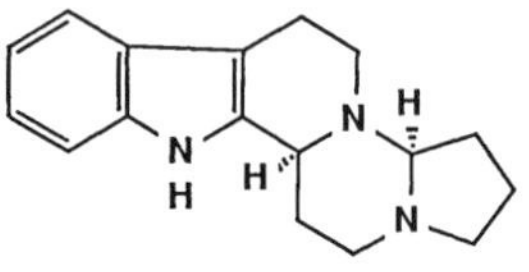 elaeocarpidine **16**

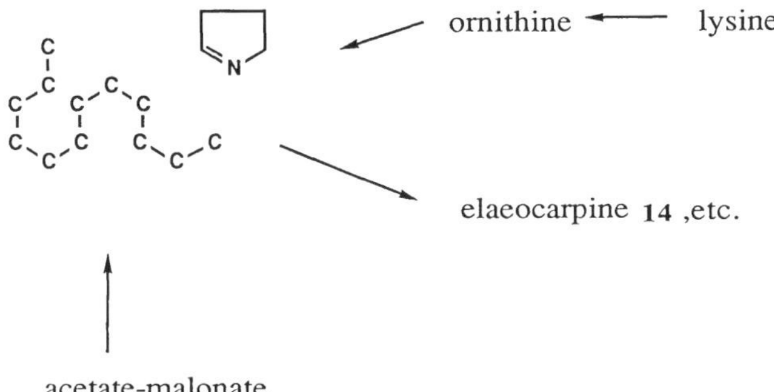

Fig. 5.2.2. Proposed biosynthetic route to Elaeocarpus alkaloids

elaeocarpidine 16
etc.

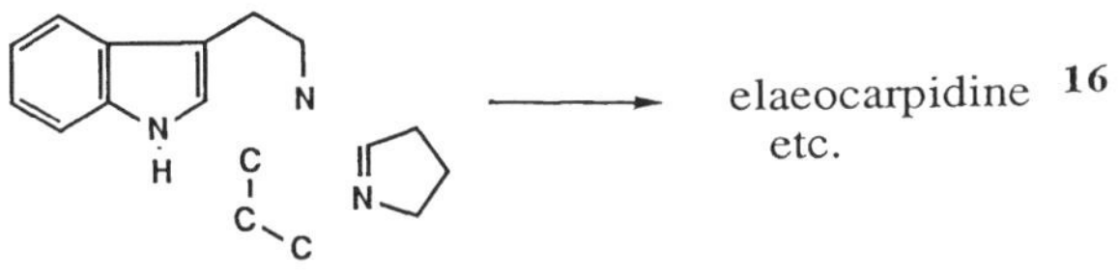

18

Fig. 5.2.3. Formation of elaeocarpidine

Elaeocarpine was obtained from *Elaeocarpus polydactylus* and other species in the racemic form. Elaeocarpiline, which was isolated from *Elaeocarpus sphaericus*, has a partially saturated aromatic ring. An alkaloid, elaeocarpidine, having an indole moiety, was isolated from *Elaeocarpus dolichostylis* and other species.

Biosynthesis of *Elaeocarpus* alkaloids has not been fully clarified. A possible route in the formation of elaeocarpine entails combination of a C_{12} polyketide chain and pyrrolidine derived from ornithine (Fig. 5.2.2).

Formation of elaeocarpidine is explained by a similar participation of ornithine (Fig. 5.2.3).

5.2.2.2 Alkaloids from Lysine

5.2.2.2.1 Punica Alkaloids

The simple piperidine alkaloids, pelletierine, isopelletierine, *N*-methylpelletierine, and pseudopelletierine, were isolated (106, 108) from the root bark of *Punica granatum* (Punicaceae), which had been used as an anthelmintic. Later, it was found that (−)-pelletierine possessed the *R*-configuration, and isopelletierine was *racemic* pelletierine (12, 51).

Biogenetically, pelletierine is derived from lysine via piperidine and a C_3 unit (acetate), and pseudopelletierine is formed by oxidative cyclization of *N*-methylpelletierine (Fig. 5.2.4).

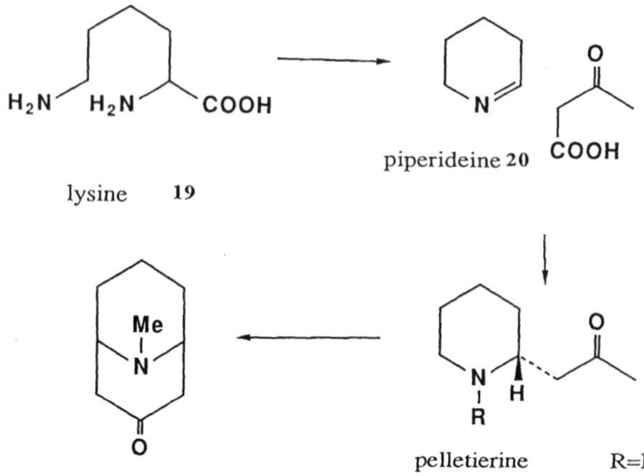

H_2N H_2N COOH → piperideine **20** COOH

lysine **19**

pseudopelletierine **23**

pelletierine R=H **21**

N-methylpelletierine R=Me **22**

Fig. 5.2.4. Formation of pelletierine and pseudopelletierine

5.2.2.2.2 Lythraceae Alkaloids

An interesting type of quinolizidine alkaloids occurs in the Lythraceae family and is not known in other plants (52). The alkaloids of woody plants of the genus *Lagerstroemia* are described here.

Lagerstroemine and lagerine were isolated as major alkaloids together with decinine, decamine, dihydroverticillatine, and methyllagerine from *Lagerstroemia indica.* These alkaloids are concentrated in the seed pods, with trace amounts in

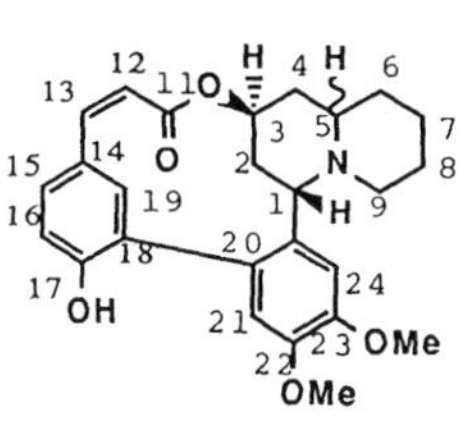

lythrine, 5 β-H, **27**

vertine, 5 α-H, **28** (cryogenine)

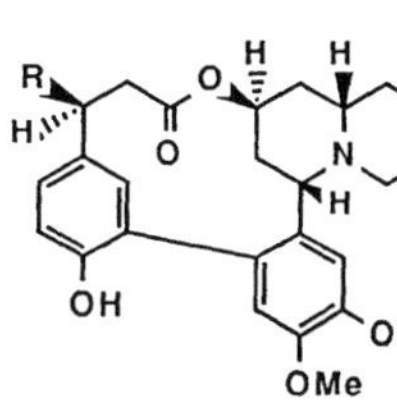

decinine, R=H, **29**

sinine, R=OH, **30** (lythridine)

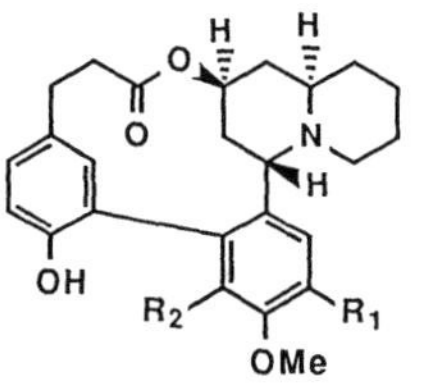

decamine, R_1=OMe, R_2=H, **31**

dihydroverticillatine R_1=H, R_2=OH, **32**

lagerstroemine R_1=H, R_2=OMe, **33**

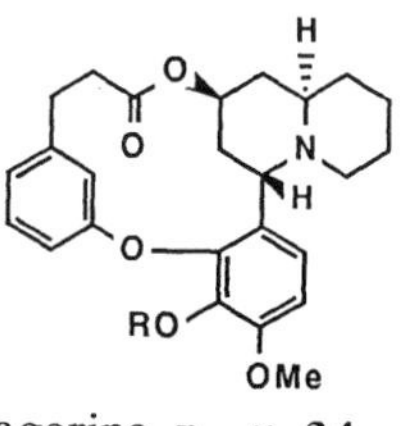

lagerine, R=H, **34**

methyllagerine, R=Me, **35**

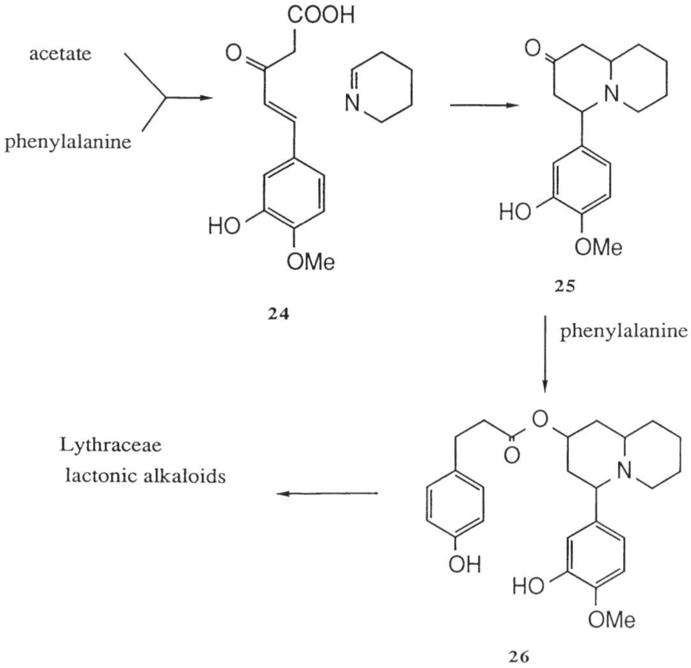

Fig. 5.2.5. Formation of Lythraceae lactonic alkaloids

the leaves and stems (43). They possess a biphenyl of biphenylether lactone structure (12- or 14-membered ring) linked between C-1 and C-3 positions of quinolizidine ring. The related alkaloids – lythrine, vertine (cryogenine), and sinine (lythridine) – were isolated from the leaves of *Lagerstroemia faurier* (47). Lysine and two C_6-C_3 units derived from phenylalanine are incorporated into the lactonic alkaloids by biogenetic studies using *Heimia* and *Decodon* genera (Fig. 5.2.5).

The phenylquinolizidine alkaloids lasubine I and lasubine II were obtained from the leaves of *Lagerstroemia subcostata* as major alkaloids, together with the ester alkaloids subcosine I and subcosine II. These alkaloids are considered to be intermediates in the biogenetic route of the lactonic alkaloids. Thus, lysine, phenylalanine, and a C_2 unit (acetate) produce the phenylquinolizidine alkaloids. These are then esterified with another phenylalanine unit to the ester alkaloids, followed by conversion to the lactonic alkaloids by oxidative coupling.

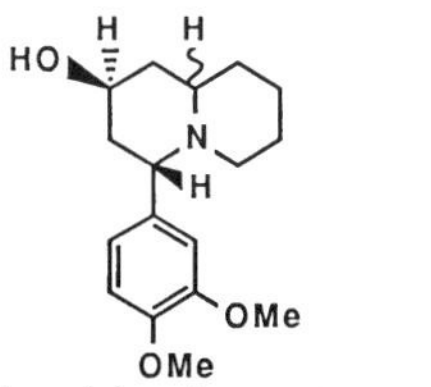

lasubine I 5α-H, **36**
lasubine II 5β-H, **37**

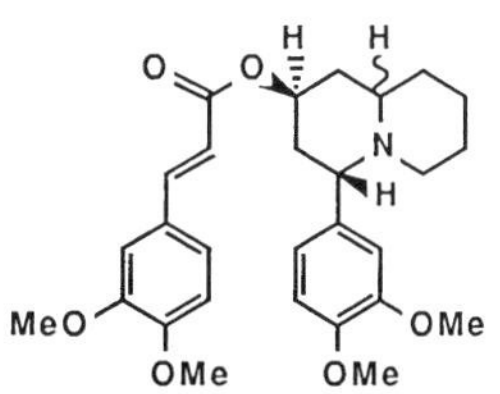

subcosine I 5α-H, **38**
subcosine II 5β-H, **39**

5.2.2.2.3 Securinega Alkaloids

The securinega alkaloids occur in several species of the *Securinega* and *Phyllanthus* genera (150). Securinine was first isolated from the leaves and stems of *Securinega suffruticosa* (Euphorbiaceae). Securinine nitrate is a central nervous system stimulant similar to strychnine but less toxic. It was found to be useful in treatment of paresis and paralysis following infectious diseases and psychic disorders. The related alkaloids were obtained from *Securinega suffruticosa*: allosecurinine, dihydrosecurinine, securinol A, and securinol B from the leaves, and securitinine from the roots.

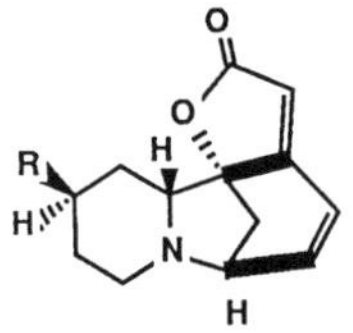

securinine, R=H ,40

phyllanthine, R=OMe ,41

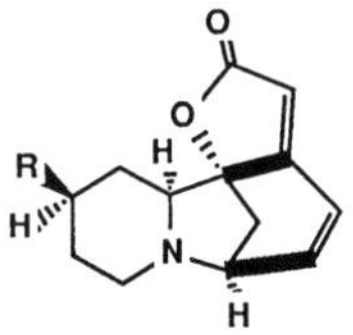

allosecurinine, R=H ,42

securitinine, R=OMe,43

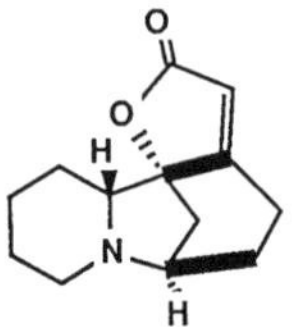

dihydrosecurinine 44

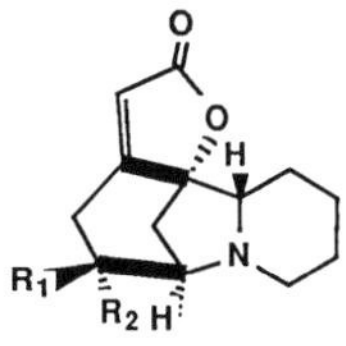

securinol A, R_1=OH , R_2=H ,45

securinol B, R_1=H , R_2=OH ,46

Virosecurinine and viroallosecurinine obtained from the leaves of *Securinega virosa* were found to be the enantiomers of securine and allosecurinine, respectively. *Securinega* alkaloids obtained from the roots of *Phyllanthus discoides* are securinine, allosecurinine, and phyllantine. These alkaloids are biogenetically produced from lysine and a C_6-C_3 unit derived from tyrosine (Fig. 5.2.6).

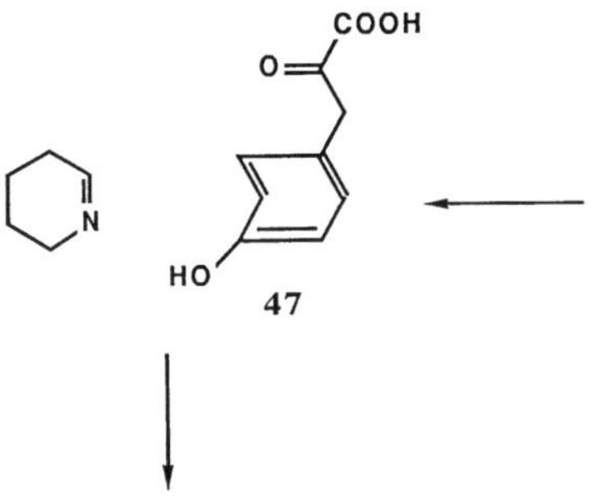

COOH

HO

47

tyrosine

Securinega alkaloids

Fig. 5.2.6. Biogenetic route to Securinega alkaloids

5.2.2.2.4 Cytisus Alkaloids

Cytisine, methylcytisine, and anagyrine were obtained from *Cytisus laburnum* (Leguminosae). All parts of the plant are poisonous because of the cytisine. It is one of the Cytisus alkaloids (95, 96).

The most common tetracyclic alkaloid, (−)-sparteine, was isolated from *Cytisus scoparius* with (−)-isosparteine and (+)-lupanine. Sparteine sulfate has been used as an oxytocic.

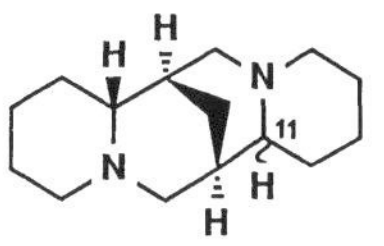

(-)-sparteine, **11** α**-H,48**

(-)-α-isosparteine, **11** α**-H,49**

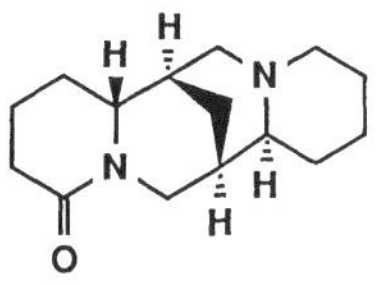

(+)-lupanine, **50**

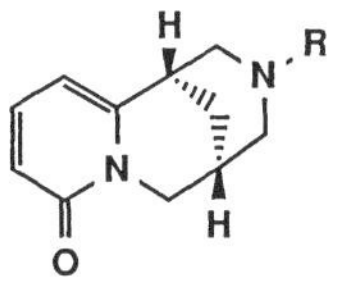

cytisine, **R=H,51**

methylcytisine, **R=Me,52**

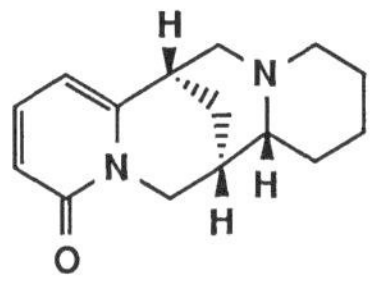

(-)-anagyrine, **53**

These tetracyclic and tricyclic quinolizidine alkaloids are commonly known as lupine alkaloids. The simple bicyclic alkaloids such as lupinine are produced from two lysine units via cadaverine. Tricyclic alkaloids such as cytisine are considered to be formed from tetracyclic alkaloids through anagyrine-type alkaloids (Fig. 5.2.7).

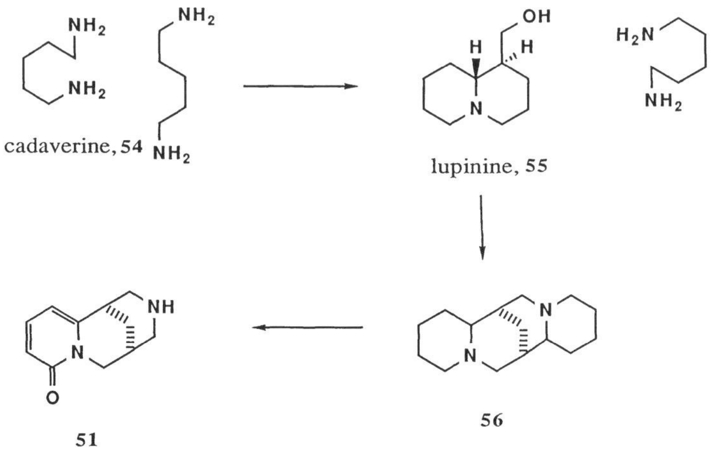

cadaverine, **54**

lupinine, **55**

51

56

Fig. 5.2.7. Formation of tricyclic Cytisus alkaloids

5.2.2.3 Alkaloids from Anthranilic Acid

5.2.2.3.1 Quinoline and Furoquinoline Alkaloids

Quinoline and furoquinoline alkaloids (57, 119) are widespread within the Rutaceae family. Many of these alkaloids have oxygen functions (hydroxy, methoxy, and methylenedioxy groups) at C-6, C-7, and/or C-8 of the quinoline ring.

Simple quinoline alkaloids have been isolated from several genera – e.g., 4-methoxy-1-2-quinolone from *Fagara boniensis*, and edulitine from *Casimiroa edulis.*

Balfourodendron riedelianum is a small tree indigenous to Brazil and Argentina where it has found popular use for the treatment of stomach and intestinal ailments. The plant is interesting because of its variety of alkaloid types (124). From the bark of *Balfourodendron riedelianum*, various types of alkaloids have been obtained: quinoline (1-methyl-2-phenyl-4-quinolone), dihydrofuroquinoline (balfourodine, *O*-methylbarfourodinium salt), dihydropyranoquinoline (isobalfourodine), furoquinoline (dictamnine, maculosidine, kokusaginine, skimmianine).

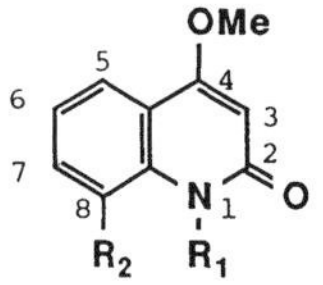

R_1=Me,R_2=H,58

edulitine, R_1=H,R_2=OMe,59

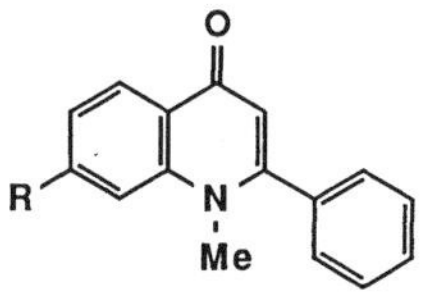

R=H,60

edulein, R=OMe,61

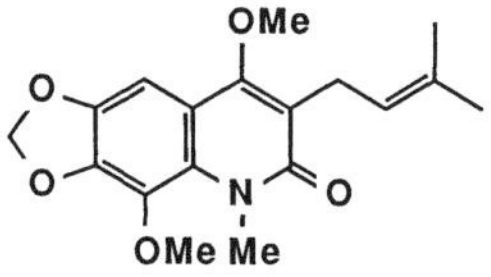

pterecortine, 62

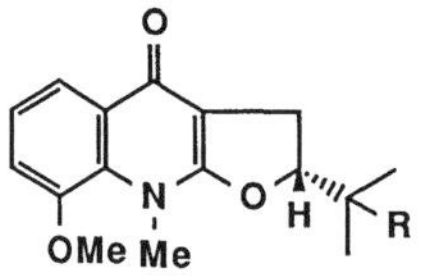

lunacrine, R=H,63

balfourodine, R=OH,64

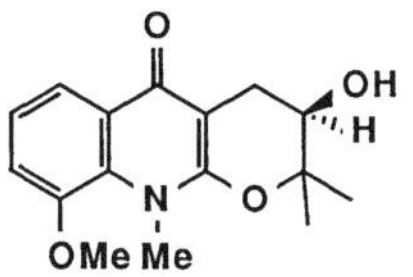

isobalfourodine, 65

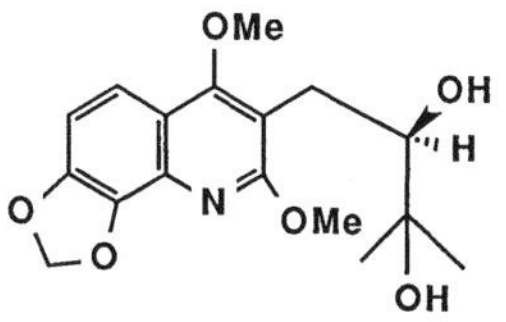

orixine, 66

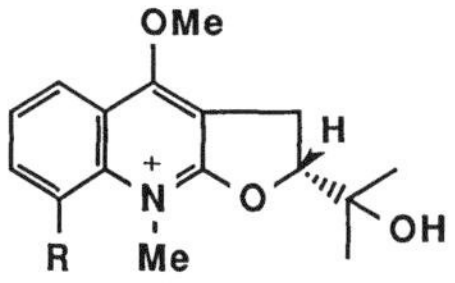

N-methylplatydesminium salt, R=H,67

pteleatinium salt, R=OH,68

O-methylbalfourodinium salt, R=OMe,69

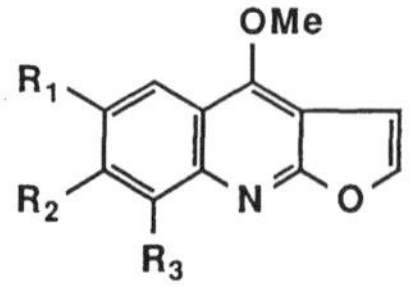

dictamnine, R_1=R_2=R_3=H,70

maculosidine, R_1=R_3=OMe,R_2=H,71

kokusaginine, R_1=R_2=OMe,R_3=H,72

skimmianine,R_1=H,R_2=R_3=OMe,77

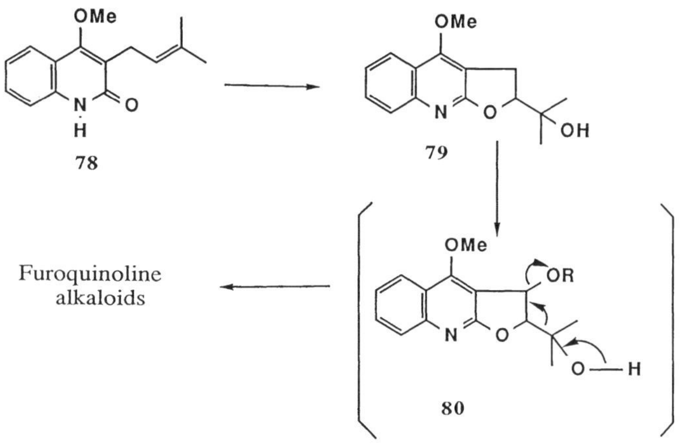

OMe
N O H
78
OMe
N O
OH
79
OMe
OR
N O
O—H
80
Furoquinoline alkaloids

Fig. 5.2.8. Formation of furoquinoline alkaloids

Related alkaloids have been obtained from other sources: *Lunasia amara, Ptelea trifoliata, Orixa japonica, Skimmia japonica*, etc.

Anthranilic acid and acetate biosynthesize the simple quinoline alkaloids, which are then prenylated with mevalonic acid at the C-3 position to 3-prenyl-2-quinolone alkaloids (e.g., pterecortine from *Ptelea trifoliata* and orixine from *Orixa japonica*). 3-Prenyl-2-quinolone alkaloids are then cyclized to dihydrofuro- or dihydropyranoquinoline alkaloids. Hydroxylation at the C-3 position of dihydrofuroquinoline alkaloids and subsequent loss of the isopropyl side chain produce the furoquinoline alkaloids (Fig. 5.2.8); dictamnine is a skeletal alkaloid and skimmianine is the most common alkaloid.

5.2.2.3.2 Acridone Alkaloids

Acridone alkaloids (49) were first isolated from the bark and leaves of some Rutaceae species in Australian tropical rain forests. The best known acridone alkaloid, acronycine, was obtained from the bark of *Acronycia bauri* together with the related alkaloids (noracronycine, de-*N*-methylacronycine, and de-*N*-methylnoracronycine) and simple acridone alkaloids (1,3-dimethoxy-*N*-methylacrydone, melicopine, melicopicine, etc.). Acronycine displayed significant activity in several tumor test systems and it was reported to be orally active.

Acridone alkaloids have been obtained from other species: evoxanthine, melicopidine, evoxanthidine, and xantheodine from *Evodia xanthoxyloides*, noracronycine, de-*N*-methylacronycine, and de-*N*-methylnoracronycine from *Glycosmis pentaphylla.*

Recently, the C-4 prenylated alkaloids glycocitrine I and glycocitrine II were found in the root and stem bark of *G. citrifolia*, along with other alkaloids.

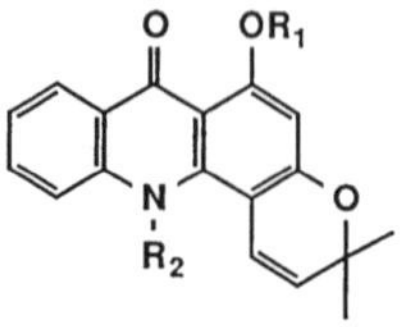

acronycine, $R_1=R_2=Me$, **81**
noracronycine, $R_1=H, R_2=Me$, **82**
de-N-methylacronycine, $R_1=Me, R_2=H$, **83**
de-N-methylnoracronycine, $R_1=R_2=H$, **84**

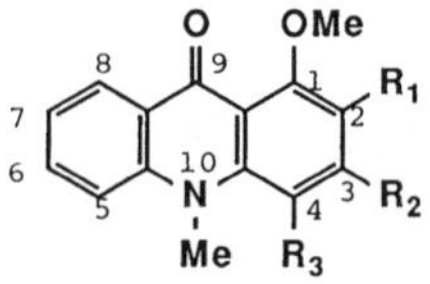

$R_1=R_3=H, R_2=OMe$, **85**
evoxanthine, $R_1,R_2=-OCH_2O-, R_3=H$, **86**
melocopine, $R_1=OMe, R_2,R_3=-OCH_2O-$, **87**
melocopidine, $R_1,R_2=-OCH_2O-, R_3=OMe$, **88**
melicopicine, $R_1=R_2=R_3=OMe$, **89**

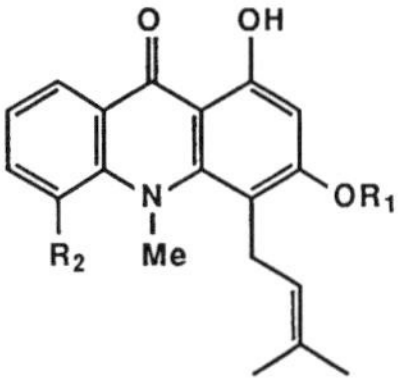

glycocitrine I $R_1=Me, R_2=OH$, **90**
glycocitrine II $R_1=R_2=H$, **91**

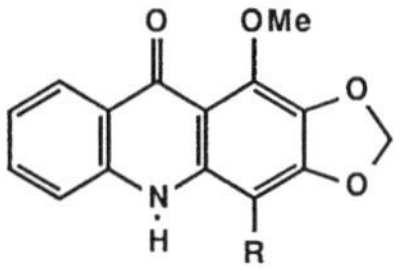

evoxanthidine, R=H, **92**
xanthevodine, R=Me, **93**

5.2.2.3.3 Evodia Alkaloids

The quinazolinocarboline alkaloids, rutaecarpine and evodiamine, were isolated from the fruit of *Evodia rutaecarpa*, which has been used in China as a drug for the treatment of headache, abdominal pain, etc. This type of alkaloid (11) occurs

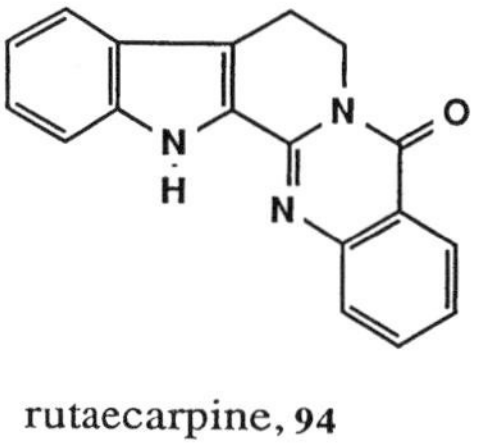

rutaecarpine, **94**

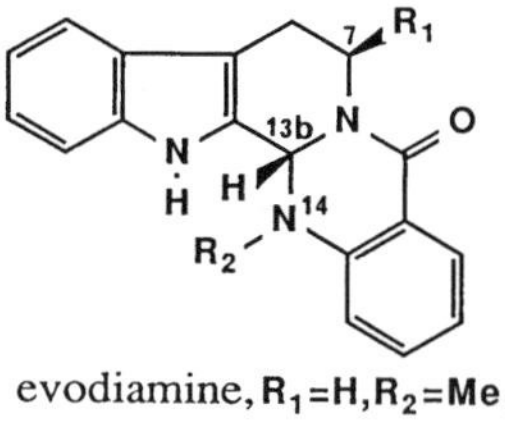

evodiamine, $R_1=H, R_2=Me$, **95**
13b,14-dihydrorutaecarpine, $R_1=R_2=H$, **96**
7-carboxyevodiamine, $R_1=COOH, R_2=Me$, **97**

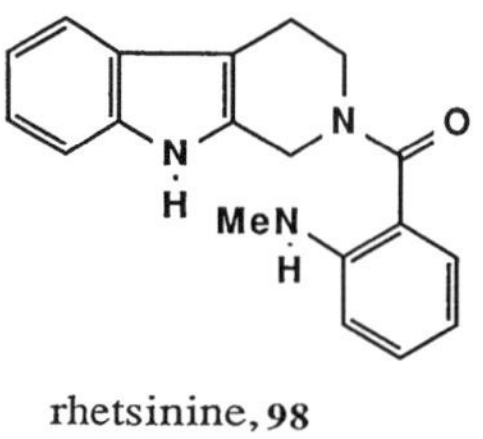

rhetsinine, **98**

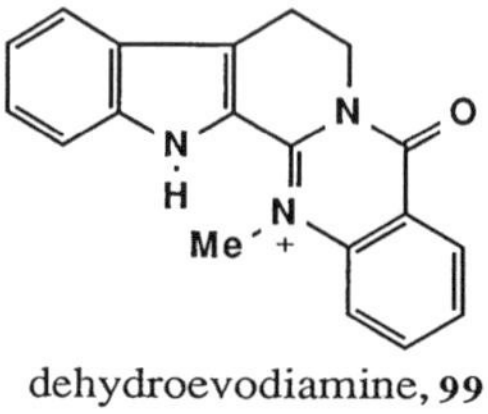

dehydroevodiamine, **99**

in a few species of the Rutaceae family; they are formed from anthranilic acid, tryptophan, and a C_1 unit. Other alkaloids have also been obtained from *Evodica rutaecarpa*; 13b,14-dihydrorutaecarpine, dehydroevodiamine, rhetsinine, and 7-carboxyevodiamine.

The absolute configuration of (+)-evodiamine had not been determined until recently, when a 13b-*S* configuration was established by correlation with (*S*)-tryptophan via (7*S*, 13b*S*)-carboxyevodiamine (31).

5.2.2.3.4 Carbazole Alkaloids

Carbazole alkaloids have been isolated from a few Rutaceae species, especially *Murraya koenigii*, the leaves of which have been used in southern India as a flavoring in food. Murrayanine, mukoeic acid, girinimbine, and mahanimbine have been isolated from the leaves. Carbazole alkaloids occur in all parts of this plant.

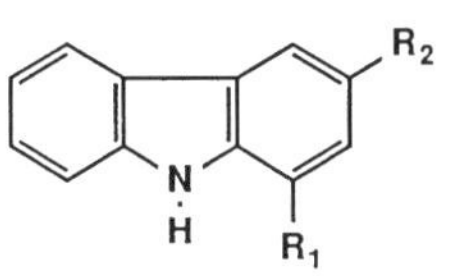

carbazole, $R_1=R_2=H$, **100**
murrayanine, $R_1=OMe, R_2=CHO$, **101**
mukoeic acid, $R_1=OMe, R_2=COOH$, **102**

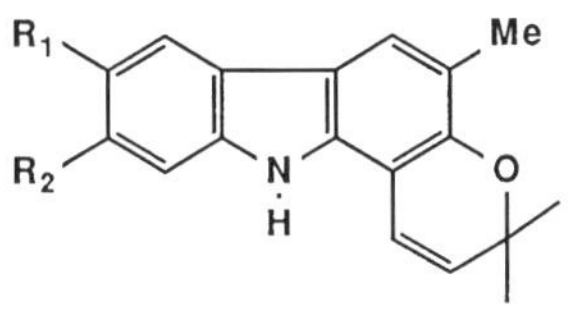

girinimbine, $R_1=R_2=H$, **103**
koenimbine, $R_1=R_2=OMe$, **104**
koenigicine, $R_1=R_2=OMe$, **105**

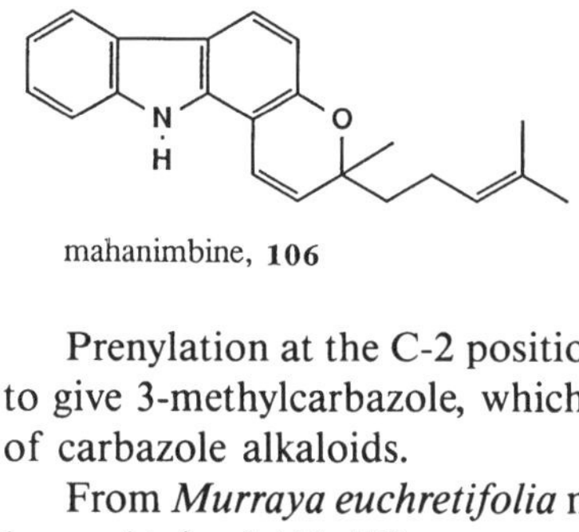

mahanimbine, **106**

Prenylation at the C-2 position of indole followed by cyclization is considered to give 3-methylcarbazole, which is a possible key intermediate in the formation of carbazole alkaloids.

From *Murraya euchretifolia* monomeric and bis-type carbazole alkaloids have been obtained (48, 76).

5.2.2.4 Alkaloids from Nicotinic Acid (Celastraceae Alkaloids)

From the poisonous European tree *Euonymus europaeus*, evonine was isolated (36). The same alkaloid was later found in *E. sieboldiana*, which grows in Asia (159). It is one of the alkaloids formed from nicotinic acid (149).

The genus *Euonymus* belongs to the family Celastraceae. Two other important alkaloid-containing genera of this family are *Maytenus* and *Catha*. Maytoline and maytine were isolated from *Maytenus ovata* collected in Ethiopia (92).

A drink made of the dried leaves of *Catha edulis* is called Khat-tea by Arabian and Ethiopian people and is used as a stimulating drink. Norpseudoephedrine has been found in this plant. From the same plant, cathidine D has been isolated.

These Celastraceae alkaloids – evonine, maytoline, maytine, and cathidine D – have nicotinate as the nitrogen-containing moiety. Biogenetic study has demonstrated that nicotinic acid is actually the precursor. It is quite evident that the remaining carbon skeleton is sesquiterpenoid. In the molecule of evonine, an additional C_5 unit is involved, but the biogenetic origin is unknown.

From *Maytenus ovata*, Kupchan isolated the well known anticancer agent maytansine (91).

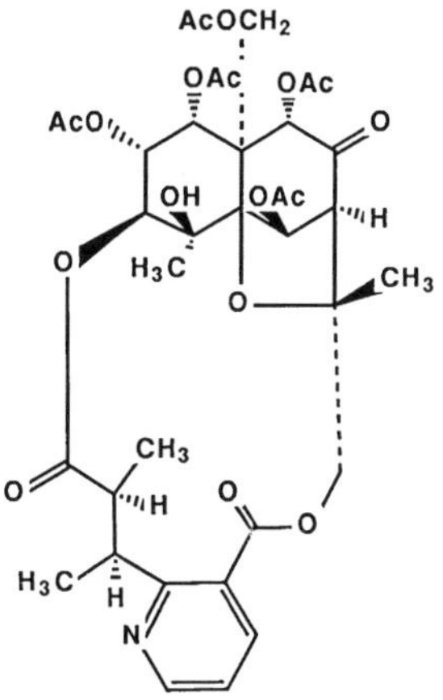

evonine, **107**

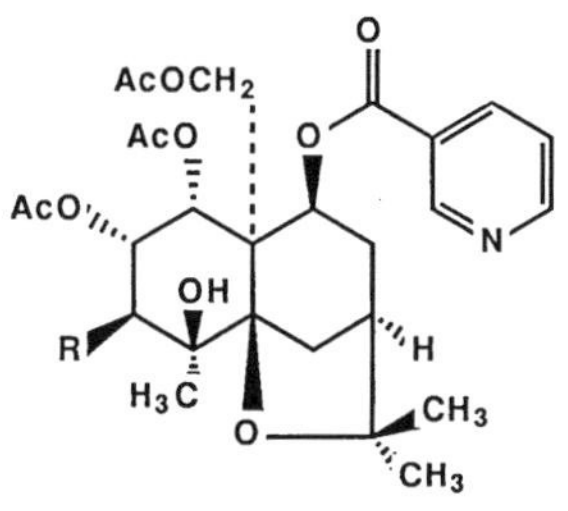

maytoline, R=OH, **108**
maytine, R=H, **109**

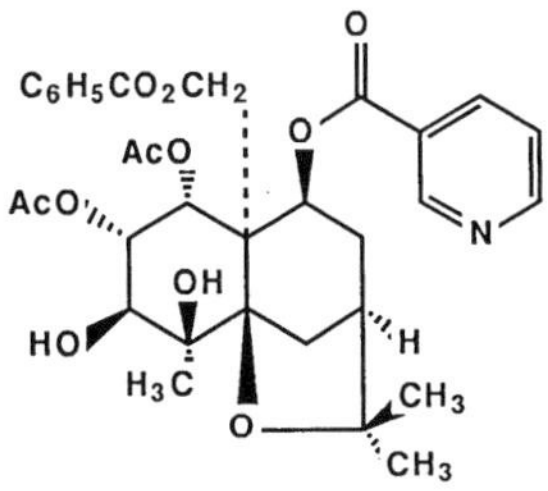

cathidine D, **110**

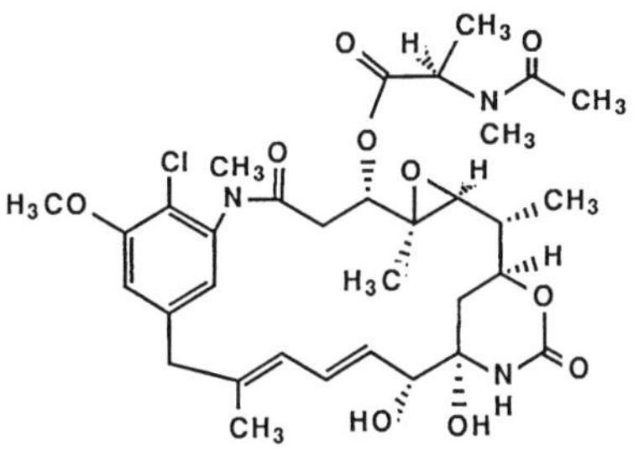

maytansine, **111**

In most cases, studies have been done on alkaloids isolated from the fruit of these Celastraceae plants, but these alkaloids seem to be present also in their wood. For example, Kupchan reported that alkaloids having anticancer activities occur in the wood of the stem of *Maytenus ovata* collected in Ethiopia and Kenya (92).

5.2.2.5 Alkaloids from Phenylalanine and Tyrosine

5.2.2.5.1 Benzylisoquinoline Alkaloids

From the bark of a Lauraceous tree, *Cinnamomum laubattii*, collected in northern Australia, a crude alkaloid mixture was obtained in 0.01% yield; this mixture was mostly composed of (+)-reticuline, a benzylisoquinoline alkaloid (37). The same alkaloid was obtained from *Cinnamomum camphor* (156).

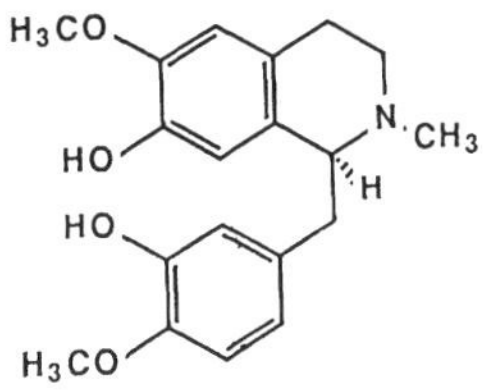

(+)-reticuline, **112**

(+)-Reticuline has been found in many plants of the families of Annonaceae, Hernandiaceae, Lauraceae, Monimiaceae, Papveraceae, and others. This alkaloid is known to serve as an important intermediate in the biosynthesis of more complicated isoquinoline alkaloids.

The antipode, (−)-reticuline, is the biogenetic intermediate of morphine. Morphine is possibly one of the most important alkaloids ever found, but we will not discuss it here because it has been found only in herbaceous plants of the Papaveraceae.

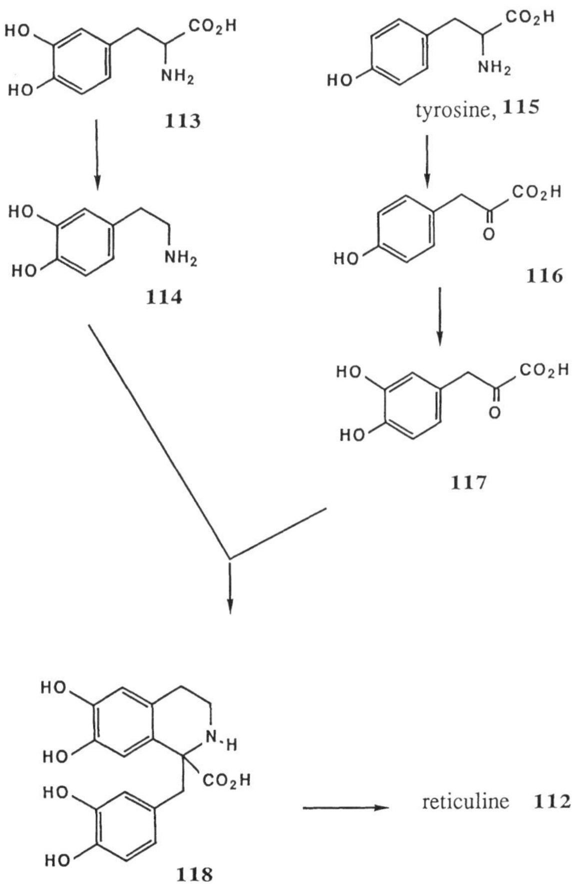

113 tyrosine, **115**

114 **116**

117

118 ⟶ reticuline **112**

Fig. 5.2.9. Biosynthesis of reticuline

Reticuline and many other structurally related alkaloids are termed benzyltetrahydroisoquinoline-type alkaloids. Sometimes they are referred to simply as benzylisoquinoline alkaloids and this term is also applied to the variety of more complex natural products that are biogenetically related to them.

The biosynthesis of reticuline is shown in Fig. 5.2.9. Tyrosine and dopa are the precursors of the fundamental skeleton. The other participating biogenetic building block is a C_1-unit that forms the *N*-methyl group.

Coclaurine was isolated from the bark and trunk (0.04%) of *Cocculus laurifolia*, a small shrub of Menispermaceae family. This alkaloid has been found in the plants of Menispermaceae (*Cocculus, Sarcopetalum*), Lauraceae (*Arseodaphne, Mechilus*), Euphorbiaceae (*Croton*), and Rhamnaceae (*Rutanilla, Zizyphus*).

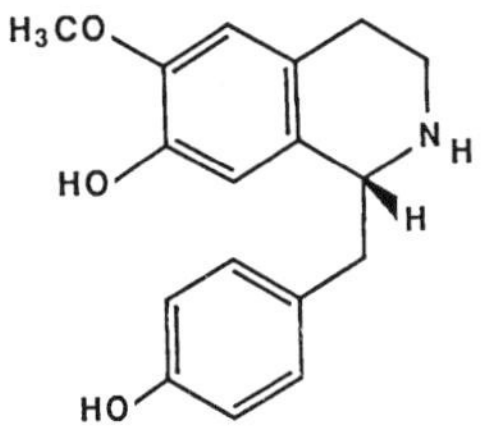

coclaurine, **119**

Coclaurine is an important alkaloid because its demethyl analog, norcoclaurine, can be regarded as the prototype building block of dimeric benzylisoquinoline alkaloid that are often referred to as biscoclaurine-type alkaloids.

The family Menispermaceae comprises over 60 genera and about 400 species, distributed mainly in the tropical and subtropical regions in the world. Most species are lianas but some are bushes or low trees. Biscoclaurine-type alkaloids occur in many of the plants of this family. From the bark and trunks of *Cocculus laurifolius*, 0.13% crude base was obtained. Trilobine (0.004%) was isolated from the base together with coclaurine (90, 157).

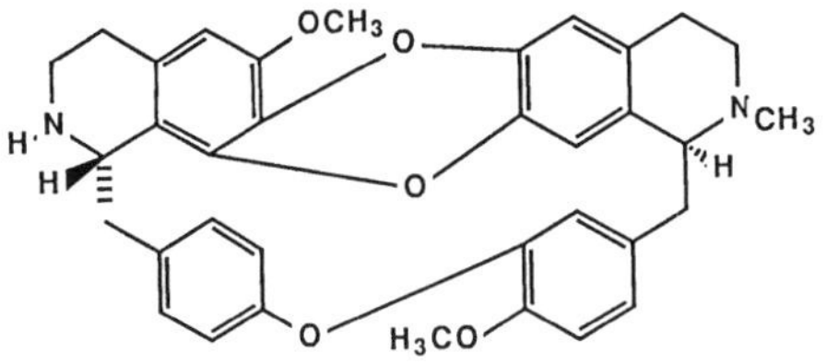

trilobine, **120**

Some other genera of Menispermaceae containing bisbenzylisoquinoline alkaloids are *Menisperm, Pycnarrhena, Stephania, Cycle, Triclisia, Limaciopsis*, and *Synchlisia* (141).

Menispermaceae is not the only family that contains these bis-type alkaloids. They are also found in Berberidaceae (*Berberis, Mahonia*), Magnoliaceae (*Michelis, Talauma*), Annonaceae (*Phaeanthus*), Monimiaceae (*Daphnandra, Doryphora*), Ranunculaceae (*Thalictrum* (herbaceous)), and others.

5.2.2.5.2 Curare Alkaloids

The South American natives in the Amazon-Orinoco basins have for centuries prepared arrow poisons called curare for hunting animals. Generally curare is classified into three groups according to the containers used for its storage. Tubocurare is stored in bamboo tubes; the active principles are from the Menispermaceous plants; this curare is discussed here. Calabash curare – stored in calabashes or gourds – is prepared from the plants of *Strychnos* species of Loganiaceae; the active principles are bis-indole alkaloids (see Sect. 5.2.2.6.11). Pot curare contains both of the above active principles, obviously prepared from a mixture of the plants of Menispermaceae and Loganiaceae.

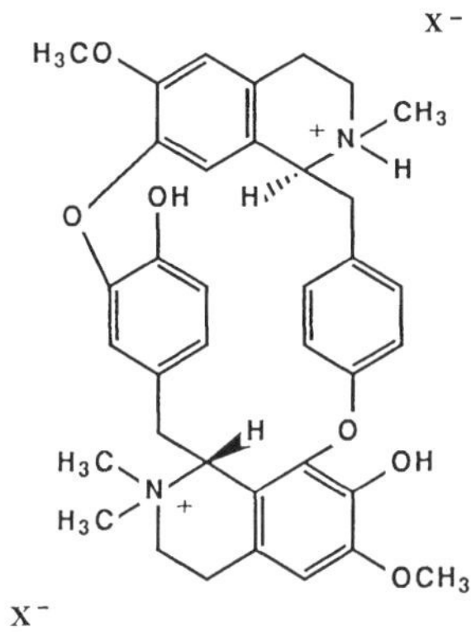

(+)-tubocurarine **121**

The most active principle of Menispermaceae curare is (+)-tubocurarine (25, 39). The origin plants of tubocurare are mainly of the *Chondodendron* species such as *C. tomentosum, C. platyphyllum*, and *C. candicans. C. tomentosum* is a liana whose bark and stem contain (+)-tubocurarine.

(+)-Tubocurarine causes muscle relaxation by blocking nerve receptors at the motor muscle by competitive binding to the acetyl choline receptor. This mode of action is generally described in the field of pharmacology as the curarizing activity.

5.2.2.5.3 Sinomenine

The stem and root of *Sinomenium acutum* are used in Chinese medicine. Sinomenine (10) was isolated from this plant and the structure, when elucidated, was found to be similar to that of morphine. The absolute configuration of the structure of sinomenine, however, is antipodal to that of morphine.

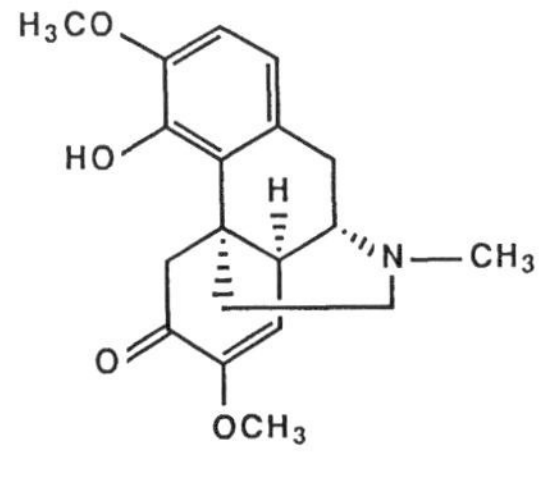

sinomenine, **122**

5.2.2.5.4 Aporphine-type Alkaloids

Magnoflorine is a quaternary alkaloid first isolated from a Magnoliaceae, *Magnolia grandiflora* (115). From the bark of this plant, magnoflorine was obtained in a yield of 0.013%. The same alkaloid was later isolated from other plants, such as *Zanthoxylum planispium* (Rutaceae), the bark and wood of which contain 0.02% magnoflorine.

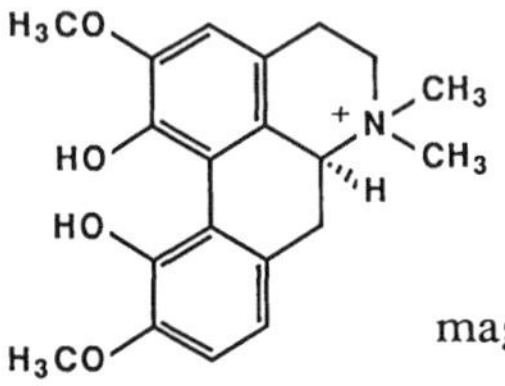

H_3CO HO HO H_3CO N^+ CH_3 CH_3 H

magnoflorine,**123**

Alkaloids having the same fundamental skeleton as magnoflorine are called aporphine-type alkaloids (58, 140). This class of alkaloids is widely distributed in nature. At least 18 families are known to contain aporphine-type alkaloids. Some of them are Annonaceae, Magnoliaceae, Lauraceae, and Hernandiaceae (Magnoliales), Ranunculaceae, Berberidaceae, and Menispermaceae (Ranunculales), Aristrochiaceae (Aristrochiales), Papaveraceae (Fumariaceae) (Papaverales), Rutaceae (Rutales), and Rhamnaceae (Rhamnales).

5.2.2.5.5 Berberine

Berberine (79, 103) occurs in the bark of a Rutaceae tree, *Phellodendron amurense*. Bark, 2–4 mm thick, can be ripped from the wood of an old tree. Cuttings of the bark are used as a crude drug in Chinese medicine as a bitter stomachic. Berberine is a quaternary base with a yellow color. The bark of *P. amurense* is also yellow because of the presence of over 1% berberine. The alkaloid exhibits antibacterial activity.

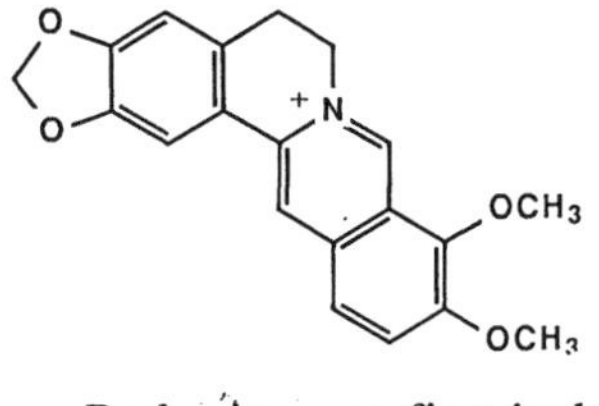

O O N^+ OCH_3 OCH_3

berberine,**124**

Berberine was first isolated from *Berberis vulgaris*, a shrub of Berberidaceae (21). Since then this alkaloid has been isolated from many plants, namely Annonaceae, Ranunculaceae (mostly herbaceous), Berberidaceae, Menispermaceae, and Papaveraceae (Fumariaceae) (mostly herbaceous).

Berberine has one extra carbon in the molecule (C-ring) in addition to a benzyl isoquinoline framework. It has been fully confirmed that this carbon "berberine bridge" originates from *N*-methyl carbon of the benzyl isoquinoline precursors (Fig. 5.2.10).

The general name protoberberine alkaloid is given to quaternary alkaloids possessing the same skeleton as berberine. The positions of the oxygen functions

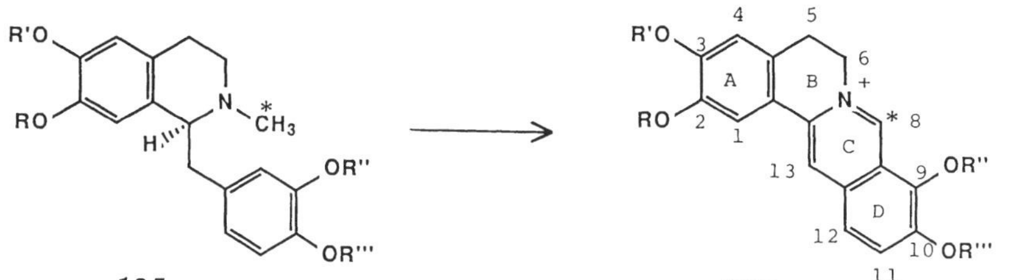

Fig. 5.2.10. Biosynthesis of berberine

on the D ring are at 9 and 10 in berberine. But sometimes they are at 10 and 11 in the closely related alkaloids, reflecting the direction of the biogenetic cyclization of the benzylisoquinoline precursors. The alkaloids with the saturated C ring are also often encountered in nature, they are referred to as tetrahydroprotoberberine alkaloids. These alkaloids are most frequently found in the Papaveraceae, a family whose species are mostly herbaceous.

5.2.2.5.6 Nitidine

Nitidine was isolated from the root bark and root of *Zanthoxylum nitidum*, which is a woody climber of the Rutaceae (4). The same compound has been isolated from *Zanthoxylum tsihanimposa* (root) and *Z. ailanthoides* (root). It is also present in the stem bark of *Zanthoxylum americanum* together with coumarins (44).

The aerial part of *Zanthoxylum inerme* was divided into bark and wood. From the bark, more than 0.07% of nitidine was isolated together with other related alkaloids. By contrast, none of these alkaloids were found in the wood, and 4-methoxy-1-methyl-2-quinoline was the only nitrogeneous compound occurring together with several coumarins and lignans. These results are shown in Table 5.2.2 (75). Nitidine has been shown to have cytotoxic and antitumor activities.

Nitidine and its derivatives are generally called benzo-[*c*]-phenanthridine alkaloids. This group of alkaloids is biosynthesized from the tetrahydroprotoberberine alkaloids through ring cleavage at the C–N bond and cyclization between C_6 and C_{13} (Fig. 5.2.11).

Table 5.2.2. Alkaloids of *Zanthoxylum inerme*

Alkaloid	Content in	
	Bark (%)	Wood (%)
Nitidine (as chloride)	0.071	
Avicine (as dihydro base)	0.005	
Chelrythrine ψ-cyanide	0.005	
L-(+)-armepavine methosalt	0.007	
Oxynitidine	0.010	
4-Methoxy-1-methyl-2-quinolone		0.004

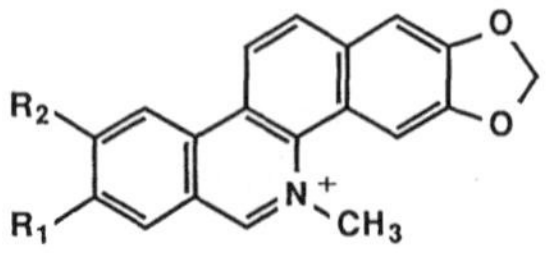

	R_1	R_2	
nitidine	OCH_3	OCH_3	**127**
avicine	-O-CH_2-O-		**128**

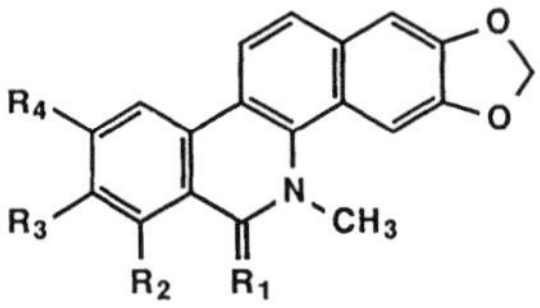

	R_1	R_2	R_3	R_4	
chelerythrine ψ-cyanide	<CN, H	OCH_3	OCH_3	H	**129**
oxynitidine	=O	H	OCH_3	OCH_3	**130**

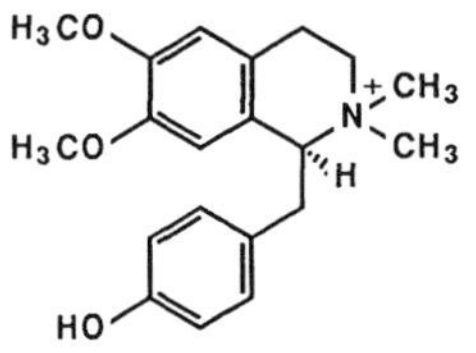

(+)-armepavine methosalt, **131**

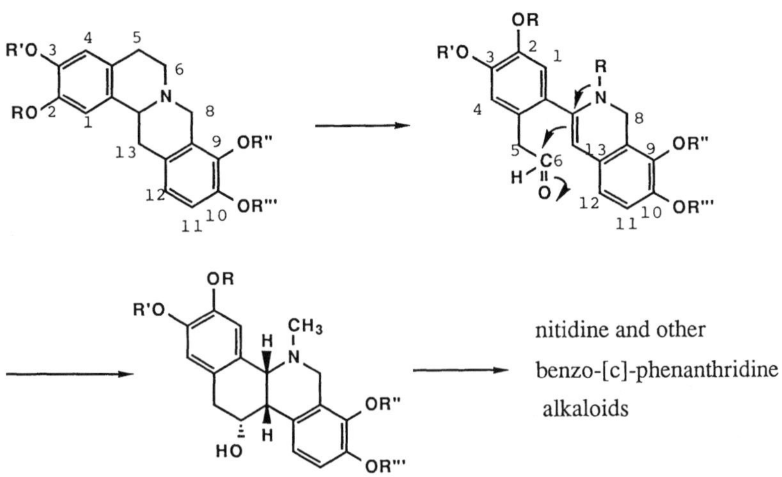

nitidine and other benzo-[c]-phenanthridine alkaloids

Fig. 5.2.11. Biosynthesis of nitidine alkaloids

5.2.2.5.7 *Erythrina Alkaloids*

From *Erythrina crystagalli* cv. *Maruba deiko*, erythraline, erythrinine, and erythratine were isolated (77). Examination by TLC demonstrated that these alkaloids are present in all heartwood, bark, and roots. Many other Erythrina alkaloids have been isolated from plants of the genus *Erythrina* (Leguminosae), but most have been obtained from leaves and fruits.

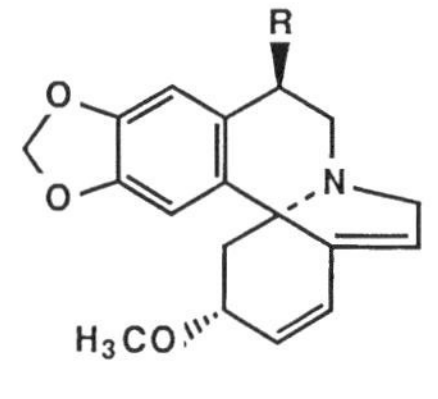

erythraline R=H **135**

erythrinine R=OH **136**

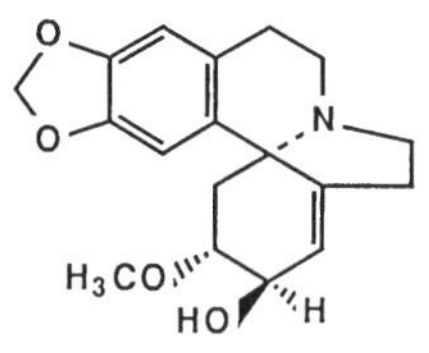

erythratine **137**

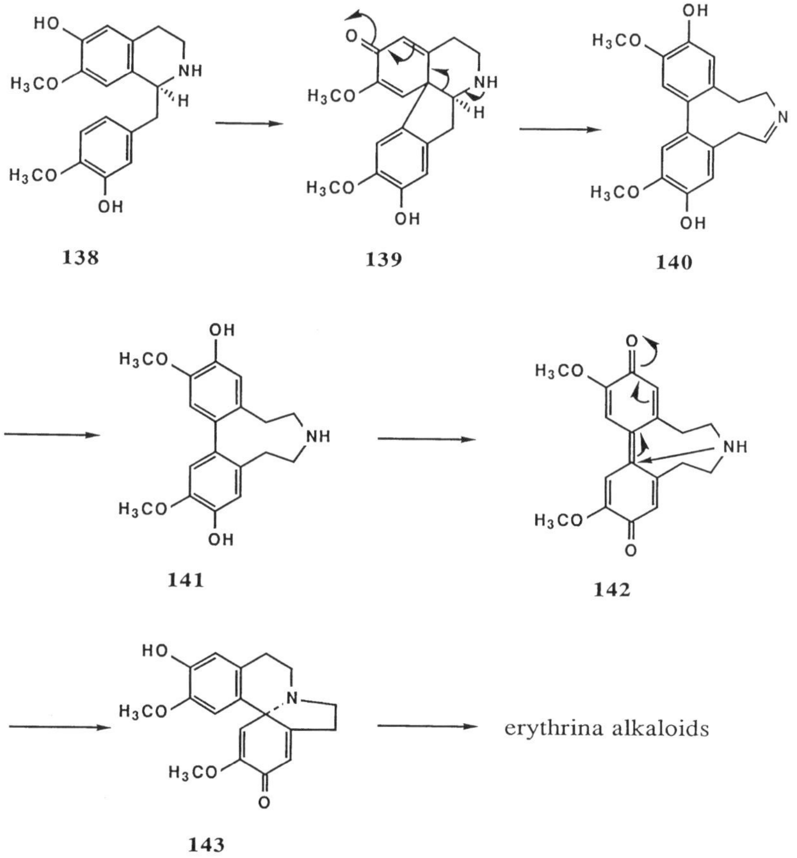

138 → **139** → **140**

→ **141** → **142**

→ **143** → erythrina alkaloids

Fig. 5.2.12. Proposed biosynthesis of Erythrina alkaloids

The biogenesis of the *Erythrina* alkaloids (15, 40, 65, 107) has been studied by using labeled precursors; a proposed general scheme is shown in Fig. 5.2.12.

It is interesting to note that a member of the *Erythrina* alkaloids, dihydroerysodine, was isolated (0.003%) from a Menispermaceae, *Cocculus laurifolius* (158).

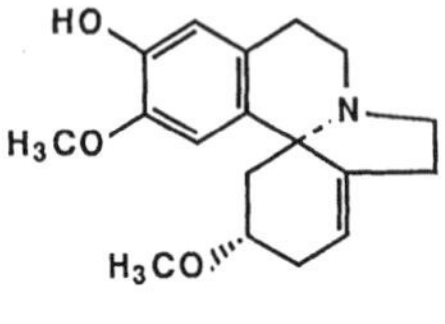

dihydroerysodine **144**

5.2.2.5.8 Cephalotaxus Alkaloids

The presence of alkaloids in the plants of *Cephalotaxus* was known for a long time but it was not until 1963 that pure alkaloids were isolated from the stems and leaves of *C. harringtonia* var. *drupacae* (121).

There are two groups of *Cephalotaxus* alkaloids: the cephalotaxine – harringtonine group and the group represented by alkaloids II and III. The latter are often called homoerythrinan alkaloids.

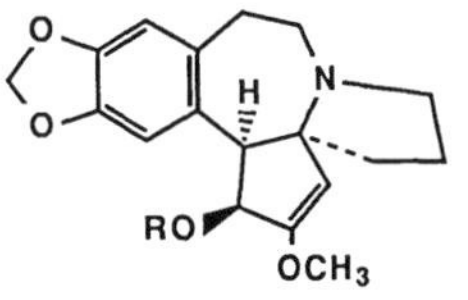

	R	
cephalotaxine	H	145
harringtonine	$-C(=O)-C(OH)(CH_2CO_2CH_3)-CH_2-CH_2-C(OH)(CH_3)-CH_3$	146
deoxyharringtonine	$-C(=O)-C(OH)(CH_2CO_2CH_3)-CH_2-CH_2-CH(CH_3)-CH_3$	147

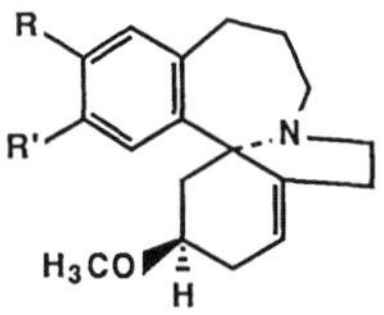

	R		R'	
alkaloid II	OCH_3 / OH	or	OH / OCH_3	148
alkaloid III	OCH_3		OCH_3	149

A possible pathway of the biosynthesis of *Cephalotaxus* alkaloids is shown in Fig. 5.2.13 (120). A similarity has been pointed out between this scheme and that of *Erythrina* alkaloids (Sect. 5.2.2.5.7). The starting open-ring intermediate in the present case is phenethylisoquinoline, instead of benzylisoquinoline for the *Erythrina* alkaloids. On this basis, homoerythrinan-type alkaloids are also called phenethylisoquinoline-type alkaloids.

A study of the constituents of *Cephalotaxus* plants was made by using a young plant grown in a controlled environment (33). The concentration of free alkaloids (homoerythrinan-type alkaloids and cephalotaxine) did not increase with age, but that of cephalotaxine esters did. Total alkaloid concentrations increased in the older leaves and decreased in the older stems. Hydrolytic cleavage

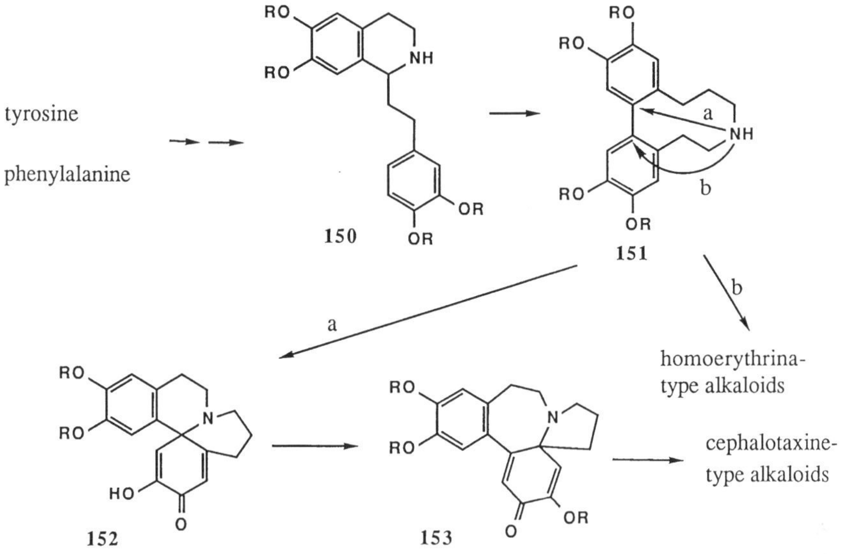

tyrosine

phenylalanine

150

151

a

b

homoerythrina-type alkaloids

152

153

cephalotaxine-type alkaloids

Fig. 5.2.13. Proposed biosynthesis of Cephalotaxus and homoerythrina-type alkaloids

of the ester part of the ester alkaloids was caused by physiological stress, such as pruning.

Much attention has been paid to the *Cephalotaxus* alkaloids, particularly in China, because of their antitumor activity.

5.2.2.5.9 Ipecacuanha and Alangium Alkaloids

Cephaelis ipecacuanha is a small tree whose original habitat is Brazil. The genus *Cephaelis* belongs to the subfamily Rubioideae of Rubiaceae. From the roots, several important alkaloids have been isolated (18, 78, 102).

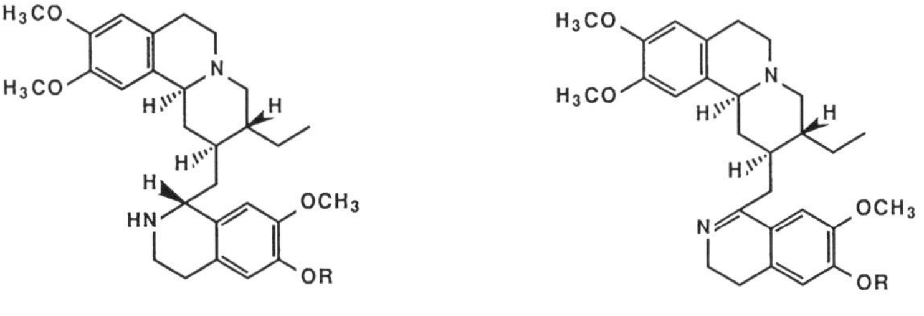

cephaeline R=H **154**
emetine R=CH_3 **155**

psychotrine R=H **156**
O-methylpsychotrine R=CH_3 **157**

Emetine and other related alkaloids are biosynthesized from dopamine and secologanin of mevalonate origin through a glycosidic intermediate, desacetylipecoside (Fig. 5.2.14) (6). Alkaloids having biogenetic importance, such as protoemetine, have also been found in the same plant (7).

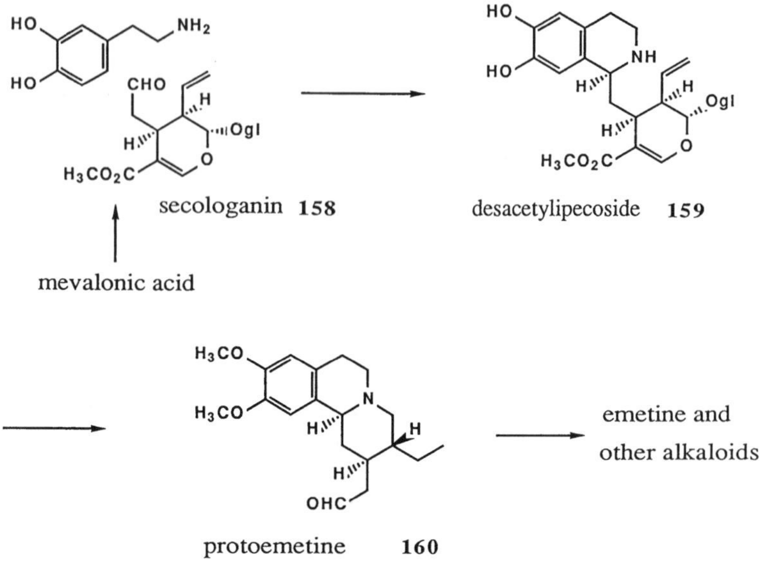

secologanin **158**

desacetylipecoside **159**

mevalonic acid

emetine and other alkaloids

protoemetine **160**

Fig. 5.2.14. Biosynthesis of emetine and related alkaloids

Emetine and other related alkaloids have also been found in the plants of *Alangium* spp. (Alangiaceae) (32). *Alangium lamarckii* is a small tree, widely distributed in India, Sri Lanka, southern China, Malaysia, and the Philippines. In traditional Indian medicine, the roots, root-bark, and bark of this plant have been used for treating various diseases, in which the usage as an emetic agent is involved. From the leaves and fruit, emetine, psychotrine, cephaeline, and other related alkaloids have been isolated. This plant also contains many tubulosine-type β-carboline alkaloids. Tubulosine was first obtained from the bark of *Pogonopus tubulosum* (Rubiaceae), a tree growing in the northern part of Argentina (16).

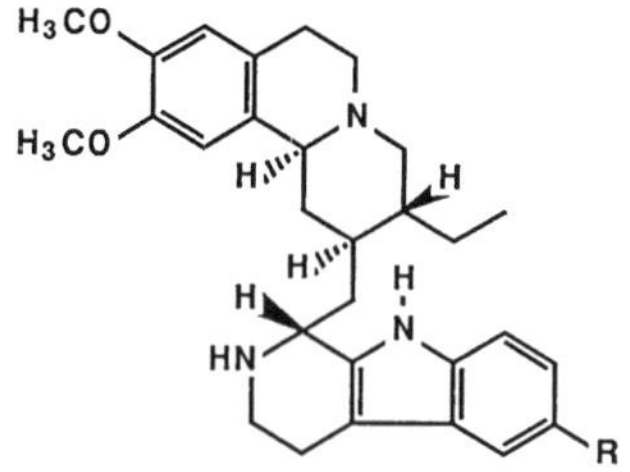

tubulosine	R=OH	**161**
deoxytubulosine	R=H	**162**

From a South African plant *Cassinopsis ilicifolia* (Icacinaceae), deoxytubulosine has been obtained (112). It is interesting to note that the closely related emetine-tubulosine alkaloids occur in three completely distinct families; namely the Rubiaceae, Alangiaceae, and Icacinaceae.

5.2.2.6 Alkaloids from Tryptophan

Alkaloids from tryptophan can be classified into two groups. One of them is a small and simple group to which belong the calycanthus, canthinone, and β-carboline alkaloids; the other is the monoterpenoid indole alkaloids.

Without exception, monoterpenoid indole alkaloids appear only in the three families of Gentianales: Apocynaceae, Loganiaceae, and Rubiaceae. Their biosyntheses have been studied for over thirty years, and a very important intermediate precursor has been found. Strictosidine, which is derived from the condensation of tryptamine and secologanin (Fig. 5.2.15), was shown to be the common precursor in the biosynthesis of the major classes of monoterpenoid indole alkaloids.

tryptamine **163** + secologanin **158** 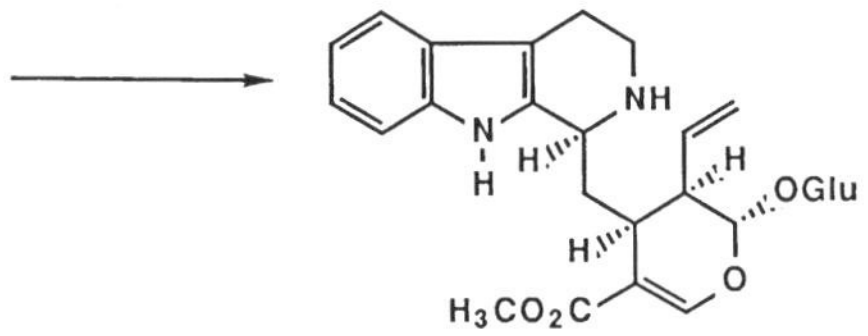 → strictosidine **164**

Fig. 5.2.15. Biosynthesis of strictosidine

Strictosidine was experimentally incorporated into a series of different alkaloids, including yohimbine-, strychnine- and gelsemine-type alkaloids (151).

Furthermore, there are two alkaloid groups whose biosyntheses are closely related to those of monoterpenoid indole alkaloids namely, quinines and camptothecins.

5.2.2.6.1 Calycanthus Alkaloids

Chimosanthus fragrans (Calycanthaceae) is a deciduous shrub of Chinese origin, commonly known as winter sweet. It bears fragrant blossoms before the leaves develop. The family is the source of *Calycanthus* alkaloids (104). From leaves of *Chimonanthus fragrans*, chimonanthine has been obtained (66). Folicanthine was obtained from the leaves of *Calycanthus floridus* and *C. occidentalis* (41). From the seeds of *Calycanthus floridus* (5) and *C. glaucus* calycanthidine was isolated. The main alkaloid in these seeds is calycanthine. It is highly toxic and can cause violent convulsions, paralysis, and cardiac depression.

These alkaloids are not of monoterpenoid origin, and it is considered that two molecules of N_b-methyl-tryptamine dimerize oxidatively to give chimonanthine or calycanthine.

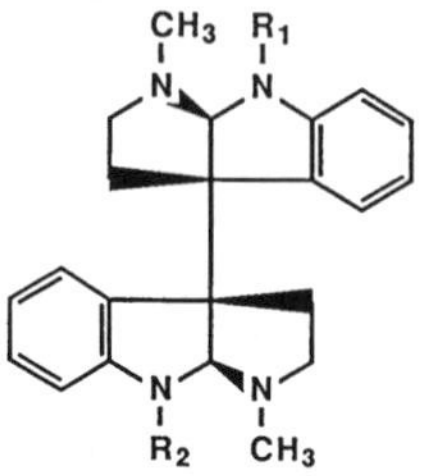

chimonanthine $R_1=R_2=H$ **165**

folicanthine $R_1=R_2=CH_3$ **166**

calycanthidine $R_1=CH_3, R_2=H$ **167**

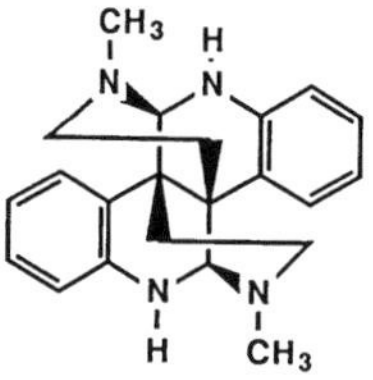

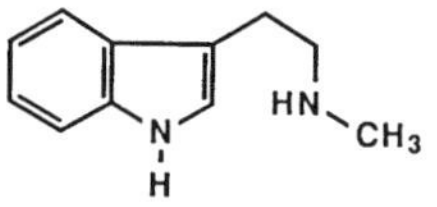

calycanthine **168**

N_b-methyltryptamine **169**

[2-^{14}C]-Tryptophan is specifically incorporated into folicanthine in *C. floridus* (116).

5.2.2.6.2 Picrasma (Pentaceras) and Carboline Alkaloids

These types of alkaloids are not of monoterpenoid origin. Plants of the Simaroubaceae family are trees that grow mainly in tropical and, to some extent, subtropical countries. *Picrasma quassioides* is a Japanese deciduous tree; its heartwood contains quassin, tannin, and nigakinone, and is used as a bitter

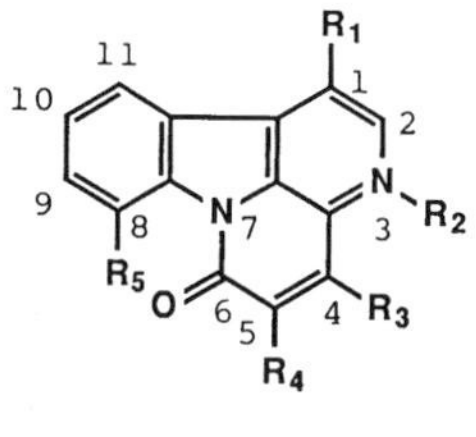

	R_1	R_2	R_3	R_4	R_5	
4,5-dimethoxycanthin-6-one	H		OMe	OMe	H	170
canthin-6-one	H		H	H	H	171
canthin-6-one-3-oxide	H	O	H	H	H	172
1-methoxycanthin-6-one	OMe		H	H	H	173
5-hydroxycanthin-6-one	H		H	OH	H	174
5-methoxycanthin-6-one	H		H	OMe	H	175
8-hydroxycanthin-6-one	H		H	H	OH	176

stomachic. 4,5-Dimethoxycanthin-6-one and related compounds have been isolated (74, 117). This plant also contains many β-carboline derivatives (117).

Ailanthus altissima is a tall deciduous tree of Chinese origin. A study of the wood extractives has revealed the presence of three alkaloids: canthin-6-one, canthin-6-one-3-oxide, and 1-methoxycanthin-6-one (118). Another member of the same family, *Simarouba amara*, contains 5-hydroxycanthine-6-one (93). The root bark of *Ailanthus excelsa* contains four canthine derivatives: canthin-6-one, 1-methoxycanthin-6-one, 5-methoxycanthin-6-one, and 8-hydroxanthin-6-one (27).

Ailanthus malabarica is a tall deciduous tree of Indian origin, cultivated in north Malaya. Its bark has been used as an incense. Several simple β-carboline derivatives occur in the bark and roots of this tree, but canthines have not been detected (84).

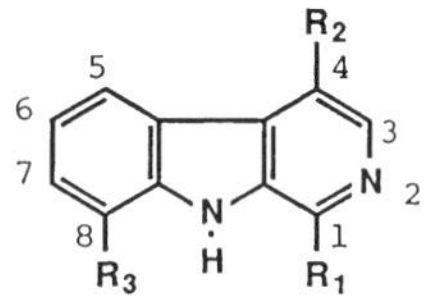

	R_1	R_2	R_3		
1-methoxycarbonyl-β-carboline	CO_2Me	H	H		177
4-methoxy-1-vinyl-β-carboline	$-CH{=}CH_2$	OMe	H		178
crenatidine	Et	OMe	OMe		179
crenatine	Et	OMe	H		180
1-acetyl-β-carboline	Ac	H	H	*	181
1-acetyl-4-methoxy-β-carboline	Ac	OMe	H	*	182
4,8-dimethoxy-1-vinyl-β-carboline	$-CH{=}CH_2$	OMe	OMe	*	183
β-carboline-1-carboxamide	$CONH_2$	H	H	*	184

* first isolated from *Ailanthus malabarica*

5.2.2.6.3 *Rauwolfia Alkaloids*

Rauwolfia serpentina (Apocynaceae) appears throughout India, Malaysia, and Thailand as small trees that grow wild in the humid forests. Now it is also cultivated in tropical countries. In India, its roots have been used as a folk medicine for snake bite and for dysentery, and as an antipyretic, among other things. The roots contain well known alkaloids: reserpine, ajmaline, ajmalicine, and serpentine.

Reserpine, which is an important antihypertensive medicine, has also been obtained from other *Rauwolfia* species such as *R. heterophylla, R. canescens, R. hirsuta, R. vomitoria, R. tetraphylla*, and *R. verticillata*.

Ajmaline has antiarrythmic activity on the heart muscle, conceivably due to its role in the sodium-carrying system, and is used clinically as a therapeutic agent in the treatment of cardiac arrythmia. Ajmaline is also used as an antihypertensive and a tranquilizer.

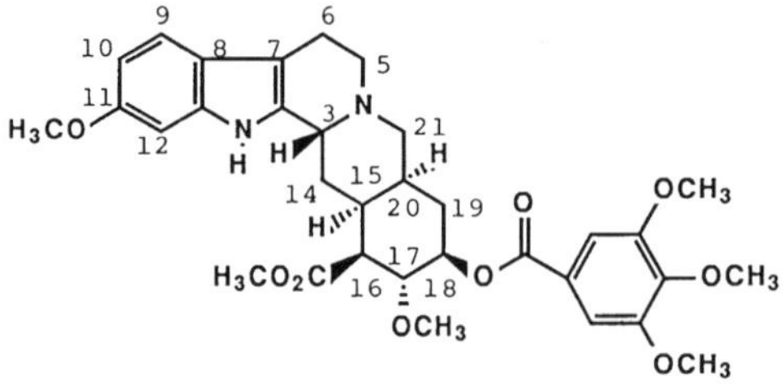

reserpine **185**

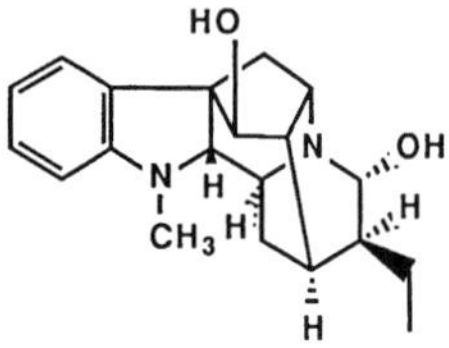

ajmaline **186** 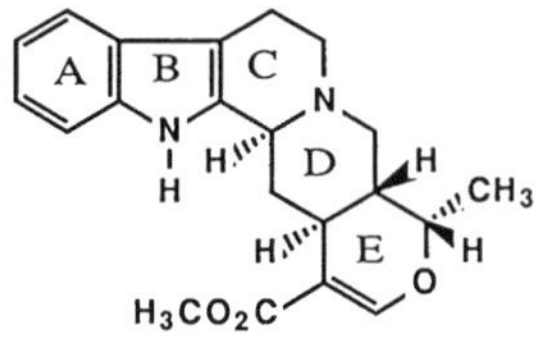 ajmalicine **187** 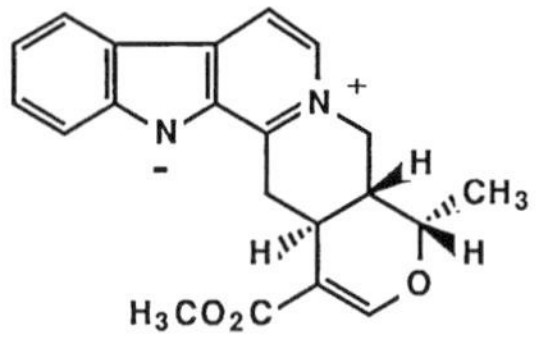 serpentine **188**

Strictosidine is incorporated into both ajmaline and ajmalicine (114). Biogenetic aromatization of ring C of ajmalicine gives rise to serpentine.

Yohimbine, which is described in Sect. 5.2.2.6.8 on Yohimbe alkaloids, has also been isolated from the roots of *Rauwolfia serpentina* (67).

The alkaloidal constituents of the stem bark of Nigerian *Rauwolfia vomitoria* have been isolated and characterized, some of which are described in Table 5.2.3 (129).

Table 5.2.3. Principal alkaloid obtained from 1.5 kg stem bark of *Rauwolfia vomitoria*

Alkaloid	Yield (%)
Reserpiline	0.04
Isoreserpiline	0.007
18-Hydroxyyohimbine	0.002
Yohimbine	0.002
Purpeline	0.002
Norpurpeline	0.002
Norseredamine	0.002
Nortetraphyllicine	0.002
Carapanaubine	0.002
Indolenine RG	0.002

The alkaloids that have been obtained from the stem bark can be divided into five groups: *E*-seco heteroyohimbines, sarpagans, dihydrolindoles, yohimbines, and heteroyohimbines (including the ψ-indoxyls and oxindoles).

It is noteworthy that *E*-seco indole alkaloids appear only in the leaves and stems, whereas dihydroindole alkaloids occur in the roots. Possible biosynthetic interrelationships of these isolated alkaloids are proposed in Fig. 5.2.16.

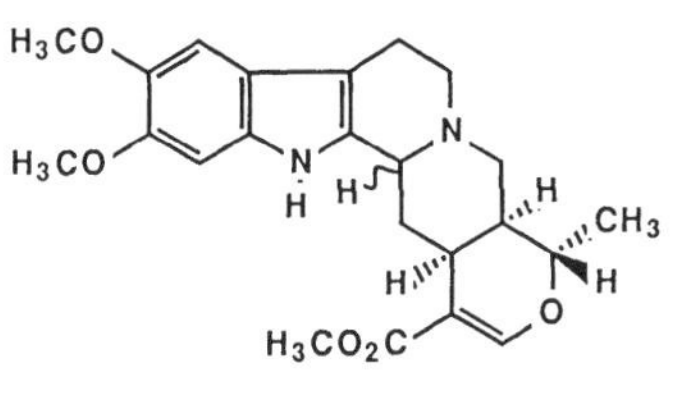

reserpiline **3H−β** **189**

idoreserpiline **3H−α** **190**

(heteroyohimbine group)

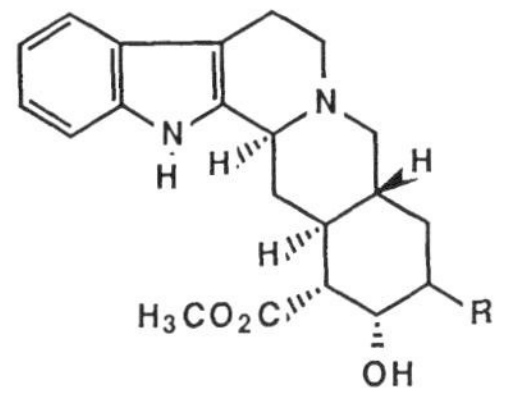

yohimbine R=H **191**

18-hydroxyyohimbine R=~OH **192**

(yohimbine group)

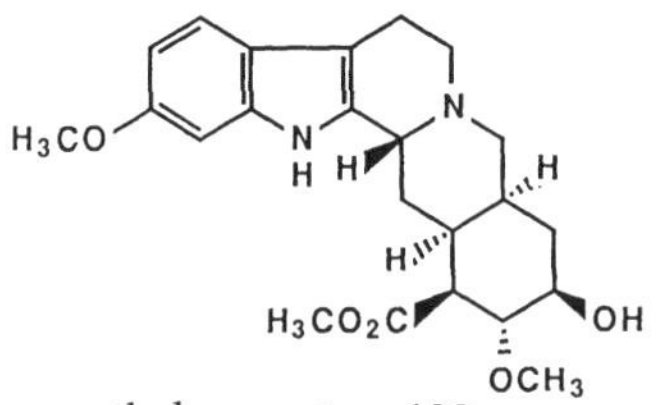

methylreserpate **193**

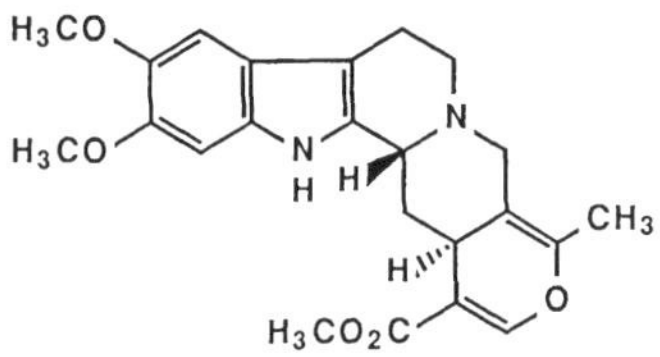

19,20-dehydroreserpiline **194**

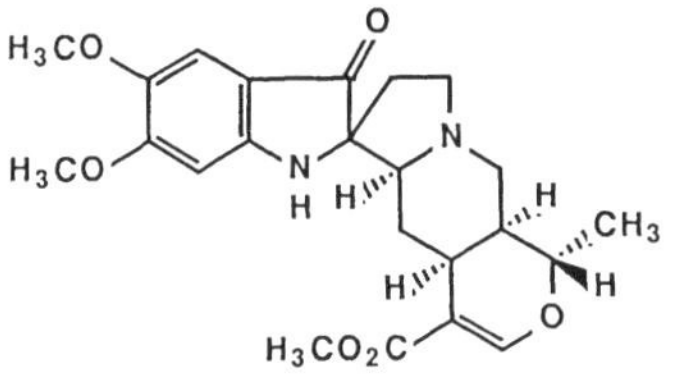

isoreserpiline-ψ-indoxyl **195**

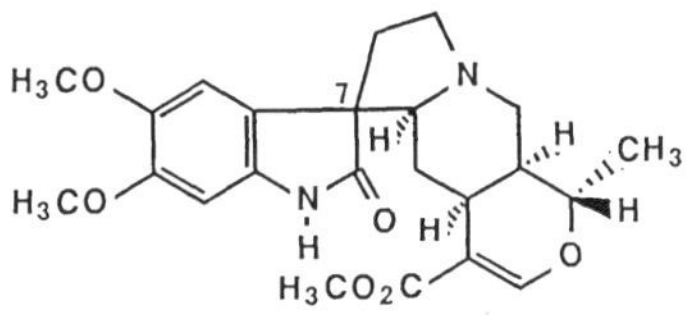

carapanaubine C7R **196**

isocarapanaubine C7S **197**

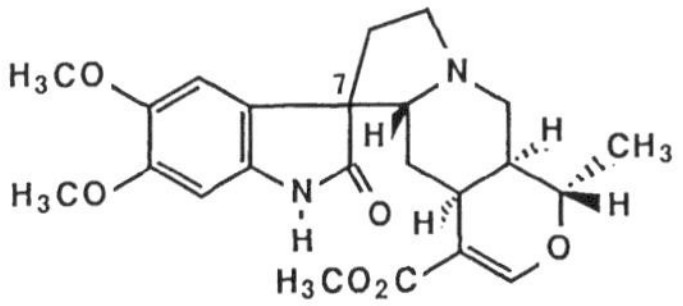

rauvoxine C7R **198**

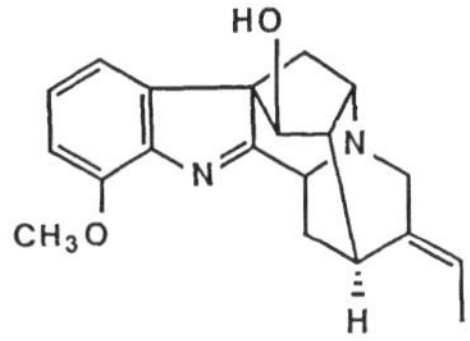

indolenine RG **199**

5.2.2.6.4 *Tabernanthe Alkaloids*

Aspidosperma quebracho blanco (Apocynaceae) is a tall tree that grows wild in Argentina, Brazil, and Chile. The bark is called "quebracho bark" or "white quebracho", and contains tannins and alkaloids. It is used in folk medicine as an antipyretic and diuretic and as a treatment for pertussis and asthma.

Many indole alkaloids have been obtained from the bark, among which aspidospermine (30.0%), yohimbine (10.0%), deacetylaspidospermine (3.0%),

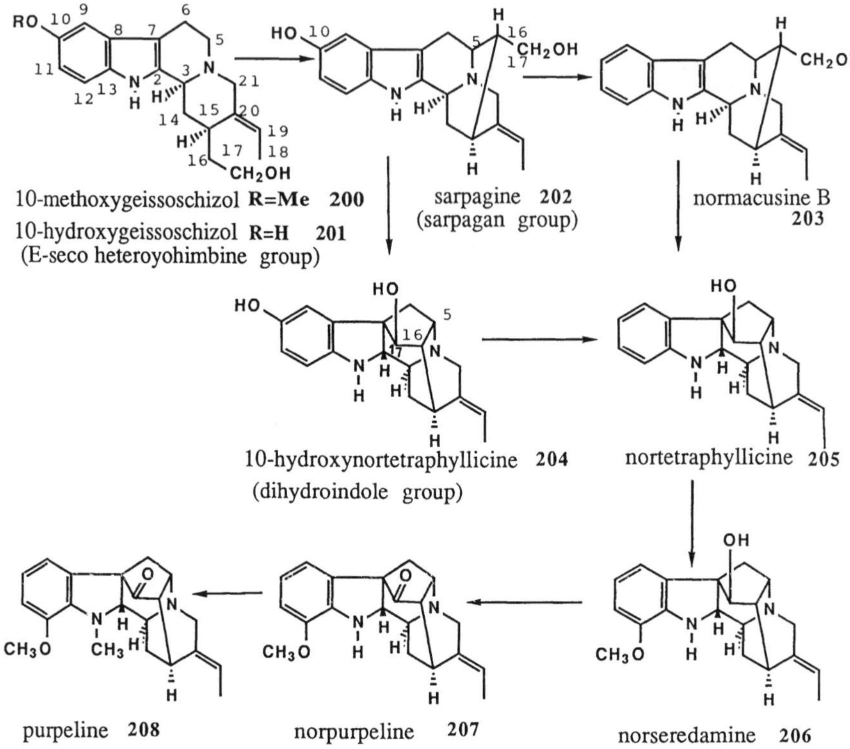

10-methoxygeissoschizol **R=Me 200**
10-hydroxygeissoschizol **R=H 201**
(E-seco heteroyohimbine group)

sarpagine **202**
(sarpagan group)

normacusine B **203**

10-hydroxynortetraphyllicine **204**
(dihydroindole group)

nortetraphyllicine **205**

purpeline **208**

norpurpeline **207**

norseredamine **206**

Fig. 5.2.16. Biosynthetic relationships of dihydroindole alkaloids

and quebrachamine (2.5%) are the main constituents (13). Eburnamenine has been also detected (139, 155).

Besides aspidospermine and deacetylaspidospermine, several indole alkaloids of these "*Aspidosperma*-type" alkaloids have also been isolated from the bark.

These "*Aspidosperma*-type" indole alkaloids occur only in Apocynaceae plants. This fact is of great interest from the standpoint of chemotaxonomy.

Tabernanthe iboga (Apocynaceae) is a shrub found in Africa. One of the major alkaloids of this plant is ibogaine, which occurs mainly in the root bark (8, 35).

Iboga extracts are said to be used by African natives to enable them to paralyze their game. Ibogaine may cause serious psychological disturbances in humans. The therapeutic category of this alkaloid is antidepressant. One of the coexisting minor alkaloids of *T. iboga* is tabernanthine, usually found in mother liquors of ibogaine. Tabernanthine is used as an analgesic and a serotonic antagonist. Tabernanthine also appears in the *Tabernaemontana, Stemmadenia,* and *Conopharingia* species.

The indole alkaloids of this type are called "iboga-type" alkaloids and are found only in Apocynaceae plants.

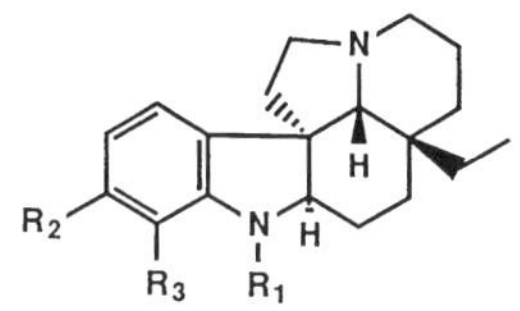

	R_1	R_2	R_3	
aspidospermine	Ac	H	OMe	209
deacetylaspidospermine	H	H	OMe	210
N_a-methyldeacetylaspidospermine	Me	H	OMe	211
(-)-pyrifolidine	Ac	OMe	OMe	212
aspidospermidine	H	H	H	213
N_a-methylaspidospermidine	Me	H	H	214

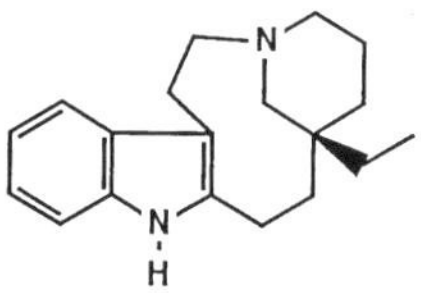

(-)-quebrachamine **215**

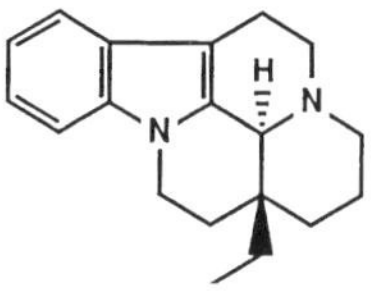

eburnamenine **217**

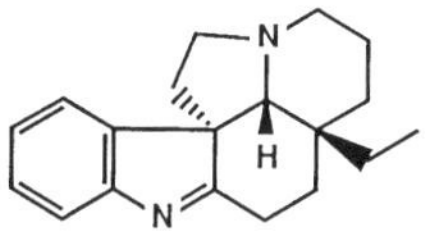

1,2-dehydroaspidospermidine **216**

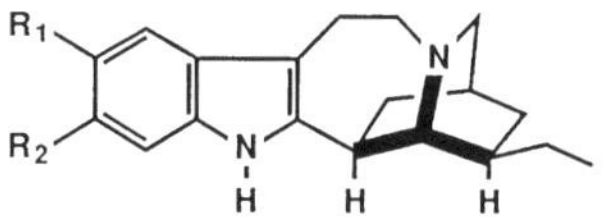

ibogaine R_1=OMe R_2=H **218**

tabernanthine R_1=H R_2=OMe **219**

5.2.2.6.5 Ochrosia Alkaloids

Ochrosia elliptica (Apocynaceae) is a small tropical evergreen tree, which contains ellipticine in its leaves and stems. Ellipticine has also been isolated from *Ochrosia sandwicensis, O. viellardii, O. silvatica* (28, 53, 85), and *Aspidosperma subincanum* (50).

Ellipticine is therapeutically used as an antineoplastic. Ellipticine and the structurally related alkaloid olivacine, which is obtained from *Aspidosperma olivaceum* (138) and other species of *Aspidosperma*, are considered to be biosynthesized via stemmadenine N-oxide by the fragmentation-addition process (69, 122).

5.2.2.6.6 Ervatamia Alkaloids

Ervatamine, 20-epiervatamine, and 19, 20-dehydroervatamine have been obtained from *Ervatamia orientalis* (Apocynaceae), a small tropical tree, and from other *Ervatamia* species (89).

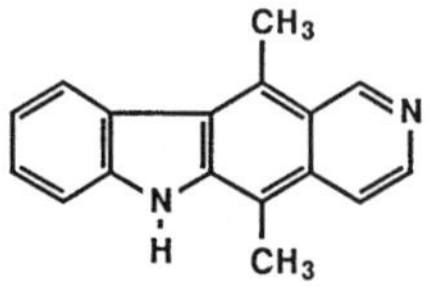

ellipticine **220**

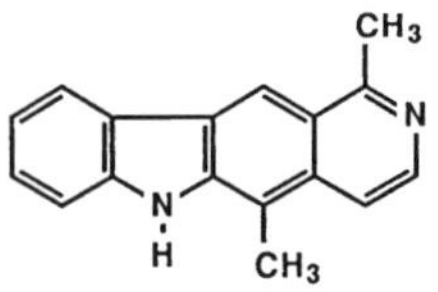

olivacine **221**

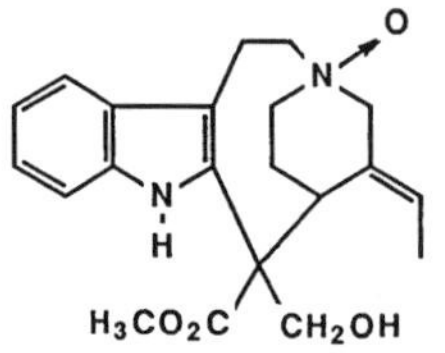

stemmadenine N-oxide **222**

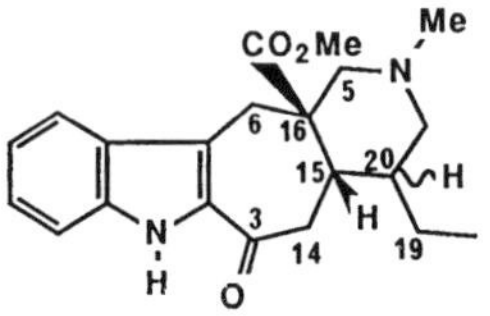

dehydroervatamine (19,20-dehydro) **223**
20-epiervatamine (20-α-ethyl) **224**
ervatamine (20-β-ethyl) **225**

These 2-acyl indole alkaloids are peculiar in that there are three carbon atoms between the indole nucleus and the N_b atom. These alkaloids are thought to be biosynthesized from vobasine-type 2-acyl indole alkaloids.

Applying the Potier-Polonovski reaction, vobasine-type indole alkaloids such as vobasine, dregamine, and tabernaemontanine could be transformed into dehydroervatamine, 20-epiervatamine, and ervatamine, respectively (70).

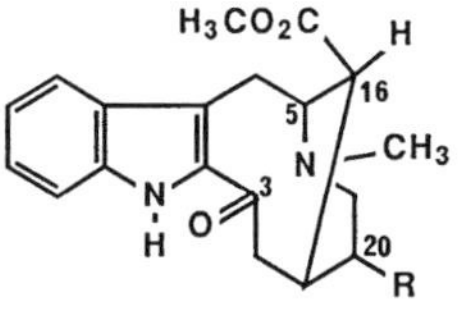

vobasine R=E-ethylidene **226**
dregamine R=α-ethyl **227**
tabernaemontanine R=β-ethyl **228**

These transformations can be related to a biogenetic process, which is supported by the fact that both types of alkaloids exist in the same plant.

5.2.2.6.7 Uncaria-Mitragyna Alkaloids

Uncaria gambier (Rubiaceae), a woody climber with recurved hooks, grows wild in eastern India, Malaysia, and Sumatra and is also cultivated in the Malaysian islands. Aqueous extracts of the leaves and stems are used in folk medicine as an astringent (a catechu called gambir). Gambir contains tannins and alkaloids. The main alkaloid, gambirine (109), has been obtained from the leaves of *U. gambier*, and dihydrocorynantheine has been isolated as a minor alkaloid. Dimeric indole alkaloids roxburghine B, C, D, and E have been also extracted from the leaves (111). They are thought to arise by combination of tryptamine and corynantheine derivatives (62, 123).

Three alkaloids, gambirtannine, dihydrogambirtannine, and oxogambirtannine, have been isolated from gambir (110). These alkaloids are considered to be generated from yohimbine-type alkaloids by dehydration and aromatization.

Uncaria rhynchophylla is a climbing vine tree growing in China, Japan, and other temperate Asian countries. Directly above its leaves are hooks into which small stems metamorphose.

Indole alkaloids hirsutine and hirsuteine are the major constituents of the root bark of *Uncaria rhynchophylla*. Along with these, minor indole alkaloids cor-

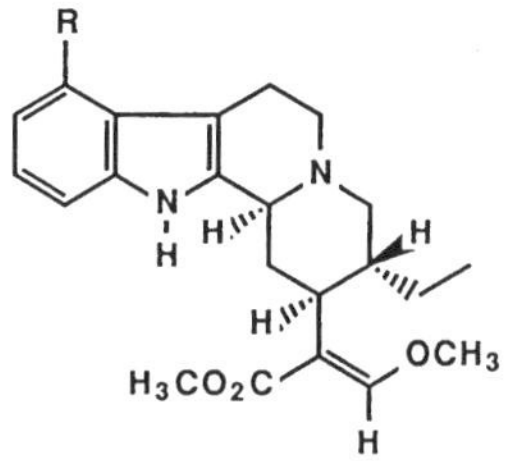

gambirine R=OH 229

dihydrocorynantheine R=H 230

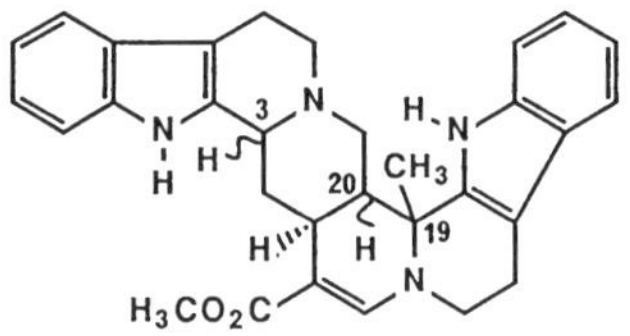

	3	20	19	
roxburghine B	β-H	α-H	β-CH_3	231
roxburghine C	α-H	β-H	α-CH_3	232
roxburghine D	β-H	β-H	α-CH_3	233
roxburghine E	β-H	β-H	β-CH_3	234

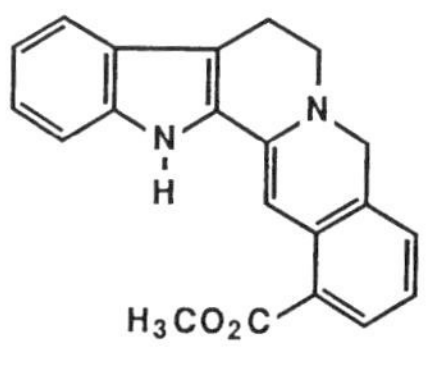

gambirtannine 235 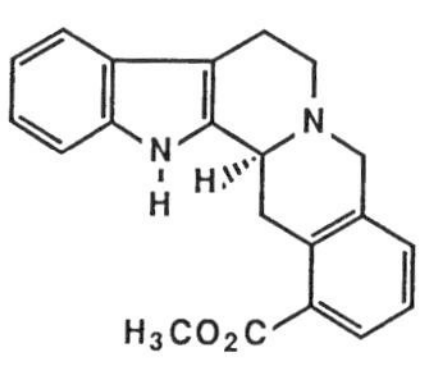 dihydrogambirtannine 236 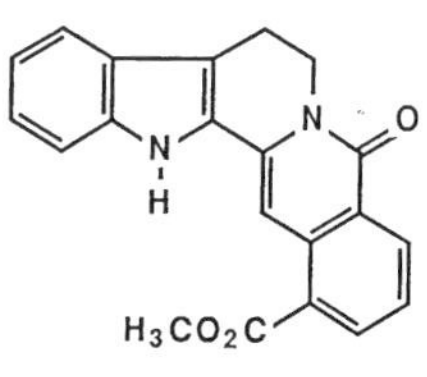 oxogambirtannine 237

ynantheine and dihydrocorynantheine have been obtained, mainly from the root bark (60). Oxindole alkaloids, such as rhynchophylline, corynoxeine, isorhyncho-phyline, and isocorynoxeinse have been isolated mainly from the aerial parts – hooks, stems, and leaves.

Liquid chromatographic studies have further confirmed these differences in alkaloidal constituents of each part of this plant (165).

From the leaves of *Uncaria rhynchophylla*, a glycoindole alkaloid, rhyn-chophine, has been isolated along with vallesiachotamine, vincoside lactam, and strictosamide (2).

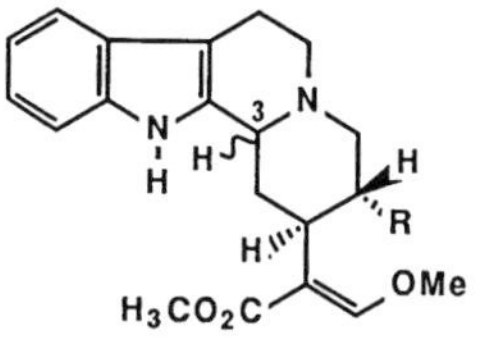

hirsutine 3β H,R=Et, pseudo 238
hirsuteine 3β H,R=-CH=CH2, pseudo 239
corynantheine 3α H,R=-CH=CH2, normal 240
dihydrocorynantheine 3α H,R=Et, normal 241

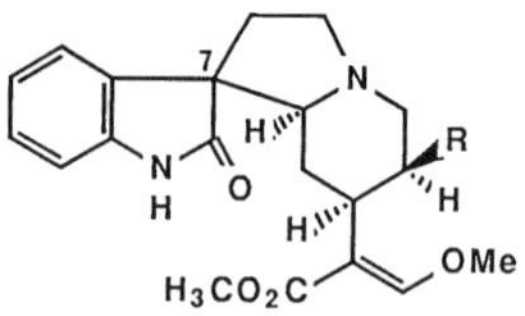

rhynchophylline C_7R ,R=Et 242
corynoxeine C_7R,R=-CH=CH2 243
isorhynchophylline C_7S ,R=Et 244
isocorynoxeine C_7S ,R=-CH=CH2 245

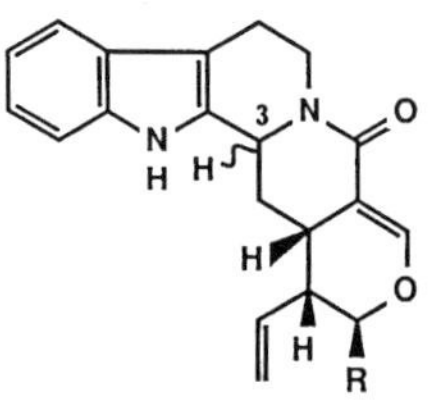

rhynchophine 3 Hβ R= (sugar–CH_2OCO–feruloyl moiety) 246

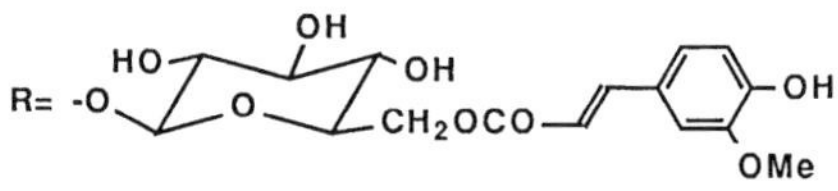

vincoside lactam 3 Hβ R= -O-β-D-glucoside 247

strictosamide 3 Hα R= -O-β-D-glucoside 248

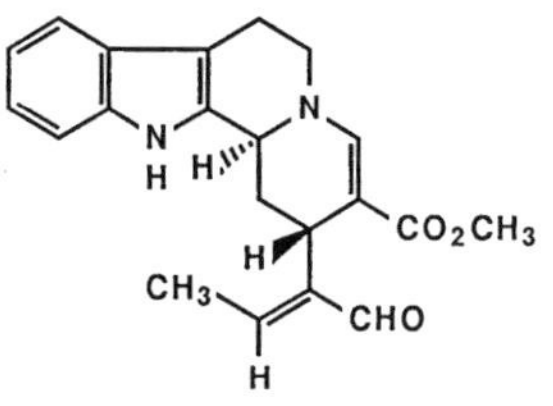

vallesiachotamine 249

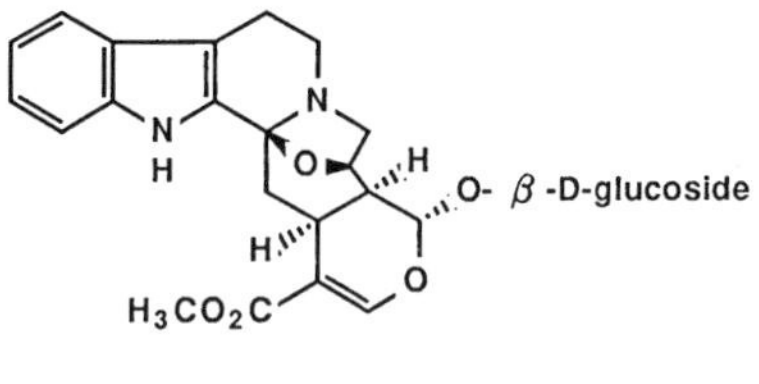

cadambine 250

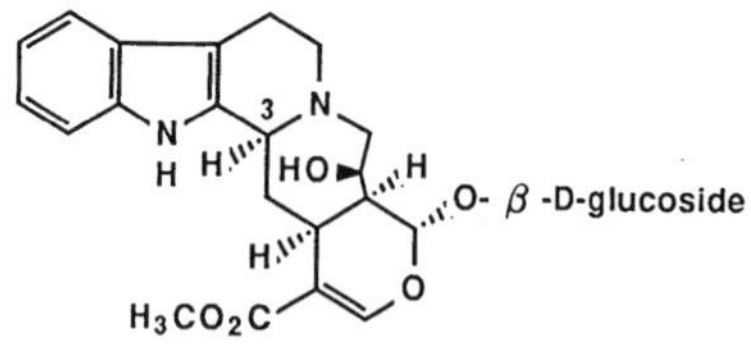

3α-dihydrocadambine 251

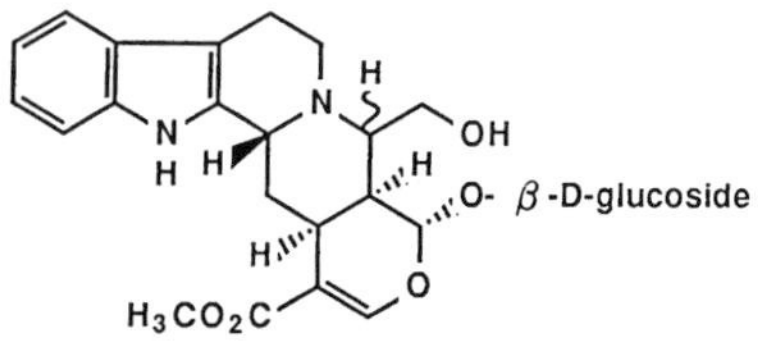

3β-isodihydrocadambine 252

The hooks of *Uncaria sinensis*, a Chinese vine tree closely resembling *U. rhynchophylla*, are used as a drug to treat infant convulsions, as an antispasmodic, and as a tranquilizer. Recently, it was found that the hypotensive principles of *Uncaria* hooks are indole-alkaloid glucosides, 3α-dihydrocadambine and 3β-isodihydrocadambine (42). Along with these glucosides, cadambine has been also obtained from the hooks.

Uncaria formosa, which also closely resembles *U. rhynchophylla*, grows wild in Formosa. Its hooks contain isoformosanine (uncarine A) and formosanine (uncarine B), and are used also in folk medicine for rheumatism and malaria.

Uncaria florida is also a Formosan climber. Pteropodine (uncarine C) and its isomers, isopteropodine (uncarine E) and speciophylline (uncarine D), have been isolated from the roots and stems of *Uncaria florida* (61). It is probable that the main alkaloid is pteropodine; it can be easily isomerized to isopteropodine under alkali extraction conditions (9).

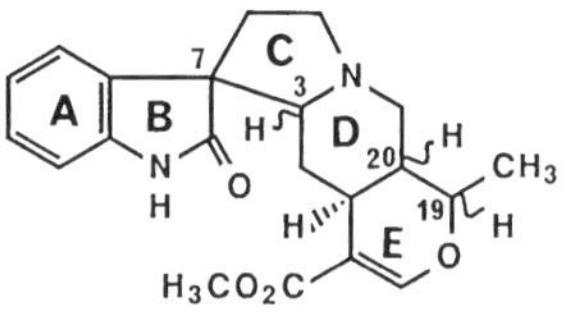

	3	7	19	20		
isoformosanine (uncarine A)	αH	S	αH	βH	normal	253
formosanine (uncarine B)	αH	R	αH	βH	normal	254
pteropodine (uncarine C)	αH	R	βH	αH	allo	255
speciophylline (uncarine D)	βH	S	βH	αH	epiallo	256
isopteropodine (uncarine E)	αH	S	βH	αH	allo	257
———— (uncarine F)	βH	R	βH	αH	epiallo	258

The distribution of alkaloids in young plants of *Mitragyna parvifolia* (Rubiaceae) grown from seeds obtained from Sri Lanka (142) and India (143) were studied separately, and the functions, movements, and biogenesis of these alkaloids were discussed.

The alkaloids present in the young leaves are the closed E-ring indole (tetrahydroalstonine and akuammigine) and oxindole (pteropodine and others) alkaloids (Fig. 5.2.17). In the mature leaves, only oxindoles were detected. The same series of oxindoles was found in the stipels and stem barks, with speciophylline the most prominent.

The *E*-seco oxindole alkaloids – rhynchophylline, isorhynchophylline, and corynoxeine – occur in the root bark. Corynoxeine is especially dominant in the root tips. In the root bark *E*-seco indole alkaloids hirsutine and hirsuteine have also been detected.

In the growing points of stems, as well as in root tips and seeds, corynoxeine is the prominent alkaloids. From this fact corynoxeine seems to have some physio-

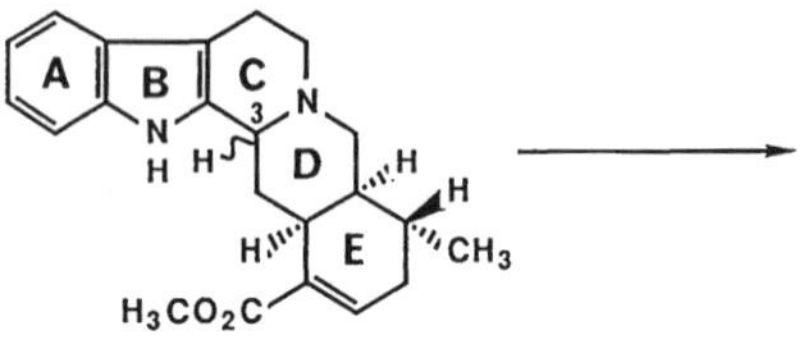

pteropodine 255
isopteropodine 257
speciophylline 256
uncarine F 258

tetrahydroalstonine (3αH, D/E cis; allo) **259**

akuammigine (3βH, D/E cis; epiallo) **260**

Fig. 5.2.17. Alkaloids present in young leaves of *Mitragyna parvifolia*

logical role. The predominance of speciophylline in the stipels suggests the protective function in the stipules.

The main difference between the constituents of the plants from Sri Lanka and those from India is the presence of mitraphylline and isomitraphylline (closed E-ring normal oxindoles) in every part of the Indian species.

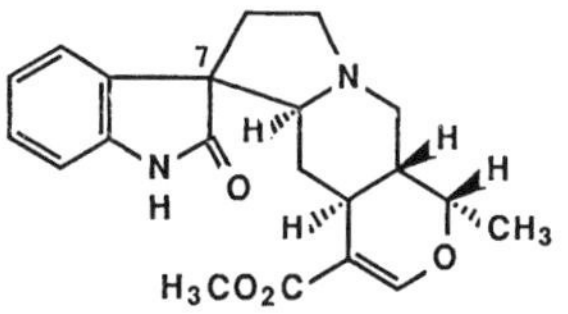

mitraphylline C7R **261**

isomitraphylline C7S **262**

5.2.2.6.8 *Yohimbe Alkaloids*

Corynanthe yohimbe (*Pausinystalia yohimbe*, Rubiaceae) is a evergreen tree native to the southern part of Africa and to Cameroon, Nigeria, and Zaire. Occasionally it is planted in tropical countries. Its bark coontains a total of 0.3% – 1.5% alkaloids. The well known indole alkaloid yohimbine has been obtained as a major alkaloid from the bark. The bark has been used traditionally as an aphrodisiac by African natives. Yohimbine hydrochloride is used as an adrenergic blocker for humans.

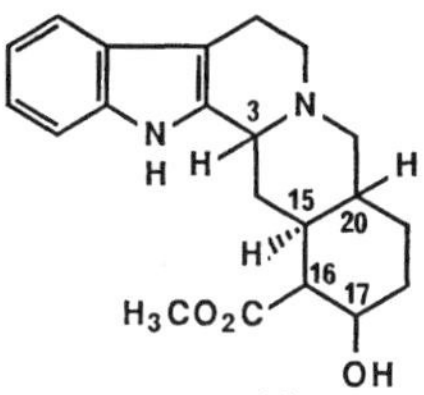

	3	16	17	20		
yohimbine	α H	α CO_2Me	α OH	β H	normal	**263**
corynanthine	α H	β CO_2Me	α OH	β H	normal	**264**
alloyohimbine	α H	α CO_2Me	α OH	α H	allo	**265**
α-yohimbine	α H	β CO_2Me	α OH	α H	allo	**266**
pseudoyohimbine	β H	α CO_2Me	α OH	β H	pseudo	**267**
β-yohimbine	α H	α CO_2Me	β OH	β H	normal	**268**

Corynantheine and several yohimbine congeners, such as corynanthine, alloyohimbine, α-yohimbine, pseudoyohimbine, and β-yohimbine, have been also isolated as minor alkaloids from the bark.

5.2.2.6.9 Cinchona Alkaloids

Cinchona succirubra (Rubiaceae) is a tall tree that grows wild in the Andes of Peru and Brazil. It is cultivated in Java and Sumatra. The bark contains 5%–8% total alkaloids. The major alkaloid, quinine, has been used as an antimalarial agent. Cinchonidine, cinchonine, and quinidine have also been obtained.

Cinchonine and cinchonidine are also used as antimalarials while quinidine is also used as a cardiac depressant (antiarrythmic).

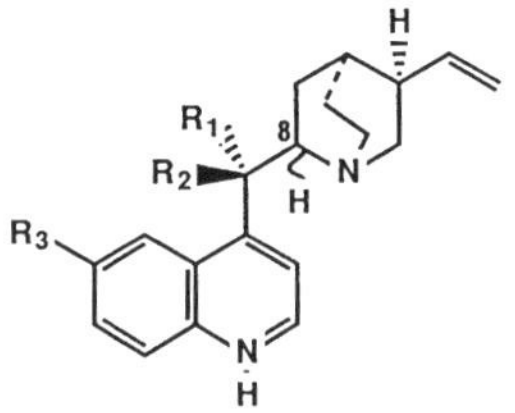

quinine 8βH R_1=H,R_2=OH,R_3=OMe 269
cinchonidine 8βH R_1=H,R_2=OH,R_3=H 270
cinchonine 8αH R_1=OH,R_2=H,R_3=H 271
quinidine 8αH R_1=OH,R_2=H,R_3=OMe 272

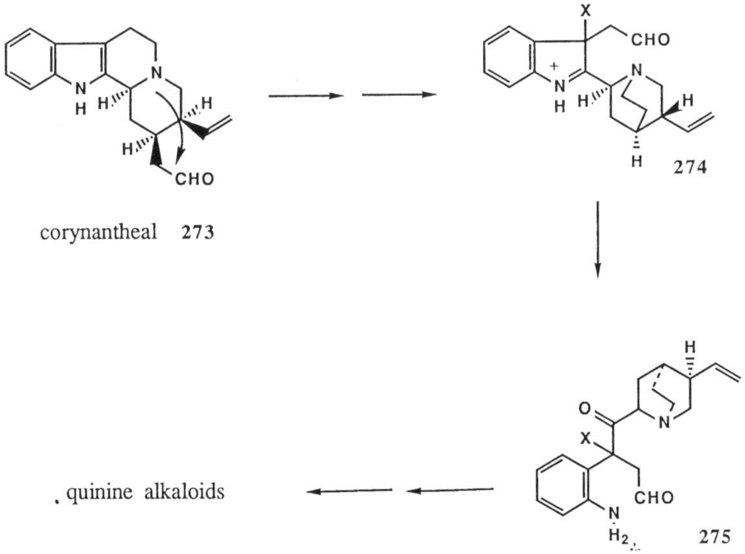

corynantheal 273

274

275

quinine alkaloids

Fig. 5.2.18. Biogenetic derivation of quinine alkaloids

Quinine alkaloids have quinoline nuclei but are considered to be derived biogenetically via strictosidine, corynantheal, and other intermediates (Fig. 5.2.18).

5.2.2.6.10 Guettarda Alkaloids

Cathenamine (20, 21-dihydroajmalicine) and 4,21-dehydrogeissoschizine have been isolated from the leaves of *Guettarda eximia* (71, 86).

Cathenamine and 4,21-dehydrogeissoschizine have both been demonstrated to be the central intermediates in the enzymatic production of heteroyohimbine-type alkaloids – ajmalicine, 19-epiajmalicine, and tetrahydroalsonine (128, 152) (Fig. 5.2.19).

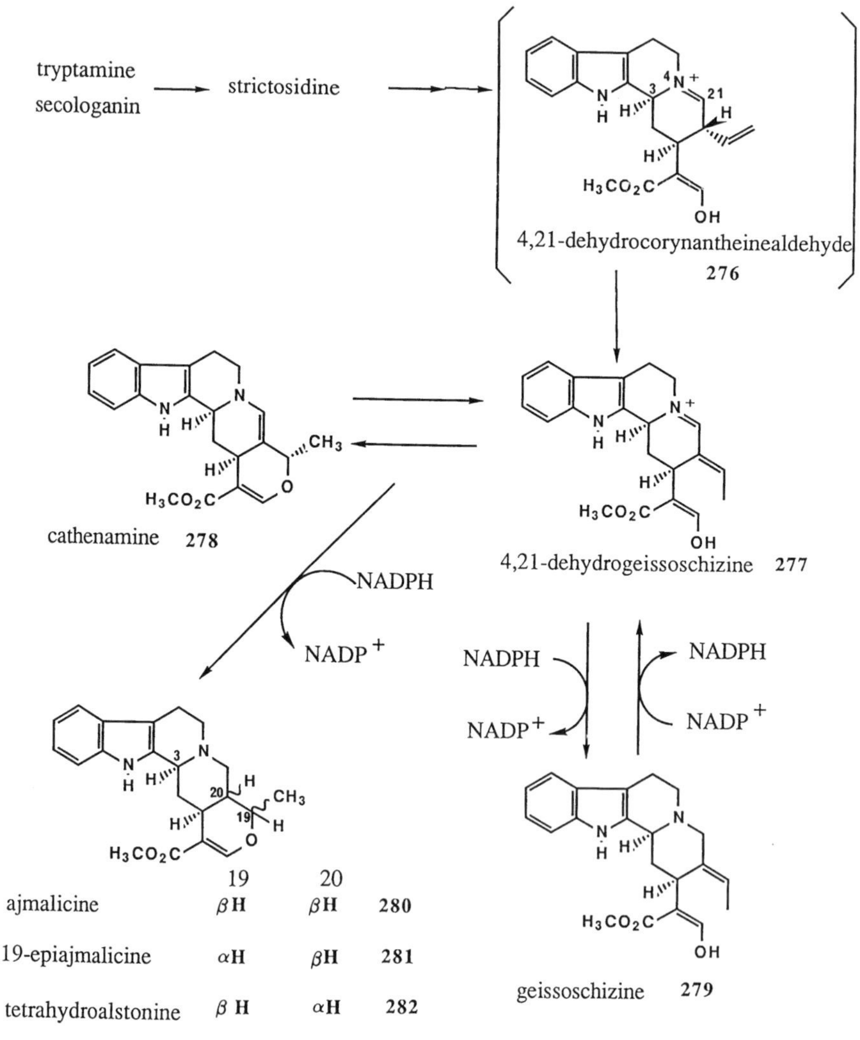

tryptamine
secologanin → strictosidine →

4,21-dehydrocorynantheinealdehyde 276

cathenamine 278

4,21-dehydrogeissoschizine 277

NADPH
$NADP^+$

geissoschizine 279

	19	20	
ajmalicine	βH	βH	280
19-epiajmalicine	αH	βH	281
tetrahydroalstonine	β H	αH	282

Fig. 5.2.19. Enzymatic production of heteroyohimbine-type alkaloids

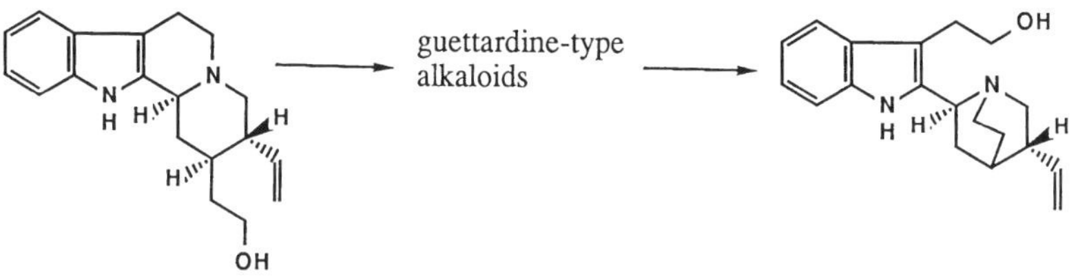

guettardine-type alkaloids

corynantheol 283

cinchonamine 284

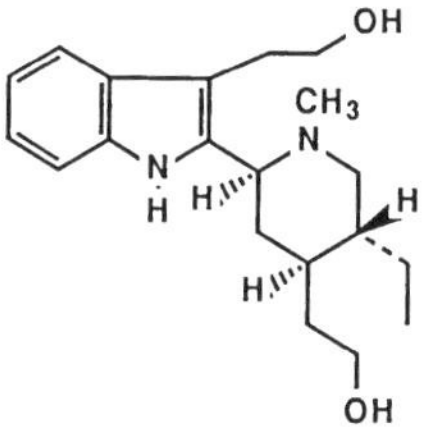

guettardine 285

Fig. 5.2.20. Biosynthesis of Guettardine and the cinchonamione-type alkaloids

4,21-Dehydrogeissoschizine was also converted into geissoschizine by chemical and enzymatic reductions.

Recently, guettardine was obtained from the bark of *Guettarda heterosepala* in a 0.025% yield (17). This alkaloid can be considered to be an intermediate between the corynanthe and the cinchonamine-type alkaloids that are precursors and congeners of quinine (Fig. 5.2.20).

5.2.2.6.11 Strychnos Alkaloids

Strychnos nux-vomica (Loganiaceae) is a woody climber that is widely distributed in India, Sri Lanka, and other Asian countries. Its seeds are virulently poisonous and contain as major alkaloids strychnine and brucine, and as minor alkaloids vomicine and novacine. Strychnine is also present in the wood and bark.

The vomica extract or tincture is made from the seeds and is used as a central nervous system stimulant. Strychnine is an antagonist of barbiturates, and for this reason, the latter are used as antidotes in strychnine poisoning.

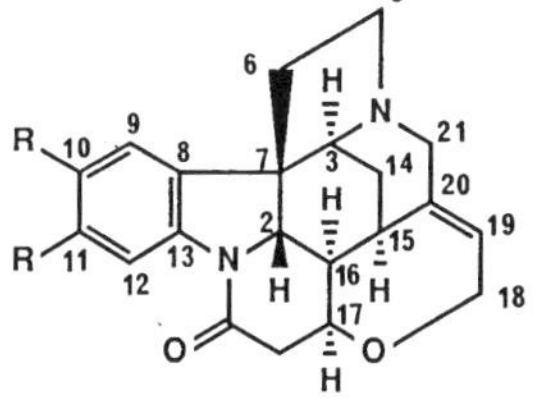

strychnine R=H 286
brucine R=OH 287

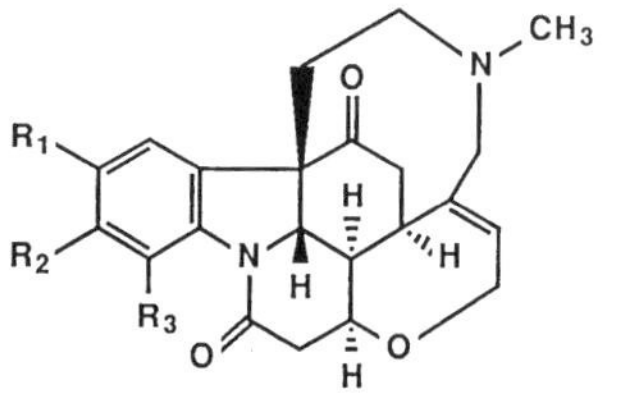

vomicine $R_1=R_2=H$, $R_3=OH$ 288
novacine $R_1=R_2=OMe$, $R_3=H$ 289

Brucine is a highly toxic alkaloid resembling strychnine and is also used as a central nervous system stimulant. In addition, it is used in analytical chemistry for separating racemic mixtures. These alkaloids possess an additional 2-carbon unit between the indolic N_a and C_{17}. It has been shown that the 2-carbon unit comes from acetate at some stage of the biogenesis (123).

Strychnos toxifera is a tree that grows wild in the Amazon basins. The stem bark and root bark contain many tertiary or quaternary bis-indole alkaloids such as C-alkaloid A, C-toxiferine I, C-calebassine, caracurine II, caracurine II matho-salt, caracurine V, and nor-dihydrotoxiferine.

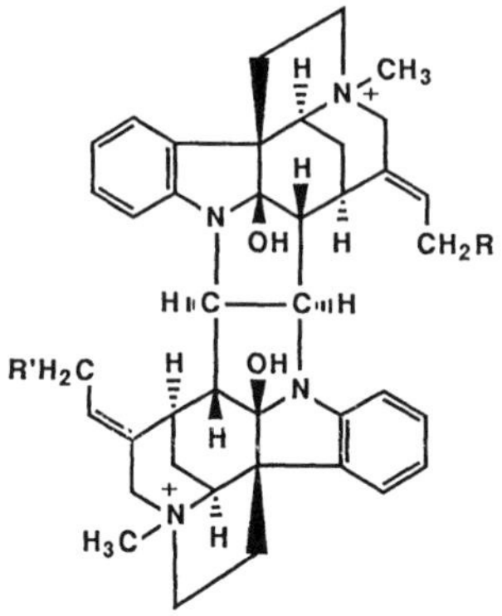

C-alkaloid A R=R'=OH 290
C-calebassine R=R'=H 291

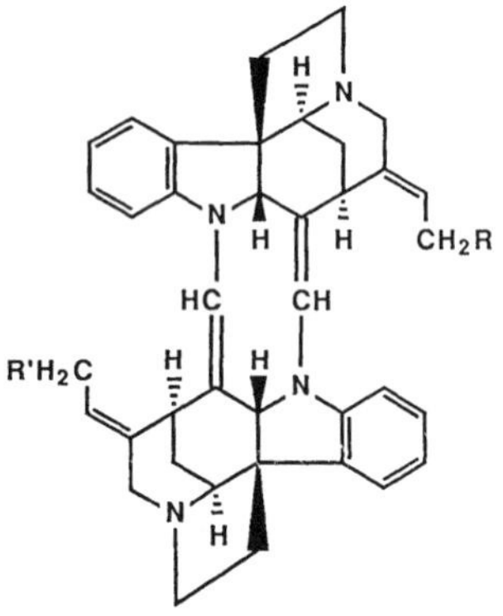

C-toxiferine I R=R'=OH N(b),N(b'):Me 292
nor-dihydrotoxiferine R=R'=H 293

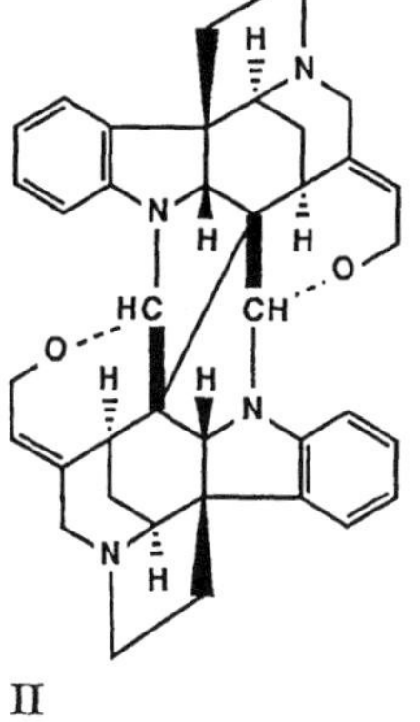

caracurine II 294
caracurine II-methosalt N(b),N(b'):Me 295

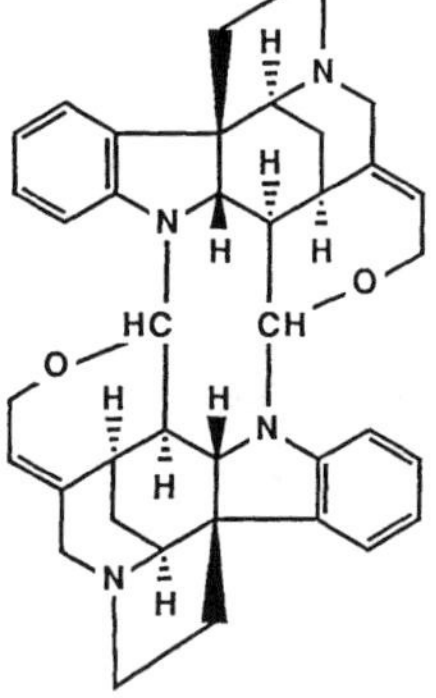

caracurine V 296

Monomeric tertiary indole alkaloids (diaboline and desacetyldiaboline (Wieland-Gumlich aldehyde)) and quaternary indole alkaloids (C-hemitoxiferine, C-fluorocurine, C-mavacurine, and macusine A, B, and C) have been also obtained.

Aqueous extracts of *S. toxifera* and other various species of *Strychnos* plants indigenous to the Amazon-Orinoco basins are called curare (calabash curare) and are used by natives for hunting (105) (Sect. 5.2.2.5.2).

The common factor in curare poisons is their smooth muscle-anesthetizing action, which blocks neuromuscular transmission. Therefore, curares are indispens-

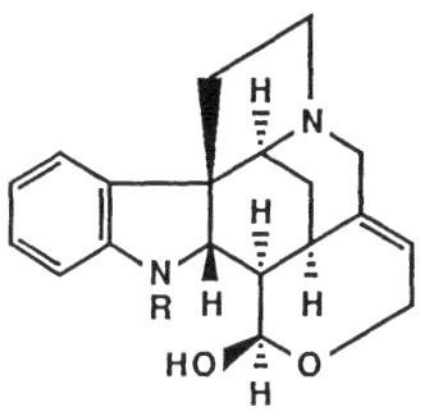

diaboline R=Ac **297**
desacetyldiaboline R=H **298**
(Wieland-Gumlich-aldehyde)

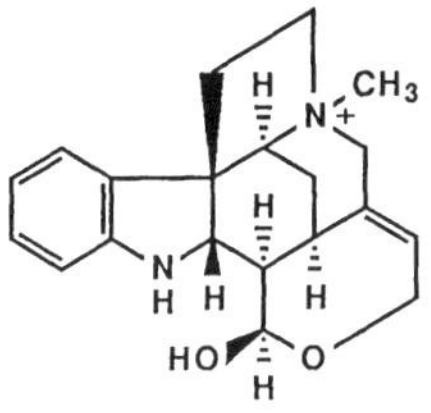

C-hemitoxiferine **299**

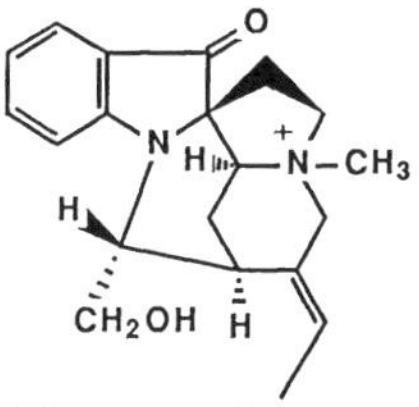

C-fluorocurine **300**

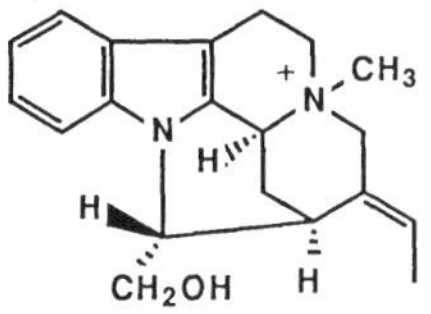

C-mavacurine **301**

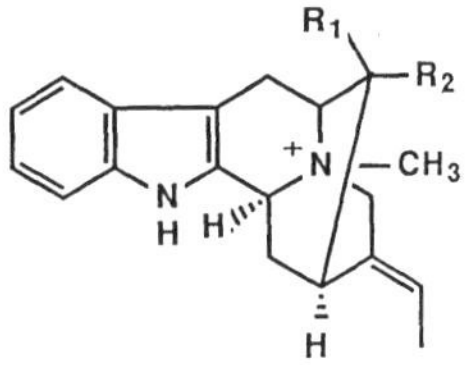

	R_1	R_2	
macusine A	CH_2OH	CO_2CH_3	**302**
macusine B	H	CH_2OH	**303**
macusine C	CO_2CH_3	CH_2OH	**304**

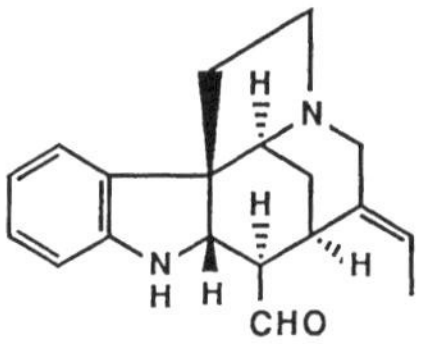

18-desoxy-Wieland-Gumlich-aldehyde **305**

able for pharmacological experiments and are used as muscle relaxants in surgical operations. Although quarternary bis-indole alkaloids were shown to be curarizing principles, and tertiary alkaloids are not curarizing, they may act as synergists for the curare activity. Indeed, curare is sometimes more active than the curarizing alkaloids at the same concentration. Curares are also used as an antidote in strychnine poisoning.

Dimeric indole alkaloids such as C-toxiferine I, nor-dihydrotoxiferine, and curarine V are derived from Wieland-Gumlich aldehyde and 18-desoxy-Wieland-Gumlich aldehyde or their corresponding N_b-methosalts (123).

5.2.2.6.12 Gelsemium Alkaloids

Three species are known as the plants of the genus *Gelsemium* (Loganiaceae); they give *Gelsemium* alkaloids (137). *Gelsemium sempervirens* (Loganiaceae) is a small vine tree that grows wild in the southeastern United States. The roots and the rhizomes are called yellow jasmine root and contain oxindole alkaloids with unique and complex ring systems: gelsemine, gelsevirine, gelsedine, gelsemicine, 14-hydroxygelsemicine, 21-oxogelsemine, and 14β-hydroxygelsedine (145). Another characteristic of these alkaloids, except for gelsemine, is the presence of an N_a-methoxy group. *Gelsemium sempervirens* is highly toxic and the yellow jasmine root has a central anesthetizing effect. Although gelsemine is the major alkaloid

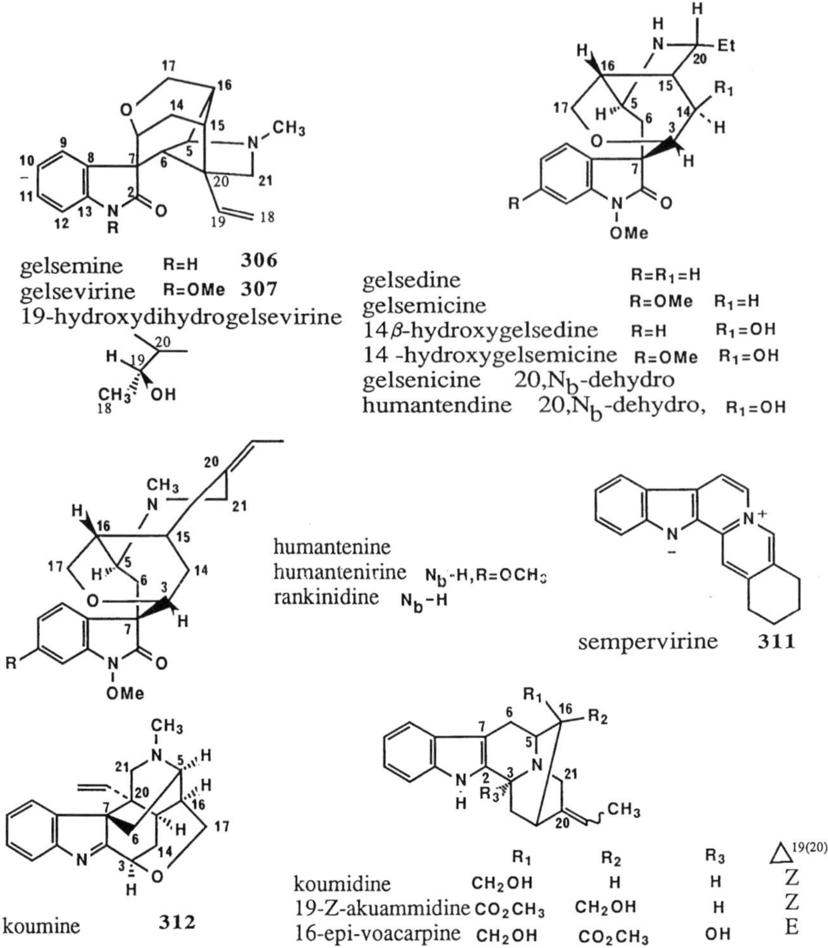

of *Gelsemium sempervirens*, gelsemicine is far more toxic. Sempervirine is another type of alkaloid of this plant.

Gelsemium elegans is a straggling shrub that is distributed in southern China and Indochina and is occasionally cultivated. Chemical studies on the alkaloids of this species have been done by several research groups. Gelsemine, gelsevirine, and 19-hydroxydihydrogelsevirine have been isolated as the alkaloids of the structure type of gelsemine (135, 137, 166). Additional gelsedine-type alkaloids, such as gelsedine, 14β-hydroxygelsedine, gelsenicine, and 14-hydroxygelsenicine, have been found (38, 145). Koumine was first isolated from this plant, and the structure was elucidated by X-ray crystallography (81, 88, 99). A chemical conversion of vobasine to koumine was done by a Chinese group (100). Independently, hydroxygardnerine, a *Gardneria* alkaloid, was converted to 11-methoxykoumine (136). Humantenine and humantenirine are the alkaloids that were first found in *Gelsemium elegans* (166, 167). Three sarpagine-type alkaloids, koumidine, 19-(*Z*)-akuammidine, and 16-epivoacarpine were also isolated from the same plant (81, 135). It is interesting to note that the ethylidene group of koumidine and 19-(*Z*)-akuammidine have *Z* configurations while that of 16-epivoachalotine has the conventional *E* configuration (135, 148).

The third species of *Gelsemium* genus is *G. rankinii* which is distributed in southeastern United States. Recent studies disclosed the presence of the following alkaloids: gelsemine, gelsevirine, 21-oxogelsevirine, humantenirine, and rankinidine (146, 147). Two of them, 21-oxogelsevirine and rankinidine have been found only in *Gelsemium rankinii*.

5.2.2.6.13 Gardneria Alkaloids

The *Gardneria* genus (Loganiaceae) has been found to be a rich source of indole alkaloids. Two species, *Gardneria nutans* and *G. multiflora*, are both climbers that grow naturally on shady, humid slopes in central to western Japan, though the distribution is not wide. Although the two species have close morphological similarities, a distinction is that *Gardneria nutans* bears one to three flowers separately on top of the stem, whereas *G. multiflora* bears the flowers densely (three to ten).

It is interesting to note that although both species contain gardneramine as the common main base, a distinct difference exists in their alkaloid constituents (132). The sarpagine-type indoles – gardnerine, gardnutine, and hydroxygardnutine – are characteristic of *Gardneria nutans*. Dimeric alkaloids gardmultine and demethoxygardmultine have been obtained from *Gardneria multiflora* (133). Other alkaloids of the gardneramine type or its oxindole form have also been isolated from *Gardneria multiflora*, but none of gardnerine-type indole alkaloids has been detected in *G. multiflora* (Table 5.2.4).

Recently, the structure of gardfloramine of *Gardneria multiflora* was determined by X-ray crystallography. This compound has three oxygen functions on its aromatic moiety; the positions of oxygen atoms were proved to be at C-9

Table 5.2.4. Alkaloidal content of two *Gardneria* species

Alkaloid	Content	
	G. nutans	*G. multiflora*
	% (based on crude base)	
Gardnerine	18	
Gardnutine	1	
Hydroxygardnutine	1.7	
Gardneramine	12	17
Gardmultine		3.84
Demethoxygardmultine		0.96
Chitosenine		0.3
18-Demethylgardneramine		2.5
Gardfloramine		0.24
Total base	0.200	0.086

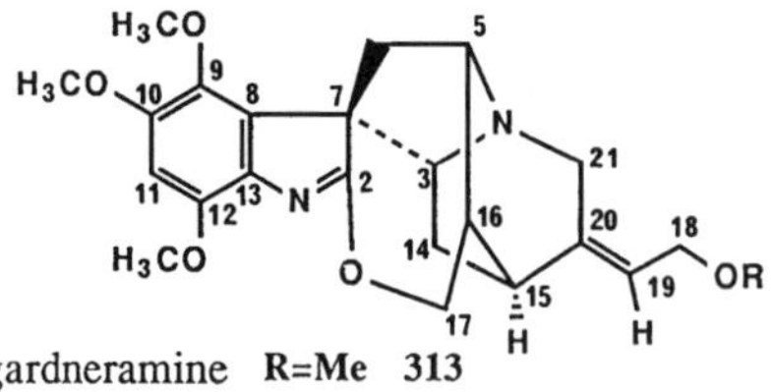

gardneramine **R=Me** **313**

18-demethylgardneramine **R=H** (19,20 Z form) **314**

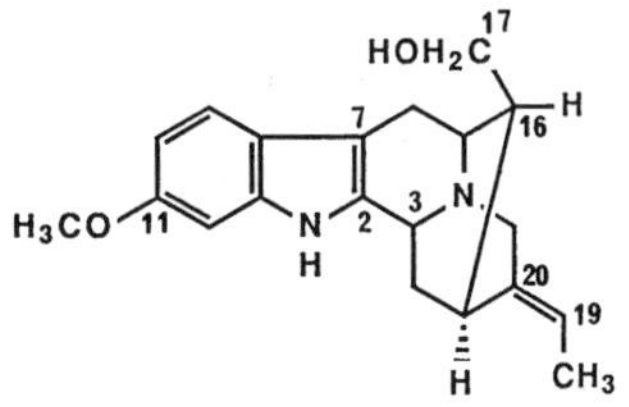

gardnerine (19,20 E form) **315**

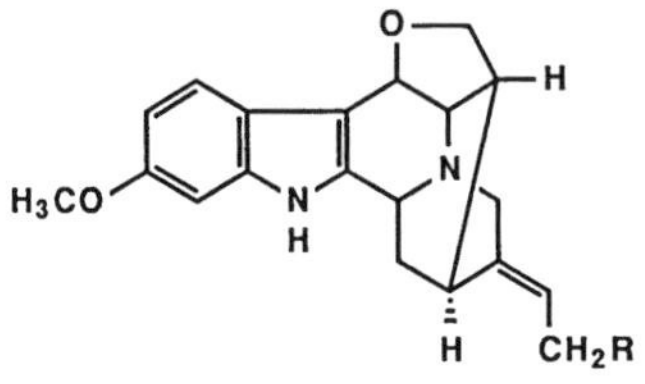

gardnutine **R=H** **316**

hydroxygardnutine **R=OH** **317**

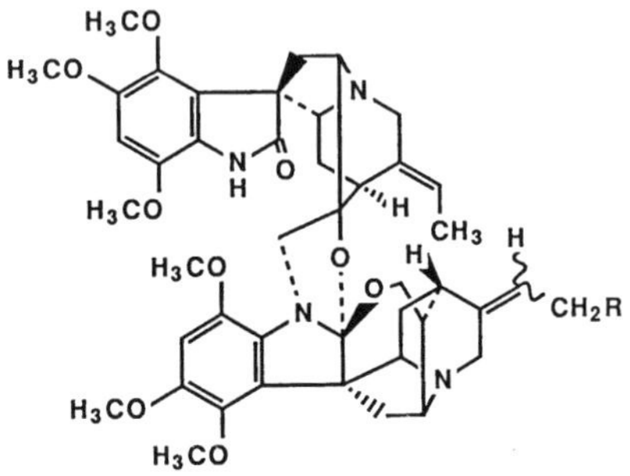

gardmultine **R=OMe** (19,20 Z form) **318**

demethoxygardmultine **R=H** (19,20 E form) **319**

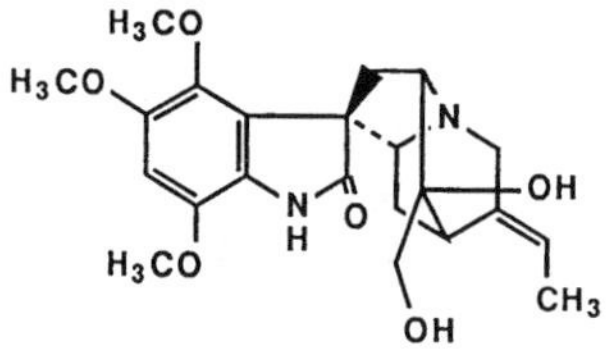

chitosenine **320**

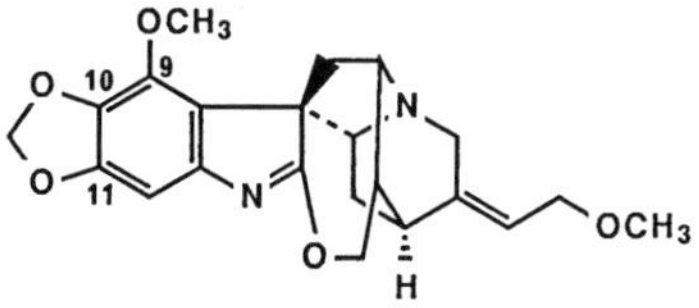

gardfloramine **321**

(methoxyl group), C-10, and C-11 (methylene dioxy group). This arrangement of oxygen functions was different from that of all the other trimethoxylated alkaloids in which methoxyl groups are located at C-9, C-10, and C-12 (134).

5.2.2.6.14 Camptothecins

Camptotheca acuminata (Nyssaceae) is a tree that grows throughout the southern part of China. From its stem wood, camptothecin has been obtained (160). Camptothecin is notable because of its remarkable antitumor and antileukemia activity. The other minor camptothecin analogues that have been obtained are 10-methoxycamptothecin (161), and 20-deoxycamptothecin (1). 11-Hydroxycamptothecin and 11-methoxycamptothecin have been obtained from the fruit (68, 98).

Camptothecin and its analogues (22) have been also found in other plants. 9-Methoxycamptothecin, camptothecin, and mappicine have been isolated from

	R	R'	
camptothecin	H	OH	322
10-hydroxycamptothecin	10-OH	OH	323
10-methoxycamptothecin	10-OMe	OH	324
20-deoxycamptothecin	H	H	325
11-hydroxycamptothecin	11-OH	OH	326
11-methoxycamptothecin	11-OMe	OH	327
9-methoxycamptothecin	9-OMe	OH	328

mappicine **329**

Nothapodytes foetida (Icacinaceae) (55, 56). From *Ophiorrhiza mungos* (Rubiaceae), camptothecin and 10-methoxycamptothecin have been obtained, and from *Ervatamia heyneana* (Apocynaceae), camptothecin and 9-methoxycamptothecin (59). From *Merrilliodendron megacarpum* (Icacinaceae), camptothecin and 9-methoxycamptothecin have been isolated (3). Table 5.2.5 lists yields from various plants.

The content of camptothecin in different parts of *C. acuminata* has been investigated, when it was found that the content in the fruit or root bark is higher than that in the stem bark (22, 97). The content of camptothecin in *N. foetida* was higher than that of any other plants.

Table 5.2.5. Yields of camptothecin and 9-methoxycamptothecin from the plants

Source		Alkaloid	
Genus/species	Plant part	Camptothecin (%)	9-Methoxycamptothecin (%)
Camptotheca acuminata	stem wood	0.005	–
Nothapodytes foetida	whole plant	0.075	0.0015
Ophiorrhiza mungos	leaf	0.00015	–
Ervatamia heyneana	stem wood/stem bark	0.00013	0.0004
Merrilliodendron megacarpum	stem wood/stem bark	0.053	0.017

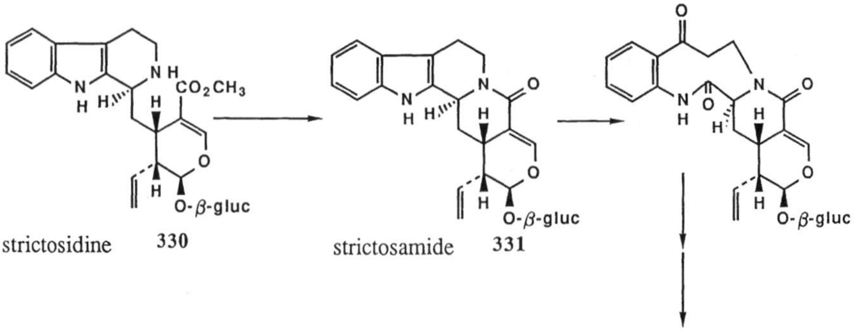

strictosidine 330 strictosamide 331

camptothecin 332

Fig. 5.2.21. Biosynthesis of camptothecin

Many feeding experiments (72, 144) have established that, as monoterpenoid indole alkaloids, camptothecin is also generated from tryptamine and secologanin. From strictosidine, a nine-membered ring intermediate can be derived via strictosamide (73) by oxidative cleavage of the C_2-C_7 double bond. A C_2-C_6 condensation occurs and, finally, camptothecin, is generated (Fig. 5.2.21).

5.2.3 Pseudoalkaloids

5.2.3.1 Alkaloids from Polyketides

5.2.3.1.1 Pinidine

Very few alkaloids of polyketide origin are known in nature, particularly in woody plants. Pinidine is one of them. This alkaloid has been isolated from the needles of the Pinaceae species *Pinus jeffreyi, P. subiniana*, and *P. torreyana* (153, 154).

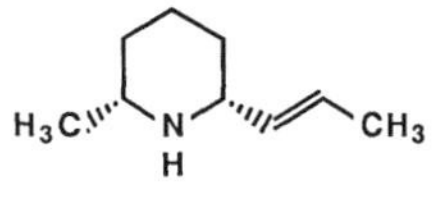

pinidine 333

It is interesting to note that these three alkaloid-containing species are also characteristic in that their terpentines are devoid of α- and β-pinenes, which are the most widely spread bicyclic monoterpenes in the ordinary *Pinus* species.

Through feeding experiments, Leete suggested the biosynthetic scheme (94) shown in Fig. 5.2.22.

5.2.3.1.2 Galbulimima Alkaloids

Galbulimima belgraveana (Himanthandraceae) is a large tree that reaches a height of 130 feet and is distributed in New Guinea, the Moluccas, and northern Australia (14). This plant contains over 25 alkaloids, mainly in the bark, albeit

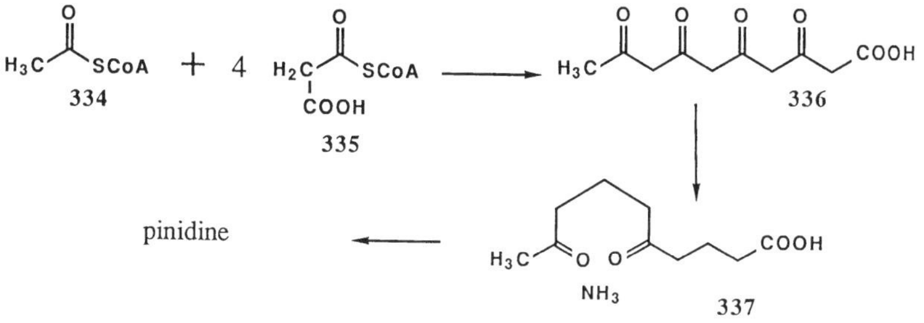

$H_3C-CO-SCoA$ (334) + 4 $H_2C(COOH)-CO-SCoA$ (335) → $H_3C-CO-CH_2-CO-CH_2-CO-CH_2-CO-CH_2-COOH$ (336) → 337 (NH_3) → pinidine

Fig. 5.2.22. Biosynthetic route to pinidine

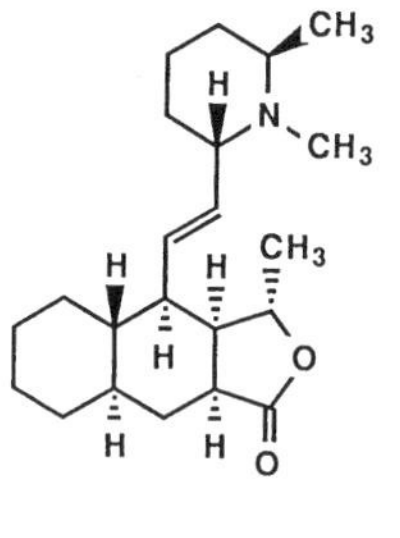

himbacine 338 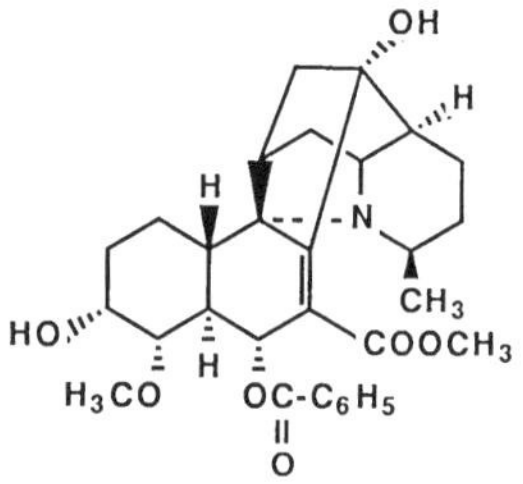   himandridine 339

with minor amounts in the leaves and wood. The main bases are himbacine and himandridine.

Their biogenesis has not been clarified but a hypothesis has been proposed (Fig. 5.2.23) that formation of various types of Galbulimina alkaloids (125, 126) is best explained by nine acetate units plus one pyruvate unit (126).

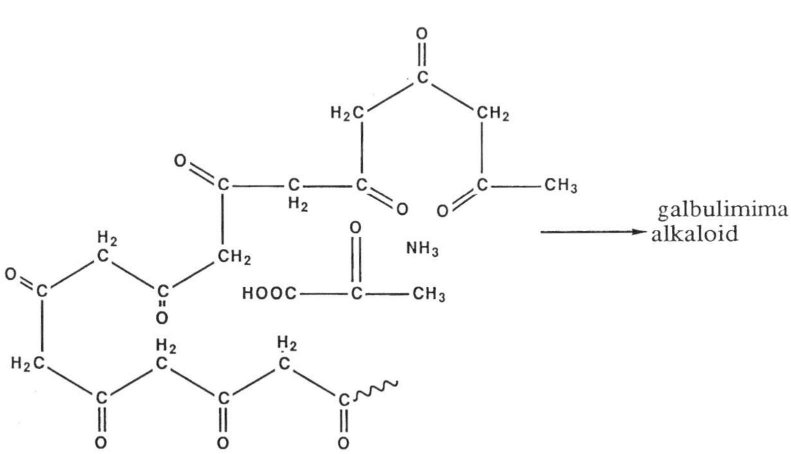

$HOOC-CO-CH_3$ + NH_3 → galbulimima alkaloid

Fig. 5.2.23. Proposed formation of various types of Galbulimima alkaloids

5.2.3.2 Alkaloids from Mevalonate

5.2.3.2.1 Spiraea Alkaloids

The presence of alkaloids in *Spiraea japonica* was reported by a Russian research group in 1964 (46). This plant is a small shrub, about 1 m high, belonging to the Rosaceae. It must be noted that the occurrence of alkaloids in Rosaceae plants is quite rare. Several alkaloids were isolated by Japanese researchers from the plant; the structures of spiradine A and spiradine D are shown (54).

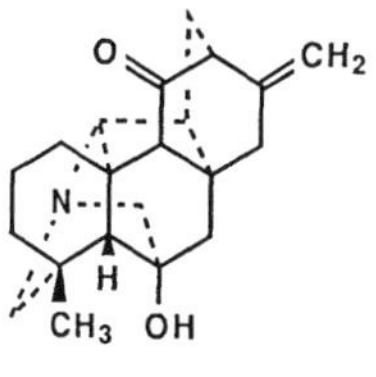

spiradine A **340**

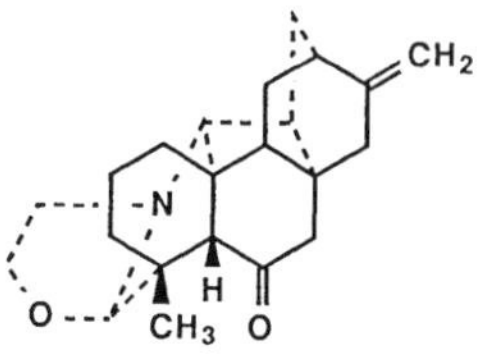

spiradine D **341**

Similar types of diterpene alkaloids have been found in Garryaceae plants. Garryine and veatchine were isolated from the bark of *Garrya veatchii*, a tree distributed in the area from the southwestern United States to Mexico and Guatemala (162).

Anopterus (Escalloniaceae) is another genus known to contain similar diterpene alkaloids (34).

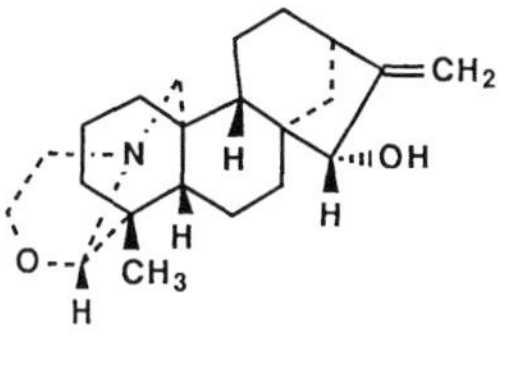

garryine **342**

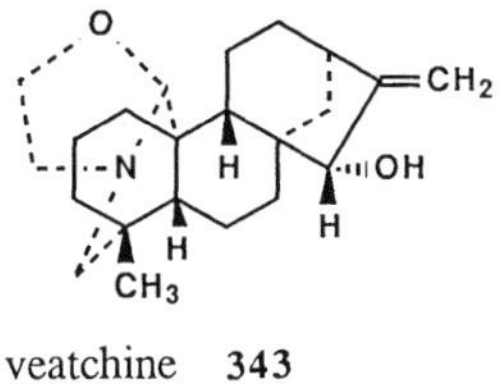

veatchine **343**

It is well known that a great many structurally related diterpene alkaloids have been found in the herbaceous plants of the Ranunculaceae (*Aconitum* and *Delphinium*) and Compositae (*Inula*). Some of them have strong toxicity or therapeutic activities.

5.2.3.2.2 Erythrophleum Alkaloids

Cassaine was isolated from the bark of *Erythrophleum guineense*, a Leguminosae (Caesalpinioideae) tree (29). This plant grows in western and central parts of equatorial Africa. The bark has been used by the natives as an arrow poison, as an ordeal drug, and as a therapeutic drug. Cassaine and other related alkaloids (30, 113) have such pharmacological effects as digitalis-like cardiac activity, paralyzing activity of the respiratory center, and long-lasting local anesthetic activity.

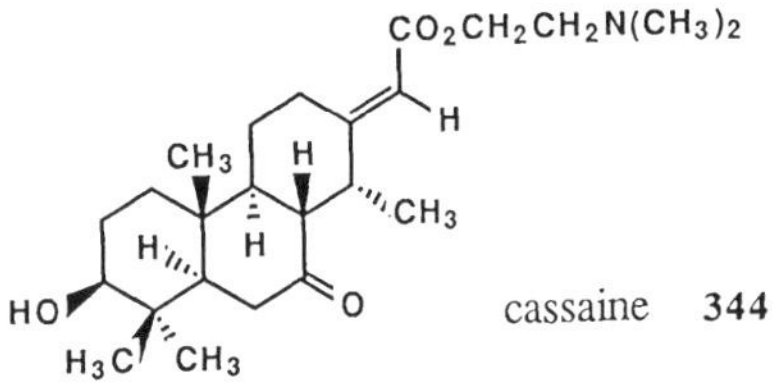

cassaine **344**

From the bark of *Erythrophleum chlorostachya*, a tall Australian tree, norerythrostachaldine and other highly cytotoxic alkaloids have been isolated (101).

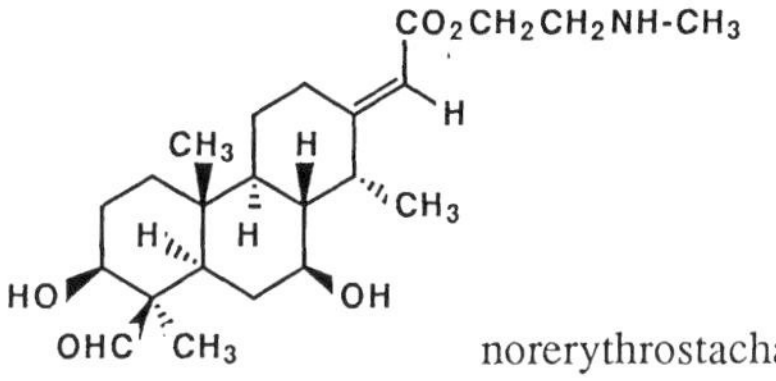

norerythrostachaldine **345**

5.2.3.2.3 Daphniphyllum Alkaloids

From *Daphniphyllum macropodum*, a group of alkaloids (164) with complicated structures has been isolated (131). This plant is large evergreen tree growing in the central to southwestern parts of Japan. Taxonomically this genus is classified in Euphorbiaceae but there is an argument that this genus should be treated as an independent family, Daphniphyllaceae.

The main *Daphniphyllum* alkaloid is daphniphylline, the structure of which was clarified by X-ray crystallography (130). Several other alkaloids have been isolated, including secodaphniphylline and yuzurimine.

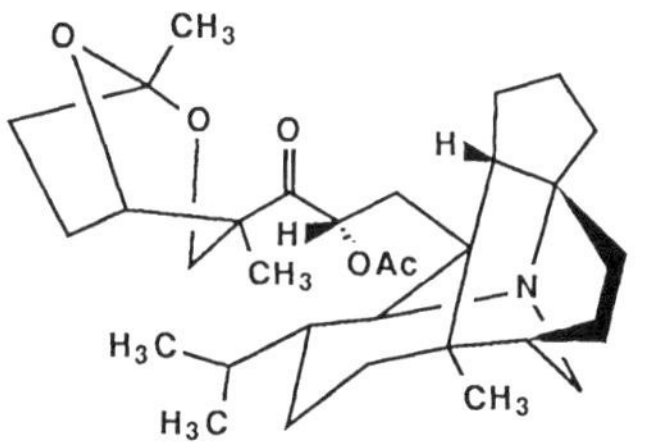

daphniphylline **346**

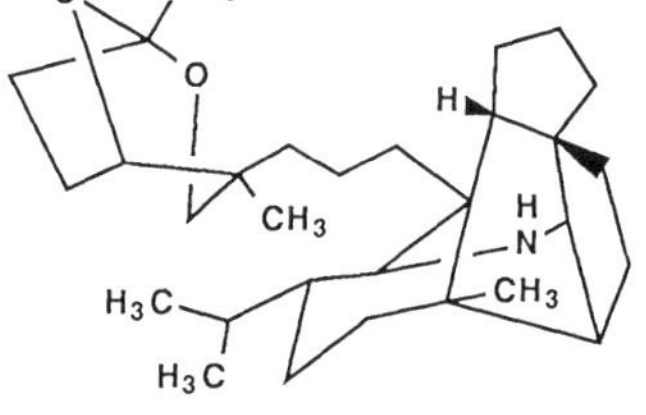

secodaphniphylline **347**

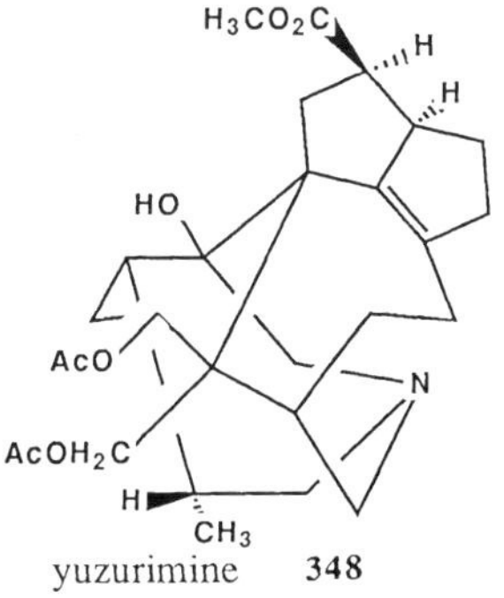

yuzurimine **348**

A careful inspection of the molecular structures suggested that the skeletal structure was derived from triterpenoids. Subsequent tracer experiments supported this view.

5.2.3.2.4 *Apocynaceae Steroidal Alkaloids*

Steroidal alkaloids of the group of pseudoalkaloids have been found in the bark and fruit of *Holarrhena* spp. (Apocynaceae), distributed in Asia and Africa (23). They are sometimes referred to as Kurchi alkaloids (80).

The family Apocynaceae is divided into two subfamilies, Plumerioideae and Apocynoideae. The genus *Holarrhena* belongs to the subfamily Apocynoideae and is taxonomically close to the genus *Strophanthus*, which is well known for its cardiac steroidal glycosides, which include conessine and holarrhimine.

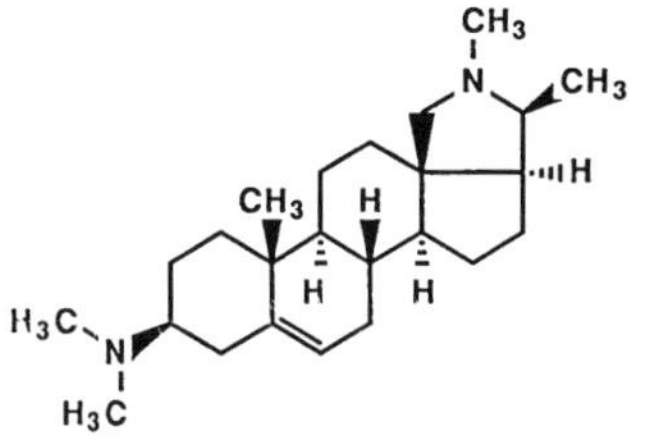

conessine **349**

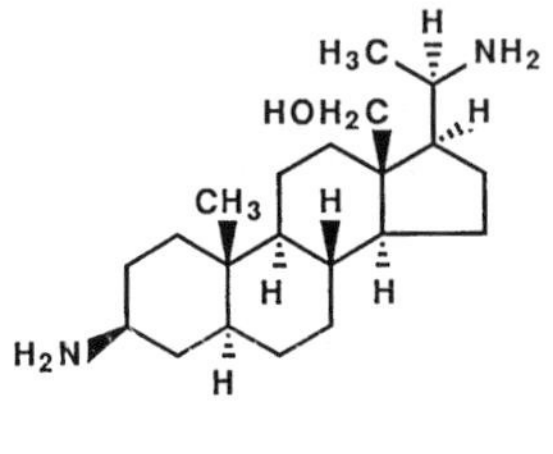

holarrhimine **350**

Similar types of alkaloids have been isolated from the plants of the closely related species such as *Funtumia* and *Malouetia*. It is interesting to note that an alkaloid isolated from the wood of an African *Malouetia* was reported to possess a curarizing effect (87).

5.2.3.2.5 *Buxaceae Steroidal Alkaloids*

The common box wood, *Buxas sempervirens*, has long been known for its medicinal properties; extracts of the plant (23) were used against various kinds of diseases such as malaria and venereal diseases. The extract was reported to possess an antitubercular property. Steroidal amines were isolated from the plant and their structures, including cyclobuxine D and cyclomicophylline A were clarified (19, 20).

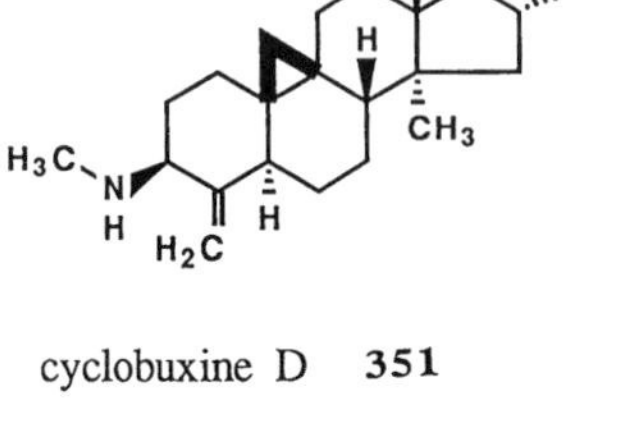

cyclobuxine D **351**

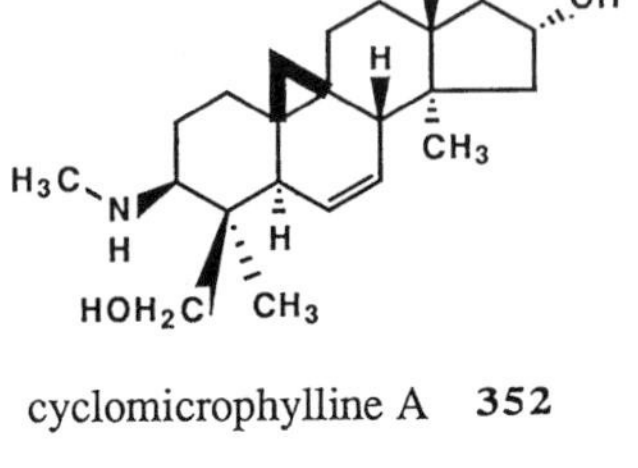

cyclomicrophylline A **352**

References

1 Adamovics J A, Cina J A, Hutchinson C R 1979 Minor alkaloids of Camptotheca acuminata. Phytochemistry 18:1085–1086
2 Aimi N, Shito T, Fukushima K, Itai Y, Aoyama C, Kunisawa K, Sakai S, Haginiwa J, Yamasaki K 1982 Studies on plants containing indole alkaloid glycosides and other constituents of the leaves of Uncaria rhynchophylla Miq. Chem Pharm Bull 30:4046–4051
3 Arisawa M, Gunasekera S P, Cordell G A, Farnsworth N R 1981 Plant anticancer agents. XXI. Constituents of Merrilliodendron megacarpum. Planta Med 47:404–407
4 Arther H R, Hui W H, Ng Y L 1959 An examination of the Rutaceae of Hong Kong. II. The alkaloids nitidine and oxynitidine, from Xanthoxylum nitidum. J Chem Soc 1840–1845
5 Barger G, Jacob A, Madinaveitia J 1938 Calycanthidine, a new simple indole alkaloid. Rec Trav Chim Pay-Bas 57:548–554
6 Battersby A R, Davidson G C, Harper B J T 1959 Ipecacuanha alkaloids. Part I. Fractionation of two new alkaloids. J Chem Soc 1744–1748
7 Battersby A R, Harper B J T 1959 Ipecacuanha alkaloids. Part II. The structure of protoemetine and a partial synthesis of (–)-emetine. J Chem Soc 1748–1753
8 Bartlett M F, Dickel D F, Taylor W I 1958 The alkaloids of Tabernanthe iboga. Part IV. The structures of ibogamine, ibogaine, tabernanthine, and voacangine. J Am Chem Soc 80:126–136
9 Beecham A F, Hert N K, Johns S R, Lamberton J A 1968 The stereochemistry of oxindole alkaloids: Uncarines A, B (formosanine), C (pteropodine), D (speciophylline), E (isopteropodine), and F. Aust J Chem 21:491–504
10 Bentley K W 1978 The alkaloids of the morphine groups. In: Coffey S (ed) Rodd's chemistry of carbon compounds. Vol. IVG. Elsevier New York, 267–321
11 Bergman J 1983 The quniazolinocarboline alkaloids. In: Brossi A (ed) The alkaloids. XIX. Academic Press New York, 29–54
12 Beyerman H C, Maat L, Visser J P 1967 Optical rotatory dispersion of pelletierine, confirmation of the absolute configuration. Rec Trav Chim Pay-Bas 86:80–84
13 Biemann K, Friedmann-Spiteller M, Spiteller G 1961 An investigation by mass spectrometry of the alkaloids of Aspidosperma quebracho-blanco. Tetrahedron Lett 485–492
14 Binns S V, Dunstan P J, Guise G B, Holder G M, Hollis A F, McCredie R S, Pinhey J T, Prager R H, Rasmussen M, Ritchie E, Taylor W C 1965 The chemical constituents of Galbulimima species. Aust J Chem 18:569–573
15 Boekelheide V 1960 The Erythrina alkaloids. In: Manske R H F, Holmes H L (eds) The alkaloids. VII. Academic Press New York, 201–228
16 Brauchli P, Deulofeu V, Budzikiewicz H, Djerassi C 1964 The structure of tubulosine, a novel alkaloid from Pogonopus tubulosus (DC) Schumann. J Am Chem Soc 86:1895–1896
17 Brillanceae M H, Kan-Fan C, Kan S K, Husson H-P 1984 Guettardine, a possible biogenetic intermediate in the formation of corynanthe-cinchona alkaloids. Tetrahedon Lett 2767–2770
18 Brossi A, Teitel S, Parry G V 1971 The ipecac alkaloids. In: Manske R H F, Holmes H L (eds) The alkaloids. XIII. Academic Press New York, 189–212
19 Brown K S Jr, Kupchan S M 1962 The structure of cyclobuxine. J Am Chem Soc 84:4590–4591
20 Brown K S Jr, Kupchan S M 1962 The configuration of cyclobuxine and its interrelation with cycloeucalenol. J Am Chem Soc 84:4592–4594
21 Buchner J A, Buchner C A 1837 Das Berberin. Justus Liebig's Ann Chem 24:228–238
22 Cai J-C, Hutchinson C R 1983 Camptothecin. In: Brossi A (ed) The alkaloids. XIX. Academic Press New York, 101–137
23 Cerny V, Sorm F 1967 Steroid alkaloids: Alkaloids of Apocynaceae and Buxaceae. In: Manske R H F, Holmes H L (eds) The alkaloids. IX. Academic Press New York, 305–426
24 Clarke R L 1977 The tropane alkaloids. In: Manske R H F, Holmes H L (eds) The alkaloids. XVI. Academic Press New York, 83–180
25 Codding P W, James M N 1972 Molecular conformation of (+)-tubocurarine chloride, a monoquaternary curare alkaloid. J Chem Soc Chem Commun 1174–1175
26 Cordell G A 1981 Introduction to alkaloids, a biogenetic approach. John Wiley New York, 1055 pp
27 Cordell G A, Ogura M, Farnsworth N R 1978 Alkaloid constituents of Ailanthus excelsa (Simaroubaceae). Lloydia 41:166–168

28 Cosson J P, Schmidt M 1970 Les alcaloides d'Ochrosia silvatica (Apocynacées). Phytochemistry 9:1353–1354

29 Dalma G 1939 Zur Kenntnis der Erythrophleum Alkaloide. (1). Cassain, ein kristallisiertes Alkaloid aus der Rinde von Erythrophleum guineense (G Don). Helv Chim Acta 22:1497–1512

30 Dalma G 1954 The Erythrophleum alkaloids. In: Manske R H F, Holmes H L (eds) The alkaloids. IV. Academic Press New York, 265–273

31 Danieli B, Lesma G, Palmisano G 1982 The configuration of (+)-evodiamine: A long-standing problem in the chemistry of indole alkaloids. J Chem Soc Chem Commun 1092–1093

32 Dasgupta B 1965 Chemical investigations of Alangium lamarckii. I. Isolation of a new alkaloid ankorine from the leaves. J Pharm Sci 54:481–483

33 Delfel N E 1980 Alkaloid distribution and catabolism in Cephalotaxus harringtonia. Phytochemistry 19:403–408

34 Denne W A, Johns S R, Lamberton J A, Mathieson A McL, Suares H 1972 The absolute structure of anopterine. Tetrahedron Lett 2727–2730

35 Dickel D F, Holden C L, Maxfield R C, Paszek L E, Taylor W I 1958 The alkaloids of Tabernanthe iboga, Part III. Isolation studies. J Am Chem Soc 80:123–125

36 Doebel K, Reichstein T 1949 Isolierung von drei kristallisierten Alkaloiden aus dem Pfaffenhütchen, Euonymus europaea L. Helv Chim Acta 32:592–597

37 Deulofeu V, Comin J, Vernengo M J 1968 The benzylisoquinoline alkaloids. In: Manske R H F, Holmes H L (eds) The alkaloids. X. Academic Press New York, 401–461

38 Du X-B, Dai Y-H, Zhang C-G, Lu S-L, Liu Z-G 1982 Studies on the Gelsemium alkaloids. I. The structure of gelsemicine. Huaxue Xuebao 40:1137–1141

39 Dutcher J D 1946 Curare alkaloids from Chondrodendron tomentosum Ruiz and Pavon. J Am Chem Soc 68:419–424

40 Dyke S F, Quessy S N 1981 Erythrina and related alkaloids. In: Manske R H F, Rodrigo R G A (eds) The alkaloids. In: Manske R H F, Rodrigo R G A (eds) The alkaloids. XVIII. Academic Press New York, 1–98

41 Eiter K, Svierak O 1951 Folicanthine, a new alkaloid from the leaves of Calycanthus floridus. Monatschrift 82:186–188

42 Endo K, Oshima Y, Kikuchi H, Koshihara Y, Hikino H 1983 Hypotensive principles of Uncaria hooks. Planta Med 49:188–190

43 Ferris J P, Briner R C, Boyce C B 1971 Lythraceae alkaloids. IX. The isolation and structure elucidation of the alkaloids of Lagerstroemia indica L. J Am Chem Soc 93:2958–2962

44 Fish F, Gray A I, Waterman P G, Donachie F 1975 Alkaloids and coumarins from North American Xanthoxyllum species. Lloydia 38:268–270

45 Fodor G 1971 The tropane alkaloids. In: Manske R H F, Holmes H L (eds) The alkaloids. XIII. Academic Press New York, 351–396

46 Frolova V I, Ban'kovskii A I, Kuzovkov A D, Molodozhnikov M M 1964 Med Prom SSSR 18:19–21

47 Fuji K, Yamada T, Fujita E, Murata H 1978 Lythraceous alkaloids. X. Alkaloids of Lagerstroemia subcostata and L. fauriei: A contribution to the chemotaxonomy. Chem Pharm Bull 26:2515–2521

48 Furukawa H, Ito C, Yago M, Wu T-S 1986 Structures of murrayastine, murrayaline, and pyrayafoline: three new carbazole alkaloids from Murraya eucrestifolia. Chem Pharm Bull 34:2672–2675

49 Gerzon K, Svoboda G H 1983 Acridone alkaloids: Experimental antitumor activity of acconycine. In: Brossi A (ed) The alkaloids. XIX. Academic Press New York, 1–28

50 Gilbert B 1965 The alkaloids of Aspidosperma, Diplorrhynchus, Kopsia, Ochrosia, Pleiocarpa, and related genera. In: Manske R H F (ed) The alkaloids. VIII. Academic Press New York, 335–513

51 Gilman R E, Marion L 1961 La pelletierine de tanret. Bull Soc Chim Fr 1993–1995

52 Golebiewski W, Wrobe J T 1981 The Lythraceae alkaloids. In: Manske R H F, Rodrigo R G A (eds) The alkaloids. XVIII. Academic Press New York, 263–322

53 Goodwin S, Smith A F, Horning E C 1959 Alkaloids of Ochrosia elliptica Labill. J Am Chem Soc 81:1903–1908

54 Goto G, Sasaki K, Sakabe N, Hirata Y 1968 The alkaloids obtained from Spiraea japonica L. fil. Tetrahedron Lett 1369–1373

55 Govindachari T R, Ravindranath K R, Viswanathan N 1974 Mappicine, a minor alkaloid from Mappia foetica Miers. J Chem Soc Perkin I 1215–1217
56 Govindachari T R, Viswanathan N 1972 Alkaloids of Mappia foetida. Phytochemistry 11:3529–3531
57 Grundon M F 1979 Quinoline alkaloids related to anthranilic acid. In: Manske R H F, Rodrigo R G A (eds) The alkaloids. XVII. Academic Press New York 105–198
58 Guinaudeau H, Leboeuf M, Cave A 1983 Aporphinoid alkaloids. J Nat Prod 46:761–835
59 Gunasekera S P, Badawi M M, Cordell G A, Farnsworth N R, Chitnis M 1979 Plant anticancer agents. X. Isolation of camptothecin and 9-methoxycamptothecin from Ervatamia heyneana. J Nat Prod 42:475–477
60 Haginiwa J, Sakai S, Aimi N, Yamanaka E, Shinma N 1973 Studies of plants containing indole alkaloids. (2) On the alkaloids of Uncaria rhynchophylla Miq. Yakugaku Zasshi 93:448–452
61 Haginiwa J, Sakai S, Takahashi K, Taguchi M, Seo S 1971 Studies of plants containing indole alkaloids. I. Alkaloids in Uncaria genus. Yakugaku Zasshi 91:575–578
62 Hemingway S R, Phillipson J D 1980 Alkaloids of the Rubiaceae. In: Phillipson J D, Zenk M H (eds) Indole and biogenetically related alkaloids. Academic Press New York, 63–90
63 Herbert R B 1977 Biosynthesis. In: Grundon M F (ed) Specialist periodical reports: Alkaloids. VII. Chemical Society London, 6–8
64 Hesse M 1978 Alkaloidchemie. Thieme Stuttgart, 231 pp
65 Hill R K 1967 The Erythrina alkaloids. In: Manske R H F, Holmes H L (eds) The alkaloids. IX. Academic Press New York, 483–516
66 Hodson H F, Robinson B, Smith G F 1961 Chimonanthine, a new Calycanthaceous alkaloid. Proc Chem Soc 465–466
67 Hofmann A 1954 Die Isolierung weiterer Alkaloide aus Rauwolfia serpentina Benth. 3. Mitteilung über Rauwolfia Alkaloide. Helv Chim Acta 37:849–865
68 Hsu J-S, Chao T-Y, Lin L-T, Hsu C-F 1977 Chemical constituents of the anticancer plant Camptotheca acuminata Decne. II. Chemical constituents of the fruits of Camptotheca acuminata Decne. Hua Hsueh Hsueh Pao 35:193–200
69 Husson H-P 1980 Iminium salts as intermediates in the biomimetic syntheses of indole alkaloids. In: Phillipson J D, Zenk M H (eds) Indole and biogenetically related alkaloids. Academic Press New York, 185–200
70 Husson A, Langlois Y, Riche C, Husson H-P, Potier P 1973 Études en serie indolique. VI. Transformation des alcaloides du type vobasine en alcaloides du type dehydroervatamine. Analyse aux rayons X de l'ervatamine. Tetrahedron 28:3095–3098
71 Husson H-P, Kan-Fan C, Sevenet T, Vidal J-P 1977 Structure de la cathenamine, intermediaire de la biosynthese des alcaloides indoliques. Tetrahedron Lett 1889–1892
72 Hutchinson C R, Heckendorf A H, Daddona P E, Hagaman E, Wenkert E 1974 Biosynthesis of Camptothecin. I. Definition of the overall pathway assisted by carbon-13 nuclear magnetic resonance analysis. J Am Chem Soc 96:5609–5611
73 Hutchinson C R, Heckendorf A H, Straughn J L, Daddona P E, Cane D E 1979 Biosynthesis of camptothecin. 3. Definition of strictosamide as the penultimate biosynthetic precursor assisted by ^{13}C and ^{2}H NMR spectroscopy. J Am Chem Soc 101:3358–3369
74 Inamoto N, Masuda S, Shimamura O, Tsuyuki T 1961 4,5-Dimethoxycanthin-6-one and 2,6-dimethoxy-p-benzoquinone from Picrasma ailanthoides Planchon. Bull Chem Soc Japan 34:888–889
75 Ishii H, Ohida H, Haginiwa J 1972 The chemical constituents of Xanthoxylum inerme Koidz, (Fagara boniensis Koidz). Isolation of the chemical constituents from bark and wood. Yakugaku Zasshi 92:118–128
76 Ito C, Wu T-S, Furukawa H 1987 Three new carbazole alkaloids from Murraya euchrestifolia. Chem Pharm Bull 35:450–452
77 Ito K, Furukawa H, Haruna M, Ito M 1973 Alkaloids of Erythrina crystagalli (L) cv. Marubadeiko H Murata. Yakugaku Zasshi 93:1674–1678
78 Janot M M 1953 The ipecac alkaloids. In: Manske R H F, Holmes H L (eds) The alkaloids. III. Academic Press New York, 363–394
79 Jeffs P W 1967 Protoberberine alkaloids. In: Manske R H F, Holmes H L (eds) The alkaloids. IX. Academic Press New York, 41–116

80 Jeger O, Prelog V 1960 Steroid alkaloids: The Holarrhena group. In: Manske R H F, Holmes H L (eds) The alkaloids. VII. Academic Press, New York, 319–342
81 Jin H-L, Xu R-S 1982 Studies on the alkaloids of Gelsemium elegans Benth. The structure of koumidine. Acta Chimica Sinica 40:1129–1135
82 Johns S R, Lamberton J 1973 Elaeocarpus alkaloids. In: Manske R H F (ed) The alkaloids, XIV. Academic Press New York, 325–346
83 Johns S R, Lamberton J A, Sioumis A A 1977 Tropine 3,4,5-trimethoxycinnamate, a new alkaloid from Erythroxylum ellipticum (Erythroxylaceae). Aust J Chem 23:421–422
84 Joshi B S, Kamat V N, Gawad D H 1977 Some carboline alkaloids of Ailanthus malabarica DC. Heterocycles 7:193–200
85 Kan-Fan C, Das B C, Potier P, Schmid M 1970 Alcaloides des feuilles d'Ochrosia vieilardii (Apocynacées). Phytochemistry 9:1351–1352
86 Kan-Fan C, Husson H-P 1979 Isolation and biomimetic conversion of 4,21-dehydrogeissoschizine. J Chem Soc Chem Commun 1015–1016
87 Khuong-Huu-Laine F, Bisset N G, Goutarel R 1965 Malouetia bequaertiana alkaloids with a review of the Malouetia genus and of guachamaca, a Venezuelan curarizing drug. Ann Pharm France 23:395–409
88 Khuong-Huu F, Chiaroni A, Riche C 1981 Structure of koumine, an alkaloid from Gelsemium elegasns Benth. Tetrahedron Lett 22:733–734
89 Knox J R, Slobbe J 1971 Three novel alkaloids from Ervatamia orientalis. Tetrahedron Lett 2149–2151
90 Kondo H, Nakazato T 1924 [Investigations on Sinomenin and Cocculus-alkaloids. III. On Trilobin, an Alkaloid of C. trilobus D.C.] [in Japanese]. Yakugaku Zasshi 511:691–697
91 Kupchan S M, Komoda Y, Branfman A R, Sneden A T, Court W A, Thomas G J, Nagao Y, Daley R G, Zimmerly V A, Sumner W C 1977 The maytansinoids, isolation, structural elucidation and chemical interrelation of novel ansa macrolides. J Org Chem 42:2349–2357
92 Kupchan S M, Smith R M, Bryan R F 1970 Maytoline, a nicotinoyl sesquiterpene alkaloid prototype from Maytenus ovatus. J Am Chem Soc 92:6667–6668
93 Lassak E V, Polonsky J, Jacquemin 1977 5-Hydroxycanthin-6-one from Simarouba amara. Phytochemistry 16:1126–1127
94 Leete E, Lechleriter J C, Carver R A 1975 Determination of the 'starter' acetate unit in the biosynthesis of pinidine. Tetrahedron Lett 3779–3782
95 Leonard N J 1953 Lupin alkaloids. In: Manske R H F, Folmes H L (eds) The alkaloids. III. Academic Press New York, 120–199
96 Leonard N J 1967 Lupin alkaloids. In: Manske R H F (ed) The alkaloids. IX. Academic Press New York, 175–221
97 Lin L-T, Chao T-Y, Hsu J-S 1977 Chemical constituents of the anticancer plant Camptotheca acuminata Decne. I. Chemical constituents of the roots of Camptotheca acuminata Decne. Hua Hsueh Hsueh Pao 35:227–231
98 Lin L-T, Sung C-C, Hsu J-S 1979 A new anticancer alkaloid 11-hydroxycamptothecin. K'o Hsueh Tung Pao 24:478–479
99 Liu C-T, Wang Q-W 1981 Structure of koumine. J Am Chem Soc 103:4634–4635
100 Liu Z-J, Yu Q-S 1986 A partial synthesis of koumine. Youzi Huaxu 1:36–37
101 Loder J W, Nearn R H 1975 Structure of norerythrostachaldine, a cytotoxic alkaloid from Erythrophleum chlorostachys (Leguminosae). Aust J Chem 28:651–656
102 Manske R H F 1960 The ipecac alkaloids. In: Manske R H F, Holmes H L (eds) The alkaloids. VII. Academic Press New York, 419–422
103 Manske R H F, Ashford W R 1954 The protoberberine alkaloids. In: Manske R H F, Holmes H L (eds) The alkaloids. In: Manske R H F, Holmes H L (eds) The alkaloids. IV. Academic Press New York, 77–118
104 Manske R H F 1965 The alkaloids of Calycanthaceae. In: Manske R H F, Holmes H L (eds) The alkaloids. VIII. Academic Press New York, 581–591
105 Marini Bettolo G B 1981 Recent advances in the research on curare. Udit (Verhandelingen van de Koninglijke Academie voor Geneeskunde van Belgie) XLIII 3:185–212
106 Marion L 1950 The alkaloids of the pomegranate root bark. In: Manske R H F, Holmes H L (eds) The alkaloids. I. Academic Press New York, 176–189

107 Marion L 1952 The Erythrina alkaloids. In: Manske R H F, Holmes H L (eds) The alkaloids. II. Academic Press New York, 499–512
108 Marion L 1960 The pyridine alkaloids. In: Manske R H F (ed) The alkaloids. VI. Academic Press New York London, 125–126
109 Merlini L, Mondelli R, Nasini G, Hesse M 1967 Gambirine, a new indole alkaloid from Uncaria gambier Roxb. Tetrahedron Lett 1571–1574
110 Merlini L, Mondelli R, Nasini G, Hesse M 1967 Indole alkaloids from gambir. Structure of gambirtannine, oxogambirtannine and dihydrogambirtannine. Tetrahedron 23:3129–3145
111 Merlini L, Mondelli R, Nasini G, Hesse M 1970 The structure of roxburghines A–E, new indole alkaloids from an Uncaria sp. Tetrahedron 26:2259–2279
112 Monteiro H, Budzikiewicz H, Djerassi C, Arndt R R, Baarschers W H 1965 Alkaloid studies. Part LIV. Structure of deoxytubulosine and interconversion with tubulosine. J Chem Soc Chem Commun 317–318
113 Morin R B 1968 Erythrophleum alkaloids. In: Manske R H F, Holmes H L (eds) The alkaloids. X. Academic Press New York, 287–303
114 Nagakura N, Ruffer M, Zenk M H 1979 The biosynthesis of monoterpenoid indole alkaloids from strictosidine. J Chem Soc Perkin I 2308–2312
115 Nakano T 1954 Studies on the alkaloids of Magnoliaceous plants. XIV. Alkaloids of Magnolia grandiflora L. (3). Structure of magnoflorine. Chem Pharm Bull 2:329–334
116 O'Donovan D G, Koegh M F 1966 The biosynthesis of folicanthine. J Chem Soc (C) 1570–1572
117 Ohmoto T, Koike K 1982 Studies on the constituents of Picrasma quassioides Bennet. I. Chem Pharm Bull 30:1204–1209
118 Ohmoto T, Tanaka R, Nikaido T 1976 Studies on the constituents of Ailanthus altissima Swingle. On the alkaloidal constituents. Chem Pharm Bull 24:1532–1536
119 Openshaw H T 1967 Quinoline alkaloids, other than those of Cinchona. In: Manske R H F (ed) The alkaloids. IX. Academic Press New York London, 223–267
120 Parry R J, Chang M N T, Schwab J M, Foxman B M 1980 Biosynthesis of the Cephalotaxus alkaloids. Investigation of the early and late stages of cephalotaxine biosynthesis. J Am Chem Soc 102:1099–1111
121 Paudler W W, Kerley G I, McKay J 1963 The alkaloids of Cephalotaxus drupaceae and Cephalotaxus fortunei. J Org Chem 28:2194–2197
122 Potier P, Janot M M 1973 Biogenesis of ellipticine indole alkaloids. C R Acad Sci Paris 276:1727–1729
123 Rahman A-u, Basha A 1983 Biosynthesis of indole alkaloids. Clarendon Press Oxford, 270 pp
124 Rapoport H, Holden K G 1959 Isolation of alkaloids from Balfourodendron riedelianum. J Am Chem Soc 81:3738–3743
125 Ritchie E, Taylor W C 1967 The galbulimima alkaloids. In: Manske R H F, Holmes H L (eds) The alkaloids. IX. Academic Press New York, 529–544
126 Ritchie E, Taylor W C 1971 The Galbulimima alkaloids. In: Manske R H F, Holmes H L (eds) The alkaloids. XIII. Academic Press New York, 227–271
127 Robinson T 1968 The biochemistry of alkaloids. Springer Berlin Heidelberg New York, 149 pp
128 Rueffer M, Kan-Fan C, Husson H-P, Stockigt J, Zenk M H 1979 4,21-Dehydrogeissoschizine, an intermediate in heteroyohimbine alkaloid biosynthesis. J Chem Soc Chem Commun 1016–1018
129 Sabri N N, Court W E 1978 Stem alkaloids of Rauwolfia vomitoria. Phytochemistry 17:2023–2026
130 Sakabe N, Hirata Y 1966 X-Ray structure determination of a new type of alkaloid daphniphylline hydrobromide. Tetrahedron Lett 965–968
131 Sakabe N, Irikawa H, Sakurai H, Hirata Y 1966 Isolation of three new alkaloids from Daphniphyllum macropodium Miquel. Tetrahedron Lett 963–964
132 Sakai S 1976 The indole alkaloids of Japanese plants; structures and reactions. Heterocycles 4:131–168
133 Sakai S, Aimi N, Yamaguchi K, Yamanaka E, Haginiwa J 1982 Gardneria alkaloids. Part 13. Structure of gardmultine and demethoxygardmultine; bis-type indole alkaloids of Gardneria multiflora Makino. J Chem Soc Perkin I 1257–1262
134 Sakai S, Aimi N, Yamaguchi K, Ogata K, Haginiwa J 1987 Gardneria alkaloids. Part 14. The structure of gardfloramine and 18-demethoxygardfloramine. Chem Pharm Bull 35:453–455

135 Sakai S, Wongseripipatana S, Ponglux D, Yokota M, Ogata K, Takayama H, Aimi N 1987 Indole alkaloids isolated from Gelsemium elegans (Thailand) 19-(Z)-akuamidine, 16-epivoacarpine, 19-hydroxydihydrogelsevirine, and the revised structure of koumidine. Chem Pharm Bull 35:4668–4671

136 Sakai S, Yamanaka E, Kitajima M, Yokota M, Aimi N, Wongserpipatana S, Ponglux D 1986 Biomimetic synthesis of koumine skeleton; partial synthesis of 11-methoxykoumine (Gelsemium-type alkaloid) from 18-hydroxygardnerine. Tetrahedron Lett 27:4585–4588

137 Saxton J E 1965 Alkaloids of Gelsemium species. In: Manske R H F (ed) The alkaloids. VIII. Academic Press New York London, 93–117

138 Schmutz J, Hunziker F 1958 Alkaloids of Aspidosperma olivaceum. Pharm Acta Helv 33:341–347

139 Schnoes H K, Burlingame A L, Biemann K 1962 Application of mass spectrometry to structure problems. The occurrence of eburnamenine and related alkaloids in Rhazya stricta and Aspidosperma quebracho blanco. Tetrahedron Lett 993–999

140 Shamma M 1967 The aporphine alkaloids. In: Manske R H F, Holmes H L (eds) The alkaloids. IX. Academic Press New York, 1–40

141 Shamma M, Maniot J L 1976 The systematic classification of bisbenzylisoquinolines. Heterocycles 4:1817–1824

142 Shellard E J, Houghton P J 1972 The Mitragyna species of Asia. XXI. The distribution of alkaloids in young plants of Mitragyna parvifolia grown from seed obtained from Ceylon. Planta Med 21:382–392

143 Shellard E J, Houghton P J 1973 The Mitragyna species of Asia. XXII. The distribution of alkaloids in young plants of Mitragyna parvifolia grown from seeds obtained from Uttar Pradesh State of India. Planta Med 22:97–102

144 Sheriha G M, Rapoport H 1976 Biosynthesis of Camptotheca acuminata alkaloids. Phytochemistry 15:505–508

145 Shun Y, Cordell G A 1985 14β-Hydroxygelsedine, a new oxindole alkaloid from Gelsemium sempervirens. J Nat Prod 48:788–791

146 Shun Y, Cordell G A, Garland M 1986 21-Oxogelsevirine, a new alkaloid from Gelsemium rankinii. J Nat Prod 49:483–487

147 Shun Y, Cordell G A 1986 Rankinidine, a new indole alkaloid from Gelsemium rankinii. J Nat Prod 49:806–808

148 Shun Y, Cordell G A 1987 Revision of the stereochemistry of koumidine. Phytochemistry 26:2875–2876

149 Smith R M 1977 The Celastraceae alkaloids. In: Manske R H F, Holmes H L (eds) The alkaloids. XVI. Academic Press New York, 215–248

150 Snieckus V 1973 The Securinega alkaloids. In: Manske R H F (ed) The alkaloids. XIV. Academic Press New York, 425–506

151 Stockigt J 1980 The biosynthesis of heteroyohimbine-type alkaloids. In: Phillipson J D, Zenk M H (eds) Indole and biogenetically related alkaloids. Academic Press New York, 113–141

152 Stockigt J, Husson H-P, Kan-Fan C, Zenk M H 1977 Cathenamine, a central intermediate in the cell-free biosynthesis of ajmalicine and related indole alkaloids. J Chem Soc Chem Commun 164–166

153 Tallent W H, Horning E C 1956 The structure of pinidine. J Am Chem Soc 78:4467–4469

154 Tallent W H, Stronberg V L, Horning E C 1955 Pinus alkaloids. The alkaloids of P. sabiniana Dougl. and related species. J Am Chem Soc 77:6361–6364

155 Taylor W I 1965 The Pentaceras and the eburnamine (Hunteria)-vincamine alkaloids. In: Manske R H F (ed) The alkaloids. VIII. Academic Press New York, 249–267

156 Tomita M, Kozuka M 1964 On the alkaloids of Cinnamomum camphora. Yakugaku Zasshi 84:365–367

157 Tomita M, Kusuda F 1953 Studies on the alkaloids of Menispermaceous plants. CI. Alkaloids of Cocculus laurifolius DC (Suppl III). Pharm Bull 1:1–5

158 Tomita M, Yamaguchi H 1956 Alkaloids of Cocculus laurifolius D C (Suppl 7). Isolation of dihydroerysodine. Chem Pharm Bull 4:225–229

159 Wada H, Shizuri Y, Yamada K, Hirata Y 1971 Evonine, an alkaloid obtained from Euonymus sieboldiana Blume I. Tetrahedron Lett 2655–2658

160 Wall M E, Wani M C, Cook C E, Palmer K H, McPhail A T, Sim G A 1966 Plant antitumor agents. I. The isolation and structure of camptothecin, a novel alkaloidal leukemia and tumor inhibitor from Camptotheca acuminata. J Am Chem Soc 88:3888–3890

161 Wani M C, Wall M E 1969 Plant antitumor agents. II. The structure of two new alkaloids from Camptotheca acuminata. J Org Chem 34:1364–1367

162 Wiesner K, Figdor S K, Bartlett M F, Henderson D R 1952 Garrya alkaloids. I. The structure of garryine and veatchine. Can J Chem 30:608–626

163 Wood H C S, Wrigglesworth R 1977 Tropane alkaloids. In: Coffey S (ed) Rodd's chemistry of carbon compounds. IVB. Elsevier New York, 201–235

164 Yamamura S, Hirata Y 1975 The Daphniphyllum alkaloids. In: Manske R H F, Holmes H L (eds) The alkaloids. XV. Academic Press New York, 41–82

165 Yamanaka E, Kimizuka Y, Aimi N, Sakai S, Haginiwa J 1983 Studies on plants containing indole alkaloids. IX. Quantitative analysis on the tertiary alkaloids in various parts of Uncaria rhynchophylla Miq. Yakugaku Zasshi 103:1028–1033

166 Yang J-S, Chen Y-W 1983 Chemical studies on the alkaloids of Hu Man Teng (Gelsemium elegans Benth). I. Isolation of the alkaloids and the structure of humantenmine. Yaoxue Xuebao 18:104–112

167 Yang J-S, Chen Y-W 1985 Elucidation of the structures of humantenine and humantenirine. Acta Pharm Sin 19:686–690

Chapter 6

Aliphatic and Alicyclic Extractives

The aliphatic and alicyclic extractives of woody tissue mostly fall into two groups. The first group are the compounds involved in intermediary metabolism and their close relatives. All plants have the same general metabolic cycles, and the intermediates are thus ubiquitous in woody tissue. These intermediates are usually transient in nature, although occasionally one or more will accumulate. Others of these highly oxygenated extractives are closely related to metabolic intermediates, but here their occurrence can be quite species specific.

The second major group are the compounds derived by the linear polymerization of acetate to yield straight-chain aliphatic compounds with relatively few oxygens. Prominent among these are the fatty acids and their esters with glycerol, sterols, and other alcohols. These fats mostly serve as energy reserves just as the same compounds do in humans. Further polymerization leads to the longer-chain wax acids and their esters so important in providing a barrier to water loss.

6.1 Simple Organic Acids

Y. OHTA

6.1.1 Introduction

This section is mainly concerned with the non-volatile, non-nitrogen-containing organic acids. It is restricted to the aliphatic acids and alicyclic acids (except for fatty acids, which are covered in Sect. 6.4.2); the aromatic acids, except for shikimic and quinic acid, are covered in Chap. 7. Since only a limited number of acids have been reported to occur in woody tissues such as bark and heartwood, the acids present in other parts, such as leaves and fruit, are also included. Those found only in herbs are excluded. No attempt has been made to search all the woody plants for the occurrence of these compounds. In the following, a brief introductory explanation about the bioformation and function in plants is described for the respective organic acids (10, 33, 43, 61, 120, 121), and this is followed by tables listing the occurrence of the acid in woody plants, if any.

6.1.2 Organic Acids in the TCA and Glyoxylate Cycles

The tricarboxylic acid (TCA) cycle is an essential metabolic pathway in plants. This cycle is amphibolic and functions not only in catabolism to supply energy as ATP, but also to generate precursors for anabolic pathways. Certain intermediates of the cycle, particularly α-ketoglutaric acid and oxaloacetic acid, are utilized

Table 6.1.1. Sources of simple organic acids

Acid	Family	Genus and species	Source	References
Aconitic acid	Rhamnaceae	*Helinus ovatus*	Leaves	34
	Rutaceae	*Citrus sinnensis*	Fruit	106
Isocitric acid	Lecythidaceae	*Couroupita guianensis*	Fruit	81
	Oleaceae	*Olea europaea*		24
	Rosaceae	*Pyrus malus*	Fruit	14, 15
		Rubus fruticosus		19
	Rutaceae	*Citrus sinnensis*	Peel	106
	Theaceae	*Camellia sinnensis*	Leaves	104
α-Ketoglutaric acid	Euphorbiaceae	*Phyllanthus simplex*		23
	Leguminosae	*Albizzia lebbeck*	Seedling	76
			Leaves	77
		Bauhinia purpurea		78
	Rosaceae	*Pyrus malus*	Fruit	47
	Rutaceae	*Citrus sinnensis*	Peel	106
Fumaric acid	Aceraceae	*Acer saccharum*	Sap, syrup	95
	Berberidaceae	*Nandina domestica*		88
	Eucommiaceae	*Eucommia ulmoides*	Leaves, stem	38, 130
	Rosaceae	*Prunus cerasus*		57
	Vitaceae	*Vitis vinifera*	Vegetative part	16, 129
Oxaloacetic acid	Leguminosae	*Bauhinia purpurea*		78
		Albizzia lebbeck	Seedling	76
			Leaves	77
	Rosaceae	*Pyrus malus*	Fruit	47
Glyoxylic acid	Cornaceae	*Cornus mas*	Fruit	108
	Leguminosae	*Albizzia lebbeck*	Seedling	76
		Bauhinia purpurea	Seedling	78
	Rosaceae	*Prunus domestica*	Fruit	12
		Pyrus malus		12
		Ribes sp.	Fruit	12
	Vitaceae	*Vitis vinifera*		39

for the biosynthesis of amino acids. Citric acid and succinyl-CoA, an biologically active form of succinic acid, are also withdrawn from the cycle for various biosynthetic reactions. The glyoxylate cycle, a modified form of the TCA cycle, is especially prominent in the plant tissues where fats are converted into carbohydrates through acetyl-CoA. A turn of the cycle can generate one molecule of succinate from two molecules of acetyl-CoA. Among the members involved in these central metabolic pathways, citric acid, malic acid, and succinic acid are encountered very frequently in higher plants, including woody plants. Since they appear to be ubiquitous in higher plants, the occurrences of them are not listed. Although the rest of the members are also expected to be commonly found ubiquitously in plant tissues, reports on their occurrence have been rather limited, as shown in the list for each compound (Table 6.1.1).

6.1.2.1 Citric Acid

The major pathway for the formation of citric acid is through the condensation of oxaloacetate and acetyl-CoA catalyzed by citrate synthase, the first enzyme in

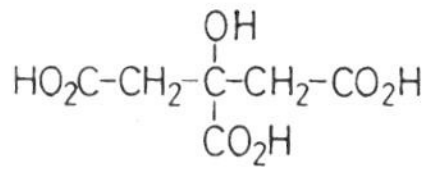

$$\begin{array}{c} \quad\quad\quad\quad OH \\ HO_2C\text{-}CH_2\text{-}\overset{|}{\underset{|}{C}}\text{-}CH_2\text{-}CO_2H \\ \quad\quad\quad\quad CO_2H \end{array}$$

1 citric acid

the TCA cycle. In the glyoxylate cycle, citric acid is also formed from acetyl-CoA arising from fatty acid breakdown.

6.1.2.2 Aconitic Acid

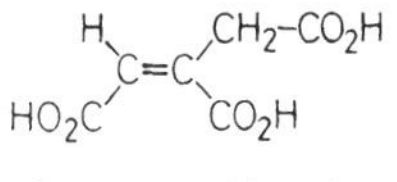

$$(H)(HO_2C)C{=}C(CH_2\text{-}CO_2H)(CO_2H)$$

2 cis-aconitic acid

Aconitic acid involved in the TCA and glyoxylate cycles and the acid commonly occurring in nature has the *cis*-configuration. The *trans*-isomer has also been isolated from some plant materials – for example, sugarcane (*Saccharum officinarum*) juice (17), tomato (*Lycopersicon esculentum*) (56) or moss (Bryophyta) (34). However, some of the occurrences might be artifacts of the isolation procedures, for an interconversion between two isomers of aconitic acid has been reported (14, 81). In both cycles, *cis*-aconitic acid is formed upon dehydration of citric acid catalyzed by aconitase (aconitate hydratase) which also catalyzes the rehydration of *cis*-aconitate to isocitric acid.

6.1.2.3 Isocitric Acid

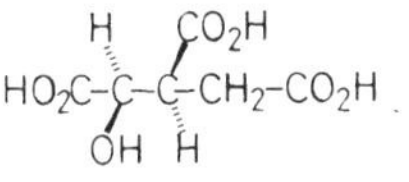

$$HO_2C\text{-}CH(OH)\text{-}CH(CO_2H)\text{-}CH_2\text{-}CO_2H$$

3 isocitric acid

Of the four stereoisomers, *erythro*-D_S- and L_S-, and *threo*-D_S- and L_S-isocitric acid, it is the *threo*-D_S-isomer that occurs in nature. The formation of isocitric acid from *cis*-aconitic acid is catalyzed by aconitase. This acid is also accumulated in Crassulacean plants as a result of dark CO_2 fixation.

6.1.2.4 α-Ketoglutaric Acid

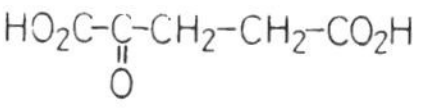

$$HO_2C\text{-}\underset{\|}{C}\text{-}CH_2\text{-}CH_2\text{-}CO_2H \\ O$$

4 α-ketoglutaric acid

α-Ketogultaric acid is formed through oxidative decarboxylation of *threo*-D_S-isocitric acid in the TCA cycle under the action of NAD-linked isocitrate dehydrogenase. *α*-Ketoglutaric acid plays an important role in the biosynthesis of all amino acids. Ammonium ion is primarily assimilated into glutamate to yield glutamine by the action of glutamine synthetase. Transamination between glutamine and α-ketoglutaric acid catalyzed by glutamine: 2-oxoglutarate aminotransferase (glutamate synthase) generates two molecules of glutamate. Glutamate is also formed, under certain conditions, from α-ketoglutaric acid by reductive amination catalyzed by glutamate dehydrogenase. Subsequently the amino group of glutamate is transferred to various α-keto acids to yield the corresponding α-amino acids.

6.1.2.5 Succinic Acid

$HO_2C\text{-}CH_2\text{-}CH_2\text{-}CO_2H$

5 succinic acid

In the TCA cycle, succinic acid is formed from α-ketoglutaric acid via succinyl-CoA. The oxidation of α-ketoglutaric acid to succinyl-CoA is carried out by α-ketoglutarate dehydrogenase in the presence of NAD and CoA. The coenzyme-A group is eliminated from succinyl-CoA by an energy-conserving reaction with guanosine diphosphate (GDP) and phosphate under the action of succinyl-CoA synthetase. In the glyoxylate cycle, this acid is produced by direct cleavage of isocitric acid by the action of isocitrate lyase. This is the key reaction in the glyoxylate cycle in providing the precursor for gluconeogenesis.

6.1.2.6 Fumaric Acid

H CO₂H / C=C / HO₂C H

6 fumaric acid

Fumaric acid, an intermediate in the TCA cycle, is formed by dehydrogenation of succinic acid under the action of the flavoprotein succinate dehydrogenase. In the catabolism of amino acids in higher plants, fumaric acid is formed as one of the end products of degradation of tyrosine (72). In contrast, degradation of phenylalanine to fumarate and acetoacetate, as in mammal cells, does not occur to any great extent in plant tissues. Phenylalanine appears to be mainly metabolized toward formation of various secondary metabolites and lignin (84, 116). Fumaric acid is also formed from adenylosuccinate during conversion of inosine 5′-phosphate into adenosine 5′-monophosphate in the biosynthesis of purines.

6.1.2.7 Malic Acid

Malic acid is known in two active isomeric forms; the naturally occurring isomer has the L-(−)-configuration. This dibasic acid as a member of the TCA cycle is formed by hydration of fumaric acid catalyzed by fumarase. In the glyoxylate

$$HO_2C-\underset{\overset{\equiv}{OH}}{\overset{H}{C}}-CH_2-CO_2H$$

7 malic acid

cycle, the condensation of glyoxylic acid and acetyl-CoA to form malic acid is carried out by malate synthase. In addition, malic acid is also formed by the reduction of oxaloacetic acid which is the initial product of CO_2 fixation in C_4 plants and Crassulacean plants.

6.1.2.8 Oxaloacetic Acid

$$HO_2C-\underset{\overset{\|}{O}}{C}-CH_2-CO_2H$$

8 oxaloacetic acid

Oxaloacetic acid, the starter compound of both the TCA and the glyoxylate cycles is regenerated from malic acid under the action of malate dehydrogenase. In green plants, oxaloacetic acid can normally be supplied in any amount needed for operation of the cycles from phosphoenol pyruvate. Oxaloacetic acid is the initial product of CO_2 fixation in C_4-photosynthetic and Crassulacean acid pathways. In these pathways, oxaloacetic acid is formed by β-carboxylation of phosphoenol pyruvate catalyzed by phosphoenol pyruvate carboxylase.

6.1.2.9 Glyoxylic Acid

HO_2C-CHO

9 glyoxylic acid

Glyoxylic acid as an intermediate in the glyoxylate cycle is formed by the cleavage of isocitric acid catalyzed by isocitrate lyase. This acid is also an oxidation product of glycolic acid during photorespiration. This reaction is carried out by glycolate oxidase, which is localized in peroxisomes.

6.1.3 Other Metabolically Important Organic Acids

A listing of some major acids and their occurrences in woody plants is found in Table 6.1.2.

6.1.3.1 Glycolic Acid

HO_2C-CH_2OH

10 glycolic acid

Table 6.1.2. Sources for other metabolically important acids

Acid	Family	Genus and species	Source	References
Glycolic acid	Caprifoliaceae	*Sambucus nigra*	Leaf	5
	Combretaceae	*Combretum micranthum*	Leaf	5
	Dichapetalaceae	*Dichapetalum cymosum*		70
	Ericaceae	*Arbutus unedo*	Leaf	5
		Erica multiflora	Twig	5
	Labiatae	*Rosmarinus* spp.	Twig	5
	Myrtaceae	*Psidium guajava*		128
	Pinaceae	*Juniperus communis*	Berry	5
	Rosaceae	*Prunus cerasus*		57
	Vitaceae	*Ampelopsis hederaceae*	Leaves	35
Glyceric acid	Rosaceae	*Prunus cerasus*		57
	Vitaceae	*Vitis vinifera*		16, 129
Pyruvic acid	Leguminosae	*Bauhinia purpurea*		78
	Rutaceae	*Citrus unshiu*		20
	Sapindaceae	*Litchia chinensis*	Leaves	96
Malonic acid	Leguminosae	*Sophora* sp.		8
	Rhamnaceae	*Ceanothus americanus*		65
	Rosaceae	*Prunus cerasus*		57
	Rutaceae	*Citrus sinnensis*	Peel	106
		C. paradisi	Fruit	75
	Sapindaceae	*Litchia chinensis*	Leaves	96
Shikimic acid and quinic acid	Anacardiaceae	*Rhus* spp.	Leaves	41
		Illicium religiosum	Sapwood	44
	Casuarinaceae	*Casuarina equisetifolia*		68
	Combretaceae	*Anogeissus tatifolia*	Branches	58
	Euphorbiaceae	*Sapium japonicum*	Leaves	41
		S. siebiferum	Leaves	41
		Aleurites cordata	Leaves	41
	Fagaceae	*Quercus stenophylla*	Branches	50
		Q. pedunculata	Leaves	11
	Ginkgonaceae	*Ginkgo biloba*	Leaves	42, 131
	Guttiferae	*Mammea americana*	Leaves	93
	Hamamelidaceae	*Hamamelis* spp.		92
	Meliaceae	*Melia azedarach*	Heartwood	6
	Pinaceae	*Abies alba*	Leaves	63
		Juniperus communis	Leaves	63
		Larix spp.		63, 90
		Picea abies	Needles	63
		Pinus spp.		18, 63
		Pseudotsuga menziesii	Bark	45
		Taxus baccata		63
	Saxifragaceae	*Ribes aureum*		94
	Salicaceae	*Populus trichocarpa*	Whole plant	37
	Vitaceae	*Vitis vinifera*	Whole plant	16
			Root	115

Glycolic acid is a substrate of photorespiration, which mainly occurs in C_3 plants. This light-enhanced respiration is recognized as a wasteful process in preventing plants from attaining maximum yield in photosynthesis. A variety of mechanisms have been presented for the formation of glycolic acid from intermediates of photosynthetic carbon metabolism. The most important pathway for the formation of glycolic acid is the oxidative cleavage of ribulose *bis*-phosphate catalyzed by RuBP carboxylase/oxidase. The acid can also be formed reductively from glyoxylic acid, or oxidatively from glycolaldehyde or sugar phosphate.

6.1.3.2 Glyceric Acid

$HO_2C-C(H)(OH)-CH_2OH$

11 glyceric acid

Glyceric acid is formed as an intermediate of glycolate pathway. It is also obtained by dephosphorylation of 3-phosphoglyceric acid which is an intermediate of carbohydrate metabolism, the initial product of photosynthetic CO_2 fixation via the Calvin cycle, or an oxygenation product of ribulose bis-phosphate in photorespiration.

6.1.3.3 Pyruvic Acid

$H_3C-C(=O)-CO_2H$

12 pyruvic acid

Pyruvic acid is important as an intermediate in sugar metabolism. This acid is formed as the final product of the glycolytic pathway from 3-phosphoglycerate via phosphoenolpyruvate. Pyruvic acid is also formed through the oxidative pentose phosphate cycle. The degradation of glucose in the cycle yields the C_3 product glyceraldehyde 3-phosphate, which can be oxidized to pyruvate. Pyruvic acid is the principal precursor for the biosynthesis of amino acids such as alanine, as well as leucine and valine.

6.1.3.4 Malonic Acid

$HO_2C-CH_2-CO_2H$

13 malonic acid

The biologically active form of malonic acid, malonyl-CoA, is essential for the biosynthesis of fatty acids and for the elongation of carbon chains including the formation of a variety of aromatic secondary metabolites. Malonyl-CoA is formed from acetyl-CoA and bicarbonate in the cytosol by the action of acetyl-CoA carboxylase.

6.1.3.5 Shikimic Acid and Quinic Acid

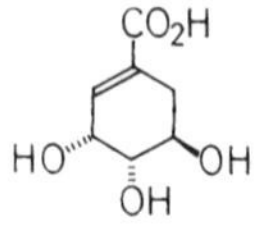

14 shikimic acid 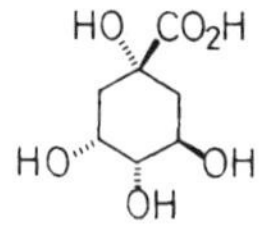 15 quinic acid

Shikimic acid plays an important role in the biosynthesis of a variety of aromatic compounds, such as aromatic amino acids, flavonoids, or lignins. This acid had been isolated from the fruit of *Illicium religiosum* (Japanese name: shikimi-no-ki), in which the acid constitutes approximately 20% of the dry matter (25). Shikimic acid has subsequently been detected in a wide variety of plants throughout the plant kingdom (124). The biosynthesis of shikimic acid is known to occur through condensation of phosphoenol pyruvate and D-erythrose-4-phosphate, which are general products of carbohydrate metabolism, yielding 3-deoxy-D-*arabino*-heptulosonic acid 7-phosphate. Oxidative cyclization of the condensation product affords 3-dehydroquinic acid which, in turn, gives 3-dehydroshikimic acid upon dehydration. Shikimic acid is formed by the reduction of dehydroshikimic acid by the action of NADPH-linked shikimate dehydrogenase, also known as dehydroshikimate reductase (124).

Quinic acid is also found throughout the plant kingdom, usually with shikimic acid. It occurs as the free acid in many plants and its phenolic esters are also very common. The simplest assumption for the biosynthesis of quinic acid is through the reduction of 3-dehydroquinic acid, catalyzed by quinate dehydrogenase. However, because of the rare occurrence of this enzyme, the biosynthesis of quinic acid still remains to be explained.

6.1.4 Organic Acids of an End-Product Nature

Several organic acids are known to occur as end products of metabolism in plant cells. These organic acids are deposited mainly in vacuoles to prevent lowering the pH of the cytoplasm to a physiologically dangerous extent. These acids and their sources are listed in Table 6.1.3.

6.1.4.1 Lactic Acid

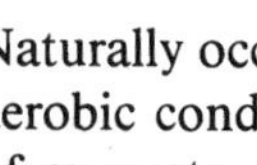

16 L-lactic acid

Naturally occurring lactic acid has the L-configuration. It accumulates under anaerobic conditions in a number of plants. L-Lactic acid is formed by reduction of pyruvate, which is an intermediary breakdown product of carbohydrate via the glycolysis pathway.

Table 6.1.3. Sources of other organic acids

Acid	Family	Genus and species	Source	References
Lactic acid	Ericaceae	*Vaccinium myrtillus*		48
	Fagaceae	*Fagus sylvatica*	Nut	52
	Myricaceae	*Myrica rubra*	Fruit	54
	Rosaceae	*Prunus cerasus*		57
		Pyrus spp.		29
		Rubus spp.		31, 109, 110
	Rutaceae	*Citrus sinnensis*		106
Oxalic acid	Aceraceae	*Acer saccharum*	Sap	123
	Anacardiaceae	*Mangifera indica*		125
	Bombacaceae	*Adansonia digitata*	Fruit	1
	Caprifoliaceae	*Sambucus nigra*	Leaves	7
	Combretaceae	*Terminalia* spp.	Bark	9
	Cornaceae	*Cornus sericea*	Fruit pulp	117
	Ericaceae	*Vaccinium myrtillus*		48
	Fagaceae	*Fagus sylvatica*		55
	Hippocastanaceae	*Aesculus hippocastanum*	Leaves	7
	Juglandaceae	*Carya illinoensis*		32
	Myrtaceae	*Eucalyptus* spp.	Bark	114
	Pinaceae	*Larix decidua*	Cambial sap	26
		L. sibirica	Needles	90
	Punicaceae	*Punica granatum*	Bark	82
	Rhamnaceae	*Ceanothus americanus*		65
		Rhamnus spp.		103
		Ziziphus jujuba		49
	Rosaceae	*Prunus* spp.		27, 79
		Pyrus spp.		29, 83
		Rubus spp.		30, 125
	Rutaceae	*Citrus* spp.		75, 106, 112, 113
	Sapindaceae	*Litchia chinensis*	Leaves	96
	Saxifragaceae	*Ribes* spp.		13, 28, 125
	Sterculiaceae	*Sterculia platanifolia*		3
		Theobroma cacao		36, 125
	Theaceae	*Camellia sinensis*		36, 46, 104
	Vitaceae	*Vitis vinifera*	Leaves	22, 129
Tartaric acid	Aceraceae	*Acer saccharum*		80
	Bombacaceae	*Adansonia digitata*		1
	Caprifoliaceae	*Viburnum nudum*	Fruit	64
	Cornaceae	*Cornus* spp.		111
	Ericaceae	*Vaccinium corymbosum*		40
	Eucommiaceae	*Eucommia ulmoides*	Leaves	38
	Guttiferae	*Garcinia gambogia*	Fruit rind	59
	Leguminosae	*Cassia obtusifolia*		71
	Moraceae	*Morus indica*	Leaves	49
	Myrtaceae	*Psidium guajava*		105
	Polygonaceae	*Polygonum reynoutria*	Stem	119
	Rhamnaceae	*Ziziphus jujuba*	Leaves	2
	Rosaceae	*Prunus* spp.		107, 118
		Pyrus spp.		99, 118
		Rubus spp.		21

Table 6.1.3 (continued)

Acid	Family	Genus and species	Source	References
	Rutaceae	*Citrus decumana*	Fruit	74
	Sapindaceae	*Litchia chinensis*	Leaves	96
	Saxifragaceae	*Ribes* spp.		62, 100
	Vitaceae	*Vitis vinifera*		4, 22, 53, 129
Chelidonic acid	Hippocastanaceae	*Aesculus flava*	Leaves	98
		A. hippocastanum		97
	Rhamnaceae	*Paliurus aculeatus*	Pedicel	98
		Rhamnella franguloides	Leaves	98
		Rhamnus spp.		98
		Segeretia minutifolia	Leaves	98
		Zizyphus lotus		98
	Rubiaceae	*Uragoga ipecacuanha*		60
	Thymelaeaceae	*Daphne* spp.		98
		Gnidia spp.		98
		Passerina spp.		98
		Pimelea flava	Leaves	98
		Stellera chamaejasme	Leaves	98
		Thymelaea spp.		98
		Wikstroemia spp.		98
Fluoroacetic acid	Dichapetallaceae	*Dichapetalum cymosum*		70
		D. toxicarium	Seed	91
		D. barteri		86
	Euphorbiaceae	*Spondianthus preussii* var.*glaber*		51
	Leguminosae	*Acacia georginae*		87
		Gastrolobium grandiflorum		66, 67
	Rubiaceae	*Palicourea marcgravii*		89

6.1.4.2 Oxalic Acid

$HO_2C\text{-}CO_2H$

17 oxalic acid

Oxalic acid is formed by oxidation of glyoxylate in peroxisomes. Another pathway leading to oxalic acid is through oxidative cleavage of L-ascorbic acid (85). Oxalic acid is one of the most widely occurring organic acids in plants. It accumulates as potassium or calcium salt in the cells of plants, particularly in those of *Oxalis* and *Rumex*. In *Begonia*, this acid exists as the free acid form. Oxalic acid is generally regarded as a metabolic end product, and is not utilized further by plants.

6.1.4.3 Tartaric Acid

The naturally occurring tartaric acid has the L-configuration. Isolation of D-isomer or DL-forms have also been reported for some plants. Tartaric acid is widely

$$HO_2C-\overset{OH}{\underset{H}{C}}-\overset{H}{\underset{OH}{C}}-CO_2H$$

18 tartaric acid

distributed in plants, particulary in fruit, as a free acid and as a potassium or calcium salt (121).

Two independent pathways starting from L-ascorbic acid have been suggested for the biosynthesis of L-tartaric acid in *Vitis vinifera* and in *Pelargonium*, respectively (126, 127). In *V. vinifera*, ascorbic acid is successively converted to 5-keto-L-idonic acid (5k-IA) through 2-keto-L-idonic acid and L-idonic acid. Cleavage of 5k-IA yields L-tartaric acid (101). In *Pelargonium* (102), ascorbic acid is oxidatively cleaved to oxalic acid and L-threonic acid which gives tartaric acid. In this plant, another minor pathway leading to the formation of tartaric acid has been proved in which D-glucose is metabolized to 5k-IA via D-gluconic acid.

6.1.4.4 Chelidonic Acid

19 chelidonic acid

Chelidonic acid has been found in many species of Amarylidaceae, Liliaceae, and also in Papaveraceae – particularly *Chelidonium majus*. It has been proposed that chelidonic acid is biosynthesized from 3-phospho-glyceric acid and a C_4 unit such as erythrose-4-phosphate, which are products of the reductive photosynthetic pathway, in *C. majus* and *Narcissus incomparibilis* (69). It is clear that chelidonic acid is not derived from aromatic amino acids, from members of the TCA cycle, or from polyketide intermediates (49).

6.1.4.5 Fluoroacetic Acid

$$FH_2C-CO_2H$$

20 fluoroacetic acid

Fluoroacetic acid was identified as the toxic principle of *Dichapetalum cymosum*, which is one of the most poisonous plants growing in South Africa (70). This acid was subsequently found in many plants belonging to several families in Africa and Australia. The seeds of *D. toxicarium* contain, besides a small amount of fluoroacetic acid, ω-fluorooleic acid as the major toxic compound (91). The presence of ω-fluoropalmitic, ω-fluorocapric, and ω-fluoromyristic acids have also been reported (122). Biosynthesis of fluoroacetic acid has been proposed to occur by a nucleophilic incorporation of a fluoride ion into a C_3 unit linked to pyridoxal phosphate. Fluoroacetic acid may be formed from this fluoro-(C_3-pyridoxal phosphate)unit by (a) transamination followed by release of fluoropyruvic acid and its oxidative decarboxylation or (b) decarboxylation, transamination, and oxidation of fluoroacetaldehyde released by hydrolysis (73).

References

1 Airan J W, Desai R M 1954 Sugars and organic acids in *Adansonia digitata* (Linn.) fruit. Univ Bombay J 22:23–27

2 Akhmedov U A, Khalmatov Kh Kh 1968 Comparative phytochemical study of various strains of jujube. Polez Dikorastushchie Rast Uzb Ikh Ispol'z 154–158

3 Alexandrow W G, Timofeev A S 1926 Solution of calcium oxalate crystals in plants. Bot Arch 15:279–293

4 Amerine M A, Winkler A J 1942 Maturity studies with California grapes. II. The titratable acidity, pH, and organic acid content. Proc Am Soc Hortic Sci 40:313–324

5 Balansard J 1951 A study of the hepato-renal diuretics. V. The presence of glycolic acid in various drugs used as diuretics. Med Trop 11:638–639

6 Baqui S, Shah J J, Pandalai R C, Kothari I L 1979 Histochemical changes during transition from sapwood to heartwood in *Melia azedarach* L. Indian J Exp Biol 17:1032–1037

7 Bau A 1921 The oxalic acid content of early spring leaves, and some observations concerning this acid. Z Tech Biol 8:151–155

8 Bentley L E 1952 Occurrence of malonic acid in plants. Nature 170:847–848

9 Bhatia K, Lal J, Swaleh M 1981 Storage of *Terminalia* barks in the open: Its effect on oxalic acid and tannin content. Indian For 107:519–523

10 Bonner J, Varner J E (eds) 1976 Plant biochemistry, 3rd ed. Academic Press New York, 925 pp

11 Boudet A, Gadal P, Alibert G, Marigo G 1967 Biosynthesis and metabolism of aromatic compounds in higher plants. Demonstration of the alicyclic precursors: quinic, 5-dehydroquinic, shikimic, and 5-dehydroshikimic acids in *Quercus pedunculata*. C R Acad Sci 265:119–122

12 Brunner H, Chuard E 1886 Phytochemical studies. Ber Deutsch Chem Ges 19:595–622

13 Bryan J D 1947 The organic acids of some common fruits. Bristol Univ, Agr Hortic Res Sta Ann Rpt 138–140

14 Bryant F, Overell B T 1951 Displacement chromatography on ion exchange columns of the carboxylic acids in plant tissue extracts. Nature 167:361–362

15 Bryant F, Overell B T 1953 Quantitative chromatographic analysis of organic acids in plant tissue extracts. Biochem Biophys Acta 10:471–476

16 Carles J, Alquier-Bouffard A 1962 The organic acids of grape vines and their gradients. CR Acad Sci 254:925–927

17 Coic Y, Lesaint Ch, Le Roux F 1961 The transformation of *cis*-aconitic acid to the *trans* isomer during chromatography on silica. CR Acad Sci 253:1124–1126

18 Cranswick A M, Zabkiewicz J A 1979 Quantitative analysis of monosaccharides, cyclitols, sucrose, quinic and shikimic acids in *Pinus radiata* extracts on a glass support-coated open tubular capillary column by automated gas chromatography. J Chromatogr 171:233–242

19 Curl A L, Nelson E K 1943 The occurrence of citric and isocitric acid in blackberries and in dewberry hybrids. J Agr Res 67:301–303

20 Daito H, Tominaga S 1981 Constituents of sugars, organic acids and amino acids in the fruit juice at various locations within canopies of differenty trained Satsuma mandarin (*Citrus unshiu* cultivar Sugiyama) trees. J Jpn Soc Hortic Sci 50:143–156

21 Daughters M R 1918 The loganberry and the acid content of its juice. J Ind Eng Chem 10:30

22 Delmas J, Poitou N, Levadou B 1963 Chromatographic separation of the principal organic acids in grape leaves. Chim Anal (Paris) 45:63–65

23 Dogra J V V, Sinha S K P 1979 Observations on the age related changes in the level of α-ketoglutaric acid in leaves of *Phyllanthus simplex* (Retz.). Comp Physiol Ecol 4:35–37

24 Donaire J P, Sanchez-Raya A J, Lopez-Gorge J, Recalde L 1977 Physiological and biochemical studies on olives. I. Variations of the concentrations of some metabolites during the development cycle. Agrochimica 21:311–321
25 Eijkman J F 1885 *Illicium religiosum.* Rec Trav Chim Pays-Bas 4:32–54
26 Franz M, Meier H 1969 Organic acids in the cambial sap of *Larix decidua.* Planta 85:202–208
27 Franzen H, Helwert F 1922 The chemical constituents of green plants. XX. The acids of the cherry (*Prunus avium*). Hoppe-Seylers Z Physiol Chem 122:46–85
28 Franzen H, Helwert F 1922 The chemical constituents of green plants. XXII. The presence of succinic acid and of oxalic acid in the currant (*Ribes rubrum*). Hoppe-Seylers Z Physiol Chem 124:65–74
29 Franzen H, Helwert F 1923 Chemical constituents of green plants. XXV. Acids of the apple (*Pirus malus*). Hoppe-Seylers Z Physiol Chem 127:14–38
30 Franzen H, Keyssner E 1923 The chemical constituents of green plants. XXIX. Several water-soluble constituents of blackberry leaves (*Rubus fructicosus*). Hoppe-Seylers Z Physiol Chem 129:309–319
31 Franzen H, Stern E 1922 The chemical constituents of green plants. XIX. Occurrence of lactic and succinic acids in the leaves of raspberry (*Rubus idaeus*). Hoppe-Seylers Z Physiol Chem 121:195–220
32 Gallager R N, Jones J B 1976 Total, extractable and oxalate calcium and other elements in normal and mouse ear pecan tree tissues. J Am Soc Hortic Sci 101:692–696
33 Gibbs M, Latzko E (eds) 1979 Encyclopedia of plant physiology. New series 6. Photosynthesis II. Springer, Berlin Heidelberg New York Tokyo, 578 pp
34 Goodson J A 1920 Constituents of the leaves of *Helinus ovatus.* J Chem Soc 117:140–144
35 von Gorup-Besanez E 1872 Chemical composition of leaves of *Ampelopsis hederaceae.* Justus Liebigs Ann Chem 161:225–232
36 Goudswaard A 1934 Oxalate-containing plants and oxaluria. Pharm Weekbl 71:114–119
37 Grabovskis J, Kreicberga Z 1964 The distribution of quinic and shikimic acids in poplars. Lalvijas PSR Zinatnu Akad Vestis Kim Ser 113–117
38 Guseva A R 1952 Chemical composition of leaves of *Eucommia ulmoides.* Akad Nauk SSSR Dok 82:757–760
39 Haagen-Smit A J, Hirosawa F N, Wang T H 1949 Chemical studies on grapes and wines. I. Volatile constituents of zinfandel grapes (*Vitis vinifera* var. *zinfandel*). Food Res 14:472–480
40 Harris C H, Thrams W D 1916 The fruit of *Vaccinium corymbosum* (huckelberry). Chem News 114:73
41 Hattori S, Matsuda H 1964 Phenolic compounds in the leaves of *Rhus* and related plants in Japan. Symp Phytochem [Proc Meeting Univ Hong Kong 1961] 164–165
42 Hattori S, Yoshida S, Hasegawa M 1954 Occurrence of shikimic acid in the leaves of Gymnosperms. Physiol Plant 7:283–289
43 Hegnauer R 1964, 1966, 1969, 1973 Chemotaxonomie der Pflanzen, vol 3, 4, 5, 6. Birkhäuser Basel
44 Higuchi T, Fukazawa K, Nakashima S 1964 Mechanism of heartwood formation. I. Histochemistry of wood tissue. Mokuzai Gakkaishi 10:235–241
45 Holmes G W, Kurth E F 1961 The chemical composition of the newly formed inner bark of Douglas-fir. Tappi 44:893–898
46 Hoover A A, Karunairatnam M C 1945 Oxalate content of some leafy green vegetables and its relation to oxaluria and calcium utilization. Biochem J 39:237–238
47 Hulme A C 1954 Organic acids in the apple fruit. Cong Int Bot Paris, Raps Commun, Sect 11, 12, 8:394–398
48 Kaiser H 1925 Acids of the whortleberry and tamarind. Süddeut Apoth Ztg 65:48–49
49 Kalyankar G D, Krishnaswamy P R, Sreenivasaya M 1952 Papyrographic characterization and estimation of organic acids in plants. Current Sci (India) 21:220–222
50 Kamano Y, Tachi Y, Otake T, Sawada J 1977 Studies on the constituents of Quercus spp. VIII. Constituents of *Quercus stenophylla* Makino. Yakugaku Zasshi 97:1131–1133
51 Kamgue R T, Sylla O, Pousset J L, Laurens A, Brunet J C, Sere A 1979 Isolation and characterization of toxic principles from *Spondianthus preussii* var. *glaber* Engler. Plant Med Phytother 13:252–259
52 van Kampen G B 1927 Lactic acid in phanerogams. Biochem Z 187:180–182

53 Klein G, Werner O 1925 Micro- and histo-chemical demonstration of free and bound oxalic, succinic, malic, tartaric, and citric acids. Hoppe-Seylers Z Physiol Chem 143:141–153
54 Komatsu S, Nodzu R 1925 Chemistry of Japanese plants. VII. Phytochemical study of yamamomo fruit. Kyoto Univ Col Sci Mem, Ser A 8:223–229
55 Krauze S, Dziedzianowicz W 1959 Toxicity of beech seeds (*Fagus silvatica*). Nahrung 3:213–227
56 Krebs H A, Eggleston L V 1944 Microdetermination of isocitric acid and *cis*-aconitic acids in biological material. Biochem J 38:426–437
57 Krishna Das S, Markakis P, Bedford C L 1965 Nonvolatile acids of red tart cherries. Mich State Univ, Agr Expt Sta, Quart Bull 48:81–88
58 Krishna Reddy K, Sastry K N S, Rajadurai S, Nayudamma Y 1961 Studies in biosynthesis of tannins in indigenous plants. III. Polyols, plant acids, and amino acids of dhava. Bull Central Leather Res Inst Madras 8:64–66
59 Kuriyan K I, Pandya K C 1931 A note on the main constituents of the dried rind of the fruit of *Garcinia gambogia*. J Indian Chem Soc 8:469–470
60 Kwasniewski V 1953 Occurrence and detection of chelidonic acid in plants. Pharm Zh 92:5–7
61 Lehninger A L 1975 Biochemistry, 2nd ed. Worth New York, 1104 pp
62 Linder K, Jurics E W 1965 Direct densitometric methods for evaluating paper chromatograms of malic, tartaric, and citric acid of fruits. Z Lebensm-Unters Forsch 128:65–70
63 Linder W, Grill D 1978 Acids in conifer needles. Phyton (Horn, Australia) 18:137–144
64 Lott R H 1909 The fruit of *Viburnum nudum*. Chem News 99:169–171
65 Lynch T A, Miya T S, Jelleff Carr C 1958 Blood-coagulating principles from *Ceanothus americanus*. J Am Pharm Assoc 47:816–819
66 McEwan T 1964 Isolation and identification of the toxic principle of *Gastrolobium grandiflorum*. Nature 201:827
67 McEwan T 1964 Isolation and identification of the toxic principle of *Gastrolobium grandiflorum*. Queensland J Agr Sci 21:1–14
68 Madhusudanamma W, Sastry K N S, Rao V, Sundara S, Santappa M 1978 Studies on casuarina (*Casuarina equisetifolia*). I. Isolation and characterization of alicyclic acids, polyols and amino acids from the different parts of casuarina. Leather Sci 25:369–371
69 Malcom M J, Gear J R 1971 Biosynthesis of chelidonic acid. Can J Biochem 49:412–416
70 Marais J S C 1944 Monofluoracetic acid, the toxic principle of "gifblaar" *Dichapetalum cymosum* (Hook) Engl. Onderstepoort J Vet Sci Anim Ind 20:67–73
71 Matuura S, Yoshioka S, Iinuma M 1978 Studies on the constituents of the useful plants: VII. The constituents of the leaves of *Cassia obtusifolia* L. Yakugaku Zasshi 98:1288–1291
72 Mazelis M 1980 Amino acid catabolism. In: Miflin B J (ed) The biochemistry of plants. 5. Amino acids and derivatives. Academic Press New York, 541–567
73 Mead R J, Segal W 1972 Fluoroacetic acid biosynthesis. Proposed mechanism. Aust J Biol Sci 25:327–333
74 Menchikowsky F, Popper S 1932 Organic acids in Palestinian grapefruits. Hadar 5:181–183
75 Monselise S P, Galily D 1979 Organic acids in grapefruit (*Citrus paradisi*) fruit tissues. J Am Soc Hortic Sci 104:895–897
76 Mukherjee D 1976 Seed storage, germination and metabolic changes in *Albizzia lebbeck*. Physiol Sex Reprod Flowering Plants, Int Symp 300–305
77 Mukherjee D 1977 Sulfur amino acids and keto acids in leaves and flowers of *Albizzia lebbeck*. Plant Biochem J 4:34–38
78 Mukherjee D, Laloraya M M 1980 Changes in the levels of keto acids in different parts of seedlings of *Bauhinia purpurea* L. during growth. Plant Biochem J 7:120–125
79 Nelson E K 1924 Non-volatile acids of the dried apricot. J Am Chem Soc 46:2506–2507
80 Nelson E K 1928 Acids of maple sugar "sand". J Am Chem Soc 50:2028–2031
81 Nelson E K, Wheeler D H 1937 Constituents of the cannonball fruit (*Couroupita guianensis* Aubl.). J Am Chem Soc 59:2499–2500
82 Netolitzky F 1925 Something theoretical and practical about calcium oxalate. Chem Ztg 49:397, 80
83 Nithammer A 1930 Biochemistry and histochemistry of plant fruits and grains. I. Biochem Zbl 220:348–357
84 Nozzolillo C, Paul K B, Godin C 1971 The fate of L-phenylalanine fed to germinating pea seeds, *Pisum sativum* (L.) var. *Alaska*, during imbibition. Plant Physiol 47:119–123

85 Nuss R A, Loewus F A 1978 Further studies on oxalic acid biosynthesis in oxalate accumulating plants. Plant Physiol 61:590–592
86 Nwude N, Parsons L E, Adaudi A O 1977 Acute toxicity of the leaves and extracts of *Dichapetalum barteri* (Engl.) in mice, rabbits and goats. Toxicology 7:23–29
87 Oelrichs P B, McEwan T 1962 The toxic principle of *Acacia georginae*. Queensland J Agr Sci 19:1–16
88 Ohta T, Miyazaki T 1953 Isolation of fumaric acid from the seed of *Nandina domestica*. Tokyo Yakka Daigaku Kenkyu Nempu 7:104–105
89 de Oliveira M M 1963 Chromatographic isolation of monofluoroacetic acid from *Palicourea marcgravii*. Experientia 19:586–587
90 Osipov V I 1973 Organic acids of the tissues of the Siberian larch. Metabol Khvoinykh svyazi Period Ikh Rosta 125–133
91 Peters R A, Hall R J, Ward P F V 1960 Chemical nature of the toxic compounds containing fluorine in the seeds of *Dichapetalum toxicarium*. Biochem J 77:17–22
92 Plouvier V 1961 Quinic and shikimic acids, bergenine and heterosides in some Hamamelidaceae. CR Acad Sci 252:599–601
93 Plouvier V 1964 L-Inositol, L-quebrachitol, and D-pinitol in some botanical groups. The presence of shikimic acid in *Mammea americana*. CR Acad Sci 258:2921–2924
94 Plouvier V 1965 Heterosides of certain Saxifragaceae; deutzioside, a new compound isolated from *Deutzia*; presence of skimmine in Hydrangea. CR Acad Sci 261:4268–4271
95 Porter W L, Buch M L, Willits C O 1951 Maple syrup. III. Preliminary study of the non-volatile acid fraction. Food Res 16:338–341
96 Prasad U S, Jha O P, Mishra A 1979 Qualitative changes in sugars, free organic acids, free amino acids in leaves and fruit parts of *Litchia chinensis*. J Indian Bot Soc 58:114–119
97 Ramstad E 1945 Distribution of chelidonic acid in drugs. Pharm Acta Helv 20:145–154
98 Ramstad E 1953 The presence and distribution of chelidonic acid in some plant families. Pharm Acta Helv 28:45–57
99 Reed B B 1909 The fruit of *Pyrus arbutifolia*. Chem News 99:302–303
100 Sah K C, Gupta D R 1960 Papyrographic characterization of non-volatile organic acids in some medicinal plants of Naini Tal. Agra Univ J Res Sci 9:175–176
101 Saito K, Kasai Z 1984 Synthesis of L-(+)-tartaric acid from L-ascorbic acid via 5-keto-gluconic acid in grapes. Plant Physiol 76:170–174
102 Saito K, Kasai Z 1984 Synthesis of L-(+)-tartaric acid from 5-keto-gluconic acid in *Pelargonium*. Proc Ann Meeting Symp Jpn Soc Plant Physiol 71
103 Salgues R 1962 New chemical and toxicological studies on the genus *Rhamnus*. Qual Plant Mater Veg 9:15–32
104 Sanderson G W, Selvendran R R 1965 Organic acids in tea plants. Nonvolatile organic acids separated on silica gel. J Sci Food Agr 16:251–258
105 Santini R Jr 1953 Polybasic organic acids present in West Indian cherries and in three varieties of guava. Puerto Rico Univ J Agr 37:195–198
106 Sasson A, Monselise S P 1977 Organic acid composition of "Schamouti" oranges at harvest and during prolonged postharvest storage. J Am Soc Hortic Sci 102:331–336
107 Schenker H H, Rieman W 1953 Determination of malic and citric acids in fruit by ion exchange chromatography. Anal Chem 25:1637–1639
108 Schindelmeiser J 1907 The fruit of *Cornus mas*. Apoth Ztg 22:482
109 Schneider A 1939 The occurrence of lactic acid in higher plants. Planta 29:747–749
110 Schneider A 1941 The formation of lactic acid in higher plants; especially during germination. Planta 32:234
111 Sheets G 1911 The fruit of *Cornus paniculatum*. Chem News 103:172–173
112 Sinclair W B, Eny D M 1947 Ether-soluble organic acids of mature valencia orange leaves. Plant Physiol 22:257–269
113 Sinclair W B, Eny D M 1947 Ether-soluble organic acids and buffer properties of citrus peels. Bot Gaz 108:398–407
114 Smith H G 1905 Occurrence of calcium oxalate in the bark of the *Eucalyptus*. Proc R Soc N S Wales 39:23–32
115 Stankova N V, Smirnova T A 1975 Phenolcarboxylic acids of grape roots. Khim Prir Soedin 22:508–509

116 Stewart C R, Beevers H 1967 Gluconeogenesis from amino acids in germinating castor bean endosperm and its role in transport to the embryo. Plant Physiol 42:1587–1595
117 Stockton E, Eldredge C G 1908 Fruits of *Caulophyllum thalictroides* and *Cornus sericea*. Chem News 98:190–191
118 Swaby R J 1943 Extraction of citrates and tartarates from fruits. Agr Gaz N S Wales 54:571–573
119 Takemoto T, Koike H 1953 Acid components of *Polygonum reynoutria*. Yakugaku Zasshi 73:100
120 Trebst A, Avron M (eds) 1977 Encyclopedia of plant physiology. New series 5. Photosynthesis I. Springer, Berlin Heidelberg New York Tokyo
121 Wang D 1973 Nonvolatile organic acids. In: Miller L P (ed) Phytochemistry III. Van Nostrand Reinhold New York, 74–111
122 Ward P F V, Hall R J, Peters R A 1964 Fluoro fatty acids in the seeds of *Dichapetalum toxicarium*. Nature 201:611–612
123 Warren W H 1911 "Sugar sand" from maple sap. Source of malic acid. J Am Chem Soc 33:1205–1211
124 Weiss U, Edwards J M 1980 In: The biosynthesis of aromatic compounds. John Wiley New York, 40–62
125 Widmark E M P, Ahldin G 1933 Oxalic acid content of vegetable foodstuffs. Biochem Z 265:241–244
126 Williams M, Loewus F A 1978 Biosynthesis of (+)-tartaric acid from L-[4-^{14}C]ascorbic acid in grape and *Geranium*. Plant Physiol 61:672–674
127 Williams M, Saito K, Loewus F A 1979 Ascorbic acid metabolism in geranium and grape. Phytochemistry 18:953–956
128 Wilson C W, Shaw P E, Campbell C W 1982 Determination of organic acids and sugars in guava (*Pisidium guajava*) cultivars by high-performance liquid chromatography. J Sci Food Agr 33:777–780
129 Wormall A 1924 Constituents of the sap of the vine (*Vitis vinifera* L.). Biochem J 18:1187–1202
130 Wu T-S 1977 Constituents of *Cinnamomum osmophloeum* Kanehira. Tai-wan Yao Hsueh Tsa Chih 29:15–18
131 Yamashita T, Sato F 1930 Shikimic acid from the leaves of *Ginkgo biloba*. Yakugaku Zasshi 50:113–117

6.2 Complex Aliphatic and Alicyclic Extractives

Y. OHTA

6.2.1 Introduction

This section is a medley of rather small groups of compounds with diverse chemical nature found in wood extractives. The occurrence of each group of compounds is restricted to a few closely related genera of plants. For example, γ-lactones are mainly found in Lauraceae and cyclopentenyl cyanogenic glycosides are included in Passifloraceae and related families. Most of these compounds have no particular metabolic or physiological function in the plants except for cyanogenic glycosides. Cyanogenic glycosides which liberate hydrogen cyanide enzymically are important to humans and livestock, since a number of important food plants contain these toxic compounds. The function of the cyanogens can be considered as a defense substance, usually against herbivores.

These compounds will be described in the order of γ- and δ-lactones, cyanogenic glycosides, cyclohexane epoxides, cyclohexane diols, polycyclic compounds, and miscellaneous.

6.2.2 γ-Lactones

Aliphatic lactones are important as aroma components in foods and flavors. Among them, γ-lactones preferentially occur in plants and δ-lactones are mainly found in animal products. The details on naturally-occurring lactones as aroma components are described in the literature (87). It is interesting that quercus lactone-a (**1**) and -b (**2**), two diastereomeric branched nonlactones, have been detected as ingredients in aged whiskeys (88, 134). These lactones were isolated from woods of some *Quercus* species which are widely used to make barrels for aging spirits such as whisky and brandy (72, 73).

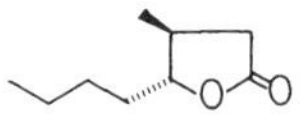

1 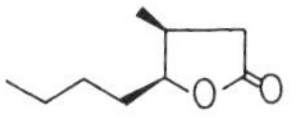 2

In 1972, Takeda et al. isolated six exocyclic α,β'-unsaturated γ-lactones from the roots of *Litsea japonica* (138) (Table 6.2.1). These lactones, named litsenolide

Table 6.2.1. Litsenolides from *Litsea japonica* (138)

Compound	R^1	R^2	mp (°C)	$[\alpha]_D$	Yield (%)
Litsenolide A_1 (**3**)	H	$(CH_2)_9CH=CH_2$	oil	−2.4	0.25
Litsenolide A_2 (**4**)	$(CH_2)_9CH=CH_2$	H	oil	−40.4	0.40
Litsenolide B_1 (**5**)	H	$(CH_2)_9C\equiv CH$	40–41	−5.9	0.15
Litsenolide B_2 (**6**)	$(CH_2)_9C\equiv CH$	H	42–43	−44.9	0.27
Litsenolide C_1 (**7**)	H	$(CH_2)_{12}CH_3$	60–62	−9.4	0.03
Litsenolide C_2 (**8**)	$(CH_2)_{12}CH_3$	H	44–45	−45.2	0.10

A_1, A_2, B_1, B_2, C_1, C_2 (**3**–**8**), are three pairs of 2-alkylidene-3-hydroxy-4-methyl butanolides. The alkyl side chains are terminated with an ethyl (C-series), ethynyl (A-series), or ethynyl group (B-series). The two components in each pair are geometrical *cis*- and *trans*-isomers at the double bond conjugated with the lactone ring. It should be remembered that the carbon numbers of litsenolide C_1 and C_2 are two units larger than those of litsenolide A_1, A_2, B_1 and B_2.

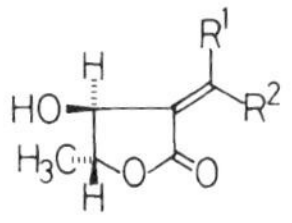

HO H R^1 R^2 H_3C H O O

3-8

A large group of γ-lactones with a similar structural relationship to litsenolides were found in the trunk wood of *Clinostemon mahuba* (70, 71). Those are seven dihydromahuba lactones of series A (**9**–**15**) (Table 6.2.2), seven dihydromahuba lactones of series B (**16**–**22**) (Table 6.2.3), and eight mahuba lactones (**23**–**30**) (Table 6.2.4).

Table 6.2.2. Dihydromahuba lactones of series A (70, 71)

Compound	R^1	R^2	mp (°C)	$[\alpha]_D$	Yield (%)
Dihydromahubenolide A (**9**)	H	$(CH_2)_{13}CH=CH_2$	48 – 50	+8.5°	0.6
Isodihydromahubenolide A (**10**)	$(CH_2)_{13}CH=CH_2$	H	45 – 46	+37.1°	3.3
Dihydromahubynolide A (**11**)	H	$(CH_2)_{13}C{\equiv}CH$	61 – 63	+9.0°	0.6
Isodihydromahubynolide A (**12**)	$(CH_2)_{13}C{\equiv}CH$	H	64 – 65	+42.4°	4.6
Dihydromahubanolide A[a] (**13**)	H	$(CH_2)_{14}CH_3$	–	–	0.1
Isodihydromahubanolide A (**14**)	$(CH_2)_{14}CH_3$	H	49 – 50	+36.6°	0.6
Isodihydromahubanolide-23 A (**15**)	–	–	–	–	0.1

[a] Stereochemistry at C-3 and C-4 is not known.

Table 6.2.3. Dihydromahuba lactones of series B (70, 71)

Compound	R^1	R^2	mp (°C)	$[\alpha]_D$	Yield (%)
Dihydromahubenolide B (**16**)	H	$(CH_2)_{13}CH=CH_2$	46 – 47	−35.1°	0.5
Isodihydromahubenolide B (**17**)	$(CH_2)_{13}CH=CH_2$	H	66 – 67	−96.5°	1.9
Dihydromahubynolide B (**18**)	H	$(CH_2)_{13}C{\equiv}CH$	55 – 56	−32.6°	0.2
Isodihydromahubynolide B (**19**)	$(CH_2)_{13}C{\equiv}CH$	H	79 – 80	−90.2°	1.1
Dihydromahubanolide B[a] (**20**)	H	$(CH_2)_{14}CH_3$	–	–	0.1
Isodihydromahubanolide B (**21**)	$(CH_2)_{14}CH_3$	H	70 – 71	−93.3°	0.4
Isodihydromahubanolide-23 B (**22**)	–	–	–	–	0.1

[a] Stereochemistry at C-3 and C-4 is not known.

Table 6.2.4. Mahuba lactones (70, 71)

Compound	R^1	R^2	mp (°C)	$[\alpha]_D$	Yield (%)
Mahubenolide (**23**)	H	$(CH_2)_{13}CH = CH_2$	oil	+6.2°	1.1
Isomahubenolide (**24**)	$(CH_2)_{13}CH = CH_2$	H	oil	+22.0°	11.6
Mahubynolide (**25**)	H	$(CH_2)_{13}C{\equiv}CH$	oil	+8.0°	0.7
Isomahubynolide (**26**)	$(CH_2)_{13}C{\equiv}CH$	H	oil	+23.1°	3.5
Mahubanolide[a] (**27**)	H	$(CH_2)_{14}CH_3$	–	–	0.1
Isomahubanolide[a] (**28**)	$(CH_2)_{14}CH_3$	H	–	–	0.1
Mahubanolide-23[a] (**29**)	H	$(CH_2)_{16}CH_3$	–	–	0.1
Isomahubanolide-23 (**30**)	$(CH_2)_{16}CH_3$	H	oil	+18.9°	2.1

[a] Stereochemistry at C-3 is not known.

Table 6.2.5. Obtusilactones in *Lindera obtusiloba* (83, 84)

Compound	R^1	R^2	mp (°C)	$[\alpha]_D$	Yield (%)
Obtusilactone (**31**)	H	$(CH_2)_9CH = CH$	oil	−53°	0.094
Isoobtusilactone (**32**)	$(CH_2)_9CH = CH_2$	H	oil	−56°	0.002
Obtusilactone A (**33**)	H	$(CH_2)_{12}CH_3$	oil	−46°	0.017
Isoobtusilactone A (**34**)	$(CH_2)_{12}CH_3$	H	oil	−54°	0.007
Obtusilactone B (**35**)	H	$(CH_2)_5CH = CH(CH_2)_7CH_3$	oil	–	0.001
Isoobtusilactone B (**36**)	$(CH_2)_5CH = CH(CH_2)_7CH_3$	H	oil	–	0.002

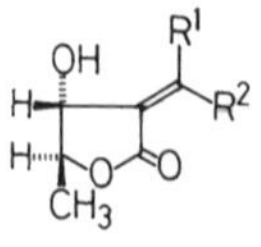

9-15 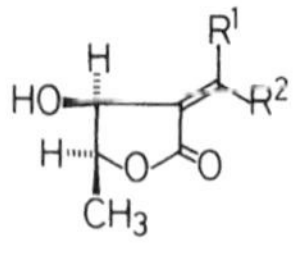 16-22 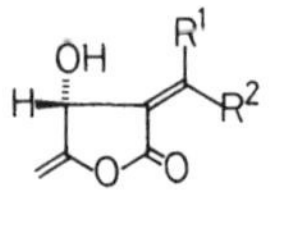 23-30

The relationship of the β,γ-substituents in dihydromahuba lactones of series A was determined to be *threo* and that in the series B as *erythro* by the LIS ^{1}H-NMR techniques (71). Mahuba lactones carry an exomethylene group at C-5 of the butanolide ring. The long alkyl side chains in these lactones are in the same alkane-alkene-alkyne-type relationship as in litsenolides. Isodihydromahubanolide 23-A **(15)** and -B **(22)**, mahubanolide-23 **(29)**, and isomahubanolide-23 **(30)** are larger than the rest by two carbon units.

Obtusilactones **(31–34)** isolated from leaves of *Lindera obtusiloba* (Lauraceae) are also isomeric butanolides with alkane-alkene-type side chains (83, 84) (Table 6.2.5). Obtusilactone-B **(35)** and isoobtusilactone-B **(36)** with an endo-olefinic alkyl substituent are geometrical isomers at the trisubstituted double bond conjugated with the carbonyl group of the lactone ring.

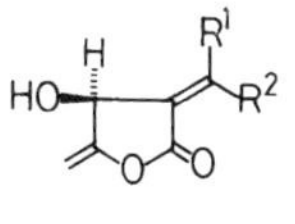

31-36

L. obtusiloba also contains two lactone dimers, C_{17}-obtusilactone dimer **(37)** (mp 57–58 °C, $[\alpha]_D$ +49.3°) and C_{19}-obtusilactone dimer **(38)** (mp 62–63 °C, $[\alpha]_D$ +42.7°) (85, 86). These lactones, found in *L. obtusiloba*, are reported to be cytotoxic (85, 86). Again, C_n compounds with an unsaturated terminals **(31, 32)** co-occur with C_{n+2} lactones having alkyl side chain terminated with ethyl group **(33, 34)**.

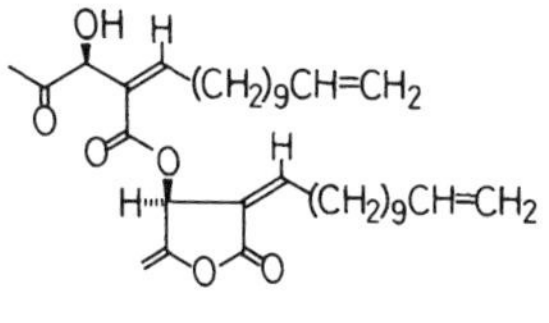

37 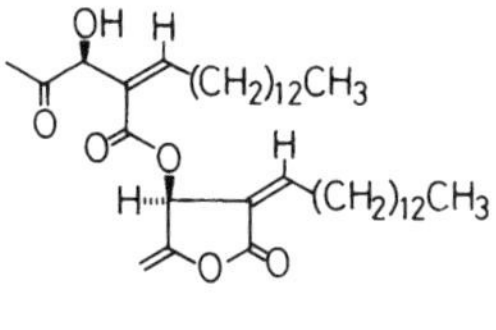 38

Rubrenolide **(39)** (mp 100 °C, $[\alpha]_D$ +21°) and rubrynolide **(40)** (mp 88 °C, $[\alpha]_D$ +21°) from *Nectandra rubra* (Lauraceae) are another type of γ-lactones of alkene-alkyne pair (39). In these compounds, the alkene or alkyne substituent is attached to C-4 of the butanolide ring.

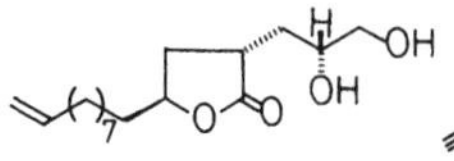

39 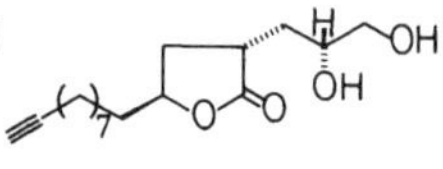 40

Trunk wood of *Mezilaurus synandra*, one of the species of *Mezilaurus* belonging to the same subtribe as *Clinostemon*, was shown to contain a similar γ-lactone, (2*R*,3*R*,4*S*)-2-dodec-ω-enyl-3-hydroxy-4-methylbutanolide (**41**) (mp 72–76 °C, $[\alpha]_D$ −7.5°) (115). The butenolide (**42**) from the same source is considered to be an artificial dehydration product of the 3-hydroxy-lactone (**41**). All of the above aliphatic γ-lactones were isolated from Lauraceae; 2-hydroxy-4-(heptadec-8′-enyl)butanolide (**43**) obtained as an oil from the bark of *Garcinia mannii* (Guttiferae) is the only compound of this type found in other families (53).

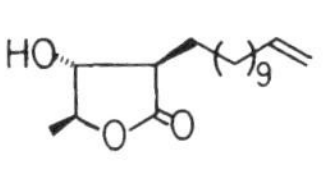

41 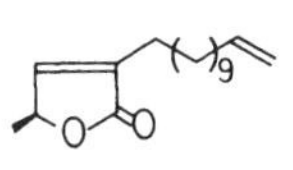 42 $H_3C(CH_2)_7CH \overset{Z}{=} CH(CH_2)_7$ 43

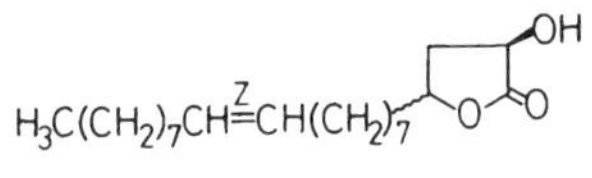

The biogenesis of these γ-lactones can be explained reasonably according to Fig. 6.2.1 (13, 15).

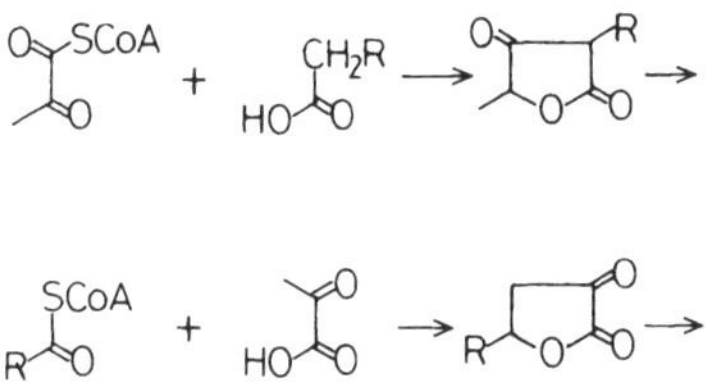

Fig. 6.2.1. Biogenesis of γ-lactones

This scheme involves the interaction of α-methylene of fatty acid and carboxyl of pyruvate, which will lead to compounds with a tetronic acid nucleus and also to litsenolides, mahubalactones, obtusilactones, and rubre(y)nolides. The interaction of the carboxyl of fatty acid and methyl of pyruvate, on the other hand, will afford the lactone (**43**) in *G. mannii.* Gottlieb has postulated another candidate for the biogenesis of these lactones, particularly to explain the formation of rubre(y)nolides of which carbon skeletons show ramification (47). In his scheme, the extrusion of a C-atom from a normal polyketide precursor is reasonably explained through a Favorsky-type rearrangement. However, the former pathway explains more satisfactorily the co-occurrence of C_n compounds carrying side chains terminated with an ethenyl or ethynyl group and C_{n+2} compounds with ethyl terminals as observed in litsenolides, mahubalactones, obtusilactones, or

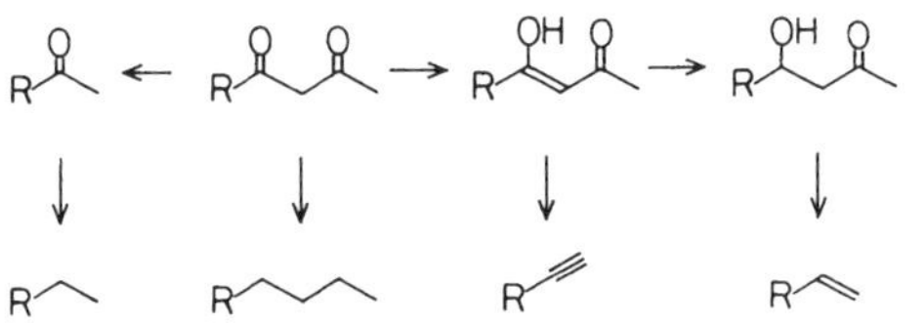

Fig. 6.2.2. Biogenesis of terminal ethyl, ethenyl, or ethynyl groups

rubre(y)nolides (15). The formation of terminal ethyl, ethenyl, or ethynyl groups in the side chain of these lactones is illustrated in Fig. 6.2.2 (15).

6.2.3 δ-Lactones (2-Pyrones)

A number of 2-pyrone derivatives with aromatic substituents are widely distributed in several plant taxa, such as Lauraceae, Annonaceae, Leguminosae, or Piperaceae. Compounds of chemotaxonomic importance are described in the literature (47) and elsewhere in this book. In contrast, aliphatic δ-lactones are rather uncommon in plants and only a limited number of them have been isolated from woody plants.

The simplest example of this type of compound, parasorbic acid (**44**), has long been known to occur only in the fruits of *Sorbus aucuparia* (Rosaceae) (30) before Cardellina and Meinwald isolated it from *Vaccinium macrocarpon* (Ericaceae) in 1980 (21). Parasorbic acid is known to inhibit fungal growth (16) and seed germination (17, 76). This may be responsible for the allelochemical nature of *V. macrocarpon* (21).

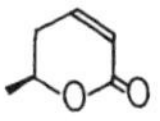

44

Massoilactone (**45**) is the major component of the bark oil of *Cryptocarya massoia* (Lauraceae) (1, 74). The absolute configuration of (−)-massoilactone is (R) at C-5, which is enantiomeric to parasorbic acid (29, 78). An analogous lactone (**46**) was also isolated from the same plant (22). Massoi bark has been used for many centuries as a constituent of native medicines. The sedative action expected from the structural similarity between massoilactone and active Kawa-type lactones is dubious (9). (−)-Massoilactone has been isolated from cane molasses as one of, and the most characteristic, "sugary-flavor" substances (48). This compound was also obtained from two species of formicine ants of the genus *Camponotus* (22). This powerful skin-irritating compound might be responsible for the defensive factor secreted by the ants.

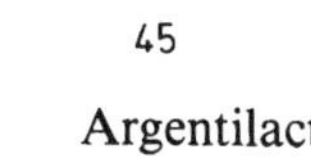

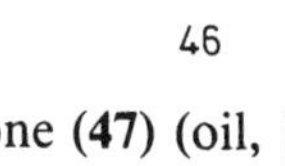

n-C_5H_{11} 45 n-C_7H_{15} 46

Argentilactone (**47**) (oil, $[\alpha]_D$ −21.1°), also a skin irritating compound, was obtained from the rhizomes of *Aristolochia argentina* (Aristolochiaceae) in 2.4% yield (99). Some species of Labiatae contain closely related δ-lactones, which have oxygenated substituents in the side chains. They are hyptolide (**48**) (mp 88.5 °C, $[\alpha]_D$ +7.4°) in *Hyptis pectinata* (11), boronolide (**49**) (mp 90 °C, $[\alpha]_d$ +25°), in *Tetradenia fruticosa* (40), and umuravumbolide (**50**), ($[\alpha]_D$ 0°), deacetylumuravumbolide (**51**) (mp 45.5 °C, $[\alpha]_D$ 0°), deacetylboronolide (**52**) (mp 105.5 °C, $[\alpha]_D$ 0°) and 1′,2′-dideacetylboronolide (**53**) (mp 80.5 °C), in *Iboza riparia* (100,

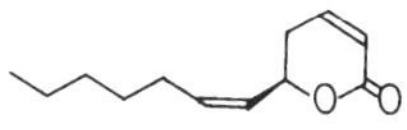

47 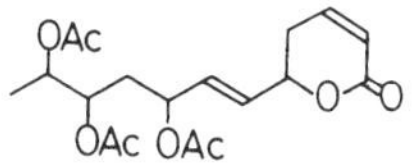 48 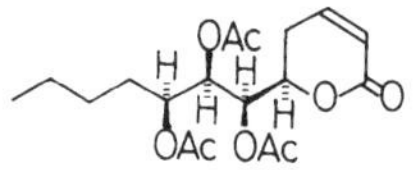 49 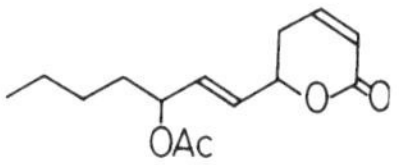 50

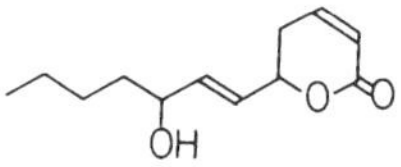

51 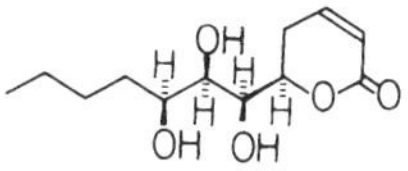 52 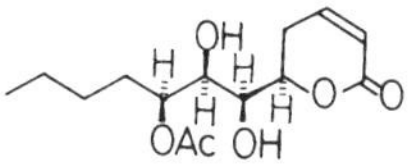 53

101). The absolute stereochemistry has only been confirmed for boronolide as 6*R*-(1*R*,2*R*,3*R*)-5,6-dihydro-6-[1,2,3-*tris*(acetoxy)-heptyl]-2*H*-pyran-2-one (62).

Conrauana lactone (**54**), 4-hydroxy-6-pentadecyl-2-pyrone, is one of the major secondary metabolites in the bark of *Garcinia conrauana* (Guttiferae) (145). According to Hussain and Waterman (53), the other three species of *Garcinia* growing in a forest reserve of West Cameroun contain xanthones, biflavonoids, and benzophenones as characteristic metabolites. In contrast, *G. conrauana* is atypical in producing large amounts of conrauana lactone. Another compound, 3-(3″,3″-dimethylallyl)-conrauana lactone (**55**) was also isolated from the stem bark of *G. conrauana* (53). *G. mannii* contained γ-lactone (**43**) as a minor component as described above.

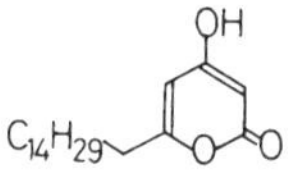

54 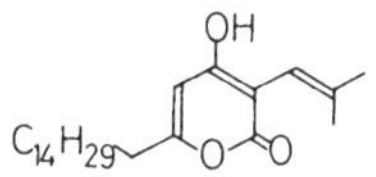 55

Altholactone (**56**), found in the bark of unnamed *Polyalthia* (Annonaceae), has a unique tetrahydrofuro-pyrone structure (69). This compound is biogenetically related to a number of α-pyrones with aromatic substituents, found in families of Lauraceae, Annonaceae, or Piperaceae. It can reasonably be regarded as an oxygenated and cyclized derivative of goniothalamin (**57**), distributed in several species of *Goniothalamus* (Annonaceae) (56, 69).

Elenolide (**58**) is another example of a δ-lactone with uncommon structure. This compound was isolated from the bark, leaves, and fruit of the olive tree, *Olea europa* (10).

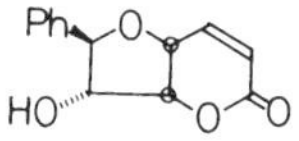

56 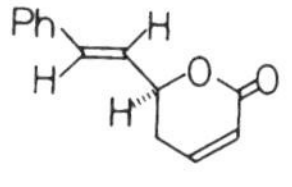 57 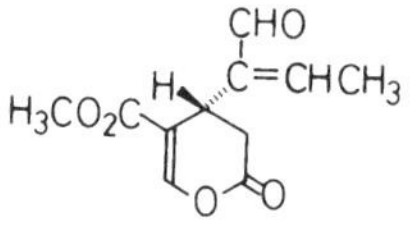 58

6.2.4 Cyanogenic Glycosides and Related Compounds

Many plants accumulate compounds capable of liberating hydrogen cyanide upon acid or enzymatic hydrolysis. Two types of compounds, cyanogenic glycosides and cyanolipids, are responsible for this cyanogenetic ability. Cyanolipids have been isolated mainly from seed oils of the family Sapindaceae (109). At least 800 species of plants representing 70 to 80 families are known to contain cyanogenic glycosides, and about 30 such compounds have been isolated for structure determination (36, 110). Such cyanogenetic substances have been recognized in a number of important food plants, such as apricots (*Prunus armeniaca*), peaches (*P. persica*), almonds (*P. amygdalus*), or cassava (*Manihot* spp.) (110). Generally, the fruit, seeds, or tubers of these plants have long been known to be poisonous for man and livestock. However, it was found rather recently that all parts of the plants including leaves, stems, and bark are also cyanogenic in some species, particularly the members of Passifloraceae or Sapindaceae (45, 112, 122, 123). Among these cyanogenic glucosides, about one-third contain an aromatic ring in the molecule. The occurrence and chemical natures of the aromatic cyanoglycosides are described in reviews (e.g., 36, 110).

Linamarin (**59**) has been known as the principal cyanogenic glucoside of manioc (*Manihot esculenta*, Euphorbiaceae) and flax (*Linum usitatissimum*) seedlings (24). Until recently, the nature of the glucose linkage had not been rigorously established, but Clapp et al. determined the linkage to be β by synthesizing both α- and β-anomers (24). Lotaustralin (**60**) frequently co-occurs with linamarin in Compositae, Leguminosae, Euphorbiaceae, Linaceae, and Papaveraceae (110). Its content in *Manihot* is generally much smaller compared to linamarin; for example, the ratio of the two is 4:96 in *M. cartheginensis* (18) and 7:93 in *M. utilissima* (82), whereas that in the green tissues of *Linum usitatissimum* is approximately 1:1 (19).

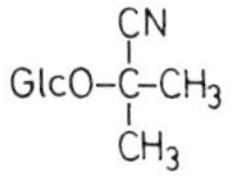

GlcO–C(CN)(CH$_3$)–CH$_3$ GlcO–C(CN)(CH$_3$)–C$_2$H$_5$

59 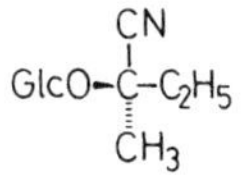   60

Cardiospermin (**61**) was isolated from the vegetative portion of the woody vine *Cardiospermum hirsutum* (Sapindaceae) (112). Later, cardiospermin-5-sulfate (**62**) was obtained from the same plant as the first example of sulfur containing cyanogenic glycoside (52). Acacipetalin (**63**) was found in two South African species of Acacia (*A. sieberiana* var. *woodii* and *A. hebeclada* (Leguminosae) (35, 102, 133). The structure of acacipetalin (**63**) originally assigned by Rimington (102) had been incorrectly revised as 2-(β-D-glucopyranosyloxy)-3-methyl-

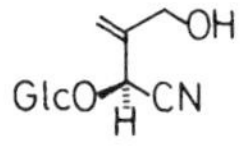

GlcO, CN, H, OH   GlcO, CN, H, O–SO$_3^-$

61 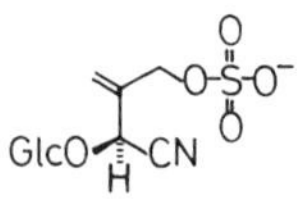   62

GlcO, CN   GlcO, H, CN   NC, H, OGlc

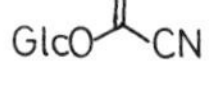

63 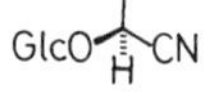   64 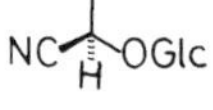   65

but-3-enenitrile (20), but in 1977, Ettlinger et al. reconfirmed the former structure for acacipetalin (35). Furthermore, they isolated another cyanogenic constituent from these two species of *Acacia* and named it proacacipetalin (**64**). It is the (*S*)-isomer of 2-(β-D-glucosyl)-3-methylbut-3-enenitrile (which was previously incorrectly assigned to acacipetalin except for stereochemistry). Proacacipetalin rearranges to acacipetalin under relatively strong basic conditions as a result of deprotonation at α-position and slower reprotonation at γ-position (35). The (*R*)-isomer of proacacipetalin, epiproacacipetalin (**65**) described by Ettlinger et al. (35), was later detected in *A. globulifera* (111).

Linustatin (**66**) and neolinustatin (**67**) are disaccharide derivatives extracted from seeds of *Linum usitatissimum* (119). Their structural features are closely related to linamarin and lotaustralin, respectively.

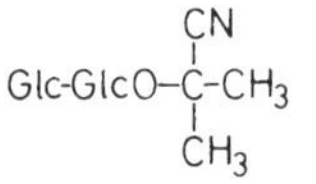

Glc-GlcO–C(CN)(CH$_3$)–CH$_3$ Glc-GlcO–C(CN)(CH$_3$)–C$_2$H$_5$

66 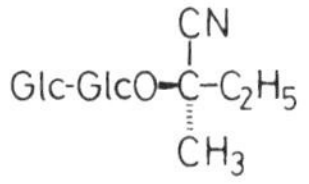   67

Biosynthesis of linamarin and lotaustralin has been studied by Conn et al. (19, 149). On the basis of highly effective incorporation of ^{14}C-labelled L-valine into linamarin and of L-isoleucine into lotaustralin, they concluded that these amino acids are precursors of the two cyanoglucosides. Since the glucosyl transferase (UDP-glucose:ketone cyanohydrin-β-D-glucosyltransferase) was shown to be unable to distinguish between the two possible enantiomers – (*R*,*S*)-2-hydroxy-2-methyl-butyro-nitriles – L-isoleucine ([3*S*]-L$_S$-isoleucine), which supplies the *R*-enantiomer, must be the sole precursor for naturally occurring (*R*)-lotaustralin (149). It is quite possible that L-leucine is the precursor of acacipetalin, proacacipetalin, epiproacacipetalin, and cardiospermin.

Another type of cyanogenic glycoside contains a cyclopentene ring in the molecule. To date, 15 cyclopentenoid cyanogens have been isolated and identified. Gynocardin (**68**) was first isolated in 1904 from the seed of *Gynocardia odorata* (Flacourtiaceae) (98). Its structure was determined in 1966 as O-β-D-glucosyl-1-cyano-4,5-dihydroxycyclopent-2-ene (27), and its stereochemistry was established later by X-ray crystallography (61). Gynocardin is typically found in members of Flacourtiaceae (27, 61, 131). While this compound has been reported from several species of Passifloraceae (*P. adenopoda*, *P. allardii*, *P. caelurea*, *P. suberosa*, and *Adenia lobata*) (139), later work has proved that *P. incarnata* is the only species of Passifloraceae that contains gynocardin (37, 113, 124). The occurrence of epigynocardin (**69**) is also described for *Pangium edule* (Flacourtiaceae) (129).

Deidaclin (**70**), first called daidamin, was obtained from *Deidamia clematoides* (Passifloraceae) (139). The structure of deidaclin was established as

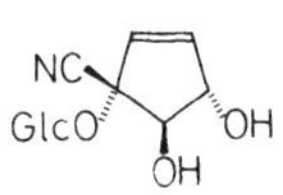

68 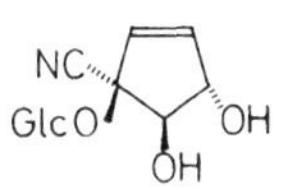   69    70 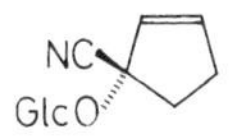   71 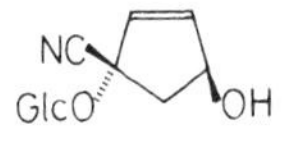   72

the β-D-glucopyranoside of an enantiomer of 2-cyclopenten-1-one cyanohydrin by identification of the aglucone obtained upon hydrolysis with gynocardase (25). Since then, deidaclin has been detected in several species of Passifloraceae (113, 130) as well as in *Turnera ulmifolia* (Turneraceae) (121). The occurrence of this cyanoglucoside in *T. ulmifolia* provided chemotaxonomic evidence in relating Turneraceae with Passifloraceae and Flacourtiaceae (121).

Tetraphyllin A (**71**) (mp 116° to 118 °C, $[\alpha]_D$ −14.0°) was first isolated in a crystalline form from *Tetrapathaea tetrandra* (Passifloraceae) (104). The structure of this compound, determined from spectroscopic evidence derived from itself and its derivatives, was similar to that of deidaclin (**70**). These two cyanogens were shown to be epimers to each other by NMR spectral studies (110, 113). The stereochemistry at the chiral centers of these compounds was determined by X-ray chrystallography to be (*R*) in deidaclin and (*S*) in tetraphyllin A, respectively (54).

Tetraphyllin B (**72**) seems to be the most widely distributed cyclopentenoid glucoside in the family of Passifloraceae (45, 104, 113, 123). Tetraphyllin B (mp 169–170 °C, $[\alpha]_D$ −35.6°) was first isolated from *Tetrapathaea tetrandra* together with tetraphyllin A (104). Its absolute configuration was determined by X-ray diffraction and enzymatic hydrolysis to be (1*S*,4*S*)-1-cyano-4-hydroxypent-2-en-β-D-glucoside (42). Isolation of epitetraphyllin B (**73**) was first reported from *Adenia volkensii*, an African shrub used to prepare arrow poisons, and its structure was proposed to be epimeric with tetraphyllin B at the chiral carbon bearing the nitrile and glucose moieties and not at the allylic hydroxyl group (45). However, a recent work has proved that epi-tetraphyllin B differs from tetraphyllin B at both asymmetric centers of the aglucone (55). The difference in configuration of the cyanohydrin carbon is apparent from the NMR spectroscopic evidence. Hydroxyenones obtained from both cyanogenic glucosides upon hydrolysis with β-glucosidase exhibited opposite CD Cotton effects to each other. Thus, these two cyanogens are glucosides of enantiomeric, not epimeric, cyclopentene cyanohydrins and the authors proposed a new trivial name, volkenin (**74**), for epitetraphyllin B (55).

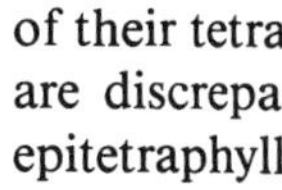

NC GlcO OH — 73

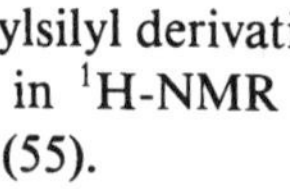

NC GlcO OH — 74

Barterin, another cyanogenic glucoside isolated from *Barteria fistulosa* (Passifloraceae) (89) and thought to be identical with tetraphyllin B (113), was proved to be identical with epitetraphyllin B by comparison of ^{1}H-NMR spectra of their tetramethylsilyl derivatives (125). However, this is still obscure, since there are discrepancies in ^{1}H-NMR spectra and rotation data between barterin and epitetraphyllin B (55).

Passibiflorin (**75**) and epipassibiflorin (**76**) are another epimeric pair of cyclopentenoid cyanogenic glycosides isolated from *P. biflora* and *P. talamancensis* (127). The sugar moieties of these compounds are β-(1→6)-rhamnosyl-glucose as identified by HPLC and TLC. Passitrifasciatin (**77**), isolated from *P. trifascia*, is also a rhamnoglucoside of 4-hydroxycyclopent-2-en-1-nitrile except that the glycoside linkage is β-(1→4) (127).

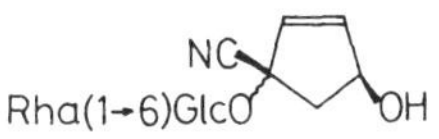
NC
Rha(1→6)GlcO
OH

75 passibiflorin

76 epipassibiflorin

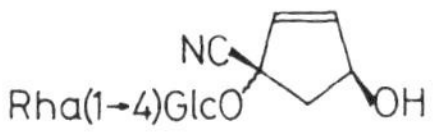
NC
Rha(1→4)GlcO
OH

77 pessitrifasciatin

Very polar cyanogenic compounds, tetraphyllin B-4-sulfate (**78**) and epitetraphyllin B-4-sulfate (**79**), were isolated as a mixture from *P. caerulea* and *P. alto-caerulea* (113) and later from some other members of Passifloraceae (128). Although these two sulfated cyanogens have been described as epimers at the chiral centers bearing nitrile and glucose moieties, as are tetraphyllin B and epitetraphyllin B (113), the stereochemistry of these compounds should be reexamined, since the latter pair of compounds has been proved to be in the enantiomeric relationship (55). A similar sulfated cyclopentenoid cyanogenic glycoside, passicoccin (**80**), is included in *P. coccinea* (128). The glycosyl moiety of passicoccin is 6-O-β-D-rhamnopyranosyl-β-D-glucopyranose. The stereochemistry of this cyanogen has not been elucidated and passococcin may exist as a mixture of its epimer, epipassicoccin (**81**), as are tetraphyllin B sulfates (128).

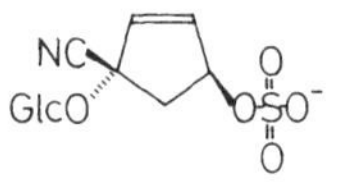
NC
GlcO
O
OSO⁻
O

78

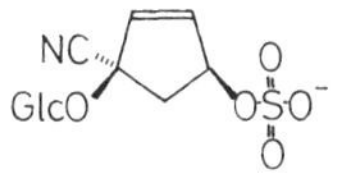
NC
GlcO
O
OSO⁻
O

79

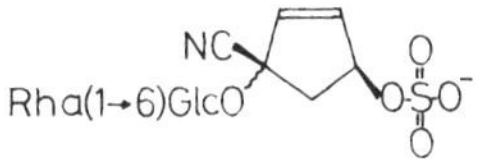
NC
Rha(1→6)GlcO
O
OSO⁻
O

80 passicoccin

81 epipassicoccin

Passicapsin (**82**) (mp 116–119 °C) was isolated from *P. capsularis*, and its structure was determined as 4-boivinosyl-(epi)tetraphyllin B except for the stereochemistry. Dideoxy sugars such as boivinose, 2,6-dideoxy-xylohexopyranose, have not been previously reported from Passifloraceae and related families (37).

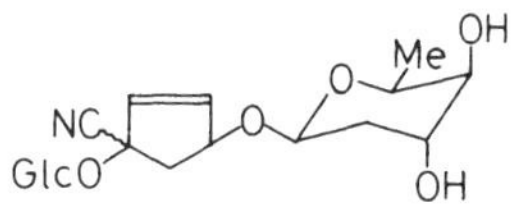
NC
GlcO
O
O
Me
OH
OH

82

The occurrence of these cyanogenic glycosides, which possess aglycones with a cyclopentenyl structure, appears to be restricted to the Passifloraceae, Turneraceae, Flacourtiaceae, and Malesherbiaceae (45, 121, 122, 123, 126, 129, 130, 132, 143). These cyclopentenyl-glycosides may be chemotaxonomic markers in these plants.

The biosynthetic origin for the aglycones in these cyanogens can be considered to be the corresponding amino acids, cyclopentenylglycines. Indeed, 2-cyclopentenyl-glycine, a possible precursor of deidaclin (**70**) and tetraphyllin A (**71**), has been found in seeds of *Hydnocarpus anthelminthica* and leaves of *Caloncoba echinata* (Flacourtiaceae) (28, 149). This amino acid was also proved to occur in leaves and seedlings of *Turnera ulmifolia* in fairly large amounts (143). When

[2-^{14}C]cyclo-pentenylglycine was fed to seedlings of *T. ulmifolia*, a specific incorporation of the label into the nitrile group of deidaclin and tetraphyllin A, which are the major cyanogenic glycosides of this plant, was observed (143). However, the incorporation of the label into epitetraphyllin B and tetraphyllin B, which also co-occur in the seedlings in small amounts, was very low. These results suggest that the hydroxylation at C-4 of the cyclopentenyl skeleton may take place during the biosynthesis of the amino acid itself (143).

Another small group of cyanoglucosides comprise those with a methylene cyclohexene ring. Simmondsin (**83**) was found in the seeds of *Simmondsia californica* (jojoba plant, Buxaceae) (32). Jojoba meal in the diet caused rats to refuse to eat. Simmondsin was isolated from the ethyl acetate extract of ground seeds as the active substance that inhibits feeding (32). Careful analysis of the coupling pattern of NMR signals led to the structure (**83**) for this compound in which the glucose-bearing oxygen adopts the unfavorable axial configuration. This conformation may be stabilized to some extent through association of the equatorial hydroxyl group at C-3 with the cyano group on the *exo*-double bond (32). Jojoba has also been reported to contain a mixture of *cis*- and *trans*-simmondsin-2′-ferulate in which feruloyl group was attached to the position 2 of the glucosyl unit (33).

Lithospermoside (**84**) was isolated from roots of *Lithospermum purpureocaeruleum* and *L. officinale* (Borginaceae) (120), and *Thalictrum rugosum* and *T. dasycarpum* (Ranunculaceae) (148). Griffonin, isolated from root of *Griffonia simplicifolia* (Caesalpinaceae) (31), was shown to be identical with lithospermoside (148). Lithospermoside (griffonin) afforded the aglycone griffonilide (**85**), upon enzymatic hydrolysis with emulsin (31, 148). X-ray crystallography of griffonilide (46) and analysis of exciton split CD Cotton effect of its dibenzoate (148) proved the absolute configuration of lithospermoside to be (4*R*,5*S*,6*S*)-6-(*β*-D-glucosyloxy)-4,5-dihydroxy-2-cyclohexene-$\Delta^{1,\alpha}$-acetonitrile. Dasycarponin (**86**) obtained from *T. dasycarpum* is the C-4 epimer of lithospermoside (148). Enzymatic hydrolysis of dasycaraponin gave D-glucose and dasycarponilide (**87**) as the aglycone.

Menisdaurin, (*Z*)-6(*S*)-(*β*-D-glucopyranosyloxy)-4(*R*)-hydroxy-2-cyclohexene-$\Delta^{1,\alpha}$-acetonitrile (**88**), was isolated from vines of *Menispermum dauricum*

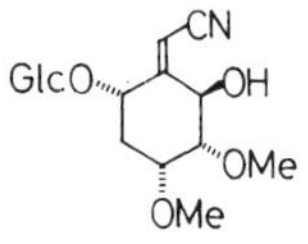

GlcO CN OH OMe OMe
83

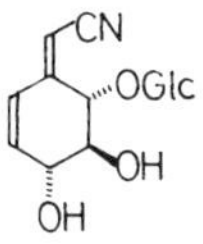

CN OGlc OH OH
84

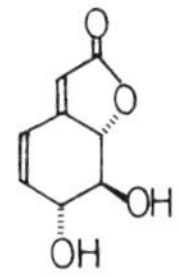

O O OH OH
85

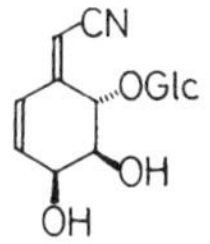

CN OGlc OH OH
86

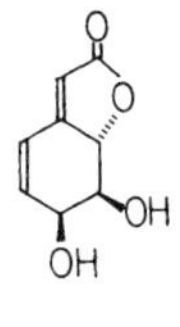

O O OH OH
87

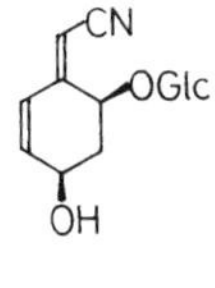

CN OGlc OH
88

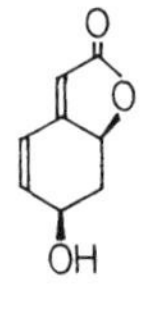

O O OH
89

(Menispermaceae) (136). Acid hydrolysis of menisdaurin afforded an aglycone, menisdaurilide (**89**), and D-glucose, as did lithospermoside.

These four cyanoglucosides, none of which are cyanogenetic, occur in members of completely separated families. This is in clear contrast to the other two groups of aliphatic and alicyclic cyanogenic glucosides, which have been found in closely related families.

As for the role of these cyanogenic compounds in plants, there exist reports supporting their defensive nature usually against herbivores (26, 80) or the role as resources for developing seeds (68). When a seed germinates, the defensive secondary metabolites, such as cyanogenic glycosides, are expected to be transferred intact to growing seedlings and to offer protection against herbivores by liberation of hydrogen cyanide from the injured tissue. The translocation of intact linamarin from the seed (cotyledons) to the developing seedlings has been demonstrated in lima beans (*Phaseolus lunatus*) (26). While the amount of the cyanogenic precursor linamarin per seedling and the cyanogenic potential of the plant remained constant during seedling development for at least 26 days, the distribution of linamarin present initially in the seed changed dramatically with time. Hydrogen cyanide liberated from the cyanogenic glycosides can also be expected to act as allelopathic agents during the germination of seeds (146).

The role of the cyanogenic glycosides as resources for developing seedlings has been observed in several plants (12, 68, 81). During seedling development of *Hevea brasiliensis* (Euphorbiaceae), cyanogenic potential of the entire seedling decreased to 15% within 19 days and the cyanogenic precursors, such as linamarin, were metabolized to form noncyanogenic substances to recycle the nitrogen (68). Hydrogen cyanide liberated in the plant tissue first reacts with cysteine to form β-cyanoalanine catalyzed by β-alanine synthase (38, 75). This amino acid is further metabolized to asparagine, which can eventually be incorporated into the general metabolism of the plant. Other metabolic pathways catalyzed by rhodanase (23, 144) or formamide hydrolyase (114) are also candidates for hydrogen cyanide metabolism, but few studies have been concerned with higher plants.

6.2.5 Highly Oxygenated Cyclohexanes

The highly oxygenated cyclohexane derivatives listed in Table 6.2.6 were found in several plants belonging to different orders: Magnoliales, Geraniales, and Piperales. Some of them are constituents of fruits or leaves.

In 1968, Holland et al. isolated senepoxide (**90**) and related seneol (**91**) from *Uvaria catocarpa* (Annonaceae, Magnoliales) (51). The fruit of this plant, so-called senasena in Madagascar, has been used as a medicine, particularly for children.

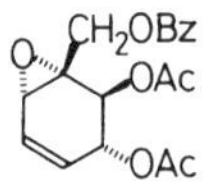

90 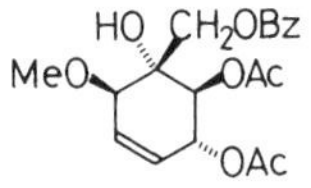 91 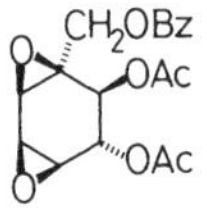 92

Table 6.2.6. Highly oxygenated cyclohexanes

Compound	mp (°C)	$[\alpha]_D$	Yield (%)	Source	Reference
Senepoxide **(90)**	85	−197	−	*Uvaria catocarpa*	51
Seneol **(91)**	113−114	−128	−	*U. catocarpa*	51
Crotepoxide **(92)**	151−153	+74	0.43	*Croton macrostachys*	65
(Futoxide)	153	−	0.1	*Piper futokadzura*	135
	148−149	+72	0.03	*P. hookeri*	117
	148−149	−	0.05	*P. brachystachyum*	116
Pipoxide **(93)**	152−154	+25	0.2	*P. hookeri*	118
	152	+53	0.15	*U. purpurea*	50
β-Senepoxide **(94)**	72−73	+62	−	*U. ferruginea*	63
Tingtanoxide **(95)**	−	−306	−	*U. ferruginea*	57
Ferrudiol **(96)**	191−192	−141	−	*U. ferruginea*	107
Zeylenol **(97)**	144−145	−116	0.02	*U. zeylanica*	57

However, the biological activity of these two compounds was not examined in their report (51).

Soon after, Kupchan et al. obtained a similar compound, crotepoxide **(92)**, from the fruit of *Croton macrostachys* (Euphorbiaceae, Geraniales) during their search for tumor inhibitors from plant sources (65, 66). Crotepoxide showed a significant tumor-inhibitory activity against Lewis lung carcinoma in mice and Walker intramuscular carcinosarcoma in rats (66). They attributed this activity to the diepoxide functionality in the molecule. Futoxide, which has been shown to be identical with crotepoxide, was obtained from the stem of *Piper futokadzura* (Piperales) (135). Crotepoxide was also present in two other species of Piperaceae, *P. hookeri* (117) and *P. brachystachyum* (116). It is interesting to note that the same crotepoxide was also isolated (64) from *C. macrostachys* when the extraction of the active principle was monitored by insect antifeedant assays against the African armyworm *Spodoptera exempta*; it is quite common that cytotoxic assays and insect antifeedant assays lead to the isolation of the same compound (Nakanishi, personal communication).

Pipoxide **(93)** was found in *P. hookeri* (118) as well as in *Uvaria purpurea* (50). The structure of pipoxide was determined to be **(93)** by NMR spectroscopy and by X-ray crystallography (50).

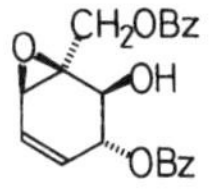

O CH$_2$OBz OH OBz
93

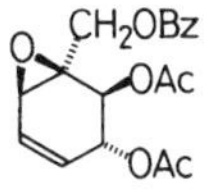

O CH$_2$OBz OAc OAc
94

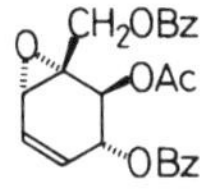

O CH$_2$OBz OAc OBz
95

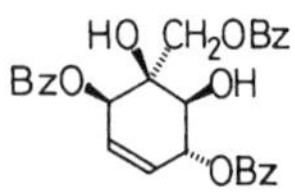

HO CH$_2$OBz BzO OH OBz
96

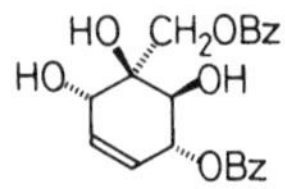

HO CH$_2$OBz HO OH OBz
97

Two other cyclohexane epoxide derivatives, β-senepoxide (**94**) and tingtanoxide (**95**), were isolated from *Uvaria ferruginea* (63). β-Senepoxide is an epimeric isomer of senepoxide at the epoxide ring. Ferrudiol (**96**), isolated from *U. ferruginea* (107), zeylenol (**97**), from *U. zeylanica* (57), and seneol (**91**) are similarly highly oxidized cyclohexanes and would be formed by opening of the oxirane ring. However, in ferrudiol, an unusual *cis*-hydrolysis of the epoxide ring is necessary (107).

These compounds are not aromatic, because the aromatic ring has been oxidized to an epoxide (67). The biogenesis of these cyclohexane-epoxides remains to be solved, although hypothetical pathways involving an arene oxide have been proposed (43, 57).

Ganem and Holbert suggested that isochorismic acid (**98**) (Fig. 6.2.3), the least stable seven-carbon acid in the shikimic acid pathway, would serve as the starting material for the formation of the arene oxide (**99**). The enol pyruvate in isochorismate molecule could act as a stable leaving group leading to the formation of the oxirane ring (43). Instead, Jolad et al. postulated benzyl benzoate (**100**) (Fig. 6.2.3) to be the precursor for the arene oxide (57). Although Ganem et al. reported that the synthesized arene oxide is unexpectedly stable (43, 44), such a highly reactive intermediate has never been detected in nature.

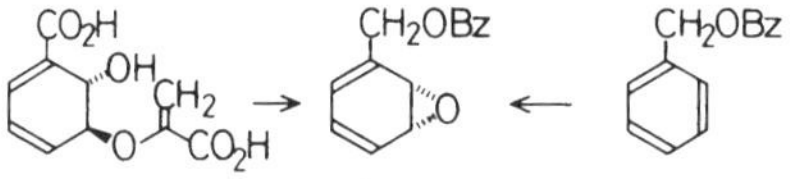

CO_2H OH CH_2 O CO_2H → CH_2OBz O ← CH_2OBz

98 99 100

Fig. 6.2.3. Biogenesis of the arene oxide (**99**)

The recent success in the isolation of three 1,6-desoxy derivatives **101, 102, 103** from *U. purpurea* (108) and *U. ferruginea* (63), which are key intermediates between the arene oxide and the natural compounds **90** through **97**, strongly support the proposed biosynthetic pathway involving the arene oxide.

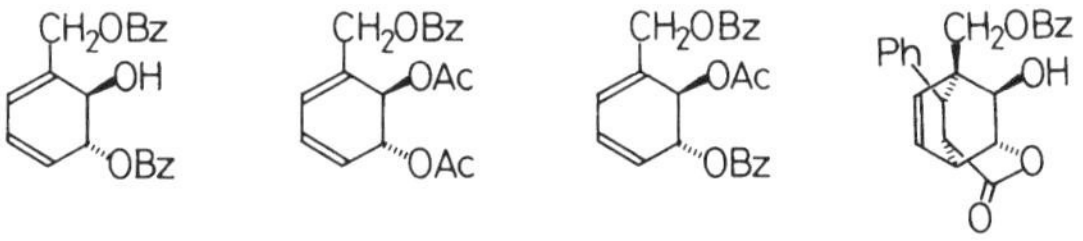

CH_2OBz OH OBz CH_2OBz OAc OAc CH_2OBz OAc OBz Ph CH_2OBz OH O O

101 102 103 104

Also interesting is the isolation of zeylena (**104**) from *U. zeylanica* (57). This compound would be formed by the addition of cinnamic acid to the arene oxide intermediate followed by an enzyme-catalyzed intramolecular Diels-Alder reaction in the intermediate triene (57).

6.2.6 Cyclohexane Diols

Aspen and cotton wood and some species of willow that grow to tree sizes are used for veneer, pulpwood, boxes, and crates. These trees are in the *Populus* and

Salix genera of the family Salicaceae. The most characteristic and most widely studied chemical constituents common to these two genera are phenolic glycosides (49). Salicin (**105**) is one of the most frequently encountered phenolic glycosides in these genera. Details on the phenolic glycosides in Salicaceae are described in the literature (49, 103, 141).

Together with these phenolic glucosides, *p*-coumaroyl or caffeoyl esters of *cis*-cyclohexanediol glucosides have been isolated from the hot water extract of the bark of Salicaceae species. Grandidentatin (**106**) was obtained from *P. grandidentata*, big-tooth aspen (90). The aglucone of grandidentatin, *cis*-1,2-cyclohexane diol, has been isolated as crystals from *P. trichocarpa* (90, 97).

Purpurein (**107**), isolated from the bark of *Salix purpurea*, was established as a diastereomer of grandidentatin by comparison of the de-*p*-coumaroylated purpurein with grandidentin, a product of alkali hydrolysis of grandidentatin (92). The absolute stereochemistry of these two coumaroyl esters of 1,2-*cis*-hexane diol remains to be solved. Grandidentoside (**108**), another cyclohexane diol glucoside isolated from *P. grandidentata*, was determined to be a caffeic acid ester of the similar glucoside (34).

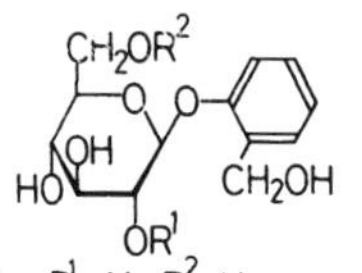

105 R^1=H, R^2=H
113 R^1=Benzoyl, R^2=H
114 R^1=H, R^2=Benzoyl

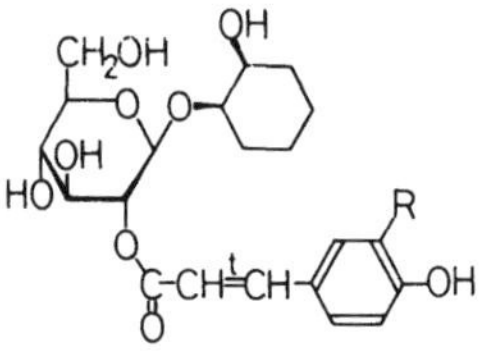

106,107 R = H
108 R = OH

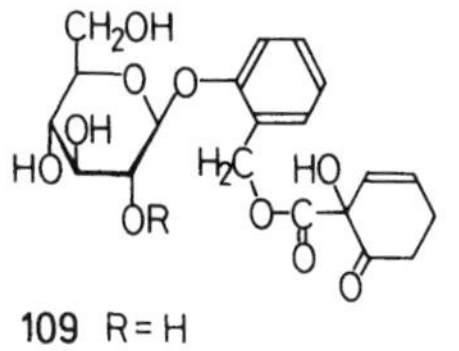

109 R=H
110 R=Benzoyl

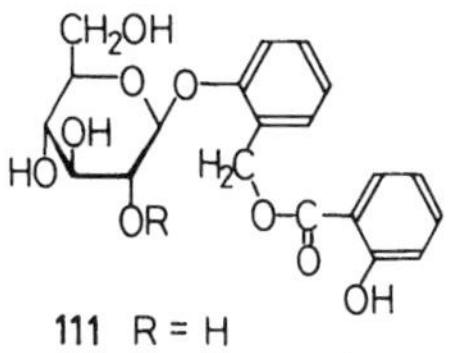

111 R = H
112 R = Benzoyl

The direct application of polyamide column chromatography to the bark extractives of Salicaceae led to the isolation of two labile complex glucosides. Salicortin (**109**) was first isolated from *Salix purpurea* in 1964 (140) and was shown later to occur in quantity in the bark of a number of *Populus* and *Salix* species (91, 93–95). The structure of salicortin has been esablished as ω-(1-hydroxy-6-oxo-2-cyclo-hexane-1-caroxylic acid ester) of salicin (**105**) (96). Tremulacin (**110**), *O*-benzoylated salicortin at C-2 of the glucose moiety, is also commonly distributed in Salicaceae species (94–96, 142). Studies on these two labile glucosides suggest that many of the phenolic glucosides found in the barks of *Populus* and *Salix* species are artifacts formed during isolation (96). Acid treatment of salicortin or tremulacin yield salicyloyl salicin (**111**) or its benzoyl ester, salicyloyltremuloidin (**112**), respectively. The latter two, in turn, will yield salicin (**105**), tremuloidin (**113**), or populin (**114**) upon hydrolysis. Furthermore, pyrocatechol and/or salicylic acid usually found during the processing of the plant materials might also originate from the alicyclic ring of these two labile glucosides (96).

6.2.7 Polycyclic Compounds

Endiandra introrsa (Lauraceae) is a large tree growing in northern New South Wales and southern Queensland, Australia. Endiandric acid A (**115**) was obtained from leaves and leaf stems of this plant in a relatively high yield (0.45%) (2). The structure of this compound was determined as 2-(6′-phenyltetracyclo[5,4,2,$0^{3,13}$, $0^{10,12}$]-trideca-4′,8′-dien-11′-yl)acetic acid by X-ray crystallographic analysis (2, 5).

This structure is unusual in that it contains a new 13-membered tetracyclic framework.

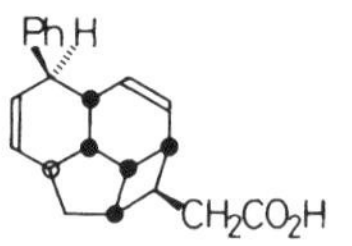

115 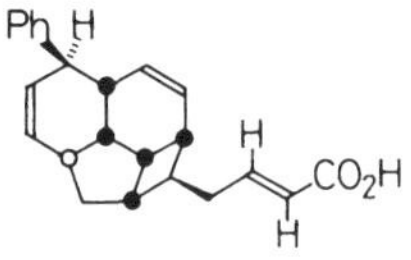 116 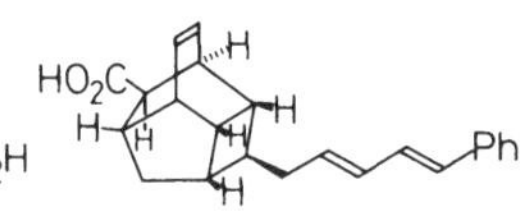 117

Two other related compounds were isolated from a different colony of the same species growing in a separate district. Endiandric acid B, (*E*)-4-(6′-phenyl-tetracyclo-[5,4,2,$0^{3,13}$,$0^{10,12}$]trideca-4′,6′-dien-11′-yl)but-2-enoic acid (**116**), was purified from the small amount of co-existing endiandric acid A (**115**) by esterification with diazomethane, separation, and subsequent hydrolysis (4). Chemical and spectroscopic evidence proved that endiandric acid B was a vinylog of endiandric acid A. The structure of the other compound, abbreviated as acid C, was determined by ^{1}H-NMR spectroscopy and also by X-ray crystallography on a Diels-Alder adduct with *N*-methyl-maleimide as 4-[(*E*,*E*)-5′-phenylpenta-2′,4′-dien-1′-yl]tetracyclo-[5,4,0,$0^{2,5}$,$0^{3,9}$]undec-10-ene-8-carboxylic acid (**117**) (6).

All three compounds are racemic in spite of the presence of eight chiral centers in each structure. Examination of the structures of these racemic compounds led to a postulation of a biosynthetic pathway including three successive electrocyclic reactions starting from non-chiral precursors (3) (Fig. 6.2.4).

Recently, Banfield et al. succeeded in isolating one of the predicted intermediates in the biogenetic pathway (8). This unstable compound, endiandric acid D (**118**), was obtained from the fiesh plant material of *E. introrsa*, which contains endiandric acid A as the major acid component; the structure of endiandric acid D was determined by means of a combination of one- and two-dimensional NMR techniques as 2-(8′-[(*E*,*E*)-5″-phenylpenta-2″,4″-dien-1″-yl]bicyclo[4,2,0]octa-2′,4′-dien-7′-yl)acetic acid. This structure was confirmed later by X-ray crystallographic analysis of an adduct of the acid D with 4-phenyl-1,2,4-triazoline-3,5-dione (7).

It should be noted that an Indian group has reported the isolation of two olefinic carboxylic acids, cryptocaryic acid (**119**) and amygdalinic acid (**120**), car-

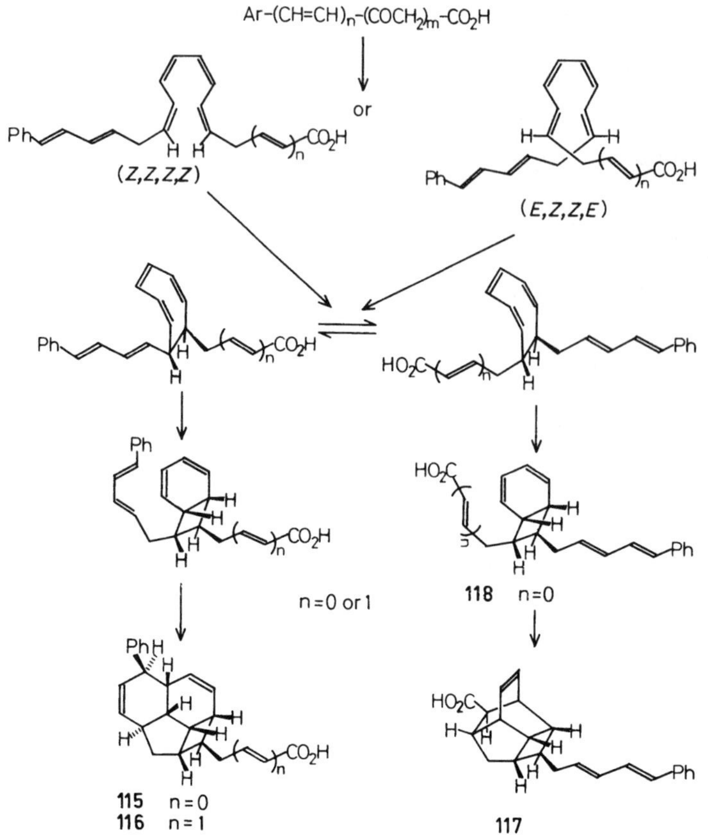

Fig. 6.2.4. Biogenesis of endiandric acids A, B, C

rying a $[4,3,0^{2,4},0^{3,7}]$-tetracyclic moiety from the bark of *Cryptocarya amygdalina* (Lauraceae) (14).

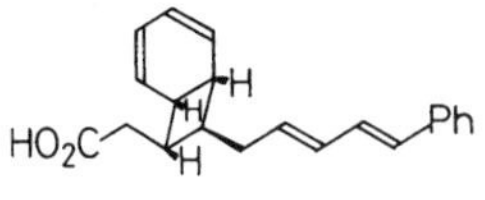

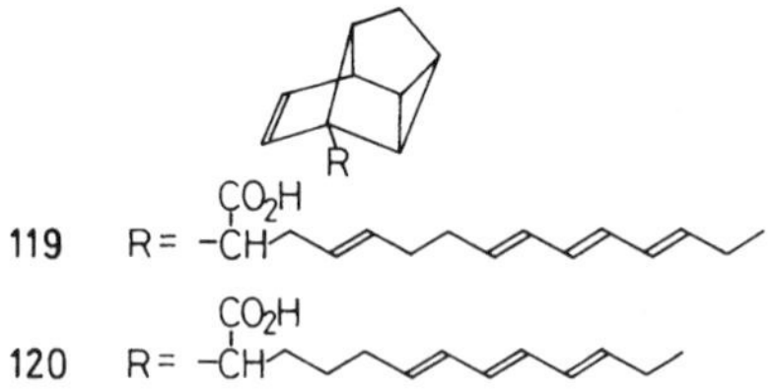

6.2.8 Miscellaneous

Dendropanax trifidus (Araliaceae) has been used as an antiseptic agent. Silica gel chromatography of an ethyl acetate extract of the fresh leaves of this plant gave (16*R*)-*cis*-9,17-octadecadiene-12,14-diyne-1,16-diol **(121)** ($[\alpha]_D$ −2.1°) and *R*-(−)-16-hydroxy-*cis*-9,17-octadecadiene-12,14-diynoic acid **(122)** ($[\alpha]_D$ −0.3°), both as a colorless oil, in yields of 0.034% and 0.23%, respectively (60). Conidium germination of *Cochliobus miyabeanus* was completely inhibited at concentrations of 12.5 μg/ml of **121** and 50.0 μg/ml of **122**.

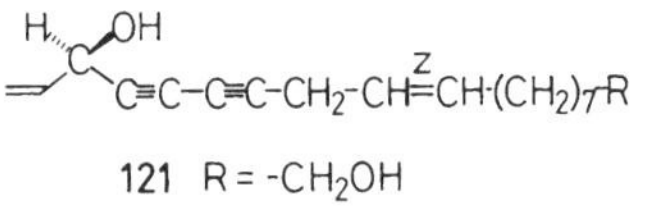

121 R = $-CH_2OH$
122 R = $-CO_2H$

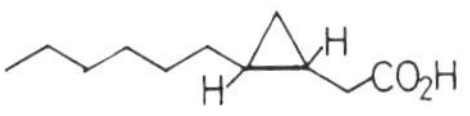

123

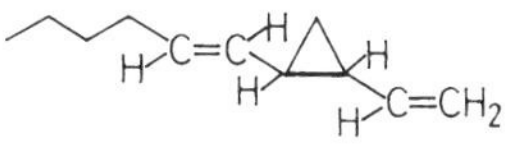

124

The bark of *Croton eluteria* (Euphorbiaceae) yields, on steam distillation, cascarilla essential oil which is used as a tonic. The main component of the acid fraction of this oil is cascarillic acid ($[\alpha]_D$ −10.5°, methyl ester), a C_{11} cyclopropane carboxylic acid (**123**) (79). The stereochemistry of the cyclopropane ring was established as *trans* by comparing ^{1}H-NMR spectra and retention times on gas chromatograms of methyl esters of the synthetic *trans*- and *cis*-isomers with those of the natural compound (147). It is interesting that compounds with closely related structures have been found in an algae and in an insect. Dictyopterene A (**124**) was isolated from the essential oil of algae of the genus *Dictyopteris* (77) and (*Z*)-3-decenoic acid from carpet beetle, *Anthrenus flavipes*, as a sex pheromone (41).

The hexane extract of shavings of the heartwood of a sesquiterpene tree, *Cinnamomum camphora*, yielded a cyclopentenone, 5-dodecanyl-4-hydroxy-4-methyl-2-cyclopentenone (**125**) (mp 42–43 °C, $[\alpha]_D$ −1.14°) (137).

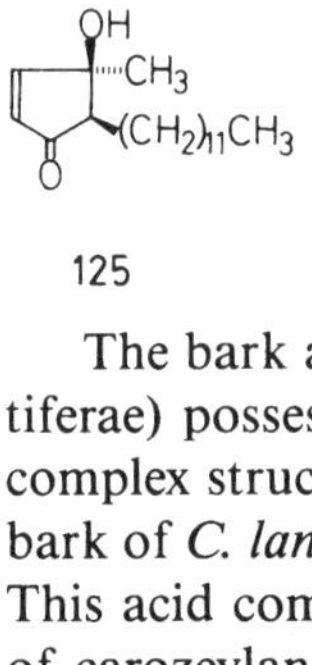

125

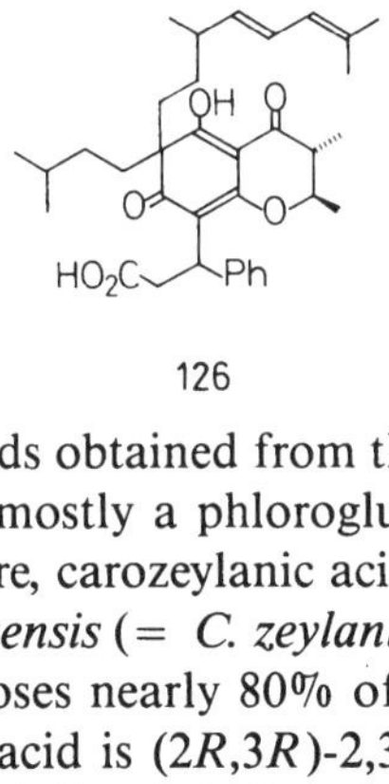

126

The bark acids obtained from the various parts of *Calophyllum* species (Guttiferae) possess mostly a phloroglucinol ring system (106). An acid with a very complex structure, carozeylanic acid (**126**) (gum, $[\alpha]_D$ +12.6°), was found in the bark of *C. lankaensis* (= *C. zeylanicum*), *C. thwaitesii*, and *C. walkeri* (105, 106). This acid composes nearly 80% of the acid fraction in each case. The structure of carozeylanic acid is (2*R*,3*R*)-2,3-dimethyl-5-hydroxy-6-(3-methylbutyl)-6-(3,7-dimethyl-octa-3,6-dienyl)-8-(2-carboxyl-1-phenylethyl)-2,3,6,7-tetrahydrobenzopyran-4,7-dione (**126**), a cyclohexadienone system fused to a 2,3-dimethylchromone ring.

Dithiolanes have been found in stem and bark of *Brugiera conjugata* and *B. cylindrica* (Rhizophoraceae) (58, 59). Brugierol (**127**) (mp 84 °C) and isobrugierol

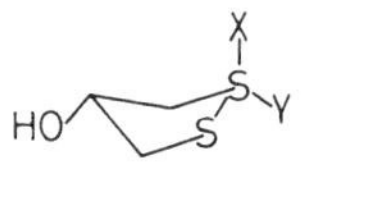

127 X = O, Y = lone pair
128 X = lone pair, Y = O

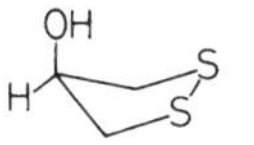

129

(**128**) (oil) are *trans*- and *cis*-isomers of 4-hydroxy-1,2-dithiolane-1-oxide, respectively, obtained from *B. conjugata* (58). Another 1,2-dithiolane (**129**) was isolated from *B. cylindrica* together with brugierol and isobrugierol (59).

References

1 Abe S 1936 Massoy oil. I. J Chem Soc Jpn 58:246–251

2 Bandaranayake W M, Banfield J E 1980 Endiandric acid, a novel carboxylic acid from *Endiandra introrsa* (Lauraceae): X-Ray structure determination. J Chem Soc Chem Commun 162–163

3 Bandaranayake W M, Banfield J E, Black D St C 1980 Postulated electrocyclic reactions leading to endiandric acid and related natural products. J Chem Soc Chem Commun 902–903

4 Bandaranayake W M, Banfield J E, Black D St C 1982 Constituents of *Endiandra* species. II. (*E*)-4-(6′-Phenyltetracyclo[5,4,2,$0^{3,13}$,$0^{10,12}$]-trideca-4′,8′-dien-11′-yl)but-2-enoic acid from *Endiandra introrsa* (Lauraceae). Aust J Chem 35:557–565

5 Bandaranayake W M, Banfield J E, Black D St C, Fallon G D, Gatehouse B M 1981 Constituents of *Endiandra* species. I. Endiandric acid, a novel carboxylic acid from *Endiandra introrsa* (Lauraceae), and a derived lactone. Aust J Chem 34:1655–1667

6 Bandaranayake W M, Banfield J E, Black D St C, Fallon G D, Gatehouse B M 1982 Constituents of *Endiandra* species. III. 4-[(*E*,*E*)-5′-Phenylpenta-2′,4′-dien-1′-yl]tetracyclo-[5,4,0,$0^{2,5}$,$0^{3,9}$]-undec-10-ene-8-carboxylic acid from *Endiandra introrsa* (Lauraceae). Aust J Chem 35:567–579

7 Banfield J E, Black D St C, Fallon G D, Gatehouse B M 1983 Constituents of *Endiandra* species. V. 2-[3′,5′-Dioxo-4′-phenyl-10′-(*E*,*E*)-5″-phenylpenta-2″,4″-dien-1″-yl-2′,4′,6′-triaza-tetracyclo-[5,4,2,$0^{2,6}$,$0^{8,11}$]tridec-12′-en-9′-yl]acetic acid derived from *Endiandra introrsa* (Lauraceae). Aust J Chem 36:627–632

8 Banfield J E, Black D St C, Johns S R, Willing R I 1982 Constituents of *Endiandra* species. IV. Isolation of 2-(8′-[(*E*,*E*)-5″-phenylpenta-2″,4″-dien-1″-yl]bicyclo[4,2,0]octa-2′,4′-dien-7′-yl)-acetic acid, a biogenetically predicted metabolite of *Endiandra introrsa* (Lauraceae) and its structure determination by means of 1D and 2D NMR spectroscopy. Aust J Chem 35:2247–2256

9 Benoni H, Hardebeck K 1964 Zur Isolierung und pharmakologischen Untersuchung von natürlichem linksdrehenden Massoia-Lacton. Arzneim Forsch 14:40–42

10 Beyerman H C, Van Dijck L A, Levisalles J, Melera A, Veer W L C 1961 Structure de l'elenolide. Bull Soc Chim Fr 1812–1820

11 Birch A J, Butler D N 1964 The structure of hyptolide. J Chem Soc 4167–4168

12 Bough W A, Gander J E 1971 Exogenous L-tyrosine metabolism and dhurrin turnover in sorghum seedlings. Phytochemistry 10:67–77

13 Bohlman F, Grenz M 1971 Neue Betenolid-Derivate aus Umbelliferen. Tetrahedron Lett 3623–3626

14 Borthakur N, Mahata P K, Rastogi R C 1981 Alkaloids and olefinic acids from *Cryptocarya amygdalina*. Phytochemistry 20:501–504

15 Braz Filho R, Diaz D P P, Gottlieb O R 1980 Tetronic acid and diarylpropanes from *Iryanthera elliptica*. Phytochemistry 19:455–459

16 Buston H W, Roy S K 1949 The physiological activity of some simple unsaturated lactones. I. Effect on the growth of certain microorganisms. Arch Biochem 22:1–7

17 Buston H W, Roy S K, Hatcher E S J, Rawes M R 1949 The physiological activity of some simple unsaturated lactones. II. Effect on certain tissues of higher plants. Arch Biochem 22:269–274

18 Butler G W 1965 The distribution of the cyanoglucosides linamarin and lotaustralin in higher plants. Phytochemistry 4:127–131

19 Butler G W, Conn E E 1964 Biosynthesis of the cyanogenic glucosides, linamarin and lotaustralin. I. Labeling studies in vivo with *Linum usitatissimum*. J Biol Chem 239:1674–1679

20 Butterfield C S, Conn E E, Seigler D S 1975 Elucidation of structure and biosynthesis of acacipetalin. Phytochemistry 14:993–997

21 Cardellina II J H, Meinwald J 1980 Isolation of parasorbic acid from the cranberry plant, *Vaccinium macrocarpon*. Phytochemistry 19:2199–2200

22 Cavill G W K, Clark D V, Whitefield F B 1968 Insect venoms, attractants, and repellents. XI. Massoilactone from two species of formicine ants, and some observations on constituents of the bark oil of *Cryptocarya massoia*. Aust J Chem 21:1819–1823
23 Chew M Y 1973 Rhodanase in higher plants. Phytochemistry 12:2365–2367
24 Clapp R C, Bissett F H, Coburn R A, Long L Jr 1966 Cyanogenesis in manioc: Linamarin and isolinamarin. Phytochemistry 5:1323–1326
25 Clapp R C, Ettlinger M G, Long L Jr 1970 Deidaclin: A natural glucoside of cyclopentenone cyanohydrin. J Am Chem Soc 92:6378–6379
26 Clegg D O, Conn E E, Janzen D H 1979 Developmental fate of the cyanogenic glucoside linamarin in Costa Rican wild lima bean seeds. Nature 278:343–344
27 Coburn R A, Long L Jr 1966 Gynocardin. J Org Chem 31:4312–4314
28 Cramer U, Rehfeldt A G, Spener F 1980 Isolation and biosynthesis of cyclopentenylglycine, a novel nonproteinogenic amino acid in Flacourtiaceae. Biochem J 19:3074–3080
29 Crombie L, Firth P A 1968 Biosynthesis of parasorbic acid (hex-2-en-5-olide) by the Rowan berry (*Sorbus aucuparia* L.). J Chem Soc (C):2852–2856
30 Doebner O 1894 Über das flüchtige Öl der Vogelbeeren, die Parasorbinsäure und ihre Isomerie mit der Sorbinsäure. Ber Deutsch Chem Ges 27:344
31 Dwuma-Badu D, Watson W H, Gopalakrishna E M, Okarter T U, Knapp J E, Schiff P L Jr, Slatkin D J 1976 Constituents of West African medicinal plants. XVI. Griffonin and griffonilide, novel constituents of *Griffonia simplicifolia*. Lloydia 39:385–390
32 Elliger C A, Waiss A C Jr, Lundin R E 1973 Simmondsin, an unusual 2-cyanomethylenecyclohexyl glucoside from *Simmondsia californica*. J Chem Soc Perkin Trans I 2209–2212
33 Elliger C A, Waiss A C Jr, Lundin R E 1974 Cyanomethylenecyclohexyl glucosides from *Simmondsia californica*. Phytochemistry 13:2319–2320
34 Erickson R L, Pearl I A, Darling S F 1970 Populoside and grandidentoside from the bark of *Populus grandidentata*. Phytochemistry 9:857–863
35 Ettlinger M, Jaroszewski J W, Jensen S R, Nielsen B J, Narty F 1977 Proacacipetalin and acacipetalin. J Chem Soc Chem Commun 952–953
36 Evjolfsson R 1970 Recent advances in the chemistry of cyanogenic glycosides. In: Progress in the chemistry of organic natural products, vol 28. Springer Berlin Heidelberg New York Tokyo, 74–108
37 Fischer F C, Fung S Y, Lankhorst P P 1982 Cyanogenesis in Passifloraceae. II. Cyanogenic compounds from *Passiflora capsularis*, *P. warmingii* and *P. perfoliata*. Planta Med 45:42–45
38 Floss H G, Hadwiger L, Conn E E 1965 Enzymatic formation of β-cyanoalanine from cyanide. Nature 208:1207–1208
39 Franca N C, Gottlieb O R, Coxon D T 1977 Rubrenolide and rubrynolide: An alkene-alkyne pair from *Nectandra rubra*. Phytochemistry 16:257–262
40 Franca N C, Polonsky J 1971 Sur la structure du boronolide, isole du *Tetradenia fruticosa* Benth. C R Acad Sci 273:439–441
41 Fukui H, Matsumura F, Ma M C, Burkholder W E 1974 Identification of the sex pheromone of the furniture carpet beetle, *Anthrenus flavipes* LeConte. Tetrahedron Lett 3563–3566
42 Gainsford G J, Russel G B, Reay P F 1984 Absolute configuration of tetraphyllin B, a cyanoglucoside from *Tetrapathaea tetrandra*. Phytochemistry 23:2527–2529
43 Ganem B, Holbert G W 1977 Arene oxides in biosynthesis. On the origin of crotepoxide, senepoxide, and pipoxide. Bioorg Chem 6:393–396
44 Ganem B, Holbert G W, Weiss L B, Ishizumi K 1978 A new approach to substituted arene oxides. Total synthesis of senepoxide and seneol. J Am Chem Soc 100:6483–6491
45 Gondwe A T D, Seigler D S, Dunn J E 1978 Two cyanogenic glucosides, tetraphyllin B and epi-tetraphyllin B, from *Adenia volkensii*. Phytochemistry 17:271–274
46 Gopalakrishna E M, Watson W H, Dwuma-Badu D, Okarter T U, Knapp J E, Schiff P L Jr, Slatkin D J 1976 Griffonilide, $C_8H_8O_4$. Cryst Struc Commun 5:779–783
47 Gottlieb O R 1972 Chemosystematics of the Lauraceae. Phytochemistry 11:1537–1570
48 Hashizume T, Kikuchi N, Sasaki Y, Sakata I 1968 Constituents of cane molasses. Part III. Isolation and identification of (–)-2-deceno-5-lactone (massoilactone). Agr Biol Chem 32:1306–1309
49 Hegnauer R 1973 Chemotaxonomie der Pflanzen. Birkhäuser Basel, 241–258
50 Holbert G W, Ganem B, Van Engen D, Clardy J, Borsub L, Chantrapromma K, Sadavongvivad C, Thebtaranonth Y 1979 Shikimate-derived metabolites. Revised structure and total synthesis of pipoxide. Tetrahedron Lett 715–718

51 Hollands R, Becher D, Gaudemer A, Polonsky J 1968 Étude des constituants des fruits d'*Uvaria catocarpa* (Annonacee). Structure de senepoxyde et du seneol. Tetrahedron 24:1633–1650
52 Hübel W, Nahrstedt A 1979 Cardiosperminsulfate – A sulphur-containing cyanogenic glucoside from *Cardiospermum grandiflorum*. Tetrahedron Lett 4395–4396
53 Hussain R A, Waterman P G 1982 Lactones, flavonoids and benzophenones from *Garcinia conrauana* and *Garcinia manii*. Phytochemistry 21:1393–1396
54 Jarozewski J W, Jensen B 1985 Deidaclin and tetraphyllin A, epimeric glucosides of 2-cyclopentenone cyanohydrin, in *Adenia globosa* ssp. *globosa* Engl. (Passifloraceae). Crystal structure of deidaclin tetraacetate. Acta Chem Scand B 39:867–875
55 Jaroszewski J W, Olafsdottir E S 1986 Natural glycosides of cyclopentenone cyanohydrins: Revised structure of so-called epitetraphyllin B. Tetrahedron Lett 27:5297–5300
56 Jewers K, Davis J B, Dougan J, Manchanda A H, Blunden G, Wetchapinan S 1972 Goniothalamin and its distribution in four *Goniothalamus* species. Phytochemistry 11:2025–2030
57 Jolad S D, Hoffmann J J, Schram K H, Cole J R, Tempesta M S, Bates R B 1981 Structures of zeylenol and zeylena, constituents of *Uvaria zeylanica* (Annonaceae). J Org Chem 46:4267–4272
58 Kato A, Numata M 1972 Brugierol and isobrugierol, *trans*- and *cis*-1,2-dithiolane-1-oxide, from *Brugiera conjugata*. Tetrahedron Lett 203–206
59 Kato A, Takahashi J 1976 A new naturally occurring 1,2-dithiolane from *Brugiera cylindrica*. Phytochemistry 15:220–221
60 Kawazu K, Noguchi H, Fujishita K, Iwasa J, Egawa H 1973 Two new antifungal compounds from *Dendropanax trifidus*. Tetrahedron Lett 3131–3132
61 Kim H S, Jeffrey G A, Panke D, Clapp R C, Coburn R A, Long L Jr 1970 The X-ray crystallographic determination of the structure of gynocardin. J Chem Soc Chem Commun 381–382
62 Kjaer A, Norrestam R, Polonsky J 1985 Boronolide: Structure and stereochemistry (X-ray analysis). Acta Chem Scand B 39:745–749
63 Kodpinid M, Sadavongvivad C, Thebtaranonth C, Thebtaranonth Y 1983 Structures of β-senepoxide, tingtanoxide, and their diene precursors. Constituents of *Uvaria ferruginea*. Tetrahedron Lett 24:2019–2022
64 Kubo I, Nakanishi K 1977 Insect antifeedants and repellents from African plants. I. In: Hedin P (ed) Host plant resistance to pests. Am Chem Soc Symp Series 62:165
65 Kupchan S M, Hemingway R J, Coggon P, McPhail A T, Sim G A 1968 Crotepoxide, a novel cyclohexane diepoxide tumor inhibitor from *Croton macrostachys*. J Am Chem Soc 90:2982–2983
66 Kupchan S M, Hemingway R J, Smith R M 1969 Tumor inhibitors. XLV. Crotepoxide, a novel cyclohexane diepoxide tumor inhibitor from *Croton macrostachys*. J Org Chem 34:3898–3902
67 Leboeuf M, Cave A, Bhaumik P K, Mukherjee B, Mukherjee R 1982 The phytochemistry of the Annonaceae. Phytochemistry 21:2783–2813
68 Lieberei R, Nahrstedt A, Selmar D, Gasparotto L 1986 Occurrence of lotaustralin in the genus Hevea and changes of HCN-potential in developing organs of *Hevea brasiliensis*. Phytochemistry 25:1573–1578
69 Loder J W, Nearn R H 1977 Altholactone, a novel tetrahydrofuro[3,2b]pyran-5-one from a *Polyalthia* species (Annonaceae). Heterocycles 7:113–118
70 Marinez V J C, Yoshida M, Gottlieb O R 1979 Six groups of ω-ethyl-, ω-ethenyl- and ω-ethynyl-α-alkylidene-γ-lactones (1). Tetrahedron Lett 1021–1024
71 Martinez V J C, Yoshida M, Gottlieb O R 1981 ω-Ethyl, ω-ethenyl, and ω-ethynyl-α-alkylidene-γ-lactones from *Clinostemon mahuba*. Phytochemistry 20:459–464
72 Masuda M, Nishimura K 1981 Branched nonalactones from some *Quercus* species. Phytochemistry 10:1401–1402
73 Masuda M, Nishimura K 1981 Absolute configurations of quercus lactones, (3*R*,4*R*)- and (3*S*,4*S*)-3-methyl-4-octanolide, from oak wood and chiroptical properties of mono-cyclic γ-lactones. Chem Lett 1333–1336
74 Meyer Th M 1940 Essential oil of massoy bark. Rec Trav Chim 59:191–201
75 Miller J M, Conn E E 1980 Metabolism of hydrogen cyanide by higher plants. Plant Physiol 65:1199–1202
76 Moewus F, Schader E 1951 The effect of coumarin and parasorbic acid upon the germination of potato tubers. Z Naturforsch 6b:112–115
77 Moore R E, Pettus J A Jr, Doty M S 1968 Dictyopterene A, an odoriferous constituent from algae of the genus *Dictyopteris*. Tetrahedron Lett 4787–4790

78 Mori K 1976 Absolute configuration of (−)-massoilactone as confirmed by a synthesis of its (*S*)-(+)-isomer. Agr Biol Chem 40:1617−1619
79 Motl O, Amin M, Sednera P 1972 The structure of cascarillic acid from cascarilla essential oil. Phytochemistry 11:407−408
80 Nahrstedt A 1985 Cyanogenic compounds as protecting agents for organisms. Plant Syst Evol 150:35−47
81 Nahrstedt A, Kant J-D, Hösel W 1984 Aspects on the biosynthesis of the cyanogenic glucoside triglochinin in *Triglochin maritima*. Planta Med 394−398
82 Narty F 1968 Studies on cassava, *Manihot utilissima* Pohl I. Cyanogenesis: The biosynthesis of linamarin and lotaustralin in etiolated seedlings. Phytochemistry 7:1307−1312
83 Niwa M, Iguchi M, Yamamura S 1975 The isolation and structure of obtusilactone. Tetrahedron Lett 1539−1542
84 Niwa M, Iguchi M, Yamamura S 1975 Three new obtusilactones from *Lindera obtusiloba* Blume. Chem Lett 655−658
85 Niwa M, Iguchi M, Yamamura S 1975 The structures of C_{17}-obtusilactone dimer and two C_{21}-obtusilactones. Tetrahedron Lett 4395−4398
86 Niwa M, Iguchi M, Yamamura S 1977 The isolation and structure of C_{19}-obtusilactone dimer. Chem Lett 581−582
87 Ohloff G 1978 Recent developments in the field of naturally-occurring aroma components. In: Herz A, Griesbach H, Kirby G W (eds) Progress in the chemistry of organic natural products. Vol 35. Springer Berlin Heidelberg New York Tokyo, 447−454
88 Otsuka K, Zenibayashi Y, Itoh M 1974 Presence and significance of two diastereomers of β-methyl-γ-octalactone in aged distilled liquors. Agr Biol Chem 38:485−490
89 Paris M, Bouquet A, Paris R-R 1969 Biochemie vegetale. Sur le barterioside, novel heteroside cyanogenetique des ecorces de racine du *Barteria fistulosa* Mast. C R Acad Sci D 268:2804−2806
90 Pearl I A, Darling S F 1962 Studies on the barks of the family Salicaceae. V. Grandidentatin, a new glucoside from the bark of *Populus grandidentata*. J Org Chem 27:1806−1809
91 Pearl I A, Darling S F 1969 Investigation of the hot water extractives of *Populus balsamifera* bark. Phytochemistry 8:2393−2396
92 Pearl I A, Darling S F 1970 Purpurein, a new glucoside from the bark of *Salix purpurea*. Phytochemistry 9:853−856
93 Pearl I A, Darling S F 1970 Phenolic extractives of *Salix purpurea* bark. Phytochemistry 9:1277−1281
94 Pearl I A, Darling S F 1971 Hot water extractives of the bark and leaves of diploid *Populus tremuloides*. Phytochemistry 10:483−484
95 Pearl I A, Darling S F 1971 Studies on the hot water extractives of the bark and leaves of *Populus deltoides* Bartr. Can J Chem 49:49−55
96 Pearl I A, Darling S F 1971 The structure of salicortin and tremulacin. Phytochemistry 10:3161−3166
97 Pearl I A, Darling S F, DeHaas H, Loving B A, Scott D A, Turley R H, Werth R E 1961 Studies on the barks of the family Salicaceae. IV. Preliminary evaluation for glucosides of barks of several species of the genus *Populus*. Tappi 44:475−478
98 Power F B, Gornall F H 1904 Gynocardin, a new cyanogenetic glucoside. Preliminary note. Proc Chem Soc (London) 20:137
99 Priestap H A, Bonafede J D, Ruveda E A 1977 Argentilactone, a novel 5-hydroxyacid lactone from *Aristolochia argentina*. Phytochemistry 16:1579−1582
100 van Puyvelde L, De Kimpe N, Dube S, Chagnon-Dube M, Boily Y, Borremans F, Schamp N, Anteunis M J O 1981 1′,2′-Dideacetylboronolide, an α-pyrone from *Iboza riparia*. Phytochemistry 20:2753−2755
101 van Puyvelde L, Dube S, Uwimana E, Uwera C, Dommisse R A, Esmans E L, van Schoor O, Vlietinck A J 1979 New α-pyrones from *Iboza riparia*. Phytochemistry 18:1215−1218
102 Rimington C 1935 The occurrence of cyanogenetic glucosides in South African species of *Acacia*. II. Determination of the chemical constitution of acacipetalin. Its isolation from *Acacia stolonifera* Burch. Onderstepoort J Vet Sci 5:445−464
103 Rowe J W, Conner A H 1979 Extractives in eastern hardwoods − A review. U S For Serv Gen Tech Rep FPL-18, 39−44

104 Russell G B, Reay P F 1971 The structures of tetraphyllin A and B, two new cyanoglucosides from *Tetrapathaea tetrandra*. Phytochemistry 10:1373–1377
105 Samaraweera U, Sotheeswaran S, Sultanbawa M U S 1981 Calozeylanic acid, a new bark acid from three *Calophyllum* species (Guttiferae). Tetrahedron Lett 22:5083–5086
106 Samaraweera U, Sotheeswaran S, Sultanbawa M U S, Balasubramaniam S 1983 Bark acids of seven *Calophyllum* species (Guttiferae). J Chem Soc Perkin Trans I:703–706
107 Schulte G R, Ganem B, Chantrapromma K, Kodpinid M, Sudsuansri K 1982 The structure of ferrudiol. A highly oxidized constituent of *Uvaria ferruginea*. Tetrahedron Lett 23:289–292
108 Schulte G R, Kodpinid M, Thebtaranonth C, Thebtaranonth Y 1982 Studies on highly oxidized cyclohexanes. Constitution of a new key metabolic intermediate. Tetrahedron Lett 23:4303–4304
109 Seigler D S 1974 Determination of cyanolipids in seed oils of the Sapindaceae by means of their NMR spectra. Phytochemistry 13:841–843
110 Seigler D S 1975 Isolation and characterization of naturally occurring cyanogenic compounds. Phytochemistry 14:9–29
111 Seigler D S, Dunn J E, Conn E E, Pereira J F 1983 Cyanogenic glycosides from four Latin American species of *Acacia*. Biochem Syst Ecol 11:15–16
112 Seigler D S, Eggerding C, Butterfield C 1974 A new cyanogenic glycoside from *Cardiospermum hirsutum*. Phytochemistry 13:2330–2332
113 Seigler D S, Spencer K C, Statler W S, Conn E E, Dunn J E 1982 Tetraphyllin B and epitetraphyllin B sulphates. Novel cyanogenic glucosides from *Passiflora caerulea* and *P. alto-caerulea*. Phytochemistry 21:2277–2285
114 Shirai R 1978 Study of cyanide metabolizing activity in mesocarp of Rosaceae. J Coll Arts Chiba Univ B-11
115 Silva R, Nagem T J, Mequita A A L, Gottlieb O R 1983 γ-Lactones from *Mezilaurus synandra*. Phytochemistry 22:772–773
116 Singh J, Atal C K 1969 Studies of the genus *Piper*. VIII. Chemical constituents of *Piper brachystachyum* Wall. Indian J Pharm 31:129–130
117 Singh J, Dhar K L, Atal C K 1969 Crotepoxide, an antitumour principle from *Piper hookeri* stems. Curr Sci 471
118 Singh J, Dhar K L, Atal C K 1970 Studies on the genus *Piper*. X. Structure of pipoxide. A new cyclohexene epoxide from *P. hookeri* Linn. Tetrahedron 26:4403–4406
119 Smith C R Jr, Weisleder D, Miller R W, Palmer I S, Olson O E 1980 Linustatin and neolinustatin: Cyanogenic glycosides of linseed meal that protect animals against selenium toxicity. J Org Chem 45:507–510
120 Sosa A, Winternitz F, Wylde R, Pavia A A 1977 Structure of a cyanoglucoside of *Lithospermum purpureo-caeruleum*. Phytochemistry 16:707–709
121 Spencer K C, Seigler D S 1980 Deidaclin from *Turnera ulmifolia*. Phytochemistry 19:1863–1864
122 Spencer K C, Seigler D S 1982 Tetraphyllin B and epitetraphyllin B from *Adenia glauca* Shinz. Onderstepoort J Vet Res 49:137–138
123 Spencer K C, Seigler D S 1982 Tetraphyllin B from *Adenia digitata*. Phytochemistry 21:653–655
124 Spencer K C, Seigler D S 1984 Gynocardin from *Passiflora*. Planta Med 50:356–357
125 Spencer K C, Seigler D S 1984 The cyanogenic glycoside barterin from *Barteria fistulosa* is epitetraphyllin B. Phytochemistry 23:2365
126 Spencer K C, Seigler D S 1985 Cyanogenic glycosides of *Malesherbia*. Biochem Syst Ecol 13:23–24
127 Spencer K C, Seigler D S 1985 Passibiflorin, epipassibiflorin and passitrifasciatin: Cyclopentenoid cyanogenic glycoside from *Passiflora*. Phytochemistry 24:981–986
128 Spencer K C, Seigler D S 1985 Passicoccin: A sulphated cyanogenic glycoside from *Passiflora coccinea*. Phytochemistry 24:2615–2617
129 Spencer K C, Seigler D S 1985 Cyanogenic glycoside and the systematics of the Flacourtiaceae. Biochem Syst Ecol 13:421–431
130 Spencer K C, Seigler D S, Domingo J L 1983 Tetraphyllins A and B, deidaclin and epitetraphyllin B from *Tetrapathaea tetrandra* (Passifloraceae). Phytochemistry 22:1815–1816
131 Spencer K C, Seigler D S, Fikenscher L H, Hegnauer R 1982 Gynocardin and tetraphyllin B from *Carpotroche brasiliensis*. Planta Med 44:28–29
132 Spencer K C, Seigler D S, Fraley S W 1985 Cyanogenic glycosides of the Turneraceae. Biochem Syst Ecol 13:433–435

133 Steyn D G, Rimington C 1935 The occurrence of cyanogenetic glucosides in South African species of Acacia. I. Onderstepoort J Vet Sci 4:51–63
134 Suomalainen H, Nykänen L 1970 Composition of whisky flavor. Process Biochem 5:13–18
135 Takahashi S 1969 The presence of the tumor inhibitor crotepoxide (futoxide) in *Piper futokadzura*. Phytochemistry 8:321–322
136 Takahashi K, Matsuzawa S, Takani M 1978 Studies on the constituents of medicinal plants. XX. The constituents of the vines of *Menispermum dauricum* DC. Chem Pharm Bull 26:1677–1681
137 Takaoka D, Imooka M, Hiroi M 1979 A novel cyclopentenone, 5-dodecanyl-4-hydroxy-4-methyl-2-cyclopentenone from *Cinnamomum camphora*. Phytochemistry 18:488–489
138 Takeda K, Sakurai K, Ishii H 1972 Components of the Lauraceae family. I. New lactonic compounds from *Litsea japonica*. Tetrahedron 28:3757–3766
139 Tantisewie B, Ruijgrok H W L, Hegnauer R 1969 Die Verbreitung der Blausäure bei den Cormophyten. 5. Mitteilung: Über Cyanogene Verbindungen bei den Parietales und bei einigen weiteren Sippen. Pharm Weekbl 104:1341–1355
140 von Thieme H 1964 Isolierung eines neuen Phenolglucosids aus *Salix purpurea* L. Pharmazie 19:725
141 von Thieme H 1967 Über die Phenolglykoside der Gattung *Populus*. Planta Med 15:35–40
142 von Thieme H, Richter R 1966 Isolierung eines neuen Phenolglykosids aus *Populus tremula* L. Pharmazie 21:251
143 Tober I, Conn E E 1985 Cyclopentenylglycine, a precursor of deidaclin in *Turnera ulmifolia*. Phytochemistry 24:1215–1218
144 Tomati U, Federici G, Cannella C 1972 Rhodanase activity in chloroplasts. Physiol Chem Physics 4:193–196
145 Waterman P G, Crichton E G 1980 Pyrones from the bark of *Garcinia conrauana*: Conrauanalactone, a novel C_{20} derived α-pyrone. Phytochemistry 19:1187–1189
146 Weiss M 1960 HCN in Apfelembryonen. Flora 149:386
147 Wilson S R, Prodan K A 1976 The synthesis and stereochemistry of cascarillic acid. Tetrahedron Lett 4231–4234
148 Wu J, Fairchild E H, Beal J L, Tomimatsu T, Doskotch R W 1979 Lithospermoside and dasycarponin, cyanoglucosides from *Thalictrum*. J Nat Prod 42:500–511
149 Zilg H, Conn E E 1974 Stereochemical aspects of lotaustralin biosynthesis. J Biol Chem 249:3112–3115

6.3 Fats and Fatty Acids

D. F. ZINKEL

6.3.1 Introduction

The previous sections of this chapter have examined the organic plant acids arising from the tricarboxylic acid (Krebs) and glyoxylate cycles and related conversions of intermediate metabolism. Although this section (6.3) addresses a part of the general subject of metabolism, the focus is on presence of fats and fatty acids in woody tissue as food reserves. Also included is a short review on the lower molecular weight acids. Excluded, however, are the fatty acids in waxes, as this subject is covered in Sect. 6.4.

6.3.2 Fats as Food Reserves

The most prevalent fatty acids are those found in natural fats and oils. Both fats and oils and the derived fatty acids are important industrial and food materials.

Although tree fruits (olives, palm) and seeds (coconuts, palm, and tung) are important sources of commodity oils, the only commercial source of fatty acids derived from woody tissue is tall oil, the byproduct of the kraft pulping process (Sect. 10.1). Nevertheless, fats are widely distributed in woody tissue as part of the food and energy reserves. Indeed, tree genera have been classified based on the types of food reserves (11). Genera classed as predominately "fat" are the hardwoods *Aesculus*, *Betula* (some species), *Catalpa*, *Cornus* (some), *Juglans* (some), *Populus* (some), *Tilia*, *Rhus* (some), and *Viburnum* (some), and the softwoods *Pinus*, *Picea*, *Pseudotsuga*, *Tsuga*, and *Taxus*. Other genera are classified as either mixed fatty acids – starch or predominately starch. The fatty acids are found in the parenchyma cells. In some hardwood families – e.g., the Lauraceae – enlarged parenchyma cells called oil cells contain materials (3) which, though not all characterized, are likely of terpenic (such as the camphor in the oil cells of *Cinnamomum camphora* or the wood of *Laurus nobilis* (17)) rather than of fatty nature.

The fatty acids occur under usual conditions predominately as triglycerides. For pine wood, the ratio of triglycerides to free fatty acids is about 10:1. This ratio and the total amounts of fatty acids, however, can change with factors such as species, position in the tree, season, and stress. For example, total fatty acids are highest in conifers; wood from southern pine species (*Pinus elliottii*, *P. taeda*, *P. palustris*, *P. echinata*, and *P. virginiana*) contains 10 to 20 mg of total fatty acids per gram of oven-dried wood (40, 41). Hardwoods, on the other hand, usually have only one-tenth as much [e.g., 1.9 mg/g in sweetgum, *Liquidambar styraciflua*, (6)]. However, the trunk and root woods of *Tilia cordata*, a member of one of the "fat" genera, contain as much as 20 to 30 mg per gram of dry wood (15). Total lipids increase from the periphery to the older rings of the sapwood in both conifers (23) and hardwoods (14). This increase can be 2 to 3 fold (23) (Zinkel DF, 1976, unpublished data for *P. elliottii*). There does not appear to be a similar effect vertically in the stem, as indicated by data for the first 20 m in height of Norway spruce (*Picea abies*) (28). Because fatty acids serve as a food reserve, seasonal variation in quantity would be expected. However, neither the earlier literature (23) nor more recent publications are consistent in verifying this a priori expectation. In the recent literature, a rather dramatic decrease in total fatty acids (decreased triglycerides and elevated free fatty acids) was reported for *Tilia cordata* samples taken in spring (15), but seasonal changes have not been observed for the conifers – e.g., pine (26) or Norway spruce (25, 27). Hydrolysis of the triglycerides to free fatty acids occurs in stress situations such as the paraquat-induced lightwood formation in pines (42), in harvested logs or wood chips, and in heartwood formation (12 p 86).

6.3.3 Aliphatic Monocarboxylic Acids

Saturated straight-chain monocarboxylic acids of the series $C_nH_{2n+1}COOH$ and the mono- and poly-unsaturated derivatives are ubiquitous to all life forms. Those acids from C_4 to about C_{24} (and somewhat longer chain length for unsaturated) are found in fats and oils and are conventionally called fatty acids; those acids, mostly saturated, C_{24} and above are usually called wax acids, indicating their usual derivation (see Sect. 6.4). In practice, the term fatty acids is more broadly used to include all aliphatic monocarboxylic acids including acetic acid. Indeed, this practice is justified in that acetic acid has been reported to occur in the triglycerides of the parasitic animal nematode, *Ascaris lumbricoides* (5).

6.3.3.1 Volatile Fatty Acids

The lower, saturated members of the series often are called volatile fatty acids. The lowest member of the series, formic acid, has not been found to occur naturally in woody plants, although it is found as a degradation product in kraft pulping liquors. The next higher homolog, acetic acid, is a universal component of intermediate metabolism and contributes to the acidity of the woody tissue. Most of the acetic acid, however, does not occur as such in living tissue, but is present as acetate of hemicelluloses. Total acetic acid is usually in the 3% to 5% range (dry material basis) in hardwoods and about one-half that amount in softwoods (29, 35). At one time, destructive distillation of hardwoods was an important source of acetic acid (30). Autohydrolysis of the acetates of hemicellulose in wood can have a detrimental influence on wood utilization, being a major contributor in the corrosion of metal (31) such as fasteners and woodworking machinery (13). This autohydrolysis can be responsible for the greater acidity of some heartwoods (12 p 115).

Volatile fatty acid homologs higher than acetic acid have not been found as natural products of woody plants, although such acids and their esters are important components that contribute to the organoleptic properties of fruits. Nevertheless, propionic, butyric, valeric, and caproic acids occur in wetwood, in addition to elevated levels of acetic acid, as anaerobic fermentation products of starch (34, 36, 37, 43). For example, the concentration of acetic, propionic, and butyric acids of 0 to 3 mM in normal sapwood of white fir increases to as much as 38, 55, and 23 mM, respectively, in the wetwood of some fir species (38). The volatile fatty acids in gum turpentine and in the low wine (the aqueous phase of turpentine distillation) and turpentine tailings (24) also are likely the product of anaerobic fermentation. However, it should be noted that the oleoresin turpentine from *Pinus sabiniana* consists primarily of *n*-heptane (22), and the turpentine from *P. jeffreyi* contains a significant proportion of *n*-heptane (22) and smaller amounts of *n*-pentane, nonane, and undecane (39). The *n*-heptane is derived not through the mevalonate pathway but rather by decarboxylation of octanoic acid (32). Presumably, the other hydrocarbons are also formed by decarboxylation.

6.3.3.2 Constituent Fatty Acids of Fats

Much of the published literature concerning the detailed analysis of fatty acids of woody tissue is for a limited, but important, number of pulping species. Nevertheless, a large variety of saturated (normal, iso-, and antiiso-) and unsaturated fatty acids has been reported. In most cases, the predominating fatty acids are usually 18-carbon acids, of which the mono-unsaturated oleic and di-unsaturated linoleic acid are the major components. There is considerably more linoleic acid than oleic acid in hardwoods as a rule [e.g., *Tilia cordata* and *Robinia pseudoacacia* (14), *Betula verrucosa* (1, 2, 4, 14, 33), *Liquidambar syraciflua* (6) and several North American species (2)] but the proportion of oleic acid increases greatly in conifers (e.g., 16, 26, 27, 40, 41), particularly pines. Linoleic acid (the all *cis*-9,12-diene) is accompanied in conifers by a small proportion of the all *cis*-5,9-derivative (16, 19). The trienoic acid, linolenic acid (all *cis*-9,12,15), is a component of both hardwoods and softwoods but comprises only a few percent of most softwood fatty acids. However, the isomeric all *cis*-5,9,12-linoleic acid (21) appears to be unique to the Pinaceae (1). It is only a relatively minor component (on the order of 1% – 2%) in American southern pines (41) but is a significant component of Scandinavian Pinales [e.g., 10% to 25% of the fatty acids of *Pinus sylvestris* (19, 26), *Abies sibirica* (1), *Picea abies* (19, 27) = *Picea excelsa* (1), *Larix russica* (1) = *Larix sibirica*, and *Cedrus libani* (10)]. An analogous series of three 20-carbon unsaturated acids (20, 41) is also present in conifers (e.g., 19, 41) but to a much smaller extent. The all *cis*-5,11,14-eicosatrienoic acid (18) is the major component of the three at a level of about 5%. The much higher content of the 5,11,14-eicosatrienoic acid in Scandinavian conifers would indicate a trend of increasing unsaturation with more northerly latitudes, as observed for other plants. Indeed, oleic and linoleic acid concentrations increase in pine seedlings on exposure to low temperatures (7), and the amount of unsaturated fatty acids increases in colder climates for both hardwoods (9) and conifers such as *Pinus sylvestris* (8).

Although the total amount of saturated fatty acids is usually much less than that of unsaturated fatty acids, some hardwoods have been reported to contain as much as 30% saturated acids (14). Palmitic acid is the predominant constituent in all cases.

References

1 Assarsson A, Akerlund G 1966 Studies on wood resin, especially the change in chemical composition during seasoning of the wood. Svensk Papperstidn 69:517 – 525

2 Clermont L P 1961 The fatty acids of aspen poplar, basswood, yellow birch and white birch. Pulp Pap Mag Can 62:T511 – 514

3 Dadswell H E, Hillis W E 1962 Wood. In: Hillis W E (ed) Wood extractives and their significance to the pulp and paper industries. Academic Press New York, 21 – 22

4 Ekman R, Pensar G 1973 Studies on components in wood. 7. Identification of total fatty acid in birch (*Betula verrucosa*) by gas chromatography-mass spectrometry. Suom Kemist Tied 82:104 – 113

5 Fairbairn D 1955 Lipids of the female reproductive organs in *Ascaris lumbricoides*. Can J Biochem Physiol 33:31 – 37

6 Foster D O, Zinkel D F 1984 Tall oil precursors in sweetgum. Wood Fiber Sci 16:298 – 301

7 Fuksman I L, Komshilov N F 1979 Change in the acid composition of resins and lipids of pinewood exposed to various temperatures. Khim Drev (4):86 – 92, Chem Abst 91:207484k (1979)

8 Fuksman I L, Komshilov N F 1981 Effect of the geographical latitude of tree stock growth on the composition and content of resins and lipids in pinewood (*Pinus sylvestris* L.). Khim Drev (2):96 – 102

9 Fuksman I L, Letonmyaki M N, Komshilov N F 1979 Changes in the resin and lipid content of trees of some forest-forming species in relation to climatic conditions. Rastit Resur 15:446–451, Chem Abst 91:137142w (1979)
10 Hafizoglu H 1987 Studies on the chemistry of *Cedrus libani* A. Rich. I. Wood extractives of *Cedrus libani*. Holzforschung 41:27–38
11 Hillis W E 1962 Distribution and formation of polyphenols. In: Hillis W E (ed) Wood extractives and their significance to the pulp and paper industries. Academic Press New York, 96–99
12 Hillis W E 1987 Heartwood and tree extractives. Springer Berlin Heidelberg New York Tokyo
13 Hillis W E, McKenzie W M 1964 Chemical attack as a factor in the wear of woodworking cutters. For Prod J 14:310–312
14 Holl W, Poschenrieder G 1975 Radial distribution and partial characterization of lipids in the trunk of three hardwoods. Holzforschung 29:119–123
15 Holl W, Priebe S 1985 Storage lipids in the trunk- and rootwood of *Tilia cordata* Mill. from the dormant to the growing period. Holzforschung 39:7–10
16 Holmbom B, Ekman R 1978 Tall oil precursors of scots pine and common spruce and their change during sulphate pulping. Acta Acad Abo, Ser B 38(3):11 pp
17 Kekelidze N A 1987 Essential oils of the bark and wood of the stems of *Laurus nobilis*. Khim Prir Soedin (3):458–459
18 Lehtinen T, Elomaa E, Alhojarvi J 1963 Fatty acid composition of tall oil. II. *cis*-5,11,14-Eicosatrienoic acid. Suom Kemistil B36:124–125
19 Lehtinen T, Elomaa E, Alhojarvi J 1963 Fatty acid composition of tall oil. III. *cis*-5,9-Octadecadienoic acid. Suom Kemistil B36:154–155
20 Lehtinen T, Elomaa E, Alhojarvi J 1964 Investigations into the fatty acid composition of tall oil. IV. *cis*-11-Eicosenoic acid and conjugated octadecadienoic acids. Suom Kemistil B37:27–30
21 Lehtinen O, Karkkainen V J, Antila M 1962 5,9,12-Octadecatrienoic acid in Finnish pine wood and tall oil. Suom Kemistil B35:179–180
22 Mirov N T 1961 Composition of gum turpentines of pines. U S Dep Agr Tech Bull 1239:62, 66
23 Mutton D B 1962 Wood resins. In: Hillis W E (ed) Wood extractives and their significance in the pulp and paper industries. Academic Press New York, 357–358
24 Parker E D, Goldblatt L A 1952 Acid composition of gum spirits of turpentine and low wines. Ind Eng Chem 44:2211–2213
25 Pensar G 1968 Distribution and composition of extractives in wood. II. Ether extractives of earlywood and latewood in pine (*Pinus silvestris*). Acta Acad Abo, Math Phys 28(1):33 pp
26 Pensar G 1970 Distribution and composition of wood extract components. 5. Radial distribution of resin components in pine trunk section at different times of the year. Suom Kemist Tied 79(1):4–8
27 Pensar G 1970 Distribution and composition of extractives in wood. 6. Radial distribution of resin components in cross sections of spruce during different seasons of the year. Acta Acad Abo, Math Phys 30(3):5 pp
28 Pensar G, Ekman R, Peltonen C 1981 On the distribution and seasonal variation of lipophilic nonvolatile extractives in stems of Norway spruce (*Picea abies*). Ekman-Days 1981, Int Symp Wood Pulp Chem Stockholm 1:52–54
29 Pettersen R C 1984 The chemical composition of wood. In: Rowell R (ed) The chemistry of solid wood. Adv Chem 207:114–116
30 Riegel E R 1942 Distillation of hardwood for charcoal and by-products. In: Industrial chemistry. Reinhold New York, 283–285
31 Sandermann W, Gerhardt U, Weissmann G 1970 Investigations on volatile organic acids in various wood species. Holz Roh-Werkst 28(2):59–67
32 Sandermann W, Schweers W, Beinhoff O 1960 The biosynthesis of *n*-heptane in *Pinus jeffreyi*. Chem Ber 93:2266–2271
33 Selby L 1960 Birch wood constituents. II. The ether extract. Svensk Papperstidn 63:81–85
34 Terashima N 1975 Odor problems. Mokuzai Kogyo 30:508–510
35 Timell T E 1964 Wood hemicelluloses. Part I. Adv Carbohyd Chem 19:250, 274
36 Ward J C, Hann R A, Baltes R C, Bulgrin B H 1972 Honeycomb and ring failure in bacterially infected red oak timber after kiln drying. USDA For Serv Res Pap FPL 165, 36 pp
37 Ward J C, Pong W Y 1980 Wetwood in trees: A timber resource problem. USDA For Serv Gen Tech Rep PNW-112, 56 pp

38 Worrall J J, Parmeter J R Jr 1982 Formation and properties of wetwood in white fir. Physiol Biochem 72:1209–1212
39 Zavarin E, Wong Y, Zinkel D F 1978 Lightwood as a source of unusual naval stores chemicals. Proc Lightwood Res Coord Counc 5th Ann. Meeting. USDA For Serv SE For Exp Sta Asheville NC, 19–30
40 Zinkel D F 1975 Tall oil precursors of loblolly pine. Tappi 58(2):118–121
41 Zinkel D F, Foster D O 1980 Tall oil precursors in the sapwood of four southern pines. Tappi 63(5):137–139
42 Zinkel D F, McKibben C R 1978 Chemistry of naval stores from pine lightwood – A critical review. Proc Lightwood Res Coord Counc 5th Ann. Meeting. USDA For Serv SE For Exp Sta Asheville NC, 133–156
43 Zinkel D F, Ward J C, Kukachka B F 1969 Odor problems from some plywoods. For Prod J 19(12):60

6.4 Chemistry, Biochemistry, and Function of Suberin and Associated Waxes

P. E. KOLATTUKUDY and K. E. ESPELIE

6.4.1 Introduction

In 1665, Robert Hooke was examining suberized cork cells from the bark of *Quercus suber* when he gave the first description of a cell. Von Höhnel was also examining cork cells from *Q. suber* more than 200 years later when he described the lamellar structure characteristic of suberin (466). It has become quite clear that suberized walls contain an insoluble polymeric material called suberin, which is associated with a complex mixture of nonpolar compounds collectively called waxes. Electron microscopic examination of suberized walls usually reveals a lamellar structure in which the light and dark bands are probably composed of the soluble waxes and the polymer, respectively. Since suberin is laid down as an insoluble polymer in the cell wall it is not possible to isolate suberin as a pure polymer. It can be isolated only as a suberin-enriched wall fraction containing about 50% cell wall carbohydrate material. The insolubility of this material and the complexity of its structure and composition make chemical and biochemical studies of suberin extremely difficult. From the limited number of studies conducted so far we have some knowledge about the location, composition, structure, biosynthesis, biodegradation, and function of suberized cell walls.

6.4.2 Waxes

6.4.2.1 Analysis of Plant Waxes

Complex mixtures of aliphatic compounds, collectively called waxes, are deposited in and on the cuticle and in the periderm in association with the polymer suberin. With the availability of modern chromatographic techniques, especially combined gas chromatography/mass-spectrometry, it has become possible to isolate and identify the components of such complex waxes. The majority of these components fall into a few major classes – including hydrocarbons and derivatives, wax esters, fatty alcohols, and fatty acids – but many other classes can also be found. Within each class there are several homologous series of components. The variations in the types of classes found and the distribution of components within each class result in a wax that is characteristic of a given plant (253, 292, 460). It must be pointed out that wax composition can vary with the anatomical region, the physiological state, and environmental conditions. Table 6.4.1 lists the most common, as well as some of the rare, components found in plant waxes, and gives their chain length range. The coverage is intended to give a general picture and is not necessarily a comprehensive list.

Table 6.4.1. Cutin- and suberin-associated wax components[1]

Compound type	General structure	Chain length range	Comments
Hydrocarbons n-alkanes	$CH_3(CH_2)_nCH_3$	C_{21} to C_{35}	Most common wax component. High proportion of even chain length in suberin-associated waxes
iso-alkanes	$(CH_3)_2CHR$	C_{25} to C_{35}	Not as widespread as n-alkanes; odd chain length
anteiso-alkanes	$CH_3CH_2CH(CH_3)R$	C_{24} to C_{36}	Even chain length
Alkenes	$R_1CH = CHR_2$	C_{17} to C_{33}	In algae dienes and trienes also found. Branched alkenes might also be minor components
Terpenoid hydrocarbons	Farnasene, pristine, and cyclic hydrocarbons, phyllocladene, isophyllocladene (+) kaurene, and isokaurene, etc.		
Ketones	R_1COR_2	C_{25} to C_{33}	Not as common as alkanes
α-Ketols	$R_1CH(OH)COR_2$		Minor and rare component
Secondary alcohols	$R_1CH(OH)R_2$	C_9 to C_{33}	Hydroxyl usually near the middle but toward the end also found; 2-ols found in suberin-associated waxes
β-Diketones	$R_1COCH_2COR_2$	C_{29}, C_{31}, C_{33}	Carbonyls at 12,14-, 14,16-, 16,18-positions; in some cases major wax components
Hydroxy- or oxo-β-diketones	$R_1CH(OH)(CH_2)_nCOCH_2COR_2$	C_{31}	Usually minor components
Monoesters	R_1COOR_2	C_{30} to C_{60}	Composed of even chain-saturated acids and alcohols. Esters of aromatic acids found in carnauba
Phenolic esters	Ferulic acid esters of primary alcohols	C_{18} to C_{28}	Commonly found in suberin-associated wax

Table 6.4.1 (continued)

Compound type	General structure	Chain length range	Comments
Polyesters	O ‖ $C(CH_2)_nCH_2)O$ O \ \\// $O(CH_2(CH_2)_mC)_n$		Cyclic and noncyclic. Major components only in gymnosperms
Primary alcohols	RCH_2OH	C_{12} to C_{34}	Very common; even chains predominate
Aldehydes	RCHO	C_{14} to C_{34}	Not as common as alcohols, polymeric forms also found. Even chain and saturated
Acids			
alkanoic acids	RCOOH	C_{12} to C_{36}	Very common. Even chain-saturated chains predominate
dicarboxylic acids	$HOOC(CH_2)_nCOOH$	C_{16} to C_{30}	Found in suberin-associated wax
ω-hydroxy acids	$CH_2(OH)(CH_2)_nCOOH$	C_{10} to C_{34}	Found in suberin-associated wax
Diols			
α,ω-diols	$CH_2(OH)(CH_2)_nCH_2OH$	C_{20} to C_{32}	Probably occur as esters, not very common in waxes
α,β-diols	$CH_2(OH)CH(OH)(CH_2)_nCH_3$	C_{14} to C_{32}	Rare and minor
terpenes	E.g., friedelin, cerin, betulin, serrantendiols, cycloartenol lactones, α- and β-amyrin	C_{15} to C_{30}	Triterpenes are major wax components of bark from many trees and shrubs

[1] Adapted from Kolattukudy (225)

Waxes are usually isolated by extracting the tissue with a nonpolar solvent such as chloroform or hexane. Cuticular waxes can be extracted by a quick dip into the solvent at room temperature, but suberin-associated waxes are more difficult to remove because they are embedded in the suberized cell wall (232). Wax from the suberized cells of bark has to be isolated by Soxhlet extraction of dried and powdered tissue to assure its complete removal. Isolated wax can be subjected to GC/MS analysis either after separation into various classes by thin-layer or column chromatography or directly after derivitization of the functional groups (232, 253, 459, 460).

6.4.2.2 *Composition of Suberin-Associated Waxes*

Suberin-associated waxes have not been studied in great detail. Since it is impossible to isolate pure suberin, it is difficult to isolate waxes associated with the polymer without contamination from other cellular lipids. This problem is especially relevant in relation to the suberin-associated waxes in the bark because large amounts of extractives can be obtained from barks, but much of it may not be components of the wax (292). Although suberin-associated waxes from only a few plant species have been analyzed, it seems that cuticular and suberin waxes share the same major components (233). However, several features of suberin-associated waxes make them distinct from cuticular waxes.

6.4.2.2.1 Hydrocarbons in Suberin-Associated Waxes

A survey of the suberin-associated waxes from the periderm of the underground storage organs of seven plants has indicated that the hydrocarbons from suberin-associated waxes have some characteristic features (116). The hydrocarbons seem to have a much higher proportion of even chain-length components and the average chain length appears to be shorter. Additionally the hydrocarbons have a broader distribution of chain-length components. This feature is in contrast to hydrocarbons in cuticular waxes, which are usually dominated by one or two odd chain-length components, usually C_{29} or C_{31} (232). The three differences mentioned above are illustrated in Fig. 6.4.1, where the distribution of chain lengths of the hydrocarbons from the suberin-associated wax is compared to that of a cuticular wax from the same plant. There are several reports of hydrocarbon com-

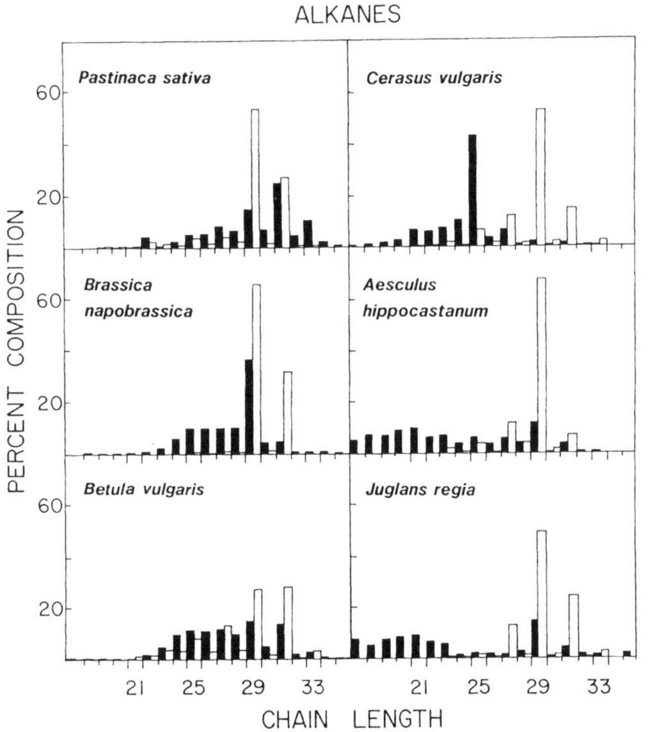

Fig. 6.4.1. Chain-length distribution of the alkanes in the wax from periderm (solid bars) and leaf cuticle (open bars). The periderm wax was extracted from underground storage organs of vegetables (116) and from the bark of trees (440)

position from bark wax where one or more of these features hold true (23, 26, 160, 200, 387, 389, 438, 440). For example, the major hydrocarbons from the bark wax of *Pinus contorta* were odd-chain, but were relatively short (C_{23} and C_{21}), and 31% of the entire alkane fraction had even chain lengths (389). Similarly the major hydrocarbon in the bark wax of *Fagus sylvatica* was C_{23}, and 48% of the hydrocarbon fraction had even chain lengths (438). In the suberin-associated bark wax of *Saraca indica*, although C_{29} was the major alkane, 46% of the hydrocarbon fraction had even chain lengths (23). The suberin-associated wax from the periderm of the underground storage organ of *Solanum tuberosum* had a hydrocarbon fraction with predominantly odd chain length, but the chains were relatively short (C_{25}, C_{23}, C_{21}) and the distribution was broad (94).

6.4.2.2.2 Wax Esters

Wax esters usually constitute a minor component in suberin-associated waxes. In the periderms of underground storage organs they comprised 1% to 7% of the total wax (116), and there were no detectable wax esters in the suberin-associated wax from the periderm surrounding the crystal idioblasts of *Agave americana* (117). In many cases where wax esters have been reported to be components of bark waxes, they comprised an unknown or low percentage of wax – e.g., *Pinus monticola* (7%) (82), *Pinus contorta* (9%) (389) and *Pinus banksiana* (7%) (387).

6.4.2.2.3 Free Fatty Alcohols

Free fatty alcohols are perhaps the most common components of cuticular waxes and often are also a major constituent of suberin-associated waxes (232). They comprised between 10% and 45% of the waxes from the periderm of underground storage organs (116). The dominant chain lengths of free fatty alcohols of cuticular waxes are C_{26} and C_{28} (232, 253, 292, 460), but the free fatty alcohols of suberin-associated waxes usually have a shorter chain length. For example, the dominant alcohols in the leaf cuticular wax of *Agave americana* were C_{26} (36% of the alcohols) and C_{28} (62%), whereas the dominant component in the fatty alcohols of the suberin-associated wax from the periderm layer surrounding the crystal idioblasts within the same leaf was C_{22} (88%) (117). The most common fatty alcohols reported as components of bark wax are C_{24} and C_{22}.

6.4.2.2.4 Free Fatty Acids

Free fatty acids are major components of many suberin-associated waxes, whereas they are usually minor components of cuticular waxes (245). For example, free fatty acids were the major class of components in the suberin-associated wax from the storage organs of *Daucus carota* (50% of the wax) and *Brassica napobrassica* (49%) (116, 219), and the bark waxes of *Pinus contorta* (43%) (389) and *Ailanthus glandulosa* (59%) (76). The major fatty acids in many bark waxes are C_{20}, C_{22}, C_{24} or a combination of these acids. For example, C_{22} and C_{24} were the two dominant components (23% and 34%, respectively) of the free fatty acid fraction that comprised 34% of the wax from the bark of *Picea abies* (484). In many analyses of bark wax the proportion of free fatty acids is not known because the waxes were saponified before analysis. It does seem clear, however, that free fatty acids constitute a characteristic component of many suberin-associated waxes.

6.4.2.2.5 Polar Wax Components

Since the suberin-associated waxes are obtained by rigorous extraction, the polar components present in such extracts could also represent cellular lipids in addition to the true suberin waxes. In spite of such uncertainties, some of the components found in the polar fraction of suberin-associated waxes obviously arise from suberized layers. Approximately 50% of the chloroform extract from the periderm layer of the underground storage organ of seven plants was too polar to move from the origin of a thin-layer chromatogram (116). A partial analysis of these polar components revealed that they contained long-chain (C_{16} to C_{28}) esterified fatty acids and fatty alcohols as well as ω-hydroxy acids and dicarboxylic acids (C_{16} to C_{20}). Dicarboxylic acids ($C_{18:1}$, C_{16}, and C_{20}) have also been reported in the bark wax of *Thuja plicata* (130). Both ω-hydroxy acids (C_{16} to C_{24}) and dicarboxylic acids (C_{16} to C_{24}) were present as esters in the wax from the bark of *Pseudotsuga menziesii* (262, 283, 284). Likewise, both ω-hydroxy acids and dicarboxylic acids were present in the suberin-associated bark wax of *Fagus sylvatica* (437, 439) and *Abies concolor* (262). In all of the cases listed above these acids comprised a small portion ($<5\%$) of the wax, but they seem to be characteristic components in many suberin-associated waxes.

Terpenes, especially triterpenes, are major components of extractives from barks (104, 292, 388). Among the most commonly reported major triterpenes are α- and β-amyrin, betulin, friedelin, cycloartenol lactones, ursolic acid, oleanolic acid, serrantendiols, and sterols (see Chap. 8).

6.4.2.2.6 Ferulic Acid Esters

Ferulic acid esters have been reported in the bark wax from several trees since they were first identified as a component of the aerial and root barks of *Phylocladus glaucus* (58). Ferulic acid esters comprised 2% of the benzene extract from the bark of *Pinus banksiana* (387). Among gymnosperms they have also been found in the bark wax from *Abies grandis* (384), *Picea abies* (329), *Pinus radiata* (502), *P. sylvestris* (10, 330), and *Pseudotsuga menziesii* (2). Ferulic acid esters were also found in the bark of hardwood trees *Erythrina variegata*, *E. stricta*, and *E. blakei* (414–416). In most of the above cases, C_{22} and C_{24} fatty alcohols were the dominant components esterified to ferulic acid with smaller amounts of C_{16}, C_{18}, C_{20}, C_{26}, and C_{28}. Ferulates appear to be a common and characteristic component of suberin-associated wax, although they may comprise only a small part of the wax. Esters of other aromatic acids such as vanillic acid and *p*-coumaric acid have also been found in bark waxes such as those from *Pinus monticola* (324) and *Leptospermum scoparium* (84). It is likely that such esters will be found as common minor components of suberin-associated waxes.

6.4.2.2.7 Tabular Survey of Bark Wax Components

The major types of components reported in the bark wax of trees are shown in Table 6.4.2; a list of compounds found in plant waxes can be found elsewhere (225, 232, 292). The plants from which bark has been studied are listed according to family, with gymnosperms given first. Some of the results shown here have been reviewed previously (72, 177, 206, 260, 270, 388, 432, 434).

Table 6.4.2. Wax components isolated from bark of various trees

Family	Genus/species	Wax components[1]	Reference(s)
Cupressaceae	*Chamaecyparis lawsoniana*	Wax esters, free fatty acids	421
	Juniperus horizontalis	Wax esters (C_{20} fatty acid)	321
	Libocedrus decurrens	Wax esters, free fatty acids	421
Pinaceae	*Abies concolor*	Wax esters (C_{22} fatty acid), free fatty acids (C_{22}), fatty alcohols (C_{24}), ω-hydroxy acids (C_{16}, C_{22}, C_{20}), dicarboxylic acids (C_{16})	181, 262
	Abies grandis	Hydrocarbons (C_{31}), wax esters (C_{24}, C_{22} alcohol and acid), ferulate esters (C_{24}, C_{22} alcohol), fatty alcohol (C_{22})	266, 384
	Abies magnifica	Wax esters (C_{24} alcohol), free fatty acids (C_{22})	21
	Picea abies	Hydrocarbons (50% even-chain length), wax esters (acids $C_{12}-C_{24}$, alcohols $C_{18}-C_{26}$), free fatty acids (C_{20}, C_{22}, C_{24}), free fatty alcohols ($C_{18}-C_{26}$), ferulate esters (C_{22}, C_{24} alcohol)	10, 329
	Picea glauca	Wax esters, free fatty acids (C_{24}, C_{22}), fatty alcohols (C_{24})	38
	Pinus banksiana	Hydrocarbons (C_{23}, C_{25}, C_{27}), Fatty acids (C_{24}, C_{26}), fatty alcohols (C_{24}, C_{22}), ferulate esters (C_{24}, C_{22} alcohol)	387
	Pinus caribaea	Fatty acids (C_{24}, C_{20}), fatty alcohol (C_{24})	357–359
	Pinus contorta	Hydrocarbons (C_{23}, C_{21}), wax esters (C_{22} alcohol and acid), free fatty acids, fatty alcohols (C_{22}, C_{24}, C_{20})	389
	Pinus halepensis	Wax esters (C_{24} acid)	98
	Pinus monticola	Wax esters (C_{24}, C_{22} acids), free fatty acids (C_{24}, C_{22}, C_{20}), fatty alcohols (C_{18}, C_{17}), vanillic acid esters	82, 324
	Pinus ponderosa	Free fatty acids, fatty alcohols	264
	Pinus radiata	Fatty acids (C_{22}, C_{24}), ferulate esters (C_{24}, C_{22})	165, 502

Table 6.4.2 (continued)

Family	Genus/species	Wax components[1]	Reference(s)
	Pinus roxburghii	Fatty alcohol (C_{26})	24
	Pinus sylvestris	Wax esters, free fatty acids (C_{20}, C_{22}, C_{24}), free fatty alcohols (C_{22}, C_{24}, ferulate esters (C_{22}, C_{24} alcohol) hydrocarbons (C_{21}),	10
		free fatty acids (C_{22}, C_{24}, C_{20}) fatty alcohols (C_{20}), ferulic acid esters	330, 440, 443
	Pinus taeda	Wax esters, free fatty acid (C_{20}), fatty alcohols (C_{24}, C_{20}) wax esters ($C_{38}-C_{48}$)	357–360
	Pseudotsuga mensziesii	Free fatty acids (C_{24}, C_{22}),	2, 262, 265, 269
		fatty alcohols (C_{24}, C_{22}),	270, 283, 284, 299
		ferulate esters (C_{22}, C_{24}), ω-hydroxy acids (C_{16}, C_{22}, C_{20}, C_{18}, C_{24}), dicarboxylic acids (C_{16}, $C_{18:1}$, C_{18}, C_{20})	486
	Thuja plicata	Free fatty acids (C_{22}, C_{24}), dicarboxylic acids (C_{16}, $C_{18:1}$, C_{20})	130
	Tsuga heterophylla	Wax esters, ferulate esters	131, 182, 183
	Tsuga mertensiana	Fatty acids (C_{24}), fatty alcohols (C_{24})	261, 262
Anacar-diaceae	*Schinus molle*	Wax esters ($C_{18:1}$, C_{24} acids)	98
Betulaceae	*Alnus rubra*	Fatty acids (C_{18})	263
	Betula pendula	Hydrocarbons (C_{25})	440
	Betula platyphylla	Hydrocarbons (C_{16}, C_{18}, C_{20}, C_{22}, C_{24})	333
Compositae	*Petasites hybridus*	Hydrocarbons (C_{27})	200
Cornaceae	*Conrus stolonifera*	Free fatty acids (C_{24}, C_{26})	318
Fagaceae	*Fagus sylvatica*	Hydrocarbons (C_{23}, C_{22}, C_{24}, C_{25})	176, 437, 438, 439
		fatty acids (C_{22}, C_{24}), fatty alcohols (C_{22}, C_{24}), ω-hydroxy acids (C_{24}, C_{26}, C_{22}), dicarboxylic acids (C_{22}, C_{20})	
Fagaceae	*Quercus suber*	Hydrocarbons ($C_{22}-C_{32}$, 49% even-chain length), fatty alcohols (C_{22}, C_{24})	25, 26, 429
Guttiferae	*Kielmeyera coriacea*	Hydrocarbons ($C_{18}-C_{21}$)	160
Hippocasta-naceae	*Aesculus hippocastanum*	Hydrocarbons (C_{29})	440
Juglandaceae	*Juglans regia*	Hydrocarbons (C_{29})	440
Leguminosea	*Erythrina blakei*	Wax esters, fatty acids, fatty alcohols, ferulate esters,	415

Table 6.4.2 (continued)

Family	Genus/species	Wax components[1]	Reference(s)
	Erythrina stricta	Hydrocarbons (C_{18}, C_{27}), fatty acids, fatty alcohols (C_{32}, C_{36}), ferulate esters (C_{32}, C_{24} alcohols)	416
	Pithecolobium dulce	Wax esters (C_{30} acid, C_{16} alcohol)	328
	Saraca indica	Hydrocarbons (C_{29}, 46% even-chain length), wax esters (C_{26} acid and alcohol), free fatty alcohols (C_{28}, C_{26})	23
Malvaceae	*Hibiscus rosa-sinensis*	8-nonynoic acid, 9-decynoic acid	319
Myrtaceae	*Eucalyptus tereticornis*	Hydrocarbons (C_{28}, C_{24})	449
	Leptospermum scoparium	Wax esters (C_{26}, C_{28}, C_{24} acid, C_{26}, C_{24} alcohol), *p*-coumarate esters (C_{24}, C_{22} alcohol)	84
Oleaceae	*Syringa vulgaris*	Hydrocarbons (C_{23})	440
Platanaceae	*Platanus hybrida*	Hydrocarbons (C_{21})	440
Podocarpaceae	*Phyllocladus glaucus*	Ferulate esters (C_{22}, C_{24} alcohols)	58
Rosaceae	*Cerasus vulgaris*	Hydrocarbons (C_{25})	440
	Prunus cornuta	α,ω-diols (C_{22}, C_{24})	16
	Prunus puddum	Fatty acids (C_{22})	16
Salicaceae	*Populus balsamifera*	Wax esters (C_{26}, C_{28} alcohol, C_{24}, C_{26}, acid), ferulate esters	1
	Populus nigra	Hydrocarbons (C_{27})	440
	Populus tremuloides	Fatty acids (C_{24}), fatty alcohols (C_{26})	193
Simaroubaceae	*Ailanthus glandulosa*	Fatty acids (C_{22})	76
Tamaricaceae	*Tamarix pentandra*	Hydrocarbons (C_{29}, C_{31}), Fatty acid (C_{18})	494

[1] Chain lengths reported are listed in order of dominance, beginning with the most prevalent.

6.4.2.3 Biosynthesis of Wax Components

6.4.2.3.1 Very Long Fatty Acids

The very long aliphatic chains (C_{20} to C_{30}) characteristic of plant waxes are formed by chain elongation of preformed fatty acids. Chain elongation was demonstrated by the incorporation of C_2 to C_{24} acids into long-chain fatty acids in several different plant tissues (233, 253). Cell-free preparations from different plants were also shown to catalyze chain elongation (66, 237, 275, 285). These studies demonstrated that microsomal enzymes catalyze elongation of fatty acids with malonyl-CoA, NADPH, and an endogenous acyl moiety as the cosubstrates. It was reported that this endogenous acyl moiety may be CoA ester rather than an acyl carrier protein derivative (65). Recent results indicate that the elongation system is located in the cytoplasmic side of the endoplasmic reticulum (42). The wide variety of chain-length classes found in plant waxes may be due to chain-elongating enzyme systems with different specificities. This possibility was suggested when inhibitors showed differential effects on the synthesis of the different chain length classes (59, 167, 229, 236, 237, 241, 282). Similar evidence with other plant tissues and inhibitors (65, 468), as well as genetic evidence (467), was subsequently presented to support this hypothesis. The differences in chain lengths of components found in various waxes are probably due to such a difference in specificity of the elongating systems.

6.4.2.3.2 Fatty Alcohols

For many years it was assumed that fatty alcohols were formed by the reduction of fatty acids. The incorporation of exogenous labeled C_2 to C_{20} fatty acids into very long fatty alcohols in plant tissue suggested that these fatty acids were elongated and then reduced (62, 220, 232, 298). A cell-free preparation that catalyzed the reduction of fatty acid to fatty alcohol was first isolated from *Euglena gracilis* (227). The NADH-dependent acyl-CoA reductase, located in the microsomal membranes, is coupled to an aldehyde reductase so that free aldehydes are not found, although the aldehyde intermediate can be trapped (216, 227). Acyl-CoA reductase together with acyl-CoA hexadecanol transacylase was solubilized from the microsomes but the enzyme could not be purified because it could not be kept in solution without high salt concentration (218). An enzyme fraction that catalyzed the reduction of acyl-CoA to aldehyde with NADH as the prefered reductant was also isolated from the leaves of *Brassica oleracea* (228). A second protein fraction from the same leaves catalyzed an NADPH-dependent reduction of aldehyde to alcohol. Neither enzyme has been purified from any plant. It is possible that in most cases a microsomal reductase(s) reduces the long-chain acyl-CoA synthesized by the elongating enzyme in a coupled manner, as suggested for *E. gracilis* (227).

6.4.2.3.3 Wax Esters

Wax ester biosynthesis probably involves an acyl transfer mechanism. The high thioesterase activity found in crude plant extracts makes it difficult to demonstrate acyl-CoA involvement in wax ester synthesis. However, partial purification of an acetone powder extract from the leaves of *B. oleracea* gave a protein fraction that catalyzed an acyl-CoA-dependent esterification of fatty alcohols (222). Additionally the acetone powder extract from *B. oleracea* leaves appeared to catalyze the direct transfer of acyl moieties from phospholipids to fatty alcohols. The leaf extract also catalyzed under appropriate conditions the esterification of fatty alcohols to free fatty acids. The transacylase mechanism is likely to be the main mechanism of wax ester synthesis in vivo. The fact that labeled wax esters were synthesized by a membrane-bound microsomal fraction from *Hordeum vulgare* leaves following incubation with radioactive alcohols, but not after incubation with free fatty acids (17), is consistent with the proposed acyl transfer mechanism. In *E. gracilis* the acyl-CoA reductase is functionally coupled to the acyl transferase (227). Both of these activities were solubilized from the microsomes

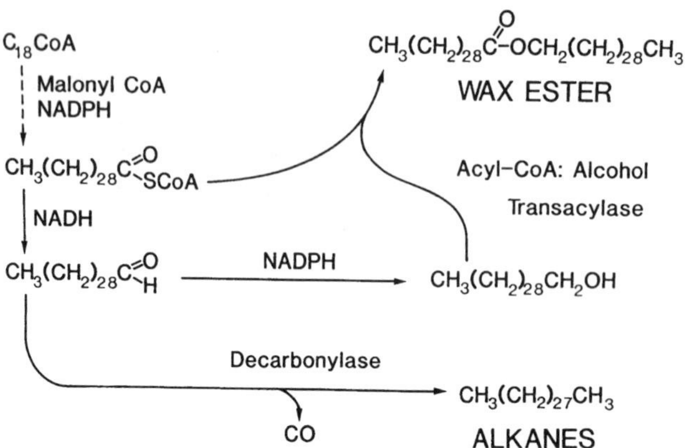

Fig. 6.4.2. Biosynthetic pathway for very long acids, aldehydes, alcohols, wax esters, and hydrocarbons

and partially resolved. It is highly likely that in higher plants the reductase and the transacylase are located in the microsomes. Figure 6.4.2 summarizes the current knowledge concerning the biosynthesis of wax esters.

6.4.2.3.4 Hydrocarbons and Derivatives

For many years it was assumed that the long-chain hydrocarbons found in plant waxes were formed by a head-to-head condensation of two fatty acids with the decarboxylation of one acid to yield a mid-chain ketone, which would subsequently be reduced, dehydrated, and reduced to yield the final hydrocarbon. Experiments with specifically and doubly labeled C_{16} and C_{18} acids showed that the entire carbon chain of the acids was incorporated without decarboxylation into C_{29} and C_{31} hydrocarbons in *B. oleracea* and in *Spinacea oleracea* (220, 224). C_{18} fatty acid was found to be incorporated more rapidly than C_{16} acid into C_{29} alkane (220), and the carbonyl carbon of nonacosane-15-one was shown to be derived from C-2 and C-4 of exogenous C_{16} and C_{18} acids, respectively (246, 247). Such experimental evidence ruled out the head-to-head condensation mechanism.

On the basis of the experimental evidence indicated above and other evidence, it was proposed that long-chain hydrocarbons are generated by an elongation-decarboxylation mechanism (221). According to this hypothesis, hexadecanoic acid, produced by fatty acid synthetase, is elongated by the stepwise addition of C_2 units until an acid of the appropriate size is obtained, which would then be decarboxylated to yield a hydrocarbon (Fig. 6.4.2). A large amount of experimental evidence obtained in this laboratory supports this hypothesis (223, 232, 241, 253) and similar results have been subsequently reported by others working not only with plants (62, 64, 468) but also insects (201) and animals (63). The most direct evidence for the proposed mechanism was that exogenous long-chain acids were converted directly to alkanes with one less carbon than the precursor fatty acid (64, 238, 240). In direct support of this mechanism, cell-free preparations from *Pisum sativum* catalyzed the conversion of [9-, 10-, 11-^{3}H]C_{32} acid into C_{31} and C_{30} alkanes (217). Ascorbate and oxygen were required for this conversion, which was inhibited by thiol compounds such as dithioerythreitol, as previously observed with tissue slices (59).

The mechanism of conversion of fatty acids to alkanes remained elusive until recently. Cell-free preparations generated alkanes from n-C_{18}, n-C_{22}, n-C_{24}, and n-C_{32} acids at comparable rates (42, 217). The alkanes thus generated in vitro contained two carbon atoms fewer than the parent acid. This result suggested that the cell-free preparation contained an enzyme that first converted the acid to an intermediate containing one less carbon, and that this intermediate probably lost another carbon to generate the alkane. Analysis of the products generated by the *P. sativum* cell-free preparation from exogenous acids showed that the only metabolite containing one carbon less than the parent acid was the aldehyde, suggesting that the aldehyde might be the precursor of the alkane (42). In fact, exogenous aldehydes gave rise to alkanes containing one less carbon. The mechanism of conversion of the aldehyde to the alkane was shown to be a decarbonylation that was catalyzed by a cuticular membrane-containing particulate preparation isolated by density-gradient centrifugation (74). This preparation catalyzed the conversion of octadecanal to heptadecane and CO. The aldehydic hydrogen was retained in the alkane just as previously observed in non-enzymatic decarbonylation catalyzed by Ru-porphyrin complex (99). The decarbonylase also probably involved a metal ion as the enzyme was inhibited by metal ion chelators. The aldehydes used as substrates for the decarbonylation arise from the reduction of acyl-CoA. Fatty acid elongase and acyl-CoA reductase are localized in the endoplasmic reticulum, and the very-long-chain aldehydes are presumably exported to the extracellular location, where the decarbonylase generates alkanes. Thus the species-specific composition of the alkanes is determined by the elongating system rather than by the specificity of the decarbonylase, which in vitro failed to show any chain length specificity. The decarbonylase has not been purified and characterized and, therefore, the details of the mechanism of decarbonylation are not understood.

The oxygenated derivatives such as ketones and secondary alcohols, which are often found in plant waxes, were originally thought to be intermediates in the biosynthesis of hydrocarbons. With the finding that alkane biosynthesis does not involve such intermediates, the possibility that the oxygenated compounds are derived from alkanes was considered. Exogenous labeled nonacosane was shown to be converted directly to nonacosan-15-ol and nonacosan-15-one, while exogenous labelede nonacosan-15-ol was converted to only nonacosan-15-one in *Brassica oleracea* (250). The obvious conclusion that the alkane was hydroxylated and the resulting secondary alcohol was subsequently oxidized to the ketone was also supported by double labeling experiments with [1-^{14}C,2-^{3}H]hexadecanoic acid (226). In further support of the above conclusion, 4-oxooctadecanoic acid was not preferentially converted to the ketone and [4-^{14}C]octadecanoic acid gave rise to nonacosan-15-one with C-4 of the precursor at the carbonyl carbon of the ketone (247). Chemical degradation of the secondary alcohol derived from [4-^{14}C]octadecanoic acid indicated that the intact C_{18} chain was incorporated into nonacosan-14-ol and nonacosan-15-ol, both of which were identified as natural wax components of *B. oleracea* (239). Exogenous [G-^{3}H]nonacosane was incorporated into nonacosan-14-ol and nonacosan-15-ol in a ratio similar to that found in the natural wax. When a randomly tritiated natural mixture of secondary alcohols was incorporated into ketones by *B. oleracea* it was found that the nonacosan-15-ol was preferentially oxidized to the ketone. These results were consistent with the finding that the naturally occurring ketone was mainly the 15-one. Since the conversion of the exogenous nonacosane to the secondary alcohol required O_2 and was inhibited by metal-ion chelating agents, it was concluded that a mixed-function oxidase hydroxylates the alkane and the resulting secondary alcohol is oxidized to the ketone by an appropriate dehydrogenase (239). It is highly likely that a similar mechanism is involved in the biosynthesis of other secondary alcohols and ketones although the position of hydroxylation and the preference for further oxidation to the ketone might depend on the species and tissue.

6.4.3 Suberin

6.4.3.1 Ultrastructure

6.4.3.1.1 Ultrastructural Characterization

Initial electron microscopic studies of suberized walls impregnated with OsO_4 revealed that the suberin polymer appears as a lamellar structure of alternating light and dark bands (417, 418). Cork cells from the bark of *Quercus suber* and *Acacia seyal* exhibited this lamellar pattern as did the periderm layers from the tuber of *S. tuberosum* and from the velamen of *Vanilla planifolia* (20, 95, 120, 375, 419, 420). It was proposed that the lamellar appearance of suberin was due to alternating layers of wax (light layer) and polymer (dark) (229, 304, 417, 420). Experimental evidence for this hypothesis was provided by the finding that selective inhibition of wax synthesis by trichloroacetate (TCA) prevented the development of these light layers in the wound periderm of *S. tuberosum* (428). Slices of *S. tuberosum* that were wound healed in 2 mM TCA developed periderm layers that exhibited discontinuous and irregularly spaced light layers, as seen in the electron microscope. Tissue wound-healed in the presence of higher concentrations of TCA (8 mM) had almost no light layers in the periderm. Although TCA treatment did not inhibit the deposition of polymeric suberin (as measured by reductive release of $C_{18:1}$ diol), the TCA treatment did inhibit the deposition of waxes into the periderm layer and prevented the development of diffusion resistance (428). This combined ultrastructural and chemical study indicated that the suberin-associated waxes are located in the light layers of the periderm layer and that they are necessary for the development of diffusion resistance. Based on the observation that chloroform extraction of the periderm layer of *S. tuberosum* did not change the ultrastructural appearance, it was proposed that the lamellar appearance is due to alternating layers of suberin polymers differing in polarity (399). However, this conclusion does not take into consideration the possibility that removal of waxes may not lead to collapse of the polymer matrix, and therefore the lucent layers might appear the same in the electron micrograph whether or not the waxes are present. In the case of *S. tuberosum* suberin, which has very little diversity in the aliphatic components, it might be especially difficult to generate polymers differing widely in polarity. In any case, there is no reliable evidence to refute the hypothesis that the lucent layers of suberin are composed of waxes associated with the suberin polymer. Although in most of the cases so far examined suberized walls showed lamellar appearance, such an appearance should not be expected of all suberized cell walls.

Suberized layers in plant roots have been ultrastructurally characterized in detail (457). The characteristic lamellar appearance has been observed in the endodermal cell walls of the roots of several plants (286, 309, 381, 403, 495). Similarly, suberin layers have been observed in hypodermal and epidermal cells of plant roots (77, 78, 122, 174, 334, 365, 366, 380, 498). Lamellar suberin has also been ultrastructurally characterized at various internal locations in the plant such as the bundle sheath of grasses (51, 118, 173, 331, 332), the chalazal region of the seedcoat (111), and in the cell wall of crystal idioblasts (441, 479, 480) (Fig. 6.4.3).

There have been very few combined ultrastructural and chemical analyses of suberin polymers (234). Since a pure polymer cannot be isolated, it is difficult to correlate the amount of lamellar material with the amount of aliphatic and aromatic monomers released by depolymerization. Reduction of suberin-enriched preparations from the periderm layer of roots and tubers released 3% to 10% of the mass as aliphatic material (249) and the chalazal region of the inner seed coat of *Citrus paradisi* released 5% of its mass as aliphatic monomers upon depolymerization; such layers also exhibited a high degree of suberization under the electron microscope (111). Whereas the most suberized cells have 2 to 10 lamellae in the wall, the cell walls of the *Citrus* seed coat were entirely suberized with approximately 40 dark layers in the lamellar array (Fig. 6.4.3). Growth of *Zea mays* under magnesium-deficient conditions gave plants whose roots released 3.5 times as much aliphatic suberin monomers as did roots of control plants (372). Ultrastructural analysis of the hypodermis of these roots showed that the magnesium-deficient roots had 8 to 10 lamellar layers (Fig. 6.4.3) compared to 2 to 3 layers for the control roots. Suberin was shown by both ultrastructural and chemical characterization to be deposited on the surface of cultured protoplasts isolated from the fruit of *Lycopersicon esculentum* (376, 377). Chemical methods used to measure suberin may not detect even a heavily suberized region if the suberin is highly localized and only a very limited portion of the tissue is involved. In such cases ultrastructural approaches could possibly detect suberized regions.

6.4.3.1.2 Ultrastructural Identification of Suberin in Bark

Bark is composed of three layers: an inner bark or phloem, a periderm layer, and an outer bark or rhytidome (110). Each of these layers can vary widely in appearance and thickness depending upon the tree, its age, and environmental conditions. The periderm layer consists of three cell types. The phellogen is the meristem that produces the phelloderm, a parenchymatous layer of cells produced internally, and the phellem, which is formed on the external side. The phellem cells are the cork cells that lay down a layer of suberin over the primary cell wall and seal off the cell, leading to eventual cell death. The anatomy of cork cells in bark has been reviewed by several authors (71, 73, 195, 295, 301, 312, 315, 320, 339, 344, 386, 432, 489, 490), and many of the woody plants whose periderm lay-

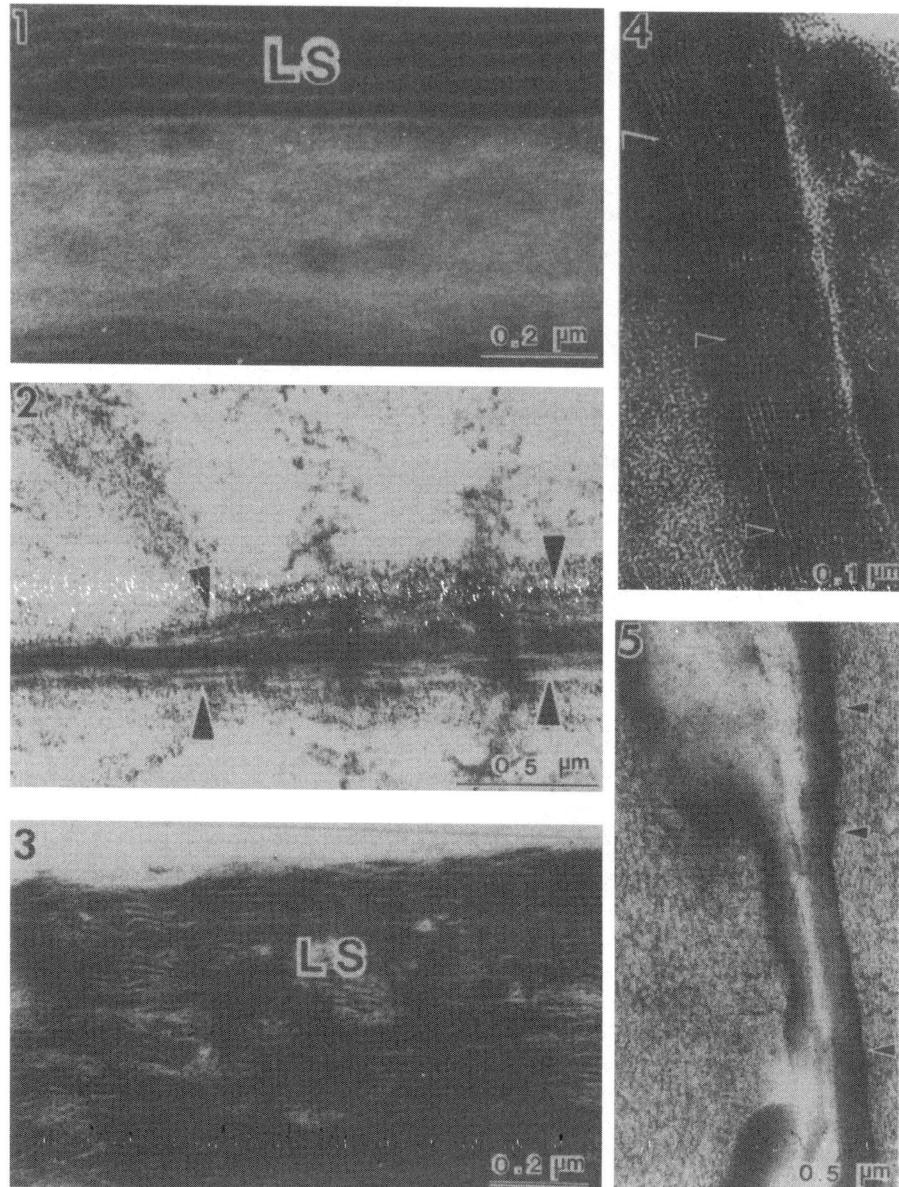

Fig. 6.4.3. Electron micrographs of suberized walls of *Solanum tuberosum* tuber periderm (1), bundle sheaths of *Zea mays* (courtesy of Dr. T.P. O'Brien) (2), the chalazal region of the inner seed coat of *Citrus paradisi* (3), hypodermis in *Zea mays* root grown in Mg^2-deficient medium (4), and idioblast from *Agave americana* leaf (5). Arrowheads, LS = lamellar suberin

Table 6.4.3. Partial listing by family of woody plants whose bark periderm layer[1] has been examined microscopically

Family	Genus/species	Microscopy method[2]	Comment(s)	Reference(s)
Araucariaceae	*Agathis australis*	LM	Crystals in phelloderm cell walls	69
Cupressaceae	*Juniperus occidentalis*	LM	Wound periderm formation	202
	Libocedrus bidwillii	LM	Thin-walled phellem cells	68
Pinaceae	*Abies alba*	LM	Periderm development	142
	Abies balsamea	LM, TEM	Lamellar suberin in phellem cell walls	140, 142, 154, 452
	Abies grandis	LM	Three types of periderm based on fluorescence microscopy	316
	Abies lasiocarpa	LM	Many layers of phellem cells	308
	Abies procera	LM	Suberin in endodermis	492
	Larix decidua	TEM	Lamellar suberin in the walls of CaC_2O_4 crystal cells	476, 482
	Larix laricinia	LM	Wound periderm formation	296, 446
	Picea abies	TEM, SEM	Fungal degradation of bark; lamellar suberin in CaC_2O_4 crystal cells	343, 349, 482
	Picea sitchensis	LM, TEM	Lamellar suberin in endodermis of root	287
	Picea glauca	LM, TEM, SEM	Lamellar suberin in phellem cell walls	138 – 140, 154
	Pinus brutia	LM	Bark cells have high phenolic content	485
	Pinus echinata	LM	Periderm structurally weakest point in bark	293, 294
	Pinus halepensis	LM	Suberization of root cell walls	274
	Pinus nigra	LM	Interlocking phellem cells	352
	Pinus palustris	LM, SEM	Phellem cells extensively interlocked	293, 294
	Pinus pinea	LM	Periderm development	279
	Pinus radiata	LM, TEM	Phellem cells rich in phenolics, lamellar suberin	18, 127, 352, 397
	Pinus resinosa	LM	Periderm development	49
	Pinus rigida	LM	Flat periderm layer	293
	Pinus serotina	LM	Flat periderm layer	293, 294
	Pinus strobus	LM	Broad periderm layer	294
	Pinus sylvestris	SEM	Thin-walled cork cells	121
	Pinus taeda	LM, SEM	Thick-walled phellem cells irregular in shape	293, 294
	Pseudotsuga menziesii	LM, TEM	Lamellar suberin adjacent to wax layer	100, 147, 257, 280, 281, 385
	Tsuga canadensis	LM, TEM	Lamellar suberin in phellem cell walls	140, 153, 154, 452, 487

[1] Studies of periderm layers from regions other than aerial bark are noted.
[2] Method of microscopy: FM, fluorescence microscopy; LM, light microscopy; TEM, transmission electron microscopy; SEM, scanning electron microscopy.

Table 6.4.3 (continued)

Family	Genus/species	Microscopy method[2]	Comment(s)	Reference(s)
	Tsuga heterophylla	LM	Variation in thickness of periderm layers	54, 305, 316
Taxodiaceae	*Sequoia sempervirens*	LM	Thin layer of phellem cells	199
Aceraceae	*Acer saccharum*	LM	Multilayered periderm	294, 374
Anacardiaceae	*Pachycormus discolor*	LM	Thick suberized walls	136
	Rhus typhina	LM	Phellem cells are alternately narrow and wide	110
Araceae	*Acorus calamus*	TEM	Lamellar suberin in oil cell wall	9
Asdepiadaceae	*Hoya carnosa*	LM, TEM, SEM	Lamellar suberin in root exodermis	334
Betulaceae	*Betula papyrifera*	LM	Suberin formed after wounding	29
	Betula pendula	LM, TEM, SEM	Periderm is fairly permeable to water	27, 205, 402
	Betula populifolia	LM	Phellem with thick cell walls	110
Bombacaceae	*Adansonia digitata*	LM	Wound periderm formation	124
Burseraceae	*Bursera coallifera*	LM	Thick-walled periderm cells	143
Celastraceae	*Euonymus alatus*	LM	Development of cork wings	53
Ericaceae	*Rhododendron maximum*	LM	Phellem cells vary in size	110
	Vaccinium corymbosum	LM	Phellem cells vary in form	110
Euphorbiaceae	*Hevea brasiliensis*	LM, TEM	Fungal penetration of lamellar suberin	325–327
Fagaceae	*Fagus grandifolia*	LM	Correlate periderm anatomy to bark pattern	336
	Fagus sylvatica	TEM, SEM	Fungal degradation of bark	60, 348, 350
	Nothofagus menziesii	LM	Thick phellem cell walls	353
	Quercus alba	LM	Layers of cork cells alternate with layers of lignified cells	195, 196
	Quercus falcata	LM	Thick phellem cell walls	195, 196
	Quercus robur	LM, TEM	Lamellar suberin, formation of wound suberin	342, 354, 355, 491
	Quercus rubra	LM	Thin-walled phellem cells	294
	Quercus suber	LM, TEM, SEM	Lamellar suberin in cork cell walls	205, 276, 345, 417–420
	Quercus velutina	LM	Periderm layer not pronounced	294
Fouquieriaceae	*Fouquieria diguetii*	LM	Thick periderm	178
	Fouquieria splendens	LM	Thick-walled cork cells	322
Hippocastanaceae	*Aesculus hippocastanum*	LM	Accumulation of CaC_2O_4 crystal cells in bark	105
Juglandaceae	*Juglans nigra*	LM	Wound periderm formation	11
	Juglans regia	LM	Thick suberin layer presents strong diffusion barrier	18
Lauraceae	*Laurus nobilis*	TEM	Lamellar suberin in wall of oil cells	291
Lecythidaceae	*Grias cauliflora*	LM, TEM	Suberin in crystal cells	346
Leguminosae	*Acacia raddiana*	LM	Periderm development	14
	Acacia senegal	TEM	Lamellar suberin deposition in bark of young twigs	347, 477, 478, 481

Table 6.4.3 (continued)

Family	Genus/species	Microscopy method[2]	Comment(s)	Reference(s)
	Acacia seyal	TEM	Lamellar suberin in cork cell wall	120
	Gleditsia triacanthos	LM, SEM	CaC_2O_4 in suberized cells	48
	Laburnum vossii	LM	Periderm layer	288
	Prosopis spicigera	LM	Wound periderm formation	259
	Robinia pseudoacacia	LM	Periderm development	13, 49, 469
Liliaceae	*Cordyline terminalis*	LM	Periderm has storied crok	110
Magnoliaceae	*Liriodendron tulpifera*	LM	Periderm cells before and after wax extraction	337
Meliaceae	*Entandrophragma condollei*	TEM	Lamellar cell wall of bark sclereids	340
Myristicaceae	*Dialyanthera otoba*	LM	Periderm development	277
Myrtaceae	*Eucalyptus elaeophora*	LM	Periderm development	373
	Eucalyptus fastigata	LM	Suberin in endodermis of roots	278
	Eucalyptus marginata	LM	Suberin formed in response to fungal invasion	455
	Eucalyptus obliqua	LM, TEM	Lamellar suberin in root cell walls	454
	Eucalyptus st. johnii	LM, TEM	Lamellar suberin in root cell walls	454
	Eugenia capensis	LM	Thin-walled phellem cells	463
	Melaleuca quinquenervia	LM, SEM	Many periderm layers	75
	Tristania conferta	LM	Thick-walled phellem cells	19
Oleaceae	*Fraxinus pennsylvanica*	LM	Periderm development	49
Orchidaceae	*Vanilla planifolia*	TEM	Lamellar suberin in aerial root cell walls	120
Podocarpaceae	*Phyllocladus alpinus*	SEM	No rhytidome	258
Proteaceae	*Hakea suaveolens*	LM	Periderm layer	288
Rhizophoraceae	*Bruguiera gymnorrhiza*	TEM	Suberin in roots	272
	Rhizophora mangle	LM	Periderm development	214
Rosaceae	*Fragaria chiloensis*	LM	Polyderm in root	323
	Malus pumila	LM, TEM, SEM	Lamellar suberin in periderm formed during fruit russeting and wound-healing	97, 256, 413
	Prunus persica	FM, LM, TEM	Lamellar suberin formed after wounding	28, 30, 36, 37
	Rubus crataegifolius	LM	Wound periderm formation	302, 496
Rutaceae	*Citrus limonia*	LM	Periderm development	400
	Citrus reticulata	LM, TEM	Salt stress causes suberization	470
Salicaceae	*Populus grandidentata*	LM	Alternating layers of thick- and thin-walled phellem cells	294
	Populus tremuloides	LM	CaC_2O_4 in periderm	194, 314
	Populus trichocarpa	LM	Periderm acts as fungal barrier	41
Salvadoraceae	*Azima tetracantha*	LM	Thin-walled periderm cells	338

Table 6.4.3 (continued)

Family	Genus/species	Microscopy method[2]	Comment(s)	Reference(s)
Simarou-baceae	*Ailanthus altissima*	LM	Periderm development	49
Tamaricaceae	*Tamarix aphylla*	LM, TEM	Lamellar suberin in salt glands	50
Tiliaceae	*Tilia americana*	LM	Suberin formed after wounding	29
	Tilia cordata	LM, TEM	Lamellar suberin in endodermis of root	287
Ulmaceae	*Ulmus americana*	LM	Periderm has alternating layers of thick- and thin-walled cells	294
Verbenaceae	*Tectona grandis*	LM, SEM	Broad periderm layers	144
Vitaceae	*Vitis vinifera*	LM	Suberin protects plant from freezing damage	303, 351
Zygophyll-aceae	*Balanites aegyptiaca*	LM	Thin-walled phellem cells	341
	Larrea tridentata	TEM, SEM	Lamellar suberin adjacent to trichome	451

ers have been examined microscopically are listed by family in Table 6.4.3. Suberin layers detected in bark with the light microscope have usually been characterized by histochemical stains (204).

Recently periderm layers from the bark of several species of trees were ultrastructurally characterized. Suberin deposition appeared to immediately precede the deposition of a tertiary carbohydrate layer during the formation of cork cells in the bark of *Acacia senegal* (477). A tertiary carbohydrate layer overlaying the suberin lamellae was also seen in the periderm of *Acorus calamus*, but not in the suberized periderms of *Larix decidua* or *Picea abies* (478, 482). It is possible that

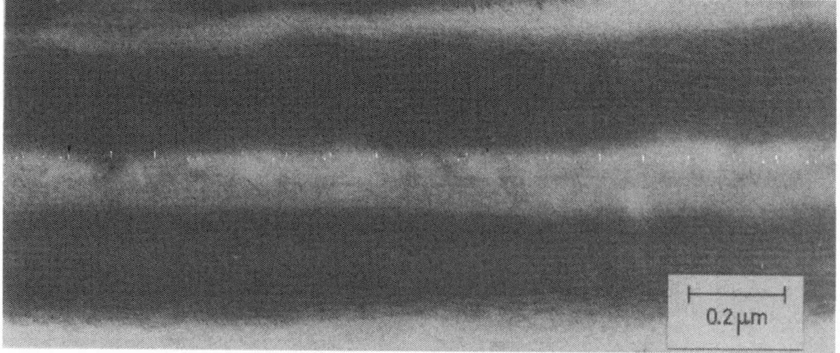

Fig. 6.4.4. Electron micrograph of lamellar suberin from periderm in the bark of *Pseudotsuga menziesii* (courtesy of R.W. Davis)

some suberized gymnosperm cell walls may have a lower carbohydrate content and may have very different permeability characteristics. No tertiary wall is seen in the periderm layer of *Pseudotsuga menziesii* (Fig. 6.4.4) where the suberized layer comprised the majority of the cell wall and an extractable wax layer appeared to be deposited inside the suberin (281). The fact that the wall of *P. menziesii* is so highly suberized may explain the low proportion of carbohydrate seen in the ^{13}C-NMR spectrum of this suberin (see Sect. 6.4.3.2.3). A detailed analysis of the phellem cells from the bark of *Tsuga canadensis* also indicated that suberized lamellae comprised the major part of the cell wall and that there was no tertiary carbohydrate deposited on the suberin layer (153). Phellem cells from the bark of *Picea glauca* and *Abies balsamea* as well as *Tsuga canadensis* all had suberized walls with a thin, lucent, apparently wax-like, layer deposited over the suberin (140). There is, as yet, no known correlation between bark morphology and ultrastructural appearance of the suberized layers of periderms, nor is there any definite correlation between chemical composition of the polymer and its ultrastructural appearance.

6.4.3.2 Chemical Composition and Structure of the Polymer

6.4.3.2.1 Composition of the Aliphatic Portion

Chemical analysis of suberin from the cork of *Quercus suber* began in the nineteenth century when it was shown that base hydrolysis of the polymer released aliphatic acids (292). The discovery that many of these acids were hydroxy fatty acids suggested that the structure of suberin might be similar to that of the epidermal polymer cutin in that suberin has domains of interesterified fatty acids. Research done in the 1950s on suberin from the bark of *Q. suber*, *Betula verrucosa* and *Pseudotsuga menziesii* began to give a general idea about the aliphatic com-

Table 6.4.4. Trivial and systematic names of *n*-aliphatic acids found in suberin

Chain length	Trivial name	Systematic name
C_{16}	Palmitic	Hexadecanoic
	Juniperic	16-hydroxyhexadecanoic
	Thapsic	Hexadecane-1,16-dioic
C_{18}	Stearic	Octadecanoic
	Suberinic	18-hydroxyoctade-9-enoic
	Phloionolic	9,10,18-trihydroxyoctadecanoic
	Phloionic	9,10-dihydroxyoctadecane-1,18-dioic
C_{20}	Arachadic	Eicosanoic
C_{22}	Behenic	Docosanoic
	Phellonic	22-hydroxydocosanoic
	Phellogenic	Docosane-1,22-dioic
C_{24}	Lignoceric	Tetracosanoic
C_{26}	Cerotic	Hexacosanoic
C_{28}	Montanic	Octacosanoic

position of suberin (180, 207–210, 262). The structures of several of the hydroxylated acids were determined over a period of several years (101, 102, 157, 158). Several of the polymer components were given trivial names prior to complete characterization. Some of these components are listed together with their systematic names in Table 6.4.4.

The advent of combined gas chromatography/mass spectrometry has made the characterization of the aliphatic portion of the suberin polymer relatively straightforward. It is important that the initial physical isolation of the suberized tissue be done with care as this step determines the degree of suberin-enrichment in the final fraction (235, 243). Treatment of the tissue with hydrolytic enzymes, such as pectinase and cellulase, may remove a considerable portion of the cell wall carbohydrate. Strongly acidic, basic, or oxidizing conditions should be avoided to prevent chemical changes in the suberin components. The final residue is dried,

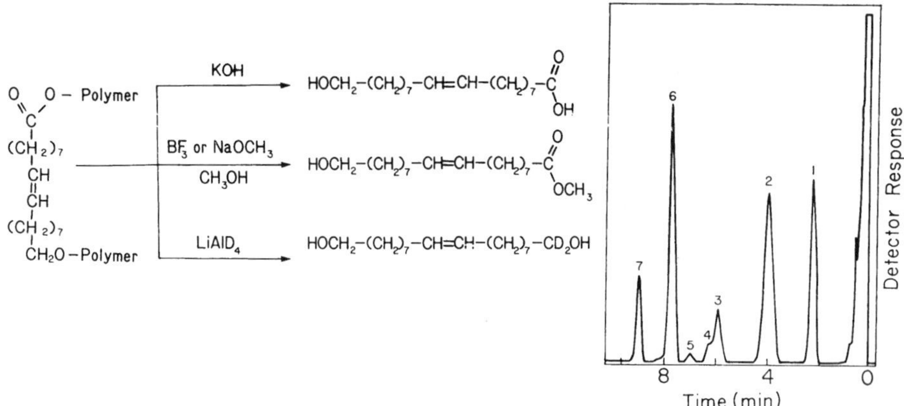

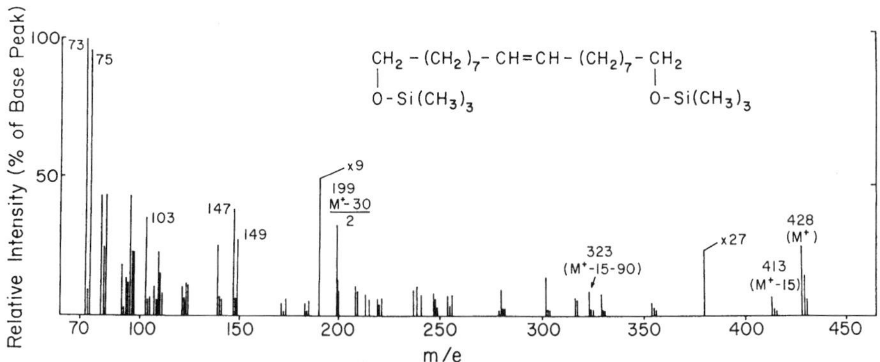

Fig. 6.4.5. Methods used to analyze the aliphatic components of suberin. Top left: chemical methods used to depolymerize suberin. Top right: gas-liquid chromatogram of the mixture of monomers generated by $LiAlD_4$ treatment of suberin from the chalazal region of the inner seed coat of *Citrus paradisi*; the components are trimethylsilyl ethers of 1,16-dihydroxyhexadecane (1), 1,18-dihydroxyoctadecene (2), 1,9,18-trihydroxyoctadecene (3), 1,9,18-trihydroxyoctadecane (4), 1,9,10,18-tetrahydroxyoctadecane (5), 1,22-dihydroxydocosane (6), 1,24-dihydroxytetracosane (7). Bottom: mass spectrum of component 2 from gas chromatogram (111)

ground to a fine powder, and then thoroughly extracted with organic solvents to remove all non-polymeric components. Since it is extremely difficult to remove all of the wax components trapped in the polymer matrix, the efficiency of the extensive solvent extraction should be verified before the final residue is subjected to chemical studies.

The aliphatic components of the polymer can be released by a variety of depolymerization techniques that rely on cleavage of ester bonds. These methods, which have been employed to depolymerize either suberin or cutin, include alkaline hydrolysis (88, 106, 297), transesterification with sodium methoxide/methanol (190, 235, 249, 254) and hydrogenolysis with $LiAlH_4$ in tetrahydrofuran (212, 227, 252, 472). The use of $LiAlD_4$ permits identification of the original carboxyl carbon by deuterium labeling (472). This technique also allows identification of other moieties such as ketones and epoxides by the incorporation of a deuterium during reduction. Such deuterium labeling led to the first recognition that 9,10-epoxy-18-hydroxyoctadecanoic acid was a major component in some lipid-derived plant polymers (254). The commonly used techniques are illustrated in Fig. 6.4.5.

Monomers released from suberin by depolymerization can be identified by gas chromatography/mass spectrometry with or without initial separation of the various classes of monomers by thin-layer chromatography. Suberin from the periderm layer of several underground root and tuber crops was analyzed by combined gas-liquid chromatography/mass spectrometry (55, 56, 235, 249, 382, 383). On the basis of these studies, certain characteristic features of suberin were recognized and these features were proposed to be useful in distinguishing suberin from cutin (235, 249). Thus suberin characteristically has high proportions of long-chain ($C_{16}-C_{24}$) ω-hydroxy acids and dicarboxylic acids with smaller amounts of very-long-chain ($C_{20}-C_{30}$) fatty acids and fatty alcohols (Table 6.4.5). This generalization held true when the compositions of the aliphatic components of suberin from a variety of anatomical locations were subsequently determined (231, 245). Very often mono-unsaturated C_{18} monomers are the dominant aliphatic components. This is true for suberin in the periderm of the tuber of *Solanum tuberosum* (Table 6.4.5), the root of *Vicia faba* (112), the root of *Manihot esculenta*, the root of *Zingiber officinale*, and the velamen of *Philodendron pertusum* (unpublished results). When the aliphatic portion of the polymer is comprised of monomers of various chain lengths, ω-hydroxy $C_{18:1}$ acid and $C_{18:1}$ diacid are often the major monomers with saturated C_{16}, C_{20}, C_{22}, and C_{24} ω-hydroxy acids and diacids comprising a smaller proportion of the polymer. There are exceptions, however. The aliphatic portion of the suberin polymer found in the fiber of green cotton (*Gossypium hirsutum*) consists of primarily two monomers, ω-hydroxydocosanoic acid (70%) and docosanedioic acid (25%) (393, 501).

It has been reported that the major aliphatic monomers of suberin polymers from several tree barks are fatty acids with both ω- and mid-chain oxygenation (192). These monomers include 9,10-epoxy-18-hydroxyoctadecanoic acid and 9,10,18-trihydroxyoctadecanoic acid, both of which are common components of cutin polymers (232). For example, 9,10-epoxy-18-hydroxyoctadecanoic acid was reported to be the major monomer (31%) of *Quercus ilex* bark suberin and

Table 6.4.5. Composition of the aliphatic components of suberin from the periderm of the underground storage organs of three plants

Component	Monomer as % of aliphatic component[1]	Chain-length composition								
		C_{16}	$C_{18:1}$	C_{18}	C_{20}	C_{22}	C_{24}	C_{26}	C_{28}	C_{30}
Fatty acid										
in *S. tuberosum*	8	4.2	7.0	–	3.5	14.2	16.0	16.1	31.7	–
P. sativa	14	7.1	6.8	21.0	18.0	15.4	22.5	2.4	3.9	–
I. batatas	9	0.9	3.9	1.1	0.9	3.3	4.7	16.6	20.1	34.3
Fatty alcohol										
in *S. tuberosum*	6	3.1	4.8	–	9.8	26.5	12.9	18.1	24.6	–
P. sativa	4	–	–	5.1	7.1	48.8	36.6	2.4	–	–
I. batatas	4	–	–	68.3	15.5	6.1	3.0	3.0	0.6	0.1
ω-Hydroxy acid										
in *S. tuberosum*	61	0.6	95.5	–	–	1.1	1.7	0.8	0.3	
P. sativa	30	31.7	53.3	8.3	4.2	1.7	0.2	–	–	
I. batatas	36	4.9	90.7	1.3	0.3	0.3	0.5	0.9	0.7	
Dicarboxylic acid										
in *S. tuberosum*	24	4.3	95.6	–	–	–	–	–	–	
P. sativa	27	23.8	45.1	16.9	5.0	5.7	0.4	–	–	
I. batatas	21	6.6	80.5	9.1	0.2	–	–	–	–	

[1] Ref. 235, 249.

9,10,18-trihydroxyoctadecanoic acid was found to be the major monomer (32%) of *Castanea sativa* bark suberin (192). Since the bark samples analyzed constituted ill-defined mixtures of anatomical portions that were not properly examined, the possibility exists that the tissue analyzed contained some proportion of a cutin polymer in addition to the suberin. An ultrastructural examination of suberized cork cells in the bark of young twigs of *Acacia senegal* revealed that the outermost layer of cells was an epidermis with a distinct cuticle (477). It is certainly possible that there are suberin polymers with high proportions of mid-chain oxygenation in the aliphatic monomers. Such polymers probably have extensive aliphatic polyester domains similar to cutin, and these domains serve unique functions. Table 6.4.6 lists the trees and shrubs from which the suberin polymer in the bark has been chemically analyzed and also lists the major monomers reported for each polymer.

6.4.3.2.2 Phenolic Composition

The phenolic domain of the suberin polymer has not been well characterized. Early chemical studies on suberin from tree bark indicated that the polymer contains phenolics. Suberin from the cork of *Pseudotsuga menziesii* bark released phenolic acids (43% of extractive-free cork) and aliphatic acids (35%) upon saponification (180). Suberin, which was subjected to a limited saponification (2% KOH at room temperature) and successive solvent extraction with hexane,

Table 6.4.6. Major components released by the depolymerization of suberin polymers from the bark of various trees and shrubs.

Family	Genus/species	Suberin content (%)	Depolymerization products		Reference(s)
			Component	Content (%)	
Cupressaceae	*Chamaecyparis nootkatensis*	1.4	C_{18} diacid	–	444
			ferulic acid	–	
	Cupressus leylandii	28	ω-hydroxy C_{16},	13	192
			ω-hydroxy $C_{18:1}$,	18	
			$C_{18:1}$ diacid	17	
Ginkgoaceae	*Gingko biloba*	11	ω-hydroxy C_{22},	–	170
			C_{22} diacid	–	
Pinaceae	*Abies amabilis*	1.8	ω-hydroxy C_{18},	–	444
			C_{18} diacid,	–	
			ferulic acid	–	
	Abies concolor	–	Hydroxy fatty acid,	–	119, 179
			ferulic acid	–	
	Picea sitchensis	1	ω-hydroxy C_{18},	–	444
			C_{18} diacid,	–	
			ferulic acid	–	
	Pinus densiflora	3	ω-hydroxy C_{22},	–	170
			C_{22} diacid	–	
	Pinus pinaster	1	C_{22} diacid	–	464
	Pinus radiata	–	ω-hydroxy C_{16},	11	129
			ω-hydroxy $C_{18:1}$,	9	
			C_{18},	8	
			C_{16} diacid,	6	
			dihydroxy C_{16}	–	
	Pinus taeda	2	–	–	359
	Pinus virginiana	1	–	–	267
	Pseudotsuga menziesii	21	ω-hydroxy C_{16},	2	179, 180, 292, 444
			ω-hydroxy C_{22},	–	
			ferulic acid	–	

Table 6.4.6 (continued)

Family	Genus/species	Suberin content (%)	Depolymerization products		Reference(s)
			Component	Content (%)	
	Thuja plicata	4.5	ω-hydroxy C_{18},	–	
			C_{18} diacid,	–	443, 444
			ferulic acid	–	444
	Tsuga heterophylla	7	ω-hydroxy C_{18},	–	
			C_{18} diacid,	–	444
			ferulic acid	–	
	Tsuga mertensiana	8	ω-hydroxy C_{18},	–	
			C_{18} diacid,	–	444
			ferulic acid	–	
Aceraceae	*Acer griseum*	26	ω-hydroxy $C_{18:1}$,	7	192
			ω-hydroxy C_{22},	24	
			9,10-dihydroxy C_{18} diacid,	17	
			9,10,18-trihydroxy C_{18}	11	
	Acer pseudoplatanus	27	ω-hydroxy C_{16},	13	192
			ω-hydroxy C_{22},	12	
			C_{16} diacid	12	
Betulaceae	*Betula pendula*	43	ω-hydroxy C_{22},	16	192
			9,10,18-trihydroxy C_{18}	43	
	Betula platyphylla	–	alkane diols,	–	1
			alkane triols	–	
	Betula verrucosa	30	ω-hydroxy C_{22},	14	43, 44, 107
			9,10-epoxy-18-hydroxy C_{18}	36	
Caprifoliceae	*Sambucus nigra*	22	ω-hydroxy $C_{18:1}$,	22	
			ω-hydroxy C_{16} diacid,	9	192
			$C_{18:1}$ diacid	11	
Celastraceae	*Euonymus alatus*	8	ω-hydroxy $C_{18:1}$,	22	192
			ω-hydroxy C_{22},	18	
			ω-hydroxy C_{24}	14	
Fagaceae	*Castanea sativa*	43	9,10-dihydroxy C_{18},	5	192
			9,10,18-trihydroxy C_{18}	32	
			9,10-epoxy-18-hydroxy C_{18}	13	

	Fagus sylvatica	48	ω-hydroxy C_{22},	9	170, 192
			9,10,18-trihydroxy C_{18},	26	
			9,10-epoxy-18-hydroxy C_{18}	29	
	Quercus acutissima	10	ω-hydroxy C_{22},		170
			C_{22} diacid		
	Quercus robur	25	ω-hydroxy $C_{18:1}$,	7	192
			9,10,18-trihydroxy C_{18},	17	
			9,10-epoxy-18-hydroxy C_{18}	31	
	Quercus ilex	40	ω-hydroxy $C_{18:1}$,	8	192
			9,10,18-trihydroxy C_{18}	25	
			9,10-epoxy-18-hydroxy C_{18}	23	
	Quercus suber	38	ω-hydroxy $C_{18:1}$,	12	6–8, 12, 179, 188
			ω-hydroxy C_{22}	25	203, 362, 363,
					382, 383
			9,10-dihydroxy C_{18} diacid	15	406, 407, 431
Guttiferae	*Kielmeyera coriacea*	46	C_{16} diacid,		160
			9,10-dihydroxy C_{18} diacid,		
			C_{28} fatty acid		
Juglandaceae	*Carya ovata*	7	–	166	
Leguminosae	*Labrunum anagyroides*	62	ω-hydroxy C_{16},	15	192
			ω-hydroxy $C_{18:1}$,	25	
			C_{16} diacid,	9	
			dihydroxy C_{16}	9	
Malvaceae	*Gossypium hirsutum*	–	C_{16} diacid,	13	392
			$C_{18:1}$	35	
Oleaceae	*Fraxinus excelsior*	22	ω-hydroxy $C_{18:1}$ diacid,	15	192
			$C_{18:1}$,	12	
			9,10-dihydroxy C_{18} diacid	14	
Rosaceae	*Malus pumila*	31	ω-hydroxy C_{22},	8	191
			9,10-dihydroxy C_{18} diacid,	9	
			9,10,18-trihydroxy C_{18}	39	
	Rubus idaeus	–	ω-hydroxy $C_{18:1}$,	19	123
			$C_{18:1}$ diacid,	12	
			dihydroxy C_{16}	7	
Salicaceae	*Populus tremula*	38	ω-hydroxy C_{16},	22	192
			ω-hydroxy $C_{18:1}$,	21	

Table 6.4.6 (continued)

Family	Genus/species	Suberin content (%)	Depolymerization products		Reference(s)
			Component	Content (%)	
			C_{16} diacid,	13	
			$C_{18:1}$ diacid	11	
Saxifragaceae	*Ribes americanum*	30	ω-hydroxy C_{16},	24	189
			C_{16} diacid	15	
	Ribes davidii	31	ω-hydroxy C_{16},	15	189
			ω-hydroxy $C_{18:1}$,	21	
			$C_{18:1}$ diacid,	12	
			C_{20} α,ω-diol	6	
	Ribes futurum	31	ω-hydroxy C_{16},	19	189
			ω-hydroxy $C_{18:1}$,	15	
			$C_{18:1}$ diacid,	10	
			C_{20} α,ω-diol	7	
	Ribes grossularia	16	ω-hydroxy C_{16},	11	189
			ω-hydroxy $C_{18:1}$,	26	
			$C_{18:1}$ diacid,	15	
			C_{20} α,ω-diol	7	
	Ribes houghtonianum	29	ω-hydroxy C_{16},	19	189
			ω-hydroxy $C_{18:1}$,	17	
			$C_{18:1}$ diacid,	12	
			C_{20} α,ω-diol	8	
	Ribes nigram	23	ω-hydroxy $C_{18:1}$,	14	189
			ω-hydroxy C_{20},	15	
			C_{20} α,ω-diol	6	

[1] Terazawa M, Miyake M 1979 Lipid polymers in plant surface. [Presented, 177th Am Chem Soc Meet Honolulu] ACS Washington DC

benzene, and ethyl ether yielded a soluble fraction (22% of the extractive-free cork) that was partially characterized as a phenolic acid-hydroxy acid ester. Saponification of this ester gave a fatty acid characterized by mixed melting point as 11-hydroxy lauric acid. Although recent examination of the suberin of *P. menziesii* bark by reductive depolymerization and GC/MS analysis revealed that the major aliphatic monomers are C_{16} and $C_{18:1}$ ω-hydroxy acids and diacids (unpublished results), it seems quite likely that a phenolic acid-hydroxy acid ester (or mixture of esters) was isolated by the above procedure. Such an ester was isolated after partial saponification of the cork suberin from *Quercus suber* (159). This component was shown to be ferulic acid esterified to ω-hydroxydocosanoic acid. Ferulic acid was also shown to be a component (6%) of suberin from the bark of *Abies concolor* (179). A 1% HCl/dioxane extract of *Quercus suber* cork yielded a waxy fraction (55% of the extracted cork) thought to be a mixture of phenolic acids esterified to hydroxy aliphatic acids (179). Saponification of this fraction released hydroxy aliphatic acids (65%), uncharacterized phenolic acids (30%), and 3,5-dimethoxy-4-hydroxycinnamic acid (5%). Depolymerization of *Thuja plicata* bark suberin with KOH/ethanol yielded ethyl esters of 3,4-dimethoxycinnamic acid and ferulic acid, as well as ω-hydroxyoctadecanoic acid (443). Similar treatment released ferulic acid and hydroxylated fatty acids from the bark suberin of *Tsuga mertensiana*, *T. heterophylla*, *Abies asabilis*, *Chamaecyparis nootkatensis*, and *Picea sitchensis* (444). Whether these acids were part of the unextracted wax or were covalently attached to the suberin polymer remains to be established. More recently small amounts of ferulic acid were shown to be esterified to the suberin polymer in the periderm layer of the underground storage organ of *Ipomoea batatas* and in the periderm of storage organs of several other plants (379). Depolymerization of the suberin polymer from all sources results in the release of colored, water-soluble phenolic components and leaves behind an insoluble residue that contains phenolic components (188, 233, 235, 249). The nature of these phenolic components derived from suberin is still unknown.

Alkaline nitrobenzene oxidation, similar to that used in the study of lignin structure (70, 164, 311), was recently used in an effort to release aromatic components from the suberin polymer. Nitrobenzene oxidation of suberized tissue from wound-healed slices of *Solanum tuberosum* released *p*-hydroxybenzaldehyde and vanillin (85). Nitrobenzene oxidation of suberin from the hypodermis and endodermis of *Zea mays* roots (372) and from the calcium oxalate crystal idioblasts of *Agave americana* (117) gave *p*-hydroxybenzaldehyde and vanillin, whereas oxidation of the lignin-containing vascular tissue from *Agave* produced a large amount of syringaldehyde (Fig. 6.4.6). It seems that a characteristic of a suberin polymer may be that it generates higher proportions of *p*-hydroxybenzaldehyde and vanillin, whereas lignin polymers give larger amounts of syringaldehyde. Thus nitrobenzene oxidation of bark from *Populus trimuloides* yielded vanillin and syringaldehyde in a 1:1 ratio whereas the ratio from oxidation of wood of the same tree was 1:3 (79). Similarly, the oxidation of bark samples from *Pinus monticola* (168) and *P. thunbergii* (184) released higher levels of vanillin and *p*-hydroxybenzaldehyde relative to syringaldehyde when compared to oxidation of wood lignin from the same source. Both the natural and wound-healed periderm

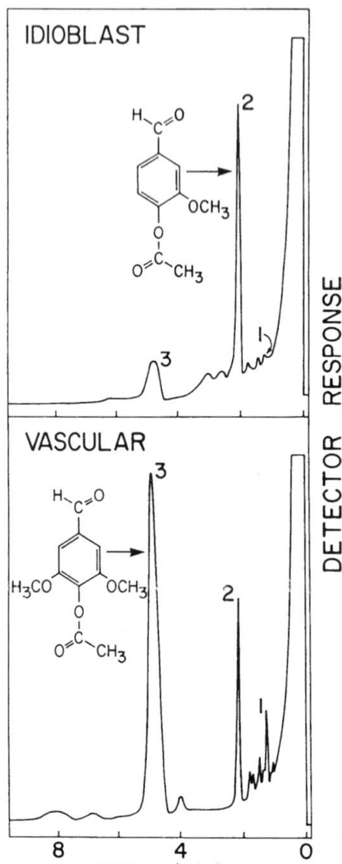

Fig. 6.4.6. Gas-liquid chromatograms of acetylated aldehydes generated by alkaline nitrobenzene oxidation of suberized idioblast cell walls and lignified vascular tissue from *Agave americana* leaves. Compounds 1, 2, and 3 were identified by their mass spectra to be acetylated *p*-hydroxybenzaldehyde, vanillin, and syringaldehyde, respectively (117)

of *Ipomoea batatas* gave primarily vanillin and *p*-hydroxybenzaldehyde upon oxidation (471). Higher levels of vanillin and *p*-hydroxybenzaldehyde were also produced from polymers generated by wound-healing tissue of *Raphanus sativa* (15) and *Lycopersicon esculentum* (137), whereas in both cases oxidation of the normal tissue gave higher levels of syringaldehyde. In these two cases and in many others (133, 462), the polymeric material laid down by the plant was designated "wound lignin," but it has been suggested that in many instances these polymers may in fact be suberin (233). Chemical analysis of both the aliphatic and aromatic components of these polymers would help to resolve this question.

A fraction analogous to Björkman lignin (39) was obtained when a portion of finely-powdered suberin polymer from the periderm of *S. tuberosum* was solubilized with dioxane. This soluble fraction, however, was not enriched in either aromatic or aliphatic components over the insoluble residue (unpublished results). Other procedures from lignin chemistry – including refluxing in HCl/dioxane or HCl/dimethylformamide, and dioxane treatment at 160 °C and high pressure (268, 396) – resulted in 20% to 50% solubilization of the suberin preparation, but with each method the insoluble material contained the majority of both the aromatic and aliphatic components of the polymer. In the past, treatment with HCl/dioxane has been used to solubilize portions of bark from *Abies concolor* (179), *Quercus crispula* (169), *Pinus densiflora* (424), and *Pseudotsuga menziesii* (180). Examination of these solubilized fractions with modern analytical techniques should yield additional information regarding the aromatic portion of the suberin polymer.

Histochemical and ultrastructural studies also indicate that suberized layers contain aromatic components. Phenolic components were detected histochemically in suberized regions (epidermis and endodermis) of *Eucalyptus fastigata* (278), *E. obliqua*, and *E. St. johnii* (476). Ferric chloride has been utilized as a phenolic-specific stain suitable for light microscopy and scanning and transmission electron microscopy (57). With this stain, phenolics have been shown to be present in the suberin layers in the hypodermis of the roots of *Allium cepa* (370), the endodermis of *Ranunculus acris* roots (404), and the Casparian band of *Tulipa kaufmanniana* roots (499). These results coupled with the chemical studies described above clearly indicate that aromatic components comprise a significant portion of the suberin polymer.

6.4.3.2.3 Structure

Based on the information available, a tentative model has been proposed for the suberin structure (229, 231, 233) (Fig. 6.4.7). According to this model a phenolic matrix somewhat similar to lignin is covalently attached to the cell wall. The aliphatic components, which cannot form an extensive polymer, are probably attached to the phenolic polymer. The dicarboxylic acids and ω-hydroxy acids may serve to cross-link the aromatic matrix, and the ω-hydroxy acids may form linear polyester domains. Suberin polymers with high proportions of mid-chain oxygenated aliphatic acids may have larger polyester domains similar to cutin. Since the very-long-chain fatty acids and fatty alcohols have only one functional group, these components must be esterified to the polymer in terminal positions. As shown in Fig. 6.4.7, the aliphatic portion of the polymer may interact with the non-polar wax layers, which are thought to alternate with polymer layers to generate the characteristic lamellar appearance of the polymer.

Initial studies, primarily done on suberin from the periderm of *S. tuberosum*, give additional indication of the secondary structure of the polymer. Enriched preparations of suberin always contain high levels of carbohydrate (as much as 50%). Obviously the suberin-enriched preparations contain covalently attached cell wall carbohydrates and the linkages between suberin and carbohydrate may be similar to those proposed for the attachment of lignin to carbohydrate (132, 268). Chemical studies on the polymer have shown that very few of the hydroxyl groups of ω-hydroxy acids are free and that the polymer has very few ($<5\%$) free aliphatic carboxyl moieties (230). Fractionation of *S. tuberosum* suberin by partial solubilization with 1% HCl/dioxane has indicated that the aliphatic components may be in separate domains, for polymeric fractions that contained a larger proportion of fatty acids and fatty alcohols but a lower proportion of ω-hydroxy acids and dicarboxylic acids have been isolated. These fatty acids and fatty alco-

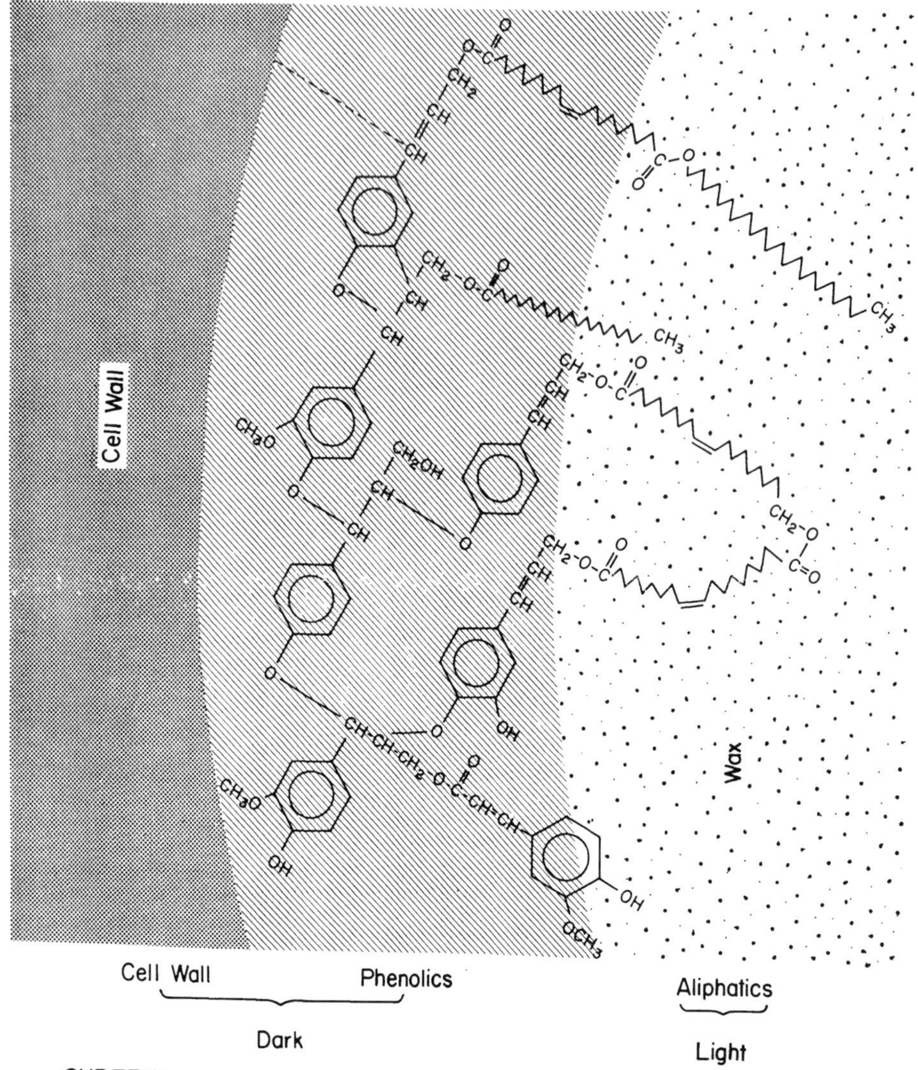

SUBERIN

Major Monomers

$CH_3(CH_2)_mCOOH$

$CH_3(CH_2)_mCH_2OH$

$\underset{\displaystyle OH}{\underset{|}{C}}H_2(CH_2)_nCOOH$

$HOOC(CH_2)_nCOOH$

Phenolics

(m = 18 – 30 ; n = 14 – 20)

Fig. 6.4.7. The major monomers of suberin and a proposed model for the polymer. Dark and light refer to the lamellar layers seen in electron micrographs of suberin

hols may be terminally attached to a distinct region of the phenolic matrix that can be fractionated by this treatment (unpublished results).

The insoluble nature of suberin prevents its analysis by conventional NMR spectroscopy. The development of solid sample ^{13}C-NMR spectroscopy with cross polarization/magic-angle spinning provides an opportunity to obtain spectra of the intact polymer. The NMR spectrum of the cutin polymer from the fruit of *Malus pumila* indicates, as expected, that the majority of carbons are methylenes (CH_2, 20–30 ppm). The spectrum of suberin from *S. tuberosum* showed the presence of a high proportion of aliphatic CH_2 but also had a large amount of CHOH (60 to 80 ppm) carbon, probably from contaminating cell wall carbohydrates. This presumed carbohydrate material was still present in the residue left after $LiAlH_4$ reduction, whereas the majority of aliphatic CH_2 had been removed by this treatment (Fig. 6.4.8). This result shows that not much of the

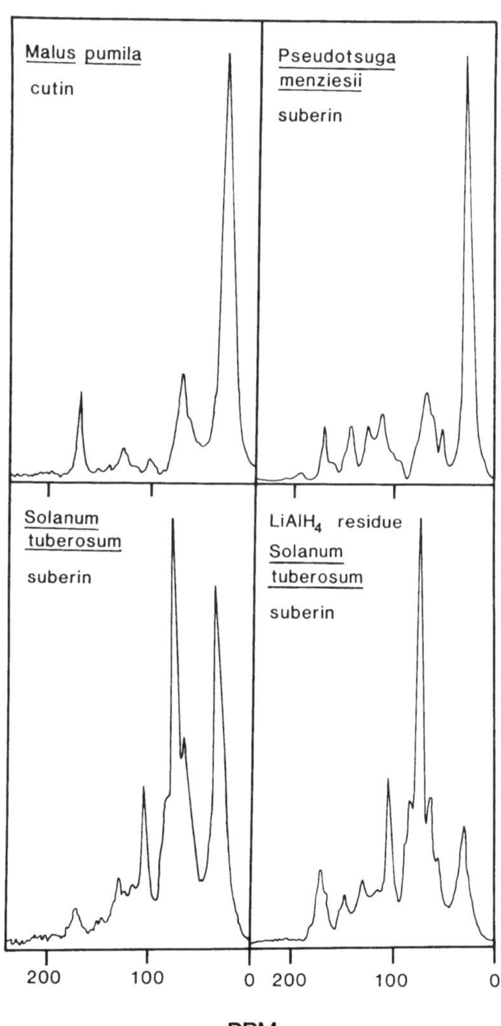

Fig. 6.4.8. ^{13}C Cross-polarization/magic-angle spinning NMR spectra of cutin from the fruit peel of *Malus pumila*, suberin from the bark of *Pseudotsuga menziesii*, suberin from the periderm of the tuber of *Solanum tuberosum*, and the residue left after $LiAlH_4$ depolymerization of *S. tuberosum* suberin. (Spectra courtesy of Regional NMR Center, Fort Collins, Colorado.)

aliphatic components of suberin can be bound by linkages not susceptible to $LiAlH_4$ cleavage. The *S. tuberosum* samples seem to have a low amount of aromatic carbons (120 to160 ppm). The spectrum of the sample from *Pseudotsuga menziesii* bark indicated that the proportion of aromatic carbon was much higher in this suberin polymer, whereas the amount of carbohydrate appeared to be quite low (Fig. 6.4.8). An early H-NMR study of an HCl/dioxane extract of *Thuja plicata* bark indicated that there were few aromatic protons, a high proportion of aliphatic protons, and fewer O-CH_3 moieties than were found in a comparable extract from wood of the same plant (442). Although a more recent ^{13}C-NMR study of extracts from suberin-enriched preparations from *Rubus idaeus*, *Quercus suber*, and *S. tuberosum* found very low levels of aromatic components (506), it is inlikely that either the aromatic or aliphatic domains of the polymer were solubilized under the conditions used, and no test was performed to determine whether any characteristic suberin component was present in the extract examined. It is highly likely that true suberin was not even included in the sample studied. In any case the limited array of NMR results so far available indicates that the aromatic matrix of suberin is more highly cross-linked than that in lignin and this conclusion is in agreement with the observation that only low levels of aromatic aldehydes were released by nitrobenzene oxidation of suberin samples (85, 117, 372). Similar low levels of aromatic aldehydes were also released from an aromatic matrix generated from phenolic acids by the isoperoxidase of isolated cell walls (40).

The following observations support a proposed model for the structure of suberin: 1) Depolymerization techniques that cleave ester bonds release aliphatic monomers from suberin (235, 249), and the composition of these monomers suggestes that they cannot by themselves form an extensive polymer; 2) Treatment of suberin from several sources with nitrobenzene generated vanillin and *p*-hydroxybenzaldehyde (85, 117, 372, 412) rather than syringaldehyde, usually generated by oxidation of lignin; 3) Suberized cell walls stain positively for phenolics with indications that suberin contains monohydroxy aromatic rings (404) and has fewer *O*-methoxy groups than does lignin (233); 4) Inability to solubilize the aromatic components of suberin-enriched preparations by the methods used for lignin suggests that suberin structure is distinctly different from that of lignin, possibly due to aliphatic cross-linking and a higher degree of condensation (233); 5) Phenolic acids (213) and aliphatic acids (94) are both involved in the biosynthesis of suberin, and phenolic acids are not synthesized in tissue slices that do not undergo suberization (47, 213). Inhibition of the synthesis of the aromatic matrix by inhibitors of phenylalaline ammonia lyase inhibited deposition of the aliphatic components and prevented development of diffusion resistance (W. Cottle and P.E. Kolattukudy, unpublished). Inhibition of synthesis of the peroxidase by iron deficiency prevented deposition of aliphatic components of suberin (412); 6) The time course of deposition of aromatic monomers into the polymer laid down by suberizing tissue slices indicated that the phenolic matrix is deposited simultaneously with or slightly before the aliphatic components (85). The appearance of a specific anionic peroxidase followed this same time course of appearance in suberizing tissue slices (47) and in abscisic acid-induced suberizing tissue culture (86). Immunocytochemical study of this specific peroxidase showed it to be localized in suberizing cell walls (113). The hypothetical model for suberin presented here is based on limited knowledge and is intended primarily as an aid in devising further experimental tests directed towards elucidation of the structure.

6.4.3.3 Suberin Biosynthesis

6.4.3.3.1 Biosynthesis of the Aliphatic Monomers

6.4.3.3.1.1 ω-Hydroxylation of Fatty Acids

ω-Hydroxylation of long-chain fatty acids is the first unique step in the biosynthesis of the aliphatic components of cutin and suberin. An endoplasmic reticulum fraction from embryonic shoots of *Vicia faba* catalyzed ω-hydroxylation of hexadecanoic acid with NADPH and O_2 as the required cofactors (426). Free fatty acid seemed to be the substrate for ω-hydroxylation in this system since all attempts to detect the possible involvement of a CoA ester of the fatty acid were unsuccessful. Microsomal preparations from the excised epidermis of expanding leaves and from root tips of *V. faba* also catalyzed an ω-hydroxylation of octadecanoic acid and other fatty acids. Mixed function oxidase inhibitors such as sodium azide, metal-ion chelators (*o*-phenanthroline, 8-hydroxyquinoline, and α,α-bipyridyl), and thiol-directed reagents all inhibited the ω-hydroxylation activity of the microsomal preparation. Inhibition of the ω-hydroxylation by CO suggested that a cytochrome P_{450}-type protein might be involved in the reaction. However, the ω-hydroxylase was unusually sensitive to CO, with 30% CO causing complete inhibition. Light at 420 to 460 nm failed to reverse the inhibition even when a low level (10%) of CO was used. If a cytochrome P_{450}-type protein is involved in this hydroxylation, it must have an unusually high affinity for CO.

6.4.3.3.1.2 Oxidation of ω-Hydroxy Acids

A high content of dicarboxylic acid is the most characteristic feature of the composition of aliphatic components of suberin (231). The generation of dicarboxylic acids by oxidation of endogenous ω-hydroxy fatty acids has been demonstrated in cell-free preparations from the excised epidermis of *Vicia faba* leaves (242). This dehydrogenase activity, which showed a strong preference for NADP, was located in the 100000-g supernatant. Modification of the substrate, ω-hydroxyhexadecanoic acid, by removal or esterification of the carboxyl or incorporation of a hydroxyl moiety at C-10, rendered it a poor substrate. ω-Oxohexadecanoic acid could be trapped by dinitrophenylhydrazine, indicating that the ω-oxo acid was probably an intermediate in the reaction. Additionally, synthetic ω-oxohexadecanoic acid was converted to C_{16} dicarboxylic acid by the enzyme preparation.

Wound-healing slices of *Solanum tuberosum* were utilized to isolate the enzymes involved in dicarboxylic acid generation since they were known to develop a suberin polymer under controlled conditions (243). Extracts of wound-healing slices of *S. tuberosum* catalyzed a two-step oxidation of ω-hydroxy acid to the dicarboxylic acid with the ω-oxo acid as an intermediate (3). Both steps were catalyzed by separate dehydrogenases (4). The ω-hydroxy acid dehydrogenase was induced by wounding and the time course of appearance of the enzyme correlated with suberization, whereas the ω-oxo acid dehydrogenase was not induced. With the use of protein and nucleic acid synthesis inhibitors, the transcriptional and translational processes involved in the synthesis of the dehydrogenases were shown to take place 72 hours after wounding the potato tubers (4). The wound-induced ω-hydroxy acid dehydrogenase was purified to homogeneity from extracts of the acetone powder prepared from wound-healing tissue slices of potato tuber.

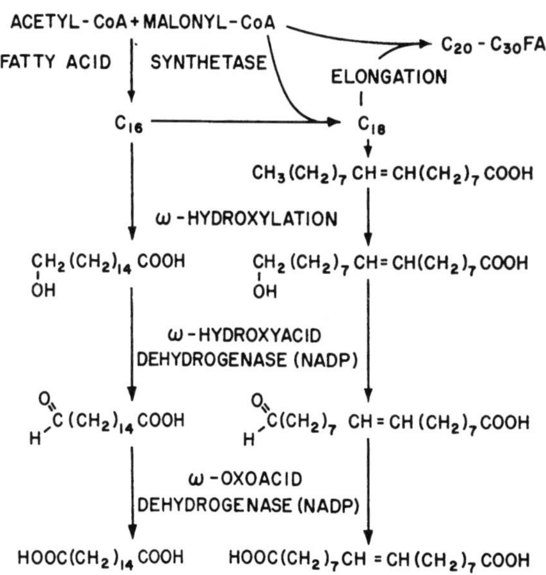

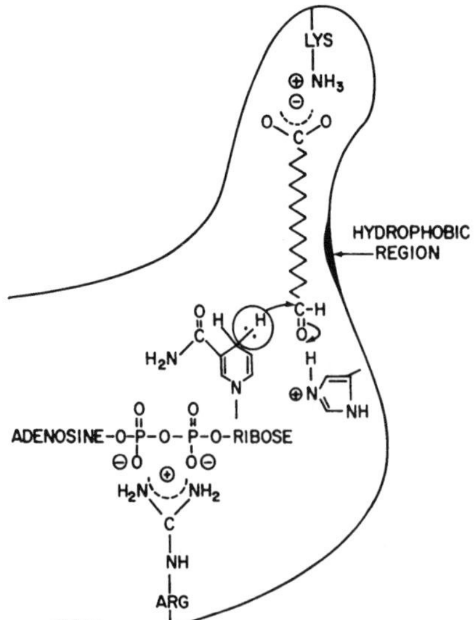

Fig. 6.4.9. Biosynthetic pathways for the major aliphatic components of suberin and some essential elements of the active site of ω-hydroxy fatty acid dehydrogenase, a key enzyme in suberin biosynthesis (5)

It was found to be a dimer composed of two 30000 MW monomers. The apparent K_m values for NADP, ω-hydroxyhexadecanoic acid, NADPH, and ω-oxohexadecanoic acid were 100, 20, 5, and 7 μM, respectively. Surprisingly, hydride from the A side of NADPH was added to ω-oxohexadecanoic acid by the purified enzyme at almost the same rate as that from the B side.

Initial velocity and product inhibition studies suggested an ordered sequential mechanism for ω-hydroxy fatty acid dehydrogenase (5). These studies suggested that NADPH was added first, followed by the ω-oxo fatty acid and that the ω-hydroxy acid was released prior to NADP. Chemical modification studies showed that an arginine residue was involved in binding NADPH and that the imidazole moiety of a histidine was involved in catalysis. The ε-amino group of a lysine residue was shown to be involved in binding the carboxyl moiety of the ω-oxo fatty acid substrate. This lysine residue was postulated to be located at the bottom of the hydrophobic pocket that participates in the substrate binding. The key features of this mechanism and the pathway of dicarboxylic acid biosynthesis are shown in Fig. 6.4.9. With the use of aldehydes of varying chain lengths, it was suggested that this hydrophobic pocket accommodated substrates with chain lengths up to C_{20}. Longer substrates did not bind effectively. This binding specificity is in accordance with the observation that although a suberin polymer may have a high proportion of long-chain ($>C_{20}$) ω-hydroxy acids, the major dicarboxylic acids are shorter than C_{20} (234, 244).

6.4.3.3.1.3 Biosynthesis of Mid-Chain Oxygenated Suberin Monomers

The mid-chain oxygenated fatty acids that are found in some suberin polymers are generated by the biosynthetic pathways shown in Fig. 6.4.10. Although these pathways have been demonstrated by studies with cell-free preparations of cutin-synthesizing systems, similar reactions are most probably involved in the biosynthesis of suberin monomers.

One of the most commonly found mid-chain hydroxylated components is dihydroxyhexadecanoic acid, which has hydroxyl moieties at C-10, C-9, C-8 or C-7, and on C-16 (232, 243, 244). A crude cell-free preparation from excised epidermis of *V. faba* catalyzed C-10 hydroxylation of 16-hydroxyhexadecanoic acid (473). This hydroxylation reaction was also catalyzed by the endoplasmic reticulum fraction from the embryonic shoots of *V. faba*. This preparation required O_2 and NADPH to catalyze mid-chain hydroxylation and the activity was inhibited by the typical mixed-function oxidase inhibitors and also by CO (427). The inhibition by CO was photoreversible, as expected of a cytochrome P_{450} hydroxylase.

Based on labeling studies, the members of the C_{18} family of mid-chain oxygenated acids were postulated to be derived from ω-hydroxyoctadecanoic acid as shown in Fig. 6.4.10 (255). The observed direct conversion of 18-hydroxy[18-^{3}H]-octadecanoic acid into 9,10-epoxy-18-hydroxyoctadecanoic acid and 9,10,18-trihydroxyoctadecanoic acid in plant tissue supported this predicted pathway (89). A 3000-g particulate fraction from a cell-free preparation of *Spinacia oleracea* leaves catalyzed the formation of *cis*-9,10-epoxy-18-hydroxyoctadecanoic acid from 18-hydroxy[18-^{3}H]octadecanoic acid (90). This epoxidation reaction re-

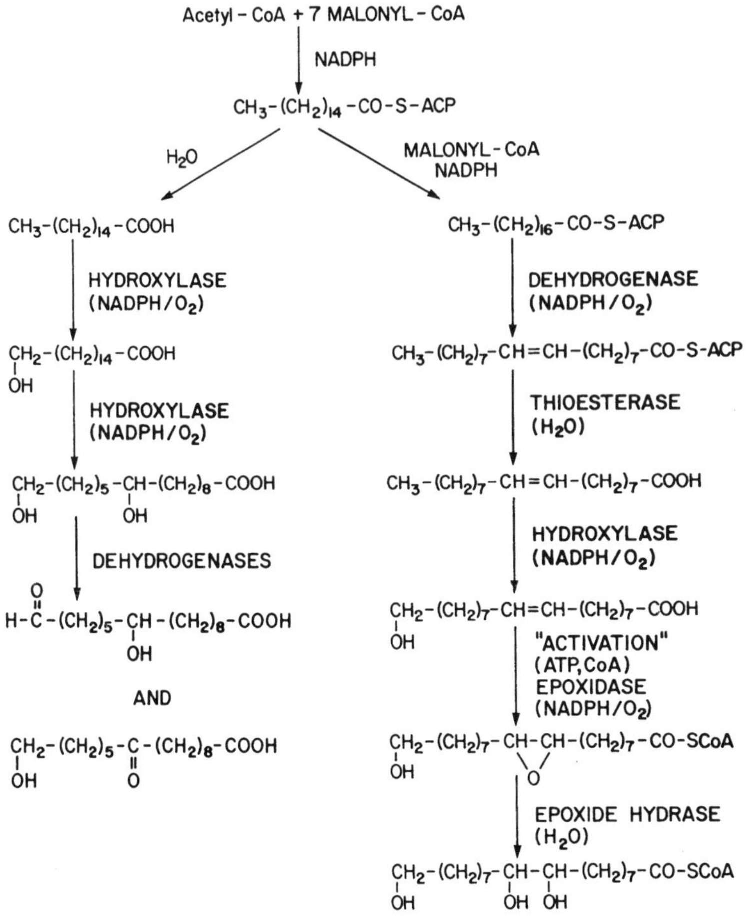

Fig. 6.4.10. Biosynthetic pathways for mid-chain oxygenated aliphatic hydroxy acids

quired O_2, NADPH, ATP, and CoA as cofactors, suggesting that this mixed-function oxidase required a thioester of the hydroxy acid as the substrate. This epoxidase was inhibited by CO and this inhibition was reversed by light (~450 nm), strongly suggesting that a cytochrome P_{450}-type enzyme catalyzed this epoxidation.

9,10,18-Trihydroxyoctadecanoic acid was generated from 9,10-epoxy-18-hydroxyoctadecanoic acid by a 3000-g particulate fraction obtained from the skin of the young fruit of *Malus pumila* (91). This enzyme activity did not require any cofactors and showed fairly stringent substrate specificity. It catalyzed the hydration of the *cis*-epoxide to the *threo*-product and this stereospecificity is consistent with the natural occurrence of the *threo*-product in cutin. The trihydroxy C_{18} acid thus generated probably is the biosynthetic precursor of 9,10-dihydroxyoctadecanedioic acid, which has been found in several suberin polymers (244).

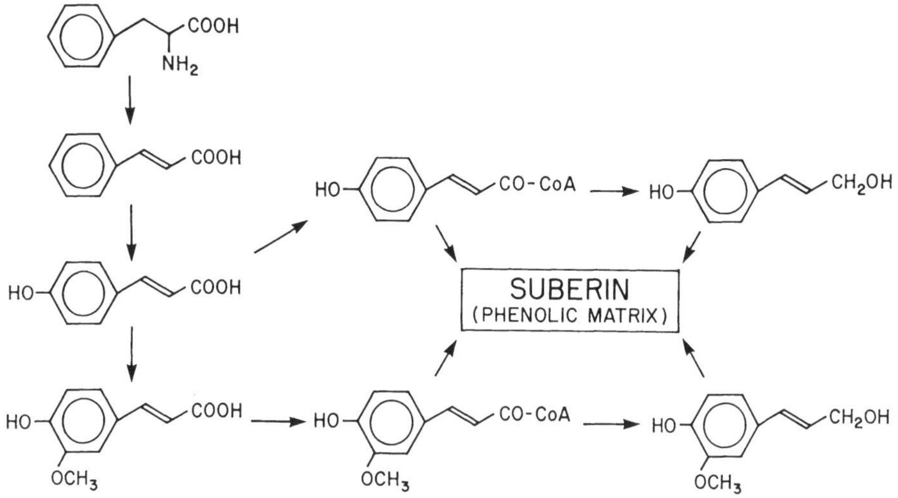

Fig. 6.4.11. Biosynthetic pathway for the aromatic domain of suberin

6.4.3.3.2 Biosynthesis of the Aromatic Components of Suberin

The phenolic metabolism of wounded plant tissues has been studied extensively (378), and metabolic events that give rise to the aromatic acids are well understood. Although these studies were concerned with metabolic changes that occur 24 to 48 hours after wounding, somewhat earlier than the peak of deposition of suberin, the biosynthetic pathways elucidated by such studies are clearly relevant to suberization and thus the reactions involved in the synthesis of aromatic monomers can be depicted as shown in Fig. 6.4.11. The individual reactions shown in this pathway have been studied with cell-free preparations and, in some cases, with purified enzymes (149, 163, 378, 505).

The time course of deposition of aromatic monomers into the phenolic portion of suberin in wound-healing slices of *Solanum tuberosum* tubers was determined using the amount of *p*-hydroxybenzaldehyde and vanillin generated by alkaline nitrobenzene oxidation as a measure of aromatic deposition (85). The deposition of such phenolics into the polymer exhibited a lag period of about three days after wounding followed by several days when the deposition of phenolics increased rapidly, and subsequently the process ceased. Exogenous L-[U-^{14}C]-phenylalanine and [U-^{14}C]cinnamic acid were incorporated into the insoluble polymeric material by wound-healing slices of *S. tuberosum* (85). Nitrobenzene oxidation of the polymeric material derived from labeled cinnamic acid released labeled *p*-hydroxybenzaldehyde and vanillin. The time course of incorporation of phenolics into the suberin polymer correlated with the time course for the deposition of aliphatic monomers into the polymer, the deposition of suberin-associated waxes into the periderm layer and the development of diffusion resistance (Fig. 6.4.12) (86).

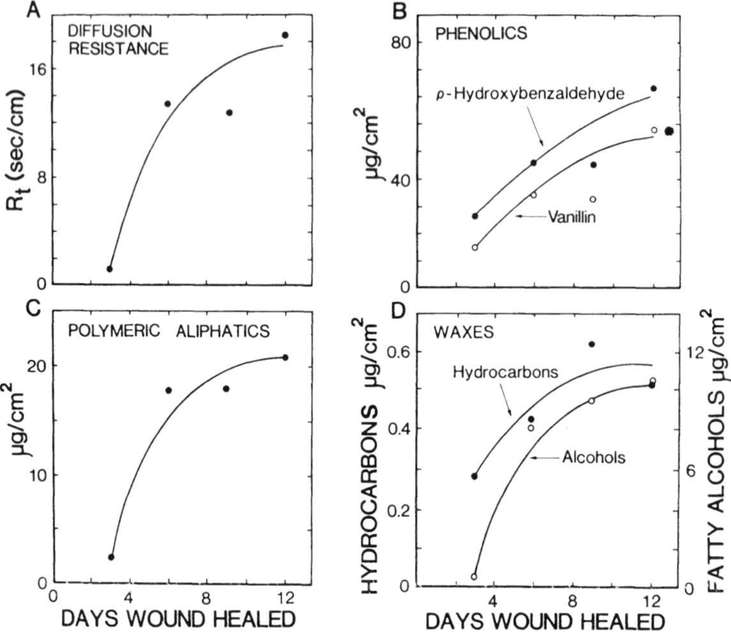

Fig. 6.4.12. Time-course of development of diffusion resistance and deposition of suberin-associated components during the wound healing of *Solanum tuberosum* tuber tissue slices. **A** Diffusion resistance; **B** phenolics measured as aromatic aldehydes generated by alkaline nitrobenzene oxidation; **C** polymeric aliphatics measured as α,ω-diols generated by $LiAlH_4$ reduction; **D** waxes recovered by $CHCl_3$ extraction. (W. Cottle and P. E. Kolattukudy, unpublished results)

6.4.3.3.3 Biosynthesis of Suberin from Aliphatic and Aromatic Monomers

No direct experimental evidence concerning the mechanism by which suberin polymer is generated from monomers is available. The aromatic monomers may polymerize in a manner similar to that proposed for lignin formation (148, 149). Peroxidase can catalyze the polymerization of phenolic monomers as well as the covalent attachzment of a phenolic polymer to the carbohydrates in the cell wall (148, 149, 488). The following observations suggest that the synthesis of the phenolic matrix of the suberin polymer is probably catalyzed by a peroxidase. A specific anionic peroxidase was induced during the wound-healing of *S. tuberosum* tissue slices, and this isoperoxidase was found to be localized in the suberizing cells (46). Both the time course of appearance and the tissue distribution of this wound-induced peroxidase activity correlated with that of the suberization process (47) (Chung and Kolattukudy, unpublished results). With immunocytochemical techniques it was shown that the anionic isoperoxidase appeared only in suberizing cells and only during suberization and was located only on the walls of the suberizing cells (113). The anionic peroxidases induced by fungal infection of wound-healing tuber slices of *S. tuberosum* (456) and *Ipomea batatas* (215, 461) were also probably involved in suberization. The time course of abscisic acid-induced suberization in callus tissue culture of *S. tuberosum* was the same as that of the induction of the specific isoperoxidase by the hormone (86). Plant roots appear to have peroxidase activity localized in epidermal and endodermal cell walls (135, 211, 313, 411, 422), both of which are areas known to undergo suberization (234). Parenchyma tissue of *S. tuberosum* tubers treated with *p*-coumaric and ferulic acids generated polymeric material that appeared to be as highly condensed as suberin phenolics since nitrobenzene oxidation of both materials generated only small amounts of *p*-hydroxybenzaldehyde and vanillin (40, 85). An increased production of peroxidase is associated with the wound-healing lesions in the fruit of *Lycopersicon esculentum* (125, 126) and wound-healing of this tissue is known to involve suberization (93). Growth of bean seedlings in iron-deficient medium resulted in decreased suberization. The suppression of suberization was not because of a lack of hydroxylated fatty acids but due to a decrease in the level of anionic isoperoxidase (412).

The cDNA for the anionic peroxidase involved in suberization in potato tuber has been cloned and sequenced (381a). With this cDNA as the probe the peroxidase gene from tomato was cloned and sequenced. Two highly homologous tandemly organized genes were found (381b). The gene transcripts were induced in wound-healing potato tubers and tomato fruits. In both potato and tomato tissue cultures abscisic acid induced transcription of this anionic peroxidase gene.

The hydrogen peroxide necessary for the peroxidase-catalyzed polymer formation might be generated by the same method proposed for lignification (150). Malate dehydrogenase associated with the cell wall might generate the NADH necessary for peroxidase-catalyzed H_2O_2 synthesis (109, 141, 151, 152). This isoperoxidase may be distinct from the one involved in polymerization of the phenolic monomers (289). The high concentrations of H_2O_2 found in the bark of *Populus gelrica* (394) may be utilized for polymerization of the aromatic matrix of suberin.

The aliphatic components of suberin form polyester domains that are attached to the phenolics as indicated above (see Fig. 6.4.7). Activated ω-hydroxy acids and dicarboxylic acids are probably involved in the esterification as shown in cutin biosynthesis (232). It was proposed that the long-chain fatty alcohols found in suberin might be incorporated into the suberin polymer by peroxidase-catalyzed polymerization of phenylpropanoic acid esters of these alcohols (232), such as the ferulic acid esters of C_{18}-C_{28} alcohols frequently found in suberin-associated waxes (Table 6.4.2).

6.4.3.4 Function of Suberin and Associated Waxes

6.4.3.4.1 Prevention of Water Loss

Suberin and the related polymer, cutin, function primarily as the structural components of barrier layers, which always have waxes associated with them (231, 232). Studies with isolated cuticular layers showed that the wax provides the major barrier to moisture diffusion: Removal of the wax resulted in a 300- to 2000-fold increase in permeability (292, 398, 401). Similar studies with cuticular layers from the leaves of *Citrus aurantium* demonstrated that wax provides the major barrier to gas diffusion (273). Wax probably plays a similar role in suberized layers, but the location of suberin within the cell wall prohibits analogous studies of isolated polymers. However, the observation that specific inhibition of wax biosynthesis with trichloroacetate – without affecting suberin polymer synthesis – prevented development of diffusion resistance in wound-healing *S. tuberosum* tissue slices strongly suggested that waxes constitute the major diffusion barrier in suberized cell walls (428). Subsequently it was shown that the extraction of suberin-associated waxes from the periderm of *S. tuberosum* and *Ipomoea batatas* storage organs caused a dramatic decrease in diffusion resistance (116). It was claimed, however, that wax did not provide the major barrier to diffusion in the periderm from the bark of mature *Betula pendula* trees (402). A subsequent publication from the same laboratory (465) confirmed the original conclusion of Soliday et al. (428) that wax does provide the major barrier to diffusion in suberized layers. It is probable that the efficiency of various suberized layers to act as a diffusion barrier varies a great deal depending on many factors, including species, anatomical site of occurrence of the layer, environment, etc.

6.4.3.4.2 Suberization in Wound-Healing

Suberin deposited during the wound-healing of underground storage organs has the same composition as the natural periderm polymer, and, together with the associated waxes, wound suberin also restores diffusion resistance (93, 243). Wounding causes suberization not only on tubers, which are normally protected by suberin, but also on fruits and leaves, which are normally protected by cutin, leading to the conclusion that suberization is a universal response to wounding in all plant organs (93). Trees respond to wounding by depositing a series of "barrier zones" (31–35, 197, 300, 306, 356, 410) of which some have been characterized as suberized layers. Periderms with heavily suberized cell walls were laid down in response to wounding of the roots of *Abies balsamea* and *Tsuga canadensis* (452). A barrier zone laid down in *Quercus robur* in response to wounding was shown both chemically and ultrastructurally to be suberized (354). Periderm was formed in response to wounding in the leaves of *Magnolia grandiflora*, *Michelia figo* (458), and *Nicotiana glutinosa* (500) and in the fruit of *Carica papaya* (433). Wound periderm was also formed in *Juniperus occidentalis* and *Larix laricina* stems in response to plant parasite attack (202, 446) and in the stems of *Juglans nigra* and *Rubus crataegifolius* in response to insect attack (11, 302).

6.4.3.4.3 Suberization in Response to Stress

Stress seems to affect suberization. Salt stress due to mineral deficiencies can result in changes in suberization. For example, magnesium deficiency caused the hypodermis (Fig. 6.4.3) and endodermis of *Zea mays* roots to have more heavily suberized cell walls (372), whereas iron deficiency in *Phaseolus vulgaris* led to a drastic decrease in suberin deposition in the roots (412). Mechanical stress in the form of physical impedance to root growth caused increased suberization in the walls of endodermal cells of *Hordeum vulgare* roots (497).

Suberin also plays a role in cold stress. The periderm layers in the bark serve as insulation for the plant (110). Thus suberin deposition in the bark of *Vitis vinifera* was shown to protect the plant from freezing damage (303, 351). The leaves of *Secale cereale* were shown by both ultrastructural and chemical analysis to deposit increased levels of aliphatic components in the epidermis and in the mestome sheath in response to growth at low temperatures (146).

Suberin can serve as a nearly impermeable chemical and physical barrier to fungal attack. There are indications that soluble compounds associated with the suberin polymer, such as phenolics or wax components, may act as antifungal agents (67, 92, 128, 145, 156, 185, 187, 307, 310, 390, 435, 447). Since the outer layer of the very young roots is not usually suberized (unpublished results) such regions are susceptible to microbial ingress, whereas regions that are encased in suberized walls, such as older roots and endodermis, usually provide an effective barrier (22, 83). Not only does existing suberin act as a barrier, but there are several reports of fungal or viral attack eliciting the deposition of suberin in cell walls, which serves to limit infection (80, 96, 175, 186, 430, 448, 453). Barrier zones laid down in both roots and stems of *Eucalyptus marginata* as a result of fungal invasion (453) might also be suberized. Fungal infection also appeared to induce the deposition of suberin in the sapwood of *Quercus robur*, and then this suberin served as a barrier to further spread of the fungus (355). Suberin deposited in the inner cell walls at the site of wounding of stems of *Pachysandra terminalis* led to increased resistance to fungal attack (198). Recently it was reported that *Verticillium alboatrum* induced a cell-wall coating response in tomato (*Lycopersicon esculentum*) xylem parenchyma cells, adjacent intercellular spaces, and vessels (436). Although in the absence of chemical evidence, the barrier layers induced by fungal infection have been variously designated as lignin, suberin, lignin-suberin-like, etc., it was proposed that suberization might be a general response to fungal infection (231). Recent chemical analysis of the depolymerization products of the wall coating induced by *V. alboatrum* in tomatoes showed that the wall coating was suberin (P.E. Kolattukudy and J. Robb, unpublished). Furthermore, the amount of aliphatic suberin components deposited on the walls of resistant tomatoes was many-fold higher than that in the susceptible near-isolines. Thus, degree of suberization could be related to resistance.

6.4.3.4.4 Suberization as a Means of Compartmentalization

Suberin is laid down in various internal locations in order to seal off specific regions of the plant (234). In the primary development of plant roots, suberin is deposited in the Casparian band of endodermal cells (45, 103, 162, 317, 364, 367, 369, 474, 483). The presence of suberin in the Casparian band of the endodermis of *Sorghum bicolor* was shown chemically by depolymerization (244). Suberized layers are also found in the mestome sheath and bundle sheath cell walls of grasses (52, 61, 108, 118, 171, 172, 331, 368), and these layers have been chemically characterized in *Zea mays* and *Secale cereale* (114, 146).

Ultrastructural studies indicated that suberin is deposited in seeds, when they reach maturity, as a means of sealing off the seed where vascular tissue was attached. Thus suberin was shown to be deposited in the pigment strand of *Triticum aestivum* (423, 503, 504) and *Oryza sativa* (335). Suberin was also shown ultrastructurally to be laid down in the chalazal region of the seeds of *Pennisetum americanum* (134), *Hordeum vulgare* (81), and *Citrus paradisi* (110). The only case in which both ultrastructural and chemical studies showed heavy suberization is in the seeds of *Citrus* (Fig. 6.4.3).

Suberin is laid down in the cell wall of calcium oxalate crystal idioblasts. This deposition has been characterized in crystal cells in the bark of *Larix decidua*, *Acacia senegal*, and *Picea abies* (481, 482). The suberin in crystal idioblast cell walls from *Agave americana* was characterized both ultrastructurally (Fig. 6.4.3) and chemically, and the major aliphatic component was shown to be octadocenedioic acid (25%) (117). The basal walls of leaf trichomes are suberized as shown in *Abutilon megapotamicum* (161), *Ananas comosus* (395), and *Larrea tridentata* (451). The walls of oil cells found in *Acorus calamus* (9) and *Laurus nobilis* (291) also appear ultrastructurally to be suberized. Suberized cells encircle silica cells in the stem of *Lolium temulentum* (271). In all of the cases listed above, the suberin may serve to protect the rest of the plant from potentially toxic components.

6.4.3.5 Regulation of Suberization

Whenever a physiological change or a stress requires the erecting of a diffusion barrier in the plant, suberization results. For example, wounding causes the formation of a suberized periderm irrespective of the nature of the organ and its natural protective layer. The molecular mechanism by which suberization is induced is not understood. Thorough washing of tuber slices of *S. tuberosum* inhibited the wound-healing process, suggesting that some component produced at the wound might be responsible for initiating suberization (425). This washing removed abscisic acid (ABA) and the washed tissue partially regained the ability to wound-heal when ABA was added, suggesting that this hormone might play a role in initiating suberization. This suggestion was supported by the fact that 10^{-4} M ABA induced suberization in *S. tuberosum* tissue culture. This induced suberization was demonstrated by the fact that treatment of the tissue culture with ABA resulted in a four-fold increase in the deposition of polymeric aromatic components, a three-fold increase in aliphatic polymeric material, and a five-fold increase in suberin-associated waxes (86). Suberization was inhibited by washing wound-healing slices of *S. tuberosum* as late as three days after wounding and, since ABA was not removed by this washing, the role of this hormone appeared to be indirect (425). Experiments with the inhibitors actinomycin D and cycloheximide suggested that the transcriptional and translational processes directly affecting the biosynthesis of the aliphatic domains of suberin take place between 72 and 96 hours after wounding (3). Suberization was inhibited when the tissue was washed any time before this 72- to 96-hour period. It seems that ABA is generated within 24 hours of wounding and that it initiates a process that causes the formation of a suberization-inducing factor, which, in turn, causes the induction of enzymes involved in suberization (Fig. 6.4.13).

The ABA induction of two key enzymes involved in suberization – ω-hydroxy fatty acid dehydrogenase and a suberization-linked peroxidase – was studied in

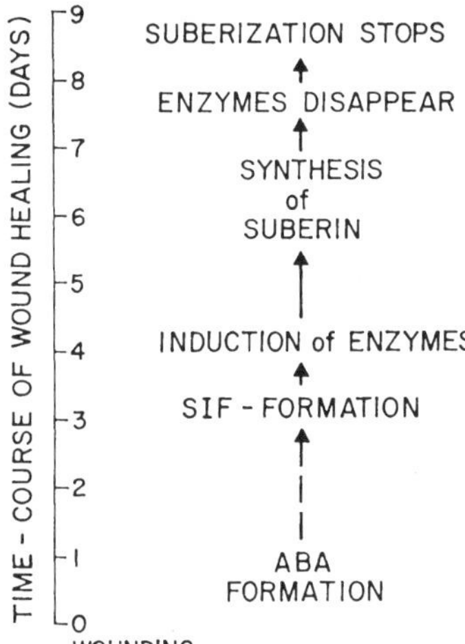

Fig. 6.4.13. A tentative scheme showing the time-course of the processes involved in suberization in wound-healing slices of *Solanum tuberosum*. SIF = a hypothetical suberin-inducing factor

S. tuberosum tissue culture (86). Although the dehydrogenase is induced during wound-healing in tissue slices (5), the initial tissue culture had fairly high levels of this enzyme, and ABA treatment caused only a small increase in this enzyme activity. However, ABA treatment caused a dramatic increase in the anionic isoperoxidase associated with suberization (86). This isoperoxidase was isolated and purified from ABA-treated tissue culture and wound-healing slices of *S. tuberosum*. The enzyme from both sources appeared to be very similar immunologically (115). ABA at 10^{-4} M induced dormancy in roots of *Picea sitchensis*, which was characterized by increased suberization (371). It seems quite likely that some of the effects that ABA has on root growth and development, such as inhibition of ion uptake (87, 408), might be due to ABA-induced suberization.

6.4.3.6 Enzymatic Degradation of Suberin

Although suberin serves as an efficient barrier to some fungal attack, it is obvious that the polymer must be subject to biological breakdown. Whether this occurs before or only after plant death may depend on the source and location of the polymer. When bark from *Picea abies* and *Fagus sylvatica* was buried in soil and allowed to decay naturally, the suberized walls were the last component of the bark to show signs of degradation (349). After three years of decay the suberized walls appeared to be partially degraded, as shown by ultrastructural analysis. Earlier studies of bark buried in soil had indicated that suberin is very resistant to degradation under these conditions (405, 493). There is considerable evidence that fungi can penetrate suberized walls. Early studies with *Armillaria mellea* indicated that it was able to penetrate the cork layer of *Juglans regia*, *J. hindsii*, *Prunus persica*, and *Pyrus communis* (450). Light microscopic examination showed that penetration by the fungi appeared to involve degradation of suberized walls rather than mechanical rupture. A subsequent study of *A. mellea* degradation of *Brachyotegia spiciformis* bark indicated that a 10-month growth period of fungus on bark led to a 60% reduction of what was called suberin content, while cellulose and lignin degradation were only 14% and 9%, respectively (445). More recently the same fungus was shown by electron microscopy to penetrate the cork walls of *Pinus sylvestris* bark by degradation of the suberized lamellae (391). Similarly, electron microscopy indicated that *Fomes annosus* penetrated suberized walls in the root bark of *Picea abies*, generating holes large enough to allow fungal penetration (361), and that *Gaeumannomyces graminis* was able to penetrate suberized walls in the roots of *Triticum aestivum* (290). The penetration of a fungus through a suberized cell wall was shown by scanning electron microscopy in the case of *Pisolithus tinctorius* penetrating the roots of *Pinus taeda* (475). The openings generated in the root cell wall conformed to the size and shape of the fungal hyphae, possibly suggesting again that penetration was enzymatic rather than mechanical. Similar penetration holes through the suberized cell walls of the bark of *Fagus sylvatica* were shown by transmission electron microscopy (60). This bark was infected with a cork-inhabiting fungus, *Ascodichaena rugosa*, which limits its growth to cork cells, passing from one suberized cell to another, while inducing the host to generate greater quantities of suberized cells. Ultrastructural studies also indicated that the fungus *Botrytis cinera* was able to penetrate the periderm layer of *Daucus carota* by enzymatic degradation of suberin (409) and that *Phellinus noxius* and *Rigidoporus lignosus* could penetrate the suberized walls in the roots of *Hevea brasiliensis* (325, 326).

Degradation of suberin was demonstrated by chemical analysis in the case of attack on the periderm of *Rubus idaeus* by *Thomasiniana theobaldi* (155). After enzymatic attack by larvae-released enzymes, the periderm was shown to have 52% fewer aliphatic suberin components, and hemicellulose and cellulose were decreased by 28% and 54%, respectively. The cell wall degradation caused by this insect attack was thought to provide subsequent access for fungal infection.

No direct evidence for enzymatic degradation of suberin was available until recently. To study the enzymology of suberin degradation, it was necessary to find a fungus that would grow well on suberin; many fungi were tested for their ability to grow on suberin-enriched cell wall preparations from *S. tuberosum* tuber periderm as the sole source of carbon. *Helminthosporium solani*, *H. sativum*, *Leptosphaeria coniothyrium*, *Ascodichaena rugosa*, *Rizoclonia corcorium*, *Cyanthus stercoreus*, and *F. solani pisi* grew under such conditions (Kolattukudy, unpublished results). Because *F. solani pisi* showed the most rapid growth, the extracellular fluid from the culture of this organism was examined for enzymes that degrade suberin.

Since suberin contains both aliphatic and aromatic domains, attempts were made to detect enzymes that degrade either domain. For the aromatic domains, *S. tuberosum* wound-periderm suberin, biosynthetically labeled by incorporation of [^{3}H]cinnamate into wound-healing *S. tuberosum* slices, was used, and the release of soluble radioactivity was used as a measure of such activity. To measure the activity of the enzyme that hydrolytically releases the esterified aliphatic acids, either *S. tuberosum* periderm suberin, biosynthetically labeled with [^{3}H]oleic acid, or radioactive cutin was used as the substrate. The extracellular fluid from suberin-grown *F. solani pisi* contained both types of enzyme activities; these could be resolved from each other by ion exchange chromatography (234, 248, 251). The enzyme that released esterified aliphatic acids was purified to homogeneity and characterized (123). This enzyme was found to be identical in all its properties with cutinase produced by the same organism when grown on cutin. The esterase released aliphatic monomers from the suberin polymer isolated from the periderm of *Rubus idaeus* when incubated for 16 hours in vitro. Analysis of the products revealed that the purified enzyme released all types of aliphatic monomers present in the suberin (123). The enzyme(s) responsible for releasing the aromatic components of suberin have not been purified. Even exhaustive treatment of the labeled *S. tuberosum* suberin released only <10% of the total label contained in the polymer derived from [^{3}H]cinnamate. HPLC analysis of the aromatic components released by this enzyme showed the presence mainly of labeled ferulic acid and coumaric acid, strongly suggesting that the enzyme probably released the esterified phenolic acids (Woloshuk and Kolattukudy, unpublished results). The aromatic components held by the more refractory linkages such as ether bonds were not released by the extracellular enzyme at measurable levels. Such components are probably degraded by mechanisms similar to those used in lignin degradation.

Acknowledgement. This work was supported in part by a grant from the National Science Foundation.

References

1 Abramovitch R A, Coutts R T, Knaus E E 1967 Some extractives from the bark of *Populus balsamifera*. Can J Pharm Sci 2:71–74

2 Adamovics J A, Johnson G, Stermitz F R 1977 Ferulates from cork layers of *Solanum tuberosum* and *Pseudotsuga menziesii*. Phytochemistry 16:1989–1990

3 Agrawal V P, Kolattukudy P E 1977 Biochemistry of suberization. ω-Hydroxy acid oxidation in enzyme preparations from suberizing potato tuber disks. Plant Physiol 59:667–672

4 Agrawal V P, Kolattukudy P E 1978 Purification and characterization of a wound-induced ω-hydroxy fatty acid: NADP oxidoreductase from potato tuber disks (*Solanum tuberosum* L.). Arch Biochem Biophys 191:452–465

5 Agrawal V P, Kolattukudy P E 1978 Mechanism of action of a wound-induced ω-hydroxy fatty acid: NADP oxidoreductase isolated from potato tubers (*Solanum tuberosum* L.). Arch Biochem Biophys 191:466–478

6 Agullô C, Collar C, Seone E 1984 Estudio comparativo de la cutina de las hojas y de la suberina de la corteza del *Quercus suber*. Ann Quim 80C:20–24

7 Agullô C, Seone E 1981 Free hydroxyl groups in the cork suberin. Chem Ind 30:608–609

8 Agullô C, Seone E 1982 Hidrogenolisis de la suberina del corcho con $LiBH_4$ grupos carboxilo libres. Ann Quim 78C:389–393

9 Amelunxen F, Gronau G 1969 Elektronenmikroskopische Untersuchungen an den Ölzellen von *Acorus calamus* L. Z Pflanzenphysiol 60:156–168

10 Ånäs E, Ekman R, Holmbom B 1983 Composition of nonpolar extractives in bark of Norway spruce and Scots pine. J Wood Chem Technol 3:119–130

11 Armstrong J E, Kearby W H, McGinnes E A Jr 1979 Anatomical responses and recovery of twigs of *Juglans nigra* following oviposition injury inflicted by the two-spotted treehopper, *Enchenopa biontata*. Wood Fiber 11:29–37

12 Arno M, Serra M C, Seone E 1981 Metanolisis de la suberina del corcho. Identification y estimacion de sus components acidos, como esteres metilicos. Ann Quim 77C:82–86

13 Arzee T, Lipschitz N, Waisel Y 1968 The origin and development of the phellogen in *Robinia pseudoacacia* L. New Phytol 67:87–93
14 Arzee T, Waisel Y, Lipschitz N 1970 Periderm development and phellogen activity in the shoots of *Acacia raddiana savi.* New Phytol 69:395–398
15 Asada Y, Okaguchi T, Matsumoto I 1960 Biosynthesis of lignin in Japanese radish root infected by downy mildew fungus. In: Tomiyama K, Daly J M, Uritani I, Oku H, Ouchi S (eds) Biochemistry and cytology of plant-parasite interactions. Elsevier Amsterdam, 200–212
16 Austin P W, Seshadri T R, Sood M S 1969 Chemical study of *Prunus puddum* (stem bark) and *Prunus cornuta* (stem bark and wood). Ind J Chem 7:44–48
17 Avato P 1984 Synthesis of wax esters by a cell-free system from barley (*Hordeum vulgare* L.). Planta 162:487–494
18 Backhaus R A, Sachs R M, Paul J 1980 Morphactin transport and metabolism in tree bark tissues: comparative studies in vitro. Physiol Plant 50:131–136
19 Bamber R K 1962 The anatomy of the barks of Leptospermoideae. Aust J Bot 10:25–54
20 Barckhausen R 1978 Ultrastructural changes in wounded plant storage tissue cells. In: Kahl G (ed) Biochemistry of wounded plant tissues. de Gruyter Berlin, 1–42
21 Becker E S, Kurth E F 1958 The chemical nature of the extractives from the bark of red fir. Tappi 41(7):380–384
22 Beckman C H, Talboys P W 1981 Anatomy of resistance. In: Mace M E, Bell A A, Beckman C H (eds) Fungal wilt diseases of plants. Academic Press New York, 587–621
23 Behari M, Andhiwal C K, Streibl M 1977 Hydrocarbons, esters and free alcohols in the bark of *Saraca indica* L. Coll Czech Chem Commun 42:1385–1388
24 Beri R M 1970 Chemical constituents of the bark of *Pinus roxburghii* Sargent. Ind J Chem 8:469–470
25 Bescansa-López J L, Gil Curbera G, Ribas-Marqués I 1966 Quimica del corcho. XV. Alcoholes parafinicos del corcho. Ann Rev Soc Esp Fis Quim 62:865–869
26 Bescansa-López J L, Ribas-Marqués I 1966 Quimica del corcho. XVI. Hidrocarburos parafinicos en el corcho. Ann Rev Soc Esp Fis Quim 62:871–874
27 Bhat K M 1982 Anatomy, basic density and shrinkage of birch bark. IAWA Bull 3:207–213
28 Biggs A R 1984 Boundary-zone formation in peach bark in response to wounds and *Cytospora leucostoma* infection. Can J Bot 62:2814–2821
29 Biggs A R 1984 Intracellular suberin: occurrence and detection in tree bark. IAWA Bull 5:243–248
30 Biggs A R 1985 Detection of impervious tissue in tree bark with selective histochemistry and fluorescence microscopy. Stain Tech 60:299–304
31 Biggs A R 1986 Phellogen regeneration in injured peach tree bark. Ann Bot 57:463–470
32 Biggs A R 1986 Prediction of lignin and suberin deposition in boundary zone tissue of wounded tree bark using accumulated degree days. J Amer Soc Hort Sci 111:757–760
33 Biggs A R 1986 Wound age and infection of peach bark by *Cytospora leucostoma*. Can J Bot 64:2319–2321
34 Biggs A R, Merrill W, Davis D D 1984 Discussion: response of bark tissues to injury and infection. Can J For Res 14:351–356
35 Biggs A R, Miles N W 1985 Suberin deposition as a measure of wound response in peach bark. Hortscience 20:903–905
36 Biggs A R, Northover J 1985 Formation of the primary protective layer and phellogen after leaf abscission in peach. Can J Bot 63:1547–1550
37 Biggs A R, Stobbs L W 1986 Fine structure of the suberized cell walls in the boundary zone and necrophylactic periderm in wounded peach bark. Can J Bot 64:1606–1610
38 Bishop C T, Harwood V D, Purves C B 1950 Separation of some chemical constituents from white spruce bark. Pulp Pap Mag Can 51(1):90–92
39 Björkman A 1956 Studies on finely divided wood. Part I. Extraction of lignin with neutral solvents. Svensk Papperstidn 59:477–485
40 Bland D E, Logan A F 1965 The properties of syringyl, guaiacyl and *p*-hydroxyphenyl artificial lignins. Biochem J 95:515–520
41 Bloomberg W J 1962 *Cytospora* canker of poplars: the moisture relations and anatomy of the host. Can J Bot 40:1281–1292
42 Bognar A L, Paliyath G, Rogers L, Kolattukudy P E 1985 Biosynthesis of alkanes by particulate and solubilized enzyme preparations from pea leaves (*Pisum sativum*). Arch Biochem Biophys 235:8–17

43 Bolmgren I 1980 Lipid chemistry with special reference to lipids from the outer bark of birch. Ph. D. Thesis. Royal Institute of Technology Stockholm, 74 pp
44 Bolmgren I, Norin T 1981 Direct methylation of salts of acids in hydrolysis mixtures of natural products by phase transfer catalysis. An investigation of the suberin fraction of birch bark (*Betula verrucosa* Erh.). Acta Chem Scand 20B:742
45 Bonnet H T Jr 1968 The root endodermis: fine structure and function. J Cell Biol 37:199–205
46 Borchert R 1974 Isoperoxidases as markers of the wound-induced differentiation pattern in potato tuber. Dev Biol 36:391–399
47 Borchert R 1978 Time course and spatial distribution of phenylalanine ammonia-lyase and peroxidase activity in wounded potato tuber tissue. Plant Physiol 62:789–793
48 Borchert R 1984 Functional anatomy of the calcium-excreting system of *Gleditsia triacanthos* L. Bot Gaz 145:474–482
49 Borger G A, Kozlowski T T 1972 Early periderm ontogeny in *Fraxinus pennsylvanica*, *Ailanthus altissima*, *Robinia pseudoacacia*, and *Pinus resinosa* seedlings. Can J For Res 2:135–143
50 Bosabalidis A M, Thomson W W 1984 Ultrastructural differentiation of an unusual structure lining the anticlinal walls of the inner secretory cells in *Tamarix* salt glands. Bot Gaz 145:427–435
51 Botha C E J, Evert R F 1986 Free-space marker studies on the leaves of *Saccharum officinarum* and *Bromus unioloides*. S Afr J Bot 52:335–342
52 Botha C E J, Evert R F, Cross R H M, Marshall D M 1982 The suberin lamella, a possible barrier to water movement from the veins to the mesophyll of *Themeda triandra* Forsk. Protoplasma 112:1–8
53 Bowen W R 1963 Origin and development of winged cork in *Euonymus alatus*. Bot Gaz 124:256–261
54 Bramhall A E, Kellog R M, Meyer R W, Warren W G 1977 Bark-tissue thickness of coastal western hemlock in British Columbia. Wood Fiber 9:184–190
55 Brieskorn C H, Binnemann P H 1972 Über den chemischen Aufbau des Kartoffelkorkes. Tetrahedron Lett 12:1127–1130
56 Brieskorn C H, Binnemann P H 1975 Carbonsäuren und Alkanole des Cutins und Suberins von *Solanum tuberosum*. Phytochemistry 14:1363–1367
57 Brisson J D, Peterson R L, Robb J, Rauser W E, Ellis B E 1977 Correlated phenolic histochemistry using light, transmission, and scanning electron microscopy, with examples taken from phytopathological problems. Scan Electr Microsc II:667–676
58 Brooker E G 1959 Constituents of *Phyllocladus glaucus*. N Z J Sci 2:212–214
59 Buckner J S, Kolattukudy P E 1973 Specific inhibition of alkane synthesis with accumulation of very long chain compounds by dithioerythritol, dithiothreitol, and mercaptoethanol in *Pisum sativum*. Arch Biochem Biophys 156:34–45
60 Butin H, Parameswaran N 1980 Ultrastructure of *Ascodichaena rugosa* on beech bark. Arch Microbiol 126:87–95
61 Canny M J 1986 Water pathways in wheat leaves. III. The passage of the mestome sheath and the function of the suberized lamellae. Physiol Plant 66:637–647
62 Cassagne C 1970 Les hydrocarbures végétaux: biosynthesis et localisation cellulaire. Ph.D. Thesis. Univ Bordeaux France, 142 pp
63 Cassagne C, Darriet D, Bourre J M 1977 Evidence of alkane synthesis by the sciatic nerve of the rabbit. FEBS Lett 82:51–54
64 Cassagne C, Lessire R 1974 Studies on alkane biosynthesis in epidermis of *Allium porrum* L. leaves. Arch Biochem Biophys 165:274–280
65 Cassagne C, Lessire R 1978 Biosynthesis of saturated very long chain fatty acids by purified membrane fractions from leek epidermal cells. Arch Biochem Biophys 191:146–152
66 Cassagne C, Lessire R 1979 Biosynthesis of the very long chain fatty acids in higher plants from exogenous substrates. In: Appelqvist L-A, Liljenberg C (eds) Advances in the biochemistry and physiology of plant lipids. Elsevier New York, 383–398
67 Cerny G 1973 Hemmung einiger Ektoenzyme von *Fomes annosus* (Fr.) Cke. durch Holz- und Bastextrakte aus *Picea abies* (Karst.). Eur J For Pathol 3:214–220
68 Chan L-L 1985 The anatomy of the bark of *Libocedrus* in New Zealand. IAWA Bull 6:23–34
69 Chan L-L 1986 The anatomy of the bark of *Agathis* in New Zealand. IAWA Bull 7:229–241
70 Chang H-M, Allan G G 1971 Oxidation. In: Sarkanen K V, Ludwig C H (eds) Lignins: Occurrence, formation, structure and reactions. Wiley-Interscience New York, 433–485

71 Chang Y-P 1954 Bark structure of North American conifers. USDA For Serv Tech Bull 1095, 86 pp
72 Chang Y-P, Mitchell R L 1955 Chemical composition of common North American pulpwood barks. Tappi 38(5):315–320
73 Chattaway M M 1953 The anatomy of bark. I. The genus *Eucalyptus*. Aust J Bot 1:402–433
74 Cheesbrough T M, Kolattukudy P E 1984 Alkane biosynthesis by decarbonylation of aldehydes catalyzed by a particulate preparation from *Pisum sativum*. Proc Natl Acad Sci USA 81:6613–6617
75 Chiang S T, Wang S 1984 The structure and formation of *Melaleuca* bark. Wood Fiber Sci 16:357–373
76 Chiarlo B, Tacchino E 1965 Ricerche sui costituenti della corteccia di *Ailanthus glandulosa* Desf. subspontaneo della *Liguria*. II. Contenuto in acidi grassi totali fino a C_{22}. Riv Ital Sost Grass 42:122–124
77 Clarke K J, McCully M E, Miki N K 1979 A developmental study of the epidermis of young roots of *Zea mays* L. Protoplasma 98:283–309
78 Clarkson D T, Robards A W, Stephens J E, Stark M 1987 Suberin lamellae in the hypodermis of maize (*Zea mays*) roots; development and factors affecting the permeability of hypodermal layers. Plant Cell Environ 10:83–93
79 Clermont L P 1970 Study of lignin from stone cells of aspen poplar inner bark. Tappi 53(1):52–57
80 Cline M N, Neely D 1983 Wound-healing process in geranium cuttings in relationship to basal stem rot caused by *Pythium ultimum*. Plant Dis 67:636–638
81 Cochrane M P 1983 Morphology of the crease region in relation to assimilate uptake and water loss during caryopsis development in barley and wheat. Aust J Plant Physiol 10:473–491
82 Conner A H, Nagasampagi B A, Rowe J W 1980 Terpenoid and other extractives of western white pine bark. Phytochemistry 19:1121–1131
83 Cooper R M 1981 Pathogen-induced changes in host ultrastructure. In: Staples R C, Toenniessen G H (eds) Plant disease control. Resistance and susceptibility. Wiley-Interscience New York, 105–142
84 Corbett R E, McDowall M A, Wyllie S G 1964 Extractives from the New Zealand Myrtaceae. Part VII. Neutral and phenolic compounds from the bark of *Leptospermum scoparium*. J Chem Soc 1283–1287
85 Cottle W, Kolattukudy P E 1982 Biosynthesis, deposition, and partial characterization of potato suberin phenolics. Plant Physiol 69:393–399
86 Cottle W, Kolattukudy P E 1982 Abscisic acid stimulation of suberization. Induction of enzymes and deposition of polymeric components and associated waxes in tissue cultures of potato tuber. Plant Physiol 70:775–780
87 Cram W J, Pitman M G 1972 The action of abscisic acid on ion uptake and water flow in plant roots. Aust J Biol Sci 25:1125–1132
88 Crisp C E 1965 The biopolymer cutin. Ph.D. Thesis. University of California Davis, 225 pp
89 Croteau R, Kolattukudy P E 1974 Direct evidence for the involvement of epoxide intermediates in the biosynthesis of the C_{18} family of cutin acids. Arch Biochem Biophys 162:471–480
90 Croteau R, Kolattukudy P E 1975 Biosynthesis of hydroxy fatty acid polymers. Enzymatic epoxidation of 18-hydroxy oleic acid to 18-hydroxy-cis-9,10-epoxystearic acid by a particulate preparation from spinach (*Spinacia oleracea*). Arch Biochem Biophys 170:61–72
91 Croteau R, Kolattukudy P E 1975 Biosynthesis of hydroxy fatty acid polymers. Enzymatic hydration of 18-hydroxy-cis-9,10-epoxystearic acid to threo-9,10,18-trihydroxystearic acid by a particulate preparation from apple (*Malus pumila*). Arch Biochem Biophys 170:73–81
92 Davies W P, Lewis B G 1981 Antifungal activity in carrot roots in relation to storage infection by *Mycocentrospora acerina* (Hartig) Deighton. New Phytol 88:109–119
93 Dean B B, Kolattukudy P E 1976 Synthesis of suberin during wound-healing in jade leaves, tomato fruit, and bean pods. Plant Physiol 58:411–416
94 Dean B B, Kolattukudy P E 1977 Biochemistry of suberization. Incorporation of [1-^{14}C]oleic acid and [1-^{14}C]acetate into the aliphatic components of suberin in potato tuber disks (*Solanum tuberosum*). Plant Physiol 59:48–54
95 Dean B B, Kolattukudy P E, Davis R W 1977 Chemical composition and ultrastructure of suberin from hollow heart tissue of potato tubers (*Solanum tuberosum*). Plant Physiol 59:1008–1010

96 DeLeeuw G T N 1985 Deposition of lignin, suberin and callose in relation to the restriction of infection by *Botrytis cinerea* in ghost spots of tomato fruits. Phytopath Z 112:143–152
97 De Vries H A M A 1968 Development of the structure of the russeted apple skin. Acta Bot Neerl 17:405–415
98 Diamantoglou S, Kull U 1981 Das jahresperiodische Verhalten der Fettsäuren in Rinden und Blättern von *Pinus halepensis* Mill. und *Schinus molle* L. Z Pflanzenphysiol 103:157–164
99 Domazetis G, James B R, Tarpey B, Dolphin D 1981 Decarbonylation of aldehydes using ruthenium (II) porphyrin catalysts. In: Ford P C (ed) Catalytic activation of carbon monoxide. Am Chem Soc Washington DC, 243–252
100 Dougal E F, Krahmer R L 1984 Ultrastructural observations of parenchyma and sclereids in Douglas-fir (*Pseudotsuga menziesii* (Mirb.) Franco) bark. Wood Fiber Sci 16:246–256
101 Duhamel L 1963 Etudes sur l'acide phellonique. Ann Chim 3:315–346
102 Duhamel L 1965 Etudes sur les produits cireux du liège: les hydroxyacides en C_{18}. Bull Soc Chim Fr 399–402
103 Dupont E M, Leonard R T 1977 The use of lanthanum to study the functional development of the Casparian strip in corn roots. Protoplasma 91:315–323
104 Eckerman C, Ekman R 1985 Comparison of solvents for extraction and crystallisation of betulinol from birch bark waste. Pap Puu 67:100–106
105 Eckstein D, Leise W, Parameswaran N 1976 On the structural changes in wood and bark of a salt-damaged horsechestnut tree. Holzforschung 30:173–178
106 Eglinton G, Hunneman D H 1968 Gas chromatographic-mass spectrometric studies of long chain hydroxy acids. I. The constituent cutin acids of apple cuticle. Phytochemistry 7:313–322
107 Ekman R 1983 The suberin monomers and triterpenoids from the outer bark of *Betula verrucosa* Ehrh. Holzforschung 37:205–211
108 Eleftheriou E P, Tsekos I 1979 Development of mestome sheath cells in leaves of *Aegilops comosa* var. *thessalica*. Protoplasma 100:139–153
109 Elstner E F, Heupel A 1976 Formation of hydrogen peroxide by isolated cell walls from horseradish (*Armoracia lapathifolia* Gilib.). Planta 130:175–180
110 Esau K 1977 Anatomy of seed plants. 2nd ed. John Wiley New York, 550 pp
111 Espelie K E, Davis R W, Kolattukudy P E 1980 Composition, ultrastructure and function of the cutin- and suberin-containing layers in the leaf, fruit peel, juice-sac and inner seed coat of grapefruit (*Citrus paradisi* Macfed.). Planta 149:498–511
112 Espelie K E, Dean B B, Kolattukudy P E 1979 Composition of lipid-derived polymers from different anatomical regions of several plant species. Plant Physiol 64:1089–1093
113 Espelie K E, Franceschi V R, Kolattukudy P E 1986 Immunocytochemical localization and time course of appearance of an anionic peroxidase associated with suberization in wound-healing potato tuber tissue. Plant Physiol 81:487–492
114 Espelie K E, Kolattukudy P E 1979 Composition of the aliphatic components of 'suberin' from the bundle sheaths of *Zea mays* leaves. Plant Sci Lett 15:225–230
115 Espelie K E, Kolattukudy P E 1985 Purification and characterization of an abscisic acid-inducible anionic peroxidase associated with suberization in potato (*Solanum tuberosum*). Arch Biochem Biophys 240(2):539–545
116 Espelie K E, Sadek N Z, Kolattukudy P E 1980 Composition of suberin-associated waxes from the subterranean storage organs of seven plants: parsnip, carrot, rutabaga, turnip, red beet, sweet potato and potato. Planta 148:468–476
117 Espelie K E, Wattendorff J, Kolattukudy P E 1982 Composition and ultrastructure of the suberized cell wall of isolated crystal idioblasts from *Agave americana* L. leaves. Planta 155:166–175
118 Evert R F, Botha C E J, Mierzwa R J 1985 Free-space marker studies on the leaf of *Zea mays* L. Protoplasma 126:62–73
119 Fahey M D, Kurth E F 1957 The chemical nature of the phenolic acid in the cork from white fir bark. Tappi 40(7):506–512
120 Falk H, El-Hadidi M N 1961 Der Feinbau der Suberinschichten verkorkter Zellwände. Z Naturforsch 16b:134–137
121 Fengel D, Wegener G 1984 Wood. Chemistry, ultrastructure, reactions. de Gruyter Berlin New York, 613 pp
122 Ferguson I B, Clarkson D T 1976 Ion uptake in relation to the development of a root hypodermis. New Phytol 77:11–14

123 Fernando G, Zimmermann W, Kolattukudy P E 1984 Suberin-grown *Fusarium solani* f. sp. *pisi* generates a cutinase-like esterase which depolymerizes the aliphatic components of suberin. Physiol Plant Pathol 24:143–155

124 Fisher J B 1981 Wound healing by exposed secondary xylem in *Adansonia* (Bombaceae). IAWA Bull 2:193–199

125 Fleuriet A, Deloire A 1982 Aspects histochimiques et biochimiques de la cicatrisation des fruits des tomates blessés. Z Pflanzenphysiol 107:259–268

126 Fleuriet A, Macheix J-J 1984 Orientation nouvelle du métabolisme des acides hydroxycinnamiques dans les fruits des tomates blessés (*Lycopersicon esculentum*). Physiol Plant 61: 64–68

127 Foster R C, Rovira A D, Cock T W 1983 Ultrastructure of the root-soil interface. Am Phytopathol Soc St. Paul, 157 pp

128 Franich R A, Gadgil P D, Shain L 1983 Fungistatic effects of *Pinus radiata* needle epicuticular fatty and resin acids on *Dothistroma pini*. Physiol Plant Pathol 23:183–195

129 Franich R A, Volkman J K 1982 Constituent acids of *Pinus radiata* stem cutin. Phytochemistry 21:2687–2689

130 Fraser H S, Swan E P 1978 The chemistry of western red cedar bark. Can For Serv For Tech Publ 27, 17 pp

131 Fraser H S, Swan E P 1979 Phenolic character of sequential solvent extracts from western hemlock and white spruce barks. Can J For Res 9:495–500

132 Freudenberg K 1968 The constitution and biosynthesis of lignin. In: Freudenberg K, Neish A C (eds) Constitution and biosynthesis of lignin. Springer New York, 47–122

133 Friend J 1976 Lignification in infected tissue. In: Friend J, Threfall R (eds) Biochemical aspects of plant-parasite relationships. Academic Press New York, 291–303

134 Fussel L K, Dwarte D M 1980 Structural changes of the grain associated with black region formation in *Pennisetum americanum*. J Exp Bot 31:645, 654

135 Geiger J P, Nicole M, Nandris D, Rio B 1986 Root rot diseases of *Hevea brasiliensis*. I. Physiological and biochemical aspects of host aggression. Eur J For Path 16:22–37

136 Gibson A C 1981 Vegetative anatomy of *Pachycormus* (Anacardiaceae). J Linn Soc London Bot 83:273–284

137 Glazener J A 1982 Accumulation of phenolic compounds in cells and formation of lignin-like polymers in cell walls of young tomato fruits after inoculation with *Botrytis cinerea*. Physiol Plant Pathol 20:11–125

138 Godkin S E, Grozdits G A, Keith C T 1977 A lipid-dense layer in the periderm of *Picea glauca* (Moench) Voss. Proc Microsc Soc Can IV:60–61

139 Godkin S E, Grozdits G A, Keith C T 1978 Structure of the thick-walled phellem cells in eastern white spruce periderm. Proc Microsc Soc Can V:62–63

140 Godkin S E, Grozdits G A, Keith C T 1983 The periderms of three North American conifers. Part 2: Fine structure. Wood Sci Technol 17:13–30

141 Goldberg R, Lê T, Catesson A M 1985 Localization and properties of cell wall enzyme activities related to the final stages of lignin biosynthesis. J Exp Bot 36:503–510

142 Golinowski W O 1971 The anatomical structure of the common fir (*Abies alba* Mill.) bark. I. Development of bark tissues. Acta Soc Bot Pol 40:149–181

143 Gómez-Vazquez B G, Engleman E M 1984 Bark anatomy of *Bursera longipes* (Rose) Standley and *Bursera copallifera* (Sessé and Moc.) Bullock. IAWA Bull 5:335–340

144 Gottwald H, Parameswaran N 1980 Anatomy of wood and bark of *Tectona* (Verbenaceae) in relation to taxonomy. Bot Jahrb Syst 101:363–384

145 Grambow H J, Grambow G E 1978 The involvement of epicuticular and cell wall phenols of the host plant in the in vitro development of *Puccinia graminis* f. sp. *tritici*. Z Pflanzenphysiol 91:1–9

146 Griffith M, Huner N P A, Espelie K E, Kolattukudy P E 1985 Lipid polymers accumulate on the epidermis and in mestome sheath cell walls during low temperature development of winter rye leaves. Protoplasma 125:53–64

147 Grillos S J, Smith F H 1959 The secondary phloem of Douglas-fir. For Sci 5:377–388

148 Grisebach H 1981 Lignins. In: Conn E E (ed) The biochemistry of plants. Vol 7. Academic Press New York, 301–316

149 Gross G G 1977 Biosynthesis of lignin and related monomers. In: Loewus F A, Runeckles V C (eds) Recent advances in phytochemistry. Vol 11. The structure, biosynthesis, and degradation of wood. Plenum New York, 141–184
150 Gross G G 1980 The biochemistry of lignification. Adv Bot Res 8:25–63
151 Gross G G, Janse C 1977 Formation of NADH and hydrogen peroxide by cell wall-associated enzymes from *Forsythia* xylem. Z Pflanzenphysiol 84:447–452
152 Gross G G, Janse C, Elstner E F 1977 Involvement of malate, monophenols, and the superoxide radical in hydrogen peroxide formation by isolated cell walls from horseradish (*Armoracia lapathifolia* Gilib.). Planta 136:271–276
153 Grozdits G A 1982 Microstructure of sequent periderms and ultrastructure of periderm cell walls in *Tsuga canadensis* (L.) Carr. Wood Sci 15:110–118
154 Grozdits G A, Godkin S E, Keith C T 1982 The periderms of three North American conifers. Part 1: Anatomy. Wood Sci Technol 16:305–316
155 Grünwald J, Seemüller E 1979 Zerstörung der Resistenzeigenschaften des Himbeerrutenperiderms als Folge des Abbaus von Suberin und Zellwandpolysacchariden durch die Himbeerrutengallmücke *Thomasiniana theobaldi* Barnes (Dipt., Cecidomyiidae). Z Pflanzenkr Pflanzenschutz 86:305–314
156 Grzywacz A, Rosochacka J 1977 Attempt at elucidation of the role of fatty acids in the resistance of *Pinus silvestris* L. seeds to infection by damping-off fungi in dependence on the colour of their seed shells. Acta Soc Bot Pol 46:569–575
157 Guillemonat A, Cesaire G 1949 Sur la constitution chimique du liège. 1er mèmoire. L'acide phloionique (octodecane diol-9-10 diolique). Bull Soc Chim Fr 792
158 Guillemonat A, Strich A 1950 Sur la constitution chimique du liège. 2e mèmoire. (1) Structure des acides phellonique et phellongènique. Bull Soc Chim Fr 860–861
159 Guillemonat A, Traynard J-C 1963 Sur le constitution chimique du liège. IVe mèmoire: Structure de la phellochryséine. Bull Soc Chim Fr 142–144
160 Guillemonat A, Triaca M 1968 Sur la constitution chimique du liège. V^{e} mèmoire. Étude préliminaire du liège de *Kielmeyera coriacea*. Bull Soc Chim Fr 950–952
161 Gunning B E S, Hughes J 1976 Quantitative assessment of symplastic transport of prenectar into the trichomes of *Abutilon* nectaries. Aust J Plant Physiol 3:619–637
162 Haas D L, Carothers Z B, Robbins R R 1976 Observations on the phi-thickenings and Casparian strips in *Pelargonium* roots. Am J Bot 63:863–867
163 Hanson K R, Havir E A 1979 An introduction to the enzymology of phenylpropanoid biosynthesis. Recent Adv Phytochem 12:91–137
164 Hartley R D 1971 Improved methods for the estimation by gas-liquid chromatography of lignin degradation products from plants. J Chromatogr 54:335–344
165 Hartman L, Weenink R O 1967 A note on the fatty acid composition of the lipids from the bark of *Pinus radiata*. N Z J Sci 10:636–638
166 Harun J, Labosky P Jr 1985 Chemical constituents of five northeastern barks. Wood Fiber Sci 17:274–280
167 Harwood J L, Stumpf P K 1971 Fat metabolism in higher plants. XLIII. Control of fatty acid synthesis in germinating seeds. Arch Biochem Biophys 142:281–291
168 Hata K, Schubert W J, Nord F F 1966 Fungal degradation of the lignin and phenolic acids of the bark of western pine. Arch Biochem Biophys 113:250–256
169 Hata K, Sogo M 1960 Chemical studies on the bark. VII. On the lignin of the outer bark of "Mizunara" (*Quercus crispula* Blume). Mokuzai Gakkaishi 6:71–75
170 Hata K, Sogo M, Fukuhara T, Hochi M 1969 On the suberin in the outer bark of some Japanese tree species. Tech Bull Fac Agr Kagawa Univ 20(2):112–119
171 Hattersley P W, Browning A J 1981 Occurrence of the suberized lamella in leaves of grasses of different photosynthetic types. I. In parenchymatous bundle sheaths and PCR ("Kranz") sheaths. Protoplasma 109:371–401
172 Hattersley P W, Perry S 1984 Occurrence of the suberized lamella in leaves of grasses of different photosynthetic types. II. In herbarium material. Aust J Bot 32:465–473
173 Hattersley P W, Wong S-C, Perry S, Roksandic Z 1986 Comparative ultrastructure and gas exchange characteristics of the C_3-C_4 intermediate *Neurachne minor* S. T. Blake (Poaceae). Plant Cell Environ 9:217–233

174 Hay M J M, Dunlop J 1982 Anatomy and development of the superficial layers in stolons of white clover (*Trifolium repens* L.). N Z J Bot 20:315–324

175 Heale J B, Dodd K S, Gahan P B 1982 The induced resistance response of carrot root slices to heat-killed conidia and cell-free germination fluid of *Botrytis cinerea* Pers. ex Pers. I. The possible role of cell death. Ann Bot 49:847–857

176 Hejno K, Jarolím V, Sorm F 1965 Über einige Inhaltsstoffe des weissen Teiles der Birkenrinde. Coll Czech Chem Commun 30:1009–1015

177 Hemingway R W 1981 Bark: its chemistry and prospects for chemical utilization. In: Goldstein I S (ed) Organic chemicals from biomass. CRC Press Boca Raton, 189–248

178 Henrickson J 1969 Anatomy of periderm and cortex of Fouquieriaceae. Aliso 7(1):97–126

179 Hergert H L 1958 Chemical composition of cork from white fir bark. For Prod J 8:335–339

180 Hergert H L, Kurth E F 1952 The chemical nature of the cork from Douglas-fir bark. Tappi 35(2):59–66

181 Hergert H L, Kurth E F 1953 The chemical nature of the extractives from white fir bark. Tappi 36:137–144

182 Hergert H L, von Blaricom L E, Steinberg J C, Gray K R 1965 Isolation and properties of dispersants from western hemlock bark. For Prod J 15:485–491

183 Herrick F W 1980 Chemistry and utilization of western hemlock bark extractives. J Agric Food Chem 28:228–237

184 Higuchi T, Ito Y, Shimada M, Kawamura I 1967 Chemical properties of bark lignins. Cellul Chem Technol 1:585–589

185 Hillis W E, Inoue T 1968 The formation of polyphenols in trees. IV. The polyphenol formed in *Pinus radiata* after *Sirex* attack. Phytochemistry 7:13–22

186 Hock W K, Klarman W L 1967 The function of the endodermis in resistance of Virginia pine seedlings to damping off. For Sci 13:108–112

187 Hoitink H A J 1980 Composted bark, a lightweight growth medium with fungicidal properties. Plant Dis 64:142–147

188 Holloway P J 1972 The composition of suberin from the corks of *Quercus suber* L. and *Betula pendula* Roth. Chem Phys Lipids 9:158–170

189 Holloway P J 1972 The suberin composition of the cork layers from some *Ribes* species. Chem Phys Lipids 9:171–179

190 Holloway P J 1974 Intracuticular lipids of spinach leaves. Phytochemistry 13:2201–2207

191 Holloway P J 1982 Suberins of *Malus pumila* stem and root corks. Phytochemistry 21:2517–2522

192 Holloway P J 1983 Some variations in the composition of suberin from the cork layers of higher plants. Phytochemistry 22:495–502

193 Hossfeld R L, Hunter W T 1958 The petroleum ether extractives of aspen bark. Tappi 41(7):359–362

194 Hossfeld R L, Kaufert F H 1957 Structure and composition of aspen bark. For Prod J 7:437–439

195 Howard E T 1971 Bark structure of the southern pines. Wood Sci 3:134–148

196 Howard E T 1977 Bark structure of southern upland oaks. Wood Fiber 9:172–183

197 Hudler G W 1984 Wound healing in bark of woody plants. J Arboric 10:241–245

198 Hudler G W, Banik M T 1986 Wound repair and disease resistance in stems of *Pachysandra terminalis*. Can J Bot 64:2406–2410

199 Isenberg I H 1943 The anatomy of redwood bark. Madrono 7:85–91

200 Ismailov A Ya, Stránský K, Streibl M 1975 On some lipidic components from the root of *Petasites hybridus* (L.) G.M. Sch. Coll Czech Chem Commun 40:3731–3733

201 Jackson L L, Blomquist G J 1976 Insect waxes. In: Kolattukudy P E (ed) Chemistry and biochemistry of natural waxes. Elsevier Amsterdam, 94–146

202 Jaramillo A E 1980 Wound periderm formation in *Juniperus occidentalis* in response to *Arceuthobium campylopodum*. Can J Bot 58:400–403

203 Jensen W 1952 Studies on suberin. IV. Isolation of phloionic acid from suberin of *Quercus suber*. Pap Puu 34:468–470

204 Jensen W 1962 Botanical histochemistry. Freeman New York London, 523 pp

205 Jensen W 1972 A study of the outer bark of birch (*Betula verrucosa*) and cork oak (*Quercus suber*) by scanning electron microscopy. Ann Rev Soc Esp Fis Quim 68:871–878

206 Jensen W, Fremer K E, Sierilä P, Wartiovaara V 1963 The chemistry of bark. In: Browning B L (ed) The chemistry of wood. Wiley-Interscience New York London, 587–666
207 Jensen W, Ihalo P, Varsa K 1957 Studies on suberin. VIII. Separation of the acids of suberin by means of countercurrent distribution of their methyl esters. Pap Puu 39:237–242
208 Jensen W, Rinne P 1953 Studies on suberin. V. The constitution of phloionic acid from the suberin of *Quercus suber*. Pap Puu 35:407–408
209 Jensen W, Rinne P 1954 Studies on suberin. VI. Isolation and constitution of phloionolic acid from the suberins of *Betula verrucosa* and *Quercus suber*. Pap Puu 36:32–34
210 Jensen W, Tinnis W 1957 Studies on suberin. IX. Identification of unsaturated acids in suberin of *Quercus suber* and *Betula verrucosa*. Pap Puu 39:261–264
211 Johnson-Flanagan A M, Owens J N 1985 Peroxidase activity in relation to suberization and respiration in white spruce (*Picea glauca* [Moench] Voss) seedling roots. Plant Physiol 79:103–107
212 Jones J H 1978 Chemical changes in cutin obtained from cuticles isolated by the zinc chloride-hydrochloric acid method. Plant Physiol 62:831–832
213 Kahl G 1974 Metabolism in plant storage tissue slices. Bot Rev 40:263–314
214 Karstedt P, Parameswaran N 1976 Beitrag zur Anatomie und Systematik der atlantischen Rhizophora-Arten. Bot Jahrb Syst 97:317–338
215 Kawashima N, Uritani I 1963 Occurrence of peroxidases in sweet potato infected by the black rot fungus. Agr Biol Chem 47:409–417
216 Khan A A, Kolattukudy P E 1973 Control of synthesis and distribution of acyl moieties in etiolated *Euglena gracilis*. Biochemistry 12:1939–1948
217 Khan A A, Kolattukudy P E 1974 Decarboxylation of long chain fatty acids to alkanes by cell-free preparations of pea leaves (*Pisum sativum*). Biochem Biophys Res Commun 61:1379–1386
218 Khan A A, Kolattukudy P E 1975 Solubilization of fatty acid synthetase, acyl-CoA reductase, and fatty acyl-CoA alcohol transacylase from the microsomes of *Euglena gracilis*. Arch Biochem Biophys 170:400–408
219 Knowles L O, Flore J A 1983 Quantitative and qualitative characterization of carrot root periderm during development. J Am Soc Hort Sci 108:923–928
220 Kolattukudy P E 1966 Biosynthesis of wax in *Brassica oleracea*. Relation of fatty acids to wax. Biochemistry 5:2265–2275
221 Kolattukudy P E 1967 Biosynthesis of paraffins in *Brassica oleracea*: fatty acid elongation-decarboxylation as a plausible pathway. Phytochemistry 6:963–975
222 Kolattukudy P E 1967 Mechanisms of syntheses of waxy esters in broccoli (*Brassica oleracea*). Biochemistry 6:2705–2717
223 Kolattukudy P E 1968 Biosynthesis of surface lipids. Science 159:498–505
224 Kolattukudy P E 1968 Tests whether a head to head condensation mechanism occurs in the biosynthesis of *n*-hentriacontane, the paraffin of spinach and pea leaves. Plant Physiol 43:1466–1470
225 Kolattukudy P E 1970 Biosynthesis of cuticular lipids. Ann Rev Plant Physiol 21:163–192
226 Kolattukudy P E 1970 Biosynthetic relationships among very long chain hydrocarbons, ketones, and secondary alcohols and the noninvolvement of alkenyl glyceryl ethers in their biosynthesis. Arch Biochem Biophys 141:381–383
227 Kolattukudy P E 1970 Reduction of fatty acids to alcohols by cell-free preparations of *Euglena gracilis*. Biochemistry 9:1095–1102
228 Kolattukudy P E 1971 Enzymatic synthesis of fatty alcohols in *Brassica oleracea*. Arch Biochem Biophys 142:701–709
229 Kolattukudy P E 1977 Lipid polymers and associated phenols, their chemistry, biosynthesis, and role in pathogenesis. In: Loewus F A, Runeckles V C (eds) Recent advances in phytochemistry. Vol 11. The structure, biosynthesis, and degradation of wood. Plenum New York, 185–246
230 Kolattukudy P E 1978 Chemistry and biochemistry of the aliphatic components of suberin. In: Kahl G (ed) Biochemistry of wounded plant tissues. de Gruyter Berlin New York, 43–84
231 Kolattukudy P E 1980 Biopolyester membranes of plants: cutin and suberin. Science 208:990–1000
232 Kolattukudy P E 1980 Cutin, suberin and waxes. In: Stumpf P K, Conn E E (eds) The biochemistry of plants. Vol 4. Academic Press New York, 541–645
233 Kolattukudy P E 1981 Structure, biosynthesis, and biodegradation of cutin and suberin. Ann Rev Plant Physiol 32:539–567

234 Kolattukudy P E 1984 Biochemistry and function of cutin and suberin. Can J Bot 62:2918–2933
235 Kolattukudy P E, Agrawal V P 1974 Structure and composition of aliphatic constituents of potato tuber skin (suberin). Lipids 9:682–691
236 Kolattukudy P E, Brown L 1974 Inhibition of cuticular lipid biosynthesis in *Pisum sativum* by thiocarbamates. Plant Physiol 53:903–906
237 Kolattukudy P E, Buckner J S 1972 Chain elongation of fatty acids by cell-free extracts of epidermis from pea leaves (*Pisum sativum*). Biochem Biophys Res Commun 46:801–807
238 Kolattukudy P E, Buckner J S, Brown L 1972 Direct evidence for a decarboxylation mechanism in the biosynthesis of alkanes in *B. oleracea*. Biochem Biophys Res Commun 47:1306–1313
239 Kolattukudy P E, Buckner J S, Liu T Y J 1973 Biosynthesis of secondary alcohols and ketones from alkanes. Arch Biochem Biophys 156:613–620
240 Kolattukudy P E, Croteau R, Brown L 1974 Hydroxylation of palmitic acid and decarboxylation of C_{28}, C_{30} and C_{32} acids in *Vicia faba* flowers. Plant Physiol 54:670–677
241 Kolattukudy P E, Croteau R, Buckner J S 1976 The biochemistry of plant waxes. In: Kolattukudy P E (ed) The chemistry and biochemistry of natural waxes. Elsevier New York, 289–347
242 Kolattukudy P E, Croteau R, Walton T J 1975 Biosynthesis of cutin. Enzymatic conversion of ω-hydroxy fatty acids to dicarboxylic acids by cell-free extracts of *Vicia faba* epidermis. Plant Physiol 55:875–880
243 Kolattukudy P E, Dean B B 1974 Structure, gas chromatographic measurement, and function of suberin synthesized by potato tuber tissue slices. Plant Physiol 54:116–121
244 Kolattukudy P E, Espelie K E 1985 Biosynthesis of cutin, suberin and associated waxes. In: Higuchi T (ed) The biochemistry of wood. Academic Press New York, 161–207
245 Kolattukudy P E, Espelie K E, Soliday C L 1981 Hydrophobic layers attached to cell walls: cutin, suberin and associated waxes. In: Loewus F A, Tanner W (eds) Encyclopedia of plant physiology. Plant carbohydrates. 13B:225–254
246 Kolattukudy P E, Jaeger R H, Robinson R 1968 Biogenesis of nonacosan-15-one in *Brassica oleracea*: dual mechanisms in the synthesis of long-chain compounds. Nature 219:1038–1040
247 Kolattukudy P E, Jaeger R H, Robinson R 1971 Biogenesis of nonacosan-15-one in *Brassica oleracea*. Phytochemistry 10:3047–3051
248 Kolattukudy P E, Köller W 1983 Fungal penetration of the first line defensive barriers of plants. In: Callow J A (ed) Biochemical plant pathology. John Wiley New York, 79–100
249 Kolattukudy P E, Kronman K, Poulose A J 1975 Determination of structure and composition of suberin from the roots of carrot, parsnip, rutabaga, turnip, red beet, and sweet potato by combined gas-liquid chromatography and mass spectrometry. Plant Physiol 55:567–573
250 Kolattukudy P E, Liu T Y J 1970 Direct evidence for biosynthetic relationships among hydrocarbons, secondary alcohols and ketones in *Brassica oleracea*. Biochem Biophys Res Commun 41:1369–1374
251 Kolattukudy P E, Soliday C L 1985 Effects of stress on the defensive barriers of plants. In: Key J L, Kosuge T (eds) Cellular and molecular biology of plant stress. Alan R Liss New York, 381–400
252 Kolattukudy P E, Walton T J 1972 Structure and biosynthesis of the hydroxy fatty acids of cutin in *Vicia faba* leaves. Biochemistry 11:1897–1907
253 Kolattukudy P E, Walton T J 1973 The biochemistry of plant cuticular lipids. Prog Chem Fats Other Lipids 13:119–175
254 Kolattukudy P E, Walton T J, Kushwaha R 1971 Epoxy acids in the lipid polymer cutin and their role in the biosynthesis of cutin. Biochem Biophys Res Commun 47:1306–1313
255 Kolattukudy P E, Walton T J, Kushwaha R P S 1973 Biosynthesis of the C_{18} family of cutin acids: ω-hydroxyoleic acid, ω-hydroxy-9,10-epoxystearic acid, 9,10,18-trihydroxystearic acid, and the Δ^{12}-unsaturated analogs. Biochemistry 12:4488–4498
256 Krähmer H 1980 Wundreaktionen von Apfelbäumen und ihr Einfluss auf Infektionen mit *Nectria galligena*. Z Pflanzenkr Pflanzenschutz 87:97–112
257 Krahmer R L, Wellons J D 1973 Some anatomical and chemical characteristics of Douglas-fir cork. Wood Sci 6:97–105
258 Kucera L J, Butterfield B G 1977 Resin canals in the bark of *Phyllocladus* species indigenous to New Zealand. N Z J Bot 15:657–663
259 Kumar A 1969 Bark morphology of desert plants. I. *Prosopis spicigera* L. Ind Bot Soc J 48:76–84

260 Kurth E F 1947 The chemical composition of barks. Chem Rev 40:33–49
261 Kurth E F 1958 Chemical analyses of mountain hemlock. Tappi 41(12):733–734
262 Kurth E F 1967 The chemical composition of conifer bark waxes and corks. Tappi 50(6):253–258
263 Kurth E F, Becker E L 1953 The chemical nature of the extractives from red alder. Tappi 36(10):461–466
264 Kurth E F, Hubbard J K 1951 Extractives from ponderosa pine bark. Ind Eng Chem 43:896–900
265 Kurth E F, Kiefer H J 1950 Wax from Douglas-fir bark. Tappi 33(4):183–186
266 Kurth E F, Tokos G M 1953 The chemical composition of grand fir bark. Tappi 36(7):301–304
267 Labosky P 1979 Chemical constituents of four southern pine barks. Wood Sci 12:80–85
268 Lai Y Z, Sarkanen K V 1971 Isolation and structural studies. In: Sarkanen K V, Ludwig C H (eds) Lignins. Occurrence, formation, structure and reactions. Wiley-Interscience New York, 165–240
269 Laver M L, Fang H H-L 1986 Douglas-fir bark. III. Sterol and wax esters of the *n*-hexane wax. Wood Fiber Sci 18:553–564
270 Laver M L, Loveland P M, Chen C-H, Fang H H-L, Liu Y-C L 1977 Chemical constituents of Douglas-fir bark: a review of more recent literature. Wood Sci 10:85–92
271 Lawton J R 1980 Observations on the structure of epidermal cells, particularly the cork and silica cells, from the flowering stem internode of *Lolium emulentum* L. (Gramineae). Bot J Linn Soc 80:161–177
272 Lawton J R, Todd A, Naidoo D K 1981 Preliminary investigations into the structure of the roots of the mangroves, *Avicennia marina* and *Bruguiera gymnorrhiza*, in relation to ion uptake. New Phytol 88:713–722
273 Lendzian K J 1982 Gas permeability of plant cuticles. Oxygen permeability. Planta 155:310–315
274 Leshem B 1970 Resting roots of *Pinus halepensis*: structure, function and reaction to water stress. Bot Gaz 131:99–104
275 Lessire R 1973 Étude de la biosynthèse des acides gras a tres longue chaine et leurs relations avec les alcanes dans la cellule d'epiderme d'*Allium porrum*. Ph.D. Thesis. Univ Bordeaux France, 118 pp
276 Liese W, Günzerodt H, Parameswaran N 1983 Biological alterations of cork quality affecting its utilization. Cortica-Boletim do Instituto dos Produtos Florestais 541:287–299
277 Liese W, Parameswaran N 1972 On the variation of cell length within the bark of some tropical hardwood species. In: Ghouse A K M, Yanas M (eds) Research trends in plant anatomy. Tata McGraw-Hill Bombay, 83–90
278 Ling-Lee M, Chilvers G A, Ashford A E 1977 A histochemical study of phenolic materials in mycorrhizal and uninfected roots of *Eucalyptus fastigata* Deane and Maiden. New Phytol 78:313–328
279 Lipschitz N, Lev-Yadan S, Rosen E, Waisel Y 1984 The annual rhythm of activity of the lateral meristems (cambium and phellogen) in *Pinus halepensis* Mill. and *Pinus pinea* L. IAWA Bull 5:263–274
280 Litvay J D, Krahmer R L 1976 The presence of callose in cork cells. Wood Fiber 8:146–151
281 Litvay J D, Krahmer R L 1977 Wall layering in Douglas-fir cork cells. Wood Sci 9:167–173
282 Liu T Y J 1972 The biosynthetic relationships among long chain fatty acids, alkanes, secondary alcohols and ketones in plants. M.Sc. Thesis. Washington State University Pullman, 82 pp
283 Loveland P M, Laver M L 1972 Monocarboxylic and dicarboxylic acids from *Pseudotsuga menziesii* bark. Phytochemistry 11:430–432
284 Loveland P M, Laver M L 1972 ω-Hydroxy fatty acids and fatty alcohols from *Pseudotsuga menziesii* bark. Phytochemistry 11:3080–3081
285 Macey M J K, Stumpf P K 1968 Fat metabolism in higher plants. 36: Long chain fatty acid synthesis in germinating peas. Plant Physiol 43:1637–1647
286 MacKenzie K A D 1979 The development of the endodermis and phi layer of apple roots. Protoplasma 100:21–32
287 MacKenzie K A D 1983 Some aspects of the development of the endodermis and cortex of *Tilia cordata* and *Picea sitchensis*. Plant Soil 71:147–153
288 Mader H 1954 Untersuchungen an Korkmembranen. Planta 43:163–181
289 Mäder M, Ungemach J, Schloss P 1980 The role of peroxidase isoenzyme groups of *Nicotiana tabacum* in hydrogen peroxide formation. Planta 174:467–470
290 Manners J G, Myers S 1975 The effect of fungi (particularly obligate pathogens) on the physiology of higher plants. Symp Soc Exp Bot 29:279–296

291 Maron R, Fahn A 1979 Ultrastructure and development of oil cells in *Laurus nobilis* L. leaves. Bot J Linn Soc 78:31–40
292 Martin J T, Juniper B E 1970 The cuticles of plants. St Martins New York, 347 pp
293 Martin R E 1969 Characterization of southern pine barks. For Prod J 19(8):23–30
294 Martin R E, Crist J B 1968 Selected physical-mechanical properties of eastern tree barks. For Prod J 18(11):54–60
295 Martin R E, Crist J B 1970 Elements of bark structure and terminology. Wood Fiber 2:269–279
296 Mater J 1967 Bark utilization. A review and projections. For Prod J 17(12):15–20
297 Matic M 1956 The chemistry of plant cuticle. A study of cutin from *Agave americana* L. Biochem J 63:168–178
298 Mazliak P 1963 La cire cuticulaire des pommes (*Malus pumila* L.). Ph.D. Thesis. Univ Paris, 115 pp
299 McDonald E C, Howard J, Bennett B 1983 Chemicals from forest products by supercritical fluid extraction. Fluid Phase Equil 10:337–344
300 McGinnes E A Jr, Phelps J E, Szopa P S, Shigo A L 1977 Wood anatomy after tree injury – a pictorial study. Univ Missouri College Agric Res Bull 1025, 35 pp
301 McMillin C W, Manwiller F G 1980 The wood and bark of hardwoods growing on southern pine sites – a pictorial atlas. USDA For Ser Gen Tech Rep SO-29, 58 pp
302 McNicol R J, Williamson B, Jennings D L, Woodford J A T 1983 Resistance to raspberry cane midge (*Resseliella theobaldi*) and its association with wound periderm in *Rubus crataegifolius* and its red raspberry derivatives. Ann Appl Biol 103:489–495
303 Meiering A G, Paroschy J H, Peterson R L, Hostetter G, Neff A 1980 Mechanical freezing injury in grapevine trunks. Am J Enol Vitic 31:81–89
304 Meyer M 1938 Die submikroskopische Struktur der rutinisierten Zellmembranen. Protoplasma 29:552–586
305 Meyer R W, Kellogg R M, Warren W G 1981 Relative density, equilibrium moisture content and dimensional stability of western hemlock bark. Wood Fiber 13:86–96
306 Middleton G E, Bostock R M 1985 Histopathology of wounded almond bark in relation to infection by *Ceratocystis fimbriata*. Phytopathology 75:1374
307 Mizicko J, Livingston C H, Johnson G 1974 The effects of dihydroquercitin on the cut surface of seed potatoes. Am Potato J 51:216–222
308 Mogensen H L 1968 Studies on the bark of the cork bark fir: *Abies lasiocarpa* var. *arizonica* (Merriam) Lemmon. I. Periderm ontogeny. J Arizona Acad Sci 5:36–40
309 Moon G J, Peterson C A, Peterson R L 1984 Structural, chemical and permeability changes following wounding in onion roots. Can J Bot 62:2253–2259
310 Moore K E 1978 Barrier-zone formation in wounded stems of sweetgum. Can J Bot 8:389–397
311 Morohoshi N, Glasser W G 1979 The structure of lignins in pulps. Part 4: Comparative evaluation of five lignin depolymerization techniques. Wood Sci Technol 13:165–178
312 Moss E H, Gorham A L 1953 Interxylary cork and fission of stems and roots. Phytomorphology 3:285–294
313 Mueller W C, Beckman C H 1978 Ultrastructural localization of polyphenol oxidase and peroxidase in roots and hypocotyls of cotton seedlings. Can J Bot 56:1579–1587
314 Muhammad A F, Micko M M 1984 Accumulation of calcium crystals in the decayed wood of aspen attacked by *Fomes ignarious*. IAWA Bull 5:237–241
315 Mullick D B 1977 The non-specific nature of defense in bark and wood during wounding, insect and pathogen attack. In: Loewus F A Runeckles V C (eds) Recent advances in phytochemistry. Vol 11. The structure, biosynthesis, and degradation of wood. Plenum New York, 395–441
316 Mullick D B, Jensen G D 1973 Cryofixation reveals uniqueness of reddish-purple sequent periderm and equivalence between brown first and brown sequent periderms of three conifers. Can J Bot 51:135–143
317 Nagahashi G, Thomson W W, Leonard R T 1974 The Casparian strip as a barrier to the movement of lanthanum in corn roots. Science 183:670–671
318 Nair G V, von Rudloff E 1960 Isolation of hyperin from red-osier dogwood (*Cornus stolonifera* Michx.). Can J Chem 38:2531–2533
319 Nakatani M, Yamachika T, Tanoue T, Hase T 1985 Structures and synthesis of seed-germination inhibitors from *Hibiscus rosa-sinensis*. Phytochemistry 24:39–42
320 Nanko H, Côté W A 1980 Bark structure of hardwoods grown on southern pine sites. Syracuse University Press New York, 56 pp

321 Narasimhachari N, von Rudloff E 1961 The chemical composition of the wood and bark extractives of *Juniperus horizontalis* Moench. Can J Chem 39:2572–2581
322 Nedoff J A, Ting I P, Lord E M 1985 Structure and function of the green stem tissue in Octillo (*Fouquieria splendens*). Am J Bot 72:143–151
323 Nelson P E, Wilhelm S 1957 Some aspects of the strawberry root. Hilgardia 26:631–642
324 Nickles W C, Rowe J W 1962 Chemistry of western white pine bark. For Prod J 12:374–376
325 Nicole M, Geiger J P, Nandris D 1982 Interactions hôte-parasite entre *Hevea brasiliensis* et les agents de pourriture racinaire *Phellinus noxius* et *Rigidoporous lignosis*: étude physio-pathologique comparée. Phytopath Z 105:311–326
326 Nicole M, Geiger J P, Nandris D 1986 Penetration and degradation of suberized cells of *Hevea brasiliensis* infected with root rot fungi. Physiol Mol Plant Pathol 28:181–185
327 Nicole M, Geiger J P, Nandris D 1986 Root rot diseases of *Hevea brasiliensis*. II. Some host reactions. Eur J For Path 16:37–55
328 Nigam S K, Mitra C R 1967 *Pithecolobium dulce*: Part III – minor constituents of trunk bark. Ind J Chem 5:395
329 Norin T, Winell B 1972 Extractives from the bark of common spruce, *Picea abies* L. Karst. Acta Chem Scand 26:2289–2296
330 Norin T, Winell B 1972 Extractives from the bark of Scots pine, *Pinus silvestris* L. Acta Chem Scand 26:2297–2304
331 O'Brien T P, Carr D J 1970 A suberized layer in the cell walls of the bundle sheath of grasses. Aust J Biol Sci 23:275–287
332 O'Brien T P, Kuo J 1975 Development of the suberized lamella in the mestome sheath of wheat leaves. Aust J Bot 23:783–794
333 Ohara S, Yatagai M, Hayashi Y 1986 Utilization of wood extractives. I. Extractives from the bark of *Betula platyphylla* Sukatchev var. *japonica* Hara. Mokuzai Gakkaishi 32:266–273
334 Olesen P 1978 Studies on the physiological sheaths in roots. I. Ultrastructure of the exodermis in *Hoya carnosa* L. Protoplasma 94:325–340
335 Oparka K J, Gates P J 1982 Ultrastructure of the developing pigment strand of rice (*Oryza sativa* L.) in relation to its role in solute transport. Protoplasma 113:33–43
336 Ostrofsky W D, Blanchard R O 1984 Variation in bark characteristics of American beech (*Fagus grandifolia*). Can J Bot 62:1564–1566
337 Ottone S P, Baldwin R C 1981 The relationship of extractive content to particle size distribution in milled yellow-poplar (*Liriodendron tulipifera* L.) bark. Wood Fiber 13:74–85
338 Outer R W D, van Veenendaal W L H 1981 Wood and bark anatomy of *Azima tetracantha* Lam. (Salvadoraceae) with descriptions of its included phloem. Acta Bot Neerl 30:199–207
339 Outer R W D, Vooren A P 1980 Bark anatomy of some Sarcolaenaceae and Rhopalocarpaceae and their systematic position. Meded Landbouwhogeschool Wageningen 80(6):1–15
340 Parameswaran N 1975 Zur Wandstruktur von Sklereiden in einigen Baumrinden. Protoplasma 85:305–314
341 Parameswaran N, Conrad H 1982 Wood and bark anatomy of *Balanites aegyptiaca* in relation to ecology and taxonomy. IAWA Bull 3:75–88
342 Parameswaran N, Knigge H, Liese W 1985 Electron microscopic demonstration of a suberised layer in the tylosis wall of beech and oak. IAWA Bull 6:269–271
343 Parameswaran N, Kruse J, Liese W 1976 Aufbau und Feinstruktur von Periderm und Lenticellen der Fichtenrinde. Z Pflanzenphysiol 77:212–221
344 Parameswaran N, Liese W 1970 Mikroskopie der Rinde tropischer Holzarten. In: Freund H (ed) Handbuch der Mikroskopie in der Technik. Band V. Mikroskopie des Holzes und des Papiers. Teil 1. Mikroskopie des Rohholzes und der Rinden. Umschau Frankfurt, 227–306
345 Parameswaran N, Liese W, Günzerodt H 1981 Characterization of wetcork in *Quercus suber* L. Holzforschung 35:195–199
346 Parameswaran N, Richter H-G 1984 The ultrastructure of crystalliferous cells in some Lecythidaceae with a discussion of their terminology. IAWA Bull 5:229–236
347 Parameswaran N, Schultze R 1974 Fine structure of the chambered crystalliferous cells in the bark of *Acacia senegal*. Z Pflanzenphysiol 71:90–93
348 Parameswaran N, Sinner M 1979 Topochemical studies on the wall of beech bark sclereids by enzymatic and acidic degradation. Protoplasma 101:197–215

349 Parameswaren N, Wilhelm G E 1979 Micromorphology of naturally degraded beech and spruce barks. Eur J For Pathol 9:103–112
350 Parameswaran N, Wilhelm G E, Liese W 1976 Ultrastructural aspects of beech bark degradation by fungi. Eur J For Pathol 5:274–286
351 Paroschy J H, Meiering A G, Peterson R L, Hostetter G, Neff A 1980 Mechanical winter injury in grapevine trunks. Am J Enol Vitic 31:227–232
352 Patel R N 1975 Bark anatomy of radiata pine, Corsican pine, and Douglas-fir grown in New Zealand. N Z J Bot 13:149–167
353 Patel R N, Shand J E 1985 Bark anatomy of *Nothofagus* species indigenous to New Zealand. N Z J Bot 23:511–532
354 Pearce R B, Holloway P J 1984 Suberin in the sapwood of oak (*Quercus robur* L.): its composition from a compartmentalization barrier and its occurrence in tyloses in undecayed wood. Physiol Plant Pathol 24:71–81
355 Pearce R B, Rutherford J 1981 A wound-associated suberized barrier to the spread of decay in the sapwood of oak (*Quercus robur* L.). Physiol Plant Pathol 19:359–369
356 Pearce R B, Woodward S 1986 Compartmentalization and reaction zone barriers at the margin of decayed sapwood in *Acer saccharinum* L. Physiol Mol Plant Pathol 29:197–216
357 Pearl I A 1975 The water-soluble and petroleum ether-soluble extractives of loblolly and slash pine barks. Tappi 58:142–145
358 Pearl I A 1975 The ether-soluble and ethanol-soluble extractives of loblolly and slash pine barks. Tappi 58:135–137
359 Pearl I A 1975 Variations of loblolly and slash pine bark extractive components and wood turpentine components on a monthly basis. Tappi 58:146–149
360 Pearl I A, Buchanan M A 1976 A study of the inner and outer barks of loblolly pine. Tappi 59:136–139
361 Peek R-D, Liese W, Parameswaran N 1972 Infektion und Abbau der Wurzelrinde von Fichte durch *Fomes annosus*. Eur J For Pathol 2:104–115
362 Pereira H 1981 Quimica da cortica. IV. Determinacao da suberina em cortica virgem e em cortica de reproducao de *Quercus suber* L. Cortica. Bol Inst Prod Florestais 516:233–235
363 Pereira H 1982 Studies on the chemical composition of virgin and reproduction cork of *Quercus suber* L. Anals Inst Superior Agron Lisboa 40:17–25
364 Perumalla C J, Peterson C A 1986 Deposition of Casparian bands and suberin lamellae in the exodermis and endodermis of young corn and onion roots. Can J Bot 64:1873–1878
365 Peterson C A, Emanuel M E, Humphreys G B 1981 Pathway of movement of apoplastic fluorescent dye tracers through the endodermis at the site of secondary root formation in corn (*Zea mays*) and broad bean (*Vicia faba*). Can J Bot 59:618–625
366 Peterson C A, Emanuel M E, Weerdenburg C A 1981 The permeability of phi thickenings in apple (*Pyrus malus*) and geranium (*Pelargonium hortorum*) roots to an apoplastic fluorescent dye tracer. Can J Bot 59:1107–1110
367 Peterson C A, Emanuel M E, Wilson C 1982 Identification of a Casparian band in the hypodermis of onion and corn roots. Can J Bot 60:1529–1535
368 Peterson C A, Griffith M, Huner N P A 1985 Permeability of the suberized mestome sheath in winter rye. Plant Physiol 77:157–161
369 Peterson C A, Perumalla C A 1984 Development of the hypodermal Casparian band in corn and onion roots. J Exp Bot 35:51–57
370 Peterson C A, Peterson R L, Robards A W 1978 A correlated histochemical and ultrastructural study of the epidermis and hypodermis of onion roots. Protoplasma 96:1–21
371 Philipson J J, Coutts M P 1979 The induction of root dormancy in *Picea sitchensis* (Bong.) Carr. by abscisic acid. J Exp Bot 30:371–380
372 Pozuelo J M, Espelie K E, Kolattukudy P E 1984 Magnesium deficiency results in increased suberization in endodermis and hypodermis of corn roots. Plant Physiol 74:256–260
373 Pryor L D, Chattaway M M, Kloot N H 1956 The inheritance of wood and bark characters in *Eucalyptus*. Aust J Bot 4:216–239
374 Rademacher P, Bauch J, Shigo A L 1984 Characteristics of xylem formed after wounding in *Acer*, *Betula*, and *Fagus*. IAWA Bull 5:141–151
375 Rainbow A, White D J B 1972 Preliminary observations on the ultrastructure of maturing cork-cells from tubers of *Solanum tuberosum* L. New Phytol 71:899–902

376 Rao G S R L, Willison J H M, Ratnayake W M N 1984 Suberin production by isolated tomato fruit protoplasts. Plant Physiol 75:716–719
377 Rao G S R L, Willison J H M, Ratnayake W M N 1985 Suberization of tomato (*Lycopersicon esculentum*) locule tissue. Can J Bot 63:2177–2180
378 Rhodes M J C, Wooltorton L S C 1978 The biosynthesis of phenolic compounds in wounded plant storage tissues. In: Kahl G (ed) Biochemistry of wounded plant tissue. de Gruyter Berlin, 243–286
379 Riley R G, Kolattukudy P E 1975 Evidence for covalently attached *p*-coumaric acid and ferulic acid in cutins and suberins. Plant Physiol 56:650–654
380 Robards A W, Clarkson D T, Sanderson J 1979 Structure and permeability of the epidermal/hypodermal layers of the sand sedge (*Carex arenaria* L.). Protoplasma 101:331–347
381 Robards A W, Jackson S M, Clarkson D T, Sanderson J 1973 The structure of barley roots in relation to the transport of ions into the stele. Protoplasma 77:291–311
381a Roberts E, Kutchan T, Kolattukudy P E 1988 Cloning and sequencing of cDNA for a highly anionic peroxidase from potato and the induction of its mRNA in suberizing potato tubers and tomato fruits. Plant Mol Biol 11:15–26
381b Roberts E, Kolattukudy P E 1989 Molecular cloning, nucleotide sequence and abscisic acid induction of a suberization-associated highly anionic peroxidase. Mol Gen Genetics 217:223–232
382 Rodríguez-Miguéns B, Ribas-Marqués I 1972 Investigaciones quimicas sobre el corcho de *Solanum tuberosum* L. (patata). Ann Rev Soc Esp Fis Quim 68:303–308
383 Rodríguez-Miguéns B, Ribas-Marqués I 1972 Contribucion a la estructura quimica de la suberina. Ann Rev Soc Esp Fis Quim 68:1301–1306
384 Rogers I H, Grierson D 1972 Extractives from grand fir [*Abies grandis* (Dougl.) Lindl.] bark. Wood Fiber 4:33–37
385 Ross W D, Krahmer R L 1971 Some sources of variation in structural characteristics of Douglas-fir bark. Wood Fiber 3:35–46
386 Roth I 1981 Structural patterns of tropical barks. Gebrüder Borntraeger Berlin Stuttgart, 609 pp
387 Rowe J W, Bower C L, Wagner E R 1969 Extractives of jack pine bark: occurrence of *cis*- and *trans*-pinosylvin dimethyl ether and ferulic acid esters. Phytochemistry 8:235–241
388 Rowe J W, Conner A H 1979 Extractives in eastern hardwoods: a review. USDA For Ser Gen Tech Rep FPL-18, 67 pp
389 Rowe J W, Scroggins J H 1964 Benzene extractives of lodgepole pine bark. Isolation of new diterpenes. J Org Chem 29:1554–1562
390 Russin J S, Shain L 1984 Colonization of chestnut blight cankers by *Ceratocystis microspora* and *C. eucastaneae*. Phytopathology 74:1257–1261
391 Rykowski K 1975 Modalité d'infection des *Pinus sylvestris* par L' *Armillariella mellea* (Vahl.) Karst dans les cultures forestières. Eur J For Pathol 5:65–82
392 Ryser U, Holloway P J 1985 Ultrastructure and chemistry of soluble and polymeric lipids in cell walls from seed coat fibres of *Gossypium* species. Planta 163:151–163
393 Ryser U, Meier H, Holloway P J 1983 Identification and localization of suberin in the cell walls of green cotton fibres (*Gossypium hirsutum* L., var. green lint). Protoplasma 117:196–205
394 Sagisaka S 1976 The occurrence of peroxide in a perennial plant, *Populus gelrica*. Plant Physiol 57:308–309
395 Sakai W S, Sanford W G 1980 Ultrastructure of the water-absorbing trichomes of pineapple (*Ananas comosus*, Bromeliaceae). Ann Bot 46:7–11
396 Sakakibara A, Nakayama N 1962 Hydrolysis of lignin with dioxane and water. II. Identification of hydrolysis products. Mokuzai Gakkaishi 8:153–162
397 Sands R 1975 Radiata pine bark – aspects of morphology, anatomy and chemistry. N Z J For Sci 5:74–86
398 Schmidt H W, Mérida T, Schönherr J 1981 Water permeability and fine structure of cuticular membranes isolated enzymatically from leaves of *Clivia miniata* Reg. Z Pflanzenphysiol 105:41–51
399 Schmidt H W, Schönherr J 1982 Fine structure of isolated and non-isolated potato tuber periderm. Planta 154:76–80
400 Schneider H 1955 Ontogeny of lemon tree bark. Am J Bot 42:893–905
401 Schönherr J 1976 Water permeability of isolated cuticular membranes: the effect of cuticular waxes on diffusion of water. Planta 131:159–164

402 Schönherr J, Ziegler H 1980 Water permeability of *Betula* periderm. Planta 147:345–354
403 Scott M G, Peterson R C 1979 The root endodermis in *Ranunculus acris*. I. Structure and ontogeny. Can J Bot 57:1040–1062
404 Scott M G, Peterson R C 1979 The root endodermis in *Ranunculus acris*. II. Histochemistry of the endodermis and the synthesis of phenolic compounds in roots. Can J Bot 57:1063–1077
405 Sen J 1961 The nature of cork in ancient buried wood with special reference to normal present-day representatives. Riv Ital Paleont 67:77–88
406 Seone E, Arno M 1977 Total synthesis and stereochemistry of phloionolic acids. Ann Rev Soc Esp Fis Quim 73:1336–1339
407 Seone E, Serra M C, Agulló C 1977 Two new epoxy-acids from the cork of *Quercus suber*. Chem Ind 662
408 Shaner D L, Bertz S M Jr, Arntzen C J 1975 Inhibition of ion accumulation in maize roots by abscisic acid. Planta 122:79–90
409 Sharman S, Heale J B 1977 Penetration of carrot roots by the grey mould fungus *Botrytis cinera* Pers. ex Pers. Physiol Plant Path 10:63–71
410 Shigo A L, Marx H G 1977 Compartmentalization of decay in trees. USDA For Serv Ag Inform Bull 405, 73 pp
411 Sijmons P C 1986 Cellular localization and organ specifity of the suberin-associated isoperoxidase. In: Greppin H, Penel C, Gaspar T (eds) Molecular and physiological aspects of plant peroxidase. Université de Genève, Centre de Botanique Genève, 221–229
412 Sijmons P C, Kolattukudy P E, Bienfait H F 1985 Iron deficiency decreases suberization in bean roots through a decrease in suberin-specific peroxidase activity. Plant Physiol 78:115–120
413 Simons R K, Chu M C 1978 Periderm morphology of mature "golden delicious" apple with special reference to russeting. Sci Hort 8:333–340
414 Singh H, Chawla A S, Jindal A K, Conner A H, Rowe J W 1975 Investigation of *Erythrina* spp. VII. Chemical constituents of *Erythrina variegata* var. *orientalis* bark. Lloydia 38:97–100
415 Singh H, Chawla A S, Kapoor V K, Kumar J 1981 Investigations of *Erythrina* spp. X. Chemical constituents of *Erythrina blakei* bark. Planta Med 41:100–103
416 Singh H, Chawla A S, Kapoor V K, Kumar J, Piatak D M, Nowicki W 1981 Investigation of *Erythrina* spp. IX. Chemical constituents of *Erythrina stricta* bark. J Nat Prod 44:526–529
417 Sitte P 1955 Der Feinbau verkorkter Zellwände. Mikroskopie 10:178–200
418 Sitte P 1957 Der Feinbau der Kork-Zellwände. In: Treiber E (ed) Die Chemie der Pflanzenzellwand. Springer Berlin Göttingen Heidelberg, 421–432
419 Sitte P 1962 Zum Feinbau der Suberinschichten im Flaschenkork. Protoplasma 54:555–559
420 Sitte P 1975 Die Bedeutung der molekularen Lamellenbauweise von Korkzellwänden. Biochem Physiol Pflanz 168:287–297
421 Smith J E, Kurth E F 1953 The chemical nature of cedar barks. Tappi 36(2):71–78
422 Smith M M, O'Brien T P 1979 Distribution of autofluorescence and esterase and peroxidase activities in the epidermis of wheat roots. Aust J Plant Physiol 6:201–219
423 Sofield I, Wardlaw I F, Evans L T, Zee S Y 1977 Nitrogen, phosphorus and water contents during grain development and maturation in wheat. Aust J Plant Physiol 4:799–810
424 Sogo M, Ishihara T, Hata K 1966 Chemical studies on the bark. XIII. On the hydrogenolysis of the outer bark lignin of *Pinus densiflora*. Mokuzai Gakkaishi 112:96–101
425 Soliday C L, Dean B B, Kolattukudy P E 1978 Suberization: inhibition by washing and stimulation by abscisic acid in potato disks and tissue culture. Plant Physiol 61:170–174
426 Soliday C L, Kolattukudy P E 1977 Biosynthesis of cutin. ω-Hydroxylation of fatty acids by a microsomal preparation from germinating *Vicia faba*. Plant Physiol 59:1116–1121
427 Soliday C L, Kolattukudy P E 1978 Midchain hydroxylation of 16-hydroxypalmitic acid by the endoplasmic reticulum fraction from germinating *Vicia faba*. Arch Biochem Biophys 188:338–347
428 Soliday C L, Kolattukudy P E, Davis R W 1979 Chemical and ultrastructural evidence that waxes associated with the suberin polymer constitute the major diffusion barrier to water vapor in potato tuber (*Solanum tuberosum* L.). Planta 146:607–614
429 Sorm F, Bazant V 1950 On the neutral components of oak wax. Coll Czech Chem Commun 15:73–81
430 Speakman J B, Lewis B G 1978 Limitation of *Gaeumannomyces graminis* by wheat root responses to *Phialophora radicicola*. New Phytol 80:373–380

431 Speer E O 1984 *Melophia ophiospora* (Lev.) Sacc., *Mutatis characteribus*, champignon phellophage sur le chêne-liège. Mycotaxon 21:235–240
432 Srivastava L M 1964 Anatomy, chemistry, and physiology of bark. Int Rev For Res 1:206–277
433 Stanghellini M E, Aragaki M 1966 Relation of periderm formation and callose deposition to anthracnose resistance in papaya fruit. Phytopathology 56:444–450
434 Stanley R G 1969 Extractives of wood, bark, and needles of the southern pines. For Prod J 19(11):50–56
435 Stewart A, Mansfield J W 1985 The composition of wall alterations and appositions (reaction material) and their role in the resistance of onion bulb scale epidermis to colonization by *Botrytis allii*. Plant Pathol 34:25–37
436 Street P F S, Roble J, Ellis B E 1986 Secretion of vascular coating components by xylem parenchyma cells of tomatoes infected with *Verticillium alboatrum*. Protoplasma 132:1–11
437 Streibl M, Konecný K, Mahdalík 1969 Chemical composition of beech bark. I. Investigation of lipids. Cellul Chem Technol 3:653–656
438 Streibl M, Mahdalík M 1971 Chemical composition of some hydrocarbons of the lipoid fraction from the beech bark. Holzforsch Holzverwert 23:32–35
439 Streibl M, Mahdalík M, Konecný K 1970 Composition of the lipid portion of the beech bark. Holzforsch Holzverwert 22:1–3
440 Streibl M, Stránský K, Herout V 1978 *n*-Alkanes of some tree barks. Coll Czech Chem Commun 43:320–326
441 Sutherland J M, Sprent J I 1984 Calcium-oxalate crystals and crystal cells in determinate root nodules of legumes. Planta 161:193–200
442 Swan E P 1966 A study of western red cedar bark lignin. Pulp Pap Mag Can 67:T456–T460
443 Swan E P 1968 Alkaline ethanolysis of extractive-free western red cedar bark. Tappi 51(7):301–304
444 Swan E P, Naylor A F S 1969 Alkaline ethanolysis of eastern conifer barks. Dep Fish For Can Bi-Mon Res Notes 25(4):32–33
445 Swift M J 1965 Loss of suberin from bark tissue rotted by *Armillaria mellea*. Nature 207:436–437
446 Tainter F H, French D W 1971 The role of wound periderm in the resistance of eastern larch and jack pine to dwarf mistletoe. Can J Bot 49:501–504
447 Tattar T A, Rich A E 1973 Extractable phenols in clear, discolored, and decayed woody tissues and bark of sugar maple and red maple. Phytopathology 63:167–169
448 Taylor M S, Wu J H, Daniel R S 1974 Fine structure of virus lesion periderm formation in *Nicotiana glutinosa*. J Ultrastruct Res 48:173–174
449 Theagarajan K S, Parbha V V 1983 Studies on the chemical constituents of *Eucalyptus* hybrid bark. Ind J For 6:238–239
450 Thomas H E 1934 Studies on *Armillaria mellea* (Vahl.) Quel., infection, parasitism, and host resistance. J Agric Res 48:187–218
451 Thomson W W, Platt-Aloia K, Koller D 1979 Ultrastructure and development of the trichomes of *Larrea* (creosote bush). Bot Gaz 140:249–260
452 Tippett J T, Bogle A L, Shigo A L 1982 Response of balsam fir and hemlock roots to injuries. Eur J For Pathol 12:357–364
453 Tippett J T, Hill T C 1984 Role of periderm in resistance of *Eucalyptus marginata* roots against *Phytophthora cinnamomi*. Eur J For Pathol 14:431–439
454 Tippett J T, O'Brien T P 1976 The structure of Eucalypt roots. Aust J Bot 24:619–632
455 Tippett J T, Shea S R, Hill T C, Shearer B L 1983 Development of lesions caused by *Phytophthora cinnamomi* in the secondary phloem of *Eucalyptus marginata*. Aust J Bot 31:197–210
456 Tomiyama K, Stahmann M A 1964 Alteration of oxidative enzymes in potato tuber tissue by infection with *Phytophthora infestans*. Plant Physiol 39:483–490
457 Torrey J G, Clarkson D T (eds) 1975 The development and function of roots. Academic Press New York, 683 pp
458 Tucker S C 1975 Wound regeneration in the lamina of Magnoliaceous leaves. Can J Bot 53:1352–1364
459 Tulloch A P 1975 Chromatographic analysis of natural waxes. J Chromatograph Sci 13:403–407

460 Tulloch A P 1976 Chemistry of waxes of higher plants. In: Kolattukudy P E (ed) Chemistry and biochemistry of natural waxes. Elsevier New York, 235–287

461 Uritani I, Oba K 1978 The tissue slice system as a model for studies of host-parasite relationships. In: Kahl G (ed) Biochemistry of wounded plant tissue. de Gruyter Berlin New York, 287–308

462 Vance C P, Kirk T K, Sherwood R T 1980 Lignification as a mechanism of disease resistance. Ann Rev Phytopathol 18:259–288

463 Van Wyk A E 1985 The genus Eugenia (Myrtaceae) in southern Africa; structure and taxonomic value of bark. S Afr J Bot 51:156–180

464 Vázquez G, Antorrena G, Parajó J C 1987 Studies on the utilization of *Pinus pinaster* bark. Part 1: chemical constituents. Wood Sci Technol 21:65–74

465 Vogt E, Schönherr J, Schmidt H W 1983 Water permeability of periderm membranes isolated enzymatically from potato tubers (*Solanum tuberosum* L.). Planta 158:294–301

466 von Höhnel F 1877 Über den Kork und verkorkte Gewebe überhaupt. Sitz-Ber Wiener Akad Wiss 76:507–562

467 von Wettstein-Knowles P 1974 Gene mutation in barley inhibiting the production and use of C_{26} chains in epicuticular wax formation. FEBS Lett 42:187–191

468 von Wettstein-Knowles P 1979 Genetics and biosynthesis of plant epicuticular waxes. In: Applequist L-A, Liljenberg C (eds) Advances in the biochemistry and physiology of plant lipids. Elsevier New York, 1–26

469 Waisel Y, Lipschitz N, Arzee T 1967 Phellogen activity in *Robinia pseudoacacia* L. New Phytol 66:331–335

470 Walker R R, Sedgley M, Belsing M A, Douglas T J 1984 Anatomy, ultrastructure and assimilate concentrations of roots of *Citrus* genotypes differing in ability for salt exclusion. J Exp Bot 35:1481–1494

471 Walter W M Jr, Schadel W E 1983 Structure and composition of normal skin (periderm) and wound tissue from cured sweet potatoes. J Am Soc Hort Sci 108:909–914

472 Walton T J, Kolattukudy P E 1972 Determination of the structures of cutin monomers by a novel depolymerization procedure and combined gas chromatography and mass spectrometry. Biochemistry 11:1885–1897

473 Walton T J, Kolattukudy P E 1972 Enzymatic conversion of 16-hydroxypalmitic acid into 10,16-dihydroxypalmitic acid in *Vicia faba* epidermal extracts. Biochem Biophys Res Commun 46:16–21

474 Warmbrodt R D 1985 Studies on the root of *Hordeum vulgare* L. – ultrastructure of the seminal root with special references to the phloem. Am J Bot 72:414–432

475 Warrington S J, Black H D, Coons L B 1981 Entry of *Pisolithus tinctorius* hyphae into *Pinus taeda* roots. Can J Bot 59:2135–2139

476 Wattendorff J 1969 Feinbau und Entwicklung der verkorkten Calciumoxalat-Kristallzellen in der Rinde von *Larix decidua* Mill. Z Pflanzenphysiol 60:307–347

477 Wattendorff J 1974 The formation of cork cells in the periderm of *Acacia senegal* Willd. and their ultrastructure during suberin deposition. Z Pflanzenphysiol 72:119–134

478 Wattendorff J 1974 Ultrahistochemical reactions of the suberized cell walls in *Acorus*, *Acacia* and *Larix*. Z Pflanzenphysiol 73:214–225

479 Wattendorff J 1976 Ultrastructure of the suberized styloid cells in *Agave* leaves. Planta 128:163–165

480 Wattendorff J 1976 A third type of raphide crystal in the plant kingdom: six-sided raphides with laminated sheaths in *Agave americana* L. Planta 130:303–311

481 Wattendorff J 1978 Feinbau und Entwicklung der Calciumoxalat-Kristallzellen mit suberinähnlichen Kristallscheiden in der Rinde und im sekundären Holz von *Acacia senegal* Willd. Protoplasma 95:193–206

482 Wattendorff J, Schmid H 1973 Prüfung auf perjodarteaktive Feinstrukturen in den suberinisierten Kristallzell-Wänden der Rinde von *Larix* und *Picea*. Z Pflanzenphysiol 68:422–431

483 Weerdenburg C A, Peterson C A 1984 Effect of secondary growth on the conformation and permeability of the endodermis of broad bean (*Vicia faba*), sunflower (*Helianthus annus*), and garden balsam (*Impatiens balsamina*). Can J Bot 62:907–910

484 Weissmann G 1977 Rindenwachse von Kiefer und Fichte. Seifen Öle Fette Wachse 103:399–403

485 Weissmann G, Ayla C 1980 Untersuchung der Rindenextrakte von *Pinus brutia* Ten. Holz Roh-Werkst 38:307–312
486 Weissmann G, Ayla C 1984 Untersuchung der Rinde von *Pseudotsuga menziesii* (Mirb.) Franco. Holz Roh-Werkst 42:203–207
487 White M S 1979 Water vapor diffusion through eastern hemlock periderm. Wood Fiber 11:171–178
488 Whitmore F W 1978 Lignin-carbohydrate complex formed in isolated cell walls by peroxidases of *Pinus elliottii*. Phytochemistry 15:375–378
489 Whitmore T C 1962 Studies in systematic bark morphology. I. Bark morphology in Dipterocarpaceae. New Phytol 61:191–207
490 Whitmore T C 1962 Studies in systematic bark morphology. II. General features of bark construction in Dipterocarpaceae. New Phytol 61:208–220
491 Whitmore T C 1963 Studies in systematic bark morphology. IV. The bark of beech, oak and sweet chestnut. New Phytol 62:161–169
492 Wilcox H 1954 Primary organization of active and dormant roots of noble fir, *Abies procera*. Am J Bot 41:812–821
493 Wilhelm G E 1976 Über chemische Veränderungen von Buchen- und Fichtenrinde beim Abbau durch Mikroorganismen. Holzforschung 30:202–212
494 Wilkinson R E 1980 Ecotypic variation of *Tamarix pentandra* epicuticular wax and possible relationship with herbicide sensitivity. Weed Sci 28:110–113
495 Williams R R, Taji A M, Bolton J A 1984 Suberization and adventitious rooting in Australian plants. Aust J Bot 32:363–366
496 Williamson B 1984 Polyderm, a barrier to infection of red raspberry buds by *Didymella applanta* and *Botrytis cinerea*. Ann Bot 53:83–89
497 Wilson A J, Robards A W 1978 The ultrastructural development of mechanically impeded barley roots. Effects on the endodermis and pericycle. Protoplasma 95:225–265
498 Wilson A J, Robards A W 1980 Observations of the pattern of secondary wall development in the hypodermis of onion (*Allium cepa*) roots. Protoplasma 104:149–156
499 Wilson C A, Peterson C A 1983 Chemical composition of the epidermal, hypodermal, endodermal and intervening cortical cell walls of various plant roots. Ann Bot 51:759–769
500 Wu J H 1973 Wound-healing as a factor in limiting the size of lesions in *Nicotiana glutinosa* leaves infected by the very mild strain of tobacco mosaic virus (TMV-VM). Virology 51:474–484
501 Yatsu L Y, Espelie K E, Kolattukudy P E 1983 Ultrastructural and chemical evidence that the cell wall of green cotton fiber is suberized. Plant Physiol 73:521–524
502 Yazaki Y, Hillis W E 1977 Polyphenolic extractives of *Pinus radiata* bark. Holzforschung 31:20–25
503 Zee S-Y 1975 Accumulation of an adcrusting substance in the pigment strand of a diploid and tetraploid wheat. Can J Bot 53:2246–2250
504 Zee S-Y, O'Brien T P 1970 Studies on the ontogeny of the pigment strand in the caryopsis of wheat. Aust J Biol Sci 23:1153–1171
505 Zenk M H 1979 Recent work on cinnamoyl CoA derivatives. In: Swain T, Harborne J B, van Sumere C (eds) Recent advances in phytochemistry. Vol. 12. Biochemistry of plant phenolics. Plenum New York, 139–176
506 Zimmermann W, Nimz H, Seemüller E 1985 ^{1}H and ^{13}C NMR spectroscopic study of extracts from corks of *Rubus ideaeus*, *Solanum tuberosum*, and *Quercus suber*. Holzforschung 39:45–49

Chapter 7

Benzenoid Extractives

Whereas two-thirds of wood consists of polysaccharides, one-third of wood consists of lignin, a phenylpropane polymer formed via the shikimic acid pathway. Therefore, it is perhaps not surprising that wood extractives are extraordinarily rich in a wide array of benzenoid natural products, many of which are derived from the same phenylpropane monomers.

A few benzenoid compounds are isoprenoids and are discussed in Chap. 8. Many of the alkaloids (Chap. 5) have benzenoid rings based on benzenoid amino acid precursors. However, the compounds most directly derived from phenylpropanes are the simple monoaryl compounds (Sect. 7.1). The extraordinarily complex hydrolyzable tannins (Sect. 7.2) are rather elaborate derivatives of glucose combined with gallic acid from the shikimic acid pathway. The lignans (Sect. 7.3) are dimeric phenylpropanes and are related to the interesting compounds in Sect. 7.4. The flavonoids (Sect. 7.5) together with their low-molecular weight polymers (Sect. 7.6) and their high-molecular weight polymers, the condensed tannins (Sect. 7.7), are based on a phenylpropane to which has been added an acetyl-CoA triketide that subsequently aromatizes to give the A ring of the basic $C_6-C_3-C_6$ skeleton of this group.

The vast array of natural products produced from these simple building blocks can only be viewed with considerable awe. The chemistry involved in isolation, proof of structure, synthesis, and tracing the biogenesis includes some of the most impressive chemistry of this century. But we also need to stand in awe of Nature wherein natural selection has led to the elaboration of ever more complex natural products to protect the tree and promote survival. These extractives have a long history of serving humankind where no synthetic substitute existed.

7.1 Monoaryl Natural Products

O. THEANDER and L.N. LUNDGREN

7.1.1 Introduction

In this brief review we will generally restrict ourselves to monoaryl compounds found in woody plants of commercial interest in forestry and agriculture. Because similar monoaryl compounds are often found both in wood and in bark and because there is growing interest in whole-tree utilization, as well as in the use of plant phenols and other extractives as chemicals and other technical products, we will also include bark. Monoaryl compounds in leaves, needles, or other parts of trees – in particular when they illustrate interesting structural features – will be noted to some extent in relation to the chemistry of wood to which they may be translocated.

Even with these botanical restrictions, however, one is left with a wide choice as to which compounds and species – and, in particular, which references – to include of the enormous number of reports from various botanical species and fractions. Some of the compounds presented in this chapter are obvious native intermediates in the biosynthesis of lignin and other polyaryl structures. On the other hand, some reported monoaryl compounds occurring after high temperature extractions and other treatments of wood might be artifacts arising through incipient lignin degradation. We have tried to present structures, occurrence, and, to some extent, properties and taxonomic significance of such natural compounds from the different classes as seem to be particularly abundant or interesting. References are also given to more general reviews of plant phenols of these types. Karrer's monograph is, of course, classical in the field of organic plant compounds (89).

In the following division into classes we present the monoaryl compounds in the order of increasing number of carbon atoms in the side-chain.

7.1.2 Simple Phenols (C_6)

Phenol (**1**) has long been reported as occurring in various parts of *Pinus sylvestris* (89–91); the wood contains 0.03%–0.12%. More recently, it was identified in *Picea abies* wood by GC/MS of the trimethylsilylethers of the acetone extract (48).

In spite of the fact that dihydroxy- and trihydroxybenzene units are very common in more complex plant phenols, they have very rarely been found and, if so, in only small amounts in the free state. Catechol (**2**) has been found in the bark of two *Populus* species (53, 130) and resorcinol (**3**) has been identified in the heartwood of five *Morus* (46) species as well as in *Quercus rubra* wood (152).The 4-methylcatechol, which inhibits fungal growth, was isolated in the reaction zone in sapwood of *Picea abies* attacked by *Fomes annosus* (133).

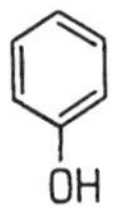

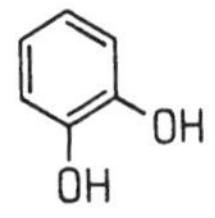

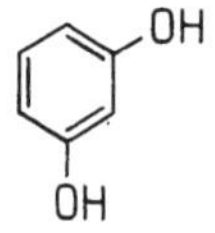

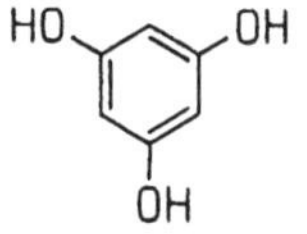

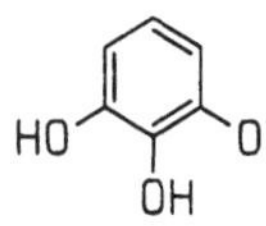

1 phenol 2 catechol 3 resorcinol 4 hydroquinone 5 phloroglucinol 6 pyrogallol

Hydroquinone (**4**) is more widely distributed in plants. It occurs in Ericaceae, Rosaceae, and Compositae, and is the only C_6-phenol that is of some systematic interest (71). This compound is found in leaf and bark of the pear tree, *Pyrus communis* (71), and in trace amounts in the heartwood of *Pinus resinosa* (142) and *P. radiata* (80). Its mono-*O*-β-D-glucoside, arbutin, quite common in many plants, is found in the stems and leaves of some *Pyrus* species (31). Phloroglucinol (**5**), among several other monoaryl compounds, has been identified by two-dimensional paper chromatography in bark extracts from various conifers (78). This phenol and pyrogallol (**6**) have been reported to occur in dried cones of *Sequoia sempervirens* and *Sequoiadendron giganteum* (98).

7.1.3 Phenolic Acids, Salicins and Other C_6-C_1 Compounds

7.1.3.1 Benzoic Acids and Related Compounds

Benzoic acid (**7**) has been isolated from the bark of *Prunus serotina* (134), the wood of *Aniba firmula* (66), and the needles of *Pinus sylvestris* (119). Benzyl benzoate has been found in *Aniba* species ("trunk wood") (55) and in the essential oil (0.6% of dry weight) from *Aniba riparia* (56). The ester constitutes 74%, and benzaldehyde 10%, of the oil. *p*-Hydroxybenzoic (**8**), protocatechuic (**9**), vanillic (**10**), and syringic acids (**11**), with their corresponding aldehydes, seem to be widely present in low concentrations in plants (12, 72, 77, 141). The acids are present as glycosides, esters, or in the free state. Protocatechuic acid, together with the aldehyde, was reported to be present in the bark of several conifers (78) and also in heartwood of *Metasequoia glyptostroboides* (148). Erdtman and Harmatha (50) found the isoprenoid cumic acid (**12**) and several hydroxy derivatives of **12** in *Libocedrus yateensis* wood. Two glycosides of 2,5-dihydroxybenzoic acid (gentisic acid) are found in *Prunus yedoensis* wood and bark (60). Gallic acid (**13**) and its dimer ellagic acid are mainly present in combined form in the commercially important hydrolyzable tannins (see Sect. 7.2). Hydroxybenzoic acids and their esters in plant have been reviewed by Herrmann (79).

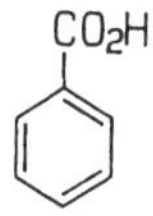

CO_2H

7 benzoic acid

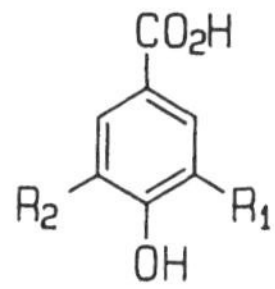

CO_2H, R_2, R_1, OH

8	*p*-hydroxybenzoic	($R_1=R_2=H$)
9	protocatechuic	($R_1=H$, $R_2=OH$)
10	vanillic	($R_1=H$, $R_2=OMe$)
11	syringic	($R_1=R_2=OMe$)
13	gallic	($R_1=R_2=OH$)

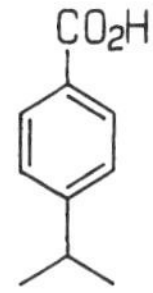

CO_2H

12 cumic acid

7.1.3.2 Salicins and Related Compounds

The glycosides of salicyl and gentisyl alcohols in *Salix* spp. and *Populus* spp., genera of the Salicaceae family, have been extensively studied by Pearl et al. (124–130) and Thieme et al. (168–171) (see summary and references in ref. 123). Salicin (**14**), the glucoside of salicyl alcohol, occurs characteristically in this family of plants. This compound is found in the bark of all species in the family, but often in small amounts. Another characteristic compound of the many salicin-related compounds occurring in most *Salix* species is salicortin (**15**), in which the

Table 7.1.1. Structures of salicins and related compounds

Name	Compound	Substitutions	Ref.
salicyl alcohol		$R = R_1 = H$	127, 128
populin		$R = H$, R_1 = glu-6-benzoate	162
tremuloidin		$R = H$, R_1 = glu-2-benzoate	125, 162
fragilin		$R = H$, R_1 = glu-6-acetate	168
trichocarposide	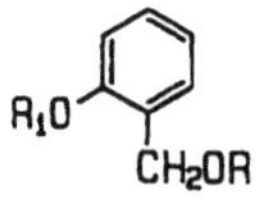 R_1O … CH_2OR	$R = H$, R_1 = glu-6-coumarate	130
salicyloylsalicin		R = salicyloyl, R_1 = glu	128, 130, 162
salicyloyltremuloidin		R = salicyloyl, R_1 = glu-2-benzoate	128
populoside		R = caffeoyl, R_1 = glu	51, 130
tremulacin	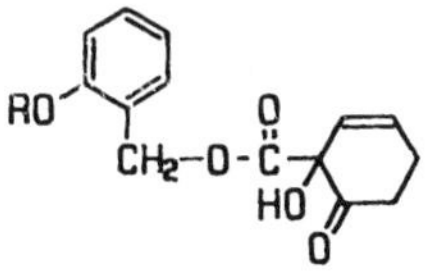 RO … CH_2–O–C(=O) … HO … O	R = glu-2-benzoate	129, 162
xylosmacin	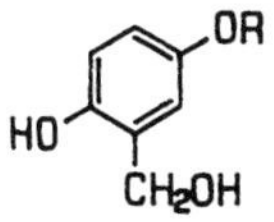 HO … OR … CH_2OH	R = glu-6-benzoate	40

gluO … CH_2OH

14 salicin

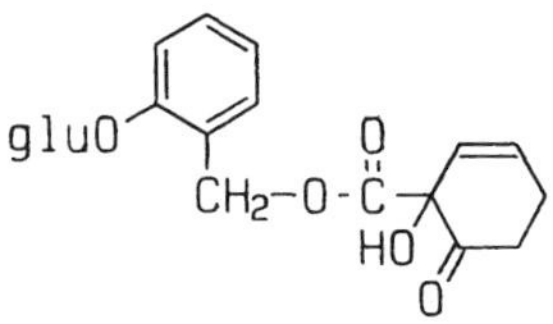

gluO … CH_2–O–C(=O) … HO … O

15 salicortin

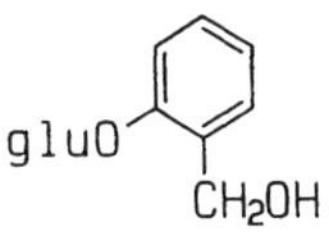

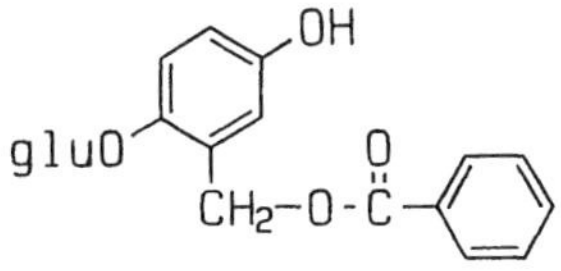

gluO … OH … CH_2–O–C(=O)

16 salireposide

gluO … OH … C(=O)–O–CH_2

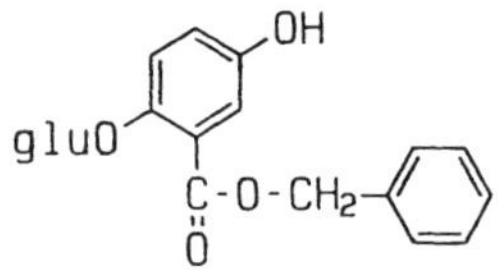

17 trichocarpin

glu = ß-D-glucopyranose

hydroxymethyl group is esterified. Also present in many *Salix* species is the benzoylated glucoside of gentisyl alcohol, salireposide (**16**), but several other salicyl and gentisyl alcohol derivatives have also been found. For a more complete list of reports of these types of compounds in woody sources, see Table 7.1.1.

Salicin compounds have a commercial potential as pharmaceuticals (124). The chemistry of *Salix* is also characterized by some differences between female and male plants. Thieme (171) has found interesting differences in the pattern of the characteristic glycosides in bark of trees of different sexes. Differences have also been reported among tissues (127) and even within the same tissue. For example, the concentration of trichocarpin (**17**) in twig bark of *Populus balsamifera* is twice as high as in trunk bark, while the salicin content is seven times as high in the twig bark. The concentration of phenolic glycosides in bark of *Salix* and *Populus* species generally increases toward the winter and the lowest values are observed during September and October (170).

7.1.4 Acetophenones and Other C_6-C_2 Compounds

Picein (**18**) and its aglycone 4-hydroxyacetophenone are found in various botanical fractions in *Picea, Pinus,* and *Salix* species (12, 72, 77, 141). Particularly high contents of this glycoside and the related glucoside, pungenin, (3,4-dihydroxyacetophenone having the glucose in the 3 position) are found in *Picea* needles. Pungenin has been isolated in 5% of dry weight from *Picea pungens* needles (118). Some other acetophenones have been reported from the bark of *Cedrelopsis grevei* (149) and from bark of *Prunus domestica* (114), and acetovanillone (together with vanillin and vanillyl alcohol) has been isolated from the wood of *Sitka* spruce (97).

p-Ethylphenol has been isolated from the oleoresin of *Picea abies* (94). Notably, nitro compounds have also been found, namely 1-nitro-2-phenylethane in wood and bark of *Ocotea pretiosa* and *Aniba canelilla* (65), and thalictoside (**19**) and its glucose 6-caffeoate in the bark of *Parabenzoin praecox* (157). A glucoside, salidroside, of 4-hydroxy-hydroxyethylbenzene has been isolated from the bark of *Salix triandra* (169).

Phenylacetic acids have rarely been reported in woody plants. Thus, 4-hydroxy-3-methoxyphenylacetic acid was indicated by paper chromatography in the bark of *Pinus sylvestris* (121). Recently, we isolated the taxifolin 3′-*O*-*β*-D-(6″-*O*-phenylacetyl)-glucopyranoside from the needles of *Pinus massoniana* (154). A carbohydrate substituted with phenylacetic acid (**20**) does not appear to have been reported previously.

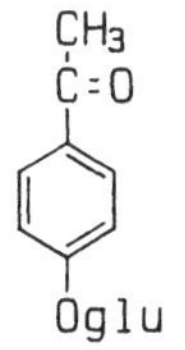

18 picein 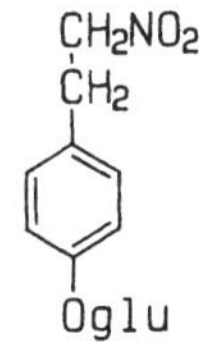 **19** thalictoside 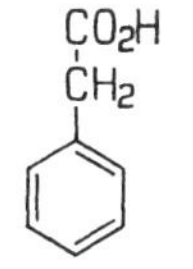 **20** phenylacetic acid 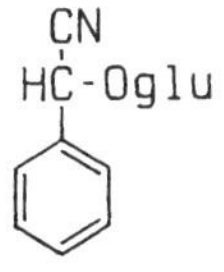 **21** prunasin

Cyanogenic glycosides constitute an important group of phenols in plant life. Prunasin (**21**) was isolated from the bark of *Prunus* species as far back as the beginning of this century (89). Recent studies have dealt with the presence of prunasin in *Acacia* species (150). For a review of cyanogenic glycosides found in the bark, leaves, or fruit stones of several woody species of the families Rosaceae, Leguminosae, and Myrtaceae, see Seigler (151).

7.1.5 Cinnamic Acids, Coumarins and Other Phenylpropanoids (C_6-C_3)

7.1.5.1 Cinnamic Acids

One or more of the four cinnamic acids – *p*-coumaric (**22**), caffeic (**23**), ferulic (**24**), and sinapic (**25**) – seem to occur, mainly in combined form, in most higher plants. The *o*-hydroxycinnamic acids are found more rarely, probably because they are readily lactonized to coumarins (see below). The natural cinnamic acids are *trans* isomers, but equilibrium mixtures of the *trans* and *cis* isomers are formed in the presence of light. Free unsubstituted cinnamic acid has been isolated from the bark of *Phebalium nudum* (23); *p*-coumaric acid was identified in the bark of *Populus trichocarpa* (53) and, together with caffeic acid, in the bark of *Pinus sylvestris* (121), and ferulic acid in wood of *Pinus radiata* (80). Coumaric, ferulic, and caffeic acids, together with vanillic acid, were recently found in the stalk of jute (*Corchorus capsularis* and *C. olitorius*) both in free and bound form (110). The extensive reports of cinnamic acid derivatives (esters, glycosides, and amides) in woody plants are exemplified in Table 7.1.2. The esters are more common than the glycosides and the esters with glucose and quinic acid are the best studied. It is probable that the elucidation of their structures has often been complicated by acyl migration during isolation, and some reports on structures must be treated with caution. Many bound forms of hydroxycinnamic acids other than esters and glycosides of carbohydrates are known. The ferulate esters of fatty alcohol-fatty acid complexes, which are major components of waxes and suberins in tree barks (Chap. 6.4) form a special group of esters. Certain cinnamoyl esters are intermediates in the biosynthesis of lignin and other high-molecular phenolic compounds. This will be discussed elsewhere in this series. For a recent, more complete review of hydroxycinnamic acids and their esters, see Herrmann (79).

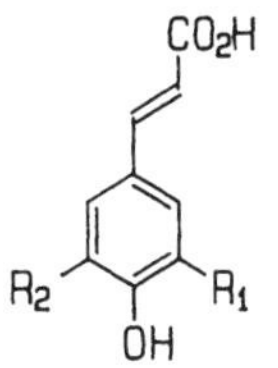

CO_2H

R_2 R_1

OH

22	*p*-coumaric	($R_1=R_2=H$)
23	caffeic	($R_1=H$, $R_2=OH$)
24	ferulic	($R_1=H$, $R_2=OMe$)
25	sinapic	($R_1=R_2=OMe$)

7.1.5.2 Coumarins

Coumarins are lactones that are formally derived from *o*-hydroxycinnamic acids by ring closure between the phenol and carboxyl groups. The chemistry of natu-

Table 7.1.2. Structures of esters, glycosides, and amides of cinnamic acids

Name	Compound	Substitutions	Ref.
----------		$R=Me, R_1=R_2=H$	8
----------		$R=R_1=Me, R_2=H$	8
----------		$R=$ shikimic acid, $R_1=R_2=H$	62
chlorogenic acid		$R=$ quinic acid, $R_1=H, R_2=OH$	31, 96
----------		$R=$ shikimic acid, $R_1=H, R_2=OH$	62
----------	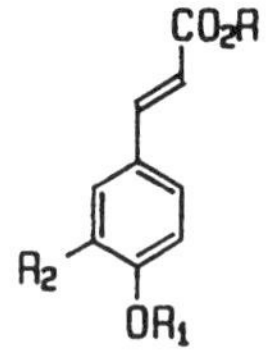	$R=$ quinic acid, $R_1=H, R_2=OMe$	62
----------		$R=$ shikimic acid, $R_1=H, R_2=OMe$	62
----------		$R=$ glu-6-caffeate, $R_1=R_2=H$	156
----------		$R=$ glu-6-caffeate, $R_1=H, R_2=OH$	156
eicosanyl ferulate		$R=n\text{-}C_{20}H_{41}, R_1=H, R_2=OMe$	117
tetracosyl ferulate		$R=n\text{-}C_{24}H_{49}, R_1=H, R_2=OMe$	39
hexacosyl ferulate		$R=n\text{-}C_{26}H_{53}, R_1=H, R_2=OMe$	36
octacosanyl ferulate		$R=n\text{-}C_{28}H_{57}, R_1=H, R_2=OMe$	88
octacosanylacetoxy ferulate		$R=n\text{-}C_{28}H_{57}, R_1=Ac, R_2=OMe$	88
----------	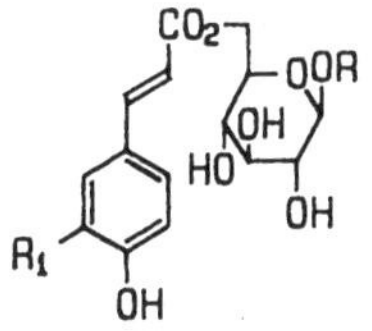	$R=R_1=H$	156
----------		$R=H, R_1=OH$	155
----------	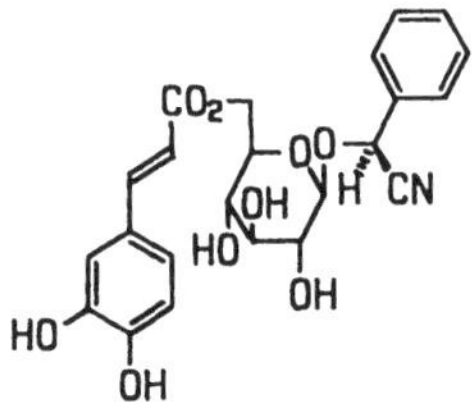		156
grandidentatin	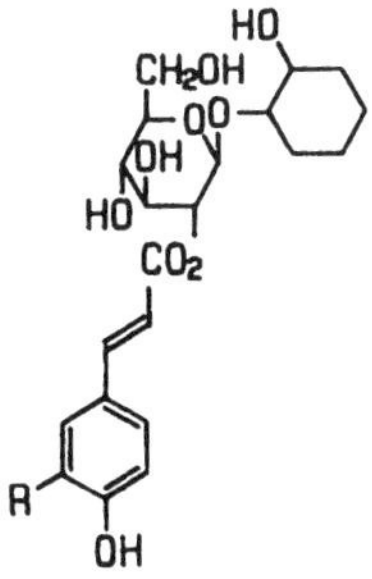	$R=H$	126, 162
grandidentoside		$R=OH$	51

Table 7.1.2 (continued)

Name	Compound	Substitutions	Ref.
specioside	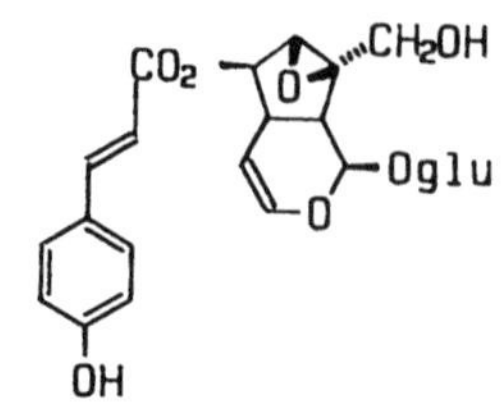		45
grandifloroside	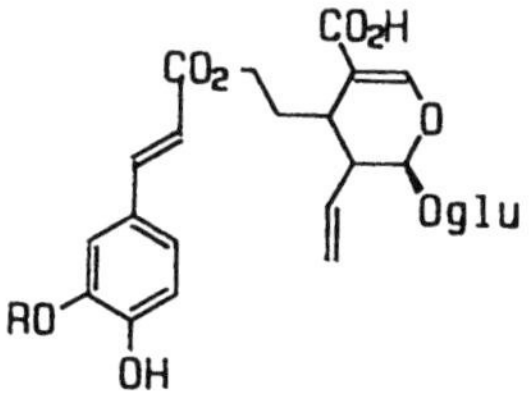	R=H	34
methylgrandifloroside		R=Me	34
10-caffeoyldeacetyldaphylloside	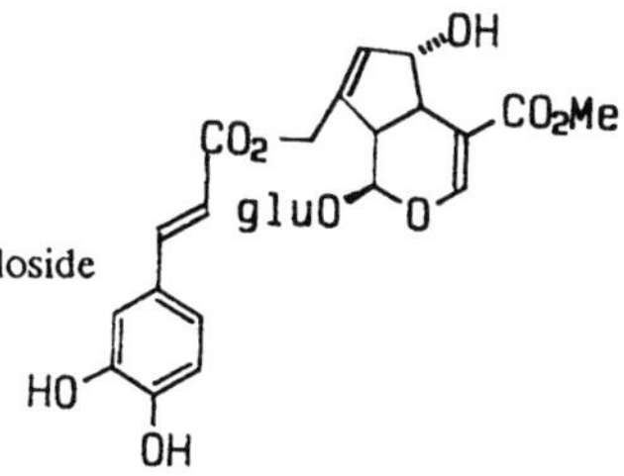		144
fagaramid	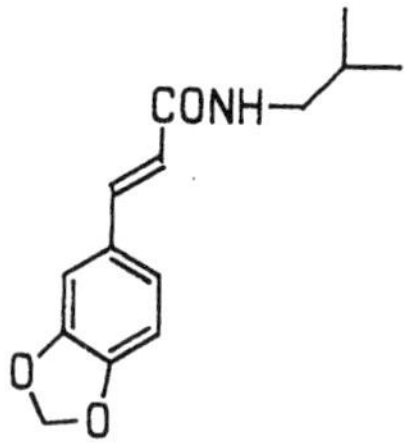		99, 120
aegeline	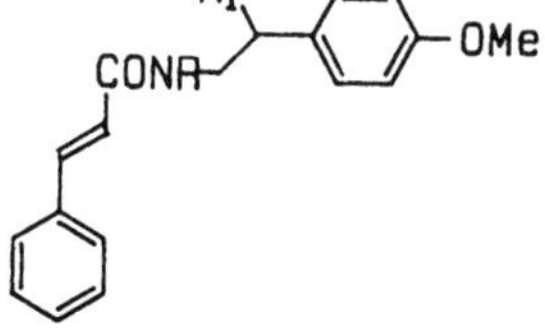	R=H, R_1=OH	104
herclavin		R=Me, R_1=H	43

Table 7.1.3.A. Structures of simple coumarins

Name	R_1	R_2	R_3	R_4	Ref.
skimmin	H	H	Oglu	H	13
floribin	OH	OMe	H	H	115
isoscopoletin	H	OH	OMe	H	173
scoparone	H	OMe	OMe	H	4, 42, 74, 95, 122, 173
fraxidin	H	OMe	OMe	OH	160
----------	H	OMe	OMe	Oglu	140
isofraxidin	H	OMe	OH	OMe	160
fraxinol	OMe	OH	OMe	H	115, 161
mandschurin	OMe	Oglu	OMe	H	140
----------	H	OMe	OMe	OMe	74, 95
osthol	H	H	OMe		81, 138
7-demethyl suberosin	H		OH	H	13, 20, 21, 63, 95, 176
suberosin	H		OMe	H	54, 95
cedrelopsin	H	OMe	OH		52, 108
coumurrayin	OMe	H	OMe		41
brosiparin	H		OMe	OH	19, 20, 21, 63
brosiprenin			OMe	OH	20, 21, 63
swietenol	H	OH	OH	H	112
geijerin	H	O	OMe	H	101

Table 7.1.3.A (continued)

Name	R_1	R_2	R_3	R_4	Ref.
mexoticin	OMe	H	OMe	OH OH	29
aculeatin	OMe	O	OMe	H	47
----------	H	OMe	OH	OH	108
O-prenyl brosiparin	H		OMe	O	63
brayleyanin	H	OMe	O		138
casegravol	H	H	OMe	OH CH_2OH	165
collinin	H	H	O	OMe	138
marmin	H	H	O OH OH	H	35, 37
----------	H	H	OMe	13 CH_2OH	158

rally occurring plant coumarins (about 600 are described) has been reviewed by Murray (113). The coumarins in Rutaceae have more recently been treated by Gray (68) and Gray and Waterman (69).

Coumarin, together with *o*-hydroxycinnamic acid and other compounds, has been reported in wood and barks of infected stems of *Prunus yedoensis* (60). The simple coumarins (Table 7.1.3.A) vary in their substitution of the aromatic ring. Prenylation (i.e., dimethylallyl substitution) also contributes to the great diversity. Umbelliferone (**26**), aesculetin (**27**), and scopoletin (**28**) – corresponding in structure to *p*-coumaric, caffeic, and ferulic acid – and various types of derivatives, are the three most common hydroxycoumarins in plants, including many species

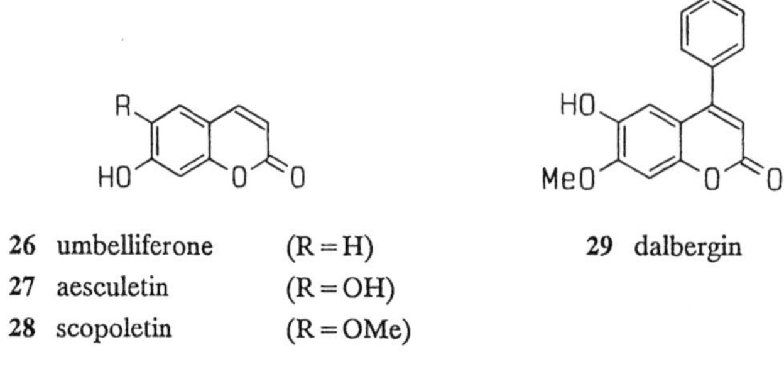

26 umbelliferone (R = H)
27 aesculetin (R = OH)
28 scopoletin (R = OMe)

29 dalbergin

Table 7.1.3.B. Structures of furanocoumarins

Name	R_1	R_2	Ref.
	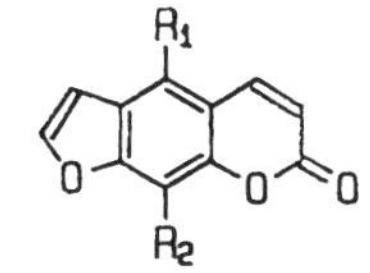		
psoralen	H	H	63, 95
bergapten	OMe	H	20, 63
xanthotoxin	H	OMe	3
----------	OMe	OH	23
isopimpinellin	OMe	OMe	14, 138
imperatorin	H	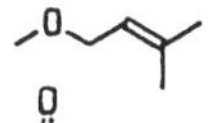	178
swietenone	H	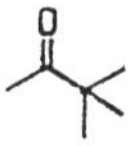	14, 111
swietenocoumarin A	H	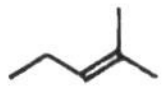	14
swietenocoumarin B	OMe	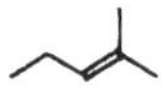	14
swietenocoumarin C	H	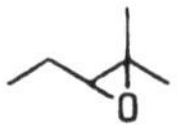	14
swietenocoumarin D	OMe	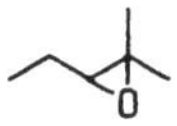	14
swietenocoumarin E	H	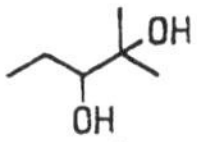	14
swietenocoumarin F	OMe	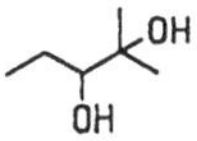	14
swietenocoumarin G	OMe	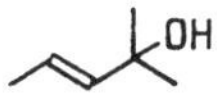	135
swietenocoumarin H	H	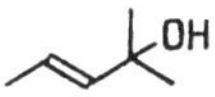	135
halfordin	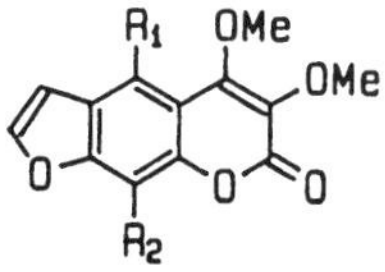	R_1 = OMe, R_2 = H	76, 100
isohalfordin		R_1 = H, R_2 = OMe	76, 100
xylotenin	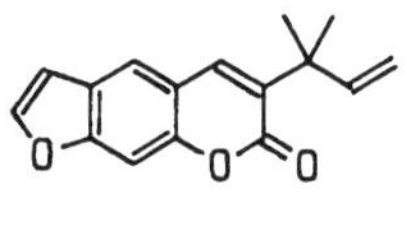		14

Table 7.1.3.C. Structures of miscellaneous coumarins

Name	Compound	Substitutions	Ref.
----------	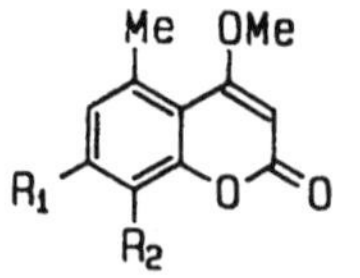	R_1 = OH, R_2 = H	122
----------		R_1 = Oglu, R_2 = H	122
siderin		R_1 = OMe, R_2 = H	116
----------		R_1 = OH, R_2 = OMe	122
mammea B/BB	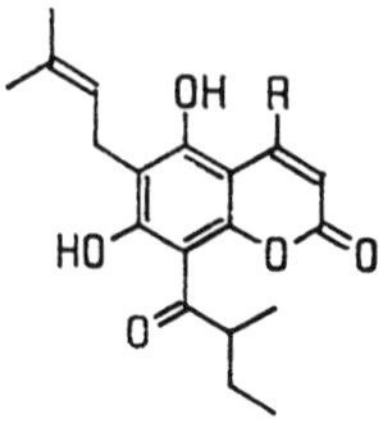	R = propyl	26
ferruol A		R = 2-methylpropyl	67
MAB 2	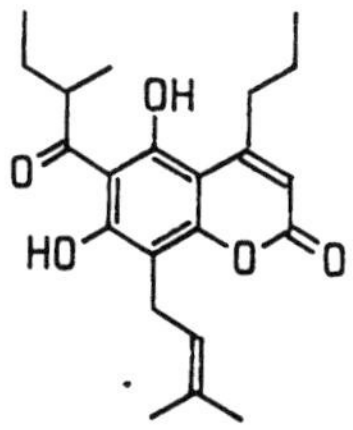		26
----------	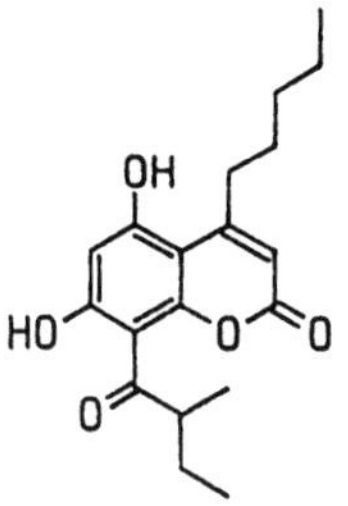		26
swietenocoumarin I	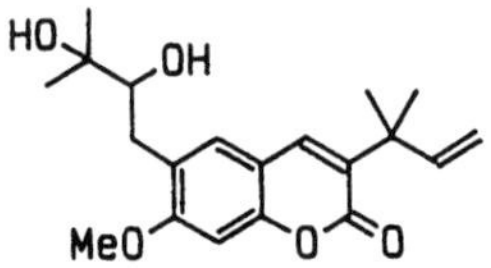		135

Table 7.1.3.C (continued)

Name	Compound	Substitutions	Ref.
marmesin	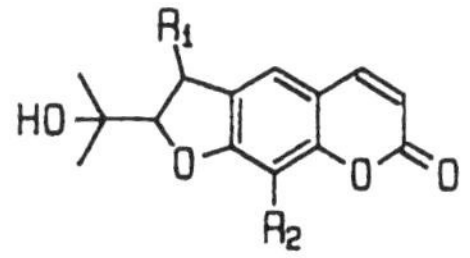	$R_1 = R_2 = H$	38
xanthoarnol		$R_1 = OH$, $R_2 = H$	86
8-prenylnodakenetin		$R_1 = H$, $R_2 =$ 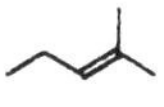	14
chalepin	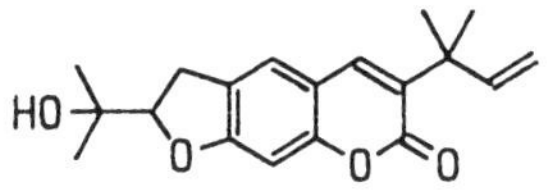		14
micromelin	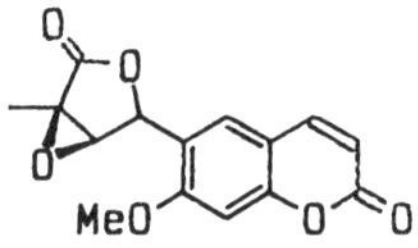		165
----------	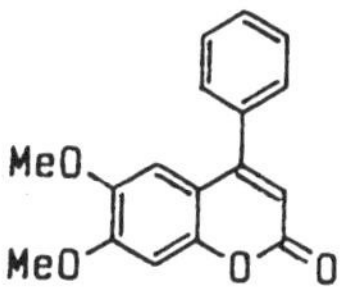		2
----------	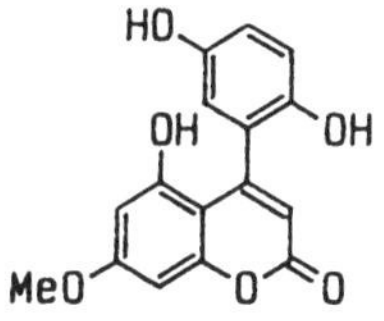		137
mammeisin	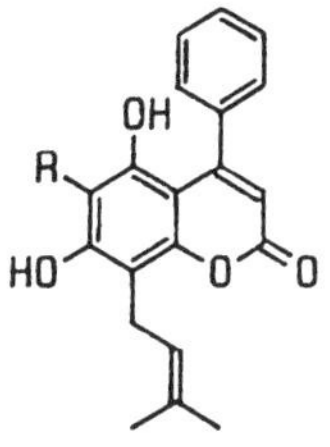	R= 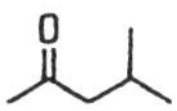	26
MAB 1		R= 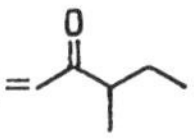	26
arnottianin	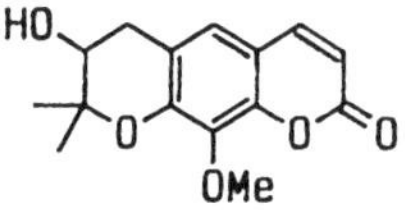		85

Table 7.1.3.C (continued)

Name	Compound	Substitutions	Ref.
xanthyletin	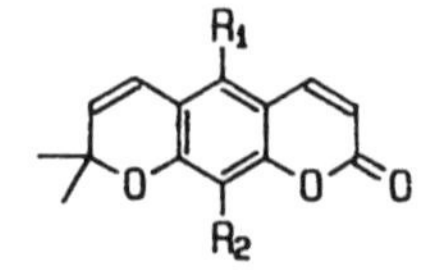	$R_1 = R_2 = H$	19, 20, 21, 63, 95, 112,
demethylluvangetin		$R_1 = H$, $R_2 = OH$	14
xanthoxyletin		$R_1 = OMe$, $R_2 = H$	76, 95, 112, 176
luvangetin		$R_1 = H$, $R_2 = OMe$	14, 20, 21, 63
----------	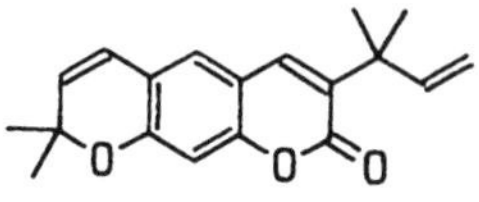		164
MAB 4	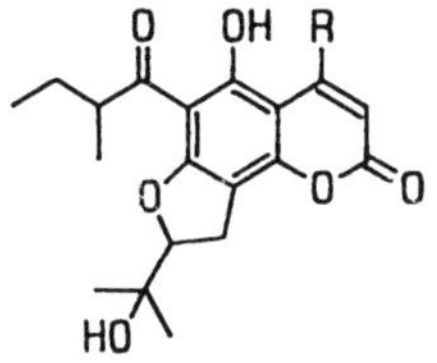	R = propyl	26
MAB 3		R = phenyl	26
seselin	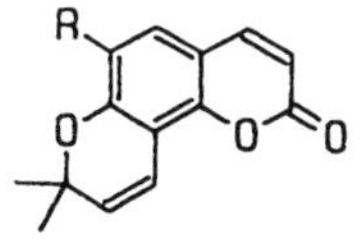	R = H	138
braylin		R = OMe	138
MAB 6	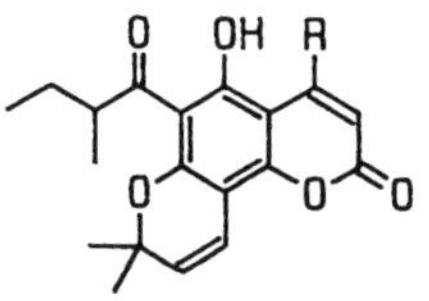	R = propyl	26
MAB 5		R = phenyl	26
alloxanthoxyletin	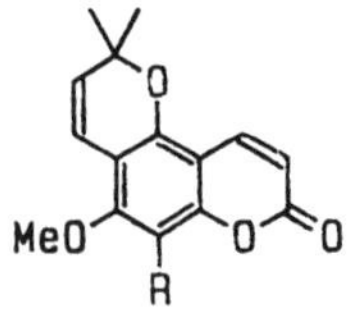	R = H	112
ceylantin		R = OMe	3
edgeworthin	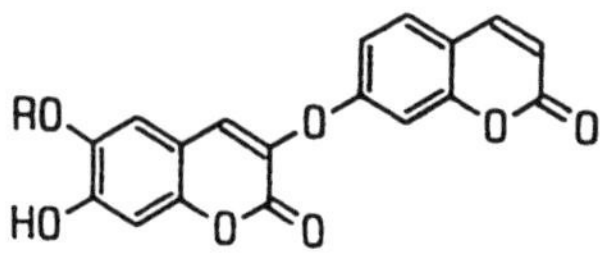	R = H	103
daphnoretin		R = Me	103

Table 7.1.3.C (continued)

Name	Compound	Substitutions	Ref.
mexolide	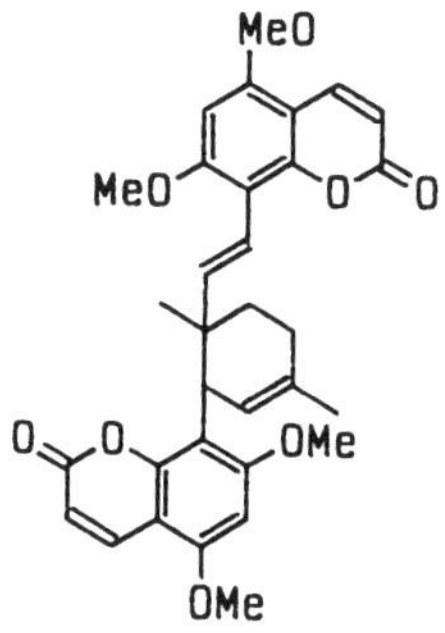		30
cleomiscosin A	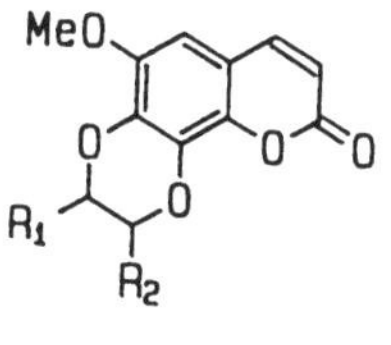	R_1 = 4-hydroxy-3-methoxyphenyl, $R_2 = CH_2OH$	10
cleomiscosin B		$R_1 = CH_2OH$, R_2 = 4-hydroxy-3-methoxyphenyl	166
propacin (a or b)		a) R_1 = 4-hydroxy-3-methoxyphenyl, R_2 = Me	180
		b) R_1 = Me, R_2 = 4-hydroxy-3-methoxyphenyl	180

of forest trees (69, 89–91). The 4-phenylcoumarins are coumarins of another type, and a typical representative, dalbergin (**29**), is found in the heartwood of *Dalbergia* (Leguminosae) trees (2). Glycosides of coumarins are not often reported. Many coumarins contain a furan ring fused onto the benzene ring, and these are collectively known as furanocoumarins (see Table 7.1.3.B). Other phenylcoumarins, pyranocoumarins, coumarinolignans, and other miscellaneous coumarins are given in Table 7.1.3.C.

Coumarins as many other phenolics have the important biological properties of being fungistatic and bacteriostatic and of causing inhibition of seed germination. There are many examples known, where they act as allelopathic chemicals, that is inhibiting the growth of competing plants.

7.1.5.3 Other Phenylpropanoids

Apart from cinnamoyl esters, aromatic amino acids (see Chap. 5.1) and the alcohols corresponding to the common hydroxycinnamic acids, as well as some glycerolaryl compounds, are of interest in connection with the biosynthesis of lignin and other polycyclic phenolic compounds.

The alcohols corresponding to the *p*-coumaric, ferulic, and sinapic acids – in particular those bound as phenolic glucosides, such as the *p*-coumaryl alcohol 4-*O*-glucosides, coniferin (**30**) and syringin (**31**) – seem to be widespread lignin-precursors in the cambial layer of woody plants (89, 94). Harmatha has isolated the *cis* isomer of coniferyl alcohol 3′-glucoside (73) from the bark of *Fagus silvatica* and more recently the corresponding *trans* compound was isolated in this laboratory from needles of *Pinus sylvestris* (6). An isomer (triandrin) of the *p*-coumaryl alcohol, with the glucose linked to the hydroxyl in the side-chain has been isolated from bark of *Salix triandra* (168). Savidge (147) has recently isolated dihydroconiferyl alcohol, a probable growth hormone, from *Pinus contorta* wood. Coniferyl aldehyde is reported from wood of spruce (59), western hemlock (62), and oak (17, 39). Compounds related to coniferyl alcohol are isoeugenol, isolated from the bark of *Phebalium nudum* (23), and its isomer, eugenol, with the double bond at the 1,2 position in the side chain. The latter and its derivative saffrol (**32**) are reported as extractives in the bark of *Nemuaron humboldtii* (28), with the latter as a predominant component. Saffrol is also a component, together with the related compounds myristicin and elemicin, of the essential oil of the "Nepal sassafras" or "Nepal camphor" tree (131).

Propiovanillone has been identified in *Quercus rubra* wood (152) and some phenylglycerol compounds are also found in various woody species. Thus, guaiacylglycerol (**33**) has been found in the sapwood of western hemlock (11) and the heartwood of *Pinus resinosa* (142). Further, the 1-*O*-glucoside and the 1-*O*- and 3-*O*-glucosides of *p*-hydroxyphenylglycerol have been reported from the inner bark of *Larix leptolepis* (105) and *Picea glehnii* (145), respectively, and an isoprene derivative of the 3-(4-hydroxyphenyl)-propan-1-ol has been isolated from wood and bark of *Zanthoxylum cuspidatum* (86).

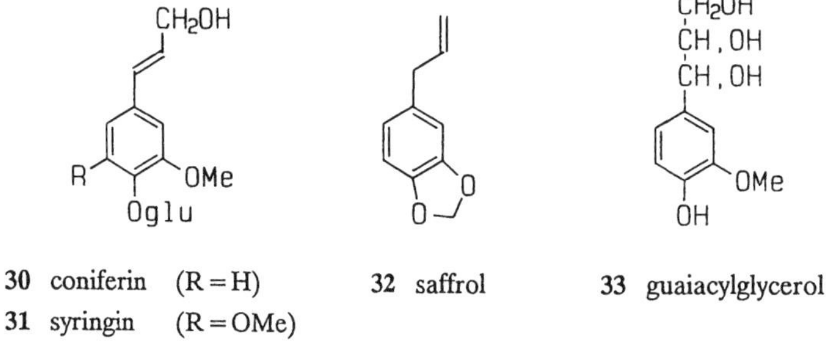

30 coniferin (R = H)
31 syringin (R = OMe)

32 saffrol

33 guaiacylglycerol

We have found traces of glucosides of both guaiacyl- and-*p*-hydroxyphenylglycerol in bark and wood of *Pinus sylvestris* and *Picea abies*, and have isolated free aglycones and glucosides, with the sugar linked to the 1 and 2 positions of the side chain of both the aglycones, in about 0.1% total yields from the needles of the corresponding trees (102, 167). It is notable that not only *threo* and *erythro* forms of the compounds were isolated, but our results also indicate the presence of both *D* and *L* forms of the respective aglycones in the plant material.

Table 7.1.4. Structures of chromanes and chromones

Name	Compound	Substitutions	Ref.
alloevodionol	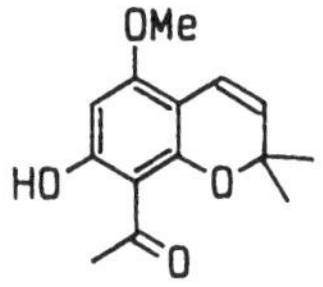		149
----------	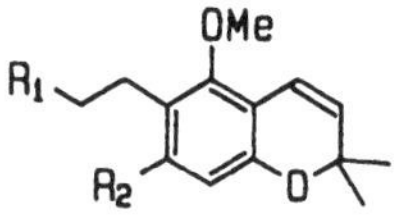	$R_1 = CH_2OH$, $R_2 = H$	42
----------		$R_1 = CH_2OH$, $R_2 = OMe$	42
----------		$R_1 = CO_2Me$, $R_2 = H$	41, 42
----------		$R_1 = CO_2H$, $R_2 = OMe$	41, 42
----------		$R_1 = CO_2Me$, $R_2 = OMe$	41, 42
----------	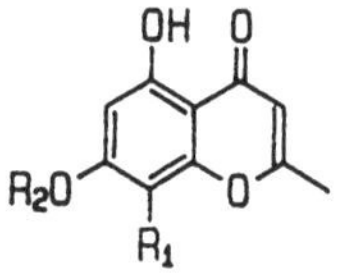	$R_1 = H$, $R_2 = glu$	70
----------		$R_1 = OMe$, $R_2 = Me$	22
----------	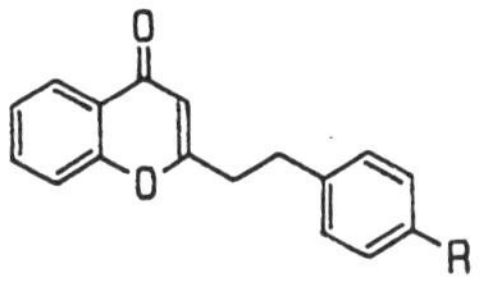	R = H	75
----------		R = OMe	75
----------	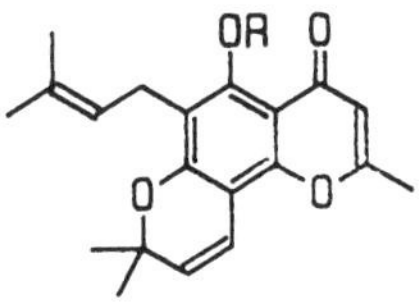	R = H	25
----------		R = Me	25
alloptaeroxylin	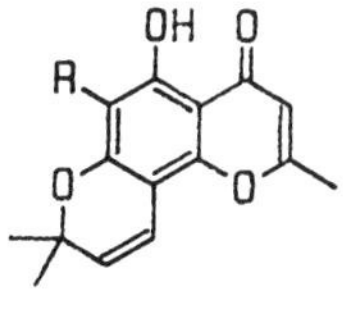	R = H	108
sorbifolin		R 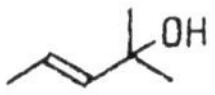 = [structure] OH	32
----------		R = [structure] 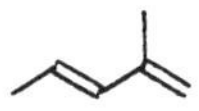	108

Table 7.1.4 (continued)

Name	Compund	Substitutions	Ref.
----------	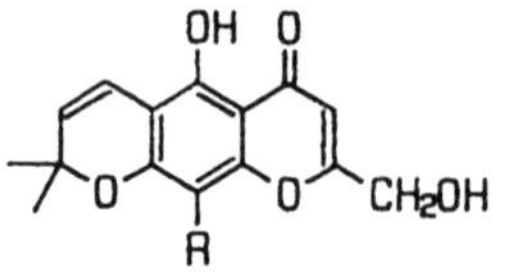	R= 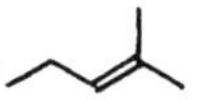	108
----------		R= 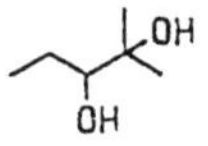	108
ptaeroxylin	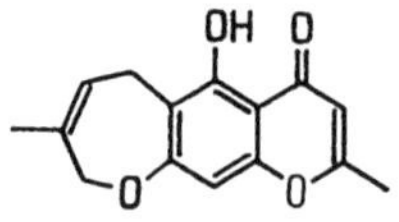		52
----------	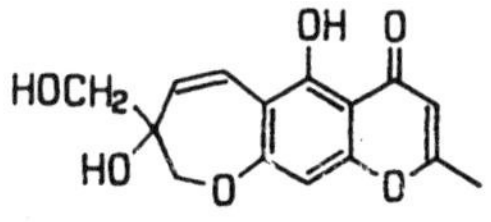		108

Table 7.1.5. Structures of non-coumarin lactones

Name	Compound	Substitutions	Ref.
djalonensin	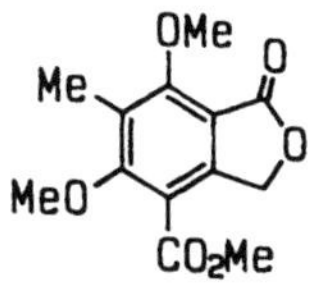		120
----------	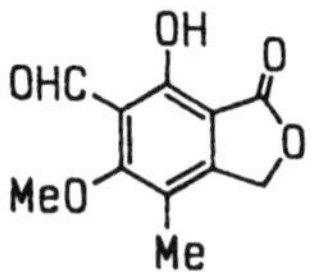		5
catalpalactone	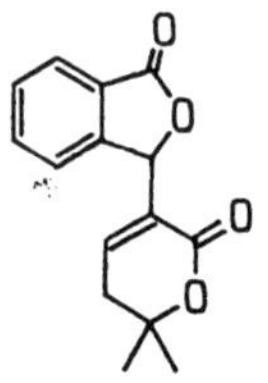		84

Table 7.1.5 (continued)

Name	Compound	Substitutions			Ref.
erythrocentaurine	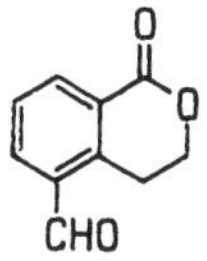				33
		R_1	R_2	R_3	
----------	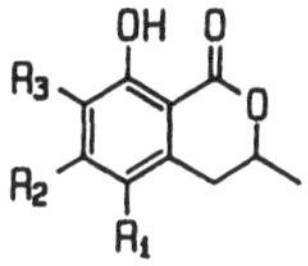	Me	H	H	5, 22, 27
----------		H	H	Me	5
----------		CHO	H	H	5
----------		CO_2H	H	H	5
----------		OH	H	H	5
----------		OMe	H	H	5
----------		Me	OMe	H	5
----------		Me	OMe	OMe	5
----------	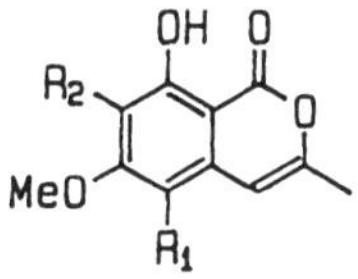	$R_1 = H$, $R_2 = Cl$			22
----------		$R_1 = Cl$, $R_2 = H$			22
4-methoxyparacotoin	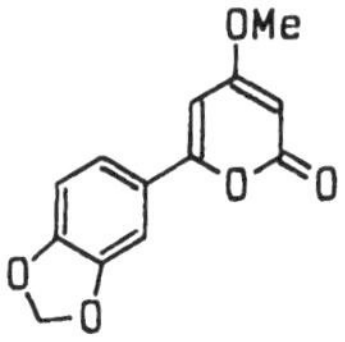				16, 109
dihydrokawain	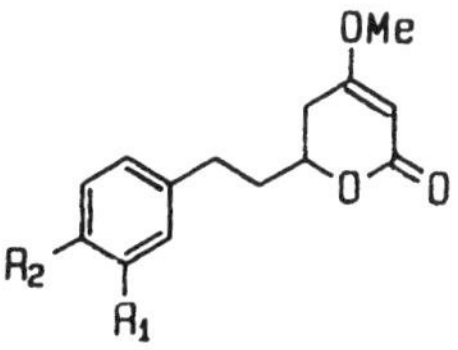	$R_1 = R_2 = H$			58
tetrahydroiangonin		$R_1 = H$, $R_2 = OMe$			58
----------		$R_1 = R_2 = OMe$			58

Table 7.1.5 (continued)

Name	Compound	Substitutions	Ref.
dihydromethysticin	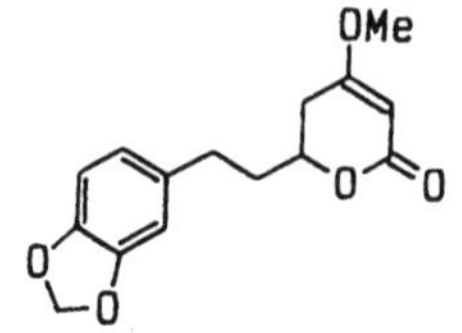		58, 109
----------	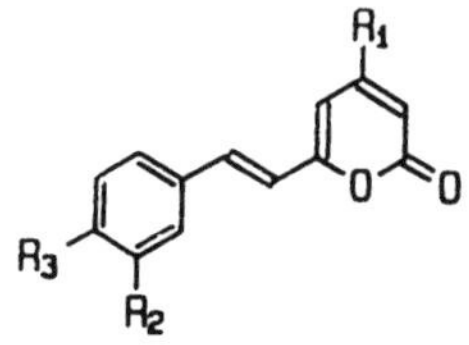	$R_1 = R_2 = R_3 = H$	16
----------		$R_1 = H$, $R_2 = OMe$, $R_3 = OH$	16, 44
----------		$R_1 = H$, $R_2 = R_3 = OMe$	44
----------	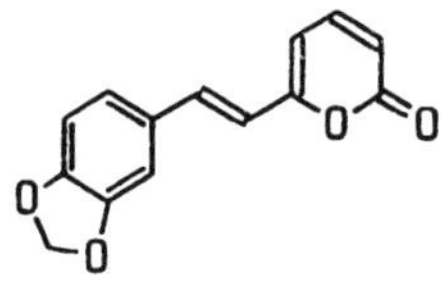		16, 44
iriyelliptin	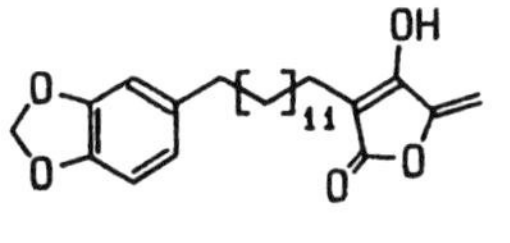		18
juruenolide	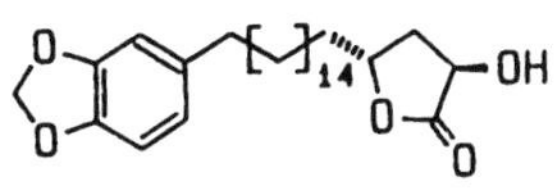		57
juruenolide-B	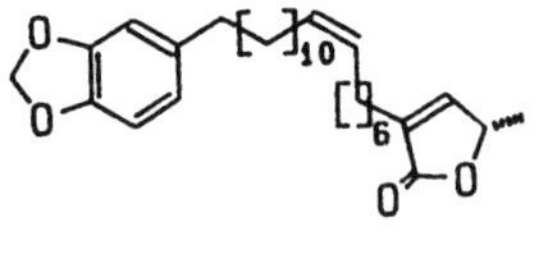		175
bergenin	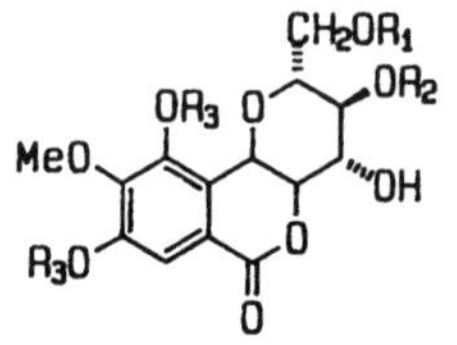	$R_1 = R_2 = R_3 = H$	136
----------		$R_1 = R_2 = H$, $R_3 = Me$	136
----------		$R_1 = R_3 = Me$, $R_2 = H$	136
----------		$R_1 = H$, $R_2 = R_3 = Me$	136

Table 7.1.6. Structures of miscellaneous monoaryl compounds

Name	Compound	Substitutions	Ref.
----------	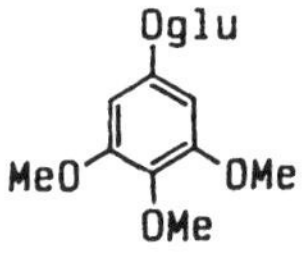		157
asaraldehyde	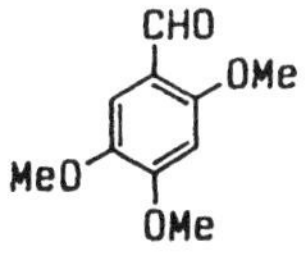		49
piperonylic acid	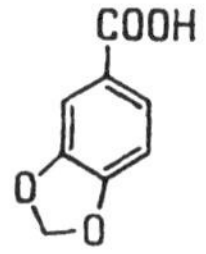		64
6-O-veratryl catalposide	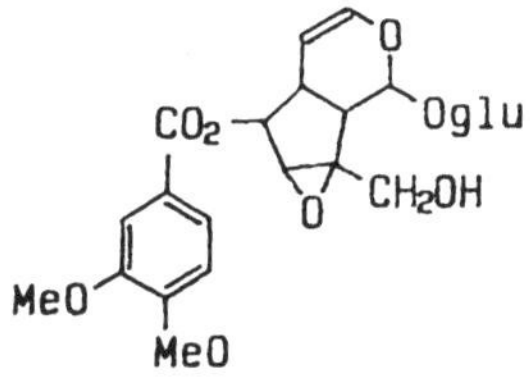		87
tembamide	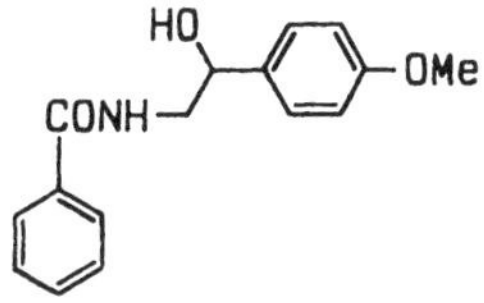		104
salidrosid	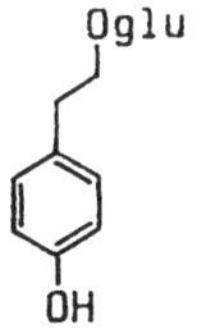		169
10-hydroxyligstroside	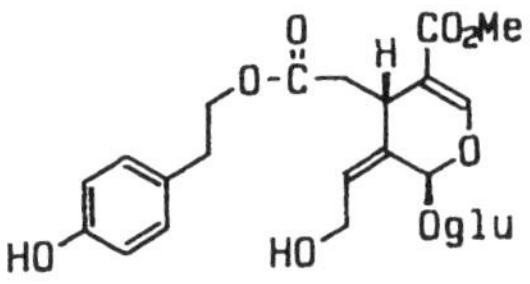		140
----------	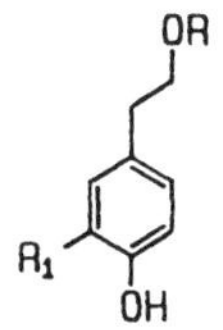	R=glu, R_1=OH	155
----------		R=glu-6-caffeate, R_1=H	155
----------		R=glu-6-caffeate, R_1=OH	155

Table 7.1.6 (continued)

Name	Compound	Substitutions				Ref.
2,4,5-trimethoxystyrene	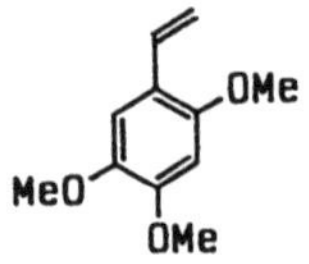					177
myristicin	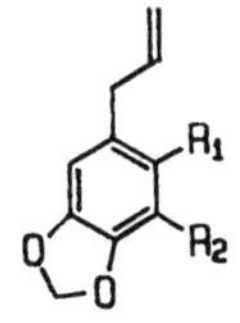	$R_1 = H, R_2 = OMe$				131
dillapiol		$R_1 = R_2 = OMe$				7, 61
		R_1	R_2	R_3	R_4	
isoeugenol	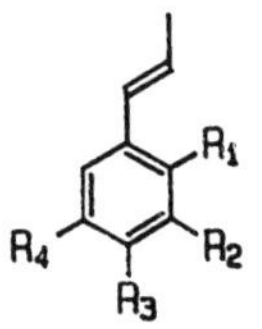	H	OMe	OH	H	23
----------		Me	H	OMe	OMe	49
isoelemicin		H	OMe	OMe	OMe	49
isomyristicin		OMe	OMe	OMe	OMe	49
patagonaldehyde	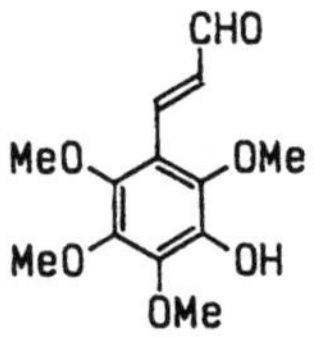					106
----------	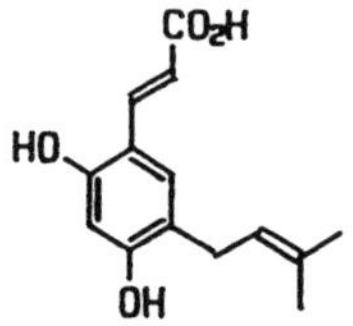					112
zanthoxylol	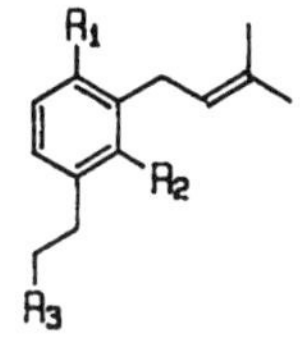	$R_1 = OH, R_2 = H, R_3 = CH_2OH$				52
----------		$R_1 = R_2 = OMe, R_3 = CO_2Me$				41
acronylin	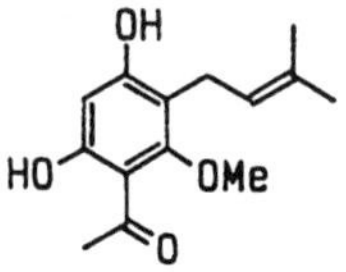					15

Table 7.1.6 (continued)

Name	Compound	Substitutions	Ref.
himasecolone	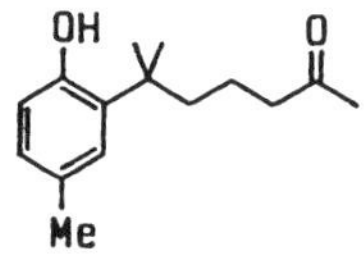		1
alliodorin	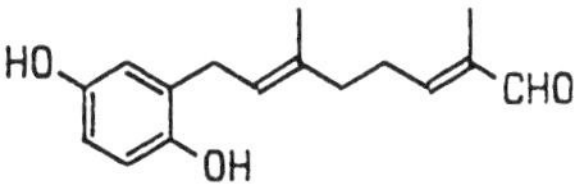		163
----------	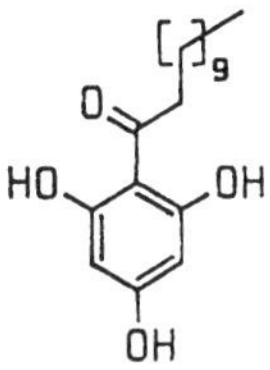		172
grevillol	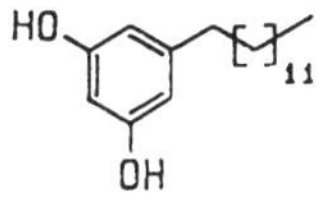		139
(-)-7-hydroxycalamenene	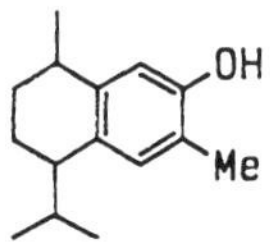		24
----------	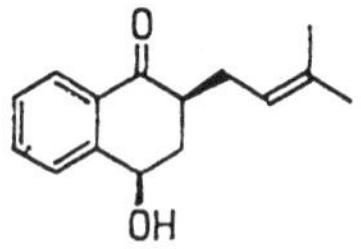		83
leucocordiachrome H	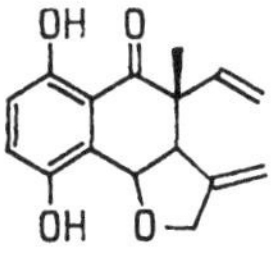		107
----------	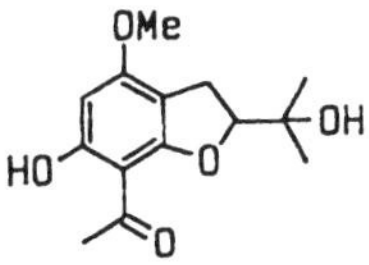		149
virolaflorine	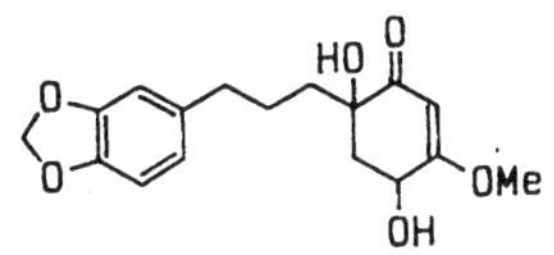		92

Table 7.1.6 (continued)

Name	Compound	Substitutions	Ref.
----------	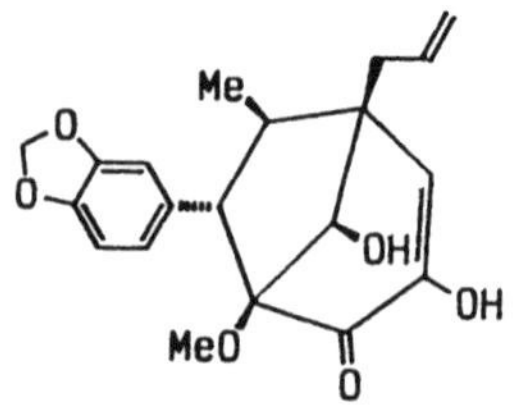		44
piperlongumin	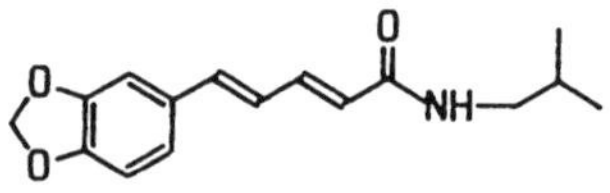		99
----------	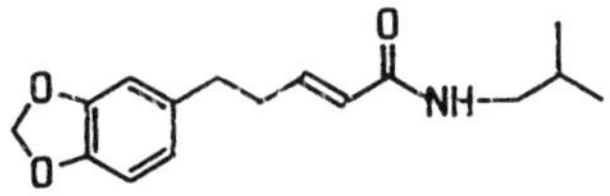		99

7.1.6 Miscellaneous Monoaryl Compounds

Several other types of monoaryl compounds have more than three carbons in the side-chain, including phenolic compounds of isoprenoidic type or more high-molecular compounds, in which the monoaryl unit is part of a larger, complicated structure. It is outside the scope of this section to treat these different classes more extensively. Apart from a few examples in the text below, we therefore refer the reader to Tables 7.1.4 (chromenes and chromones), 7.1.5 (non-coumarin lactones), and 7.1.6 (other miscellaneous monoaryl compounds) giving a brief selection of this large heterogenous group of monoaryl compounds. Reviews on plant chromones have recently been written by Gray (68) and by Saengchantara and Wallace (143).

An interesting C_6-C_4 compound is 4-(*p*-hydroxyphenyl)-butan-2-ol, originally isolated from *Rhododendron chrysanthum* (9) and called rhododendrol (**34**). Its glucoside, called rhododendrin, was also isolated. Both compounds were later isolated from the bark of *Betula alba* (159) and called betuligenol and betuloside. They were later shown to be identical with the *Rhododendron* compounds (93). More recently, the distribution of rhododendrin has been studied in *Betula* species (146) and also in Aceraceae plants (82). For long-chain non-isoprenoid phenols of plant origin, we recommend the comprehensive review by Tyman (174). The types of phenolic lipids discussed there are of 'polyketide' origin, having no branched chains. The isoprenoid phenolic lipids as well as more low-molecular isoprenoid monoaryl compounds, biogenetically derived from mevalonate, will be

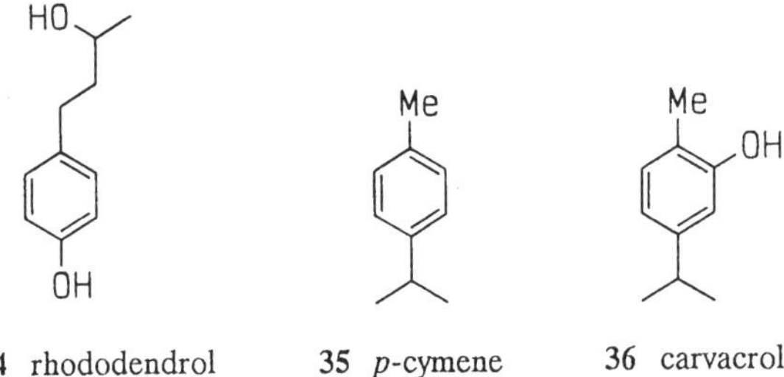

34 rhododendrol 35 *p*-cymene 36 carvacrol

treated in Chap. 8. Examples of the latter are *p*-cymene (**35**), isolated from several woods, and its 2-hydroxy derivative, carvacrol (**36**) (132, 153, 179).

References

1 Agrawal P K, Rastogi R P 1981 Terpenoids from *Cedrus deodara*. Phytochemistry 20:1319–1321

2 Ahluwalia V K, Seshadri T R 1957 Constitution of dalbergin. Part. II. J Chem Soc 970–972

3 Ahmed J, Shamsuddin K M, Zaman A 1984 A pyranocoumarin from *Atlantica ceylanica*. Phytochemistry 23:2098–2099

4 Aiba J C, Gottlieb O R, Maia J G S, Pagliosa F M, Yoshida M 1978 Benzofuranoid neolignans from *Licaria armeniaca*. Phytochemistry 17:2038–2039

5 Alvarenga M A et al. (21 authors) 1978 Dihydroisocoumarins and phthalide from wood samples infested by fungi. Phytochemistry 17:511–516

6 Andersson R, Lundgren L N 1988 Monoaryl and cyclohexenone glycosides from needles of *Pinus sylvestris*. Phytochemistry 27:559–562

7 Andrade C H S, Braz Filho R, Gottlieb O R 1980 Neolignans from *Aniba ferrea*. Phytochemistry 19:1191–1194

8 Anjaneyulu A S R, Rao A M, Rao V K, Row L R 1975 The lignans of *Gmelina asiatica*. Phytochemistry 14:824

9 Archangelsky K 1901 Über Rhododendrol, Rhododendrin und Andromedotoxin. Arch Exp Pathol Pharmacol 46:313

10 Arisawa M, Handa S S, McPherson D D, Lankin D C, Cordell G A, Fong H H S, Farnsworth N R 1984 Cleosmine A from *Simaba multiflora, Soulamea soulameoides*, and *Matayba arborescens*. J Nat Prod 47:300–307

11 Barton G M 1968 Significance of Western hemlock phenolic extractives in pulping and lumber. Forest Prod J 18(5):76–80

12 Bate-Smith E C 1962 The simple polyphenolic constituents of plants. In: W E Hillis (ed) Wood extractives. Academic Press New York, 133–158

13 Bhattacharjee S R, Mullick D K 1960 Isolation of scopoletin, skimmin and umbelliferone from the trunk-bark of *Skimmia laureola* Hook. J Indian Chem Soc 37:420–422

14 Bhide K S, Mujumdar R B, Rama Rao A V 1977 Phenolics from the bark of *Chloroxylon swietenia* D C. Indian J Chem 15B:440–444

15 Biswas G K, Chatterjee A 1970 Isolation and structure of acronylin: a new phenolic compound from *Acronychia laurifolia*. Chem Ind (London) 20:654–655

16 Bittencourt A M, Gottlieb O R, Mors W B, Magalhaes M T, Mageswaran S, Ollis W D, Sutherland I O 1971 The natural occurrence of 6-steryl-2-pyrones and their synthesis. Tetrahedron 27:1043–1048

17 Black R A, Rosen A A, Adams S L 1953 The chromatographic separation of hardwood extractive components giving color reactions with phloroglucinol. J Am Chem Soc 75: 5344–5346

18 Braz Filho R, Diaz D P P, Gottlieb O R 1980 Tetronic acid and diarylpropanes from *Iryanthera elliptica*. Phytochemistry 19: 455–459

19 Braz Filho R, Magalhaes A F, Gottlieb O R 1970 New coumarins from *Brosimum paraense*. An Acad Brasil Cienc 42:139–142

20 Braz Filho R, Magalhaes A F, Gottlieb O R 1971 Brosiparin and other coumarins from *Brosimum rubescens*. An Acad Brasil Cienc 43: 585–586

21 Braz Filho R, Magalhaes A F, Gottlieb O R 1972 Coumarins from *Brosimum rubescens*.Phytochemistry 11:3307–3310

22 Braz Filho R, De Moraes M P L, Gottlieb O R 1980 Pterocarpans from *Swartzia laevicarpa*.Phytochemistry 19:2003–2006

23 Briggs L H, Cambie R C 1958 The constituents of *Phebalium nudum* Hook. I. The bark. Tetrahedron 2:256–270

24 Burden R S, Kemp M S 1983 (–)-7-Hydroxycalamenene, a phytoalexin from *Tilia europea*. Phytochemistry 22:1039–1040

25 Campos A M, Khac D D, Fetizon M 1987 Chromones from *Dictyoloma incandescens*.Phytochemistry 26:2819–2823

26 Carpenter I, McGarry E J, Scheinmann F 1971 The isolation and structure of nine coumarins from the bark of *Mammea africana* G. Don. J Chem Soc (C):3783–3790

27 Carpenter R C, Sotheeswaran S, Sultanbawa M U S, Balasubramaniam S 1980 (–)-5-Methylmellein and catechol derivatives from four *Semecarpus* species. Phytochemistry 19:445–447

28 Chabeau E 1938 Chemical composition of the bark of *Nemuaron humboldtii*. Bull Acad Roy Med Belg 3:46–66

29 Chakraborty D P, Chowdhury B K 1967 Mexoticin, a new coumarin from *Murraya exotica*. Tetrahedron Lett 36:3471–3473

30 Chakraborty D P, Roy S, Chakraborty A, Mandal A K, Chowdhury B K 1980 Structure and synthesis of mexolide: a new antibiotic dicoumarin from *Murraya exotica*. Tetrahedron 36:3563–3564

31 Challice J S 1973 Distribution of phenols amongst the various tissues of the *Pyrus* stem. J Sci Food Agric 24:285–293

32 Chan W R, Taylor D R, Willis C R 1967 The structure of sorbifolin, a chromone from *Spathelia sorbifolia* L. J Chem Soc (C) 2540–2542

33 Chapelle J P 1973 Isolement de derivés seco-iridoides d'*Anthocleista zambesiaca*. Phytochemistry 12:1191–1192

34 Chapelle J P 1976 Grandifloroside et methylgrandifloroside, glucosides iridoides nouveaux d'*Anthocleista grandiflora*. Phytochemistry 15:1305–1307

35 Chatterjee A, Bhattacharya A 1959 The isolation and constitution of marmin, a new coumarin from *Aegle marmelos* Correa. J Chem Soc 1922–1924

36 Chatterjee A, Dhara K P, Rej R N, Ghosh P C 1977 Hexacosylferulate, a phenolic constituent of *Pinus roxburghii*. Phytochemistry 16:397–398

37 Chatterjee A, Dutta C P, Bhattacharyya S, Audier H E, Das B C 1967 The structure of marmin. Tetrahedron Lett 5:471–473

38 Chatterjee A, Roy S K 1959 Constituents of the heartwood of *Aegle marmelos*. J Indian Chem Soc 36:267–269

39 Chen C L 1970 Constituents of *Quercus alba*. Phytochemistry 9:1149

40 Cordell G A, Chang P T O, Fong H H S, Farnsworth N R 1977 Xylosmacin, a new phenolic glucoside ester from *Xylosma velutina*. Lloydia 40:340–343

41 Correa D de B, Gottlieb O R, de Padua A P 1975 Dihydrocinnamic acids from *Hortia badinii*. Phytochemistry 14:2059–2060

42 Correa D de B, Gottlieb O R, de Padua A P 1979 Dihydrocinnamyl alcohols from *Hortia badinii*. Phytochemistry 18:351

43 Crombie L 1955 Structure and stereochemistry of neoherculin. J Chem Soc 995–998

44 De Diaz A M P, Gottlieb O R, Magalhaes A F, Magalhaes E G, Maia J G S, Santos C C 1977 The chemistry of Brazilian Lauraceae. XLVI. Notes on *Aniba* species. Acta Amazonica 7:41–43

45 De Silva L B, Hearth W H M W, Jennings R C, Mahendran M, Wannigama G P 1982 Isolation of specioside from *Stereospermum personatum*. Indian J Chem 21B: 703–704

46 Deshpande V H, Srinivasan R, Rama Rao A V 1975 Phenolics of the heartwood of five *Morus* species. Ind J Chem 13:453–457

47 Dutta P 1942 Constituents of natural coumarins of *Toddalia aculeata*. J Indian Chem Soc 19:425–428

48 Ekman R 1976 Analysis of lignans in Norway spruce by combined gas chromatography-mass spectrometry. Holzforschung 30:79–85
49 Enquiquez R G, Chavez M A, Jauregui F 1980 Propenylbenzenes from *Guatteria gaumeri*. Phytochemistry 19:2024–2025
50 Erdtman H, Harmatha J 1979 Phenolic and terpenoid heartwood constituents of *Libocedrus yateensis*. Phytochemistry 18:1495–1500
51 Erickson R L, Pearl I A, Darling S F 1970 Populoside and grandidentoside from the bark of *Populus grandidentata*. Phytochemistry 9:857–863
52 Eshiett I T, Taylor D A H 1968 The isolation and structure elucidation of some derivatives of dimethyl-allyl-coumarin, chromone, quinoline, and phenol from *Fagara* species, and from *Cedrelopsis grevei*. J Chem Soc (C) 481–484
53 Estes T K, Pearl I A 1967 Hot-water extractives of the green bark of *Populus trichocarpa*. Tappi 50:318–323
54 Ewing J, Hughes G K, Ritchie E 1950 The chemical constituents of Australian *Zanthoxylum* species. I. Suberosin, 6-(3,3-dimethylallyl)-7-methoxycoumarin. Aust J Sci Res 3A:342–345
55 Fernandes J B, Gottlieb O R 1976 Neolignans from *Aniba* species. Phytochemistry 15:1033–1036
56 Franca N C, Gottlieb O R, Magalhaes M T, Mendes P H, Maia J G S, Da Silva M L, Gottlieb H E 1976 Tri-O-methylgalangin from *Aniba riparia*. Phytochemistry 15:572–573
57 Franca N C, Gottlieb O R, Rosa B P 1975 Juruenolide: A γ-lactone from *Iryanthera juruensis*. Phytochemistry 14:590–591
58 Franca N C, Gottlieb O R, Suarez A M P 1973 6-Phenylethyl-5,6-dihydro-2-pyrones from *Aniba gigantifolia*. Phytochemistry 12:1182–1184
59 Freudenberg K, Harkin J M 1963 The glycosides of cambial sap of spruce. Phytochemistry 2:189–193
60 Fujii S, Aoki H, Komoto M, Munakata K 1971 Fluorescent glycosides accumulated in *Taphrina wiesneri*-infected cherry stems. Agric Biol Chem 35:1526–1541
61 Giesbrecht A M, Franca N C, Gottlieb O R 1974 The neolignans of *Licaria canella*. Phytochemistry 13:2285–2293
62 Goldschmid O, Hergert H L 1961 Examination of Western hemlock for lignin precursors. Tappi 44:858–870
63 Gottlieb O R, Da Silva M L, Maia J G S 1972 Distribution of coumarins in Amazonian *Brosimum* species. Phytochemistry 11:3479–3480
64 Gottlieb O R, Magalhaes M T 1958 Isolation of piperonylic acid from *Ocotea pretiosa*. Nature 182:742–743
65 Gottlieb O R, Magalhaes M T 1959 Occurrence of 1-nitro-2-phenylethane in *Ocotea pretiosa* and *Aniba canelilla*. J Org Chem 24:2070–2071
66 Gottlieb O R, Mors W B 1959 Isolation of 5,6-dehydrokavain and 4-methoxyparacotoin from *Aniba firmula* Mez. J Org Chem 24:17–18
67 Govindachari T R, Pai B R, Subramaniam P S, Ramdas Rao U, Muthukumaraswamy N 1967 Ferruol A, a new 4-alkylcoumarin. Tetrahedron 23:4161–4165
68 Gray A I 1983 Structural diversity and distribution of coumarins and chromones in the Rutales. Phytochem Soc Eur Ser 22:97–146
69 Gray A I, Waterman P G 1978 Coumarins in the Rutaceae. Phytochemistry 17:845–864
70 Gujaral V K, Gupta S R, Verma K S 1979 A new chromone glucoside from *Tecomella undulata*. Phytochemistry 18:181–182
71 Harborne J B 1964 Biochemistry of phenolic compounds. Academic Press London, 618 pp
72 Harborne J B, Simmonds N W 1974 The natural distribution of the phenolic aglycones. In: Harborne J B (ed) Biochemistry of phenolic compounds. Academic Press New York
73 Harmatha J, Lubke H, Rybarik I, Mahdalik M 1978 *cis*-Coniferyl alcohol and its glucoside from the bark of beech (*Fagus silvatica* L.). Collect Czech Chem Commun 43:774–780
74 Hasegawa M, Miyoshi T 1960 Minor elements contained in the wood of *Fagaria ailanthoides*. J Jap For Soc 42:222–223
75 Hashimoto K, Nakahara S, Inoue T, Sumida Y, Takahashi M, Masada Y 1985 A new chromone from agarwood and pyrolysis products of chromone derivatives. Chem Pharm Bull 33:5088–5091
76 Hegarty M P, Lahey F N 1956 The coumarins of *Halfordia scleroxyla*. Aust J Chem 9:120–131
77 Hegnauer R 1960–1973 Chemotaxonomie der Pflanzen, Vol I–VI. Birkhäuser Basel

78 Hergert H L 1960 Chemical composition of tannins and polyphenols from conifer wood and bark. Forest Prod J 10:610–617
79 Herrmann K 1978 Hydroxyzimtsäure und Hydroxybenzoesäuren enthaltende Naturstoffe in Pflanzen. Fortsch Chem Org Naturst 35:73–132
80 Hillis W E, Inoue T 1968 The polyphenols formed in *Pinus radiata* after *Sirex* attack. Phytochemistry 7:13–22
81 Hollis A F, Prager R H, Ritchie E, Taylor W C 1961 The constituents *Flindersia pubescens* Bail. and *F. schottiana* F. Muell. Aust J Chem 14:100–105
82 Inoue T, Ishidate Y, Fujita M, Kubo M, Fukushima M, Nagai M 1978 Constituents in the leaves and the stem bark of *Acer nikoense* Maxim. Yakugaku Zasshi 98:41–46
83 Inouye H, Okuda T, Hayashi T 1971 Naphtochinonderivate aus *Catalpa ovata* G. Don. Tetrahedron Lett 39:3615–3618
84 Inouye H, Okuda T, Hirata Y, Nagakura N, Yoshizaki M 1967 Structure of catalpalactone, a new phthalide from catalpa wood. Chem Pharm Bull 15:786–792
85 Ishii H, Hosoya K 1972 Arnottianin: A new dihydropyranocoumarin. Chem Pharm Bull 20:860
86 Ishii H, Ishikawa T, Sekiguchi H, Hosoya K 1973 Xanthoarnol. New dihydrofuranocoumarin. Chem Pharm Bull 21:2346–2348
87 Joshi K C, Prakash L, Singh L B 1975 6-O-Veratryl catalposide, a new iridoid glucoside from *Tecomella undulata*. Phytochemistry 14:1441–1442
88 Joshi K C, Sharma A K, Singh P 1986 A new ferulic ester from *Tecomella undulata*. Planta Med 71–72
89 Karrer W 1958 Konstitution und Vorkommen von organischen Pflanzenstoffen. Birkhäuser Basel, 1207 pp
90 Karrer W, Cherbuliez E, Eugster C H 1977 Konstitution und Vorkommen von organischen Pflanzenstoffen. Ergänzungsband 1. Birkhäuser Basel, 1308 pp
91 Karrer W, Hürlimann H, Cherbuliez E 1981 Konstitution und Vorkommen von organischen Pflanzenstoffen. Ergänzungsband 2. Birkhäuser Basel, 939 pp
92 Kijjoa A, Giesbrecht A M, Gottlieb O R, Gottlieb H E 1981 1,3-Diaryl-propanes and propan-2-ols from *Virola* species. Phytochemistry 20:1385–1388
93 Kim Ki-U 1942 J Pharm Soc Japan 62:455
94 Kimland B, Norin T 1972 Wood extractives of common spruce, *Picea abies* (L.) Karst. Svensk Papperstidn 10:403–409
95 King F E, Housley J R, King T J 1954 Coumarin constituents of *Fagara macrophylla, Zanthoxylum flavum,* and *Chloroxylon swietenia*. J Chem Soc 1392–1399
96 Klimczak M, Kahl W, Grodzinska-Zachwieja Z 1972 Phenolic acids in poplar *Populus*. Diss Pharm Pharmacol 24:181–185
97 Kohlbrenner P J, Schverch C 1959 Benzene-alcohol-soluble extractives of Sitka spruce. J Org Chem 24:166–172
98 Kritchevsky G, Anderson A B 1955 The cone solid of coast redwood (*Sequoia sempervirens*) and giant sequoia (*Sequoia gigantea*). J Org Chem 20:1402–1406
99 Kubo I, Matsumoto T, Klocke J A, Kamikawa T 1984 Molluscicidal and insecticidal activities of isobutylamides isolated from *Fagara macrophylla*. Experientia 40:340–341
100 Lahey F N, MacLeod J K 1968 Halfordin and isohalfordin – revised structures. Tetrahedron Lett 4:447–451
101 Lahey F N, Wluka D J 1955 Geijerin, a new coumarin from the bark of *Geijera salicifolia*. Aust J Chem 8:125–128
102 Lundgren L N, Popoff T, Theander O 1982 Acrylglycerol glucosides from *Pinus sylvestris*. Acta Chem Scand B36:695–699
103 Majumder P L, Sengupta G C, Dinda B N, Chatterjee A 1974 Edgeworthin, a new bis-coumarin from *Edgeworthia gardneri*. Phytochemistry 13:1929–1931
104 Marcano D D C, Hasegawa M, Castaldi A 1972 Neutral compounds and alkaloides of *Zantoxylum ocumarense*. Phytochemistry 11:1531–1532
105 Miki K, Sasaya T 1979 Glycerol derivatives in the inner bark of *Larix leptolepis* Gord. Mokuzai Gakkaishi 25:361–366
106 Moir M, Thomson R H 1973 A new cinnamaldehyde from *Patagonula americana*. Phytochemistry 12:2501–2503

107 Moir M, Thomson R H 1973 Cordiachromes from *Patagonula americana* L. J Chem Soc Perkin Trans I 1556–1561
108 Mondon A, Callsen H 1975 Chromone und Cumarine aus *Cneorum pulverulentum*. Chem Ber 108:2005–2020
109 Mors W B, Gottlieb O R, Djerassi C 1957 The chemistry of rosewood. Isolation and structure of anibine and 4-methoxyparacotoin. J Am Chem Soc 79:4507–4511
110 Mosihuzzaman M, Hoque M F, Chowdhury T A, Theander O, Lundgren L N 1988 Phenolic acids in fresh and wretted jute plants (*Corchorus capsularis* and *Corchorus olitorius*). J Sci Food Agric 42:141–147
111 Mujumdar R B, Rama Rao A V, Rathi S S, Venkataraman K 1975 Swietenone, the first natural *t*-butyl ketone, from *Chloroxylon swietenia*. Tetrahedron Lett 11:867–868
112 Mujumdar R B, Rathi S S, Rama Rao A V 1977 Heartwood constituents of *Chloroxylon swietenia* D C. Indian J Chem 15B:200
113 Murray R D H 1978 Naturally occurring plant coumarins. Fortschr Chem Org Naturst 35:199–429
114 Nagarajan G R, Parmar V S 1977 Phloroacetophenone derivatives in *Prunus domestica*. Phytochemistry 16:614–615
115 Nagarajan G R, Rani U, Parmar V S 1983 Floribin – a new coumarin from *Fraxinus floribunda* Wall. Pharmazie 38:72–73
116 Nagasampagi B A, Sriraman M C, Yankov L, Dev S 1975 Siderin from *Cedrela toona*. Phytochemistry 14:1673
117 Nair G V, Rudloff E v 1959 The chemical composition of the heartwood extractives of tamarack (*Larix laricina* K. Koch). Can J Chem 37:1608–1613
118 Neish A C 1957 Pungenin: A glucoside found in leaves of *Picea pungens* (Colorado spruce). Can J Biochem Physiol 35:161–167
119 Norin T, Sundin S, Theander O 1971 Dehydropinifolic acid, a diterpene acid from the needles of *Pinus sylvestris* L. Acta Chem Scand 25:607–610
120 Okorie D A 1976 A new phthalide and xanthones from *Anthocleista djalonensis* and *A. vogelli*. Phytochemistry 15:1799–1800
121 Oksanen H 1960 Paper chromatography of phenolic constituents in the bark of pine. Suom Kemistil 34B:91–92
122 Oliveira A B de, Fonseca e Silva L G, Gottlieb O R 1972 Flavanoids and coumarins from *Platymiscium praecox*. Phytochemistry 11:3515–3519
123 Palo T R 1984 Distribution of birch, willow, and poplar secondary metabolites and their potential role as chemical defense against herbivores. J Chem Ecol 10:499–520
124 Pearl I A 1969 Water extractives of American *Populus* pulpwood species bark – a review. Tappi 52:428–431
125 Pearl I A, Darling S F 1959 Salireposide from the bark of *Populus tremuloides*. J Org Chem 24:1616
126 Pearl I A, Darling S F 1962 Grandidentatin, a new glucoside from the bark of *Populus grandidentata*. J Org Chem 27:1806–1809
127 Pearl I A, Darling S F 1968 Continued studies on the hot water extractives of *Populus balsamifera* bark. Phytochemistry 7:1851–1853
128 Pearl I A, Darling S F 1970 Phenolic extractives of *Salix purpurea* bark. Phytochemistry 9:1277–1281
129 Pearl I A, Darling S F 1970 The structure of salicortin and tremulacin. Tetrahedron Lett 44:3827–3830
130 Pearl I A, Darling S F 1971 Studies of the hot water extractives of the bark and leaves of *Populus deltoides* Bartr. Can J Chem 49:49–55
131 Pickles S S 1912 The essential oil of the "Nepal sassafras" or "Nepal camphor" tree. J Chem Soc 101:1433–1443
132 Pilo C, Runeberg J 1960 Heartwood constituents of *Juniperus chinensis*. Acta Chem Scand 14:353–358
133 Popoff T, Theander O, Johansson M 1975 Changes in sapwood of roots of Norway spruce, attacked by *Fomes annosus*. Physiol Plant 34:347–356
134 Power F B, Moore C W 1909 The constituents of the bark of *Prunus serotina*. Isolation of 1-mandelonitrile glucoside. J Chem Soc 95:243–261

135 Rama Rao A V, Bhide K S, Mujumdar R B 1980 Isolation of swietenocoumarins G, H and I. Indian J Chem 19B:1046–1048
136 Ramaiah P A, Row L R, Reddy D S, Anjaneyulu A S R, Ward R S, Pelter A 1979 Isolation and characterization of bergenin derivatives from *Macaranga peltata*. J Chem Soc Perkin Trans I: 2313–2316
137 Reher G, Kraus L, Sinnwell V, König W A 1983 A neoflavanoid from *Coutarea hexandra* (Rubiaceae). Phytochemistry 22:1524–1525
138 Ritchie E 1964 Chemistry of *Flindersia* species. Rev Pure Appl Chem 14:47–57
139 Ritchie E, Taylor W C, Vautin T K 1965 Chemical studies of the Proteaceae. I: *Grevillea robusta* A. Cunn and *Orites excelsa* R. Br. Aust J Chem 18:2015–2020
140 Rosendal Jensen S, Juhl Nielsen B 1976 A new coumarin, fraxidin 8-*O*-β-D-glucoside and 10-hydroxy-ligstroside from bark of *Fraxinus excelsior*. Phytochemistry 15:221–223
141 Rowe J W, Conner A H 1979 Extractives in eastern hardwoods. USDA For Serv Gen Tech Rep FPL-18, 67 pp
142 Rudloff E v 1965 Guaiacylglycerol and quinol in the heartwood of *Pinus resinosa* Ait. Chem Ind 4:180–181
143 Saengchantara S T, Wallace T W 1986 Chromanols, chromanones and chromones. Nat Prod Rep 465–475
144 Sainty D, Delaveau P, Bailleul F, Moretti C 1982 New iridoid from *Randia formosa:* 10-caffeoyl-deacetyldaphylloside. J Nat Prod 45:486–488
145 Sano Y, Sakakibara A 1978 Studies on chemical components of akaezomatsu (*Picea glehnii*) bark. 3. Isolation of two glycosides of *p*-hydroxyphenylglycerol. Hokkaido Daigaku Nogakubu Enshurin Kenkyu Hokoku 35:157–163
146 Santamour F S Jr, Vettel H E 1978 The distribution of rhododendrin in birch (*Betula*) species. Biochem Syst Ecol 6:107–108
147 Savidge R A 1987 Dihydroconiferyl alcohol in developing xylem of *Pinus contorta*. Phytochemistry 26:93–94
148 Sato A, Senda M, Kakutani T, Watanabe Y, Kiato K 1966 Extractives from heartwood of *Metasequoia glyptostroboides*. Mokuzai Kenkya 39:13–21
149 Schulte K E, Ruecker G, Klewe U 1973 Some constituents of the bark of *Cedrelopsis grevei*. Arch Pharm 306:857–865
150 Secor J B, Conn J E, Dunn J E, Seigler D S 1976 Detection and identification of cyanogenic glucosides in six species of *Acacia*. Phytochemistry 15:1703–1706
151 Seigler D S 1975 Isolation and characterization of naturally occurring cyanogenic compounds. Phytochemistry 14:9–29
152 Seikel M K, Hostettler F D, Niemann G J 1971 Phenolics of *Quercus rubra* wood. Phytochemistry 10:2249–2251
153 Senter P, Zavarin E, Zedler P H 1975 Volatile constituents of *Cupressus stephensonii* heartwood. Phytochemistry 14:2233–2235
154 Shen Z, Theander O 1985 Flavonoid glycosides from needles of *Pinus massoniana*. Phytochemistry 24:155–158
155 Shimomura H, Sashida Y, Adachi T 1987 Phenolic glucosides from *Prunus grayana*. Phytochemistry 26:249–251
156 Shimomura H, Sashida Y, Adachi T 1988 Phenylpropanoid glucose esters from *Prunus burgeriana*. Phytochemistry 27:641–644
157 Shimomura H, Sashida Y, Oohara M, Tenma H 1988 Phenolic glucosides from *Parabenzoin praecox*. Phytochemistry 27:644–646
158 Singh H, Chawla A S, Kapoor V K, Kumar N, Piatak D M, Nowicki W 1981 Chemical constituents of *Erythrina stricta* bark. J Nat Prod 44:526–529
159 Sosa A, Sosa-Boudouil C 1936 Variations dans la composition du Boulean (*Betula alba* L.) au cours de la végétation d'une année. Bull Soc Chim Biol 18:918
160 Späth E, Jerzmanowska-Sienkiewiczowa Z 1937 Über Fraxidin und Isofraxidin. Ber Deutsch Chem Ges 1019–1020
161 Späth E, Jerzmanowska-Sienkiewiczowa Z 1937 Über Fraxinol, einen neuen Inhaltsstoff der Eschenrinde. Ber Deutsch Chem Ges 698–702
162 Steel J W, Weitzel P F, Audette R C S 1972 Investigation of the bark of *Salix peteolaris*. J Chromatogr 71:435–441

163 Stevens K L, Jurd L, Manners G 1973 Alliodorin, a phenolic terpenoid from *Cordia alliodora*. Tetrahedron Lett 31:2955–2958
164 Subba Rao G S R, Raj K, Kumar V P S 1981 Chemical examination of *Clausena willdenovii* W & A: Isolation of 3-(1,1-dimethylallyl)xanthyletin from the root and bark. Indian J Chem 20B:88–89
165 Talapatra S K, Ganguly N C, Goswami S, Talapatra B 1983 Chemical constituents of *Casearia graveolens*: some novel reactions and the preferred molecular conformation of the major coumarin, micromelin. J Nat Prod 46:401–408
166 Tanaka H, Kato I, Ichino K, Ito K 1986 Coumarinolignoids, cleomiscosin A and cleomiscosin B from *Aesculus turbinata*. J Nat Prod 49:366
167 Theander O 1982 Hydrophilic extractives from the needles of Scots pine and Norway spruce. Svensk Papperstidn 9:R64–R68
168 Thieme H 1963 Isolierung eines neuen Phenolglycosids aus *Salix triandra* L. Die Naturwissenschaften 50:571
169 Thieme H 1964 Zur Konstitution des Salidrosides, eines Phenolglycosids aus *Salix triandra* L. Die Naturwissenschaften 51:360
170 Thieme H 1965a Die Phenolglycoside der Salicaceen: Untersuchungen über jahrzeitlich bedingte Veränderungen der Glycosidkonzentrationen, über die Abhängigkeit des Glycosidgehalts von der Tageszeit und von alten Pflanzenorganen. Pharmazie 20:688–691
171 Thieme H 1965b Die Phenolglycosidspectren und der Glycosidgehalt der mitteldeutschen *Salix*-arten. Pharmazie 20:570–574
172 Tillekeratne L M V, Jayamanne D T, Weerasuria K D V, Gunatilaka A A L 1982 Lignans of *Horsfieldia iryaghedhi*. Phytochemistry 21:476–478
173 Tsukamoto H, Hisada S, Nishibe S, Roux D G, Rourke J P 1984 Coumarins from *Olea africana* and *Olea capensis*. Phytochemistry 23:699–700
174 Tyman J H P 1979 Non-isoprenoid long-chain phenols. Chem Soc Rev 8:499–537
175 Viera P C, Yoshida M, Gottlieb O R, Filho H F P, Nagem T J, Braz Filho R 1983 Lactones from *Iryanthera* species. Phytochemistry 22:711–713
176 Vrkoc J, Sedmera P 1972 Extractives of *Chloroxylon swietenia*. Phytochemistry 11:2647–2648
177 Waterman P G 1976a 2,4,5-Trimethoxystyrene from *Pachypodanthium standtii*. Phytochemistry 15:347
178 Waterman P G 1976b Alkaloids and coumarins from *Zanthoxylum flavum*. Phytochemistry 15:578–579
179 Zavarin E, Anderson A B 1955 Extractive components from incence-cedar heartwood (*Libocedrus decurrens* Torrey). I. Occurrence of carvacrol, hydrothymoquinone and thymoquinone. J Org Chem 20:82–88
180 Zoghbi M D G B, Roque N F, Gottlieb O R 1981 Propacin, a coumarinolignoid from *Protium opacum*. Phytochemistry 20:180

7.2 Gallic Acid Derivatives and Hydrolyzable Tannins

E. Haslam

7.2.1 Introduction

Gallic acid and its derivatives as they occur in extracts of oak-galls constitute a chemical reagent of considerable antiquity. The blue-black color produced when an aqueous infusion is treated with salts of iron was first described by Pliny; its use in the analysis of mineral waters and as a component of invisible ink were noted as early as the sixteenth century. The eminent Swedish chemist Scheele first isolated gallic acid from oak-galls toward the end of the eighteenth century and the acid was named *gallic* by Braconnot in reference to its origin in plant galls.

One of the first published preparations of gallic acid, however, is that of Sir Humphrey Davy (79) in the Journal of the Royal Institute (1803); he followed a directive from the trustees and governors to present a series of lectures on 'the chemical principles of the art of tanning.' This incident vividly illustrates the intimate relationship, for both scientific and historical reasons, that studies of the chemistry and biochemistry of gallic acid and its derivatives have with those of the vegetable tannins and ultimately the proanthocyanidins (142).

Tanning is a process whereby a raw animal hide or skin is converted into leather; in earlier times this was achieved, almost exclusively, with plant extracts such as those from oak bark and quebracho (142). In the botanical sense, the word 'tannin' has acquired a very wide usage, which has occasioned inevitable unfortunate misunderstandings. Indeed the reproach that "anything that gives a blue-black color with iron salts has been termed a tannin" often seems merited (36). The relationship to the practice of vegetable tanning, however, stimulated a great deal of interest among many eminent chemists in the early years of the twentieth century to discover the nature of the substances responsible for these distinctive properties. Several important substances were isolated in crystalline form – notably, chebulinic acid (121, 123, 124), acer tannin (114), hamameli tannin (70), (+)-catechin, and (−)-epicatechin. The initial enthusiasm, however, gradually waned as the great complexity of many plant extracts was eventually realized. Studies of the vegetable tannins became one of the untidy corners of organic chemistry. It was the advent of new methods of isolation and separation, and structure determination that gave a new impetus to research in this field and made possible the comparatively recent expansion in our knowledge (36, 40–42).

7.2.2 Metabolism of Gallic Acid – General Observations

Although the metabolism of gallic acid is but one facet of general phenolic metabolism in higher plants, it is characterized by several distinctive and possibly unique features. When these are coupled with the properties that many gallic acid derivatives display, these features inevitably prompt several questions: What are the biological function and purpose of this form of secondary metabolism? What selectionary advantages may it have conferred during the course of evolution? Before such questions can be properly approached – let alone, answered – a great deal more must be learned concerning the general biochemical matrix from which per se secondary metabolism derives, its enzymology, and its control. Until such time, these characteristic features of gallic acid metabolism can only be commented upon and, where desirable, highlighted. The debate concerning their possible role in the life of the plant is likely to remain a tantalizingly inconclusive one.

In his pioneering study of the phenolic constituents of plants and their taxonomic significance, Bate-Smith (2) suggested that plant phenols "...because they occupy a metabolic cul-de-sac, may provide a plane of reference for vascular plants generally to which other aspects of plant chemistry may be referred."

He noted that the more common phenolic constituents of dicotyledons are eight in number (Fig. 7.2.1). The majority of the other phenolic constituents encountered in plants Bate-Smith formulated as derived from these groups by a lim-

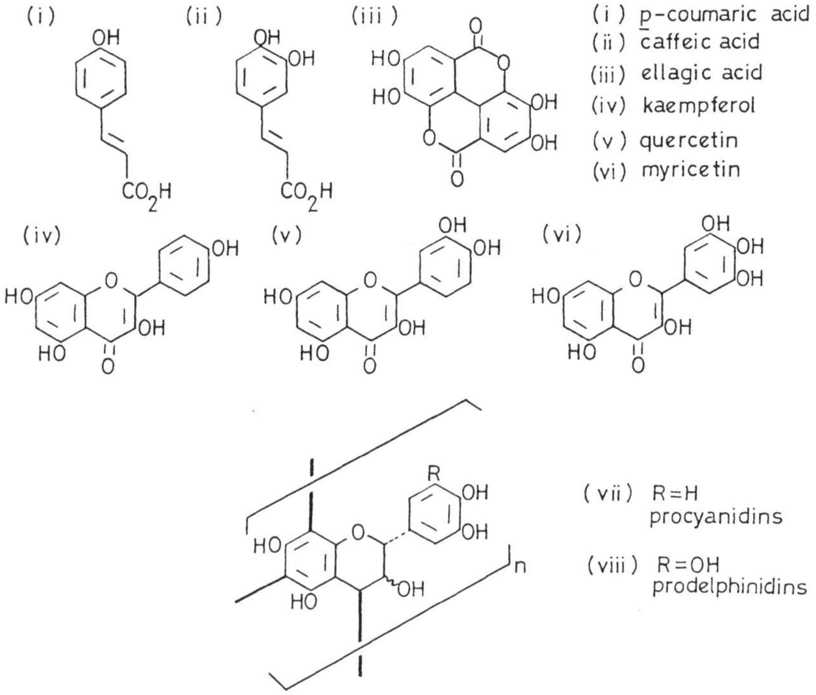

(i) p-coumaric acid
(ii) caffeic acid
(iii) ellagic acid
(iv) kaempferol
(v) quercetin
(vi) myricetin

(vii) R=H procyanidins
(viii) R=OH prodelphinidins

Fig. 7.2.1. The more common constituents of dicotyledons

ited number of unit transformations. Bate-Smith drew particular attention to the fact that 3,4,5-trihydroxycinnamic acid **(6)**, except as its various *O*-methyl ethers, was not present in plants but that hexahydroxydiphenic acid **(4)** – determined as its stable dilactone ellagic acid **(5)** after acid hydrolysis – was widely distributed in plants in ester form **(3)**. This evidence and various phylogenetic considerations persuaded Bate-Smith to suggest that hexahydroxydiphenic acid **(4)** was the taxonomic equivalent of the 'missing' 3,4,5-trihydroxycinnamic acid **(6)**. However, Schmidt and Mayer (132) in their concomitant studies of the ellagitannins had postulated that hexahydroxydiphenoyl[1] esters are formed in vivo by oxidative coupling of suitably disposed galloyl ester groups **(2)** in a precursor substrate (Fig. 7.2.2); and, although direct experimental evidence to reinforce this opinion is still awaited, there is strong circumstantial evidence to support this idea. Accepting this view, it follows that gallic acid **(1)** is now perhaps more correctly considered as the ultimate taxonomic equivalent of Bate-Smith's 'missing acid', **(6)**.

Parenthetically it should perhaps be noted that the absence of 3,4,5-trihydroxycinnamic acid **(6)** from plant tissues is now, in contrast to the early 1960s, more readily explicable. Thus if one accepts that the occurrence of the hydroxycinnamic acids and their various conjugates in plants derive from their formation

[1] This nomenclature is used as an abbreviation for the 6,6-dicarbonyl-2,2′,3,3′,4,4′-hexahydroxybiphenyl radical.

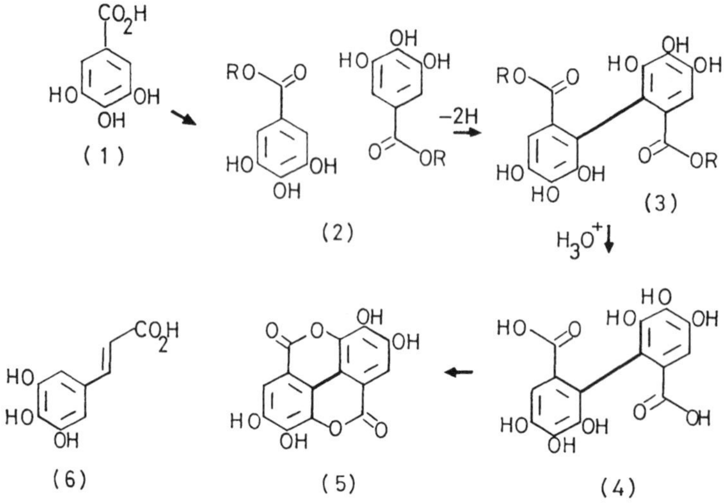

Fig. 7.2.2. Formation of hexahydroxydiphenoyl esters

as shunt metabolites from the corresponding hydroxycinnamoyl coenzyme A ester intermediates in lignin biosynthesis (30), then derivatives of **6** would not be expected to be plant metabolites since this acid is not a recognized intermediate in the pathway to lignins. In this context it is also important to underline that the consensus of experimental evidence and opinion now favors the concept of successive aryl hydroxylation reactions in the B-ring of a C_{15} (possibly chalcone) intermediate as the *modus vivendi* to flavonoids such as myricetin and the prodelphinidins (Fig. 7.2.1), which possess a 3′,4′-5′-trihydroxy substituted B-ring (32).

Gallic acid and its derivatives thus occupy a particular niche in taxonomic terms; other noteworthy features of gallic acid metabolism in plants worthy of emphasis relate to the patterns and modes of occurrence of other natural phenolic esters. Thus esters of *o*- and *p*-hydroxybenzoic, resorcylic, and protocatechuic acids are relatively rarely encountered in plants. Their occurrence (for example, that of salicylic acid in the Salicaceae) is often something of a taxonomic speciality. In contrast, as Bate-Smith (2) demonstrated, gallic acid and its derivatives are metabolized fairly widely by plants. Similarly, although esters of the hydroxycinnamic acids (*p*-coumaric, caffeic, ferulic, and sinapic) are almost universally distributed in plants, their recorded occurrence is as mono- and very occasionally bis-esters with polyols (e.g., chlorogenic acid and isochlorogenic acid) and sugars (51). In contradistinction, gallic acid is found in a variety of ester forms that range in complexity from simple mono-esters such as *β*-D-glucogallin (**7**) and theogallin (**8**) to the complex polyesters with D-glucose, whose composition gives molecular weights extending to at least 2000 (41, 42). As far as the author is aware, this diversity and complexity of ester forms is unique in the plant kingdom.

Many of their characteristic properties (in particular, their complexation with proteins, polysaccharides, and various nitrogenous metabolites) draws attention once again to the distinctive position of gallic acid as a secondary plant product.

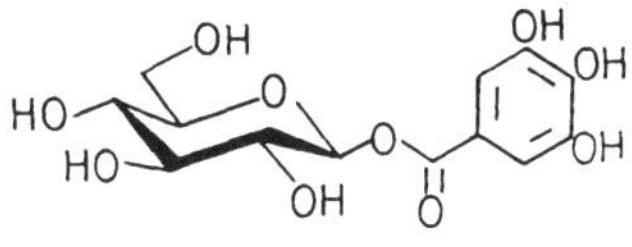

(7)

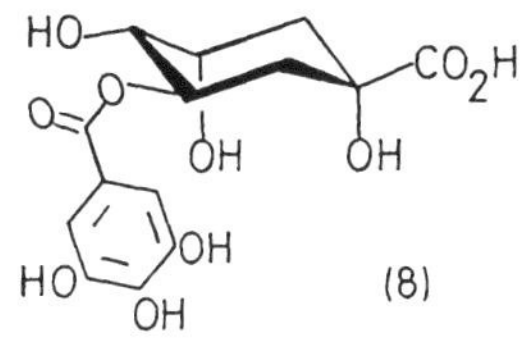

(8)

Together with the proanthocyanidins, these polyphenolic esters constitute the so-called vegetable tannins of the earlier chemical and botanical literature (79). Bate-Smith observed that the capacity of plants to metabolize polyphenols of this type is a primitive characteristic that has tended, with phylogenetic specialization, to become lost and not regained. Bate-Smith (2) thus accorded the presence of proanthocyanidins the character *a*, and the capability to synthesize the vic-trihydroxy aromatic ring (1,2,3-OH or pyrogallol type) – such as is found in gallic acid, myricetin and the prodelphinidins – the character *b*. Those that did not metabolize these phenolic constituents were designated a_0 and b_0, respectively. Plants were broadly divided by Bate-Smith into the four groups ab, a_0b, ab_0, and a_0b_0; the thread of evolution from the earliest emergence of the dicotyledons was assumed always to be in the direction $ab \rightarrow a_0b_0$.

Later work by Bate-Smith (3–8) led to an amplification and modification of some of these views. Proanthocyanidins are present in the most primitive of vascular plants, such as ferns and gymnosperms, but the galloyl and hexahydroxydiphenoyl esters are not. These latter two groups appear for the first time in dicotyledons and Bate-Smith was led to propose a phylogenetic sequence: proanthocyanidins → hexahydroxydiphenoyl esters (ellagitannins) → galloyl esters (gallotannins). In an examination of the chemistry and taxonomy of *Ribes* (4) he further speculated that ellagitannins appear to be produced at the expense of prodelphinidins, and he suggested that their presence in plant tissues was a derived and not a primitive character. Further hints of this type of relationship are revealed by an examination of the polyphenolic metabolism in several plant families such as Aceraceae, Ericaceae, Fagaceae, and Rosaceae, all of which retain the capacity to synthesize complex polyphenols (Haslam E, unpublished observations). Thus in the Ericaceae, species of the genera *Calluna, Erica,* and *Rhododendron* are all rich sources of proanthocyanidins whereas *Arctostaphylos uva-ursi* has only a minimal capacity for proanthocyanidin synthesis and combines with this a very high level of gallic acid (gallotannin) metabolism. This distinctive relationship is more than matched in the Aceraceae (35).

Additional indications of a possible metabolic link come from a consideration of phenolic synthesis in plants such as *Camellia sinensis* (tea), which metabolizes

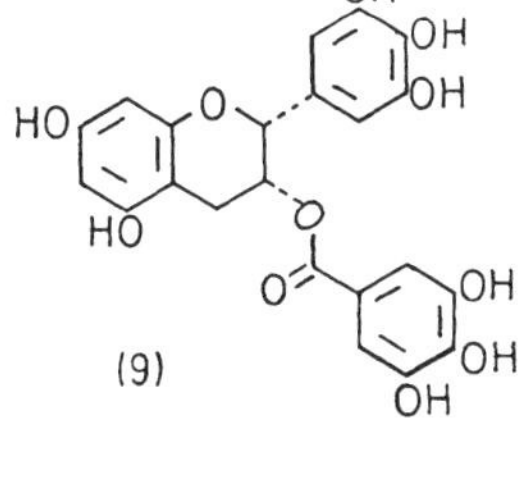

(9)

substantial levels of flavan-3-ols (normally associated with proanthocyanidin synthesis) but principally as their 3-galloyl esters (e.g., (**9**). Only minimal levels of proanthocyanidins are found in *Camellia sinensis*. Whether these observations are a reflection of a biosynthetic link (metabolic balance within species), or a phylogenetic situation involving two widely different streams, it is not yet possible to decide.

7.2.3 Biosynthesis of Gallic Acid

Gallic acid thus occupies a distinctive position in the overall phenolic metabolism of plants. In these circumstances it is perhaps surprising that some ambiguity still surrounds the actual pathway of biosynthesis of gallic acid (41, 42). Three pathways have been formulated. To some extent they reflect the different experimental approaches adopted and, as such, underline one of the weaknesses of the isotopic tracer method when applied to whole organisms. Zenk (144) formulated a conventional pathway from L-phenylalanine to the 'missing' 3,3,4-trihydroxycinnamic acid (**6**) followed by β-oxidation to give gallic acid (**1**) (Fig. 7.2.3). This conclusion was based on feeding studies utilizing ^{14}C-labelled L-phenylalanine, L-tyrosine, and benzoic acid in *Rhus typhina* (sumac, Anacardiaceae) and is analogous to that proposed for the formation of other hydroxybenzoic acids in higher plants (145). The mechanism of β-oxidation is presumed to proceed via the intermediacy of coenzyme-A esters and significantly predicts the initial formation of the activated form of gallic acid – its coenzyme A thiol ester. Neish and Towers (77) favored a variation of this scheme (Fig. 7.2.3 (b)), in which β-oxidation ensues at the caffeic acid stage to give protocatechuic acid, which is then further hydroxylated to give gallic acid (**1**). Both of these pathways therefore postulate that gallic acid is derived, overall, by oxidative degradation – shortening of the C_3 side chain and hydroxylation of the aryl ring – of one of the end products (L-phenylalanine) of the shikimate pathway.

In contrast, the third pathway for which experimental evidence has been obtained – first with the mould *Phycomyces blakesleeanus* and later with *Rhus typhina* and plants of *Acer* and *Geranium* sp. and *Camellia sinensis* (16, 17, 20, 60, 120) – suggests that gallic acid is a shunt or overflow metabolite of the shikimate pathway, derived by direct dehydrogenation of the intermediate 3-dehydroshikimate (**11**). This pathway predicts that not only the carbon skeleton of gallic acid but also the three phenolic groups originate directly from **11**. Parenthetically, it is interesting to note that both shikimic acid (**12**) and the closely related quinic acid (**13**) are often found in substantial amounts in many plant tissues. The relationship of quinic acid (**13**) to 3-dehydroquinic acid (**10**) suggests that it is a shunt metabolite derived by reduction of **10**; by analogy, the accumulation of shikimic acid (**12**), a key intermediate in the pathway, similarly points to some overall rate-limiting process in the shikimate pathway that occurs after the intermediate (**12**). An extrapolation of this idea suggests that the third route to gallic acid (Fig. 7.2.3 (c)) may operate as a shunt mechanism to relieve this metabolic block in somewhat different circumstances. In support of this type of mechanism is the observation (Amrhein, private communication) that glyphosate (**14**), which

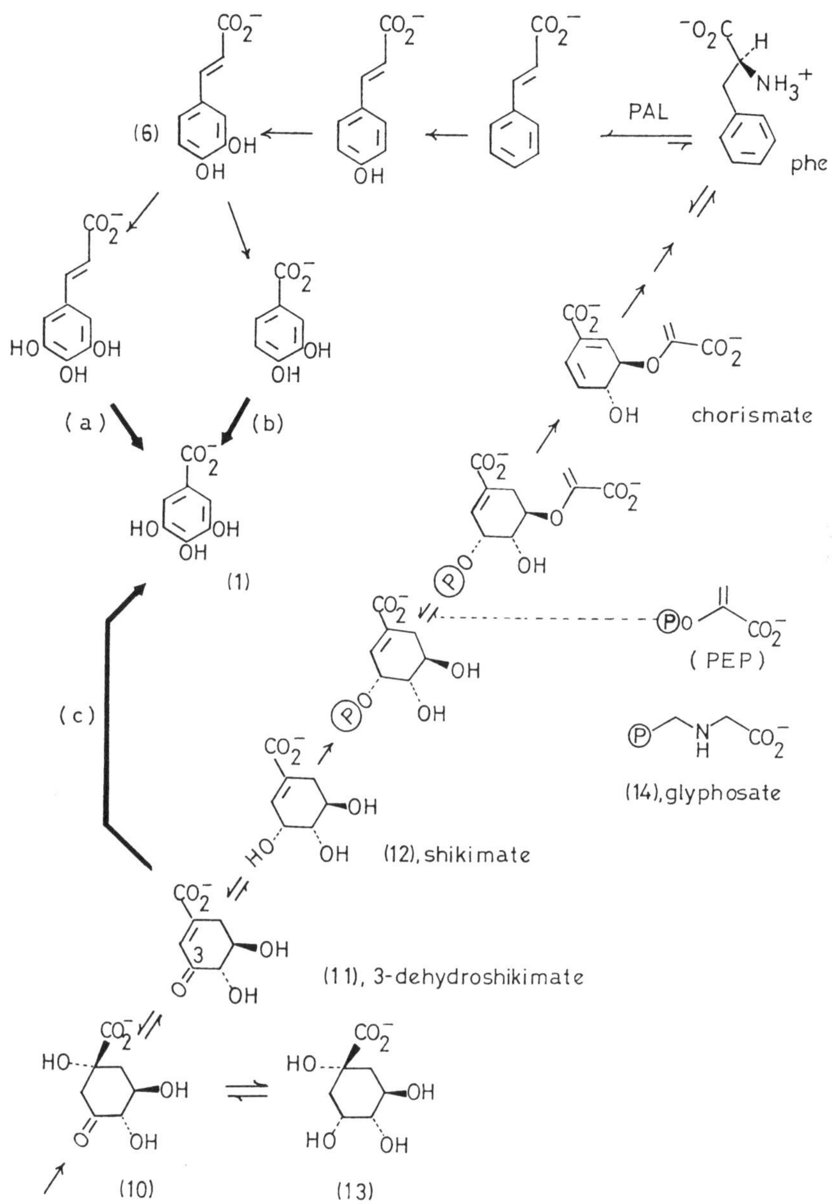

Fig. 7.2.3. Three biogenetic routes to gallic acid (**1**)

acts as a competitive inhibitor of the enzyme 5-ESP-synthase (Fig. 7.2.3), leads to an increase in the production of gallic acid (**1**) when it is administered in limiting amounts to plant tissues that metabolize gallic acid.

Comparatively little is yet known concerning the biosynthesis of the various derivatives of gallic acid, such as the esters and depsides that occur in plants. Biogenetic relationships have been proposed to link these groups of substances and they appear to bring considerable clarification and order to a very wide range of apparently tenuously related structures (41). However it is important to stress that, valuable as these are in present circumstances, they are based almost entirely on data relating to their normal occurrence and distribution and upon the chemical intuition common to all biogenetic hypotheses. A study of the enzymology of biosynthesis of esters of gallic acid was recently commenced by Gross (26–29). Cell-free extracts of fresh oak leaves (*Quercus robur*) were found to catalyze the esterification of gallic acid (^{14}C labelled) and UDP-glucose. The product of the reaction was identified as β-D-glucogallin (**7**). This first intermediate is then thought to undergo sequential substitution by further molecules of gallic acid (possibly via the participation of a carboxyl-activated species such as galloyl coenzyme-A), leading finally to β-pentagalloyl-D-glucose (**15**) which, it is believed, plays a central role in the biosynthesis of both the gallotannins and the ellagitannins (41, 44, 132, 137).

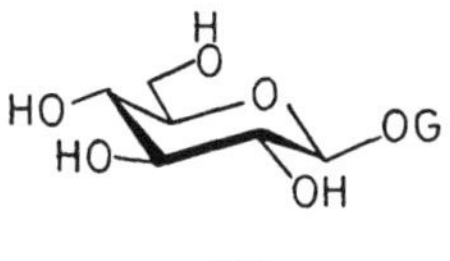

(7) 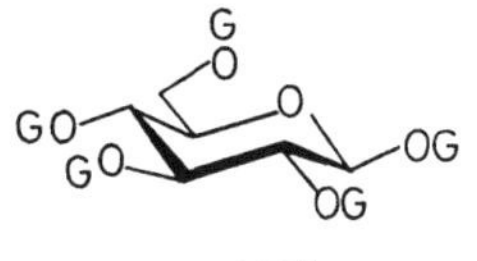 (15)

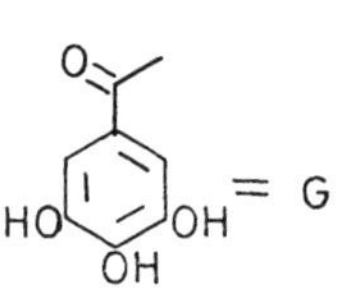

7.2.4 Metabolites of Gallic Acid

Because of their importance to commerce in earlier days, and particularly to the trade in leather, plants that metabolize substantial quantities of the polyesters of gallic acid and hexahydroxydiphenic acid have been the subject of the most intensive chemical study and analysis. A watershed in this work was the investigations of Emil Fischer (22, 23) on the constituents of Chinese and Aleppo galls. Between 1910 and 1940 several distinguished chemists – Fischer, Karl Freudenberg, and Paul Karrer among them – made substantial contributions to this field. Notable work in this particular area of plant chemistry has been carried out more recently by Schmidt and Mayer in Heidelberg in their chemical work on the ellagitannins

Table 7.2.1. Sources of complex esters of gallic acid and hexahydroxydiphenic acid of commercial importance

Family	Genus, species	Part	Common name
Anacardiaceae	*Rhus typhina*		
	R. coriaria	Leaves	Sumac
	R. semialata	Galls	Chinese sumac
Leguminosae	*Caesalpinia coriaria*	Fruit pods	Divi-divi
	C. spinosa	Fruit, fruit pods	Tara
	C. brevifolia	Fruit, fruit pods	Algarobilla
Fagaceae	*Castanea* spp.	Wood, bark, leaves	Chestnut
	Quercus spp.	Wood, bark	Oak
	Q. aegilops	Acorn cups	Valonea
	Q. infectoria	Galls	Turkish, Aleppo
Combretaceae	*Terminalia chebula*	Fruit	Myrabolans

(Table 7.2.1) and by Japanese investigators, such as Okuda and Nishioka, interested in the potential pharmacological value of polyphenols.

Bate-Smith's original observations (2) on the phenolic constituents of plants and their taxonomic significance (Sect. 7.2.2) were based on surveys of plants from nearly 200 plant families. The inevitable attention that chemists have given to individual plants, such as those listed in Table 7.2.1, whose esters are of potential commercial value, has undoubtedly circumscribed our knowledge of the wider role that gallic acid plays in phenolic biosynthesis throughout the plant kingdom. In the present context, so far as is practicable, the subject is presented in an order based on biogenetic considerations. These appear, to the author, to provide a logical and unifying framework for such a discussion (41–46).

7.2.4.1 Simple Esters – Occurrence and Detection

Gallic acid is almost invariably found in plant tissues in ester form. Various simple esters of gallic acid have been described from plant sources (Table 7.2.2, Table 7.2.3) and these metabolites are in many ways analogous to the hydroxycinnamoyl esters occurring in plants. These esters are thus formed by association with sugars,

Table 7.2.2. Naturally occurring galloyl esters of phenols and phenolic glycosides

Ester	Genus, species	Family	References
(–)-Epicatechin-3-gallate	*Camellia sinensis*	Theaceae	11
(–)-Epigallocatechin-3-gallate	*C. sinensis*	Theaceae	11
(+)-Catechin-3-gallate	*Bergenia crassifolia*	Saxifragaceae	38
	B. cordifolia	Saxifragaceae	38
(–)-Epicatechin-3,5-digallate	*Camellia sinensis*	Theaceae	18
(–)-Epigallocatechin-3,5-digallate	*C. sinensis*	Theaceae	18
2-*O*-Galloylquinol-*β*-D-glucoside	*Bergenia crassifolia*	Saxifragaceae	12
6-*O*-Galloylquinol-*β*-D-glucoside	*B. cordifolia*	Saxifragaceae	12
p-Galloyloxyphenyl-*β*-D-glucoside	*Arctostaphylos uva-ursi*	Ericaceae	12
2′,6′-Dihydroxy-4′-galloyloxymethyl-phenyl-*β*-D-glucoside (cretanin)	*Castanea crenata*	Fagaceae	111
6-Acetyl-2′,6′-dihydroxy-4′-galloyl oxymethyl-phenyl-*β*-D-glucoside	*C. crenata*	Fagaceae	113
4-(4′-Hydroxyphenyl)-2-butanon-4′-*O*-*β*-D-(6″-*O*-galloyl)-glucoside (lindleyin)	*Rhei rhizoma*	Polygonaceae	87
	Aconium lindleyi	Crassulaceae	25
Quercetin-3-*O*-(6″-*O*-galloyl)-*β*-D-glucoside	*Tellima grandiflora*	Saxifragaceae	10
Quercetin-3-*O*-(6″-*O*-galloyl)-*β*-D-galactoside	*Euphorbia* sp.	Euphorbiaceae	115
Quercetin-3-*O*-(2″-*O*-galloyl)-*β*-D-galactoside	*Euphorbia* sp.	Euphorbiaceae	115
Myricetin-3-*O*-(6″-*O*-galloyl)-*β*-D-glucoside	*Tolmiea menziesii*	Saxifragaceae	9
Naringenin-7-*O*-(6″-*O*-galloyl)-*β*-D-glucoside	*Acacia farnesiana*	Leguminosae	141
Digalloyl procyanidin B-2 monogalloyl procyanidin B2	*Rhei rhizoma*	Polygonaceae	87
5-*O*-galloylquinic acid (theogallin)	*Camellia sinensis*	Theaceae	117
2-*O*-Cinnamoyl-1-*O*-galloyl-*β*-D-glucoside	*Rhei rhizoma*	Polygonaceae	74
Glycerol-1-*O*-gallate	*R. rhizoma*	Polygonaceae	86

Table 7.2.3. Naturally occurring galloyl esters of β-D-glucose and related polyols

Galloyl ester	Genus, species	Part	Family	Reference(s)
β-1-*O*-galloyl-D-glucose (β-D-glucogallin)	*Rheum officinale*		Polygonaceae	24, 84, 87
β-1,6-di-*O*-galloyl-D-glucose	*Quercus infectoria*	Galls	Fagaceae	45
β-1,2,6-tri-*O*-galloyl-D-glucose	*Nuphar japonicum*	Leaves	Nymphaeaceae	88
	Rubus fructicosus	Leaves	Rosaceae	45
	R. idaeus	Leaves	Rosaceae	45
	Rosa canina	Leaves	Rosaceae	45
	Fuchsia sp.	Leaves	Onagraceae	45
	Epilobium angustifolium	Leaves	Onagraceae	45
	Rhei rhizoma		Polygonaceae	86, 87
β-1,3,6-tri-*O*-galloyl-D-glucose	*Terminalia chebula*	Fruit	Combretaceae	137
β-1,2,3,6-tetra-*O*-galloyl-D-glucose	*Fuchsia* sp.	Leaves	Onagraceae	45
	Epilobium angustifolium	Leaves	Onagraceae	45
	Quercus infectoria	Galls	Fagaceae	45
	Ceratonia siliqua	Pods	Leguminosae	45
β-1,2,4,6-tetra-*O*-galloyl-D-glucose	*Bergenia crassifolia* *B. cordifolia*	Roots and rhizomes	Saxifragaceae	45
β-1,3,4,6-tetra-*O*-galloyl-D-glucose	*Caesalpinia brevifolia*	Pods	Leguminosae	129
β-penta-*O*-galloyl-D-glucose	*Nuphar japonicum*	Leaves	Nymphaeaceae	88
	Terminalia chebula	Fruit	Combretaceae	137
	Acer platanoides	Leaves	Aceraceae	45
	A. campestre	Leaves	Aceraceae	45
	Quercus infectoria	Galls	Fagaceae	45
	Q. borealis	Leaves	Fagaceae	45
	Rhus semialata	Galls	Anacardiaceae	45
	R. coriaria	Leaves	Anacardiaceae	45
	R. typhina	Leaves	Anacardiaceae	45
	Cotinus coggyria	Leaves	Anacardiaceae	45
	Fuchsia sp.	Leaves	Onagraceae	45
	Epilobium angustifolium	Leaves	Onagraceae	45
	Rubus idaeus	Leaves	Rosaceae	45
	R. fructicosus	Leaves	Rosaceae	45
	Rosa canina	Leaves	Rosaceae	45
	Geranium robertanium	Leaves	Geraniaceae	45
2,6-di-*O*-galloyl-1,5-anhydro-D-glucitol (aceritannin)	*Acer ginnale*	Leaves	Aceraceae	43, 78, 114
	A. saccharinum	Leaves	Aceraceae	45
	A. tartaricum	Leaves	Aceraceae	45
2′,5-di-*O*-galloyl-D-hamamelose (Hamameli tannin)	*Hamamelis virginiana*	Bark	Hamamelidaceae	71
	Quercus rubra	Bark	Fagaceae	71
	Castanea sativa	Bark	Fagaceae	71

polyols, glycosides, and other phenols. In contrast to the hydroxycinnamic acids however, fully authenticated examples of the occurrence of *N*-acyl derivatives (with amines, amino acids, and alkaloids) of gallic acid have not yet been described in the literature (51).

The examination of plant extracts to determine the presence of esters of gallic acid is considerably facilitated by HPLC and by two-dimensional paper chromatography, which gives a 'fingerprint' of the plant's phenolic metabolism. Galloyl esters are detected by their violet (mono, bis) to dark blue (tris, etc.) absorption in UV light, which is enhanced by fuming with ammonia, and by applying various chromogenic sprays. Specific for galloyl esters is KIO_3 solution, which gives a rose-pink color and reacts with gallic acid to form the characteristic orange of purpurogallin carboxylic acid. R_F values in the butanol-acetic acid-water system are broadly comparable for a whole range of galloyl glucose esters (0.30 – 0.50) but R_F values in aqueous acetic acid are very sensitive to the degree of esterification and decrease with increasing galloylation.

Although derivatization (methyl ether, acetate) is usually a prerequisite to obtaining micro-analytical data, structural information is generally most easily obtained with the free phenolic forms of galloyl esters. Hydrolysis with the adaptive enzyme tannase (obtained by growth of *Aspergillus niger* on tannic acid) (47) is a specific means of cleavage of natural esters to yield gallic acid and the polyol (usually D-glucose). With appropriate controls the enzyme may be used to quantitatively determine the ratio of gallic acid to the polyol. ^{1}H and ^{13}C NMR spectroscopy with spin-decoupling usually provides a complete structural analysis for natural galloyl-D-glucopyranose derivatives. When the anomeric center is acylated, the galloyl group almost invariably adopts the β-configuration and the sugar usually adopts the C-1 ($^4C_1$) conformation (21, 57). Immediate recognition of these two features is given by the characteristic low field doublet for H-1 (δ 6.39 to 5.7 ppm TMS, J = 9.5 Hz) in the ^{1}H NMR spectrum. The position of this signal in the ^{1}H NMR spectrum is remarkably sensitive to the presence of other galloyl ester groups in the molecule and is a measure of the index of galloylation in the D-glucose series.

Galloyl groups are distinguished by a series of 2-proton singlets (d_6 acetone, δ 6.95 to 7.25 ppm TMS); similar information derives from ^{13}C NMR measurements. The position of the ^{13}C resonances due to the galloyl ester carbonyl groups is solvent sensitive and in prototropic media these resonances occur further downfield from TMS relative to the corresponding signals of spectra measured in d_6 acetone. The relative intensities of these signals and the multiplicity provide a good measure of the extent of esterification of the glucose nucleus by gallic acid. The ^{13}C NMR signals of the individual carbon atoms in the D-glucose residue accord generally with the relative positions of the ^{13}C signals for the carbon atoms in β-D-glucose itself. In all cases the anomeric carbon occurs at lowest field (δ 93.4 to 95.5 ppm TMS) and C-6 at highest field (δ 62.9 to 63.8 ppm TMS) in acetone and methanol. ^{13}C resonances of the carbon atoms of the aryl rings of the galloyl ester groups are also readily recognized and identified. Assignments are made relative to those of methyl gallate. In polygalloyl esters – e.g., β-penta-*O*-galloyl-D-glucose (**10**) – the ^{13}C resonances due to C-2 and C-3 in the individual galloyl groups are not resolved at 25.15 MHz and almost invariably occur as singlets at δ 109.8 to 110.6 and δ 145.9 to 146.3 ppm TMS, respectively. The ^{13}C resonances due to C-1 and C-4 of the various galloyl groups are however generally resolved at 25.15 MHz and, in an analogous manner to the multiplicity of signals due to the ester carbonyl group carbon atoms, they provide a convenient index of the number of galloyl ester groups attached to the D-glucose core.

A further means for ready identification of a galloyl ester is by mass spectroscopy of the methylated (diazomethane) ester. The mass spectral fragmentation pattern usually displays the ion **16** as base peak, m/e 195 (37, 112).

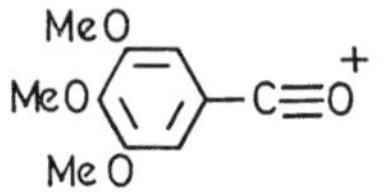

$$\text{(MeO)}_3\text{C}_6\text{H}_2\text{–C}\equiv\text{O}^+ \quad (16),\ m/e = 195$$

Structure analysis of galloyl D-glucose derivatives in which the anomeric hydroxyl group (C-1 position) is unacylated is, by comparison with those derivatives in which this hydroxyl group is acylated, much more difficult to carry out. In solution these esters not only set up an equilibrium mixture of the α- and β-anomeric forms, but they may assume alternative conformations of the glucopyranose ring to the normally preferred C-1 ($^4C_1$) conformation (52, 57). Derivatization (e.g., acetylation) similarly usually produces a mixture of α and β derivatives and unless these are separable, spectroscopic identification (particularly ^{1}H NMR) is complicated by the doubling of signals in the spectrum.

β-D-Glucogallin (β-1-*O*-galloyl-D-glucose) was first isolated as its crystalline hydrate from the roots of Chinese rhubarb (*Rheum officinale*) by Gilson (24). Although in the interim several galloyl-D-glucose derivatives were obtained by degradation of natural products, it was half a century later before further simple galloyl esters of D-glucose were isolated and described as natural products in their own right. Schmidt and his colleagues then described (137) the isolation of β-penta-*O*-galloyl-D-glucose (**15**) and β-1,3,6-tri-*O*-galloyl-D-glucose from myrabolans (*Terminalia chebula*) fruit and later (133) β-1,3,4,6-tetra-*O*-galloyl-D-glucose from algarobilla (*Caesalpinia brevifolia*) fruit pods. In more recent work, β-penta-*O*-galloyl-D-glucose (**15**) has been isolated and identified in a range of plants (Table 7.2.3) where it appears to act as a key intermediate in the further metabolism of gallic acid (vide infra). The various, simple, unmodified, naturally occurring esters of gallic acid with D-glucose (and other sugars), which have now been recognized in higher plants, are shown in Table 7.2.3. It is interesting to note, among the partially esterified derivatives, the predisposition for the ester groups to be located at positions 1, 2 and 6 on the D-glucose core. (This feature may also be noted in Table 7.2.2). Also included in Table 7.2.3 are *Acer* tannin (structure revised to 2,6-di-*O*-galloyl-1,5-anhydro-D-glucitol)(44, 78) and Hamameli tannin, which have been recognized for many years as natural products. However it should be stressed that the association of the word tannin with these esters is unfortunate since it implies an ability to readily precipitate proteins from solution – an ability that these compounds do not possess. For the esters of gallic acid and D-glucose, this ability to precipitate proteins only appears to any significant extent at the stage of β-1,3,6-tri-*O*-galloyl-D-glucose (39).

Using phenolic biosynthesis in the leaves of a plant as a point of reference, the metabolism of gallic acid for many plants begins and ends with their ability to synthesize esters of gallic acid with D-glucose (and occasionally other sugars), phenols, and phenolic glycosides. For a great many other plants, however, this is not the limit of the range of metabolites associated with gallic acid; in this particular context, those plants may be noted that are able to biosynthesize β-penta-

O-galloyl-D-glucose (**15**). This intermediate appears to mark a biosynthetic watershed for many plants, and from it other biosynthetic pathways subsequently diverge to give the gallotannins and various ellagitannins (41, 45).

In the context of this particular biosynthetic proposal and supporting the key position that *β*-penta-*O*-galloyl-D-glucose (**15**) is deemed to occupy in many plants, it has been observed that callus tissue of *Quercus robur* metabolizes the same ester (**15**) in very small amounts, along with gallic acid and pyrogallol but metabolizes none of the many other derivatives of gallic and hexahydroxydiphenic acids that leaves of the fully differentiated plant *Quercus robur* synthesize. The evidence suggests that under the conditions of growth of this tissue cell culture only a part of the genetic information for secondary metabolite production is expressed (E. Haslam and M. Fowler, unpublished observations).

The patterns of further metabolism of **15**, which have been discerned in the leaves of plants (46), are broadly divisible into at least three groups, labelled A, B, and C (Fig. 7.2.4). Attention is directed in this review to these three major patterns but it should be noted that this scheme probably does not yet represent an all-embracing classification of gallic acid metabolism in higher plants. We previously noted those plant species that provided extracts of commercial importance in history: whereas several points of contact between the phenolic metabolites of these particular plants and Fig. 7.2.4 are self-evident (vide infra) others – notably *Terminalia chebula, Caesalpinia coriaria*, and *C. brevifolia* – quite clearly produce a range of phenolic metabolites which, on the basis of the available evidence,

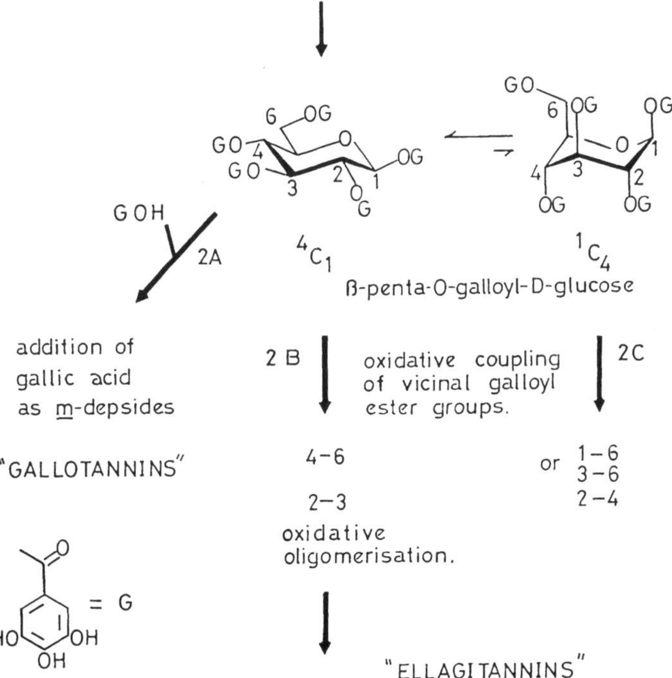

Fig. 7.2.4. Conversion of *β*-penta-*O*-galloyl-D-glucose (**15**) to gallo- and ellagitannins

are extensions of categories B and C. Additionally it is also clear that specialized organs of a particular plant (e.g. , fruit, fruit pod, bark) may well metabolize other phenols derived from one (or more) of the 'primary' phenolic metabolites formed by routes A, B, or C.

7.2.4.2 Depside Metabolites – Group 2A

The ability to produce metabolites in which additional gallic acid molecules are esterified to a pre-existing galloyl ester in the form of depsides is limited to a comparatively few plant families. Many of the products of this form of gallic acid metabolism were grouped together in the earlier literature under the generic term 'gallotannin' and are referred to here as Group 2A. The depsidically linked galloyl ester residues are specifically cleaved under very mild conditions (methanol, r.t., pH 6.0 or refluxing methanol), leaving other galloyl esters unaffected. This methanolysis reaction is undoubtedly facilitated by participation of the phenolic group adjacent to the depside link (Fig. 7.2.5). The reaction provided Haworth and his collaborators (48) with a ready means of degradation and structural analysis for this type of molecule (13, 37, 59, 116). Combined with HPLC and paper chromatography, it undoubtedly provides the most ready means of identification of this type of metabolite in plant extracts.

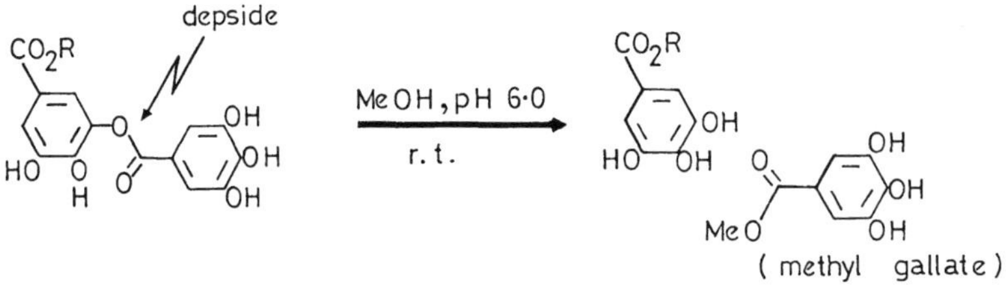

Fig. 7.2.5. Methanolysis of the depside link

The structure of the complex polyphenol derived from the twig galls of *Rhus semialata* (Chinese gallotannin – tannic acid) as a core of **15** to which up to five other galloyl ester groups are linked in depside fashion was first proposed by Fischer and Freudenberg (22) and later refined by Haworth and Haslam (1, 37). Molecules with similar polyester structures based on **15** have now been obtained from several additional plants (Table 7.2.4). The presence of depsidically linked galloyl ester groups is indicated by methanolysis, ^{1}H NMR, and ^{13}C NMR. Methanolysis splits the depsidically linked galloyl ester groups from the core of **15** and number average value of such groups may be determined by quantitative HPLC to give the molar ratio of **15** to methyl gallate (Tab. 7.2.4). The position of linkage of the depsidically bound galloyl ester groups on the β-penta-*O*-galloyl-D-glucose core of these metabolites has proved difficult to define and this has, over the years, led to a number of different formulations.

Table 7.2.4. Polygalloyl esters based upon β-penta-*O*-galloyl-D-glucose (**15**)

Family	Genus, species	Part	Molar ratio methyl gallate: β-penta-*O*-galloyl-D-glucose[1]
Anacardiaceae	*Rhus semialata*	Galls	1.8
	R. coriaria	Leaf	2.0
	R. typhina	Leaf	1.7
	Cotinus coggyria	Leaf	1.6
Aceraceae	*Acer platanoides*	Leaf	1.2
	A. campestre	Leaf	1.1
	A. rubrum	Leaf	1.2
Ericaceae	*Arctostaphylos uva-ursi*	Leaf	1.1
Geraniaceae	*Pelargonium* sp.	Leaf	1.3
Paeoniceae	*Paeonia officinalis*	Leaf	1.3
Hamamelidaceae	*Hamamelis mollis*	Leaf	1.5
	Parrottia persica	Leaf	1.5

[1] Determined after methanolysis by HPLC (45).

A novel method of analysis based on the use of ^{13}C NMR was recently discovered independently by Nishioka (80) and by Haslam and his colleagues (45). This work shows that the typical heterogeneity of the gallotannins is based on the biosynthesis in the plant of a series of related polygalloyl-D-glucose derivatives. These range from β-penta-*O*-galloyl-D-glucose (**15**) itself to possibly deca- and undecagalloyl-D-glucose compounds. The proportion of each type determines the final composition of the gallotannin. The hexa- to undecagalloyl-D-glucose metabolites have structures derived by a 'partly random' addition of further gallic acid molecules, by esterification as *m*-depsides, to **15**. Using ^{13}C NMR Haslam and his colleagues (45) determined the position of the depside linkages as being predominantly to the galloyl ester groups attached to C-3, C-4, and C-6 (**17a**). Nishioka and co-workers (80) favored, on similar evidence, the groups at C-2, C-3, and C-4 (*17b*). This work therefore supports the view, originally expressed by Emil

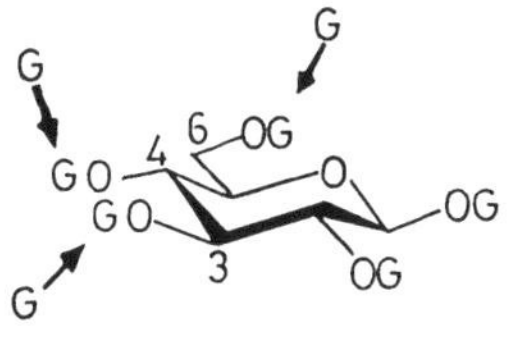

(17a)

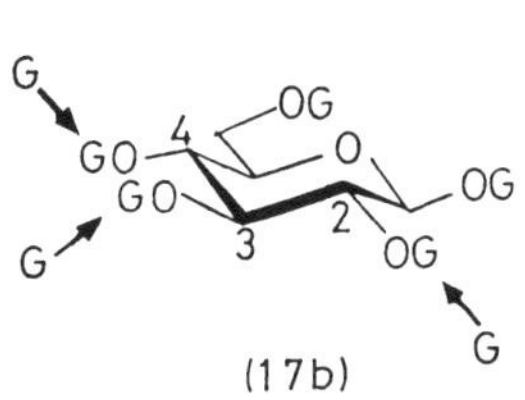

(17b)

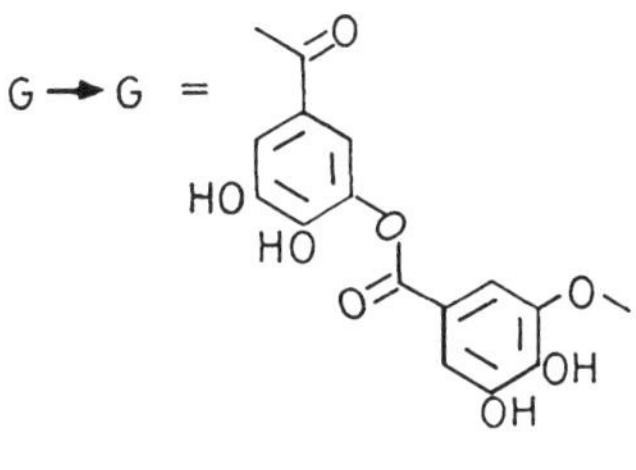

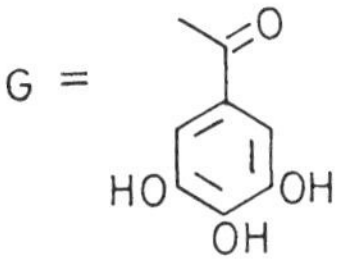

G → G =

G =

Fischer (22), that the gallotannins are not only mixtures of isomers but also of substances of differing empirical formulae.

To date, the ester **17a,b** is the most widely encountered in the plant kingdom and it is worth noting that in plants that metabolize this phenol it overwhelmingly dominates the phenolic extract. The ester **17a,b** has R_F values very similar to those of *β*-penta-*O*-galloyl-D-glucose **(15)** and is distinguished by preparative chromatography on Sephadex LH-20. Table 7.2.4 shows plants in which this complex phenol has been identified.

Although the principal form in which gallic acid is found in depside linkage is the ester **17a,b,** other esters of this type have been described. The galls of various oak species (*Quercus infectoria, Q. lusitanica*) thus yield an ester **(18)** in which the depsidically linked galloyl groups are attached to a *β*-1,3,4,6-tetra-*O*-galloyl-D-glucose core and the fruit pods of *Caesalpinia spinosa* yield the ester **(19)**, which is based on a 3,4,5-tri-*O*-galloyl quinic acid core. More recently a mixture of the two tri-*O*-galloyl-1,5-anhydro-D-glucitols **(20, 21)** has been obtained from the leaves of *Acer saccharinum* (45).

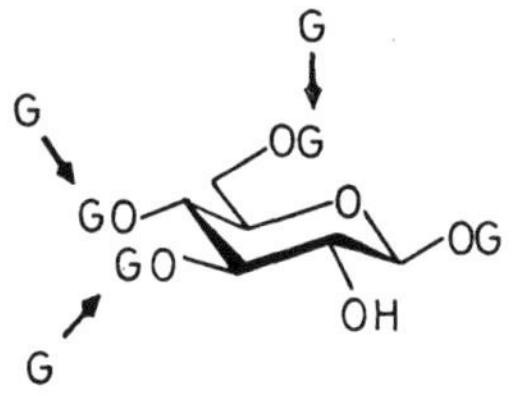

(18) 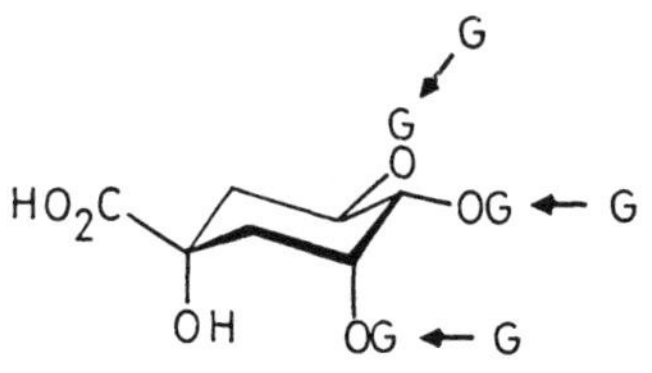 (19)

(G and G→G as in (17)a,b)

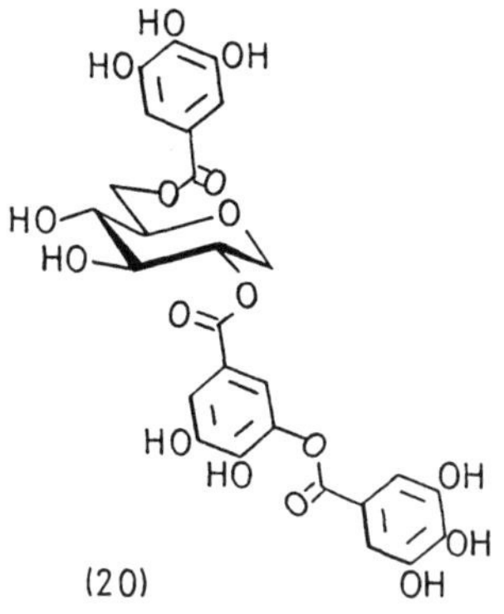

(20) 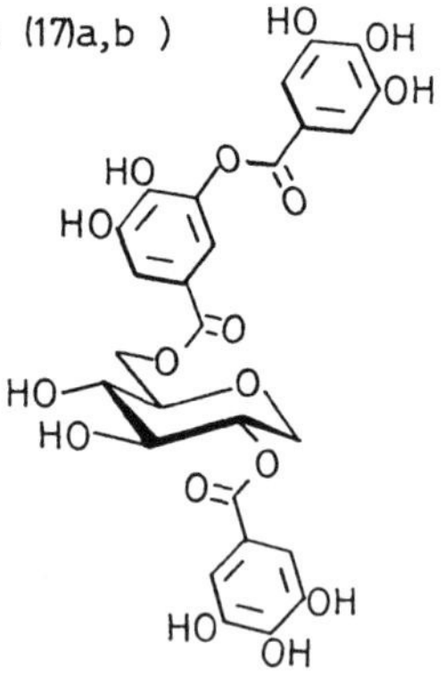 (21)

There is, on the present recorded evidence, a very close association of this form of metabolism of gallic acid with the Rhoideae tribe in the Anacardiaceae. This ability to metabolize depsides of gallic acid may be used quite successfully as a taxonomic marker and hence as a guide to interrelationships in particular plant families. It has been used successfully in this way in the Aceraceae, Ericaceae, Geraniaceae and Hamamelidaceae (36, 47).

7.2.4.3 Metabolites Formed by Oxidative Coupling of Galloyl Esters – Groups 2B and 2C, Ellagitannins

In his taxonomic surveys of the 1950s and 1960s, Bate-Smith (2) drew particular attention to the occurrence of ellagic acid (**5**) in the acid hydrolysates of dicotyledonous plant extracts. From this he inferred that its presumed precursor – hexahydroxydiphenic acid, bound in ester form (3) – was widely distributed in the plant kingdom. Subsequent work has confirmed the correctness of this assumption. Notable work in this particular area of plant chemistry has been carried out by Schmidt (121 – 124) and Mayer (63, 132) in their classical work on the ellagitannins. The following discussion will emphasize the fundamental knowledge that has emerged from these crucial observations but it will also seek to place the distinguished contributions from the Heidelberg school and more recent contributions from Japan and elsewhere in the context of the emerging, although as yet incomplete, biogenetic framework.

In this work, numerous complex polyphenols have been isolated, whose structures are composed of a range of phenolic acids related in structure to gallic acid (**1**). The formal structural and the implied biogenetic relationships (oxidative coupling: C-C and C-O bond formation, hydrolytic ring fission, oxidation) of these to gallic acid are shown in Table 7.2.5 and in Figs. 7.2.6 and 7.2.7. Some of the principal structures of substances that have been isolated and that contain structural elements based upon these various additional phenolic acids are shown in later tables.

Table 7.2.5. Formal relationships between gallic acid and some naturally occurring phenolcarboxylic acids

Number of gallic acid molecules	Formal chemical relationship	Phenolcarboxylic acid	Hydrolysis products of esters	Reference(s)
2	$-2H$	Hexahydroxydiphenic acid	Ellagic acid	122, 123, 132
2	$-2H$, $+2H_2O$	'Hydrated' chebulic acid	Chebulic acid (split acid)	49, 50, 124
2	$-4H$	Dehydrohexahydroxy-diphenic acid	Chlorellagic acid (HCl)	133, 134
	$+H_2O$, $-4H$	Brevifolin carboxylic acid	Brevifolin carboxylic acid	123, 125
	$-CO_2$		Brevifolin	
2	$-2H$	Dehydrodigallic acid	Dehydrogallic acid	62, 110, 111
3	$-4H$	Valoneic acid	Valoneic acid dilactone	130
3	$-4H$	Sanguisorbic acid	Sanguisorbyl lactone	43, 81
3	$-4H$	Nonahydroxytriphenic acid	Flavogallol	64, 70, 73
3	$-4H$, $+2H_2O$	Trilloic acid	Trilloic acid trilactone	72
4	$-6H$, $-2H_2O$	Gallagic acid	Dehydrodiellagic acid	68

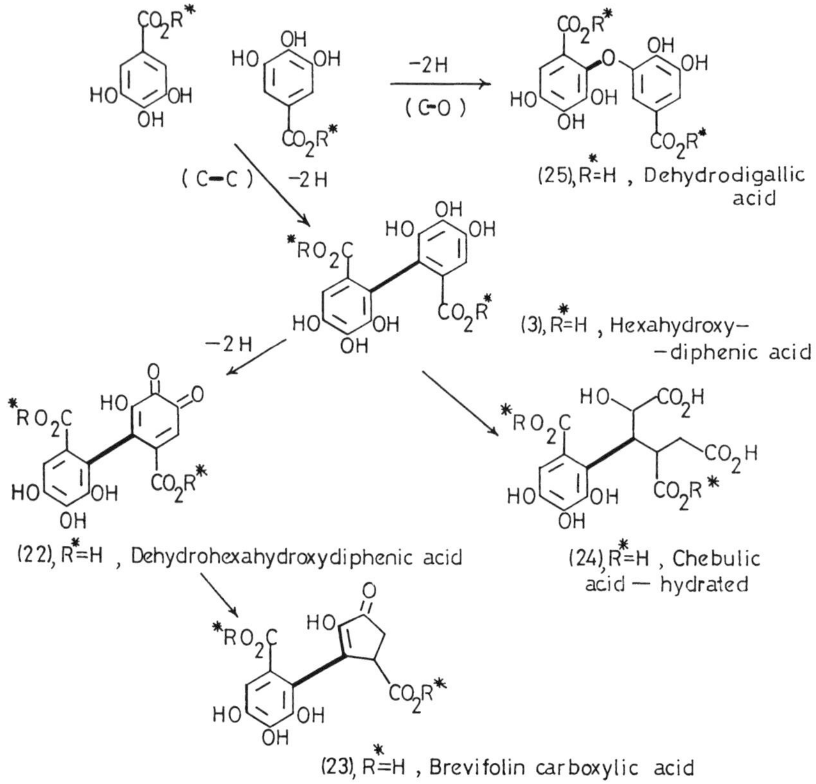

Fig. 7.2.6. Relationship of several polyphenols (**3**, **22**, **23**, **24**, **25**) to gallic acid (**1**)

7.2.4.3.1 Hexahydroxydiphenoyl Esters

Hexahydroxydiphenic acid in a bound form (**3**) is readily detected in a plant extract by an old but very distinctive color test – the Procter-Paessler reaction. Esters of type **3** react with nitrous acid to give a carmine or rose red, changing to brown-green, then purple, and finally indigo blue. The chemical basis of this reaction has not been discussed but it has been employed by Bate-Smith (6, 7) for the quantitative determination of **3** in plant extracts, and it forms the basis of a very useful spray reagent for the detection of **3** by paper chromatography (6, 7, 43).

Although there is no firm experimental proof, it is generally assumed, following the hypothesis of Schmidt and Mayer (130), that esters of type **3** are formed by oxidative coupling of suitably placed galloyl ester groups in a suitable polygalloyl ester precursor. One specific chirality is imposed on the biphenyl system of the hexahydroxydiphenoyl ester group as it is formed and this chirality is determined by the need of the hexahydroxydiphenoyl group to bridge particular positions in the alcohol portion of the ester. In all cases so far discussed by Schmidt and Mayer and by other workers the hexahydroxydiphenoyl group bridges positions within the D-glucose molecule. It has been demonstrated that the hexahydroxydiphenic acid that bridges the 3,6 positions of D-glucopyranose is the

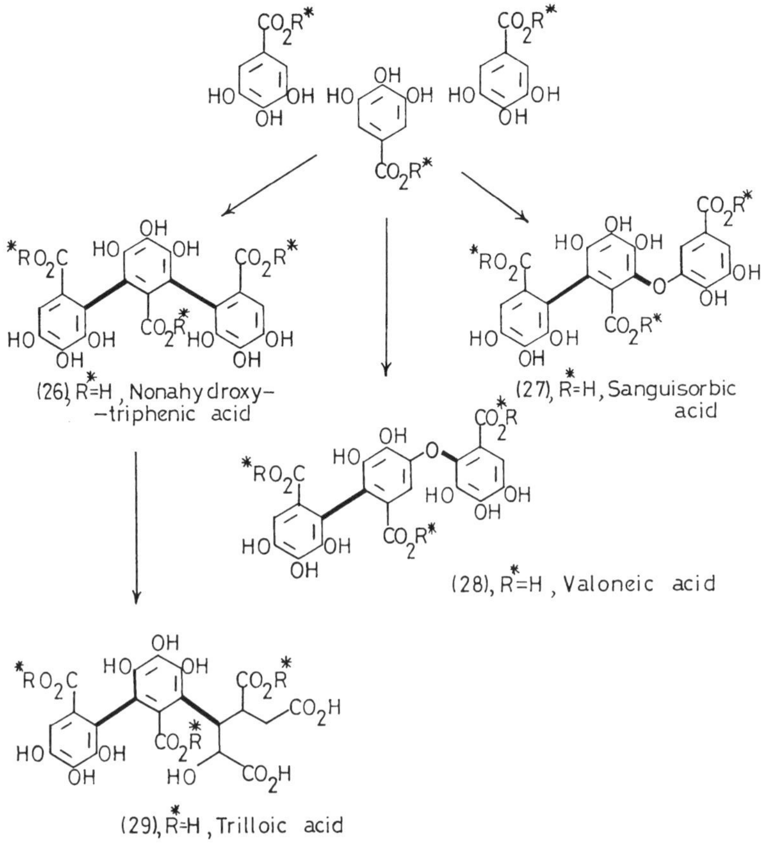

(26), R*=H, Nonahydroxy-triphenic acid

(27), R*=H, Sanguisorbic acid

(28), R*=H, Valoneic acid

(29), R*=H, Trilloic acid

Fig. 7.2.7. Relationship of several polyphenols (**26, 27, 28, 29**) to gallic acid (**1**)

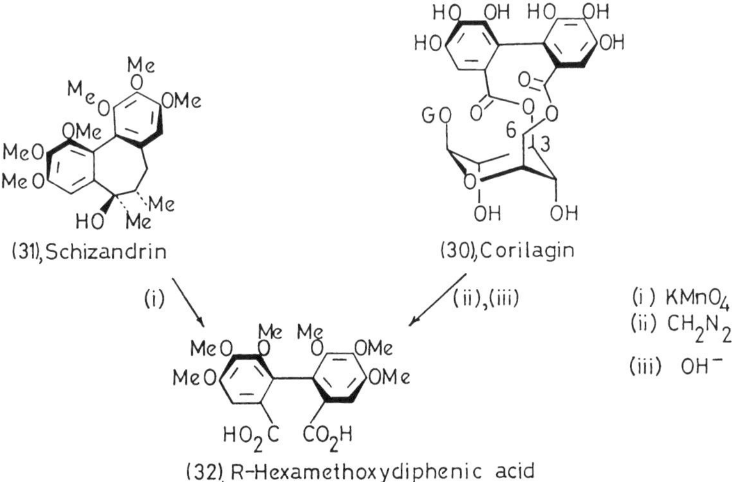

Fig. 7.2.8. (+)-Hexamethoxydiphenic acid derived from corilagin and schizandrin

dextrorotary form, while that which bridges the 2,3 or 4,6 positions of D-glucopyranose is the laevorotatory isomer (127, 128).

Okuda (89) has recently correlated (+)-hexamethoxydiphenic acid (**32**) from corilagin (**30**) with the same acid derived from the dibenzocyclooctane lignan, schizandrin (**31**), whose chirality had been shown to be R by comparison with the X-ray-derived molecular structure of the lignan, gomisin D (54) (Fig. 7.2.8).

It is interesting to note that the same conclusions regarding the chirality of hexahydroxydiphenoyl groups bound to D-glucopyranose follow from theoretical considerations. Two assumptions are necessary:

1) that the ester carbonyl groups of the hexahydroxydiphenoyl residue have, like other ester groups (19), a preference for the eclipsed conformation of the ester carbonyl oxygen and the adjacent hydrogen atom on the glucose ring (**33**) and
2) that in the hexahydroxydiphenoyl group there is a preference for the anti or opposed arrangement of the two carbonyl groups with the C-O dipoles aligned antiparallel (**34**).

If these two features dominate the oxidative coupling reactions of galloyl esters of D-glucose, then the geometrical restrictions imposed by the sugar ring lead directly to the conclusion that the chirality of the derived hexahydroxydiphenoyl group is determined by the sugar. Molecular models show that the 3,6-bridges (e.g., corilagin (**30**)) have the *R*-configuration and 4,6- or 2,3-hexahydroxydiphenoyl linkages (e.g., pedunculagin) have the alternative *S* configuration. Although hexahydroxydiphenoyl esters linked 2,4 have not yet been isolated from natural sources, this type of argument suggests that they may have the *S* or *R* configuration. Reasoning along these lines, one infers therefore that the chirality of bridging hexahydroxydiphenoyl ester groups in ellagitannins is induced directly by the chirality of the D-glucose residue and is not the result of enzymic control of the coupling reaction.

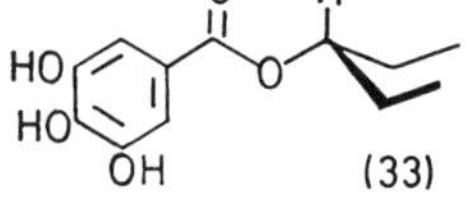

(33) 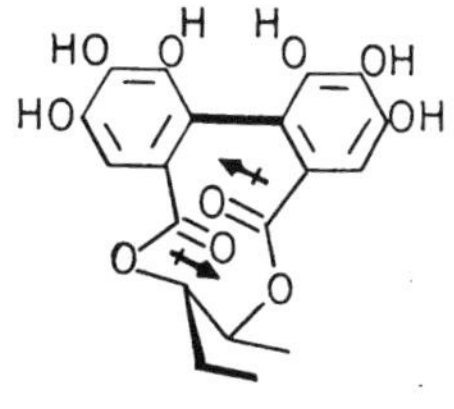 (34)

The presence of a hexahydroxydiphenoyl group in a polyphenolic metabolite is readily defined by spectroscopic means (41, 43, 103, 104). The chirality (and hence their probable position of attachment to the D-glucose core) may be determined by cd spectra (43, 103, 104).

The formation of a diphenoyl ester bridge in a galloyl D-glucose derivative gives the molecule much greater rigidity and this may result in constraints on the D-glucose residue to adopt an unfavorable shape or conformation. Schmidt and Jochims have demonstrated this feature in ^{1}H NMR studies. Thus, for example, corilagin (**30**) (135) may adopt a flexible boat conformation or the chair form (**30**), in which the bulky substituents are all axially disposed. The conformational restraints that result from the presence of hexahydroxydiphenoyl ester groups in a molecule usually result in changes in the ^{13}C chemical shift of the D-glucose carbon atom signals relative to the related galloyl-D-glucose derivative.

7.2.4.3.2 Dehydrohexahydroxydiphenoyl Esters

In chemical terms a perhaps unexpected but biosynthetically interesting variant of the hexahydroxydiphenoyl ester group first noted by Schmidt is its dehydro-derivative (**22**). It was identified by Schmidt as a structural component of brevilagin 1 and 2, polyphenols from the fruit pods of *Caesalpinia brevifolia* (133, 134). Subsequent work has shown the ester to be present in the structure of other plant polyphenols, notably geraniin (**51**) in plants of the Geraniaceae (Table 7.2.9).

Spectroscopically (by ^{1}H NMR and ^{13}C NMR), the presence of the dehydrohexahydroxydiphenoyl ester group (**22**) in a molecule leads to a complexity of signals that is both characteristic and distinctive but difficult to interpret satisfactorily in structural terms. The isomerization occurs in solution and leads to an equilibrium mixture of isomers and duplication of signals. The attainment of the position of equilibrium is accelerated in hydroxylic media. Although in the ^{1}H NMR spectra the most obvious changes occur in the protons of the ester group **22**, transformations in the protons of the associated D-glucose are also observed. In recent work Okuda and others (44, 89, 94, 100, 101) have put forward the suggestion that in solution the ester group **22** is present as an equilibrium mixture of two hemi-acetal structures (**22a** and **22b**). This to some extent rationalizes the very obvious equilibration of substrates containing **22**, which takes place in hydroxylic media. It also provides a rationale for the unexpected stability of the formal *ortho*-quinone grouping present in **22**.

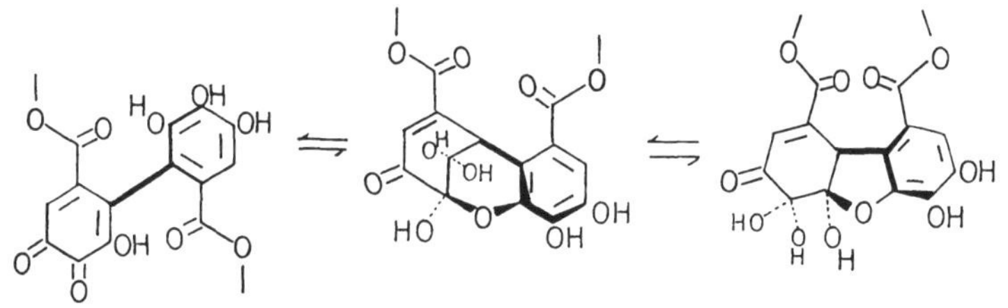

(22) (22a) (22b)

7.2.4.3.3 Group 2B Metabolites

Two-dimensional paper chromatographic analysis of plant phenolic extracts has proved to be an invaluable aid in the delineation of the various patterns of metabolism of proanthocyanidins and now of hexahydroxydiphenoyl esters and their derivatives in the plant kingdom. A very widely distributed metabolic fingerprint is that in which β-penta-*O*-galloyl-D-glucose (**15**) (it is assumed) is further transformed by oxidative coupling of pairs of adjacent galloyl ester groups (2,3 and 4,6) on the β-D-glucopyranose core. Some details of this pattern of metabolism were hinted at in earlier work by Hillis and Seikel (52), Jurd (58), and Wilkins and Bohm (143). Schmidt et al. (139) earlier isolated and identified pedunculagin (**39**) (2,3:4,6-bishexahydroxydiphenoyl-D-glucose) from oak galls. This ester (**39**)

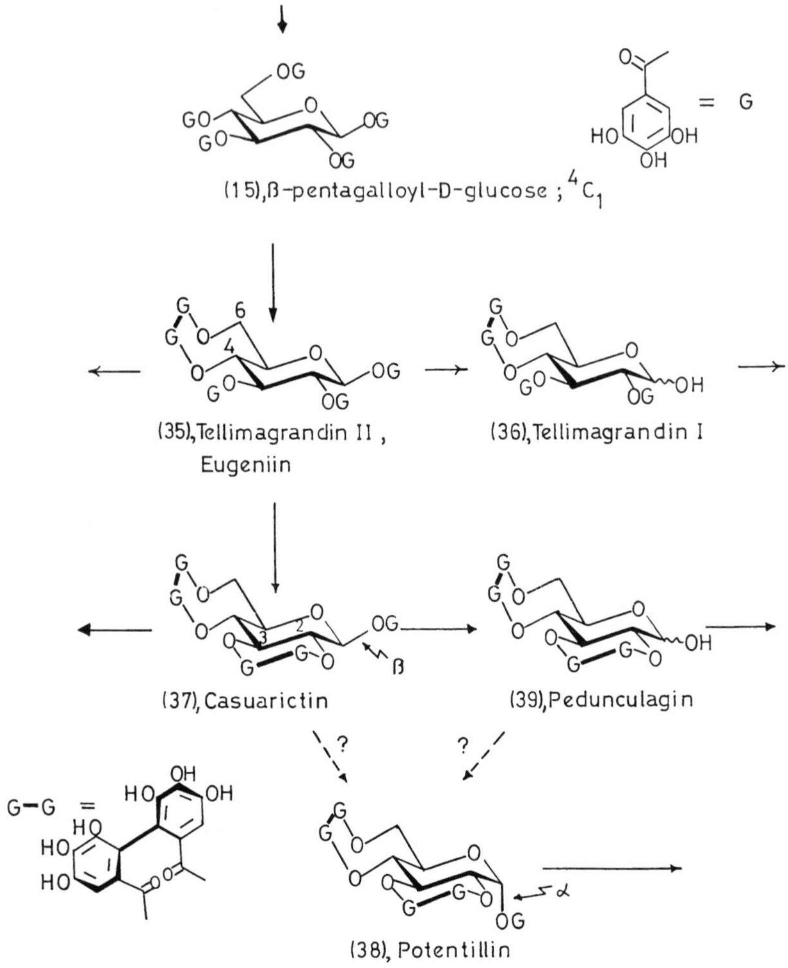

(15),β-pentagalloyl-D-glucose ; $^4C_1$

(35),Tellimagrandin II, Eugeniin

(36),Tellimagrandin I

(37),Casuarictin

(39),Pedunculagin

(38), Potentillin

Fig. 7.2.9. Formation of Group 2B products from galloyl-galloyl coupling of β-penta-galloyl-D-glucose

is a key metabolite in the paper chromatographic phenolic fingerprint of this group (Fig. 7.2.9).

Plants whose phenolic metabolism places them within this category furnish, along with β-penta-*O*-galloyl-D-glucose **(15)**, one or more of the hexahydroxydiphenoyl esters – tellimagrandin II **(35)**, tellimagrandin I **(36)**, casuarictin **(37)**, potentillin **(38)**, and pedunculagin **(39)** (Fig. 7.2.9) – as key phenolic metabolites (43, 46). The ratio of these metabolites varies markedly from plant to plant, even within the same family. Parallel with this variation is a concomitant change in the nature of higher molecular-weight polyphenolic metabolites formed by intramolecular C-O oxidative coupling of the metabolites **35** through **39**.

It is noticable that the major biosynthetic thrust in these plants is towards the formation of these 'dimers' and 'trimers' (Table 7.2.4). Thus, for example, in the plant family Rosaceae, the leaves of *Rosa canina* and *Filipendula ulmaria* metabolize substantial quantities of the phenolic esters **35** and **36**; however, the principal phenolic metabolites of these tissues are the 'dimers' rugosin D (**40**), and rugosin E (**41**) (43, 46) formed presumably by oxidative C–O coupling of the precursors tellimagrandins I and II (**35** and **36**) (143). It is noteworthy that the two glucose residues are linked by the formation of a valoneic acid residue (**28**) (Fig. 7.2.7). Correspondingly in the same plant family *Rubus idaeus, fructicosus, Geum rivale,* and *Potentilla* sp. metabolize substantial quantities of the 'dimer'(sanguin H-6, **42**) (45, 47) alongside its presumed precursors, casuarictin (**37**) and potentillin (**38**). In this polyphenol the two glucose residues are linked by a sanguisorbic acid residue (**27**) (Fig. 7.2.7) – that is, isomeric with valoneic acid and formed similarly, it is presumed, by C-O oxidative coupling (41, 81). Finally among polyphenols of the 'dimeric' form may be noted gemin A (**43**) from *Geum japonicum* (104). In this structure, units of tellimagrandin II (**35**) and potentillin (**38**) are linked by a third type of structural fragment, a dehydrodigallic acid unit (**25**) (62).

Further examples of this type of polyphenolic structure in which molecular complexity and size are increased by oxidative coupling of two or more glucose derivatives (**35** – **39**) by formation of new C-O bonds are shown in Table 7.2.6.

Table 7.2.6. Hexahydroxydiphenoyl ester derivatives, Group 2B: The structure of some 'dimeric' and 'trimeric' C-O oxidatively coupled metabolites

Compound	Source	Structure ('Monomer' units)	Linking group	Reference(s)
Agrimoniin	*Agrimonia pilosa*	Two units of potentillin (**38**)	Dehydrodigallic acid (**25**)	107
Gemin A (**43**)	*Geum japonicum*	One unit of potentillin (**38**) one unit of tellimagrandin II (**35**)	Dehydrodigallic acid (**25**)	109
Gemin B	*G. japonicum*	One unit of tellimagrandin II (**35**) and β-1-*o*-galloyl-2,3-hexahydroxydiphenoyl-D-glucose	Dehydrodigallic acid (**25**)	108
Gemin C	*G. japonicum*	One unit of potentillin (**38**) and β-1-*o*-galloyl-2,3-digalloyl-D-glucose	Dehydrodigallic acid (**25**)	108
Rugosin D (T_{2B}) (**40**)	*Rosa* sp. *Filipendula ulmaria*	Two units of tellimagrandin II (**35**)	Valoneic acid (**28**)	93
Rugosin E	*Rosa* sp.	One unit of tellimagrandin II (**35**)	Valoneic acid (**28**)	93
(T_{2A}) (**41**)	*Filipendula ulmaria*	One unit of tellimagrandin I (**36**)		
Rugosin F	*Rosa* sp.	One unit of tellimagrandin II (**35**) one unit of casuarictin (**37**)	Valoneic acid (**28**)	93
Rugosin G	*Rosa* sp.	Three units of tellimagrandin II (**35**)	Valoneic acid (**28**)	93
Sanguin H-6	*Rubus* sp.	One unit of potentillin (**38**)	Sanguisorbic acid (**27**)	82
(T_1) (**42**)	*Sanguisorbia officinalis*	one unit of casuarictin (**37**)		43

Table 7.2.7. Hexahydroxydiphenoyl ester derivatives, Group 2B: The structure of some principal metabolites and intermediates

Compound	Source	Structure	Reference(s)
Tellimagrandin II (eugeniin) (**35**)	*Tellima grandiflora* *Eugenia caryophyllata* *Rosa* sp. *Quercus* sp. *Fuchsia* sp. *Cornus* sp.	*β*-1-*O*-galloyl; 2,3-bisgalloyl-hexahydroxydiphenic acid linked 4,6; D-glucopyranose conformation-1C	43, 84, 143
Tellimagrandin I (**36**)	*Tellima grandiflora* *Rosa* sp. *Quercus* sp. *Fuchsia* sp. *Camellia japonica* *Geum japonicum* *Casuarina stricta* *Cornus* sp.	2,3-bisgalloyl-hexahydroxydiphenic acid linked 2,3 and 4,6; D-glucopyranose conformation-1C (predominantly)	43, 102, 143
Casuarictin (**37**)	*Casuarina stricta* *Quercus* sp. *Rubus* sp.	*β*-1-*O*-galloyl; (*S*)-hexahydroxy diphenic acid linked 2,3 and 4,6; D-glucopyranose conformation-1C	43, 107
Potentillin (**38**)	*Potentilla* sp. *Quercus* sp. *Rubus* sp.	*β*-1-*O*-galloyl; (*S*)-hexahydroxydiphenic acid linked 2,3 and 4,6; D-glucopyranose conformation-1C	43, 107
Pedunculagin (**39**)	*Quercus* sp. *Rubus* sp. *Casuarina stricta* *Stachyurus praecox* *Camellia japonica* *Juglans* sp. *Potentilla* sp.	(*S*)-hexahydroxydiphenic acid linked 2,3 and 4,6; D-glucopyranose conformation C-1	43, 52, 139
Alnusiin	*Alnus sieboldiana*	(*S*)-hexahydroxydiphenic acid linked 2,3; valoneic acid linked 4,6	99
Isoterchebin (trapain)	*Cornus* sp. *Trapa japonica*	*β*-1-*O*-galloyl; 2,3-bisgalloyl; dehydrohexahydroxydiphenic acid linked 2,3; D-glucopyranose conformation-1C	94
Casuariin (**44**)	*Casuarina stricta* *Stachyurus praecox* *Psidium guajava*	(*S*)-hexahydroxydiphenic acid linked 2,3 and 4,6; *C*-glycosidic link to ester group at position 2; open chain form of D-glucose; (*S*)-configuration at C-1	102, 106
Casuarinin (**45**)	*Casuarina stricta* *Stachyurus praecox* *Psidium guajava*	Casuariin (**44**) with galloyl group at C-5	90
Stachyurin (**48**)	*Casuarina stricta* *Stachyurus praecox* *Psidium guajava*	Isomeric casuarinin (**45**); *R*-configuration at C-1	102, 106
Castalin (**47**)	*Castanea sativa*	Based on nonahydroxytriphenic acid (**26**) linked 2,3,5 with C-glycosidic link to C-1 of open chain form of D-glucose; (*S*)-configuration at C-1	67, 73
Vescalin (**50**)	*Quercus sessiflora*	Isomeric with castalin (**46**); (*R*)-configuration at C-1	70
Castalagin (**46**)	*Castanea sativa*	Castalin (**47**) with (*S*)-hexahydroxydiphenic acid linked 4,6	64, 73
Vescalagin (**49**)	*Quercus sessiflora*	Vescalin (**50**) with (*S*)-hexahydroxydiphenic acid linked 4,6	75

Table 7.2.7
(continued)

Compound	Source	Structure	Reference(s)
Valolaginic acid	*Q. valonea*	Based on trilloic acid (**29**) linked 2,3,5 with C-glysidic link to C-1 of open chain form of D-glucose; (*S*)-hexahydroxydiphenic acid linked 4,6; (*S*)-configuration at C-1	69
Isovalolaginic acid	*Q. valonea*	Isomeric with valolaginic acid at anomeric centre C-1; (*R*)-configuration	66
Castavalonenic acid	*Q. valonea*	Castalin (**47**) with valoneic acid bridging 4,6 positions	65
Rugosin C	*Rosa* sp.	β-1-*O*-galloyl; (*S*)-hexahydroxydiphenic acid bridging 2,3; valoneic acid-4,6; conformation of D-glucopyranose-1C	97
Rugosin A	*R. rugosa*	β-1-*O*-galloyl; 2,3-digalloyl; 4,6-valoneic acid; conformation of D-glucopyranose-1C	97
Rugosin B	*R. rugosa*	2,3-digalloyl; valoneic acid-4,6; attached to D-glucopyranose	93
Praecoxin A	*Stachyurus praecox*	(*S*)-hexahydroxydiphenic acid linked 2,3; valoneic acid-4,6; attached to D-glucopyranose	96

With the isolation in recent years of more and more polyphenolic structures of this class (Tables 7.2.6 and 7.2.7), it is apparent that this particular pathway of progressive oxidative metabolism of β-penta-*O*-galloyl-D-glucose (**15**) – C-C coupling to form hexahydroxydiphenoyl ester groups bridging the 4,6 and 2,3 positions, followed by C-O coupling to form 'dimers' and very probably higher molecular weight species (41, 89) – is a very common one among certain plant families. Although it is not yet clear what factors determine the detailed phenolic profile or fingerprint in such plants, it is nonetheless apparent that these accumulated biosynthetic end products may be utilized for taxonomic purposes. It is noteworthy that whereas all plants in this category (Group 2B) are presumed to possess enzymes to facilitate C-C coupling of the 4 and 6 galloyl ester groups in **15** to give **35** (Fig. 7.2.9), many do not possess the means to bring about the subsequent coupling of the 2 and 3 galloyl ester groups. Members of this class thus form dimers from species **35** and **36**, whereas those that posses the additional facility of oxidative coupling of the ester groups at 2 and 3 may form dimers from all the potential precursors (**35** – **39**).

One further, and as yet unexplained, item of note is the presence of the bis-hexahydroxydiphenoyl ester, potentillin (**38**), with the α-configuration at C-1 of the glucose residue. This metabolite is almost invariably found alongside the β-anomer, casuarictin (**37**), and is presumably of some crucial biogenetic significance. It is the only example, to date, of an α-glucoside detected among the 'monomeric' galloyl and hexahydroxydiphenoyl ester metabolites.

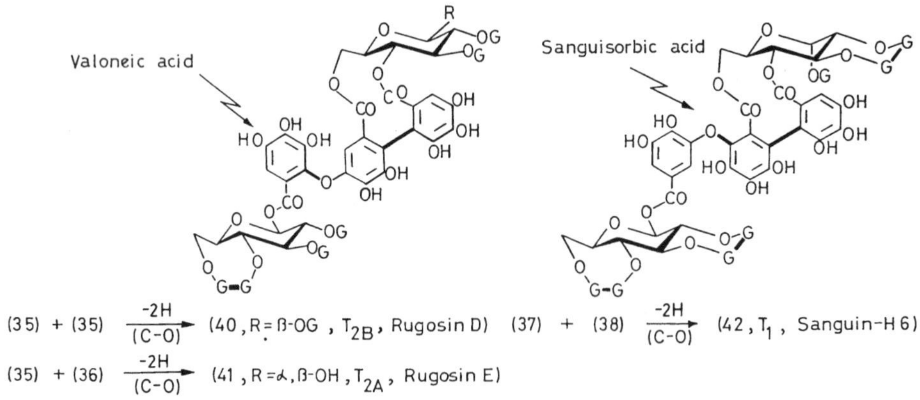

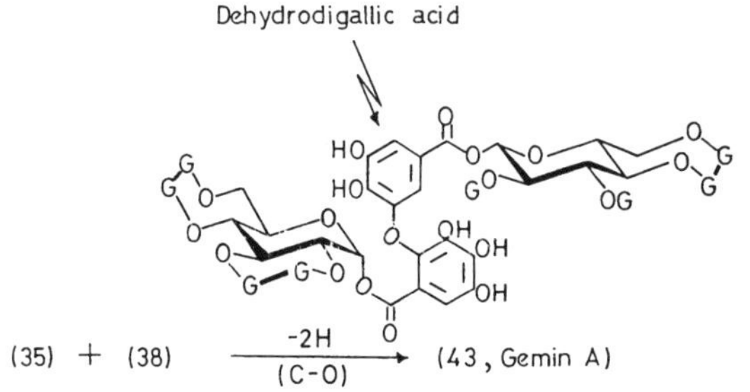

(35) + (35) $\xrightarrow[(C-O)]{-2H}$ (40, R = β-OG, T_{2B}, Rugosin D)

(35) + (36) $\xrightarrow[(C-O)]{-2H}$ (41, R = α,β-OH, T_{2A}, Rugosin E)

(37) + (38) $\xrightarrow[(C-O)]{-2H}$ (42, T_1, Sanguin-H 6)

(35) + (38) $\xrightarrow[(C-O)]{-2H}$ (43, Gemin A)

Fig. 7.2.10. Products formed by C-O oxidative coupling

A very interesting series of variations on the patterns of metabolism discussed above occurs in members of the plant families Fagaceae, Casuarinaceae, Stachyuraceae, Juglandaceae, Myrtaceae (64–67, 69, 70, 73, 75, 102, 105), and others. The ability to bring about both 4,6- and 2,3-oxidative coupling of galloyl ester groups in **15** is retained by plants in these families. However the principal phenolic metabolites are *not* those of additional C-O oxidative coupling as outlined (Fig. 7.2.10) but are 'monomeric' products, formed presumably from pedunculagin **(39)** or casuarictin **(37)**, by opening of the pyranose ring of the sugar and by formation of a C-glycosidic link to the hexahydroxydiphenoyl group bridging the 2,3-positions and galloylation at C-5 (Fig. 7.2.11) (Table 7.2.7). The most highly condensed polyphenols of this type are vescalagin **(49)** and castalagin **(46)**, in which an additional intramolecular carbon-carbon bond to the galloyl group at

Table 7.2.8. Hexahydroxydiphenoyl ester metabolites: Group 2B, occurrence in plant families

Order	Family	Order	Family
Fagales	Betulaceae	Myrtales	Combretaceae
	Fagaceae	Hamamelidae	Hamamelidaceae
Rosales	Rosaceae	Casuarinales	Casuarinaceae
	Saxifragaceae	Cornales	Cornaceae
Theales	Theaceae	Proteales	Elaeganaceae
Myrtales	Myrtaceae	Juglandales	Juglandaceae
	Onagraceae		

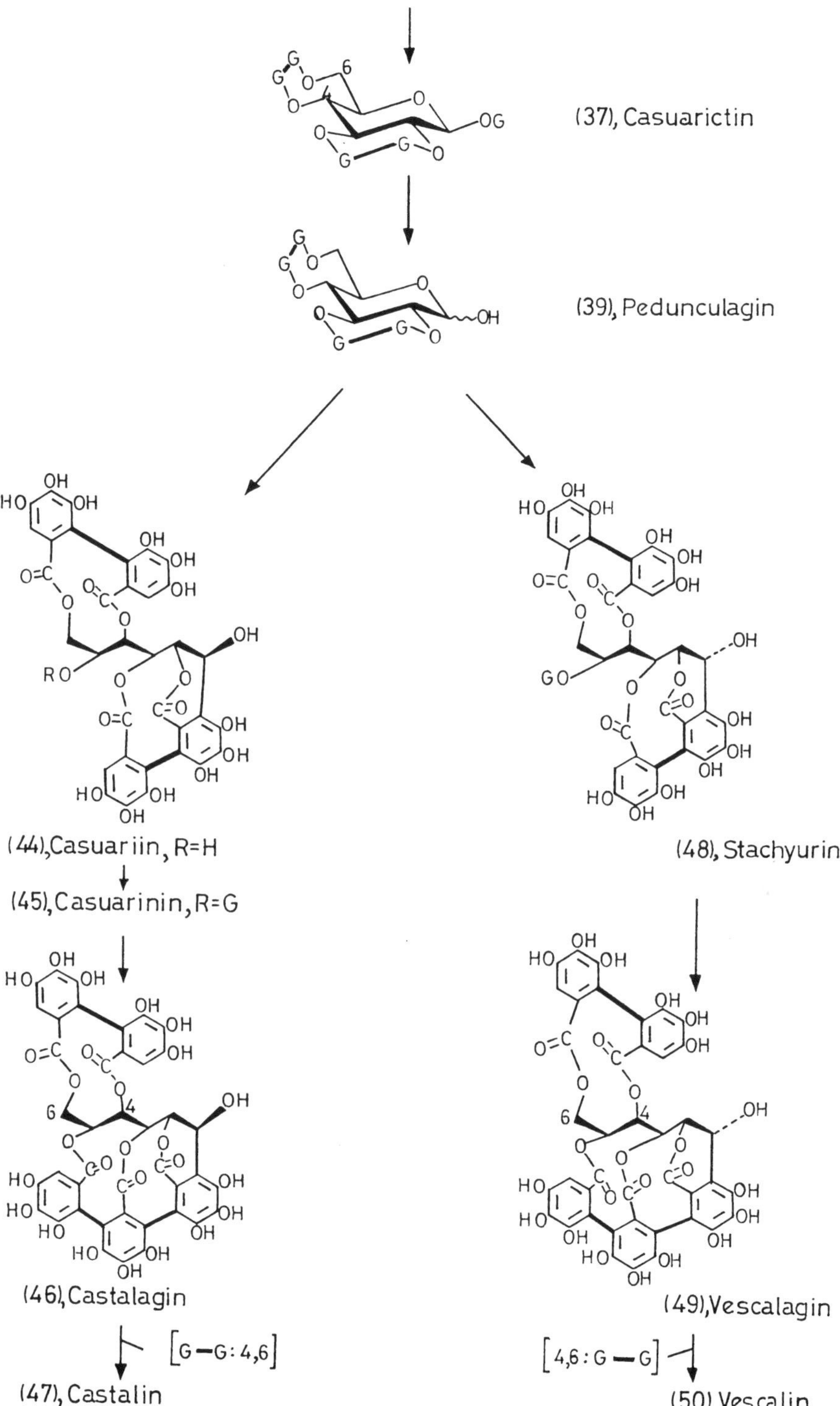

Fig. 7.2.11. C-glycosidic links formed by opening of the pyranose ring of the sugar

C-5 has been formed. The creation in these molecules of three C-C biphenyl linkages and a C-glycosidic bond results in the formation of relatively inflexible propeller-shaped species that contrast quite markedly with their presumed biosynthetic precursor, β-penta-*O*-galloyl-D-glucose (**15**), which has a perfectly flexible disc-like shape. This indeed appears to be a common feature of all the polyphenolic metabolites – grouped under the heading 2B – namely, the progressive formation of more highly condensed and often less conformationally mobile polyphenolic structures.

Plants whose phenolic metabolism places them within the broad category 2B, as outlined in the preceeding discussion, are shown in Table 7.2.8, see p. 424.

7.2.4.3.4 Group 2C Metabolites

According to present evidence (46) a rather smaller group of plants adopts a further metabolic variation in which oxidative coupling of adjacent galloyl ester groups occurs 1 – 3 to form both *R*- and *S*-hexahydroxydiphenoyl esters in a D-glucopyranose precursor, which itself adopts the less favorable $^1C_4$ (C-1) or an intermediate skew-boat conformation. An additionally significant feature of this last form of metabolism is that one or more of the hexahydroxydiphenoyl ester groups, in key metabolites, may be further dehydrogenated to give derivatives of the dehydrohexahydroxydiphenoyl ester group (**22**). Phenolic metabolites of this class (Group 2C) have been observed in members of the plant families Cercidiphyllaceae, Ericaceae, Onagraceae, Combretaceae, Nyssaceae, Aceraceae, Punicaceae, Simaroubaceae, and Geraniaceae (44, 46).

For its routine metabolic processes, Nature appears to prefer the conformationally most stable sugars; among the D-hexoses, the three most stable ones – glucose, mannose, and galactose – are very widely distributed. Those with higher free energies occur rarely, if ever. It is a point of some curiosity, therefore, that in this particular form of oxidative metabolism of galloyl esters of D-glucose transformations apparently occur with the galloyl-D-glucopyranose precursor in an energetically unfavorable chair ($^1C_4$, C-1) or related skew-boat conformation. The difference in free energy between the $^1C_4$ (C-1) and $^4C_1$ (1-C) forms of D-glucopyranose is calculated as + 5.95 kcal mol^{-1} (24.8 kJ mol^{-1}) and this difference may in part explain why oxidative coupling of galloyl ester groups via the $^4C_1$ (C-1) form (Class B) of the D-glucopyranose precursor is much more widely encountered in plants than that via the alternative $^1C_4$ (1-C) or skew-boat forms (Class C). As yet only one plant, *Fuchsia* sp. (Onagraceae), has been detected in which both classes of metabolite co-exist (44).

It was shown earlier that the chirality of a hexahydroxydiphenoyl ester group attached to D-glucopyranose ($^4C_1$, 1-C) could be predicted assuming that it was biosynthetically derived by oxidative coupling of two vicinal galloyl ester groups. Using a similar approach it may be predicted that 3,6-coupling in a galloyl D-glucopyranose ($^1C_4$, C-1) will preferentially form the *R* configuration and 1,6-coupling will form the corresponding *S* configuration in the derived hexahydroxydiphenoyl ester group. This prediction is confirmed for the 3,6-coupled esters by cd measurements, which show the chirality of the hexahydroxydiphenoyl ester group in corilagin (**30**) to be *R*. In contrast both *R* and *S* configurations ap-

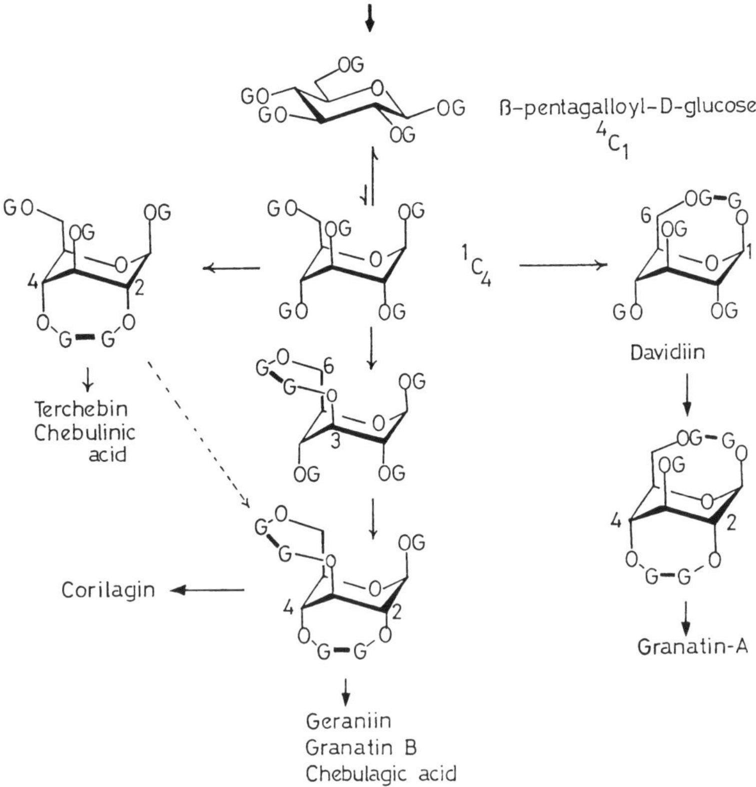

Fig. 7.2.12.
Formation of Group 2C products from galloyl-galloyl coupling of β-pentagalloyl-D-glucose

pear equally accessible in 2,4-oxidative coupling. However in both instances the two aroyl rings adopt an almost orthogonal relationship that leads (vide infra) to facile hydrolysis.

In contrast to the Group 2B polyphenolic metabolites, it has not yet been possible to discern a characteristic 'fingerprint' of Group 2C metabolites. Table 7.2.9 lists polyphenols isolated from plants that are deemed to fall within the 2C category; Fig. 7.2.12 outlines suggested pathways of metabolism from β-penta-*O*-galloyl-D-glucose (**15**) that lead to their formation. A key metabolite in a number of plants is the yellow crystalline compound, geraniin (**51**), first isolated by Okuda (95, 100, 101) from *Geranium* and *Euphorbia* species and from *Acer* and *Cercidiphyllum* species by Haslam and his collaborators (**44**). The dehydrohexahydroxydiphenoyl ester group (**22**), present in the structure of geraniin, is often another characteristic, but not invariable, structural feature in metabolites of this class.

Schmidt first isolated and described the unusual dehydrohexahydroxydiphenoyl ester group in two compounds, brevilagin 1 and 2, from the dried fruit pods of *Caesalpinia brevifolia* (Algarobilla, Leguminosae) (137, 138). A distinctive characteristic of the dehydrohexahydroxydiphenoyl ester group in geraniin and the brevilagins is its isomerization in hydroxylic media. This produces in the ^{1}H and

^{13}C NMR spectra a complex pattern of signals due to the various isomeric forms present in solution of **22**, **22a**, and **22b**. Hydroxy-*o*-benzoquinones are normally reactive systems but in **22**, stabilization is achieved by intramolecular hemi-acetal formation. The two aromatic nuclei of a hexahydroxydiphenoyl ester group when it bridges the 2,4-positions of D-glucopyranose are constrained to an almost orthogonal relationship. This conformation facilitates hydrolysis by neighboring group participation and thus $t_{1/2}$ (H_2O, 90 °C) for the 2,4-linked hexahydroxydiphenoyl ester group is ≃0.5 hr. Dehydrogenation of one phenolic nucleus and stabilization by internal hemi-acetal formation converts the hexahydroxydiphenoyl ester into a very stable grouping ($t_{1/2}$ (H_2O, 90 °C) ≃24 hr). If molecules such as geraniin have pertinent biological functions, then this device may be a simple means to ensure their longer term survival in the plant tissue.

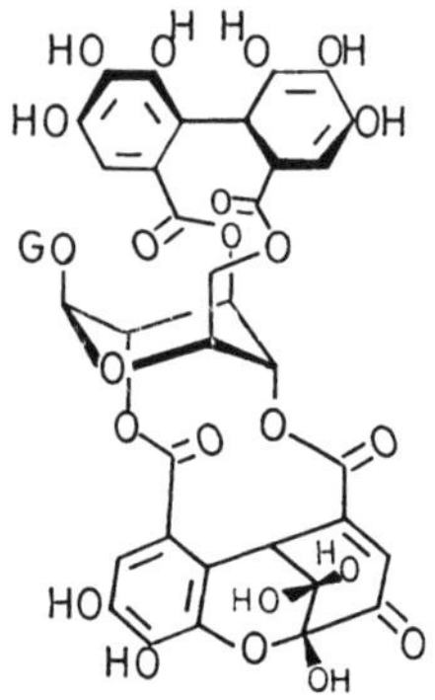

(51), Geraniin

No evidence has yet been obtained to show that 'dimeric' products are formed by oxidative coupling of the 'monomeric' species such as was noted in Class 2B (vide supra). However, brief comment should be made on two phenolic metabolites, isolated early in the studies of the ellagitannins: chebulinic acid (91, 140) and chebulagic acid (91) (Table 7.2.9). It is clear from biogenetic considerations that these are phenolic metabolites very probably derived by further oxidative ring fission of one aryl ring in a precursor such as geraniin **(51)**.

It should perhaps be noted that neither the dehydrohexahydroxydiphenoyl ester group **(22)** nor the chirality of the hexahydroxydiphenoyl ester group is definitive for either of the groups 2B or 2C. Thus, although the 3,6-hexahydroxydiphenoyl ester group is characteristic of a number of metabolites of Group 2C (corilagin, geraniin, chebulagic acid, granatin B) and has the *R* configuration, the alternative 1,6-linkage has the alternative *S* configuration. This emphasizes that this particular facet of polyphenolic metabolism may well be controlled and directed by the chirality of the D-glucose molecule, central to all these structures. Similarly both Schmidt (123) and Okuda (89) have described metabolites containing the dehydrohexahydroxydiphenoyl ester group but with the $^4C_1$ conformation of the D-glucose and the bridging patterns typical of members of the Group 2B.

Table 7.2.9. Hexahydroxydiphenoyl ester derivatives, Group 2C: The structure of some principal metabolites

Compound	Source	Structure	Reference(s)
Corilagin (**30**)	*Terminalia chebula* *Acer* sp. *Geranium* sp.	β-1-*O*-galloyl; (*R*)-hexahydroxydiphenic acid linked 3,6; D-glucopyranose conformation-1B	126, 127, 131, 135, 136; 101
Geraniin (**51**)	*Geranium* sp. *Euphorbia* sp. *Acer* sp. *Cercidiphyllum* sp.	β-1-*O*-galloyl; (*R*)-hexahydroxydiphenic acid linked 3,6; dehydrohexahydroxydiphenic acid (**22**) linked 2,4	100 44
Granatin B	*Punica granatum*	β-1-*O*-galloyl; (*R*)-hexahydroxydiphenic acid linked 3,6; dehydrohexahydroxydiphenic acid (**22**) linked 2,4; isomeric with geraniin	92
Granatin A	*Punica granatum* *Davidia involucrata*	β-1,6(*S*)-hexahydroxydiphenoyl-D-glucose; dehydrohexahydroxydiphenic acid (**22**) linked 2,4	44, 89
Davidiin	*D. involucrata*	2,3,4-trigalloyl-D-glucose; (*S*)-hexahydroxydiphenic acid linked β-1,6	44
Mallotinic acid	*Mallotus japonicus*	β-1-*O*-galloyl, valoneic acid (**28**) linked 3,6 to D-glucopyranose	98
Mallotusonic acid	*M. japonicus*	Dehydrohexahydroxydiphenic acid linked 2,4 to mallotinic acid	98
Chebulinic acid	*Terminalia chebula*	β-1,3,6-trigalloyl; chebulic acid linked 2,4 to D-glucopyranose	140, 91
Chebulagic acid	*T. chebula*	β-galloyl; (*R*)-hexahydroxydiphenic acid linked 3,6; chebulic acid (**24**) linked 2,4 to D-glucopyranose	91, 140
Terchebin	*T. chebula*	β-1,3,6-trigalloyl; dehydrohexahydroxydiphenoyl ester linked 2,4 to D-glucopyranose	92, 94, 138

7.2.4.3.5 Postscript

In the majority of cases, the sources of plant materials employed in the isolation of these polyphenolic compounds by early workers were the traditional sources of economically important ellagitannins – for example, myrabolans, algarobilla, valonea, oak and chestnut bark and oak galls. It is clear from paper chromatographic analysis that several of these particular plant materials are specialized sources and in certain cases a direct link cannot yet be discerned with the patterns of gallic acid metabolism found generally in the plants discussed above. Where a relationship can be discerned (e.g., chebulinic acid and chebulagic acid in myrabolans), this usually involves an extension of one of the basic patterns of gallic acid metabolism (A, B, or C) outlined above.

A number of metabolites isolated from various plant sources do not appear to fall clearly into either of the classes 2B or 2C outlined. Their structures are noted in several references (68, 83, 85, 86, 138, 139).

7.2.5 The Interaction of Proteins with Metabolites of Gallic Acid

The study of the interaction of polyphenols with proteins has a long history, stemming principally from its intimate connection with the art of leather manufacture. One of the first scientific papers on this subject is that of Sir Humphrey Davy in 1803 (79); early work such as that demonstrated some of the macroscopic features of the complexation process. They likewise permitted several empirical definitions of the term 'vegetable tannin' to be advanced. More recent work (53, 61, 76, 118, 119) has delineated the principal means – hydrogen bonding, hydrophobic and ionic interactions – whereby polyphenols may associate with proteins at the molecular level. However, comprehensive studies that relate the structure of polyphenols to their ability to bind to proteins have, in general, not been possible because of the lack of polyphenols with clearly defined structures. Work in the area of gallic acid metabolites over the past decade has now made available polyphenolic molecules with clearly defined structural parameters. In turn the study of 'structure-activity' relationships in the complexation of polyphenols proteins is feasible. A brief discussion of some facets of this work (E. Haslam and T.H. Lilly, unpublished observations) is given here. The results are used to comment on the nature of the protein-polyphenol interactions and on the putative role of polyphenols in the evolution of defense mechanisms of plants.

For almost 200 years studies of the chemistry of natural products have constituted a dominant theme of organic chemistry. Studies of gallic acid metabolites, such as outlined above, are a continuation of that historical thread to the present day. With the flowering of biochemistry in the twentieth century, there came the realization that for some natural products (e.g., fatty acids, amino acids, and nucleotides) a distinctive role in the life of all organisms could be assigned. Natural products such as these occur in broadly similar patterns and the pathways by which they are synthesized are similar, if not identical, in most organisms (33). They are frequently referred to as primary or intermediate metabolites. In contrast, an infinitely greater body of natural products, principally of plant and microbial origin, such as alkaloids, terpenes, polyacetylenes, phenols, and mycotoxins, occur sporadically in Nature. Moreover they appear to have no explicit function in the economy of the producing organism. These substances have, for this reason, often been collectively designated as secondary metabolites (14).

Nonetheless the question of the general role of secondary metabolism in plants and microorganisms is one that has excited continued interest and speculation, particularly over the last two decades. Several propositions center on the suggestion that it is the processes of secondary metabolism and not, in the general case, the secondary metabolites themselves that are of importance to the organism (14, 15). However one alternative theory that has gained increasing prominence in some circles is based on a theory of plant and animal co-evolution (34, 55, 56). It focuses on the secondary metabolites and the premise that plants have evolved the stratagem of a chemical armory appropriate to the environmental needs and pressures they face. As one exponent of this hypothesis suggests (55): ".... natural selection serves as a mechanism by which a population of herbivores may 'call forth de novo' the evolution of a biosynthetic pathway producing compounds toxic to the herbivore....".

In this context the purported role of polyphenols (≡ vegetable tannins) is frequently cited. The relevant biological property of polyphenols is considered to be their astringency, which derives from their ability to complex with proteinaceous materials (3). This may render tissues unpalatable to a predator by precipitating salivary proteins, may decrease their nutritional value or, via immobilizing enzymes, may impede the invasion of host tissues by predators and parasites. A phylogenetic subdivision of plants is, as outlined above, now possible on the basis of the products of gallic acid metabolism (46). Quantitative studies, moreover, show that the principal biosynthetic thrust is invariably towards the elaboration of the condensed and high-molecular weight polyphenols – end products of the various biosynthetic pathways (2A, B, C) – rather than the accumulation of intermediates along the pathways. Thus, for example, the following polyphenols overwhelmingly predominate in the tissue of particular plants: **17a, b** (MW 1250) in *Rhus* sp., *Pelargonium* sp., *Cotinus* sp., and some *Acer* sp.; **40** (rugosin D, T_{2B}) (MW 1874) in *Rosa* sp. and *Filipendula ulmaria*; **42** (sanguin H-6, T_1) (MW 1870) in *Rubus* sp., *Potentilla* sp., and *Geum* sp.; **49** (vescalagin) and its C-1 diastereoisomer **46** (castalagin) (MW 934) in various *Quercus* sp. It is therefore assumed that the astringency of these tissues is attributable, in large part, to the particular polyphenols indicated. These and related molecules and their presumed biosynthetic precursor, *β*-penta-*O*-galloyl-D-glucose (**15**), have been studied to determine the strength of their association with the protein Bovine Serum Albumin (BSA) using equilibrium dialysis and microcalorimetric analytical procedures.

The results show a clear ranking order among the polyphenols examined in terms of their ability to complex with the protein. Molecular size is important. Thus the efficacy of binding to BSA increases incrementally in the galloyl-D-glucose series with the addition of each galloyl ester group (di→tri→tetra→penta). It reaches a maximum in the flexible disclike structure of **15**. Acretion of further depsidically linked galloyl ester groups to **15** as in the gallotannins (Group 2A) (**17a, b**) (Table 7.2.5), although it increases molecular weight (940→ 1250: penta →hepta-octagalloyl-D-glucose), does not lead to an increase in protein binding capacity. Presumably, in this case, the increase in molecular weight, compared to **15**, does not bring with it a concomitant increase in the number of groups (phenolic hydroxyl and aryl rings) on the surface of the molecule available to bond with the protein. The importance of molecular size is, however, reflected once again in the dimer **40** (rugosin D, T_{2B}), which, of the various metabolites of gallic acid encountered to date in higher plants, has proven to be the most efficient in its complexation with protein.

Equally significant as molecular size, however, is conformational mobility and flexibility in the polyphenol substrate. This is succinctly demonstrated among the galloyl ester metabolites of Group 2B (Figs. 7.2.9, 7.2.10, 7.2.11), whose biosynthesis proceeds by oxidative coupling of adjacent (4,6 and 2,3) galloyl ester groups in the precursor (**15**). Thus as galloyl ester groups in **15** are constrained by oxidative coupling and the formation of intramolecular biphenyl linkages (e.g., **35, 37, 42, 46,** and **49**) (Fig. 7.2.13) then the reduced conformational mobility is reflected in a reduced capacity to bind to BSA. The apotheosis of this effect is seen in the case of castalagin (**46**) and vescalagin (**49**). These propeller-shaped and virtually inflexible molecules are in a sense analogues of **15**, but on a molar basis they are

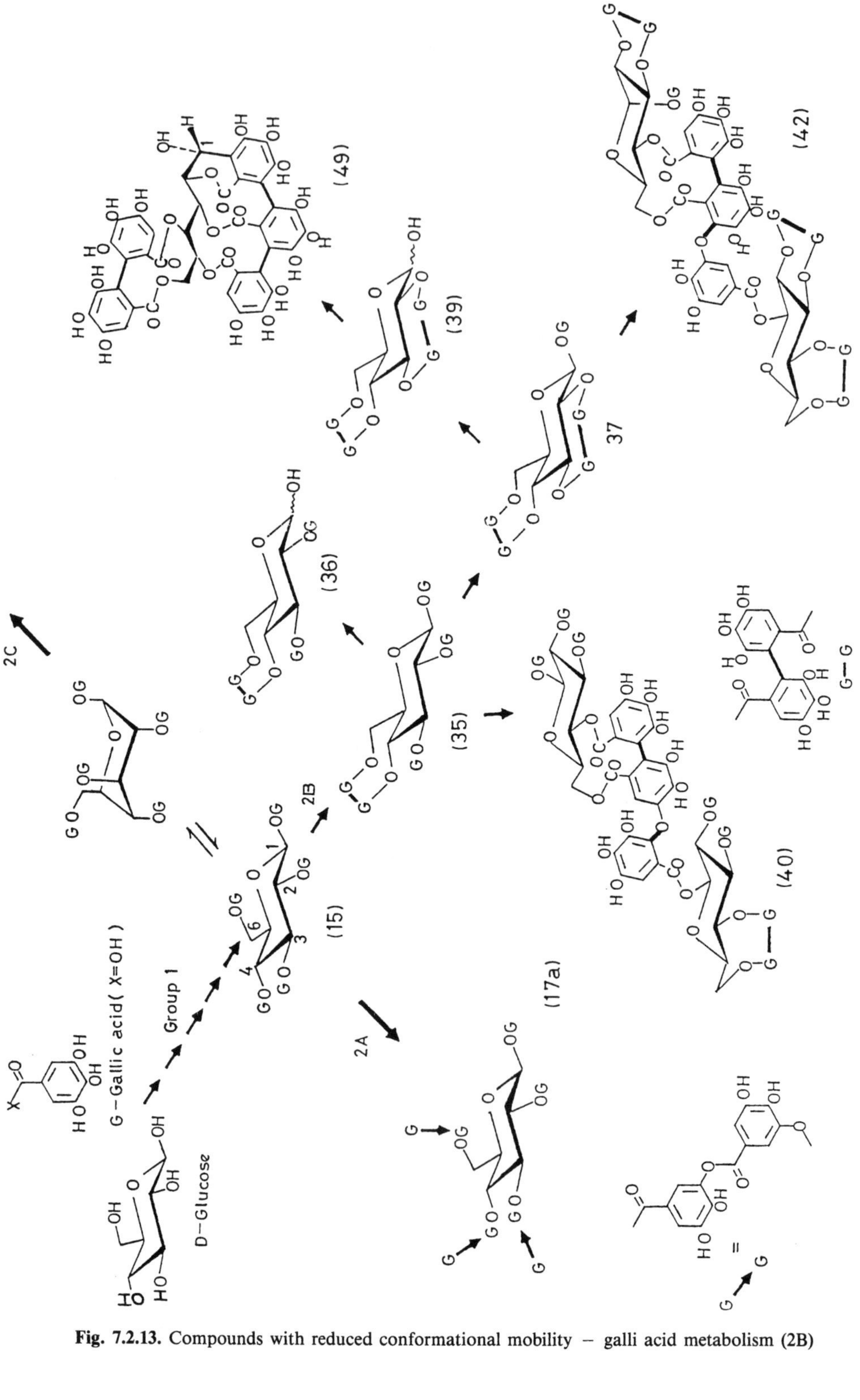

Fig. 7.2.13. Compounds with reduced conformational mobility – galli acid metabolism (2B)

substantially less effective in binding to BSA. Similarly, comparison of the 'dimeric' species **40** and **42** shows that the introduction of the two additional biphenyl linkages in **42** leads once again to a large decrease in protein binding capacity. All the species (**35**, **37**, **42**, **46**, **49**) bind less effectively to protein than their precursor (**15**) and this suggests that there is a strong degree of cooperativity in the association mechanism.

Collectively these results complement those of Hagerman and Butler (31) who, in a separate study, showed that conformationally mobile proteins – as opposed to those with a tightly coiled secondary structure – have high affinities for polyphenols. Complimentarity between the polydentate ligand (polyphenol) and receptor (protein) is thus maximized by conformational flexibility in both components. In this context the comparatively lower astringency of the proanthocyanidins (3) may, in part, be explicable in terms of the conformational restraints imposed by restricted rotation about the repeating inter-flavan bonds (40).

The same patterns of results have also derived from two separate but related investigations (E. Haslam, T.H. Lilley, and K. Hostettman, unpublished observations). Polyphenols act as inhibitors of many enzymes and the overall trends in enzyme inhibition by particular substrates are broadly related to their capacity to bind to protein. Similarly the molluscicidal activity of various polyphenolic substrates against the schistosomiasis-transmitting snail *Biomphalaria glabrata* appears to be closely parallel to their ability to complex with proteins.

These observations add substantially to the understanding of the mechanism of association of polyphenols with proteins. They also permit of some comments on the theory of plant herbivore co-evolution and hence of the possible function of polyphenol synthesis in higher plants. The data show a wide variability in the protein binding capabilities of polyphenols. Frequently, but not invariably, among metabolites of gallic acid this ability is substantially impaired when compared to that of the central biosynthetic intermediate β-penta-*O*-galloyl-D-glucose (**15**) from which they are presumed to be derived. This evidence suggests that although the retention of polyphenolic synthesis may confer an advantage on the plant and may well be the basis of selective pressure, such protection bestows a secondary benefit on the plant. In particular, this evidence does not clearly support the idea that the prime cause and purpose of polyphenolic synthesis in plants is as a general defense against predators.

Polyphenols constitute a group of natural products of great structural diversity and wide phylogenetic distribution (2, 40, 41). Such proliferation and structural diversity may be alternatively viewed, as Bu'Lock has succinctly remarked in the related context of microbial secondary metabolism (15), ".... as a result of there being very little selection pressure on their identity; i.e., the products bring no great advantage or disadvantage per se."

References

1 Armitage R, Bayliss G S, Gramshaw J G, Haslam E, Haworth R D, Jones K, Searle T 1961 Gallotannins, Part 3. The Constitution of Chinese,Turkish, sumac and tara tannins. J Chem Soc 1842–1853

2 Bate-Smith E C 1962 The phenolic constituents of plants and their taxonomic significance. J Linn Soc London Bot 95–173
3 Bate-Smith E C 1973 Haemanalysis of tannins: The concept of relative astringency. Phytochemistry 12: 907–912
4 Bate-Smith E C 1976 Chemistry and taxonomy of *Ribes*. Biochem Syst Ecol 4: 13–18
5 Bate-Smith E C 1977 Chemistry and taxonomy of the Cunoniaceae. Biochem Syst Ecol 5: 95–105
6 Bate-Smith E C 1977 Astringent tannins of *Acer* species. Phytochemistry 16: 1421–1426
7 Bate-Smith E C 1978 Systematic aspects of the astringent tannins of *Acer* species. Phytochemistry 17: 1945–1948
8 Bate-Smith E C 1981 Astringent tannins of the leaves of *Geranium* species. Phytochemistry 20: 211–216
9 Bohm B A 1979 Flavonoids of *Tolmiea menziesii*. Phytochemistry 18: 1079–1080
10 Bohm B A, Collins F W, Wilkins C K 1975 The flavonoid glycoside gallates from *Tellima grandiflora*. Phytochemistry 14: 1099–1102
11 Bradfield A E, Penney M 1948 The catechins of green tea. Part 2. J Chem Soc 224–2254
12 Britton G, Haslam E 1965 Gallotannins. Part 12. Phenolic constituents *Arctostaphylos uva-ursi*. J Chem Soc 7312–7319
13 Britton G, Haslam E, Crabtree P W, Stangroom J E 1966 Gallotannins. Part 13. The structure of Chinese gallotannin. Evidence for a polygalloyl chain. J Chem Soc (C) 783–790
14 Bu'Lock J D 1965 The biosynthesis of natural products. McGraw-Hill London, 149
15 Bu'Lock J D 1980 Mycotoxins as secondary metabolites. In: Steyn P S (ed) The biosynthesis of mycotoxins. Academic Press London New York, 1–16
16 Conn E E, Swain T 1961 Biosynthesis of gallic acid in higher plants. Chem Ind 592–593
17 Cornthwaite D C,Haslam E 1965 Gallotannins. Part 9. The biosynthesis of gallic acid in *Rhus typhina*. J Chem Soc 3008–3011
18 Coxon D T, Holmes A, Ollis W D, Vora V C, Grant M S, Tee J L 1972 Flavonol digallates in green tea leaf. Tetrahedron 28: 2819–2826
19 Culvenor C J 1966 The conformation of esters and the 'acylation shift'. NMR evidence from pyrrolizidine alkaloids. Tetrahedron Lett 1091–1099
20 Dewick P M, Haslam E 1969 Phenol biosynthesis in higher plants – gallic acid. Biochem J 113: 537–542
21 Ferrier R J, Collins P M 1972 Monosaccharide chemistry. Penguin London, 40
22 Fischer E 1919 Untersuchungen über Depside und Gerbstoffe. Springer Berlin, 541
23 Fischer E, Bergmann K 1918 Über das tannin und die synthese ähnlicher stoffe. V. Ber Deutsche Chem Ges 51: 1760–1804
24 Gilson E R 1903 Sur deux nouveaux glucotannoides. CR Acad Sci 136: 385
25 Gonzalez A G, Francisco G G, Friere R, Hernandez R, Salazar J, Suarez E 1976 Lindleyin, a new phenolic gallyl glucoside from *Aconium lindleyi*. Phytochemistry 15: 344–346
26 Gross G G 1982 Synthesis of galloyl coenzyme A thiol ester. Z Naturforsch 37C: 778–783
27 Gross G G 1982 Synthesis of β-glucogallin from UDP glucose and gallic acid by an enzyme from oak leaves. Fed Eur Biochem Lett 148: 67–70
28 Gross G G 1983 Synthesis of mono-, di- and trigalloyl-β-D-glucose by β-glucogallin dependent galloyl transferases from oak leaves. Z Naturforsch 38C: 519–523
29 Gross G G 1983 Partial purification and properties of UDP-glucose: Vanillate 1-*O*-glucosyl transferase from oak leaves. Phytochemistry 22: 2179–2182
30 Gross S R 1978 Recent advances in the chemistry and biochemistry of lignin. In: Swain T, Harborne J B, van Sumere C (eds) Recent advances in phytochemistry – Plant phenolics, vol 12. Plenum Press New York, 177–220
31 Hagerman A E, Butler L G 1981 The specificity of proanthocyanidin-protein interactions. J Biol Chem 256: 4494–4497
32 Hahlbrock K 1981 Flavonoids. In: Conn E E (ed) The biochemistry of plants, vol 7. Academic Press London New York, 425–456
33 Harborne J B 1967 Comparative biochemistry of the flavonoids. Academic Press London New York, 383
34 Harborne J B 1977 Introduction to ecological biochemistry. Academic Press London New York, 2
35 Haslam E 1965 Galloyl esters in the Aceraceae. Phytochemistry 4: 495–498
36 Haslam E 1966 Chemistry of vegetable tannins. Academic Press London New York, 179

37 Haslam E 1967 Gallotannins. Part 14. Structure of the gallotannins. J Chem Soc (C) 1734–1738
38 Haslam E 1969 (+) Catechin-3-gallate and a polymeric procyanidin from *Bergenia* sp. J Chem Soc (C) 1824–1828
39 Haslam E 1974 Polyphenol-protein interactions. Biochem J 139: 285–288
40 Haslam E 1977 Symmetry and promiscuity in procyanidin biochemistry. Phytochemistry 16: 1625–1640
41 Haslam E 1982 The metabolism of gallic and hexahydroxydiphenic acid in higher plants. Fortschr Chem Org Naturst 41: 1–46
42 Haslam E 1982 Vegetable tannins. In: Conn E E (ed) The biochemistry of plants, vol 7. Academic Press London New York, 527–556
43 Haslam E, Gupta R K, Al-Shafi S M K, Layden K 1982 The metabolism of gallic acid and hexahydroxydiphenic acid in plants. Part 2. Esters of (S)-hexahydroxydiphenic acid with D-glucopyranose ($^4C_1$). J Chem Soc Perkin I 2525–2534
44 Haslam E, Haddock E A, Gupta R K 1982 The metabolism of gallic acid and hexahydroxydiphenic acid in plants. Part 3. Esters of (*R*)- and (*S*)-hexahydroxydiphenic acid and dehydrohexahydroxydiphenic acid with D-glucopyranose ($^1C_4$ and related conformations). J Chem Soc Perkin I 2535–2545
45 Haslam E, Haddock E A, Gupta R K, Al-Shafi S M K, Magnolato D 1982 The metabolism of gallic acid and hexahydroxydiphenic acid in plants. Part 1. Naturally occurring galloyl esters. J Chem Soc Perkin I 2515–2524
46 Haslam E, Haddock E A, Gupta R K, Al-Shafi S M K, Layden K, Magnolato D 1982 The metabolism of gallic and hexahydroxydiphenic acid in plants: Biogenetic and molecular taxonomic considerations. Phytochemistry 21: 1049–1062
47 Haslam E, Stangroom J E 1966 The esterase and depsidase activities of tannase. Biochem J 990: 28–31
48 Haworth R D 1961 Some problems in the chemistry of the gallotannins: Pedler Lecture. Proc Chem Soc 401–410
49 Haworth R D, da Silva L B 1951 Chebulinic acid. Part 1. J Chem Soc 3511
50 Haworth R D, da Silva L B 1954 Chebulinic acid. Part 2. J Chem Soc 3611–3617
51 Herrmann K 1978 Hydroxyzimtsäure und Hydroxybenzosäuren enthaltende Naturstoffe in Pflanzen. Fortschr Chem Org Naturst 35: 73–132
52 Hillis W E, Seikel M 1970 Hydrolysable tannins of *Eucalyptus delegatensis* wood. Phytochemistry 9: 1115–1128
53 Hoff J E, Oh H I, Armstrong G S, Hoff C A 1980 Hydrophobic interactions in tannic acid – protein complexes. J Agr Food Chem 28: 394–398
54 Ikeya Y, Taguchi H, Yosioka I, Kobayashi H 1979 The constituents of *Schizandra chinensis* Baile: Isolation and structure determination of five new lignans, gomisin A, B, C, F and G and the absolute structure of schizandrin. Chem Pharm Bull 27: 1383–1394
55 Janzen D H 1969 Coevolution. Science 164: 415
56 Janzen D H 1979 New horizons in the biology of plant defences. In: Rosenthal G A, Janzen D H (eds) Herbivores – their interactions with secondary plant metabolites. Academic Press New York, 331–350
57 Jochims J C, Taigel G, Schmidt O Th 1968 Protonenresonanze-spektrum und Konformationsbestimmung einiger natürlicher Gerbstoffe. Justus Liebigs Ann Chem 717: 169–185
58 Jurd L 1958 Plant polyphenols. III. The isolation of a new ellagitannin from the pellicle of the walnut. J Am Chem Soc 80: 2249–2252
59 Keen P C, Haworth R D, Haslam E 1962 Gallotannins. Part 7. Tara gallotannins. J Chem Soc 3814–3818
60 Knowles P F, Haworth R D, Haslam E 1961 Gallotannins. Part 4. The biosynthesis of gallic acid. J Chem Soc 1854–1859
61 Loomis W D 1974 Overcoming problems of phenolics and quinones in the isolation of plant enzymes and organelles. Methods Enzymol 31: 528–544
62 Mayer W 1954 Dehydro-digallusäure. Justus Liebigs Ann Chem 578: 34–44
63 Mayer W 1973 Otto Theodor Schmidt. Justus Liebigs Ann Chem 1759–1776
64 Mayer W, Bilzer B, Schauerte K 1971 Isolierung von Castalagin und Vescalagin aus Valoneagerbstoffen. Justus Liebigs Ann Chem 754: 149–151

65 Mayer W, Bilzer W, Schilling G 1976 Castavaloninsäure, Isolierung und Strukturmittelung. Justus Liebigs Ann Chem 876–881
66 Mayer W, Busath H, Schick H 1976 Isovalolaginsäure. Justus Liebigs Ann Chem 2169–2177
67 Mayer W, Einwiller A, Jochims J C 1967 Die Struktur des Castalins. Justus Liebigs Ann Chem 707: 182–189
68 Mayer W, Gorner A, Andra K 1977 Punicalagin und Punicalin, zwei Gerbstoffe aus den Schalen der Granatäpfel. Justus Liebigs Ann Chem 1976–1986
69 Mayer W, Gunther A, Busath H, Bilzer W, Schilling G 1976 Valolaginsäure. Justus Liebigs Ann Chem 987–995
70 Mayer W, Kullman F, Schilling G 1971 Die Struktur des Vescalins. Justus Liebigs Ann Chem 747: 51–59
71 Mayer W, Kunz N, Loebich F 1965 Die Struktur des Hamamelitannins. Justus Liebigs Ann Chem 688: 232–238
72 Mayer W, Schick H, Schilling G 1976 Trillosäure, eine neue Phenolcarbonsäure aus Valoneagerbstoffen. Justus Liebigs Ann Chem 2178–2184
73 Mayer W, Seitz H, Jochims J C 1969 Die Struktur des Castalins. Justus Liebigs Ann Chem 707: 182–189
74 Mayer W, Schultz G, Wrede S, Schilling F G 1975 2-O-Cinnamoyl-1-*O*-galloyl-β-D-glucopyranose aus *Rhizoma rhei.* Justus Liebigs Ann Chem 946–952
75 Mayer W, Seitz H Y, Jochims J C, Schauerte K, Schilling G 1971. Struktur des Vescalagins. Justus Liebigs Ann Chem 751: 60–68
76 McManus J P, Davis K G, Lilley T H, Haslam E 1981 The association of phenols with proteins. J Chem Soc Comm 309–311
77 Neish A C, Towers G H N, Chen D, El-Basyouni S Z, Ibrahim R K 1964 The biosynthesis of hydroxybenzoic acids in higher plants. Phytochemistry 3: 485–492
78 Nielsen B J, Lacour N F, Jensen S R, Bock K 1980 The structure of acer tannin. Phytochemistry 19: 2033
79 Nierenstein M 1932 Incunabula of tannin chemistry. Edward Arnold London, 116
80 Nishioka I, Nonaka G, Nishizawa M, Yamagishi T 1982 Tannins and related compounds. Part 5. Structure of Chinese gallotannin. J Chem Soc Perkin I 2963–2968
81 Nishioka I, Nonaka G, Tanaka T 1982 Tannins and related compounds. Part 3. A new phenolic acid, sanguisorbic acid dilactone and three new ellagitannins sanguins H-1, H-2 and H-3 from *Sanguisorba officinalis.* J Chem Soc Perkin I 1067–1073
82 Nishioka I, Nonaka G, Tanaka T, Nita M 1982 A dimeric hydrolysable tannin sanguin H-6 from *Sanguisorba officinalis.* Chem Pharm Bull 30: 2255–2257
83 Nishira H, Joslyn M A 1968 The galloyl glucose compounds in green carob pods (*Ceratonia siliqua*). Phytochemistry 7: 2147–2156
84 Nonaka G, Harada M, Nishioka I 1980 Eugeniin, a new ellagitannin from cloves. Chem Pharm Bull 28: 685–687
85 Nonaka G, Matsumoto Y, Nishioka I 1981 Trapain, a new hydrolysable tannin from *Trapa japonica* Fleroy. Chem Pharm Bull 29: 1184–1186
86 Nonaka G, Nishioka I 1983 Tannins and related compounds: Rhubarb. Isolation and structure of a glycerol gallate, gallic acid glucoside gallates, galloyl glucoses and isolindleyin. Chem Pharm Bull 31: 1652–1658
87 Nonaka G, Nishioka I, Nagasawa T, Oura H 1981 Tannins and related compounds: Rhubarb. Chem Pharm Bull 29: 2862–2870
88 Nonaka G, Nishizawa N, Yamagishi T, Nishioka I, Baudo H 1982 Novel hydrolysable tannins from *Nuphar japonicum.* Chem Pharm Bull 30: 1094–1097
89 Okuda T 1981 Tannins of medicinal plants and drugs. Heterocycles 15: 1323–1342
90 Okuda T, Ashida M, Yoshida T 1981 Casuarictin and casuarinin – two new ellagitannins from *Casuarina stricta.* Heterocycles 16:1681–1685
91 Okuda T, Fujii R, Yoshida T 1980 Revised structures of chebulinic and chebulagic acid. Chem Pharm Bull 28: 3713–3715
92 Okuda T, Hatano T, Nitta H, Fujii R 1980 Hydrolysable tannins having enantiomeric dehydrohexahydroxydiphenoyl group: Revised structure of terchebin and granatin B. Tetrahedron Lett 21: 4361–4364

93 Okuda T, Hatano T, Ogawa N 1982 Rugosin D, E, F, and G. Dimeric and trimeric hydrolysable tannins. Chem Pharm Bull 30: 4234–4237

94 Okuda T, Hatano T, Yasui T 1981 Revised structure of isoterchebin isolated from *Cornus officinalis*. Heterocycles 16: 1321–1325

95 Okuda T, Hatano T, Yazaki K 1982 Dehydrogeraniin, furosinin and furosin, dehydroellagitannins from *Geranium thunbergii*. Chem Pharm Bull 30: 1113–1116

96 Okuda T, Hatano T, Yazaki K 1983 Praecoxin B, C, D and E. Novel ellagitannins from *Stachyurus praecox*. Chem Pharm Bull 31: 333–336

97 Okuda T, Hatano T, Yazaki K, Ogawa N 1982 Rugosin A, B and C and praecoxin A, tannins having a valoneoyl group. Chem Pharm Bull 30: 4230–4233

98 Okuda T, Seno K 1978 Mallotusinic acid and mallotinic acid, new hydrolysable tannins from *Mallolotus japonicus*. Tetrahedron Lett 139–142

99 Okuda T, Usman-Memon M, Yoshida T 1981 Alnusiin, a novel ellagitannin from *Alnus sieboldiana* fruits. Heterocycles 16: 1085–1087

100 Okuda T, Yoshida T, Hatano T 1982 Constituents of *Geranium thunbergii* Sieb et Zudd. Part 12. Hydrated stereostructure and equilibration of geraniin. J Chem Soc Perkin I 9–14

101 Okuda T, Yoshida T 1980 ^{13}C Nuclear magnetic resonance spectra of corilagin and geraniin. Heterocycles 14: 1743–1747

102 Okuda T, Yoshida T, Ashida M, Yazaki K 1982 Casuariin, stachyurin and strictinin. New Ellagitannins from *Casuarina stricta*. Chem Pharm Bull 30: 766–769

103 Okuda T, Yoshida T, Hatano T, Koga T, Toh N, Kuriyama N 1982 Circular dichroism of hydrolysable tannins. I. Ellagitannins and gallotannins. Tetrahedron Lett 23: 3937–3940

104 Okuda T, Yoshida T, Hatano T, Koga T, Toh N, Kuriyama N 1982 Circular dichroism of hydrolysable tannins. II. Ellagitannins and gallotannins. Tetrahedron Lett 23: 3941–3944

105 Okuda T, Yoshida T, Hatano T, Yazaki K, Ashida M 1982 Ellagitannins of the Casuarinaceae, Stachyuraceae, Myrtaceae. Phytochemistry 21: 2871–2874

106 Okuda T, Yoshida T, Koga T, Toh N 1982 Absolute configurations of chebulic, chebulinic and chebulagic acids. Chem Pharm Bull 30: 2655–2658

107 Okuda T, Yoshida T, Kuwahara M, Usman-Memon M, Shingu T 1982 Agrimoniin and potentillin, an ellagitannin dimer and monomer having an α-glucose core. J Chem Soc Chem Comm 163–164

108 Okuda T, Yoshida T, Maruyama Y, Usman-Memon M, Shingu T 1982 Gemin B and C, dimeric ellagitannins from *Geum japonicum*. Chem Pharm Bull 30: 4245–4248

109 Okuda T, Yoshida T, Usman-Memon M, Shingu T 1982 Structure of gemin A, a new dimeric ellagitannin having α- and β-glucose cores. J Chem Soc Chem Comm 351–353

110 Ozawa T, Arai N, Takino Y 1978 Structure of a new phenolic glycoside chesnatin from chestnut galls. Agr Biol Chem 42: 1907–1910

111 Ozawa T, Haga K, Arai N, Takino Y 1978 Structure of a new phenolic glycoside from chestnut galls. Agr Biol Chem 42: 1511–1514

112 Ozawa T, Kobayashi D, Takino Y 1977 Structure of the new phenolic glycosides MP-2 and MP-10 from chestnut galls. Agr Biol Chem 41: 1257–1262

113 Ozawa T, Odaira Y, Imegawa H, Takino Y 1980 A new phenolic glycoside acetylcretanin and flavonoids from chestnut galls. Agr Biol Chem 44: 581–587

114 Perkin A G, Uyeda Y 1922 Occurrence of a crystalline tannin in the leaves of *Acer ginnala*. J Chem Soc 66–76

115 Pohl R, Nahrstedt A, Dumkow K, Janistyn B 1974 Quercetin-galactoside gallate in Euphorbiaceae. Tetrahedron Lett 559–562

116 Reddy K K, Rajadurai S, Sastry S K N, Nayudamma Y 1964 Studies of the Dhava tannins. 1. The isolation and constitution of a gallotannin from Dhava (*Anogeissus latifolia*) 17: Aust J Chem 238–245

117 Roberts E A H, Myers M 1958 Theogallin, a polyphenol occurring in tea. II Identification as a galloyl quinic acid. J Sci Food Agric 9: 701–705

118 Robinson W B, Calderon P, van Buren J 1968 Factors influencing the formation of precipitates and hazes by gelatin and condensed and hydrolysable tannins. J Agr Food Chem 16: 479–482

119 Robinson W B, van Buren J 1969 Formation of complexes between protein and tannic acid. J Agr Food Chem 17: 772–777

120 Saijo R 1983 Pathway of gallic acid biosynthesis and its esterification with catechins in young tea shoots. Agr Biol Chem 47: 455–460
121 Schmidt O Th 1954 Ellagengerbstoffe. Leder 5: 129–134
122 Schmidt O Th 1955 Natürliche Gerbstoffe. In: Peach K, Tracey M V (eds) Moderne Methoden der Pflanzenanalyse, vol III. Springer Berlin, 517–548
123 Schmidt O Th 1956 Gallotannine und Ellagen-Gerbstoffe. Fortschr Chem Org Naturst 13: 71–136
124 Schmidt O Th 1957 Über Chebulagsäure und Chebulinsäure. Leder 8: 106
125 Schmidt O Th, Bernauer K 1954 Brevifolin und Brevifolin-Carbonsäure. Justus Liebigs Ann Chem 588: 211–230
126 Schmidt O Th, Blin F, Lademann R 1952 Über die Bindung der Ellagsäure in Corilagin und Chebulagsäure. Justus Liebigs Ann Chem 576: 75–93
127 Schmidt O Th, Demmler K 1952 Optisch aktiv 2,3,4,2′,3′,4′-hexamethoxydiphenylcarbonsäure-6,6′. Justus Liebigs Ann Chem 576: 85–93
128 Schmidt O Th, Demmler K 1954 Racemische und optisch aktive 2,3,4,2′,3′,4′-Hexaoxydiphenylcarbonsäure-6,6′. Justus Liebigs Ann Chem 586: 170–193
129 Schmidt O Th, Ebert W, Koff M 1969 1,3,4,6-Tetragalloyl-β-D-glucose aus Algarobilla. Justus Liebigs Ann Chem 729: 251–252
130 Schmidt O Th, Komarek E 1955 Valoneasäure. Justus Liebigs Ann Chem 591: 156–176
131 Schmidt O Th, Lademann R 1951 Corilagin, ein weiterer kristallisierter Gerbstoff aus Divi-Divi. Justus Liebigs Ann Chem 571: 232–238
132 Schmidt O Th, Mayer W 1956 Natürliche Gerbstoffe. Angew. Chem 68: 103–106
133 Schmidt O Th, Schanz R, Eckert R, Wurmb R 1967 Brevilagin 1. Justus Liebigs Ann Chem 706: 131–153
134 Schmidt O Th, Schanz R, Wurmb R, Groebke W 1967 Brevilagin 2. Justus Liebigs Ann Chem 706: 154–168
135 Schmidt O Th, Schmidt D M 1953 Über das Vorkommen von Corilagin in Myrabolanen. Justus Liebigs Ann Chem 578: 31–34
136 Schmidt O Th, Schmidt D M, Herok J 1954 Die Konstitution und Konfiguration des Corilagins. Justus Liebigs Ann Chem 587: 67–80
137 Schmidt O Th, Schulz J, Feisser H 1967 Die Gerbstoffe der Myrabolanen. Justus Liebigs Ann Chem 706: 187–212
138 Schmidt O Th, Schulz J, Wurmb R 1967 Terchebin. Justus Liebigs Ann Chem 706: 169–179
139 Schmidt O Th, Wurtele L, Harreus A 1965 Penduculagin, eine 2,3:4,6-Di-*S*-(–)hexahydroxydiphenoyl-D-glucose aus Knoppern. Justus Liebigs Ann Chem 690: 150–162
140 Uddin M, Haslam E 1967 Gallotannins. Part 15. Some observations on the structure of chebulinic acid and its derivatives. J Chem Soc (C) 2381–2384
141 Wagner H, Iyengar M A, Seligmann O, El-Sissi H I, Saleh N A M, El-Negoumy S I 1974 Prunin-O-6′-gallate aus *Acacia farnesiana*. Phytochemistry 13: 2843–2844
142 White T 1957 Tannins, their occurrence and significance. J Sci Food Agr 8: 377–385
143 Wilkins C K, Bohm B A 1976 Ellagitannins from *Tellima grandiflora*. Phytochemistry 15: 211–214
144 Zenk M H 1964 Zur Frage der Biosynthese von Gallusäure. Z Naturforschung 19B: 83–84
145 Zenk M H 1978 Recent work on cinnamoyl CoA derivatives. In: Swain T, Harborne J B, van Sumere C (eds) Recent advances in phytochemistry, vol 12. Biochemistry of plant phenolics. Plenum Press London, 139–176

7.3 Lignans

O. R. GOTTLIEB and M. YOSHIDA

7.3.1 Introduction

Interest in lignans and neolignans has risen sharply in the last few years. This is due chiefly to their powerful, varied biological activities (Table 7.3.1) exemplified by three of the most noteworthy representatives. First, the lignan podophyllotoxin (**1**), in the form of several semisynthetic derivatives, is one of the very few natural products from higher plants used clinically in cancer chemotherapy (30). Next kadsurenone (**2**), also isolated from an angiosperm species has been described as

Table 7.3.1. Bioactivity of lignans and neolignans (56, 101, 103)

Lignans and neolignans as	Are active on
Chromosome damagers	Nucleic acids
Inhibitors of nucleoside uptake	
Inhibitors of cAMP phosphodiesterase	Enzymes
Inhibitors of fungal enzymes	
Inhibitors of mitochondrial respiration	
Inhibitors of oxidases	
Antimitotics	Cells
Cytotoxic agents	
Tubulin binders	
Antitumorals	Tumors
Cytostatics	
Cytotoxins	
Inhibitors of viral release	Viruses
Inhibitors of growth	Bacteria
Cercaricides	Helminths
Inhibitors of growth	Fungi
Toxins	Fish
Antifeedants	Insects
Inhibitors of larval growth	
Juvenile hormones	
Synergists of insecticides	
Inhibitors of germination	Plants
Allergens	Mammals
Antihepatotoxins	
Antileukemic agents	
Antipyretic agents	
Antitumorals	
Cathartic agents	
CNS stimulants	
Narcotics	
Neuroleptic agents	
Muscular relaxants	
Psychotrophic agents	
Stress reducers	
Toxins	
Urinary antiseptics	

1 podophyllotoxin

2 kadsurenone

3 enterolactone

4 matairesinol

5 prestegane A

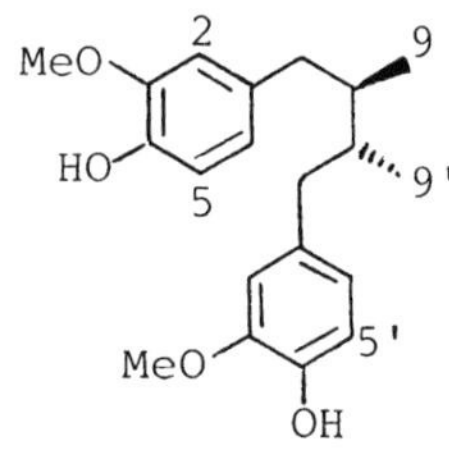

6 dihydroguaiaretic ac

7 burchellin

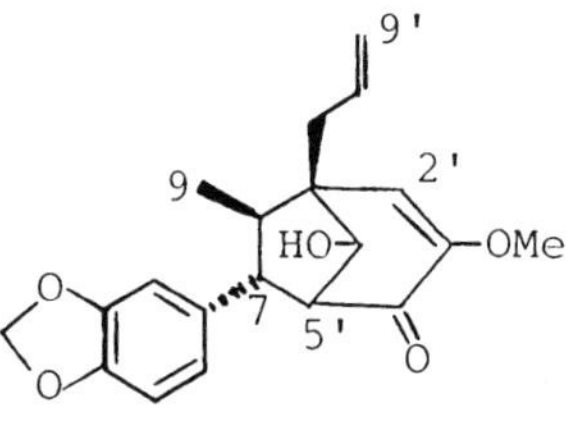

8 guianin

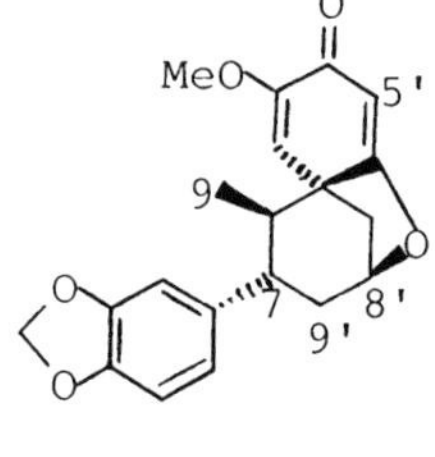

9 futoenone

10 lancifolin B

11 carinatone

12

13 dehydrodieugenol

14 isoasatone

15a Δ7 isoeugenol

15b Δ8 eugenol

a potent antagonist of PAF, the mediator of inflammatory diseases such as asthma, allergy, and thrombosis (28). Finally enterolactone (**3**), identified in biological fluids of several mammals, including humans, was shown to possess endogenous digitalis-like activity (19) and may even be an example of a new type of hormone-controlling cell growth (161).

The structures of the lignoids **1**, **2** and **3** illustrate some common and some different features. All three compounds comprise two C_6-C_3-moieties, numbered 1 to 9 and 1′ to 9′, and hence their monomers are presumably metabolites of the shikimate pathway. Since this pathway is restricted to plants, the biosynthesis of **3** is a mystery. Could enterolactone be a bacterial degradation product of lignins ingested with plant food, or is it synthesized ex novo (133)? If so, what are its precursors – a pertinent question since enterolactone lacks oxy-functions at the 4- and 4′-positions of the aryls, a structural particularity not shared with any other naturally occurring C_6-C_3-dimer known. Indeed free *p*-hydroxyls are essential for the oxidative coupling of the monomers regardless of the mechanism of the reaction.

7.3.2 Nomenclature and Numbering

Application of systematic nomenclature to oxidative C_6-C_3-dimers results in designations that are not internally comparable, such is the diversity of skeletons and functions. A structural definition separating lignans [products involving an 8.8′-bridge with (e.g. **1**) or without (e.g. **4**, **5**) other bridges] from neolignans (products such as **2** and **6** through **14** involving bridges connecting any locations except 8.8′) (51, 52) is still thought by several authors to be preferable. Indeed synthesis of representatives belonging to both groups does not require any basically different strategies (161). Besides, the biosynthetic origins of lignans and neolignans are chiefly matters of hypotheses, as little experimental work has been reported in this area (see, however, 69, 70). Hence, it was suggested that it would be wiser (164) to base a classification on structure (51, 52) than on biogenesis (53, 54).

However, by analogy with other groups of natural products, it is to be expected that, in time, the structural definitions will be abandoned in favor of biogenetic ones. Although, as stated above, direct biochemical evidence is still scarce, circumstantial evidence can already be adduced to separate lignans (products such as **1**, **4** and **5** of oxidative coupling involving cinnamyl alcohols and/or cinnamic acid precursors) and neolignans (products of oxidative coupling of propenylphenol *plus* propenylphenol (e.g. **6**) and propenylphenol *plus* allylphenol (e.g. **7** through **12**), as well as allylphenol *plus* allylphenol (e.g. **13**, **14**)).

First and most conspicuously, spectra and reactivity of lignans and neolignans, classified according to this biogenetic criterion, are different. Chiefly, the many interesting rearrangements in the allyl-substituted compounds can, of course, be observed only in the neolignans. Secondly, lignans appear sporadically throughout the gymnosperms (Table 7.3.2) and angiosperms (Table 7.3.3), while their structural analogues – the 8.8′-neolignans (e.g. **6**) – are practically restricted to Magnoliiflorae and Piperales *sensu* Dahlgren (32), botanical groups in

Table 7.3.2. Lignan-containing gymnospermous and pteridophytous families (common termination-aceae) and species

Family	Genus/species	Code
Araucari	*Araucaria angustifolia*	AraAraang
Cupress	*Callitris drummondi*	CupCaldru
Cupress	*Callitris collumellaris*	CupCalcol
Cupress	*Calocedrus formosana*	CupCalfor
Cupress	*Chamaecyparis obtusa*	CupChaob1
Cupress	*Chamaecyparis obtusa* var. *breviramea*	CupChaob2
Cupress	*Fitzroya cupressoides*	CupFitcup
Cupress	*Juniperus bermudiana*	CupJunber
Cupress	*Juniperus formosana*	CupJunfor
Cupress	*Juniperus phoenica*	CupJunpho
Cupress	*Juniperus sabina*	CupJunsab
Cupress	*Libocedrus bidwillii*	CupLibbid
Cupress	*Libocedrus yateensis*	CupLibyat
Cupress	*Thuja plicata*	CupThupli
Cupress	*Thuja standishi*	CupThusta
Ephedr	*Ephedra alata*	EphEphala
Pin	*Abies nephrolepis*	PinAbinep
Pin	*Cedrus deodara*	PinCeddeo
Pin	*Larix decidua*	PinLardec
Pin	*Larix leptolepis*	PinLarlep
Pin	*Picea abies*	PinPicabi
Pin	*Picea ajanensis*	PinPicaja
Pin	*Picea excelsa*	PinPicexc
Pin	*Picea jezoensis*	PinPicjez
Pin	*Picea obovata*	PinPicobo
Pin	*Picea vulgaris*	PinPicvul
Pin	*Pinus laricio*	PinPinlar
Pin	*Pinus massoniana*	PinPinmas
Pin	*Pinus sibirica*	PinPinsib
Pin	*Tsuga chinensis* var. *formosana*	PinTsuchi
Pin	*Tsuga heterophylla*	PinTsuhet
Pin	*Tsuga sieboldii*	PinTsusie
Podocarp	*Dacrydium intermedium*	PodDacint
Podocarp	*Podocarpus spicatus*	PodPodspi
Pterid	*Pteris vittata*	PtePtevit
Tax	*Taxus baccata*	TaxTaxbac
Tax	*Taxus cuspidata*	TaxTaxcus
Tax	*Taxus wallichiana*	TaxTaxwal
Taxodi	*Taiwania cryptomerioides*	TaxTaicry

which they co-occur with neolignans of many other structural types (Table 7.3.3). Furthermore, neolignans of *different* carbon-skeletons (e.g. **7**, **8**, **9**) very probably belong to the *same* biogenetic type. All three are clearly derived from identical propenylphenol (**15a**) *plus* allylphenol (**15b**) precursors, and occur together in the same or in related plant species. At least in phytochemical terms, it would be misleading indeed to consider them in different structurally oriented groups. Finally, neolignans of *identical* carbon skeletons (e.g. **10**, **11**, **12**) clearly belong to *different* biogenetic types and it would be misleading indeed to consider them in the

Table 7.3.3. Lignan- and neolignan-containing angiospermous superorders (common termination -iflorae), orders (-ales), families (-aceae) and species

Superorder	Order	Family	Genus/species	Code
Magnoli	Illici	Schisandr	*Kadsura coccinea*	SchKadcoc
Magnoli	Illici	Schisandr	*Kadsura japonica*	SchKadjap
Magnoli	Illici	Schisandr	*Kadsura longipedunculata*	SchKadlon
Magnoli	Illici	Schisandr	*Schisandra chinensis*	SchSchchi
Magnoli	Illici	Schisandr	*Schisandra henryi*	SchSchhen
Magnoli	Illici	Schisandr	*Schisandra rubriflora*	SchSchrub
Magnoli	Illici	Schisandr	*Schisandra* sp.	SchSchsph
Magnoli	Illici	Schisandr	*Schisandra sphenanthera*	SchSchsp
Magnoli	Magnoli	Magnoli	*Liriodendron tulipifera*	MagLirtul
Magnoli	Magnoli	Magnoli	*Magnolia acuminata*	MagMagacu
Magnoli	Magnoli	Magnoli	*Magnolia denudata*	MagMagden
Magnoli	Magnoli	Magnoli	*Magnolia fargesii*	MagMagfar
Magnoli	Magnoli	Magnoli	*Magnolia grandiflora*	MagMaggra
Magnoli	Magnoli	Magnoli	*Magnolia kachirachirai*	MagMagkac
Magnoli	Magnoli	Magnoli	*Magnolia kobus*	MagMagkob
Magnoli	Magnoli	Magnoli	*Magnolia liliflora*	MagMaglil
Magnoli	Magnoli	Magnoli	*Magnolia officinale*	MagMagoff
Magnoli	Magnoli	Magnoli	*Magnolia salicifolia*	MagMagsal
Magnoli	Magnoli	Magnoli	*Magnolia stellata*	MagMagste
Magnoli	Magnoli	Magnoli	*Talauma hodgsoni*	MagTalhod
Magnoli	Magnoli	Himantandr	*Himantandra baccata*	HimHimbac
Magnoli	Magnoli	Himantandr	*Himantandra belgraveana*	HimHimbel
Magnoli	Annon	Myristic	*Horsfieldia iryaghedi*	MyrHoriry
Magnoli	Annon	Myristic	*Iryanthera grandis*	MyrIrygra
Magnoli	Annon	Myristic	*Myristica cagayanensis*	MyrMyrcag
Magnoli	Annon	Myristic	*Myristica fragrans*	MyrMyrfra
Magnoli	Annon	Myristic	*Myristica malabarica*	MyrMyrmal
Magnoli	Annon	Myristic	*Myristica otoba*	MyrMyroto
Magnoli	Annon	Myristic	*Myristica simarum*	MyrMyrsim
Magnoli	Annon	Myristic	*Osteophloeum platyspermum*	MyrOstpla
Magnoli	Annon	Myristic	*Virola calophylloidea*	MyrVircal
Magnoli	Annon	Myristic	*Virola carinata*	MyrVircar
Magnoli	Annon	Myristic	*Virola cuspidata*	MyrVircus
Magnoli	Annon	Myristic	*Virola elongata*	MyrVirelo
Magnoli	Annon	Myristic	*Virola guggenheimii*	MyrVirgug
Magnoli	Annon	Myristic	*Virola pavonis*	MyrVirpav
Magnoli	Annon	Myristic	*Virola sebifera*	MyrVirseb
Magnoli	Annon	Myristic	*Virola surinamensis*	MyrVirsur
Magnoli	Annon	Cannel	*Cinnamosma madagascarinensis*	CanCinmad
Magnoli	Annon	Eupomati	*Eupomatia laurina*	EupEuplau
Magnoli	Laur	Trimeni	*Piptocalix moorei*	TriPipmoo
Magnoli	Laur	Trimeni	*Trimenia papuana*	TriTripap
Magnoli	Laur	Austrobailey	*Austrobaileya scandens*	AusAussca
Magnoli	Laur	Monimi	*Laurelia novae-zelandiae*	MonLaunov
Magnoli	Laur	Laur	*Aniba affinis*	LauAniaff
Magnoli	Laur	Laur	*Aniba burchellii*	LauAnibur
Magnoli	Laur	Laur	*Aniba citrifolia*	LauAnicit
Magnoli	Laur	Laur	*Aniba ferrea*	LauAnifer
Magnoli	Laur	Laur	*Aniba guianensis*	LauAnigui
Magnoli	Laur	Laur	*Aniba lancifolia*	LauAnilan
Magnoli	Laur	Laur	*Aniba megaphylla*	LauAnimeg
Magnoli	Laur	Laur	*Aniba simulans*	LauAnisim

Table 7.3.3 (continued)

Superorder	Order	Family	Genus/species	Code
Magnoli	Laur	Laur	*Aniba* sp.	LauAnisp
Magnoli	Laur	Laur	*Aniba* sp. 41	LauAni41
Magnoli	Laur	Laur	*Aniba terminalis*	LauAniter
Magnoli	Laur	Laur	*Cinnamomum camphora*	LauCincam
Magnoli	Laur	Laur	*Cinnamomum* sp.	LauCinsp
Magnoli	Laur	Laur	*Eusideroxylon zwageri*	LauEuszwa
Magnoli	Laur	Laur	*Licaria aritu*	LauLicari
Magnoli	Laur	Laur	*Licaria armeniaca*	LauLicarm
Magnoli	Laur	Laur	*Licaria canella*	LauLiccan
Magnoli	Laur	Laur	*Licaria chrysophylla*	LauLicchr
Magnoli	Laur	Laur	*Licaria macrophylla*	LauLicmac
Magnoli	Laur	Laur	*Licaria rigida*	LauLicrig
Magnoli	Laur	Laur	*Licaria* sp. 04	LauLic04
Magnoli	Laur	Laur	*Licaria* sp. 41	LauLic41
Magnoli	Laur	Laur	*Licaria* sp. 63	LauLic63
Magnoli	Laur	Laur	*Litsea gracilipes*	LauLitgr1
Magnoli	Laur	Laur	*Litsea grandis*	LauLitgr2
Magnoli	Laur	Laur	*Litsea turfosa*	LauLittur
Magnoli	Laur	Laur	*Machilus edulis*	LauMacedu
Magnoli	Laur	Laur	*Machilus japonica*	LauMacjap
Magnoli	Laur	Laur	*Machilus thumbergii*	LauMacthu
Magnoli	Laur	Laur	*Machilus zuihoensis* var. *long*	LauMaczui
Magnoli	Laur	Laur	*Mezilaurus itauba*	LauMezita
Magnoli	Laur	Laur	*Nectandra miranda*	LauNecmir
Magnoli	Laur	Laur	*Nectandra polita*	LauNecpol
Magnoli	Laur	Laur	*Nectandra puberula*	LauNecpub
Magnoli	Laur	Laur	*Nectandra rigida*	LauNecrig
Magnoli	Laur	Laur	*Nectandra* sp.	LauNecsp
Magnoli	Laur	Laur	*Ocotea aciphylla*	LauOcoaci
Magnoli	Laur	Laur	*Ocotea catharinensis*	LauOcocat
Magnoli	Laur	Laur	*Ocotea cymbarum*	LauOcocym
Magnoli	Laur	Laur	*Ocotea porosa*	LauOcopor
Magnoli	Laur	Laur	*Ocotea* sp.	LauOcosp
Magnoli	Laur	Laur	*Ocotea veraguensis*	LauOcover
Magnoli	Laur	Laur	*Parabenzoin trilobum*	LauPartri
Magnoli	Laur	Laur	*Sassafras randaiense*	LauSasran
Magnoli	Laur	Laur	*Urbanodendron verruscosum*	LauUrbver
Magnoli	Laur	Hernandi	*Hernandia cordigera*	HerHercor
Magnoli	Laur	Hernandi	*Hernandia ovigera*	HerHerovi
Magnoli	Laur	Hernandi	*Hernandia peltata*	HerHerpel
Magnoli	Aristolochi	Aristolochi	*Aristolochia albida*	AriArialb
Magnoli	Aristolochi	Aristolochi	*Aristolochia indica*	AriAriind
Magnoli	Aristolochi	Aristolochi	*Aristolochia* sp.	AriArisp
Magnoli	Aristolochi	Aristolochi	*Aristolochia taliscana*	AriArital
Magnoli	Aristolochi	Aristolochi	*Aristolochia triangularis*	AriAritri
Magnoli	Aristolochi	Aristolochi	*Asarum sieboldi*	AriAsasie
Magnoli	Aristolochi	Aristolochi	*Asarum taitonense*	AriAsatai
Magnoli	Aristolochi	Aristolochi	*Heterotropa takahoi*	AriHettak
Nympha	Piper	Saurur	*Saururus cernuus*	SauSaucer
Nympha	Piper	Piper	*Macropiper excelsum*	PipMacexc
Nympha	Piper	Piper	*Piper brachystachyum*	PipPipbra
Nympha	Piper	Piper	*Piper clusii*	PipPipclu
Nympha	Piper	Piper	*Piper cubeba*	PipPipcup

Table 7.3.3 (continued)

Superorder	Order	Family	Genus/species	Code
Nympha	Piper	Piper	*Piper futokadsura*	PipPipfut
Nympha	Piper	Piper	*Piper guineense*	PipPipgui
Nympha	Piper	Piper	*Piper lacunosum*	PipPiplac
Nympha	Piper	Piper	*Piper peepuloides*	PipPippee
Nympha	Piper	Piper	*Piper sylvaticum*	PipPipsyl
Ranuncul	Ranuncul	Berberid	*Berberis buxifolia*	BerBerbux
Ranuncul	Ranuncul	Berberid	*Berberis chilensis*	BerBerchi
Ranuncul	Ranuncul	Berberid	*Diphylleia grayi*	BerDipgra
Ranuncul	Ranuncul	Berberid	*Epimedium grandiflorum*	BerEpigra
Ranuncul	Ranuncul	Berberid	*Podophyllum emodi*	BerPodem1
Ranuncul	Ranuncul	Berberid	*Podophyllum emodi* var. *hexandrum*	BerPodem2
Ranuncul	Ranuncul	Berberid	*Podophyllum hexandrum*	BerPodhex
Ranuncul	Ranuncul	Berberid	*Podophyllum peltatum*	BerPodpel
Ranuncul	Ranuncul	Berberid	*Podophyllum pleianthum*	BerPodple
Ranuncul	Ranuncul	Berberid	*Podophyllum sikkimensis*	BerPodsik
Ranuncul	Ranuncul	Menisperm	*Tinospora cordifolia*	MenTincor
Malv	Euphorbi	Euphorbi	*Amanoa oblongifolia*	EupAmaobl
Malv	Euphorbi	Euphorbi	*Cleistanthus collinus*	EupClecol
Malv	Euphorbi	Euphorbi	*Cleistanthus patulus*	EupClepat
Malv	Euphorbi	Euphorbi	*Jatropha gossypifolia*	EupJatgos
Malv	Euphorbi	Euphorbi	*Phyllanthus niruri*	EupPhynir
Malv	Urtic	Ulmac	*Ulmus thomasii*	UlmUlmtho
Malv	Urtic	Urticac	*Boehmeria tricuspis*	UrtBoetri
Malv	Thymelae	Thymelae	*Daphne oleoides*	ThyDapole
Malv	Thymelae	Thymelae	*Daphne tangutica*	ThyDaptan
Malv	Thymelae	Thymelae	*Dirca occidentalis*	ThyDapocc
Malv	Thymelae	Thymelae	*Gnidia latifolia*	ThyGnilat
Malv	Thymelae	Thymelae	*Stellera chamaejasme*	ThyStecha
Malv	Thymelae	Thymelae	*Wikstroemia foetida* var. *oahuensis*	ThyWikfoe
Malv	Thymelae	Thymelae	*Wikstroemia indica*	ThyWikind
Malv	Thymelae	Thymelae	*Wikstroemia uva-ursi*	ThyWikuva
Viol	Viol	Curcubit	*Herpestospermum caudigerium*	CurHercau
Primul	Eben	Eben	*Diospyrus kaki*	EbeDiokak
Ros	Fag	Betul	*Alnus glutinosa*	BetAlnglu
Ros	Ros	Rosac	*Prinsepia utilis*	RosPriuti
Ros	Ros	Rosac	*Pygeum acuminatum*	RosPygacu
Myrt	Myrt	Combret	*Conocarpus erectus*	ComConere
Myrt	Myrt	Combret	*Eucalyptus hemiploia*	ComEuchem
Rut	Rut	Rut	*Acronychia muelleri*	RutAcrmue
Rut	Rut	Rut	*Fagara boninensis*	RutFagbon
Rut	Rut	Rut	*Fagara xanthoxyloides*	RutFagxan
Rut	Rut	Rut	*Haplophyllum dauricum*	RutHapdau
Rut	Rut	Rut	*Haplophyllum hispanicum*	RutHaphis
Rut	Rut	Rut	*Haplophyllum obtusifolium*	RutHapobt
Rut	Rut	Rut	*Haplophyllum perforatum*	RutHapper
Rut	Rut	Rut	*Haplophyllum popovii*	RutHappop
Rut	Rut	Rut	*Haplophyllum tuberculatum*	RutHaptub
Rut	Rut	Rut	*Haplophyllum versicolor*	RutHapver
Rut	Rut	Rut	*Haplophyllum vulcanicum*	RutHapvul
Rut	Rut	Rut	*Ruta graveolens*	RutRutgra
Rut	Rut	Rut	*Ruta microcarpa*	RutRutmic
Rut	Rut	Rut	*Ruta pinnata*	RutRutpin

Table 7.3.3 (continued)

Superorder	Order	Family	Genus/species	Code
Rut	Rut	Rut	*Zanthoxylum acanthopodium*	RutZanaca
Rut	Rut	Rut	*Zanthoxylum alanthoides*	RutZanala
Rut	Rut	Rut	*Zanthoxylum alatum*	RutZanala
Rut	Rut	Rut	*Zanthoxylum arnottianum*	RutZanarn
Rut	Rut	Rut	*Zanthoxylum capense*	RutZancap
Rut	Rut	Rut	*Zanthoxylum chalybeum*	RutZancha
Rut	Rut	Rut	*Zanthoxylum dinklagei*	RutZandin
Rut	Rut	Rut	*Zanthoxylum inerme*	RutZanine
Rut	Rut	Rut	*Zanthoxylum oxophyllum*	RutZanoxo
Rut	Rut	Rut	*Zanthoxylum piperitum*	RutZanpip
Rut	Rut	Rut	*Zanthoxylum pluvitile*	RutZanplu
Rut	Rut	Rut	*Zanthoxylum podocarpum*	RutZanpod
Rut	Rut	Rut	*Zanthoxylum senegalense*	RutZansen
Rut	Rut	Rut	*Zanthoxylum* sp.	RutZansp
Rut	Rut	Rut	*Zanthoxylum viride*	RutZanvir
Rut	Rut	Burser	*Bursera ariensis*	BurBurari
Rut	Rut	Burser	*Bursera fagaroides*	BurBurfag
Rut	Rut	Burser	*Bursera microphylla*	BurBurmic
Rut	Rut	Burser	*Bursera morelensis*	BurBurmor
Rut	Rut	Burser	*Commiphora incisa*	BurCominc
Rut	Gerani	Zygophyll	*Guaiacum officinale*	ZygGuaoff
Rut	Gerani	Zygophyll	*Larrea cuneifolia*	ZygLarcun
Rut	Gerani	Zygophyll	*Larrea divaricata*	ZygLardiv
Rut	Gerani	Zygophyll	*Larrea nitida*	ZygLarnit
Rut	Polygal	Krameri	*Krameria cistisoides*	KraKracis
Rut	Polygal	Krameri	*Krameria ramosissima*	KraKraram
Rut	Polygal	Krameri	*Krameria triandra*	KraKratri
Rut	Polygal	Polygal	*Polygala chinensis*	PolPolchi
Rut	Polygal	Polygal	*Polygala polygama*	PolPolpol
Rut	Polygal	Krameri	*Ratanhiae radix*	KraRatrad
Santal	Santal	Visc	*Viscum album* var. *coloratum*	VisVisalb
Arali	Arali	Arali	*Acanthopanax chiisanensis*	AraAcachi
Arali	Arali	Arali	*Acanthopanax koreanum*	AraAcakor
Arali	Arali	Api	*Anthriscus nemerosa*	ApiAntnem
Arali	Arali	Api	*Anthriscus sylvestris*	ApiAntsyl
Arali	Arali	Api	*Bupleurum frutiscescens*	ApiBupfru
Arali	Arali	Api	*Chaerophyllum maculatum*	ApiChamac
Arali	Arali	Api	*Steganotaenia araliacea*	ApiSteara
Aster	Aster	Aster	*Anacyclus pyrethrum*	AstAnapyr
Aster	Aster	Aster	*Arctium lappa*	AstArclap
Aster	Aster	Aster	*Arctium leiospermum*	AstArclei
Aster	Aster	Aster	*Artemisia absinthium*	AstArtabs
Aster	Aster	Aster	*Helichrysum bracteatum*	AstHelbra
Aster	Aster	Aster	*Heliopsis buphthalmoides*	AstHelbup
Aster	Aster	Aster	*Heliopsis scabra*	AstHelsca
Aster	Aster	Aster	*Lappa major*	AstLapmaj
Aster	Aster	Aster	*Lappa minor*	AstLapmin
Aster	Aster	Aster	*Lappa tomentosa*	AstLaptom
Aster	Aster	Aster	*Onopordon alexandrinum*	AstOnoale
Corn	Eucommi	Eucommi	*Eucommia ulmoides*	EucEuculm
Corn	Eric	Eric	*Lyonia ovatifolia* var. *elliptica*	EriLyoova
Corn	Corn	Hydrange	*Forsythia arten*	HydForart
Corn	Corn	Hydrange	*Forsythia suspensa*	HydForsus

Table 7.3.3 (continued)

Superorder	Order	Family	Genus/species	Code
Corn	Corn	Hydrange	*Forsythia verdissima*	HydForver
Corn	Corn	Hydrange	*Forsythia coreana*	HydForcor
Gentian	Gentian	Rubi	*Hedyotis lawsoniae*	RubHedlaw
Gentian	Gentian	Apocyn	*Carissa edulis*	ApoCaredu
Gentian	Gentian	Apocyn	*Trachelospermum asiaticum* var. *intermedium*	ApoTraasi
Gentian	Gentian	Apocyn	*Trachelospermum foetidum*	ApoTrafoe
Gentian	Gentian	Apocyn	*Trachelospermum liukiuense*	ApoTraliu
Gentian	Gentian	Logani	*Buddleja davidii*	LogBuddav
Gentian	Gentian	Ole	*Fraxinus japonica*	OleFrajap
Gentian	Gentian	Ole	*Fraxinus mandshurica*	OleFraman
Gentian	Gentian	Ole	*Lingstrum japonicum*	OleLinjap
Gentian	Gentian	Ole	*Olea africana*	OleOleafr
Gentian	Gentian	Ole	*Olea capensis*	OleOlecap
Gentian	Gentian	Ole	*Olea cunninghamii*	OleOlecun
Gentian	Gentian	Ole	*Olea europaea*	OleOleeur
Gentian	Dipsac	Caprifoli	*Lonicera hypoleuca*	CapLonhyp
Lami	Scrophulari	Scrophulari	*Aptosimum spinescens*	ScrAptspi
Lami	Scrophulari	Scrophulari	*Lancea tibetica*	ScrLantib
Lami	Scrophulari	Scrophulari	*Paulownia tomentosa*	ScrPautom
Lami	Scrophulari	Globulari	*Globularia alypum*	GloGloaly
Lami	Scrophulari	Pedali	*Sesamum angolense*	PedSesang
Lami	Scrophulari	Acanth	*Justicia extensa*	AcaJusext
Lami	Scrophulari	Acanth	*Justicia flava*	AcaJusfla
Lami	Scrophulari	Acanth	*Justicia hayatay* var. *decumbens*	AcaJushay
Lami	Scrophulari	Acanth	*Justicia procumbens*	AcaJuspr1
Lami	Scrophulari	Acanth	*Justicia procumbens* var. *leucantha*	AcaJuspr2
Lami	Scrophulari	Acanth	*Justicia prostata*	AcaJuspr3
Lami	Scrophulari	Acanth	*Justicia simplex*	AcaJussim
Lami	Scrophulari	Acanth	*Playlopsis falcisepala*	AcaPlafal
Lami	Scrophulari	Bignoni	*Dolichandrone crispa*	BigDolcri
Lami	Scrophulari	Bignoni	*Kigelia pinnata*	BigKigpin
Lami	Scrophulari	Bignoni	*Stereospermum kunthianum*	BigStekun
Lami	Lami	Verben	*Clerodendron inerme*	VerCleine
Lami	Lami	Verben	*Gmelina arborea*	VerGmearb
Lami	Lami	Verben	*Gmelina asiatica*	VerGmeasi
Lami	Lami	Verben	*Gmelina leichhardtii*	VerGmelei
Lami	Lami	Lami	*Sideritis canariensis*	LamSidcan
Lami	Lami	Lami	*Thymus longiflorus*	LamThylon
Ar	Ar	Ar	*Acorus calamus*	AraAcocal

same structurally oriented group. As suggested by the oxygenation pattern, the formation of **10** (bridgeheads 8.3′) and of **11** (bridgeheads 8.5′) should involve two different types of direct oxidative coupling of **15a** and **15b**.

At this stage, the inclusion of the C_6-C_3-dimers in their putative biogenetic class is much more important than a decision concerning the lignan/neolignan controversy. Indeed it has already been pointed out (54) that there is no need to add the prefix "neo" – in complete designations. Such designations are in themselves sufficient to discriminate between, for example, (8*S*,8′*R*)-4′-hydroxy-

3,4,3′-trimethoxy-8.8′,9.O.9′-lignan (**5**) and (8*R*,8′*R*)-4,4′-dihydroxy-3,3′-dimethoxy-8.8′-lignan (**6**), the former a lignan by virtue of the presence of oxygen on both the 9 and the 9′ positions, and the latter a neolignan by virtue of the absence of oxygen from at least one of the 9 or 9′ positions.

The construction of a comprehensive system of nomenclature requires some additional points to be considered. Numbering of the carbons of the two C_6-C_3 units from 1 to 9 and from 1′ to 9′ has been mentioned. In neolignans, whenever possible or meaningful, the unprimed positions are assigned to the propenylphenol-derived C_6-C_3 unit coupled *via* a C-8 bridgehead to any of the primed positions of the additional C_6-C_3 unit. For the purpose of numbering, the putative precursors (e.g. **15a, 15b**) are numbered first. These numbers are then transferred to the analogous positions of the dimers (or oligomers) (e.g. **10, 11**). The advantage of this procedure over the previous one, suggesting the assignment of the smallest possible numerals for bridgeheads (54), is exemplified by **10, 11** and **12**. These structures show identical carbon skeletons but clearly do not belong to the same biogenetic type. As suggested by the oxygenation pattern, the formation of **10** (bridgeheads 8.3′) and of **11** (bridgeheads 8.5′) represent two different types of direct oxidative coupling of **15a** and **15b**, while the formation of **12** requires additionally the rearrangement of the allyl group from 1′ to 5′ (i.e., compound **12** belongs to type 8.1′, allyl 1′→5′).

The graphic representation of structural types and subtypes of lignoids is formulated in Figs. 7.3.1 through 7.3.7 to impress upon the reader the considerable number of structural variants of the C_6-C_3-dimers in nature. The same information, completed by stereochemical details, is given in Tables 7.3.4 and 7.3.5. In these tables, all lignoids are classified according to structural types, defined by the positions of the biogenetically important bridgeheads. In the definitions of types only C−C bonds between the monomeric units are considered. C−O−C bridges characterize types only if forming the sole link between the two units. Usually, however, C−O−C bridges define subtypes and then only if incorporated in a cycle.

It is highly probable that this proposal concerning nomenclature and numbering of lignoids will not meet everybody's approval. Indeed skepticism is justified at least on one point. We have made some previous proposals in quest of the perfect system and thus may have involuntarily caused confusion in the minds of our readers. Nevertheless, the long pathway to the present system was well worthwhile. As here proposed (Tables 7.3.4 and 7.3.5), the system is totally exempt of trivial designations and other non-systematic items; the symbols used in the description of any compound are familiar to all organic chemists; the designation of any compound is formed by a linear and usually brief string of symbols; and the coding and decoding processes yield unique results. Most importantly the system is "closed" (55) – i.e. applicable to *any* lignoid, including previously

Fig. 7.3.1. Skeletons of 8.8′-lignans and neolignans and of the rearrangement products of the latter. In this and the following figures the presently known numbers of lignans (L) and neolignans (N) are registered for each structural subtype

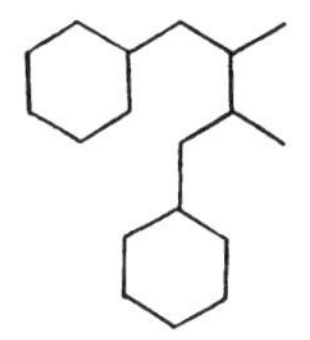

8.8'
L13 N18

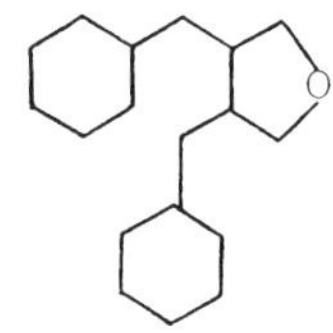

8.8',9.O.9'
L58

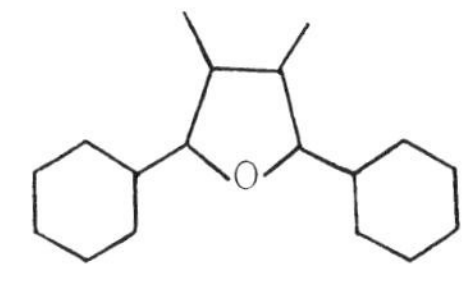

8.8',7.O.7'
L2 N15

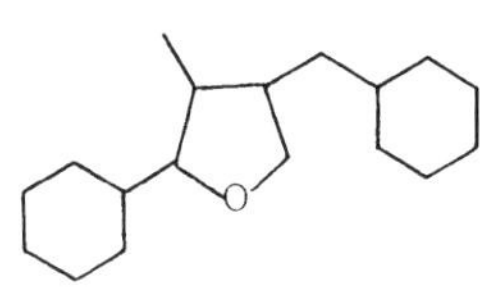

8.8',7.O.9'
L16 N2

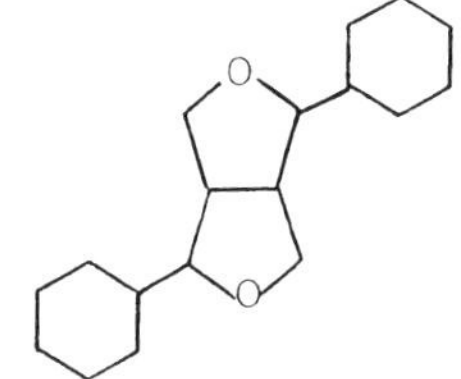

8.8',7.O.9',9.O.7'
L70

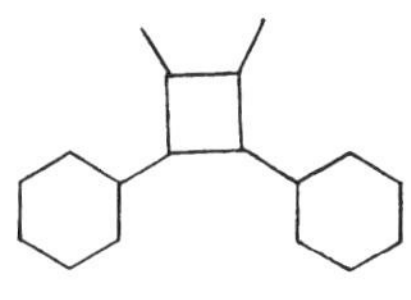

8.8',7.7'
N3

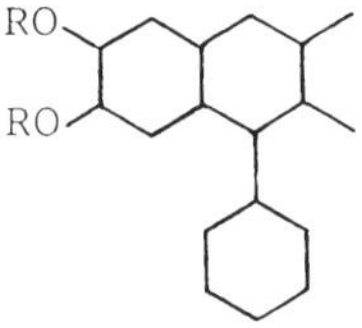

RO
RO

8.8',6.7'
L14 N40

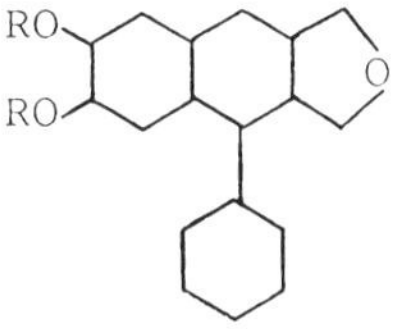

RO
RO

8.8',6.7',9.O.9'
L68

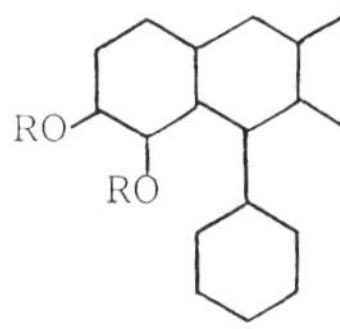

8.8',2.7'
L1 N2

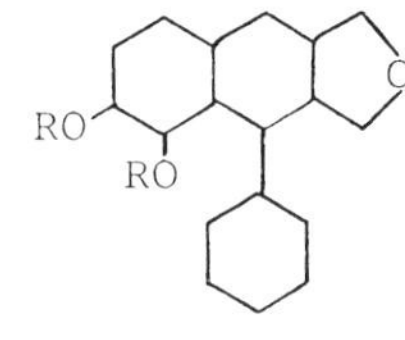

8.8',2.7',9.O.9'
L5

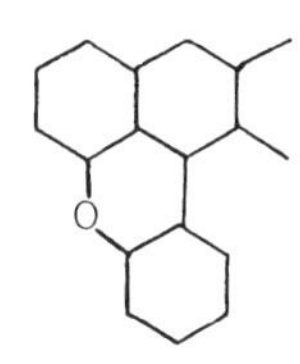

8.8',2.7',3.O.6'
N1

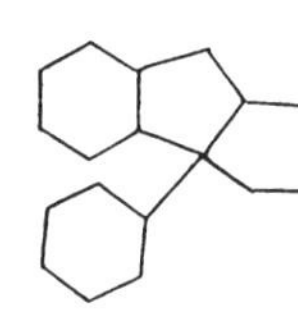

8.8',2.7',8:8'→7'
N2

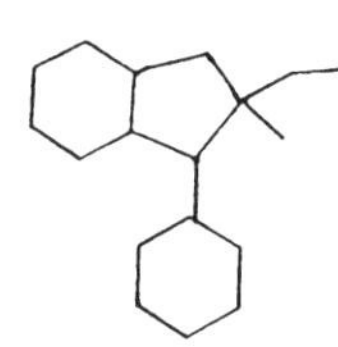

8.8',2.7',7:8→8'
N1

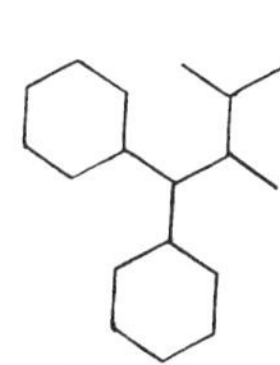

8.8',6.7', seco 1.7
N1

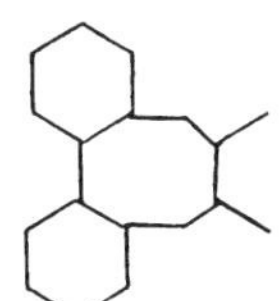

8'8',2.2'
N47

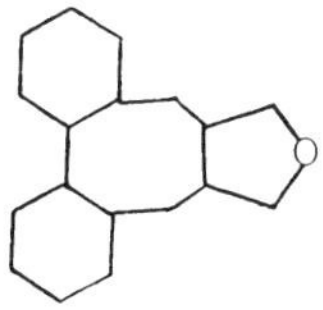

8.8',2.2',9.O.9'
L4

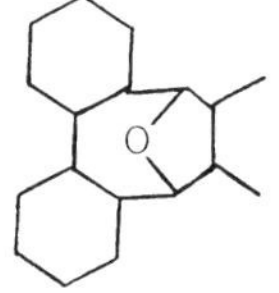

8.8',2.2',7.O.7'
N3

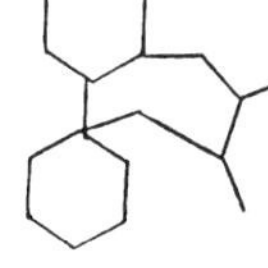

8.8',2.1'
N3

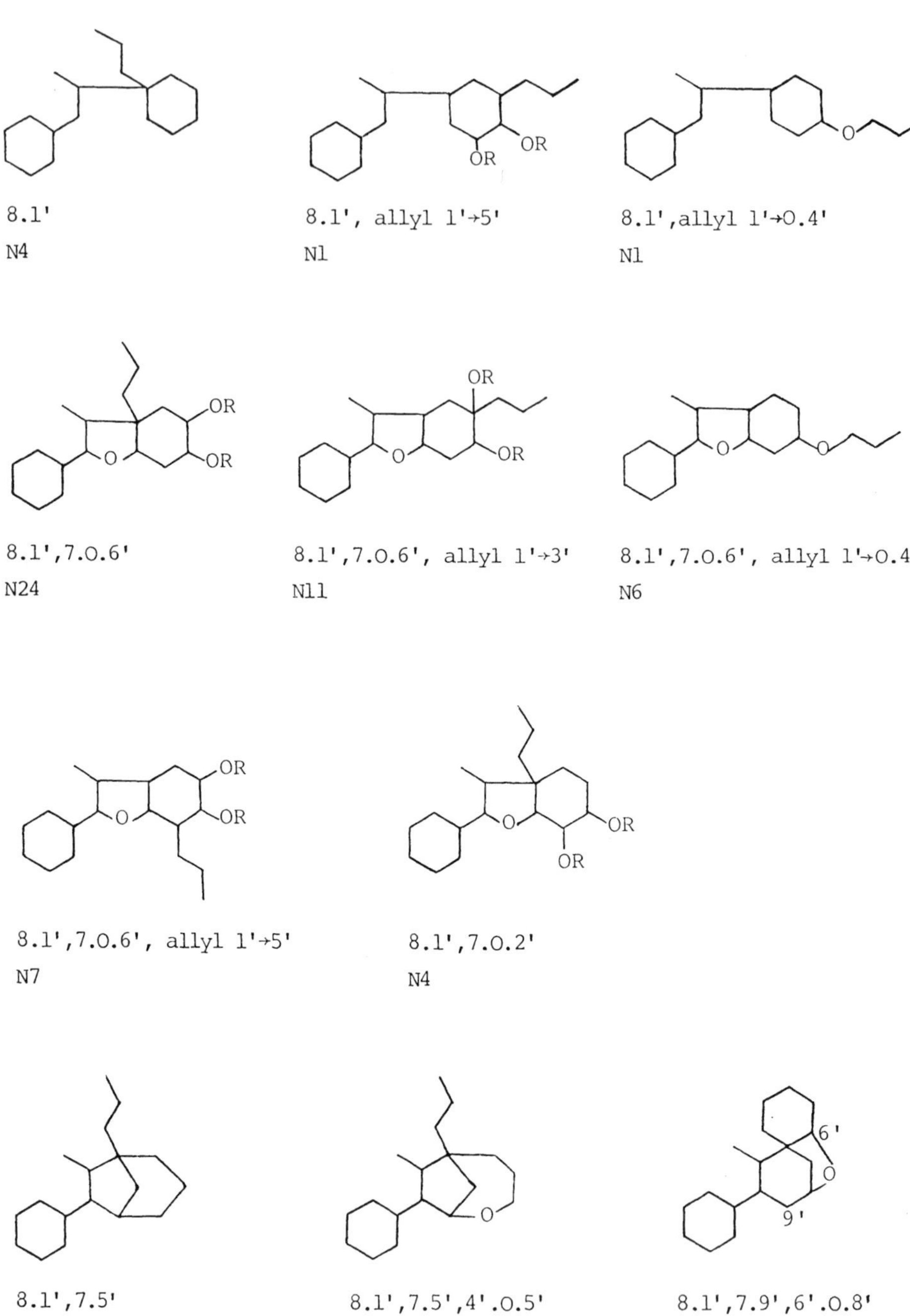

Fig. 7.3.2. Skeletons of 8.1′-neolignans and of their rearrangement products

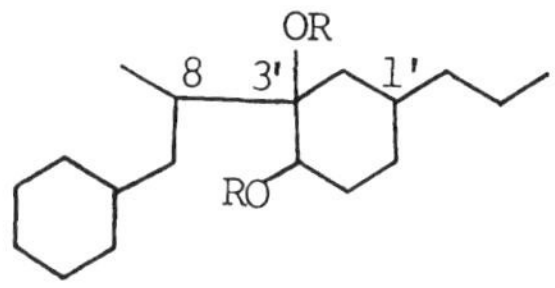

8.3'

N7

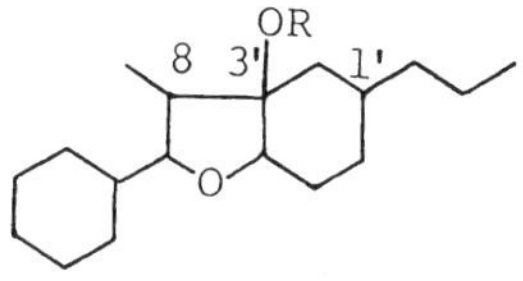

8.3',7.O.4'

N9

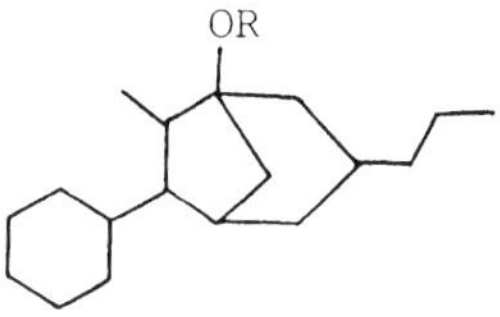

8.3',7.5'

N3

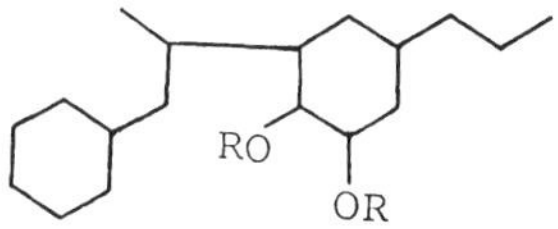

8.5'

N2

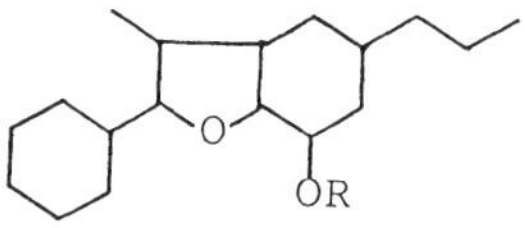

8.5',7.O.4'

L2 N26

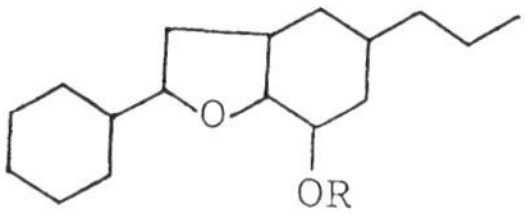

8.5',7.O.4',9-nor

N7

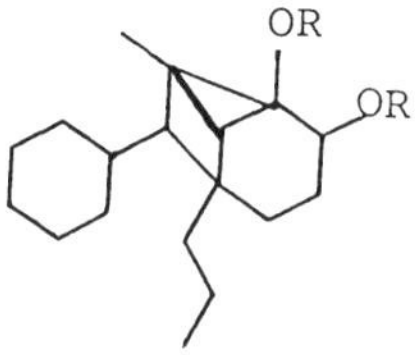

8.3',8.2',7.1'

N1

Fig. 7.3.3. Skeletons of 8.3'- and 8.5'-neolignans

7.1',8.O.2'

N4

7.1',8.O.2',allyl 1'→3'

N2

7.1',8.O.2',allyl 1'→O.4'

N1

8.7',7.O.8'

N1

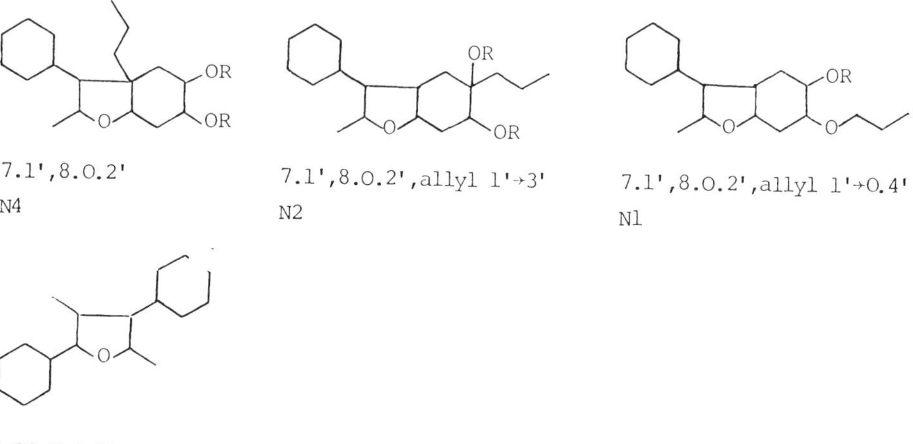

Fig. 7.3.4. Skeletons of some rarer 7.1'- and 8.7'-neolignans

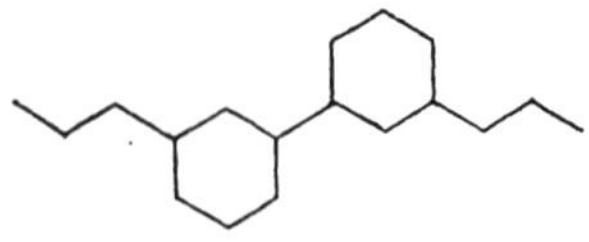

5.5'

N9

Fig. 7.3.5. Skeleton of 5.5′-neolignans

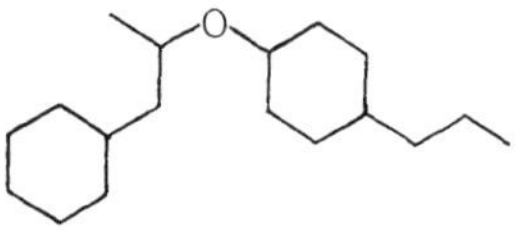

O

8.O.4'

N11

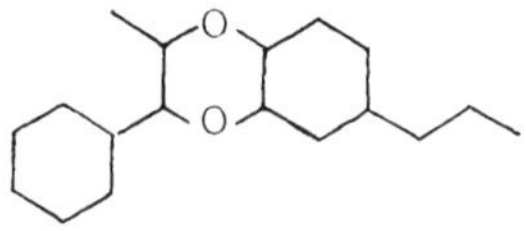

O

O

8.O.4',7.O.3'

N9

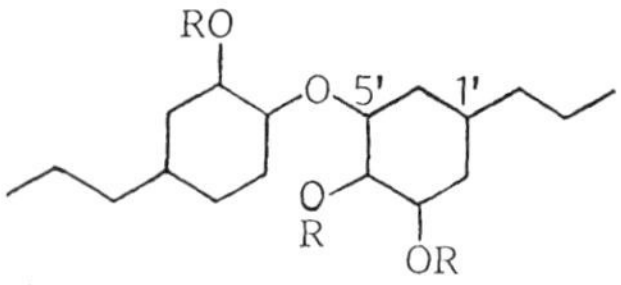

RO

O 5' 1'

O

R

OR

4.O.5'

N2

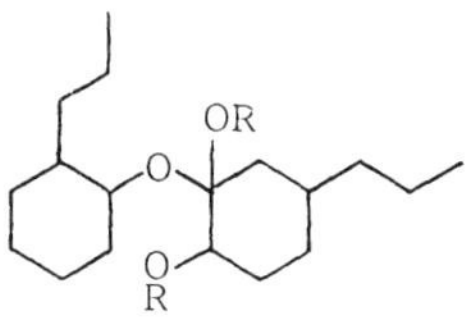

OR

O

O

R

2.O.3'

N1

Fig. 7.3.6. Skeletons of 8.O.4′- and of other neolignans with C–O–C bridges

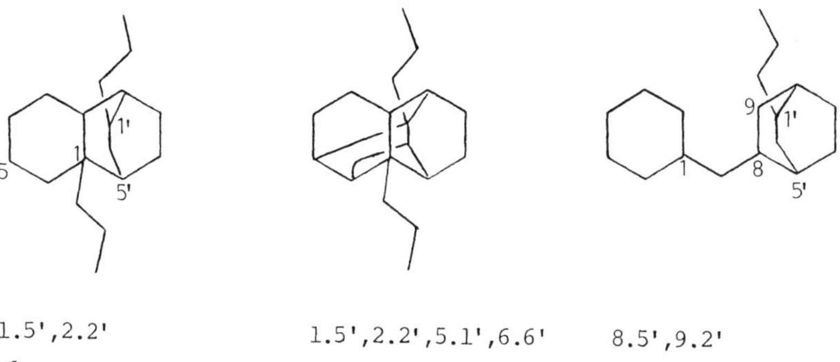

1.5',2.2'

N1

1.5',2.2',5.1',6.6'

N1

8.5',9.2'

N2

Fig. 7.3.7. Skeletons of 1.5′,2.2′- and of 8.5′,9.2′-neolignans

unknown, unusual, and complicated ones. For example, previously devised conventions of lignoid nomenclature are hardly applicable to the case of isofutoquinol A, while in our proposal the skeleton is conveyed with utmost ease by the short expression 8.3′, 8.2′, 7.1′ (Fig. 7.3.3). The location of functional groups on this skeleton is of course no problem (see type 8.3′, subtype 8.3′, 8.2′, 7.1′ in Table 7.3.5).

Table 7.3.4. Lignan structures classified according to skeletal types and subtypes

Type	Subtype	Compound	Trivial name	Occurrence in species	Ref.
8.8′	8.8′	8*R*,8′*R*; OH: 9,9′; O_2CH_2: 3,4,3′,4′; *Δ*: 1,3,5,1′,3′,5′	dihydrocubebin	MyrHoriry PipPipclu PipPipcub	148 164 118
		8*R*,8′*R*; OH: 4,9,4′,9′; OMe: 3,3′; *Δ*: 1,3,5,1′,3′,5′	secoisolariciresinol	ApoCaredu RutZanala PinLarlep	164 66 142
		OH: 4,9,4′,9′; OMe: 3,3′; *Δ*: 1,3,5,1′,3′,5′	mesosecoisolari-ciresinol	AriAriind PinCeddeo	1 5
		OH: 9,4′,9′; OMe: 3,4,3′; *Δ*: 1,3,5,1′,3′,5′	mesosecioisolari-ciresinol monomethylether	AraAraang	164
		8*S*,8′*S*; OH: 4,9,4′,8′,9′; OMe: 3,3′; *Δ*: 1,3,5,1′,3′,5′	carinol	ApoCaredu	164
		8*S*,8′*S*; O_2CH_2: 3,4; OMe: 5,9,3′,4′,9′; *Δ*: 1,3,5,1′,3′,5′	(+)-niranthin	EupPhynir	122
		8*S*,8′*S*; OMe: 3,4,9,3′,4′,9′; *Δ*: 1,3,5,1′,3′,5′	phyllanthin	EupPhynir	122
		8*R*,8′*R*; OAc: 9,9′; O_2CH_2: 3,4,3′,4′; *Δ*: 1,3,5,1′,3′,5′	ariensin	BurBurari	60
		8*R*,8′*R*; OH: 9,9′; OMe: 3′,4′,5′; O_2CH_2: 3,4; *Δ*: 1,3,5,1′,3′,5′	dihydroclusin	PipPipcub	122
		OMe; 9,9′; O_2CH_2: 3,4,3′,4′; oxo: 9,9′; *Δ*: 1,3,5,1′,3′,5′	heliobuphthalmin	BerPodhex AstHelbup	69 17
		O_2CH_2: 3,4,3′,4′; oxo: 7,7′; *Δ*: 1,3,5,8(8′),1′,3′,5′	zuihonin D	LauMaczui	164
		OMe: 9,9′; oxo: 9,9′; O_2CH_2: 3,4,3′,4′; *Δ*: 1,3,5,7,1′,3′,5′	dehydrohelio-buphthalmin	AstHelbup	17, 73
		OH: 8; OMe: 9,9′; oxo: 9,9′; O_2CH_2: 3,4,3′,4′; *Δ*: 1,3,5,1′,3′,5′	8-hydroxyhelio-buphthalmin	AstHelbup	73
	8.8′, 9.O.9′	O_2CH_2: 3,4,3′,4′; oxo: 9′; *Δ*: 1,3,5,7,1′,3′,5′,7′	taiwanin A	PipPiplac	156

Table 7.3.4 (continued)

Type	Subtype	Compound	Trivial name	Occurrence in species	Ref.
		8*R*; O_2CH_2: 3,4,3,',4'; oxo: 9'; Δ: 1,3,5,1',3',5',7'	(−)-savinin = hibalactone	ApiAntnem CupChaob1 CupJunsab	164 122 122
		8*S*,8'*R*; O_2CH_2: 3,4,3',4'; oxo: 9'; Δ: 1,3,5,1',3',5'	(−)-hinokinin	PipPipcub AriAriind CupCalcol CupChaob1	118 1 122 122
		7*R*,8*R*,8'*R*; OH: 7; O_2CH_2: 3,4,3',4'; oxo: 9'; Δ: 1,3,5,1',3',5'	parabenzlactone	LauPartri	122
		OH: 7'; O_2CH_2: 3,4,3',4'; oxo: 9'; Δ: 1,3,5,1',3',5'	sventenin	RutRutmic	122
		8*R*,8'*R*; OH: 4'; OMe: 3'; O_2CH_2: 3,4; oxo: 9'; Δ: 1,3,5,1',3',5'	pluviatolide	RutZanpod	122
		8*R*,8'*R*; OH: 4,4'; OMe: 3,3'; oxo: 9'; Δ: 1,3,5,1',3',5'	matairesinol	ThyStecha HydForsus ApoTraasi PinAbinep PinPicaja PinPicobo PodPodspi	114, 146 83 122 122 122 122 122
		8*R*,8'*R*; OH: 4,4',5'; OMe: 3,3'; oxo: 9'; Δ: 1,3,5,1',3',5'	thujaplicatin	CupThusta	122
		8*R*,8'*R*; OH: 4,4'; OMe: 3,3',5'; oxo: 9'; Δ: 1,3,5,1',3',5'	thujaplicatin methylether	CupThusta	122
		8*R*,8'*S*; OH: 4,4',8'; OMe: 3,3',5' oxo: 9'; Δ: 1,3,5,1',3',5'	hydroxythuja-plicatin methylether	CupThupli CupThusta	122
		8*R*,8'*S*; OH: 4,8,4',8'; OMe: 3,3',5'; oxo: 9'; Δ: 1,3,5,1',3',5'	dihydroxythuja-plicatin methylether	CupThusta	122
		8*R*,8'*R*; OMe: 3,4,3',4',5'; oxo: 9' Δ: 1,3,5,1',3',5'	di-O-methylthuja-plicatin methylether	PipPipcub	10

Table 7.3.4 (continued)

Type	Subtype	Compound	Trivial name	Occurrence in species	Ref.
		7*R*,8*R*,8'*R*; oxo: 9'; OH: 4,7,4'; OMe: 3,3'; *Δ*: 1,3,5,1',3',5'	hydroxymatai-resinol	PinAbinep PinPicabi PinPicaja PinPicjez TaxTaxwal	122 122 122 122 164
		8*R*,8'*S*; OMe: 3,3'; oxo: 9'; OH: 4,4',8'; *Δ*: 1,3,5,1',3',5'	nortrachelogenin = wikstromol	ApoCaredu ThyWikind ApoTraaso	164 74 122
		8*R*,8'*S*; OH: 4',8'; OMe: 3,4,3'; oxo: 9'; *Δ*: 1,3,5,1',3',5'	trachelogenin	ApoTraasi	122
		8*R*,8'*S*; OH: 4,8,4',5',8'; OMe: 3,3'; oxo: 9'; *Δ*: 1,3,5,1',3',5'	dihydroxythuja-plicatin	CupThusta	122
		8*R*,8'*R*; OMe: 3,4; O_2CH_2: 3',4'; oxo: 9'; *Δ*: 1,3,5,1',3',5'	kusunokinin	RutHapvul	57
		8*S*,8'*S*; OH: 4'; OMe: 3,4,3'; oxo: 9'; *Δ*: 1,3,5,1',3',5'	arctigenin	ApoTraasi	122
		AcO: 7'; O_2CH_2: 3,4,3',4'; oxo: 9'; *Δ*: 1,3,5,1',3',5'	sventeninacetate	RutRutpin	122
		8*R*,8'*R*; OH: 4'; OMe: 3,4,3'; oxo: 9'; *Δ*: 1,3,5,1',3',5'	(−)-arctigenin	ApoTraasi HydForsus ThyWikind	122 83 164
		8*R*,8'*R*; *O*-*β*-*D*-Glupyr: 4'; OMe: 3,4,3'; oxo: 9'; *Δ*: 1,3,5,1',3',5'	(+)-arctiin	ApoTrafoe ApoTraliu AstLaptom AstOnoale HydForart HydForsus	122 122 122 116 122 122
		8*R*,8'*R*; OH: 4,2',4'; OMe: 3,3'; oxo: 9'; *Δ*: 1,3,5,1',3',5'	gnidifolin	Thygnilat	20
		OMe: 3',4',5'; O_2CH_2: 3,4; oxo: 9'; *Δ*: 1,3,5,1',3',5'	isoanthricin	ApiAntsyl	84

Table 7.3.4 (continued)

Type	Subtype	Compound	Trivial name	Occurrence in species	Ref.
		8*R*; OMe: 3′,4′; O_2CH_2: 3,4; oxo: 9′; Δ: 1,3,5,1′,3′,5′,7′	(−)-kaerophyllin	ApiChamac	110
		8*R*,8′*R*; OMe: 3,4,5,3′,4′; oxo: 9′; Δ: 1,3,5,1′,3′,5′	(−)-lignanolide 1	LauCincam	164
		8*R*,8′*R*; OH: 4′; OMe: 3,4,5,3′; oxo: 9′; Δ: 1,3,5,1′,3′,5′	(−)-lignanolide 2	ApoTrafoe	164
		OH: 4,4′; OMe: 3,3′; oxo: 9′; O-pOHBenzoyl: 7′; Δ: 1,3,5,1′,3′,5′	lignanolide-A	PinAbinep	122
		OH: 4,4′; OMe: 3,3′; OVan: 7′; oxo: 9′; Δ: 1,3,5,1′,3′,5′	lignanolide-B	PinAbinep	122
		8*R*,8′*R*; OH: 4,4′,8′; OMe: 3,3′; oxo: 9′; Δ: 1,3,5,1′,3′,5′		ApoCaredu	164
		8*R*,8′*R*; OH: 3′; OMe: 3,4,4′; oxo: 9′; Δ: 1,3,5,1′,3′,5′	prestegane A	ApiSteara	138
		8*R*; OMe: 3,4; O_2CH_2: 3′,4′; oxo: 9′ Δ: 1,3,5,1′,3′,5′,7′	suchilactone	PolPolchi	122
		8*R*,8′*S*; OH: 4,4′,8′; OMe: 3,3′; oxo: 9′; Δ: 1,3,5,1′,3′,5′	(+)-wikstromol	ThyWikfoe ThyWikuva	149 149
		8*R*,8′*R*; OMe: 3′,4′,5′; O_2CH_2: 3,4; oxo: 9′; Δ: 1,3,5,1′,3′,5′	(−)-yatein	CupLibyat HerHerovi HerHercor PipPipcub	39 26 127 10
		8*R*,8′*R*; OMe: 3,4,5; O_2CH_2: 3′,4′; oxo: 9′; Δ: 1,3,5,1′,3′,5′	(−)-isoyatein	PipPipcub	10
		OMe: 5,3′,4′,5′; O_2CH_2: 3,4; oxo: 9′; Δ: 1,3,5,1′,3′,5′	5-methoxyyatein	HerHercor	126
		8*R*,8′*R*; OMe: 3,4,5,5′; O_2CH_2: 3′,4′; oxo: 9′; Δ: 1,3,5,1′,3′,5′	(−)-cubebinone	PipPipcub	10

Table 7.3.4 (continued)

Type	Subtype	Compound	Trivial name	Occurrence in species	Ref.
		8*R*,8′*R*; OMe: 3,4,5,3′,4′,5′; oxo: 9′; *Δ*: 1,3,5,1′,3′,5′	cordigerin = cubebininolide	HerHercor PipPipcub	127 10
		8*R*,8′*R*; OH: 9′; O_2CH_2: 3,4,3′,4′; *Δ*: 1,3,5,1′,3′,5′	(−)-cubebin	AriAriind AriAritri PipPipclu PipPipcub PipPiplac	1 164 164 118 156
		8*R*,8′*R*; OH: 9′; OMe: 3′,4′; O_2CH_2: 3,4; *Δ*: 1,3,5,1′,3′,5′	3′,4′-dimethoxy- 3′,4′-demethylene- dioxycubebin	AriAritri	164
		8*R*,8′*R*; OH: 9′; OMe: 3,4′ O_2CH_2: 3′,4′ *Δ*: 1,3,5,1′,3′,5′	3,4-dimethoxy- 3,4-demethylene- dioxycubebin	AriAritri	164
		8*R*,8′*R*; OH: 9′; OMe: 3,4,5,3′,4′,5′ *Δ*: 1,3,5,1′,3′,5′	(−)-cubebinin	PipPipcub	118
		8*R*,8′*R*; OH: 4,7,4′,7′; OMe: 3,3′; *Δ*: 1,3,5,1′,3′,5′	(−)-liovil	PinAbinep PinPicaja	122 122
		8*R*,8′*R*; OH: 8; OMe: 3,4,3′,4′; *Δ*: 1,3,5,1′,3′,5′	(−)-kusunokinol	LauCincam	122
		8*R*,8′*R*; OH: 4,4′,8′,9′; OMe: 3,3′; *Δ*: 1,3,5,1′,3′,5′	(−)-carissanol	ApoCaredu	164
		8*R*,8′*R*; OH: 9; OMe: 3′,4′,5′; O_2CH_2: 3,4′; *Δ*: 1,3,5,1′,3′,5′	(−)-clusin	PipPipclu PipPipcub	164 118
		OH: 4,7,4′; OMe: 3,3′; *Δ*: 1,3,5,1′,3′,5′		MenTincor	59
		OFer: 4′; OMe: 3′; *Δ*: 1,3,5,1′,3′,5′	(−)-divanillyl- tetrahydrofuran ferulate	EbeDiokak	108
		8*R*,8′*R*; OH: 3,3′; OMe: 4,4′; oxo: 9′; *Δ*: 1,3,5,1′,3′,5′	prestegane B	ApiSteara	138
		8*R*; O_2CH_2: 3,4,3′,4′; oxo: 9′; *Δ*: 1,3,5,1′,3′,5′,7′	(−)-hibalactone	CupCalfor	42

Table 7.3.4 (continued)

Type	Subtype	Compound	Trivial name	Occurrence in species	Ref.
		8*S*,8′*S*; O_2CH_2: 3,4,3′,4′; oxo: 9′ Δ: 1,3,5,1′,3′,5′	heliobuphthalmin lactone	AstHelbup BigKigpin	73 164
		8*R*,8′*R*; O_2CH_2: 3,4,3′,4′; oxo: 9′; Δ: 1,3,5,1′,3′,5′,7′	(+)-calocendrin	CupCalfor	42
		8*S*,8′*R*; OMe: 3,4; O_2CH_2: 3′,4′; oxo: 9′; Δ: 1,3,5,1′,3′,5′		MyrVirseb	164
		8*R*,8′*R*; OMe: 3,4; O_2CH_2: 3′,4′; oxo: 9′; Δ: 1,3,5,1′,3′,5′		MyrVirseb	164
		8*R*,8′*R*; OMe: 3,4,3′,4′; oxo: 9; Δ: 1,3,5,1′,3′,5′		MyrVirseb	164
		O_2CH_2: 3,4,3′,4′; oxo: 9′; Δ: 1,3,5,7,1′,3′,5′,7′	7,7′-didehydro-heliobuphthalmin-lactone	AstHelbup	73
	8.8′, 7.O.7′	7*S*,8*S*,7′*S*,8′*S*; OH: 4,9,4′,9′ OMe: 3,3′; Δ: 1,3,5,1′,3′,5′	neo-olivil	LamThylon	164
	8.8′, 7.O.9′	8*R*,8′*S*; OH: 4,9,4′,8′; OMe: 3,3′; Δ: 1,3,5,1′,3′,5′	olivil	ApoCaredu BigStekun EucEuculm OleFrajap OleFraman	164 31 33 153 153
		7*S*,8*R*,8′*R*; OH: 4,9,4′; OMe: 3,3′; Δ: 1,3,5,1′,3′,5′	(+)-lariciresinol	ApoCaredu PinLarlep	164 142
		7*R*,8*S*,8′*S*; OH: 4,9,4′; OMe: 3,3′; Δ: 1,3,5,1′,3′,5′	(−)-lariciresinol	ThyDirocc PinLardec ThyDaptan	9 122 94
		7*S*,8*R*,8′*R*; OH: 4,9; OMe: 3,3′,4′; Δ: 1,3,5,1′,3′,5′	(+)-lariciresinol-4-methylether	AraAraang	164

Table 7.3.4 (continued)

Type	Subtype	Compound	Trivial name	Occurrence in species	Ref.
		7*S*,8*R*,8′*R*; *O*-β-D-Glu: 9; OH: 4,4′; OMe: 3,3′; Δ: 1,3,5,1′,3′,5′	(+)-lariciresinol-9-*O*-β-D-glucoside	PtePtevit	131
		OH: 4,9,4′; OMe: 3,5,3′,5′; oxo: 7′; Δ: 1,3,5,1′,3′,5′	3-hydroxymethyl-2-syringa-4-syringoyl-tetrahydrofuran	PtePtevit	131
		7*R*,8*S*,8*S*′; OH: 9; O_2CH_2: 3,4,3′,4′; Δ: 1,3,5,1′,3′,5′	(−)-dihydroses-amin	ThyDaptan	94, 95
		7*S*,8*R*,8′*S*; *O*-β-D-Glupyr: 4′; OH: 4,9,8′; OMe: 3,3′; Δ: 1,3,5,1′,3′,5′	(−)-olivil-*O*-β-D-glucopyranoside	OleLinjap	164
		7*S*,8*R*,8′*R*; OH: 9; OMe: 5,3′,4′,5′; O_2CH_2: 3′,4′; Δ: 1,3,5,1′,3′,5′	dihydrosesartemin	MyrVirelo	104
		OH: 9; OMe: 3,4,5,2′,4′,5′; Δ: 1,3,5,1′,3′,5′	β-dihydroyan-gambin	MyrVirelo	104
		7*R*,8*S*,8′*R*; OH: 4,9,4′,8′; OMe: 3,3′; Δ: 1,3,5,1′,3′,5′	(−)-massoniresinol	PinPinmas	134
		7*R*,8*R*,8′*S*; OH: 9; OMe: 3,4,3′,4′,5′ oxo: 7′; Δ: 1,3,5,1′,3′,5′	(+)-sylvone	PipPipsyl	11
		OH: 4,4′; OMe: 3,3′; OVan: 9; Δ: 1,3,5,1′,3′,5′		PinAbinep	122
		OH: 4,4′,8′; OMe: 3,3′; OCou: 9; Δ: 1,3,5,1′,3′,5′		PinAbinep	122
		OH: 4,4′,8′; OMe: 3,3′; OFer: 9; Δ: 1,3,5,1′,3′,5′		PinAbinep	122
	8.8′, 7.O.9′, 9.O.7′	OMe: 3,4,5,3′,4′,5′; Δ: 1,3,5,1′,3′,5′	diayangambin	PipMacexc	122
		7*S*,7′*R*; OMe: 3,4,5,3′,4′,5′; Δ: 1,3,5,1′,3′,5′	epiyangambin	MyrVirelo	102

Table 7.3.4 (continued)

Type	Subtype	Compound	Trivial name	Occurrence in species	Ref.
		7*S*,7′*S*; OMe: 3,4,5,3′,4′,5′; *Δ*: 1,3,5,1′,3′,5′	yangambin = dimethyllirioresinol B	MonLaunov	4
				MyrVirelo	102
		7*S*,7′*R*; O_2CH_2: 3,4,3′,4′; *Δ*: 1,3,5,1′,3′,5′	asarinin = episesamin	AcaJussim	164
				MyrHoriry	148
		O_2CH_2: 3,4,3′,4′; *Δ*: 1,3,5,1′,3′,5′	(±)-asarinin	AcaJussim	164
				RutZanala	66
				RutZansp	157
		7*S*,7′*S*; O_2CH_2: 3,4,3′,4′; *Δ*: 1,3,5,1′,3′,5′	(+)-sesamin	EupAmaobl	45
				EupClecol	164
				LamSidcan	122
				RutZancap	122
				RutZancha	122
				RutZandin	122
				RutZanpip	122
				RutZansen	122
				RutZanvir	122
				ScrAptspi	122
				AstAnapyr	122
				MagTalhod	144
				RutZanaca	122
		7*R*,7′*R*; O_2CH_2: 3,4,3′,4′; *Δ*: 1,3,5,1′,3′,5′	(−)-sesamin	AriAsaste	122
				RutZanarn	122
				RutZanpod	122
				RutZanp	157
				AcaJussim	164
				MagMagkob	65
				MagMagste	164
				PedSesang	122
				ThyDapole	147
		O_2CH_2: 3,4,3′,4′; *Δ*: 1,3,5,1′,3′,5′	(±)-fagarol	RutFagxan	122
		OH: 9′; O_2CH_2: 3,4,3′,4′; *Δ*: 1,3,5,1′,3′,5′	cinnamonol	LauCincam	122
		7*S*,7′*R*; OH: 8′; O_2CH_2: 3,4,3′,4′; *Δ*: 1,3,5,1′,3′,5′	paulownin	BigDodCri	119
				EupAmaobl	45
				EupClecol	164
				PinAbinep	122
				VerGmeasi	122
				ScrPautom	122
		7*S*,7′*S*; OH: 8,8′; O_2CH_2: 3,4,3′,4′; *Δ*: 1,3,5,1′,3′,5′	(−)-wodeshiol	EupClecol	164
				PinAbinep	122

Table 7.3.4 (continued)

Type	Subtype	Compound	Trivial name	Occurrence in species	Ref.
		7*S*,7'*R*; OH: 9,8'; O_2CH_2: 3,4,3',4'; Δ: 1,3,5,1',3',5'	gummadiol	PinPicaja PinAbinep VerGmearb	122 122 122
		7*S*,7'*S*; OH: 4,4'; OMe: 3,3'; Δ: 1,3,5,1',3',5'	(+)-pinoresinol	PinPicabi PinPicexc PinPicvul PinLarlep RutZanala SchSchhen ScrAptspi ThyDapole ThyStecha	122 122 122 142 66 97, 163 122 147 114, 146
		OH: 4,4'; OMe: 3,3'; Δ: 1,3,5,1',3',5'	(±)-pinoresinol	HydForsus HydForcor PinPicaja	164 83, 163 122
		7*R*,7'*S*; OH: 4,4'; OMe: 3,3'; Δ: 1,3,5,1',3',5'	(−)-epipinoresinol = symplocosigenol	RutZanala	66
		OH: 4,4'; OMe: 3,3'; Δ: 1,3,5,1',3',5'	(±)-epipinoresinol	PinPicexc	122
		7*R*,7'*S*; OAc: 8; *O*-β-D-Glu: 4; OH: 4'; OMe: 3,3'; Δ: 1,3,5,1',3',5'	(+)-acetoxypinoresinol-4-*O*-β-D-glucoside	OleOleeur	164
		7*S*,7'S; OH: 4,4'; OMe: 3,3',5'; Δ: 1,3,5,1',3',5'	(+)-medioresinol	ThyDirocc	9
		7*S*,7'*S*; OMe: 3,4,3',4'; Δ: 1,3,5,1',3',5'	(+)-pinoresinol dimethylether = eudesmin	AraAraang MagTalhod MonLaunov ScrAptspi	122 144 4 122
		7*S*,7'*S*; OH: 4'; OMe: 3,4,3'; Δ: 1,3,5,1',3',5'	(+)-pinoresinol methylether	AraAraang ScrAptspi	122 122
		7*S*,7'*S*; *O*-β-D-Glupyr: 4; OH: 4'; OMe: 3,3'; Δ: 1,3,5,1',3',5'	(+)-pinoresinol-*O*-β-D-glucopyranoside	OleLinjap	164
		7*S*,7'*R*; *O*-β-D-Glupyr: 4'; OH: 4,8'; OMe: 3,3'; Δ: 1,3,5,1',3',5'	(+)-8'-hydroxy-pinoresinol-4'-*O*-β-D-glucopyranoside	EucEuculm	33
		7*S*,7'*S*; *O*-β-D-Glupyr: 4,4'; OMe: 3,3',5'; Δ: 1,3,5,1',3',5'	(+)-medioresinol-di-*O*-β-D-glucopyranoside	EucEuculm	33

Table 7.3.4 (continued)

Type	Subtype	Compound	Trivial name	Occurrence in species	Ref.
		7*S*,7′*S*; OMe: 5,5′; O_2CH_2: 3,4,3′,4′; Δ: 1,3,5,1′,3′,5′	(+)-excelsin	PipMacexc	122
		7*S*,7′*R*; OMe: 5,5′; O_2CH_2: 3,4,3′,4′; Δ: 1,3,5,1′,3′,5′	(+)-epiexcelsin	PipMacexc	122
		7*S*,7′*S*; OMe: 5′; O_2CH_2: 3,4,3′,4′; Δ: 1,3,5,1′,3′,5′	demethoxyexcelsin	PipMacexc	122
		7*S*,7′*S*; OH: 7′,8′; O_2CH_2: 3,4,3′,4′; Δ: 1,3,5,1′,3′,5′	arboreol	VerGmearb	122
		7*R*,7′*S*; OH: 4′; OMe: 3,4,3′; Δ: 1,3,5,1′,3′,5′	(+)-phillygenol	MagMagkob	65
		7*S*,7′*R*; OMe: 3′,4′,5′; O_2CH_2: 3,4; Δ: 1,3,5,1′,3′,5′	(+)-epiaschantin	HerHerovi	26
		7*S*,7′*S*; OMe: 3,4,3′,4′; Δ: 1,3,5,1′,3′,5′	(+)-eudesmin	AcaPlafal	3
				ComEuchem	122
				MagMagkob	65
				MagMagste	164
				RutZanaca	122
				RutZanala	122
				RutZanoxo	122
		7*R*,7′*R*; OMe: 3,4,3′,4′; Δ: 1,3,5,1′,3′,5′	diaeudesmin	PipPippee	122
		7*S*,7′*R*; OMe: 3,4,3′,4′; Δ: 1,3,5,1′,3′,5′	(+)-epieudesmin	MagMagkob	65
				RutZanaca	122
				RutZanala	122
				RutZanoxo	122
				VisVisalb	158
		7*R*,7′*R*; OH: 8′; OMe: 3,4,3′,4′; Δ: 1,3,5,1′,3′,5′	gmelinol	VerGmearb	122
				VerGmeasi	122
				VerGmelei	122
		7*S*,7′*S*; OH: 4,4′; OMe: 3,5,3′,5′; Δ: 1,3,5,1′,3′,5′	(+)-syringaresinol = lirioresinol B	RutZanala	66
				RutZanarn	122
				RutZanoxo	122
				SchSchsph	96
				ThyWikfoe	149
				ThyWikuva	164
		7*R*,7′*R*; OH: 4,4′; OMe: 3,5,3′,5′; Δ: 1,3,5,1′,3′,5′	(−)-syringaresinol	BerBerbux	41
				BerBerchi	150
				EphEphala	112
				PolPolpol	62
				RutFagbon	67, 122
				RutZanine	122
				VisVisalb	158

Table 7.3.4 (continued)

Type	Subtype	Compound	Trivial name	Occurrence in species	Ref.
		7*R*,7′*S*; OH: 4,4′; OMe: 3,5,3′,5′; *Δ*: 1,3,5,1′,3′,5′	(+)-episyringaresi-nol = lirioresinol A	MagLirtul	122
		OH: 4,4′; OMe: 3,5,3′,5′; *Δ*: 1,3,5,1′,3′,5′	diasyringaresinol	MagLirtul	122
		7*S*,7′*S*; OMe: 3,4,3′,4′,5′; *Δ*: 1,3,5,1′,3′,5′	magnolin	MagMaggra	122
		7*R*,7′*S*; OMe: 3,4,3′,4′,5′; *Δ*: 1,3,5,1′,3′,5′	epimagnolin	HerHerovi	26
		7*S*,7′*S*; *O*-*β*-D-Glupyr: 4,4′; OMe: 3,5,3′,5′; *Δ*: 1,3,5,1′,3′,5′	syringaresinol-bis-*O*-*β*-D-glucopyra-noside = liriodendrin	OleLinjap	164
		7*S*,7′*S*; *O*-*β*-D-Glu: 4; OH: 4′; OMe: 3,5,3′,5′; *Δ*: 1,3,5,1′,3′,5′	syringaresinol mono-*O*-*β*-D-gluco-side	VisVisalb	158
		7*S*,7′*S*; OH: 9; O_2CH_2: 3,4,3′,4′; *Δ*: 1,3,5,1′,3′,5′	(+)-9-hydroxyses-amin	BerPodple MagMagkob PipPipcub	72 65 10
		7*R*,7′*R*; *O*-*β*-D-Glu: 4,4′; OMe: 3,5,3′,5′; *Δ*: 1,3,5,1′,3′,5′	acanthoside D = eleutheroside E	AraAcachi	81
		7*R*,7′*R*; OMe: 3,4,5,5′; O_2CH_2: 3′,4′; *Δ*: 1,3,5,1′,3′,5′	(+)-diasesartemin	AstArtabs	164
		7*S*,7′*R*; OMe: 3,4,5,5′; O_2CH_2: 3′,4′; *Δ*: 1,3,5,1′,3′,5′	(+)-episesarte-min A	AstArtabs	164
		7*R*,7′*S*; OMe: 3,4,5,5′; O_2CH_2: 3′,4′; *Δ*: 1,3,5,1′,3′,5′	(+)-episesarte-min B	AstArtabs	164
		7*S*,7′*R*; OMe: 3′,4′; O_2CH_2: 3,4; *Δ*: 1,3,5,1′,3′,5′	(+)-fargesin	BerBerbux MagMaggra MagTalhod RutZanaca RutZanala	40 122 144 122 122

Table 7.3.4 (continued)

Type	Subtype	Compound	Trivial name	Occurrence in species	Ref.
		7*S*,7′*S*; OH: 2; O_2CH_2: 4,5,3′,4′; *Δ*: 1,3,5,1′,3′,5′	(+)-justisolin	AcaJussim	164
		7*R*,7′*R*; OH: 8,8′; O_2CH_2: 3,4,3′,4′; *Δ*: 1,3,5,1′,3′,5′	(−)-kigeliol	BigKigpin	164
		7*S*,7′*S*; OMe: 3′,4′; O_2CH_2: 3,4; *Δ*: 1,3,5,1′,3′,5′	(+)-kobusin	MagMagkob	65
		7*S*,7′*S*; *O*-*β*-D-Glu: 4,4′; OMe: 3,5,3′,5′; *Δ*: 1,3,5,1′,3′,5′	liriodendrin	GloGloaly	29
		7*S*,7′*S*; OH: 4′; OMe: 3′; O_2CH_2: 3,4; *Δ*: 1,3,5,1′,3′,5′	(+)-piperitol	AstHelbra MagMagste RutZanpip ScrAptspi	164 122 122 122
		7*S*,7′*S*; OMe: 3′; OCH_2CHCMe_2: 4′; O_2CH_2: 3,4; *Δ*: 1,3,5,1′,3′,5′	(+)-piperitol-dimethylallylether	RutZanpip	122
		7*S*,7′*S*; *O*-*β*-D-Glu: 4′; OMe: 3′; O_2CH_2: 3,4; *Δ*: 1,3,5,1′,3′,5′	(+)-piperitol-*O*-*β*-D-glucoside	AstHelbra	164
		7*R*,7′*S*; OH: 4; OMe: 3; O_2CH_2: 3′,4′; *Δ*: 1,3,5,1′,3′,5′	(−)-pluviatilol	RutZanplu	122
		7*R*,7′*R*; OH: 4,8,4′,8′; OMe: 3,3′; *Δ*: 1,3,5,1′,3′,5′	(−)-prinsepiol	RosPriuti	80
		7*S*,7′*S*; OMe: 3,4,5,5′; O_2CH_2: 3′,4′; *Δ*: 1,3,5,1′,3′,5′	(+)-sesartemin	AstArtabs	164
		7*S*,7′*S*; *O*-*β*-D-Glupyr: 4; OMe: 3; O_2CH_2: 3′,4′; *Δ*: 1,3,5,1′,3′,5′	(−)-simplexoside	AcaJussim	164
		7*S*,7′*S*; OMe: 3′,4′; O_2CH_2: 3,4; *Δ*: 1,3,5,1′,3′,5′	(+)-spinescin	ScrAptspi	122
		7*R*,7′*R*; *O*-*β*-D-Glupyr: 4,4′; OMe: 3,5,3′,5′; *Δ*: 1,3,5,1′,3′,5′	eleutheroside E	AraAcakor VisVisalb	82 158
		7*S*,7′*R*; *O*-L-Rham-(1-6)-*β*-D-Glu: 4′; OH: 4; OMe: 3,3′; *Δ*: 1,3,5,1′,3′,5′	versicoside	RutHapver	14

Table 7.3.4 (continued)

Type	Subtype	Compound	Trivial name	Occurrence in species	Ref.
		7*S*,7′*R*; OMe: 3,3′,4′; Β-D-Xyl-(1 – 6)-Β-D-Glu: 4; *Δ*: 1,3,5,1′,3′,5′	lantibeside	ScrLantib	122
		7*R*,7′*S*; O_2CH_2: 3,4,3′,4′; *Δ*: 1,3,5,1′,3′,5′	(–)-asarinin	AstArtabs RutZanala RutZanarn RutZanpod	122 66 122 125
		7*R*,7′*S*; OMe: 3; O_2CH_2: 3′,4′; OCH_2CHCMe_2: 4; *Δ*: 1,3,5,1′,3′,5′	dimethylallylether of pluviatilol	RutZanpod	125
	8.8′, 6.7′	8*R*,7′*S*,8′*R*; OH: 4,9,3′,4′,9′; OMe: 3; *Δ*: 1,3,5,1′,3′,5′	(+)-isotaxiresinol	CupFitcup TaxTaxbac TaxTaxcus	122 122 165
		8*R*,7′*S*,8′*R*; OH: 9,3′,4′,9′; OMe: 3,4; *Δ*: 1,3,5,1′,3′,5′	(+)-isotaxiresinol-6-methylether	CupFitcup	122
		8*R*,7′*S*,8′*S*; OH: 4,8,9,4′,5′,8′,9′; OMe: 3,3′; oxo: 9′; *Δ*: 1,3,5,1′,3′,5′	plicatic acid	CupThupli	122
		7′*R*,8′*S*; OH: 4,9,4′,9′; OMe: 3,5,3′,5′; oxo: 9; *Δ*: 1,3,5,7,1′,3′,5′	thomasic acid	UlmUlmtho	122
		8*R*,7′*S*,8′*R*; OH: 4,9,4′,9′; OMe: 3,3′; *Δ*: 1,3,5,1′,3′,5′	(+)-isolariciresinol	AcaJusfla AraAraang CupFitcup PinPicaja PinPicexc TaxTaxcus	115 122, 163 122 122 122 165
		8*R*,7′*S*,8′*R*; OH: 4,9,4′,9′; OMe: 3,5,3′,5′; *Δ*: 1,3,5,1′,3′,5′	dimethoxyisolariciresinol = lyoniresinol	BetAlnglu EriLyoova UlmUlmtho	122 122 122
		8*S*,7′*R*,8′*S*; OH: 4,9,4′; OMe: 3,3′,5′; *O*-*β*-D-Xylpyr: 9′; *Δ*: 1,3,5,1′,3′,5′	5-methoxy-9-*O*-*β*-D-xylopyranosyl-(–)-isolariciresinol	CanCinmad	164
		8*R*,7′*S*,8′*R*; OH: 4,9,9′; OMe: 3,3′,4′; *Δ*: 1,3,5,1′,3′,5′	(+)-isolariciresinol-4′-methylether	AraAraang	164

Table 7.3.4 (continued)

Type	Subtype	Compound	Trivial name	Occurrence in species	Ref.
		8*S*,7′*S*,8′*S*; OH: 4,8,9,4′,9′; OMe: 3,3′; *Δ*: 1,3,5,1′,3′,5′	cycloolivil = iso-olivil	BigStekun EucEuculm OleFrajap OleFraman OleOleafr OleOlecun OleOleeur	31 33 153 153 154 122 154
		8*R*,7′*S*,8′*R*; OMe: 5,9,3′,4′,9′; O_2CH_2: 3,4; *Δ*: 1,3,5,1′,3′,5′	nirtetralin	EupPhynir	122
		8*S*,7′*R*,8′*S*; OMe: 3,4,9,3′,4′,9′; *Δ*: 1,3,5,1′,3′,5′	phyltetralin	EupPhynir	122
		8*S*,7′*S*,8′*R*; OMe: 5,9,3′,4′,9′; O_2CH_2: 3,4; *Δ*: 1,3,5,1′,3′,5′	(+)-hypophyl-lanthin	EupPhynir	48, 122 132, 164
		8*R*,7′*S*,8′*R*; OMe: 3,4,9,9′; O_2CH_2: 3′,4′; *Δ*: 1,3,5,1′,3′,5′	lintetralin	EupPhynir	164
		8′*S*,7′*R*,8′*S*; OH: 4,4′,9′; OMe: 3,5,3′,5′; *O*-*β*-D-Xyl: 9; *Δ*: 1,3,5,1′,3′,5′	(−)-pygeoside	RosPygacu	27
	8.8′, 6.7′, 9.O.9′	8*S*,9*S*,7′*S*,8′*S*; OH: 4,8,9,4′; OMe: 3,3′; *Δ*: 1,3,5,1′,3′,5′	(+)-africanal	OleOleafr	164
		8*R*,9*R*,7′*S*,8′*R*; OAc: 4,4′; OMe: 3,9,3′; *Δ*: 1,3,5,1′,3′,5′	(−)-epoxycyclolig-nan 1	PodDacint	22, 23
		8*R*,9*S*,7′*S*,8′*R*; OAc: 4,4′; OMe: 3,9,3′; *Δ*: 1,3,5,1′,3′,5′	(+)-epoxycyclolig-nan 1	PodDacint	23
		8*R*,9*S*,7′*S*,8′*R*; OAc: 4,4′; OEt: 9; OMe: 3,3′; *Δ*: 1,3,5,1′,3′,5′	(+)-epoxycyclolig-nan 2	PodDacint PodPodsp	23 122
		8*R*,9*S*,7′*S*,8′*R*; OEt: 9; OH: 4,4′; OMe: 3,3′; *Δ*: 1,3,5,1′,3′,5′	(+)-ethyl-*β*-coni-dendral	PodDacint	23

Table 7.3.4 (continued)

Type	Subtype	Compound	Trivial name	Occurrence in species	Ref.
		8*R*,9*R*,7′*S*,8′*R*; OH: 4,4′; OMe: 3,9,3′; Δ: 1,3,5,1′,3′,5′	(−)-methyl-α-conidendral = formosanol	PodDacint	23
		8*R*,9*S*,7′*S*,8′*R*; OH: 4,4′; OMe: 3,9,3′; Δ: 1,3,5,1′,3′,5′	(+)-methyl-β-conidendral	PodDacint	23
		8*R*,9*R*,7′*S*,8′*R*; OH: 4,4′; OMe: 3,9,3′; Δ: 1,3,5,1′,3′,5′	formosanol	CupJunfor	85
		8*R*,9*R*,7′*S*,8′*R*; OH: 4,4′; OMe: 3,9,3′; Δ: 1,3,5,1′,3′,5′	tsugacetal	AusAussca PinTsuchi	122 43
		8*R*,7′*R*,8′*R*; OMe: 5′; O_2CH_2: 3,4,3′,4′; oxo: 9′; Δ: 1,3,5,1′,3′,5′	austrobailignan-1	AusAussca	122
		8*R*,7′*R*,8′*S*; OMe: 5′; O_2CH_2: 3′,4′; oxo: 9′; Δ: 1,3,5,1′,3′,5′	austrobailignan-2	BerDipgra BerPodem1 BerPodem2 CupCalfor CupJunsab AusAussca	122 122 122 122 122 122
		7*R*,8*R*,7′*R*,8′*R*; OH: 7; OMe: 3′,4′,5′; O_2CH_2: 3,4; oxo: 9′; Δ: 1,3,5,1′,3′,5′	(−)-podophyllotoxin	BerPodhex BerPodple PolPolpol ApiBupfru BerPodpel	71, 122 72, 122 122 122 71
		8*R*,7′*R*,8′*R*; OMe: 3′,5′; OH: 4′; O_2CH_2: 3,4; oxo: 9′; Δ: 1,3,5,1′,3′,5′	(−)-demethyldehydroxypodophyllotoxin	BerPodem1 PolPolpol	122 62, 122
		7*R*,8*R*,7′*R*,8′*R*; OMe: 3′,5′; OH: 7,4′; O_2CH_2: 3,4; oxo: 9′; Δ: 1,3,5,1′,3′,5′	(−)-demethylpodophyllotoxin	CanCinmad PolPolpol	164 62

Table 7.3.4 (continued)

Type	Subtype	Compound	Trivial name	Occurrence in species	Ref.
		8*R*,7′*R*,8′*R*; OMe: 3′,4′,5′; O_2CH_2: 3,4; oxo: 9′; *Δ*: 1,3,5,1′,3′,5′	(−)-dehydroxy-podophyllotoxin = anthricin = sili-colin = hernandian	CanCinmad BerPodhex BerPodpel BerPodple CupJunpho PolPolpol	164 71, 164 71 72 21 63
		OH: 7; OMe: 3′,4′,5′; O_2CH_2: 3,4; oxo: 9′; *Δ*: 1,3,5,7,1′,3′,5′,7′	dehydropodophyl-lotoxin	ApiAntsyl BerPodhex BerPodple BurBurmic BurBurmor CupCaldru CupJunber CupJunpho CupLibbid PolPolpol	122 71, 164 72, 122 122 122 122 122 21 122 63
		8*R*,7′*R*,8′*R*; OMe: 3′,4′,5′; O_2CH_2: 3,4; oxo: 7,9′; *Δ*: 1,3,5,1′,3′,5′	(−)-podophyllo-toxone	BerPodhex	71
		8*R*,7′*R*,8′*R*; OH: 2,4′; OMe: 3′,5′; O_2CH_2: 3,4; oxo: 9′; *Δ*: 1,3,5,1′,3′,5′	(−)-*α*-peltatin	BerPodhex BerPodpel	71 122
		8*R*,7′*R*,8′*R*; OH: 2; OMe: 3′,4′,5′; O_2CH_2: 3,4; oxo: 9′; *Δ*: 1,3,5,1′,3′,5′	(−)-*β*-peltatin	CupJunber	164
		8*R*,7′*R*,8′*R*; OMe: 2,3′,4′,5′; O_2CH_2: 3,4; oxo: 9′; *Δ*: 1,3,5,1′,3′,5′	(−)-*β*-peltatin-methylether	CupJunpho CupLibbid	21 122
		7*R*,8*R*,7′*R*,8′*R*; OH: 7; OMe: 3′,4′,5′; O_2CH_2: 3,4; oxo: 9′; *Δ*: 1,3,5,1′,3′,5′	(−)-*β*-peltatin-A-methylether	BerPodhex BerPodple CupJunpho	71 72, 122 21
		8*S*,7′*R*,8′*R*; OMe: 3′,4′,5′; O_2CH_2: 3,4; oxo: 7,9′; *Δ*: 1,3,5,1′,3′,5′	(−)-isopicropodo-phyllone	AcaJussim BerPodple	164 72
		7*R*,8*R*,7′*R*,8′*S*; OH: 7; OMe: 3′,4′,5′; O_2CH_2: 3,4; oxo: 9′; *Δ*: 1,3,5,1′,3′,5′	(+)-picropodo-phyllin	BerDipgra BerPodem2 BerPodhex BerPodpel	122 122 71 71

Table 7.3.4 (continued)

Type	Subtype	Compound	Trivial name	Occurrence in species	Ref.
		OH: 7; OMe: 3,4,3′,4′,5′; oxo: 9′; Δ: 1,3,5,7,1′,3′,5′	(−)-sikkimotoxin	BerPodsik	122
		OH: 4,7,3′,4′; OMe: 3,5′; oxo: 9′; Δ: 1,3,5,1′,3′,5′,7′	plicatinaphtol	CupThupli	122
		OMe: 7; O_2CH_2: 3,4,3′,4′; oxo: 9′; Δ: 1,3,5,7,1′,3′,5′,7′	taiwanin-E-methyl-ether (Justicidin F)	AcaJuspr2	122
		O_2CH_2: 3,4,3′,4′; oxo: 9′; Δ: 1,3,5,1′,3′,5′,7′	dihydrotaiwanin C	EupClecol	164
		OH: 7,3′,4′; OMe: 3,4; oxo: 9′; Δ: 1,3,5,7,1′,3′,5′,7′	taiwanin H	BurBurfra TaxTaicry	122 86
		O_2CH_2: 3,4,3′,4′; oxo: 9′; Δ: 1,3,5,7,1′,3′,5′,7′	taiwanin C = justicidin E	EupClecol TaxTaicry	164 122
		OH: 7; O_2CH_2: 3,4,3′,4′; oxo: 9′; Δ: 1,3,5,7,1′,3′,5′,7′	taiwanin E	EupClecol TaxTaicry	164 90, 122
		O_2CH_2: 3,4,3′,4′; 3,4-di-*O*-Me-D-Xyl: 7; oxo: 9′; Δ: 1,3,5,7,1′,3′,5′,7′	taiwanin E-3,4-di-*O*-methyl-*D*-xylo-side	EupClecol	164
		OMe: 3,4; O_2CH_2: 3′,4′; oxo: 9′; Δ: 1,3,5,7,1′,3′,5′,7′	justicidin-B	AcaJushay AcaJuspr1 AcaJuspr2 RutHapper RutHappop RutHapver	122 122 122 122 124 14
		OMe: 3,4,7; O_2CH_2: 3′,4′; oxo: 9′; Δ: 1,3,5,7,1′,3′,5′,7′	justicidin-A	AcaJushay AcaJuspr1 AcaJuspr2 RutHaptub	122 122 122 77
		OMe: 7; O_2CH_2: 3,4,3′,4′; oxo: 9; Δ: 1,3,5,7,1′,3′,5′,7′	justicidin-D	AcaJusext	164
		OMe: 3′,4′; O_2CH_2: 3,4; oxo: 9′; Δ: 1,3,5,7,1′,3′,5′,7′	chinensin	PolPolchi	122
		OH: 7; OMe: 3′,4′; O_2CH_2: 3,4; oxo: 9′; Δ: 1,3,5,7,1′,3′,5′,7′	chinensinaphtol = isodiphyllin	ApiBupfru	122

Table 7.3.4 (continued)

Type	Subtype	Compound	Trivial name	Occurrence in species	Ref.
		OMe: 7,3′,4′; O_2CH_2: 3,4; oxo: 9′; Δ: 1,3,5,7,1′,3′,5′,7′	chinensinaphtol methylether	PolPolchi	122
		OH: 7; OMe: 3,4; O_2CH_2: 3′,4′; oxo: 9′; Δ: 1,3,5,7,1′,3′,5′,7′	diphyllin	Acajuspr1	122
				Acajuspr2	122
				BerDipgra	122
				EupClecol	7, 122, 164
				EupClepat	164
				RutHapdau	13
				RutHaphis	49
				RutHapobt	13
				RutHapvul	57
				TaxTaicry	90
		8*R*,7′*S*,8′*R*; OH: 4,4′; OMe: 3,3′; oxo: 9′; Δ: 1,3,5,1′,3′,5′	(−)-conidendrin	PolPolpol	63
				TaxTaxwal	164
		8*R*,7′*R*,8′*R*; O_2CH_2: 3,4,3′,4′; oxo: 9′; Δ: 1,3,5,1′,3′,5′	(−)-polygamain	PolPolpol	63
				BurCominc	120
		8*R*,7′*R*,8′*S*: O_2CH_2: 3,4,3′,4′; oxo: 9′; Δ: 1,3,5,1′,3′,5′	picropolygamain	BurCominc	120
		OMe: 3,4; O_2CH_2: 3′,4′; 3,4-di-*O*-Me-Xyl: 7; oxo: 9′; Δ: 1,3,5,7,1′,3′,5′,7′	cleistanthin A	EupClecol	7, 8
				EupClepat	164
		OMe: 3,4; O_2CH_2: 3′,4′; 2,3-di-*O*-Me-Xylpyr-1,4-*O*-*β*-D-Glupyr: 7; oxo: 9′; Δ: 1,3,5,7,1′,3′,5′,7′	(−)-cleistanthin C	EupClecol	164
				EupClepat	164
		OMe: 3,4; O_2CH_2: 3′,4′; 2,3,5-tri-OMe-D-Xyl: 7; oxo: 9′; Δ: 1,3,5,7,1′,3′,5′,7′	cleistanthin D	EupClecol	164
		OMe: 3,4; O_2CH_2: 3′,4′;	cleistanthin E	EupClecol	164

Table 7.3.4 (continued)

Type	Subtype	Compound	Trivial name	Occurrence in species	Ref.
		2,3,5-tri-OMe-D-Xyl-fur-2,3-di-OMe-D-Xyl-pyr-1,4-*O*-*β*-D-Glu-pyr: 7; oxo: 9′; *Δ*: 1,3,5,7,1′,3′,5′,7′			
		OMe: 3,4; O_2CH_2: 3′,4′; *O*-*β*-D-Glupyr-(1 – 2)-3,4-di-OMe-D-Xyl-pyr: 7; oxo: 9′; *Δ*: 1,3,5,7,1′,3′,5′,7′	cleistanthoside A	EucClepat	164
		OH: 3; OMe: 4; O_2CH_2: 3′,4′; oxo: 9′; *Δ*: 1,3,5,7,1′,3′,5′,7′	daurinol	RutHapdau	13
		OMe: 3,4; O_2CH_2: 3′,4′; Xyl: 7; oxo: 9′; *Δ*: 1,3,5,7,1′,3′,5′,7′	diphyllinin	RutHaphis	49
		OMe: 3,4; O_2CH_2: 3′,4′; 5-CroXyl: 7; oxo: 9′; *Δ*: 1,3,5,7,1′,3′,5′,7′	diphyllinin crotonate	RutHaphis	49
		OMe: 3,4; O_2CH_2: 3′,4′; 5-AceXyl: 7; oxo: 9′; *Δ*: 1,3,5,7,1′,3′,5′,7′	diphyllinin monoacetate	RutHaphis	49
		OH: 4,4′; OMe: 3,3′; oxo: 9; *Δ*: 1,3,5,1′,3′,5′	conidendrin	PinPicexc PinTsuhet PinTsuste TaxTaicry	122 122 122 122
		OMe: 7; O_2CH_2: 3,4,3′,4′; oxo: 9; *Δ*: 1,3,5,7,1′,3′,5′,7′	justicidin-D = neo-justicidin-A	AcaJuspr1 AcaJuspr2	122 122
		OMe: 3,4,7; O_2CH_2: 3′,4′; oxo: 9; *Δ*: 1,3,5,7,1′,3′,5′,7′	justicidin-C = neo-justicidin-B	AcaJuspr1 AcaJuspr2	122 122
		O_2CH_2: 3,4,3′,4′; oxo: 9; *Δ*: 1,3,5,7,1′,3′,5′,7′	justicidin-E	AcaJuspr1 AcaJussim	122 130
		OMe: 3′,4′; O_2CH_2: 3,4; oxo: 9; *Δ*: 1,3,5,7,1′,3′,5′,7′	deoxyisodiphyllin	ApiBupfru	122

Table 7.3.4 (continued)

Type	Subtype	Compound	Trivial name	Occurrence in species	Ref.
		OH: 4,4′; OMe: 3,3′; oxo: 9; Δ: 1,3,5,7,1′,3′,5′,7′	detetrahydroconi-dendrin	CupJunfor	88
		OH: 6′; OMe: 2; O_2CH_2: 3,4,3′,4′; oxo: 9; Δ: 1,3,5,7,1′,3′,5′,7′	prostalidin A	AcaJuspr1	164
		OMe: 2,6′; O_2CH_2: 3,4,3′,4′; oxo: 9; Δ: 1,3,5,7,1′,3′,5′,7′	prostalidin B	AcaJuspr1	164
		OH: 6′; O_2CH_2: 3,4,3′,4′; oxo: 9; Δ: 1,3,5,7,1′,3′,5′,7′	prostalidin C	AcaJuspr1	164
		OMe: 3′,4′; O_2CH_2: 3,4; oxo: 9; Δ: 1,3,5,7,1′,3′,5′,7′	retrochinensin	AcaJuspr1	164
	8.8′, 2.7′	OMe: 5,9,3′,4′,9′; O_2CH_2: 3,4; Δ: 1,3,5,1′,3′,5′	hypophyllanthin	EupPhynir	122
	8.8′, 2.7′, 9.O.9′	OH: 5′; O_2CH_2: 3,4,3′,4′; oxo: 9; Δ: 1,3,5,7,1′,3′,5′,7′	justicinol	AcaJusfla	164
		7′*R*; OMe: 3,4; O_2CH_2: 3′,4′; oxo: 9′; Δ: 1,3,5,8,1′,3′,5′,8′	(−)-β-apopolyga-main	PolPolpol	63
		8*R*,7′*R*,8′*R*; OMe: 3,4; O_2CH_2: 3′,4′; oxo: 9′; Δ: 1,3,5,1′,3′,5′	(−)-polygamatin	PolPolpol	63
		O_2CH_2: 3,4,3′,4′; oxo: 9′; Δ: 1,3,5,7,1′,3′,5′,7′	helioxanthin	AcaJusfla AcaJussim AstHelbup RutRutgra	164 130 122 122
		OMe: 3,7; OH: 4; O_2CH_2: 3′,4′; oxo: 9′; Δ: 1,3,5,7,1′,3′,5′,7′	orosunol	AcaJusfla	115
	8.8′, 2.2′, 9.O.9′	OMe: 4,5,3′,4′; OAng: 7′; oxo: 9′; Δ: 1,3,5,1′,3′,5′	steganolide B	ApiSteara	128
		OMe: 4,5,3′,4′; OTig: 7; oxo: 9′; Δ: 1,3,5,1′,3′,5′	steganolide C	ApiSteara	128

Table 7.3.4 (continued)

Type	Subtype	Compound	Trivial name	Occurrence in species	Ref.
		8*R*,8′*R*; OMe: 4,5,3′,4′,5′; oxo: 9′; *Δ*: 1,3,5,1′,3′,5′	neoisostegane	ApiSteara	61, 164
		7*S*,8*R*,8′*S*; OMe: 4,5; O_2CH_2: 4′,5′; OAng: 7; oxo: 9′; *Δ*: 1,3,5,1′,3′,5′	araliangine	ApiSteara	164
8.5′	8.5′, 7.O.4′	OH: 4; OMe: 3,3′; OFer: 9,9′; *Δ*: 1,3,5,1′,3′,5′	boehmenan	UrtBoetri	122
		OH: 9,3′,9′; OMe: 3; Polyol: 4; *Δ*: 1,3,5,1′,3′,5′		PinPinsib	109
8.8′, 9.O.9′, 5′.8″		OH: 4,4′,4″,7″,9″; OMe: 3,3′,3″; oxo: 9; *Δ*: 1,3,5,1′,3′,5′, 1″,3″,5″	lappaol C	AstArclap	164
		OH: 4′,4″,7″,9″; OMe: 3,4,3′,3″; oxo: 9′; *Δ*: 1,3,5,1′,3′,5′, 1″,3″,5″	lappaol D	AstArclap	164
8.8′, 9.O.9′, 4′.O.8″		OH: 4,4″,7″,9″; OMe: 3,3′,3″; oxo: 9; *Δ*: 1,3,5,1′,3′,5′,1″,3″,5″	lappaol E	AstArclap	164
8.8′, 9.O.9′, 4′.O.7″, 5′.8″		OH:4,4″,9″; OMe: 3,3′,3″; oxo: 9; *Δ*: 1,3,5,1′,3′,5′,1″,3″,5″	lappaol A	AstArclap	164
		OH: 4″,9″; OMe: 3,4,3′,3″; oxo: 9′; *Δ*: 1,3,5,1′,3′,5′, 1″,3″,5″	lappaol B	AstArclap	164
8.8′, 7.O.9′, 9.O.7′, 5′.8″		OH: 4,4′,4″; OMe: 3,3′,3″; *Δ*: 1,3,5,1′,3′,5′, 1″,3″,5″; oxo: 7,2′	herpetrione	CurHercau	164
8.8′, 7.O.9′, 9.O.7′, 4′.O.8″		OH: 4,4″,7″,9″; OMe: 3,5,3′,5′,3″; *Δ*: 1,3,5,1′,3′,5′, 1″,3″,5″	buddlenol C	LogBuddav	64
		OH: 4,4″,7″,9″; OMe: 3,5,3′,5′,3″,5″; *Δ*: 1,3,5,1′,3′,5′, 1″,3″,5″	buddlenol D	LogBuddav	64

Table 7.3.4 (continued)

Type	Subtype	Compound	Trivial name	Occurrence in species	Ref.
		OH: 4,4″,7″,9″; OMe: 3,3′,5′,3″; Δ: 1,3,5,1′,3′,5′,1″,3″,5″	buddlenol E	LogBuddav	64
		OH: 4,4″,7″,9″; OMe: 3,3′,5′,3″,5″; Δ: 1,3,5,1′,3′,5′,1″,3″,5″	buddlenol F	LogBuddav	64
8.8′, 7.O.9′, 9.O.7′, 3′.8″, 4′.O.7″,		*7S,8R,7′S,8′R*; OH: 4,4″,9″; OMe: 3,5′,3″; Δ: 1,3,5,1′,3′,5′,1″,3″,5″	ledyotol A	RutHedlaw	79
8.O.4′, 8′.O.4″		OH: 4,7,9,7′,9′,9″; OMe: 3,3′,3″; Δ: 1,3,5,1′,3′,5′, 1″,3″,5″	leptolepisol B	PinLarlep	164
8.O.4′; 8.5″, 7′.O.4″		OH: 4,7,9,9′,9″; OMe: 3,3′,3″; Δ: 1,3,5,1′,3′,5′, 1″,3″,5″	leptolepisol A	PinLarlep	164
		OH: 4,7,9,9′; OMe: 3,3′,5′,3″; oxo: 9″; Δ: 1,3,5,1′,3′,5′, 1″,3″,5″	buddlenol A	LogBuddav	64
		OH: 4,7,9,9′,9″; OMe: 3,3′,5′,3″; Δ: 1,3,5,1′,3′,5′, 1″,3″,5″,7″	buddlenol B	LogBuddav	64
8.5′, 8.8″, 9′.O.9″, 5″.8‴		OH: 4,7,9,4′,4″, 4‴,7‴,9‴ OMe: 3,3′,3″3‴; oxo: 9′ Δ: 1,3,5,1′,3′,5′, 1″,3″,5″,1‴3‴,5‴	lappaol H	AstArclap	164
8.5′, 7.O.4′, 8′.8″, 9′.O.9″, 5″.8‴, 4″.O.7‴		OH: 4,9,4‴,9‴; OMe: 3,3′,3″,3‴; oxo: 9′ Δ: 1,3,5,1′,3′,5′,1″,3″, 5″,1‴,3‴,5‴	lappaol F	AstArclap	164
8.O.4′, 7′.O.9″, 8′.8″, 9′.O.7″, 4″.O.8‴		OH: 4,7,9,4‴,7‴,9‴; OMe: 3,3′,5′,3″,5″,3‴; Δ: 1,3,5,1′,3′,5′,1″,3″, 5″,1‴,3‴,5‴	hedyotisol A	RutHedlaw	107

Table 7.3.5. Neolignan structures classified according to skeletal types and subtypes

Type	Subtype	Compound	Trivial name	Occurrences in species	Ref.
8.8′	8.8′	8′*R*; OH: 4,4′; OMe: 3,3′; Δ: 1,3,5,7,1′,3′,5′	(−)-guaiaretic acid	ZygGuaoff	164
		8*R*,8′*R*; OH: 4,4′; OMe: 3,3′; Δ: 1,3,5,1′,3′,5′	(−)-dihydroguaiaretic acid	EupEuplau ZygGuaoff MyrOstpla	54 164 19
		8*R*,8′*R*; OH: 4′; OMe: 3′,O_2CH_2: 3,4; Δ: 1,3,5,1′,3′,5′	(−)-6-austrobailignan	AusAussca	54
		8*R*,8′*R*; O_2CH_2: 3,4,3′,4′; Δ: 1,3,5,1′,3′,5′	(−)-5-austrobailignan	AusAussca	54
		OMe: 3,4,3′,4′; Δ: 1,3,5,1′,3′,5′		MyrMyroto	113
		8*R*,8′*S*; OH: 4′; OMe: 3′; O_2CH_2: 3,4; Δ: 1,3,5,1′,3′,5′	dl-anwulignan	SchSchsph	96
		8*R*,8*S*,8′*S*; OH: 7′; OMe: 3,4,3′,4′; Δ: 1,3,5,1′,3′,5′		LauNecpub	111
		8*R*,8′*S*; OMe: 3,4,3′,4′; oxo: 7′; Δ: 1,3,5,1′,3′,5′		LauNecpub	111
		8*S*,8′*R*; OH: 4,4′; Δ: 1,3,5,1′,3′,5′	conocarpol	ComConere	54
		8*S*,8′*R*; OH: 3,4,3′,4′; Δ: 1,3,5,1′,3′,5′	*meso*-nordihydroguaiaretic acid (NDGA)	ZygLarcun ZygLardiv ZygLarnit	54 54 54, 164
		8*S*,8′*R*; OH: 3,4,4′; OMe: 3′; Δ: 1,3,5,1′,3′,5′		ZygLardiv	54
		8*S*,8′*R*; OH: 4,4′; OMe: 3,3′; Δ: 1,3,5,1′,3′,5′	*meso*-dihydroguaiaretic acid	LauLiccan LauMacedu ZygLarcun	54 54 54
		8*R*,8′*S*; OH: 7′; OMe: 3,4,3′,4′; Δ: 1,3,5,1′,3′,5′		MyrVirseb	99
		8*S*,8′*R*; O_2CH_2: 3,4,3′,4′; Δ: 1,3,5,1′,3′,5′	machilin A	LauMacthu	135
		8*R*,8′*R*; OMe: 3,4; oxo: 7,7′; O_2CH_2: 3′,4′; Δ: 1,3,5,1′,3′,5′		MyrVirseb	99
		8*R*,8′*S*; OH: 4; OMe:3; oxo: 7′; O_2CH_2: 3′,4′; Δ: 1,3,5,1′,3′,5′		MyrVirseb LauNecpub	99 111

Table 7.3.5 (continued)

Type	Subtype	Compound	Trivial name	Occurrence in species	Ref.
		8*R*,8′*S*; oxo: 7′; O_2CH_2: 3,4,3′,4′; Δ: 1,3,5,1′,3′,5′		LauNecpub	111
		OH: 3,3′; OMe: 4,5,4′,5′; Δ: 1,3,5,1′,3′,5′	pregomisin	SchSchchi	164
	8.8′, 7.O.7′	7*R*,8*S*,7′*S*,8′*R*; OH: 4,4′; OMe: 3,3′; Δ: 1,3,5,1′,3′,5′	tetrahydrofuro-guaiacin B	ZygGuaoff	164
		7*S*,8*S*,7′*R*,8′*R*; OMe: 3,4,3′,4′; Δ: 1,3,5,1′,3′,5′	galgravin	MagMagacu HimHimbel LauNecrig	54 54 89
		7*S*,8*S*,7′*R*,8′*R*; OH: 4,4′; OMe: 3,3′; Δ: 1,3,5,1′,3′,5′	tetrahydrofuro-guaiacin A	ZygGuaoff LauNecrig MyrMyrmal	164 89 121
		7*S*,8*S*,7′*R*,8′*R*; OH: 4; OMe: 3,3′,4′; Δ: 1,3,5,1′,3′,5′	nectandrin A	LauNecrig	89
		7*R*,8*S*,7′*R*,8′*R*; OMe: 3,4; O_2CH_2: 3′,4′; Δ: 1,3,5,1′,3′,5′	machilusin	LauMacjap	164
		7*S*,8*S*,7′*S*,8′*S*; O_2CH_2: 3,4,3′,4′; Δ: 1,3,5,1′,3′,5′	galbacin	HimHimbac HimHimbel LauEuszwa	164 164 54
		7*S*,8*S*,7′*S*,8′*S*; OMe: 3,4,3′,4′; Δ: 1,3,5,1′,3′,5′	galbegin	LauMacjap	164
		7*S*,8*S*,7′*S*,8′*S*; OMe: 3,4,5,3′,4′,5′; Δ: 1,3,5,1′,3′,5′	(−)-grandisin	LauLitgr1 LauLitgr2	54 54
		7*R*,8*S*,7′*S*,8′*S*; OH: 4; OMe: 3; O_2CH_2: 3′,4′; Δ: 1,3,5,1′,3′,5′	(+)-austrobailig-nan-7	LauUrbver AusAussca	164 54
		7*R*,8*S*,7′*S*,8′*S*; OMe: 3,4; O_2CH_2: 3′,4′; Δ: 1,3,5,1′,3′,5′	(+)-calopiptin	LauUrbver MagMagacu TriPipmoo	164 54 54
		7*R*,8*S*,7′*S*,8′*S*; OMe: 3,4,3′,4′; Δ: 1,3,5,1′,3′,5′	veraguensin	LauOcosp MagMagden MagMaglil MyrVirsur TriPipmoo	54 164 142 164 54
		7*R*,8*S*,7′*S*,8′*S*; OH: 4,4′; OMe: 3,3′; Δ: 1,3,5,1′,3′,5′	verrucosin	MagMagden	164

Table 7.3.5 (continued)

Type	Subtype	Compound	Trivial name	Occurrence in species	Ref.
		OH: 4,4′; OMe: 3,3′; Δ: 1,3,5,7,1′,3′,5′,7′	furoguaiacin	ZygGuaoff	54,164
		OH: 4; OMe: 3,3′,4′; Δ: 1,3,5,7,1′,3′,5′,7′	methylfuroguaiacin	ZygGuaoff	164
	8.8′, 7.O.9′	7*S*,8*S*,8′*R*; OH: 7′; OMe: 3′,4′; O_2CH_2: 3,4; Δ: 1,3,5,1′,3′,5′	magnostellin-C	MyrVirelo	75
		7*S*,8*S*,8′*R*; OH: 7′; OMe: 3,4,3′,4′; Δ: 1,3,5,1′,3′,5′	magnostellin-A	MagMagste	164
	8.8′, 7.7′	7*R*,8*S*,7′*S*,8′*R*; OMe: 3,4,6,3′,4′,6′; Δ: 1,3,5,1′,3′,5′	heterotropan	AriHettak	164
		OMe: 3,4,6,3′,4′,6′; Δ: 1,3,5,1′,3′,5′	acoradin	AraAcocal	117
		7*R*,8*R*,7′*R*,8′*R*; OMe: 3,4,6,3′,4′,6′; Δ: 1,3,5,1′,3′,5′	magnosalin	MagMagsal	164
	8.8′, 6.7′	8*R*,7′*S*,8′*S*; OH: 4,4′; OMe: 3,3′; Δ: 1,3,5,1′,3′,5′	(+)-guaiacin	MagMagkac ZygGuaoff LauMacedu MyrOstpla	68 164 54 19
		8*R*,7′*S*,8′*S*; OH: 4′; OMe: 3′; O_2CH_2: 3,4; Δ: 1,3,5,1′,3′,5′	(−)-iso-otoba-phenol	MyrVircar	54
		8*R*,7′*S*,8′*S*; OH: 4; OMe: 3; O_2CH_2: 3′,4′; Δ: 1,3,5,1′,3′,5′	(+)-otobaphenol	MyrMyroto MyrOstpla	54 19
		8*R*,7′*S*,8′*S*; OMe: 3,4; O_2CH_2: 3′,4′; Δ: 1,3,5,1′,3′,5′	(+)-isogalcatin or isootobain	MyrMyroto	54, 113
		8*R*,7′*S*,8′*S*; OMe: 3′,4′; O_2CH_2: 3,4; Δ: 1,3,5,1′,3′,5′	(−)-galcatin	MyrVircar MyrMyroto HimHimbac	54 113 54
		8*R*,7′*S*,8′*S*; OMe: 3,4,3′,4′; Δ: 1,3,5,1′,3′,5′	(−)-galbulin	MyrMyroto HimHimbac	113 54
		8*R*,7′*R*,8′*R*; OH: 3,4,3′; OMe: 4′; Δ: 1,3,5,1′,3′,5′	(−)-nor-isoguaiacin	ZygLardiv	54
		8*R*,7′*R*,8′*R*; OH: 4,4′; OMe: 3,3′; Δ: 1,3,5,1′,3′,5′	(−)-isoguaiacin	ZygGuaoff	164

Table 7.3.5 (continued)

Type	Subtype	Compound	Trivial name	Occurrence in species	Ref.
		8*R*,7′*R*,8′*R*; OMe: 3,4,3′,4′; Δ: 1,3,5,1′,3′,5′		LauNecpub MyrVirseb	111 99
		8*R*,7′*R*,8′*R*; OH: 4,4′; OMe: 3; Δ: 1,3,5,1′,3′,5′	3′-demethoxy-guaiacin	ZygLardiv	54
		8*R*,7′*R*,8′*S*; OMe: 3′; O_2CH_2: 3,4,4′,5′; Δ: 1,3,5,1′,3′,5′	(−)-austrobailig-nan-3	AusAussca	54
		8*R*,7′*R*,8′*S*; OMe: 3′,4′,5′; O_2CH_2: 3,4; Δ: 1,3,5,1′,3′,5′	(−)-austrobailig-nan-4	AusAussca	54
		8*R*,7′*R*,8′*S*; OH: 7′; OMe: 2; O_2CH_2: 4,5,3′,4′; Δ: 1,3,5,1′,3′,5′		MyrVircal	105
		8*R*,7′*S*,8′*R*; OH: 7′; OMe: 3′; O_2CH_2: 4,5,4′,4′; Δ: 1,3,5,1′,3′,5′		MyrVircal	105
		8*R*,7′*R*,8′*S*; OMe: 3,4; oxo: 7; O_2CH_2: 3′,4′; Δ: 1,3,5,1′,3′,5′		MyrVirseb	164
		8*R*,7′*S*,8′*R*; OH: 7′; OMe: 3,4; oxo: 7; O_2CH_2: 3′,4′; Δ: 1,3,5,1′,3′,5′		MyrVircal	105
		OH: 8,7′; OMe: 3,4; oxo: 7; O_2CH_2: 3′,4′; Δ: 1,3,5,1′,3′,5′		MyrVirseb	98
		OH: 8,7′; OMe: 3′,4′; oxo: 7; O_2CH_2: 3,4; Δ: 1,3,5,1′,3′,5′		MyrVirseb	98
		OH: 8; OMe: 3,4; oxo: 7; O_2CH_2: 3′,4′; Δ: 1,3,5,1′,3′,5′,7′		MyrVirseb	98
		OH: 7′; OMe: 3,4; oxo: 7; O_2CH_2: 3′,4′; Δ: 1,3,5,8,1′,3′,5′		MyrVirseb	98
		8*S*,7′*R*,8′*S*; OH: 3; OMe: 4,3′,4′; oxo: 7; Δ: 1,3,5,1′,3′,5′	schisandrone	SchSchsph	91

Table 7.3.5 (continued)

Type	Subtype	Compound	Trivial name	Occurrence in species	Ref.
		8*S*,8'*S*; OH: 4'; OMe: 3'; oxo: 7; O_2CH_2: 3,4; *Δ*: 1,3,5,1',3',5'	enshicine	SchSchhen	97
		8*R*,7'*R*,8'*S*; OMe: 3,4,3',4'; *Δ*: 1,3,5,1',3',5'		MyrMyroto	113
		8*R*,7'*R*,8'*S*; OMe: 3,4; O_2CH_2: 3',4'; *Δ*: 1,3,5,1',3',5'		MyrMyroto	113
		8*R*,7'*S*,8'*S*; O_2CH_2: 3,4,3',4' *Δ*: 1,3,5,1',3',5'	cagayanin	MyrMyrcag	87, 93
		8*S*,7'*S*,8'*S* OMe: 3,4,3',4'; *Δ*: 1,3,5,1',3',5'		MyrVirseb	99
		8*R*,7'*S*,8'*S*; OH: 4,4'; *Δ*: 1,3,5,1',3',5'		MyrIrigra	164
		8*R*; OMe: 2; O_2CH_2: 4,5,3',4'; *Δ*: 1,3,5,1',3',5',7'		MyrVircal	105
		8*R*; O_2CH_2: 3,4,3',4'; *Δ*: 1,3,5,1',3',5',7'	otobaene	MyrVircus MyrMyroto	54 54, 161
		OH: 4,4'; OMe: 3,3'; *Δ*: 1,3,5,7,1',3',5',7'	dehydroguaiaretic acid	ZygGuaoff	164
		OH: 9'; OMe: 3,4; oxo: 9'; O_2CH_2: 3',4'; *Δ*: 1,3,5,7,1',3',5',7'		MyrVirseb	99
		OMe: 3,4; oxo: 9'; O_2CH_2: 3',4'; *Δ*: 1,3,5,7,1',3',5',7'		MyrVirseb	99
	8.8', 2.7'	8*S*,7'*S*,8'*S*; oxo: 7; O_2CH_2: 3,4,3',4'; *Δ*: 1,3,5,1',3',5'	otobanone	MyrMyrsim	54
		8*S*,7'*R*,8'*R*; O_2CH_2: 3,4,3',4'; *Δ*: 1,3,5,1',3',5'	otobain	MyrMyroto	54
		8*S*,7'*R*,8'*S*; OH: 7'; oxo: 7; O_2CH_2: 3,4,3',4'; *Δ*: 1,3,5,1',3',5'	hydroxyoxootobain	MyrOstpla	19

Table 7.3.5 (continued)

Type	Subtype	Compound	Trivial name	Occurrence in species	Ref.
		8*S*,7′*S*,8′*R*; OH: 7; O_2CH_2: 3,4,3′,4′; *Δ*: 1,3,5,1′,3′,5′	hydroxyotobain	MyrOstpla	19
	8.8′, 2.7′, 3.O.6′	2*R*,8*R*,7′*S*,8′*R*; oxo: 6; O_2CH_2: 3,4,3′,4′; *Δ*: 1(8),4,1′,3′,5′	carpanone	LauCinsp	54
	8.8′, 2.7′, 8: 8′→7′	8*R*,7′*S*; OH: 7′; OMe: 3′,4′; oxo: 7,8′; O_2CH_2: 4,5; *Δ*: 1,3,5,1′,3′,5′		MyrVirseb	98
		OMe: 3′,4′; oxo: 7; O_2CH_2: 4,5; *Δ*: 1,3,5,1′,3′,5′,7′		MyrVirseb	98
	8.8′, 2.7′, 7: 8→8′	8*S*,7′*R*; OH: 8; OMe: 3′,4′; oxo: 7,8′; O_2CH_2: 4,5; *Δ*: 1,3,5,1′,3′,5′		MyrVirseb	98
	8.8′, 6.7′, seco 1.7	8*S*,7′*R*,8′*R*; OH: 7; OMe: 3,4; *Δ*: 1,3,5,1′,3′,5′		MyrMyrseb	99
	8.8′, 2.2′	8*S*,8′*R*; OMe: 3,3′,4′,5′; O_2CH_2: 4,5; *Δ*: 1,3,5,1′,3′,5′	gomisin N	SchSchchi	164
		8*S*,8′*R*; OH: 5,5′; OMe: 3,4,3′,4′; *Δ*: 1,3,5,1′,3′,5′	gomisin J	SchSchchi	139, 164
		8*S*,8′*R*; OH: 5; OMe: 3,4,3′,4′,5′ *Δ*: 1,3,5,1′,3′,5′	gomisin K_1	SchSchchi	164
		8*S*,8′*R*; OH: 3; OMe: 4,5,3′; O_2CH_2: 4′,5′; *Δ*: 1,3,5,1′,3′,5′	gomisin M_1	SchKadlon	145
		8*S*,8′*R*: OMe: 3,4,5,3′; O_2CH_2: 4′,5′; *Δ*: 1,3,5,1′,3′,5′	kadsuranin	Schkadlon	145
		8*S*,8′*R*; OMe: 3,3′; O_2CH_2: 4,5,4′,5′; *Δ*: 1,3,5,1′,3′,5′	wuweizisu C	Schkadlon SchSchchi SchSchrub	145 164 159

Table 7.3.5 (continued)

Type	Subtype	Compound	Trivial name	Occurrence in species	Ref.
		8*S*,8′*R*; OH: 3; OMe: 3′,4′,5′; O_2CH_2: 4,5; *Δ*: 1,3,5,1′,3′,5′		SchKadcoc	92
		8*S*,8′*R*; OMe: 3,3′,4′,5′; O_2CH_2: 4,5; *Δ*: 1,3,5,1′,3′,5′	isokadsuranin	SchKadcoc	92
		8*S*,8′*R*; OH: 3′; OMe: 3,4,5; O_2CH_2: 4′,5′; *Δ*: 1,3,5,1′,3′,5′	gomisin M_2	SchKadcoc	92
		8*S*,8′*R*; OMe: 3,4,5,3′,4′,5′; *Δ*: 1,3,5,1′,3′,5′	deoxyschisandrin	SchKadcoc	92
		8*S*,8′*R*; Ang: 3; OMe: 3′,4′,5′; O_2CH_2: 4,5; *Δ*: 1,3,5,1′,3′,5′	kadsutherin	SchKadcoc	92
		8*S*,8′*R*; Ang: 3; OMe: 4,5,3′; O_2CH_2: 4′,5′; *Δ*: 1,3,5,1′,3′,5′	(+)-angeloylgomi-sin M_1	SchKadlon	145
		OH: 4,4′,8′; OMe: 3,5,3′,5′; *Δ*: 1,3,5,1′,3′,5′	schisandrol	SchSchchi	54
		8*S*,8′*S*; OH: 8′; OMe: 3,4,5,3′,4′,5′ *Δ*: 1,3,5,1′,3′,5′	schisandrin	SchSchchi	54, 164
		8*S*,8′*S*; Ang: 3; OH: 8′; OMe: 4,5,3′,4′,5′; *Δ*: 1,3,5,1′,3′,5′	angeloylgomisin H	SchSchchi	164
		8*S*,8′*S*; Tig: 3; OH: 8′; OMe: 4,5,3′,4′,5′; *Δ*: 1,3,5,1′,3′,5′	tigloylgomisin H	SchSchchi	164
		8*S*,8′*S*; Ben: 3; OH: 8′; OMe: 4,5,3′,4′,5′; *Δ*: 1,3,5,1′,3′,5′	benzoylgomisin H	SchSchchi	164
		8*S*,8′*S*; OH: 3,8′; OMe: 4,5,3′,4′,5′; *Δ*: 1,3,5,1′,3′,5′	gomisin H	SchSchchi	164

Table 7.3.5 (continued)

Type	Subtype	Compound	Trivial name	Occurrence in species	Ref.
		8*S*,7′*R*,8′*S*; OH: 7′; OMe: 3,3′,4′,5′; O_2CH_2: 4,5; Δ: 1,3,5,1′,3′,5′	gomisin O	SchSchchi SchSchrub	164 159
		8*S*,7′*S*,8′*S*; OH: 7′; OMe: 3,3′,4′,5′; O_2CH_2: 4,5; Δ: 1,3,5,1′,3′,5′	epigomisin O	SchSchchi	164
		8*S*,7′*R*,8′*S*; OMe: 3′,4′,5′; O_2CH_2: 4,5; $OCH_2CH(CH_3)C$ $(CH_3)(OH)$-COO: 3 − 7′ Δ: 1,3,5,1′,3′,5′	gomisin E	SchSchchi SchKadcoc	164 92
		8*S*,7′*R*,8′*S*; OH: 7′; OMe: 3,4,5,3′; O_2CH_2: 4′,5′; Δ: 1,3,5,1′,3′,5′		SchKadcoc	92
		8*S*,7′*R*,8′*S*; Ben: 7′; OMe: 3,4,5,3′; O_2CH_2: 4′,5′; Δ: 1,3,5,1′,3′,5′	benzoyliso-gomisin C	SchKadcoc	92
		8*S*,7′*R*,8′*S*; Ang: 7′; OMe: 3,3′; O_2CH_2: 4,5,4′,5′; Δ: 1,3,5,1′,3′,5′	angeloylgomisin R	SchKadlon	145
		8*S*,7′*R*,8′*R*; Tig: 7′; OH: 8′ OMe: 3,3′,4′,5′; O_2CH_2: 4,5; Δ: 1,3,5,1′,3′,5′	tigloylgomisin D	SchSchchi	164
		7*R*,8*R*,8′*R*; OAc: 7; OMe: 3,3′,4′,5′; O_2CH_2: 4,5; Δ: 1,3,5,1′,3′,5′	kadsurin	SchKadjap	164
		8*R*,8′*S*; OMe: 3,4,5,5′; O_2CH_2: 3′,4′; Δ: 1,3,5,1′,3′,5′	γ-schizandrin	SchSchchi	54
		8*R*,8′*S*: OH: 3′; OMe: 3,4,5,4′,5′; Δ: 1,3,5,1′,3′,5′	schisanhenol or (+)-gomisin K_3	SchSchchi SchSchhen SchSchrub	164 164 159
		8*S*,8′*S*; OH: 8′; OMe: 3,3′,4′,5′ O_2CH_2: 4,5; Δ: 1,3,5,1′,3′,5′	gomisin A	SchSchchi	54, 164

Table 7.3.5 (continued)

Type	Subtype	Compound	Trivial name	Occurrence in species	Ref.
		8*R*,8′*R*; OMe: 3,4,5,3′,4′,5′; *Δ*: 1,3,5,1′,3′,5′	deoxyschisandrin	SchSchchi SchSchrub	54 159
		8*R*,8′*R*; OMe: 3,4,5,5′; O_2CH_2: 3′,4′ *Δ*: 1,3,5,1′,3′,5′	pseudo-*γ*-schisan-drin	SchSchchi	54
		8*S*,7′*S*,8′*S*; Ang: 7′; OH: 8′; OMe: 3,3′,4′,5′; O_2CH_2: 4,5; *Δ*: 1,3,5,1′,3′,5′	gomisin B	SchSchchi	54, 164, 140
		8*S*,7′*S*,8′*S*; Ben: 7′; OH: 8′; OMe: 3,3′,4′,5′; O_2CH_2: 4,5; *Δ*: 1,3,5,1′,3′,5′	gomisin C	SchSchchi	54, 140, 164
		8*S*,7′*S*,8′*S*; Ang: 7′; OH: 8′; OMe: 3,4,5,3′; O_2CH_2: 4′,5′; *Δ*: 1,3,5,1′,3′,5′	gomisin F	SchSchchi	54, 140, 164
		8*S*,7′*S*,8′*S*; Ben: 7′; OH: 8′; OMe: 3,4,5,3′; O_2CH_2: 4′,5′; *Δ*: 1,3,5,1′,3′,5′	gomisin G	SchSchchi	54, 140, 164
		8*S*,7′*S*,8′*S*; Ben: 7′; OH: 8′; OMe: 5,3′,4′,5′; O_2CH_2: 3,4; *Δ*: 1,3,5,1′,3′,5′	schisantherin A	SchSchsph	44, 164
		8*S*,7′*S*,8′*S*; Ang: 7′; OH: 8′; OMe: 5,3′,4′,5′; O_2CH_2: 3,4; *Δ*: 1,3,5,1′,3′,5′	schisantherin B	SchSchsph	164
		8*S*,7′*S*,8′*S*; Tig: 7′; OH: 8′; OMe: 5,3′,4′,5′; O_2CH_2: 3,4; *Δ*: 1,3,5,1′,3′,5′	schisantherin C	SchSchsph	164
		8*S*,7′*S*,8′*S*; Ben: 7′; OH: 8′; OMe: 3,3′; O_2CH_2: 4,5,4′,5′; *Δ*: 1,3,5,1′,3′,5′	schisantherin D	SchSchchi SchSchsph	164 164

Table 7.3.5 (continued)

Type	Subtype	Compound	Trivial name	Occurrence in species	Ref.
		8*S*,7′*S*,8′*S*; Ben: 7′; OH: 5,8′; OMe: 3,4,3′,4′,5′; *Δ*: 1,3,5,1′,3′,5′	schisantherin E	SchSchsph	164
		8*S*,7′*S*,8′*S*; Ang: 7′; OH: 8′; OMe: 3,4,5,3′,4′,5′ *Δ*: 1,3,5,1′,3′,5′	angeloylgomisin Q	SchSchchi	164
		8*S*,7′*S*,8′*S*; OH: 8′; OMe: 3′,4′,5′; O_2CH_2: 4,5; $OCH_2CH(CH_3)C(CH_3)(OH)$-COO: 3 – 7′; *Δ*: 1,3,5,1′,3′,5′	gomisin D	SchSchchi	54, 137
		7*R*,8*S*,7′*S*,8′*S*; Ang: 7′; OAc: 7; OH: 8′; OMe: 3,3′,4′,5′; O_2CH_2: 4,5; *Δ*: 1,3,5,1′,3′,5′	kadsuranin	SchKadcoc SchKadjap	92 164
		8*S*,7′*R*,8′*S*; OH: 7; OMe: 3,3′; O_2CH_2: 4,5,4′,5′; *Δ*: 1,3,5,1′,3′,5′	gomisin R	SchSchchi	164
		8*S*,7′*R*,8′*R*; Ang: 7′; OH: 8′; OMe: 3,3′,4′,5′ O_2CH_2: 4,5; *Δ*: 1,3,5,1′,3′,5′	angeloylgomisin P	SchSchchi	164
		8*S*,7′*R*,8′*R*; Tig: 7′; OH: 8′; OMe: 3,3′,4′,5′; O_2CH_2: 4,5; *Δ*: 1,3,5,1′,3′,5′	tigloylgomisin P	SchSchchi	164
	8.8′, 2.2′, 7.O.7′	OMe: 3,3′; O_2CH_2: 4,5,4′,5′; *Δ*: 1,3,5,1′,3′,5′		VerCleine	136
		OMe: 3,3′,4′,5′; O_2CH_2: 4,5; *Δ*: 1,3,5,1′,3′,5′		VerCleine	136
		OMe: 3,4,5,3′; O_2CH_2: 4′,5′; *Δ*: 1,3,5,1′,3′,5′		VerCleine	136
	8.8′, 2.1′	7*R*,8*S*,8′*R*; OAc: 7; OMe: 3,4,5,3′,5′; oxo: 4′; *Δ*: 1,3,5,2′,5′	eupodienone-1	EupEuplau	164

Table 7.3.5 (continued)

Type	Subtype	Compound	Trivial name	Occurrence in species	Ref.
		7*R*,8*S*,8′*R*; OH: 7; OMe: 3,4,5,3′,5′; oxo: 4′; *Δ*: 1,3,5,2′,5′	eupodienone-2	EupEuplau	164
		7*R*,8*S*,8′*R*; OCOPh: 7; OMe: 3,4,5,3′,5′; oxo: 4′; *Δ*: 1,3,5,2′,5′	eupodienone-3	EupEuplau	164
8.1′	8.1′	7*R*,8*S*,1′*R*,3′*R*; OH: 7; OMe: 3,4,5,3′; oxo: 6′; *Δ*: 1,3,5,4′,8′	megaphone	LauAnimeg	164
		7*R*,8*S*,1′*R*,3′*R*; OAc: 7; OMe: 3,4,5,3′; oxo: 6′; *Δ*: 1,3,5,4′,8′	megaphone acetate	LauAnimeg	164
		7*R*,8*S*,1′*R*,3′*R*; OMe: 3,3′; oxo: 6′; O_2CH_2: 4,5; *Δ*: 1,3,5,4′,8′	megaphyllone	LauAnimeg	164
		7*R*,8*S*,1′*R*,3′*R*; OH: 7; OMe: 3,4,3′; oxo: 6′; *Δ*: 1,3,5,4′,8′		LauOcopor	34
	8.1′, allyl: 1′→5′	OH: 9,4′; OMe: 3,4,3′; oxo: 7; *Δ*: 1,3,5,1′,3′,5′,8′	carinatonol	MyrVircar	164
	8.1′, allyl: 1′→O.4′	OMe: 3,4,5,3′,5′; *Δ*: 1,3,5,1′,3′,5′,8′	aurein	LauLic63	54
	8.1′, 7.O.6′	7*S*,8*S*,1′*R*; OMe: 3′; oxo: 4′; O_2CH_2: 3,4; *Δ*: 1,3,5,2′,5′,8′	burchellin	LauAniaff LauAnibur LauAnisp LauAniter MagMagden MagMaglil	54 54 106 54 164 164
		7*S*,8*S*,1′*R*; OMe: 3,4,3′; oxo: 4′; *Δ*: 1,3,5,2′,5′,8′		LauAniaff LauAnibur	54 164
		7*S*,8*S*,1′*R*; OMe: 3,4,5,3′; oxo: 4′; *Δ*: 1,3,5,2′,5′,8′		LauNecsp	164
		7*S*,8*S*,1′*R*; OMe: 3,4,3′,5′; oxo: 4′; *Δ*: 1,3,5,2′,5′,8′		LauAniaff	54

Table 7.3.5 (continued)

Type	Subtype	Compound	Trivial name	Occurrence in species	Ref.
		7*S*,8*S*,1′*R*; OMe: 3′,5′; oxo: 4′; O_2CH_2: 3,4; Δ: 1,3,5,2′,5′,8′		LauOcover	78
		7*S*,8*S*,1′*S*; OMe: 3′; oxo: 4′; O_2CH_2: 3,4; Δ: 1,3,5,2′,5′,8′		LauAnisp LauAniter	106 54
		7*S*,8*S*,1′*S*; OMe: 3,4,5,3′; oxo: 4′; Δ: 1,3,5,2′,5′,8′		LauAnisim LauNecmir	54 54
		7*R*,8*S*,1′*S*; OMe: 3,4,5,3′; oxo: 4′; Δ: 1,3,5,2′,5′,8′		LauAnisim	54
		7*R*,8*S*,1′*S*; OMe: 3′; oxo: 4′; O_2CH_2: 3,4; Δ: 1,3,5,2′,5′,8′		LauAnisp LauAniter LauOcocat	106 54, 164 164
		7*R*,8*S*,1′*S*; OMe: 3,3′; oxo: 4′; O_2CH_2: 4,5; O_2CH_2: 4,5; Δ: 1,3,5,2′,5′,8′		LauAnisim LauAniter	54 54,164
		7*R*,8*S*,1′*S*; OMe: 3,4,5,3′; oxo: 4′; Δ: 1,3,5,2′,5′,8′		LauAnisim LauAniter	54 54,164
		7*R*,8*S*,1′*S*; OMe: 3′,5′; oxo: 4′; O_2CH_2: 3,4; Δ: 1,3,5,2′,5′,8′		LauOcocat	164
		7*R*,8*S*,1′*R*,3′*R*; OMe: 3,4,3′; oxo: 4′; Δ: 1,3,5,5′,8′	porosin	LauOcopor LauUrbver	34, 54 164
		7*R*,8*S*,1′*R*,3′*R*; OMe: 3′; oxo: 4′; O_2CH_2: 3,4; Δ: 1,3,5,5′,8′	porosin-B	LauOcopor LauUrbver	13 164
		7*R*,8*S*,1′*S*,3′*S*; OMe: 3′; oxo: 4′; O_2CH_2: 3,4; Δ: 1,3,5,5′,8′		LauOcopor	34
		7*S*,8*R*,1′*S*,3′*S*; OMe: 3,3′,5′; oxo: 4′; O_2CH_2: 4,5; Δ: 1,3,5,5′,8′		LauAnifer	164
		7*S*,8*R*,1′*S*,3′*S*; OMe: 3′,5′; oxo: 4′; O_2CH_2: 3,4; Δ: 1,3,5,5′,8′		LauAnifer	164

Table 7.3.5 (continued)

Type	Subtype	Compound	Trivial name	Occurrence in species	Ref.
		$7R,8R,1'R,3'R$; OMe: 3′,5′; oxo: 4′; O_2CH_2: 3,4; Δ: 1,3,5,5′,8′	canellin B	LauLiccan	54
		$7R,8R,1'R,3'R$; OMe: 3′; oxo: 4′; O_2CH_2: 3,4; Δ: 1,3,5,5′,8′		LauOcopor	34
		$7S,8S,1'R,3'R$; OMe: 3,3′; oxo: 4′; O_2CH_2: 4,5; Δ: 1,3,5,5′,8′	armenin C	LauAnisp	152
		$7R,8R,1'R,3'R$; OMe: 3,4,5,3′; oxo: 4′; Δ: 1,3,5,5′,8′	canellin D	LauAnisp	152
		$7R,8R,1'R,3'R$; OMe: 3,3′,5′; oxo: 4′; O_2CH_2: 4,5 Δ: 1,3,5,5′,8′	canellin E	LauAnisp	152
		$7R,8S,1'R,3'R,4'R$, $6'S$; OH: 4′,6′; OMe: 3,4,3′; Δ: 1,3,5,8′		LauOcopor	34
		$7R,8S,1'R,3'R,4'S$, $6'S$; OH: 4′,6′; OMe: 3,4; Δ: 1,3,5,8′		LauOcopor	34
	8.1′, 7.O.6′, allyl: 1′→3′	$7S,8S,3'S$; OMe: 3′; oxo: 4′; O_2CH_2: 3,4,3′; Δ: 1,3,5,1′,5′,8′		LauAnibur LauAnisp LauAniter	164 106 54
		$7S,8S,3'S$; OMe: 3,4,3′; oxo: 4′; Δ: 1,3,5,1′,5′,8′		LauAniaff LauAnibur	54 164
		$7S,8S,3'R$; OMe: 3,3′; oxo: 4′; O_2CH_2: 4,5; Δ: 1,3,5,1′,5′,8′		LauAnisim	54
		$7S,8S,3'R$; OMe: 3,3′,5′; oxo: 4′; O_2CH_2: 4,5; Δ: 1,3,5,1′,5′,8′		LauAnisim	54
		$7S,8S,3'R$; OMe: 3,4,5,3′; oxo: 4′; Δ: 1,3,5,1′,5′,8′		LauAnisim LauNecmir LauNecsp	54 54 164

Table 7.3.5 (continued)

Type	Subtype	Compound	Trivial name	Occurrence in species	Ref.
		7*S*,8*S*,3′*R*; OMe: 3,4,5,3′; oxo: 4′; *Δ*: 1,3,5,1′,5′,8′		LauAnicit	151
		7*S*,8*S*,3′*R*; OMe: 3,4,5,3′,5′; oxo: 4′; *Δ*: 1,3,5,1′,5′,8′		LauAnisim	54
		7*S*,8*S*,3′*R*; OMe: 3′,5′; oxo: 4′; O_2CH_2: 3,4; *Δ*: 1,3,5,1′,5′,8′		LauOcocat LauOcover	164 78
		7*S*,8*S*,3′*R*; OMe: 3′; oxo: 4′; O_2CH_2: 3,4; *Δ*: 1,3,5,1′,5′,8′		LauAnisp LauAniter LauOcover	106 54 78
		7*R*,8*S*,3′*R*; OMe: 3′; oxo: 4′; O_2CH_2: 3,4; *Δ*: 1,3,5,1′,5′,8′		LauAniter	54
		7*R*,8*S*,3′*R*; OMe: 3,3′; oxo: 4′; O_2CH_2: 4,5; *Δ*: 1,3,5,1′,5′,8′		LauAnisim	54
	8.1′, 7.O.6′, allyl: 1′→O.4′	7*S*,8*S*; OMe: 3′; O_2CH_2: 3,4; *Δ*: 1,3,5,1′,3′,5′,8′		LauAnibur LauAniter	164 54
		7*S*,8*S*; OMe: 3,3′; O_2CH_2: 4,5; *Δ*: 1,3,5,1′,3′,5′,8′		LauAnisim	54
		7*S*,8*S*; OMe: 3,4,5,3′; *Δ*: 1,3,5,1′,3′,5′,8′		LauAnisim LauNecmir	54 54
		7*S*,8*S*; OMe: 3,4,5,3′,5′; *Δ*: 1,3,5,1′,3′,5′,8′		LauAnisim	54
		7*S*,8*S*; OMe: 3′,5′; O_2CH_2: 3,4; *Δ*: 1,3,5,1′,3′,5′,8′		LauOcover	78
		7*R*,8*S*; OMe: 3,3′; O_2CH_2: 4,5; *Δ*: 1,3,5,1′,3′,5′,8′		LauAnisim	54
	8.1′, 7.O.6′, allyl: 1′→5′	7*S*,8*S*; OH: 4′; OMe: 3′; O_2CH_2: 3,4; *Δ*: 1,3,5,1′,3′,5′,8′		LauAniter	54
		7*S*,8*S*; OH: 4′; OMe: 3,3′; O_2CH_2: 4,5; *Δ*: 1,3,5,1′,3′,5′,8′		LauAnisim	54

Table 7.3.5 (continued)

Type	Subtype	Compound	Trivial name	Occurrence in species	Ref.
		7*R*,8*S*; OH: 4′; OMe: 3′; O_2CH_2: 3,4; Δ: 1,3,5,1′,3′,5′,8′		LauAniter	54
		7*R*,8*S*; OH: 4′; OMe: 3,3′; O_2CH_2: 3,4; Δ: 1,3,5,1′,3′,5′,8′		LauAnisim	54
		OH: 4′; OMe: 3′; O_2CH_2: 3,4; Δ: 1,3,5,7,1′,3′,5′,8′		LauAniter	54
		OH: 4′; OMe: 3,3′; O_2CH_2: 4,5; Δ: 1,3,5,7,1′,3′,5′,8′		LauAnisim	54
		OH: 4′; OMe: 3,4,5,3′; Δ: 1,3,5,7,1′,3′,5′,8′		LauNecmir	54
	8.1′, 7.O.2′	7*R*,8*S*,1′*S*; OH: 2′; OMe: 3; O_2CH_2: 4,5; oxo: 3′; Δ: 1,3,5,4′,8′	ferrearin-A	LauAnifer	164
		7*R*,8*S*,1′*S*; OH: 2′; oxo: 3′; O_2CH_2: 3,4; Δ: 1,3,5,4′,8′	ferrearin-B	LauAnifer	164
		7*S*,8*S*,1′*S*; OH: 2′; oxo: 3′; O_2CH_2: 3,4; Δ: 1,3,5,4′,8′	ferrearin-C	LauOcoaci	129
		7*S*,8*S*,1′*S*; OH: 3′; OMe: 2′; O_2CH_2: 3,4; Δ: 1,3,5,4′,8′		LauOcopor	34
	8.1′, 7.5′	7*R*,8*R*,1′*R*,5′*R*,6′*S*; OH: 6′; OMe: 3′; oxo: 4′; O_2CH_2: 3,4; Δ: 1,3,5,2′,8′	guianin	LauAniaff LauAnibur LauAnigui	54 164 54
		7*S*,8*R*,1′*S*,5′*R*,6′*S*; OH: 6′; OMe: 3,3′,5′; oxo: 4′; O_2CH_2: 4,5; Δ: 1,3,5,2′,8′		LauAnicit LauAnifer	151 164
		7*S*,8*R*,1′*R*,5′*S*,6′*R*; OH: 6′; OMe: 3′,5′; oxo: 4′; O_2CH_2: 3,4; Δ: 1,3,5,2′,8′	5-methoxyguianin	LauAniaff LauAnifer LauAnigui LauOcoaci	54 164 54 129
		7*S*,8*R*,1′*R*,5′*S*;6′*R*; OAc: 6′; OMe: 3′,5′; oxo: 4′; O_2CH_2: 3,4; Δ: 1,3,5,2′,8′		LauAniaff	54
		7*S*,8*R*,1′*R*,5′*S*,6′*R*; OAc: 6′; OMe: 3′,5′; oxo: 4′; O_2CH_2: 3,4; Δ: 1,3,5,2′,8′		LauLicarm	164

Table 7.3.5 (continued)

Type	Subtype	Compound	Trivial name	Occurrence in species	Ref.
		7*S*,8*R*,1'*R*,5'*S*,6'*R*; OAc: 6'; OH: 3'; OMe: 5'; oxo: 4'; O_2CH_2: 3,4; Δ: 1,3,5,2',8'		LauAniaff	54
		7*R*,8*R*,1'*S*,5'*S*,6'*R*; OH: 3',6'; OMe: 3; oxo: 4'; O_2CH_2: 4,5; Δ: 1,3,5,2',8'		LauAnisim	164
		7*S*,8*R*,1'*S*,5'*R*,6'*S*; OH: 3',6'; OMe: 3,5'; oxo: 4'; O_2CH_2: 4,5; Δ: 1,3,5,2',8'		LauAnisim	164
		7*R*,8*S*,1'*R*,5'*S*,6'*R*: OH: 3',6'; OMe: 5'; oxo: 4'; O_2CH_2: 3,4; Δ: 1,3,5,2',8'		LauOcocat	164
		7*R*,8*R*,1'*S*,5'*S*; OMe: 3'; oxo: 4',6'; O_2CH_2: 3,4; Δ: 1,3,5,2',8'		LauAniaff	164
		7*R*,8*R*,1'S,5'*S*; OH: 4; OMe: 3,3'; oxo: 4',6'; Δ: 1,3,5,2',8'		LauAnibur LauAnisp	164 106
		7*R*,8*S*,1'*S*,4'*S*,5'*R*,6'*S*; OH: 4',6'; OMe: 3',5'; O_2CH_2: 3,4; Δ: 1,3,5,2',8'		LauOcoaci LauOcocat	129 164
		7*R*,8*S*,1'*S*,4'*R*,5'*R*,6'*S*; OH: 4',6'; OMe: 3,3'; O_2CH_2: 4,5; Δ: 1,3,5,2',8'		LauOcocat	164
		7*S*,8*R*,1'*R*,4'*S*,5'*S*,6'*R*; OAc: 4'; OH: 6'; OMe: 3',5'; O_2CH_2: 3,4; Δ: 1,3,5,2',8'		LauOcosp	164
		7*S*,8*R*,1'*R*,4'*S*,5'*S*,6'*R*; OAc: 4'; OH: 6'; OMe: 3,3',5'; O_2CH_2: 4,5; Δ: 1,3,5,2',8'		LauOcosp	164
		7*R*,8*S*,1'*R*,3'*R*,4'*S*, 5'*R*,6'*S*; OH: 4',6'; OMe: 3',5'; O_2CH_2: 3,4; Δ: 1,3,5,8'	canellin A	LauLiccan LauLicrig LauOcoaci	54 19 164

Table 7.3.5 (continued)

Type	Subtype	Compound	Trivial name	Occurrence in species	Ref.
		7R,8S,1'R,3'R,4'S, 5'R,6'S; OH: 4',6'; OMe: 3,3',5'; O_2CH_2: 4,5; Δ: 1,3,5,8'	methoxycanellin A	LauAnifer LauAnisp	164 164
		7R,8S,1'R,4'R,5'R,6'S; OH: 4',6'; OMe: 5'; oxo: 3'; O_2CH_2: 3,4; Δ: 1,3,5,8'	canellin C	LauLiccan LauLicrig LauOcocat	54 164 164
		7S,8S,1'R,4'R,5'S,6'R; OH: 4',6'; OMe: 3; oxo: 3'; O_2CH_2: 4,5; Δ: 1,3,5,8'		LauAnisim	54
		7R,8S,1'R,4'R,5'R,6'S; OH: 4',6'; OMe: 3,5'; oxo: 3'; O_2CH_2: 4,5; Δ: 1,3,5,8'		LauAnisim LauOcocat	54 164
		7R,8R,1'R,4'S,5'S,6'R; OH: 6'; OMe: 3,4'; oxo: 3'; O_2CH_2: 4,5; Δ: 1,3,5,8'		LauAnisim LauAnisp	54 106
		7S,8R,1'R,4'S,5'R,6'S; OH: 6'; OMe: 3,4',5'; oxo: 3'; O_2CH_2: 4,5; Δ: 1,3,5,8'		LauAnisim	54
		7R,8R,1'R,5'S,6'R; OH: 4',6'; oxo: 3'; O_2CH_2: 3,4; Δ: 1,3,5,8'		LauAnisp	106
		7R,8R,1'R,5'S,6'R; OH: 6'; OMe: 4'; oxo: 3'; O_2CH_2: 3,4; Δ: 1,3,5,8'		LauAnisp	106
		7S,8R,1'S,4'R,5'S,6'R; OAc: 4'; OH: 6'; OMe: 3,5'; oxo: 3'; O_2CH_2: 4,5; Δ: 1,3,5,8'		LauOcosp	164
		7S,8R,1'S,4'R,5'S,6'R; OAc: 4'; OH: 6'; OMe: 5'; oxo: 3'; O_2CH_2: 3,4; Δ: 1,3,5,8'		LauOcosp	164
		7S,8R,1'R,3'S,4'S,5'S; OH: 4'; OMe: 3,3',5'; oxo: 6'; O_2CH_2: 4,5; Δ: 1,3,5,8'		LauAnisp	164

Table 7.3.5 (continued)

Type	Subtype	Compound	Trivial name	Occurrence in species	Ref.
		7*S*,8*R*,1′*R*,3′*S*,4′*S*, 5′*S*; OH: 4′; OMe: 3,4,5,3′,5′; oxo: 6′; Δ: 1,3,5,8′		LauAnisp	164
		7*S*,8*R*,1′*R*,4′*R*,5′*R*; OH: 4′; OMe: 3,4,5,3′,5′; oxo: 6′; Δ: 1,3,5,2′,8′		LauOcosp	164
		7*S*,8*R*,1′*S*,4′*S*,5′*S*; OH: 4,4′; OMe: 3,3′,5′; oxo: 6′; Δ: 1,3,5,2′,8′		LauAnisp	106
		7*R*,8*R*,1′*R*,5′*S*; OH: 4; OMe: 3,3′; oxo: 4′,6′; Δ: 1,3,5,2′,8′		LauAnisp	106
	8.1′, 7.5′, 4′.O.5′	7*R*,8*S*; OH: 6; OMe: 3′; oxo: 4′; O_2CH_2: 3,4; Δ: 1,3,5,2′,8′	oxamethoxyguianin	LauOcoaci	46
	8.1′, 7.9′, 6′.O.8′	7*S*,8*R*; OMe: 3,4,5,3′; oxo: 4′; Δ: 1,3,5,2′,5′	maglifloenone or denudatone	MagMagden MagMaglil	164 143, 164
		7*S*,8*R*; OMe: 3′; oxo: 4′; O_2CH_2: 3,4; Δ: 1,3,5,2′,5′	futoenone	MagMagden MagMaglil PipPipfut	164 143, 164 54
8.3′	8.3′	8*R*,3′*S*; OH: 4; OMe: 3,3′,4′; oxo: 6′; Δ: 1,3,5,1′,4′,8′	lancifolin A	LauAnilan	164
		8*R*,3′*R*; OH: 4; OMe: 3,3′,4′; oxo: 6′; Δ: 1,3,5,1′,4′,8′	lancifolin B	LauAnilan	164
		8*R*,3′*S*; OMe: 3,4,3′,4′; oxo: 6′; Δ: 1,3,5,1′,4′,8′	lancifolin C	LauAnilan	164
		8*R*,3′*R*; OMe: 3,4,3′,4′; oxo: 6′; Δ: 1,3,5,1′,4′,8′	lancifolin D	LauAnilan	164
		8*R*,3′*S*; OMe: 3′,4′; oxo: 6′; O_2CH_2: 3,4; Δ: 1,3,5,1′,4′,8′	lancifolin E	LauAnilan	164
		8*R*,3′*R*; OMe: 3′,4′; oxo: 6′; O_2CH_2: 3,4; Δ: 1,3,5,1′,4′,8′	lancifolin F	LauAnilan	164
		OMe: 3′,4′; oxo: 6′; O_2CH_2: 3,4; Δ: 1,3,5,7,1′,4′,8′	futoquinol	PipPipfut	164

Table 7.3.5 (continued)

Type	Subtype	Compound	Trivial name	Occurrence in species	Ref.
	8.3′, 7.O.4′	7*S*,8*R*,3′*S*; OMe: 3,4,5,3′; oxo: 6′; *Δ*: 1,3,5,1′,4′,8′	mirandin A	LauNecmir LauNecsp	54 164
		7*S*,8*R*,3′*R*; OMe: 3,4,5,3′; oxo: 6′; *Δ*: 1,3,5,1′,4′,8′	mirandin B	LauNecmir	54
		7*S*,8*R*,3′*R*; OMe: 3′; oxo: 6′; O_2CH_2: 3,4 *Δ*: 1,3,5,1′,4′,8′	denudatin A	MagMagden MagMaglil	164 164
		7*S*,8*R*,3′*R*; OMe: 3,4,3′; oxo: 6′; *Δ*: 1,3,5,1′,4′,8′	denudatin B	MagMagden MagMaglil	164 164
		7*S*,8*R*,3′*S*,4′*R*; OMe: 3,4,3′,4′; oxo: 6′; *Δ*: 1,3,5,1′,8′	piperenone	MagMaglil PipPipfut	164 54
		7*S*,8*R*,3′*S*,4′*R*; OH: 4; OMe: 3,3′,4′; oxo: 6′; *Δ*: 1,3,5,1′,8′	liliflone	MagMaglil	164
		7*R*,8*S*,3′*R*; OMe: 3,4,3′; oxo: 6′; *Δ*: 1,3,5,1′,4′,8′	kadsurenone	PipPipfut	28
		7*R*,8*S*,3′*R*,4′*S*; OMe: 3′,4′; oxo: 6′; O_2CH_2: 3,4; *Δ*: 1,3,5,1′,8′	kadsurin A	PipPipfut	28
		7*R*,8*S*,3′*R*,4′*S*; OH: 6′; OMe: 3′,4′; O_2CH_2: 3,4; *Δ*: 1,3,5,1′,8′	kadsurin B	PipPipfut	28
	8.3′, 7.5	7*S*,8*R*,3′*R*,4′*S*,5′*R*; OH: 4′; OMe: 3,3′,5′; oxo: 6′; O_2CH_2: 4,5; *Δ*: 1,3,5,1′,8′	macrophyllin	LauLicmac LauNecsp	54 164
		7*R*,8*R*; OH: 6′; OMe: 3,4,5,3′; oxo: 4′; *Δ*: 1,3,5,1′,8′		LauNecsp	164
		7*R*,8*R*; OH: 4; OMe: 3,3′; oxo: 4′,6′; *Δ*: 1,3,5,1′,8′	liliflodione	MagMaglil	164
	8.3′, 8.2′, 7.1′	OMe: 4′; oxo: 6′; O_2CH_2: 3,4; *Δ*: 1,3,5,4′,8′	isofutoquinol A	PipPipfut	164
8.5′	8.5′	OH: 7,4′; OMe: 3,4,3′; *Δ*: 1,3,5,1′,3′,5′,8′	carinatol	MyrVircar	164

Table 7.3.5 (continued)

Type	Subtype	Compound	Trivial name	Occurrence in species	Ref.
		OH: 4′; OMe: 3,4,3′; Δ: 1,3,5,1′,3′,5′,8′	carinatone	MyrVircar	164
	8.5′, 7.O.4′	7*S*,8*S*; OH: 4; OMe: 3,3′; Δ: 1,3,5,1′,3′,5′,7′	licarin A	LauAnisim	54
				LauLicari	54
				LauNecrig	89
				LauUrbver	164
				MagMagkac	68
				MyrMyrfra	54
		7*S*,8*S*; OMe: 3′; O_2CH_2: 3,4; Δ: 1,3,5,1′,3′,5′,7′	licarin B or (+)-eupomatenoid 8	AriArital	164
				EupEuplau	54
				LauLicari	54
				LauMacjap	164
				LauOcopor	34
				MagMagkac	68
		7*S*,8*S*; OMe: 3,4,3′; Δ: 1,3,5,1′,3′,5′,7′	(−)-acuminatin	MagMagkac	38, 68
		7*S*,8*S*; OH: 3,4; OMe: 3′; Δ: 1,3,5,1′,3′,5′,7′	kachirachirol B	MagMagkac	68
		7*R*,8*R*; OMe: 3,4,3′; Δ: 1,3,5,1′,3′,5′,7′	licarin-D	LauOcoaci	46
				LauUrbver	164
		7*R*,8*R*; OMe: 3,4,5,3′; Δ: 1,3,5,1′,3′,5′,7′	licarin-C	LauNecmir	54
		7*R*,8*R*; OH: 4; Δ: 1,3,5,1′,3′,5′,7′	conocarpin	ComConere	54
		7*R*,8*R*; OH: 9′; OMe: 3′; O_2CH_2: 3,4; Δ: 1,3,5,1′,3′,5′,7′	machilin B	LauMacthu	135
		OMe: 3′; O_2CH_2: 3,4; Δ: 1,3,5,7,1′,3′,5′,7′	eupomatenoid-1 or eupomatene	MagMagkac	68
		OH: 4; OMe: 3,3′; Δ: 1,3,5,7,1′,3′,5′,7′	eupomatenoid-7	EupEuplau	54
				MagMagkac	68
		OMe: 3,4,3′; Δ: 1,3,5,7,1′,3′,5′,7′	eupomatenoid-12	EupEuplau	54
		OH: 4; OMe: 3′; Δ: 1,3,5,7,1′,3′,5′,7′	kachirachirol A or eupomatenoid-13	EupEuplau	164
				MagMagkac	68
		OH: 4; ORha: 2; Δ: 1,3,5,7,1′,3′,5′,7′	eupomatenoid-2	EupEuplau	54
		O_2CH_2: 3,4; Δ: 1,3,5,7,1′,3′,5′,7′	eupomatenoid-3	EupEuplau	54
		OMe: 3,4; Δ: 1,3,5,7,1′,3′,5′,7′	eupomatenoid-4	EupEuplau	54
		OH: 4; OMe: 3 Δ: 1,3,5,7,1′,3′,5′,7′	eupomatenoid-5	EupEuplau	54
		OH: 4; Δ: 1,3,5,7,1′,3′,5′,7′	eupomatenoid-6 or ratanhiaphenol II	PolRatrad	164
				EupEuplau	54
		OH: 4,7′,8′; OMe: 3; Δ: 1,3,5,7,1′,3′,5′	eupomatenoid-9	EupEuplau	54

Table 7.3.5 (continued)

Type	Subtype	Compound	Trivial name	Occurrence in species	Ref.
		8′*R*; OH: 8′; O_2CH_2: 3,4; *Δ*: 1,3,5,7,1′,3′,5′	eupomatenoid-11	EupEuplau	54
		OH: 4; OMe: 3; oxo: 7′; *Δ*: 1,3,5,7,1′,3′,5′	eupomatenoid-10	EupEuplau	54
		7*S*,8*S*; OH: 4; OMe: 3,3′; *Δ*: 1,3,5,1′,3′,5′,8′		MyrVircar	164
		OH: 4; OMe: 3,3′; *Δ*: 1,3,5,7,1′,3′,5′,8′	carinatidin	MyrVircar	164
		OMe: 3,4,3′; *Δ*: 1,3,5,7,1′,3′,5′,8′	carinatin	MyrVircar	164
		OH: 9; OMe: 3,4,3′; *Δ*: 1,3,5,1′,3′,5′,8′	dihydrocarinatinol	MyrVircar	164
		OH: 4; OMe: 3,5,3′; *Δ*: 1,3,5,1′,3′,5′,7′	5-methoxydehydro-diisoeugenol	MyrMyrfra	54
	8.5′, 7.O.4′, 9-nor	OH: 2; OMe: 4; *Δ*: 1,3,5,7,1′,3′,5′,7′	ratanhiaphenol I	KraRatrad	164
		OMe: 2,3,4,5,6; *Δ*: 1,3,5,7,1′,3′,5′,7′		KraKraram	2
		OMe: 2,3,4,6; *Δ*: 1,3,5,7,1′,3′,5′,7′		KraKraram	2
		OMe: 2,4,6; *Δ*: 1,3,5,7,1′,3′,5′,7′		KraKraram	2
		OH: 2; OMe: 4,6; *Δ*: 1,3,5,7,1′,3′,5′,7′		KraKraram	2
		OH: 2,6; OMe: 4; *Δ*: 1,3,5,7,1′,3′,5′,7′		KraKraram	2
		OMe: 2,4; *Δ*: 1,3,5,7,1′,3′,5′,7′		KraKraram	2
7.1′	7.1′, 8.O.2′	7*R*,8*S*,1′*R*,5′*R*; OMe: 3,3′,5′; oxo: 4′; O_2CH_2: 4,5; D: 1,3,5,2′,8′		LauLicchr	47
		7*R*,8*S*,1′*R*,5′*R*; OMe: 3,4,5,3′,5′; oxo: 4′; D: 1,3,5,2′,8′		LauLicchr	47
		7*R*,8*S*,1′*S*; OMe: 3,4,5,3′,4′,5′; oxo: 4′; D: 1,3,5,2′,5′,8′	chrysophyllon IB	LauLicchr	100

Table 7.3.5 (continued)

Type	Subtype	Compound	Trivial name	Occurrence in species	Ref.
		7*R*,8*S*,1′*S*; OMe: 3,3′,5′; oxo: 4′; O_2CH_2: 4,5; D: 1,3,5,2′,5′,8′	chrysophyllon IA	LauLicchr	100
	7.1′, 8.O.2′, allyl: 1′→3′	7*R*,8*S*,3′*R*; OMe: 3,4,5,3′,5′; oxo: 4′; D: 1,3,5,1′,5′,8′	chrysophyllon IIB	LauLicchr	100
		7*R*,8*S*,3′*R*; OMe: 3,3′,5′; oxo: 4′; O_2CH_2: 4,5; D: 1,3,5,1′,5′,8′	chrysophyllon IIA	LauLicchr	100
	7.1′, 8.O.2′, allyl: 1′→0.4′	7*R*,8*S*; OMe: 3,4,5,3′,5′; Δ: 1,3,5,1′,3′,5′,8′	chrysophyllon IIIB	LauLicchr	100
		7*R*,8*S*; OMe: 3,3′,5′; O_2CH_2: 4,5 Δ: 1,3,5,1′,3′,5′,8′	chrysophyllon IIIA		
8.7′	8.7′, 7.O.8′	7*S*,8*S*,7′*R*,8′*S*; OMe: 3,4,6,3′,4′,6′ Δ: 1,3,5,1′,3′,5′	magnosalicin	MagMagsal	155
5.5′	5.5′	OH: 4,4′; Δ: 1,3,5,8,1′,3′,5′,8′	magnolol	MagMagoff LauSasran	54 37, 169
		OH: 4,4′,9′; Δ: 1,3,5,8,1′,3′,5′,7′	randainol	LauSasran	37
		OH: 4,4′; oxo: 9′; Δ: 1,3,5,8,1′,3′,5′,7′	randainal	LauSasran	164
		OH: 4,1′,4′; Δ: 1,3,5,8,1′,3′,5′	randaiol	LauSasran	164
		OH: 4,4′; OMe: 3,3′; Δ: 1,3,5,8,1′,3′,5′,8′	dehydrodieugenol	LauLittur LauNecpol LauOcocym MyrVircar	54 164 36 76
		OH: 4; OMe: 3,3′,4′; Δ: 1,3,5,8,1′,3′,5′,8′	dehydrodieugenol monomethylether	LauNecpol LauOcocym MyrVircar	164 36 76
		OMe: 3,4,3′,4′; Δ: 1,3,5,8,1′,3′,5′,8′	di-*O*-methyl-dehydrodieugenol	LauNecpol	164
		OH: 4,2′; Δ: 1,3,5,8,1′,3′,5′,8′	honokiol	MagMagoff	162
		OH: 4; OMe: 2′; Δ: 1,3,5,8,1′,3′,5′,8′	honokiol mono-*O*-methylether	MagMaggra	123
8.O.4′	8.O.4′	OH: 7; OMe: 3,4,5,3′; Δ: 1,3,5,1′,3′,5′,7′	surinamensin	MyrVirsur	164

Table 7.3.5 (continued)

Type	Subtype	Compound	Trivial name	Occurrence in species	Ref.
		8*R*; OH: 4,7; OMe: 3,5,3′,5′; Δ: 1,3,5,1′,3′,5′,7′		MyrVircar	25
		OMe: 3,4,3′; Δ: 1,3,5,1′,3′,5′,7′	virolin	MyrVirsur	164
		8*R*; OH: 4; OMe: 3,5,3′,5′; Δ: 1,3,5,1′,3′,5′,7′		MyrVircar MyrVirpav	25 24
		OMe: 3,4,5,3′,5′; Δ: 1,3,5,1′,3′,5′,7′		MyrVirelon	104
		OMe: 3,3′,5′; O_2CH_2: 4,5; Δ: 1,3,5,1′,3′,5′,8′		LauEuszwa MyrVircar	164 25
		OH: 7,9′; OMe: 3,4,3′; Δ: 1,3,5,1′,3′,5′,8′	carinatidiol	MyrVircar	164
		7,8 erythro; OH: 4,7; OMe: 3,3′; Δ: 1,3,5,1′,3′,5′,7′	machilin-C	LauMacthu	135
		7,8 threo; OH: 4,7; OMe: 3,3′; Δ: 1,3,5,1′,3′,5′,7′	machilin-D	LauMacthu	135
		OAc: 7; OH: 9′; OMe: 3′; O_2CH_2: 3,4; Δ: 1,3,5,1′,3′,5′,7′	machilin-E	LauMacthu	135
	8.O.4′, 7.O.3′	7*R*,8*R*; OMe: 3,4,5,5′; Δ: 1,3,5,1′,3′,5′,8′	eusiderin-A	LauAni41 LauEuszwa LauLic04 LauLic41 LauLicrig MyrVircar MyrVirelo MyrVirgug MyrVirpav SchSchsph	164 54 54 54 164 25 104 164 24 91
		7*R*,8*R*; OMe: 5′; O_2CH_2: 3,4; Δ: 1,3,5,1′,3′,5′,8′	eusiderin-B	LauLic41 LauLicrig MyrVircus	54 164 164
		7*R*,8*R*; OH: 4; OMe: 3,5,5′ Δ: 1,3,5,1′,3′,5′,7′	eusiderin-E	MyrVircar	25
		7*R*,8*R*; OH: 5′; OMe: 3,4,5; Δ: 1,3,5,1′,3′,5′,8′	eusiderin-H	MyrVirpav	24
		OMe: 3,5′; O_2CH_2: 4,5; Δ: 1,3,4,1′,3′,5′,8′		LauEuszwa	164
		7*S*,8*R*; OMe: 3,4,5,5′; Δ: 1,3,5,1′,3′,5′,8′	eusiderin-C	MyrVirpav	24, 164

Table 7.3.5 (continued)

Type	Subtype	Compound	Trivial name	Occurrence in species	Ref.
		7*S*,8*R*; OMe: 3,4,5′; Δ: 1,3,5,1′,3′,5′,8′	eusiderin-D	MyrVirpav	164
		7*S*,8*R*; OH: 7; OMe: 3,4,5,5′; Δ: 1,3,5,1′,3′,5′,8′	eusiderin-F	LauAnisp	35
		7*S*,8*R*; OMe: 3,4,5,5′; oxo: 9′; Δ: 1,3,5,1′,3′,5′,7′	eusiderin-G	LauAnisp	35
4.O.5′	4.O.5′	OH: 4′; OMe: 3,3′; D: 1,3,5,8,1′,3′,5′,8′	dihydro- dieugenol B	MyrVircar LauOcocym	164 36
		Δ: 1,3,5,8,1′,3′,5′,8′	isomagnolol	LauSasran	164
2.O.3′	2.O.3′	OMe: 4,5,3′,4′; oxo: 6′; Δ: 1,3,5,8,4′,8′	lancilin	LauAnilan	164
1.5′	1.5′,2.2′	OMe: 3,3,5,3′,3′,5′; oxo: 4,4′; Δ: 8,8′	(−)-isoasatone	AriAsatai AriHettak	54 164
	1.5′,2.2′ 5.1′,6.6′	OMe: 3,3,5,3′,5′,5′; oxo: 4,4′; Δ: 8.8′	(±)-isoasatone	AriAsatai	54
	1.5′,2.2′ 8′.5″, 9′.2″	8′*R*; OMe: 3,3,5,3′,3′, 5′,3″,3″,5″; oxo: 4,4′,4″; Δ: 5,8,8″	heterotropatrione	AriHettak	164
		8′*S*; OMe: 3,3,5,3′,3′, 5′,3″,3″,5″; oxo: 4,4′,4″; Δ: 5,8,8″	isoheterotropatri- one	AriHettak	164
8.5′,9.2′		8*R*; OMe: 3,4,5,3′,3′,5′; oxo: 4′; Δ: 1,3,5,8′	heterotropanone 1	AriHettak	164
		8*S*; OMe: 3,4,5,3′,3′,5′; oxo: 4′; Δ: 1,3,5,8′	heterotropanone 2	AriHettak	164
8.8′,7.O.7′, 4′.O.8″		OH: 4,7″; OMe: 3,3′,3″,4″; Δ: 1,3,5,1′,3′,5′, 1″,3″,5″	saucerneol	SauSaucer	164
8.O.4′,8.8′, 7′.O.7″,4″.O.8‴		OH: 7,7‴; OMe: 3,4, 3′,3″,3‴,4‴; Δ: 1,3,5,1′,3′,5′, 1″,3″,5″,1‴,3‴,5‴	manassantin A	SauSaucer	164
		OH: 7,7‴; OMe: 3′; 3″,3‴,4‴; O_2CH_2: 3,4; Δ: 1,3,5,1′,3′,5′,1″,3″, 5″,1‴,3‴,5‴	manassantin B	SauSaucer	164

8.8',9.O.9',5'.8"
L1

8.8',9.O.9',4'.O.8"
L1

8.8',9.O.9',4'.O.7",5'.8"
L2

8.8',7.O.9',9.O.7',5'.8"
L1

8.8',7.O.9',9.O.7',4'.O.8"
L4

8.8',7.O.9',9.O.7',3'.8",4'.O.7"
L1

8.8',7.O.7',4'.O.8"
N1

8.O.4',8'.O.4"
L1

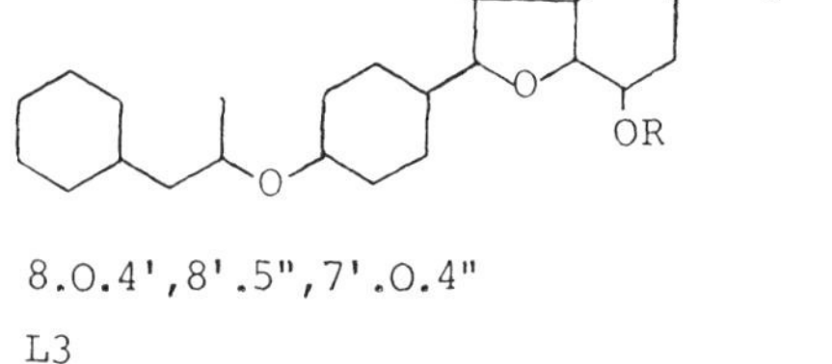

8.O.4',8'.5",7'.O.4"
L3

1.5',2,2',8'.5",9',2"
N2

Fig. 7.3.8. Skeletons of sesquilignans and sesquineolignans

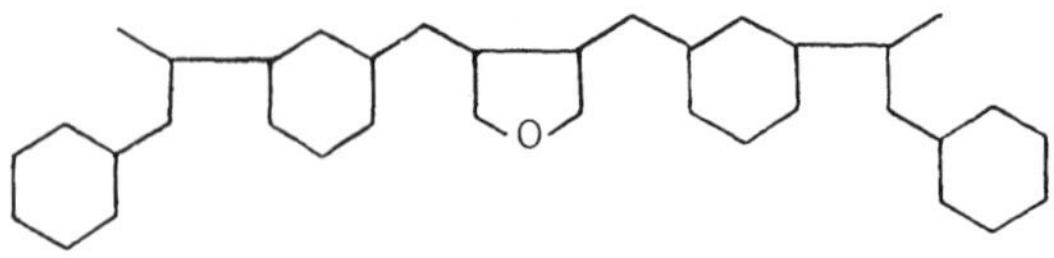

8.5',8'.8",9'.O.9",5".8'''

L1

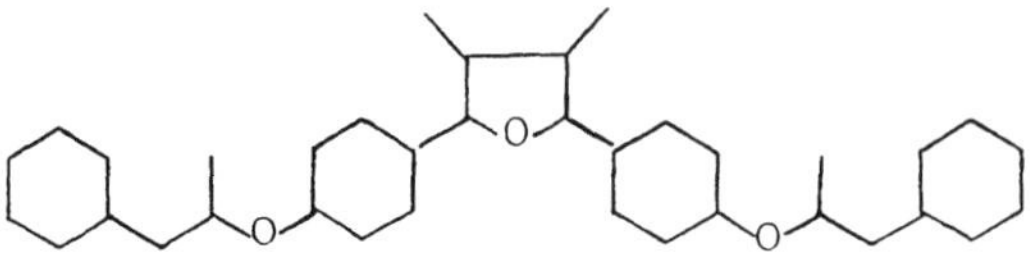

8.O.4',8'.8",7'.O.7",4".O.8'''

N2

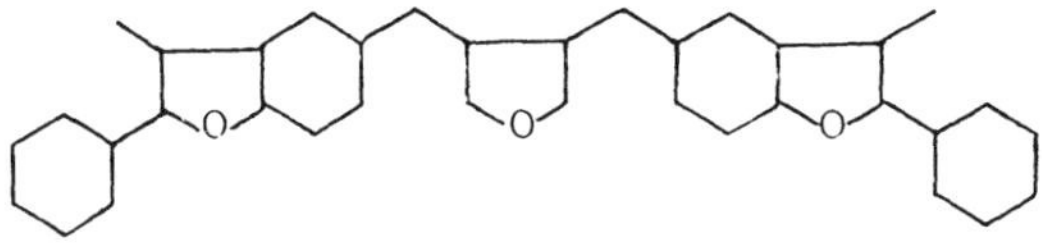

8.5',7.O.4',8'.8",9'.O.9",5".8''',4".O.7'''

L1

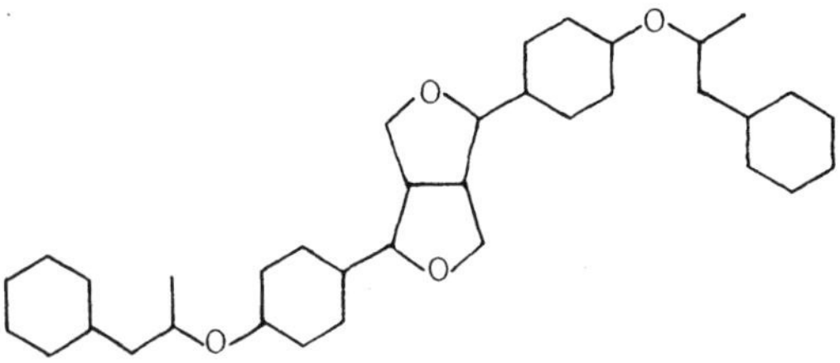

8.O.4',8'.8",7'.O.9",9'.O.7",4".O.8'''

L1

Fig. 7.3.9. Skeletons of dilignans and dineolignans

Last, but by no means least, our system of nomenclature and numbering can be applied not only to lignans and neolignans (i.e. C_6-C_3-dimers) but also to C_6-C_3-oligomers without any change of concept. Several trimers (Fig. 7.3.8) and tetramers (Fig. 7.3.9) have already been isolated and it is easy to foresee that many more will be described in the future.

Thus although based on biogenetic reasoning, supported at times by biomimetic interconversions or even only by mechanistic speculations, we consider the nomenclature and numbering proposed in the present work to constitute the best classificatory criterion in the lignan-neolignan domain of secondary metabolites.

7.3.3 Chemistry

There is at present little need for reviews on chemistry and synthesis of lignans and neolignans. An adequate chemical coverage up to mid-1976 (122) was subsequently complemented up to the end of 1983 (56, 164). Synthesis was covered in a very substantial paper published in 1982 (161). This does not mean, of course, that after the indicated dates science has lost interest in lignoids. Indeed, precisely the opposite has happened due to the powerful stimulus provided by the discovery of new and potentially exciting applications. Concomitantly, synthetic procedures became more varied (16) and analytical methods became more sophisticated.

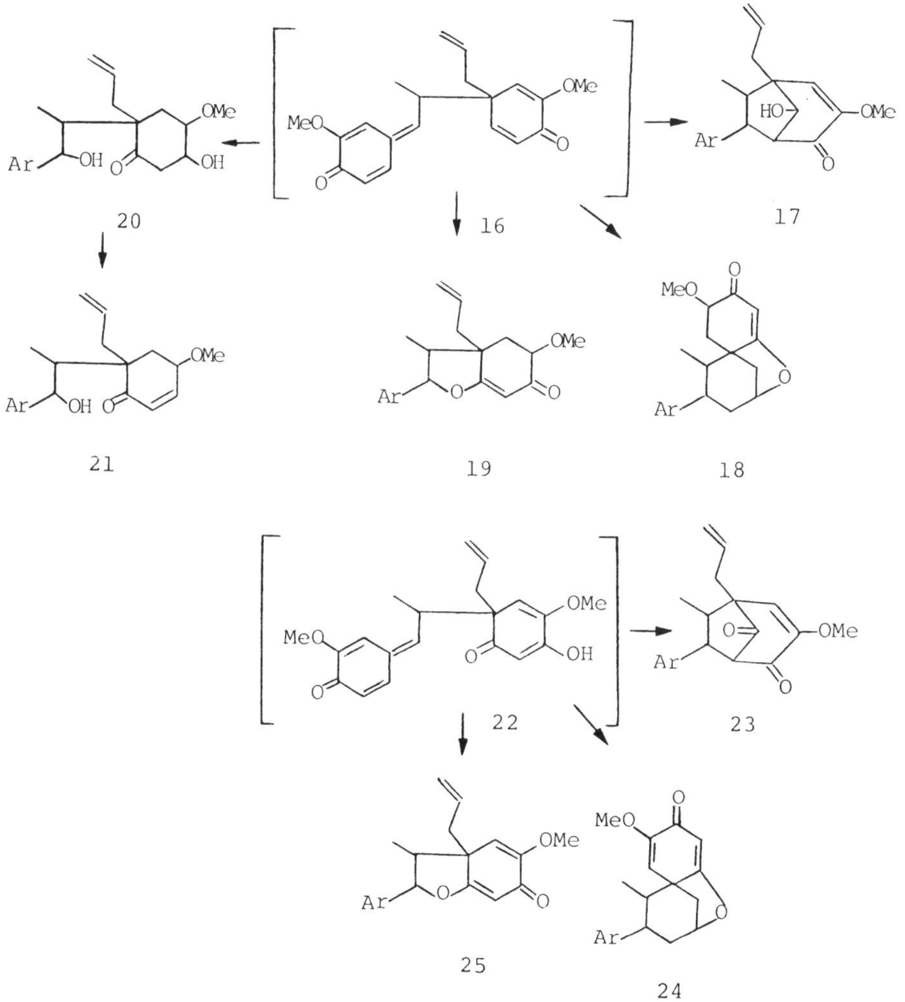

Fig. 7.3.10. Biogenesis of neolignans belonging to the bicyclo[3.2.1]octanoid (**17, 23**), futoenone (**18, 24**), porosin (**19**), burchellin (**25**), and megaphone (**20, 21**) groups. Stereochemical indications are omitted in order to maintain the information as generally valid as possible

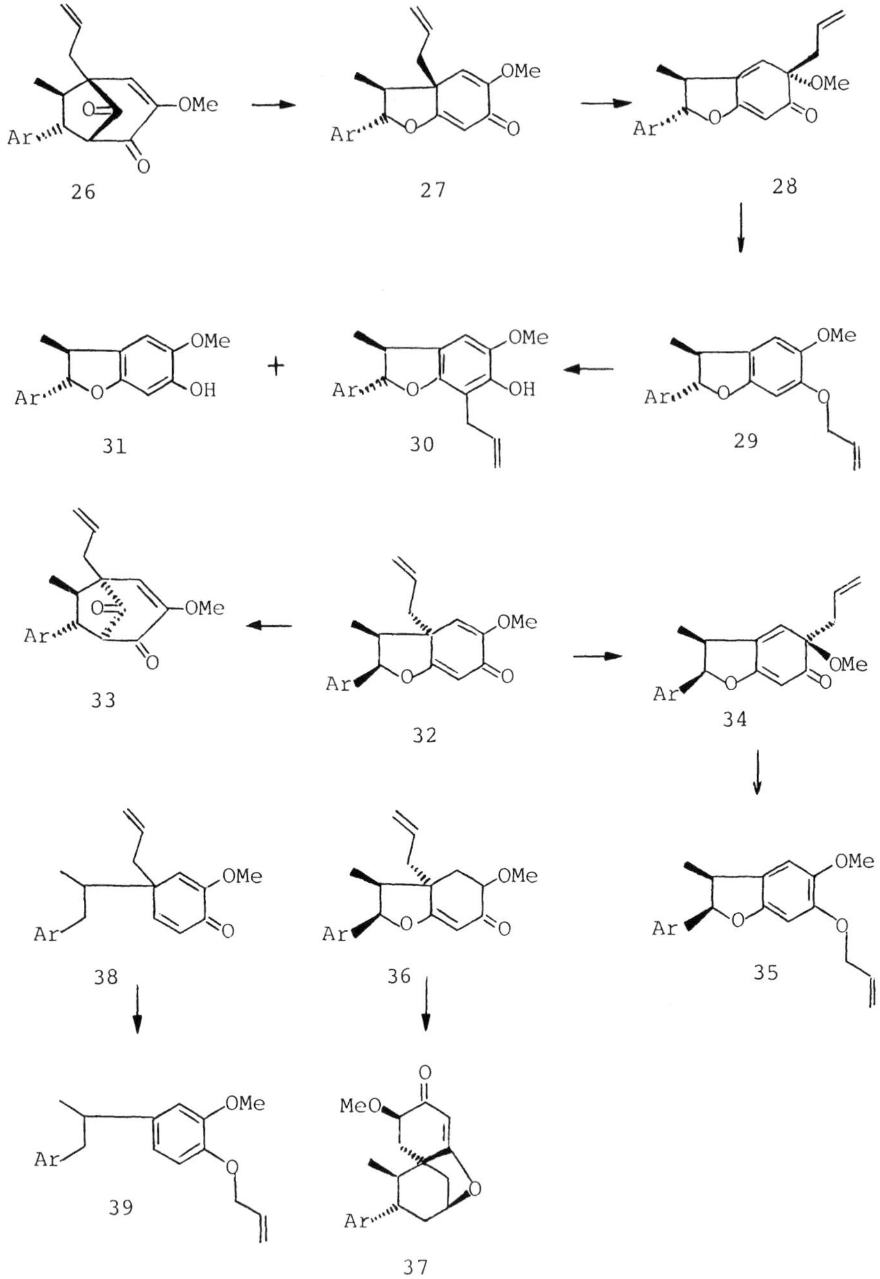

Fig. 7.3.11. Biogenetic and biomimetric relationships of neolignans belonging to the bicyclo[3.2.1]-octanoid (**27**, **28**, 32, **34**), hexahydrobenzofuranoid (**36**), and futoenone (**37**) groups. Also represented is the putative precursor (**38**) of aurein (**39**)

Thus conformational preferences determined by NMR (163) and computer-modelling were used in the verification of structure-activity relationships (15).

The chemical properties of lignans have been under scrutiny for many decades and, hence, are well known. In comparison the chemistry of neolignans has been described only recently. We focus on four different examples here. Two are of fundamental importance for the understanding of the adopted system of nomenclature or, in other words, the classification of lignoids. Whatever the mechanism, leading by oxidative coupling from **15a** + **15b** or from more highly oxygenated analogues to neolignans, the intermediary role of **16** and **22** seems to be assured. Different subsequent reactions lead from **16** or **22** as precursors to three of the most common types of neolignan skeletons (Fig. 7.3.10). Neolignans of such structural types frequently co-occur in the same or closely related plant groups. Although we have known this fact for several years we continue to marvel that so many different structural types can derive from just one precursor. Seemingly, whatever may happen to the intermediates **16** and **22** does happen, an exceptional phenomenon in the enzymatically catalyzed production of natural products.

The second example regards a series of rearrangements (6) (Fig. 7.3.11). Again the particular neolignans co-occur in the same or in closely related plant groups. The indicated biogenetic sequences were all reproduced in vitro. Transformations to and correlations with 2-aryldihydrobenzofuran neolignans were especially important, since the absolute configurations and the chiroptical properties of 2-aryldihydrobenzofuran neoflavonoids are known (58). Again neolignans are exceptional natural products combining an unusually large number of chiral centers per sp^3-carbons of such small molecules.

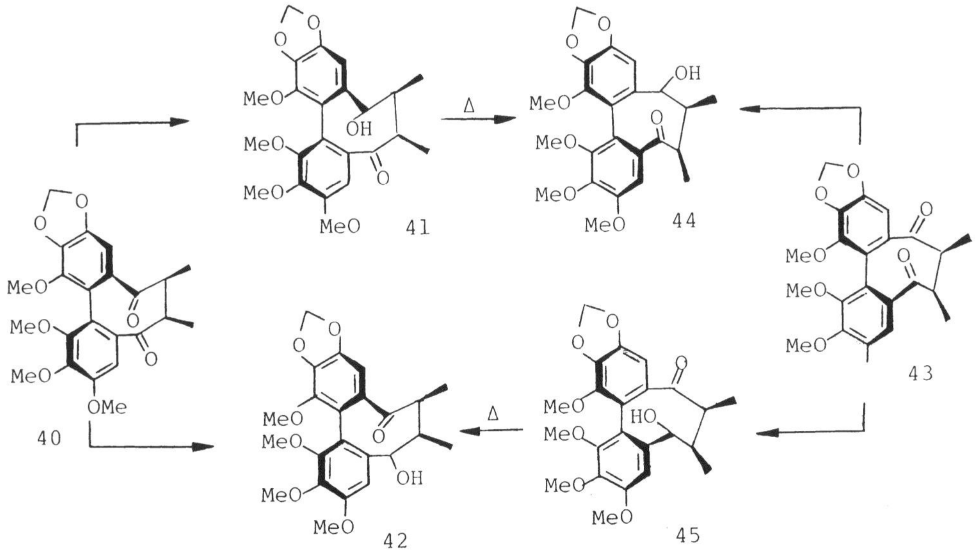

Fig. 7.3.12. Stereoselective transformation involving kadsurin (acetate of **41**). Reagents: $LiAl(tBu)_3H$ (**40**→**41**, **43**→**45**), H_2/Pd (**40**→**42**, **43**→**44**), alumina (**41**→**44**, **45**→**42**)

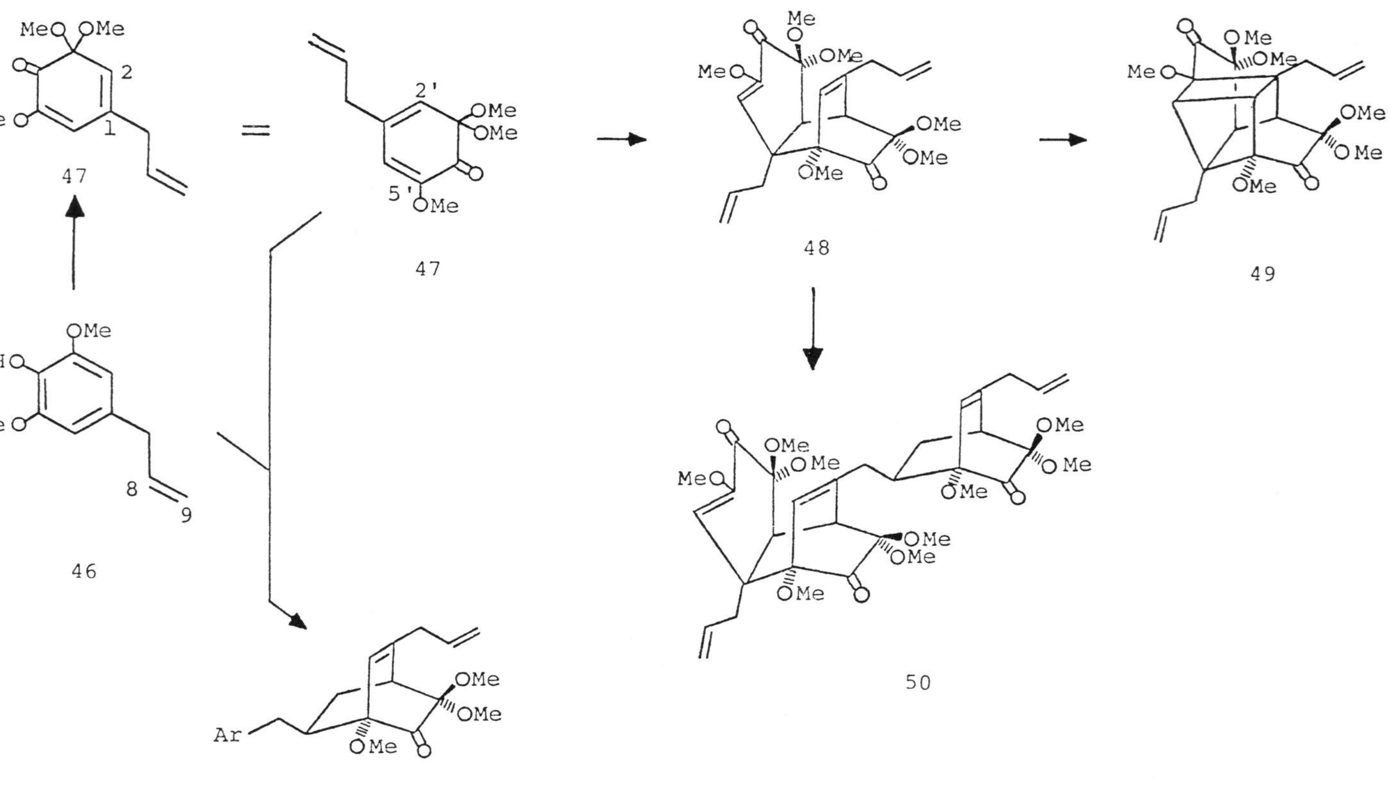

Fig. 7.3.13. Biogenetic relationships in the asatone (**48**) group of neolignans

Third, during the development of a synthetic approach to kadsurin, the acetate of the dibenzocyclo-octadiene neolignan **41**, a high stereoselectivity in reactions involving the eight-membered ring of some intermediates was observed. Moreover, it was found that some of these intermediates, which are isomeric by hindered biaryl rotation, have very different stabilities, and as a result one rotamer can be converted thermally into another. These properties are undoubtedly related to well-defined conformations of the octacycle. An explanation of these findings was based on the self-consistent conformational analysis summarized in Fig. 7.3.12 (50).

These three examples of neolignan chemistry refer to groups of compounds generated, at least formally, by oxidative coupling of propenylphenols *plus* allylphenols (Figs. 7.3.10 and 7.3.11) and of propenylphenols *plus* propenylphenols (Fig. 7.3.12).

The fourth and last example selected refers to a structurally surprising group of neolignans generated by the coupling of allylphenols *plus* allylphenols (Fig. 7.3.13). The process is initiated by oxidative methoxylation of the precursor **46** to **47**. Diels-Alder additions of **47** (as dienophile) + **47** (as diene) or of **46** (as dienophile) + **47** (as diene) lead respectively to **48** (asatone), and hence to **49**, and to **51** (164). Biosynthesis of the sesquineolignan **50** involves both these alternative cycloadditions. Biomimetic synthesis by anodic oxidation of **46** in methanol led to asatone (161).

7.3.4 Oligomeric Lignoids

Perusal of the literature on lignans leads to the conclusion that most if not all natural lignan skeletons have been described. The opposite is true for neolignans. The isolation of neolignans from plant extracts is frequently a tedious job since many representatives are difficult to crystallize. It is possible, consequently, that they have been frequently overlooked in fractionation work and that the compounds described in the present review are merely representative of a vast class of natural products. It is noteworthy in this sense, that most neolignans do not bear free phenolic hydroxyls. Since, however, such groups are a prerequisite for the oxidative coupling reaction that initiates their biosynthetic process, etherification must, of necessity, be the last step of this process. This implies the existence, even if transitory, of phenolic neolignans, potential substrates for oxidative oligomerization reactions. A search for naturally occurring oligomers of propenyl- and allylphenols in the viscous masses that form large parts of the extractives from tropical woods may prove of interest. Indeed, some sesquineolignans (Fig. 7.3.8) and dineolignans (Fig. 7.3.9) have already been found.

References

1 Achari B, Bandyopadhyay S, Saha C R, Pakrashi S C 1983 A phenanthroid lactone, steroid and lignans from *Aristolochia indica*. Heterocycles 20:771–774

2 Achenbach H, Grob J, Domingues X A, Star J V, Salgado F 1987 Ramosissin and other methoxylated nor-neolignans from *Krameria ramosissima*. Phytochemistry 26:2041–2043

3 Adesomoju A A, Okogun J I 1985 Phytochemical investigation of *Playlopsis falcisepala*. Fitoterapia 56:279–280
4 Adjaye A E, Dobberstein R H, Venton D L, Fong H H S 1984 Phytochemical investigation of *Laurelia novae-zelandiae*. J Nat Prod 47:553–554
5 Agrawal P K, Rastogi R P 1982 Two lignans from *Cedrus deodara*. Phytochemistry 21:1459–1461
6 de Alvarenga M A, Brocksom U, Gottlieb O R, Yoshida M, Braz-Filho R, Figliuolo R 1978 Hydrobenzofuranoid-bicyclo [3.2.1] octanoid neolignan rearrangement. J Chem Soc Chem Comm 831–833
7 Annapoorani K S, Damodaran C, Sekharan P C 1984 Solid-state fluorodensitometric quantitation of arylnaphthalene lignan lactones of *Cleistanthus collinus*. J Chromatogr 303:296–305
8 Annapoorani K S, Damodaran C, Sekharan P C 1985 High pressure liquid chromatographic separation of arylnaphthalide lignan lactones. J Liq Chromatogr 8:1173–1194
9 Badawi M M, Handa S S, Kinghorn A D, Cordell G A, Farnsworth N R 1983 Plant anticancer agents XXVII: Antileukemic and cytotoxic constituents of *Dirca occidentalis* (Thymelaeaceae). J Pharm Sci 72:1285–1287
10 Badheka L P, Prabhu B R, Mulchandani N B 1986 Lignans from *Piper cubeba*. Part 2. Dibenzylbutyrolactone lignans from *Piper cubeba*. Phytochemistry 25:487–489
11 Banerji A, Sarkar M, Ghosal T, Pal S C, Shoolery J N 1984 Sylvone, a new furanoid lignan of *Piper sylvaticum*. Tetrahedron 40:5047–5052
12 Banerji J, Das B, Chatterjee A, Shoolery J N 1984 Gadain, a lignan from *Jatropha gossypifolia*. Phytochemistry 23:2323–2327
13 Batirov E Kh, Matkarimov A D, Batsuren D, Malikov V M 1981 New coumarins and lignans from *Haplophyllum obtusifolium* and *H. dahuricum*. In: Atanasova B (ed) Inf Conf Chem Biotech Biol Act Nat Prod. Vol 1. Bulg Acad Sci Sofia Bulgaria
14 Batirov E Kh, Matkarimov A D, Malikov V M, Yugudaev M R 1985 Versicoside – a novel lignan glycoside from *Haplophyllum versicolor*. Khim Prir Soedin 624–628
15 Beard A R, Drew M G B, Hilgard P, Hudson B D, Mann J, Neidle S, Wong L T F 1987 Podophyllotoxin analogues: Synthesis and computer-modelling investigation of structure-activity relationships. Anti-Cancer Drug Des 2:247–252
16 Beard A R, Drew M G B, Mann J, Wong L T F 1987 Synthesis of analogues of 4-deoxypodophyllotoxin. Tetrahedron 43:4207–4215
17 Bohlmann F, Lonitz M, Knoll K 1978 Neue Lignan-Derivate aus der Tribus Heliantheae. Phytochemistry 17:330–331
18 Braquet P, Senn N, Fagoo M, Garay R, Robin J-P, Esanu A, Chabrier E, Godfraind T 1986 Lignanes de mammifères: une possible implication dans l'activité digitale endogène. C R Acad Sci Paris. Serie III 302:443–448
19 Braz-Filho R, de Carvalho M G, Gottlieb O R 1984 The chemistry of Brazilian Myristicaceae. XVIII. Eperudiendiol, glycerides and neolignans from fruits of *Osteophloeum platyspermum*. Planta Med 53–55
20 Bryan R F, Shen M S 1978 Gnidifolin (*trans*-2-(2,4-dihydroxy-3-methoxybenzyl)-3-(4'-hydroxy-3'-methoxybenzyl)-butyrolactone). Acta Cryst B34:327–329
21 Cairnes D A, Ekundayo O, Kingston D G I 1980 Plant anticancer agents. X. Lignans from *Juniperus phoenicea*. J Nat Prod 43:495–497
22 Cambie R C, Clark G R, Craw P A, Jones T C, Rutledge P S, Woodgate P D 1985 Chemistry of the Podocarpaceae. LXIX. Further lignans from the wood of *Dacrydium intermedium*. Aust J Chem 38:1631–1645
23 Cambie R C, Parnell J C, Rodrigo R 1979 New lignans from *Dacrydium intermedium*: PMR and CMR spectroscopy of phenyltetralin lactones and acetals. Tetrahedron Lett 1085–1088
24 Carneiro M O, Yoshida M, Gottlieb O R 1983 Algumas neolignanas de *Virola pavonis*. Cienc Cult 35(Suppl):457
25 Cavalcante S de H, Yoshida M, Gottlieb O R 1985 The chemistry of Brazilian Myristicaceae. XXV. 8.O.4'-Neolignans from *Virola carinata* fruit. Phytochemistry 24:1051–1055
26 Chalandre M C, Pareyre C, Bruneton J 1984 Lignans and alkaloids of *Hernandia ovigera* L. var. *mascarenensis* Meissn. Ann Pharm Fr 42:317–321
27 Chandel R S 1980 Pygeoside, a new lignan xyloside from *Pygeum acuminatum*. Indian J Chem 19B:279–282

28 Chang M N, Han G-Q, Arison B H, Springer J P, Hwang S-B, Shen T Y 1985 Neolignans from *Piper futokadsura*. Phytochemistry 24:2079–2082
29 Chaudhuri R K, Sticher O 1981 1. New iridoid glucosides and a lignan diglucoside from *Globularia alypum* L. Helv Chim Acta 64:3–15
30 Chemistry in Britain 1981 Tumor drug. From warts to cancer. Chem Brit 17:457
31 Chogomu-Tin R, Bodo B, Nyasse B, Sondengam B L 1985 Isolation and identification of (–)-olivil and (+)-cycloolivil from *Stereospermum kunthianum*. Planta Med 464
32 Dahlgren R 1980 A revised system of classification of the angiosperms. Bot J Linn Soc 80:91–124
33 Deyama T, Ikawa T, Kitagawa S, Nishibe S 1986 The constituents of *Eucommia ulmoides* Oliv. III. Isolation and structure of a new lignan glycoside. Chem Pharm Bull 34:523–527
34 Dias D A, Yoshida M, Gottlieb O R 1986 The chemistry of Brazilian Lauraceae. LXX. Further neolignans from *Ocotea porosa*. Phytochemistry 25:2613–2616
35 Dias S M C, Fernandes J B, Maia J G S, Gottlieb O R, Gottlieb H E 1986 The chemistry of Brazilian Lauraceae. LXXVII. Eusiderins and other neolignans from an *Aniba* species. Phytochemistry 25:213–217
36 de Diaz A M P, Gottlieb H E, Gottlieb O R 1980 The chemistry of Brazilian Lauraceae. LIX. Dehydrodieugenols from *Ocotea cymbarum*. Phytochemistry 19:681–682
37 El-Feraly F S 1984 Randainol: a neolignan from *Sassafras randainense*. Phytochemistry 23:2329–2331
38 El-Feraly F S, Cheatham S F, Hufford C D, Li W-S 1982 Optical resolution of (+/–)-dehydrodiisoeugenol: structure revision of acuminatin. Phytochemistry 21:1133–1135
39 Erdtman H, Harmatha J 1979 Phenolic and terpenoid heartwood constituents of *Libocedrus yateensis*. Phytochemistry 18:1495–1500
40 Fagbule M O, Olatunji G A 1984 Isolation and characterization of the lignan fargesin. Cellul Chem Technol 18:293–296
41 Fajardo V, Urzua O A, Torres R, Cassels B K 1979 Secondary metabolites of *Berberis buxifolia*. Rev Latinoamer Quim 10:131–134
42 Fang J-M, Jan S-T, Cheng Y-S 1985 (+)-Calocendrin, a lignan dihydroanhydride from *Calocedrus formosana*. Phytochemistry 24:1863–1864
43 Fang J-M, Wei K-M, Cheng Y-S, Cheng M-C, Wang Y 1985 (+)-Tsugacetal, a lignan acetal from Taiwan hemlock. Phytochemistry 24:1363–1365
44 Fang S-D, Huang M-F, Liu J-S, Gao Y-L, Hsu J-S 1975 Studies on the constituents of Hua-Zong Wu-Wei-Zi (*Schisandra sphenanthera* Rehd. et Wills.). I. The isolation and structure of schisantherin A. Hua Hsueh Hsueh Pao 33:57–60
45 Fang X, Manayakkara N P D, Phoebe C H J, Pezzuto J M, Kinghorn A D, Farnsworth N R 1985 Plant anticancer agents. XXXVII. Constituents of *Amanoa oblongifolia*. Planta Med 346–347
46 Felicio J D'A, Motidome M, Yoshida M, Gottlieb O R 1986 Further neolignans from *Ocotea aciphylla*. Phytochemistry 25:1707–1710
47 Ferreira S Z, Roque N F, Gottlieb O R, Gottlieb H E 1982 The chemistry of Brazilian Lauraceae. LXVIII. An unusual porosin type neolignan from *Licaria chrysophylla*. Phytochemistry 21:2756–2758
48 Ganeshpure P A, Schneiders G E, Stevenson R 1981 Structure and synthesis of hypophyllanthin, nirtetralin, phyltetralin and lintetralin. Tetrahedron Lett 22:393–396
49 Gonzalez A G, Darias V, Alonso G 1979 Cytostatic lignans isolated from *Haplophyllum hispanicum*. Planta Med 200–203
50 Gottlieb H E, Mervič M, Ghera E, Frolow F 1982 Conformational study of dibenzocyclo-octadiene systems related to the schisandrin-type lignans. J Chem Soc Perkin Trans I 2353–2358
51 Gottlieb O R 1972 Chemosystematics of the Lauraceae. Phytochemistry 11:1537–1570
52 Gottlieb O R 1974 Lignans and neolignans. Rev Latinoamer Quim 5:1–10
53 Gottlieb O R 1977 Chemistry of neolignans with potential biological activity. In: Wagner H, Wolff P (eds) New natural products and plant drugs with pharmacological, biological or therapeutical activity. Springer Heidelberg, 227–248
54 Gottlieb O R 1978 Neolignans. Progr Chem Org Nat Prod 35:1–72
55 Gottlieb O R, Kaplan M A C 1986 Replacement-nodal-subtractive nomenclature and codes of chemical compounds. J Chem Inf Comp Sci 26:1–3
56 Gottlieb O R, Yoshida M 1984 Lignóides, com atenção especial a química das neolignanas. Química Nova (SBQ, São Paulo) 7:250–273

57 Gozler T, Gozler B, Patra A, Leet J E, Freyer A J, Shamma M 1984 Konyanin: a new lignan from *Haplophyllum vulcanicum*. Tetrahedron 40:1145–1150

58 Gregson M, Ollis W D, Redman B T, Sutherland I O, Dietrichs H H, Gottlieb O R 1978 The neoflavonoid group of natural products. 8. Obtusastyrene and obtustyrene, cinnamylphenols from *Dalbergia retusa*. Phytochemistry 17:1395–1400

59 Hanuman J B, Mishra A K, Sabata B 1986 A natural phenolic lignan from *Tinospora cordifolia* Miers. J Chem Soc Perkin Trans I 1181–1185

60 Hernandez J D, Roman L U, Espineira J, Joseph-Nathan P 1983 Ariensin, a new lignan from *Bursera ariensis*. Planta Med 215–217

61 Hicks R P, Sneden A T 1983 Neoisostegane, a new bisbenzocyclooctadiene lignan lactone from *Steganotaenia araliacea* Hochst. Tetrahedron Lett 24:2987–2990

62 Hokanson G C 1978 Podophyllotoxin and 4′-demethylpodophyllotoxin from *Polygala polygama* (Polygalaceae). Lloydia 41:497–498

63 Hokanson G C 1979 The lignans of *Polygala polygama* (Polygalaceae). Deoxypodophyllotoxin and three new lignan lactones. J Nat Prod 42:378–384

64 Houghton P J 1985 Lignans and neolignans from *Buddleja davidii*. Phytochemistry 24:819–826

65 Iida T, Nakano M, Ito K 1982 Hydroperoxysesquiterpene and lignan constituents of *Magnolia kobus*. Phytochemistry 21:673–675

66 Ishii H, Ishikawa T, Mihara M, Akaike M 1983 Studies on the chemical constituents of rutaceous plants. XLVIII. The chemical constituents of *Zanthoxylum alanthoides* Sieb. et Zucc. (*Fagara alanthoides*) bark. Yakugaku Zasshi 103:279–292

67 Ishii H, Murakami K, Takeishi K, Ishikawa T, Haginiwa J 1981 Studies on the chemical constituents of rutaceous plants. XLVIII. The chemical constituents of *Fagara boninensis* Koidz. root. Yakugaku Zasshi 101:504–514

68 Ito K, Ichino K, Iida T, Lai J 1984 Neolignans from *Magnolia kachirachirai*. Phytochemistry 23:2643–2645

69 Jackson D E, Dewick P M 1984 Biosynthesis of *Podophyllum* lignans. I. Cinnamic acid precursors of podophyllotoxin in *Podophyllum hexandrum*. Phytochemistry 23:1029–1035

70 Jackson D E, Dewick P M 1984 Biosynthesis of *Podophyllum* lignans. II. Interconversions of aryltetralin lignans in *Podophyllum hexandrum*. Phytochemistry 23:1037–1042

71 Jackson D E, Dewick P M 1984 Aryltetralin lignans from *Podophyllum hexandrum* and *Podophyllum peltatum*. Phytochemistry 23:1147–1152

72 Jackson D E, Dewick P M 1985 Tumor-inhibitory aryltetralin lignans from *Podophyllum pleianthum*. Phytochemistry 24:2407–2409

73 Jakupovic J, Schuster A, Bohlmann F, King R M, Robinson H 1986 New lignans and isobutylamides from *Heliopsis buphthalmoides*. Planta Med 18–20

74 Kato A, Hashimoto Y, Kidokoro M 1979 (+)-Nortrachelogenin, a new pharmacologically active lignan from *Wikstroemia indica*. J Nat Prod 42:159–162

75 Kato M J, Paulino-Filho H F, Yoshida M, Gottlieb O R 1986 The chemistry of Brazilian Myristicaceae. XXIX. Neolignans from *Virola elongata*. Phytochemistry 25:279–280

76 Kawanishi K, Hashimoto Y 1981 Dehydrodieugenol and its monomethyl ether from *Virola carinata*. Phytochemistry 20:1166

77 Khalid S A, Waterman P G 1981 Alkaloid, lignan and flavonoid constituents of *Haplophyllum tuberculatum* from Sudan. Planta Med 148–152

78 Khan M R, Gray A I, Waterman P G 1987 Neolignans from the stem bark of *Ocotea veraguensis*. Phytochemistry 26:1155–1158

79 Kikuchi T, Matsuda S, Kadota S, Tai T 1985 Studies on the constituents of medicinal and related plants in Sri Lanka. III. Novel sesquilignans from *Hedyotis lawsoniae*. Chem Pharm Bull 33:1444–1451

80 Kilidhar S B, Parthasarathy M R, Sharma P 1982 Prinsepiol, a lignan from stems of *Prinsepia utilis*. Phytochemistry 21:796–797

81 Kim J H, Hahn D R 1981 Studies on the chemical constituents of *Acanthopanax chiisanensis* Nakai roots. Arch Pharmacol Res 4:59–62

82 Kim Y H, Chung B S, Kim H J 1985 Studies on the constituents of *Acanthopanax koreanum* Nakai (I). Saengyak Hakhoechi 16:151–154

83 Kitagawa S, Hisada S, Nishibe S 1984 Phenolic compounds from *Forsythia* leaves. Phytochemistry 23:1635–1636

84 Kosawa M, Morita N, Hata K 1978 Chemical constituents of the roots of *Anthriscus sylvestris* Hoffm. I. Structures of an acyloxycarboxylic acid and a new ester, anthriscusin. Yakugaku Zasshi 98:1486–1490
85 Kuo Y-H, Chen W-C, Wu T-R 1985 A new lignan formosanol, from *Juniperus formosana*. J Chin Chem Soc (Taipei) 32:377–379
86 Kuo Y-H, Lin Y-T 1985 Taiwanin H, a new lignan from the barks of *Taiwania cryptomerioides* Hayata. J Chin Chem Soc (Taipei) 32:381–382
87 Kuo Y-H, Wu R-E 1985 A new lignan, cagayanin, from nutmeg of *Myristica cagayanensis* Merr. J Chin Chem Soc (Taipei) 32:177–178
88 Kuo Y-H, Wu T-R, Lin Y-T 1982 A new lignan detetrahydroconidendrin from *Juniperus formosana* Hayata. J Chin Chem Soc (Taipei) 29:213–215
89 LeQuesne P W, Larrahondo J E, Raffauf R F 1980 Antitumor plants. X. Constituents of *Nectandra rigida*. J Nat Prod 43:353–359
90 Lee C L 1984 Three lignans from the bark of *Taiwania cryptomerioides*. Kuo Li T'ai-wan Ta Hsuch Yuan Yen Chiu Pao Kao 24:41–45
91 Li L, Xue H 1985 Schisandrone, a new 4-aryltetralone lignan from *Schisandra sphenanthera*. Planta Med 217–219
92 Li L, Xue H, Tan R 1985 Dibenzocyclooctadiene lignans from roots and stems of *Kadsura coccinea*. Planta Med 297–300
93 Lin Y-T, Kuo Y-H, Kao S-T 1971 Extractive components from the seed of *Myristica cagayanensis*. J Chin Chem Soc (Taiwan) 18:45–49
94 Lin-gen Z, Seligmann O, Jurcic K, Wagner H 1982 Inhaltsstoffe von *Daphne tangutica*. Planta Med 172–176
95 Lin-gen Z, Seligmann O, Lotter H, Wagner H 1983 (–)-Dihydrosesamin, a lignan from *Daphne tangutica*. Phytochemistry 22:265–267
96 Liu J, Huang M 1984 Constituents of *Schisandra sphenanthera* Redh. et Wils. IV. Structures of anwuweizic acid and dl-anwulignan, and the absolute configurations of d-epigalbacin. Huaxue Xuebao 42:264–270
97 Liu J-S, Huang M-F, Ayer W A, Nakashima T T 1984 Structure of enshicine from *Schisandra henryi*. Phytochemistry 23:1143–1145
98 Lopes L M X, Yoshida M, Gottlieb O R 1984 The chemistry of Brazilian Myristicaceae. XXI. Aryltetralone and arylindanone neolignans from *Virola sebifera*. Phytochemistry 23:2021–2024
99 Lopes L M X, Yoshida M, Gottlieb O R 1984 The chemistry of Brazilian Myristicaceae. XXII. Further lignoids from *Virola sebifera*. Phytochemistry 23:2647–2652
100 Lopes M N, da Silva M S, Barbosa-Filho J M, Ferreira Z S, Yoshida M, Gottlieb O R 1978 The chemistry of Brazilian Lauraceae. LXXXI. Unusual benzofuranoid neolignans from *Licaria chrysophylla*. Phytochemistry 25:2609–2612
101 Mabry T J, Ulubelen A 1980 Chemistry and utilization of phenylpropanoids including flavonoids, coumarins and lignans. J Agric Food Chem 28:188–196
102 MacRae W D, Towers G H N 1984 An ethnopharmacological examination of *Virola elongata* bark: a South American arrow poison. J Ethnopharmacol 12:75–92
103 MacRae W D, Towers G H N 1984 Biological activities of lignans. Phytochemistry 23:1207–1220
104 MacRae W D, Towers G H N 1985 Non-alkaloidal constituents of *Virola elongata* bark. Phytochemistry 24:561–566
105 Martinez-V J C, Cuca-S L E, Yoshida M, Gottlieb O R 1985 The chemistry of Colombian Myristicaceae. IV. Neolignans from *Virola calophylloidea*. Phytochemistry 24:1867–1868
106 Martinez-V J C, Maia J G S, Yoshida M, Gottlieb O R 1980 The chemistry of Brazilian Lauraceae. LV. Neolignans from an *Aniba* species. Phytochemistry 19:474–476
107 Matsuda S, Kadota S, Tai T, Kikuchi T 1984 Isolation and structure of hedyotisol-A, -B and -C, novel dilignans from *Hedyotis lawsoniae*. Chem Pharm Bull 32:5066–5069
108 Matsuura S, Iinuma M 1985 Lignan from calyces of *Diospyros kaki*. Phytochemistry 24:626–628
109 Medvedeva S A, Ivanova S Z, Babkin V A, Rezvukhin A I 1985 Volkova IV. Lignan glycoside from needles of *Pinus sibirica*. Khim Prir Soedin 560
110 Mikaya G A, Turabelidze D G, Kemertelidze E P, Wulfson N S 1981 Kaerophyllin, a new lignan from *Chaerophyllum maculatum*. Planta Med 43:378–380
111 Moro J C, Fernandes J B, Vieira P C, Yoshida M, Gottlieb O R, Gottlieb H E 1987 Neolignans from *Nectandra puberula*. Phytochemistry 26:269–272

112 Nawwar M A M, Barakat H H, Buddrus J, Linscheid M 1985 Alkaloidal, lignan and phenolic constituents of *Ephedra alata*. Phytochemistry 24:878–879
113 Nemethy E K, Lago R, Hawkins D, Calvin M 1986 Lignans of *Myristica otoba*. Phytochemistry 25:959–960
114 Niwa M, Tatematsu H, Liu G, Chem X, Hirata Y 1985 Constituents of *Stellera chamaejasme* L. Tennen Yuki Kagobutsu Toronkai Koen Yoshishu 27:734–741
115 Olaniyi A A 1982 Two new arylnaphthalide lignans from *Justicia flava* roots. Planta Med 154–156
116 Omar A A 1978 Arctiin, a lignan glucoside from *Onopordon alexandrinum*. Lloydia 41:638–639
117 Patra A, Mitra A K 1979 Constituents of *Acorus calamus* Linn. Indian J Chem 17B:412–414
118 Prabhu B R, Mulchandani N B 1985 Lignans from *Piper cubeba*. Phytochemistry 24:329–331
119 Prakash L, Singh R 1980 Chemical constituents of the stem bark and stem heartwood of *Dolichandrone crispa* Seem. Pharmazie 35:122–123
120 Provan G J, Waterman P C 1985 Picropolygamain, a new lignan from *Commiophora incisa* resin. Planta Med 271–272
121 Purushothaman K K, Sarada A, Connolly J D 1984 Structure of malabaricanol – a lignan from the aril of *Myristica malabarica* Lam. Indian J Chem 23B:46–48
122 Rao C B S (ed) 1978 Chemistry of lignans. Andhra University Press Waltair, 377 pp
123 Rao K V, Davis T L 1982 Constituents of *Magnolia grandiflora*. I. Mono-O-methylhinokiol. Planta Med 57–59
124 Razakova D M, Bessanova I A 1981 Lignans from *Haplophyllum popovii*. Khim Prir Soedin 516–517
125 Ren L, Xie F, Feng J, Xuez Z 1984 Studies on the chemical constituents of *Zanthoxylum podocarpum* Hemsl. Yaoxue Xuebao 19:268–273
126 Richomme P, Bruneton J, Adrien C 1985 Study of Hernandiaceae. VII. 5′-Methoxyyatein and 5′-methoxypodorhizol. New lignans isolated from *Hernandia cordigera* Viell. Heterocycles 23:309–312
127 Richomme P, Bruneton J, Cabalion P, Debray M M 1984 Study of Hernandiaceae. IX. Lignans from two Melanesian *Hernandia*. J Nat Prod 47:879–881
128 Robin J P, Davoust D, Taafrout M 1986 Steganolides B and C, new lignans analogous to episteganacin and isolated from *Steganotaenia araliacea* Hochst. Carbon-13-proton correlation of steganacin using two-dimensional NMR. Tetrahedron Lett 27:2871–2874
129 Romoff P, Yoshida M, Gottlieb O R 1984 The chemistry of Brazilian Lauraceae. LXXVI. Neolignans from *Ocotea aciphylla*. Phytochemistry 23:2101–2104
130 Sastry K V, Rao E V, Pelter A, Ward R S 1979 4-Aryl-2,3-naphthalide lignans from *Justicia simplex*. Indian J Chem 17B:415–416
131 Satake T, Murakami T, Saiki Y, Chen C-M 1978 Chemische Untersuchungen der Inhaltsstoffe von *Pteris vittata* L. Chem Pharm Bull 26:1619–1622
132 Schneiders G E, Stevenson R 1982 Structure and synthesis of the aryltetralin lignans hypophyllanthin and nirtetralin. J Chem Soc Perkin Trans I 999–1003
133 Setchel K D R, Borriello S P, Gordon H, Lawson A M, Harkness R, Morgan D M L, Kirk D N, Adlercreutz H, Anderson L C, Arelson M 1981 Lignan formation in man. Microbial involvement and possible roles in relation to cancer. Lancet July 4:4–7
134 Shen Z, Theander O 1985 (–)-Massoniresinol, a lignan from *Pinus massoniana*. Phytochemistry 24:364–365
135 Shimomura H, Sashida Y, Oohara M 1987 Lignans from *Machilus thunbergii*. Phytochemistry 26:1513–1515
136 Spencer G F, Flippen-Anderson J L 1981 Isolation and X-ray structure determination of a neolignan from *Clerodendron inerme* seeds. Phytochemistry 20:2757–2759
137 Syafii W, Yoshimoto T, Samejima M 1985 Neolignans from the heartwood of ulin (*Eusideroxylon zwageri*). Mokuzai Gakkaishi 31:952–955
138 Taafrout M, Rouessac F, Robin J 1983 Prestegane A, from *Steganotaenia araliacea* Hochst.: The first natural dibenzylbutanolide with a meta-phenol. Tetrahedron Lett 24:3237–3238
139 Taguchi H, Iketami Y 1979 Gomisin J. Jpn Kokai Tokkyo Koho 79/32, 462 (Cl.C07C43/22)
140 Taguchi H, Ikeya Y 1977 The constituents of *Schisandra chinensis* Baill. The structures of two new lignans, gomisin F and G, and the absolute structures of gomisin A, B and C. Chem Pharm Bull 25:364–366

141 Takeda S, Maemura S, Sudo K, Kase Y, Arai I, Ohkura Y, Funo S, Fujii Y, Aburada M, Hosoya E 1986 Effects of gomisin A, a lignan component of *Schisandra* fruits, on experimental liver injuries and liver microsomal drug-metabolizing enzymes. Nippon Yakurigaku Zasshi 87:169–187
142 Takehara T, Sasaya T 1979 Studies on the extractives of larch. Phenolic constituents from sapwood of *Larix leptolepis* Gork. Hokkaido Daigaku Nogakubu Enshurin 36:681–693
143 Talapatra B, Chaudhuri P K, Talapatra S K 1982 (–)-Maglifloenone, a novel spirocyclohexadienone neolignan and other constituents from *Magnolia liliflora*. Phytochemistry 21:747–750
144 Talapatra S K, Mukhopadhyay S K, Talapatra B 1977 Lignan and alkaloids of *Talauma hodgsoni*. J Indian Chem 54:790–791
145 Tan R, Li L, Fan Q 1984 Studies on the chemical constituents of *Kadsura longipedunculata*: Isolation and structure elucidation of five new lignans. Planta Med 414–417
146 Tatematsu H, Hurokawa M, Niwa N, Hirata Y 1984 Piscicidal constituents of *Stellera chamaejasme* L. II. Chem Pharm Bull 32:1612–1613
147 Thusoo A, Raina N, Minhaj N, Ahmed S R, Zaman A 1981 Crystalline constituents from *Daphne oleoides*. Indian J Chem 20B:937–938
148 Tillekertne L M V, Jayamanne D T, Weerasuria K D V, Guntatilaka A A L 1982 Lignans from *Horsfieldia iryaghedhi*. Phytochemistry 21:476–478
149 Torrance S J, Hoffmann J J, Cole J R 1979 Wikstromol. Antitumor lignan from *Wikstroemia foetida* var *oahuensis* Gray and *Wikstroemia uva-ursi* Gray (Thymelaeaceae). J Pharm Sci 68:664–665
150 Torres R, Delle Monache F, Marini-Bettolo G B 1979 Biogenetic relationships between lignans and alkaloids in *Berberis* genus. Lignans and berbamine from *Berberis chilensis*. Planta Med 32–36
151 Trevisan L M V, Yoshida M, Gottlieb O R 1984 The chemistry of Brazilian Lauraceae. LXXIV. Neolignans from *Aniba citrifolia*. Phytochemistry 23:701
152 Trevisan L M V, Yoshida M, Gottlieb O R 1984 The chemistry of Brazilian Lauraceae. LXXV. Hexahydrobenzofuranoid neolignans from an *Aniba* species. Phytochemistry 23:661–665
153 Tsukamoto H, Hisada S, Nishibe S 1984 Lignans from bark of *Fraxinus mandshurica* var. *japonica* and *F. japonica*. Chem Pharm Bull 32:4482–4489
154 Tsukamoto H, Hisada S, Nishibe S 1984 Lignans from bark of the olea plants. I. Chem Pharm Bull 32:2730–2735
155 Tsuruga T, Ebizuka Y, Nakajima J, Chun Yui-To, Noguchi H, Iitaka Y, Sankawa U 1984 Isolation of a new neolignan, magnosalicin from *Magnolia salicifolia*. Tetrahedron Lett 25:4129–4132
156 Vanegas T M, Cabezas F F, Gottlieb O R, Sandoval L, Montero E O 1984 Isolation and structure of the lignan cubebin from *Piper lacunosum*. Rev Latinoamer Quim 14:139–140
157 Vaquette J, Cave A, Waterman P G 1979 Alkaloids, triterpenes and lignans from *Zanthoxylum sp.* Sevenet 1183-a species indigenous to New Caledonia. Planta Med 42–47
158 Wagner H, Feil B, Seligmann O, Petricic J, Kalogjera Z 1986 Cardioactive drugs. V. Phenylpropane and lignans of *Viscum album*. Planta Med 102–104
159 Wang H, Chen Y 1985 Lignans from *Schisandra rubriflora* Rhed. et Wils. Yaoxue Xuebao 20:832–841
160 Wang Y, Cheng M-C, Lee J-S, Chen F-C 1982 Molecular and crystal structure of magnolol-$C_{18}H_{18}O_2$. J Chin Chem Soc (Taipei) 30:215–221
161 Ward R S 1982 The synthesis of lignans and neolignans. Chem Rev 11:75–125
162 Watanabe K, Watanabe H, Goto Y, Yamaguchi M, Yamamoto N, Hagino K 1983 Pharmacological properties of magnolol and hinokiol extracted from *Magnolia officinalis*: Central depressant effects. Planta Med 103–108
163 Wenkert E, Gottlieb H E, Gottlieb O R, Pereira M O da S, Formiga M D 1976 ^{13}C NMR spectroscopy of neolignans. Phytochemistry 15:1547–1551
164 Whiting D A 1985 Lignans and neolignans. Nat Prod Rep 2(3):191–211
165 Zhou Y C Y 1982 Studies on the lignans of Zhi Shan (*Taxus cuspidata* Sieb. et Zucc.). Zhongcaoyao 13:1–2

7.4 Stilbenes, Conioids, and Other Polyaryl Natural Products

T. Norin

7.4.1 Introduction

The various types of wood extractives that will be discussed in this section are structurally very different from one another and are not closely related from a biogenetic point of view. The only common feature is that their structures incorporate two or more aryl groups and that they do not belong to any of the other classes of compounds described in this book. The classes of natural products to be presented are:

stilbenes $[C_6-C_2-C_6]$, including structurally related compounds (e.g., dihydrostilbenes);

conioids (norlignans) $[C_6-C_3(C_2)-C_6]$, including condensed and structurally related compounds;

aucuparins $[C_6-C_6]$, and structurally related biphenyls;

diarylheptanoids $[C_6-C_7-C_6]$, structurally related diarylalkanoids $[C_6-C_n-C_6]$ and bridged biphenyls (cyclophanes), (dihydrostilbenes $[C_6-C_2-C_6]$ and wood extractives of $C_6-C_3-C_6$-type are discussed in Sects. 7.4.2 and 7.5, respectively);

miscellaneous diaryl and polyaromatic compounds.

Many of the compounds are rather common and are widely distributed among woody plants. However, they rarely occur in abundant quantities. Some of the compounds possess significant biological properties – e.g. they contribute to the resistance of plants to disease, pest insects, or grazing and foraging animals. The roles that many of the compounds play in the plants are, however, to a large extent unknown. None of the compounds included in this section has commercial use, but some of them contribute to the properties of the wood and influence the properties and the processing of forest products.

7.4.2 Stilbenes and Structurally Related Compounds

The stilbene class of plant phenolics is a small but important group of natural products found in the wood, bark, and leaves of gymnosperms and angiosperms (14, 38, 85). Although the biosynthesis of the stilbenes of woody plants is not verified by labelling experiments in vivo, the oxygenation patterns of the compounds strongly indicate that they are formed from a cinnamic acid (C_6-C_3) moiety and an extended triacetic (acetogenin) moiety (Fig. 7.4.1) (see also Chap. 3). Support for this hypothesis has been presented by Rupprich and Kindl (101), who have isolated an enzyme from *Rheum rhaponticum* (rhubarb) rhizomes, that converts *p*-coumaroyl co-enzyme A and malonyl co-enzyme A into resveratrol (**1**) (Fig. 7.4.1). Further support for this biosynthetic pathway comes from the fact that the

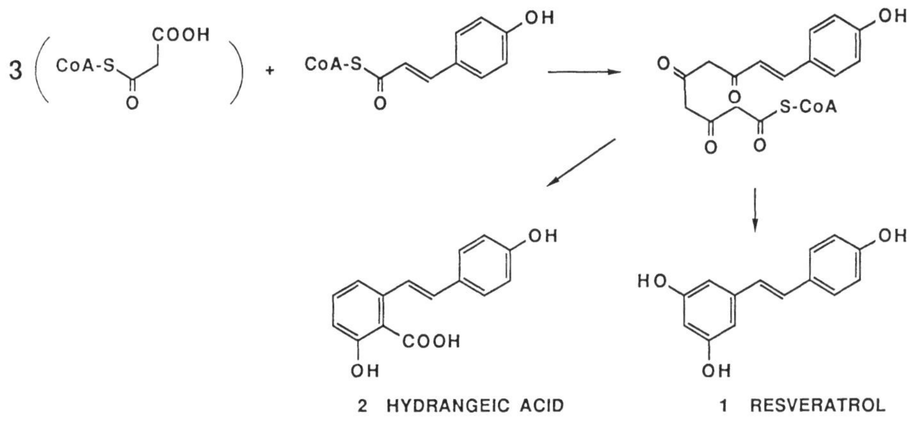

Fig. 7.4.1. The biosynthesis of stilbenes

intermediate stilbene carboxylic acid, hydrangeic acid (**2**) (cf., Fig. 7.4.1), either as such or in the form of the corresponding isocoumarin derivative, is naturally occurring (*Hydrangea* spp., Rosales, Saxifragaceae) (7).

The stilbenes are widely distributed among plants. They often co-occur with flavanoids, which are related to the stilbenes on biogenetic grounds. The structures and distribution of the stilbenes in woody plants are presented in Table 7.4.1.

A common oxygenation pattern of the natural stilbenes is the 3,5-dioxy substitution. Thus, the first natural stilbenes to be isolated from wood – pinosylvin and its monomethyl and dimethyl ethers-carry 3,5-dioxy substituents (35).

Although the natural stilbenes are widely distributed among plants, they appear to be significant only to certain genera and seem to be expressive taxonomic markers (38). Thus pinosylvin and related stilbenes are characteristic heartwood constituents of *Pinus* spp., and the heartwood of many *Eucalyptus* spp. contain resveratrol. Further investigations on species belonging to the Moraceae and Leguminosae families may reveal the presence of stilbenes and biogenetically related compounds.

Some natural sources are particularly rich in phenolic compounds of stilbene type. The heartwood of several *Pinus* spp. contains considerable amounts of pinosylvin. 3,4,3′,4′-Tetrahydroxystilbene (piceatannol) (23, 50) and its 3-methyl ether are significant constituents of the bark of Norway spruce (*Picea abies*) (23, 50; K. Sunnerheim and O. Theander, personal communication). The heartwood of laburnum (*Laburnum alpinum*) is remarkable in its content of stilbenes since it contains up to 4% by dry weight of the 3,4,3′,5′-tetrahydroxystilbene (37).

Most stilbenes isolated from natural sources are of *trans* (*E*) configuration. However, Rowe et al. (98) have isolated the *cis* (*Z*) isomer of pinosylvin dimethyl ether from the bark of *Pinus banksiana*. The bark contains a mixture of the *cis/trans* isomers with the *trans* isomer as the major compound.

An interesting derivative that recently has been reported is the maackiasin (**3**), which is a 3,4,3′,5′-tetrahydroxystilbene linked to the isoflavone retusin by a diether bond. The compound is found in the heartwood of *Maackia amurensis* (Leguminosae) (74).

Table 7.4.1. Stilbenes and related compounds in woody plants

Compound	Genus, species (family)	Reference(s)
trans-Stilbene	*Alnus firma* (Betulaceae)	8
4-Hydroxystilbene	*Pinus excelsa* (Pinaceae)	78
3,4'-Dihydroxystilbene	*Morus laevigata* (Moraceae)	118
Pinosylvin, 3,5-dihydroxystilbene	*Pinus* spp. (Pinaceae)	38, 85
	Alnus sieboldiana (Betulaceae)	9
Pinosylvin monomethyl ether; 3-OH, 5-OCH_3	*Pinus* spp. (Pinaceae)	38, 85
	Alnus sieboldiana (Betulaceae)	9
trans-Pinosylvin dimethyl ether; 3,5-di-OCH_3	*Pinus* spp. (Pinaceae)	38, 85
	Alnus sieboldiana (Betulaceae)	9
cis-Pinosylvin dimethyl ether	*Pinus banksiana* (Pinaceae)	98
3,5-Dimethoxy-4-isopentenylstilbene	*Derris rariflora* (Leguminosae)	46
2,4,3'-Trihydroxystilbene	*Morus laevigata* (Moraceae)	118
3-Methoxy-4,5-dihydroxystilbene	*Alnus viridis* (Betulaceae)	43
Resveratrol; 3,5,4'-trihydroxystilbene (**1**)	*Pinus taeda* (Pinaceae)	57
	Picea spp. (Pinaceae)	49, 80, 104
	Eucalyptus spp. (Myrtaceae)	55, 60, 62
	Maackia amurensis (Leguminosae)	79
Pterostilbene; 3,5-dimethoxy-4'-hydroxy	*Pterocarpus* spp. (Leguminosae)	71
3,4,3',5'-Tetrahydroxystilbene	*Picea* spp. (Pinaceae)	9, 48, 50
(piceatannol; astringenin)	*Pinus radiata* (Pinaceae)	129
	Laburnum alpinum (Leguminosae)	37
	Vouacapoua spp. (Leguminosae)	73
	Centrolobium spp. (Leguminosae)	1
Piceatannol 3'-glucoside (astringin)	*Picea* spp. (Pinaceae)	80, 85
Piceatannol 4'-glucoside	*Pinus radiata* (Pinaceae)	129
Piceatannol 3'-methoxy	*Picea* spp. (Pinaceae)	80, 104
Chlorophorin; 2,4,3',5'-tetrahydroxy-4'(3,7-dimethyl-2,6-octadienyl)-stilbene	*Chlorophora excelsa* (Moraceae)	72
3,4,5,3',5'-Pentahydroxystilbene	*Vouacapoua* spp. (Leguminosae)	73
Dihydrostilbenes	*Pinus* spp. (sub-group Haploxylon; Pinaceae)	13
	Morus spp. (Moraceae)	13
Benzils (2,4,2'-trihydroxy-4-methoxy)	*Zollernia paraensis* (Leguminosae)	45
Resveratrol oligomers	*Dipterocarpaceae* spp.	
	Citaceae spp.	65
	Guataceae spp.	
Hopeaphenol (**4**)	*Hopea* spp. (Dipterocarpaceae)	
Copalliferol A (**5**) } (Stemonoporol (**6**) }	*Hopea* spp. (Dipterocarpaceae)	110
	Shorea stipularis (Dipterocarpaceae)	110
	Vateria copallifera (Dipterocarpaceae)	110
Pallidol (**7**)	*Cissus pallida* (Vitaceae)	65
Vaticaffinol (for structure see Ref. (111))	*Vatica affinis* (Dipterocarpaceae)	111
ε-viniferin (**8**)		

Dihydrostilbenes have been found in *Pinus* spp. (only in the Haploxylon subgroup) and in *Morus* spp. (13). They are, together with the stilbenes, useful taxonomic markers. It is noticeable that some dihydrostilbenes (e.g. 2,3'-dihydroxy-5'-methoxy) are identified as plant-growth inhibitors (52).

A natural benzil derivative, which appears to be biogenetically related to the stilbenes, has been isolated from the Leguminosae species, *Zollernia paraensis*, which contains 2,4,2'-trihydroxy-4'-methoxybenzil (45).

Oligomers of resveratrol are significant constituents of some woody plants (Table 7.4.1). Dimers, trimers, and tetramers are known. They exhibit a rather complex stereochemistry. The structure of the tetramer hopeaphenol (**4**) has been determined by X-ray methods, whereas those of other members are mainly based on NMR evidence (110). The absolute configurations are not known.

The resveratrol oligomers appear to be formed via oxidative coupling reactions, which are enzymatically controlled to yield optically active products. Many polyphenols of the stilbene type (e.g. resveratrol) are sensitive to oxidation in air and polymerization. The dark-colored products thus formed should be structural-

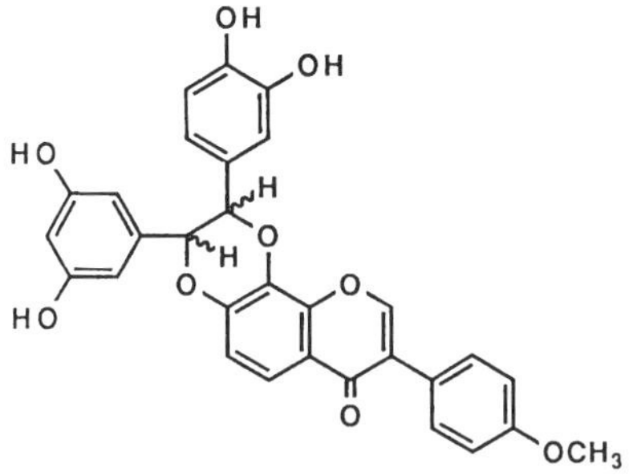

3 MAACKIASIN

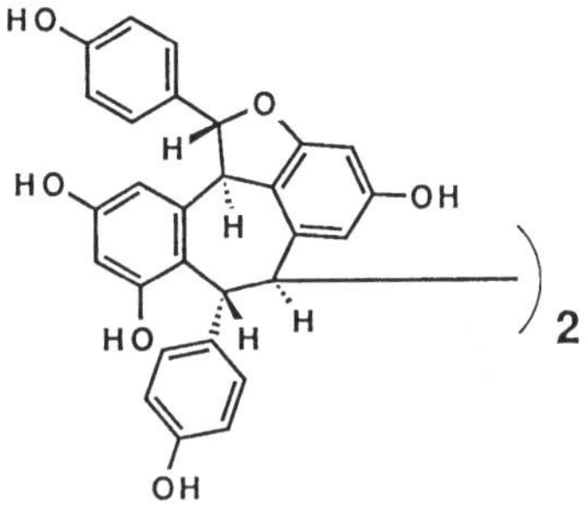

4 HOPEAPHENOL

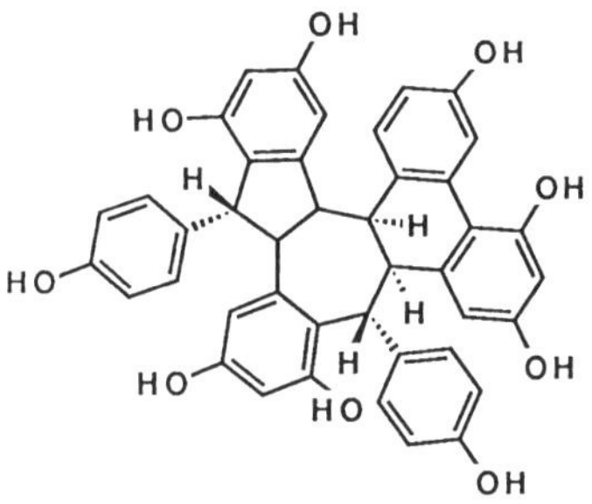

5 COPALLIFEROL A

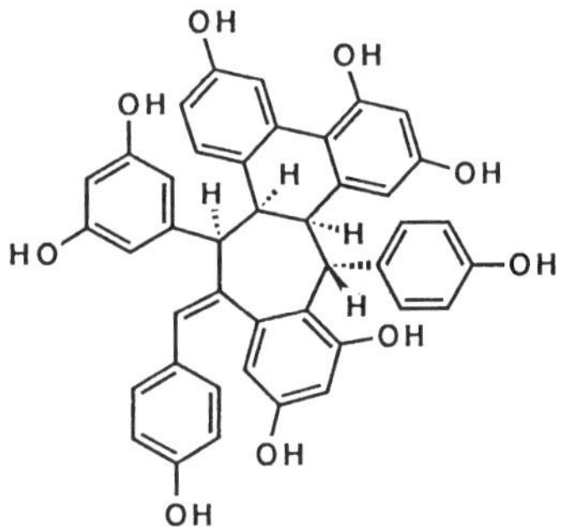

6 STEMONOPOROL

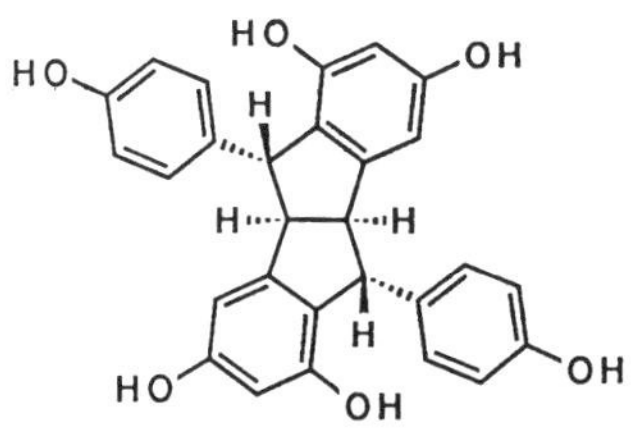

7 PALLIDOL

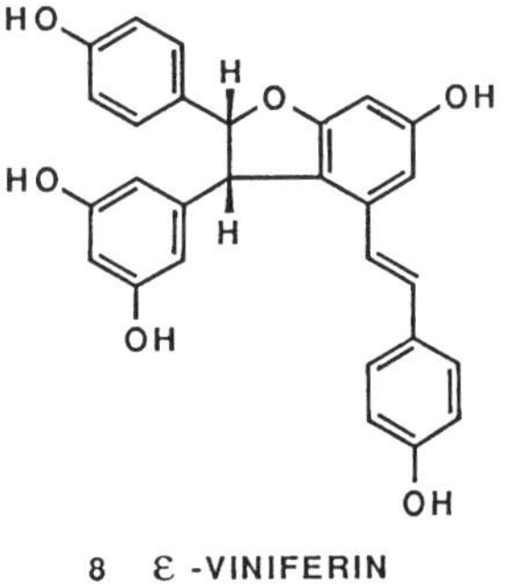

8 ε -VINIFERIN

ly related to the naturally occurring oligomers, but are, of course, racemic. Such racemic polymers may also be present in woody plants.

The presence of hydroxystilbenes and related compounds is considered to contribute to the durability of the wood. Thus pinosylvin and related stilbenes exhibit fungistatic properties when tested on cultures in vitro (51). Similarly oligomeric products from resveratrol show antifungal properties (65). These oligomers are reported to be antifungal phytoalexins, which are formed in infected grape vine leaves (90). However, it is notable that no significant protection was observed when pinosylvin, pinosylvin monomethyl ether, or resveratrol were applied by impregnation to decay-susceptible wood (51, 77). Further studies are needed in order to establish the relationships between natural fungicides and the durability of wood.

Pinosylvin and pinosylvin monomethyl ether are formed in the reaction zone in *Pinus* spp. as a response to fungus infections (57, 61). There is also an interesting correlation between the stilbene production in *Pinus radiata* and the presence of ethylene (plant hormone) in the wood (105, 106). The processes involved in heartwood formation and the formation of phenolic compounds in reaction zones have been extensively investigated and reviewed by Hillis (59).

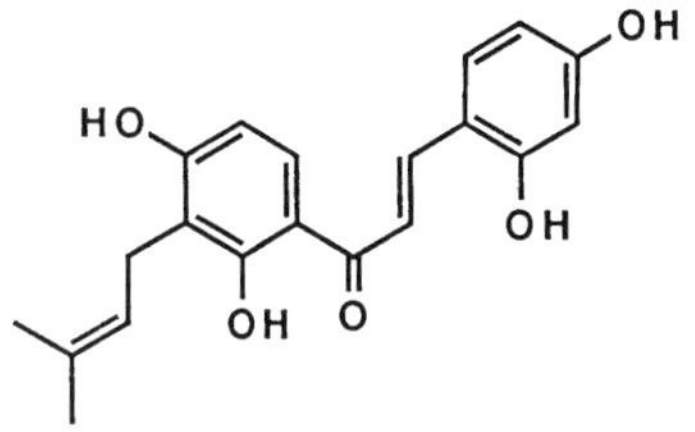

9 DEHYDROMORACIN C

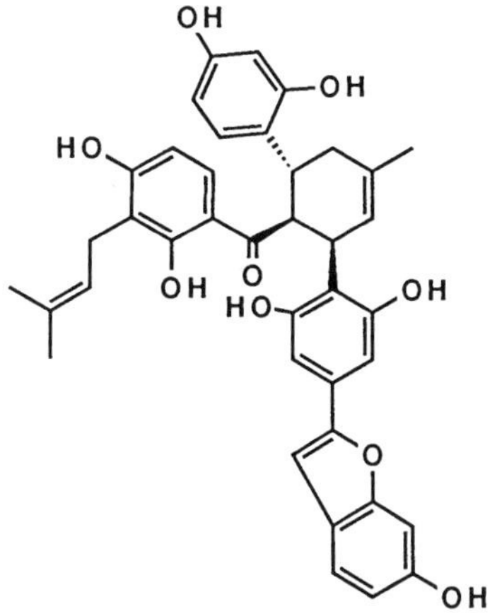

10 CHALCOMORACIN

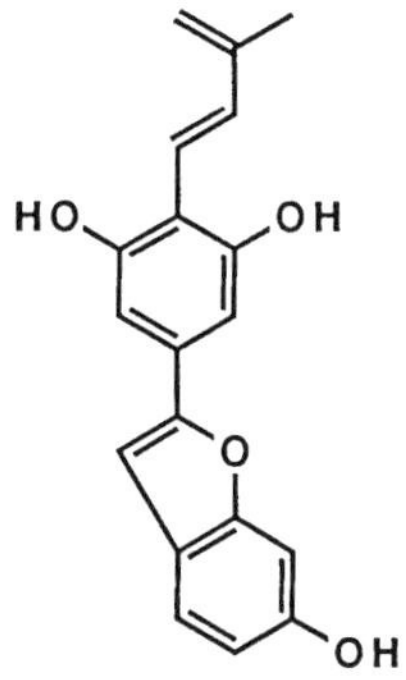

11 MORACHALCONE

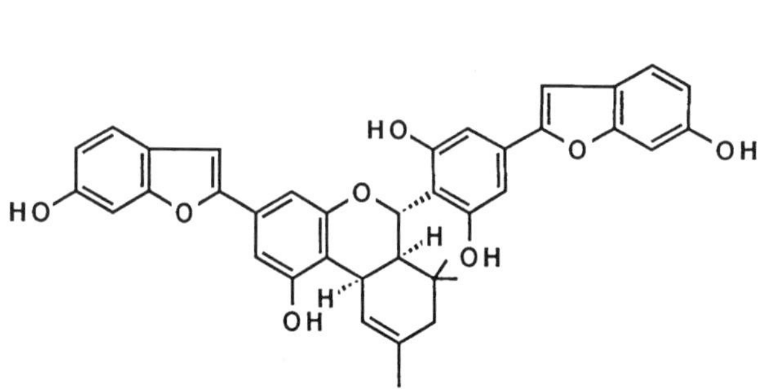

12 DIMORACIN

Some interesting plant phenolics of stilbene and 2-phenylbenzofuran types have been identified as phytoalexins of the mulberry tree (*Morus alba*) (19, 107, 112, 113). Of particular interest are the oligomeric compounds chalcomoracin (**10**) and dimoracin (**12**), which may be considered to be Diels-Alder adducts of morachalcone (**11**) and dehydromoracin (**9**). However, the adducts are optically active, unlike some other natural products that are proposed to be Diels-Alder adducts.

Plant phenolics of stilbene type are also reported to contribute to the insect resitance of the wood. Thus wood samples impregnated with pinosylvin monomethyl ether are resistant to the attack of the West Indian termite *Cryptotermes brevis* (123, 124). Pinosylvin dimethyl ether and dihydropinosylvin were less effective. Sandermann and co-workers (103) have also found that pinosylvin and related hydroxystilbenes as well as some other phenolic wood extractives contribute to the termite and insect resistance of wood. However, our knowledge on this topic is still limited, and further studies shall reveal useful information.

Some hydroxystilbenes – e.g. pinosylvin – interfere with the delignification process during acid-sulfite pulping (36). The pinosylvin competes with the nucleophilic hydrogensulfite anion and inhibits the formation of soluble lignosulfonates. Instead, black-colored insoluble products are formed. The polyhydroxy stilbenes also contribute to the color of wood since they are easily oxidized to dark-colored products.

Chlorophorin is a prenylated tetrahydroxystilbene of the West African hardwood iroko (*Chlorophora excelsa*) (72). This wood is durable and resembles teak or oak. It is used for both outdoor and indoor constructions. The wood cannot be varnished or surface treated with lacquer since the chlorophorin, like many other 2- and 4-hydroxysubstituted stilbenes, prevents proper hardening of the polyester. Workers exposed to saw dust of iroko wood develop strong allergic reactions that have been ascribed to the chlorophorin content of the wood (56) (see also Chap. 9.5).

7.4.3 Conioids (Norlignans), Including Condensed and Structurally Related Compounds

In 1960 Erdtman and Vorbrüggen (42) isolated a new C_{17}-phenolic compound with a backbone of $C_6-C_3(C_2)-C_6$ from the Tasmanian conifer *Athrotaxis selaginoides* (Cupressales, Taxodiaceae). The compound was named athrotaxin and was accompanied by a number of easily oxidizable phenols, some of which were later obtained in a pure crystalline state by careful chromatography (24). They include the agatharesinol (**13a**), hinokiresinol (**14**), and sugiresinol (**15a**). These three compounds, as well as athrotaxin (**16a**), belong to a new family of phenols having a unique carbon skeleton. The compounds have been referred to as norlignans by Whiting (121), although the biogenetic origin is not known and has only been postulated based on structural relationships of co-occurring wood constituents (41, 121). The C_{17}-phenolic compounds of this type have so far only been found in conifers of the orders Cupressales and Araucariales. The name conioids has therefore been coined for this family of plant phenolics by Erdtman and co-workers.

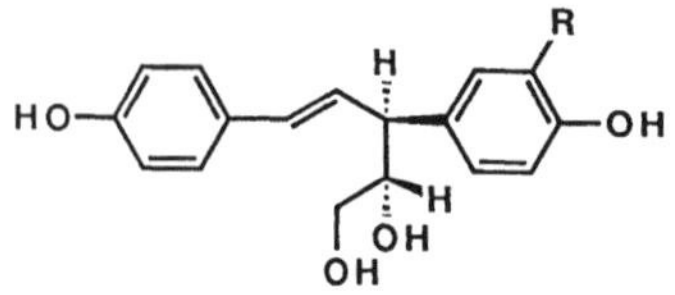

13 a AGATHARESINOL R=H
b SEQUIRIN R=OH

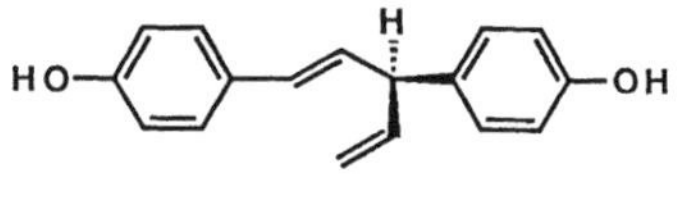

14 HINOKIRESINOL

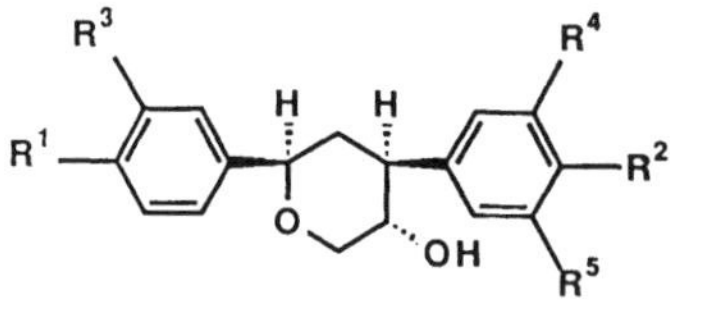

15 a SUGIRESINOL $R^1=R^2=OH$; $R^3=R^4=R^5=H$
b HYDROXYSUGIRESINOL $R^1=R^2=R^4=OH$; $R^3=R^5=H$
c SEQUIRIN-E $R^1=R^2=R^3=OH$; $R^4=R^5=H$
d SEQUIRIN-F $R^1=R^2=R^3=R^4=OH$; $R^5=H$
e SEQUIRIN-G $R^1=R^2=R^3=R^4=R^5=OH$

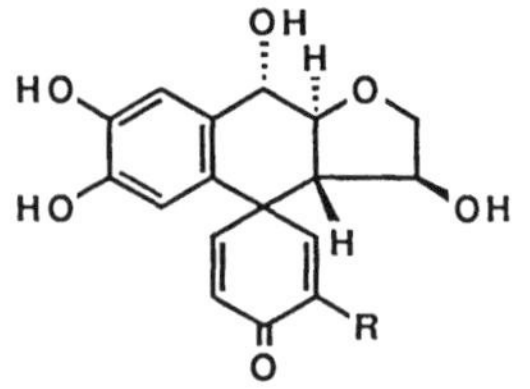

16 a ATHROTAXIN R=H
b HYDROXYATHROTAXIN R=OH

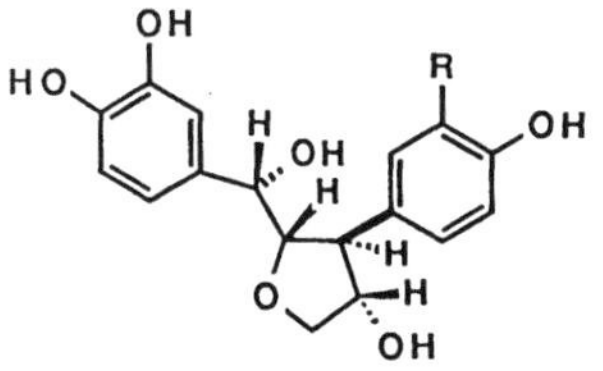

17 a METASEQUIRIN-A R=H
b HYDROXYMETASEQUIRIN R=OH

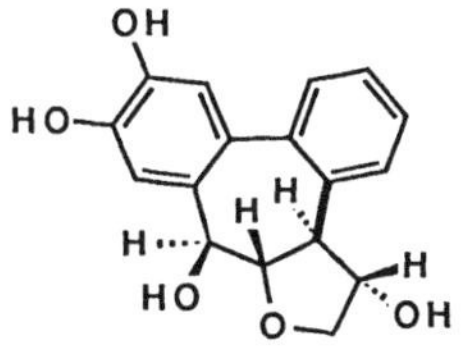

18 METASEQUIRIN-B

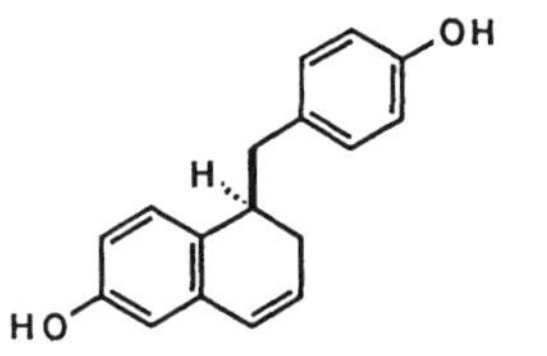

19 SEQUIRIN-D

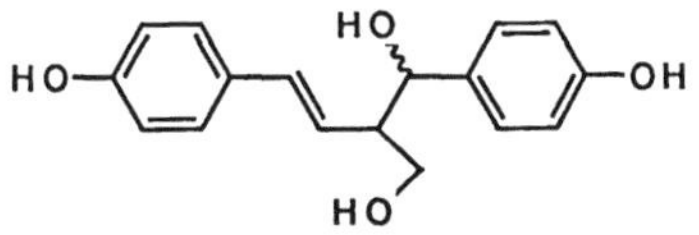

20 YATERESINOL

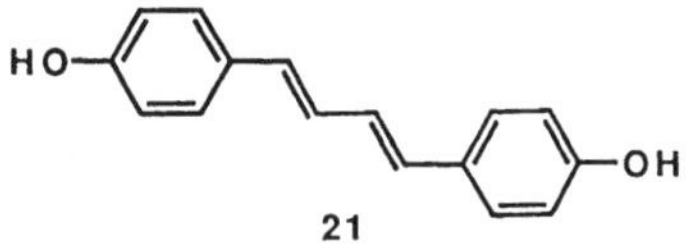

21

The various members (**13** to **18**) of the conioid family are presented in Table 7.4.2 (for a recent review see also Ref. (121)). It is of interest to note the close structural relationships between agatharesinol (**13a**) and sugiresinol (**15a**) and between sequirin (**13b**) and hydroxysugiresinol (**15b**). Acid treatment of agatharesinol yields sugiresinol and similarly sequirin can be converted to hydroxysugi-

Table 7.4.2. Coniods of wood

Compound	Structure		Source	Reference(s)
	No.	Reference(s)	Genus, species	
Agatharesinol	**13a**	33, 34	*Agathis australis*	32, 33
			Athrotaxis selaginoides	24
			Sequoia sempervirens	41
			Sequoiadendron giganteum	
Sequirin (sequirin C)	**13b**	70	*Sequoia sempervirens*	10
			Sequoiadendron giganteum	58
Hinokiresinol	**14**	34, 63	*Chamaecyparis obtusa*	63
			Athrotaxis selaginoides	24
Sugiresinol (sequirin A)	**15a**	34, 67, 93	*Athrotaxis selaginoides*	24
			Cryptomeria japonica	66
			Sequoia sempervirens	93
			Sequoiadendron giganteum	58
			Metasequoia glyptostroboides	41
			Tacodium dubicum	67[1]
Hydroxysugiresinol (sequirin B)	**15b**	53, 67, 93	*Cyptomeria japonica*	53, 66
			Sequoia sempervirens	93
			Sequoiadendron giganteum	58
			Metasequoia glyptostroboides	41
			Taxodium dubicum	67[1]
Sequirin E	**15c**			
Sequirin F	**15d**	58	*Sequoia giganteum*	58
Sequirin G	**15e**			
Athrotaxin	**16a**	24	*Athrotaxis selaginoides*	24, 42
Hydroxyathrotaxin	**16b**	30, 31	*Metasequoia glyptostroboides*	30, 31
Metasequirin	**17a**	30, 31	*Metasequoia glyptostroboides*	30, 31
Hydroxymetasequirin	**17b**			

[1] Hergert H, 1969, personal communication

resinol (93). Sugiresinol and hydroxysugiresinol have therefore been postulated to be artifacts. In this context it is notable that the epimeric cyclization products have not yet been detected.

The absolute configurations of agatharesinol (**13a**), hinokiresinol (**14**), and sugiresinol (**15a**), have been assigned on the bases of a comparison of spectropolarimetric data (ORD and $[\alpha]_D$) (34). The absolute configurations of the other members of the family have not yet been studied but the C_3-configuration should be related. The structure of athrotaxin (**16a**) indicates that it could be a dehydrogenation product of dihydroxysugiresinol.

Sequirin D (**19**) from *Sequoia sempervirens* (12, 54) and yateresinol (**20**) of *Libocedrus yateensis* (41) are two compounds that may be mentioned in this context although they do not possess the characteristic C_{17} backbone of the conioids. A diphenylbutadiene (**21**) has also been isolated from the heartwood of *L. yateensis* (41). Possible biogenetic relationships between the common lignans, yateresinol (**20**), the conioids, and the diphenylbutadiene (**21**) have been discussed by Erdtman and co-workers (41) and by Whiting (121).

The lability of several conioids suggests that natural oligomers and polymers may occur. Polymerization of sequirin under acidic conditions results in a material having an infrared spectrum nearly identical with that of the water-soluble polymers ("tannin" and "phlobaphen") from redwood (*Sequoia sempervirens*) (93). Milder acid treatment of sequirin and also of hydroxysugiresinol yields isosequirin, the structure of which has been postulated mainly on the basis of a mechanism of its formation (93).

7.4.4 Aucuparins and Structurally Related Biphenyls

Aucuparin (**22a**) was first isolated from the heartwood of mountain ash (*Sorbus aucuparia*, Rosaceae) where it co-occurs with methoxyaucuparin (**22b**). They constitute up to 7% (dry-weight) of the wood (39, 40, 83). Their structures have been firmly established by syntheses (83). The compounds have later been found to be common wood constitutents of many *Sorbus* spp. (H. Erdtman, T. Norin, E. von Rudloff, unpublished results). 4-Hydroxyaucuparin (**22c**) has been isolated from *S. aucuparia*. Chromatographic studies have shown that this compound is also present in many other *Sorbus* spp. (B. Kimland, T. Norin, unpublished results).

Aucuparin has also been isolated from the wood of *Kielmeyera* spp. (Guttiferae), where it occurs together with xanthones. This observation has led Gottlieb (47) to propose that the biosyntheses of the two types of compounds are related.

Aucuparin has been found to be a fungicidic phytoalexin of the loquat (*Eriobotry japonica*, Rosaceae) (119) and the compound also occurs in the sapwood of the apple tree (*Malus pumila*, Rosaceae) when infected with the fungus that causes silver leaf disease (*Chondrostereum purpureum*) (70). The infected wood of the closely related pear tree (*Pyrus communis*, Rosaceae) (69) and the infected wood of *Cotoneaster lactea* (Rosaceae) (20) contain phytoalexins of the biogenetically related dibenzofuran type (α-, β-, and γ-pyrufuran, and cotonefuran, **23a–d**).

The aucuparins appear to be significant constituents of only a limited number of genera and families and are therefore useful taxonomic markers. So far they have only been found in species belonging to the Rosaceae family. They are part of the defense system and are formed in response to infections or as a consequence of heartwood formation. Further members of this class of plant phenolics will certainly be found in the future.

Plant phenolics of biphenyl type but of different biogenetic origin to those of the aucuparins are constituents of the sap of the lac tree (*Rhus vernicifera*, Anacardiaceae). The sap is the raw material of the Japan lacquer. It dries into a tough and brilliant film and has been used in the Orient as a coating material for thousands of years (75) (see also Chap. 1.1). The major constituent of the sap is urushiol (**24**). The drying process is believed to be an enzymatic oxidative coupling of urushiol under the influence of oxidoreductases – e.g. laccase to form biphenyls (e.g. **25**), dibenzofurans (e.g. **26**), oligomers, and polymers (88).

A common phenolic plant constituent that is of the biphenyl type is ellagic acid (**27**). It occurs in some wood species (47) and is the significant compound

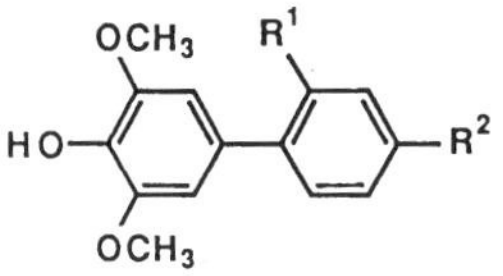

22		R^1, R^2
a	AUCUPARIN	$R^1=R^2=H$
b	METHOXYAUCUPARIN	$R^1=OCH_3$; $R^2=H$
c	p-HYDROXYAUCUPARIN	$R^1=H$; $R^2=OH$

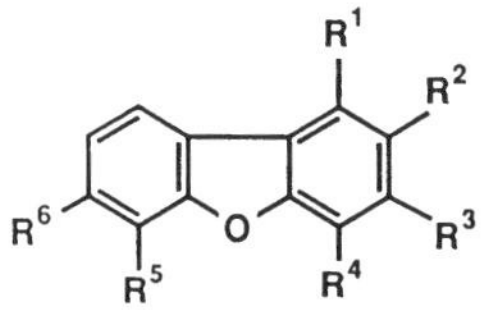

23		R^1	R^2	R^3	R^4	R^5	R^6
a	α-PYRUFURAN	OCH_3	OH	OCH_3	OCH_3	H	H
b	β-	OCH_3	OCH_3	OH	OCH_3	H	H
c	γ-	OCH_3	OH	OCH_3	OCH_3	H	OH
d	COTONEFURAN	OCH_3	OH	OCH_3	OCH_3	OCH_3	OH

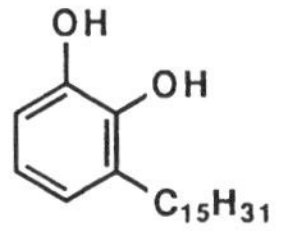

24 URUSHIOL

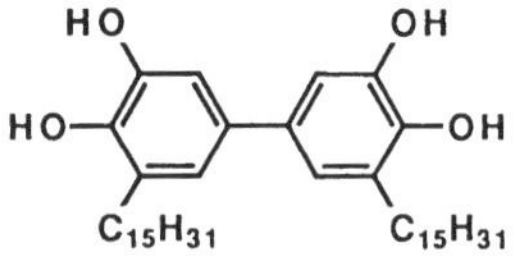

25

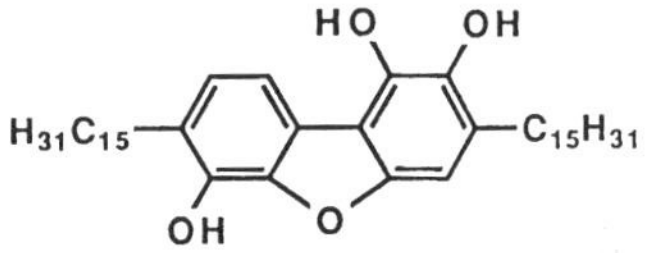

26 DIBENZOFURAN

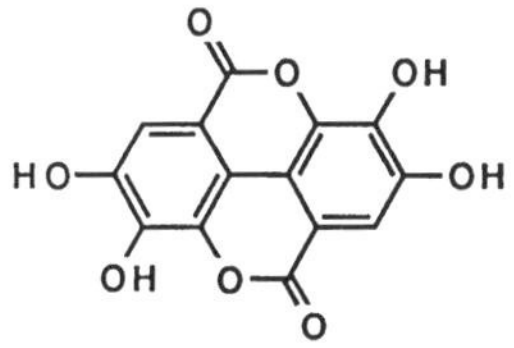

27 ELLAGIC ACID

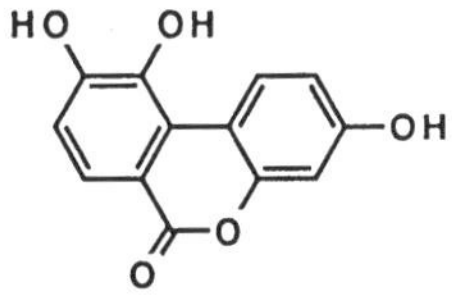

28 FASCICULIFEROL

in tannins (see Sect. 7.2). A related compound is fasciculiferol (**28**), which occurs together with a number of flavonoids and related compounds in the heartwood of *Acacia* spp. (Mimosoideae) (17).

7.4.5 Diarylheptanoids, Structurally Related Diarylalkanoids, and Bridged Biphenyls (Cyclophanes)

Natural phenolic compounds of diarylalkanoid ($C_6-C_n-C_6$) or related bridged biphenyl types have been isolated from the wood, bark, or leaves of many hardwoods including those of Betulaceae, Leguminosae, Myricaceae, and Proteaceae. Although only a limited number of compounds of these types are known, it seems that they are common among certain plant families and can serve as taxonomic markers.

Several diarylheptanoids ($C_6-C_7-C_6$) and related bridged biphenyls, [7,0] metacyclophanes, have been isolated (**29** to **36**). Diarylalkanoids with shorter

Table 7.4.3. Diarylalkanoids and bridged biphenyls (cyclophanes) of woody plants

General type	Compound	Structure no.	Source Genus, species (family)	Reference(s)
$C_6-C_4-C_6$ with C_1 branch	Agrimolide	**29**	*Agrimonia pilosa* L. (Rosaceae)	125, 126
$C_6-C_5-C_6$	1,5-diphenylpentan-3-ol	**30a**	*Flindersia* spp. (Flindersiaceae)	18
	(−)-erythro-1,5-diphenylpentan-1,3-diol	**30b**		
$C_6-C_7-C_6$	(+)-centrolobol	**31a**		
	(−)-centrolobol	**31b**		
	(+)-centrolobin	**32a**	*Centrolobium* spp. (Leguminosae)	1
	(−)-centrolobin	**32a**; 2*R*, 6*S*		
	(+)-de-*O*-methylcentrolobin	**32b**		
	(−)-de-*O*-methylcentrolobin	**32b**; 2*R*, 6*S*		
	Platyphyllonol (glucoside; platyphyllin)	**31c**	*Betula platyphylla* (Betulaceae)	114
	Hirsutanonol	**31d**	*Alnus hirsuta* (Betulaceae)	115
	Hirsutenone	**31e**		
	Oregonin	**31i**	*Alnus rubra* (Betulaceae)	68
	Dicoumaroylmethane	**31f**		
	Coumaroylferuloylmethane	**31g**	*Curcuma* spp. (Zingiberaceae)	68, 76, 97
	Curcurmin	**31h**		
	Yashabushiketol	**31j**	*Alnus firma* (Betulaceae)	8
	Dihydroyashabushiketol (1,7-diphenyl-5-hydroxy-3-heptanone)	**31k**		
C_6-C_7 \ / C_7	Myricanol	**35a**	*Myrica nagi* (Myricaceae)	11
	Myricanone	**33b**		
	Alnusdiol	**33c**	*Alnus japonica*	84
	Asadanin	**33d**		
	Epiasadanol	**33e**	*Ostrya japonica* (Carpinaceae)	127, 128
	Isoasadanol	**33f**		
$C_6-C_{14}-C_6$	Striatol	**34**	*Grevillea striata* (Proteaceae)	91
	Robustol	**35**	*Grevillea striata* (Proteaceae)	21
C_6-C_6 \ / C_{14}	Turriane and derivatives	**36a** **37b,c,d**	*Grevillea striata* (Proteaceae)	92

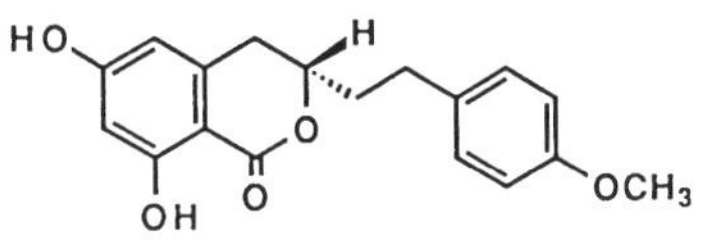

29 AGRIMOLIDE

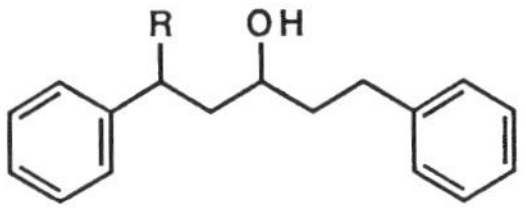

30 a R=H
b R=OH (erythro)

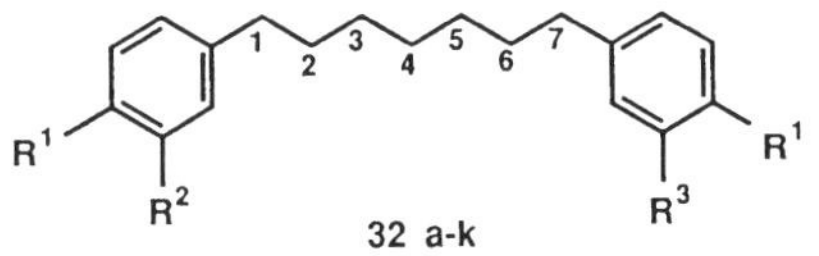

32 a-k

	R^1	R^2	R^3	alkyl subst.	name
31 a	OH	H	H	3R-hydroxy	(+)-CENTROLOBOL
b	OH	H	H	3S-hydroxy	(-)-CENTROLOBOL
c	OH	H	H	3-keto-5-hydroxy	PLATYPHYLLYLONOL
d	OH	OH	OH	3-keto-5-hydroxy	HIRSUTANONOL
e	OH	OH	OH	3-keto-4-ene	HIRSUTENONE
f	OH	H	H	3,5-diketo-1,6-diene	DICOUMAROYLMETHANE
g	OH	OCH_3	OCH_3	3,5-diketo-1,6-diene	COUMAROYLFERULOYLMETHANE
h	OH	OCH_3	OCH_3	3,5-diketo-1,6-diene	CURCURMIN
i	OH	OH	OH	3-keto-5-O-xyloside	OREGONIN
j	H	H	H	3-keto-5-hydroxy-1-ene	YASHABUSHIKETOL
k	H	H	H	3-keto-5-hydroxy	DIHYDROYASHABUSHIKETOL

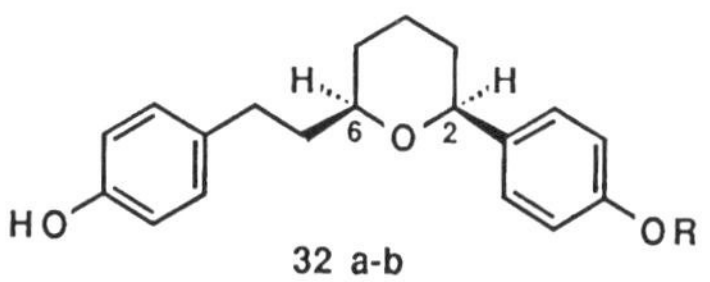

32 a-b

32 a (+)-CENTROLOBIN $R=CH_3$
b (+)-DE-O-METHYLCENTROLOBIN R=H

(C_5) or longer (C_{14}) alkyl chains are also known (Table 7.4.3). The biosyntheses of the various compounds of these types are not fully understood (11, 68, 97). The bridged analogues, however, should arise from the corresponding diarylalkanoids by oxidative coupling (84).

The occurrence of opposite enantiomers of diarylheptanoids in closely related *Centrolobium* spp. and possibly even in variants of the same species (*C. robustum* var. *macrochaete* Mart. and *C. robustum* var. *microchaete* Mart.) (1) is most interesting from biogenetic and phytochemical points of view. Diarylheptanoids of the same chiral relationships – e.g. (+)centrolobin (**32a**), (+)-de-*O*-methyl centrolobin (**32b**), and (+)-centrolobol (**31a**) – co-occur and should therefore be biogenetically related (1).

Diarylheptanoids and related compounds exhibit antibiotic effects (11) and appear to be part of the defense system of the plants. It is also notable that the stem bark of the Indian tree *Myrica nagi* is used as a source of fish poison and that the bridged biphenyls myricanol (**33a**) and myricanone (**33b**) may be responsible for this biological effect (11).

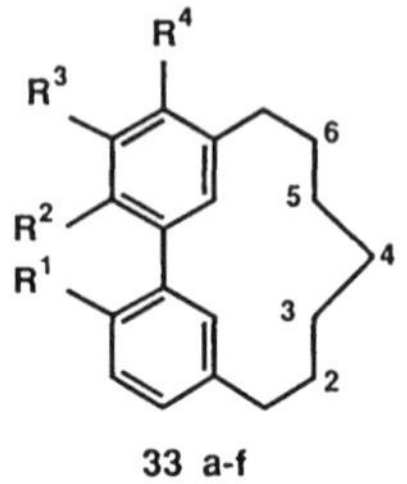

33 a-f

	R^1	R^2	R^3	R^4	alkyl subst.	name
33 a	H	OCH_3	OCH_3	OH	3R-hydroxyl	MYRICANOL
b	H	OCH_3	OCH_3	OH	3-keto	MYRICANONE
c	OH	OH	H	H	3,5-dihydroxy	ALNUSDIOL
d	OH	OH	H	H	$2\beta,3\beta,5\alpha$-trihydroxy-4-keto	ASADANIN
e	OH	OH	H	H	$2\beta,3\beta,4\alpha,5\alpha$-tetrahydroxy	EPIASADANOL
f	OH	OH	H	H	diastereomer of 33e	ISOASADANOL

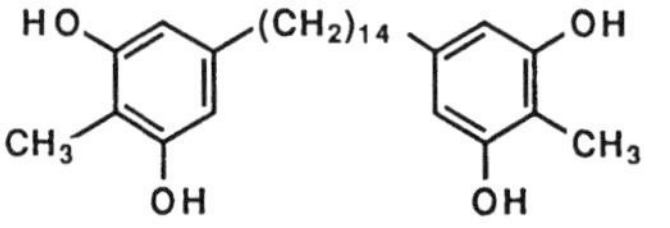

34 STRIATOL

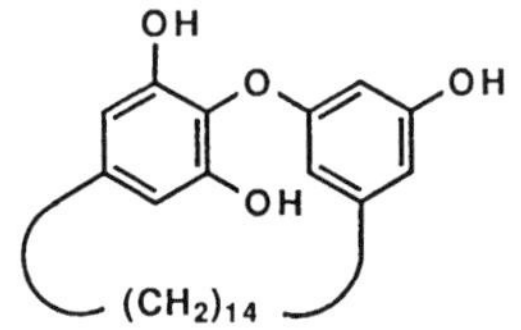

35 ROBUSTOL

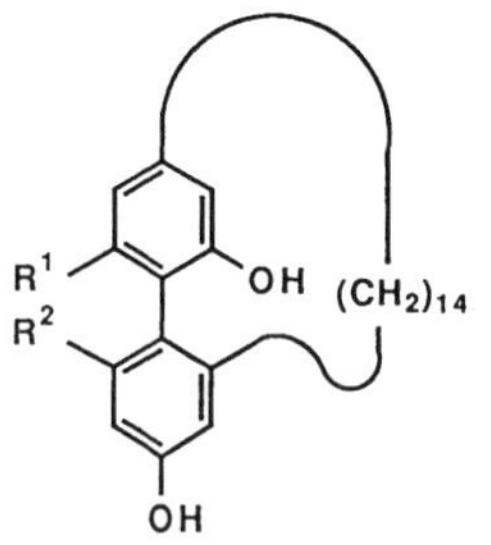

36 a-d

	R^1	R^2	alkyl subst.	name
36 a	OH	OCH_3	-	TURRIANE
b	OCH_3	OCH_3	-	
c	OH	OCH_3	cis-4-ene	
d	OCH_3	OCH_3	cis-4-ene	

The red-orange staining that occurs on freshly cut wood and bark of various alder species (*Alnus* spp.) appears to be due to the oxidation of diarylheptanoids (68). Investigations of the possible precursors of the staining in red-alder wood (Oregon alder, *Alnus rubra*) revealed the presence of the diarylheptanoid xyloside oregonin (**31i**) (68).

7.4.6 Miscellaneous Diaryl and Polyaromatic Compounds

There are several other wood constituents that should be discussed within the frame of this section. They are not frequently found and will therefore only be briefly discussed. Some of the compounds are representatives of important classes of plant products – e.g. xanthones and the related chromane type of compounds, naphthoquinones, and anthraquinones.

Xanthones and the related chromane type of compounds are important pigments in certain hardwoods (29) – e.g. in *Calophyllum* spp. (Guttiferae) (26) and in *Toromita* spp. (Guttiferae) (86). They also exhibit interesting biological and pharmacological properties. Thus, extract of the bark of osage orange, which comes from a North American tree (*Maclura pomifera*, Moraceae), exhibits anti-termitic properties and is toxic to fish. These properties are attributed to the pre-nylated xanthone derivatives macluroxanthone (**37**) and osajaxanthone (**38**), respectively (99). Other examples of prenylated xanthones from wood are 6-deoxy-jacareubin (**39**) and pyramojacareubin (**40**) from Guttiferae species (89, 120).

Chromones and related compounds are also found in wood. A comprehensive review of this class of compounds (chromanols, chromanones, and chromones) has recently been published (102).

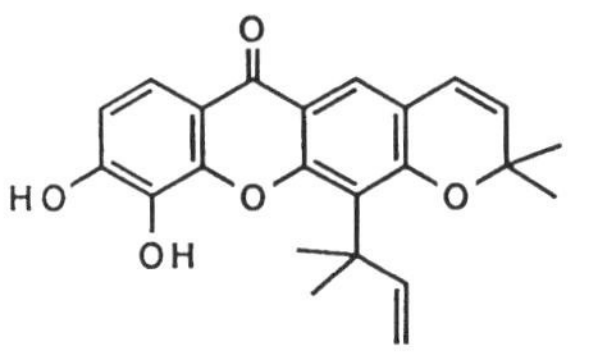

37 MACLUROXANTHONE

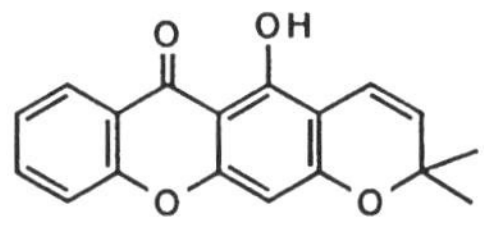

38 OSAJAXANTHONE

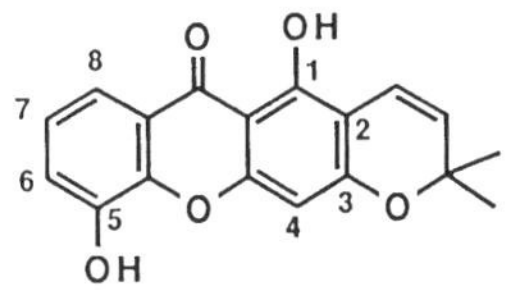

39 6-DEOXYJACAREUBIN

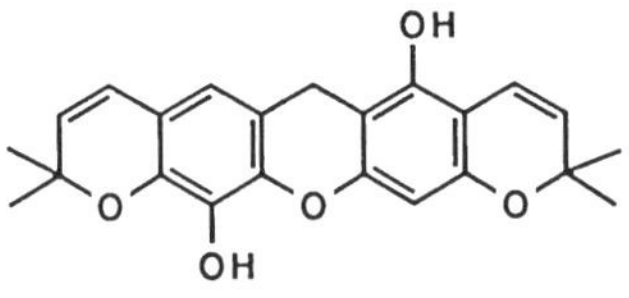

40 PYRANOJACAREUBIN

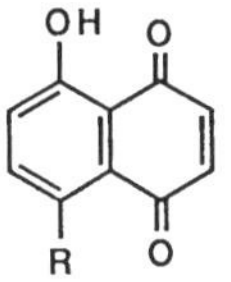

41 a JUGLONE R=H
b 7-METHYLJUGLONE R=CH_3

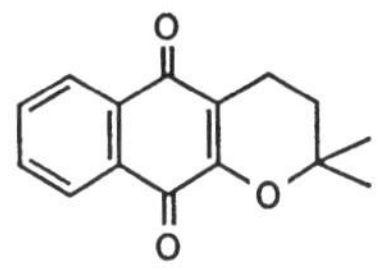

42 α-LAPACHON

Several naturally occurring quinones have been isolated from woody plants (116). Apart from being characteristic pigments, many of the quinones possess significant biological and pharmacological properties. Thus juglone (**41a**), which occurs in the seed hull of black walnut (*Juglans nigra*, Juglandaceae) and the bark of shagbark hickory (*Carya ovata*, Juglandaceae), is an insect repellent and exhibits pharmacological effects including those of a fungicide and antibiotic. The related 7-methyljuglone (**41b**), which occurs in the wood of the persimmon (*Diospyros virginiana*, Ebenaceae), exhibits termiticidal properties (22).

Other interesting phenolic wood constituents of napthoquinone type are the prenylated derivatives (**42**) in the durable wood of *Catalpa ovata* (Bignoniaceae) (64). Sesquiterpenes of naphthalene type (cadalene type e.g. **43**), which occur in

Ulmus species (Ulmaceae) (99, 100) may also be mentioned in this context. Some of these naphthalene derivatives cause staining of painted surfaces on the wood of slippery elm (*Ulmus rubra*) (99). Related pigments of wood is mansonones – e.g. mansonone F (**44**) – that have been isolated from *Mansonia altissima* Chev. (Sterculiaceae) (81). The wood of this species is used for furnitures. Several allergic symptoms and heart troubles have been reported for workers handling this wood (81).

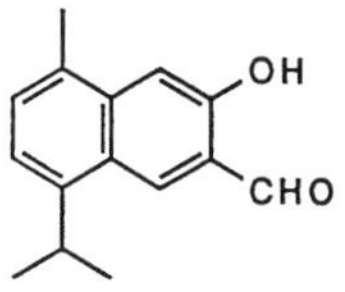

43 7-HYDROXYCADALENAL

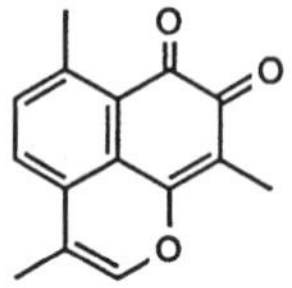

44 MANSONONE F

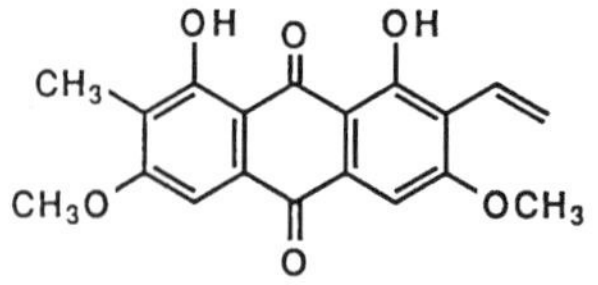

45

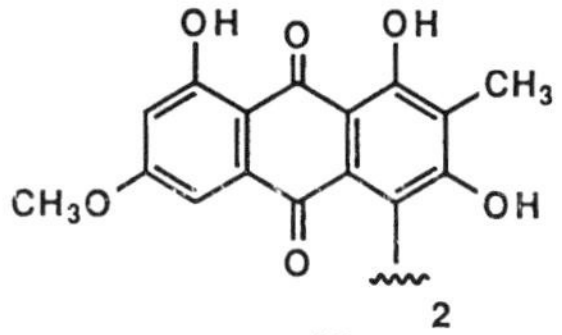

46

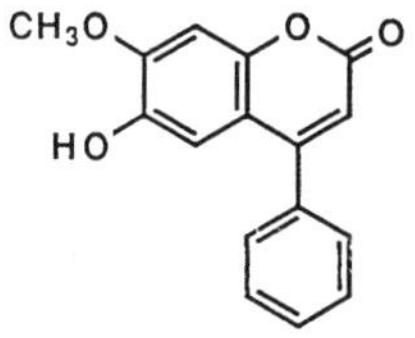

47 DALBERGIN

Anthraquinones are plant pigments that occur in some woods (e.g. Rubiaceae, Leguminosae, Rhamnaceae, and Polygonaceae) (22, 122). They sometimes are present as glycosides. In the Rubiaceae family, the sugar moiety is often primeveroside (glucose-xylose) (122). *Cassia* species of the Leguminosae family are rich sources of anthraquinones (e.g. **45**) (25, 109, 117). Some new anthraquinones (25) including a bianthraquinone (**46**) (109) have recently been isolated from this genus.

Flavonoids of wood are covered separately in Sect. 7.5. Neoflavanoids as well as other rearranged or condensed flavanoids and related compounds are also significant constituents of some wood species. They will only briefly be discussed in this section. They contribute to the properties of wood and can serve as taxonomic markers. Representative neoflavanoids are dalbergin (**47**) and related compounds of *Dalbergia* and *Machaerium* spp. (Leguminosae) (27, 28, 87).

The "arrested metabolic pool" (17) of the heartwood of *Acacia* spp. (Mimosoideae) is very rich in phenolic compounds of different structural types (14, 15, 16, 17). Several of these compounds are flavanoids or related compounds.

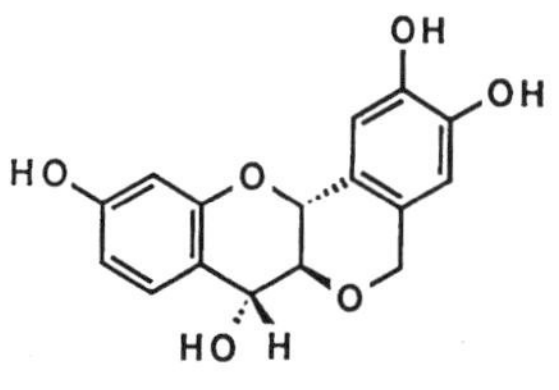

48 (+)-PELTOGYNOL

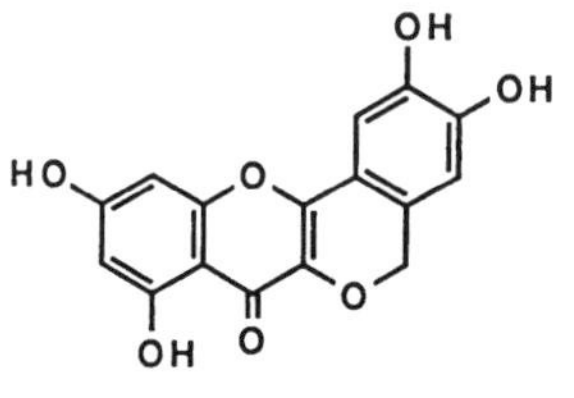

49 β-PHOTOMETHYLQUERCETIN

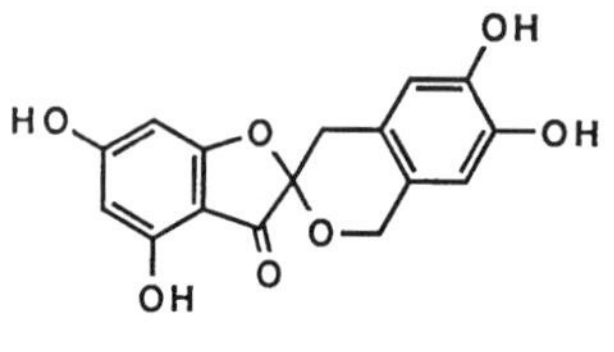

50 CROMBENIN

The peltogynoids belong to such a class of compounds. Structurally they can be classified as resorcinol and phloroglucinol analogues of (+)-peltogynol (**48**). Some additional representatives of this class are β-photomethylquercetin (**49**) and crombenin (**50**), the latter being a spiropeltogynoid (14, 15, 16, 17).

Some pigments of wood have also structural relationships to flavonoid compounds. Dracorubin (**51**) is a constituent of "Dragon's blood", which is a commercially available resin from a palm tree, *Daemonorops draco* (Palmae), from Southeast Asia (82, 94). This resin also contains flavans, biflavanoids, anthocyaninidins, triflavonoids, and chalcones. An unusual constituent is a secobiflavonoid (**52**) (82).

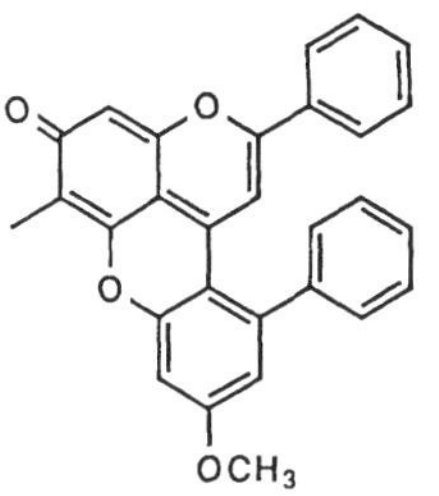

51 DRACORUBIN

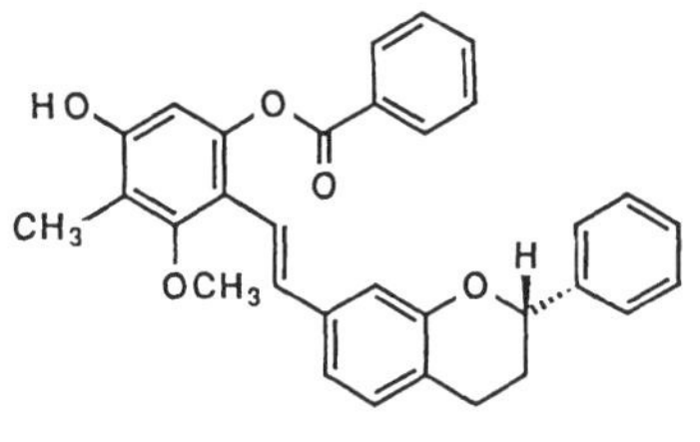

52 SECOBIFLAVANOID

The chemistry of the pigments of the "insoluble red woods" has attracted the attention of many chemists for more than 150 years (3, 4, 5, 6, 95, 96). These woods form a group of dye woods from *Pterocarpus* spp. (*P. santalinus*, red sandalwood of Southeast Asia) and *Baphia* spp. (*B. nitida*, cam wood of West Africa) of the Leguminosae family. The structures of some of the pigments – e.g.,

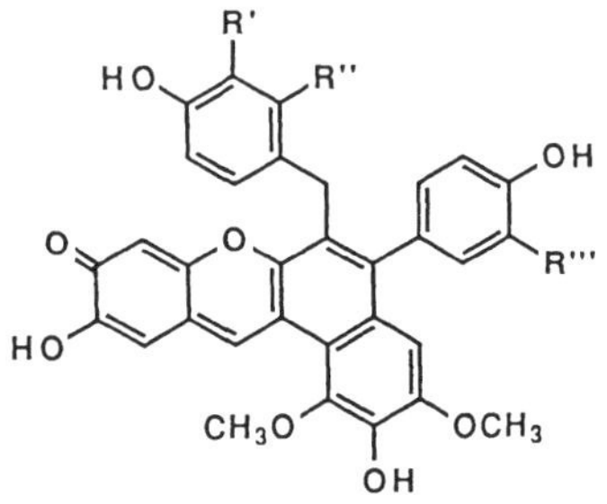

53 a-d

53					
	a	SANTALIN	A	R'=OH	R"=R'''=H
	b		B	R'=OCH_3	R"=R'''=H
	c	SANTARUBIN	A	R'=H	R"=R'''=OCH_3
	d		B	R'=H	R"=OCH_3; R'''=OH

santalins A (**53a**) and B (**53b**), and santarubins A (**53c**) and B (**53d**) – have been established mainly by NMR methods (3, 4, 5, 6). Recently the structures have been confirmed by X-ray structure analysis of a parent compound (44).

7.4.7 Concluding Remarks

This section has dealt with a diversity of compounds of very different structural types and of different biogenetic origins. Some types of compounds have been investigated in more detail than others. This situation does not reflect the relative importance of the compounds but rather the academic interest of natural product chemists. The literature is, of course, not covered completely since the intention has been to emphasize common constituents or constituents of general interest. Additional references to recent work on some of the classes of compounds, discussed in this section, can be obtained from a comprehensive review on benzenoid and polycyclic aromatic natural products by Simpson (108).

References

1 Agagao Craveizo A, da Costa Prado A, Gottlieb O R, Wellerson de Albuquerque P C 1970 Diarylheptanoids of *Centrolobium* species. Phytochemistry 9:1869–1875

2 Arakawa H, Torimoto N, Masui Y 1968 Die absolute Konfiguration des (−)-β-Tetralols und des Agrimonolides. Tetrahedron Lett 4115–4117

3 Arnone A, Camarda L, Merlini L, Nasini G 1975 Structure of the red sandalwood pigments santalins A and B. J Chem Soc Perkin Trans I 186–194

4 Arnone A, Camarda L, Merlini L, Nasini G, Taylor D A H 1977 Coloring matters of the west African redwoods *Pterocarpus osun* and *P. soyauxii*. Structures of santarubins A and B. J Chem Soc Perkin Trans I 2116–2118

5 Arnone A, Camarda L, Merlini L, Nasini G 1977 13C Nuclear magnetic resonance spectral analysis of santalin and santarubin permethyl ethers. J Chem Soc Perkin Trans I 2118–2122

6 Arnone A, Camarda L, Merlini L, Nasini G, Taylor D A H 1981 Isoflavonoid constituents of the west African redwood *Baphia nitida*. Phytochemistry 20:799–801

7 Asahina Y, Asano J 1930 Über die Konstitution von Hydrangenol und Phyllodulcin (II. Mitteil.). Ber Deutsch Chem Ges 63:429–437.

8 Asakawa Y 1970 A new ketol, 1,7-dimethyl-5-hydroxy-3-heptanone, and trans-stilbene from *Alnus firma* Sieb. et Zucc. Bull Chem Soc Jpn 43:575

9 Asakawa Y 1971 Chemical constituents of *Alnus sieboldiana* (Betulaceae). II. The isolation and structure of flavanoids and stilbenes. Bull Chem Soc Jpn 44:2761

10 Balogh B, Anderson A B 1965 Chemistry of the genus *Sequoia*. II. Isolation of sequirins, new phenolic compounds from the coast redwood (*Sequoia sempervirens*). Phytochemistry 4: 569–575

11 Begley MJ, Campbell R V M, Crombie L, Tuck B, Whiting D A 1971 Constitution and absolute configuration of meta, meta-bridged, strained biphenyls from *Myrica nagi*; X-ray analysis of 16-bromomyricanol. J Chem Soc (C) 3634–3642

12 Begley M J, Davies R V, Henley-Smith P, Whiting D A 1978 The constitution of (1*R*)-sequirin-D (*Sequoia sempervirens*), a biogenetically novel norlignan, by direct X-ray analysis. J Chem Soc Perkin Trans I 750–754.

13 Billeck G 1964 Stilbene im Pflanzenreich. Fortschr Chem Org Naturst 22:115–152.

14 Brandt E V, Roux D G 1979 Metabolites from the purple heartwood of Mimosoideae. 1. *Acacia peuce* F. Muell. The first natural 2,3-*cis* peltogynoids. J Chem Soc Perkin Trans I 777–780

15 Brandt E V, Ferreira D, Roux D G 1981 Metabolites from the purple heartwood of Mimosoideae. Part 2. Acacia carnei Maiden: Isolation, synthesis and reactions of crombeone. J Chem Soc Perkin Trans I 514–521

16 Brandt E V, Ferreira D, Roux D G 1981 Metabolites from the purple heartwood of Mimosoideae. Part 3. *Acacia combei* C. T. Wite. Stucture and partial synthesis of crombenin, a natural spiropeltogynoid. J Chem Soc Perkin Trans I 1878–1883

17 Brandt E V, van Heerden F R, Ferreira D, Roux D G 1981 Metabolites from the purple heartwood of the Mimosoideae. Part 4. *Acacia fasciculifera* F. Muell ex Benth: Fasciculiferin, fasciculiferol, and the synthesis of 7-aryl- and 7-flavanyl-peltogynoids. J Chem Soc Perkin Trans I 2483–2490
18 Breen G J W, Ritchie E, Taylor W C 1962 The chemical constituents of Australian *Flindersia* species. XVI. The constituents of the wood of *Flindersia laevicarpa* C.T. White & Francis. Austr J Chem 15:819–823
19 Brooks C J W, Watson D G 1985 Phytoalexins. Nat Prod Rep 2:427–459
20 Burden R S, Kemp M S, Wiltshire C W 1984 Isolation and structure determination of cotonefuran, an induced antifungal dibenzofuran from *Cotoneaster lactea* W.W. Sm. J Chem Soc Perkin Trans I 1445–1448
21 Cannon J R, Chow P W, Metcalf B W, Power A J, Fuller M W 1970 The structure of robustol, a novel phenol from *Grevillea robusta* A. Cunn. Tetrahedron Lett: 325–328
22 Carter F L, Garlo A M, Stanley J B 1978 Termiticidal components of wood extracts: 7-methyljuglone from *Diospyros virginiana.* J Agr Food Chem 16:869–873
23 Cunningham J, Haslam E, Haworth R D 1963 The constitution of piceatannol. J Chem Soc 2875–2883
24 Daniels P, Erdtman H, Nishimura K, Norin T, Kierkegaard P, Pilotti A M 1972 Athrotaxin, a C_{17}-phenolic constituent from *Athrotaxis selaginoides* Don. J Chem Soc Chem Commun 246–247
25 Dass A, Joshi T, Sjukla S 1984 Anthraquinones from *Cassia sophera* root bark. Phytochemistry 23:2689–2691
26 Dean F M, Kahn H, Minhaj N, Prakash S, Zaman A 1984 The structure of wightianone, the pigment of clathrate from *Calophyllum wightianum.* J Chem Soc Perkin Trans I 1755–1759
27 Donnelly B J, Donnelly D M X, O'Sullivan A M 1968 *Dalbergia* species. – VI. the occurrence of melannein in the genus *Dalbergia.* Tetrahedron 24:2617–2622
28 Donnelly D M X, O'Reilly J, Thopson J 1972 Neoflavanoids of Dalbergia cultrata. Phytochemistry 11:823–826
29 Ellis G P 1977 Chromenes, chromanones, and chromones. Wiley-Interscience New York
30 Enoki A, Takahama S, Kitao K 1977 The extractives of Metasekoia, *Metasequoia glyptostroboides* Hu et Cheng. I. The isolation of metasequirin-A, athrotaxin and agatharesinol from the heartwood. Mokuzai Gakkaishi 23:579–586
31 Enoki A, Takahama S, Kitao K 1977 The extractives of Metasekoia, *Metasequoia glyptostroboides* Hu et Cheng. II. The isolation of hydroxyathrotaxin, metasequirin-B and hydroxymetasequirin-A. Mokuzai Gakkaishi 23:587–593
32 Enzell C R, Thomas B R 1965 The chemistry of the order Araucariales. 2. The wood resin of *Agathis australis.* Acta Chem Scand 19:913–919
33 Enzell C R, Thomas B R 1966 The chemistry of the order Araucariales. 5. Agatharesinol. Tetrahedron Lett 2396–2402
34 Enzell C R, Hirose Y, Thomas B R 1967 The chemistry of the order Araucariales. 6. Absolute configurations of agatharesinol, hinokiresinol and sugiresinol. Tetrahedron Lett 793–798
35 Erdtman H 1939 Zur Kenntnis der Extraktivstoffe des Kiefernkernholzes. Naturwissenschaften 27:130–131
36 Erdtman H 1949 Compounds inhibiting the sulfite cook. Tappi 32:303–304
37 Erdtman H, Norin T 1963 Heartwood constituents of *Laburnum alpinum* Brecht. & Prest. Acta Chem Scand 17:1781–1782
38 Erdtman H 1963 Some aspects of chemotaxonomy. In: Swain T (ed) Chemical plant taxonomy. Academic Press New York, 89–125
39 Erdtman H, Eriksson G, Norin T 1961 Phenolic biphenyl derivatives from the heartwood of *Sorbus aucuparia* (L). Acta Chem Scand 15:1796
40 Erdtman H, Eriksson G, Norin T, Forsén S 1963 Aucuparin and methoxyaucuparin, two phenolic biphenyl derivatives from the heartwood of *Sorbus aucuparia* (L.). Acta Chem Scand 17:1151–1156
41 Erdtman H, Harmatha J 1979 Phenolic and terpenoid heartwood constituents of *Libocedrus yateensis.* Phytochemistry 18:1495–1500
42 Erdtman H, Vorbrüggen H 1960 Die Chemie der Natürlichen Ordnung Cupressales. XXXII. Über die Inhaltsstoffe des Kernholzes von *Athrotaxis selaginoides* Don, *Athrotaxis cupressoides* Don und *Chamaecyparis pisifera* Sieb. et Zucc. Acta Chem Scand 14:2161–2168

43 Favre-Bonvin J, Jay M, Wollenweber E 1978 A novel stilbene from bud excretion of *Alnus viridis*. Phytochemistry 17:821–822

44 Ferguson G, Ruhl B L, Whalley W B 1987 The chemistry of insoluble red wood. Part 16. The X-ray structure of 9-methoxy-6(4-methoxybenzyl)-5-(4-methoxyphenyl)benzo[*a*]xanthene. J Chem Res (S):30–31; J Chem Res (M):401–417

45 Ferrari F, de Linea R A, Marini-Bettolo G B 1984 2,4,2′-Trihydroxy-4′-methoxybenzil from *Zollernia paraensis*. Phytochemistry 23:2691–2692

46 Filho R B, Gottlieb O R, Mourao A P 1975 A stilbene and two flavones from *Derris rariflora*. Phytochemistry 14:261–263

47 Gottlieb O R 1968 Biogenetic proposals regarding aucuparins and xanthones. Phytochemistry 7:411–421

48 Gromova A S, Lutskii V I, Tyukavkina N A 1974 Hydroxystilbenes of the phloem of *Picea koraensis*. Khim Prir Soedin 10:778–779

49 Gromova A S, Lutskii V I, Tyukavkina N A 1975 Hydroxystilbenes of the inner and outer bark of *Picea ajanensis*. Khim Prir Soedin 11:82

50 Grossman W, Endres H, Pauckner W 1958 Die Konstitution des Piceatannols. Ber Deutsch Chem Ges 91:134–140

51 Hart J H, Hillis W E 1974 Inhibition of wood rotting fungi by stilbenes and other polyphenols in *Eucalyptus sideroxylon*. Phytopathology 64:939–948

52 Hashimoto T, Tajima M 1978 Structures and synthesis of the growth inhibitors botatasins IV and V, and their physiological activities. Phytochemistry 17:1179–1184

53 Hatam N A R, Whiting D A 1967 Constituents of Californian redwood. The structures of sequirin B and C. Tetrahedron Lett 781–785

54 Hatam N A R, Whiting D A 1982 Synthesis, conformation, and chirality of di-*O*-methyl sequirin D, a biogenetically novel metabolite of *Sequoia sempervirens*. J Chem Soc Perkin Trans I: 461–471

55 Hathway D E 1962 The use of hydroxystilbene compounds as taxonomic tracers in the genus *Eucalyptus*. Biochem J 83:80–84

56 Hausen B M 1973 Holzarten mit Gesundheitsschädigenden Inhaltstoffen. DRW-Verlag Stuttgart, 65–67

57 Hemingway R W, Mc Graw G W, Barras S J 1977 Polyphenols in *Ceratocystis minor*-infected *Pinus taeda*: Fungal metabolites, phloem, and xylem phenols. J Agr Food Chem 25:717–722

58 Henley-Smith P, Whiting D A 1976 New norlignans of *Sequoiadendron gigantea* phytochemical comparison with *Sequoia sempervirens*. Phytochemistry 15:1285–1287

59 Hillis W E 1977 Secondary changes in wood. In: Loewus F A, Runeckles V E (eds) The structure, biosynthesis, and degradation of wood. Vol II. Plenum New York

60 Hillis W E, Hart J H, Yazaki A 1974 Polyphenols of *Eucalyptus sideroxylon* wood. Phytochemistry 13:1591–1595

61 Hillis W E, Inoue T 1968 The formation of polyphenols in trees. IV. The polyphenols formed in *Pinus radiata* after *Sirex* attack. Phytochemistry 7:13–22

62 Hillis W E, Isoi K 1965 Variation in the chemical composition of *Eucalyptus sideroxylon*. Phytochemistry 4:541–550

63 Hirose Y, Oishi N, Nagaki H, Nakatsuka T 1965 The structure of hinokiresinol. Tetrahedron Lett 3665–3668

64 Inouye H, Okuda T, Hayashi T 1971 Naphthochinonderivative aus *Catalpa ovata* G. Don. Tetrahedron Lett 3615–3618

65 Khan M A, Nabi S G, Prakash S, Zaman A 1986 Pallidol, a resveratrol dimer from *Cissus pallida*. Phytochemistry 25:1945–1948

66 Kai Y 1965 On the phenolic constituents from *Cryptomeria japonica* D. Don. III. The structure of sugiresinol. Mokuzai Gakkaishi 11:23–26

67 Kai Y, Shimizu M 1966 The structure of sugiresinol and hydroxysugiresinol. In: Abstr 10th Symp Chem Nat Prod. Jpn Chem Soc 131–137

68 Karchesy J J, Laver M L, Barofsky D G, Barofsky E 1974 Structure of oregonin, a natural diarylheptanoid xyloside. J Chem Soc Chem Commun 649–650

69 Kemp M S, Burden R S 1984 Isolation and structure determination of γ-pyrufuran, a third induced antifungal dibenzofuran from the wood of *Pyrus communis* L. infected with *Chondrostereum purpureum* (Pers. ex Fr.) Pouzar. J Chem Soc Perkin Trans I 1441–1443

70 Kemp M S, Holloway P J, Burden R S 1985 3β,19α-Dihydro-2-oxours-12-en-28-oic acid: a pentacyclic triterpene induced in the wood of *Malus pumila* Mill. infected with *Chondrostereum purpureum* (Pers. ex Fr.) Pouzar, and a constituent of the cuticular wax of apple fruits. J Chem Res (S):154–155; J Chem Res (M):1848–1876

71 King F E, Cotterill C B, Godson D H, Jurd L, King T J 1953 The chemistry of extractives from hardwoods. Part XIII. Colourless constituents of *Pterocarpus* species. J Chem Soc 3693–3697

72 King F E, Gronden M F 1949 The constitution of chlorophorin, a constituent of iroko, the timber of Chlorophorea excelsa. Part 1. J Chem Soc 3348–3352

73 King F E, King T J, Godson D H, Manning L C 1956 The chemistry of extractives from hardwoods. Part XXVII. The occurrence of 3,4,3′,5′-tetrahydroxy- and 3,4,5,3′,5′-pentahydroxy-stilbene in *Vouacapoua* species. J Chem Soc 4477–4480

74 Krivoshchekova O E, Stepanenko L S, Maksimov O B 1986 A new isoflavone-stilbene from heartwood of *Maackia amurensis*. Khim Prir Soedin 22:39–42

75 Kumanotani J 1983 Japanese lacquer – a super durable coating. Proposed structure and expanded application. In: Carraker C D Jr, Sperling L H (eds) Polymer application of renewable-resource materials. Plenum Press New York, 225–248 (Polymer Sci Technol 17:225–248)

76 Kuroyanagi M, Natori S 1970 Curcuminoids from *Zingiberaceae* plants. Yakugaku Zasshi 90:1467–1470

77 Loman A A 1970 Bioassays of fungi isolated from *Pinus contorta* var. *latifolia* with pinosylvin, pinosylvin monomethyl ether, pinobanksin and pinocembrin. Can J Bot 48:1303–1308

78 Maheshi V B, Seshadri T R 1954 Chemical components of commercial woods and related plant materials. J Sci Ind Res India 13B:835

79 Maksimov O B, Krivoshchekova O E, Stepanenko L V 1985 Isoflavones and stilbenes of the heartwood of *Maackia amurensis*. Khim Prir Soedin 6:775–781

80 Manners G D, Swan E P 1971 Stilbenes in the barks of five Canadian *Picea* species. Phytochemistry 10:607–610

81 Marini-Bettolo G B, Casinovi C G, Galetti C 1965 A new class of quinones: sesquiterpenoid quinones of *Mansonia altissima* Chev. Tetrahedron Lett 4857–4864

82 Melini L, Nasini G 1976 Constituents of Dragon's blood. Part II. Structure and oxidative conversion of a novel secobiflavonoid. J Chem Soc Perkin Trans I 1570–1576

83 Nilsson M, Norin T 1963 Syntheses of aucuparin and methoxyaucuparin. Acta Chem Scand 17:1157–1159

84 Nomura M, Tokoroyama T 1975 Further phenolic components from *Alnus japonica* Stend. J Chem Soc Chem Commun 316–317

85 Norin T 1972 Some aspects of the chemistry of the order Pinales. Phytochemistry 11:1231–1242

86 de Oliveira W G, Mesquita A A L, de Lima R A, Gottlieb O R, Gottlieb H E 1984 Xanthones from *Toromita excelsa*. Phytochemistry 23:2390–2391

87 Ollis W D, Redman B T, Roberts R J, Sutherland I O 1968 New neoflavanoids from *Machaerium khulmannii* and *Machaerium nictitans* and the recognition of a new neoflavanoid type, the neoflavenes. J Chem Soc Chem Commun 1392–1393

88 Oshima R, Yamauchi Y, Watanabe C, Kumanotani J 1985 Enzymic oxidative coupling of urushiol in sap of the lac tree, *Rhus vernicifera*. J Org Chem 50:2613–2621

89 Owen P J, Scheinmann F 1974 Extractives from Guttiferae. Part XXVI. Isolation and structure of six xanthones, a biflavanoid, and triterpenes from the heartwood of *Pentaphalangium solomonse* Warb. J Chem Soc Perkin Trans I 1018–1021

90 Pryce R J, Longcake P 1977 α-Viniferin: an antifungal resveratrol trimer from grapevines. Phytochemistry 16:1452–1454

91 Rasmussen M, Ridley D D, Ritchie E, Taylor W C 1968 Chemical studies of the Proteaceae. III. The structure determination and synthesis of striatol, a novel phenol from *Grevillea striata* R. Br. Austr J Chem 21:2989–2999

92 Ridley D D, Ritchie E, Taylor W C 1970 Chemical studies of Proteaceae. IV. The structures of the major phenols of Grevillea striata, a group of novel cyclophanes. Austr J Chem 23:147–183

93 Riffer R, Anderson A B 1967 Chemistry of the genus *Sequoia*. IV. The structures of the C_{17} phenols from *Sequoia sempervirens*. Phytochemistry 6:1557–1562

94 Robertson A, Whalley W B, Yates J 1950 The pigment of "Dragon's Blood" resin. Part III. The constitution of dracorubin. J Chem Soc 3117–3123

95 Robertson A, Suckking C W, Whalley W B 1949 The chemistry of the insoluble red woods. Part III. The structure of santal and notes on orobol. J Chem Soc 1571–1578
96 Robertson A, Whalley W B 1954 The chemistry of the insoluble red woods. Part VI. Santalin and santarubin. J Chem Soc 2794–2801
97 Roughley P J, Whiting D A 1971 Diarylheptanoids; the problems of the biosynthesis. Tetrahedron Lett 3741–3746
98 Rowe J W, Bower C L, Wagner E R 1969 Extractives of jack pine bark: Occurrence of *cis*- and *trans*-pinosylvin dimethyl ether and ferulic acid esters. Phytochemistry 8:235–241
99 Rowe J W, Conner A H 1979 Extractives in eastern hardwoods. A review. USDA For Serv Gen Tech Rep FPL-18. 67 pp
100 Rowe J W, Seikel M, Roy D N, Jorgensen E 1972 Chemotaxonomy of *Ulmus*. Phytochemistry 11:2513–2517
101 Rupprich N, Kindl H 1978 Enzymatic synthesis of 3,5,4′-trihydroxystilbene from p-coumaroyl coenzyme A and malonyl coenzyme A. Physiol Chem 359:165–172
102 Saengchantara S T, Wallace T W 1986 Chromanols, chromanones, and chromones. Nat Prod Rep 3:465–475
103 Sandermann W, Dietrichs H H 1957 Investigations on termite-resistant wood. Holz Roh-Werkst 15:281–297
104 Sano Y, Sakakibara A 1977 Studies on chemical components of Akaezomatsu (*Picea glehnii*) bark. 2. Isolation of stilbenes. Hokkaido Daigaku Nogakubu Enshurin Kenkyu Hokoku 34:287–304
105 Shain L, Hillis W E 1972 Ethylene production in *Pinus radiata* in response to *Sirex-Amylostereum* attack. Phytopathology 62:1407–1409
106 Shain L, Hillis W E 1973 Ethylene production in xylem of *Pinus radiata* in relation to heartwood formation. Can J Bot 51:1331–1335
107 Shirata A, Takahashi K, Takasugi M, Nagao S, Ishikawa S, Ueno S, Munoz L, Masamune T 1983 Antimicrobial spectra of the compounds from mulberry. Sanshi Shikenjo Hokoku 28:793–806
108 Simpson T J 1987 Benzenoid and polycyclic aromatic natural products. Nat Prod Rep 4:639–676
109 Singh Janhavi, Singh J 1986 A bianthraquinone and a triterpenoid from the seeds of *Cassia hirsuta*. Phytochemistry 25:1985–1987
110 Sotheeswaran S, Sultanbawa M U S, Surendrakumar S, Bladon P 1983 Polyphenols from *Dipterocarp* species. Copalliferol A and stemonoporol. J Chem Soc Perkin Trans I 699–702
111 Sotheeswaran S, Sultanbawa M U S, Surendrakumar S, Balasubramaniam S, Bladon P 1985 Polyphenols from *Dipterocarp* species. Vaticaffinol and ε-viniferin. J Chem Soc Perkin Trans I 777–780
112 Takasugi M, Nagao S, Masamune T, Shirata A, Takahashi K 1980 Studies on phytoalexins of the Moraceae. 7. Chalcomoracin, a natural Diels-Alder adduct from diseased mulberry. Chem Lett 1573–1576
113 Takasugi M, Nagao S, Masamune T 1982 Studies on phytoalexins of the Moraceae. 10. Structure of dimoracin, a new natural Diels-Alder adduct from diseased mulberry. Chem Lett 1217–1220
114 Terazawa M, Koga T, Okuyama H, Miyake M 1973 Isolation of platyphyllanol, a new diarylheptanoid from the green bark of shirakanba, *Betula platyphylla* Sukatch var. *japonica* Hara. Mokuzai Gakkaishi 19:47–48
115 Terazawa M, Okuyama H, Miyake M 1973 Isolation of hirsutanonol and hirsutenone, two new diarylheptanoids from the green bark of keyamatrannoki, *Alnus hirsuta*. Mokuzai Gakkaishi 19:45–46
116 Thomson R H 1971 The naturally occurring quinones. 2nd ed. Academic Press London New York
117 Tiwari R D, Singh J 1979 A new anthraquinone digalactoside from *Cassia laevigata* pods. Phytochemistry 18:347
118 Venkataraman K 1972 Wood phenolics in the chemotaxonomy of the Moraceae. Phytochemistry 11:1571–1586
119 Watanabe L, Ishiguri Y, Nonaka F, Morita A 1982 Isolation and identification of aucuparin as a phytoalexin from *Eriobotrya japonica* L. Agr Biol Chem 46:567–568
120 Waterman P G, Crichton E G 1980 Xanthones and biflavonoids from *Garcinia densivenia*. Phytochemistry 19:2723–2726
121 Whiting D A 1987 Lignans, neolignans, and related compounds. Nat Prod Rep 4:515–525

122 Wijnsma R, Verpoorte R 1986 Anthraquinones in the Rubiaceae. In: Herz W, Griesbach H, Kirby G W, Tamm Ch (eds) Progress in the chemistry of natural products. Vol 49. Springer Wien New York, 79–149
123 Wolcott G N 1951 The termite resistance of pinosylvin and other new insecticides. J Econ Entomol 44:263–264
124 Wolcott G N 1953 Stilbenes and comparable materials for dry-wood termite control. J Econ Entomol 46:374–375
125 Yamamoto M 1958 Chemical structure of agrimonolide, a new constituent of *Agrimonia pilosa*. I. Yakugaku Zasshi 78:1086–1089
126 Yamamoto M 1959 Chemical structure of agrimonolide, a new constituent of *Agrimonia pilosa*. II. Yakugaku Zasshi 79:1067–1073
127 Yasue M, Imamura H 1966 Wood extractives XII. Relative configuration of asadanin and epiasadanol. Mokuzai Gakkaishi 12:226–231
128 Yasue M, Imamura H 1967 Wood extractives XIII. Structure of isoasadanol. Mokuzai Gakkaishi 12:213–236
129 Yazaki Y, Hillis W E 1977 Polyphenolic extractives of *Pinus radiata* bark. Holzforschung 31:20–25

7.5 Flavonoids

J. B. HARBORNE

7.5.1 Introduction

The flavonoid pigments, one of the most numerous and widespread groups of natural constituents, are of importance and interest not only because of their significant natural functions in the economy of the plant, but also because certain members of the group are physiologically active in humans. There is indeed considerable current interest in the effects flavonoids exert in animal systems (17). Flavonoids are found in all vascular plants and at least 4000 different structures have been reported to date (14, 15). One group of flavonoids, the anthocyanins, are intensely colored and their contributions to red and blue colors in flowers, fruits, and other plant tissues are well known. Such compounds have rarely been reported in woods. It is the other classes of flavonoid that lack intense color – namely the flavones, flavonols, flavanones, and flavanonols – that occur predominantly in woody tissues.

Flavonoids are known to occur widely in woody plants and are found in both hardwoods and softwoods (18). Generally, their presence is not obvious until the alcohol-soluble extractives are studied in detail. Some wood flavonoids such as the flavonol fisetin are intensely fluorescent in ultraviolet light and are thus readily detected, even when present in trace amounts. Other common wood flavonoids such as dihydroquercetin are dull absorbing in ultraviolet light and can only be detected satisfactorily by use of chromogenic reagents.

Although normally present in wood in relatively small quantities, flavonoids may be of importance in vivo in protecting the heartwood from fungal decay, along with other phenolics that may be present. Dihydroquercetin, for example, is one of the most potent fungicidal materials of Douglas-fir, a fairly decay-resistant species. In vitro, too, flavonoids may contribute to the long-term durability of timber. Economically, their presence in wood may sometimes be a disadvantage, since they are capable, even when present in small amounts, of inhibiting the sulfite pulping process (18).

The present account of wood flavonoids is limited to a consideration of the monomeric flavonoids, since biflavonoids and flavolans are dealt with in Sect. 7.6. Four major sources of information have been employed here: the two flavonoid monographs (14, 15), the 1962 review by Hergert (18), and the more recent review of free aglycones in plants by Wollenweber and Dietz (39). The available data are limited, because rather few widespread surveys of wood constituents have been attempted. We still know almost nothing of the flavonoid content of most tree species, and it is still not clear even whether flavonoids are present in most woody tissues. Detailed studies have usually been limited to individual species – e.g. *Quercus tinctoria* and *Apuleia leiocarpa* – or to individual genera – e.g. *Pinus* and *Prunus*. Some families – e.g. Leguminosae, Anacardiaceae, and Pinaceae – have been much more thoroughly investigated than most others.

Because of their ready accessibility, the leaves or needles of woody plants have been more thoroughly investigated than the heartwoods or barks, and their flavonoids have been better surveyed, notably by Bate-Smith (3). Where comparisons are possible, it is clear that wood and bark flavonoid patterns are likely to be related to those of other plant parts, but it is still not possible to predict from leaf investigations what flavonoids, if any, are likely to be found in the woody parts.

Three generalizations may be made specifically about flavonoid patterns in wood: The compounds mainly occur free and not in conjugated form as glycoside; 'reduced' flavonoids – e.g. flavanones and flavanonols – predominate; and several biosynthetically related structures representing different flavonoid classes commonly co-occur. An indication of the main structural classes found in wood is seen in Fig. 7.5.1. These classes are, of course, closely related biosynthetically, as shown in the figure. Compounds of the most common hydroxylation pattern are illustrated, but other series of hydroxylated flavonoids (e.g. 5-deoxy compounds) may also be found.

The more polar flavonoids occur in the alcohol-soluble fraction of the wood, whereas lipophilic flavonoids, if present, appear in the chloroform or ether extract. Flavonoids may crystallize directly from such extracts, but are more usually obtainable in pure form after some suitable chromatographic separation. They are then characterized by standard spectroscopic procedures (14, 15) and, where appropriate, by comparison with authentic markers. Some wood flavonoids are labile in solution, undergoing oxidation or polymerisation during handling. It is possible during the processes of extraction and purification that interconversions of one type to another (e.g. of dihydroflavonol to flavonol) may occur. Care must therefore be exercised in flavonoid analyses of wood tissues to avoid artifact production in this way. Also, some structural analyses have not been entirely unambiguous; examples will be mentioned in later sections where is has not proved possible to re-isolate and confirm the presence of a particular wood flavonoid.

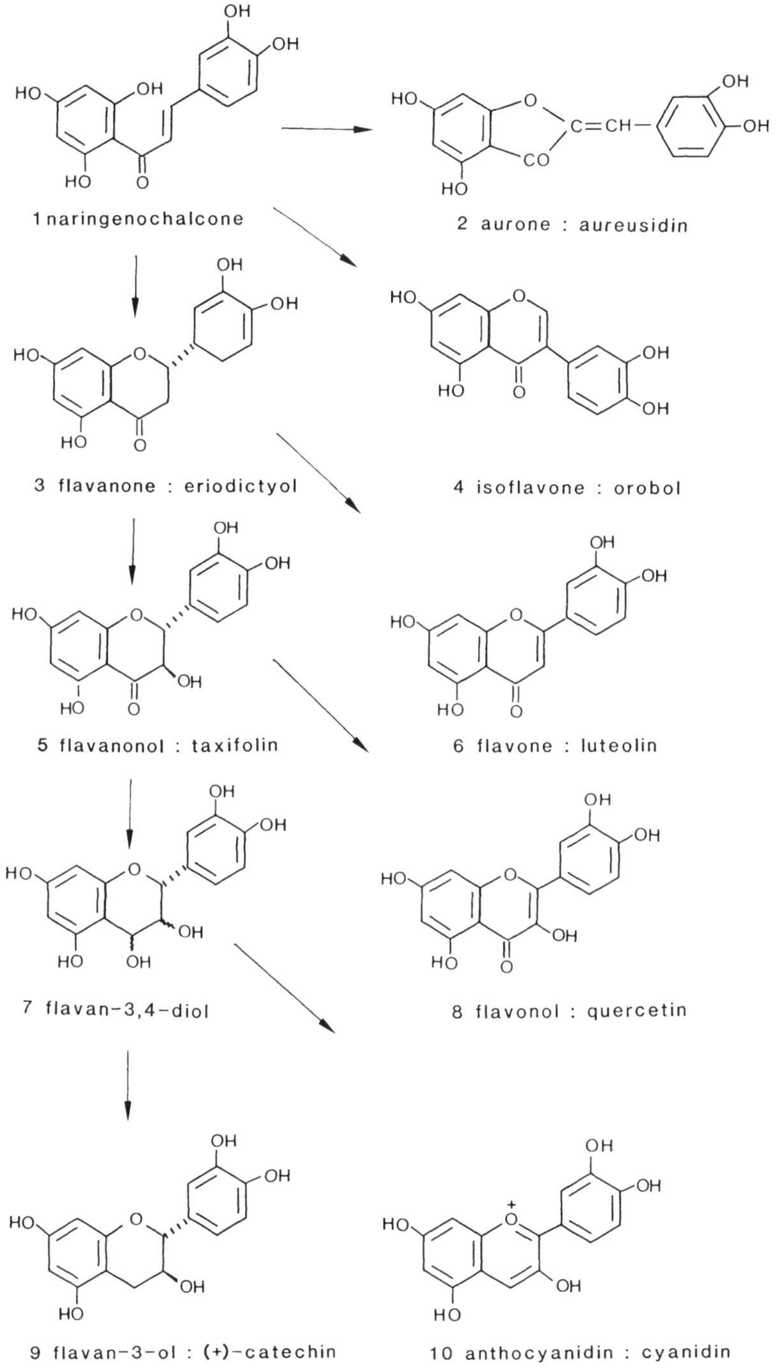

1 naringenochalcone

2 aurone : aureusidin

3 flavanone : eriodictyol

4 isoflavone : orobol

5 flavanonol : taxifolin

6 flavone : luteolin

7 flavan-3,4-diol

8 flavonol : quercetin

9 flavan-3-ol : (+)-catechin

10 anthocyanidin : cyanidin

Fig. 7.5.1. Relationship of different flavonoid classes

7.5.2 Structural Types

7.5.2.1 Flavones and Flavonols

Flavonols only differ from flavones (see Fig. 7.5.1) in the additional presence of a 3-hydroxyl group – i.e. they are flavon-3-ols – and generally the properties of the two classes are fairly similar. The distinction, however, is worth making since flavonols are biosynthetically more highly developed than flavones. In terms of natural distribution, there is a tendency for flavones to replace flavonoids within particular plant groups or vice versa so that the presence/absence of flavones and flavonols may be taxonomically significant. Chemically, the two classes differ in chromatographic and spectral properties, and in color reactions, although there is some overlap. Neither class is as common in wood tissues as their dihydro derivatives, the flavanones and the flavanonols.

The various flavones that have been reported in wood or bark of higher plants are listed in Table 7.5.1 under five headings: simple flavones, *C*-methylflavones, isoprenylated flavones, flavone *O*-glycosides, and *C*-glycosylflavones. Only a small number of simple hydroxyflavones (or their *O*-methyl ethers) are recorded. They are only regularly reported as heartwood constituents of *Pinus* and *Prunus*.

Table 7.5.1. Flavones reported to occur in wood or bark

Type Structure (name)	Genus, species (family)	Location[a]
Simple flavones		
5,6-dimethoxy	*Casimiroa edulis* (Rutaceae)	B
5,7-dihydroxy (chrysin)	*Pinus* subg. *Haploxylon* (Pinaceae)	HW
	Prunus spp. (Rosaceae)	HW
5-hydroxy-7-methoxy (tectochrysin)	*Pinus* subg. *Haploxylon* (Pinaceae)	HW
	Prunus spp. (Rosaceae)	HW
5,6,2′-trimethoxy	*Casimiroa edulis* (Rutaceae)	B
5,7-dihydroxy-6-methoxy (oroxylon)	*Oroxylum indicum* (Bignoniaceae)	B
5-hydroxy-7,8-dimethoxy	*Uvaria rufas* (Annonaceae)	B
5,7,8-trimethoxy	*Actinodaphne madraspatane* (Lauraceae)	HW
5,7,4′-trihydroxy (apigenin)	*Garcinia multiflora* (Hypericaceae)	HW
5,7-dihydroxy-6,4′-dimethoxy (pectolinarigenin)	*Cassia reniger* (Leguminosae)	B
5,7,2′,4′-tetrahydroxy (noroartocarpetin)	*Artocarpus* spp. (Moraceae)	HW
5,2′,4′-trihydroxy-7-methoxy (artocarpetin)	*Artocarpus* spp. (Moraceae)	HW
5,7,3′,4′-tetrahydroxy (luteolin)	*Prunus ssiori* (Rosaceae)	HW
5,4′-dihydroxy-7,3′-dimethoxy (velutin)	*Ceanothus velutinus* (Rhamnaceae)	B
5,7,8,3′,4′,5′-hexamethoxy	*Aniba* sp. (Lauraceae)	HW
C-Methyl flavones		
5,7-dihydroxy-6-*C*-methyl (strobochrysin)	*Pinus strobus* (Pinaceae)	HW
5-hydroxy-7,4′-dimethoxy-6-*C*-methyl	*Eucalyptus torelliana* (Myrtaceae)	HW
5-hydroxy-7,4′-dimethoxy-6,8-di-*C*-methyl (eucalyptin)	*E. sideroxylin* (Myrtaceae)	HW

Table 7.5.1 (continued)

Type Structure (name)	Genus, species (family)	Location[a]
Isoprenylated flavones		
5,7,2′,4′-tetrahydroxy derivatives		
artocarpesin, cycloartocarpesin, oxydihydroartocarpesin, norartocarpin, cycloartocarpin, chaplashin, integrin	*Artocarpus* spp. (Moraceae)	HW
mulberrin, mulberrochromene, cyclomulberrin, cyclomulberrochromene, mulberranol, rubroflavones A – D	*Morus* spp. (Moraceae)	B
5-allyloxy-6,7,4′-trimethoxy	*Tinospora malabarica* (Menispermaceae)	HW
5,7,2′,4′,5′-pentahydroxy derivatives		
cycloheterophyllin, isocycloheterophyllin, heterophyllin	*Artocarpus* spp. (Moraceaea)	B
5,6,7,2′,4′,5′-hexahydroxy derivatives artobilichromene	*Artocarpus nobilis* (Moraceae)	B
7,8-furanyl-3′,4′-methylenedioxy (pongaglabrone)	*Pongamia glabra* (Leguminosae)	HW
Flavone *O*-glycosides		
chrysin 5-glucoside	*Docyniopsis tschonoski* (Rosaceae)	B
chrysin 7-glucoside	*Prunus aequinoctialis* (Rosaceae)	W
3′,4′-dihydroxyflavone 4′-glucoside	*Rhus* spp. (Anacardiaceae)	HW
genkwanin 5-glucoside	*Prunus serrata* (Rosaceae)	B
genkwanin 4′-glucoside	*P. puddum* (Rosaceae)	B
luteolin 7-glucoside	*Prunus ssiori* (Rosaceae)	B
5,7,3′-trihydroxy-6,8,4′-trimethoxy-flavone 5-(6″-acetylglucoside)	*Vitex negundo* (Verbenaceae)	B
Flavone C-glycosides		
apigenin 8-glucoside (vitexin), apigenin 6-glucoside (isovitexin), agipenin 6,8-diglycosides (vicenins 1 – 3), luteolin 6,8-diglycosides (lucenins 1 – 3)	*V. lucens* (Verbenaceae)	HW
bayin (5-deoxyvitexin)	*Castanospermum australe* (Leguminosae)	W

[a] B = bark; HW = heartwood; W = wood

The occurrence of flavones lacking B-ring hydroxylation – e.g. of chrysin – is noteworthy. A wider range of flavones occur in *Prunus* than in *Pinus*, with luteolin being repordes in *Prunus ssiori*. It is interesting that the barks of *Prunus* species appear to contain some of the 5- or 7-glucosides of those flavones that occur free in the heartwood.

The most unusual substitution pattern among the simple flavones is that of 5,6,2′-trimethoxyflavone, reported in the bark of *Casimiroa edulis*. Two further flavones, zapotin (5,6,2′,6′-tetramethoxy) and zapotinin (5-hydroxy-6,2′,6′-trimethoxy), have been isolated from the same source, but these structures are still uncertain, since they have not been confirmed by subsequent synthetic experiments.

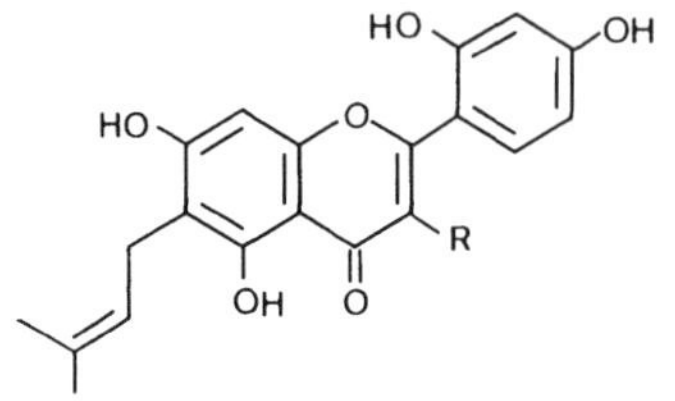

11 artocarpesin R=H

12 norartocarpin R=$CH_2CH{=}CMe_2$

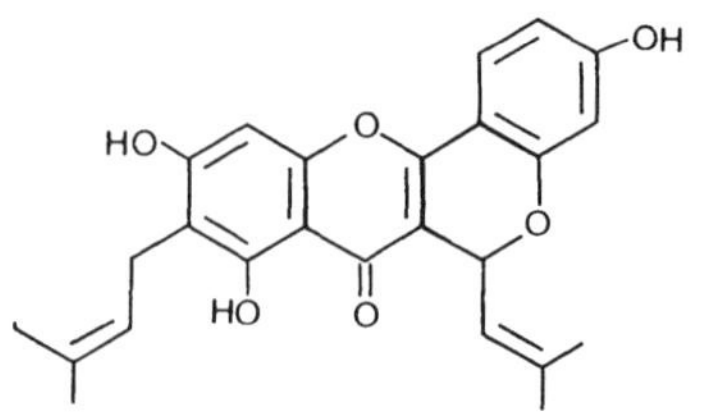

13 cycloartocarpin

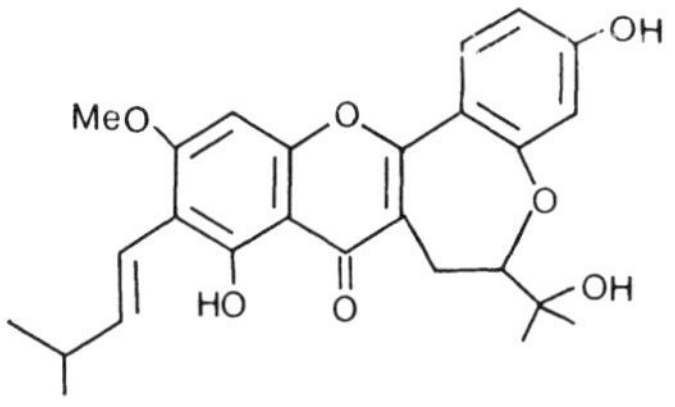

14 chaplashin

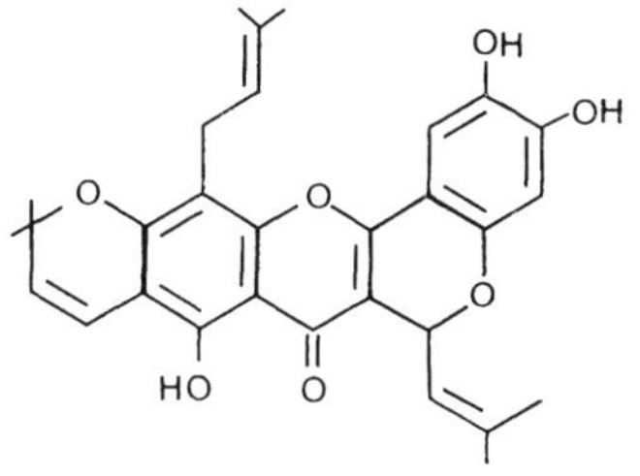

15 cycloheterophyllin

Isopentenyl-substituted flavones mainly characterize two genera of the Moraceae, *Artocarpus* and *Morus*. *Artocarpus* heartwoods have yielded some 12 isoprenylated flavones, based on a 5,7,2′,4′-tetrahydroxylated structure. One of the simplest is artocarpesin (**11**) with a single isoprene residue at C-6. It occurs with the related oxydihydro derivative and with cycloartocarpesin, in which the isoprene residue has cyclized with the 7-hydroxyl group. Disubstitution with isoprene residues at C-3 and C-6 gives rise to norartocarpin, (**12**) and its 7-methyl ether, artocarpin. Further cyclization of the C-3 isoprene residue in these structures with the 2′-hydroxyl produces either cycloartocarpin (**13**) with a new pyran ring or chaplashin (**14**) with a seven-membered oxepine ring. Finally, in *Artocarpus* bark, there are three more complex flavones trisubstituted with isoprenyl groups, heterophyllin, cycloheterophillin (**15**) and isocycloheterophyllin. These are all colored compounds, due to the extended conjugated systems present. Cycloheterophyllin has one free isopentenyl at C-8, whereas those at C-3 and C-6 are cyclized, producing a pentacyclic system (38).

A related series of isoprenylated flavones occur in the bark of *Morus* species (Table 7.5.2) one of which, mulberrin, only differs from norartocarpin in the position of one the exocyclic double bonds. Again, cyclized derivatives of mulberrin occur, the tetracyclic cyclomulberrin and mulberrochromene and the pentacyclic

Table 7.5.2. Flavonols reported to occur in wood or bark

Type Structure (name)	Genus, species (family)	Location[a]
Simple flavonols		
3,7-dihydroxy	*Platymiscium praecox* (Leguminosae)	W
3,5,7-trihydroxy (galangin)	*Pinus griffithii* (Pinaceae)	W
5,7-dihydroxy-3-methoxy	*P. griffithii* (Pinaceae)	W
3,5-dihydroxy-7-methoxy	*P. griffithii* (Pinaceae)	W
	P. excelsa (Pinaceae)	W
	Aniba riparia (Lauraceae)	W
3,5,7-trimethoxy	*A. riparia* (Pinaceae)	W
3,7,4′-trihydroxy	*Rhus javanica* (Anacardiaceae)	HW
	Schinopsis spp. (Anacardiaceae)	HW
	Platymiscium praecox (Leguminosae)	W
3,5,7,4′-tetrahydroxy (kaempferol)	*Pinus* spp. (Pinaceae)	B
5,7,4′-trihydroxy-3-methoxy	*Didierea madagascariensis* (Didiereaceae)	B
3,5,7-trihydroxy-4′-methoxy	*Prunus mume* (Rosaceae)	W
3,7,8,4′-tetrahydroxy	*Acacia* spp. (Leguminosae)	HW
7,8,4′-trihydroxy-3-methoxy	*Acacia* spp. (Leguminosae)	HW
3,7,3′,4′-tetrahydroxy (fisetin)	*Mangifera indica* (Anacardiaceae)	B
	Pistacia chinensis	HW
	Rhus javanica	HW
	Schinopsis spp.	HW
	Acacia mearnsii (Leguminosae)	B
	Robinia pseudacacia	HW
	Gleditsia triacanthos	HW
7,3′,4′-trihydroxy-3-methoxy	*Acacia mearnsii* (Leguminosae)	B
3,7,3′-trihydroxy-4′-methoxy	*Schinopsis lorentsii* (Leguminosae)	W
3,7,3′,4′-tetramethoxy	*Pongamia glabra* (Leguminosae)	B
3,5,7-trihydroxy-8,4′-dimethoxy	*Prunus domestica* (Rosaceae)	HW
3,5,7,2′,4′-pentahydroxy (morin)	*Chlorophora tinctoria* (Moraceae)	W
	Morus spp. (Moraceae)	W
3,7,2′,3′,4′-pentahydroxy	*Jacaranda acutifolia* (Bignoniaceae)	B
3,5,7,3′,4′-pentahydroxy (quercetin)	*Abies* spp. (Pinaceae)	B
	Picea spp. (Pinaceae)	B
	Pseudotsuga spp. (Pinaceae)	B, W
	Taxus spp. (Taxaceae)	B
	Taxodium spp. (Taxodiaceae)	B
	Tsuga spp. (Pinaceae)	B
	Podocarpus spp. (Podocarpaceae)	W
	Nothofagus spp. (Fagaceae)	W
	Soymida febrifuga (Meliaceae)	HW
7,3′,4′-trihydroxy-3,5-dimethoxy (caryatin)	*Carya pecan* (Juglandaceae)	B
5,3′-dihydroxy-3,7,4′-trimethoxy	*Apuleia leiocarpa* (Leguminosae)	W
	Distemonanthus benthamianus (Leguminosae)	HW
5-hydroxy-3,7,3′,4′-tetramethoxy	*Sterculia foetida* (Sterculiaceae)	B
3-hydroxy-5,7,3′,4′-tetramethoxy	*S. foetida* (Sterculiaceae)	B
3,7,8,3′,4′-pentahydroxy (melanoxetin)	*Acacia* spp. (Leguminosae)	HW
	Albizia spp. (Leguminosae)	HW
7,8,3′,4′-tetrahydroxy-3-methoxy	*Acacia* spp. (Leguminosae)	HW
3,7,3′,4′-tetrahydroxy-8-methoxy	*A. kempeana* (Leguminosae)	HW
7,3′,4′-trihydroxy-3,8-dimethoxy	*A. kempeana* (Leguminosae)	HW
3,7,3′,4′,5′-pentahydroxy (robinetin)	*Acacia* spp. (Leguminosae)	HW

Table 7.5.2 (continued)

Type	Structure (name)	Genus, species (family)	Location[a]
		Butea spp. (Leguminosae)	HW
		Gleditsia spp. (Leguminosae)	HW
		Millettia spp. (Leguminosae)	HW
		Robinia spp. (Leguminosae)	HW
	5,6,3′-trihydroxy-3,7,4′-trimethoxy (oxyayanin B)	*Distemonanthus benthamianus* (Leguminosae)	W
		Apuleia leiocarpa (Leguminosae)	W
	5,4′-dihydroxy-3,7,8,3′-tetramethoxy	*Melicope mantellii* (Rutaceae)	B
	5,2′,3′-trihydroxy-3,7,4′-trimethoxy (apuleidin)	*Apuleia leiocarpa* (Leguminosae)	W
	5,7,2′,5′-tetrahydroxy-3,7,4′-trimethoxy (oxyayanin A)	*A. leiocarpa* (Leguminosae)	W
		Distemonanthus benthamianus (Leguminosae)	
	2′,5′-dihydroxy-3,6,7,4′-tetramethoxy	*Apuleia leiocarpa* (Leguminosae)	
	3,5,7,3′,4′,5′-hexahydroxy (myricetin)	*Pinus* spp. (Pinaceae)	B
		Taxus spp. (Taxaceae)	B
		Soymida febrifuga (Meliaceae)	HW
	3,5,7,3′,5′-pentahydroxy-4′-methoxy	*Alluaudia* spp. (Didiereaceae)	B
	3,5,7,4′-tetrahydroxy-3′,5′-dimethoxy (syringetin)	*Soymida febrifuga* (Meliaceae)	HW
	5-hydroxy-3,7,2′,4′,5′-pentamethoxy	*Distemonanthus benthamianus* (Leguminosae)	HW
	2′-hydroxy-3,5,7,4′,5′-pentamethoxy	*Distemonanthus benthamianus* (Leguminosae)	HW
	3,5,7,2′,4′,5′-hexamethoxy	*Distemonanthus benthamianus* (Leguminosae)	HW
	3,5,6,7,2′,3′,4′-heptamethoxy	*Distemonanthus benthamianus* (Leguminosae)	HW
	6,2′-dihydroxy-3,5,7,4′,5′-pentamethoxy	*Distemonanthus benthamianus* (Leguminosae)	HW
	5,6-dihydroxy-3,7,2′,4′,5′-pentamethoxy	*Distemonanthus benthamianus* (Leguminosae)	HW
	3,5,6,7,2′,4′,5′-heptamethoxy	*Distemonanthus benthamianus* (Leguminosae)	HW
	5,6,2′,3′-tetrahydroxy-3,7,4′-trimethoxy (apuleisin)	*Apuleia leiocarpa* (Leguminosae)	HW
	3,5,6,7,4′-pentahydroxy-2′,5′-dimethoxy (apulein)	*A. leiocarpa* (Leguminosae)	HW
	5,7,3′,4′,5′-pentahydroxy-3,6-dimethoxy	*Alluaudia ascendens* (Didiereaceae)	B
	5,6,7,3′,5′-pentahydroxy-3,4′-dimethoxy	*A. ascendens* (Didiereaceae)	B
	3,5,7,3′,5′-pentahydroxy-6,4′-dimethoxy	*A. ascendens* (Didiereaceae)	B
	5,7,3′,5′-tetrahydroxy-3,6,4′-trimethoxy	*A. ascendens* (Didiereaceae)	B
	5,7,4′,5′-tetrahydroxy-3,6,3′-trimethoxy	*Decarya madagascariensis* (Didiereaceae)	B
	5,6,5′-trihydroxy-3,7,3′,4′-tetramethoxy (apuleitrin)	*Apuleia leiocarpa* (Didiereaceae)	HW
	6,4′-dihydroxy-3,5,7,3′,5′-pentamethoxy (apuleirin)	*A. leiocarpa* (Didiereaceae)	HW
C-Methyl flavonols			
	3,5,7,3′,4′-pentahydroxy-6-*C*-methyl (pinoquercetin)	*Pinus ponderosa* (Pinaceae)	B
	3,5,7,3′,4′,5′-hexahydroxy-6-*C*-methyl (pinomyricetin)	*P. ponderosa* (Pinaceae)	B
	5,7,3′,4′,5′-pentahydroxy-3-methoxy-6-*C*-methyl	*Alluaudia dumosa* (Didiereaceae)	B
		A. humbertii (Didiereaceae)	B
Methylenedioxy flavonols			
	3,5-dimethoxy-6,7-O_2CH_2-3′,4′-O_2CH_2 (meliternatin)	*Melicope* spp. (Rutaceae)	W
	3,7-dimethoxy-3′,4′-O_2CH_2 (demethoxykanugin)	*Pongamia glabra* (Leguminosae)	HW

Table 7.5.2 (continued)

Type	Structure (name)	Genus, species (family)	Location[a]
	3,7,3′-trimethoxy-4′,5′-O_2CH_2	*P. glabra* (Leguminosae)	HW
	5-hydroxy-3,6,7-trimethoxy-3′,4′-O_2CH_2 (melisimplin)	*Melicope simplex* (Rutaceae)	W
	3,5,6,7-tetramothoxy-3′,4′-O_2CH_2 (melisimplexin)	*M. simplex* (Rutaceae) *M. broadbentiana* (Rutaceae)	W W
	3,5,7,8-tetramethoxy-3′,4′-O_2CH_2 (meliternin)	*M. ternata* (Rutaceae)	W
	5,3′,4′-trihydroxy-3-methoxy-7,8-O_2CH_2 (wharangin)	*M. ternata* (Rutaceae)	B
	3,5,6,7,8-pentamethoxy-3′,4′-O_2CH_2 (melibentin)	*M. broadbentiana* (Rutaceae)	W
Flavonol glycosides			
	kaempferol 3-rhamnoside	*Afzelia* sp. (Leguminosae)	HW
	kaempferol 7-methyl ether 3-dirhamnosylgalactoside	*Rhamnus alaternus* (Rhamnaceae)	B
	kaempferol 4′-methyl ether 7-glucoside (mumenin)	*Prunus mume* (Rosaceae)	W
	quercetin 3-rhamnoside	*Quercus tinctoria* (Fagaceae)	B
	quercetin 7-glucoside	*Prunus emarginata* (Rosaceae)	B
	quercetin 7-methyl ether 3-rhamninoside	*Rhamnus cathartica* (Rhamnaceae)	B
	quercetin 7,4′-dimethyl ether 3-rhamninoside	*Rhamnus* spp. (Rhamnaceae)	B
	myricetin 3-rhamnoside	*Myrica rubra* (Myricaceae)	B
	myricetin 4′-methyl ether 3-digalactoside	*Vitex negundo* (Verbenaceae)	B
	3,7,2′,3′,4′-pentahydroxyflavone 3-neohesperidoside	*Jacaranda acutifolia* (Bignoniaceae)	B
	6-*C*-glucosyl kaempferol 6-*C*-glucosyl quercetin 6-*C*-glucosyl kaempferol 7-methyl ether 6-*C*-glucosyl quercetin 7-methyl ether	*Zelkowa serrata* (Ulmaceae)	W

[a] B = bark; HW = heartwood; W = wood

cyclomulberrochromene. *Morus* barks have also yielded four flavones substituted at C-3 with a C_{10}-geranyl sidechain. These are named rubraflavones A through D and differ in having additional isopentenyl groups at C-6 or C-8, which may or may not be cyclized to the 7-hydroxyl (38).

The few flavone glycosides reported from wood or bark (Table 7.5.1) are mainly based on glucose. A noteworthy source of *C*-glycosylflavones is the wood of *Vitex lucens*, since one of the most common of all such substances, vitexin or apigenin 8-glucoside, was named after this plant. The wood additionally contains isovitexin (apigenin 6-glucoside) and a series of di-*C*-substituted derivatives of both apigenin and luteolin **(6)** (32).

The flavonols reported in woody tissues are listed in Table 7.5.2. Some typical structures are **16** through **23**. The most common flavonol of higher plants is quercetin and this substance is reported as a very frequent constituent of gymnosperm woods and barks. In the angiosperm woods, the flavonols most frequently reported include fisetin (5-deoxyquercetin) (**16**) characteristic of both the Anacardiaceae and Leguminosae, robinetin (**89**) of *Robinia pseudacacia* heartwood, and morin (**17**) first isolated from old fustic, the bark of *Chlorophora tinctoria*. Perhaps the most complex mixture of flavonols reported from a single source are those described from *Apuleia leiocarpa* (5).

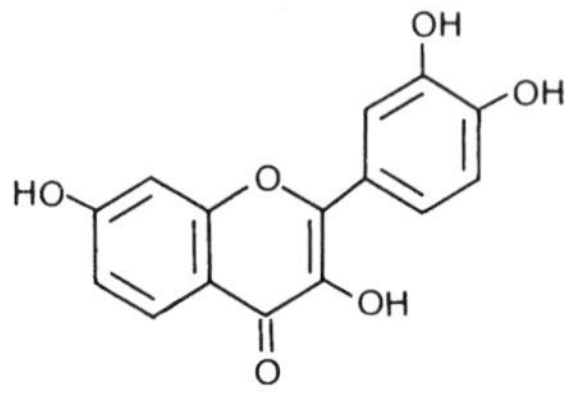

16 fisetin

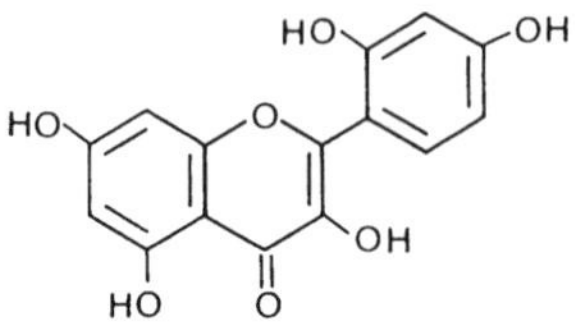

17 morin

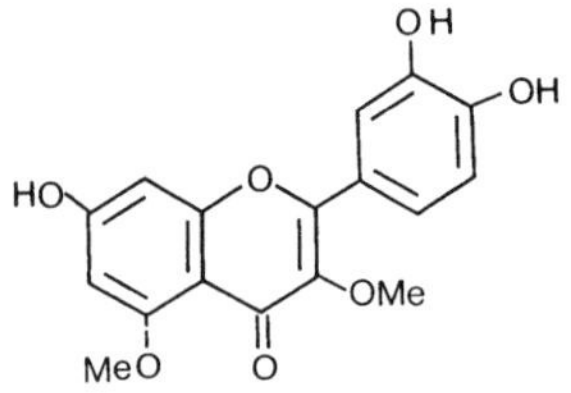

18 caryatin

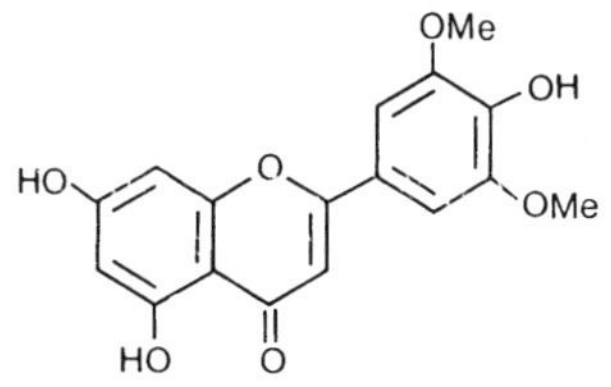

19 syringetin

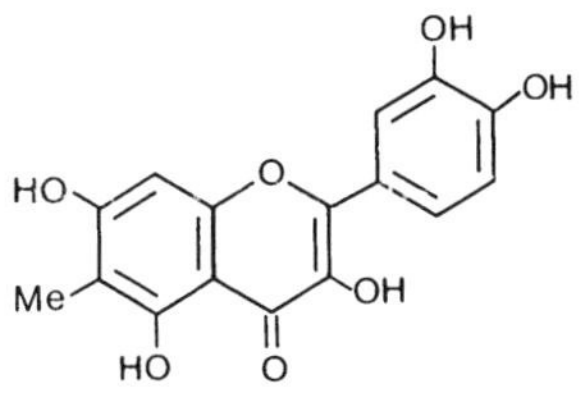

20 pinoquercetin

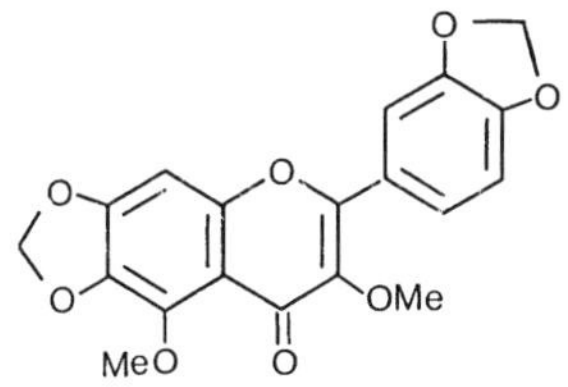

21 meliternatin

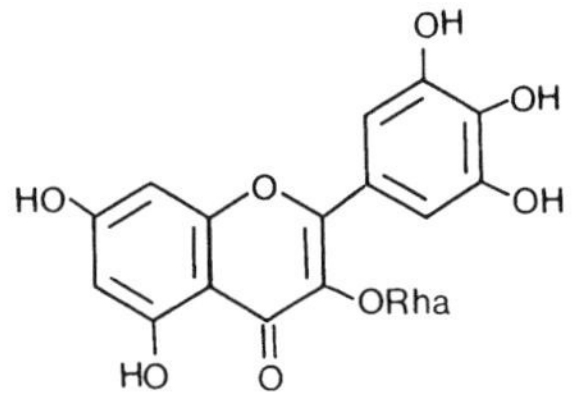

22 myricitrin

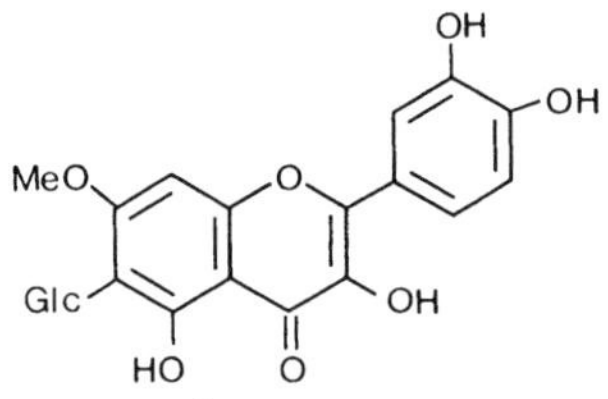

23 keyakinin

C-methylflavonols are uncommon in wood, having been reported so far only from *Pinus* in the gymnosperms and *Alluaudia* in the angiosperms. Methylenedioxyflavonols are also rare, being characteristic of a single rutaceous genus, *Melicope*.

Glycosides are relatively infrequently reported in this series, and are more typically found in bark than in heartwood. Three common flavonol 3-rhamnosides were first characterized from tree barks: afzelichin (kaempferol 3-rhamnoside) from *Afzelia*, quercitrin (querctin 3-rhamnoside) from *Quercus*, and myricitrin (**22**) (myricetin 3-rhamnoside) from *Myrica*. Finally, it may be noted that the only known source of *C*-glycosylflavonols, as distinct from the much more common *C*-glycosylflavones, is the woody tissue of *Zelkowa serrata*. One such glycoflavonol is keyakinin (**23**) the 6-*C*-glucoside of kaempferol 7-methyl ether.

7.5.2.2 Flavanones and Flavanonols

Flavanones are 2,3-dihydroflavones, while flavanonols are 2,3-dihydroflavonols. Indeed, flavanols are also termed dihydroflavonols and individual members may be named both from the plant source and from the derived flavonol. The most common flavanonol, for example, is known both as dihydroquercetin and as taxifolin. Reduction of the 2,3-double bond introduces an asymmetric carbon atom at C-2 and flavanones are capable of existing in two stereoisomeric forms. In fact, most if not all naturally occurring flavanones are laevorotatory and probably belong to the same (2*S*) configurational series. This has been proved rigorously for (−)-liquiritigenin (**24**).

Dihydroflavonols, which have different substituents at both C-2 and C-3, can exist in both *cis*- and *trans*-forms. (+)-Dihydroquercetin (**25**), has been established to have the 2*R*:3*R* stereochemistry and the majority of other flavanonols that have been studied also possess this configuration. However, a few dihydroflavonols have been isolated in different optically active forms. Notably, both (+)-fustin (**26a**) and (-)-fustin (**26b**) are known, the first occurring in *Acacia mollissima* heartwood and the second in *Rhus cotinus* wood. These compounds have opposite configurations.

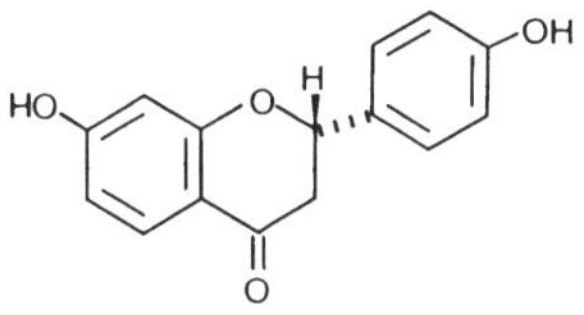

24 (−)(2S)-liquiritigenin

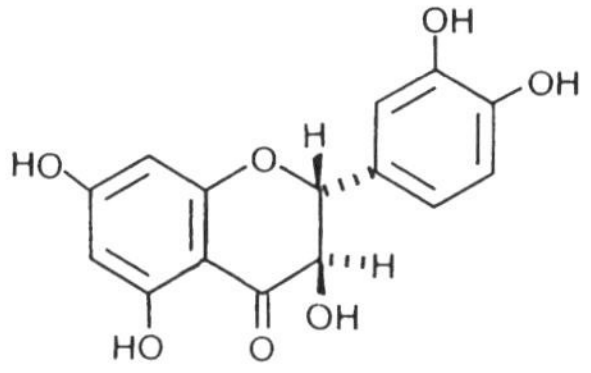

25 (+)-(2R,3R)-dihydroquercetin

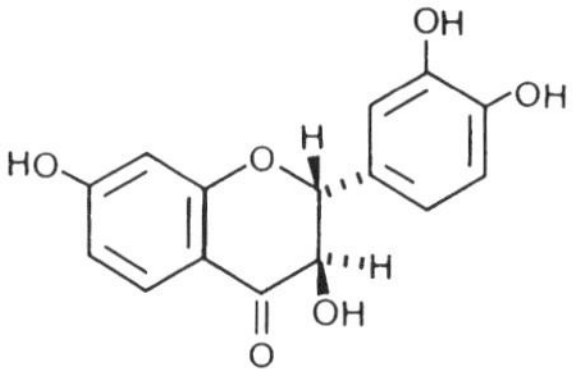

26a (+)-(2R,3R)-fustin

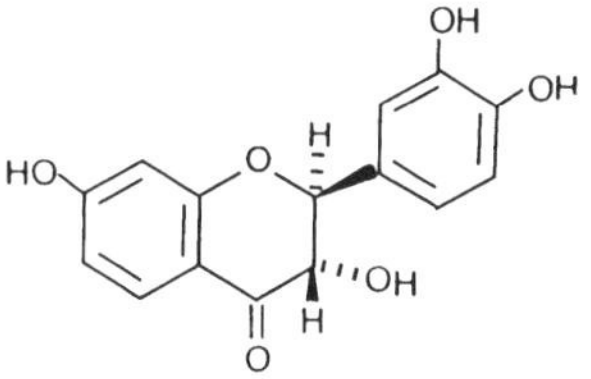

26b (−)-(2S,3S)-fustin

The flavanones isolated from different species of trees are listed in Table 7.5.3. It will be seen that flavanones lacking B-ring hydroxyls (e.g. pinocembrin) (**71**) or having *C*-methyl substitution are represented, occurring notably in *Pinus* along with the related flavones (see Table 7.5.1). 5-Deoxyflavanones – e.g. butin are expectedly found in a variety of legume heartwoods. Naringenin, the flavanone corresponding to apigenin, and its methyl ethers are present in many sources. Naringenin is more widely occurring than is apparent from the Table 7.5.3. Its simple *O*-glucosides have also been reported, principally from bark of different *Prunus* species.

The known wood flavanonols (Table 7.5.4) mainly correspond in substitution pattern with the flavanones listed in Table 7.5.3. Related pairs of flavanones and

Table 7.5.3. Flavanones reported to occur in wood and bark

Type Structure (name)	Genus, species (family)	Location[a]
Simple flavanones		
5,7-dihydroxy (pinocembrin)	*Pinus* spp. (Pinaceae)	HW
	Aniba rosaeodora (Lauraceae)	W
	Prunus spp. (Rosaceae)	HW
5-hydroxy-7-methoxy (pinostrobin)	*Larix dahurica* (Pinaceae)	HW
	Pinus spp. (Pinaeceae)	HW
	Prunus spp. (Rosaceae)	HW
5,7-dimethoxy	*Aniba riparia* (Lauraceae)	W
7,4′-dihydroxy (liquiritigenin)	*Dalbergia sericea* (Leguminosae)	B
	Diplotropis purpurea (Leguminosae)	W
7-hydroxy-4′-methoxy	*Dalbergia retusa* (Leguminosae)	HW
5,7-dihydroxy-8-methoxy (dihydrowogonin)	*Prunus* spp. (Rosaceae)	HW
5,7,4′-trihydroxy (naringenin)	*Larix dahurica* (Pinaceae)	HW
	Nothofagus spp. (Fagaceae)	W
5,4′-dihydroxy-7-methoxy (sakuranetin)	*Juglans regia* (Juglandaceae)	B
	Prunus spp. (Rosaceae)	W
5,7-dihydroxy-4′-methoxy (isosakuranetin)	*Eucalyptus maculata* (Myrtaceae)	B
	Prunus spp. (Rosaceae)	W
7,8,4′-trihydroxy	*Acacia rhodoxylon* (Leguminosae)	W
7,3′,4′-trihydroxy (butin)	*Acacia* spp. (Leguminosae)	W
	Gliricidia spp. (Leguminosae)	
	Robinia spp. (Leguminosae)	
5,7,2′,4′-tetrahydroxy	*Morus alba* (Moraceae)	HW
5,2′,4′-trihydroxy-7-methoxy (artocarpanone)	*Artocarpus integrifolia* (Moraceae)	HW
7-hydroxy-5,2′,4′-trimethoxy (cerasinone)	*Prunus cerasus* (Rosaceae)	HW
5,4′-dihydroxy-7,3′-dimethoxy	*Melicope sarcocolla* (Rutaceae)	B
5,3′-dihydroxy-7,4′-dimethoxy (persicogenin)	*Prunus persica* (Rosaceae)	B
6,7,3′,4′-tetrahydroxy (plathymenin)	*Plathymenia reticulata* (Leguminosae)	HW
7,8,3′,4′-tetrahydroxy (iso-okanin)	*Acacia* spp. (Leguminosae)	HW
7,3′,4′,5′-tetrahydroxy (robtin)	*Robinia pseudacacia* (Leguminosae)	HW
C-Methyl flavanones		
5,7-dihydroxy-6-*C*-methyl (strobopinin)	*Pinus* subg. *Haploxylon* (Pinaceae)	HW
5-hydroxy-7-methoxy-6-*C*-methyl (cryptostrobin)	*Pinus* subg. *Haploxylon* (Pinaceae)	HW
5-hydroxy-7-methoxy-6,8-di-*C*-methyl (desmethoxymatteucinol)	*Pinus krempfii* (Pinaceae)	HW
8-*C*- methylnaringenin	*Qualea labouriauana* (Vochysiaceae)	W
6,8-di-*C*-methyleriodictyol 3′-methyl ether	*Qualea paraensis* (Vochysiaceae)	W
Isopentenyl flavanone		
6-prenylpinocembrin	*Chlorophora tinctoria* (Moraceae)	HW
Flavanone glycosides		
pinocembrin 5-glucoside	*Prunus* spp. (Rosaceae)	B
dihydrowogonin 7-glucoside	*Prunus cerasus* (Rosaceae)	B
naringenin 7-glucoside (prunin)	*Prunus* spp. (Rosaceae)	B
naringenin 5-glucoside	*Salix* spp. (Salicaceae)	B
isosakuranetin 7-glucoside	*Prunus* spp. (Rosaceae)	B
sakuranetin 5-glucoside	*Prunus* spp. (Rosaceae)	B
	Juglans nigra (Juglandaceae)	HW, B
persicogenin 5-glucoside	*Prunus* spp. (Rosaceae)	B
5-hydroxy-6,7,3′,4′,5′-pentamethoxy-flavanone 5-rhamnoside	*Cassia renigera* (Leguminosae)	B

[a] B = bark; HW = heartwood; W = wood

Table 7.5.4. Flavanonols (dihydroflavonols) reported to occur in wood or bark

Type Structure (name)	Genus, species (family)	Location[a]
Simple flavanonols		
3.7-dihydroxy	*Platymiscium praecox* (Leguminosae)	W
3,5,7-trihydroxy (pinobanksin)	*Pinus* spp. (Pinaceae)	B
	Platanus vulgaris (Platanaceae)	B, HW
	Tilia platyfolia (Tiliaceae)	W
3,5-dihydroxy-7-methoxy (alpinone)	*Pinus excelsa* (Pinaceae)	HW
	P. clausa (Pinaceae)	
	P. strobus (Pinaceae)	
3-hydroxy-5,7-dimethoxy	*Aniba riparia* (Lauraceae)	W
3,7-dihydroxy-6-methoxy	*Dalbergia ecastophyllum* (Leguminosae)	W
3,7,4′-trihydroxy (garbanzol)	*Toxicodendron succedaneum* (Anacardiaceae)	HW
3,5,7,4′-tetrahydroxy (aromadendrin, dihydrokaempferol)	Many sources incl. *Larix* (Pinaceae)	
	Nothofagus (Fagaceae)	
	Pinus (Pinaceae)	
	Prunus (Rosaceae)	
3,5,4′-trihydroxy-7-methoxy	*Prunus avium* (Rosaceae)	HW
3,5,7-trihydroxy-4′-methoxy	*Prunus domestica* (Rosaceae)	HW
3,5-dihydroxy-7,4′-dimethoxy	*Cephalanthus spathelliferus* (Rubiaceae)	B
3,7,3′,4′-tetrahydroxy (fustin)	*Acacia* (Leguminosae)	HW
	Mimosa Leguminosae)	HW
	Pistacia (Anacardiaceae)	HW
	Rhus (Anacardiaceae)	HW
3,5,7,3′,4′-pentahydroxy (taxifolin, dihydroquercetin)	Many sources	
3,5,3′,4′-tetrahydroxy-7-methoxy	*Prunus puddum* (Rosaceae)	W
3,7,8,3′,4′-pentahydroxy	*Acacia kempeana* (Leguminosae)	W
(dihydromelanoxetin)	*Albizia adianthifolia* (Leguminosae)	W
3,7,3′,4′-tetrahydroxy-8-methoxy	*Acacia kempeana* (Leguminosae)	W
3,7,3′,4′,5′-pentahydroxy	*Robinia pseudacacia* (Leguminosae)	HW
(dihydrorobinetin)	*Sequoia gigantea* (Taxodiaceae)	HW
3,7,3′,5′-tetrahydroxy-4′-methoxy	*Gliricidia sepum* (Leguminosae)	HW
3,5,7,8,3′,4′-hexahydroxy	*Acacia nigrescens* (Leguminosae)	HW
3,5,7,3′,4′,5′-hexahydroxy	*Pinus* spp. (Pinaceae)	B, W
(dihydromyricetin)	*Adenanthera pavonina* (Leguminosae)	W
3,5,7,4′-tetrahydroxy-3′,5′-dimethoxy	*Soymida febrifuga* (Meliaceae)	HW
C-Methyl flavanonols		
3,5,7-trihydroxy-6-Me (strobobanksin)	*Pinus* subg. *Haploxylon* (Pinaceae)	HW
3,7,4′-trihydroxy-5-methoxy-6-Me	*Qualea labouriauana* (Vochysiaceae)	W
3,5,7,3′,4′-pentahydroxy-6-Me	*Populus deltoides* (Salicaceae)	B
(cedeodarin)	*Cedrus deodara* (Pinaceae)	W
3,5,7,3′,4′,5′-hexahydroxy-6-Me (cedrin)	*C. deodara* (Pinaceae)	W
Flavanonol glycosides		
dihydrokaempferol 3-rhamnoside	*Nothofagus* spp. (Fagaceae)	HW
	Engelhardtia formosana (Juglandaceae)	B
dihydroquercetin 3-rhamnoside	*Quintinnia serrata* (Saxifragaceae)	B
dihydroquercetin 3′-glucoside	*Larix* spp. (Pinaceae)	B
	Pseudotsuga spp. (Pinaceae)	B
cedrin 3′-glucoside	*Cedrus deodara* (Pinaceae)	W

[a] B = bark; HW = heartwood; W = wood

flavanonols are often found together in the same wood extract. Flavanonols are probably more widespread than flavanones, and dihydroquercetin, in particular, has been recorded from a multitude of sources. Dihydrokaempferol and dihydromyricetin are also quite common.

Three *C*-methylflavanonols have been reported (Table 7.5.4), namely the 6-methyl derivatives of pinobanksin, dihydroquercetin, and dihydromyricetin. Additionally, the 8-methyl isomer of dihydroquercetin, named deodarin, was thought to occur in cedar wood, *Cedrus deodara*. This could not be confirmed by later workers, who found instead the 6-methyl isomer, together with 6-methyl dihydromyricetin and dihydroquercetin itself (2). Only a very few glycosides of flavanonols have been described so far (Table 7.5.4). The rare substitution of glucose at the 3′-position of dihydroquercetin is distinctive, and this compound is characteristically present in many members of the Pinaceae (see Sect. 7.5.3.2).

7.5.2.3 Chalcones and Aurones

The two classes of yellow phenolic anthochlor pigments – the chalcones and aurones – are of restricted distribution in the plant kingdom (14, 15). Chalcones are reported from 25 families and aurones from about 10. The two classes are related in that aurones are formed from chalcones by an oxidative process and related chalcone-aurone pairs tend to be found together in the same plant source. The

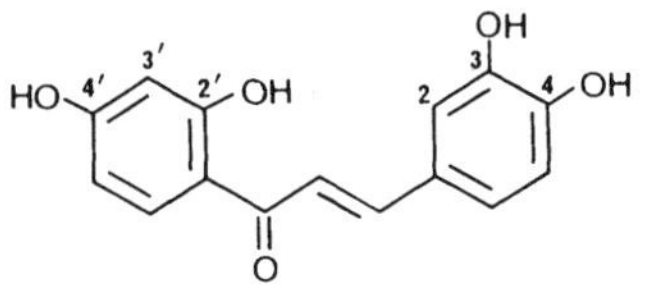

27 butein

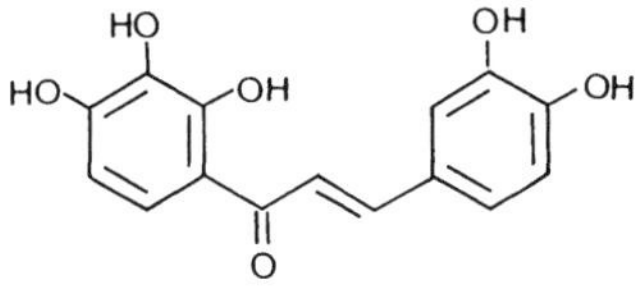

28 okanin

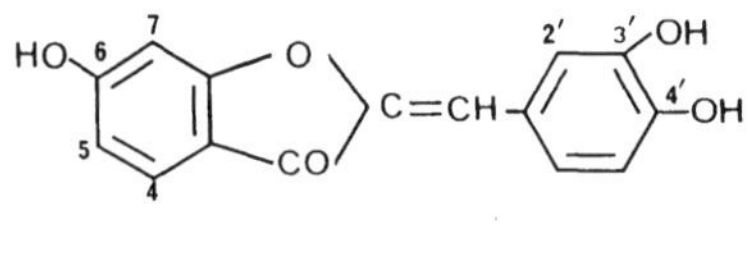

29 sulphuretin

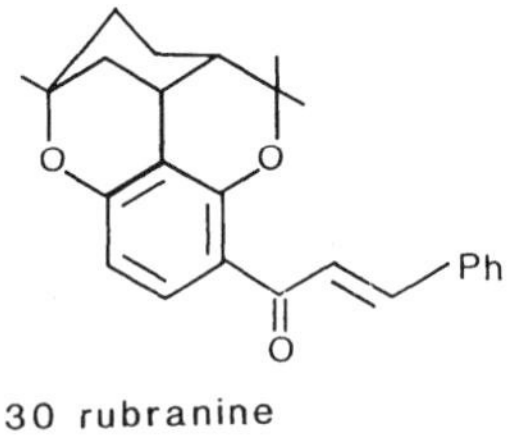

30 rubranine

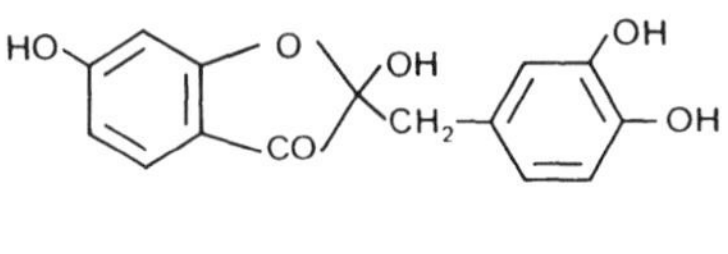

31 2,6,3′,4′-tetrahydroxy-2-benzylcoumaranone

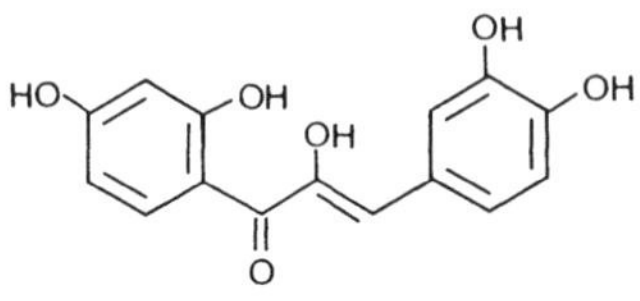

32 2′,4′,3,4,α-pentahydroxychalcone

Table 7.5.5. Chalcones and aurones in wood and bark

Type Structure (name)	Genus, species (family)	Location[1]
Chalcones[2]		
2′-hydroxy-4′,6′-dimethoxy (flavokawin B)	*Aniba riparia* (Lauraceae)	TW
2′,4′,4-trihydroxy (isoliquiritigenin)	*Acacia* spp. (Leguminosae)	HW
	Cyclolobium sp. (Leguminosae)	HW
	Machaerium villosum (Leguminosae)	HW
2-hydroxy-4,6-dimethoxy	*Pinus excelsa* (Pinaceae)	HW
2′,4′,4-trihydroxy-3′-methoxy	*Acacia* spp. (Leguminosae)	HW
2′,4′,3,4-tetrahydroxy (butein)	*Rhus javanica* (Anacardiaceae)	HW
	Acacia spp. (Leguminosae)	HW
2′,4′,3-trihydroxy-4-methoxy (homobutein)	*Acacia* spp. (Leguminosae)	HW
2′,3′,4′,3,4-pentahydroxy (okanin)	*Acacia* spp. (Leguminosae)	HW
	Albizia adianthifolia (Leguminosae)	HW
	Cylicodiscus gabunensis (Leguminosae)	HW
3′,4′,2,4,5-pentahydroxy (neoplathymenin)	*Plathymenia reticulata* (Leguminosae)	HW
2′,4′,3,4,α-pentahydroxy	*Trachylobium verrucosum* (Leguminosae)	HW
2′,4′,3,4,5-pentahydroxy (robtein)	*Robinia pseudacacia* (Leguminosae)	HW
2′,4′-dihydroxy-6′,3,5-trimethoxy	*Prunus cerasus* (Rosaceae)	HW
2′-hydroxy-4′,6′,3,5-tetramethoxy	*P. cerasus* (Rosaceae)	HW
2′-hydroxy-3′,4′,5′,6′,3,4-hexamethoxy	*Macaranga peltata* (Euphorbiaceae)	B
Chalcone glycosides		
butein 2′,4′-diglucoside	*Amphipterigium adstrigens* (Anacardiaceae)	HW
2′,4′,6′,4-tetrahydroxy 2′-glucoside (isosalipurposide)	*Salix* spp. (Salicaceae)	B
2′,4,6′-trihydroxy-4′-methoxy-2′-glucoside (neosaturanin)	*Prunus* spp. (Rosaceae)	HW
isoliquiritigenin 3′-*C*-glucoside	*Cladrastis platycarpa* (Leguminosae)	W
Isoprenoidchalcone		
rubranine (see **30**)	*Aniba rosaedora* (Lauraceae)	W
Aurones[2]		
6,3′,4′-trihydroxy (sulphuretin)	*Rhus* spp. (Anacardiaceae)	HW
4,6,3′,4′-tetrahydroxy (aureusidin) 6,3′,4′-trihydroxy-4-methoxy	*Melanorrhoea aptera* (Anacardiaceae)	HW
Aurone glycoside		
sulphuretin 6,3′-diglucoside	*Rhus* spp. (Anacardiaceae)	HW

[1] B = bark; HW = heartwood; TW = trunkwood; W = wood
[2] Numbering of chalcone and aurone ring systems is indicated in formulae **27** and **29**.

main occurrence of chalcones and aurones are in the floral tissues of members of the Compositae, where they are responsible for the yellow color in certain tribes and genera – e.g. in *Coreopsis*.

As wood constituents, chalcones have been reported as heartwood components in Anacardiaceae, Lauraceae, Leguminosae, Pinaceae, and Rosaceae, and as bark components in Salicaceae. Chalcones usually occur in the *trans*-form, but both *cis*- and *trans*-isomers of okanin (**28**) have been found together in *Cylico-*

discus gabunensis heartwood. In all, some 12 chalcone aglycones and four glycosides have been reported in woody tissues (Table 7.5.5). Rarely, chalcones may have an additional hydroxyl attached to the central C_3-portion of the molecule in the α-position. One such compound, **32**, has been reported from the wood of *Trachylobium*.

Structurally, the only unusual chalcone of woody tissues is rubranine (**30**) from *Aniba rosaedora*, which is a citral-chalcone condensation product. The possibility of rubranine being an artifact of the isolation procedure was considered by Montero and Winternitz (28), who synthesized it by condensing the essential oil citral with pinocembrin in the presence of the alkaloid anibine, which also occurs naturally in *Aniba* wood. However, De Alleluia and coworkers (7) later confirmed that it was a natural product from the evidence that it was optically active, and that it could be isolated by very brief washing of the wood with hexane.

Aurones found in woody tissues are few in number (Table 7.5.5) and are so far restricted to members of the Anacardiaceae. Co-occurring with the aurones in some genera are the related but colorless 2-hydroxy-2-benzylcoumaranones. The sulphuretin analogue (**31**) occurs, for example, in the wood of *Schinopsis* species. The taxonomic significance of aurone occurrence in the Anacardiaceae is considered further in Sect. 7.5.3.3.

7.5.2.4 Isoflavonoids and Neoflavonoids

Isoflavonoids share a common chalcone precursor in their biosynthesis with the flavonoids described earlier and they resemble the flavonoids in many of their properties (e.g. phenolic nature, etc.). The key difference is in the placement of the side phenyl group in the 3- instead of in the 2-position, so that isoflavonoids are all derivatives of 3-phenylchroman. Just as there are flavones, flavanones, and flavans at different levels of oxidation in the central pyran ring, so there are isoflavones, isoflavanones, and isoflavans. There are also several new classes of isoflavonoid not directly represented in the flavonoid series, such as pterocarpans and coumestans. Isoflavones are by far the most numerous group among the isoflavonoids. The neoflavonoids are yet a further group of flavonoid-related oxygen heterocycles that can by formally derived from a 4-phenylchroman skeleton. Related open-chain structures, corresponding to chalcones, are known, as well as compounds that are 1,4-quinones. The neoflavonoids as a group can also be regarded as coumarin-derived – i.e. they are 4-phenylcoumarins or related structures.

In terms of their distribution in the plant kingdom, both the isoflavonoids and the neoflavonoids are much more restricted in their occurrences than other flavonoids. Both groups occur together in the Leguminosae, notably in the genus *Dalbergia* and related members of the tribe Dalbergieae. Otherwise, isoflavonoids as occasional constituents of a few other families, notably the Myristicaceae and Rosaceae, and the neoflavonoids have also been reported in Guttiferae, Passifloraceae, and Rubiaceae.

Recent reviews of the isoflavonoids include those of Wong (in (15)), Dewick (in (14)), and Ingham (21). According to the latter two reviews, the number of known structures is now at least 630, and what is significant here is that about half these compounds have been isolated from woody tissues of the Leguminosae. To list all these structures in the space available here is not possible. Fortunately, the excellent listing of known isoflavonoids up to 1981 by Ingham (21) has the particular merit of indicating the part of the plant from which the different structures have been obtained. Here we can give only an indication of the range of structural variants that have been encountered in wood and bark of trees. Similarly, we provide an abbreviated review of the known naturally occurring neoflavonoids isolated from trees.

Among the simple hydroxyisoflavones of legume woods, there are two main series, those with and those without 5-hydroxylation. Two of the commonest heartwood isoflavones are genistein (with a 5-hydroxyl) (**34**) and daidzein (without) (**33**). Genistein or 5,7,4′-trihydroxyisoflavone is found, for example, in the heartwood of *Laburnum alpinum*, the sapwood of L. *anagyroides*, in the trunkwood and trunkbark of *Erythrina cristo-gallii* and in twigs of *Piptanthus nepalensis*. Daidzein is similarly recorded in a variety of sources, including heartwood of *Dalbergia stevensonii* and of *Machaerium villosum*. All three possible monomethyl ethers of genistein are known: the 5-methyl ether (isoprunetin) in *Ormosia excelsa* wood, the 4′-methyl ether (biochanin A) in *Andira parviflora*

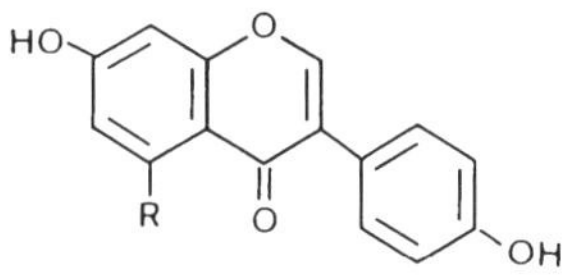

33 daidzein, R=H
34 genistein, R=OH

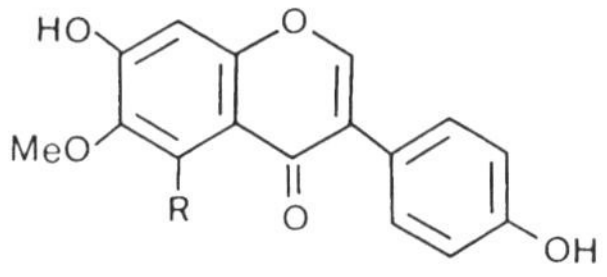

35 glycitein, R=H
36 tectorigenin, R=OH

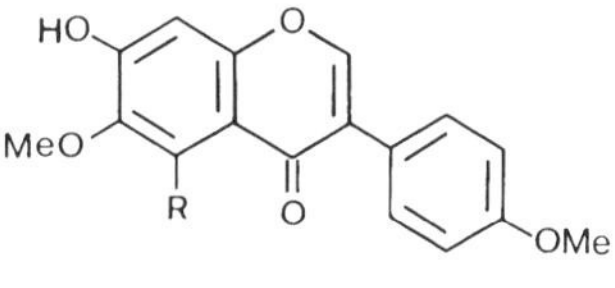

37 afrormosin, R=H
38 5-methoxyafrormosin, R=OMe

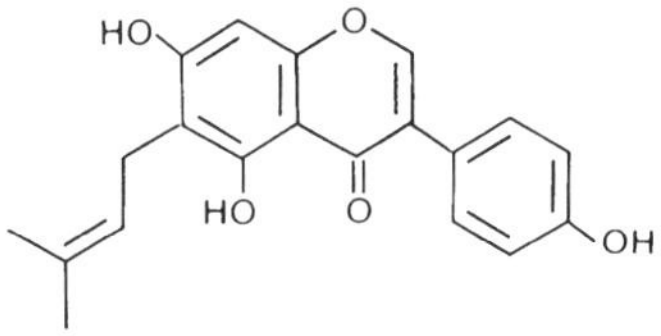

39 wighteone

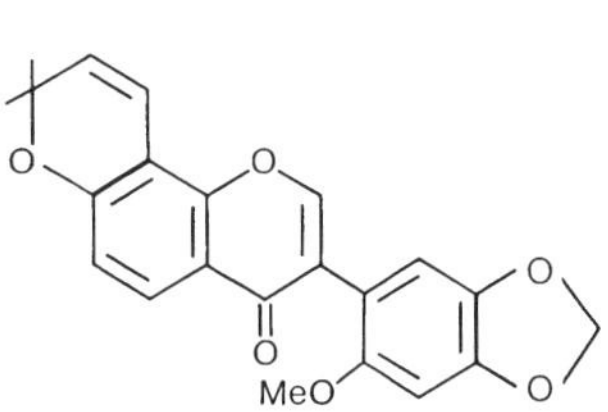

40 jamaicin

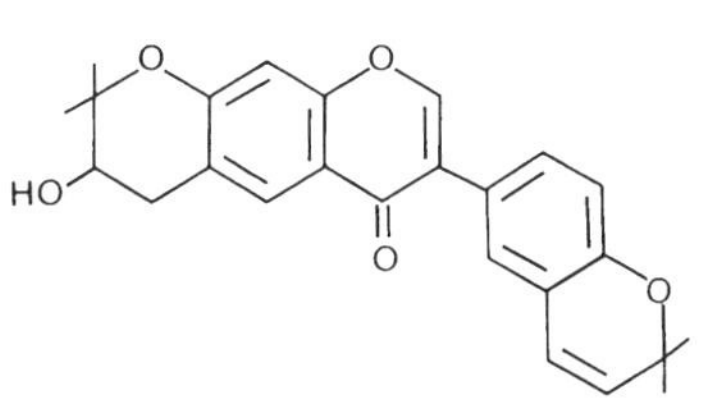

41 mundulone

trunkwood, and the 7-methyl ether (prunetin) in *Pterocarpus angolensis* heartwood. Prunetin is one of the few isoflavones also to occur outside the Leguminosae, since it is found in wood and bark of several *Prunus* species (Rosaceae). A common structural substitution besides *O*-methylation is the methylenedioxy bridge between adjacent phenolic groups; the simplest example among isoflavones is pseudobaptigenin (7-hydroxy-3′,4′-methylenedioxyisoflavone), known to occur in the heartwood of several *Dalbergia* species.

Most other heartwood isoflavones are derived from either genistein or daidzein by addition of extra hydroxyl (and/or methoxyl) groups to either aromatic ring. Four of many possible examples are glycitein (6-methoxydaidzein) (**35**) from heartwood of *Mildbraedeodendron excelsa*, tectorigenin (6-methoxygenistein) (**36**) from heartwood of *Dalbergia riparia*, afrormosin (6-methoxydaidzein 4′-methyl ether) (**37**) first isolated from heartwood of *Afrormosia elata* and subsequently found in eight related legume genera, and 5-methoxyafrormosin (**38**) from bark, heartwood, and sapwood of *Cladrastis platycarpa*. One might also mention here 2′-hydroxygenistein 4′-methyl ether or 2′-hydroxybiochanin A, which occurs uniquely outside the Leguminosae, in the trunkwood of two *Virola* species, *V. caducifolia* and *V. multinerva* (Myristicaceae). Three other isoflavones in *Virola* wood are formononetin, biochanin A, and 2′-methoxybiochanin A. Finally, there are several isoflavones in legume woods that have isopentenyl substituents. Three typical examples are wighteone (**39**) from the bark of *Erythrina variegata*, jamaicin, **40**, from the bark of *Piscidia erythrina*, and mundulone, **41**, from the bark of *Mundulea sericea*. The known isoprenylated isoflavones, of which there are over 50, are rarely recorded in heartwood, being found more usually in bark (as in the above examples), in seeds, or in roots.

Other related isoflavonoid types are known that may co-occur with one or more isoflavones in legume heartwoods. Related pterocarpans and isoflavans, for example, accompany isoflavones in the wood of various *Dalbergia* and *Machaerium* species (see Sect. 7.5.3.3). Again, the isoflavone quinone bowdichione (**42**) and the pterocarpan homopterocarpin, (**46**) occur with the isoflavones genistein (**34**) and calycosin (3′-hydroformononetin) (**111**) in the heartwood of *Bowdichia nitida*.

Bowdichione (**42**) is colored, unlike most isoflavonoids, which are colorless. It is also unique in that it remains the only quinonoid derivative in this series. More numerous are the isoflavanones, of which 39 structures are known. Only some of these occur in woody tissues, such as dihydrodaidzein (**43**) in *Pericopsis montana* and dalbergioidin (**44**) in *Lespedeza cyrtobotrya*. Isoflavanones have an optically active center at C-3, as do the pterocarpens (e.g. **45**–**48**). Racemization may occur during isolation so that optically inactive forms are sometimes described in the literature. In fact, the absolute configurations of most isoflavanones and pterocarpans have still to be determined.

Although there are optically active centers at both the 6a and 11a positions in pterocarpans, considerations of steric hindrance mean that only two of the four possible configurations are likely in nature, namely the 6a*R*:11a*R* and the 6a*S*:11a*S* structures. Note that the pterocarpan numbering differs from that in the isoflavones or isoflavanones.

There are two series of pterocarpans in legume woods, those without and those with a hydroxyl at C-6a. Respective examples are medicarpin (**45**) from the

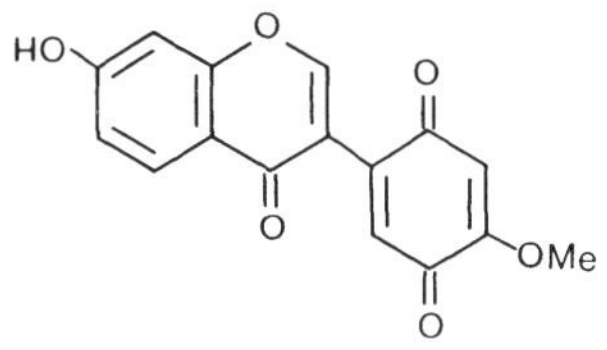

42 bowdichione

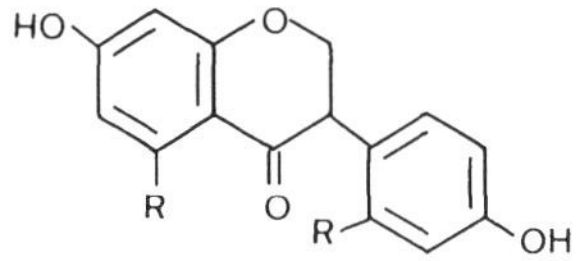

43 dihydrodaidzein, R=H
44 dalbergioidin, R=OH

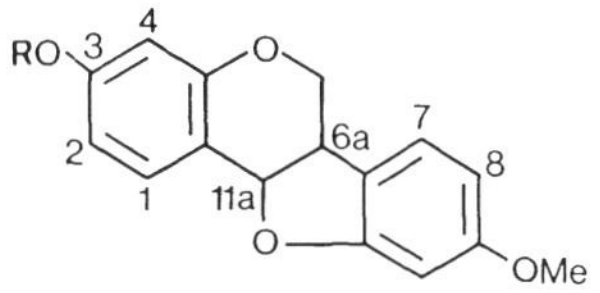

45 medicarpin, R=H
46 homopterocarpin, R=Me

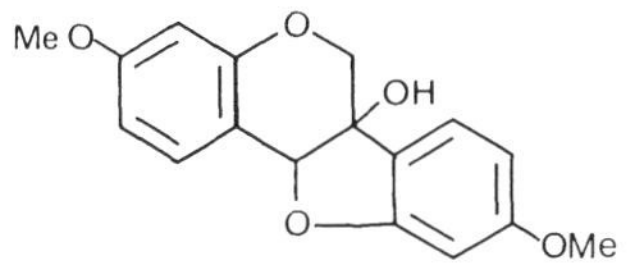

47 variabilin

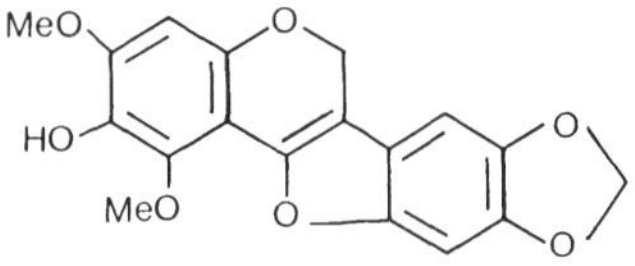

48 leiocalycin

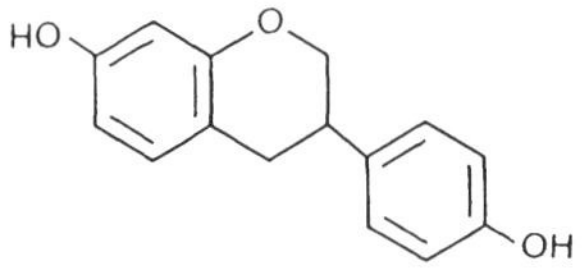

49 equol

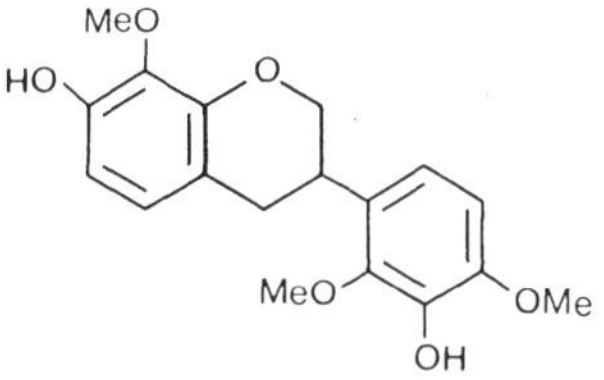

50 (−)-duartin

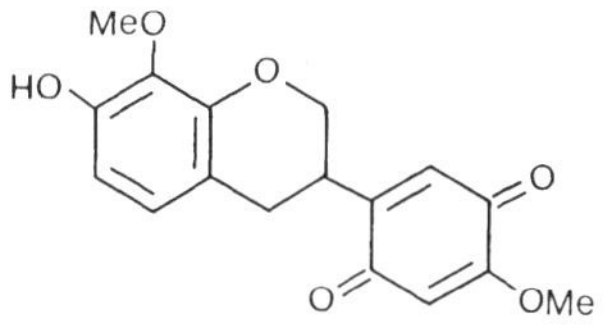

51 mucroquinone

trunkwood of *Aldina heterophylla* and variabilin (**47**) from the wood of *Dalbergia variabilis.* Loss of a molecule of water across the 6a-11a bond in 6a-hydroxypterocarpans produces the related pterocarpene. Twelve such compounds are known, mainly from heartwoods, one example being leiocalycin (**48**) from *Swartzia leiocalycina.*

While flavans are quite rare in nature, the isomeric isoflavans are more numerous (there are 39 of them), and they occur fairly widely, at least in members of the Leguminosae. Isoflavans are optically active and compounds with both the *R*- and *S*- configuration at C-3 have been described. Two typical isoflavans are equol (**49**) which occurs in the heartwood of *Millettia pendula* but is better known as a urinary metabolite in mammals of the isoflavone formononetin (**109**) and (−)-duartin (**50**) present in the heartwood of several *Machaerium* species. Isoflavans (and pterocarpans) occur in a different context than wood chemistry in the

Leguminosae; they are also commonly formed in leaf tissue as phytoalexins, in response to microbial infection (21). Two structural types closely related to the isoflavans are the isoflavan quinones and the isoflavenes. Examples are mucroquinone **(51)** from woods of *Machaerium mucronulatum* and *Cyclolobium clausseni*, and haginins B and D **(52a, 52b)** from wood of *Lespedeza cyrtobotrya.*

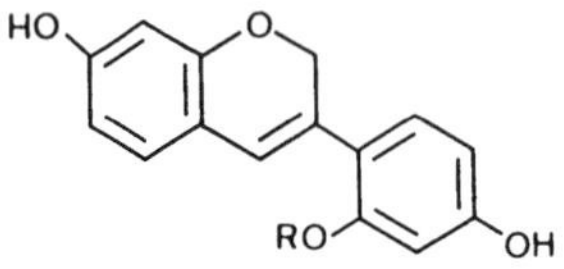

52a haginin B, R=Me

52b haginin D, R=H

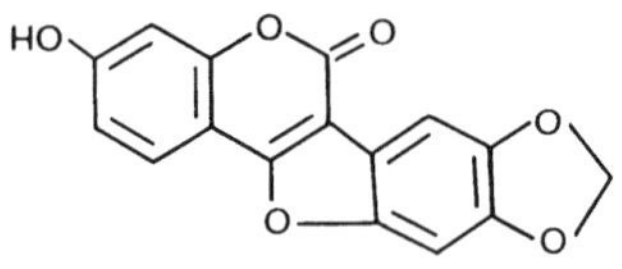

53 medicagol

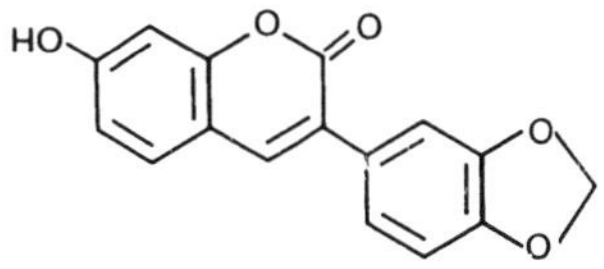

54

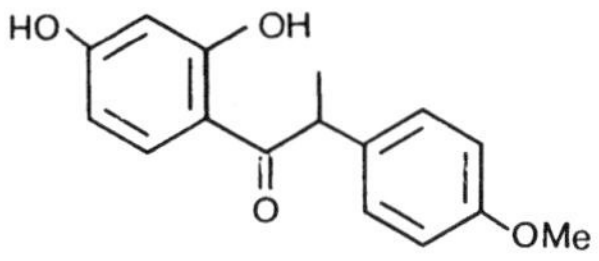

55 angolensin

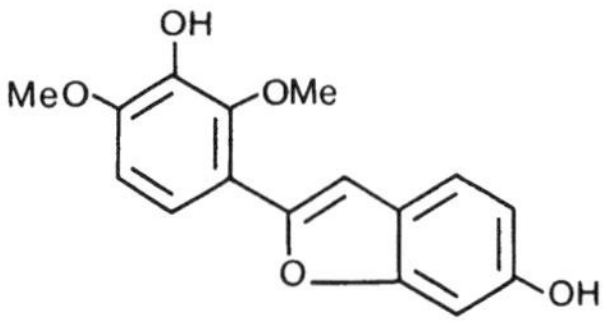

56 pterofuran

The coumestans and 3-arylcoumarins are two further structural types, which can be seen as the 2-keto derivatives of pterocarpenes and isoflavenes, respectively. Again, they are found occasionally in heartwoods, accompanying other isoflavonoid types. Examples are the coumestan medicagol **(53)** and the 3-arylcoumarin, 7-hydroxy-3′,4′-methylenedioxy-3-phenylcoumarin **(54)**, which occur together in heartwood of *Dalbergia oliveri.* Two more structural types, found rarely in legume woods, are α-methyldeoxybenzoins, an open-chain form of isoflavanone, and 2-arylbenzofurans, which are formally related although their skeletons have one less carbon atom than in other isoflavonoids. Examples are angolensin **(55)** and pterofuran **(56)** both from the heartwood of *Pterocarpus indicus.*

The insecticidal rotenoids, which are another class of isoflavonoid characteristically occurring in the Leguminosae, have yet to be reported from any heartwood. Rotenoids are almost entirely formed in roots or seeds of more shrubby members of the family – e.g. members of the genus *Derris*. They occasionally appear in root bark or stem bark extracts.

Isoflavone *O*-glycosides are also not usually encountered in heartwood extracts, but they do appear in bark extracts. Thus daidzein 7-glucoside has been reported from bark of *Erythrina crista-galli*, formononetin 7-glucoside from bark of *Wisteria floribunda*, and biochanin A 7-glucoside from bark of *Dalbergia paniculata*. Isoflavone *C*-glycosides are rare, compared to *O*-glycosides, but they

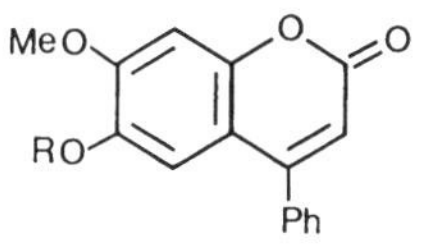

57 dalbergin, R=H
58 methyldalbergin, R=Me

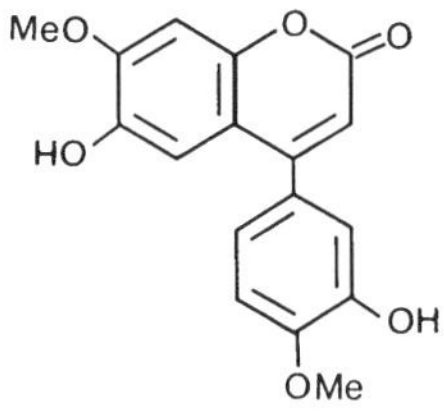

59 melannein

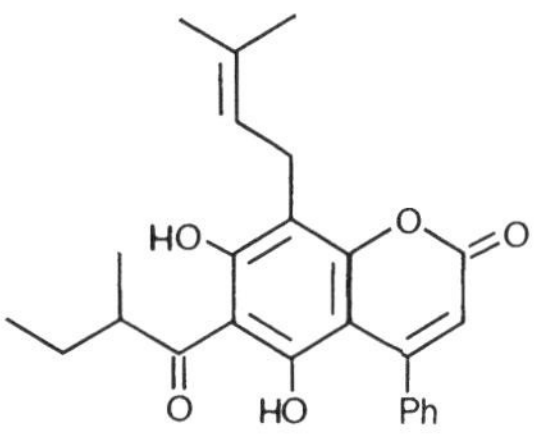

60 mammea A/AB

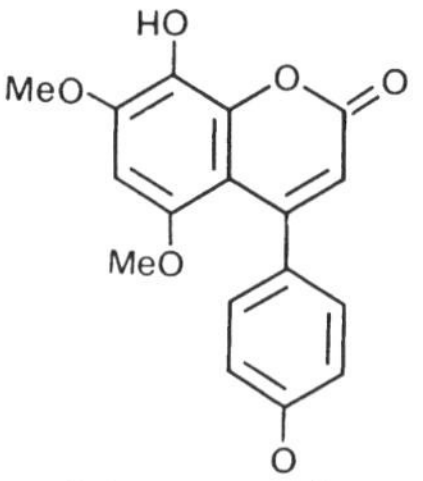

61 exostemin

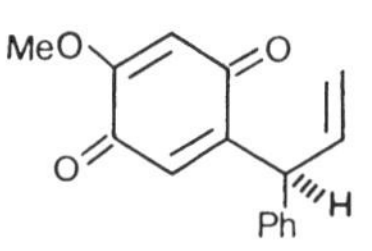

62a (S)-4-methoxy-dalbergione

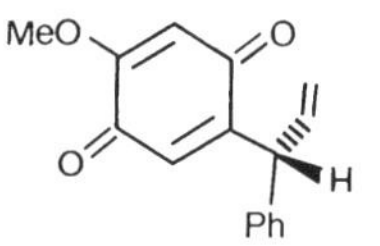

62b (R)-4-methoxy-dalbergione

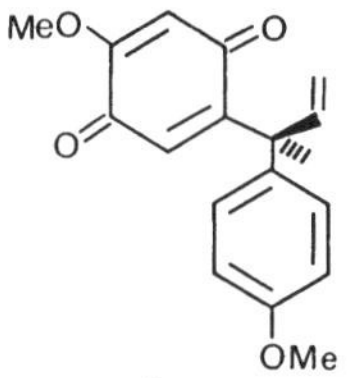

63 (S)-4,4′-dimethoxydalbergione

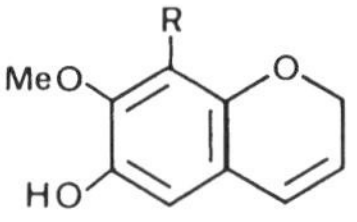

neoflavenes
64a R=H
64b R=OMe

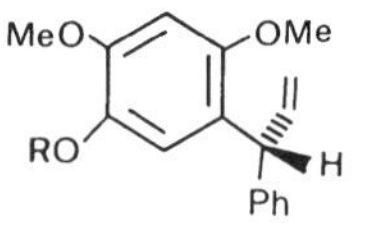

65 latifolin, R=H
66 5-methyllatifolin, R=Me

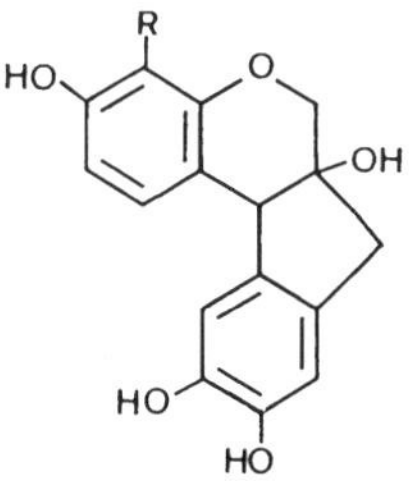

67 brazilin, R=H
68 haematoxylin, R=OH

have been found in the barks of *Dalbergia* species – e.g. prunetin 8-*C*-glucoside from *D. paniculata* and orobol (5,7,3′,4′-tetrahydroxyisoflavone) 8-*C*-glucoside from *D. nitidula*.

Representative of those naturally occurring neoflavonoids that are found in woody tissues are structures **57** through **68**. The largest group are probably the 4-arylcoumarins. The simplest compounds of this class occur widely in heartwood, sapwood, or bark of *Dalbergia* and *Machaerium* species, examples being dalbergin (**57**) and methyldalbergin (**58**) in which the 4-phenyl is unsubstituted, and melannein (**59**) from *D. melanoxylon*, in which there is hydroxymethoxy substitution in the side phenyl group. There is a single report of a 4-phenylcoumarin in the unrelated Rubiaceae, with the occurrence of exostemin in wood of *Exostemma caribaerum*. Finally, there are some isoprenylated 4-arylcoumarins known in the bark and other tissues of *Mammea*, *Mesua*, and *Calophyllum* species of the Guttiferae. One typical structure here is mammea A/AB (**60**), which occurs in the bark of *Mammea africana*.

Three other classes of neoflavonoid have been reported in legume woods, along with the 4-phenylcoumarins; these are the dalbergiones, the dalbergiquinols, and the neoflavenes. The dalbergiones, which are open-chain-substituted benzoquinones, exist in different isomeric forms. 4-Methoxydalbergione (**62**), for example, has been isolated in both the *R* and *S* configurations. More interestingly, (*R*)-4-methoxydalbergione (**62b**) and (*S*)-4,4′-dimethoxydalbergione (**63**) are present in *Dalbergia nigra* wood as a quasi-racemic mixture. Most of the related dalbergiquinols – e.g. latifolin and its 5-methyl ether (**65**, **66**) are known only in the *R*-form. Isomerization of dalbergiones in vivo can give rise to neoflavenes, but so far only two compounds of this type are known: 6-hydroxy-7-methoxyneoflavene (**64a**) from bark of *D. latifolia* and *D. sisoo* and the related 8-methoxy compound (**64b**) from heartwood of *M. nictitans*, *M. kuhlmannii* and *M. pedicellatum*.

A few other wood constituents with related structures have been included under the general heading of neoflavonoid (see, for example, Donnelly (8, 15), but the only other two compounds that must be mentioned here are the 4-arylchromans, brazilin (**67**) and haematoxylin (**68**). These compounds have long been studied, brazilin for example first having been isolated from brazil wood, *Caesalpinia echinata*, by Chevreul in 1808. Details of the structural analysis and synthesis of these two well known dyeing materials were given in a review by Robinson in 1962 (31). Haematoxylin is an extractive of logwood, *Haematoxylon campeachianum*, a tree in the same tribe as *Caesalpinia*.

7.5.3 Distribution

7.5.3.1 Distribution within the Plant

Flavonoids, when present in trees, are likely to be found equally in sapwood, heartwood, and bark, but the amounts in the different tissues may vary considerably. Heartwood is usually a better source than sapwood, the amounts present reaching up to 0.5% to 1% dry weight. Bark tissues may contain larger amounts.

For example, Douglas-fir bark, *Pseudotsuga menziesii*, contains up to 5% dihydroquercetin (**85**) whereas bark of *Pinus contorta* contains between 0.7% and 2% myricetin (**83**). Angiosperm barks seem to have rather lesser amounts; the recovery of quercetin (**8**) from bark or *Quercus tinctoria* of *Q. phellus* does not usually exceed 0.5%.

69 chrysin, R=H
70 tectochrysin, R=Me

71 pinocembrin, R=H
72 pinostrobin, R=Me

73 pinobanksin, R_1=R_2=H
74 alpinone R_1=Me, R_2=H
75 strobobanksin, R_1=R_2=Me

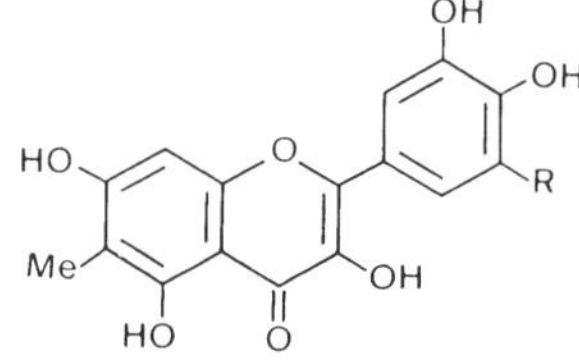

76 strobopinin R_1=R_2=H
77 cryptostrobin R_1=Me,R_2=H
78 6,8-dimethylpinocembrin, R_1=R_2=Me

20 pinoquercetin, R=H
79 pinomyricetin, R=OH

80 galangin, R=H
81 izalpinin, R=Me

82

8 quercetin, R=H
83 myricetin, R=OH

84 dihydrokaempferol, R_1 R_2=H
85 dihydroquercetin, R_1=OH, R_2=H
86 dihydromyricetin, R_1=R_2=OH

The compounds found in the different parts of the wood also show significant differences in those cases in which comparisons have been made. While the pattern in the heartwood usually mirrors that in the sapwood, that of the bark often diverges. In the genus *Morus*, for example, the heartwood contains mixtures of dihydroflavonols (up to 0.03% dry wt) and flavonols (about 0.01% total), while the bark contains isopentenylflavones in yield up to 0.2% total (38). In *Prunus*, there is some overlap between heartwood and bark components, but the heartwood is generally richer in structural types. In *Prunus avium*, for example, there are 21 flavonoids recorded in the heartwood, only two of which – dihydrokaempferol and dihydroquercetin (**84**, **85**) – are also present in the bark. On the other hand, *Prunus avium* bark contains the dihydroflavonol pinobanksin (**73**), which is not found in the heartwood. Again in the cedar tree, *Cedrus deodara*, the wood is known to contain dihydroquercetin and its 3′-glucoside, cedeodarin, cedrin, dihydromyricetin (**86**) and cedrinoside, whereas the bark has only dihydroquercetin and cedeodarin. Another important difference between heartwood and bark constituents, which has already been mentioned, is that glycosides are much more likely to be found in bark than in heartwood.

Heartwood flavonoids are usually present in the free state, though it is possible that they may be loosely associated with other wood components. In the case of *Artocarpus* heartwood, it appears that at least some of the flavonol morin is present as a calcium chalate (29). Since most heartwood flavonoids have the ability to chelate metals, this may happen in other cases too.

7.5.3.2 Patterns in Gymnosperm Woods

The available data on flavonoid distribution in the wood of gymnosperms are summarized in Table 7.5.6. As can be seen, the overall pattern is fairly similar in that dihydroflavonols are of regular occurrence throughout. The 3′-glucoside of dihydrokaempferol is regularly present within the Pinales, but, because it is not known in any other gymnosperm group, it may be a marker for the Pinales. The occurrence of isoflavones in the wood is very rare, the only report being in *Podocarpus spicatus*. Whether other *Podocarpus* species also contain isoflavones in the wood is yet to be determined.

The single genus that has been thoroughly investigated for its heartwood and bark flavonoids is *Pinus*. Flavones, flavanones, and flavanonols have been reported widely in heartwood and there is a clear-cut correlation between flavonoid pattern and subgeneric classification (9). Thus species of the subgenus *Haploxylon* have a complex mixture of flavones (e.g. chrysrin (**69**)), flavanones (e.g. pinocembrin (**71**)) and flavanonols (e.g. pinobanksin (**73**)). By contrast, species of the *Diploxylon* subgenus lack most of these constituents, containing only the flavanone, namely pinocembrin, and the flavanonol pinobanksin. The situation is slightly more complex within the *Haploxylon* group in that there are further minor variations that correlate with subsectional classification. In particular, three components, strobochrysin, strobopinin (**76**), and strobobanksin (**75**), are confined to the two subsections *Strobi* and *Gerardianae*. Finally, two species of subsection *Gerardianae* do not conform with the other *Haploxylon* pines since they lack flavones that are otherwise so characteristic of the subgenus.

Certain individual pine species have additional components superimposed on the patterns already discussed. Thus the heartwood of *Pinus griffithii* contains a chalcone with 2-hydroxy-4,6-dimethoxy substitution (**82**) as well as the flavonols galangin (**80**) and izalpinin (**81**). These latter two are related to the flavanonols pinobanksin and alpinone (**26**). Again, *P. krempfii* (also classified as *Ducampopinus krempfii*) contains a further flavanone 6,8-dimethylpinocembrin (**78**) in addition to the other *Haploxylon* constituents (10). Morphologically, this species

Table 7.5.6. Flavonoid patterns in gymnospermous woods

Order Genus, species	Location[1]	Compounds reported
Pinales		
Abies amabilis } *A. concolor* }	HW, B	dihydrokaempferol and its 3′-glucoside
Cedrus atlantica	W, B	dihydrokaempferol and its 3′-glucoside
C. deodara	W	dihydrokaempferol and its 3′-glucoside, cedeodarin, cedrin, cedrinoside and dihydromyricetin
Larix (8 spp.)	HW	dihydrokaempferol, dihydroquercetin
Picea sitchensis	HW, B	dihydrokaempferol and its 3′-glucoside
Pinus subgenus *Haploxylon*[2] (17 spp.)	HW[3]	pinocembrin, pinobanksin, chrysin, tectochrysin, strobobanksin, strobopinin, cryptostrobin
Pinus subgenus *Diploxylon* (35 spp.)	HW	pinocembrin, pinobanksin
Pseudotsuga spp.	HW, B	dihydrokaempferol, dihydroquercetin, quercetin, pinocembrin, pinobanksin
Tsuga heterophylla	W, B	dihydrokaempferol 3′-glucoside
Podocarpales		
Podocarpus spicatus	W	quercetin, genistein, podospicatin
Cupressales		
Biota	W	dihydrokaempferol, dihydroquercetin
Libocedrus	W	dihydroquercetin
Sequoia	W, B	dihydroquercetin
Taxodium	B	dihydroquercetin
Thuja	B	dihydroquercetin
Taxales		
Taxus baccata	B	quercetin, myricetin

[1] B = bark; HW = heartwood; W = wood
[2] *P. griffithii* HW also contains 2-hydroxy, 4,6-dimethoxychalcone, galangin, and izalpinin
[3] Barks of *Pinus* spp. contain the same flavones as heartwood in lesser amount; exceptionally, pinoquercetin and pinomyricetin occur in *P. ponderosa* bark

is rather anomalous, but the chemical resemblance in heartwood flavonoids places it unambiguously in the genus *Pinus*.

Other tissues of *Pinus* trees also contain flavonoids though usually in lesser quantities. Erdtman (9) writes "small amounts of the heartwood constituents have been found in the cambium and sapwood of many pine species ... and it is obvious that the heartwood constituents are formed in the cambium and transported to the heartwood via the rays of the sapwood." Bark constituents, by contrast, differ somewhat from those of the wood so that there must be some separation at the cambium in the pathways of biosynthesis. Pine barks, for example, may contain the common dihydroflavonols and flavonols in addition to the flavanones and flavanonols of the wood. *Pinus contorta* bark, for example, contains sufficient myricetin to make it a useful source of this flavonol, while *P. pinaster* bark is a good source of quercetin. In addition, bark of *Pinus ponderosa* contains two *C*-methylated flavonols, pinoquercetin **(20)** and pinomyricetin **(79)**, that have not been reported anywhere else.

Much less is known of the comparative biochemistry of the wood flavonoids in other members of the Pinales. The wood and bark of two of the four known

Cedrus species have been investigated recently; the deodar tree *Cedrus deodara* has yielded a richer mixture of dihydroflavonols than has the atlas cedar *C. atlantica* (Table 7.5.6) (2). Needle constituents of these two species, as well as of *C. libani*, have also been examined, and there are again considerable differences between species.

The flavonoids in needles of the Pinales have been more extensively surveyed than those of the wood or bark (25, 30); generally there is little correspondence between needle and wood patterns. More highly oxidized flavonoids predominate in the green tissues, with myricetin, quercetin, and kaempferol glycosides being common, accompanied in *Abies*, *Larix*, *Keeteleeria* and *Tsuga* by glycosylflavones. Dihydroflavonols have been occasionally recorded in needles – e.g. in *Pinus* – whereas dihydroquercetin 3′-glucoside, the most characteristic of the heartwood constituents, has been reported in needles of *Picea sitchensis*, *Pseudotsuga menziesii*, and *Tsuga heterophylla* (19). There are two interesting parallels in *Pinus* between wood and needle components. 6-*C*-Methylation, which is such a common feature of the heartwood flavanones, is represented in the leaf needle chemistry of *Pinus contorta* in the synthesis of 6-*C*-methylkaempferol 3-glucoside (20). Again, 8-*C*-methylation, a minor feature of the wood chemistry (cf., 6,8-dimethylpinocembrin (**78**)) is represented in the needle chemistry of *P. sylvestris* by the production of sylpin (5,6,4′-trihydroxy-3-methoxy-8-*C*-methylflavone) (27).

Outside the Pinales, little comparative work has been attempted on heartwood flavonoids (see Table 7.5.6). Only limited information is available; the large group of trees within the Cupressales, for example, have hardly been studied at all. The only point of note is the apparent absence of biflavonoids from the heartwood on any conifer. Biflavonoids are so characteristic in leaf extracts and they have been recorded now in leaves of every gymnosperm order, apart from the Pinales (Geiger and Quinn, in (14)).

7.5.3.3 Patterns in Angiosperm Woods

Of the 400 or more angiosperm families recognized by taxonomists, some 60 are more or less completely arborescent. However, many other families have woody as well as herbaceous members. Bate-Smith, in a limited survey of the angiosperms, listed over 68 families with a primitive 'woody' phenolic pattern in the leaves (4). Leaf flavonoids are more widely investigated generally than those of the wood. Considering only the 60 arborescent families, our knowledge of flavonoid patterns in the wood and bark is quite fragmentary. For only about two families – the Anacardiaceae and Leguminosae – are there sufficient data from which patterns can be discerned. More limited information is available for a few other groups such as the Fagaceae (*Quercus*), Lauraceae (*Aniba*), Moraceae (*Artocarpus*, *Morus*), Myrtaceae (*Eucalyptus*), Rhamnaceae (*Rhamnus*), Rosaceae (*Prunus*) and Rutaceae (*Casimiroa*, *Melicope*).

With such a limited data base, it is difficult to draw too many comparisons between flavonoid patterns in angiosperms vis-à-vis the gymnosperms. However, there is a general trend in flavonoid synthesis within the vascular plants toward complexity and elaboration of the basic C_{15} flavonoid skeleton, and this can be observed in the wood constituents. Angiosperm woods tend to contain more complex flavonoids than gymnosperm woods. An elaborate series of isopentenyl-substituted flavones (**11** through **15**) has been uncovered, for example, in the bark of *Artocarpus* species (Moraceae). Again, in the Leguminosae tree extractives, additional oxygenation of both A- and B-rings of the flavones and flavonols is common; for example, a series of ten highly substituted flavonols have been encountered in the heartwood of *Apuleia leiocarpa* (see Sect. 7.5.3.3.2).

Another sign of biosynthetic complexity is the accumulation of increasing numbers of flavonoid types in a given heartwood. In gymnosperm woods, dihydroflavonols are the main class of flavonoid, occasionally accompanied by flavanones. In angiosperm woods, more highly oxidized types, notably flavonols, accompany the reduced skeletons. Thus in heartwoods of both Anacardiaceae and Leguminosae, it is not uncommon to find representatives of flavones, flavonols, dihydroflavonols, flavanones, and chalcones occurring together. In Leguminosae, an additional group of oxygen heterocycles, the isoflavonoids, are regular wood constituents.

In the gymnosperms, isoflavones have so far only been recorded in wood of *Podocarpus spicata* (Podocarpaceae). They have actually been reported also in *Juniperus macropoda* (Cupressaceae), but in the leaves, not in the wood. In the angiosperms, by contrast, isoflavones are widely present in legume woods, having been recorded in about 100 genera. In addition, they have been observed in woody tissues of *Prunus* (Rosaceae) and of *Osteophleum* and *Virola* (Myristicaceae). Furthermore, the isoflavone character is not specifically associated with the woody habit in angiosperms, since there are recent records of isoflavones being found in a herbaceous family, the Compositae, namely in the genus *Wyethia* (21).

In this account of angiosperm wood flavonoids, attention will be focused on the two families that have been the subject of detailed investigation.

7.5.3.3.1 Heartwood Flavonoids of the Anacardiaceae

The Anacardiaceae, containing some 600 species and 77 genera of trees, shrubs and lianas, is mainly tropical and subtropical in habitat. Familiar members of economic importance include the pistachio, *Pistacia vera*, the mango, *Mangifera indica*, and the quebracho tree, *Schinopsis quebracho-colorada*. Wood flavonoids have now been investigated in representatives of about 17 genera. Most attention was given in the earlier literature to the genera *Rhus* and *Schinopsis*. *Rhus succedeana* (≡ *Toxicodendron succedeana*) has been intensively investigated and fisetin (**16**), fisetin 7-glucoside, fustin (**93**), garbanzol (**92**), sulphuretin (**29**), and 3,7,4′-trihydroxyflavone (**88**) have been reported (24). The yellow coloring matter of young fustic wood, *Rhus cotinus* (≡ *Cotinus coggyria*), was identified as sulphuretin (**29**) and another aurone, rengasin (**87**) was observed in *Melannorrhoea* (23). Additionally, the quebracho tree heartwood was found to contain three dihydroflavonol-flavonol pairs: fustin (**93**) and fisetin (**16**), fustin 4′-methyl ether and fisetin 4′-methyl ether (**90**), dihydrorobinetin and robinetin (**89**).

Already from these varied investigations, a distinctive chemistry was apparent in the heartwood of these trees (**16, 27, 29, 87–93**). Three notable features of flavonoid synthesis are: presence of 5-deoxyflavonoids, a characteristic also of members of the Leguminosae; presence of aurones, otherwise rare as wood constituents; and presence of several related dihydroflavonol-flavonol pairs.

More recent investigations have centered on the use of heartwood flavonoids to illuminate infrageneric relationships in *Rhus* (41) and to assist in the reclassification of the family Julianaceae as part of the Anacardiaceae (40).

Heartwoods of 23 taxa of *Rhus* were screened and 14 flavonoids detected. All uniformly contained eight deoxyflavonoids: sulphuretin (**29**) and its 6,3′-diglucoside, fisetin (**16**), butein (**27**), 3,7,4′-trihydroxyflavone (**88**), 3′,4′-dihydroxyflavone (**91**), fustin (**93**), and garbanzol (**92**). Among the new compounds reported is

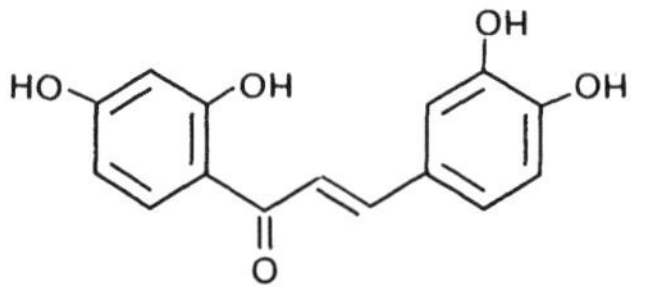

chalcone : 27 butein

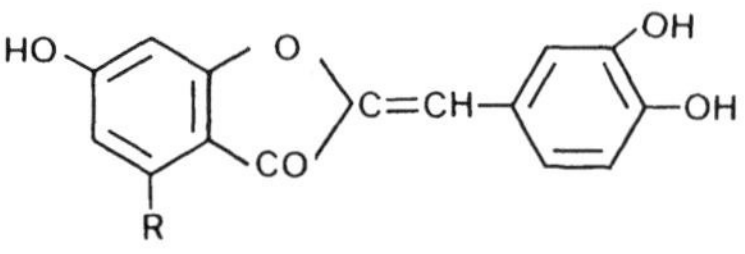

aurones : 29 sulphuretin, R=H
87 rengasin, R=OMe

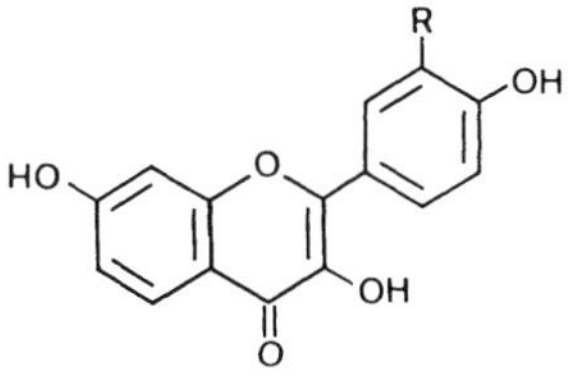

Flavonols : 88 3,7,4′ triOH, R=H
16 fisetin , R=OH

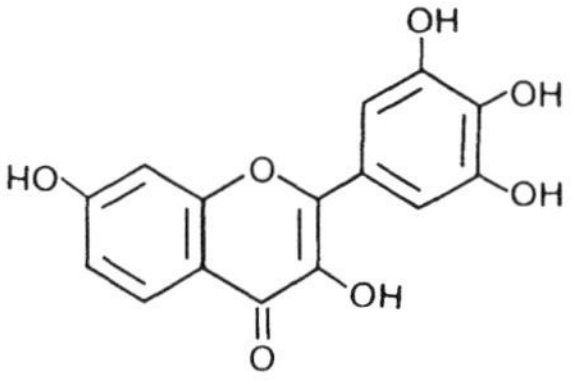

Flavonol : 89 robinetin

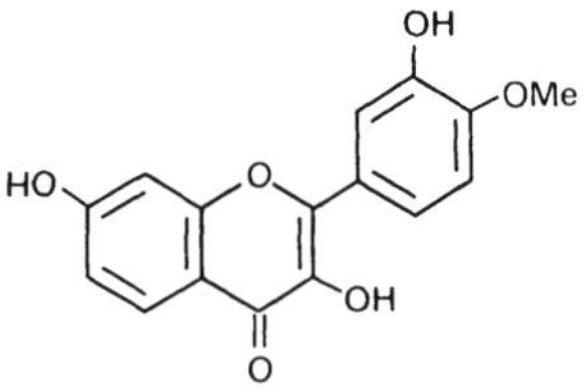

Flavonol : 90 fisetin 4′-methyl ether

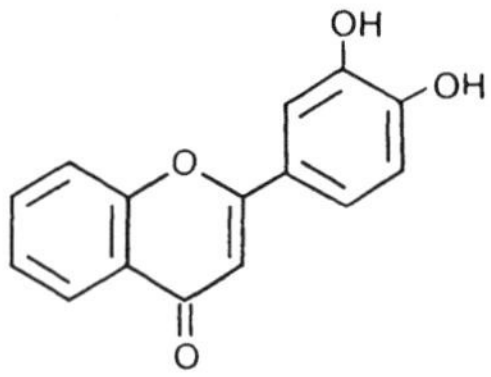

Flavone : 91 3′,4′-dihydroxy

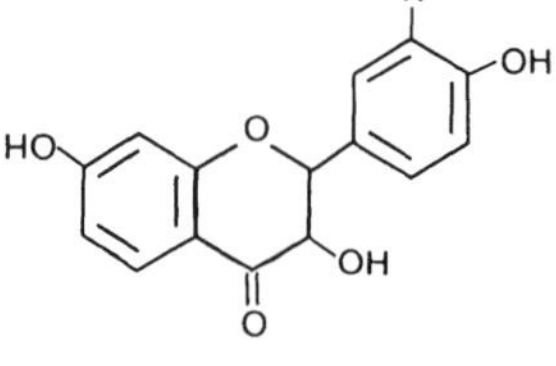

Flavanonols : 92 garbanzol, R=H
93 fustin, R=OH

Table 7.5.7. Distribution of 5-deoxyflavonoids and of rengasin in heartwood extracts of various species of Anacardiaceae

Tribe	Genus, species	Compound							
		butein glucoside	sulphuretin	sulphuretin glucoside	fisetin	fisetin 7-glucoside	fustin	3,7,4′-trihy-droxyflavone	rengasin
Anacardiaeae	*Anacardium excelsum*	–	+	–	+	–	–	+	+
	Mangifera altissima	–	+	+	+	–	–	+	–
	Swintonia luzonensis	–	+	–	+	–	–	+	+
Rhoeae	*Amphipterygium adstrigens*	+	+	+	+	+	+	+	+
	Actinocheita filicina	–	+	–	+	–	–	+	?
	Astronium graveolens	–	–	–	+	–	–	+	–
	Cetinus obovatus	–	+	+	+	+	+	–	?
	Loxopterygium sagottii	–	+	–	+	–	–	+	+
	Malosma laurina	–	+	+	+	+	+	+	–
	Metopium brownei	+	+	+	+	+	+	+	+
	M. toxiferum	–	+	+	+	–	–	+	+
	Pistacia texana	–	+	–	+	+	–	+	+
	Rhus chondroloma	+	+	+	+	+	+	+	+
	R. copallina	–	+	+	+	–	+	+	+
	R. glabra	–	+	+	+	–	+	+	+
	R. microphylla	–	+	+	+	+	+	+	+
	R. rubifolia	–	+	+	+	+	+	+	+
	R. typhina	–	+	–	+	–	+	+	+
	R. virens	+	+	+	+	+	+	+	+
	Toxicodendron diversilobum	–	+	+	+	+	+	–	+
	Trichoscypha arborea	–	+	–	+	–	–	–	+
Spondiaeae	*Dracontomelon edule*	–	+	–	–	–	–	–	–
	Spondias mombin	–	+	+	+	–	–	–	+
	Tapirira guanensis	–	–	–	+	–	–	–	–

3′,4′-dihydroxyflavone, previously only known as a leaf constituent in *Primula*. The chemical patterns in *Rhus* correlated with subgeneric classification in that 14 species of the subgenus *Lobadium* contained glucosides of fisetin, butein, and 3′,4′-dihydroxyflavone, in addition to the aglycones listed above, whereas the four species of subgenus *Rhus* studied lacked them. The chemical data in toto indicate a close relationship between all these taxa, and support a conservative view of the taxonomy – i.e. that all these trees should be regarded as members of a single genus rather than split up, as proposed by some taxonomists, into several segregate genera.

A related study of the heartwood of *Amphipterygium adstrigens*, a representative of the Julianaceae, a small family of five species, has shown that all the typical Anacardiaceae flavonoids are present. This family was variously placed either in association with the Juglandaceae or near the Anacardiaceae and/or Burseraceae. Chemical analysis of heartwood of representatives of these two other families failed to reveal any of the characteristic 5-deoxyflavonoids of Anacardiaceae. Hence, the chemical data on *Amphipterygium* clearly establish it as belonging in this alliance. Young (40) on this basis and that of anatomical comparisons, has proposed that the Julianaceae should be merged into the Anacardiaceae and regarded as a member of the tribe Rhoeae (Table 7.5.7).

7.5.3.3.2 Heartwood Flavonoids of the Leguminosae

More chemistry has been carried out on the heartwood flavonoids of the Leguminosae than on those of any other plant family. This is not surprising considering that there are many tree species in what is a very large family and also that these trees are used as timber, for furniture, or as a source of tannin. But it also reflects the fact that there are an enormous variety of chemical structures present, particularly of isoflavonoids. The discovery of interesting and novel structures in one tree inevitably encourages investigation of related species for further novelties.

In spite of the intensity of chemical investigation, it is still true that a completely representative survey of the family for heartwood flavonoids has still to be carried out. Chemical efforts have been concentrated in certain plant groups, notably on *Acacia* in the Australian and African floras, on *Dalbergia*, *Machaerium* and other Brazilian genera, and on some Asian taxa, such as the sandalwoods *Pterocarpus*. Our knowledge of flavonoid patterns is thus limited. Certain structural features that we assume to be characteristic of the family – e.g. the absence of 5-hydroxylation – may not in fact extend to more than a certain percentage of the family.

The Leguminosae is divided traditionally into three subfamilies: the Caesalpinioideae, Mimosoideae, and Papilionoideae. In some modern taxonomic texts, the three subfamilies are raised to family rank as the Ceasalpiniaceae, Mimosaceae, and Fabaceae. Since most legume taxonomists still regard it as a single family, this is the treatment used here. Whereas the first two subfamilies are more primitive and, hence, woodier, they are smaller in number than the Papilionoideae, the respective numbers of genera being 150, 56 and 500. It is in the larger Papilionoideae in which isoflavonoids are mainly found (they do occur rarely in the other two subfamilies), so that most chemical work has been concentrated on this subfamily. Here some mention will be made of flavonoid studies on *Acacia* (Mimosoideae), on *Apuleia* (Caesalpinioideae), and on *Dalbergia, Machaerium*, and *Pterocarpus* (all of the Papilionoideae).

One of the few legume genera that has been extensively surveyed for taxonomic purposes is *Acacia*. Early knowledge of the flavonoids was derived from those species used in the tanning industry – namely, black wattle, *A. mearnsii*, green wattle, *A. decurrens*, and golden wattle, *A. pycnantha*. More recently, several surveys have been carried out on both heartwood and bark flavonoids, and at least 400 of the approximately 750 known species have now been examined (6, 35, 36). The flavonoids present fall into four series, depending on the hydroxylation pattern: 7,4′-dihydroxy, 7,3′,4′-trihydroxy, 7,8,4′-trihydroxy, and 7,8,3′,4′-tetrahydroxy. Representative flavonols are **16, 88, 94–97**; similar substitution patterns extend to the flavononols, flavan-3,4-diols, and chalcones. Methylation of the 3- or 8-hydroxyl groups is occasionally superimposed on this pattern.

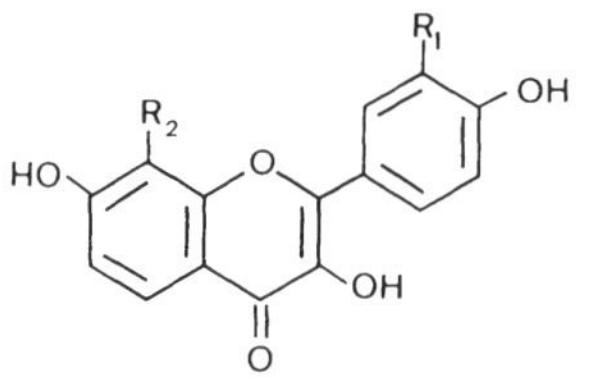

88 3,7,4′-trihydroxy, $R_1=R_2=H$
16 fisetin, $R_1=OH$, $R_2=H$
94 3,7,8,4′-tetrahydroxy, $R_1=H$, $R_2=OH$
95 3,7,8,3′,4′-pentahydroxy, $R_1=R_2=OH$

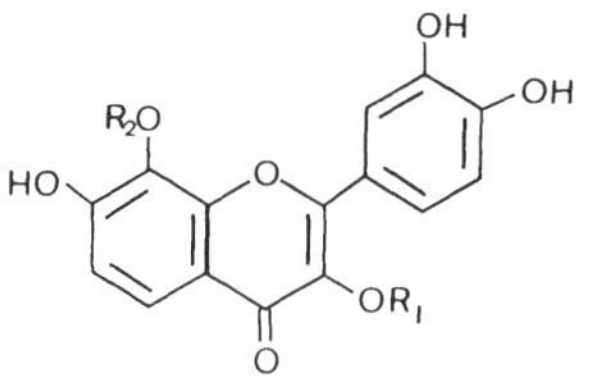

96 8-methoxyfisetin, $R_1=H$, $R_2=Me$
97 fisetin 3′-methyl ethers, $R_1=Me$, $R_2=H$

The distribution of the different hydroxyl types in the *Acacia* species is remarkably closely correlated with sectional classification and with evolutionary advancement within the genus. These patterns are summarized in Fig. 7.5.2, which is based on the work of Tindale and Roux (35). Whereas the flavonoid hydroxylation patterns encountered in *Acacia* could be regarded as being fairly common in nature, much rarer substitutions do occasionally appear in legume woods. Perhaps the most extreme example of flavonoid variation is in *Apuleia leiocarpa* (5). This is a monotypic genus of the Caesalpinioideae, the yellow heartwood of which has yielded a remarkable range of ten flavonol methyl ethers **(98–107)**. No fewer than four different B-ring substitution patterns are represented; there is also variation in 6-substitution and in which hydroxyl groups are *O*-methylated. All these flavonols have 5-oxygenation, unlike the more common types found in legumes.

These 10 flavonols are reported in both heartwood and sapwood, but the bark lacks them. Instead, the bark contains an isoprenylated pterocarpan, leiocarpin; this is again very unusual in that it is the only authentic report so far of isoflavonoids in the subfamily Caesalpinioideae. The only other legume with any of the heartwood flavonols of *Apuleia* is *Distemonanthus benthamianus*, which contains ayanin **(98)**, oxyayanin A **(100)**, and oxyayanin B **(99)**. *Distemonanthus* interestingly enough belongs to the same tribe, the Cassieae, as *Apuleia*.

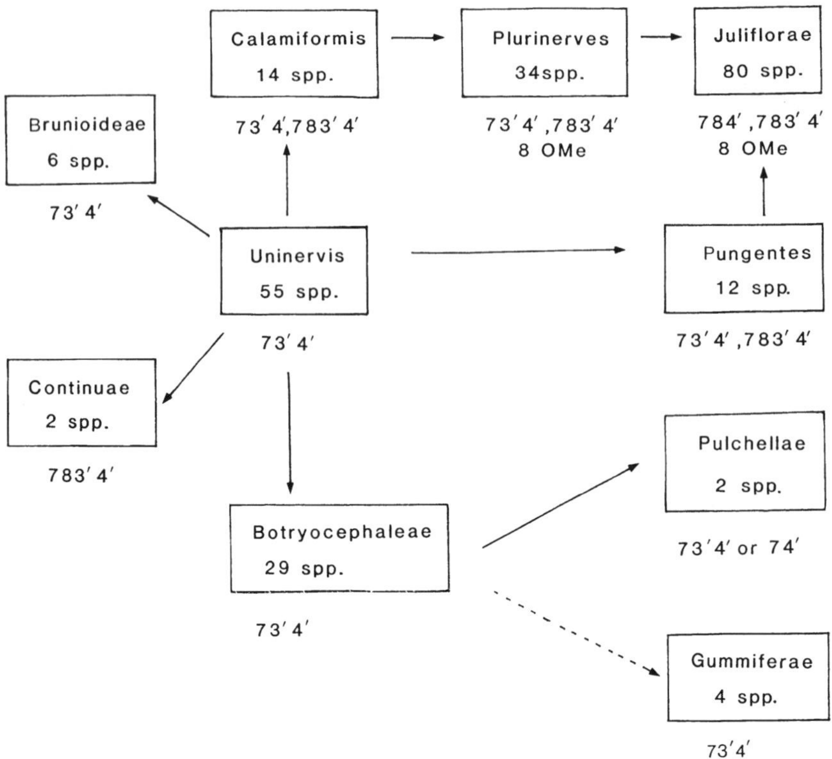

Fig. 7.5.2. Flavonoid patterns and evolutionary trends within subsections of the genus *Acacia*

The pterocarpan, leiocarpin, of *Apuleia* has also been found in an unrelated legume, in bark of *Dalbergia nitidula*, which belongs to a different subfamily, the Papilionoideae. Because of the beautiful and durable heartwoods, the genus *Dalbergia* has been extensively investigated, as well as the related *Machaerium*. Besides the isoflavonoids that are universally present in these woods, many novel neoflavonoids were first reported from these trees. The chemistry of these two genera has been reviewed by Braga de Oliviera et al. (4).

Typical compounds present are illustrated among structures **52** through **68**. The principal neoflavonoids are the dalbergiquinols, dalbergiones, neoflavenes, and dalbergins. These are related to each other in the bio-oxidative sequence: quinol → quinone → neoflavene → dalbergin. Two minor types present also are cinnamylphenols and 2,3-dihydrobenzofurans with similar substitution patterns. Most of these compounds have been isolated with either the *R*- or the *S*-configuration, so there is additional stereochemical complexity in these heartwood constituents. The isoflavonoids also form a series, which is reductive rather than oxidative: isoflavone → pterocarpan → isoflavan. In all, some 90 isoflavonoids have been recorded in *Dalbergia* woods and 19 in *Machaerium* woods.

The distribution of these phenolics in *Dalbergia* and *Machaerium* woods is indicated in Table 7.5.8. The two genera are close morphologically, and it is therefore interesting that they merge into each other chemically, with no sharp cut-off

Table 7.5.8. Numbers of structures of neoflavonoids and isoflavonoids in heartwood of *Dalbergia* and *Machaerium* species

Continent of origin	Genus, species	Neoflavonoids			Isoflavonoids		
		Dalbergin & derivs	Cinnamylphenols	Dihydro benzofurans	Isoflavones	Pterocarpans	Isoflavans
Asia	*D. sissoo*	3	–	–	3	–	–
	D. latifolia	3	–	–	–	–	–
	D. cochinsensis	3	–	–	1	–	–
	D. lanceolaria	–	–	–	–	–	–
Africa	*D. melanoxylon*	2	–	1	–	–	–
	D. baroni	3	–	–	1	–	–
America	*D. nigra*	4	–	–	1	–	–
	D. miscolobium	3	–	–	1	–	–
	D. villosa	1	2	–	1	–	–
	D. barretoana	1	–	–	2	–	–
	D. obtusa	2	3	1	1	–	–
	D. spruceana	2	–	–	5	7	–
	D. frutescens	–	–	–	2	2	3
	D. cearensis	2	–	–	2	2	1
	D. decipularis	–	–	–	1	2	–
	D. ecastophyllum	–	–	–	–	2	2
	D. volubilis	–	–	–	2	–	–
	M. nictitans	–	7	–	1	1	–
	M. kuhlmannii	1	6	1	1	1	–
	M. scleroxylon	2	3	–	1	–	–
	M. villosum	–	–	2	5	1	2
	M. acutifolium	–	–	1	–	1	2
	M. mucronulatum	–	–	3	1	1	2
	M. opacum	–	–	–	–	–	2
	M. vestitum	–	–	–	2	2	2

3′ 4′-disubstitution

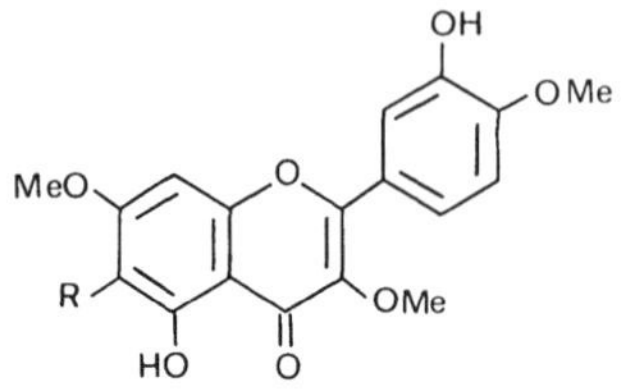

98 ayanin, R=H

99 oxyayanin B, R=OMe

2′ 4′ 5′-trisubstitution

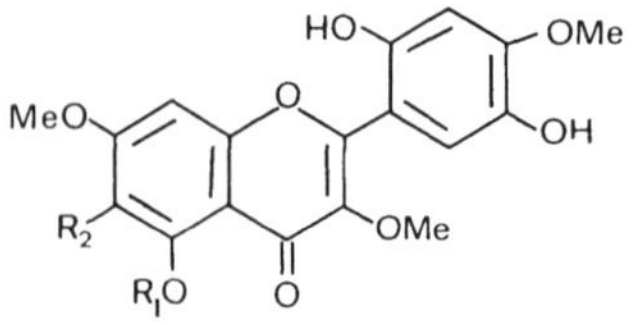

100 oxyayanin A, R_1=R_2=H

101 5-O-methyloxyayanin A, R_1=Me, R_2=H

102 5-demethylapulein, R_1=H, R_2=OMe

103 Apulein, R_1=Me, R_2=OMe

2′ 3′ 4′-trisubstitution

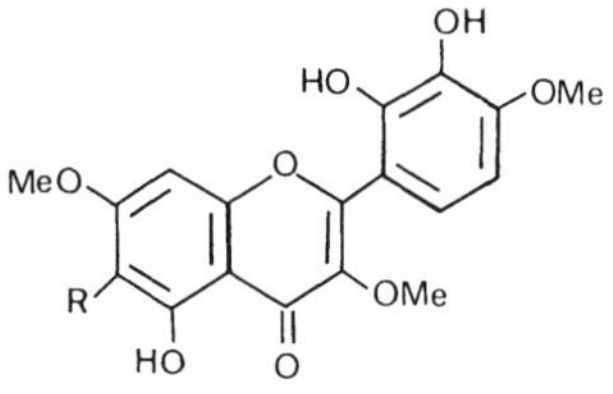

104 apuleidin, R=H

105 apuleisin, R=OH

3′ 4′ 5′- trisubstitution

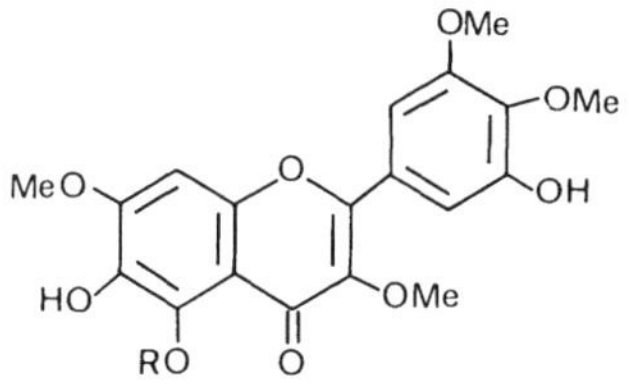

106 apuleitrin, R=H

107 apuleirin, R=Me

point between them. However, *Machaerium* species do accumulate *both* neoflavonoids and isoflavonoids, whereas *Dalbergia* species tend to have *either* neoflavonoids or isoflavonoids. Although a correlation with continent of origin is incomplete, there is a geographical trend, with Asian and African species having a simpler pattern than that in the American taxa. The chemical evidence supports the view that *Machaerium* evolved from *Dalbergia*. However, only 25 of a possible 450 species have so far been investigated, so further surveys are needed to confirm these findings.

The above analysis of isoflavonoid data on *Dalbergia*/*Machaerium* is simplistic and does not fully take into account the structural variation that is present. It is possible to consider the biosynthetic steps that separate the various constituents and relate their presence/absence to evolutionary advancement. Unfortunately, our knowledge of isoflavonoid biosynthesis in woody tissues is very limited; it is only possible to suggest chemically plausible pathways. Application of biosynthetic principles to a simpler case than *Dalbergia*, namely, species of *Pterocarpus*, is illustrated in Fig. 7.5.3. This is really only an alternative way of summarizing the known distribution of isoflavones in the heartwoods of this genus (33). Again, the sample investigated is small (eight spp) relative to the size

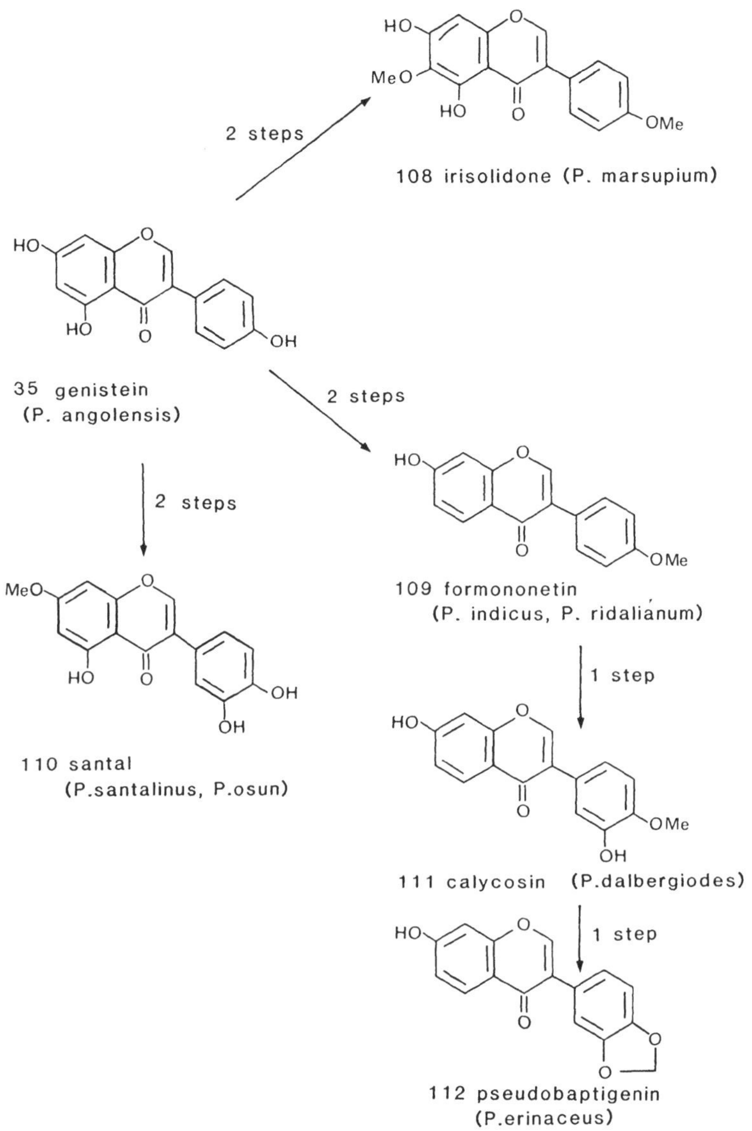

2 steps

108 irisolidone (P. marsupium)

35 genistein
(P. angolensis)

2 steps

2 steps

110 santal
(P.santalinus, P.osun)

109 formononetin
(P. indicus, P. ridalianum)

1 step

111 calycosin (P.dalbergiodes)

1 step

112 pseudobaptigenin
(P.erinaceus)

Fig. 7.5.3. Isoflavone pathways and distribution in *Pterocarpus* heartwoods

of the genus (100 spp). More sophisticated prodecures of analyzing data on legume isoflavonoids have been developed by Gottlieb (11) and applied to isoflavonoids in the Papilionoideae as a whole.

7.5.4 Properties and Function

There is as yet no definitive answer to the question: What is the role of flavonoid in heartwood, sapwood, or bark of a tree? Hergert proposed in 1962 (18) that there is probably a dual role: protection from fungal attack, and protection from wood-boring insects such as termites. Hergert discussed (18) the available evidence, which was in part conflicting and in part supporting such a dual role. Since then, much more indirect evidence has accumulated in further support.

Perhaps the most telling point in favor of a protective role against fungal attack is the recent discovery that isoflavonoids, along with certain other groups of secondary compound, are phytoalexins – compounds specifically produced in plant tissue to ward off microbial attack. Indeed, studies of phytoalexin induction in herbaceous members of the Leguminosae (e.g. *Lotus*, *Lathyrus*, *Pisum*, etc.) have shown that isoflavans and pterocarpans, first isolated as constitutive constituents in legume heartwoods, are also formed de novo in leaf tissue in response to microbial infection. For example, two of the most common phytoalexins in species of *Medicago*, *Trifolium*, and *Trigonella* are the pterocarpans medicarpin and maackiain, compounds earlier known naturally in heartwood, sapwood, or bark of a variety of legume trees. These isoflavonoid phytoalexins have been shown to be toxic to or inhibitive to the growth of a wide range of microorganisms (21). Interestingly, pterocarpans and isoflavans are more toxic on a molar basis than the related isoflavanones and isoflavones, creating a graded series of isoflavonoid structures of increasing potency (13). Structure-activity relationships have been studied (37), but it is still not possible to say exactly what the structural requirements are for maximal fungitoxicity. Although the results of using isoflavonoids as protectant fungicides on crop plants have not been entirely successful, this does not invalidate the idea that these compounds are active in vivo. The hypothesis that isoflavonoids and other phenolic constituents contribute to the natural resistance of heartwood or bark to microbial invasion now seems very reasonable.

The neoflavonoids of legume heartwoods may also have a similar role in microbial protection, although this has not yet been as well established as for the isoflavonoids. Those that are quinones are potentially more toxic than the related quinols. The dalbergiones have at least been shown to inhibit the growth of some microorganisms; they are also apparently the agents responsible for the dermatitis developed by workers handling these particular woods (15).

The insecticidal effects of heartwood flavonoids have also been studied in recent years. Again, the isoflavonoids show significant toxic effects on several insect pests. It is interesting that those isoflavonoids that are especially fungitoxic are those that show the most dramatic effects on insect mortality in laboratory experiments (16, 34).

The antifungal and insecticidal activities of flavones, flavonols, and dihydroflavonols have also been further explored since 1962, and new examples of active compounds have been described. The effects of flavonols on insect feeding are often indirect – e.g. they may seriously affect larval growth rather than cause immediate mortality (22). The possibility that heartwood constituents may be damaging to humans has been raised, following the epidemiological link between the incidence of nasal carcinoma in furniture workers and the inhalation of wood dust (1) (see also Chap. 9.5). One candidate as the carcinogen was quercetin, following the discovery that this flavonol is mutagenic in the Ames' *Salmonella* test. In fact, the evidence is now against such a view, since repeated feeding of quercetin to rats has failed to implicate it as an in vivo carcinogen. Furthermore, attempts to detect quercetin in the pine wood dust, which is the causative agent of nasal carcinoma, failed to show it occurring in any appreciable amount (12).

In his 1962 review, Hergert also considered the heartwood flavonoids from the viewpoint of their potential economic value to humans. There is no doubt that certain woods or barks are superlative sources of flavonoid materials such as quercetin, rutin, or dihydroquercetin. These may be used as such or converted simply to more marketable products. Rutin continues to be used widely as a medicinal plant product, although its efficacy is still a matter of debate. Fortunately, there is a resurgence of interest in flavonoids as potential curative agents in humans, and more determined efforts are being made to submit these products to proper clinical trial. There is also current interest in flavonoids as flavor potentiators in the food industry. Thus, the market outlook for heartwood flavonoids would seem to be more favorable today than it was in 1962.

References

1 Acheson E D 1982 Nasal cancer in furniture workers: The problem. In: Acheson E D (ed) The carcinogenicity and mutagenicity of wood dust. MRC Southampton, 1–5

2 Agrawal P K, Rastogi R P 1984 Chemistry of the true cedars. Biochem System Ecol 12:133–144

3 Bate-Smith E C 1962 The phenolic constituents of plants and their taxonomic significance. J Linn Soc Bot 58:95–172

4 Braga de Oliviera A, Gottlieb O R, Ollis W D, Rizzani C T 1971 A phylogenetic correlation of the genera *Dalbergia* and *Machaerium*. Phytochemistry 10:1893–1876

5 Braz R, Gottlieb O R 1971 The flavones of *Apuleia leiocarpa*. Phytochemistry 10:2433–2450

6 Clark-Lewis J W, Porter L J 1972 Phytochemical survey of the heartwood flavonoids of of *Acacia* species from arid zones of Australia. Aust J Chem 25:1943–1955

7 De Alleluia I B, Braz R, Gottlieb o R, Magalhaes E G, Marques R 1978 (–)-Rubranine from *Aniba rosaeodora*. Phytochemistry 17:517–521

8 Donnelly D M X 1982 Metabolites of *Dalbergia*: New structural variants of the neoflavonoids. In: Farkas L, Gabor M, Kallay R, Wagner H (eds) Flavonoids and bioflavonoids. Akademiai Kiado Budapest 263–278

9 Erdtman H 1956 Organic chemistry and conifer taxonomy. In: Todd A (ed) Perspectives in organic chemistry. Wiley-Interscience New York, 453–494

10 Erdtman H, Kimland B, Norin T 1966 Wood constituents of *Ducampopinus krempfii (Pinus krempfii)*. Phytochemistry 5:927–931

11 Gottlieb O R 1982 Micromolecular evolution, systematics and ecology. Springer Heidelberg, 170 pp

12 Harborne J B 1982 Biochemistry of hardwoods: Possible carcinogens. In: Acheson E D (ed). The carcinogenicity and mutagenicity of wood dust. MRC Southampton, 16–20

13 Harborne J B, Ingham J L 1978 Biochemical aspects of the coevolution of higher plants with their fungal parasites. In: Harborne J B (ed) Biochemical aspects of plant and animal coevolution. Academic Press London, 343–405

14 Harborne J B, Mabry T J (eds) 1982 The flavonoids: Advances in research. Chapman & Hall London, 744 pp

15 Harborne J B, Mabry T J, Mabry H (eds) 1975 The flavonoids. Chapman & Hall London, 1204 pp

16 Hart S V, Kogan M, Paxton J D 1983. Effect of soybean phytoalexins on the herbivorous insects Mexican bean beetle and soybean looper. J Chem Ecol 9:657–672

17 Havsteen B 1983 Flavonoids, a class of natural products of high pharmacological potency. Biochem Pharmacol 32:1141–1148

18 Hergert H L 1962 Economic importance of flavonoid compounds: Wood and bark. In: Geissman T A (ed) The chemistry of flavonoid compounds. Pergamon Press Oxford, 553–592

19 Hergert H L, Goldschmid O 1958 Biogenesis of heartwood and bark components. I. A new taxifolin glucoside. J Org Chem 23:700–704

20 Higuchi R, Donnelly D M X 1978 Acylated flavonol glucosides of *Pinus contorta* needles. Phytochemistry 17:787–792

21 Ingham J L 1983 Naturally occurring isoflavonoids, 1855–1981. Fortschr Chem Org Naturst 43:1–266

22 Isman M B, Duffey S S 1982 Toxicity of tomato phenolic compounds to the fruitworm *Heliothis zea.* Ent Exp & Appl 31:370–376

23 King F E, King T J, Rustidge D W 1962 The chemistry of extractives from hardwoods. 33. Extractives from *Melanorrhea* spp. J Chem Soc 1192–1194

24 King H G C, White T 1961 The colouring matter of *Rhus cotinus* wood (young fustic). J Chem Soc 3538–3539

25 Lebreton P, Sartre J 1983 Les Pinales, considerées d'un point de vue chimiotaxonomique. Can J For Res 13:145–154

26 Mahesh V B, Seshadri T R 1954 Chemical components of commercial woods. I The heartwood of *Pinus griffithii.* J Sci Ind Res India 13B:835–841

27 Medvedeva S A, Ivanova S Z, Tjukavkina N A, Zapesochnaja G G 1977 Silpin – a new *C*-methylated flavonoid from *Pinus sylvestris*. Khim Prir Soedin 650–653

28 Montero J L, Winternitz F 1973 Structure of the rubraninchalcone isolated from the rosewood *Aniba rosaeodora.* Tetrahedron 29:1243–1252

29 Mu Q, Li Q 1982 Isolation and identification of morin and morin calcium chelate from *Artocarpus pithecogallus* and *A. heterophylla.* Zhiwu Xuebao 24:147–153

30 Niemann G J 1979 Some aspects of the chemistry of Pinaceae needles. Acta Bot Neerl 28: 73–88

31 Robinson R 1962 Synthesis in the brazilin group. In: Gore T S, Joshi B S, Sunthankar S V and Tilak B D (eds) Chemistry of natural and synthetic colouring matters. Academic Press London, 1–12

32 Seikel M K, Chow J H S, Feldman L 1966 The glycosylflavonoid pigments of *Vitex lucens* wood. Phytochemistry 5:439–455

33 Seshadri T R 1972 Polyphenols of *Pterocarpus* and *Dalbergia* woods. Phytochemistry 11:881–898

34 Sutherland O R W, Russell G B, Biggs D R, Lane G A 1980 Insect feeding deterrent activity of phytoalexin isoflavonoids. Biochem System Ecol 8:73–76

35 Tindale M D, Roux D G 1974 An extended phytochemical survey of Australian species of *Acacia*: Chemotaxonomic and phylogenetic aspects. Phytochemistry 13:829–839

36 Tindale M D, Roux D G 1975 Phytochemical studies on the heartwoods and barks of African and Australian species of *Acacia.* Boissiera 24:299–305

37 Van Etten H D 1976 Antifungal activity of pterocarpans and other selected isoflavonoids. Phytochemistry 15:655–660

38 Venkataraman K 1972 Wood phenolics in the chemotaxonomy of the Moraceae. Phytochemistry 11:1571–1586

39 Wollenweber E, Dietz V H 1981 Occurrence and distribution of free flavonoid aglycones in plants. Phytochemistry 20:869–932

40 Young D A 1976 Flavonoid chemistry and the phylogenetic relationships of the Julianaceae. Syst Bot 1:149–162

41 Young D A 1979 Heartwood flavonoids and the infrageneric relationships of *Rhus* (Anacardiaceae). Am J Bot 66:502–510

7.6 Biflavonoids and Proanthocyanidins

R. W. HEMINGWAY

7.6.1 Introduction

The biflavonoids and proanthocyanidins constitute the two major classes of oligomeric flavonoids found in plants (126, 136). The biflavonoids are oxidative coupling products leading to biflavones, flavanone-flavones, and biflavanones. These compounds always carry carbonyl functions at the C-4 positions (126). In contrast, the proanthocyanidins are products of the reduction of flavanonols such as flavan-3,4-diols or oligomeric flavan-3-ols with an interflavanoid bond at C-4 being coupled to a terminating flavan-3-ol (136). Natural compounds that are crossed products between these two classes, such as a flavan-3-ol linked through C-4 to a flavanone or a flavanone oxidatively coupled to a flavan-3-ol, are extremely rare. The unusual biflavonoid larixinol, isolated from *Larix gmelini*, a condensation product of dihydrokaempferol and (−)-epiafzelechin, comes closest to a representative of the crossed flavanone-flavan-3-ol biflavanoids (336). It is likely, therefore, that this classification of oligomeric flavanoids into two categories will undoubtedly be used for some time to come.

7.6.2 Biflavonoids

Compared with the monomeric flavonoids discussed in Sect. 7.5, the biflavonoids are limited in structural variation. The biflavonoids are confined largely to oxidative coupling products of chalconaringenin. Variations in the location of the interflavonoid bond, oxidation state of the heterocyclic ring, hydroxylation and/or methoxylation, and glycosylation lead to a group of approximately 90 known natural products (125, 126). In addition to their restricted structural variation, the biflavonoids are limited to a narrow range of plants, largely those in primitive orders (125). These compounds appear in the Bryales, Psilotales, and Selaginellales. They occur in the leaves of primitive gymnosperms. Biflavonoids are found only sporadically in the angiosperms and here their association with primitive character is not so evident. Geiger and Quinn (125) have proposed that the angiosperms lost the capacity to synthesize biflavonoids in evolution, but their biogenesis was regained by a few selected families.

A number of important reviews of the biflavonoids have been prepared, and works by Baker and Ollis (13) published in 1961, Locksley (232) in 1973, and Geiger and Quinn in 1975 (125) and 1982 (126) are recommended reading for those undertaking study of these compounds. The review by Baker and Ollis (13) provides a particularly interesting account of the difficulties encountered in the elucidation of the structures of these compounds in the early years prior to modern spectroscopic methods. Excellent summaries of isolation procedures for the flavonoids are provided by Markham (237, 238) and Hostettmann and Hostettmann (163). Readers are referred to these, the review of biflavonoids by Geiger and Quinn (126), and references cited herein for details. Compilations of chromatographic properties and UV-visible spectra have been made by Dossaji et al. (78). In addition to alkaline degradation products, mass spectra (233) and ^{1}H-NMR spectra (240, 286, 291, 297) usually provide all the information necessary for structure elucidation. Proton-coupled ^{13}C-NMR spectra have been especially useful for distinguishing between various C-6 and C-8 biphenyl-linked isomers (40, 239). Further ^{13}C-NMR and CD spectral studies of *Garcinia* biflavonoids are reported by Duddeck et al. (88), and data by Liu et al. (231) and Niwa et al. (270) provide additional chemical shifts for the biflavonoids in *Stellera chamaejasme.*

Various interconversion reactions are used for the synthesis of analogs. Partial methylation and demethylation (3, 16, 130, 249–251, 263, 264), rearrangement of C-8 : C-8 to C-6 : C-8 biphenyl isomers, and dehydrogenation reactions converting flavones to flavanones have been employed (3, 16, 227, 251, 264, 293). Many biflavonoids have now been synthesized and references are compiled in Locksley's (232) and Geiger and Quinn's (125, 126) reviews.

Unfortunately, there is no commonly accepted trivial nomenclature for these compounds; their full systematic names, not to mention their often complex common names, are extremely cumbersome. Geiger and Quinn have proposed a new system (125) but it, like the trivial procyanidin nomenclature initiated by Weinges (381), requires frequent reference for understanding and has become very difficult to use as the number of new compounds has grown. Locksley (232) has proposed a system in which the position of substitution of the upper (I) and lower (II) units is defined from the usual numbering of the parent flavonoid, and each unit is also described from the parent monomer. Thus, the common biflavone amentoflavone is named apigeninyl-(I-3′,II-8)-apigenin, etc. Locksley's nomenclature is used in this chapter (232).

7.6.2.1 Structural Variations of Biflavonoids

As would be expected from their formation through oxidative coupling reactions (351), the biflavonoids exhibit great diversity in the location of the interflavonoid bond. Natural and synthetic dimers have now been described that represent nearly every possible combination of radical coupling (**1**–**10**). However, as in the hydrolyzable tannins produced by oxidative coupling of galloyl groups (Sect. 7.2), there is surprising specificity in the distribution of these compounds between different plants.

Most of the natural biflavonoids have biphenyl linkages. The amentoflavones (I-3′,II-8 biflavonoids) (**1**) occur most frequently (125, 126). All of the natural compounds in this group (Table 7.6.1) carry a 5,7-dioxy A-ring and most have a 4′-oxy B-ring, except the biflavanones isolated recently from *Semecarpus anacardium* (257). Further hydroxylation of the B-ring has been observed in a *Dicranum* sp. moss, the only known natural source of biluteolin (230), and a methylated derivative of this compound 4′,5′-di-*O*-methyl-luteolinyl-(I-3′,II-8)-luteolin or 5′-methoxybilobetin has been isolated from *Ginkgo* (180) (Sect. 7.6.2.2) (Table 7.6.1). Methylation of the C-4′ and/or C-7 hydroxyl functions of either or both of the units is common (125). As expected, methylation of the C-5 hydroxyl is

1

2

3

4

5

6

7

8

9

10

Table 7.6.1. Distribution of biflavonoids: amentoflavones and robustaflavones

Group/Class I-3′,II-8	Common name	Substitution					Occurrence[1]		Reference
		C-5	C-6	C-7	C-4′	C-5′	Coniferales	Dicotyledoneae	
Biflavone	amentoflavone	OH	H	OH	OH	H	Araucariaceae	Anacardiaceae	48, 161, 259,
		OH	H	OH	OH	H	Cephalotaxaceae	Burseraceae	269, 270, 284,
							Cupressaceae	Caprifoliaceae	285, 371
							Podocarpaceae	Clusiaceae	
							Taxodiaceae	Nandinaceae	
	sequoiaflavone	OH	H	OCH_3	OH	H	Cephalotaxaceae		252
		OH	H	OH	OH	H	Podocarpaceae		
							Taxodiaceae		
	bilobetin	OH	H	OH	OCH_3	H	Araucariaceae		12
		OH	H	OH	OH	H	Podocarpaceae		
							Taxodiaceae		
	sotetsuflavone	OH	H	OH	OH	H	Araucariaceae		12, 193
		OH	H	OCH_3	OH	H	Cupressaceae		
							Podocarpaceae		
							Taxodiaceae		
	podocarpusflavone A	OH	H	OH	OH	H	Cupressaceae	Clusiaceae	50, 255
		OH	H	OH	OCH_3	H	Podocarpaceae	Euphorbiaceae	
							Taxodiaceae		
	ginkgetin	OH	H	OCH_3	OCH_3	H	Cephalotaxaceae		262
		OH	H	OH	OH	H	Podocarpaceae		
	7,7″-di-*O*-methylamento-	OH	H	OCH_3	OH	H	Araucariaceae		169
	flavone	OH	H	OCH_3	OH	H	Taxodiaceae		
	podocarpusflavone B	OH	H	OCH_3	OH	H	Podocarpaceae	Euphorbiaceae	123, 255
		OH	H	OH	OCH_3	H			
	amentoflavone 4′,7″-	OH	H	OH	OCH_3	H	Araucariaceae		199
	dimethyl ether	OH	H	OCH_3	OH	H			
	isoginkgetin	OH	H	OH	OCH_3	H	Podocarpaceae		12
		OH	H	OH	OCH_3	H			
	amentoflavone 7″, 4″-	OH	H	OH	OH	H	Podocarpaceae		16
	dimethyl ether	OH	H	OCH_3	OCH_3	H	Taxodiaceae		
	7,4′,7″-tri-*O*-methyl-	OH	H	OCH_3	OCH_3	H	Araucariaceae		169
	amentoflavone	OH	H	OCH_3	OH	H			

	sciadopitysin	OH	H	OCH_3	OCH_3	H	Araucariaceae		190 – 194
		OH	H	OH	OCH_3	H	Cephalotaxaceae		
							Cupressaceae		
							Podocarpaceae		
							Taxodiaceae		
	heveaflavone	OH	H	OCH_3	OH	H		Euphorbiaceae	39, 235
		OH	H	OCH_3	OCH_3	H			
	kayaflavone	OH	H	OH	OCH_3	H	Araucariaceae		190 – 194
		OH	H	OCH_3	OCH_3	H	Podocarpaceae		
							Taxodiaceae		
	amentoflavone 7,7″,4′,4‴-	OH	H	OCH_3	OCH_3	H	Araucariaceae		159, 263, 296
	tetramethyl ether	OH	H	OCH_3	OCH_3	H	Podocarpaceae		
	dioonflavone	OCH_3	H	OCH_3	OCH_3	H	Dioon (Cycadales)		77, 263, 266
	5′-methoxybilobetin	OH	H	OH	OH	OCH_3	Ginkgo		180
		OH	H	OH	OH	H			
	biluteolin	OH	H	OH	OH	OH	Dicranum (Bryales)		230
		OH	H	OH	OH	OH			
	7-*O*-methyl-6-methyl	OH	CH_3	OCH_3	OH	H	Cephalotaxaceae		9
	amentoflavone	OH	H	OH	OH	H			
	strychnobiflavone (3,3″	OH	H	OH	OH	OH		Strychnos	268
	OCH_3)	OH	H	OH	OH	OH			
Flavanone-	2,3-dihydroamentoflavone	OH	H	OH	OH	H	Cycas (Cycadales)		16
Flavone		OH	H	OH	OH	H			
	2,3-dihydroamentoflav-	OH	H	OH	OH	H	Taxodiaceae		124
	one-7″,4″-dimethyl	OH	H	OCH_3	OCH_3	H			
	2,3-dihydrosciadipitysin	OH	H	OCH_3	OCH_3	H	Taxodiaceae		16
		OH	H	OH	OCH_3	H			
Biflavanone	tetrahydroamentoflavone	OH	H	OH	OH	H		Anacardiaceae	2, 171, 172
		OH	H	OH	OH	H			
		H	H	OH	OH	H		Anacardiaceae	257, 258, 260
		H	H	OH	OH	H			
		OH	H	OH	OH	H		Anacardiaceae	257, 258, 260
		H	H	OH	OH	OH			
	jeediflavanone	OH	H	OH	OH	H		Anacardiaceae	257, 258, 260
		OH	H	OH	OH	OH			
Biflavone	robustaflavone	OH	H	OH	OH	H	Araucariaceae	Anacardiaceae	49, 229, 362
		OH	–	OH	OH	H	Cupressaceae		
	tetrahydrorobustaflavone	OH	H	OH	OH	H		Anacardiaceae	171
		OH	–	OH	OH	H			

extremely rare, the exception being dioonflavone, the hexamethyl ether of amentoflavone (77, 266). Methylation of the A-ring carbons has been observed in only one natural product isolated from *Cephalotaxus* species, the 7-*O*-methyl-6-methyl-amentoflavone (9).

Flavanone-flavone analogs are found in *Metasequoia* species that contain naringeninyl-(I-3′,II-8)-7,4′-di-*O*-methyl-apigenin and 7,4′-di-*O*-methyl-naringeninyl-(I-3′,II-8)-apigenin (16). Biflavanones are now represented by a number of compounds. Naringeninyl-(I-3′,II-8)-naringenin has been found in both *Semecarpus* and *Toxicodendron* species of the Anacardiaceae (2, 94, 172). *Semecarpus anacardium* also contains three other biflavanones with different hydroxylation patterns (257). A related biflavanone is present in *Cycas revoluta* (124). Further structural modification of the amentoflavones is primarily through glycosylation. A series of C-4′ and C-7 glucosides has been isolated from five species of the ancient order of Psilotales (372). A C-glucoside has recently been obtained from nut shells of *Anacardium* (260). A methylated flavone-flavon-3-ol, abiesin, was isolated recently from *Abies webbiana* leaves (41), and a most unusual di-3-*O*-methyl-biflavon-3-ol has been isolated from *Strychnos pseudoquina* (268).

Substantial hindrance to substitution at the C-6 position is evident from the restricted occurrence of the robustaflavones (I-3′,II-6 biflavonoids) (**2**). The biapigenin, robustaflavone, and its biflavanone analog have been reported in *Araucaria* and *Juniperus* of the gymnosperms (125). Within the angiosperms, robustaflavone is present in *Rhus* species and a biflavanone has been reported in *Semecarpus* (125).

The cupressuflavones (I-8,II-8 biflavonoids) (**3**) are also a small group of compounds compared to the amentoflavones, and usually co-occur with them (Table 7.6.2). Cupressuflavone and its variously C-4′ and C-7 methylated derivatives are prominent constituents of the Araucariaceae and Cupressaceae (125). Curiously they have not yet been found in the Taxodiaceae. All of these compounds carry a 5,7,4′-substitution pattern. One flavanone-flavone dimer, mesuferrone-B (307), and a biflavanone, mesuferrone-A (308), have been described from *Mesua ferrea*.

The agathisflavones (I-6,II-8 biflavonoids) (**4**) are much more rare than their corresponding I-8,II-8 analogs with only agathisflavone and a few C-4′ and C-7 methylated derivatives being present in the Araucariaceae (125, 126). These compounds have not yet been found in the Cupressaceae, Podocarpaceae, or Taxodiaceae. As expected from the restricted substitution at the C-6 position, the I-6,II-6 biflavonoids (**5**) are not common, and, in fact, succedaneaflavone (43) is the only known natural product.

Three groups of biflavonoids linked through the C-3 of the heterocyclic ring have been isolated from plants (Table 7.6.3). The taiwaniaflavones (I-3,II-3′) (**6**) appear in only the Taxodiaceae as the biflavone taiwaniaflavone (183) and its C-4′ and C-7-*O*-methyl derivatives. The second group, the (I-3,II-8 biflavonoids) (**7**), occurs primarily in the angiosperms and as flavanone-flavone or biflavanones. *Garcinia* species contain naringeninyl-(I-3,II-8)-apigenin or volkensiflavone (150) and the 7′-*O*-glucoside, fukugiside (213), as examples of the wide range of flavanone-flavone and biflavanones present in this genus. These compounds make up one of the largest groups of biflavonoids, with the complexity of substitution ranging from bi-naringenin and its 7-*O*-glucoside through naringeninyl-

(I-3,II-8)-taxifolin to eriodictolyl-(I-3,II-8)-taxifolin (125, 126, 295) (Table 7.6.3). A flavanone-aurone derivative, zehyerin (**7A**), has been isolated from a *Phyllogeiton* species of the Rhamnaceae (371). Larixinol (**7B**), a (I-3,II-8) biflavonoid derived from unusual coupling of a dihydroflavonol with a flavan-3-ol was isolated recently from *Larix gmelini* (336).

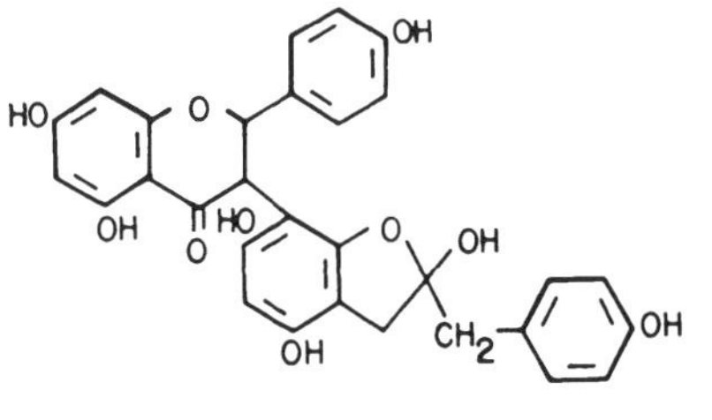

7A

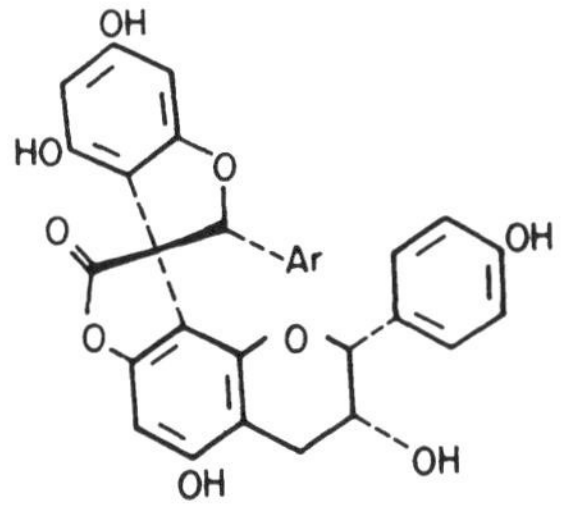

7B

The I-3,II-3 linked chamaejasmins (**8**) have received much attention recently and their number has grown to at least eight known natural products. All compounds reported to date are biflavanones, and variations in their stereochemistry account for much of the structural diversity. Chamaejasmin is reported to have the rare 2*S*,3*S* configuration (125). Seven biflavanones of this group have been isolated from *Stellera chamaejasme* in the past three years and all of these compounds are of a 2*R* stereochemistry (48, 231, 269, 270, 390).

Two groups of biflavonoids have diaryl ether linkages (Table 7.6.4). The hinokiflavones (I-4′,*O*,II-6-biflavonoids) (**9**) are often found in the Coniferales that also contain amentoflavones, but their distribution is more restricted than might be predicted from uncontrolled oxidative coupling (125, 126). As in the amentoflavones, the various C-4′- and C-7-*O*-methylated derivatives are common, particularly in the Cupressaceae and the Taxodiaceae. Only one flavanone-flavone analog has been reported, naringeninyl-(I-4′,*O*,II-6)-apigenin, and this compound was found in the primitive Cycadaceae (16). A flavanone-chalcone C-glucoside, occidentoside, has recently been isolated from the nut shells of *Anacardium occidentale* (259). Although other diaryl ether-linked biflavonoids have been synthesized (125), the only other group of natural products with this type of linkage is that of the ochnaflavones (I-4′,*O*,II-3′-biflavonoids). Ochnaflavone (**10**) and its C-4′ and C-7-*O*-methyl derivatives are found in *Ochna* species of the Theaflorae (284, 285). These biflavones vary primarily in their methoxylation patterns and now represent a comparatively large class of biflavonoids.

7.6.2.2 Distribution of Biflavonoids

A summary of the distribution of biflavonoids in woody plants is provided in Tables 7.6.1 through 7.6.4. References have been selected to demonstrate more specific distribution of biflavonoids by plant species. More detailed listings of the occurrences of these compounds in all plants as well as of known synthetic products up to 1981 are provided in the reviews prepared by Geiger and Quinn (125, 126).

Table 7.6.2. Distribution of biflavonoids: cupressuflavones, agathisflavones and succedaneaflavone

Group	Class	Common name	Substitution			Occurrence[1]		Reference
			C-5	C-7	C-4′	Coniferales	Dicotyledoneae	
I-8,II-8	Biflavone	cupressuflavone	OH	OH	OH	Cephalotaxaceae	Casuarinaceae	3, 261, 295
			OH	OH	OH	Cupressaceae		
						Podocarpaceae		
		4′-*O*-methyl-cupressuflavone	OH	OH	OCH_3	Araucariaceae		350
			OH	OH	OH			
		7-*O*-methyl-cupressuflavone	OH	OCH_3	OH	Araucariaceae		200
			OH	OH	OH	Cupressaceae		
						Podocarpaceae		
		4′-7-di-*O*-methyl-cupressuflavone	OH	OCH_3	OCH_3	Araucariaceae		169
			OH	OH	OH			
		7,7′-di-*O*-methyl-cupressuflavone	OH	OCH_3	OH	Araucariaceae		4
			OH	OCH_3	OH	Cupressuaceae		
		7,4′,7″-tri-*O*-methyl-tetramethylether	OH	OCH_3	OCH_3	Araucariaceae		199
			OH	OCH_3	OCH_3			
		cupressuflavone-tetramethylether	OH	OCH_3	OCH_3	Araucariaceae		4, 170
			OH	OCH_3	OCH_3			
	Flavanone-Flavone	mesuaferrone B	OH	OH	OH		Clusiaceae	309
			OH	OH	OH			
	Biflavanone	mesuaferrone A	OH	OH	OH		Clusiaceae	42, 46
			OH	OH	OH		Anacardiaceae	

I-6,II-8	Biflavone	agathisflavone	OH	OH	OH	Araucariaceae	Anacardiaceae	200
			OH	OH	OH			
		7-*O*-methyl agathisflavone	OH	OCH_3	OH	Araucariaceae		297
			OH	OH	OH			
		7,4″-di-*O*-methyl agathisflavone	OH	OCH_3	OH	Araucariaceae		297
			OH	OH	OCH_3			
		7,7″-di-*O*-methyl agathisflavone	OH	OCH_3	OH	Araucariaceae		200
			OH	OCH_3	OH			
		7,7″,4″-tri-*O*-methyl agathisflavone	OH	OCH_3	OH	Araucariaceae		169
			OH	OCH_3	OCH_3			
		agathisflavone-tetramethylether	OH	OCH_3	OCH_3			130
			OH	OCH_3	OCH_3			
	Flavanone-Flavone	rhusflavone	OH	OH	OH		Anacardiaceae	47
			OH	OH	OH			
	Biflavanone	rhusflavanone	OH	OH	OH		Anacardiaceae	44, 47, 228
			OH	OH	OH			
I-6,II-6	Biflavanone	succedaneaflavone	OH	OH	OH		Anacardiaceae	43
			OH	OH	OH			

Table 7.6.3. Distribution of biflavonoids: taiwaniaflavones, volkensiflavones, and chamaejasmin

Group	Class	Common name	Substitution					Occurrence[1]		Reference
			C-3	C-5	C-7	C-4′	C-3′	Coniferales	Dicotyledoneae	
I-3,II-3′	Biflavones	taiwaniaflavone	FLA	OH	OH	OH	H	Taxodiaceae		183
			H	OH	OH	OH	H			
		taiwaniaflavone-7″-methylether	FLA	OH	OH	OH	H	Taxodiaceae		183
			H	OH	OCH_3	OH	H			
		taiwaniaflavone-4′,7″-dimethylether	FLA	OH	OH	OCH_3	H	Taxodiaceae		183
			H	OH	OCH_3	OH	H			
I-3,II-8	Flavanone-Flavone	volkensiflavone	FLA	OH	OH	OH	H		Clusiaceae	150, 181, 293, 373, 374
			H	OH	OH	OH	H			
		spicatiside	FLA	OH	OH	OH	H		Clusiaceae	213
			H	OH	O-GLU	OH	H			
		fukugetin	FLA	OH	OH	OH	H		Clusiaceae	27, 91, 185, 215, 293, 373
			H	OH	OH	OH	OH			
		3-*O*-methyl fukugetin	FLA	OH	OH	OH	H		Clusiaceae	215
			H	OH	OH	OH	OCH_3			
		fukugiside	FLA	OH	OH	OH	H		Clusiaceae	213
			H	OH	O-GLU	OH	OH			
	Biflavanone	GB1a	FLA	OH	OH	OH	H		Clusiaceae	8, 14, 173
			H	OH	OH	OH	H		Liliiflorae	
		GB1a-glucoside	FLA	OH	OH	OH	H		Clusiaceae	45
			H	OH	O-GLU	OH	H			
		GB1	FLA	OH	OH	OH	H		Clusiaceae	8, 14, 173
			OH	OH	OH	OH	H			
		GB2a	FLA	OH	OH	OH	OH		Clusiaceae	8, 14, 173
			H	OH	OH	OH	OH			
		GB2	FLA	OH	OH	OH	H		Clusiaceae	8, 14, 173
			OH	OH	OH	OH	OH			
		kolaflavanone	FLA	OH	OH	OH	H		Clusiaceae	60
			OH	OH	OH	OCH_3	OH			

		xanthochymusside	FLA	OH	OH	OH	H	Clusiaceae	214
			OH	OH	OH	OCH_3	OH		
		manniflavanone	FLA	OH	OH	OH	OH	Clusiaceae	63
			OH	OH	OH	OH	OH		
I-3,II-3	Biflavanone	chamaejasmin (2*S*,3*S*;2*S*,3*S*)	FLA	OH	OH	OH	H	Thymelaeceae	125
			FLA	OH	OH	OH	H		
		7-*O*-methyl chamaejasmin	FLA	OH	OCH_3	OH	H		390
			GLA	OH	OH	OH	H		
		isochamaejasmin (2*R*,3*R*;2*R*,3*R*)	FLA	OH	OH	OH	H	Thymelaeaceae	48
			FLA	OH	OH	OH	H		
		neochamaejasmin A	FLA	OH	OH	OH	H	Thymelaeaceae	270
		(2*R*,3*S*;2*R*,3*S*)	FLA	OH	OH	OH	H		
		neochamaejasmin B (2*R*,3*S*;2*S*,3*S*)	FLA	OH	OH	OH	H	Thymelaeaceae	270
			FLA	OH	OH	OH	H		
		chamaejasmin A (2*R*,3*R*;2*S*,3*R*)	FLA	OH	OH	OCH_3	H	Thymelaeaceae	231
			FLA	OH	OH	OCH_3	H		
		chamaejasmin B (2*R*,3*S*;2*R*,3*S*)	FLA	OH	OH	OCH_3	H	Thymelaeaceae	231
			FLA	OH	OH	OCH_3	H		
		chamaejasmin C (2*R*,3*S*;2*R*,3*S*)	FLA	OH	OCH_3	OCH_3	H	Thymelaeaceae	231
			FLA	OH	OH	OCH_3	H		

Table 7.6.4. Distribution of biflavonoids: hinokiflavones and ochnaflavones

Group	Class	Common name	Substitution			Occurrence[1]		Reference
			C-5	C-7	C-4′	Coniferales	Dicotyledoneae	
I-4′-*O*-II-6	Biflavones	hinokiflavone	OH	OH	FLA	Araucariaceae	Anacardiaceae	121, 190, 194, 306
			OH	OH	OH	Cupressaceae Podocarpaceae Taxodiaceae	Casuarinaceae	
		neocryptomerin	OH	OCH_3	FLA	Podocarpaceae		251, 255
			OH	OH	OH			
		isocryptomerin	OH	OH	FLA	Cupressaceae		249, 254
			OH	OCH_3	OH	Taxodiaceae		
		cryptomerin A	OH	OH	FLA	Cuperessaceae		253
			OH	OH	OCH_3	Podocarpaceae		
		chamaecyparin	OH	OCH_3	FLA	Cupressaceae		251
			OH	OCH_3	OH			
		hinokiflavone-7,4′-dimethylether	OH	OCH_3	FLA	Cupressaceae		190
			OH	OH	OCH_3			
		cryptomerin B	OH	OH	FLA	Taxodiaceae		253, 270
			OH	OCH_3	OCH_3			
		hinokiflavone-7,7″,4″-trimethylether	OH	OCH_3	FLA	Cupressaceae		190, 250
			OH	OCH_3	OCH_3			
	Flavanone-Flavone	2,3-dihydrohinokiflavone	OH	OH	FLA	Cycadaceae		16
			OH	OH	OH			
	Flavanone-Chalcone	occidentoside	OH	OH	FLA		Anacardiaceae	259
			OH	OH	OH			
I-4′,0,II-3′	Biflavone	ochnaflavone	OH	OH	FLA		Ochnaceae	284, 285
			OH	OH	OH			
		ochnaflavone-4′-methylether	OH	OH	FLA		Ochnaceae	284, 285
			OH	OH	OCH_3			
		ochnaflavone 7-methylether	OH	OH	FLA		Ochnaceae	184
			OH	OCH_3	OH			
		ochnaflavone-7,4′-dimethylether	OH	OCH_3	FLA		Ochnaceae	284, 285
			OH	OH	OCH_3			
		ochnaflavone-7,7′,4′-trimethylether	OH	OCH_3	FLA	Ochnaceae	284, 285	
			OH	OCH_3	OCH_3			

The biflavonoids have usually been isolated from leaf tissues, although fruit and bark are also commonly good sources of these compounds. The biflavonoids are less commonly isolated from wood, but the heartwood of *Garcinia* spp. is exceptional as a source of particularly interesting compounds.

7.6.2.3 Significant Properties of Biflavonoids

By far the most interest in biflavonoids has focused on their taxonomic relationships (125, 126). However, the biflavonoids have received attention also as principal constituents of leaf extracts that have been used as medicines – for example, *Juniperus*, *Ginkgo*, and *Semecarpus* extracts. *Garcinia* extracts are further interesting examples. The nuts of *Garcinia* are habitually chewed by natives of Ghana and the nut pulp is used as an antiseptic for treatment of cuts and prevention of sore throats, and as an antidote for arrow poisons (60). Twigs and roots are used as chew sticks and the root bark is reportedly a cure for diarrhea and dysentery (63). *Garuga* leaf extracts are mixed with honey for treatment of asthma (227). Loss of the U.S. Food and Drug Administration's approval of flavonoid preparations as effective drugs has undoubtedly restricted study of the pharmacological properties of these compounds. Research continues in Europe, however, and studies reported on biological activity of the biflavonoids indicate that they often exhibit greater activity than their monomers in terms of antispasmogenic as well as peripheral vasodilation and antibradykinin activities (162, 267). A mixture of sciadopitysin, ginkgetin, and sequoiaflavone showed properties of a central nervous system depressant, and as an analgesic and antipyretic (12, 246, 326). Few of the natural biflavonoids carry a 3′,4′-dioxy B-ring generally associated with flavonoid activity on capillary permeability or ascorbate oxidation. However, because many of the biflavonoid extracts are rich sources of methylated apigenin dimers and methylated flavonoids are most effective in reducing blood cell aggregation (246), study of biflavonoids in this regard could prove beneficial. The biflavonoids have been demonstrated to inhibit cyclic GMP and AMP phosphodiesterase activity (17, 326). Because various derivatives of flavones are known as potent coronary vasodilators, heart stimulants, and as antispasmolytics (368), study of these properties on similar biflavonoid derivatives might be rewarding.

To be effective as orally ingested medicines, these compounds or their biologically active products must survive the microflora of the gastrointestinal tract. Griffith's excellent review of the mammalian metabolism of flavonoids contains no references to biflavonoids (128). Several features of these compounds such as frequent substitution at the C-8 position by either the B- or A-ring of a flavonoid unit and the common occurrence of 7-methoxy flavan units suggest that the biflavonoids may not be readily degraded by intestinal microflora. Certainly additional study of the metabolism of these compounds would be interresting from the standpoint of the fate of the biphenyl moieties. Considerable debate remains about the relationships of flavonoids to animal cell malignancy. MacGregor and Jurd (234) concluded that a 3-hydroxyl and double bond at the 2-3 positions were structural features of flavonoids required for mutagenicity. Although quercetin, for example, is known to have mutagenic activity, the biflavonoids commonly found in plant extracts do not carry a 3-hydroxyl. Biflavonoids have, in contrast, been shown to retard the development of hepatoma cells (179). There are no reports suggesting any toxicity of biflavonoids to mammals. However, podocarpusflavone-A and 7″,4″-dimethylamentoflavone inhibit the growth of insects (220). Also, since methylated flavones are noted as potent fish poisons, study of the activity of the methylated biflavonoids in this regard should be interesting (246, 363). As is true of research on most secondary plant metabolites, the past preoccupation on structure elucidation with relative neglect of the significance of these compounds needs to be rectified. Interesting opportunities are apparent in further definition of the biological properties of the biflavonoids.

7.6.3 Proanthocyanidins

Although earlier reviews by Haslam (134–136) discuss the many problems that have arisen about the systematic classification and nomenclature of proanthocyanidins, it seems useful to repeat the definition of the term proanthocyanidin as initially established by Freudenberg and Weinges (116) – "the proanthocyanidins are all the colorless substances isolated from plants which, when treated with acid, form anthocyanidins." Thus the leucoanthocyanidins (now restricted to the flavan-3,4-diols), the condensed tannins (Sect. 7.7), and any other colorless flavonoid derivatives that produce anthocyanidins on heating with acid should be grouped under this class of compounds.

The proanthocyanidins are further classified on the basis of their hydroxylation patterns (134–136). The propelargonidins (**11**), procyanidins (**12**), and prodelphinidins (**13**) are thus named for the anthocyanidin generated on treatment with acid, while their 5-deoxy analogs, the proguibourtinidins (**14**), profisetinidins (**15**), and prorobinetinidins (**16**) are named for their corresponding flavan-3,4-diols (Fig. 7.6.1). Not all known proanthocyanidins can be subdivided using this classification now that mixed procyanidin → prodelphinidin oligomers have been isolated from *Hordeum vulgare* (barley). Since mixed proanthocyanidins (containing both catechin and gallocatechin upper units for example) are now known (Sect. 7.7), this system of classification may soon become impossible to use.

The compounds within the procyanidin class were named by a trivial system introduced by Freudenberg and Weinges (116, 381) and extended by Haslam (134) (i.e., the procyanidins B1, B2, etc.). As the number of new procyanidin oligomers accumulated, this system became increasingly difficult to use. The alternative is to use the fully systematic IUPAC system (134, 316). Whereas this is fully descriptive of the absolute stereochemistry, it is difficult to use and can easily become misleading because of the large number of asymmetric centers in trimeric and tetrameric proanthocyanidins. A proposed nomenclature analogous to the IUPAC system for naming of natural products such as carbohydrates circumvents these problems (141). In this system, the basic structural units are named from their familiar monomeric flavan-3-ols for the 2*R* isomers, and for the less common 2*S* isomers the prefix *ent*- is used to define the absolute stereochemistry at C-2 and C-3 of each unit. The location of the interflavanoid bond is denoted within brack-

Fig. 7.6.1. Structural variations of proanthocyanidins and related flavan-3,4 diols

		R2a	R2b	R3a	R3b	R4a	R4b	R5	R7	R3′	R5′
(*11*)	Upper	H	PH	OH	H	H	FLA	OH	OH	H	H
	Lower	H	PH	OH	H	H	H	OH	OH	OH	H
(*12*)	Upper	H	PH	H	OH	FLA	H	OH	OH	OH	H
	Lower	H	PH	OH	H	H	H	OH	OH	OH	H
(*13*)	Upper	H	PH	H	OH	FLA	H	OH	OH	OH	OH
	Lower	H	PH	OH	H	H	H	OH	OH	OH	OH
(*14*)	Upper	PH	H	H	OH	FLA	H	H	OH	H	H
	Lower	H	PH	OH	H	H	H	OH	OH	OH	H
(*15*)	Upper	H	PH	OH	H	H	FLA	H	OH	OH	H
	Lower	H	PH	OH	H	H	H	OH	OH	OH	H
(*16*)	Upper	H	PH	OH	H	H	FLA	H	OH	OH	OH
	Lower	H	PH	OH	H	H	H	OH	OH	OH	H
(*17*)	Upper	PH	H	OH	H	H	FLA	OH	OH	OH	H
	Lower	PH	H	H	OH	H	H	OH	OH	OH	H
(*18*)		H	PH	OH	H	H	OH	H	OH	H	H
(*19*)		H	PH	OH	H	OH	H	H	OH	OH	H
(*20*)		H	PH	H	OH	H	OH	H	OH	OH	OH
(*21*)		PH	H	H	OH	OH	H	OH	OH	OH	H

ets as in the carbohydrates. The interflavanoid bond configuration at C-4 is denoted as α or β as in the IUPAC rules. Thus the familiar procyanidin B1 is named epicatechin-(4β→8)-catechin (**12**), the analogous prodelphinidin is named epigallocatechin-(4β→8)-gallocatechin (**13**), and the corresponding 2*S* enantiomer is named *ent*-epicatechin-(4α→8)-*ent*-catechin (**17**). This nomenclature is used throughout this section as well as in Sect. 7.7.

In comparison, a simplified nomenclature for the flavan-3,4-diols (Fig. 7.6.1, **18**–**21**) is not necessary. The hydroxylation patterns of these compounds are readily recognized from their common names; guibourtacacidins (**18**, 7,4′-dihydroxy), mollisacacidins (**19**, 7,3′,4′-trihydroxy), robinetinidins (**20**, 7,3′,4′,5′-tetrahydroxy), leucocyanidins (**21**, 5,7,3′,4′-tetrahydroxy), teracacidins (7,8,4′-trihydroxy), and melacacidins (7,8,3′,4′-tetrahydroxy). Because only three chiral centers need to be defined, their absolute stereochemistry is readily denoted through use of the *R* or *S* prefix. This is clearly preferable to the use of common names. For example, 2*R*,3*S*,4*R*-mollisacacidin has been variously named (+)-mollisacacidin, (+)-gleditsin and (+)-leucofisetinidin (377). Use of the *R* and *S* prefix to define the absolute stereochemistry of these compounds avoids the considerable confusion that has occurred with the informal nomenclature used in the past.

Proanthocyanidins are by far the most common oligomeric flavonoids found in plants (134–136). The low molecular weight oligomers are usually found in low concentration in plant extracts containing the condensed tannins (Sect. 7.7), and, consequently, these compounds are usually of comparatively little biological or economic significance. However, the elucidation of the structure and reactions of these oligomers has formed the basis on which most of our understanding of condensed tannin chemistry rests. Therefore, this section concentrates on the structure and reactions of these compounds and is meant to be an introduction to the review of condensed tannins in which their biosynthesis, distribution in the plant kingdom and biological significance are treated more fully in Sect. 7.7 and Chap. 10.3. The compounds reviewed in this section include the flavan-3-ols, flavan-3,4-diols and oligomeric proanthocyanidins ranging from the dimers to tetramers. Although not properly included among proanthocyanidins, the flavan-3-ols are discussed here since their chemistry is so closely linked to that of the proanthocyanidins.

7.6.3.1 Flavan-3-ols

Study of the chemistry of flavan-3-ols has historically been and undoubtedly will continue to be essential to progress in our understanding of the proanthocyanidins. The pharmacological properties of epicatechin and catechin derivatives make these compounds interesting in their own right (15, 61, 222), but this review will concentrate on their relationships with the proanthocyanidins or condensed tannins. Although the early work of Freudenberg and his colleagues (377) on acid-catalyzed condensation of catechin may not be relevant to the biosynthesis of proanthocyanidins, these reactions are descriptive of mechanisms that may occur in vitro. Similarly, oxidative coupling reactions of flavan-3-ols certainly should not be forgotten in terms of secondary enzymic dehydrogenation and autoxidation of proanthocyanidins (134, 377). As will be seen in the discussions that follow, flavan-3-ols serve as useful model compounds for study of many of the reactions of proanthocyanidins. Finally, the flavan-3-ols are, of course, important intermediates in the biogenesis of most of the proanthocyanidins since they usually serve as nucleophiles terminating their polymerization. It is appropriate then to begin our consideration of the chemistry of proanthocyanidins by first reviewing the flavan-3-ols.

Although a comprehensive review of the biosynthesis of the proanthocyanidins is provided by Hillis (154) and Porter in Sect. 7.7, brief mention must be made here of the problem of biogenesis of flavan-3-ols because this remains one of the most outstanding questions concerning these compounds. The commonly accepted route to flavan-3-ols involves the reduction of dihydroflavonols to flavan-3,4-diols followed by a second reduction to the flavan-3-ols (134, 219). The difficulty with this scheme is that nearly all natural dihydroflavonols are of a 2,3-*trans* or 2*R*,3*R* configuration (23) and yet many of the natural flavan-3-ols have a 2,3-*cis* (also 2*R*,3*R* but analogous to a 2*R*,3*S* configuration in dihydroflavonols) stereochemistry. Roux's isolation of an α-hydroxychalcone as a natural product provided a route to 2*R*,3*S*-*cis* dihydroflavonols (315). However, the fact remains that no natural dihydroflavonols of this stereochemistry had been isolated from plant tissues containing predominantly 2*R*,3*R*-proanthocyanidins until the recent isolation of (−)-2,3-*cis*-3,3′,4′,7,8-pentahydroxyflavanone from *Acacia melanoxylon* by Foo (101).

Haslam's school (175, 177, 299) also suggests a biogenetic route through an α-hydroxychalcone but, rather than formation of a 2*R*,3*S* dihydroflavonol, a flav-3-en-3-ol that would provide a symmetrical intermediate is postulated. Two stereospecific reductions could then provide the flavan-3-ols of either a 3*S* or 3*R* configuration and this could also account for the frequent occurrence of differing configurations for the C-3 hydroxyl in the procyanidin units and the terminating flavan-3-ol unit of the polymers.

Stafford (338–341) has proposed an alternative scheme in which these compounds are formed by reductions of dihydroflavonols, but a C-3 epimerase is postulated to account for production of a 2,3-*cis* (2*R*,3*S*)-dihydroflavonol. This could be held by an enzyme where it is reduced stereospecifically through two steps to give a 2,3-*cis* flavan-3-ol such as epicatechin. This scheme proposed by Stafford and that proposed by Jacques et al. (175, 177) both require two metabolic pools, one for the flavan-3,4-diols and one for the flavan-3-ols, to account for differences in the stereochemistry of the flavan-3-ols and proanthocyanidins.

Support for inversion at C-3 involving a flav-3-en-3-ol was obtained recently but, rather than its formation from an α-hydroxychalcone, a tautomeric rearrangement of a quinone methide is involved (10, 145). Reaction of polymeric procyanidins with phenylmethanethiol at ambient temperature and alkaline pH gives mono- and di-benzylsulfide derivatives of diarylpropan-2-ols (145). Loss of phenylmethanethiol and tautomeric rearrangement lead to the formation of two prop-2-one derivatives analogous to ring-opened flav-3-en-3-ols. Earlier work (10, 104, 139) shows that procyanidins are readily synthesized by reactions of flavan derivatives with good leaving groups at C-4 (i.e., 4-ols) through quinone methide intermediates. Formation of a flav-3-en-3-ol through rearrangement of a quinone methide intermediate is also consistent with the results of Haslam's labeling experiments (299) and provides for differences in stereochemistry between the flavan-3-ols and the procyanidins. However, in the most recent contribution to this question, Foo (101) demonstrated facile conversion of the 2,3-*cis* dihydroflavonol to the 2,3-*trans* isomer, and he suggested that the usual absence of the 2,3-*cis* dihydroflavonols in plant extracts is due to the extreme lability of these compounds to epimerization.

7.6.3.1.1 Structure of Flavan-3-ols

Runge first isolated "catechins" (i.e., epicatechin) **(22)** from *Acacia catechu* in 1821 but it wasn't until the 1920s that Freudenberg made progress in the elucidation of the structures of these compounds (115, 377). Their general constitution as 5,7,3′,4′-flavan-3-ols was not established until 1925 (115). The absolute stereochemistry of catechin and epicatechin was finally established by Birch et al. (22) and Hardegger et al. (131) in 1957.

Because of the importance of the shape of proanthocyanidins to many of their properties, the conformation of the heterocyclic ring is currently of interest. Engel et al. (95) examined the crystal structure of 8-bromo-5,7,3′,4′-tetra-*O*-methylcatechin and established that both the heterocyclic oxygen and C-4 lie slightly above the mean plane of the adjacent aromatic ring and that C-2 lies well above (i.e., 60 pm) and C-3 lies below (−23 pm) this plane. The heterocyclic ring in this com-

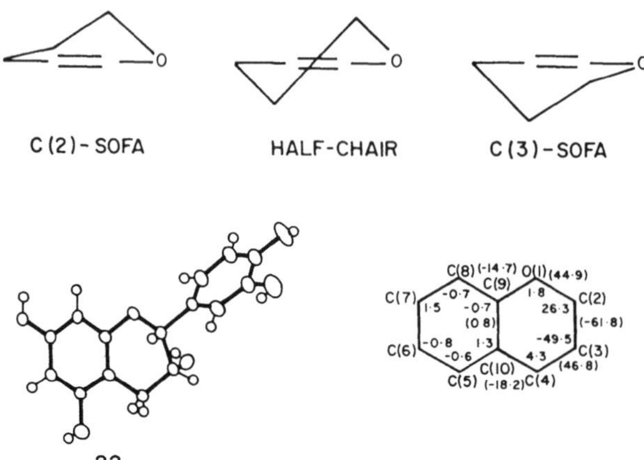

Fig. 7.6.2. Heterocyclic ring conformations found in flavan-3-ols, flavan-3,4-diols, and proanthocyanidins

pound is, therefore, between that of a C(2)-sofa and a half-chair conformation (Fig. 7.6.2). The crystal structure of 8-bromo-5,7,3′,4′-tetra-*O*-methylepicatechin has also been examined. Although details of atomic coordinates have not yet been published, Haslam (136) provides a summary account of the deviations from planarity of the atoms in the heterocyclic ring. Here the heterocyclic oxygen and C-4 atoms lie virtually coplanar with the adjacent carbons of the fused aromatic ring. The C-2 and C-3 carbons are both below this plane by 4 pm and 70 pm, respectively. The ring conformation for this compound is then described as a C(3)-sofa. Recent studies on the crystal structure of 6-bromo-tetra-*O*-methylcatechin have shown that this compound crystallizes with two heterocyclic ring conformations, one as a half-chair and the other as a sofa conformation (92).

The structure of penta-*O*-acetyl-catechin is an interesting example of problems encountered in definition of the conformation of these compounds. In the crystal state, the heterocyclic ring is in a "reverse" half-chair conformation with both the *B*-ring and 3-acetoxy substituents in axial positions (120). However, the heterocyclic ring proton coupling constants $J_{2,3} = 6.5$, $J_{3,4\alpha} = 5.0$, and $J_{3,4\beta} = 7.0$ Hz are not consistent with the crystal state conformation, indicating a change in heterocyclic ring conformation in solution. ^{1}H-NMR spectra recorded over a range of 20° to −90 °C show selective broadening of the heterocyclic ring protons with a decrease in temperature suggesting exchange between different heterocyclic ring conformations in solution (120). Porter et al. (304) have examined these properties further through molecular mechanical (MM2) studies and detailed analyses of heterocyclic ring coupling constants. These analyses also suggest that this compound exists as an equilibrium mixture and more specifically a mixture of two half-chair conformations, the *A*-conformer (B-ring axial) and *E*-conformer (B-ring equatorial). The above coupling constants can be accounted for by an equilibrium mixture of the *A*-conformer and the *E*-conformer in ratio of 52:48 in either acetone or chloroform solutions.

Fronczek et al. (119) and Spek et al. (337) have also examined the crystal and solution structure of epicatechin. The equatorial phenyl ring at C-2 and the axial hydroxyl at C-3 are confirmed, and here the heterocyclic ring is more closely described as a half-chair (Fig. 7.6.2). The heterocyclic oxygen and the C-4 lie slightly above the plane of the fused aromatic ring. The C-2 lies above the plane by 26.3 pm and C-3 lies below it by 49.5 pm. Conformational changes did not appear to occur when epicatechin dissolved in dioxane since dipole moments and their temperature dependence could be rationalized by a theoretical rotational isomeric state analysis based on the crystal structure. Likewise, the small $J_{2,3}$ coupling of the heterocyclic protons of epicatechin in solution suggests a half-chair conformation. However, the heterocyclic ring proton coupling constants had not been accurately measured because of broadening of the H-2 and H-3 signals by long-range couplings. Porter et al. (304) resolved this problem by selective decoupling experiments on a deuterium-exchanged sample of epicatechin. This showed $J_{2,3} = 1.7$, $J_{3,4\alpha} = 0.8$, and $J_{3,4\beta} = 0.6$ Hz in d_6-acetone, a $J_{2,3}$ coupling significantly larger than would be predicted for a half-chair conformation as was found in the crystal state. Line broadening was also seen for the H-2 and H-3 protons of this compound when the temperature was reduced to $-90\,°C$. The above coupling constants could be accounted for by an equilibrium between the *A*- and *E*-conformers in a ratio of 14:86, the *E*-conformer predominating to a high degree as would be expected.

Crystal structures have not yet been determined for other common flavan-3-ols such as catechin or the acetate and/or methyl ether derivatives of epicatechin. ^{1}H-NMR spectra of penta-*O*-acetyl-epicatechin showed heterocyclic ring proton coupling constants consistent with a C(2)-sofa to half-chair conformation in which the B-ring was equatorial and the C-3 acetoxy function was in an axial position (110). Dipole moment and conformational analyses of tetra-*O*-methyl-catechin and tetra-*O*-methyl-epicatechin indicated approximate half-chair conformations that were slightly puckered to permit weak hydrogen bonding between the C-3 hydroxyl and the heterocyclic ring oxygen for both compounds (241). These conclusions may have to be modified on the basis of the new approach to interpretation of heterocyclic ring proton coupling offered by Porter et al. (304). Their interpretation of the heterocyclic ring conformation of catechin is particularly informative. The coupling constants found for catechin in d_6-acetone were $J_{2,3} = 7.4$, $J_{3,4\alpha} = 5.3$, and $J_{3,4\beta} = 8.3$ Hz. If it is assumed that these coupling constants arise from a single preferred conformation, the calculated conformation would have an energy of 4.2 kcal/mole higher than that found for the half-chair forms. The observed coupling constants are most satisfactorily interpreted as an equilibrium between the *A*- and *E*-conformers in a ratio of 38:62 respectively. This approach to interpretation of the heterocyclic ring proton coupling constants, as opposed to the assumption of a single fixed C(2)- or C(3)-sofa conformation, has important implications for the interpretation of the structures of the C-4 substituted proanthocyanidins

7.6.3.1.2 Distribution of Flavan-3-ols

More than 40 flavan-3-ols are now known natural products (Table 7.6.5). The most commonly reported flavan-3-ols, catechin and epicatechin, are broadly dis-

Table 7.6.5. Natural Flavan-3-ols

Name	Substitution					Stereochemistry		Occurrence	Reference(s)
	C-5	C-7	C-3′	C-4′	C-5′	C-23	C-8		
(+)-fisetinidol	H	OH	OH	OH	H	S	R	*Afzelia, Colophospermum*	66
(−)-fisetinidol	H	OH	OH	OH	H	R	S	*Acacia*	323
(+)-epifisetinidol	H	OH	OH	OH	H	S	S	*Colophospermum*	82
(−)-robinetinidol	H	OH	OH	OH	OH	R	S	*Acacia*	321, 324, 376
(+)-afzelechin	OH	OH	H	OH	H	R	S	*Eucalyptus, Nothofagus*	155, 156
(−)-epiafzelechin	OH	OH	H	OH	H	R	R	*Afzelia*	207
(+)-epiafzelechin	OH	OH	H	OH	H	S	S	Palmae	73
(+)-catechin	OH	OH	OH	OH	OH	R	S	General distr.	22, 131, 377
(−)-epicatechin	OH	OH	OH	OH	H	R	S	General distr.	22, 131, 377
(+)-epicatechin	OH	OH	OH	OH	H	S	S	Palmae	73
3′,4′,7,8-tetrahydroxyflavan-3-ol	H	OH OH(8)	OH	OH	H	R	S	*Prosopis*	322
3,5,7,3′,5′-pentahydroxyflavan	OH	OH	OH	H	OH	R	R	*Humboldtia* sp.	328
3′-*O*-methylepicatechin	OH	OH	OCH_3	OH	H	R	R	*Symplocos*	356
5,7,3′-tri-*O*-methylepicatechin	OCH_3	OCH_3	OCH_3	OH	H	R	R	*Cinnamomum*	256
5,7,4-tri-*O*-methylcatechin	OCH_3	OCH_3	OH	OCH_3	H	R	S	*Viguiera*	71
5,7-di-*O*-methyl 3′,4′-methylenedioxyepicatechin	OCH_3	OCH_3	O	CH_2	O	R	R	*Cinnamomum*	256
3′-prenyl-3,7,4′-trihydroxy	H	OH	prenyl	OH	H	R	S	*Broussonetia*	348
(+)-catechin-3-gallate	OH	OH	OH	OH	H	R	S	*Bergenia*	132
(+)-catechin-7-gallate	OH	O-GA	OH	OH	H	R	S	*Sanguisorba*	349
(−)-epicatechin-3-gallate	OH	OH	OH	OH	H	R	R	*Camellia*	35, 62
(−)-epicatechin-3,5-digallate	O-GA	OH	OH	OH	H	R	R	*Camellia*	35, 62
(−)-catechin-3-gallate	OH	OH	OH	OH	H	S	R	*Polygonum*	274
phylloflavan	OH	OH	OH	OH	H	S	R	*Phyllocladus*	106
(+)-catechin-5-*O*-β-D-glucoside	O-GL	OH	OH	OH	H	R	S	*Rheum*	271
(+)-catechin-7-*O*-β-D-glucoside	OH	O-GL	OH	OH	H	R	S	*Rhaphiolepis*	271
catechin-3′-*O*-β-D-glucoside	OH	OH	O-GL	OH	H	R	S	*Rheum*	196
catechin-4′-*O*-β-D-glucoside	OH	OH	OH	O-GL	H	R	S	*Rheum*	196
catechin-5,3′-di-*O*-β-D-glucoside	O-GL	OH	O-GL	OH	H	R	S	*Rheum*	196
catechin-3′-di-*O*-β-D-glucoside	OH	O-GL	O-GL	OH	H	R	S	*Rheum*	196

catechin-3′,4′-di-*O*-β-D-glucoside	OH	OH	O-GL	O-GL	H	R	S	*Rheum*	196
catechin-5,4′-di-*O*-β-D-glucoside	O-GL	OH	OH	O-GL	H	R	S	*Rheum*	196
catechin-8-C-β-D-glucoside	OH	OH	OH	OH	H	R	S	*Rheum*	196
catechin-6-*C*-β-D-glucoside	OH	OH	OH	OH	H	R	S	*Rheum*	196
(+)-catechin-7-arabinoside	OH	O-GL	OH	OH	H	R	S	*Polypodium*	357, 389
(+)-catechin-7-xyloside	OH	O-GL	OH	OH	H	R	S	*Ulmus*	76
(+)-catechin-7-apioside	OH	O-GL	OH	OH	H	R	S	*Polypodium*	195
3′-*O*-methylepicatechin-7-glucoside	OH	O-GL	OCH_3	OH	H	R	R	*Symplocos*	356
(+)-gallocatechin	OH	OH	OH	OH	OH	R	S	General distr.	322, 377
(−)-epigallocatechin	OH	OH	OH	OH	OH	R	R	General distr.	322, 377
4′-*O*-methylepigallocatechin	OH	OH	OH	OCH_3	OH	R	R	Celastraceae	74
(−)-epigallocatechin-3-gallate	OH	OH	OH	OH	OH	R	R	*Camellia*	35, 62
(−)-epigallocatechin-3,5-digallate	O-GA	OH	OH	OH	OH	R	R	*Camellia*	35, 62
kopsirachin	OH	OH	OH	OH	H	R	S	*Kopsia*	160

tributed throughout the gymnosperms and angiosperms. Analogs carrying a pyrogallol B-ring, gallocatechin and epigallocatechin, are also widely distributed. In comparison, afzelechins with a 4′-hydroxy B-ring are rare. The 5-deoxy flavan-3-ols are restricted in their distribution to the Leguminosae and Anacardiaceae, and the 3,7,4′-trihydroxyflavans analogous to the quibourtacacidins have not been reported. Both fisetinidol and robinetinidol are found in *Acacia* species. The 3,5,7,3′,5′-pentahydroxyflavan has recently been isolated from the wood of *Humboldtia laurifolia* (328).

Although a number of flavan-3-ols of the 2*S* configuration are known, their distribution is quite restricted in comparison with the above compounds. Both *ent*-epiafzelechin and *ent*-epicatechin are found in several species of the Palmae (73). *Afzelia xylocarpa* and *Colophospermum mopane* contain *ent*-fisetinidol, and *ent*-epifisetinidol is also present in the latter species (66, 82, 323). The flavan-3-ols readily undergo epimerization through inversion of configuration at C-2, making precautions necessary in their isolation. The above compounds clearly are not artifacts, however.

Natural methylated flavan-3-ols are extremely rare and it has only been in the past few years that most examples have been found. The first of these compounds to be reported was 4′-*O*-methylepigallocatechin, a compound with interesting muscle-relaxant properties (74, 182, 375). Since then, 3′-*O*-methylepicatechin from *Symplocus* (356), a 4,7,3′- (or 4′)-tri-*O*-methylepicatechin and 5,7-di-*O*-methyl-3′,4′-methylenedioxyepicatechin from *Cinnamomum cassia* (256) have been reported. The most recent example is the 5,7,4′-tri-*O*-methylcatechin, obtained from *Viguiera quinqueradiata* (71). Prenylated flavan-3-ols are also rare with broussinol, 1,3′-prenyl-3,7,4′-trihydroxyflavan, only recently having been reported from *Broussonetia papyrifera* in *Fusarium solani*-infected bark (348). This compound is a true phytoalexin, for, this and related compounds do not occur in healthy tissues.

Gallate esters are found in plants containing both hydrolyzable and condensed tannins. Both catechin- and epicatechin-3-gallate were synthesized by Weinges et al. (386) and, soon thereafter, Haslam (132) reported catechin-3-gallate as a natural product from *Bergenia* species. Epicatechin-3-gallate is a major constituent of Siebel and Beaujolais grapes particularly in the early stages of their development (65). The 3-galloyl and 3,5-digalloyl esters of epicatechin and epigallocatechin are found in green tea (35, 62). Catechin-7-gallate has recently been found in *Sanguisorba officinalis* (349). *Ent*-catechin-3-*O*-gallate, one of only two C-3 sub-

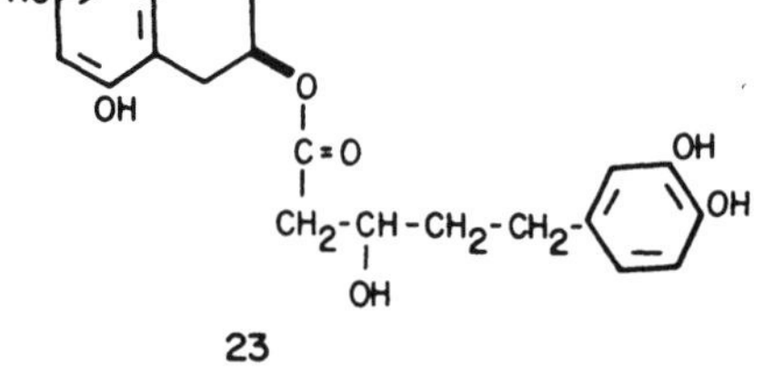

23

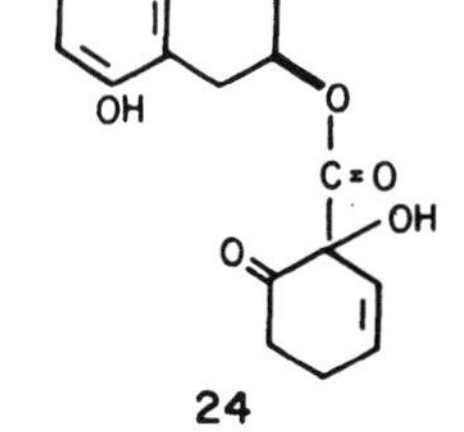

24

stituted analogs with a 2*S* configuration has been isolated from *Polygonum multiflorum* where it occurs with a mixture of procyanidin gallates (274). Foo et al. (106) have recently described another C-3 substituted flavan-3-ol derivative with a 2*S* configuration, phylloflavan – *ent*-epicatechin-3-d-(3,4-dihydroxyphenyl)-*β*-hydroxypentanoate (**23**) – isolated from *Phyllocladus alpinus*. Surprisingly, this compound co-occurred with the more usual 2*R* flavan-3-ols, catechin and epicatechin. This compound has strong hypertensive activity in rats (364). Another unusual example of acylation is in the 1,6-dihydroxy- and 1-hydroxy-6-oxo-2-cyclohexene carboxylic esters (**24**), of catechin and dimeric procyanidins recently found in the bark of *Salix sieboldiana* (166).

Flavan-3-ol glycosides are also comparatively rare. Both catechin- and epicatechin-3-*O*-glucosides were synthesized by Weinges et al. (386), but neither of these compounds has yet been found as a natural product in plants. A catechin-7-*O*-arabinoside from *Polypodium vulgare* (357, 389) and a catechin-7-*O*-xyloside from *Ulmus americana* (76) were the only two flavan-3-ol glycosides known for some time. In 1986 many additional representatives of the flavan-3-ol glycosides were found by Nishioka's group (196, 271). The (+)-catechin-5-*O*-*β*-D-glucoside with the 2*R* stereochemistry and the 2*S* *ent*-catechin-7-*O*-*β*-D-glucoside co-occur in *Rhaphiolepis umbellata* (271). Additional study of flavan-3-ol and procyanidin glycosides in *Rheum* (rhubarb) has resulted in the characterization of eight flavan-3-ol glucosides, including a series of *O*-*β*-D-glucopyranosides with substitution at the C-5, C-7, C-3′, and C-4′ positions and *C*-*β*-D-glucosides at the C-6 and C-8 positions (196). Other glycosides that have been reported recently include the 3′-*O*-methyl-epicatechin-7-*O*-glucoside, symplocoside, from *Symplocos uniflora* (356) and catechin-7-*O*-D-apioside from *Polypodium vulgare* (195).

Catechins are particularly good nucleophiles, so it is not surprising to find these compounds substituted at the A-ring with other natural products. The bark of *Quercus stenophylla* contains ellagitannins that are linked by carbon → carbon bonds to the C-6 or C-8 of catechin (277). Kopsirachin (**25**), from *Kopsia dasyrachis*, is a most interesting example of this type of catechin derivative in which both the C-6 and C-8 positions are substituted by the alkaloid skytanthine (160). Other examples of this type of compound are the C-6, C-8, and C-6 + C-8 cinnamate derivatives of catechin isolated from *Cinchona succirubra* (273, 279). It can be predicted that many more examples of this type of catechin derivative will be found in the future.

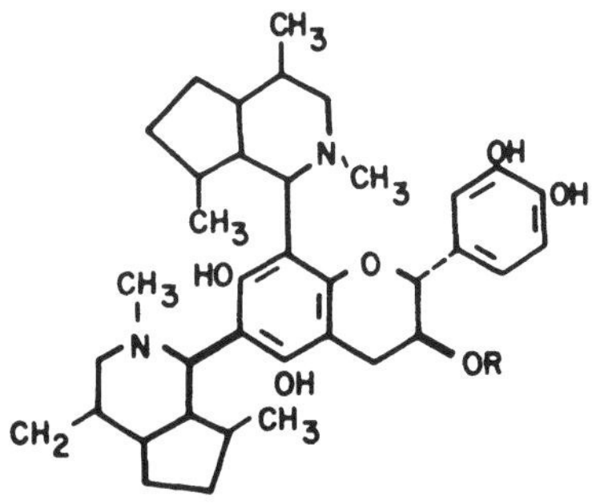

25

7.6.3.1.3 Reactions of Flavan-3-ols

Studies of the reactions of flavan-3-ols and particularly those of catechin have been central to the elucidation of the structure and development of uses for the condensed tannins. This work, initiated by Freudenberg and his colleagues at Heidelberg in the 1920s [see Weinges et al. (377) for a thorough review], continues to be an important aspect of condensed tannin chemistry. A wide range of electrophilic aromatic substitution reactions has been examined to obtain definitive evidence for the location of substitution (i.e. C-6 or C-8) of proanthocyanidins and to establish the influence of steric hindrance on the relative reactivity of these nucleophilic centers in flavan-3-ols.

Although a number of oligomeric proanthocyanidins had been isolated (352, 381) and isomers containing different interflavonoid bonds were indicated, it was only in 1978, through the work of Engle et al. (95) and Hundt and Roux (167) on products of the bromination of tetra-*O*-methylcatechin, that definitive evidence for the substitution location (i.e. C-6 or C-8) was established. In this work it was found that bromination of the methyl ether occurred essentially exclusively at the C-8 position, and reaction at C-6 did not occur until all the C-8 positions had been brominated. Bromination of the catechol B-ring did not occur unless forced by reactions with excess reagent. Since methylation increased steric restraint to substitution at C-6, McGraw and Hemingway (247) studied the bromination of catechin as the phenol, and here substitutions were observed at C-6, C-8, and C-6 + C-8 in ratios of approximately 1:2:1, respectively. Thus, steric restraint to substitution at C-6 of the 5,7-dihydroxy-flavan-3-ols was not as severe as might have been implied from the work on bromination of the methyl ethers. Similar results have recently been reported on the bromination of epicatechin (212).

Possibly the most important role of flavan-3-ols in plants is their reaction as nucleophiles terminating the polymerization of proanthocyanidins. Acid-catalyzed reactions to produce flavan-4-carbocations (either from flavan-3,4-diols or from interflavanoid bond cleavage of proanthocyanidins) that react with the A-rings of flavan-3-ols to produce dimeric proanthocyanidins and higher oligomers have been so successfully employed in recent years that they have been called biomimetic syntheses (27, 28, 133, 175, 177). Where a suitable leaving group is present at C-4 (such as a hydroxyl in flavan-3,4-diols), similar products can be made cleanly under mildly alkaline conditions through quinone methide intermediates (104, 139). Recent work by Attwood et al. (10) has shown that a quinone methide rather than a flavan-4-carbocation is the electrophilic species even under mild acidic conditions. Whether a carbocation or a quinone methide, the C-4 is as an sp^2 hybridized state with the same heterocyclic ring geometry. In either case, the flavan-3-ols serve as nucleophiles terminating the polymerization to polymeric proanthocyanidins.

The first aspect of these reactions that must be considered is the difference between kinetic control (i.e. conditions are such that products are not cleaved back to the flavan-4-carbocation or quinone methide from the chain extender units and the flavan-3-ol from the terminal unit) and thermodynamic control (i.e condensation and cleavage reactions are permitted to occur repeatedly) (177, 316).

Table 7.6.6. Regio- and stereo-selectivity in proanthocyanidin synthesis (from Botha et al. [28])

Electrophile	Nucleophile	Time (hr)	Linkage			
			[4α-8]	[4β-8]	[4α-6]	[4β-6]
mollisacacidin	fisetinidol	2	0	0	1.4	1
mollisacacidin	fisetinidol	168	0	0	1.4	1
mollisacacidin	*ent*-epifisetinidol	4.5	0	0	1.3	1.0
mollisacacidin	catechin	12	5.1	3.0	1	
mollisacacidin	catechin	48	12.0	9.6	1	0.2
teracacidin	catechin	2	4.6	0	1	–
ent-mollisacacidin	catechin	3	10.0	13.0	1	–
leucobinetinidin	catechin	2	7.0	3.4	1	
mollisacacidin	epicatechin	2	30.0	24.0	1	
leucocyanidin	catechin	0.5	3.2	0	1	0[a]

[a] Later reports [67] indicate a ratio of 10:1 for [4α→8] and [4α→6] isomers

The proanthocyanidins containing 5-deoxy upper units are known to be far more stable to acid-catalyzed cleavage than the 5,7-dihydroxy procyanidins. Therefore, provided that reaction conditions are sufficiently mild (i.e. 0.1 M HCl, 2 to 4 hours at 25 °C) – such as those employed by Botha et al. (26–29) – an assumption of kinetic control of the regioselectivity of condensation of these compounds seems appropriate. Botha et al. (26–29) have done a great deal of work to define relationships of the structures of flavan-4-carbocations or quinone methides and flavan-3-ols on the regioselectivity of their condensations in the synthesis of proanthocyanidins (Table 7.6.6).

The most obvious factor controlling the substitution at various nucleophilic centers of flavan-3-ols is the hydroxylation pattern. Thus, in reactions of mollisacacidin with fisetinidol, substitution occurs regiospecifically at the C-6 position whereas in reactions with catechin, substitution occurs predominantly at the C-8 position (Table 7.6.6) (28). Even though the C-6 position of catechin is a highly hindered site, in competing reactions of mollisacacidin with both fisetinidol and 8-(hydroxybenzyl)-catechin added as nucleophiles, substitution occurs at the C-6 of the catechin moiety. Fisetinidol is recovered largely unreacted because of the higher general nucleophilicity of the phloroglucinol ring (29).

The structure of the flavan-4-carbocation or quinone methide also exerts a significant influence on the regioselectivity of these reactions. Although reaction of catechin with carbocations or quinone methides generated from teracacidin and mollisacacidin give C-6 and C-8 substituted products in similar proportions, substitution is much more highly favored at C-8 in reactions with *ent*-mollisacacidin and also, but less so, with leucorobinetinidin (27, 28). Similarly, a change in the stereochemistry of the nucleophile also was very important since substitution at C-8 was much more highly favored in reactions of epicatechin with mollisacacidin than was observed in reactions with catechin (Table 7.6.6). Even in reactions with the 5-deoxy flavan-4-carbocations or quinone methides, changes in the ratios of products are observed if reaction time periods are extended. Reaction of catechin with mollisacacidin for 48 hours (1.0 M HCl, ambient temperature) gave products substituted more predominantly at the C-8 position than in the shorter time periods. Extended reactions of mollisacacidin with fisetinidol (7 days) still gave exclusive substitution at the C-6 position, and the ratios of the two stereoisomers at C-4 did not change.

The situation is more complex in defining the regioselectivity of procyanidin synthesis. Here differences between kinetic control and thermodynamic equilibrium ratios become particularly important because of the lability of the interflavanoid bond in these compounds and because of differences in both the relative rate of acid-catalyzed cleavage and rate of condensation for the C-6 and C-8 substituted isomers (148). Work of Haslam's group indicated that the C-8 substituted procyanidins predominated over their C-6 linked pairs by a factor of 8 to 9 to 1 (177, 352). However, Botha et al. (28) obtained catechin-(4α→8)-catechin and catechin-(4α→6)-catechin in relative yields of 3.2 to 1 from a 2-hour biomimetic synthesis with 0.1 M HCl at ambient temperature. Similarly, Hemingway et al. (143) obtained epicatechin-(4β→8)-catechin and epicatechin-(4β→6)-catechin in relative yields of about 2.5 to 1 through synthesis by reaction of *Pinus taeda* proanthocyanidins with excess catechin for 48 hours at 25 °C using HCl as a catalyst. This ratio was similar to the yield of the two isomers isolated from the phloem of *Pinus taeda*. The extreme lability of the interflavanoid bond in the procyanidins causes one to wonder if true kinetic control ratios can be obtained from acid-catalyzed reactions of the procyanidins.

Two other approaches to determination of a kinetic control ratio for substitution at the C-6 and C-8 positions of procyanidins have been attempted (139, 148). Determination of the relative rate of cleavage of C-8 and C-6 regio-isomers (2.6–1) and a determination of their relative yields at the thermodynamic equilibrium (1.3–1) permitted an estimation of a kinetic control ratio of 3.3:1 for the C-8 and C-6 linked isomers (148). The other approach was to determine the ratio of the C-8 and C-6 linked dimers after synthesis by reaction of epicatechin-(4β)-phenylsulfide and catechin at pH 9.0 and ambient temperature through quinone methide intermediates (139). Here the C-8 and C-6 linked isomers were obtained in an approximate ratio of 3.5–4:1. These later results, together with evidence for interflavanoid linkage isomerism in trimeric and polymeric procyanidins (141, 143) (Sect. 7.6.3.3), show that substitution is not as heavily favored at the C-8 position in procyanidins as had been thought.

It should be noted that the distribution of regio- and/or stereoisomers obtained by these syntheses is not always reflected in the natural products. Haslam's observations of procyanidins dominated by the C-8 linked dimers (133–136) may be due to enzymic control of the location of substitution in these plants. A similar situation exists in the dimeric proanthocyanidins isolated from *Acacia mearnsii* bark (27, 28). Here fisetinidol-(4β→8)-catechin predominates and no fisetinidol-(4α→8)-catechin was observed in the plant extracts. In contrast fisetinidol-(4α→6)-catechin was the principal product and significant amounts of the C-6 substituted isomer were also obtained from biomimetic syntheses (28).

Other electrophilic aromatic substitutions of flavan-3-ols have been examined, primarily regarding the use of proanthocyanidins in adhesives. Hillis and Urbach (157–158) studied the reactions of catechin and condensed tannins with formaldehyde early. Later studies (147) through ^{1}H-NMR spectra and GPC profiles of peracetates of catechin-formaldehyde condensation products confirmed Hillis' conclusions. ^{1}H-NMR also showed that methylol groups did not accumulate but that condensation to dimers occurred rapidly and that further condensation to higher oligomers was slower than would be expected of a homogeneous reaction. Kiatgrajai et al. (205) further examined the kinetics of these reactions and were able to isolate bis-(8-catechinyl)-methane (**26**) as the principal dimeric product. Foo and Hemingway (105) have also studied the condensation products of catechin with furfuraldehyde and have obtained 2-furyl-di(-8-catechinyl)-methane and two diasteriemers of 2-furyl-(6-catechinyl) (8-catechinyl)-methane (**27**).

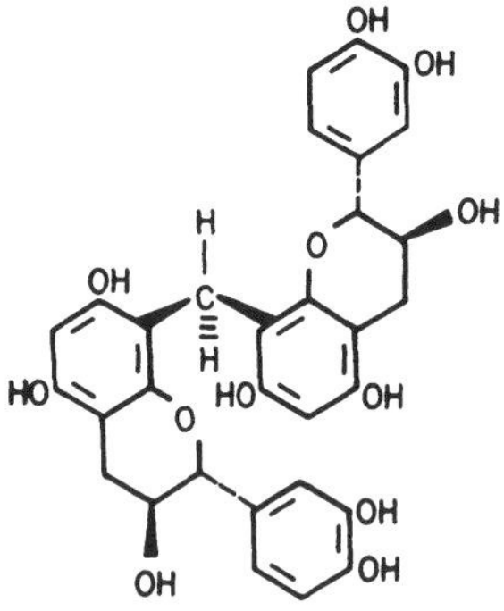

26 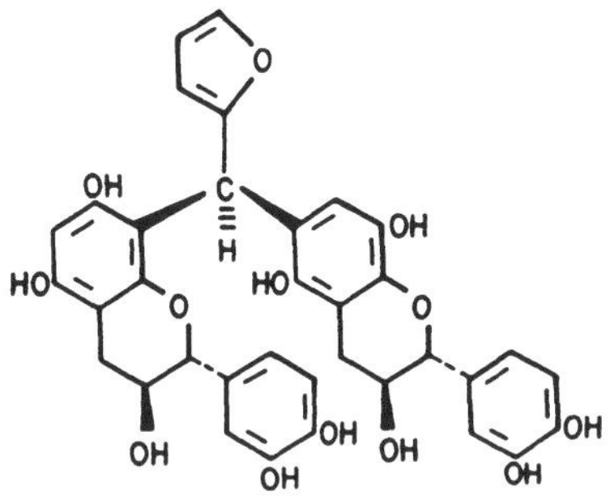 **27**

Reaction or the procyanidins with formaldehyde proceeds so rapidly that these products are usually not useful as wood adhesives. Herrick and Bock (153) advocate crosslinking these compounds with methylol-phenols rather than aldehydes. To gain further information about these reactions, both catechin and tetra-*O*-methyl catechin have been reacted with *p*- and *o*-hydroxybenzyl alcohol (142, 247). Reactions of catechin with *p*-hydroxybenzyl alcohol gave the C-6, C-8, and C-6 + C-8 disubstituted products in approximately equal mole ratios. With *o*-hydroxybenzyl alcohol, substitution at C-8 was favored, with approximate yield ratios of 1:2.5:1.7, respectively. Foo and Hemingway (105) have also examined the regioselectivity of reactions of catechin with furfuryl alcohol and have obtained 2-furyl-(6-catechinyl)-methane and 2-furyl-(8-catechinyl)-methane in approximate relative yields of 1:2.7.

A number of important reactions of flavan-3-ols (and proanthocyanidins) involve attack of the pyran ether of the heterocyclic ring. Perhaps best known and most commonly encountered is an epimerization through inversion at C-2, first recognized by Freudenberg in the 1920s (377). This reaction occurs when flavan-3-ols are heated to a high temperature in protic solutions. Epimerization occurs much more rapidly under alkaline conditions. Resistance of 5,7,3′,4′-tetra-*O*-methylcatechin to alkaline epimerization led Mehta and Whalley (248) to conclude that inversion proceeds through opening of the pyran ring with the formation of a C-2 quinone methide that recyclizes on neutralization to give either the C-2 inverted product of the starting compound. However, more recently it has been shown that catechin or epicatechin fails to epimerize if O_2 is very carefully excluded from alkaline solutions. Epimerization proceeds in the presence of oxy-

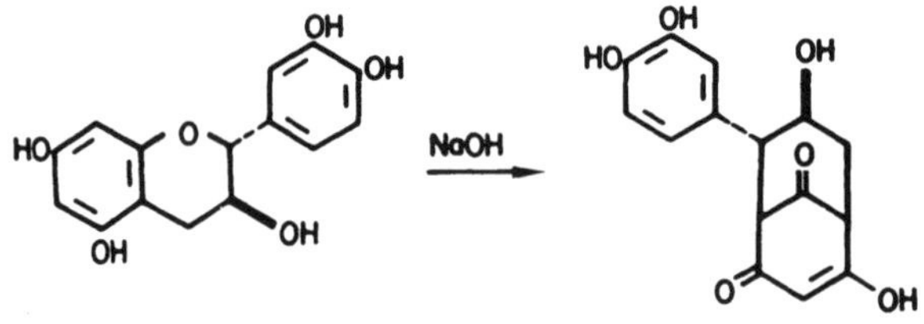

Fig. 7.6.3. Rearrangement of (+)-catechin to catechinic acid

gen, presumably through a radical anion intermediate (197). Catechin gives *ent*-epicatechin and epicatechin gives *ent*-catechin with the *trans* isomers favored by a factor of 3 to 1 (204). This reaction has been used by Foo and Porter (109, 110) to permit the synthesis of a number of procyanidin diastereoisomers (Sect. 7.6.3.3).

In addition to epimerization, a rearrangement of catechin to catechinic acid (**28**) occurs under alkaline conditions (335), but the rate of rearrangement is only about one-fifth that of epimerization (204) (Fig. 7.6.3). ESR studies of these rearrangements indicated that they occur by oxidation to the corresponding 2'- and 6'-hydroxycatechin derivatives that undergo cleavage of the pyran ring and then rearrangement to form the nonatrione ring system (178). Many industrial approaches to use of proanthocyanidins involve solution under alkaline conditions and, although it is known that the aldehyde reactivity of these preparations can be reduced substantially, only recently have these alkaline rearrangement reactions been studied for the higher molecular weight proanthocyanidins (223–225). These reactions are discussed more fully in Sect. 7.6.3.3.2.

Acid-catalyzed condensations of flavan-3-ols involving protonation of the pyran oxygen, cleavage of the heterocyclic ring and substitution at the C-2 carbocation with phenolic nucleophiles were studied extensively by Freudenberg, Mayer, and Weinges at Heidelberg (114, 117, 118, 243–245, 387, 388). Weinges et al. (377) have prepared a thorough review of this work, which should be consulted for details. Because the C-2 carbocation or quinone methide is attacked from the least hindered side (i.e. *trans*- to the 3-hydroxyl), condensation products of catechin and epicatechin with resorcinol are enantiomers of **29**. The phlorogucinol A-ring of the flavan-3-ol is involved in a dehydration to give the inverted anhydro derivatives (**30**). A similar situation exists in the reactions with phloroglucinol, but here the cyclization occurs with retention of the stereochemistry of the ring-opened products to give **31** and **32**. When catechin was dissolved in dioxane with 2N HCl for 30 hours at ambient temperature, a dimer, dicatechin, which was finally shown to be the C-8 substituted ring opened product (**33**), was obtained. The corresponding anhydro-dicatechin (**34**) was obtained when aqueous solutions of catechin were heated at 90 °C and pH 4.0 for several days.

For some time the natural condensed tannins were considered to be polymers basically of a structure similar to these two products. Although this conclusion is now known to be incorrect, condensation products of this type are known natural products. Nonaka and Nishioka (278) have isolated a number of these chalcan-

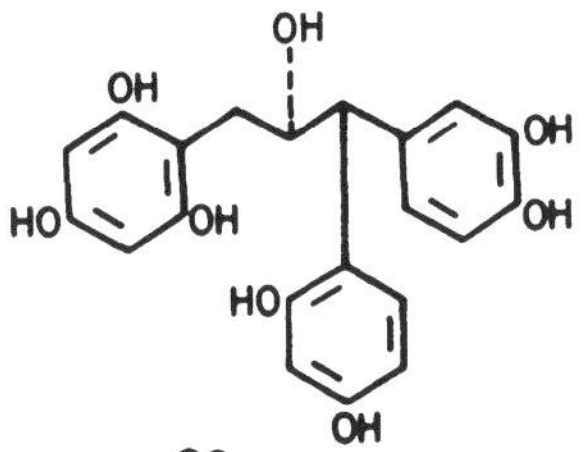

29

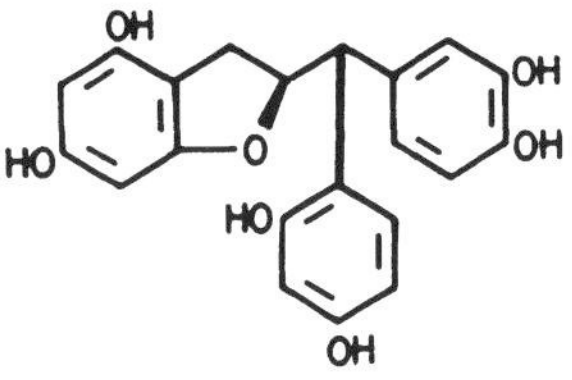

30

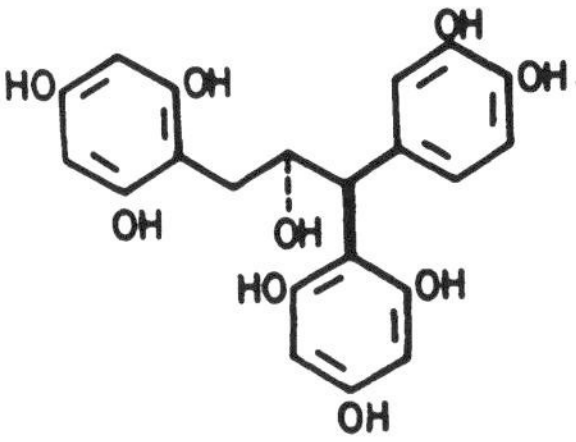

31

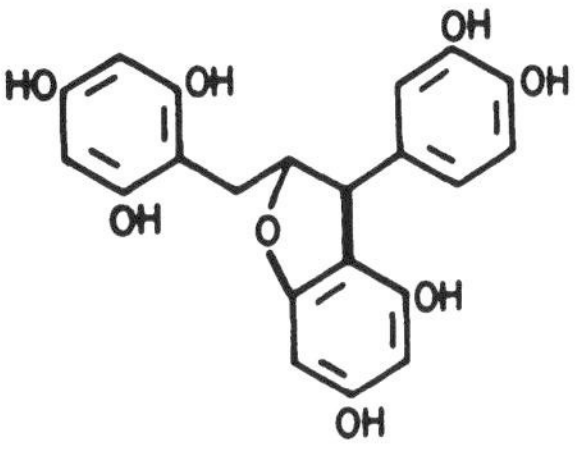

32

flavan dimers from the leaves and twigs of *Uncaria gambir.* The gambirins A_1 **(33)**, A_2 **(35)**, A_3, B_1 **(34)**, B_2, and B_3 **(36)** are the only known natural products that are clearly acid-catalyzed condensation products of flavan-3-ols of this type. Reactions of flavan-3-ols with sulfite ion as the nucleophile also are highly stereoselective. Sears (332) obtained the 2,3-*trans*-2-sulfonate derivative of catechin in reactions at pH 5.9 and 170 °C for 30 minutes.

Caution must be exercised in use of flavan-3-ols as models for the reactions of proanthocyanidins, because substitution at the C-2 or the C-4 positions of

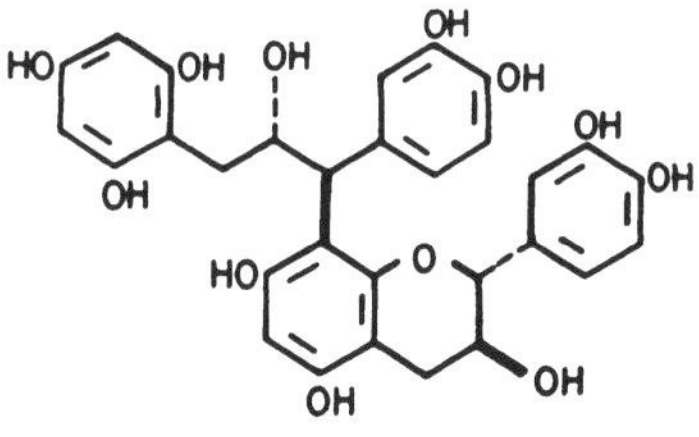

33

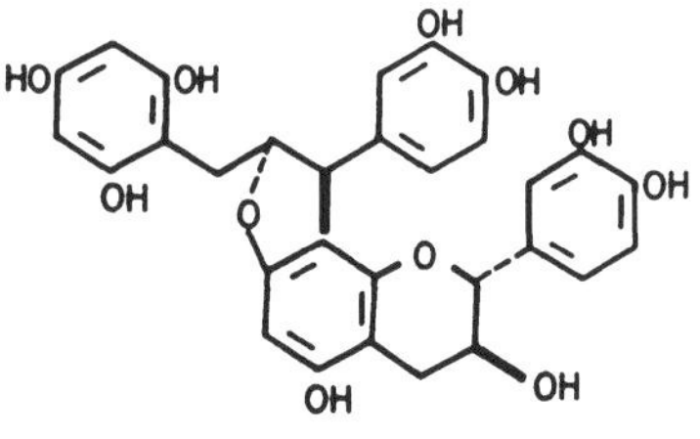

34

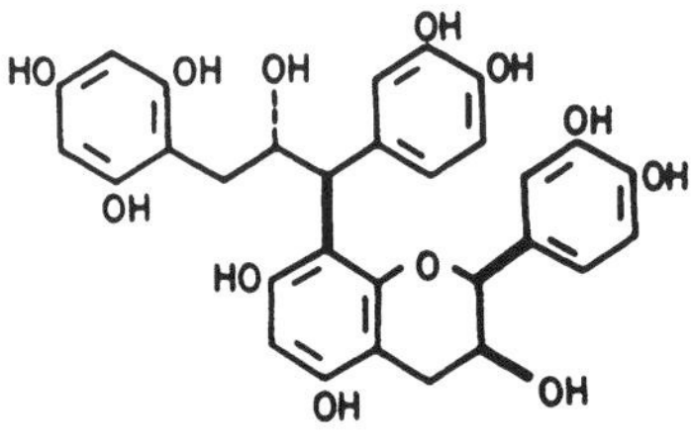

35

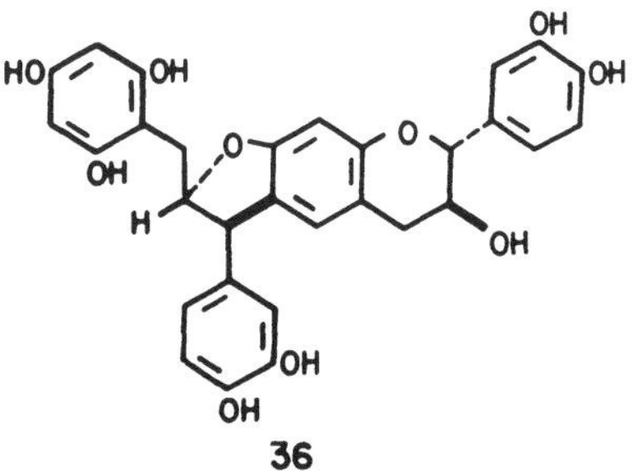

36

acid-catalyzed cleavage products of the proanthocyanidins will depend on the relative lability of the interflavanoid bond and the pyran ether (223). In reactions of polymeric procyanidins with sulfite ion, for example, the interflavanoid bond is far more labile to cleavage than the pyran ether. Hence flavan-4-sulfonates and oligomeric procyanidin-4-sulfonates are the principal reaction products (107). Similarly, reactions of polymeric procyanidins with phloroglucinol or resorcinol (31, 64, 144) give flavan-4-adducts in high or moderate yield and no evidence for substitution at C-2 has been obtained.

Where the interflavanoid bond is less labile to cleavage, as in the 5-deoxy proanthocyanidins, the relative yields of C-2 and C-4 substituted products are not as well clarified. Roux et al. (319) report that polymeric C-2 sulfonates are obtained in the reaction of wattle tannins with sulfite ion. Also, the polymeric profisetinidins from quebracho did not yield significant quantities of fisetinidol-4-sulphonates or fisetinidol-4-benzylsulfides when reacted at 135 ° to 145 °C with sodium bisulfite or phenylmethanethiol (Laks and Hemingway, unpublished results). However, solvolysis of a guibourtacacidol-(4α→8)-epiafzelchin dimer at 140 °C in the presence of excess phenylmethanethiol cleaved the interflavanoid bond with the formation of the guibourtacacidol-(4α)-benzyl sulfide and epicatechin, suggesting that the interflavanoid bond was more labile than the pyran ether in this compound (290). Additional work is needed to clarify the relative lability of the interflavanoid bonds and pyran heterocyclic ring, particularly in compounds of the 5-deoxy group.

Hydrogenolyses of the methyl ethers of catechin and epicatechin over sodium in ethanol-liquid ammonia were key reactions in establishing the stereochemistry of these compounds (22). Hydrogenolysis of flavan-3-ols (and proanthocyanidins) over Pd-C was reported to readily give 1-(3,4-dihydroxyphenyl)-3-(2,4,6-trihydroxyphenyl)-propan-2-ol and the corresponding propane derivative was reported (176). In later work, Foo (99) demonstrated that these reactions require high temperatures and hydrogen pressures. Other high temperature reactions that should be mentioned here include alkaline fusions that provide the phenol of the A-ring (i.e. resorcinol or phloroglucinol) and the carboxylic acid from the B-ring. Although a simple and well-known reaction, it is mentioned here because alkaline fusion of *Acacia mearnsii* (black wattle) heartwood can give yields of 8% of crystalline resorcinol (318). Reaction of catechin with ammonium sulfite in ammonia at 175 °C gives about 30% molar catechol together with an interesting sulfonated amino-quinoline (**37**) (Fig. 7.6.4) (221).

Hathway and Seakins (138) attracted considerable interest with their work on the polymerization of flavan-3-ols by phenoloxidase enzymes. A series of studies

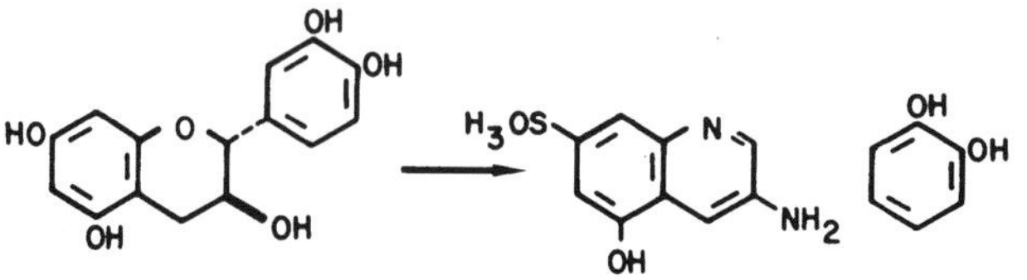

Fig. 7.6.4. Ammonium sulfite cleavage products of (+)-catechin

reviewed in detail by Hathway (137) and Weinges (377) showed that these enzymic dehydration polymers were biphenyl-linked products. These results together with the more realistic physiological conditions (i.e. pH 6–8) led to the belief that the natural condensed tannins were enzymic dehydration products, rather than polymers analogous to the acid-catalyzed products studied by Freudenberg and his colleagues (377). Although it is now known that neither of these hypotheses is correct, these oxidative coupling condensations are very important reactions of flavan-3-ols and proanthocyanidins. For example, in fungal interactions with condensed tannins it is generally believed that extracellular phenol-oxidase enzymes degrade the tannins in advance of the developing hyphae; thus, a detailed understanding of these oxidative polymerizations is clearly important (149). In addition, oxidative coupling reactions of catechin are considered responsible for darkening of thermomechanical pulps made from western hemlock wood (165). Weinges and coworkers studying the products of enzymic dehydration of catechin isolated 8-hydroxycatechin, two dimers named dehydro-dicatechin-A and -B, a trimer, and other higher oligomers (377–379, 383). Dehydro-dicatechin B (**38**) proved to be a C-6′ to C-8 biphenyl-linked dimer and the trimer was of an analogous structure. Extensive studies were made to elucidate the structure of dehydro-dicatechin-A (**39**); its structure has only recently been solved by X-ray crystallography (361, 379, 383).

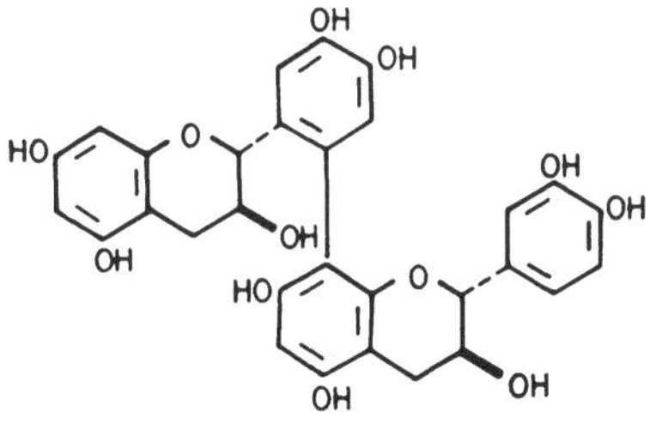

38

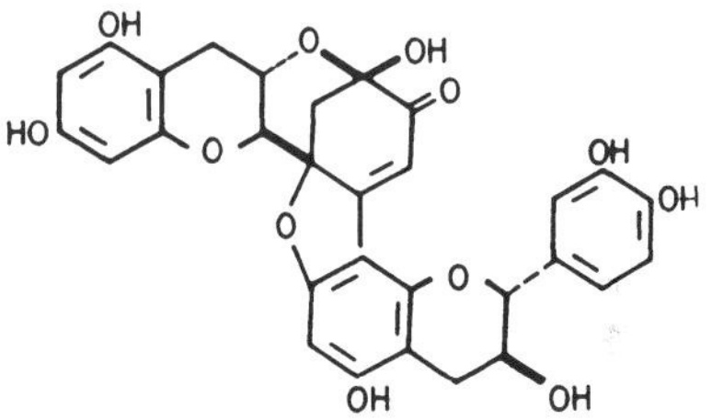

39

The theaflavins (**40**) are oxidative coupling products of epicatechin or catechin with gallocatechin or epigallocatechin. Because of the pyrogallol B-ring, these compounds condense to form purpurogallin or tropolone derivatives (35, 62, 377). In black tea, the theaflavins are also found as gallate esters and co-occur with the thearubigins that are largely polymeric proanthocyanidins and their

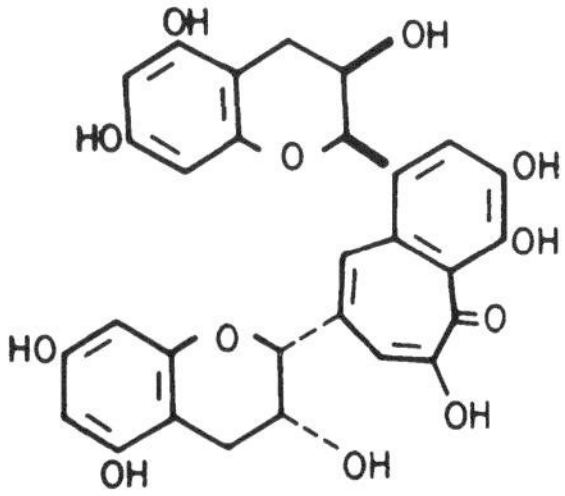

40

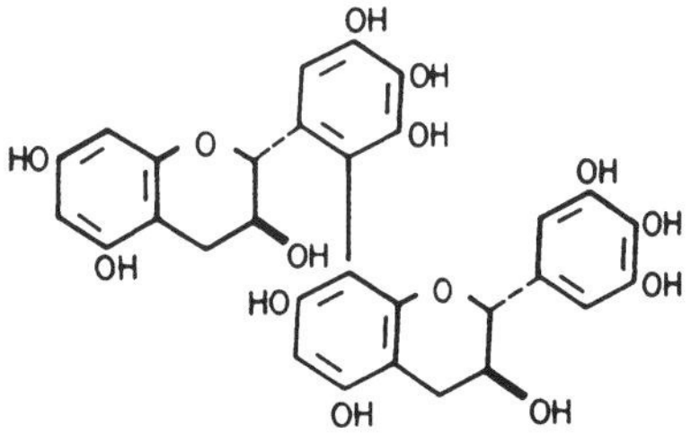

41

gallate ester derivatives. Dimers analogous to dehydro-dicatechin such as bis-flavanol (**41**) and gallate esters are present in the thearubigin fraction (358, 377).

These products are not limited to plant materials in which oxidative coupling processes are forced – e.g. as in black tea. Ahn and Gstirner (6) found mixed catechin and gallocatechin dimers, some of which were typical procyanidins and some of which were C-6′ to C-8 oxidative coupling products, in the bark of *Quercus robur.* Ahn (5) also found a catechin trimer (**42**) in which the upper linkage was C-6′ to C-8 (oxidative coupling) and the lower linkage was C-4 to C-8 (procyanidin). Further examples of oxidatively coupled flavan-3-ols have recently been found in *Prosopis glandulosa* (mesquite) in which the 2,3-*trans*-7,8,3′,4′-tetrahydroxyflavan-3-ol is oxidatively coupled through the C-5 and C-6 positions to form the biphenyl- and terphenyl-linked oligomers such as **43** (174). These compounds show that oxidative coupling of flavan-3-ols, while not common, does occur and that proanthocyanidins are labile to secondary oxidative coupling. These reactions may explain why yields of products from solvolysis of condensed tannins with thiols or phloroglucinol are sometimes comparatively low. The possibility of secondary oxidative coupling of proanthocyanidins clearly needs more attention, particularly with reference to the structures of polymers from outer bark or industrial plant extracts that may have been subjected to oxidative conditions.

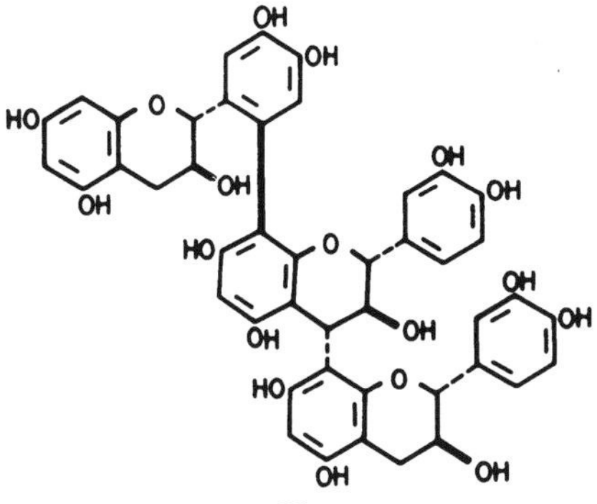

42

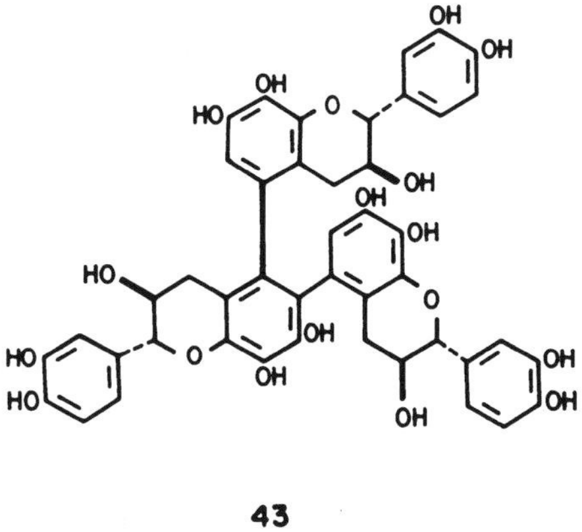

43

7.6.3.2 Flavan-3,4-diols

The significance of flavan-3,4-diols in plants rests primarily on their probable role as precursors of the polymeric proanthocyanidins. Co-occurrence of the 5-deoxy compounds – i.e., quibourtacacidins, mollisacacidins, and robinetinidins – with the related proanthocyanidins in *Acacia* species and the ready synthesis of naturally occurring proanthocyanidins from reactions of these flavan-3,4-diols with catechin under mild acidic conditions constitutes heavy but not definitive evidence for this thesis (31, 315–317).

Some inconsistencies occur in the distribution of dihydroflavonols, flavan-3,4-diols, and proanthocyanidins in plants outside the Leguminosae that have caused doubts about the central role of flavan-3,4-diols in proanthocyanidin metabolism (134–136). One concern is the fact that no 5,7-dihydroxy-substituted flavan-3,4-diol has yet been isolated from plants as a natural product. However, the 2,3-*trans*-3,4-*trans*-(2*R*,3*S*,4*R*)-3′,4′,5,7-tetrahydroxyflavan-3,4-diol can be made by sodium borohydride reduction of (+)-dihydroquercetin (301, 305). Even more significantly, the 2,3-*trans*-3,4-*cis*-(2*R*,3*S*,4*S*)-flavan-3,4-ol is produced from (+)-dihydroquercetin by reductase enzymes obtained from cell suspension cultures of Douglas-fir (339, 342). As is discussed more fully in Sect. 7.6.3.3.2, the presence of an axial hydroxyl at C-4 in the (2*R*,3*S*,4*S*)-isomer results in an especially good precursor to polymeric procyanidins. Because of the high degree of reactivity of these compounds, the failure to isolate them from plant extracts is readily explained (28, 31, 139, 305).

Another concern has been that the hydroxylation patterns of the dihydroflavonols and flavonols are often not reflected in the co-occurring proanthocyanidins. In the barks of southern pines for example, dihydromyricetin and myricetin are prominent constituents of the outer bark (151). However, the proanthocyanidins and condensed tannins have only very small proportions of prodelphinidin units (146). It is probable that the pool of dihydroflavonols used in reduction to flavan-3,4-diols and flavan-3-ols occurs prior to the hydroxylation at C-5′.

The most important concern is lack of an accounting for the preponderance of 2,3-*cis* procyanidin units in the proanthocyanidins of conifers (see Sect. 7.7). Stafford's (338, 339) proposal of a C-3 epimerase that provides a 2,3-*cis* dihydroflavonol that is held on an enzyme surface and is immediately reduced to the flavan-3,4-diol in one pool or that is reduced through two stages to the flavan-3-ol in another pool is certainly an attractive explanation. However, enzymes from cell suspension cultures of *Pseudotsuga menziesii* (Douglas-fir) or *Ginkgo biloba* gave only a 2,3-*trans* flavan-3,4-diol (338–342), and it is known that the 2,3-*cis* procyanidin units predominate over the 2,3-*trans* isomers by a factor of 3:1 in the phloem (189). Thus such an epimerase was not present in these enzyme preparations.

An explanation for the formation of 2*R*,3*R*-(2,3-*cis*)-proanthocyanidins from the 2*R*,3*S*-(2,3-*trans*)-flavan-3,4-diols could lie in a tautomeric rearrangement of quinone methide intermediates to flav-3-en-3-ols, which could then be stereospecifically converted back to either 2,3-*trans* or 2,3-*cis* quinone methides (145). Chemical evidence supporting this thesis has been obtained by the formation of diarylpropanone derivatives from the reaction of polymeric procyanidins with phenylmethanethiol under alkaline conditions (223). Enzymes controlling the quinone methide to flav-3-en-3-ol rearrangements rather than C-3 inversion of dihydroflavonols may be involved. In either case, evidence continues to mount that the flavan-3,4-diols are indeed central intermediates in the biogenesis of proanthocyanidins and that this conversion is under enzymic (genetic) control (219, 341, 342).

The 7,8-dihydroxyflavan-3,4-diols (teracacidins and melacacidins) posed another interesting question about their function in the plant. Here these flavan-3,4-diols also can readily react with phenolic nucleophiles to produce dimeric proanthocyanidins (28, 112). Until recently it was considered that the *ortho*-hydroxylation pattern reduced the nucleophilicity of the A-ring so that further condensation to polymeric proanthocyanidins did not occur. This conclusion was supported by observation of the free flavan-3,4-diols in natural plant extracts.

In *Acacia nigrescens* heartwood for example, melacacidin and isomelacacidin are prominent constituents, and this wood does not contain condensed tannins (112). Similarly, *Acacia auriculiformis* heartwoods contain a mixture of teracacidin isomers, and again no significant amounts of condensed tannins are present (85). However, Foo (102, 103) has shown that (–)-melacacidin rapidly polymerizes under mild acidic conditions to give oligomers linked through (4β→6) interflavanoid bonds and has isolated a dimeric promelacacidin, most interestingly linked through a (4α→6) interflavanoid bond (2,3-*cis*-3,4-*cis* stereochemistry) from *Acacia melanoxylon*. The woods of *Acacia* species containing melacacidins should probably be re-examined considering Foo's recent results (103).

7.6.3.2.1 Structure of Flavan-3,4-diols

King and Bottomley (206) isolated the first natural flavan-3,4-diol (–)-melacacidin from *Acacia melanoxylon* wood in 1953. This stimulated rapid progress in the chemistry of leucoanthocyanidins, particularly thanks to the work of Clark-Lewis and coworkers in Australia and Roux and coworkers in South Africa on the flavan-3,4-diols of *Acacia* species and related plants of the Leguminosae. Clark-Lewis and coworkers described the structures of a number of compounds of the melacacdin, teracacidin, and mollisacacidin groups, and defined the absolute configuration and conformation of (–)-mollisacacidin, (–)-teracacidin, and (+)-mollisacacidin (51–55). During the same period, Drewes and Roux (81, 83, 84) described the stereochemistry of (+)-mollisacacidin, (–)-leucofisetinidin, and (+)-leucorobinetinidin and related diastereoisomers, and Saayman and Roux (327) established the configurations of a number of compounds of the guibourtacacidin group.

A methylated derivative, 8-*O*-methyl-teracacidin, was first isolated from *Acacia auriculiformis* by Drewes and Roux in 1966 (85). Since then, a number of flavan-3,4-diols methylated at the aromatic hydroxyls (89, 113, 282) and, most unusually, at the aliphatic hydroxyls (113, 282) have been described. A series of (4,6)-linked oligomeric (+)-mollisacacidin derivatives, **44**–**48**, have been isolated from *Acacia mearnsii* heartwood and also synthesized (87, 370). The occurrence of these compounds in the wood of *Acacia mearnsii* together with condensed tannins of analogous structure provides powerful support for the thesis that these compounds are indeed the precursors of the condensed tannins of the profisetinidin class. These oligomers can also cyclize to produce oligomeric peltogynoids (**49**) such as those from *Acacia fasciculifera* heartwood (86, 360). The hydroxylation and methoxylation of these compounds follow distinct patterns related to the taxonomic sections of the *Acacia* genus and are therefore of chemotaxonomic interest (Table 7.6.7) (58, 354, 355).

During the same period there were numerous reports of the isolation of leucocyanidins (5,7,3′,4′-tetrahydroxyflavan-3,4-diols) as natural products from a variety of plants. Unfortunately, none of these compounds was fully characterized (i.e. no comparison of heterocyclic proton coupling constants with data from methylated synthetic racemates (11) and none of the reported melting points or optical rotations coincided with each other or, more importantly, with properties of synthetic compounds of determined structure) (301, 304, 377). These discrep-

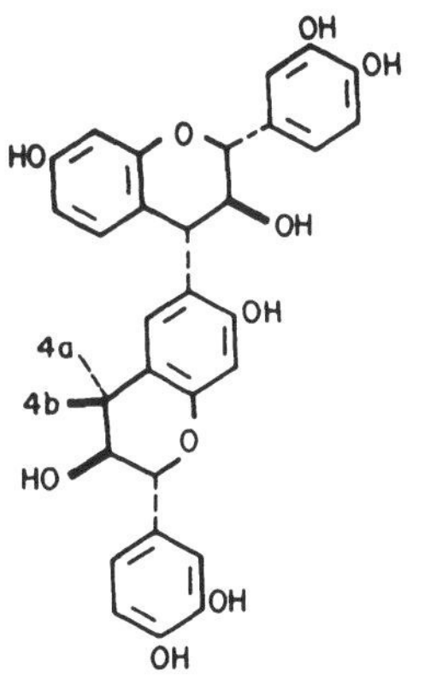

44 4a = OH
45 4b = OH

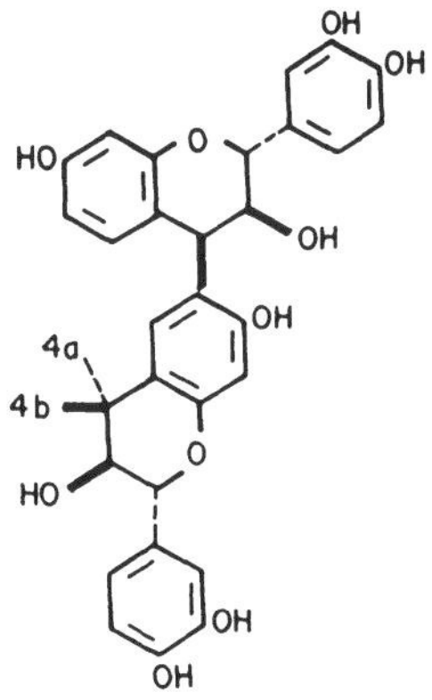

46 4a = OH
47 4b = OH

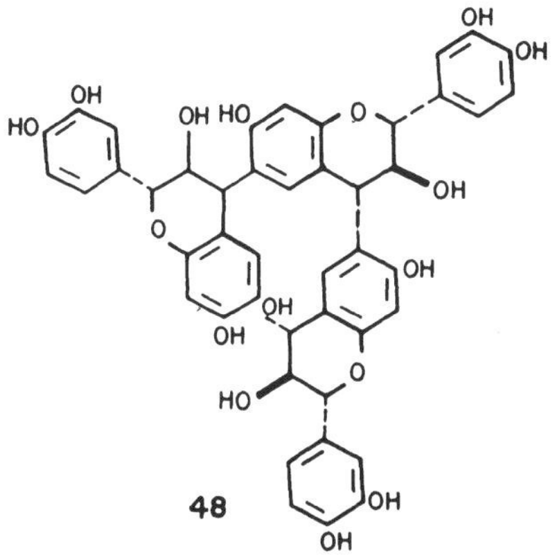

48

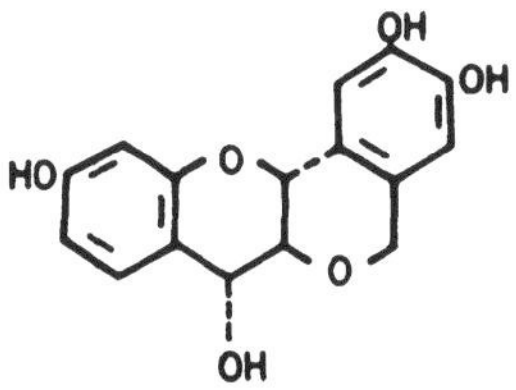

49

ancies and the extreme lability of the leucocyanidins to polymerization suggest that it is virtually certain that no leucocyanidin has been isolated from a plant (301, 304). This does not imply that leucocyanidins do not exist or are not important intermediates in the biosynthesis of procyanidins but rather that they cannot be isolated from plant tissue unless extreme precautions have been taken to prohibit their condensation with flavan-3-ols or oligomeric procyanidins that will co-occur in the plant extracts. Fletcher et al. (98) have provided evidence that the primary constituent of *Butea frondosa* gum is in fact an oligomeric procyanidin and not a (+)-leucocyanidin as proposed by Ganguly and Seshadri (122). Examples of some other leucocyanidins that have been reported as natural products, which should receive additional study to either confirm or revise their structures, include the leucocyanidin-3-glucoside and leucopelargonidin-3-rhamnoside reported from *Ficus racemosa* bark (1), and the oligomeric leucocyanidins present in the wood and bark of *Acer rubrum* (265).

Despite the extreme sensitivity of the leucocyanidins, Porter et al. (305) have recently crystallized 2*R*,3*S*,4*R*-leucocyanidin as the dihydrate from aqueous solu-

Table 7.6.7. Natural flavan-3,4-diols

Name	Substitution							Stereochemistry			Occurrence	Reference
	C-3	C-4	C-7	C-8	C-3′	C-4′	C-5′	2	3	4		
Guibourtacacidins												
(+)-guibourtacacidin	OH	OH	OH	H	H	OH	H	R	S	R	*Acacia cultriformis*	89, 313, 314, 327
	OH	OH	OH	H	H	OH	H	R	S	S	*A. cultriformis, Guibourtia coleosperma*	*89, 327*
	OH	OH	OH	H	H	OH	H	R	R	R	*G. coleosperma*	67
	OH	OH	OH	H	H	OH	H	R	R	R	*G. coleosperma*	327
Mollisacacidins												
(+)-mollisacacidin	OH	OH	OH	H	OH	OH	H	R	S	R	*Acacia mearnsii, Acacia saxatilis*	59, 81, 89, 198
(+)-gleditsin	OH	OH	OH	H	OH	OH	H	R	S	S	*Gleditsa japonica, A. cultriformis*	56, 59, 80, 81, 83, 84, 89
	OH	OH	OH	H	OH	OH	H	R	R	S	*G. coleosperma*	84
(+)-leucofisetinidin	OH	OH	OH	H	OH	OH	H	R	R	R	*G. coleosperma*	80, 83, 84
(−)-leucofisetinidin	OH	OH	OH	H	OH	OH	H	S	R	S	*Schinopsis*	59, 81, 320
	OH	OH	OH	H	OH	OH	H	S	R	R	*G. coleosperma*	84
	OH	OH	OH	H	OH	OH	H	S	S	S		80, 83
7,3′,4′-tri-*O*-methylgleditsin	OH	OH	OCH_3	H	OCH_3	OCH_3	H	R	S	S	*Neorautanenia amboensis*	282
3,4,7,3′,4′-penta-*O*-methylether	OCH_3	OCH_3	OCH_3	H	OCH_3	OCH_3	H	R	S	S	*N. amboensis*	282
Robinetinidins												
(+)-leucorobinetinidin	OH	OH	OH	H	OH	OH	OH	R	S	R	*Robinia pseudoacacia*	320, 324, 376
Teracacidins												
(−)-teracacidin	OH	OH	OH	OH	H	OH	H	R	R	R	*Acacia ovites*	52, 53, 290, 316
(−)-isoteracacidin	OH	OH	OH	OH	H	OH	H	R	R	S	*Acacia spargitalia, Acacia obtusifolia*	51 – 54, 84, 316

	OH	OH	OH	OH	H	OH	H	R	S	R		236
	OH	OH	OH	OH	H	OH	H	R	S	S		236, 316
8-*O*-methylteracacidin	OH	OH	OH	OCH_3	H	OH	H	R	S	S		85
7,8,4′-tri-*O*-methylether	OH	OH	OCH_3	OCH_3	H	OCH_3	H	R	S	R	*A. cultriformis*	89
Melacacidins												
(−)-melacacidin	OH	OH	OH	OH	OH	OH	H	R	R	R	*Acacia melanoxylon*	113, 206
(−)-isomelacacidin	OH	OH	OH	OH	OH	OH	H	R	R	S	*Acacia harpophylla*, *Acacia excelsa*	57, 89, 111, 113
	OH	OH	OH	OH	OH	OH	H	R	S	S		57, 89, 174, 206
	OH	OH	OH	OH	OH	OH	H	R	S	R		113
8-*O*-methylmelacacidin	OH	OH	OH	OCH_3	OH	OH	H	R	S	S		113
3,8-di-*O*-methylmelacacidin	OCH_3	OH	OH	OCH_3	OH	OH	H	R	S	S	*A. saxatilis*	113
7,8,3′,4′-tetra-*O*-methylether	OH	OH	OCH_3	OCH_3	OCH_3	OCH_3	H	R	S	S	*A. cultriformis*, *A. saxatilis*	89, 113

tion. X-ray crystallographic analysis showed that the conformation of the heterocyclic ring in this compound was intermediate between a half-chair and a C(3)-sofa. The pyran oxygen and C-4 are near planar. The C-2 lies 17.6 pm above and C-3 lies 51.9 pm below the plane of the A-ring. The low temperature ^{1}H-NMR spectra of this compound and its methylated and acetylated derivatives showed no selective broadening of the heterocyclic ring protons; therefore, these compounds exist in only one conformation in solution as well. Coupling constants of $J_{2,3} = 9.9$ and $J_{3,4\beta} = 7.9$ and molecular mechanical studies indicate a C(2)-sofa *E*-conformation for the phenol in solution (304).

Reduction of (+)-dihydroquercetin with enzymes from cell suspension give the 2*R*,3*S*,4*S*-(2,3-*trans*-3,4-*cis*)-flavan-3,4-diol (339, 342). ^{1}H-NMR coupling constants for the tetra-*O*-methyl ether and the tetra-*O*-methyl ether diacetate derivatives as well as molecular mechanical studies indicate that these compounds exist exclusively in half-chair *E*-conformations. The 2,3-*cis* flavan-3,4-diol methyl ethers also occur in solution predominantly in half-chair *E*-conformations (304).

7.6.3.2.2 Distribution of the Flavan-3,4-diols

All the known natural flavan-3,4-diols can be grouped in the classes of guibourtacacidins (7,4′-dihydroxy), mollisacacidins (7,3′,4′-trihydroxy), robinetinidins (7,3′,4′,5′-tetrahydroxy), teracacidins (7,8,4′-trihydroxy), and melacacidins (7,8,3′,4′-tetrahydroxy). The vast majority of these compounds are found in the wood or bark of plants of the Leguminosae and particularly in the *Acacia* species. The distribution of these compounds is summarized in Table 7.6.7.

7.6.3.2.3 Reactions of Flavan-3,4-diols

The flavan-3,4-diols are well known for their conversion to anthocyanidins on treatment with HCl in the presence of oxygen (311, 312). Long before their isolation as natural products, the flavan-3,4-diols were considered to be central precursors of the anthocyanins that are responsible for the red-blue coloration of flowers, fruits, and leaves (311–313). Despite the obvious importance of the flavan-3,4-diol or proanthocyanidin → anthocyanidin reaction (302) to both aesthetic and commercial aspects of life, little research has been directed to the in vivo reactions. Dihydroquercetin and its glycosides are readily incorporated into cyanidin and delphinidin, presumably by a flavan-3,4-diol intermediate (164, 201–203, 347). It is assumed that the vast array of anthocyanins found in plants are then produced by glycosylation, methylation, etc., of the basic anthocyanidin chromophore. Little is known, however, of the enzymes and factors controlling these reactions (164). Of all areas of proanthocyanidin chemistry, this aspect is possibly the most deserving of increasing emphasis. Health concerns regarding the use of synthetic red food colorants have further increased the importance of these reactions. If ways can be found to improve the stability of the anthocyanidin chromophore, these compounds derived either from plant extracts or from reaction of procyanidins with acid have considerable potential in the food industry (164, 168, 353). Timberlake and Bridle (353) have accomplished substantial improvement in the pH dependency of the anthocyanidin chromophore by its reaction with catechin and acetaldehyde. Further research on variations of this theme certainly needs attention. It also seems possible that anthocyanin synthesis might be accomplished through extension of Stafford's (338–342) work on the enzymes involved in reduction of dihydroflavanols to flavan-3,4-diols, with an emphasis on promoting anthocyanidin synthesis rather than the formation of condensed tannins.

The vast majority of recent works on the reactions of flavan-3,4-diols has centered on their role as precursors to flavan-4-carbocations or quinone methide electrophiles in proanthocyanidin syntheses. The regioselectivity of their condensation with flavan-3-ols or oligomeric proanthocyanidins is discussed in Sect. 7.6.3.1.2. Our concern here is with the factors governing the stability of the electrophile and the stereospecificity or stereoselectivity of its reaction with phenolic nucleophiles. Roux and coworkers have made extensive studies of these reactions and have prepared reviews of their recent results (316, 317). The intent here is to summarize some of the more important principles that have been developed.

The ease of formation or stability of the electrophile is, of course, dramatically influenced by the A-ring hydroxylation pattern. Although melacacidins and teracacidins do react with phenolic nucleophiles such as phloroglucinol or catechin under acidic conditions, because of the delocalization of the charge on the pyrogallol A-ring, these compounds would not be expected to be as reactive as the mollisacacidins. The mollisacacidins and leucorobinetinidins (resorcinolic A-ring) react smoothly with phloroglucinol or catechin in the presence of 0.1 M HCl at ambient temperature (28). In comparison, the leucocyanidins (phloroglucinolic A-ring) react much more readily at either slightly acidic (i.e. pH 5 and ambient temperature) (28, 301) or midly alkaline (i.e. pH 8–9 and ambient temperature) (139, 145) conditions. The B-ring hydroxylation pattern is also apparently important to the reactivity of these compounds. In reactions with 0.1 N HCl and catechin, (+)-leucorobinetinidin (pyrogallol B-ring) reacts approximately twice as fast as (+)-leucofisetinidin (catechol B-ring) (28).

The stereochemistry at C-3 and C-4 also has a strong bearing on the stability of the electrophile. Isomelacacidin (2*R*,3*S*,4*S*) has the 4-hydroxyl group in an axial position. It was converted to the 4-*O*-ethyl derivative by boiling in ethanol containing 1% acetic acid, whereas the isomeric (–)-melacacidin (2*R*,3*S*,4*R*) in which the 4-hydroxyl group is in an equatorial position was unreactive (57). The same principle was demonstrated in the extreme difficulty encountered in the crystallization of racemic 2,3-*cis*-3,4-*trans*-3,4-diacetyl-5,7,3′,4′-tetra-*O*-methylleucocyanidin (axial 4-acetoxyl) from ethanol solutions because of the ready solvolytic displacement of the 4-acetoxyl with an ethoxyl group at C-4 (51). Clark-Lewis and Mortimer (57) and Haslam (136) suggest that the greater reactivity of compounds with an axial leaving group at C-4 can be explained by overlapping aromatic *p* orbitals that assist in resonance stabilization of the carbocation and that the axial C-3 hydroxyl group probably further stabilizes the carbocation through formation of a protonated epoxide intermediate. Foo (111) recently suggested another explanation for the differences in reactivity of these compounds. In axial 4-O derivatives, the C–O bond orbital is favorably exposed for overlapping the *p* electrons of the aromatic A-ring and this would facilitate the formation of a quinone methide. The equatorial 4-O derivative would, by contrast, require formation of a flavan-4-carbocation, which would then rearrange to a more stable quinone methide.

In addition to marked differences in the stability of the electrophile, wide variations occur in the stereoselectivity of their condensation with flavanoid nucleophiles that depend on the relative amount of steric hindrance in the approach of the nucleophile to the C-4 position. Formation of a [4-α] or a [4-β] interflavanoid bond is largely controlled by the configuration of the C-3 hydroxyl group and to a lesser extent the C-5 hydroxyl group of the electrophile together with the A-ring substitution and configuration of the nucleophile. In reactions of 2,3-*cis* (2*R*,3*R*) electrophiles, the condensation at C-4 is invariably *trans* to the 3-hydroxyl group. For example, in reaction of (2*R*,3*R*,4*R*)-teracacidin with either resorcinol, phloroglucinol, or catechin, condensation occurs with inversion of the configuration at C-4 to give the 4*S* products exclusively (27, 29, 31). Additionally, reaction of (2*R*,3*R*,4*S*)-5,7,3′,4′-tetra-*O*-methyl-flavan-3,4-diol with mercaptoacetic acid affords stereospecific production of the 4*S*-flavan-4-thioacetic acid derivative with retention of the configuration at C-4 (36). Solvolysis of (2*R*,3*R*)-2,3-*cis* procyanidins in the presence of thiols, phloroglucinol, or flavan-3-ols have given only the 4*S* adducts (98, 143, 352). Thus the electrophile is always approached from the upper side away from the 3-hydroxyl (Fig. 7.6.5) in condensations with the 2,3-*cis* electrophiles. Note that this places the 4-substituent in a quasi-axial position and this is seen as an upfield shift of the C-2 carbon signal in ^{13}C-NMR spectra of all 2,3-*cis* procyanidins due to the γ gauche effect of the 2,4-*trans* substituents (98, 303). The consistency of the stereochemistry in the 2,3-*cis* proanthocyanidins at the C-4 position is also seen in the invariably strong positive couplet in the low wave length band of their CD spectra (15, 26, 31). Foo's (111) isolation of a 2,3-*cis*-3,4-*cis*-promelacacidin is the first natural product with this stereochemistry. Only the 2,3-*cis*-3,4-*trans* isomer is obtained by synthesis, implying enzymic control of the formation of the natural proanthocyanidins. The 2,3-*cis*-3,4-*cis* epiafzelchin-(4α)-pyrogallol synthesized from guibourtacacidol-(4β)-phloroglucinol by photolytic rearrangement constitutes a very special case (359).

In contrast, substitution of the 2,3-*trans* electrophiles is usually not stereospecific. Depending on the structures of the electrophile and the nucleophile, varying proportions of 4*R* and 4*S* isomers are produced (Table 7.6.6). Where the heterocyclic ring of the electrophile assumes a C(2)-sofa or a conformation in which the 3-hydroxyl group is in a quasi-equatorial position and thus steric hindrance to approach from either the upper or lower side does not heavily favor formation of one isomer (see Fig. 7.6.5). In reactions of (2*R*,3*S*,4*R*)-mollisacacidin with phloroglucinol the [4-α] or 3,4-*trans* isomer is favored in a ratio of 2:1 over the [4-β] or 3,4-*cis* isomer (31). In reactions with fisetinidol or catechin the [4-α] linked dimers (i.e. retention of configuration) are produced in ratios of 1.4:1 and 1.7:1 for those of the [4-β] linked dimers (i.e. inversion of configuration) (28). A similar preference for the 3,4-*trans* isomer was seen in reactions of (2*R*,3*S*,4*R*)-mollisacacidin with *ent*-epifisetinidol where the [4-α]-inked dimer was obtained in yields of 1.3 times those of the [4-β]-inked dimer and in reactions with epicatechin where the [4-α] or 3,4-*trans* isomer was obtained in yields of 1.25:1 in comparison with the [4-β] or 3,4-*cis* isomer (27–29).

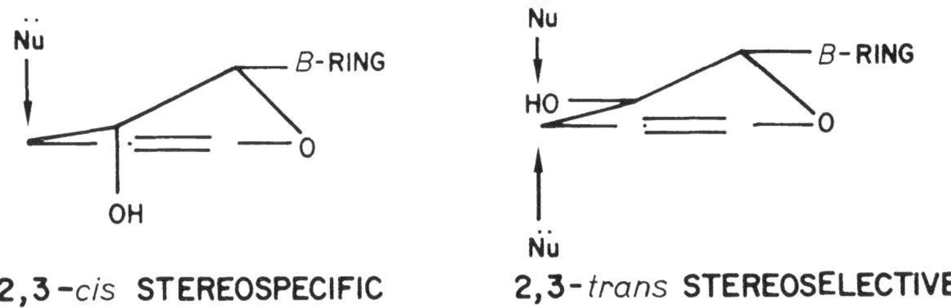

Fig. 7.6.5. Influence of stereochemistry at C-3 on substitution at C-4

Variation in the structure of the electrophile also influences the stereochemistry of these reactions. However, reaction of *ent*-mollisacacidin (2*S*,3*R*,4*S*) with phloroglucinol and catechin gave the [4-β] or 3,4-*trans*-linked dimers in a preference of 2:1 and 1.3:1 over the 3,4-*cis* isomers. Nearly the same ratios (i.e. 2:1 and 1.7:1) were obtained in reactions with (2*R*,3*S*,4*R*)-mollisacacidin. A pyrogallol B-ring in the electrophile increased the reaction rate. It also slightly increased the preference for a 3,4-*trans* substitution as reactions of (2*R*,3*S*,4*R*)-leucorobinetinidin with catechin gave the [4-α]-linked dimer in a ratio of 2:1 over the [4-β]-linked isomer, compared to a ratio of 1.7:1 for similar reactions with (2*R*,3*S*,4*R*)-mollisacacidin.

The reactions of the leucocyanidins are more highly directed to the 3,4-*trans* isomer because of greater steric restraint due to the 5-hydroxyl group. Botha et al. (28) found that reactions of leucocyanidin with phloroglucinol and resorcinol gave the 3,4-*trans*-[4-α]-linked products in a preference of 5.7:1 and 6.6:1 over their 3,4-*cis*-[4-β]-linked isomers. Porter and Foo (301) in a similar reaction of (2*R*,3*S*,4*R*)-leucocyanidin with phloroglucinol found that the 3,4-*trans*-[4-α]-linked product was obtained in preference to the 3,4-*cis* isomer by a factor of about 20. Reactions of leucocyanidin with catechin gave the [4-α]-linked dimers with no evidence for any 3,4-*cis* isomers (67). This latter result was consistent with the common occurrence of catechin-(4α→8)-catechin, catechin-(4α→8)-epicatechin, and their (4α→6) isomers in lower yield from plants containing procyanidins with absence of any reports of a catechin-(4β→6 or 8)-flavan. Recently, however, small proportions of 2,3-*trans*-3,4-*cis*-[4-β] linkages have been demonstrated in the procyanidins (68). The present evidence suggests that the synthetic 2,3-*cis* procyanidins have exclusively 3,4-*trans* stereochemistry whereas the 2,3-*trans* procyanidins have predominantly but not exclusively 3,4-*trans* stereochemistry. This contrasts markedly to the synthetic 5-deoxy proanthocyanidins such as the profisetinidins that have both 3,4-*cis* and 3,4-*trans* configurations at the interflavanoid bond as predicted from the reactions of the flavan-3,4-diols (316, 317).

7.6.3.3 Oligomeric Proanthocyanidins

Roux has commented that condensed tannins (and proanthocyanidins) have not attracted study because the weight of the research effort is invariably disproportionate to the results achieved (316). This observation is particularly descriptive of the earlier work because of the great inertia that had to be overcome to permit isolation and proof of structure of oligomeric proanthocyanidins. The advances of the past 10 to 15 years have been significant, and these compounds are now far more amenable to study. Because of this, interest in condensed tannin chemistry is growing. However, research efforts world-wide lag far behind the importance of these compounds to our production and use of foods and industrial chemicals. Despite the application of modern chromatographic methods, problems with the isolation of closely related oligomeric phenols continue to present often insurmountable problems. It is commonly necessary to resort to separations of a series of derivatives (i.e. methyl ethers, peracetates, or methyl ether peracetates) by laborious methods to obtain pure compounds. The high degree of chirality, restricted rotation about the interflavanoid bond, and conformational heterogeneity of the heterocyclic ring in these compounds vastly complicate interpretation of spectral data even with use of today's high field ^{1}H- and ^{13}C-NMR spectrometers (303, 316, 317). Fast atom bombardment mass spectrometry promises to answer some important questions about the structure and distribution of these compounds in plant extracts and use of high resolution FAB-MS has become routine for establishing elemental composition of these compounds (188). High resolution EI-MS continues to be useful in analyses of methylated derivatives at the dimer and trimer level (69). The more difficult problem is to define the absolute stereochemistry of compounds with so many asymmetric centers. New two-dimensional NMR experiments are just now beginning to be employed to solve these problems.

Nevertheless, there has been an explosive growth in our understanding of the chemistry of plant proanthocyanidins over the past 15 years. Approximately 100 oligomeric (dimers to tetramers) proanthocyanidins have now been described compared to only about 15 compounds described in the late 1960s (377). Isolation and proof of structure of pure proanthocyanidins at the tetramer level exceeds the number average molecular weight of condensed tannin isolates from wattle (*Acacia* spp.) bark and approaches the number average molecular weight of many procyanidin-based tannin isolates. The isolation and proof of structure of these higher oligomers is obviously a demanding and unending task. Since these compounds are so useful in predicting the structure and properties of condensed tannins, efforts to synthesize and prove the natural occurrence of a broader range of trimeric and tetrameric proanthocyanidins must continue. We are, however, rapidly approaching the stage of our understanding of these compounds at which much more attention must be given to the significance of these structural variations on polymer properties than to the novelty of structure. The review that follows clearly shows that, although remarkable progress has been made in definition of proanthocyanidin structure, our understanding of the properties of these compounds is extremely limited.

7.6.3.3.1 Structure and Distribution of Oligomeric Proanthocyanidins

The past ten years have seen the description of new proanthocyanidins in every class from the proguibourtacacidins through to the prodelphinidins. Much of the advance can be attributed to the synthesis of compounds in sufficient quantities to permit detailed structure analysis followed by isolation of the identical compound from plant tissue (141, 316, 317).

7.6.3.3.1.1 Proguibourtinidins

Only nine proguibourtinidins have been isolated as natural products (Table 7.6.8). A guibourtacacidol-epiafzelechin dimer was recently isolated from *Cassia fistula* sapwood for which a (4α→8) interflavanoid bond was assumed but not proven (290). *Acacia luderitzii* heartwood contains both (4α→6) and (4α→8) guibourtacacidol-catechin dimers, a (4α→6)-linked quibourtacacidol-epicatechin dimer, and a C-8 or C-6 carboxylic acid derivative of the latter compound or its (4α→8)-linked isomer (90). Guibourtacacidol-catechin and -epicatechin dimers are also present in *Julbernadia globiflora* wood (292). Now that reliable methods for distinguishing between C-4 to C-6 and C-4 to C-8 interflavanoid bonds are known, these compounds deserve further study.

The proguibourtinidins obtained from *Guibourtia coleosperma* that carry a 3,4,3′,5-tetrahydroxystilbene terminating unit are among the most interesting proanthocyanidins that have been isolated recently. As in most other 2,3-*cis* proanthocyanidins the compound **50** contains only [4-β] or 3,4-*trans* substitution. The 2,3-*trans* dimer has a [4-α] bond and, interestingly, the trimer **51** has both a 2,3-*cis* and a 2,3-*trans* proanthocyanidin unit (345). Another trimer with mixed 2,3-stereochemistry in the proguibourtinidin units was isolated, but its structure was not defined.

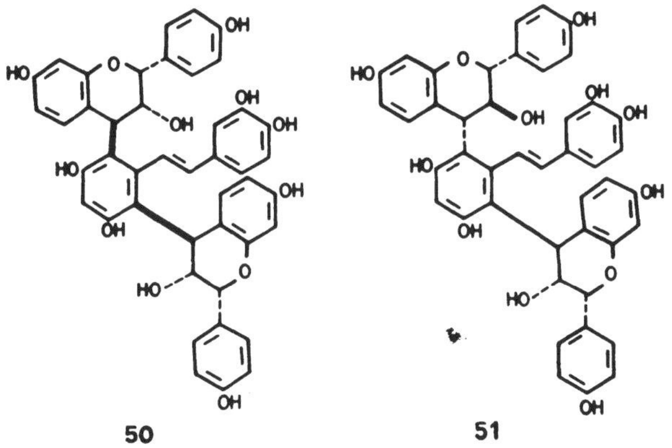

HO OH O OH OH OH HO OH OH HO O OH

50

HO OH O OH OH OH HO OH OH HO O OH

51

7.6.3.3.1.2 Profisetinidins

The profisetinidins have been studied intensively by Roux's school at The University of the Orange Free State, Bloemfontein, South Africa. The heartwoods of *Acacia mearnsii* and *Colophospermum mopane* contain a series of 2*R* isomers and these compounds have all been synthesized (Tables 7.6.9 and 7.6.10). The commercially important quebracho tannins from *Schinopsis* and *Rhus* contain a series of 2*S* isomers and these compounds have also been synthesized recently. Fifteen dimers, nine trimers, and nine tetramers have now been thoroughly de-

Table 7.6.8. Proguibourtinidin dimers

Name	Substitution						Stereochemistry[a]			Occurrence	Reference(s)
	5	6	7	8	3′	4′	2	3	4		
guibourtacacidol-(4α→8)-epiafzelhin[b]	H	H	OH	H	OH	OH	R	S		*Cassia fistula*	290
	OH	H	OH	FLA	H	OH	R	R			
guibourtacacidol-(4α→8)-catechin	H	H	OH	H	H	OH	R	S	R	*Acacia luderitzii*	90
	OH	H	OH	FLA	OH	OH	R	S			
guibourtacacidol-(4α→6)-catechin	H	H	OH	H	H	OH	R	S	R	*A. luderitzii*	90, 292
	OH	FLA	OH	H	OH	OH	R	S		*Julbernadia globiflora*	
guibourtacacidol-(4α→6)-epicatechin	H	H	OH	H	H	OH	R	S	R	*A. luderitzii*	90, 292
	OH	FLA	OH	H	OH	OH	R	R		*J. globiflora*	
guibourtacacidol-(4α→6)-8-carboxyepicatechin	H	H	OH	H	H	OH	R	S	R	*A. luderitzii*	90
	OH	FLA	OH	COOH	OH	OH	R	R			

[a] Refers to absolute configuration of the parent flavan-3,4-diol and flavan 3-ol
[b] Tentative assignment of interflavanoid bond location and location of carboxyl substitution

Table 7.6.9. Profisetinidin dimers

Name	Substitution						Stereochemistry[a]			Occurrence	Reference(s)
	5	6	7	8	3′	4′	2	3	4		
fisetinidol-(4α→6)-fisetinidol	H	H	OH	H	OH	OH	R	S	R	*Colophosphermum*	28
	H	FLA	OH	H	OH	OH	R	S		*mopane*	
fisetinidol-(4β→6)-fisetinidol	H	H	OH	H	OH	OH	R	S	S	*C. mopane*	28
	H	FLA	OH	H	OH	OH	R	S			
fisetinidol-(4α→6)-*ent*-epifisetinidol	H	H	OH	H	OH	OH	R	S	R	Synthesis only	28
	H	FLA	OH	H	OH	OH	S	S			
fisetinidol-(4β→6)-*ent*-epifisetinidol	H	H	OH	H	OH	OH	R	S	S	Syn. only	28
	H	FLA	OH	H	OH	OH	S	S			
fisetinidol-(4α→8)-catechin	H	H	OH	H	OH	OH	R	S	R	*Acacia mearnsii*	28
	OH	H	OH	FLA	OH	OH	R	S			
fisetinidol-(4α→6)-catechin	H	H	OH	H	OH	OH	R	S	R	Syn. only	28
	OH	FLA	OH	H	OH	OH	R	S			
fisetinidol-(4β→8)-catechin	H	H	OH	H	OH	OH	R	S	S	*A. mearnsii*	28
	OH	H	OH	FLA	OH	OH	R	S			
fisetinidol-(4α→8)-epicatechin	H	H	OH	H	OH	OH	R	S	R	Syn. only	28
	OH	H	OH	FLA	OH	OH	R	R			
fisetinidol-(4α→6)-epicatechin	H	H	OH	H	OH	OH	R	S	R	Syn. only	28
	OH	FLA	OH	H	OH	OH	R	R			
fisetinidol-(4β→8)-epicatechin	H	H	OH	H	OH	OH	R	S	S	Syn. only	28
	OH	H	OH	FLA	OH	OH	R	R			
ent-fisetinidol-(4β→8)-catechin	H	H	OH	H	OH	OH	S	R	S	*Schinopsis balansae,*	28
	OH	H	OH	FLA	OH	OH	R	S		*Rhus leptodictya*	369
ent-fisetinidol-(4α→8)-catechin	H	H	OH	H	OH	OH	S	R	R	*S. balansae,*	28
	OH	H	OH	FLA	OH	OH	R	S		*R. leptodictya*	
ent-fisetinidol-(4β→6)-catechin	H	H	OH	H	OH	OH	S	R	S	*S. balansae,*	369
	OH	FLA	OH	H	OH	OH	R	S		*R. leptodictya*	
ent-fisetinidol-(4α→6)-catechin	H	H	OH	H	OH	OH	S	R	R	*S. balansae,*	369
	OH	FLA	OH	H	OH	OH	R	S		*R. leptodictya*	
ent-fisetinidol-(4β→8)-epicatechin	H	H	OH	H	OH	OH	S	R	S	*R. leptodictya*	369
	OH	H	OH	FLA	OH	OH	R	R			

[a] Refers to absolute configuration of the parent flavan-3,4-diol and flavan-3-ol

Table 7.6.10. Profisetinidin trimers and tetramers

Name	Substitution						Stereochemistry[a]			Occurrence	Reference(s)
	5	6	7	8	3′	4′	2	3	4		
fisetinidol-(4β→6)-fisetinidol-(4α→6)-fisetinidol	H	H	OH	H	OH	OH	R	S	S	*Acacia mearnsii*	343
	H	FLA	OH	H	OH	OH	R	S	R		
	H	FLA	OH	H	OH	OH	R	S			
fisetinidol-(4α→8)-catechin-(6→4α)-fisetinidol	H	H	OH	H	OH	OH	R	S	R	*A. mearnsii,*	30
	OH	FLA	OH	FLA	OH	OH	R	S		*Colophospermum*	
	H	H	OH	H	OH	OH	R	S	R	*mopane*	
fisetinidol-(4α→8)-catechin-(6→4β)-fisetinidol	H	H	OH	H	OH	OH	R	S	R	*A. mearnsii,*	30
	OH	FLA	OH	FLA	OH	OH	R	S		*C. mopane*	
	H	H	OH	H	OH	OH	R	S	S		
fisetinidol-(4β→8)-catechin-(6→4α)-fisetinidol	H	H	OH	H	OH	OH	R	S	S	*A. mearnsii*	30
	OH	FLA	OH	FLA	OH	OH	R	S			
	H	H	OH	H	OH	OH	R	S	R		
fisetinidol-(4β→8)-catechin-(6→4β)-fisetinidol	H	H	OH	H	OH	OH	R	S	S	*A. mearnsii*	30
	OH	FLA	OH	FLA	OH	OH	R	S			
	H	H	OH	H	OH	OH	R	S	S		
ent-fisetinidol-(4α→8)-catechin-(6→4α)-*ent*-fisetinidol	H	H	OH	H	OH	OH	S	R	R	*Schinopsis, Rhus*	369
	OH	FLA	OH	FLA	OH	OH	R	S			
	H	H	OH	H	OH	OH	S	R	R		
ent-fisetinidol-(4α→8)-catechin-(6→4β)-*ent*-fisetinidol	H	H	OH	H	OH	OH	S	R	R	*Schinopsis, Rhus*	369
	OH	FLA	OH	FLA	OH	OH	R	S			
	H	H	OH	H	OH	OH	S	R	S		
ent-fisetinidol-(4β→8)-catechin-(6→4α)-*ent*-fisetinidol	H	H	OH	H	OH	OH	S	R	S	*Schinopsis, Rhus*	369
	OH	FLA	OH	FLA	OH	OH	R	S			
	H	H	OH	H	OH	OH	S	R	R		
ent-fisetinidol-(4β→8)-catechin-(6→4β)-fisetinidol	H	H	OH	H	OH	OH	S	R	S	*Schinopsis, Rhus*	369
	OH	FLA	OH	FLA	OH	OH	R	S			
	H	H	OH	H	OH	OH	S	R	S		
fisetinidol-(4α→8)-catechin-(6→4α)-fisetinidol, (6→4β)-fisetinidiol	H	H	OH	H	OH	OH	R	S	R	*Acacia* Syn.	343, 392
	OH	FLA	OH	FLA	OH	OH	R	S			
	H	FLA	OH	H	OH	OH	R	S	R		
	H	H	OH	H	OH	OH	R	S	S		

fisetinidol-(4β→8)-catechin-(6→4α)-fisetinidol, (6→4β)-fisetinidol	H	H	OH	H	OH	OH	R	S	S	*Acacia* Syn.	392
	OH	FLA	OH	FLA	OH	OH	R	S			
	H	FLA	OH	H	OH	OH	R	S	R		
	H	H	OH	H	OH	OH	R	S	S		
fisetinidol-(4β→6)-fisetinidol-(4α→6)-fisetinidol-(4α→8)-catechin	H	H	OH	H	OH	OH	R	S	S	*Acacia* Syn.	392
	H	FLA	OH	H	OH	OH	R	S	R		
	H	FLA	OH	H	OH	OH	R	S	R		
	OH	H	OH	FLA	OH	OH	R	S			
ent-fisetinidol-(4α→6)-*ent*-fisetinidol-(4β→6)-	H	H	OH	H	OH	OH	S	R	R	*Rhus lancea*	97
ent-fisetinidol-(4β→6)-catechin	H	FLA	OH	H	OH	OH	S	R	S		
	H	FLA	OH	H	OH	OH	S	R	S		
	OH	FLA	OH	H	OH	OH	R	S			
fisetinidol-(4β→6)-fisetinidol-(4α→8)-catechin-(6→4α)-fisetinidol	H	H	OH	H	OH	OH	R	S	S	*Acacia* Syn.	392
	H	FLA	OH	H	OH	OH	R	S	R		
	OH	FLA	OH	FLA	OH	OH	R	S			
	H	H	OH	H	OH	OH	R	S	R		
fisetinidol-(4β→6)-fisetinidol-(4β→8)-catechin-(6→4α)-fisetinidol	H	H	OH	H	OH	OH	R	S	S	*Acacia* Syn.	34, 392
	H	FLA	OH	H	OH	OH	R	S	S		
	OH	FLA	OH	FLA	OH	OH	R	S			
	H	H	OH	H	OH	OH	R	S	R		
fisetinidol-(4β→6)-fisetinidol-(4α→8)-catechin-(6→4β)-fisetinidol	H	H	OH	H	OH	OH	R	S	S	*Acacia* Syn.	392
	H	FLA	OH	H	OH	OH	R	S	R		
	OH	FLA	OH	FLA	OH	OH	R	S			
	H	H	OH	H	OH	OH	R	S	S		
ent-fisetinidol-(4α→6)-*ent*-fisetinidol-(4α→8)-catechin-(6→4β)-	H	H	OH	H	OH	OH	S	R	R	Syn. *Acacia*	34
ent-fisetinidol	H	FLA	OH	H	OH	OH	S	R	R		
	OH	FLA	OH	FLA	OH	OH	R	S			
	H	H	OH	H	OH	OH	S	R	S		
fisetinidol-(4α→8)-catechin-[(6→4β)-fisetinidol-]-	H	H	OH	H	OH	OH	R	S	R	Syn.	343
[(4α→6)-fisetinidol]	OH	FLA	OH	FLA	OH	OH	R	S			
	H	H	OH	H	OH	OH	R	S	S		
	H	FLA	OH	H	OH	OH	R	S			

[a] Refers to the absolute stereochemistry of parent flavan-3,4-diols and flavan-3-ols

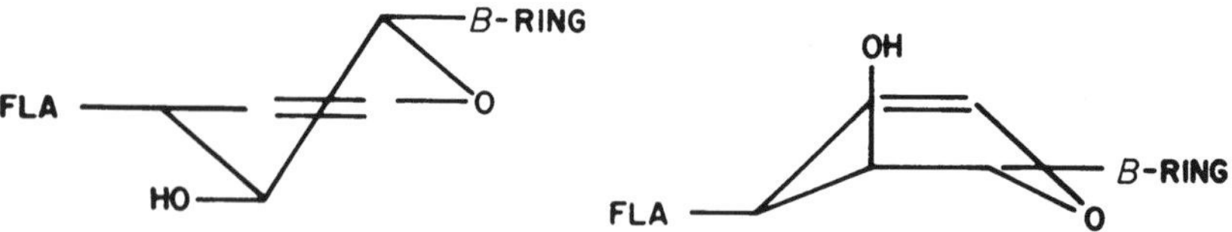

Fig. 7.6.6. Heterocyclic ring conformations of 2,3-*trans*-3,4-*trans* and 2,3-*trans*-3,4-*cis* profisetinidins

scribed (Tables 7.6.9 and 7.6.10). As discussed previously, current interest has been directed to the ratios of 3,4-*trans* and 3,4-*cis* isomers obtained from biomimetic synthesis and observed in natural plant extracts. The conformation of the heterocyclic ring in these isomers is also of great importance. Ferreira and Roux (316, 317) have assigned a half-chair heterocyclic ring conformation for the 2,3-*trans*-3,4-*trans* isomers and a skewed boat conformation for the 2,3-*trans*-3,4-*cis* isomers on the basis of ^{1}H-NMR coupling constants of the free phenols (Fig. 7.6.6). These heterocyclic ring conformations provide *quasi*-equatorial positions for the 4-flavanyl unit and the B-rings in both cases. These heterocyclic ring conformations suggest a plate-like structure for the profisetinidin-based condensed tannins. Two of the tetramers have exceptional conformational stability, resulting in unusually sharp ^{1}H-NMR spectra. Brandt et al. (34) have used nuclear Overhaeuser effect difference spectroscopy to define the conformations of these compounds. The tetramer synthesized by Steenkamp et al. (343) is interesting because it is the first true branched rather than angular proanthocyanidin known to date.

52

53

54 ---

55 —

Two dioxane-linked dimers **52** and **53**, have been isolated from the heartwood of *Acacia mearnsii*, and these compounds together with another isomeric dimer and two trimers **54** and **55**, have been recently synthesized through boron trifluoride-catalyzed condensations of mollisacacidin trimethy-ether (79, 391). The heartwood of *Dalbergia nitidula* contains a series of oligomers based on isoflavonoids. Two dimers and three trimers have now been synthesized and characterized (21, 33). Investigation of other plant tissues that contain isoflavonoids may reveal a whole new group of proanthocyanidins. Not all species of *Acacia* contain only profisetinidins in their heartwood, Foo (100) has recently shown that the heartwood of *Acacia baileyana* var. *purpurea* contains a mixture of profisetinidins, procyanidins, and prodelphinidins.

7.6.3.3.1.3 Prorobinetinidins

The bark of *Acacia mearnsii* contains a complex mixture of proanthocyanidins, including minor proportions of profisetinidins, but predominantly prorobinetinidins (316, 317). Extracts from the bark of *Acacia mearnsii* are the commercially important "wattle" or "mimosa" tannins. The prorobinetinidins carry either a catechin or gallocatechin terminal unit. Much of their structural complexity is derived from isomerism at C-4 (i.e. mixtures of 3,4-*cis*- and 3,4-*trans*-interflavanoid bonds), but the 3,4-*trans*- or [4-α]-linked oligomers predominate in the natural products. These compounds have been synthesized through reaction of (2*R*,3*S*,4*R*)-leucorobinetinidol with catechin or gallocatechin (28, 368). Because of the greater nucleophilicity of the phloroglucinol A-ring of the flavan-3-ol terminal unit as compared with the resorcinol functionality of the prorobinetinidin unit, all of the trimeric proanthocyanidins of this group are angular compounds. Of the nine prorobinetinidins that have been described, six are reported to be present in *Acacia mearnsii* bark extracts (Table 7.6.11).

7.6.3.3.1.4 Proteracacidins and Promelacacidins

Two teracacidol-catechin dimers linked (4β→8) and (4β→6) have been synthesized, but these compounds have not been reported as natural products (27, 28). A melacacidol-(4α→6)-melacacidol dimer isolated from the heartwood of *Prosopis glandulosa* (mesquite) occurs in low concentration together with the oxidatively coupled flavan-3-ols described previously (174). Foo (103) has recently isolated a melacacinidin-(4α→6)-isomelacacidin dimer from *Acacia melanoxylon*, the first natural product that has the 2*R*,3*R*,4*R*-(2,3-*cis*-3,4-*cis*)-stereochemistry. However, only its 2*R*,3*R*,4*S* (2,3-*cis*-3,4-*trans*) diastereoisomer was obtained by synthesis.

7.6.3.3.1.5 Propelargonidins

Although a number of condensed tannins containing propelargonidin oligomers have now been found (Sect. 7.7), only five propelargonidin oligomers have been described as natural products. An epiafzelechin-(4β→8)-catechin dimer (gambirin C) has been isolated from *Uncaria gambir* where it occurs in low concentration together with the C-2 substituted gambirins (278). Two epiafzelechin-4-*O*-methyl-

Table 7.6.11. Prorobinetinidins

Name	Substitution							Stereochemistry[a]			Occurrence	Reference(s)
	5	6	7	8	3′	4′	5′	2	3	4		
robinetinidol-(4α→8)-catechin	H	H	OH	H	OH	OH	OH	R	S	R	*Acacia mearnsii* bark	27, 28
	OH	H	OH	FLA	OH	OH	H	R	S			
robinetinidol-(4α→6)-catechin	H	H	OH	H	OH	OH	OH	R	S	R	Synthesis only	27, 28
	OH	FLA	OH	H	OH	OH	H	R	S			
robinetinidol-(4β→8)-catechin	H	H	OH	H	OH	OH	OH	R	S	S	Syn. only	27, 28
	OH	H	OH	FLA	OH	OH	H	R	S			
robinetinidol-(4α→8)-gallocatechin	H	H	OH	H	OH	OH	OH	R	S	R	*A. mearnsii* bark	27, 28, 368
	OH	H	OH	FLA	OH	OH	OH	R	S			
robinetinidol-(4α→8)-catechin-(6→4α)-robinetinidol	H	H	OH	H	OH	OH	OH	R	S	R	*A. mearnsii* bark	368
	OH	FLA	OH	FLA	OH	OH	H	R	S			
	H	H	OH	H	OH	OH	OH	R	S	R		
robinetinidol-(4β→8)-catechin-(6→4β)-robinetinidol	H	H	OH	H	OH	OH	OH	R	S	R	*A. mearnsii* bark	368
	OH	FLA	OH	FLA	OH	OH	OH	R	S			
	H	H	OH	H	OH	OH	OH	R	S	S		
robinetinidol-(4β→8)-catechin-(6→4β)-robinetinidol	H	H	OH	H	OH	OH	OH	R	S	S	Syn. only	368
	OH	FLA	OH	FLA	OH	OH	OH					
	H	H	OH	H	OH	OH	OH	R	S	S		
robinetinidol-(4α→8)-gallocatechin-(6→4α)-robinetinidol	H	H	OH	H	OH	OH	OH	R	S	R	*A. mearnsii* bark	368
	OH	FLA	OH	FLA	OH	OH	OH	R	S			
	H	H	OH	H	OH	OH	OH	R	S	R		
robinetinidol-(4α→8)-gallocatechin-(6→4β)-robinetinidol	H	H	OH	H	OH	OH	OH	R	S	R	*A. mearnsii* bark	368
	OH	FLA	OH	FLA	OH	OH	OH	R	S			
	H	H	OH	H	OH	OH	OH	R	S	S		

[a] Refers to absolute configuration of parent flavan-3,4-diol and flavan-3-ol

gallocatechin dimers have been isolated from *Ouratea* of the Celastraceae and have been tentatively assigned (4α→8)- and (4β→8)-interflavanoid bonds (74). The [4α→8]-interflavanoid bond with a 2,3-*cis*-3,4-*cis* stereochemistry is very unusual although Foo (103) has also recently reported a dimeric proanthocyanidin (a promelacacidin) with this stereochemistry from the wood of *Acacia melanoxylon*. The propelargonidins from *Ouratea* deserve further study to establish beyond doubt the interflavanoid bond location and stereochemistry. The only known methylated proanthocyanidins found in natural extracts are also represented by these two compounds. Two propelargonidins **56** and **57**, recently isolated from the roots of *Prunus persica* (peach) trees, are particularly interesting because of their inhibition of the growth of *Oryza sativa* (rice) seedlings and because of the stereochemistry of the interflavanoid bond (283). Although procyanidins linked with both (4β→8) and (2β→O→7) bonds are well known, these propelargonidins represent the first examples of this type of compound with a 2,3-*trans*-3,4-*cis* stereochemistry of the upper unit.

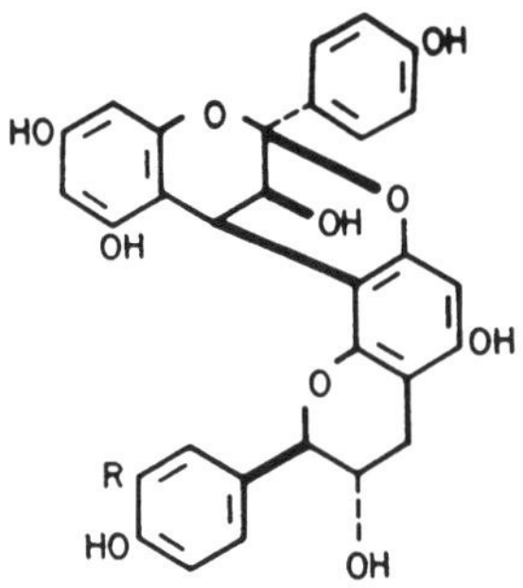

HO O OH O OH OH OH O R HO OH

56 R = H
57 R = OH

7.6.3.3.1.6 Procyanidins

The procyanidins are broadly distributed in the leaves, fruit, bark, and less commonly the wood of a wide spectrum of plants (Sect. 7.7). About 50 procyanidins ranging from dimers to pentamers have now been isolated and their structures defined. The 2*R*,3*R*-(2,3-*cis*)-procyanidins linked by (4β→8)- and/or (4β→6)-interflavanoid bonds occur most frequently. Many plants contain mixtures of 2*R*,3*R*-(2,3-*cis*) and 2*R*,3*S*-(2,3-*trans*) procyanidins but the compounds of the former stereochemistry normally predominate. The distribution of these compounds in commercially important woody plants has been extended continually from the early work of Porter (300) on *Pinus radiata* bark through the survey of Pinaceae by Samejima (329) to the recent isolation of dimers from the bark of *Juniperus communis* (94). The plants that contain predominantly 2*R*,3*S*-(2,3-*trans*) compounds are, so far, restricted to few plant genera, including fruits of *Ribes* species or the catechins of *Salix* species. All 2,3-*cis* procyanidins have [4-β]-interflavanoid bonds (i.e., 3,4-*trans* configuration). Most natural 2,3-*trans* procyanidins isolated to date have [4-α]-interflavanoid bonds (i.e. also 3,4-*trans* but opposite configuration). However, Delcour et al. (68) and Kolodziej (210, 211) have recently obtained catechin-(4β→8)-epicatechin and catechin-(4β→8)-catechin in a yield of about 10% of the corresponding catechin-(4β→8)-diastereomers by synthesis. Small proportions of 2,3-*trans*-3,4-*cis* procyanidins were recently isolated from *Potentilla erecta* so these compounds are now known as natural products as well (331).

A comparatively large group of doubly linked type-A procyanidins (i.e. 4β→8;2β→0→7 interflavanoid bonds) has now been described. Until recently, these compounds were represented only by epicatechin-(4β→8;2β→0→7)-epicatechin from *Aesculus hippocastanum* (horsechestnut) and *Persea gratissima* (avocado) (127, 242, 287, 330) and epicatechin-(4β→8;2β→0→7)-catechin from *Vaccinium vitis-idaea* (cranberries) (381). Karchesy and Hemingway (187) showed that epicatechin-(4β→8;2β→0→7)-catechin is the predominant dimeric procyanidin in *Arachis hypogea* (peanut) skins but small amounts of epicatechin-(4β→8;2β→0→7)-epicatechin and epicatechin-(4β→8)-catechin occur with it. The procyanidins in *Cinnamomum zeylanicum* (cinnamon) are most interesting. Here mixtures of (4β→8)-, (4β→6)-, and (4β→8;2β→0→7)-linked dimers are also found and higher molecular weight oligomers are found in which mixtures of these types of interflavanoid bonds co-occur in the same molecule (275). These and other dimeric procyanidins are presented in Table 7.6.12.

Most plants contain procyanidins of the 2*R* configuration but those in the Palmae in particular are noted for the occurrence of both 2*R* and 2*S* enantiomers (72, 93, 110). A number of these compounds have been synthesized but only one dimer with a 2*S* configuration has been found as a natural product.

Nearly all of the procyanidins isolated to date have either catechin or epicatechin as terminal units. It is common to find 2*R*,3*R*-(2,3-*cis*)-stereochemistry for the procyanidin extender units and a 2*R*,3*S*-(2,3-*trans*)-configuration for the terminal unit. The catechin-(4α→8)-epiafzelchin dimer isolated from *Wisteria sinensis* by Weinges (380) and the catechin-(4α→2)-phloroglucinol isolated from *Nelia meyeri* by Kolodzeij (208) are the only natural procyanidins reported to date that are not terminated with either catechin or epicatechin. *Nelia meyeri* also contains epicatechin-(4β→8)-epicatechin and epicatechin-(4β→6)-epicatechin (209).

Gallate esters have been found in *Vitis* spp. (grapes); *Rheum rhizoma* (rhubarb); *Polygonum multiflorum*, a climbing vine native to China; and in *Sanquisorba officinalis*, a Japanese plant that also contains the gambirins A1 and B3 (see Sect. 7.6.3.1.3) (274, 280, 349). The first glycosides of proanthocyanidins were described only in 1986 with the isolation of the 7-*O*-β-D-glucopyranoside of catechin-(4α→8)-catechin, both the 6- and 8-*C*-β-D-glucopyranosides of epicatechin-(4β→8)-epicatechin, and both the 6- and 8-*C*-β-D-glucopyranosides of epicatechin-(4β→8)-catechin from *Rheum* spp. (rhubarb) (196, 271). A series of 1-hydroxy-6-hydroxy- and 1-hydroxy-6-oxo-2-cyclohexene-1-carboxylic acid derivatives of 2,3-*cis* procyanidins was isolated recently from the bark of *Salix sieboldiana* (166). Other acylated proanthocyanidins will no doubt be reported in the near future because methods for their isolation and identification have improved.

None of the flavan-3-ol terminated procyanidins or their derivatives have as yet been obtained as crystals of suitable quality for crystal structure determination; difficulties exist in the interpretation of the proton spectra because of rotational isomerism and long-range couplings (98, 310). Therefore, conclusions that

Table 7.6.12. Dimeric procyanidins

Name	Substitution						Stereochemistry[a]			Occurrence	Reference(s)
	5	6	7	8	3′	4′	2	3	4		
epicatechin-(4β→8)-catechin	OH	H	OH	H	OH	OH	R	R	S	General	352, 380, 381
	OH	H	OH	FLA	OH	OH	R	S			
epicatechin-(4β→6)-catechin	OH	H	OH	H	OH	OH	R	R	S	General	143, 272, 352
	OH	FLA	OH	H	OH	OH	R	S			
epicatechin-(4β→8)-epicatechin	OH	H	OH	H	OH	OH	R	R	S	General	330, 352,
	OH	H	OH	FLA	OH	OH	R	R			380, 381
epicatechin-(4β→6)-epicatechin	OH	H	OH	H	OH	OH	R	R	S	General	352, 380, 381
	OH	FLA	OH	H	OH	OH	R	R			
catechin-(4α→8)-catechin	OH	H	OH	H	OH	OH	R	S	R	General	352, 380, 381
	OH	H	OH	FLA	OH	OH	R	S			
catechin-(4α→6)-catechin	OH	H	OH	H	OH	OH	R	S	R	General	67, 136
	OH	FLA	OH	H	OH	OH	R	S			
catechin-(4α→8)-epicatechin	OH	H	OH	H	OH	OH	R	S	R	*Ribes*	68, 352, 380
	OH	H	OH	FLA	OH	OH	R	R			
catechin-(4β→8)-epicatechin	OH	H	OH	H	OH	OH	R	S	S	Synthesis	68, 210, 211
	OH	H	OH	FLA	OH	OH	R	R			
catechin-(4α→6)-epicatechin	OH	H	OH	H	OH	OH	R	S	R	*Ribes*	68, 352
	OH	FLA	OH	H	OH	OH	R	R			
epicatechin-(4β→8)-*ent*-epicatechin	OH	H	OH	H	OH	OH	R	R	S	Syn.	109, 110
	OH	H	OH	FLA	OH	OH	S	S			
epicatechin-(4β→6)-*ent*-epicatechin	OH	H	OH	H	OH	OH	R	R	S	Syn.	109, 110
	OH	FLA	OH	H	OH	OH	S	S			
ent-epicatechin-(4α→8)-epicatechin	OH	H	OH	H	OH	OH	S	S	R	Syn.	109, 110
	OH	H	OH	FLA	OH	OH	R	R			
ent-epicatechin-(4α→8)-*ent*-epicatechin	OH	H	OH	H	OH	OH	S	S	R	*Palmaceae*, Syn.	72, 109, 110
	OH	H	OH	FLA	OH	OH	S	S			
catechin-(4α→8)-*ent*-epicatechin	OH	H	OH	H	OH	OH	R	S	R	Syn.	109, 110
	OH	H	OH	FLA	OH	OH	S	S			
catechin-(4α→8)-epiafzelechin	OH	H	OH	H	OH	OH	R	S	R	*Wisteria*	380
	OH	H	OH	FLA	H	OH	R	R			

Table 7.6.12 (continued)

Name	Substitution						Stereochemistry[a]			Occurrence	Reference(s)
	5	6	7	8	3′	4′	2	3	4		
epicatechin-(4β→8)-3-*O*-galloyl-epicatechin	OH	H	OH	H	OH	OH	R	R	S	*Vitis*	65
	OH	H	OH	FLA	OH	OH	R	R			
3-*O*-galloyl-epicatechin-(4β→8)-epicatechin	OH	H	OH	H	OH	OH	R	R	S	*Rheum*	280
	OH	H	OH	FLA	OH	OH	R	R			
3-*O*-galloyl-epicatechin-(4β→8)-catechin	OH	H	OH	H	OH	OH	R	R	S	*Rheum, Polygonum*	274, 280
	OH	H	OH	FLA	OH	OH	R	S			
3-*O*-galloyl-catechin-(4α→8)-catechin	OH	H	OH	H	OH	OH	R	S	R	*Sanguisorba*	349
	OH	H	OH	FLA	OH	OH	R	S			
3-*O*-galloyl-epicatechin-(4β→8)-3-O-galloyl-epicatechin	OH	H	OH	H	OH	OH	R	R	S	*Polygonum*	274
	OH	H	OH	FLA	OH	OH	R	S			
6-*C*-β-D-glucoepicatechin-(4β→8)-epicatechin	OH	GLU	OH	H	OH	OH	R	R	S	*Rheum*	196
	OH	H	OH	FLA	OH	OH	R	R			
8-*C*-β-D-glucoepicatechin-(4β→8)-epicatechin	OH	H	OH	GLU	OH	OH	R	R	S	*Rheum*	196
	OH	H	OH	FLA	OH	OH	R	R			
6-*C*-β-D-glucoepicatechin-(4β→8)-catechin	OH	GLU	OH	H	OH	OH	R	R	S	*Rheum*	196
	OH	H	OH	FLA	OH	OH	R	S			
8-*C*-δ-D-glucoepicatechin-(4β→8)-catechin	OH	H	OH	GLU	OH	OH	R	R	S	*Rheum*	196
	OH	H	OH	FLA	OH	OH	R	S			
7-*O*-β-D-gluco-catechin-(4α→8)-catechin	OH	H	O-GLU	OH	OH	OH	R	S	R	*Rheum*	196
	OH	H	OH	FLA	OH	OH	R	S			
epicatechin-(4β→8; 2β→*O*→7)-epicatechin	OH	H	OH	H	OH	OH	R[b]	R	S	*Aesculus Persea*	127, 174,
	OH	H	C2-*O*	FLA	OH	OH	R	R			187, 242, 287
epicatechin-(4β→8;2β→*O*→7)-catechin	OH	H	OH	H	OH	OH	R	R	S	*Arachis*	187
	OH	H	C2-*O*	FLA	OH	OH	R	S			

[a] Refers to absolute configuration of parent flavan-3,4-diol and flavan-3-ol
[b] Refers to B-ring

can be drawn from coupling constants of the heterocyclic ring protons with regard to conformation of the heterocyclic ring are limited, particularly with respect to the free phenolic forms. Porter et al. (304) have studied the heterocyclic ring conformation of the methyl ether derivative of catechin-(4α→2)-phloroglucinol in the crystal and solution state. A C(2)-sofa *E*-conformation was indicated both in the crystal state and in solution by consideration of heterocyclic ring proton coupling constants and by molecular mechanical (MM2) studies. There have been a number of attempts to apply conformational analysis techniques to more fully define the structures of oligomeric procyanidins recently. Pizzi et al. (298) used energy minimization methods to predict the conformation of procyanidins, assuming that all heterocyclic rings have the conformation that is observed in crystalline 8-bromotetra-*O*-methyl-(+)-catechin (95) and assuming that only one rotational isomer is present. However, X-ray crystallography of other catechin derivatives (119, 120, 304) and measurement of rotational isomer populations by time-resolved fluorescence (18) suggest that Pizzi's models oversimplify the structures of these compounds. Viswanadhan and Mattice (365–367) have extended these calculations to account for a variety of heterocyclic ring conformations and have considered two rotational isomers for the 16 possible dimers of catechin and epicatechin.

A number of trimeric and tetrameric procyanidins have been described (Table 7.6.13). Epicatechin-(4β→8)-epicatechin-(4β→8)-epicatechin, proposed as the major trimeric procyanidin in *Aesculus hippocastanum* (horsechestnut) seed coats (176) and in *Crataegus* (212), has now been synthesized and the locations of the interflavanoid bonds verified (141, 212). The trimeric procyanidins obtained from *Pinus taeda* (140, 141) and from *Areca catechu* (272) have both (4β→8) and (4β→6) bonds within the same molecule. Heterogeneity of the interflavanoid bond location and type [i.e. the procyanidins of *Cinnamomum zeylanicum* (cinnamon)] (275) is now well documented. Questions still exist about the presence and extent of branching in the procyanidins. An epicatechin-(4β→8)-catechin-(6→4β)-epicatechin trimer (**58**), the first compound representative of a branch point in procyanidin polymers, has been synthesized (104) but has not yet been isolated as a natural product. Absence of this compound in natural extracts supports Porter's analysis (Sect. 7.7) showing limited branching in a procyanidin from *Chaenomeles chinensis*. Two all-*trans* trimeric procyanidins, also with interflavanoid linkage heterogeneity, and a trimer with both 2,3-*trans* (4α→8)- and (4β→8)-interflavanoid bond configurations have been synthesized recently (67, 68). Two linear (epicatechin)$_3$-catechin tetramers with (4β→8) and (4β→6) bonding to the terminal catechin unit were also obtained from *Areca catechu* (272).

Jacques et al. (176) isolated two trimeric procyanidins that contained both (4β→8)- and (4β→8;2β→0→7)-interflavanoid bonds, and partially defined their structures. Nonaka et al. (275) have now unequivocally established the locations of interflavanoid bonds in a series of these compounds from *Cinnamomum zeylanicum* (cinnamon), including a trimer, two tetramers (**59**), and a pentamer. The cinchonains such as **60** and **61**, cinnamate derivatives of procyanidins are among the more unusual structural variations that have been found in the procyanidins (273). These compounds emphasize the high degree of nucleophilicity of the phloroglucinol ring in procyanidins. These no doubt are examples of a vari-

Table 7.6.13. Trimeric, tetrameric, and pentameric procyanidins

Name	Substitution						Stereochemistry[a]			Occurrence	Reference(s)
	5	6	7	8	3′	4′	2	3	4		
epicatechin-(4β→8)-epicatechin-(4β→8)-epicatechin	OH	H	OH	H	OH	OH	R	R	S	*Aesculus*	141, 166, 352
	OH	H	OH	FLA	OH	OH	R	R	S		
	OH	H	OH	FLA	OH	OH	R	R			
epicatechin-(4β→8)-epicatechin-(4β→8)-catechin	OH	H	OH	H	OH	OH	R	R	S	*Pinus taeda*,	140, 141, 272
	OH	H	OH	FLA	OH	OH	R	R	S	*Areca catechu*	
	OH	H	OH	FLA	OH	OH	R	S			
epicatechin-(4β→8)-epicatechin-(4β→6)-catechin	OH	H	OH	H	OH	OH	R	R	S	*P. taeda*	140, 141, 272
	OH	H	OH	FLA	OH	OH	R	R	S	*A. catechu*	
	OH	FLA	OH	H	OH	OH	R	S			
epicatechin-(4β→6)-epicatechin-(4β→8)-catechin	OH	H	OH	H	OH	OH	R	R	S	*P. taeda*	140, 141
	OH	FLA	OH	H	OH	OH	R	R	S		
	OH	H	OH	FLA	OH	OH	R	S			
catechin-(4β→8)-catechin-(4β→8)-catechin	OH	H	OH	H	OH	OH	R	S	R	Synthesis	67
	OH	H	OH	FLA	OH	OH	R	S	R		
	OH	H	OH	FLA	OH	OH	R	S			
catechin-(4β→8)-catechin-(4β→6)-catechin	OH	H	OH	H	OH	OH	R	S	R	Syn.	67
	OH	H	OH	FLA	OH	OH	R	S	R		
	OH	FLA	OH	H	OH	OH	R	S			
catechin-(4β→8)-catechin-(4α→8)-epicatechin	OH	H	OH	H	OH	OH	R	S	S	Syn.	68
	OH	H	OH	FLA	OH	OH	R	S	R		
	OH	H	OH	FLA	OH	OH	R	R			
catechin-(4α→8)-catechin-(4α→8)-epicatechin	OH	H	OH	H	OH	OH	R	S	R	Syn.	68
	OH	H	OH	FLA	OH	OH	R	S	R		
	OH	H	OH	FLA	OH	OH	R	R			
epicatechin-(4β→8)-catechin-(6→4β)-epicatechin	OH	H	OH	H	OH	OH	R	R	S	Syn.	104
	OH	FLA	OH	FLA	OH	OH	R	S			
	OH	H	OH	H	OH	OH	R	R	S		
epicatechin-(4β→8)-(2β→O→7)-epicatechin-(4β→8)-epicatechin	OH	H	OH	H	OH	OH	R	R	S	*Cinnamomum*	275
	OH	H	C2-O	FLA	OH	OH	R	R	S		
	OH	H	OH	FLA	OH	OH	R	R	S		

epicatechin-(4$\beta \to$8)-epicatechin-(4$\beta \to$8)-epicatechin-(4$\beta \to$8)-catechin	OH	H	OH	H	OH	OH	R	R	S	*A. catechu*	272
	OH	H	OH	FLA	OH	OH	R	R	S		
	OH	H	OH	FLA	OH	OH	R	R	S		
	OH	H	OH	FLA	OH	OH	R	S			
epicatechin-(4$\beta \to$8)-epicatechin-(4$\beta \to$8)-epicatechin-(4$\beta \to$6)-catechin	OH	H	OH	H	OH	OH	R	R	S	*A. catechu*	272
	OH	H	OH	FLA	OH	OH	R	R	S		
	OH	H	OH	FLA	OH	OH	R	R	S		
	OH	FLA	OH	H	OH	OH	R	S			
epicatechin-(4$\beta \to$8)-epicatechin-(4$\beta \to$8)-(2$\beta \to O \to$7)-epicatechin (4$\beta \to$8)-epicatechin	OH	H	OH	H	OH	OH	R	R	S	*Cinnamomum*	275
	OH	H	OH	FLA	OH	OH	R	R	S		
	OH	H	C2-O	FLA	OH	OH	R	R	S		
	OH	H	OH	FLA	OH	OH	R	R			
epicatechin-(4$\beta \to$6)-epicatechin-(4$\beta \to$8)-(2$\beta \to O \to$7)-epicatechin-(4$\beta \to$8)-epicatechin	OH	H	OH	H	OH	OH	R	R	S	*Cinnamomum*	275
	OH	FLA	OH	H	OH	OH	R	R	S		
	OH	H	C2-O	FLA	OH	OH	R	R	S		
	OH	H	OH	FLA	OH	OH	R	R			
epicatechin-(4β-8)-epicatechin-(4$\beta \to$8)-epicatechin-(4$\beta \to$8)-(2$\beta \to O \to$7)-epicatechin-(4$\beta \to$8)-epicatechin	OH	H	OH	H	OH	OH	R	R	S	*Cinnamomum*	275
	OH	H	OH	FLA	OH	OH	R	R	S		
	OH	H	OH	FLA	OH	OH	R	R	S		
	OH	H	C2-O	FLA	OH	OH	R	R	S		
	OH	H	OH	FLA	OH	OH	R	R			

[a] Refers to absolute configuration of parent flavan-3,4-diol and flavan-3-ol

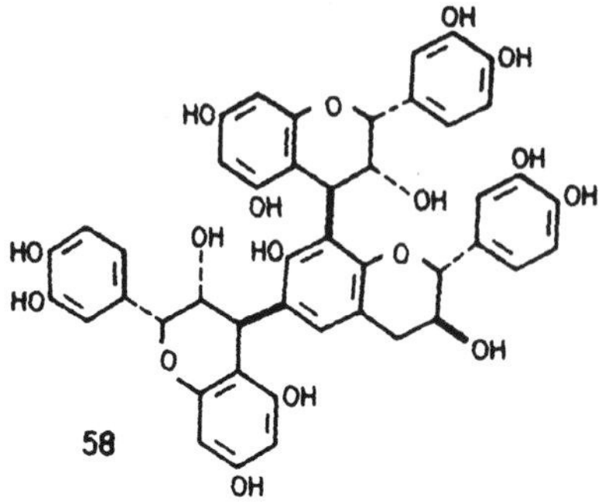

58

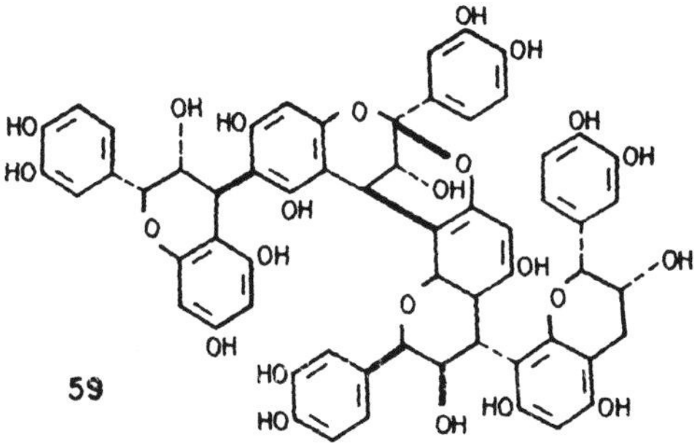

59

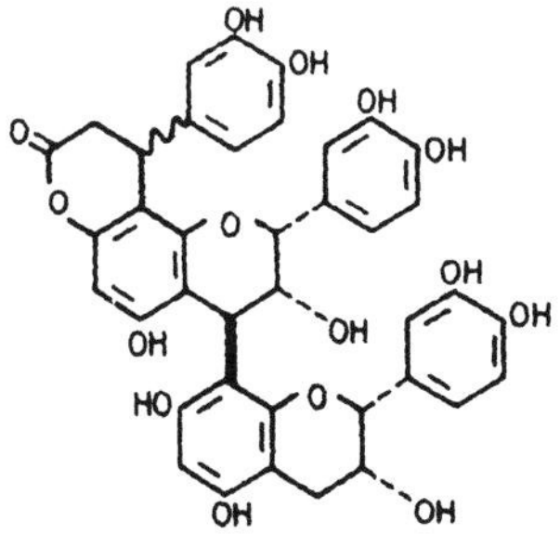

60 ---

61 —

ation in structure that will become more common as progress in isolation of procyanidins is made in the future.

^{1}H-NMR spectra of the procyanidins, like most other proanthocyanidins, show restricted rotation about the interflavanoid bond under normal temperature conditions (91, 98, 352, 382). The phenolic forms of procyanidins with a 2,3-*cis*-3,4-*trans* upper unit give broadened but first order spectra until the temperature is reduced to 0 °C where two rotational isomers become apparent (98). It is very important to establish the presence and relative proportions of rotational isomers in the free phenols at physiological temperature conditions. It is not possible to resolve these rotamers by NMR because of the comparatively slow time scale. The presence of two rotamers of the dimeric procyanidins as free phenols, and in proportions similar to those found for the locked methyl ether or acetate derivatives, has recently been shown by time-resolved fluorescence decay methods

(18). These results agree with the observation of two minimum energy states, in which the energy difference between the two minima is small enough that both rotamers may be populated to a significant extent (366, 367). Mattice and coworkers (19, 365–367) have shown that the chain conformation, particularly the amount of helical character and the right or left handedness, is determined by the rotamer population. Contrary to the conclusions of Haslam (135), in no case does the stereochemistry at C-3 have an important effect on the geometry of the helix. The presence of rotational isomerism in these compounds results in the formation of random coils rather than rigid helically-wound rods.

In the methyl ether or acetate derivatives, the rotational isomers are locked at ambient temperature and second order NMR spectra are clearly seen for dimeric procyanidin derivatives (98). Probe temperatures must be increased to 160° to 180 °C in nitrobenzene before the spectra collapse to first order. The (4β→8)-linked isomers exhibit one preferred rotational isomer whereas two rotational isomers are evident in nearly equal proportions in the (4β→6)-linked compounds. In these 2,3-*cis*-3,4-*trans* procyanidins, the lower flavan unit is in a *quasi*-axial position. Here the restricted rotation is due to interaction of the A-ring substituent(s) *ortho* to the interflavanoid bond and the C-2 hydrogen together with the π system of the A-ring of the upper unit (98).

In the 2,3-*trans*-3,4-*trans* procyanidins, which carry the lower flavan unit in a *quasi*-equatorial position, rotational energy barriers are higher. The free phenol and their derivatives all show rotational isomerism at ambient temperature. The temperature must be raised to 100 °C to obtain a first order spectrum of catechin-(4α→8)-catechin in deutero-dimethyl sulfoxide. As was observed in the 2,3-*cis* isomers, the 2,3-*trans*-4,3-*trans*-(4α→8)-linked procyanidins have one preferred rotational isomer, which dominates over the minor rotamer by a factor of about 4 to 1. However, in the deca-acetate of catechin-(4α→6)-catechin, two rotational isomers are present in approximately equal proportion. In the 2,3-*trans*-3,4-*trans* procyanidins, both the oxygen of the heterocyclic ring in the case of (4α→8) linkage and the substituent at C-7 or the substituents at both C-5 and C-7 in the case of (4α→6) linkage interact with the oxygen substituents at C-3 and C-5 of the upper unit (98).

7.6.3.3.1.7 Prodelphinidins

Even though the prodelphinidins are represented in the condensed tannins of a broad spectrum of plants (Sect. 7.7), to date only ten dimeric and four trimeric compounds have been described (Table 7.6.14). Two compounds were isolated from the male flowers of trees, epigallocatechin-(4β→8)-catechin from *Pinus sylvestris* (129) and gallocatechin-(4α→8)-catechin from *Salix caprea* (108). The tannins from the leaves of conifers are generally dominated by prodelphinidins (226), as opposed to procyanidins in the barks (329), but no oligomeric prodelphinidins have yet been isolated from conifer tree leaves. Study of the prodelphinidins in conifer tree leaves may extend the number of known compounds with mixed hydroxylation patterns. The prodelphinidins in needles of *Pinus taeda*, *P. palustris*, *P. elliottii*, and *P. echinata* (southern pines) are of particular interest because they have been shown to be sensitive indicators of tree growth

Table 7.6.14. Prodelphinidins

Name	Substitution							Stereochemistry[a]			Occurrence	Reference(s)
	5	6	7	8	3′	4′	5′	2	3	4		
epigallocatechin-(4β→8)-catechin	OH	H	OH	H	OH	OH	OH	R	R	S	*Pinus sylvestris*,	129, 218
	OH	H	OH	FLA	OH	OH	H	R	S		*Myrica nagi*	
gallocatechin-(4α→8)-catechin	OH	H	OH	H	OH	OH	OH	R	S	R	*Salix caprea*,	32, 70, 108
	OH	H	OH	FLA	OH	OH	H	R	S		*Hordeum vulgare*	
epigallocatechin-(4β→8)-epigallocatechin	OH	H	OH	H	OH	OH	OH	R	R	S	*Myrica rubra*	276
	OH	H	OH	FLA	OH	OH	OH	R	R			
epigallocatechin-(4β→6)-epigallocatechin	OH	H	OH	H	OH	OH	OH	R	R	S	*M. rubra*	276
	OH	FLA	OH	H	OH	OH	OH	R	R			
gallocatechin-(4α→8)-epigallocatechin	OH	H	OH	H	OH	OH	OH	R	S	R	*Ribes sanguineum*	108
	OH	H	OH	FLA	OH	OH	OH	R	R			
ent-epigallocatechin-(4α→8)-catechin	OH	H	OH	H	OH	OH	OH	S	S	R	Beer	70
	OH	H	OH	FLA	OH	OH	H	R	S			
epigallocatechin-(4β→8)-3-*O*-galloyl-epigallocatechin	OH	H	OH	H	OH	OH	OH	R	R	S	*M. rubra*	276
	OH	H	OH	FLA	OH	OH	OH	R	R			
3-*O*-galloyl-epigallocatechin-(4β→8)-3-*O*-galloyl-epigallo-catechin	OH	H	OH	H	OH	OH	OH	R	R	S	*M. rubra*	276
	OH	H	OH	FLA	OH	OH	OH	R	R			
3-*O*-galloyl-epigallocatechin-(4β→6)-3-*O*-galloyl-epigallo-catechin	OH	H	OH	H	OH	OH	OH	R	R	S	*M. rubra*	276
	OH	FLA	OH	H	OH	OH	OH	R	R			
3-*O*-galloyl-epigallocatechin-(4β→8)-3-*O*-galloyl-gallo-catechin	OH	H	OH	H	OH	OH	OH	R	R	S	*M. rubra*	276
	OH	H	OH	FLA	OH	OH	OH	R	S			
gallocatechin-(4α→8)-gallocatechin-(4α→8)-catechin	OH	H	OH	H	OH	OH	OH	R	S	R	*H. vulgare*	288
	OH	H	OH	FLA	OH	OH	OH	R	S	R		
	OH	H	OH	FLA	OH	OH	H	R	S			
gallocatechin-(4α→8)-catechin-(4α→8)-catechin	OH	H	OH	H	OH	OH	OH	R	S	R	*H. vulgare*	288
	OH	H	OH	FLA	OH	OH	H	R	S	R		
	OH	H	OH	FLA	OH	OH	H	R	S			
catechin-(4α→8)-gallocatechin-(4α→8)-catechin	OH	H	OH	H	OH	OH	H	R	S	R	*H. vulgare*	288
	OH	H	OH	FLA	OH	OH	OH	R	S	R		
	OH	H	OH	FLA	OH	OH	H	R	S			

rates and the phosphorous and nitrogen nutrition. The proanthocyanidin content of leaves being inversely correlated with tree vigor (Tiarks et al., unpublished results). Two dimers and four trimers have been isolated from *Hordeum vulgare* (barley). The procyanidins in barley are considered to be responsible for the chill-haze in beer that has been stored (32, 70, 288). The trimers in barley are particularly interesting because they represent the first mixed procyanidin-prodelphinidin oligomers found in natural extracts (288). Some of the prodelphinidins in beer are also derived from the hops (75). The prodelphinidins obtained from *Myrica rubra* by Nonaka et al. (276) and from *Myrica nagi* by Krishnamoorthy and Seshadri (218) are interesting because of the predominance of gallate esters. Prodelphinidin gallates are also the principal proanthocyanidins in green tea (281).

7.6.3.3.2 Reactions of Oligomeric Proanthocyanidins

The discussions of the reactions of flavan-3-ols (Sect. 7.6.3.1.3) and of the flavan-3,4-diols (Sect. 7.6.3.2.3) are, for the most part, directly applicable to those of the oligomeric proanthocyanidins and condensed tannins, the critical difference, of course, being reactions at the interflavanoid bond. Although this difference is obvious, the interflavanoid bond, particularly its lability to either acid- or base-catalyzed solvolysis (107, 144, 148, 152, 225), has not been given adequate consideration in many instances. Reactions of condensed tannins are sometimes incorrectly postulated to parallel those of the flavan-3-ols. It seems far more appropriate now to use dimeric proanthocyanidins as model compounds for study of the reactions of condensed tannins, because these compounds are easily synthesized and their spectral properties are well known.

Acid-catalyzed solvolysis is clearly the most important reaction of condensed tannin chemistry. Its application to the synthesis of oligomeric proanthocyanidins from either flavan-3,4-diols (Sect. 7.6.3.2.3) or from polymeric procyanidins has been responsible for much of the advance in our understanding of the chemistry of oligomeric proanthocyanidins. These reactions have also been central to most of the understanding of the chemistry of the condensed tannins. After Betts et al. (20) first described the reactions of these compounds with mercaptoacetic acid, Sears and Casebier (333) showed that its reaction with catechin-(4α→2)-phloroglucinol cleanly gave the catechin-(4α)- and (4β)-thioglycolates. They extended this reaction to study of the structure of *Tsuga heterophylla* (western hemlock) bark tannins (334). Brown et al. (20, 37) showed that these compounds could be cleaved to their individual flavan units with preservation of the stereochemistry at C-2 and C-3, using a number of different sulfur nucleophiles. These reactions are now routinely used in the study of the structure of polymeric procyanidins (see Sect. 7.7).

Additional structural information can be obtained through use of partial thiolytic cleavage. The demonstration of interflavanoid bond location heterogeneity in polymeric procyanidins was first established through the isolation of epicatechin-(4β→8)-epicatechin-(4β)- and epicatechin-(4β→6)-epicatechin-(4β)-phenyl sulfides from partial cleavage of *Pinus taeda* tannins (143, 186). The isolation of their corresponding benzyl sulfides from trimeric procyanidins as well as condensed tannins from *Pinus palustris* and *Photinia glabrescens* (141) cast doubt on general applicability of Haslam's helically coiled, rod-like structures for polymeric procyanidins (135).

In addition to being odorless, phloroglucinol offers significant advantages in ease of separation of products and a higher degree of stereoselectivity over the use of thiols in these solvolysis reactions (64). Because of the much higher preference for formation of the 3,4-*trans* adduct from 2,3-*trans* procyanidins, phloroglucinol is a better nucleophile than a thiol in the study of these compounds. Additionally, the separation of the phloroglucinol adducts on Sephadex LH-20 with ethanol or ethanol-water mixtures is often improved. The phloroglucinol adducts also serve as much better models for higher oligomers. For example, the interpretation of ^{13}C-NMR spectra of trimeric procyanidins was greatly aided by the synthesis of epicatechin-(4β→8)-epicatechin-(4β→2)-phloroglucinol and epicatechin-(4β→6)-epicatechin-(4β→2)-phloroglucinol by reaction of *Photinia glabrescens* tannins with phloroglucinol under conditions paralleling the thiolytic cleavage (141, 303). As would be expected from the prior discussions of the reactions of flavan-3,4-diols (Sect. 7.6.3.2.3), major differences occur in the rates of solvolysis of proanthocyanidins with variations in structure. The 2,3-*cis*-3,4-*trans* dimers epicatechin-(4β→8)-epicatechin and epicatechin-(4β→8)-catechin are especially labile to thioloytic cleavage, and both compounds carry the lower flavan unit in a *quasi*-axial position. These two compounds were cleaved 1.9 times faster than catechin-(4α→8)-catechin in which the lower flavan unit is placed approximately equatorially (148). It was not predicted that the 2,3-*cis*-[4β→6]-linked procyanidins would be far more stable than their [4β→8]-linked isomers. However, epicatechin-(4β→8)-catechin was cleaved 2.6 times faster than an epicatechin-(4β→6)-catechin and epicatechin-(4β→6)-epicatechin (148). The relative stability of the [4β→6]-linked dimers over their [4β→8]-linked counterparts has important implications to the interpretation of partial solvolysis products from polymeric procyanidins and the regioselectivity of synthesis of procyanidins under thermodynamic equilibrium conditions.

A-ring hydroxylation patterns have major influences on the rates of cleavage of the interflavanoid bond. While the procyanidins can be easily cleaved at ambient temperature in the presence of HCl or at 100 °C in the presence of acetic acid catalyst, the proanthocyanidins that carry one or two resorcinolic A-rings are much more resistant to solvolysis. Epicatechin-(4β→2)-resorcinol is virtually stable to cleavage at pH 4 and 100 °C because only very small amounts of epicatechin-(4β)-phenyl sulfide and resorcinol are obtained after 20 hours of reaction (144). Epicatechin-(4β→8)-catechin is completely cleaved within 4 hours under the same conditions. As was mentioned earlier (Sect. 7.6.3.1.3), guibourtacacidol-(4α→8)-epiafzelechin is also resistant at reflux temperature with acetic acid catalyst; but good yields of guibourtacacidol-(4α)-benzyl sulfide and epiafzelechin can be obtained in reactions at 140 °C under pressure (290). These results suggest that those interflavanoid bonds that link both resorcinolic and phloroglucinolic flavanoid units can be cleaved at elevated temperature without serious attack of the pyran ring. However, significant yields of C-4 benzyl sulfides or C-4 sulfonic acid derivatives were not obtained from the reaction of polymeric profisetinidins from quebracho with phenylmethanethiol or sodium sulfite at 135 ° to 150 °C (Laks and Hemingway, unpublished results). At the other extreme,

Brown and Shaw (37) propose an ether linkage for the condensed tannins of *Calluna vulgaris* on the basis that epicatechin-(4β)-phenyl sulfides can be obtained in good yield by refluxing the tannin with benzene thiol in ethanol-water with no added acid. This result can possibly be explained on the basis of the extreme lability of the 2,3-*cis*-3,4-*trans* procyanidins with a [4β→8] linkage (148), but their hypothesis certainly deserves further study.

Use of thiols or phloroglucinol as the nucleophile in these reactions, while very effective in study of the polymer structure, has little practical significance. An exception would be the synthesis of alkylsulfide derivatives of flavan-3-ol and low-molecular weight procyanidins obtained by reaction of condensed tannins with fatty thiols. These derivatives are extremely toxic to gram-positive bacteria and a number of wood-destroying fungi (222). Interflavanoid bond cleavage with formation of flavan-4 adducts using sulfite ion as the nucleophile to obtain low molecular weight flavan-3-ol-4-sulfonates and procyanidin-4-sulfonates from the polymers is an effective way of making intermediates for adhesive formulation (107). These compounds hold considerable promise as intermediates for the utilization of condensed tannins as cold-setting phenolic resins (217). Similarly, use of resorcinol as the nucleophile has permitted the development of wood laminating adhesives in which more than 60% of the resorcinol requirement has been replaced by tannins from southern pine bark (216). These and other approaches to the use of condensed tannins are discussed in Chap. 10.3.

Until recently, little attention has been given to the reactions of polymeric proanthocyanidins under alkaline conditions. It was known that solution of polymeric procyanidins in base caused a marked decrease in the reactivity of these polymers with aldehyde and that both carbonyl and relatively acidic functions were generated analogous to the rearrangement of catechin to catechinic acid (152, 335).

The condensed tannins from *Pinus taeda* phloem were reacted at pH 12.0 and ambient temperature in the presence of excess phenylmethanethiol to obtain information about the relative lability of the interflavanoid bond and the pyran ether under alkaline conditions (223). One stereoisomer of 1,1-(2,4,6-trihydroxyphenyl) (α-phenylmethanethio)-3,3-(3,4-dihydroxyphenyl)(β-phenylmethanethio)-propan-2-ol (Fig. 7.6.7) (**62**) was predominantly (90%) produced. The formation of one predominant isomer implies that the initial cleavage occurred at the interflavanoid bond with the formation of epicatechin-(4β)-benzylsulfide (2*R*,3*R*,4*S*-configuration). The second cleavage occurred at the pyran ether to give a quinone methide that is attacked at C-3 from the side opposite the benzyl sulfide at C-1. Small amounts of 1,1-(3,4-dihydroxyphenyl)(α-phenylmethanethio)-3-(2,4,6-trihydroxyphenyl)-propan-2-ol (**63**), were obtained and as a mixture of two stereoisomers in an approximate ratio of 3:2 from reaction of catechin cleaved from the terminal unit of the tannin. Loss of phenylmethanethiol from **62** and tautomeric rearrangement of the quinone methide thus formed resulted in the formation of the ketone, 1,1-(3,4-dihydroxyphenyl) (phenylmethanethio)-3-(2,4,6-trihydroxyphenyl)-propanone (**64**). Another ketone, 1-(2,4,6-trihydroxyphenyl)-3-(3,4-dihydroxyphenyl)-propanone (**65**), was formed by loss of phenylmethanethiol from **63** and an analogous tautomeric rearrangement of the quinone methide of the catechol ring. As has been discussed previously in Sect. 7.6.3.1.2, these ketones

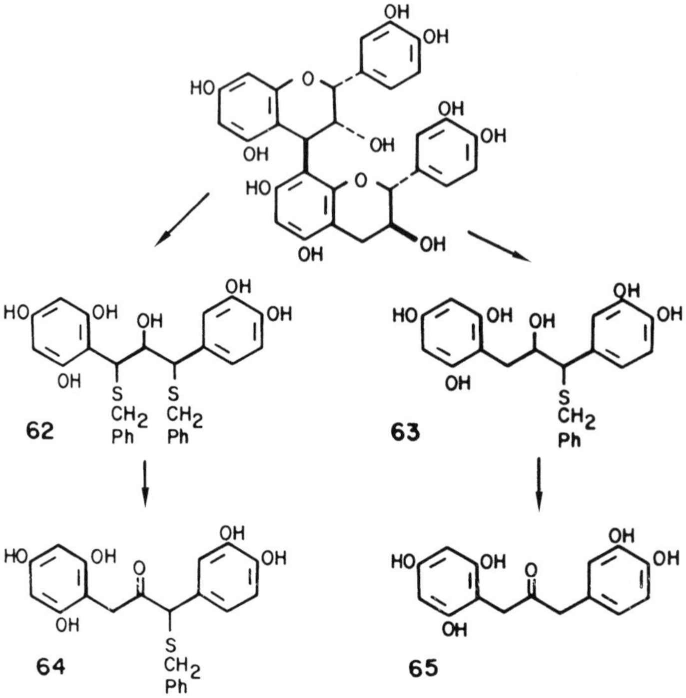

Fig. 7.6.7. Diarylpropanoids from procyanidins and phenylmethanethiol at alkaline pH

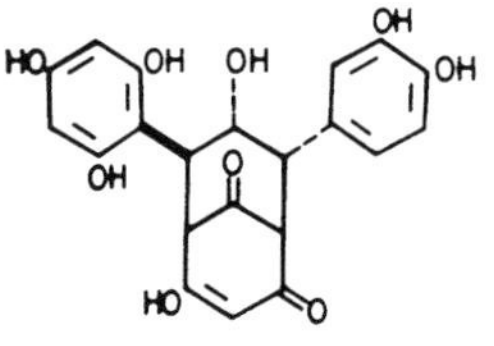

66

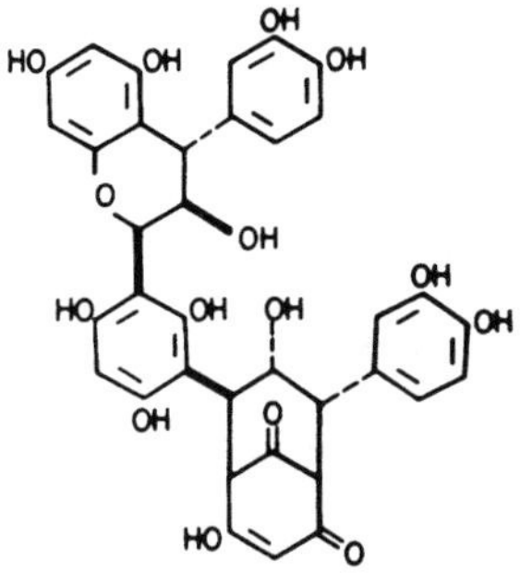

67

are analogous to the flav-3-en-3-ol proposed by Haslam (299) as an intermediate in the biosynthesis of flavan-3-ols and proanthocyanidins. An analogous rearrangement of a quinone methide obtained from a flavan-3,4-diol has been proposed as a way to explain the formation of 2*R*,3*R*-(2,3-*cis*)-procyanidins from 2*R*,3*S*-(2,3-*trans*)-flavan-3,4-diols (145). Base-catalyzed cleavage in the presence of good nucleophiles such as thiols also affords an effective way of producing low molecular weight diarylpropanoids from condensed tannins of the procyanidin or prodelphinidin class.

To learn more about the intermolecular rearrangement reactions of condensed tannins in alkaline solutions, the condensed tannins from *Pinus taeda* phloem were also reacted with phloroglucinol as the nucleophile under reactions paralleling those described above with phenylmethanethiol (225). As was evident from the above, the interflavanoid bond was particularly labile to cleavage at pH 12.0 and ambient temperature, and small amounts of epicatechin-(4β→2)-phloroglucinol were obtained. This compound rearranged to an enolic form of 8-(3,4-dihydroxyphenyl)-7-hydroxy-6-(2,4,6-trihydroxyphenyl)-bicyclo-[3,3,1]-nonane-2,4,9-trione (**66**). Incomplete interflavanoid bond cleavage resulted in the formation of a dimeric derivative, 3′-{8-(3,4-dihydroxyphenyl)-7-hydroxy-2,4,9-trioxobicyclo-[3,3,1]-nonane-6-yl}-4-(3,4-dihydroxyphenyl)-2′,3,4′,5,6′,7-hexahydroxyflavan (**67**), also in an enolic form. Cleavage of catechin from the terminal unit of the polymer and rearrangement gave catechinic acid, 6-(3,4-dihydroxyphenyl)-7-hydroxy-bicyclo-[3,3,1]-nonanetri-2,4,9-one (Fig. 7.6.3, **28**) as had been observed by Sears et al. (335). A remarkable feature of these rearrangements is the fact that both **66** and **67** are produced stereospecifically. The conformation of **66** was established by 2D COSY and HETORR NMR experiments. Roux's group has now demonstrated base-catalyzed rearrangements similar to the formation of **67** in the formation of phlobatannins from *Acacia* sp. (344, 346).

^{13}C-NMR spectra of products of condensed tannins subjected to similar alkaline conditions but without the addition of another nucleophile could be interpreted on the basis of these rearrangement reactions (224). Nearly every prominent signal could be assigned by analogy to the chemical shifts found for **66** and **67**. The loss of aldehyde reactivity and formation of carbonyl and acidic functions that had previously been observed when tannins are subjected to alkaline conditions are clearly explained by these reactions, as had been proposed by Sears et al. (335). These rearrangements are particularly important to control when attempting to use condensed tannins from conifer tree barks in wood adhesive applications in which crosslinking relies on the phloroglucinol A-ring functionality. New developments in the utilization of the abundant natural condensed tannins require additional knowledge of their reactions. The oligomeric proanthocyanidins can serve as useful model compounds for these reactions.

The forgoing discussion demonstrates that phenomenal advances have been made in our understanding of the chemistry of proanthocyanidins in terms of definitions of the structures of new compounds, broadening the range of plants in which these compounds are present, and in definition of the reactions that these compounds can be expected to undergo in the development of new uses for the condensed tannin polymers. While these compounds are important in defining the structure of the high-molecular weight polymers, their use as model compounds for understanding the reactions of these polymers is clearly most important from a practical viewpoint. For further information on the condensed tannins consult Sect. 7.7 and Chap. 10.3.

Acknowledgements. I am indebted to Linda Korb who assisted in gathering reference materials, Cleo Lane who typed a number of drafts, and Drs. L. Y. Foo, J. J. Karchesy, G. W. McGraw, L. J. Porter, and D. G. Roux who have made helpful suggestions.

References

1 Agrawal S, Misra K 1977 Leucoanthocyanidins from *Ficus racemosa* bark. Chem Scripta 12:37–39

2 Ahmad I, Ishratulla Kh, Ilyas M, Rahman W, Seligmann O, Wagner H 1981 Tetrahydroamentoflavone from nuts of *Semecarpus prainii*. Phytochemistry 20:1169–1170

3 Ahmad S, Razaq S 1971 A new approach to the synthesis of symmetrical biflavones. Tetrahedron Lett 48:4633–4636

4 Ahmad S, Razaq S 1976 New synthesis of biflavones of cupressuflavone series. Tetrahedron 32:503–506

5 Ahn B-Z 1974 Ein Catechin Trimer aus der Eichenrinde. Arch Pharmaz 307:186–197

6 Ahn B-Z, Gstirner F 1971 Über Catechin Dimer der Eichenrinde. Arch Pharmaz 304:666–673

7 Ansari F R, Ansari W H, Rahman W 1978 A biflavone from *Garuga pinnata* Roxb. (Burseraceae). Indian J Chem 16B:846

8 Ansari W H, Rahman W, Barraclough D, Maynard R, Scheinmann F 1975 Biflavonoids and a flavanone-chromone from the leaves of *Garcinia dulcis* (Roxb.) Kurz. J Chem Soc Perkin Trans I 1558–1563

9 Aqil M, Rahman W, Okigawa M, Kawano N 1976 Isolation of C-methylbiflavanone from *Cephalotaxus* and use of lanthanide shift reagent in deciding 6- or 8-C-methylflavone. Chem Ind (London) 567–568

10 Attwood M R, Brown B R, Lisseter S G, Torrero C L, Weaver P M 1984 Spectral evidence for the formation of quinone-methide intermediates from 5- and 7-hydroxyflavonoids. J Chem Soc Chem Commun 177–179

11 Baig M, Clark-Lewis J W, Thompson M J 1969 Flavan derivatives. XXVII. Synthesis of a new racemate of leucocyanidin tetramethyl ether (2,3-*cis*-3,4-*trans* isomer): NMR spectra of the four racemates of leucocyanidin tetramethyl ether (5,7,3′,4′-tetramethoxyflavan-3,4-diol). Aust J Chem 22:2645–2650

12 Baker W, Finch A C M, Ollis W D, Robinson K W 1963 The structures of naturally occurring biflavonoyls. J Chem Soc 1477–1490

13 Baker W, Ollis W D 1961 Biflavonyls. In: Ollis W D (ed) Recent developments in the chemistry of natural phenolic compounds. Pergamon London, 152–184

14 Bandaranayake W M, Selliah S S, Sultanbawa M V S, Ollis W D 1975 Biflavonoids and xanthones of *Garcinia terpenophylla* and *G. echinocarpa*. Phytochemistry 14:1878–1880

15 Barrett M W, Klyne S, Scopes P M, Fletcher A C, Porter L J, Haslam E 1979 Plant proanthocyanidins. Part 6. Chiroptical studies. Part 95. Circular dichromism of procyanidins. J Chem Soc Perkin Trans I 2375–2377

16 Beckman S, Geiger H, deGroot-Pfleiderer W 1971 Biflavone und 2,3-dihydrobiflavone aus *Metasequoia glyptostroboides*. Phytochemistry 10:2465–2474

17 Beretz A, Joly M, Stoclet J C, Anton R 1979 Inhibition of 3′,5′-AMP phosphodiesterase by biflavonoids and xanthones. Planta Med 36:193–195

18 Bergmann W R, Barkley M D, Hemingway R W, Mattice W L 1987 Heterogeneous fluorescence decay of 4→6 and 4→8 linked dimers of (+)-catechin and (−)-epicatechin as a result of rotational isomerism. J Am Chem Soc 109:6614–6619

19 Bergmann W R, Viswanadhan V N, Mattice W L 1987 Conformations of polymeric proanthocyanidins composed of (+)-catechin or (−)-epicatechin joined by 4→6 interflavan bonds. J Chem Soc Perkin Trans II:45–51

20 Betts M J, Brown B R, Shaw M R 1969 Reaction of flavonoids with mercaptoacetic acid. J Chem Soc (C) 1178–1184

21 Bezuidenhoudt B C B, Brandt E V, Roux D G 1984 Synthesis of isoflavonoid oligomers using pterocarpan as inceptive electrophile. J Chem Soc Perkin Trans I 2767–2777

22 Birch A J, Clark-Lewis J W, Robertson A V 1957 The relative and absolute configurations of catechins and epicatechins. J Chem Soc 3586–3594

23 Bohm B A 1975 Flavones and dihydroflavonols. In: Harborne J B, Mabry T J, Mabry H (eds) The flavonoids. Chapman and Hall London, 561–632

24 Bokadia M M, Brown B R, Cummings W 1960 Polymerisation of flavans. Part III. The action of lead tetra-acetate on flavans. J Chem Soc 3308–3313

25 Bonati A, Mustich G 1979 Pharmacologically active polyphenolic substances. U S Pat 4, 166, 861

26 Botha J J, Ferreira D, Roux D G 1978 Condensed tannins: Circular dichroism method of assessing the absolute configuration at C-4 of 4-arylflavan-3-ols and stereochemistry of their formation from flavan-3,4-diols. J Chem Soc Chem Commun 698–700

27 Botha J J, Ferreira D, Roux D G 1978 Condensed tannins: Direct synthesis, structure and absolute configuration of four biflavanoids from black wattle (mimosa) extract. J Chem Soc Chem Commun 700–702

28 Botha J J, Viviers P M, Ferreira D, Roux D G 1981 Synthesis of condensed tannins. Part 4. A direct biomimetic approach to (4,6) and (4,8) biflavanoids. J Chem Soc Perkin Trans I 1235–1244

29 Botha J J, Viviers P M, Ferreira D, Roux D G 1982 Condensed tannins: Competing nucleophilic centers in biomimetic condensation reactions. Phytochemistry 21:1289–1294

30 Botha J J, Viviers P M, Young D A, duPreez I C, Ferreira D, Roux D G, Hull W E 1982 Synthesis of condensed tannins. Part 5. The first angular (4,6:4,8)-triflavonoids and their natural counterparts. J Chem Soc Perkin Trans I 527–533

31 Botha J J, Young D A, Ferreira D, Roux D G 1981 Synthesis of condensed tannins. Part I. Stereoselective and stereospecific synthesis of optically pure 4-arylflavan-3-ols, and assessment of their absolute stereochemistry at C-4 by means of circular dichroism. J Chem Soc Perkin Trans I 1213–1219

32 Brandon M J, Foo L Y, Porter L J, Meredith P 1982 Proanthocyanidins of barley and sorghum; composition as a function of maturity of barley ears. Phytochemistry 21:2953–2957

33 Brandt E V, Bezuidenhoudt B C B, Roux D G 1982 Direct synthesis of the first natural bi-isoflavonoid. J Chem Soc Chem Commun 1409–1410

34 Brandt E V, Young D A, Kolodziej H, Ferreira D, Roux D G 1986 Cycloconformations of two tetraflavanoid profisetinidin condensed tannins. J Chem Soc Chem Commun 913–914

35 Brown A G, Eyton W B, Holmes A, Ollis W D 1969 The identification of the thearubigins as polymeric proanthocyanidins. Phytochemistry 8:2333–2340

36 Brown B R, MacBride J A H 1964 New methods for assignment of relative configuration to 2,3-*trans* flavan-3,4-diols. J Chem Soc 3822–3831

37 Brown B R, Shaw M R 1974 Reactions of flavanoids and condensed tannins with sulphur nucleophiles. J Chem Soc Perkin Trans I 2036–2049

38 Castro O C, Valverde V R 1985 Chamaejasmin, a biflavanone from wood of *Diphysa robinioides*. Phytochemistry 24:367–368

39 Chandramouli N, Natarjan S, Murti V V S, Seshadri T R 1971 Structure of heveaflavone. Indian J Chem 9:895–896

40 Chari V M, Ilyas M, Wagner H, Neszmelyi A, Chen F-C, Chen L-K, Lin Y C, Lin Y-M 1977 ^{13}C-NMR spectroscopy of biflavonoids. Phytochemistry 16:1273–1278

41 Chatterjii A, Kotoky J, Das K K, Bannerji J, Chakraborty T 1984 Abiesin, a biflavonoid of *Abies webbiana*. Phytochemistry 23:704–705

42 Chen F-C, Ho T-I, Lin Y-M 1975 Synthesis of hexa-*O*-methyl-8-8″-binaringenin. Heterocycles 3:833–836

43 Chen F-C, Lin Y-M 1975 Succedaneaflavanone – a new 6,6″-binargingen from *Rhus succedanea*. Phytochemistry 14:1644–1647

44 Chen F-C, Lin Y-M 1976 Rhus flavanone, a new biflavanone from the seeds of wax-tree. J Chem Soc Perkin Trans I 98–101

45 Chen F-C, Lin Y-M, Hung J-C 1975 A new biflavanone glucoside from *Garcinia multiflora*. Phytochemistry 14:818–820

46 Chen F-C, Lin Y-M, Lin Y-C 1978 Neorhusflavanone, a new biflavanone from wax-tree. Heterocycles 9:663–668

47 Chen F-C, Lin Y-M, Wu J-C 1974 Rhusflavone, a new flavano-flavone from *Rhus succedanea*. Phytochemistry 13:1571–1574

48 Chen X F, Liu G-Q, Takematsu H, Hirata Y 1984 Structure of isochamaejasmin from *Stellera chamaejasme* L. J Chem Soc Jpn Chem Lett 9:1587–1590

49 Chexal K K, Handa B K, Rahman W 1970 Thin layer chromatography of biflavonyls on silica gel; structure chromatographic behaviour correlations. J Chromatography 48:484–492

50 Chexal K, Handa B K, Rahman W, Kawano N 1970 Some optically active biflavones from *Podocarpus gracilior*. Chem Ind (London) 28

51 Clark-Lewis J W 1968 Flavan derivatives. XXI. Nuclear magnetic resonance spectra, configuration, and conformation of flavan derivatives. Aust J Chem 21:2059–2075
52 Clark-Lewis J W, Dainis I 1964 Flavan derivatives. XI. Teracacidin, melacacidin, and 7,8,4'-trihydroxy flavanol from *Acacia sparsifolia* and extractives from *Acacia ovites*. Aust J Chem 17:1170–1173
53 Clark-Lewis J W, Dainis I 1967 Flavan derivatives. XIX. Teracacidin and isoteracacidin from *Acacia obtusifolia* and *Acacia maidenii* heartwoods: Phenolic hydroxylation patterns of heartwood flavonoids characteristic of sections and subsections of the genus *Acacia*. Aust J Chem 20:2191–2198
54 Clark-Lewis J W, Katekar G F 1960 The absolute configurations of (−)-melacacidin, (−)-teracacidin, and (+)-mollisacacidin. Proc Chem Soc (London) 4502–4508
55 Clark-Lewis J W, Katekar G F, Mortimer P I 1961 Flavan derivatives. IV. Teracacidin, a new leucoanthocyanidin from *Acacia intertexta*. J Chem Soc 499–503
56 Clark-Lewis J W, Mitsuno M 1958 The identity of gleditsin and mollisacacidin. J Chem Soc 1724
57 Clark-Lewis J W, Mortimer P I 1960 Flavan derivatives. III. Melacacidin and isomelacacidin from *Acacia* species. J Chem Soc 4106–4111
58 Clark-Lewis J W, Porter L J 1972 Phytochemical survey of the heartwood flavonoids of *Acacia* species from arid zones of Australia. Aust J Chem 25:1943–1955
59 Clark-Lewis J W, Roux D G 1959 Natural occurrence of enantiomorphous leucoanthocyanidins: (+)-mollisacacidin (gleditsin) and quebracho (−)-leucofisetinidin. J Chem Soc 1402–1406
60 Cotterill P J, Scheinmann F, Stenhouse I A 1978 Extractives from Guttiferae. Part 34. Kolaflavanone, a new biflavanone from the nuts of *Garcinia kola* Heckel. Applications of 13-C nuclear magnetic resonance in elucidation of the structures of flavonoids. J Chem Soc Perkin Trans I 532–539
61 Courbat P, Valenza A 1975 Medicaments containing epicatechin-2-sulfonic acids and salts thereof. U S Pat 3, 888, 990
62 Coxon D T, Holmes A, Ollis W D, Vora V C, Grant M S, Tee J L 1972 Flavanol digallates in green tea leaf. Tetrahedron 28:2819–2826
63 Crichton E G, Waterman P G 1979 Manniflavanone, a new 3:8-linked flavanone dimer from the stem bark of *Garcinia manii*. Phytochemistry 18:1553–1556
64 Czochanska Z, Foo L Y, Newman R H, Porter L J 1980 Polymeric proanthocyanidin, stereochemistry, structural units and molecular weight. J Chem Soc Perkin Trans I 2278–2286
65 Czochanska Z, Foo L Y, Porter L J 1979 Compositional changes in lower molecular weight flavans during grape maturation. Phytochemistry 18:1819–1822
66 Dean F M, Mongkilsuk S, Podiumoung V 1965 (+)-Fisetinidol from *Afzelia xylocarpa*. J Chem Soc 828–829
67 Delcour J A, Ferreira D, Roux D G 1983 Synthesis of condensed tannins. Part 9. The condensation sequence of leucocyanidin with catechin and resultant procyanidins. J Chem Soc Perkin Trans I 1711–1717
68 Delcour J A, Serneels E J, Ferreira D, Roux D G 1985 Synthesis of condensed tannins. Part 13. The first 2,3-*trans*-3,4-*cis* procyanidins: sequence of units in a trimer of mixed stereochemistry. J Chem Soc Perkins Trans I 669–676
69 Delcour J A, Tuytens G M 1983 A mass spectrometric criterion for determining the B- and E-ring hydroxylation pattern in dimeric biflavanoids. J Chem Soc Chem Commun 1195
70 Delcour J A, Tuytens G M 1984 Structure elucidation of three dimeric proanthocyanidins isolated from a commercial Belgian pilsner beer. J Inst Brew 90:153–161
71 Delgado G, Alvarez L, Romo-de-Vivar A 1984 Terpenoids and a flavan-3-ol from *Viguiera quinqueradiata*. Phytochemistry 23:675–678
72 Delle-Monache F, Ferrari F, Marini-Bettolo G B 1971 Occurrence of (+)-epicatechin in nature. Gazz Chim Ital 101:387–395
73 Delle-Monache F, Ferrari F, Poce-Tocci A, Marini-Bettolo G B 1972 Catechins with (+)-epicatechin configuration in nature. Phytochemistry 11:2333–2335
74 Delle-Monache F, Pomponi M, Marini-Bettolo G B, D'Albuquerque I L, DeLima O G 1976 A methylated catechin and proanthocyanidin from the *Celastraceae*. Phytochemistry 15:573–574
75 Derdelinickz G, Jerumais J 1984 Separation of malt and hop proanthocyanidins on Fractogel TSK HW 40 (S). J Chromatog 285:231–234

76 Doskotch R W, Mikhail A A, Chatterjii S J 1973 Structure of the water-soluble feeding stimulant for *Scolytus multistriates*. A revision. Phytochemistry 12:1153–1155
77 Dossaji S F, Bell E A, Wallace J W 1973 Biflavones of dioon. Phytochemistry 12:371–373
78 Dossaji S F, Mabry T J, Wallace J W 1975 Chromatographic and UV-visible spectral identification of biflavonoids. Rev Latinoamer Quim 6:37–45
79 Drewes S E, Ilsley A H 1969 Dioxane-linked biflavanoid from the heartwood of *Acacia mearnsii*. J Chem Soc (C) 897–900
80 Drewes S E, Ilsley A H 1969 Isomeric leuco-fisetinidins from *Acacia mearnsii*. Phytochemistry 8:1039–1042
81 Drewes S E, Roux D G 1964 Condensed tannins. 18. Stereochemistry of flavan-3,4-diol tannin precursors: (+)-mollisacacidin, (−)-leucofisetinidin and (+)-leucorobinetinidin. Biochem J 90:343–350
82 Drewes S E, Roux D G 1965 Absolute configuration of mopanol, a new leucoanthocyanidin from *Colophospermum mopane*. J Chem Soc Chem Commun 500–501
83 Drewes S E, Roux D G 1965 Condensed tannins: Optically active diastereomers of (+)-mollisacacidin by epimerization. Biochem J 94:482–487
84 Drewes S E, Roux D G 1965 Natural and synthetic diastereoisomeric (−)-3′,4′,7-trihydroxyflavan-3,4-diols. Biochem J 96:681–687
85 Drewes S E, Roux D G 1966 A new flavan-3,4-diol from *Acacia auricultiformis* by paper ionophoresis. Biochem J 98:493–500
86 Drewes S E, Roux D G 1966 Stereochemistry and biogenesis of mopanols and peltogynols. J Chem Soc 1644–1653
87 Drewes S E, Roux D G, Eggers S H, Feeney J 1967 Three diastereomeric 4,6-linked bileucofisetinidins from the heartwood of *Acacia mearnsii*. J Chem Soc (C) 1217–1227
88 Duddeck H, Snatzke G, Yemul S S 1978 ^{13}C-NMR and CD of some 3,8″-biflavonoids from *Garcinia* species and of related flavanones. Phytochemistry 17:1369–1373
89 duPreez I C, Roux D G 1970 Novel flavan-3,4-diols from *Acacia culturiformis*. J Chem Soc (C) 1800–1804
90 duPreez I C, Rowan A C, Roux D G 1970 A biflavanoid proanthocyanidin carboxylic acid and related biflavonoids from *Acacia luderitzii* Engl. var *retinens* J Ross and Brenan. J Chem Soc Chem Commun 492–493
91 duPreez I C, Rowan A C, Roux D G, Feeny J 1971 Hindered rotation about the sp2-sp3 hybridized C–C bond between flavonoid units in condensed tannins. J Chem Soc Chem Commun 315–316
92 Einstein F W B, Kielmann E, Wolowidnyk E K 1985 Structure and nuclear magnetic resonance spectra of 6-bromo-3,3′,4′,5,7-penta-*O*-methyl-catechin. Can J Chem 63:2176–2180
93 Ellis C J, Foo L Y, Porter L J 1983 Enantiomerism: A characteristic of the proanthocyanidin chemistry of the Monocotyledonae. Phytochemistry 22:483–487
94 ElSohly M A, Craig J C, Waller C W, Turner C E 1978 Biflavonoids from fruits of poison ivy *Toxicodendron radicans*. Phytochemistry 17:2140–2141
95 Engel D W, Hattingh M, Hundt H K L, Roux D G 1978 X-ray structure, conformation and absolute configuration of 8-bromo-tetra-*O*-methyl-(+)-catechin. J Chem Soc Chem Commun 695–696
96 Engelshowe R 1983 Dimere Proanthocyanidine als Gerbstoffvorstufen in *Juniperus communis*. Planta Med 49:170–175
97 Ferreira D, Hundt H K L, Roux D G 1971 The stereochemistry of a tetraflavanoid condensed tannin from *Rhus lancea* L.f. J Chem Soc Chem Commun 1257–1259
98 Fletcher A C, Porter L J, Haslam E, Gupta R K 1977 Plant proanthocyanidins. Part 3. Conformational and configurational studies of natural procyanidins. J Chem Soc Perkin Trans I 1628–1636
99 Foo L Y 1982 Polymeric proanthocyanidins of *Photinia glaubrescens*, modification of molecular weight and nature of products from hydrogenolysis. Phytochemistry 21:1741–1746
100 Foo L Y 1984 Condensed tannins: Co-occurrence of procyanidins, prodelphinidins and profisetinidins in the heartwood of *Acacia baileyana*. Phytochemistry 23:2915–2918
101 Foo L Y 1986 The first natural 2,3-*cis*-dihydroflavonol and rationalization of the biogenesis of 2,3-*cis*-proanthocyanidins. J Chem Soc Chem Commun 675–677

102 Foo L Y 1985 Facile self-condensation of melacacidin: A demonstration of the reactivity of the pyrogallol A-ring. J Chem Soc Chem Commun 1273–1274
103 Foo L Y 1986 A novel pyrogallol A-ring proanthocyanidin dimer from *Acacia melanoxylon*. J Chem Soc Chem Commun 236–237
104 Foo L Y, Hemingway R W 1984 Condensed tannins: Synthesis of the first "branched" procyanidin trimer. J Chem Soc Chem Commun 85–86
105 Foo L Y, Hemingway R W 1985 Condensed tannins: Reactions of model compounds with furfuryl alcohol and furfuraldehyde. J Wood Chem Technol 5:135–158
106 Foo L Y, Hrstich L, Vilain C 1985 Phylloflavan, a characteristic constituent of *Phyllocladus* species. Phytochemistry 24:1495–1498
107 Foo L Y, McGraw G W, Hemingway R W 1983 Condensed tannins: Preferential substitution at the interflavanoid bond by sulfite ion. J Chem Soc Chem Commun 672–673
108 Foo L Y, Porter L J 1978 Prodelphinidin polymers: Definition of structural units. J Chem Soc Perkin Trans I 1186–1190
109 Foo L Y, Porter L J 1983 Enantiomerism in natural procyanidins polymers: Use of epicatechin as a chiral resolution reagent. J Chem Soc Chem Commun 242–243
110 Foo L Y, Porter L J 1983 Synthesis and conformation of procyanidin diastereoisomers. J Chem Soc Perkin Trans I 1535–1543
111 Foo L Y, Wong H 1986 Diastereoisomeric leucoanthocyanidins from the heartwood of *Acacia melanoxylon*. Phytochemistry 25:1961–1965
112 Fourie T G, duPreez I C, Roux D G 1972 3′,4′,7,8-Tetrahydroxyflavonoids from the heartwood of *Acacia nigrescens* and their conversion products. Phytochemistry 11:1763–1770
113 Fourie T L, Ferreira D, Roux D G 1974 8-*O*-Methyl and the first 3-*O*-methyl flavan-3,4-diol from *Acacia saxatilis*. Phytochemistry 13:2573–2581
114 Freudenberg K, Alonso deLama J M 1958 Zur Kenntnis der Catechin Gerbstoffe. Justus Liebigs Ann Chem 612:78–93
115 Freudenberg K, Fikentscher H, Harder M, Schmidt O 1925 Die Umwandlung des Cyanidins in Catechin. Justus Liebigs Ann Chem 444:135–145
116 Freudenberg K, Weinges K 1960 Systematic und Nomenclatur der Flavonoiden. Tetrahedron 8:336–349
117 Freudenberg K, Weinges K 1963 Zur Kenntnis der Flavonoid-Gerbstoffe. Justus Liebigs Ann Chem 668:92–96
118 Freudenberg K, Weinges L, Naya Y 1963 Die dimere Vorstufe der Catechin-Gerbstoffe. Naturwissenschaften 50:354
119 Fronczek F R, Gannuch G, Mattice W L, Tobiason F L, Broeker J L, Hemingway R W 1984 Dipole moment, solution conformation and solid state structure of (−)-epicatechin, a monomer of procyanidin polymers. J Chem Soc Perkin Trans II 1611–1616
120 Fronczek F R, Gannuch G, Tobiason F L, Shanafelt H A, Hemingway R W, Mattice W L 1985 Preference for occupancy of axial positions by substituents bonded to the heterocyclic ring in penta-*O*-acetyl-(+)-catechin. J Chem Soc Perkin Trans II 1383–1386
121 Fukui Y, Kawano N 1959 The structure of hinokiflavone, a new type bisflavonoid. J Am Chem Soc 81:6331
122 Ganguly A K, Seshadri T R 1959 A study of leucoanthocyanidins of plants. II. (+)-Leucocyanidin from the gum of *Butea frondosa*. Tetrahedron 6:21–23
123 Garg H S, Mitra C R 1971 Putraflavone, a new biflavonoid from *Putranjiva roxburghii*. Phytochemistry 10:2787–2791
124 Geiger H, deGroot-Pfleiderer W 1971 Über 2,3-dihydrodiflavone in *Cycas revoluta*. Phytochemistry 10:1936–1938
125 Geiger H, Quinn C 1975 Biflavonoids. In: Harborne J B, Mabry T J, Mabry H (eds) The flavonoids. Academic Press New York, 692–742
126 Geiger H, Quinn C 1982 Biflavonoids. In: Harborne J B, Mabry T J (eds) The flavonoids: Advances in research. Chapman and Hall London, 505–534
127 Geissman T A, Dittmar H F K 1965 A proanthocyanidin from avocado seed. Phytochemistry 4:359–368
128 Griffiths L A 1982 Mammalian metabolism of flavonoids. In: Harborne J B, Mabry T J (eds) The flavonoids. Advances in research. Chapman and Hall London, 681–718

129 Gupta R K, Haslam E 1981 Plant proanthocyanidins. Part 7. Prodelphinidins from *Pinus sylvestris*. J Chem Soc Trans I 1148–1150
130 Handa B K, Chexal K K, Mah T, Rahman W 1971 Some observations on partial demethylation of biflavonoyls. J Indian Chem Soc 48:177–181
131 Hardegger E, Gampler H, Zust A 1957 Die absolute Konfiguration des Catechins. Helv Chim Acta 40:1819–1822
132 Haslam E 1969 (+)-Catechin-3-gallate and a polymeric proanthocyanidin from *Bergenia* species. J Chem Soc (C) 1824–1828
133 Haslam E 1974 Biogenetically patterned synthesis of procyanidins. J Chem Soc Chem Commun 594–595
134 Haslam E 1975 Natural proanthocyanidins. In: Harborne J B, Mabry T J, Mabry H (eds) The flavonoids. Academic Press New York, 505–559
135 Haslam E 1977 Symmetry and promiscuity in procyanidin biochemistry. Phytochemistry 16:1625–1640
136 Haslam E 1982 Proanthocyanidins. In: Harborne J B, Mabry T J (eds) The flavonoids: Advances in research. Chapman and Hall London, 417–447
137 Hathway D E 1962 The condensed tannins. In: Hillis W E (ed) Wood extractives and their significance to the pulp and paper industries. Academic Press New York, 191–228
138 Hathway D E, Seakins J W T 1955 Autoxidation of catechin. Nature 176:218
139 Hemingway R W, Foo L Y 1983 Condensed tannins: Quinone methide intermediates in procyanidin synthesis. J Chem Soc Chem Commun 1035–1036
140 Hemingway R W, Foo L Y, Porter L J 1981 Polymeric proanthocyanidins: Linkage isomerism in epicatechin-4(-epicatechin-4)-catechin procyanidins. J Chem Soc Commun 320–321
141 Hemingway R W, Foo L Y, Porter L J 1982 Linkage isomerism in trimeric and polymeric 2,3-*cis* procyanidins. J Chem Soc Perkin Trans I 1209–1216
142 Hemingway R W, Karchesy J J, McGraw G W 1981 Condensation of hydroxybenzyl alcohols with catechin: A model for methylolphenols in conifer bark polyflavonoid adhesives. In: Phenolic resins, chemistry and application. Weyerhaeuser Sci Symp 2. Weyerhaeuser Tacoma WA, 33–69
143 Hemingway R W, Karchesy J J, McGraw G W, Wielesek R A 1983 Heterogeneity of interflavanoid bond location in loblolly pine bark procyanidins. Phytochemistry 22:275–281
144 Hemingway R W, Kreibich R E 1984 Condensed tannin-resorcinol adducts and their use in woodlaminating adhesives: An exploratory study. J Appl Polym Sci 40:79–90
145 Hemingway R W, Laks P E 1985 Condensed tannins: A proposed route to biogenesis of 2*R*,3*R*-(2,3-*cis*)-proanthocyanidins. J Chem Soc Chem Commun 746–747
146 Hemingway R W, McGraw G W 1976 Progress in the chemistry of shortleaf and loblolly pine bark flavonoids. J Applied Polym Sci Appl Polym Symp 28:1349–1364
147 Hemingway R W, McGraw G W 1978 Formaldehyde condensation products of model phenols for conifer bark tannins. J Liq Chrom 1:163–179
148 Hemingway R W, McGraw G W 1983 Kinetics of acid-catalyzed cleavage of procyanidins. J Wood Chem Technol 3:421–435
149 Hemingway R W, McGraw G W, Barras S J 1977 Polyphenols in *Ceratocystis minor* infected *Pinus taeda*: Fungal metabolites, phloem and xylem phenols. J Agr Food Chem 25:717–722
150 Herbin G A, Jackson B, Locksley H D, Scheinmann F, Wolstenholme W A 1970 The biflavonoids of *Garcinia volkensii* (Guttiferae). Phytochemistry 9:221–226
151 Hergert H L 1960 Chemical composition of tannins and other polyphenols from conifer wood and bark. For Prod J 10 (11):610–615
152 Herrick F W 1980 Chemistry and utilization of western hemlock bark extractives. J Agr Food Chem 28:228–237
153 Herrick F W, Bock L H 1958 Thermosetting exterior plywood type adhesives from bark extracts. For Prod J 8 (10):269–274
154 Hillis W E 1985 Biosynthesis of tannins. In: Biosynthesis and biodegradation of wood components. Academic Press New York, 325–347
155 Hillis W E, Carle A 1960 The chemistry of eucalypt kinos. III. (+)-Afzelechin, pyrogallol and (+)-catechin from *Eucalyptus calophylla* kino. Aust J Chem 13:390–395
156 Hillis W E, Inoue T 1967 The polyphenols of *Nothofagus* species. II. The heartwood of *Nothofagus fusca*. Phytochemistry 6:59–67

157 Hillis W E, Urbach G 1959 The reaction of (+)-catechin with formaldehyde. J Appl Chem 9:474–481
158 Hillis W E, Urbach G 1959 Reaction of polyphenols with formaldehyde. J Appl Chem 9:665–673
159 Hodges R 1965 5-5′-Dihydroxy-7,4′,7″,4‴-tetramethyoxy-8-3‴-biflavone from *Dacrydium cupressinum* Lamb. Aust J Chem 18:1491–1492
160 Homberger K, Hesse M 1984 Kopsirachin, ein ungewöhnliches Alkaloid aus der Apocynaceae *Kopsia dasyrachis* Ridl Chim Acta 67:237–247
161 Horhammer L, Wagner H, Reinhardt H 1965 Isolierung des Bis-(5,7,4′-trihydroxy-flavones), Amentoflavon aus der Rinde von *Viburnum prunifolium* L. Naturwissenschaften 52:161–162
162 Horhammer L, Wagner H, Reinhardt H 1967 Über neue Inhaltsstoffe aus den Rinden von *Viburnum prunifolium* L. Z Naturforsch 22B:768–776
163 Hostettman K, Hostettman M 1982 Isolation techniques for flavonoids. In: Harborne J B, Mabry T J (eds) The flavonoids: Advances in research. Chapman and Hall London, 1–18
164 Hrazdina G 1982 Anthocyanidins. In: Harborne J B, Mabry T J (eds) The flavonoids: Advances in research. Chapman and Hall London, 135–188
165 Hrutfiord B F, Luthi R, Hanover K F 1985 Color formation in western hemlock. J Wood Chem Technol 5:451–461
166 Hsu F-L, Nonaka G-I, Nishioka I 1985 Acylated flavanols and procyanidins from *Salix sieboldiana*. Phytochemistry 24:2089–2092
167 Hundt H K L, Roux D G 1978 Condensed tannins: Determination of the point of linkage in terminal (+)-catechin units and degradative bromination of 4-flavanyl-flavan-3,4-diols. J Chem Soc Chem Commun 696–698
168 Iacobucci G A, Sweeny J B 1983 The chemistry of anthocyanins, anthocyanidins, and related flavylium salts. Tetrahedron 39:3005–3038
169 Ilyas N, Ilyas M, Rahman W, Okigawa M, Kawano N 1978 Biflavones from the leaves of *Araucaria excelsa*. Phytochemistry 17:987–990
170 Ilyas M, Seligmann O, Wagner H 1977 Biflavones from the leaves of *Araucaria rulei* F. Muell. and a survey on biflavanoids of the *Araucaria* genus. Z Naturforsch 32C:206–209
171 Ishratullah Kh, Ansari W H, Rahman W, Okigawa M, Kawano N 1977 Biflavonoids from *Semecarpus anacardium* Linn. (Anacardiaceae). Indian J Chem 15B:615–619
172 Ishratullah Kh, Rahman W, Okigawa M, Kawano N 1973 Biflavones from *Taxodium mucronatum*. Phytochemistry 17:335
173 Jackson B, Locksley H D, Scheinmann F, Wolstenholme W A 1971 Extractives from Guttiferae. Part XXII. The isolation and structure of four novel biflavanones from the heartwoods of *Garcinia buchananii* Baker and *G. eugeniifolia* Wall. J Chem Soc (C):3791–3804
174 Jacobs E, Ferreira D, Roux D G 1983 Atropisomerism in a new class of condensed tannins based on biphenyl and terphenyl. Tetrahedron Lett 24:4627–4630
175 Jacques D, Haslam E 1974 Biosynthesis of plant proanthocyanidins. J Chem Soc Chem Commun 231–232
176 Jacques D, Haslam E, Bedford G R, Greatbanks D 1974 Plant proanthocyanidins. Part II. Proanthocyanidin A2 and its derivatives. J Chem Soc Perkin Trans I 2663–2671
177 Jacques D, Opie C T, Porter L J, Haslam E 1977 Plant proanthocyanidins. Part 4. Biosynthesis of procyanidins and observations on the metabolism of cyanidin in plants. J Chem Soc Perkin Trans I 1637–1643
178 Jensen O N, Pedersen J A 1983 The oxidative transformations of (+)-catechin and (−)-epicatechin as studied by ESR. Tetrahedron 39:1609–1615
179 Joly M, Beck J P, Haag-Berrurier M, Anton R 1980 Cytotoxicite de flavonoids sur cellules d'hepatome en culture. Planta Med 39:230
180 Joly M, Haag-Berrurier M, Anton R 1980 La 5′-methoxy-bilobetine une biflavone extraite du *Ginkgo biloba*. Phytochemistry 19:1999–2002
181 Joshi B S, Kamat V N, Viswananathan N 1970 The isolation and structure of two biflavones from *Garcinia talboti*. Phytochemistry 9:881–888
182 Kamal G M, Gunaherath B, Gunatilaka A A L, Sultanbawa M U S, Balasubramaniam S 1982 Dulcitol, an (−)-4′-*O*-methyl-epigallocatechin from *Kokoona zeylanica*. J Nat Prod 45:140–142
183 Kamil M, Ilyas M, Rahman W, Hasaka N, Okigawa M, Kawano N 1977 Taiwaniaflavone: a new series of naturally occurring biflavones from *Taiwania cryptomerioides*. Chem Ind (London) 160

184 Kamil M, Khan N A, Ilyas M, Rahman W 1983 Biflavones from Ochnaceae, a new biflavone from *Ochna pumillai*. Indian J Chem 22B:608

185 Karanjgaokar C G, Radhakrishanan P V, Venkataraman K 1967 Morelloflavone, a 3-(8)-flavonylflavanone from the heartwood of *Garcinia morella*. Tetrahedron Lett 3195–3198

186 Karchesy J J, Hemingway R W 1980 Loblolly pine bark polyflavanoids. J Agr Food Chem 28:222–228

187 Karchesy J J, Hemingway R W 1986 Condensed tannins: ($4\beta\rightarrow8;2\beta\rightarrow O\rightarrow7$) linked procyanidins in *Arachis hypogaea* L. J Agr Food Chem 34:966–970

188 Karchesy J J, Hemingway R W, Foo L Y, Barofsky E, Barofsky D F 1986 Sequencing procyanidin oligomers by fast atom bombardment mass spectrometry. Anal Chem 58:2563–2567

189 Karchesy J J, Loveland P M, Laver M L, Barofsky D F, Barofsky E 1976 Condensed tannins from the barks of *Alnus rubra* and *Pseudotsuga menziesii*. Phytochemistry 15:2009–2010

190 Kariyone T, Fukui Y 1960 Studies on the chemical constituents of plants of Coniferae and allied orders. Studies on the structure of hinokiflavone, a flavonoid from the leaves of *Chamaecyparis obtusa* Endlicher. Composition of hinokiflavone and its degradation in KOH solution. J Pharm Soc Jpn 80:746–749

191 Kariyone T, Sawada T 1958 Studies on flavonoids of the leaves of Coniferae and allied plants. I. On the flavonoid from the leaves of *Torreya nucifera* Sieb. et Zucc. J Pharm Soc Jpn 78:1010–1013

192 Kariyone T, Sawada T 1958 Studies on flavonoids in the leaves of Coniferae and allied plants. II. On the flavonoids from the leaves of *Cycas revoluta* Thunb. and *Cryptomeria japonica* D. Don. var. *araucarioides* Hort. J Pharm Soc Jpn 78:1013–1015

193 Kariyone T, Sawada T 1958 Studies on flavonoids in the leaves of Coniferae and allied plants. III. On the flavonoid from the leaves of *Taxus cuspidata* Sieb. et Zucc. and relation between ginkgetin, kayaflavone, sciadopitysin and sotetsuflavone. J Pharm Soc Jpn 78:1016–1019

194 Kariyone T, Sawada T 1958 Studies on flavonoids in the leaves of Coniferae and allied plants. IV. On the flavonoid from the leaves of *Chamaecyparis obtusa* Endl. J Pharm Soc Jpn 78:1020–1022

195 Karl C, Müller G, Pedersen P A 1982 Ein neues Catechin Glycosid aus *Polypodium vulgare*. Z Naturforsch 37C:148–151

196 Kashiwada Y, Nonaka G, Nishioka I 1986 Tannins and related compounds. XLV. Rhubarb (5). Isolation and characterization of flavan-3-ol and procyanidin glucosides. Chem Pharm Bull 34:3208–3222

197 Kennedy J A, Munro M H G, Powell H K J, Foo L Y 1984 The protonation reactions of catechin, epicatechin, and related compounds. Aust J Chem 37:885–892

198 Keppler H H 1975 The isolation and constitution of mollisacacidin, a new leucoanthocyanidin from the heartwood of *Acacia mollismi* Willd. J Chem Soc 2721–2724

199 Khan N U, Ansari W H, Rahman W, Okigawa M, Kawano N 1971 Two new biflavonyls from *Araucaria cunninghamii*. Chem Pharm Bull (Tokyo) 19:1500–1501

200 Khan N U, Ilyas M, Rahman W, Mashima T, Okigawa M, Kawano N 1972 Biflavones from the leaves of *Araucaria bidwillii* Hooker and *Agathis alba* Foxworthy (Araucariaceae). Tetrahedron 28:5689–5695

201 Kho K F F 1979 Biosynthese van anthocyanen in witbloeiende mutanten van *Petunia hybrida*. Pharm Weekbl 114:325–334

202 Kho K F F, Bennink G J H, Wiering H 1975 Anthocyanin synthesis in a white flowering mutant of *Petunia hybridia* by a complementation technique. Planta 127:271–279

203 Kho K F F, Bolsman-Louwen A C, Vuik J C, Bennink G J H 1977 Anthocyanin synthesis in a white flowering mutant of *Petunia hybrida*. II. Accumulation of dihydroflavonol intermediates in white flowering mutants; uptake of intermediates in isolated corollas and conversion into anthoxyanins. Planta 135:109–118

204 Kiatgrajai P, Wellons J D, Gollob L, White J D 1982 Kinetics of epimerization of (+)-catechin and its rearrangement to catechinic acid. J Org Chem 47:2910–2913

205 Kiatgrajai P, Wellons J D, Gollob L, White J D 1982 Kinetics of polymerization of (+)-catechin with formaldehyde. J Org Chem 47:2913–2917

206 King F E, Bottomley W 1954 The chemistry of extractives from heartwoods. Part XVII. The occurrence of a flavan-3,4-diol (melacacidin) in *Acacia melanoxylon*. J Chem Soc 1399–1403

207 King F E, Clark-Lewis J W, Forbes W F 1955 The chemistry of extractives from hardwoods. 25. (−)-Epiafzelechin, a new member of the catechin series. J Chem Soc 2948–2956

208 Kolodziej H 1983 The first naturally occurring 4-arylflavan-3-ol. Tetrahedron Lett 24:1825–1828

209 Kolodziej H 1984 Occurrence of procyanidins in *Nelia meyeri*. Phytochemistry 23:1745–1752

210 Kolodziej H 1985 The first 2,3-*trans*-3,4-*cis* procyanidin. Phytochemistry 24:2460–2462

211 Kolodziej H 1986 Synthesis and characterization of procyanidin dimers as their peracetates and octamethylether diacetates. Phytochemistry 25:1209–1215

212 Kolodziej H, Ferreira D, Roux D G 1984 Synthesis of condensed tannins. Direct access to (4,6) and (4,8) all 2,3-*cis* procyanidin derivatives from (−)-epicatechin: Assessment of bonding positions in oligomeric analogues from *Crataegus ocyacatha* L. J Chem Soc Perkin Trans I 343–350

213 Konoshima M, Ikeshiro Y 1970 Fukugiside, the first biflavonoid glycoside from *Garcinia spicata* Hook F. Tetrahedron Lett 1717–1720

214 Konoshima M, Ikeshiro Y, Miyahara S, Yen K-Y 1970 The constitution of biflavonoids from *Garcinia* plants. Tetrahedron Lett 4203–4206

215 Konoshima M, Ikeshiro Y, Nishinaga A, Matsuura T, Kubota T, Sakamoto H 1969 The constitution of flavonoids from *Garcinia spicata* Hook F. Tetrahedron Lett 121–124

216 Kreibich R E, Hemingway R W 1985 Condensed tannin-resorcinol adducts in laminating adhesives. For Prod J 35 (3):23–25

217 Kreibich R E, Hemingway R W 1987 Condensed tannin-sulfonates in cold-setting wood-laminating adhesives. For Prod J 37 (2):43–46

218 Krishnamoorthy V, Seshadri T R 1966 A new proanthocyanidin from the stem bark of *Myrica nagi* Thunb. Tetrahedron 2367–2371

219 Kristiansen K N 1984 Biosynthesis of proanthocyanidins in barley: Genetic control of the conversion of dihydroquercetin to catechin and procyanidins. Carlsberg Res Commun 49:503–524

220 Kubo I, Klocke J A, Matsumoto T 1983 Identification of two insect growth inhibiting biflavonoids in *Podocarpus gracilior*. Rev Latinoamer Quim 14:59–61

221 Laks P 1984 Chemical modification of bark tannins for adhesive formulation. Ph. D. Thesis. Univ British Columbia Vancouver

222 Laks P (in press) Flavonoid biocides: Phytoalexin analogues from condensed tannins. Phytochemistry

223 Laks P E, Hemingway R W 1987 Condensed tannins: Base-catalysed reactions of polymeric procyanidins with phenyl-methanethiol. Lability of the interflavanoid bond and pyran ring. J Chem Soc Perkin Trans I 465–470

224 Laks P E, Hemingway R W Condensed tannins: Structure of the "Phenolic Acids." Holzforschung (in press).

225 Laks P E, Hemingway R W, Conner A H (in press) Condensed tannins: Base-catalysed reactions of polymeric procyanidins with phloroglucinol. Intramolecular rearrangements. J Chem Soc Perkin Trans I

226 Lebreton Ph, Thivend S, Boutard B 1980 Distribution des proanthocyanidins chez les gymnosperms. Plant Med Phytother 14:105–129

227 Lin K-M, Chen F-C 1974 On the dehydrogenation and rearrangement of biflavanoids. Chem J Chin Chem Soc (Taiwan) 3:67–73

228 Lin Y-M, Chen F-C 1973 Rhusflavanone – a new biflavanone from *Rhus succedanea*. Tetrahedron Lett 4747–4750

229 Lin Y-M, Chen F-C 1974 Robustaflavone from the seed kernels of *Rhus succedanea*. Phytochemistry 13:1617–1619

230 Lindberg G, Osterdahl B-G, Nilsson E 1974 Chemical studies of Bryophytes. Chem Scripta 5:140–144

231 Liu G-Q, Tatematsu H, Kurokawa M, Niwa M, Hirata Y 1984 Novel C-3/C-3″-biflavanones from *Stellera chamaejasme* L. Chem Pharm Bull 32:362–365

232 Locksley H D 1973 The chemistry of biflavanoid compounds. Fortschr Chem Org Naturst 30:207–312

233 Mabry T J, Markham K R 1975 Mass spectrometry of flavonoids. In: Harborne J B, Mabry T J, Mabry H (eds) The flavonoids. Academic Press New York, 78–126

234 MacGregor J T, Jurd L 1978 Mutagenicity of plant flavonoids structural requirements for mutagenic activity in *Salmonella typhinurium*. Mutation Res 54:297–309

235 Madhav R 1969 Heveaflavone – a new biflavonoid from *Hevea braseliensis*. Tetrahedron Lett 2017–2019

236 Malan E, Roux D G 1975 Flavonoids and tannins of *Acacia* species. Phytochemistry 14:1835–1841

237 Markham K R 1975 Isolation techniques for flavonoids. In: Harborne J B, Mabry T J, Mabry H (eds) The flavonoids. Academic Press New York, 1–45

238 Markham K R 1982 Techniques of flavonoid identification. Academic Press New York

239 Markham K R, Chari V M, Mabry T J 1982 Carbon ^{13}C-NMR spectroscopy of flavonoids. In: Harborne J B, Mabry T J (eds) The flavonoids: Advances in research. Chapman and Hall London, 19–134

240 Markham K R, Mabry T J 1975 Resonance spectroscopy of flavonoids. In: Harborne J B, Mabry T J, Mabry H (eds) The flavonoids. Academic Press New York, 45–77

241 Mattice W L, Tobiason F L, Houglum K, Shanafelt A 1982 Conformational analysis and dipole moments of tetra-*O*-methyl-(+)-catechin and tetra-*O*-methyl-(−)-epicatechin. J Am Chem Soc 104:3359–3362

242 Mayer W, Goll L, Von Arndt E M, Mannschreck A 1966 Procyanidino-(−)-epicatechin ein zweiarmig verknüpftes kondensiertes Proanthocyanidin aus *Aesculus hippocastanum*. Tetrahedron Lett 429–435

243 Mayer W, Merger F 1961 Über Kondensationsreaktionen der Catechine. II. Die Reaktion der Catechine mit Phloroglucin. Justus Liebigs Ann Chem 644:70–77

244 Mayer W, Merger F 1961 Über Kondensationsreaktionen der Catechine. III. Die Selbstkondensation von (+)-Catechin. Justus Liebigs Ann Chem 644:79–84

245 Mayer W, Merger F, Frank G, Heyns K, Grutzmacher H F 1963 Über Kondensationsreaktionen der Catechine. IV. Die Konstitution der Kondensationsprodukte von Catechinen mit Phloroglucin. Naturwissenschaften 50:152–153

246 McClure J W 1975 Physiology and functions of flavonoids. In: Harborne J B, Mabry T J, Mabry H (eds) The flavonoids. Academic Press New York, 970–1055

247 McGraw G W, Hemingway R W 1982 Electrophilic aromatic substitution of catechins: Bromination and benzylation. J Chem Soc Perkin Trans I 973–978

248 Mehta P P, Whalley W B 1963 The stereochemistry of some catechin derivatives. J Chem Soc 5327–5332

249 Miura H, Kawano N 1967 The partial demethylation of flavones. II. Formation of isocryptomerin. Chem Pharm Bull (Tokyo) 15:232–235

250 Miura H, Kawano N 1968 On the distribution of bisflavones in the leaves of *Taxodiaceae* and *Cupressaceae* plants. J Pharm Soc Jpn 88:1459–1462

251 Miura H, Kawano N 1968 The partial demethylation of flavones. IV. Formation of new bisflavones hinokiflavone-7,7″-dimethyl ether and neocryptomerin. Chem Pharm Bull (Tokyo) 16:1838–1840

252 Miura H, Kawano N 1968 Sequoiaflavone in the leaves of *Sequoia sempervirens* and *Cunninghamia lanceolata* var. *konishii* and its formation by partial demethylation. J Pharm Soc Jpn 88:1489–1491

253 Miura H, Kawano N, Weiss A C Jr 1966 Cryptomerin A and B, hinokiflavone methyl ethers from the leaves of *Cryptomeria japonica*. Chem Pharm Bull (Tokyo) 14:1404–1408

254 Miura H, Kihara T, Kawano N 1968 New biflavones from *Podocarpus* and *Chamaecyparis* plants. Tetrahedron Lett 2339–2342

255 Miura H, Kihara T, Kawano N 1969 Studies on bisflavones in the leaves of *Podocarpus macrophylla* and *P. nagi*. Chem Pharm Bull (Tokyo) 17:150–154

256 Miyamura M, Mohara T, Tomimatsu T, Nishioka I 1983 Seven aromatic compounds from the bark of *Cinnamomum cassia*. Phytochemistry 22:215–218

257 Murthy S S N 1983 Naturally occurring biflavonoid derivatives. Part III. A new biflavanone from the nut shell of *Semecarpus anacardium* Linn. Indian J Chem 22B:1167–1168

258 Murthy S S N 1984 Partial conversions in biflavonoids: Part 5. Confirmation of the structure of jeediflavanone, a biflavanone from *Semecarpus anacardium*. Phytochemistry 23:925–927

259 Murthy S S N, Anjaneyulu A S R, Ramachandra L, Pelter A, Ward R W 1981 Occidentoside – a new biflavonoid glycoside from the nut shells of *Anacardium occidentale* Linn. Indian J Chem 20B:150–151

260 Murthy S S N, Anjaneyulu A S R, Row L R, Pelter A, Ware R S 1982 Chemical examination of *Anacardium occidentalis*. Planta Med 45:3–10

261 Murti V V S, Rahman R W, Seshadri T R 1967 Cupressuflavone, a new biflavonyl pigment. Tetrahedron 23:397–404

262 Nakazawa K 1959 Synthesis of ginkgetin tetramethyl ether. Chem Pharm Bull (Tokyo) 7:748–749

263 Nakazawa K 1962 Synthesis of nuclear-substituted flavonoids and allied compounds. IX. Synthesis of tetramethyl ether and dimethyl ether of ginkgetin. Chem Pharm Bull (Tokyo) 10:1032–1038

264 Nakazawa K 1968 Syntheses of ring substituted flavonoids and allied compounds. XI. Synthesis of hinokiflavone. Chem Pharm Bull (Tokyo) 16:2503–2511

265 Narayanan V, Seshadri T R 1969 Chemical components of *Acer rubrum* wood and bark: Occurrence of procyanidin dimer and trimer. Indian J Chem 7:213–214

266 Natarajan S, Murti V V S, Seshadri T R 1969 Biflavonyls: Part III. Mass spectrometry of biflavones. Indian J Chem 17:751–755

267 Natarajan S, Murti V V S, Seshadri T R, Ramaswumi A S 1970 Some new pharmacological properties of flavonoids and biflavonoids. Current Sci India 39:533–534

268 Nicoletti M, Goulart M F, DeLima R, Goulart A E, Delle Monache F, Marini-Bettolo G B 1984 Flavanoids and alkaloids from *Strychnos pseudoquina*. J Nat Prod 47:953–957

269 Niwa M, Liu G-Q, Tatematsu H, Hirata Y 1984 Chamaechromone, a novel rearranged biflavonoid from *Stellera chamaejasme* L. Tetrahedron Lett 25:3735–3738

270 Niwa M, Tatematsu H, Liu G-Q, Hirata Y 1984 Isolation and structures of two new C-3/C-3″-biflavanones, neochamaejasmin A and neochamaejasmin B. J Chem Soc Jpn Chem Lett 539–542

271 Nonaka G, Ezaki E, Hayashi K, Nishioka I 1983 Flavanol glucosides from rhubarb and *Rhaphiolepis umbellata*. Phytochemistry 22:1659–1661

272 Nonaka G, Hsu F-L, Nishioka I 1981 Structures of dimeric, trimeric, and tetrameric procyanidins from *Areca catechu* L. J Chem Soc Chem Commun 781–783

273 Nonaka G, Kawahara O, Nishioka I 1982 Tannins and related compounds. VIII. A new type of proanthocyanidin, cinchonains IIa and IIb from *Cinchona succirubra*. Chem Pharm Bull Tokyo 30:4277–4282

274 Nonaka G, Miura N, Nishioka I 1982 Stilbene glycoside, gallates, and proanthocyanidins from *Polygonum multiflorum*. Phytochemistry 21:429–432

275 Nonaka G, Morimoto S, Nishioka I 1983 Tannins and related compounds. Part 13. Isolation and structures of trimeric, tetrameric and pentameric proanthocyanidins from cinnamon. J Chem Soc Perkin Trans I 2139–2145

276 Nonaka G, Muta M, Nishioka I 1983 Myricatin, a galloyl-flavanol-sulfate and prodelphinidin gallates from *Myrica rubra*. Phytochemistry 22:237–241

277 Nonaka G, Nishimura H, Nishioka I 1986 Tannins and related compounds. Part 26. Isolation and structure of stenophyllanins A, B, and C, novel tannins from *Quercus stenophylla*. J Chem Soc Perkin Trans I 163–172

278 Nonaka G, Nishioka I 1980 Novel biflavonoids, chalcan-flavan dimers from gambir. Chem Pharm Bull (Tokyo) 28:3145–3149

279 Nonaka G, Nishioka I 1982 Tannins and related compounds. VII. Phenylpropanoid substituted epicatechins, cinchonains from *Cinchona succirubra*. Chem Pharm Bull (Tokyo) 30:4268–4272

280 Nonaka G, Nishioka I, Nagasawa T, Oura H 1981 Tannins and related compounds. I. Rhubarb. Chem Pharm Bull (Tokyo) 29:2862–2870

281 Nonaka G, Sakai K, Nishioka I 1984 Hydrolysable tannins and proanthocyanidins from green tea. Phytochemistry 23:1753–1755

282 Oberholzer M E, Rall G J H, Roux D G 1980 (2*R*,3*S*,4*S*)-3,4,7,3′,4′-Pentamethoxy-2,3-*trans*-3,4-*cis*-flavan, a novel flavan from *Neorautanenia amboensis*. Phytochemistry 19:2503–2504

283 Ohigashi H, Minami S, Fukui H, Koshimizu K, Mizutani F, Sugiura A, Tomana T 1982 Flavanols as plant growth inhibitors from the roots of peach, *Prunus persica* Batsh. cv. ‘Hakuto’. Agr Biol Chem 46:2555–2561

284 Okigawa M, Kawano N, Aqil M, Rahman W 1973 The structure of ochnaflavone, a new type of biflavone and the synthesis of its pentamethyl ether. Tetrahedron Lett 2003–2006

285 Okigawa M, Kawano N, Aqil M, Rahman W 1976 Ochnaflavone and its derivatives: A new series of diflavonyl ether from *Ochna squarrosa* Linn. J Chem Soc Perkin Trans I 580–583

286 Okigawa M, Khan N U, Kawano N, Rahman W 1975 Application of a lanthanide shift reagent Eu(fod)3 to the elucidation of the structures of flavones and related compounds. J Chem Soc Perkin Trans I 1563–1568
287 Otsuka H, Fujioka S, Komiya T, Mizuta E, Takemoto M 1982 Studies on anti-inflammatory agents. VI. Anti-inflammatory constituents of *Cinnamomum sieboldii* Meissn. Yakugaku Zasshi 102:162–172
288 Ottrup H 1981 Structure of prodelphinidins in barley. Proc EBC Congr 323–333
289 Ottrup H, Schaumburg K 1981 Structure elucidation of some proanthocyanidins in barley by 1H 270 MHZ NMR spectroscopy. Carlsberg Res Commun 46:43–52
290 Patil A D, Desphande V H 1982 A new dimeric proanthocyanidin from *Cassia fistula* sapwood. Indian J Chem 21B:626–628
291 Pelter A, Amenechi P I 1969 Isoflavonoid and pterocarpinoid extractives of *Lonchocarpus laxiflorus*. J Chem Soc (C) 887–896
292 Pelter A, Amenechi P I, Warren R 1969 The structures of two proanthocyanidins from *Julbernadia globiflora*. J Chem Soc (C) 2572–2579
293 Pelter A, Warren R, Chexal K K, Handa B K, Rahman W 1971 Biflavonyls for Guttiferae: *Garcinia livingstonii*. Tetrahedron Lett 27:1625–1634
294 Pelter A, Warren R, Hameed N, Khan N, Ilyas M, Rahman W 1970 Biflavonyl pigments from *Thuja orientalis* (Cupressaceae). Phytochemistry 9:1897–1898
295 Pelter A, Warren R, Handa B K, Chexal K K, Rahman W 1971 On the occurrence, isomerization, and racemization of some optically active biflavonyls. Indian J Chem 9:98–100
296 Pelter A, Warren R, Ilyas M, Usmani J N, Bhatnagar S R, Rizivi R H, Ilyas N, Rahman W 1969 The structure of W13, the first optically active biflavone of the amentoflavone series. Experientia 25:350–351
297 Pelter A, Warren R, Ilyas M, Usmani J N, Rizivi R H, Rahman W 1969 The isolation and characterization of two members of a new series of naturally occurring biflavones. Experientia 25:351–352
298 Pizzi A, Cameron F A, Eaton N J 1986 The tridimensional structure of polyflavonoid tannins by conformational analysis. J Macromol Sci Chem A23:515–540
299 Platt R V, Opie C T, Haslam E 1984 Plant proanthocyanidins. Part 8. Biosynthesis of flavan-3-ols and other secondary plant products from 2*S*-phenylalanine. Phytochemistry 23:2211–2217
300 Porter L J 1974 Extractives of *Pinus radiata* bark. Part 2. Procyanidin constituents. N Z J Sci 17:213–218
301 Porter L J, Foo L Y 1982 Leucocyanidin: Synthesis and properties of (2*R*,3*S*,4*R*)-(+)-3,4,5,7,3′,4′-hexahydroxyflavan. Phytochemistry 21:2947–2952
302 Porter L J, Hrstich L N, Chan B G 1986 The conversion of procyanidins and prodelphinidins to cyanidin and delphinidin. Phytochemistry 25:223–230
303 Porter L J, Newman R H, Foo L Y, Wong H, Hemingway R W 1982 Polymeric proanthocyanidins: ^{13}C-NMR studies of procyanidins. J Chem Soc Perkin Trans I 1217–1221
304 Porter L J, Wong R Y, Benson M, Chan B G, Vishwanadhan V N, Gandour R D, Mattice W L 1986 Conformational analysis of flavans: ^{1}H-NMR and molecular mechanical (MM2) studies of the benzpyran ring of 3′,4′,5,7-tetrahydroxyflavan-3-ols: The crystal and molecular structure of the procyanidin: (2*R*,3*SR*,4*R*)-3′,4′,5,7-tetramethoxy-4-(2,4,6-trimethoxyphenyl)-flavan-3-ol. J Chem Res (S):86–87 (M):830–880
305 Porter L J, Wong R Y, Chan B G 1985 The molecular and crystal structure of (+)-2,3-*trans*-3,4-*trans*-leucocyanidin [(2*R*,3*S*,4*R*)-(+)-3,3′,4,4′,5,7-hexahydroxyflavan] dihydrate and comparison of its heterocyclic conformation in solution and the solid state. J Chem Soc Perkin Trans I 1413–1418
306 Quasim M A, Roy S K, Asam A, Kamil M, Ilyas M 1983 Biflavones from *Zamia angustifolia*. J Sci Res (Bhopal) 5(3):179–181
307 Raju M S, Srimannarayana G, Rao N V S 1976 Structure of messuaferrone-B, a new biflavanone from the stamnes of *Mesua ferrea* Linn. Tetrahedron Lett 49:4509–4512
308 Raju M S, Srimannarayana G, Rao N V S 1978 Structure of mesuaferrone-A, a new biflavanone from the stamens of *Mesua ferrea* Linn. Indian J Chem 16B:167–168
309 Raju M S, Srimannarayana G, Rao N V S, Bala K R, Seshadri T R 1976 Structure of mesuaferrone-B. A new biflavanone from the stamens of *Mesua ferrea* Linn. Tetrahedron Lett 4509–4512

310 Rauwald H W 1982 ^{1}H-NMR-Studie zur Analytik rotamerer Procyanidinie. Planta Med 46:110–112
311 Robertson A V 1959 Mechanism of formation of anthocyanidins from flavan-3,4-diols. Can J Chem 37:1946–1954
312 Robinson G M, Robinson R 1933 A survey of anthocyanidins. Notes on the distribution of leuko-anthocyanidins. Biochem J 27:206–212
313 Roux D G 1959 Flavan-3,4-diols and leuco-anthocyanidins of *Guibourtia* spp. Nature 183:890–891
314 Roux D G, DeBruyn G C 1963 Condensed tannins. 17. Isolation of 4′,7-dihydroxyflavan-3,4-diol from *Guibourtia colesperma*. Biochem J 87:439
315 Roux D G, Ferreira D 1974 α-Hydroxychalcones as intermediates in flavonoid biogenesis: The significance of recent chemical analogies. Phytochemistry 13:2039–2048
316 Roux D G, Ferreira D 1982 The direct biomimetic synthesis structure and absolute configuration of angular and linear condensed tannins. Fortschr Chem Org Naturst 4147–4176
317 Roux D G, Ferreira D 1982 Structure and function in biomimetic synthesis of linear, angular and branched condensed tannins. Pure Appl Chem 54:2465–2578
318 Roux D G, Ferreira D, Botha J J, Garbutt D C F 1976 Heartwood extracts of the black wattle (*Acacia mearnsii*) as a possible source of resorcinol. J Appl Polym Sci Polym Symp 28:1365–1376
319 Roux D G, Ferreira D, Hundt H K L, Malan E 1975 Structure, stereochemistry and reactivity of natural condensed tannins as basis for their extended industrial application. J Appl Polym Sci Polym Symp 28:335–353
320 Roux D G, Freudenberg K 1958 Über Leuco-Robinetinidinhydrat und Leuko-Fisetinidinhydrat. Justus Liebigs Ann Chem 613:56–60
321 Roux D G, Maihs E A 1958 Black wattle catechin. Nature 182:1798
322 Roux D G, Maihs E A 1958 Condensed tannins. 3. Isolation and estimation of (−)-7,3′,4′,5′-tetrahydroxyflavan-3-ol, (+)-catechin and (+)-gallocatechin from black wattle bark extract. Biochem J 74:44–49
323 Roux D G, Paulus E 1961 Condensed tannins. 7. Isolation of 7,3′,4′-trihydroxyflavan-3-ol ((−)-fisetinidol), a naturally occurring catechin from black wattle heartwood. Biochem J 78:120–123
324 Roux D G, Paulus E 1962 Condensed tannins. 13. Interrelationships of flavonoid components from the heartwood of *Robinia pseudoacacia*. Biochem J 82:324–330
325 Roy S K, Qasim M A, Kamil M, Ilyas M, Rahman W 1983 Chemical investigation of *Ochrocarpus longifolius*. Indian J Chem 22B:609
326 Ruckstuhl M, Beretz A, Anton R, Landry Y 1979 Flavonoids are selective cyclic GMP phosphodiesterase inhibitors. Biochem Pharmacol 28:535–538
327 Saayman H M, Roux D G 1965 Configuration of quibourtacacidin and synthesis of isomeric racemates. Biochem J 96:36–42
328 Samaraweera U, Sotheeswaran S, Sultanbawa M U S 1983 3,4,7,3′,5′-Pentahydroxyflavan and 3-methoxyfriedelan from *Humboldtia laurifolia*. Phytochemistry 22:565–567
329 Samejima M, Yoshimoto T 1982 Systematic studies on the stereochemical composition of proanthocyanidins from coniferous bark. Mokuzai Gakkaishi 28:67–74
330 Schilling C, Weinges K, Müller O, Mayer W 1973 ^{13}C-NMR-spektroskopische Konstitutionsermittlung der $C_{30}H_{24}O_{12}$-Procyanidine. Justus Liebigs Ann Chem 1471–1475
331 Schleep S, Kolodziej F H 1986 The first natural procyanidin with a 3,4-*cis* configuration. J Chem Soc Chem Commun 392–393
332 Sears K D 1972 Sulfonation of catechin. J Org Chem 37:3546–3547
333 Sears K D, Casebier R L 1968 Cleavage of proanthocyanidins with thioglycolic acid. J Chem Soc Chem Commun 1437–1438
334 Sears K D, Casebier R L 1970 The reaction of thioglycolic acid with polyflavonoid bark fractions of *Tsuga heterophylla*. Phytochemistry 9:1589–1594
335 Sears K D, Casebier R L, Hergert H L, Stoudt G H, McCandlish L E 1975 The structure of catechinic acid, a base rearrangement product of catechin. J Org Chem 39:3244–3247
336 Shen Z, Falshaw C P, Haslam E, Begley M J 1985 A novel spiro-biflavonoid from *Larix gmelini*. J Chem Soc Chem Commun 1135–1136
337 Spek A L, Kojic-Podic B, Labadie R P 1984 Structure of (−)-epicatechin: (2*R*,3*R*)-2-(3,4-dihydroxyphenyl)-3,4-dihydro-2H-1-benzopyran-3,5,7-triol, $C_{15}H_{14}O_6$. Acta Cryst C40:2068–2071

338 Stafford H A 1983 Enzymic regulation of procyanidin biosynthesis: Lack of a flav-3-en-3-ol intermediate. Phytochemistry 22:2643–2646
339 Stafford H A 1984 Flavan-3-ol biosynthesis, the conversion of (+)-dihydroquercetin and flavan-3,4-*cis*-diol (leucocyanidin) to (+)-catechin by reductases extracted from cell suspension cultures of Douglas-fir. Plant Physiol 76:184–186
340 Stafford H A, Lester H H 1982 Enzymic and non-enzymic reduction of (+)-dihydroquercetin to its 3,4-diol. Plant Physiol 70:695–698
341 Stafford H A, Lester H H 1985 Flavan-3-ol biosynthesis. Plant Physiology 78:791–794
342 Stafford H A, Lester H H, Porter L J 1985 Chemical and enzymic synthesis of monomeric procyanidins (leucocyanidins or 3′,4′,5,7-tetrahydroxyflavan-3,4-diols) from (2*R*,3*R*)-dihydroquercetin. Phytochemistry 24:333–338
343 Steenkamp J A, Ferreira D, Roux D G, Hull W E 1983 Synthesis of condensed tannins. Part 8. The first branched (4,6:8:4,6)-tetraflavanoid. Coupling sequence and absolute configuration. J Chem Soc Perkin Trans I 23–28
344 Steenkamp J A, Steynberg J P, Brandt E V, Ferreira D, Roux D G 1985 Phlobatannins, a novel class of ring-isomerized condensed tannins. J Chem Soc Chem Commun 1678–1679
345 Steynberg J P, Ferreira D, Roux D G 1983 The first condensed tannins based on a stilbene. Tetrahedron Lett 24:4147–4150
346 Steynberg J P, Young D A, Burger J F W, Ferreira D, Roux D G 1986 Phlobatannins via facile ring isomerizations of profisetinidin and prorobinetinin. J Chem Soc Chem Commun 1013–1014
347 Strickland R G, Harrison B J 1977 Precursors and genetic control of pigmentation. 3. Detection and destribution of different white genotypes of bluebells (*Endymion* species). Heredity 39:327–333
348 Takasugi M, Niino N, Nagao S, Anetai M, Masamune T, Shirata A, Takahashi K 1984 Eight minor phytoalexins from diseased paper mulberry. J Chem Soc Jpn Chem Lett 689–692
349 Tanaka T, Nonaka G, Nishioka I 1983 7-*O*-Galloyl-(+)-catechin and 3-*O*-galloyl-procyanidin B3 from *Sanguisorba officinalis*. Phytochemistry 22:2576–2578
350 Taufeeq H M, Fatma W, Ilyas M, Rahman W, Kawano N 1978 Biflavones from *Cupressus lusitanica* var. *benthami*. Indian J Chem 16B:655–657
351 Taylor W I, Battersby A R 1967 Oxidative coupling of phenols. Marcel Dekker New York
352 Thompson R S, Jacques D, Haslam E, Tanner R J N 1972 Plant proanthocyanidins. Part I. Introduction, the isolation, structure, and distribution in nature of plant procyanidins. J Chem Soc Perkin Trans I 1387–1399
353 Timberlake C F, Bridle P 1977 Anthocyanins: Colour augmentation with catechin and acetaldehyde. J Sci Food Agr 28:539–544
354 Tindale M D, Roux D G 1969 Phytochemical survey of the Australian species of *Acacia*. Phytochemistry 8:1713–1727
355 Tindale M D, Roux D G 1974 An extended phytochemical survey of Australian species of *Acacia*: Chemotaxonomic and phylogenetic aspects. Phytochemistry 13:829–839
356 Tschesche R, Braum T M, VonSasson W 1980 Symplocoside, a flavanol glycoside from *Symplocos uniflora*. Phytochemistry 19:1825–1829
357 Uvarova N S, Jizba J, Herout V 1967 Plant substances. 27. The proof of structure of polydine, a glycoside of *Polypodium vulgare* L. Collect Czech Chem Commun 31:3075–3078
358 Uvatez L, Brandenberger H 1961 Plant phenols. III. Separation of fermented and black tea polyphenols by cellulose column chromatography. J Chromatogr 5:17–31
359 Van der Westhuizen J H, Ferreira D, Roux D G 1981 Synthesis of condensed tannins. Part 2. Synthesis of photolytic rearrangement, stereochemistry and circular dichromism of the first 2,3-*cis*-3,4-*cis*-4-arylflavan-3-ols. J Chem Soc Perkin Trans I 1220–1226
360 VanHeerden F R, Brandt E V, Ferreira D, Roux D G 1981 Metabolites from the purple heartwoods of the Mimosoideae. Part 4. *Acacia fasciculifera* F. Mueli ex. Benth: Fasciculiferin, fasciculiferol and the synthesis of 7-aryl- and 7-flavanyl-peltogynoids. J Chem Soc Perkin Trans I 2483–2489
361 VanSoest T L 1971 Aufklärung der Molekularstruktur des Dehydro-Dicatechins-A durch Röntgenstrukturanalyse seines Bromoheptamethyläthers. Justus Liebigs Ann Chem 754:137–138
362 Varshney A K, Rahman W, Okigawa M, Kawano N 1973 Robustaflavone – the first member of a new series of biflavones. Experientia 29:784–786

363 Venkataraman K 1975 Flavones. In: Harborne J B, Mabry T J, Mabry H (eds) The flavonoids. Academic Press New York, 268–295

364 Vilian C, Damas J, Lecompte J, Foo L Y 1985 Pharmacological properties of a novel flavan derivative isolated from the New Zealand conifer *Phyllocladus alpinus*. Cardiovascular activities of phylloflavan in rats. In: Proc 7th Hungarian Bioflavonoid Symp Szeged Hungary

365 Viswanadhan V N, Bergmann W R, Mattice W L (in press) Configurational statistics of (4→6) (β)-linked homopolymers of (+)-catechin of (−)-epicatechin. Macromolecules

366 Viswanadhan V N, Mattice W L 1986 Conformational analysis of the sixteen (4→6) and (4→8) linked dimers of (+)-catechin and (−)-epicatechin. J Comput Chem 7:711–717

367 Viswanadhan V N, Mattice W L (in press) Preferred conformations of the sixteen (4→6) and (4→8) linked dimers of (+)-catechin and (−)-epicatechin with axial or equatorial dihydroxyphenyl rings at C(2). J Chem Soc Perkin Trans II

368 Viviers P M, Botha J J, Ferreira D, Roux D G, Saayman H M 1983 Synthesis of condensed tannins. Part 7. Angular (4,6:4,8)-prorobinetinidin triflavonoids from blackwattle ("mimosa") bark extract. J Chem Soc Perkin Trans I 17–22

369 Viviers P M, Kolodziej H, Young D A, Ferreira D, Roux D G 1983 Synthesis of condensed tannins. Part II. Intramolecular enantiomerism of constituent units of tannins from the Anacardiaceae: Stoichiometric control in direct synthesis, derivation of 1H nuclear magnetic resonance parameters applicable to higher oligomers. J Chem Soc Perkin Trans I 2555–2562

370 Viviers P M, Young D A, Botha J S, Ferreira D, Roux D G, Hull W W 1982 Synthesis of condensed tannins. Part 6. The sequence of units, coupling positions and absolute configuration of the first linear (4,6:4,6) triflavanoid with terminal 3,4-diol function. J Chem Soc Perkin Trans I 535–540

371 Volsteedt F, Roux D G 1971 Zeyherin, a natural 3,8-coumaranonyl flavanone from *Phyllogeiton zeyheri* Sond. Tetrahedron Lett 20:1647–1650

372 Wallace J W, Markham K R 1978 Apigenin and amentoflavone glycosides in the Psilotaceae and their phylogenetic significance. Phytochemistry 17:1313–1317

373 Waterman D G, Crichton E G 1980 Xanthones and biflavonoids from *Garcinia densivenia* stem bark. Phytochemistry 19:2723–2726

374 Waterman D G, Hussain R A 1983 Systematic significance of xanthones, benzophenones, and biflavonoids in *Garcinia*. Biochem System Ecol 11:21–28

375 Weeratunga G, Bohlin L, Sanberg F 1984 A muscle-relaxant catechin derivative from *Elaeodendrin balae* (Celastraceae). Acta Pharm Suec 21:73–76

376 Weinges K 1958 Über Catechine und ihre Herstellung aus Leuko-Anthocyanidinhydraten. Justus Liebigs Ann Chem 615:203–209

377 Weinges K, Bahr W, Ebert W, Goritz K, Marx H-D 1969 Konstitution, Entstehung, und Bedeutung der Flavonoidgerbstoffe. Fortschr Chem Org Naturst 27:158–260

378 Weinges K, Ebert W 1968 Isolierung eines kristallisierten Dehydrierungsdimeren aus (+)-Catechin. Phytochemistry 7:153–155

379 Weinges K, Ebert W, Huthwelker D, Mattauch H, Perner J 1969 Konstitution und Bildungsmechanismus des Dehydro-Dicatechins A. Justus Liebigs Ann Chem 726:114–124

380 Weinges K, Goritz K, Nader F 1968 Zur Kenntnis der Proanthocyanidine. XI. Konfigurationsbestimmung von $C_{30}H_{26}O_{12}$-Procyanidinen und Strukturaufklärung eines neuen Procyanidins. Justus Liebigs Ann Chem 715:164–171

381 Weinges K, Kaltenhauser W, Marx H-D, Nader E, Nader F, Perner J, Seiler D 1968 Procyanidine aus Früchten. Justus Liebigs Ann Chem 711:184–204

382 Weinges K, Marx H-D, Goritz K 1970 Die Rotationsbehinderung an der C(sp2)-C(sp3)-Bindung der 4-Arylsubstituierten Polymethoxyflavane. Chem Ber 103:2336–2343

383 Weinges K, Mattauch H, Wilkins C, Frost D 1971 Spektroskopische und chemische Konstitutionsaufklärung des Dehydro-Dicatechins A. Justus Liebigs Ann Chem 754:124–136

384 Weinges K, Nagel D 1968 Massenspektroskopischer Konstitutionsbeweis des Dicatechins. Phytochemistry 7:157–159

385 Weinges J, Perner J, Marx H-D 1970 Synthese des Octamethyl-Diacetylprocyanidins B-3. Chem Ber 103:2344–2349

386 Weinges K, Seiler D 1968 Die diastereomeren Catechin-3-glucoside und -3-gallate. Justus Liebigs Ann Chem 714:193–204

387 Weinges K, Toribo F 1965 Catechin-Resorcinol Kondensate. Justus Liebigs Ann Chem 681: 161–168

388 Weinges K, Toribo F, Paulus E 1965 Die Konformation als Ursache stereoselectiver Reaktionen. Justus Liebigs Ann Chem 688:127–133
389 Weinges K, Wild R 1970 Die Konstitution des Polydins. Liebigs Ann Chem 734:46–55
390 Yang W, Xing Y, Sang M, Xu J, Huang W 1984 Separation and structural determination of 7-methoxychamaejasmin. Gaodeng Xuexiao Huaxue Xuebao 5(5):671–673
391 Young D A, Ferreira D, Roux D G 1983 Synthesis of condensed tannins. Part 10. Dioxane linked profisetinidins. J Chem Soc Perkin Trans I 2031–2035
392 Young D A, Ferreira D, Roux D G (in press) Stereochemistry and dynamic behaviour of some synthetic "angular" profisetinidin tetraflavanoid derivatives. J Polymer Sci Polym Chem

7.7 Condensed Tannins

L. J. PORTER

7.7.1 Introduction

The condensed tannins, together with hydrolyzable tannins (gallic acid metabolites, described in Sect. 7.2) constitute the vegetable tannins, which have the essential characteristic of binding proteins and, less familiarly, polysaccharides. It is this feature of their chemistry that has been of continuing fascination for scientists, and has provided much of the impetus for their study. Recent work has shown that a minor role in plants probably involves their broad-spectrum antibiotic properties (see Sect. 7.7.5) and this is also likely to involve their binding characteristics (see Sect. 7.7.2.3).

The previous section (7.6) described the chemistry of the catechins and oligomeric proanthocyanidins. The historical steps that brought Bate-Smith and Swain, Roux, and other early workers to the realization that condensed tannins are essentially polymers based on the same flavanoid units is now a familiar story (55, 57, and references therein). However, it awaited the modern chemical methodology of the 1980s to provide an unequivocal demonstration of the veracity of these predictions.

Recent advances have relied on development of methods of purification utilizing dextran (Sephadex) gels (25, 65), and application of ^{13}C NMR (25, 109), and gel permeation chromatography (GPC) (66, 124, 148) to the problem. Consequently we are now in a position in which the essential features of condensed tannin structure have been elucidated, and these are described in the following sections, particularly as they pertain to woody plants. It should be pointed out, however, that although condensed tannins do indeed occur in high concentrations in most woody plants of the Coniferae (124, 126) and Dicotyledonae (7), particularly in the bark and fruits, there is a danger that this point has been overstressed. Although they are indeed largely absent from the leaves of most of the herbaceous Dicotyledonae (7), and from many of the most common families of the largely non-woody Monocotyledonae (8), such as the Graminaceae, Amarylidaceae, and Orchidaceae, there are particularly high concentrations in the leaves and fruits of others, such as the Palmae, Musaceae, and Iridaceae (29). In fact the distinction between the occurrence of condensed tannins in woody versus herbaceous species appears now to be more quantitative than qualitative.

7.7.2 Structure and Properties

The condensed tannins may be conveniently divided into two classes on the basis of A-ring oxidation pattern. Type 1 are those containing flavanoid units with a phloroglucinol pattern A-ring: the propelargonidins (**1**, **4**), the procyanidins (**2**, **5**), and the prodelphinidins (**3**, **6**). Type 2 are those containing flavanoids units with a resorcinol pattern A-ring (and thus belong to the class of so-called 5-deoxyflavonoids): the proguibourtinidins (**7**), profisetinidins (**8**), and prorobinetinidins (**9**).

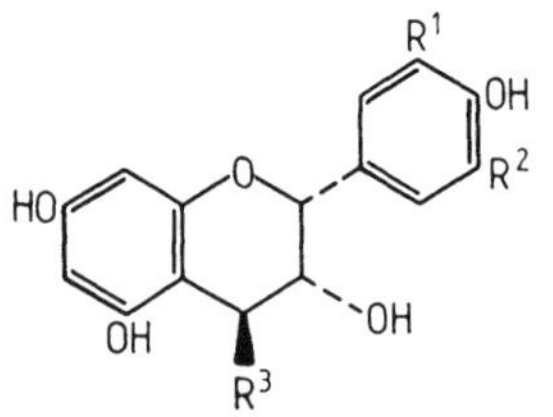

a 1-3; 10-12 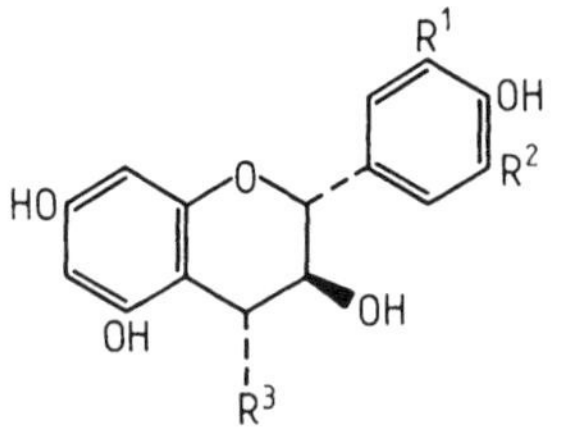 b 4-6; 13-15 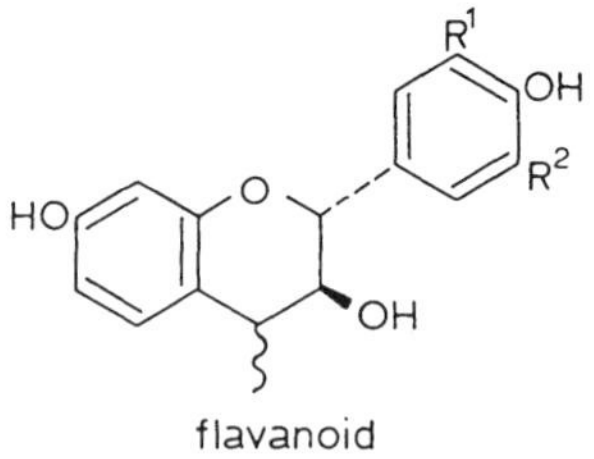 c 7; 8; 9 flavanoid

The Type 1 proanthocyanidins are distributed almost ubiquitously in the woody plants, whereas Type 2 proanthocyanidins are confined to certain families in the Leguminosae and Anacardiaceae (see Sect. 7.7.3.1), often co-existing with Type 1 proanthocyanidins either in the same or different organs of the plant. However, the Type 2 tannins are of pre-eminent importance commercially as currently the two most important sources of condensed tannins for industrial applications are wattle (*Acacia mearnsii*) bark and quebracho (*Schinopsis* spp.) wood, which are both of this type (Chap. 10.3). Our current knowledge of these tannins is almost entirely due to the efforts of David Roux and his colleagues over the past three decades.

The nomencleature for proanthocyanidin oligomers, described in Sect. 7.6.3.3, may be extended to the proanthocyanidin polymers (condensed tannins), which are separated from the oligomers somewhat artificially, for the purposes of this chapter.

7.7.2.1 Isolation and Purification

The presence of proanthocyanidins in plant tissue may be readily detected by hydrolysis with hot alcoholic acid (108) to produce the anthocyanidin pigment (e.g. cyanidin) that may be positively identified by paper chromatography or cellulose TLC in Forestal solvent (HOAc:H_2O:HCl, 30:10:3 v/v/v), or by a colorimetric method using vanillin and concentrated hydrochloric acid (17). The latter is very sensitive.

Condensed tannins have been extracted from plant material with a variety of polar solvents including water (cold or hot, the latter especially in industrial situations), acetone, methanol, and perhaps most satisfactorily with acetone-water mixtures, the latter solvent being able to break hydrogen-bonding between tannins and plant proteins and polysaccharides. Further purification and separation from non-tannin materials has been achieved by adsorption on hide powder or fractionation on paper in dilute aqueous acetic acid, and more recently by chromatography on Sephadex or polystyrene gels.

The concentrations of condensed tannin in solution has been variously estimated by ultraviolet absorbance at 280 nm (after suitable fractionation to eliminate other phenols), by hydrolysis to anthocyanidins, or formation of the vanillin condensation product and colorimetric estimation, or by the formation of colored complexes with ferric salts.

The approach we have found to be most useful for the isolation and purification of Type 1 or 2 proanthocyanidin polymers is extraction from plant tissue with acetone-water mixtures (25, 37), separation of the monomers and lower oligomers from the resulting aqueous solution with ethyl acetate, adsorption on Sephadex LH-20 of the aqueous solution diluted with an equal volume of methanol and washing with the same solvent to eliminate impurities. The polymer is then displaced with acetone-water to yield freeze-dried analytically pure material (25). In the case of procyanidins, the ethyl acetate fraction contains monomers (catechin or epicatechin), dimers, trimers, and some tetramers, whereas the LH-20 fractions contain tetramers, on up to genuinely polymeric species.

7.7.2.2 Elucidation of the Structure of Type 1 Proanthocyanidin Polymers

Once a Type 1 proanthocyanidin polymer has been isolated by the above methods, the purity of a preparation may be gauged by the vanillin-hydrochloric acid assay (17, 25) or by the $E^{1\%}$ value in the λ_{max} 270–280 region in water or methanol (25). This procedure has been shown to be effective for the isolation of polymers from a wide range of plant sources (25, 29, 37), and it is usually reasonable to assume that the constitution of the polymer so isolated is an average representation of the condensed tannins as they exist in that particular plant tissue.

The constitution of the polymers isolated in this way may be readily deduced from the ^{13}C NMR spectrum in acetone-water (25) or methanol (126), or by a combination of chemical degradation and chiroptical methods (25). The most convenient of these is a ^{13}C NMR spectrum of the proanthocyanidin, which gives a pattern of signals characteristic of both the average stereochemistry of the heterocyclic ring and B-ring oxidation pattern of the constituent flavanoid units, and also an estimate of the number-average degree of polymerization, P_n (25). The assignments for these characteristic signals are given in Table 7.7.1 for each type of proantocyanidin unit (Structures **1**–**6**), and also the chain terminating flavan-3-ol unit (Structures **10**–**15**). Assignment of the stereochemistry can be made from the chemical shifts of the heterocyclic ring carbons. In a procyanidin polymer, for example, the C-2 signal for an epicatechin-4 unit (**2**) occurs at δ 76.6, whereas that for a catechin-4 unit (**5**) is 7 ppm downfield at δ 83.6. Similarly, the stereochemistry of the chain terminating units may be readily identified: C-2 of an epicatechin unit (**11**) occurs at δ 79.1, whereas that for catechin (**14**) is at δ 81.6 (25, 109). Therefore, the stereochemistry of both the proanthocyanidin monomer units and the flavanoid chain terminating units may be readily assigned. In the case of polymers that contain units with both stereochemical fea-

Table 7.7.1. Characteristic chemical shifts (δ)[a] for proanthocyanidin polymers

Polymer	Unit											
	C2	C3	C4	C4a	C6	C8	C1′	C2′	C3′	C4′	C5′	C6′
Catechin-4	83.4	73.1	38.2	~108	97.5	~108	131.8	116.6	145.3	145.3	116.6	120.7
Epicatechin-4 (low MW polymer)	76.6	72.3	37.0	102.3	97.4	107.2	132.2	115.2	144.9	145.2	116.5	119.3
Epicatechin-4 (high MW polymer)	76.7	72.6	37.2	102.4	97.6	107.4	132.2	115.4	144.8	145.2	116.6	119.3
Gallocatechin-4	83.6	73.3	38.2	~108	97.6	~107	131.2	108.7	146.2	133.7	146.2	108.7
Epigallocatechin-4	76.6	72.6	37.1	102.5	97.5	~107	131.9	107.1	146.3	132.8	146.3	107.1
8-Catechin or -gallocatechin	81.2	67.8	29.2									
8-Epicatechin or -epigallocatechin	79.0	66.2	29.4									

[a] All chemical shifts were obtained at 20 MHz in d_6-acetone/water (1 : 1, v/v) at 30 °C with the most intense d_6-acetone signal referenced at exactly δ 30.3

tures, the ratio of diastereomers in the polymer chain may be estimated from the relative intensities of the C-2 signals (25).

As shown in Table 7.7.1, the various B-ring oxidation patterns also give rise to characteristic patterns of signals. The exact δ values for the same oxidation pattern are slightly dependent on the heterocyclic ring stereochemistry. In the case of a polymer containing both procyanidin and prodelphinidin units, the relative proportions of the two types of unit may be approximately estimated from the ratio of intensities of the signals near δ 145, which are due to the C-3′ and C-4′ of procyanidin and C-3′ and C-5′ of prodelphinidin units. The signals arising from each type of unit are separated by approximately 1 ppm.

The presence of propelargonidin units (**1**, **4**) in a polymer is also very readily detected as the 4′-hydroxy B-ring possesses a two-carbon signal at δ 129 (C-2′ and C-6′), which is quite separate from any signals from procyanidin or prodelphinidin units (29).

Some typical ^{13}C NMR spectra for the polymers are illustrated in Fig. 7.7.1. The contrasting patterns of signals arising from differences in stereochemistry and oxidation pattern between Figs. 7.7.1 a and b are quite marked, and characteristic. Other proanthocyanidin polymer spectra that are encountered are essentially these spectra with signals mixed in various portions according to B-ring oxidation and heterocyclic ring stereochemistry. An example of one such variant is shown in Fig. 7.7.1 c.

The average polymer stereochemistry may also be deduced from measurement of the specific rotation of a solution of the polymer in water (25) at 578 nm. The measurement may also be performed in methanol, methanol-water (1 : 1, v/v), or acetone-water (1 : 1, v/v), which are useful supplementary solvents for less soluble tannins. The value of $[\alpha]_{578}$ for a particular polymer is similar for any of the above solvents (L. J. Porter, unpublished observations). At 578 nm the specific rotation of pure freeze-dried epicatechin or epigallocatechin homopolymers (which

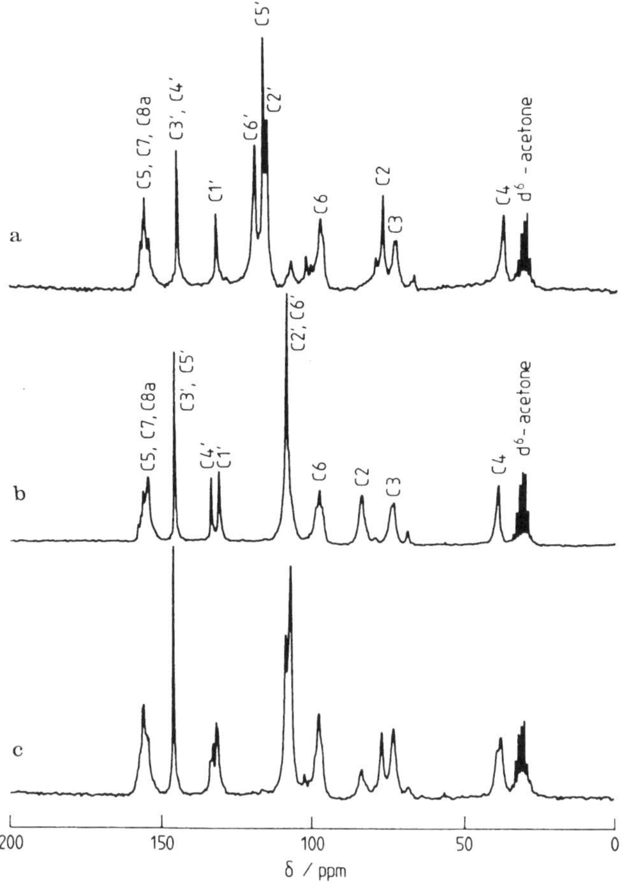

Fig. 7.7.1. ^{13}C NMR spectra of some proanthocyanidin polymers in d_6-acetone/water (1 : 1, v/v): **a.** An epicatechin (procyanidin) homopolymer, one of the most common encountered in nature, from *Chaenomeles chinensis* fruit; **b.** A gallocatechin (prodelphinidin) homopolymer from *Watsonia ardnerei* leaves; **c.** A mixed gallocatechin-4 and epigallocatechin-4 (prodelphinidin) polymer from immature flowers of *Trifolium repens*

contain 2.5 – $3H_2O$ per flavanoid unit) is approximately +160°. The specific rotation of pure catechin or gallocatechin homopolymers is approximately –320°. In the case of polymers with mixed stereochemistry (both 2,3-*cis* and 2,3-*trans* units, Fig. 7.7.1c) the fraction, X, of 2,3-*cis* units (i.e. epicatechin stereochemistry), may be calculated from the formula $X = ([\alpha] + 320)/480$, where $[\alpha]$ is the experimentally determined specific rotation for the polymer at 578 nm (25).

As was first shown by Roux and co-workers, in a detailed study of appropriate model compounds, the sign of the band near 220 nm for the circular dichroism spectra of proanthocyanidin oligomers is determined by the absolute configuration at C-4 and appears to arise largely from interactions between A-ring chromophores (16). Similarly, the sign of an intense couplet near 200 nm is determined by this interaction (5). The optical rotations (25) and circular dichroism spectra (40) of Type 1 proanthocyanidin polymers are qualitatively very similar to those of the simple 4-phloroglucinol derivatives of catechin (**14**), epicatechin (**11**), epigallocatechin (**12**), and gallocatechin (**15**), which show that the B-ring has no contribution to the magnitude of $\Delta\varepsilon$ and α. For the chiroptical properties of proanthocyanidin polymers to obey such a simple relationship to composition, and to be apparently independent of P_n, implies that either the units are behaving as a set of independent oscillators, or that the chains have a highly irregular structure (25); the latter is supported by other evidence, as will be discussed later.

Both the mean B-ring oxidation pattern and stereochemistry of polymers and oligomers have been correlated with infrared bands in the skeletal region (33). This method provides an additional, very useful device for monitoring tannin structure – especially for samples available in only small quantities. It is also very useful for deducing the presence of co-occurring hydrolyzable tannins or 3-*O*-gallate moieties (25).

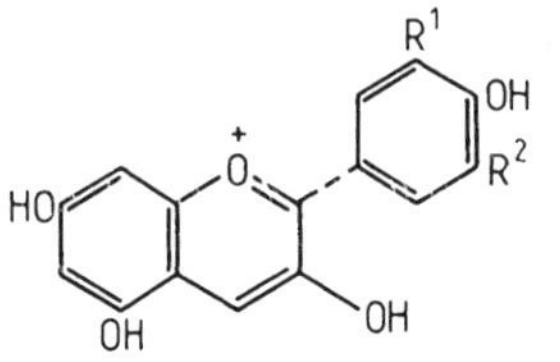

d 16–18

The B-ring oxidation pattern stoichiometry of Type 1 polymers may also be deduced from acid degradation of the polymer and estimation of the ratios of cyanidin (**16**), delphinidin (**17**), and pelargonidin (**18**) by chromatographic separation of the pigments and colorimetric estimation, or alternatively by degradation with phenylmethanethiol and acetic acid and hot ethanol to degrade the polymer to the 4-benzylsulfide derivatives of the proanthocyanidin units plus the flavan-3-ol chain terminating units (Sect. 7.6). The proportion of each product may be estimated approximately from GC analysis of the TMS ether derivates of the phenols (25) or direct HPLC analysis (125). This method is most informative, for not only are the relative proportions of the monomer units measured, but an approximate estimate of P_n may be obtained, which also provides information unobtainable from ^{13}C NMR if a polymer of mixed stereochemistry and oxidation pattern is encountered. Whereas NMR may only estimate average values for these characteristics (25), benzylsulfide degradation gives an analysis of all flavanoid units (25, 126). While the preceding uses of the thiolysis refer to exhaustive degradation with phenylmethanethiol, partial thiolysis may be used to detect the degree of linkage isomerism in polymers (62) and oligomers. Examples of the latter are given in Sect. 7.6. To illustrate the efficacy of the various methods of determining polymer structure, typical analyses for selected polymers are given in Table 7.7.2.

The above discussion showed that condensed tannins may, in general, be readily isolated and purified, and indicated the techniques from which the stoichiometry and structure of the constituent units may be deduced, but gave no

Table 7.7.2. Analysis of some typical proanthocyanidin polymers

Genus, species	Organ	$[\alpha]_{578}$ [a]	Ratio 2,3-*cis*:2,3-*trans*			Ratio PC:PD			Terminal [b] unit	Reference(s)
			From rotation	From NMR	From thiolysis	From acid hyd	From NMR	From thiolysis		
Abies firma	bark	–	–	–	39:61 [c]	70:30	–	–	c:e	126
		–	–	–	47:53 [d]	–	–	–	89:11	
Chaenomeles chinensis	fruit	+144	97:3	100:0	96:4	96:4	100:0	100:0	e, 100	37
Grevillea robusta	leaf	+35	74:26	67:33	77:23	45:55	39:61	38:62	–	25
Onobrychis viciifolia	leaf	+112	90:10	88:12	–	19:81	23:77	18:82	–	25
Pinus radiata	leaf	+125	93:7	90:10	–	24:76	18:82	–	c ~ 100	37
	phloem	+59	79:21	76:24	–	50:50	52:48	51:49	c ~ 100	25, 37
Ribes sanguineum	leaf	−230	19:81	17:83	–	13:87	9:91	10:90	egc ~ 100	25
Vaccinium corymbosum	fruit	+144	97:3	100:0	–	100:0	100:0	100:0	e, 100	37
Watsonia pyramidata	leaf	−295	5:95	0:100	–	0:100	0:100	–	g, 100	29

[a] Specific rotation in water
[b] c = catechin; e = epicatechin; egc = epigallocatechin; g = gallocatechin
[c] Procyanidin units
[d] Prodelphinidin units

indication of the nature of the interflavanoid linkages. It has been established that proanthocyanidin oligomers (Sect. 7.6) are linked by acid labile 4→6 and 4→8 bonds (55, 57, 62). It may also be readily deduced that, at least the great majority of interflavanoid linkages in the Type 1 polymers must be of the same kind as in the oligomers because of both the similarity in chemical reactivity of the bonds, and of chemical shifts at C-4 (δ 37–38) and C-6/C-8 (δ 106–108) in the ^{13}C NMR spectra, between proanthocyanidin oligomers and polymers (25, 109). Furthermore, it has been observed in naturally occurring procyanidin dimers and trimers that 4→8 linkages occur approximately three times more frequently than 4→6 linkages (Sect. 7.6.3.3.1). In contrast to this, it has been suggested that the polymers are stereoregular, and that they may possess virtually all 4→8 linkages, on the basis of molecular models and by analogy with other biopolymers (57). More recent evidence from high-field ^{13}C NMR shows, at least for the polymer system studied, that this is unlikely to be true, and that polymers have interflavanoid bond heterogeneity similar in proportion to oligomers (90).

The system studied was an epicatechin homopolymer from the fruits of *Chaenomeles chinensis* (90). The polymer may consist, theoretically, of four types of unit (Fig. 7.7.2): T ('top') units, in which an epicatechin-4 unit is linked through C-4 to C-6 or C-8 of a neighboring unit; M ('middle') units, which are doubly-linked, both through C-4, and C-6 or C-8 to two neighboring units; J ('branching' or 'junction') units, which are triply-linked through C-4, C-6 and C-8 to neighboring units; and B ('bottom') units, which are singly linked through C-6 or C-8 to neighboring units, and C-4 is unsubstituted. The *Chaenomeles* polymer has an average P_n of 14 units per chain. Chemical shifts could be identified in the ^{13}C NMR spectrum of the polymer characteristic of the various types of possible interflavanoid linkage [T(4β→6)M, T(4β→8)M, etc.]. This analysis showed that both types of bond were apparently represented throughout the chain. The most straightforward example of this was given by M(4β→6)B and M(4β→8)B. Here C-3 of a B unit occurs at δ 66.2 if 4→8 linked, and δ 66.8 if 4→6 linked. Both signals were present, and the ratio of intensities of δ 66.2 to δ 66.8 was 4:1, about the value expected.

There was also evidence that J units were present (90). If some of the chains are branched, then the population of T units will be higher than B units (see Fig. 7.7.2). The peak at δ 97.4, which contains the unsubstituted C-6/C-8 resonances, possessed a clear signal at δ 96.1, which model compounds show arises from the C-8 signal of *both* T(4β→8)M and T(4β→6)M species. Computer deconvolution shows that the intensity of the δ 96.1 peak, compared with the total C-6/C-8 peak, was approximately 20% to 30% greater than that predicted if the polymer consisted entirely of linear chains. This implied that there is approximately one branch point (J unit) per two polymer chains. These results imply a high degree of irregularity in the proanthocyanidin polymer chains, which is consistent with chiroptical observations (see earlier in this section).

It may be noted from Table 7.7.2 that a number of proanthocyanidin polymers contain units with the same B-ring oxidation but differing stereochemistry (e.g., catechin-4 and epicatechin-4), and others with common stereochemistry but differing B-ring oxidation (e.g., epicatechin-4 and epigallocatechin-4), whereas still others may have both stereochemistry and B-ring hydroxylation mixed. These ob-

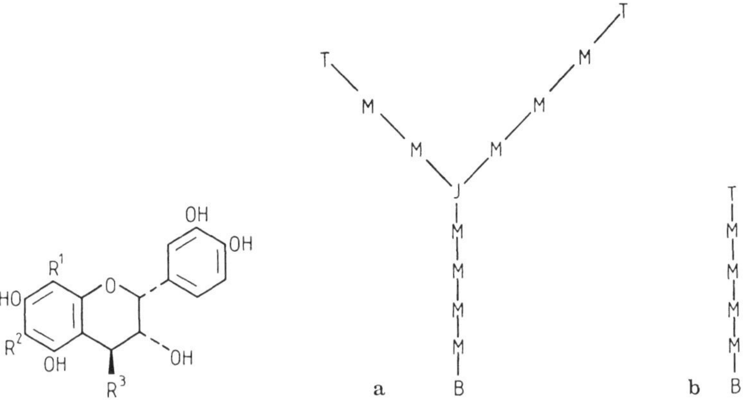

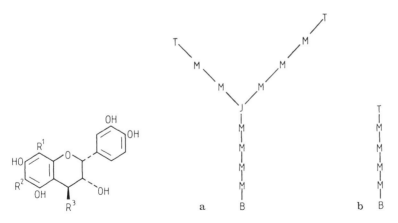

Fig. 7.7.2. Structure of an average molecule of the procyanidin polymer from *Chaenomeles chinensis* fruits as deduced from high-field ^{13}C NMR. **a** = branched chain; **b** = linear chain. T unit: $R^1 = R^2 = H$; R^3 = flavanoid. M unit: R^1 or $R^2 = H$; R^1 or $R^2 = R^3$ = flavanoid; J unit: $R^1 = R^2 = R^3$ = flavanoid; B unit: R^1 or R^2 = flavanoid; R^1 or $R^2 = R^3 = H$

servations may be interpreted two ways: Either the monomer units of differing structures occur in the same polymer chain, or else the tannin consists of mixtures of chains each containing one type of monomer. There is some evidence that the former type may prevail in nature.

Hordeum vulgare contains almost exclusively dimeric and trimeric proanthocyanidins containing only gallocatechin-4, catechin-4, and catechin units. Several groups (Sect. 7.6.3.3.1) have independently corroborated that the major trimers consist of all possible combinations of these units (i.e. gallocatechin-(4α→8)-gallocatechin-(4α→8)-catechin, gallocatechin-(4α→8)-catechin-(4α→8)-catechin, catechin-(4α→8)-gallocatechin-(4α→8)-catechin, and catechin (4α→8)-catechin-(4α→8)-catechin). Another example is the trimeric procyanidins from *Pinus taeda* phloem. The major trimers are linkage isomers of (epicatechin-4) (epicatechin-4) catechin (62). Accompanying these trimers are small amounts of catechin-(4α→8)-epicatechin-(4β→8)-catechin and epicatechin-(4β→8)-catechin-(4α→8)-catechin (Foo, Hemingway and Porter, unpublished observations). The accompanying polymer was observed to contain a small proportion of catechin-4 units (66).

Further proanthocyanidin metabolites that occur sporadically in the plant kingdom are the 3-*O*-gallates. Until recently, they were known only as acylated derivates of epicatechin or epigallocatechin, but several instances are now known of acylation of proanthocyanidin units as well (Sect. 7.7.4.1).

It was considered that proanthocyanidins may be unglycosylated as only a handful of *O*-glycosylated flavan-3-ols were known (57). Now a considerable number of *O*- or *C*-glycosylated flavan-3-ols have been isolated, and these have been accompanied by *C*- or *O*-glycosylated dimeric procyanidins in several instances (Sect. 7.6.3.1.1).

Three examples of *O*-glycosylated procyanidin polymers have also been reported (107). Unripe fruits of *Cydonia oblonga* contain an epicatechin homopolymer, but in ripe fruits, where the concentration of UDP-glucose will be high, ^{13}C NMR shows that the polymer also contains 3-*O*-β-D-glucosylpyranoside moieties largely bound to the terminal (B) epicatechin units. The other examples were from *Pinus brutia* and *Picea abies* bark, which contain glucosylated procyanidins with the sugar unit probably largely attached to the phenolic hydroxyl groups (107).

7.7.2.3 Structure of Type 2 Proanthocyanidin Polymers

Our knowledge of the structure of Type 2-condensed tannins rests largely on detailed studies of the related oligomers by Roux and his co-workers. (This work is outlined in Sect. 7.6.3.3.1.) These studies have largely concentrated on the commercially important wattle (*Acacia mearnsii*) and quebracho (*Schipnosis* spp.) tannins and show that the oligomers consist of 5-deoxyproanthocyanidin units (Structures **7**–**9**) linked to phloroglucinol pattern A-ring terminal units.

Earlier workers, especially those at the Leather Industries Research Institute, Grahamstown, South Africa, obtained evidence for polymers in these tannin sources, but their studies were largely limited to those species soluble in water and, therefore, probably underestimated the molecular weight of tannins in quebracho and wattle. Recently Foo (34) isolated the proanthocyanidins of *Acacia baileyana* heartwood and found them to consist of mixed profisetinidins (Type 2 polymers) and procyanidins and prodelphinidins (Type 1) polymers. Foo was able to separate a procyanidin-prodelphinidin polymer from the remaining proanthocyanidins on the basis of its greater water solubility. The remainder of the proanthocyanidins were subjected to Sephadex LH-20 chromatography and a profisetinidin polymer was separated from a profisetinidin-procyanidin-prodelphinidin polymer by elution with methanol-water. These results show that profisetinidins are significantly less polar than the Type 1 polymers.

These results also imply that the *Acacia baileyana* polymer is of relatively low molecular weight, as significant proportions of Type 2 polymer is retained on Sephadex LH-20 after elution with methanol-water, when wattle or quebracho tannin is subjected to similar separations (L. J. Porter, unpublished observations).

The ^{13}C NMR spectra of profisetinidin polymers are quite distinct from procyanidin polymers in several respects (34). In particular, C-4 for a fisetinidol-4 unit occurs at δ 130 and δ 110 respectively, well upfield and downfield respectively from the corresponding resonances, δ 155 and δ 97, in catechin-4 units (34). ^{13}C NMR is, however, less useful for assigning relative stereochemistry than for Type 1 polymers. Type 2 oligomers have a low degree of interflavanoid bond stereospecificity (15, 119), in contrast to Type 1 polymers (i.e. 4α or 4β), and this leads to extra spectral complexities. In addition, the B units of quebracho and wattle tannins have a phloroglucinol A-ring oxidation pattern, whereas the T and M units are of the resorcinol A-ring pattern (15).

A further feature that makes Type 2 polymers harder to characterize is the relatively high stability of the interflavanoid bond to acid cleavage compared with C-ring opening reactions (Sect. 7.6.3.3.2). Thus anthocyanidins or benzylsulfide derivatives are formed with difficulty and lower yield than occurs for the Type 1 polymers, and C-ring rearranged products, or 'phlobatannins', are formed in significant yields (150).

Detailed studies of Type 2 proanthocyanidin polymers remain to be completed.

7.7.2.4 Molecular Weight Distribution

The Type 1-condensed tannins consist of monomer units of molecular weight 272 (propelargonidin units: **1**, **4**), or 288 (procyanidin units: **2**, **5**), or 304 (prodelphinidin units: **3**, **6**), respectively (or 300 in rounded terms). The units, therefore, possess relatively large individual molecular weights compared with most natural or synthetic polymers.

Whereas ^{13}C NMR gives an often quite accurate value of P_n for an individual condensed tannin preparation, it gives no idea of the range of molecular weight (MW), or dispersivity, of the polymer chains. Typically, all synthetic and most natural polymers do not consist of a single molecular species, but contain chains of varying length – that is, they are polydisperse. To get an estimate of the dispersivity one must make an estimate of the MW distribution by a technique such as gel-permeation chromatography (GPC), or make estimates of average MW that are dependent on both size and average chain length.

GPC has not so far been successfully applied directly to the phenolic polymers because of their highly polar and strongly hydrogen bonding character. Condensed tannins will dissolve only in very polar solvents (Sect. 7.7.2.3) and the best defined GPC supports are based on polystyrene or polystyrene bonded to silica. However, several groups have derivatized condensed tannins by methylation (66) or peracetylation (36, 124, 142, 148) and performed GPC analyses on the relatively less polar derivatives in chloroform or tetrahydrofuran. The usual acetylation method is with pyridine and acetic anhydride.

All published work on the GPC of derivatized tannins has been performed on banks of μ Styragel columns, using derivatized phenols and flavonoids as lower MW standards and narrow MW distribution polystyrene as higher MW standards, using single peak position calibration. Implicit in this assumption is that the hydrodynamic volumes ($[\eta]M$) of polystyrene and the derivatized tannin are equal. This assumption may not be unreasonable, as the ratio of MW/chain length (in Å) is 83 for polystyrene and 97 for peracetylated condensed tannins (based on Dreiding models) – a 20% difference – and implying, provided that the majority of tannin chains are unbranched, that GPC will tend to underestimate MW using this method of calibration. Theoretical (MM2) calculations based on catechin or epicatechin homopolymers also support the view that procyanidin chains are more compact than polystyrene – even without branching (144). The other implicit assumption is that $[\eta]$ will vary in a similar way with chain length for the two polymers.

The few data (105) available testing the validity of these assumptions are summarized in Table 7.7.3. While some values of P_n and P_w (weight-average degree of polymerization) obtained by GPC agree reasonably well with direct measurement of P_n (by NMR and vapour pressure osmometry, VPO) and P_w (by low-

Table 7.7.3. Degree of polymerization of some proanthocyanidin polymers

Genus, species	Organ	Phenols[a]			Acetates[b]		
		P_n (NMR)	P_n (VPO)	P_w (LALLS)	P_n (VPO)	P_n (GPC)	P_w (GPC)
Aesculus × carnea	fruit	–	11.0	26.9	8.6	8.4	17.0
Chaenomeles chinensis	fruit	12.2	14.3	38.5	11.4	9.3	24.2
	fruit[c]	17.2	17.5	42.4	–	–	–
Cydonia oblonga	fruit	14.2	17.8	46.2	15.0	9.9	42.9
Crataegus oxyacantha	fruit	4.5	–	13.4	–	5.4	7.5
Photinia glabrescens	leaf	~12	–	66	–	10.8	24.0
Trifolium repens	flower	6.0	5.5	–	6.7	5.5	18.3
Watsonia ardnerei	leaf	7.5	8.0	21.4	–	5.0	12.0

[a] Measurements directly on phenolic polymers in methanol. P_n = number average degree of polymerization; P_w = weight average degree of polymerization; NMR, implies MW assessed by integration of ^{13}C NMR; VPO = vapour pressure osmometry; LALLS = low angle laser light scattering (data mostly taken from ref. [105])

[b] Measurements on peracetate derivatives synthesized by reacting the phenols with pyridine and acetic anhydride. GPC = gel permeation chromatography on μ Styragel columns

[c] Different sample harvested following year

angle laser light scattering, LALLS) (105), it does show that GPC tends to underestimate MW, as predicted above. These difficulties can be overcome by either collection of full viscosity-MW data on several tannin acetates to construct a full universal calibration curve, or by calibration with two or three broad-MW dispersion tannin acetate standards of accurately known P_n and P_w. Either or both approaches will undoubtedly be applied in the future.

However, the extant data are certainly sufficient to establish the following facts about proanthocyanidin polymers:

1) They are polydisperse, the actual range of molecular sizes being dependent on the source of plant material. The observed range of sizes varies from the procyanidins of *Rubus idaeus* leaves, which are dimeric, and those of *Hordeum vulgare* ears which are no larger than tetrameric, to other sources where there are polyflavanoid chains containing hundreds of units.

2) While some plant sources contain condensed tannins with a rather symmetrical MW distribution, others display asymmetric distributions, often with higher and lower MW maxima, suggesting perhaps two metabolic pools of proanthocyanidins in the extracted plant tissue.

3) Where relatively symmetrical MW distributions are encountered, dispersivities (P_w/P_n) of 1.5 to 3 are usually observed (105), whereas asymmetric distributions can often lead to high dispersivities – up to 44 has been reported (147).

If the growth of proanthocyanidin chains is a random process leading to linear chains, then classical theory predicts that P_w/P_n will approach a maximum of 2.0. On the other hand if chain branching occurs, achieved by double-substitution of a flavanoid A-ring (Fig. 7.7.2), then this will result in values of $P_w/P_n > 2$. The extent to which the dispersivity may exceed 2 for a particular P_n may be predicted from the classical Flory and Stockmayer theory (82).

This approach has been applied to the *Chaenomeles* polymer (82). The theory can be applied with reasonable confidence in this case as GPC shows that it possesses a symmetrical MW distribution. Measurements of P_n and P_w (Table 7.7.2) shows that the dispersivity of the polymer is 2.6. The Flory-Stockmayer theory predicts that for the chain length of this polymer the dispersivity implies that 2% of the units are branched (J units), which means that there will be 25% more T units than B units (Fig. 7.7.2), which is corroborated reasonably well by the experimental NMR value of 20% – 30% (see Sect. 7.7.2.1). Therefore, two experiments support the fact that proanthocyanidin polymer chains are branched, but that the frequency of branching is low.

The low incidence of branching apparently occurring in the phloroglucinol-pattern A-ring proanthocyanidins contrasts with the profisetinidins and prorobinetinidins found in quebracho and wattle tannins. Here the B unit is catechin, and because of the much greater nucleophilicity of the C-6 and C-8 of the phloroglucinol A-ring of catechin compared with that of the resorcinol A-ring of the proanthocyanidin units, a doubly-substituted ('angular') catechin species is the predominant trimer (15). Thus virtually 100% of the wattle tannin chains based on 5-deoxy units are branched (Sect. 7.6.3.3.1).

7.7.2.5 Conformation and Solution Properties

Porter and co-workers (111) have considered the factors that contribute to the equilibrium conformation of the heterocyclic (or C-) ring of flavan-3-ols and shown that it may be described by the equilibrium:

C(2) – Sofa		C(2) – Sofa
⇃↾		⇃↾
Half-chair	⇌ Boat ⇌	Half-chair
⇃↾		⇃↾
C(3) – Sofa		C(3) – Sofa
(E-conformers)		(A-conformers)

– where E- and A-conformers are those with the B-ring equatorial or axial respectively. The boat conformation is the high-energy transition state for the interconversion of E- and A-conformers, and the energy barrier is $5-6\ \text{kcal}\cdot\text{mole}^{-1}$ in enthalpy units. An unequal conformational energy for the E- and A-conformers is manifested by an unequal population of the two states, the one with lower energy being populated to a greater extent.

Low temperature ^1H NMR studies (111) showed that the E:A ratio for catechin (**14**) and epicatechin (**11**) were 62:38 and 86:14 respectively. Acetylation of the 3-hydroxy group stabilised the A-conformation of catechin and changed the ratio to 48:52.

Substitution at C-4 by a hydroxy or aryl substituent strongly favors the E-conformation (i.e. as in catechin-4 or epicatechin-4 units) due to minimization of 1,3-diaxial interactions and the *pseudo*-allylic or A(1, 3)-strain effect (111). It was also concluded that the favored E-conformers of catechin-4 and epicatechin-4 units would adopt C(2)-sofa and half-chair conformation, respectively as minimum energy species.

The stereochemistry of proanthocyanidins is further complicated by atropisomerism. Two such isomers occur for a proanthocyanidin dimer. For example in Figure 7.7.3, a dimer with an epicatechin-4 T unit will be 4β linked and adopt the two conformations **19** and **20**, whereas a dimer with a catechin-4 T unit will be 4α linked and adopt the conformations **21** and **22**. The atropisomers are manifested in ^1H NMR spectra of dimers by many of the resonances occurring as doublets of unequal intensity (32, 38). The ratio of atropisomers may also be estimated from fluorescence decay experiments (12).

On the assumption that catechin-4α and epicatechin-4β units will adopt a preferred conformation (**19** and **21** respectively) in a homopolymer chain, and that all the units will be 4→8 linked, Haslam and co-workers showed, on the basis of molecular models, that the chain will adopt a thread-like structure in which the central core is composed of the A-ring and C-4 of the repeating unit, with the B-rings projecting laterally from this core (i.e. E-conformers), their arrangement describing a regular helical conformation (32). Catechin homopolymers will possess a right-hand helical screw, whereas epicatechin homopolymers will be left-handed.

Mattice and co-workers (144) investigated this problem using rotational isomeric state theory. The characteristic ratios and components of the persistence

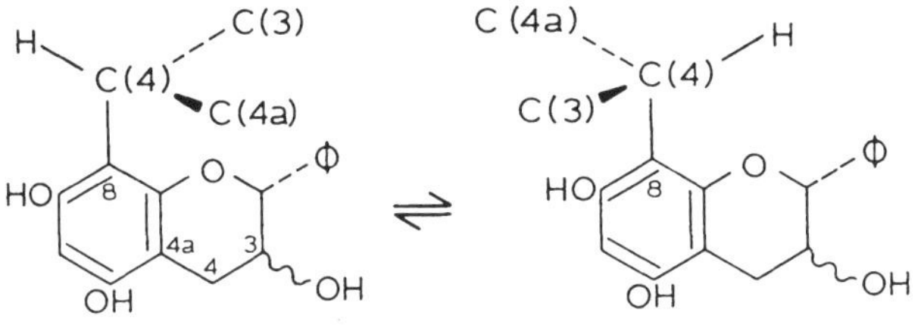

Fig. 7.7.3. Rotational isomers of 4→8 linked procyanidin dimers. ϕ = 3′,4′-Dihydroxyphenyl; H, C(3), C(4) and C(4a) are the atoms of the upper procyanidin unit linked, or adjacent to, the 4→8 interflavonoid linkage. Structures **19** and **20** are for a dimer with an upper epicatechin-4 unit, and structures **21** and **22** for a dimer with an upper catechin-4 unit

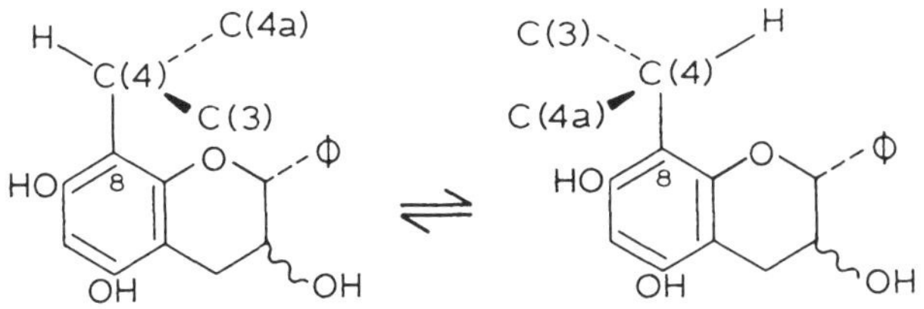

vector (i.e., degree of regularity) were evaluated for catechin or epicatechin homopolymers using structural information obtained from MM2 calculations for dimers of these molecules. These calculations showed that 4→8 linked homopolymers form random coils with unperturbed dimensions smaller than those of atactic polystyrene (agreeing with GPC results; see Sect. 7.7.2.2) if the relative population of the two rotational isomers were assigned as indicated by fluorescence decay measurements.

Current evidence also supports the view that proanthocyanidin homopolymers do not possess such regiospecificity of interflavanoid bonds, and maintain a similar ratio of 4→8/4→6 units to that observed in naturally occurring trimers, approximately 3 to 4: 1. Moreover, a proportion of chains will be branched (Sect. 7.7.2.1).

The effect of introduction of a 4→6 linkage into a 4→8 linked chain is to produce a discontinuity (or bend) in the chain. Molecular models show that this introduces increased flexibility to the chain. The effect of branching is to increase considerably the polymer-solvent associated volume of the polymer for a particular chain length, compared with a linear chain. The overall consequence of linkage isomerism and branching is to produce an average conformation in solution that is rather more globular than rod-like.

The condensed tannins are polyphenolic with a pK_1 of ~8.6, so that a 10% w/v solution in water would be expected to have a pH ~ 4.5. Proanthocyanidins may be solubilized by the addition of base through formation of phenoxide ions. Extraction of tannin sources with either sodium carbonate or sodium hydroxide

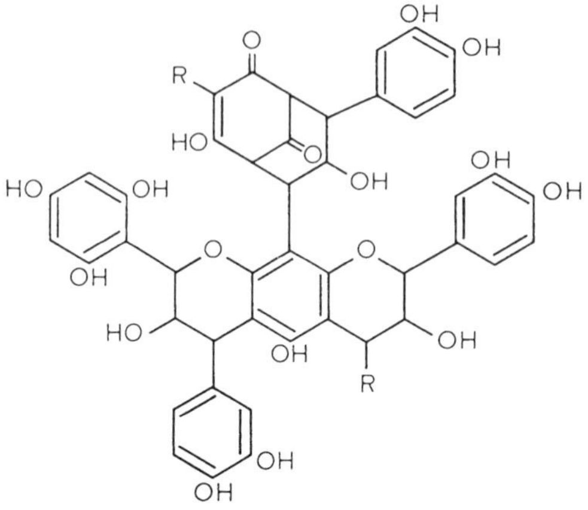

Fig. 7.7.4. Representative structure of a procyanidin treated with base. R = H or other flavanoid fragments

has been utilized commercially (Sects. 10.3.3 and 10.3.4). However, Laks and Hemingway (72) have shown that this cannot be achieved without extensive rearrangement of the A- and C- rings to form structures such as illustrated in Fig. 7.7.4. This occurs rapidly even in mildly alkaline (pH 9.0) solutions (Sect. 7.6.3.3.2).

It must also be noted that at the natural pH of proanthocyanidin solutions (pH ≈ 4.5), the interflavanoid bond of Type 1 proanthocyanidins is sufficiently labile to undergo acid-catalyzed cleavage at ambient temperatures (11, 63). Solutions of proanthocyanidins will, therefore, gradually undergo disproportionation. This stresses the need for proanthocyanidins to be isolated at relatively low temperatures and with reasonable speed for the final preparation to reflect accurately the composition in the original plant cells.

Purified tannins are usually soluble in water. However, as is usual with polymers, solubility decreases with increasing molecular weight. The optimum solvent for condensed tannins appears to be acetone-water (37), which is a hydrogen-bond breaking solvent. It enables molecular-sieving to be achieved on Dextran gels (110) by preventing self-association and hydrogen-bond formation between the tannin and the gel. It also enables solubilization of condensed tannins of relatively high MW. The optimum solvent stoichiometry on a solubility basis has been established to be 55:45 v/v acetone-water (110).

The effect of MW on solubility has been demonstrated for condensed tannins. Attempted solubilization of purified tannins from *Aesculus hippocastanum* fruits and *Vicia sativa* leaves by acetone, a moderately poor solvent for tannins, led to soluble and insoluble fractions (L. J. Porter and L. Y. Foo, unpublished observations). GPC of the peracetylated fractions showed clearly that the MW and dispersivity of the insoluble fraction were considerably higher than the soluble fraction. The point is also illustrated for the lower MW cider procyanidins which have a decreasing partition coefficient between ethyl acetate and water with increasing MW (73). A further example was the isolation of ethyl acetate, water, and acetone-water soluble polymers from *Pinus taeda* bark, which had corresponding increases in MW (66).

It has been observed many times that aqueous condensed tannin extracts, especially from pine barks, sometimes display very high viscosities. This is usually attributed to the high MW of the tannin species. Direct measurement with purified condensed tannin from *Pinus radiata* inner bark in water at 20 °C showed that the polymer had a low intrinsic viscosity ($[\eta] = 3.5$), which could not account for such behavior (L. J. Porter, unpublished observations).

The viscosity problems are usually encountered with aqueous extracts, usually performed under mildly alkaline conditions. The high observed viscosities could arise from either or both of two factors: Such conditions will solubilize pectins and hemicelluloses, which, together with the complexed tannin, will have a relatively high viscosity; also, many pine barks contain large amounts of ill-defined polymeric material that is probably derived by secondary processes from the original tannin. This material is soluble in methanol, and contains phenolic material with an apparent MW of $\approx 10^6$ (145), which is highly pigmented. Extractive free hemlock bark has also been shown to still contain tannin (128). These materials would also be expected to produce a high viscosity (see also Sect. 7.7.4.2).

7.7.2.6 Complexation

The most notable feature of hydrolyzable and condensed tannins is to bind proteins, carbohydrates, and certain nitrogen-containing organic bases (e.g. caffeine, alkaloids). The phenomenon of protein complexation has been known for centuries and forms the basis for tanning leather with plant extracts (Sect. 10.3.3). Binding with polysaccharides is a less familiar property, but may well be equally important from a biological and commercial standpoint.

Protein binding has been used in various guises as a method for analysis of tannins and tannin-containing extracts. The hide powder method has been used for decades in the leather industry. More recently, methods based on the residual absorbance of unprecipitated hemoglobin (10), radiometric titration with I^{125} labelled bovine serum albumin (BSA) (51), and the ferric chloride induced color of protein-tannin complexes resolubilized in alkali (50) have all been suggested as methods of tannin analysis.

There have been a number of studies of protein-polyphenol and -tannin binding, the results of which have been summarized by Haslam and co-workers (79) as follows:

Polyphenols (tannins) form reversible, and often insoluble, complexes with proteins. The principal mechanisms of polyphenol-protein binding are thought to be: (i) hydrogen bonding; (ii) hydrophobic interactions; (iii) ionic interactions. Our current knowledge of the structure of plant polyphenols is sufficient to eliminate (iii). The thrust of recent studies has been to try and define the relative balance between (i) and (ii) for polyphenol complexation with proteins, carbohydrates, and organic bases (the three phenomena being very closely linked).

The importance of both (i) and (ii) to complexation has been graphically illustrated by two recent studies. Nyman (97) found that protein-proanthocyanidin complexes (from *Pinus sylvestris* needle extracts) could be disassociated by addition of either polyvinylpyrrolidone (which disrupts hydrogen bonds) or the detergent Tween 80 (which disrupts hydrophobic interactions). Similarly, the *soluble* protein-proanthocyanidin complexes were adsorbed on either Polyclar AT (through hydrogen bonding) or Phenyl-Sepharose CL-4B (through hydrophobic interactions). This suggests amphiphilic capacity (simultaneous capacity for hydrophilic and hydrophobic interactions) on the part of the proanthocyanidins (97).

Haslam and co-workers (80) have taken these observations a step further by considering the crystal structures of complexes between caffeine and the polyphenols methyl gallate and potassium chlorogenate. These structures show clearly that the molecules are aligned in the lattice so that a hydrogen-bonded network is formed between the phenolic protons and caffeine nitrogen atoms. Furthermore, the aromatic rings are aligned for maximum overlap (hydrophobic interaction). Such interactions could be clearly discerned in solution by complexation-induced shifts in the ^{1}H NMR spectrum when the spectra of caffeine-polyphenol mixtures were compared with the parent compounds (41).

Further evidence that polyphenol-protein and polyphenol-nitrogen base complexation are due to similar effects was provided by Okuda and co-workers (100). They considered complexation between hemoglobin (a protein) or methylene blue (a nitrogen base) and a wide range of hydrolyzable and condensed tannins and polyphenols, spanning a mass range of 126–3100. The binding coefficients (defined relative to the hydrolyzable tannin geraniin) were positively correlated, with a correlation coefficient of 0.84 for 84 compounds or tannin fractions.

Binding efficiencies are also dependent on the MW of the tannin. Pentagalloyl glucose (MW 940) has the highest affinity, for both bovine serum albumin (79) and hemoglobin (100), of any of the monomeric hydrolyzable tannins (those with a single sugar core) but its affinity is exceeded by some dimeric tannins (79) with MW approximately double pentagalloylglucose. Similarly, proanthocyanidins display a regular increase in binding strength with hemoglobin up to a MW of 2000–2300, or 7–8 units average chain length (100, 113).

Polyphenol-protein affinities are also dependent on protein structure. Peptides with less than six residues bind only weakly (52). Tightly coiled globular proteins have much lower affinities than conformationally loose proteins (52). Maximum binding is apparently achieved with proline-rich proteins due to the formation of stronger hydrogen bonds (52). Affinities are pH dependent and maximum binding usually occurs near the isoelectric point of the protein and dissociation occurs at low pH values (79). In contrast polyphenol-polysaccharide binding is pH independent and comparative binding experiments of polyphenols on dextran gels shows that hydrophobic (non-bonding) interactions are more important than for protein binding (79).

Rahman and Richards (116) have shown that starch-proanthocyanidin complexes are precipitated with lead acetate and are regenerated by EDTA treatment. The complex coprecipitates with ethanol, whereas the proanthocyanidins alone are perfectly soluble in ethanol-water mixtures. Chromatography of the starch-proanthocyanidin complex on Sepharose or Sephadex leads to complete elution of the starch and adsorption of the proanthocyanidin, which is only partly and slowly removed by elution with concentrated urea.

Polysaccharide-proanthocyanidin binding may be very strong, as illustrated by the report of Richards and co-workers (89) of what was considered to be a covalently bound (copolymer) of proanthocyanidin and high MW glycan isolated from *Rhizophora stylosa* leaves. The complex survived several attempts at separation on Sepharose and nylon. The complex was finally split by adsorption on Sephadex LH-60 and repeated centrifugation with water, which removed the polysaccharide and allowed the polymeric proanthocyanidin glucoside to be eluted with methanol (G. N. Richards, personal communication). The proanthocyanidin glucoside was similar to those discussed earlier isolated from conifer barks (Sect. 7.7.2.2).

The *Rhizophora stylosa* polysaccharide-proanthocyanidin complex is apparently similar to the species isolated from *Larix gmelini* bark (127). The complex survived repeated chromatography on Sephadex LH-20 and the ^{13}C NMR contained signals consistent with a procyanidin polymer and polysaccharide being present.

It has been observed repeatedly that significant concentrations of proanthocyanidins remain in exhaustively extracted plant material – especially bark. It seems likely that this is largely due to strong complexes formed between proanthocyanidins and polysaccharides as demonstrated above. These are almost certainly artifacts of extraction processes, which causes mixing of cell products, or a result of post-mortem processes, which lead to cell desiccation and crushing as the bark phloem dies and forms the rhytidome. Current evidence indicates that proanthocyanidins are concentrated in vesicles that migrate and form vacuoles (152) – in line with the general observation that hydrophilic flavonoids accumulate in the vacuole, whereas lipophilic ones are secreted and accumulate in the cuticle on cell wall surfaces (139) – rather than being associated with the cell wall.

One mechanism by which proanthocyanidins and cell wall polysaccharides may become covalently bound (as either an artifactal or senescence process) has been demonstrated whereby C-glycosides may form by an acid-catalyzed reaction between C-1 of a sugar and C-6 or C-8 of a procyanidin unit (86).

7.7.3 Distribution in Plants

7.7.3.1 Chemotaxonomic and Phylogenetic Significance

Proanthocyanidins have been popular taxonomic markers because of their ready detection. Initial observations were often limited to presence/absence, or at the most to reports of the presence of specific proanthocyanidin (oxidation pattern) types (7). More recently the ready ability to isolate and purify tannins has enabled assessment of stereochemistry as an additional classification character (37).

Bate-Smith was the first to note the correlation of 'woodiness' with proanthocyanidins. In his early surveys of leaf extracts, he found, in a sample of about 800 species that included most major plant families, that 61% of the woody families had proanthocyanidins, whereas only 15% of herbaceous families were shown to contain these constituents (7). Bate-Smith (9) later formalized these observations, and his words are worth quoting here: "...but the presence or absence of leucoanthocyanins (denoted a and a_0, respectively) is highly significant taxonomically. Similarly, the presence or absence (denoted b and b_0, respectively) of the *vic*-

trihydroxy substituted flavonoids is highly significant. There is much evidence to support the view that the a and b conditions are primitive to a_0 and b_0, but the condition is also correlated with woodiness. Thus nearly all the predominantly woody families from Casuarinaceae to Ebenaceae are Type a, while nearly all the herbaceous families from Aristolochiaceae to Compositae are a_0. The b and b_0 conditions are not so directly related to woody habit, but are related to situations which are generally regarded as primitive and advanced, respectively; but since the families of predominantly woody plants are also those generally regarded as primitive, there is a tendency for such families to be of the b Type." From this it may be concluded that woody plants are expected to be largely of the ab or ab_0 type.

Another index of advancement is the fact that evolution tends to result in a decrease in the number of free hydroxyl groups on flavonoid molecules, either by loss of functionality or *O*-methylation (54). This is achieved in the proanthocyanidins either by loss of a hydroxyl from the A-ring to form the 5-deoxy flavonoids, or loss of the 3′-hydroxyl from the B-ring to form propelargonidins. Apparently *O*-methylation in these compounds is rare; there are only a few authenticated examples: 4′-*O*-methyl epigallocatechin from the Celastraceae, where it is accompanied by propelargonidins (46, 84), 3′-*O*-methyl epicatechin from *Symplocus uniflora* (143), and a range of *O*-methylated catechin and epicatechin derivatives from *Cinnamomum* and *Lindera* species (87).

The distribution of A- and B-ring deoxy proanthocyanidins are given in Table 7.7.4. It may be seen that the A-ring deoxy proanthocyanidins have a strictly limited distribution, and are confined to the Leguminosae and an Anacardiaceae, respectively. They consist of three types – the proguibourtinidins (**7**), profisetinidins (**8**), and prorobinetinidins (**9**), the profisetinidins being the most common. The genus *Acacia* is the richest source of these flavanoids, where 5-deoxyflavanoids normally occur exclusively in the heartwood, whereas they are often accompanied by procyanidins and prodelphinidins in the bark (140). Many *Acacia* species contain 8-hydroxy or -methoxy proguibourtinidins and profisetinidins. The distribution of the various flavan-3,4-diol oxidation patterns – 7,4′ (guibourtacacidin), 7,3′,4′ (mollisacacidin), 7,8,4′ (teracacidin), and 7,8,3′,4′ (melacacidin) – are valuable taxonomic criteria in the genus *Acacia* and has been used to aid division of this very large genus into sections and subsections (23, 104, 140, 141).

The profisetinidins in the Anacardiaceae are of unusual structure as they consist of flavanoid units with a 2*S* rather than the normal 2*R* configuration (see Sect. 7.6). This situation is rare in the proanthocyanidins and only a few other instances are known: the widespread occurrence of *ent*-epicatechin and *ent*-epicatechin-4 procyanidins in monocotyledonous plants (29), and the presence of *ent*-epicatechin in *Polygonum multiflorum* (93) and *Uncaria gambir* (95). *Ent*-catechin has been isolated from *Polygonum multiflorum* (93) and *Rhaphiolepsis umbellata* (91) – as the 7-*O*-glucoside and 3-*O*-gallate, respectively.

The occurrence of propelargonidins (**1**, **4**) in woody families is more widespread (see Table 7.7.4). These are normally accompanied by procyanidins and sometimes prodelphinidins. A possible tendency is for them to occur in families in the Rosidae group, the 5-deoxy proanthocyanidins occurring exclusively here. Propelargonidins, on current evidence, are absent from the Magnoliidae (the most primitive group), Caryophyllidae, and Hamamelidae.

Table 7.7.4. Occurrence of deoxyflavanoid condensed tannins in woody plants

Structure	Order (Group)[1]	Family	Species	Organ[2]	Reference
Type 1. Deoxy A-ring					
Proguibourtinidins	Fabales (R)	Leguminosae	Many *Acacia* spp.	b + w	140
			Guibourtia spp.	w	122
			Julbernadia globiflora	w	103
Profisetinidins	Fabales (R)	Leguminosae	Many *Acacia* spp.	b + w	140
			Baikiaea plurijuga	w	85
			Colophospermum mopane	w	114
			Gleditsia japonica	w	22
			Guibourtia spp.	w	122
			Neorautanenia amboensis	r	98
			Robinia pseudoacacia	w	121
	Sapindales (R)	Anacardiaceae	*Cotinus coggygria*	w	39
			Rhus lancea	w	31
			Schinopsis spp.	w	120
Prorobinetinidins	Fabales (R)	Leguminosae	Many *Acacia* spp.	b + w	140
			Robinia pseudoacacia	w	121
Type 2. Deoxy B-ring					
Propelargonidins	Dilleniales (D)	Dilleniaceae	*Davilla flexuosa*	l	49
			Doliocarpus guianensis	l	49
			D. macrocarpus	l	49
	Urticales (D)	Moraceae	*Ficus* spp.	b + l	1, 134
		Urticaceae	*Boehmeria biloba*	l	7
			Urtica dioica	l	7
	Violales (D)	Flacourtiaceae	*Casearia esculenta*	r	70
	Ebenales (D)	Sapotaceae	*Achras sapota*	f	81
		Ebenaceae	*Diospyros peregrina*	b	19
	Rosales (R)	Rosaceae	*Crataegus oxyacantha*	f	77
			Dryas octopetala	u	101
			Kerria japonica	l	7
			Malus pumila	l	76
			Neviusia alabamensis	l	7

Table 7.7.4 (continued)

Structure	Order (Group)[1]	Family	Species	Organ[2]	Reference
	Fabales (R)	Leguminosae	*Acacia excelsa*	b	140
			Albizia lebbeck	w	48
			Cassia fistula	b	88
			C. sieberiana	r	102
			Ceratonia siliqua	f	136
			Lotus tenuis	l	–[3]
			Robinia fertilis	l	–[3]
	Myrtales (R)	Rhizophoraceae	*Cassipourea verticillata*	l	7
		Myrtaceae	*Eucalyptus calophylla*	b	42
			E. diversicolor	l	–[3]
			E. maculata	b	83
	Celastrales (R)	Celastraceae	*Maytenus rigida*	b	84
			Prionostema aspera	b	84
	Rhamnales (R)	Vitaceae	*Vitis glauca*	u	118
			Vitis cv. 'Albany surprise'	l	–[3]
	Sapindales (R)	Connaraceae	*Connarus monocarpus*	r+f	3, 117
		Sabiaceae	*Meliosma oldhamii*	l	7
	Geraniales (R)	Balsaminaceae	*Impatiens balsamina*	u	60
	Polygalales (R)	Malphighiaceae	*Malphigia coccigera*	l	7
	Lamiales (A)	Verbenaceae	*Vitex pubescens*	u	60
	Rubiales (A)	Rubiaceae	*Uncaria gambir*	l	95

[1] R = Rosidae; D = Dillenidae; A = Asteridae
[2] b = bark; f = fruit; l = leaf; r = root; u = organ source unknown; w = wood
[3] L.J. Porter, unpublished results

The group for which the most detailed data is currently available for condensed tannin structures is the Coniferae. The data available (largely for bark tannins), where both the stereochemistry and the B-ring oxidation patterns are known, are summarized in Table 7.7.5. In addition, Lebreton's group have completed a survey of the needle tannins from 140 species of gymnosperms, including 130 from the conifers (74). The B-ring oxidation pattern alone is available from this survey. However, these data are sufficient to allow the conclusion that conifer tannins are entirely confined to the procyanidins and prodelphinidins. Comparison of Lebreton's data with those in Table 7.7.5 shows that the incidence and proportion of prodelphinidins is far higher in the needles than in the bark. Thus the results for *P. radiata* (Table 7.7.5) are typical. Even more extreme examples are *P. brutia* and *P. sylvestris*, which contain entirely procyanidins in the bark and almost pure prodelphinidins in the needles.

Table 7.7.5. Condensed tannin structures in the Coniferae

Family Genus, species	Organ	*cis: trans*	PC:PD	P_n	Reference
Araucariaceae					
Agathis australis	leaf	90:10	100:0	7	1
	bark		83:17	4–5	1
Cupressaceae					
Chamaecyparis obtusa	bark	40:60	100:0	7	126
C. pisifera	bark	46:54	100:0	5	126
C. nootkatensis	cone	83:17	80:20	–	1
Juniperus chinensis	bark	28:72	100:0	5	126
Thuja standishii	bark	8:92	100:0	5	126
Thujopsis dolabrata	bark	57:43	100:0	5	126
Podocarpaceae					
Podocarpus macrophyllus	bark	43:57	100:0	9	126
P. totara	leaf	70:30	100:0	8	1
Taxaceae					
Taxus cuspidata	bark	55:45	100:0	6–7	126
Taxodiaceae					
Cryptomeria japonica	bark	22:78	100:0	5	126
Sciadopitys vertillata	bark	84:16	100:0	5	126
Pinaceae					
Abies firma	bark	43:57	70:30	9	126
A. sachalinensis	bark	33:67	71:29	8	126
Larix leptolepis	bark	89:11	100:0	8	126
L. gmelini	bark	95:5	100:0	8	126
Picea glehnii	bark	94:6	98:2	10	126
P. jezoensis	bark	89:11	95:5	10	126
Pseudotsuga japonica	bark	84:16	100:0	8	126
P. menziesii	leaf	100:0	100:0	10	1
	bark	75:25	100:0	–	67
	bark	73:27	100:0	10	1
Tsuga heterophylla	bark	54:46	100:0	6	1
T. sieboldii	bark	89:11	100:0	8	126
Pinus					
Subgenus Haploxylon:					
Group Strobi:					
P. pentaphylla	bark	87:13	100:0	8	126

Table 7.7.5 (continued)

Family Genus, species	Organ	*cis: trans*	PC:PD	P_n	Reference
Subgenus Diploxylon:					
Subsection Pinaster:					
Group Laricones:					
P. densiflora	bark	93:7	100:0	9	126
P. sylvestris	bark	80:20	100:0	7	[1]
	male flower	100:0	67:33	–	47
P. thunbergii	bark	92:8	100:0	9	126
Group Australes:					
P. elliottii	bark	85:15	63:37	8	[1]
P. palustris	bark (inner)[2]	90:10	100:0	9	[1]
	bark (middle)	75:25	80:20	6	[1]
P. taeda	bark	90:10	100:0	6–7	[1]
Group Insignes:					
P. brutia	bark	–	100:0	4	[1,3]
P. contorta	bark	44:56	31:69	9	[1]
P. patula	bark (inner)[2]	75:25	35:65	8	[1]
	bark (middle)	45:55	90:10	8	[1]
P. radiata	leaf	90:10	20:80	9	[1]
	winter bud	88:12	43:57	10	[1]
	male cone	80:20	44:56	8	[1]
	bark (inner)[2]	80:20	50:50	7	[1]
	bark (middle)	42:58	85:15	6	[1]
	bark (outer)	25:75	90:10	6	[1]

[1] L. J. Porter and R. W. Hemingway, unpublished results
[2] Including phloem
[3] A glycoside

Lebreton was able to use the proanthocyanidin data to conclude that the main conifer genera *Abies*, *Picea* and probably *Pinus* evolve from a common stock in which *Picea* is the most evolved, and *Tsuga* seems to converge phytochemically towards *Picea* (74).

Considering further the data in Table 7.7.5, it may be seen that in common with the rest of the plant kingdom the most frequently occurring units have 2,3-*cis* (epicatechin) stereochemistry. Whereas there is a heavy preponderance of procyanidin units in most bark polymers (in contrast to needle polymers, above) prodelphinidins are most common in the genus *Pinus*, especially in the Australis and Insignes groups.

Haslam (57) noted earlier that two distinct and very common patterns may be distinguished in the procyanidins: (i) those based on epicatechin-4 (T and M) units with also an epicatechin terminal (B) unit, and (ii) those based on epicatechin-4 (T and M) with a contrasting catechin terminal (B) unit. Those included in Type (i) were procyanidins from *Malus pumila, Crataegus oxyacantha, Theobroma cacao, Cotoneaster* sp., *Cydonia oblonga, Prunus cerasus*, and *Aesculus hippocastanum.* We have confirmed that this holds true for the procyanidin polymers as well, and others of this type include those from *Chaenomeles chinensis, Persea americana*, and *Actinidia chinensis.*

Those Haslam noted of Type (ii) include *Vitis* spp., *Sorghum* spp., and *Vaccinium oxycoccus*. Further studies on the grape show that the leaves have a Type (i) polymer, and the Type (ii) are widespread in the Leguminosae – in *Vicia sativa, Lotus* spp., *Trifolium arvense*, and *Onobrychis viciifolia* (36). Others include *Vaccinium, Photinia* spp, and *Chaenomeles speciosa* (L. J. Porter, unpublished results) and also apparently is almost the universal type in conifer barks (*Pinus, Podocarpus, Pseudotsuga*) (66, L. J. Porter, unpublished results).

It is evident from the above that these observations do have taxonomic significance. Most of the species containing Type (i) tannins belong to the Rosaceae (*Theobroma, Aesculus, Persea* and *Actinidia* are, of course, remote). The tannins of Type (ii) are largely confined to two groups – the Coniferae and Leguminosae. It is evident from these few results that proanthocyanidins are likely to provide valuable taxonomic data as their chemistry becomes better known.

In summary, it may be concluded that the proanthocyanidins are ubiquitously distributed in the Coniferae, occurring in relatively large amounts in the bark, needles, flowers, and cones. The proanthocyanidins occur in virtually all woody dicotyledonous and monocotyledonous (viz., the palms) species. A-ring deoxy flavanoid proanthocyanidins have phylogenetically a severely circumscribed distribution in the relatively advanced Anacardiaceae and Leguminosae, whereas propelargonidins are widespread and sporadic in their distribution, usually co-occurring with PC and PD polymers, and apparently most common in orders in the group Rosidae.

7.7.3.2 Distribution and Structural Variation within Plants

Condensed tannins are usually concentrated in the epidermal tissue of plants, the leaves, bark, and bark phloem, and in the epidermi of fruit skins and seeds, although vanillin staining shows that tannins also occur in the vascular bundle inclusions in fruits such as *Malus pumila* and *Cydonia oblonga*. Condensed tannins are generally absent from sapwood and heartwood. However, there are many exceptions such as the *Eucalyptus* kinos, which have high concentrations of ellagi- and condensed tannins in the heartwood, and many Leguminous forest trees and especially the arid zone *Acacia* species of Africa and Australia, which all possess dark red heartwoods because of the high concentrations of condensed tannins and associated phenolics.

On a microscopic scale, tannins have been observed to occur in the protoplast. In the latter they often appear as inclusions in cytoplasmic vacuoles. Electron microscopy studies of tannins in cell suspension cultures of white spruce (*Picea glauca*) callus tissue showed that the tannin inclusions originated within cytoplasmic vacuoles, possibly derived from the endoplasmic reticulum, and accumulated in the central vacuole through enlargement and coallescence of those cytoplasmic vacuoles (18). Tannin accumulations in tissue culture cells of slash pine occurred in the smooth endoplasmic reticulum, in vesicles, and within cell vacuoles (4). Generalized deterioration of the cytoplasm and organelles was observed in many cells synthesizing tannin, but this apparently did not occur when they accumulated within vacuoles (4).

The tannin-containing vacuole may become very large and has been observed to occupy over 90% of the volume of *Picea glauca* cells (18) and also of tannin-containing cells in *Pinus strobus* needles (75). Tannins do not appear to be present in chloroplast cells and, in the case of Douglas-fir callus cell suspension cultures, some sort of metabolic balance appears to be struck between the chloroplast and tannin-producing cells. Too rapid production of the latter leads to death of the culture (H. A. Stafford, personal communication). Such specialized tannin cells have also been observed in *Vitis vinifera* phloem tissue and *Lotus pedunculatus* leaf and root tissues. Such a pattern may be the norm in the plant kingdom.

Condensed tannins, when present in a woody plant, may not always exhibit a constant structure throughout the plant. There are a number of examples of this phenomenon known: *Acacia* species commonly elaborate both 5-oxy and -deoxy-flavanoid tannins (for example, profisetinidins and procyanidins) in the wood (34) and bark (140), and exclusively 5-oxy-flavanoids and associated flavanoids (for example, quercetin glycosides) in the leaves (20, 138). Other examples include many *Ribes* species that commonly contain prodelphinidins. These are predominantly of the gallocatechin-4 (**6**) type in the leaves, but with the contrasting epigallocatechin-4 (**3**) stereochemistry in the fruits (37). Further, the tannins of *Pinus radiata* display contrasting structures between the bark in which 2,3-*trans* stereochemistry, catechin-4 (**5**), procyanidin polymers predominate, and tannins in the phloem, needles, male cones, and winter buds, in which a variety of mixed procyanidin and prodelphinidin polymers exist with predominantly 2,3-*cis* stereochemistry (Table 7.7.5).

The above observations tend to suggest that more than one metabolic pool for the production of proanthocyanidins may exist in a plant and these in some cases are localized in different tissues. There is some evidence for this in Stafford's observations on procyanidin biosynthesis in Douglas-fir callus tissue (130, 131). Here quite high concentrations of dimers and trimers based on catechin-4 (**5**) units are found, but the associated polymer is based almost entirely on epicatechin-4 (2) units, and evidence was obtained that the latter are formed much more rapidly than the catechin-4 oligomers (131). Thus the main biosynthetic process appears to be directed towards the synthesis of epicatechin-4 units that are diverted to polymers, and it seems likely that the two types of unit may be derived from distinct metabolic pools.

These observations may explain why, in surveys by Foo and Porter, of a wide range of plants reported to be rich sources of the dimers catechin-(4α→8)-catechin and catechin-(4α→8)-epicatechin, it was observed almost without exception that the polymers were based predominantly on epicatechin-4 units (37).

7.7.4 Metabolism

7.7.4.1 Biosynthesis

The biosynthesis of flavonoids from the malonate/shikimate level to dihydroflavonols is now well established (Fig. 7.7.5). Many of the enzyme systems responsible for the transformations have been isolated and the reactions demonstrated in cell-free media (28).

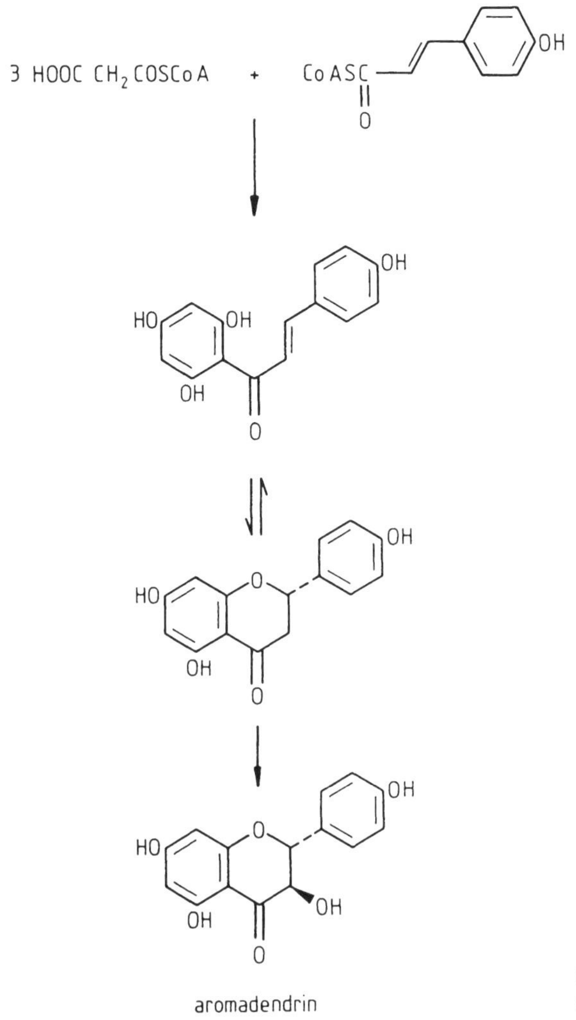

3 HOOC CH_2COSCoA + CoASC(=O)–CH=CH–C$_6$H$_4$OH

aromadendrin

Fig. 7.7.5. Early steps in the biosynthesis of flavanoids

More recently the sequence of steps leading to the proanthocyanidins has been also largely established and the current view of these reactions are summarized in Fig. 7.7.6. There were originally three major objections to a general acceptance of the dihydroflavonol → flavan-3,4-diol → flavan-3-ol sequence to proanthocyanidin formation: (i) whether or not flavan-3,4-diols with a phloroglucinol pattern A-ring were stable chemical entities. The 5-deoxy analogues were well known natural products, but attempts to establish the existence of 5-hydroxy equivalents had failed (64); (ii) doubt as to the existence of 2,3-*cis*-dihydroflavonols. It had been arqued that these were disfavored species on thermodynamic grounds (21), and until recently all naturally occurring dihydroflavonols that had been isolated were the 2,3-*trans* isomer; (iii) the fact that the proanthocyanidin (i.e. T, M, or J) units were often of contrasting stereochemistry and/or oxidation pattern to the terminal (i.e. B) flavan-3-ol units. These objections, together with certain obser-

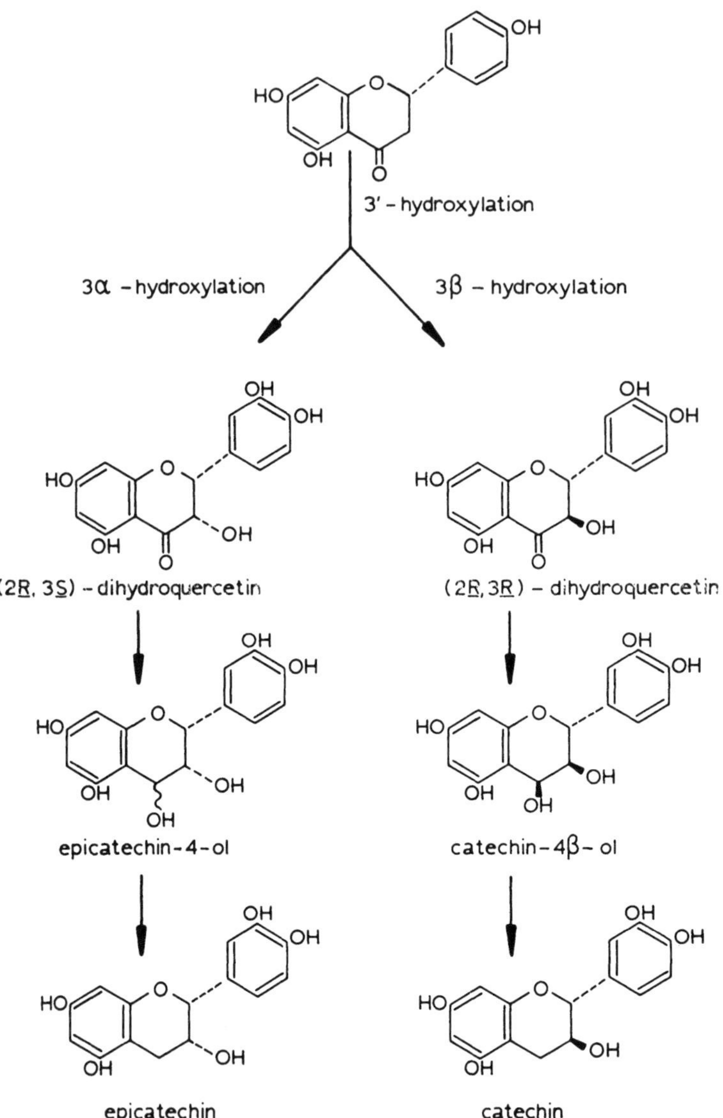

Fig. 7.7.6. Biosynthesis of procyanidins through a flavan-3,4-diol intermediate. The bottom four diol and flavan-3-ol units react together to form procyanidin oligomers and polymers

vations on the patterns of tritium loss from the incorporation of labelled cinnamic acid or *β*-phenylalanine into proanthocyanidins, led Haslam and co-workers to suggest the intermediacy of a symmetrical flav-3-en-3-ol intermediate in procyanidin biosynthesis (64).

Establishment of the sequence in Fig. 7.7.6 as the favored biosynthetic sequence for proanthocyanidins began with the synthesis of (2*R*,3*S*,4*R*)-3,3′,4,4′,5,7-hexahydroxyflavan from (2*R*,3*R*)-taxifolin by sodium borohydride reduction in ethanol (106). The structure of this product was established unequivocally by ^{1}H and ^{13}C NMR (106) and X-ray crystallography (112).

This synthesis was rapidly utilized by three groups who independently established the intermediacy of leucocyanidin in catechin (61, 71, 133) and also anthocyanin (61) formation. In particular, Stafford and co-workers (133) established that cell suspension cultures of *Pseudotsuga menziesii* callus tissue contained a NADPH-dependent reductase capable of converting (2*R*,3*R*)-dihydroquercetin to (2*R*,3*S*,4*S*)-3,3′,4,4′,5,7-hexahydroxyflavan (the 4-epimer of the above synthetic product). Further, in the presence of 2-mercaptoethanol, the extracts formed catechin in addition to the diol. This second step might be expected to involve a stoichiometric reduction by a thiol-containing hydrogen-carrying protein, analogous to the conversion of ribose to deoxyribose.

Similar results were obtained by Kristiansen (71) and Grisebach and co-workers (61) on crude enzyme preparations from *Hordeum vulgare* and *Matthiola incana*, respectively. Kristiansen (71) further showed that the enzymic product was a minor species in the sodium borohydride reduction of taxifolin, the 4*R* and 4*S* products being formed in a ratio of approximately 95 : 5. This enabled the properties of the enzymically and chemically produced 4*S* isomers to be directly compared. Stafford and Lester (132) have also shown that (2*R*,3*R*)-dihydromyricetin may be converted to 3,3′,4,4′,5,5′,7-heptahydroxy-flavan by the *Pseudotsuga menziesii* enzyme complex, thereby establishing the sequence for prodelphinidins as well.

Strong evidence that an equivalent sequence leading to 2,3-*cis* proanthocyanidins has recently been obtained by the isolation of several examples of 2,3-*cis*-dihydroflavonols: (2*R*,3*S*)-3,3′,4′,7,8-pentahydroxyflavanone from *Acacia melanoxylon*, where it is accompanied by flavon-3,4-diols with identical oxidation pattern and C-ring stereochemistry (35); (2*R*,3*S*)-dihydroquercetin 3-*O*-β-D-glucosylpyranoside from *Taxillus kaempferi* (123); 3-*O*-β-D-xylosides of (2*R*,3*S*)- and (2*S*,3*R*)-dihydroquercetin from *Thujopsis dolabrata* (92). The most important observation made in the latter study, however, was that the 2,3-*cis* aglycones resulting from cleavage of the sugar with hesperidinase in water, were perfectly stable at neutral pH and were characterized by ^{1}H NMR and circular dichroism measurements (92). The way is apparently now open to establishing the biosynthetic route to the 2,3-*cis* proanthocyanidins and flavan-3-ols.

The biosynthetic scheme suggested above contains the tacit assumption that the proanthocyanidin units and the fully reduced flavan-3-ols are in a sequential biosynthetic pathway. This view is largely derived from the fact that most workers have focused their attention on the procyanidins and the common oxidation pattern of chain extending and terminating units. However, when proanthocyanidin chemistry is considered as a whole, there are many exceptions to this. Some of the more outstanding examples are:

1) In the Celastraceae the unusual flavan-3-ol 4′-*O*-methyl-epigallocatechin occurs, and this is linked to epiafzelechin-4 units in the dimer (84).

2) In *Pinus radiata* needles the proanthocyanidin polymer consists (90%) of epigallocatechin-4 units, with a low concentration of epicatechin-4 and catechin-4 units, but the chain-terminating unit is exclusively catechin (37).

3) The young twigs of gambir (*Uncaria gambir*) contain catechin and *ent*-epicatechin and complex dimeric ring opened products based on these flavans, all containing dihydroxylated B-rings. In contrast the proanthocyanidins contain propelargonidin units, as evidenced by the isolation of epiafzelechin-(4β→8)-catechin (95).

4) The proanthocyanidins of *Hordeum vulgare* are composed predominantly of gallocatechin-4 units, with lesser amounts of catechin-4 units. The terminal unit is almost exclusively catechin (Sect. 7.6.3.3.2).

Thus there are many examples of a contrasting B-ring oxidation pattern found between the proanthocyanidin units and the terminal flavan-3-ol units. Also, as pointed out by Haslam, there are many examples of contrasting stereochemistry between these units in procyanidins (57), and our work has shown that this is equally true of prodelphinidins (37).

From this, and from the fact that feeding experiments by Haslam's and Stafford's groups have shown that the efficiency of labelling in procyanidin units is much higher than for the flavan-3-ol terminal units of dimers, it is tempting to speculate that the two types of unit may sometimes be derived from quite distinct biosynthetic pathways in the plant, so that these pathways are not always in direct sequence. The oxidation pattern and stereochemistry of the proanthocyanidin and chain-terminating groups will, therefore, not always be the same.

It has also been observed that proanthocyanidins normally accumulate in plant tissue as polymers rather than lower oligomers (dimers or trimers). This implies that the major biosynthetic route is directed toward the synthesis of proanthocyanidin units to form those polymers, in contrast to the formation of lesser amounts of monomeric flavan-3-ols. For instance, the total production of epicatechin, including that present as monomer and as a terminal (B) unit in the polymers, is only 10% that of epicatechin-4 procyanidin units, in the tannins extracted from the unripe fruits of *Chaenomeles chinensis* (L. J. Porter, unpublished observations).

There are now a number of well authenticated examples of the occurrence of gallate esters of monomeric flavan-3-ols and proanthocyanidins. Recently Haslam suggested that, in view of the apparently similar function and properties of hydrolyzable and condensed tannins, there is some form of metabolic link between these two complex polyphenols (58). Further, he observed that where gallate esters of flavan-3-ols occur, they are accompanied by only low concentrations of proanthocyanidins.

Current evidence would lead one to modify this view somewhat. There are now examples of proanthocyanidin dimers and polymers known in which both the proanthocyanidin and flavan-3-ol unit are esterified. The evidence also supports the view that there may be an even closer link between hydrolyzable tannins and the gallate esters of tannins than suggested by Haslam.

Most of the known sources of gallate esters of proanthocyanidins and flavan-3-ols also contain hydrolyzable tannins, or hydrolyzable tannin-like substances. These include the leaves of the grape (*Vitis vinifera*) (37, 68), carob pods (*Ceratonia siliqua*) (58, 136), and the roots of *Polygonum multiflorum* (93), *Rheum palmatum* (96), and *Myrica rubra* (94). Further, the acylation is almost always at the 3-position of the flavanoid unit, whether on T, M, or B units, so that conjugation occurs predominantly with an aliphatic hydroxyl group. It is tempting to suggest that a common galloyl transferase enzyme acylates both D-glucose (in the case of hydrolyzable tannins) and flavanoid units.

7.7.4.2 Seasonal Variation and Fate in Senescent Tissues

A number of useful observations have been made on the variation of extractable tannin concentrations in leaves and fruits. These show that in a species with an annual cycle of growth, tannin levels are generally low in newly formed leaves and concentrations peak in late summer/early autumn. The decrease in the level of extractable tannins is often attributed to a continuous process of increasing condensation. A number of explanations are possible:

It is still not known whether tannins are actively metabolized by the plant or are metabolic endproducts. Current evidence favors the view that flavonoids in general are turned over by the plant (6) and, if this is the case for tannins, then the decrease in concentration may merely be due to a lowering of their rate of production. Haslam has suggested that proanthocyanidin production may be limited by the reductive capacity of a plant (56). Thus, late in the season proanthocyanidin production, a net reductive process, is diverted to anthocyanidin production, which accounts in part for the autumnal reddening of leaves (56). There is some support for this as willow stems are a rich source of procyanidins in summer, but in winter only cyanidin and catechin may be extracted (L. J. Porter, unpublished observations).

Senescence in leaf tissue leads to a general desiccation of the organ and consequent breakdown of the cells. Possibly proanthocyanidins are further polymerized by the native oxidase enzymes of the leaf, and hence rendered insoluble.

Haslam has observed in *Crataegus oxyacantha* and *Rubus idaeus* fruits that once proanthocyanidin production is established there is an approximately steady-state concentration throughout the growth period, so that the amount of proanthocyanidin per fruit remains about constant, but the actual yield per weight of fruit decreases (56). Our own observations on *Crataegus oxyacantha* confirm this and moreover show that the MW of the procyanidin polymer is about the same in the early as in the late season (L. J. Porter, unpublished observations). We have observed in *Vitis* sp. fruit (grape) ripening that there are high tannin levels in the immature grapes, which decrease to an approximately steady-state concentration on ripening (26). There was also some evidence of anthocyanin production at the expense of proanthocyanidin synthesis in very ripe fruit (26).

The fate of tannin in developing bark is another aspect of this topic. Evidence from work on *Pinus radiata*, a species that develops a very thick bark, shows that the polymer formed in the bark phloem is of the normal proanthocyanidin type that can be isolated as an off-white solid; it contains largely about equal numbers of epicatechin-4 and epigallocatechin-4 together with minor amounts of catechin-4 units. During bark development, the tissue redden and the structure of the extractable tannin undergoes progressive changes across the middle and outer bark (Table 7.7.5). It is possible that such structures are a consequence of the native oxidase enzymes of the phloem cells reacting with the tannin so as to degrade or modify the more reactive units with 2,3-*cis* stereochemistry and trihydroxylated B-rings, so that the outer bark contains a polymer with largely catechin-4 units.

Hathway has shown that oak bark contains large concentrations of oxidase enzymes and demonstrated their effect on the flavan-3-ols which occur in oak bark (59). Ahn and Gstirner have reported the presence of oxidatively coupled flavan-3-ol dimers in oak bark, in addition to the normal proanthocyanidin dimers (2). Such products are consistent with the presence of oxidase/peroxidase enzymes. Such secondary processes would explain the very high dispersitivities observed for *Pinus sylvestris* tannins, compared with values for tannins isolated from living plant tissue (R. Marutzky and L. J. Porter, unpublished observations). It would also account for the extremely high MW tannin-like material isolated by Yazaki and Hillis from *Pinus radiata* bark methanol extracts by ultrafiltration (145).

More recently, Roux and co-workers (149) have isolated a complex array of oligomeric polyphenols based on (2*R*,3*S*)-3,3′,4′,7,8-pentahydroxyflavan from the heartwood of mesquite (*Prosopis glandulosa*). These include some oligomers linked by normal proanthocyanidin interflavanoid linkages (i.e. 4→6), but predominantly the form of linkage has arisen from oxidative coupling to form biphenyl and *m*-terphenyl interflavanoid bonds. These are just the sort of bonds expected to be formed in secondary processes in *any* proanthocyanidin polymer, and it is still an open question as to the extent of these processes in outer bark and heartwood.

7.7.5 Role in Plants

As pointed out recently by Zucker (153), the role of plant tannins at the ecological level is almost certainly a mixed function, involving a defense mechanism against living plant enemies and inhibition of decomposition of plant tissue when it becomes litter, either in the soil as roots, or on the soil and leaves. There is, moreover, no compelling evidence that tannins have any role in plant physiological processes.

7.7.5.1 Resistance to Insects

One of these proposed functions is to protect a plant from attack by biological agencies, particularly phytophagous insects. In support of this view is the now classical work of Feeny who showed that a decreasing tendency for lepidopterous larvae to feed on oak leaves as they matured could be correlated with a 10-fold increase in tannin concentration between young and mature leaves (30). Moreover, both Feeny and Chan and co-workers (69) showed that such larvae exhibited growth inhibition when their diet included condensed tannin.

At least five mechanisms have been suggested on how condensed tannins might act as antibiotic agents against insects; these include: reduction in the availability of dietary protein; reduction of enzymatic digestive activity; decreased availability of individual amino acids; direct toxic action; and a repellant effect diminishing palatability (69). However, in a study of the growth inhibitory effect of cotton condensed tannin on *Heliothis zea*, Klocke and Chan (69) interestingly did not find clear evidence for any of the above mechanisms! Along similar lines, Bernays has summarized the extant data available for the interaction between plant herbivores and tannins (13). She points out that the actual number of critical experiments clearly pointing to an antibiotic effect on phytophagous insects are very few, and in some of these cases there are possible alternative explanations. There are in fact a number of examples of tannins, especially hydrolyzable tannins, acting as feeding stimulants. Other work by Bernays has shown that neither hydrolyzable nor condensed tannins have any adverse effects on polyphagous Acridoidea. Recently it has also been established that the tree locust *Anacridium melanorhodon* exhibits improved growth when tannic acid is included in its diet (14).

From these results, Bernays concludes that insect feeding responses range from positive to negative with hydrolyzable tannins, depending on the species, and that condensed tannins show no or negative effects. Tannins also do affect the growth and survival of insects, but there is as yet no unequivocal proof of an antidigestive effect of tannins on insects. Therefore no generalizations may be made in this area, and it remains an almost virgin area of study.

7.7.5.2 Resistance to Decay Fungi

Zucker has recently suggested that tannins are at least to some extent tightly complexed to a variety of target molecules, such as structural protein, non-catalytic protein, starch, pectic substances, and cellulose (153). Most of these constituents are associated with organelles or in the primary and secondary layers of the cell wall, including (in some woody families such as *Acacia, Eucalyptus*, etc.) the wood itself. Zucker (153) suggests that if such associations were relatively irreversible, it would serve two important functions, namely protection of the cell wall or organelle against microbial attack and delay in decomposition of the leaf when it falls from the plant and becomes litter. Synge (135) has even gone so far as to suggest that individual species may actually modify the surrounding humus with their own brand of tannin-protein-carbohydrate mix so as to encourage the spread and well-being of that species – such as observed for the heathers (56).

As Zucker (153) points out, bacterial or fungal attack on pectins and cellulose is initiated by secretion of exo-enzymes that cleave the polysaccharide chain. If tannins form strong and specific complexes with the pectins and cellulose, such enzyme attack will be blocked or considerably slowed.

There is, however, a paucity of examples of specific anti-fungal activity by condensed tannins. There are specific studies of the inhibition of enzymes by condensed tannins (43), and they have been implicated as inhibitors of cellulases, pectinases, and amylases. One of the few examples remains that of 'black heart' disease of *Prunus armeniaca*, where Somers and Harrison (129) showed that a procyanidin was often produced in response to attack by the pathogenic fungus *Ver-*

ticillium albo-atrum, when the tannin arrested and finally killed the invading fungus over a 1-to 6-month period. Walkinshaw (147) has shown histologically that tannin production is stimulated by rust fungus attack on southern pine seedlings, but does not seem to be effective in halting gall formation. It is suggested that tannin production is a normal response by the host to a pathogen.

One fungus isolated from a rotting log of *Pinus radiata* has been shown to be capable of growth on proanthocyanidins as a sole carbon source. Grant (44) identified it as the filamentous fungus *Penicillium adametzi*. Similarly, Deschamps and Leulliette (27) have isolated a number of yeasts from decaying bark that degrade both hydrolyzable and condensed tannins.

7.7.5.3 Allelopathic Agents

Allelopathy was originally defined to be 'biochemical interactions between all plants', both deleterious and advantageous, and to include all plant phyla. However, it has become customary practice to restrict the term to higher plant-higher plant interactions (53), and it is used in this sense here.

Although most of the evidence for allelopathy is circumstantial, it is accepted today by most ecologists as a significant factor in the development of plant communities. The best known example of allelopathy is that of *Juglans* spp., which exude the toxin juglone as a glucoside, thus preventing the growth of plants in the area more or less defined by the drip-line of the tree (53). Since that time, further work has shown that the result for the walnut is rather typical of allelopathic interactions, and the active agents are generally simple organic molecules exuded from the roots or leaves. These compounds either prevent seed germination or seedling growth of competitive species, although self-species inhibition is not unknown (53).

On the basis of the above generalized findings, the proanthocyanidins are not good candidates for allelopaths. They are of high MW and, therefore, cannot act as airborne allelopaths like many terpenoids. They are, however, water-soluble, and monomeric tannin residues such as gallic acid, catechin, or epicatechin would seem to be the most likely candidates. In fact, gallic acid has been recognized as a germination inhibitor in *Arctostaphylos glandulosa* (Ericaceae) of the California chaparral and in *Euphorbia* spp. (115).

There is some evidence that tannins may act as allelopathic agents. Both hydrolyzable and condensed tannins have been shown to inhibit the gibberellin-induced growth of a number of seedlings (24) and, in contrast, appeared to enhance IAA induced growth (24), but this is in contradiction of earlier findings by Zinsmeister and Hoellmueller (151).

In a study of the biological effects of needle leachates in the ecology of *Pinus ponderosa* communities, Lodhi and Killingbeck (78) found that condensed tannins were among the factors that reduced seed germination and growth in herbaceous plant species, and also inhibited the growth of nitrification bacteria (*Nitrosomonas* sp.) when a clear toxic effect was demonstrated. In another study, Taylor and Shaw (137) showed that the condensed tannins from *Picea engelmannii* inhibited seed germination and seedling growth of a number of other conifers, including *Pinus ponderosa* and *Tsuga mertensiana*. Similarly, Yazaki and Nichols (146) showed that the proanthocyanidins from *Pinus radiata* bark displayed a phytotoxic effect toward cucumber seedlings, which may partly explain the success of this material as a ground cover and mulch. The most interesting recent example is, however, the observation that afzelechin and two propelargonidin analogues of procyanidin A2 are specific naturally occurring root growth inhibitors synthesized in the roots of mature peach trees (*Prunus persica* Batsch cv. 'Hakato') (99).

References

1 Agrawal S, Misra K 1977 Leucoanthocyanins from *Ficus racemosa* bark. Chem Scripta 12:37–39
2 Ahn B-Z, Gstirner F 1973 Über Catechin-dimere der Eichenrinde III. Arch Pharm (Weinheim) 306:338–346
3 Aiyar S N, Jain M K, Kristnamurti M, Seshadri T R 1964 Chemical components of the roots of *Connarus monocarpus*. Phytochemistry 3:335–339
4 Bann P S, Walkinshaw C H 1974 Fine structure of tannin accumulations in callus cultures of *Pinus elliotti* (slash pine). Can J Bot 52:615–623
5 Barrett M W, Klyne W, Scopes P M, Fletcher A C, Porter L J, Haslam E 1979 Plant proanthocyanidins. 6. Chiroptical studies. 95. Circular dichroism of procyanidins. J Chem Soc Perkin Trans I 2375–2377
6 Barz W, Hosel W 1975 Metabolism of flavonoids. In: Harborne J B, Mabry T J, Mabry H (eds) The flavonoids. Chapman and Hall London, 916–969
7 Bate-Smith E C 1966 The phenolic constituents of plants and their taxonomic significance. 1. Dicotyledons. J Linn Soc (Bot) 58:95–173
8 Bate-Smith E C 1968 The phenolic constituents of plants and their taxonomic significance. 2. Monocotyledons. J Linn Soc (Bot) 60:325–356
9 Bate-Smith E C 1969 Flavonoid patterns in the Monocotyledons. In: Harborne J B, Swain T (eds) Perspectives in phytochemistry. Academic Press London, 167–177
10 Bate-Smith E C 1973 Haemanalysis of tannins: The concept of relative astringency. Phytochemistry 12:907–912
11 Beart J E, Lilley T H, Haslam E 1985 Polyphenol interactions. Part 2. Covalent binding of procyanidins to proteins during acid-catalysed decomposition; observations on some polymeric proanthocyanidins. J Chem Soc Perkin Trans II 1439–1443
12 Bergmann W R, Barkley M D, Hemingway R W, Mattice W L 1987 Heterogeneous fluorescence decay of 4→6 and 4→8 linked dimers of (+)-catechin and (−)-epicatechin as a result of rotational isomerism. J Am Chem Soc 109:6614–6619
13 Bernays E A 1981 Plant tannins and insect herbivores: An appraisal. Ecol Entomol 6:353–360
14 Bernays E A, Woodhead S 1982 Plant phenols utilized as nutrients by a phytophagous insect. Science 216:201–203
15 Botha J J, Viviers P M, Young D A, du Preez I C, Ferreira D, Roux D G, Hull W E 1982 Synthesis of condensed tannins. 5. The first angular [4,6:4,8]-triflavanoids and their natural counterparts. J Chem Soc Perkin Trans I:527–533
16 Botha J J, Young D A, Ferreira D, Roux D G 1980 Synthesis of condensed tannins. 1. Stereoselective and stereospecific synthesis of optically pure 4-arylflavan-3-ols, and assessment of their absolute stereochemistry at C-4 by means of circular dichroism. J Chem Soc Perkin Trans I 1213–1219
17 Broadhurst R B, Jones W T 1978 Analysis of condensed tannins using acidified vanillin. J Sci Food Agr 29:788–794
18 Chafe S C, Durzan D J 1973 Tannin inclusions in cell suspension cultures of white spruce. Planta 113:251–262
19 Chauhan J S, Kumari G 1978 A new leucoanthocyanin from the stem of Diospyros peregrina. J Indian Chem Soc 55:1068–1070

20 Clark-Lewis J W, Dainis I 1968 Flavan derivatives. 20. A new glycoside and other extractives from *Acacia ixiophylla*: Rhamnitrin (Rhamnetin 3α-L-rhamnoside). Aust J Chem 21:415–420
21 Clark-Lewis J W, Korytnyk W 1958 Flavan derivatives. Part 1. The absolute configuration of (+)-dihydroquercetin. J Chem Soc 2367–2372
22 Clark-Lewis J W, Mitsuno M 1958 The identity of gleditsin and mollisacacidin. J Chem Soc 1724.
23 Clark-Lewis J W, Porter L J 1972 Phytochemical survey of the heartwood flavonoids of *Acacia* species from arid zones of Australia. Aust J Chem 25:1943–1955
24 Corcoran M R, Geissman T A, Phinney B O 1972 Tannins as gibberellin antagonists. Plant Physiol 49:323–328
25 Czochanska Z, Foo L Y, Newman R H, Porter L J 1980 Polymeric proanthocyanidins. Stereochemistry, structural units, and molecular weight. J Chem Soc Perkin Trans I 2278–2286
26 Czochanska Z, Foo L Y, Porter L J 1979 Compositional changes in lower molecular weight flavans during grape maturation. Phytochemistry 18:1819–1822
27 Deschamps A M, Leulliette L 1984 Tannins degradation by yeasts from decaying barks. Int Biodet 20:237–241
28 Ebel J, Hahlbrock K 1982 Biosynthesis. In: Harborne J B, Mabry T J (eds) The flavonoids: Advances in research. Chapman and Hall London, 641–679
29 Ellis C J, Foo L Y, Porter L J 1983 Enantiomerism: A characteristic of the proanthocyanidin chemistry of the Monocotyledonae. Phytochemistry 22:483–487
30 Feeny P P, Bostock H 1968 Seasonal changes in the tannin content of oak leaves. Phytochemistry 7:871–880
31 Ferreira D, Hundt H K L, Roux D G 1971 The stereochemistry of a tetraflavanoid condensed tannin from *Rhus lancea* L. F. J Chem Soc Chem Commun 1257
32 Fletcher A C, Porter L J, Haslam E, Gupta R K 1977 Plant proanthocyanidins. Part 3. Conformational and configurational studies of natural procyanidins. J Chem Soc Perkin Trans I 1628–1637
33 Foo L Y 1981 Proanthocyanidins: Gross chemical structures by infra-red spectra. Phytochemistry 20:1397–1402
34 Foo L Y 1984 Condensed tannins: Co-occurrence of procyanidins, prodelphinidins and profisetinidins in the heartwood of *Acacia baileyana*. Phytochemistry 23:2915–2918
35 Foo L Y 1987 Configuration and conformation of dihydroflavonols from *Acacia melanoxylon*. Phytochemistry 26:813–817
36 Foo L Y, Jones W T, Porter L J, Williams V M 1982 Proanthocyanidin polymers of fodder legumes. Phytochemistry 21:93–95
37 Foo L Y, Porter L J 1980 The phytochemistry of proanthocyanidin polymers. Phytochemistry 19:1747–1754
38 Foo L Y, Porter L J 1983 Synthesis and conformation of procyanidin diastereomers. J Chem Soc Perkin Trans I 1535–1543
39 Freudenberg K, Weinges K 1959 Condensation of catechin. Stereochemistry of the flavan-diol group. Chem Ind 486
40 Gaffield W, Foo L Y, Porter L J 1985 Chiroptical properties of tannins – intense split cotton effects of dimeric procyanidins at low wavelength. Proc Fed Eur Chem Soc Int Conf Circular Dichroism 6:338–346
41 Gaffney S H, Martin R, Lilley T H, Haslam E, Magnolato D 1986 The association of polyphenols with caffeine and α- and β-cyclodextrin in aqueous media. J Chem Soc Chem Commun 107–109
42 Ganguly A K, Seshadri T R 1961 Leucoanthocyanidins of plants. 3. Leucopelargonidin from *Eucalyptus calophylla* Kino. J Chem Soc 2787–2790
43 Goldstein J L, Swain T 1965 Inhibition of enzymes by tannins. Phytochemistry 4:185–192
44 Grant W D 1976 Microbial degradation of condensed tannins. Science 193:1137—1139
45 Gujer R, Magnolato D, Self R 1986 Glucosylated flavonoids and other phenolic compounds from sorghum. Phytochemistry 25:1431–1436
46 Gunaherath G M K B, Gunatilaka A A L, Sultanbawa M U S 1982 Dulcitol and (–)-4′-*O*-methylepigallocatechin from Kokoona zeylanica. J Nat Prod 45:140
47 Gupta R K, Haslam E 1981 Plant proanthocyanidins. 7. Prodelphinidins from *Pinus sylvestris*. J Chem Soc Perkin Trans I 1148–1150
48 Gupta S R, Malik K K, Seshadri T R 1966 Chemical components of *Albizzia lebbeck* heartwood. Indian J Chem 4:139–141

49 Gurni A A, Kubitzki K 1981 Flavonoid chemistry and systematics of the Dilleniaceae. Biochem System Ecol 9:109–114
50 Hagerman A E, Butler L G 1978 Protein precipitation method for the quantitative determination of tannins. J Agr Food Chem 26:809–812
51 Hagerman A E, Butler L G 1980 Determination of protein in tannin-protein precipitates. J Agr Food Chem 28:944–947
52 Hagerman A E, Butler L G 1981 The specificity of proanthocyanidin-protein interactions. J Biol Chem 256:4494–4497
53 Harborne J B 1977 Introduction to ecological biochemistry. Academic Press London, 242 pp
54 Harborne J B 1977 Flavonoids and the evolution of the angiosperms. Biochem System Ecol 5:7–22
55 Haslam E 1975 Natural proanthocyanidins. In: Harborne J B, Mabry T J, Mabry H (eds) The flavonoids. Chapman and Hall London, 505–559
56 Haslam E 1977 Symmetry and promiscuity in procyanidin chemistry. Phytochemistry 16:1625–1640
57 Haslam E 1982 Proanthocyanidins. In: Harborne J B, Mabry T J (eds) The flavonoids: Advances in research. Chapman and Hall London New York, 417–448
58 Haslam E 1982 The metabolism of gallic acid and hexahydroxydiphenic acid in higher plants. Fortschr Chem Organ Naturst 41:1–46
59 Hathway D E 1958 Oak-bark tannins. Biochem J 70:34–42
60 Hegnauer R 1962–1973 Chemotaxonomie der Pflanzen, vol 1–6. Birkhäuser Basel
61 Heller W, Forkmann G, Britsch L, Grisebach H 1985 Enzymatic reduction of (+)-dihydroflavonols to flavan-3,4-*cis*-diols with flower extracts from *Matthiola incana* and its role in anthocyanin biosynthesis. Planta 165:284–287
62 Hemingway R W, Foo L Y, Porter L J 1982 Linkage isomerism in trimeric and polymeric 2,3-*cis*-procyanidins. J Chem Soc Perkin Trans I 1209–1216
63 Hemingway R W, McGraw G W, Karchesy J J, Foo L Y, Porter L J 1983 Recent advances in the chemistry of condensed tannins. J Appl Polym Sci: Appl Polym Symp 37:967–977
64 Jacques D, Opie C T, Porter L J, Haslam E 1977 Plant proanthocyanidins. 4. Biosynthesis of procyanidins and observations on the metabolism of cyanidin in plants. J Chem Soc Perkin Trans I 1637–1643
65 Jones W T, Broadhurst R B, Lyttleton J W 1976 The condensed tannins of pasture legume species. Phytochemistry 15:1407–1409
66 Karchesy J J, Hemingway R W 1980 Loblolly pine bark polyflavanoids. J Agr Food Chem 28:222–228
67 Karchesy J J, Loveland P M, Laver M L, Barofsky D F, Barofsky E 1976 Condensed tannins from the barks of *Alnus rubra* and *Pseudotsuga menziesii*. Phytochemistry 15:2009–2010
68 Karl C, Muller G, Pedersen P A 1983 Ellagitannine aus den Blättern von *Vitis vinifera*. Z Naturforsch 38C:13–15
69 Klocke J A, Chan B G 1982 Effects of cotton condensed tannin on feeding and digestion in the cotton pest *Heliothis zea*. J Insect Physiol 28:911–915
70 Krishnan V, Rangaswami S 1965 Occurrence of leucopelargonidin in the roots of *Casearia esculenta* Roxb. Current Sci (India) 34:634–635
71 Kristiansen K N 1986 Conversion of (+)-dihydroquercetin to (+)-2,3-*trans*-3,4-*cis*-leucocyanidin and (+)-catechin with an enzyme extract from maturing grains of barley. Carlsberg Res Commun 51:51–60
72 Laks P E, Hemingway R W 1987 Condensed tannins. Structure of the 'phenolic acids'. Holzforschung 41 (in press)
73 Lea A G H, Timerlake C F 1974 The phenolics of ciders. 1. Procyanidins. J Sci Food Agr 25:1537–1545
74 Lebreton P, Thivend S, Boutard B 1980 Distribution des proanthocyanidins chez les gymnospermes. Plantes Med Phyto 14:105–129
75 Ledbetter M C, Porter K R 1970 Introduction to the fine structure of plant cells. Springer Berlin, 155
76 Lewak S, Pliszka B, Eichelberger E 1967 Leucoanthocyanidins in the leaves of apple tree. Acta Soc Bot Pol 36:251–259

77 Lewak S, Radominska A 1965 Dimeric leucoanthocyanidins of hawthorn (*Crataegus oxyacantha*). Roczniki Chem 39:1839–1845
78 Lodhi M A K, Killingbeck K T 1982 Effects of pine-produced chemicals on selected understory species in a *Pinus ponderosa* community. J Chem Ecol 8:275–283
79 McManus J P, Davis K G, Bent J E, Gaffney S H, Lilley T H, Haslam E 1985 Polyphenol interactions. Part 1. Introduction; some observations on the reversible complexation of polyphenols with proteins and polysaccharides. J Chem Soc Perkin Trans II 1429–1438
80 Martin R, Lilley T H, Bailey N A, Falshaw C P, Haslam E, Magnolato D, Begley M J 1986 Polyphenol-caffeine complexation. J Chem Soc Chem Commun 105–106
81 Mathew A G, Lakshminarayana S 1969 Polyphenols of immature sapota fruit. Phytochemistry 8:507–509
82 Mattice W L, Porter L J 1984 Molecular weight averages and ^{13}C NMR intensities provide evidence for branching in proanthocyanidin polymers. Phytochemistry 23:1309–1311
83 Mishra C S, Misra K 1980 Chemical examination of the stem bark of *Eucalyptus maculata*. Planta Med 38:169–173
84 Delle Monache F, Pomponi M, Marini-Bettolo G B, D'Albuquerque I L, Goncalves de Lima O 1976 A methylated catechin and proanthocyanidins from the Celastraceae. Phytochemistry 15:573–574
85 Morgan J W W, Orsler R J 1967 Rhodesian teak tannin. Phytochemistry 6:1007–1012
86 Morimoto S, Nonaka G, Nishioka I 1986 Tannins and related compounds. 38. Isolation and characterization of flavan-3-ol glucosides and procyanidin oligomers from cassia bark (*Cinnamonnum cassia* Blume). Chem Pharm Bull 34:633–642
87 Morimoto S, Nonaka G, Nishioka I, Ezaki N, Takizawa N 1985 Tannins and related substances. 29. Seven new methyl-derivatives of flavan-3-ols and 1,3-diaryl-propan-2-ol from *Cinnamomum cassia, C. obtusifolium* and *Lindera umbellata* var. *membranacea*. Chem Pharm Bull 33:2281–2286
88 Narayanan V, Seshadri T R 1972 Phenolics of *Cassia fistula*. Indian J Chem 10:379–381
89 Neilson MJ, Painter T J, Richards G N 1986 Flavologlycan: A novel glycoconjugate from leaves of mangrove (*Rhizophora stylosa* Griff). Carbohyd Res 147:315–324
90 Newman R H, Porter L J, Foo L Y, Johns S R, Willing R I 1987 High resolution ^{13}C NMR studies of proanthocyanidin polymers. Mag Res Chem 25:118–124
91 Nonaka G, Ezaki E, Hayashi K, Nishioka I 1983 Flavanol glucosides from rhubarb and *Rhaphiolepsis umbellata*. Phytochemistry 22:1659–1661
92 Nonaka G, Goto Y, Kinjo J, Nohara T, Nishioka I 1987 Tannins and related compounds. 52. Studies on the constituents of the leaves of *Thujopsis dolabrata* Sieb. et Zucc. Chem Pharm Bull 35:1105–1108
93 Nonaka G, Miwa N, Nishioka I 1982 Stilbene glycoside gallates and proanthocyanidins from *Polygonum multiflorum*. Phytochemistry 21:429–432
94 Nonaka G, Muta M, Nishioka I 1983 Myricatin, a galloyl flavanonol sulphate and prodelphinidin gallates from *Myrica rubra*. Phytochemistry 22:237–241
95 Nonaka G, Nishioka I 1980 Novel biflavonoids, chalcan-flavan dimers from gambir. Chem Pharm Bull 28:3145–3149
96 Nonaka G, Nishioka I, Nagasawa T, Oura H 1981 Tannins and related compounds. 1. Rhubarb (1). Chem Pharm Bull 29:2862–2870
97 Nyman B F 1985 Protein-proanthocyanidin interactions during extraction of Scots pine needles. Phytochemistry 24:2939–2944
98 Oberholzer M E, Rall G J H, Roux D G 1980 (2*R*,3*S*,4*S*)-3,4,7,3′,4′-pentamethoxy-2,3-*trans*-3,4-*cis*-flavan, a novel flavan from *Neorautanenia amboensis*. Phytochemistry 19:2503–2504
99 Ohigashi H, Minami S, Fukui H, Koshimizu K, Mizutani F, Sugiura A, Tomana T 1982 Flavanols, as plant growth inhibitors from roots of peach *Prunus persica* Batsch cv. 'Hakuto'. Agr Biol Chem 46:2555–2561
100 Okuda T, Mori K, Hatano T 1985 Relationship of the structures of tannins to the binding activities with hemoglobin and methylene blue. Chem Pharm Bull 33:1424–1433
101 Pangon J F, Jay M, Voirin B 1974 Flavonoids of *Dryas octopetala*. Phytochemistry 13: 1883–1885
102 Paris R, Etchepane S 1967 Sur les polyphenols due *Cassia sieberiana* DC. Isolement du 1-épicatéchol et du leucopélargonidol. Ann Pharm Franc 25:343–346

103 Pelter A, Amenechi P I, Warren R 1969 The structures of two proanthocyanidins from *Julbenadia globiflora*. J Chem Soc (C) 2572–2579
104 Pettigrew C J, Watson L 1975 On the classification of Australian acacias. Aust J Bot 23:833–847
105 Porter L J 1986 Number- and weight-average molecular weights for some proanthocyanidin polymers (condensed tannins). Aust J Chem 39:557–562
106 Porter L J, Foo L Y 1982 Leucocyanidin: synthesis and properties of (2*R*,3*S*,4*R*)-(+)-3,4,5,7,3′,4′-hexahydroxyflavan. Phytochemistry 21:2947–2952
107 Porter L J, Foo L Y, Furneaux R H 1985 Isolation of three naturally occurring O-β-D-glucopyranosides of procyanidin polymers. Phytochemistry 24:567–569
108 Porter L J, Hrstich L N, Chan B G 1986 The conversion of procyanidins and prodelphinidins to cyanidin and delphinidin. Phytochemistry 25:223–230
109 Porter L J, Newman R H, Foo L Y, Wong H, Hemingway R W 1982 Polymeric proanthocyanidins. ^{13}C N.M.R. studies of procyanidins. J Chem Soc Perkin Trans I 1217–1221
110 Porter L J, Wilson R D 1972 The separation of condensed tannins on Sephadex G-25 eluted with 50% aqueous acetone. J Chromatog 71:570–572
111 Porter L J, Wong R Y, Benson M, Chan B G, Viswanadhan V N, Gandour R D, Mattice W L 1986 Conformational analysis of flavans. ^1H NMR and molecular mechanical (MM2) studies of the benzpyran ring of 3′,4′,5,7-tetrahydroxyflavan-3-ols: The crystal and molecular structure of the procyanidin (2*R*,3*S*,4*R*)-3′,4′,5,7-tetramethoxy-4-(2,4,6-trimethoxyphenyl)flavan-3-ol. J Chem Res (S):86–87; (M): 0830–0880
112 Porter L J, Wong R Y, Chan B G 1985 The molecular and crystal structure of (+)-2,3-*trans*-leucocyanidin [(2*R*,3*S*,4*R*-(+)-3,3′,4,4′,5,7-hexahydroxyflavan]dihydrate, and comparison of its heterocyclic ring conformation in solution and the solid state. J Chem Soc Perkin Trans I 1413–1418
113 Porter L J, Woodruffe J 1984 Haemanalysis: The relative astringency of proanthocyanidin polymers. Phytochemistry 23:1255–1256
114 du Preez I C, Rowan A C, Roux D G 1971 Hindered rotation about the sp^2-sp^3 hybridized C–C bond between flavanoid units in condensed tannins. J Chem Soc Chem Commun 315–316
115 Putnam A R 1983 Allelopathic chemicals. Chem Eng News 61(14):34–45
116 Rahman M D, Richards G N 1988 Interactions of starch and other polysaccharides with condensed tannins in hot water extracts of ponderosa pine. J Wood Chem Tech 7 (in press)
117 Ramiah N, Prasad N B R, Nair G A 1976 Chemical investigation of the fruits (seeds, pericarp and aril) of *Connarus monocarpus*. Indian J Chem 14B:475
118 Ramiah N, Prasad N B R, Saradama P 1976 Chemical components of *Vitis glauca* Hook. (Syn. *Cissus glauca* Roxb.). J Indian Chem Soc 53:1158
119 Roux D G, Ferreira D, Botha J J 1980 Structural considerations in predicting the utilization of tannins. J Agr Food Chem 28:216–222
120 Roux D G, Ferreira D, Hundt H K L, Malan E 1975 Structure, stereochemistry, and reactivity of natural condensed tannins as a basis for their extended industrial application. J Appl Polym Sci: Appl Polym Symp 28:335–353
121 Roux D G, Paulus E 1962 Condensed tannins. 13. Interrelationships of flavonoid components from the heartwood of *Robinia pseudoacacia*. Biochem J 82:324–330
122 Saayman H M, Roux D G 1965 Configuration of guibourtacacidin, and synthesis of isomeric racemates. Biochem J 96:36–42
123 Sakurai A, Okada K, Okamura Y 1982 Chemical studies on the mistletoe. IV. The structure of isoglucodistylin, a new flavonoid glycoside isolated from *Taxillus kaempferi*. Bull Chem Soc Japan 55:3051–3052
124 Samejima M, Yoshimoto T 1981 General aspects of phenolic extractives from coniferous barks. Mokuzai Gakkaishi 27:491–497
125 Samejima M, Yoshimoto T 1981 High performance liquid chromatography of proanthocyanidins and related compounds. Mokuzai Gakkaishi 27:658–662
126 Samejima M, Yoshimoto T 1982 Systematic studies on the sterochemical composition of proanthocyanidins from coniferous bark. Mokuzai Gakkaishi 28:67–74
127 Shen Z, Haslam E, Falshaw C P, Begley M J 1986 Procyanidins and polyphenols of *Larix gmelini* bark. Phytochemistry 25:2629–2635
128 Sears K D, Casebier R L 1970 The reaction of thioglycollic acid with polyflavanoid bark fractions of *Tsuga heterophylla*. Phytochemistry 9:1589–1594

129 Somers T C, Harrison A F 1967 Wood tannins – isolation and significance in host resistance to Verticillium wilt disease. Aust J Biol Sci 20:475–479
130 Stafford H A, Cheng T Y 1980 The procyanidins of Douglas-fir seedlings, callus and cell suspension cultures derived from cotyledons. Phytochemistry 19:131–135
131 Stafford H A, Lester H H 1981 Proanthocyanidins and potential precursors in needles of Douglas-fir and in cell suspension cultures derived from seedling shoot tissues. Plant Physiol 68:1035–1040
132 Stafford H A, Lester H H 1985 Flavan-3-ol biosynthesis. Plant Physiol 78:791–794
133 Stafford H A, Lester H H, Porter L J 1985 Further studies on the chemical and enzymic synthesis of monomeric procyanidins (leucocyanidins or 3′,4′,5,7-tetrahydroxyflavan-3,4-diols) from (2*R*, 3*R*)-dihydroquercetin. Phytochemistry 24:333–338
134 Subramanian P M, Misra G S 1977 Chemical constituents of *Ficus bengalensis*. Indian J Chem 15B:762–763
135 Synge R L M 1975 Interactions of polyphenols with proteins in plants and plant products. Qual Plant 24:337–350
136 Tamir M, Nachtomi E, Alumot E 1971 Degradation of tannins from carob pods (*Ceratonia siliqua*) by thioglycollic acid. Phytochemistry 10:2769–2774
137 Taylor R J, Shaw D C 1983 Allelopathic effects of Engelmann spruce bark stilbenes and tannin-stilbene combinations on seed germination and seedling growth of selected conifers. Can J Bot 61:279–289
138 Thieme H, Khogali A 1975 Über das Vorkommen von Flavonoiden und Gerbstoffen in den Blättern einiger afrikanischer Acacia-Arten. Pharmazie 30:736–743
139 Thresh K, Ibrahimm R K 1985 Are spinach chloroplasts involved in flavonoid *O*-methylation? Z Naturforsch 40c:331–335
140 Tindale M D, Roux D G 1969 A phytochemical survey of the Australian species of *Acacia*. Phytochemistry 8:1713–1727
141 Tindale M D, Roux D G 1974 An extended phytochemical survey of Australian species of *Acacia*: Chemotaxonomic and phylogenetic aspects. Phytochemistry 13:829–839
142 Tisler V, Ayla C, Weissman G 1983 Untersuchung der Rindenextrakte von *Pinus halpensis* Mill. Holzforsch Holzverwert 35:113–116
143 Tschesche R, Braun T M, von Sassen W 1980 Symplocoside, a flavanol glycoside from *Symplocus uniflora*. Phytochemistry 19:1825–1829
144 Viswanadhan V N, Bergmann W R, Mattice W L 1987 Configurational statistics of C(4)-C(8)-linked homopolymers of (+)-catechin or (−)-epicatechin. Macromolecules 20:1539–1543
145 Yazaki Y, Hillis W E 1980 Molecular size distribution of radiata pine bark extracts and its effect on properties. Holzforschung 34:125–130
146 Yazaki Y, Nichols D 1978 Phytotoxic effects of *Pinus radiata* bark. Aust For Res 8:185–198
147 Walkinshaw C H 1978 Cell necrosis and fungus content in fusiform rust-infected loblolly, longleaf, and slash pine seedlings. Phytopathology 68:1705–1712
148 Williams V M, Porter L J, Hemingway R W 1983 Molecular weight profiles of proanthocyanidin polymers. Phytochemistry 22:569–572
149 Young D A, Brandt E V, Ferreira D, Roux D G 1986 Synthesis of condensed tannins. Part 17. Oligomeric (2*R*,3*S*)-3,3′,4′,7,8-pentahydroxy flavans: Atropisomerism and conformation of biphenyl and *m*-terphenyl analogues from *Prosopis glandulosa* ('mesquite'). J Chem Soc Perkin Trans I 1737–1749
150 Young D A, Cronje A, Botes A L, Ferreira D, Roux D G 1985 Synthesis of condensed tannins. Part 14. Biflavanoid profisetinidins as synthons. The acid induced 'phlobaphene' reaction. J Chem Soc Perkin Trans I 2521–2527
151 Zinsmeister H D, Hoellmueller W 1967 Tannins and growth. 3. Analysis of IAA-tannin antagonism in oat coleophile cylinders. Planta 73:333–338
152 Zobel A M 1985 Ontogenesis of tannin coenocytes in *Sambucus racemosa* L. 1. Development of the coenocytes from mononucleate tannin cells. Ann Bot 55:765–773
153 Zucker W V 1983 Tannins: Does structure determine function? An ecological perspective. Am Nat 121:35–365

Printed and bound by PG in the USA